T0191678

Lecture Notes in Computer Science 14223

Founding Editors

Gerhard Goos
Juris Hartmanis

The series Lecture Notes in Computer Science (LNCS), including its subseries Lecture Notes in Artificial Intelligence (LNAI) and Lecture Notes in Bioinformatics (LNBI), has established itself as a medium for the publication of new developments in computer science and information technology research, teaching, and education.

LNCS enjoys close cooperation with the computer science R & D community, the series counts many renowned academics among its volume editors and paper authors, and collaborates with prestigious societies. Its mission is to serve this international community by providing an invaluable service, mainly focused on the publication of conference and workshop proceedings and postproceedings. LNCS commenced publication in 1973.

Hayit Greenspan · Anant Madabhushi ·
Parvin Mousavi · Septimiu Salcudean ·
James Duncan · Tanveer Syeda-Mahmood ·
Russell Taylor
Editors

Medical Image Computing and Computer Assisted Intervention – MICCAI 2023

26th International Conference
Vancouver, BC, Canada, October 8–12, 2023
Proceedings, Part IV

Springer

Editors

Hayit Greenspan
Icahn School of Medicine, Mount Sinai,
NYC, NY, USA

Tel Aviv University
Tel Aviv, Israel

Parvin Mousavi
Queen's University
Kingston, ON, Canada

James Duncan ⓘ
Yale University
New Haven, CT, USA

Russell Taylor ⓘ
Johns Hopkins University
Baltimore, MD, USA

Anant Madabhushi ⓘ
Emory University
Atlanta, GA, USA

Septimiu Salcudean ⓘ
The University of British Columbia
Vancouver, BC, Canada

Tanveer Syeda-Mahmood ⓘ
IBM Research
San Jose, CA, USA

ISSN 0302-9743 ISSN 1611-3349 (electronic)
Lecture Notes in Computer Science
ISBN 978-3-031-43900-1 ISBN 978-3-031-43901-8 (eBook)
https://doi.org/10.1007/978-3-031-43901-8

This Springer imprint is published by the registered company Springer Nature Switzerland AG
The registered company address is: Gewerbestrasse 11, 6330 Cham, Switzerland

Paper in this product is recyclable.

Preface

We are pleased to present the proceedings for the 26th International Conference on Medical Image Computing and Computer-Assisted Intervention (MICCAI). After several difficult years of virtual conferences, this edition was held in a mainly in-person format with a hybrid component at the Vancouver Convention Centre, in Vancouver, BC, Canada October 8–12, 2023. The conference featured 33 physical workshops, 15 online workshops, 15 tutorials, and 29 challenges held on October 8 and October 12. Co-located with the conference was also the 3rd Conference on Clinical Translation on Medical Image Computing and Computer-Assisted Intervention (CLINICCAI) on October 10.

MICCAI 2023 received the largest number of submissions so far, with an approximately 30% increase compared to 2022. We received 2365 full submissions of which 2250 were subjected to full review. To keep the acceptance ratios around 32% as in previous years, there was a corresponding increase in accepted papers leading to 730 papers accepted, with 68 orals and the remaining presented in poster form. These papers comprise ten volumes of Lecture Notes in Computer Science (LNCS) proceedings as follows:

- Part I, LNCS Volume 14220: Machine Learning with Limited Supervision and Machine Learning – Transfer Learning
- Part II, LNCS Volume 14221: Machine Learning – Learning Strategies and Machine Learning – Explainability, Bias, and Uncertainty I
- Part III, LNCS Volume 14222: Machine Learning – Explainability, Bias, and Uncertainty II and Image Segmentation I
- Part IV, LNCS Volume 14223: Image Segmentation II
- Part V, LNCS Volume 14224: Computer-Aided Diagnosis I
- Part VI, LNCS Volume 14225: Computer-Aided Diagnosis II and Computational Pathology
- Part VII, LNCS Volume 14226: Clinical Applications – Abdomen, Clinical Applications – Breast, Clinical Applications – Cardiac, Clinical Applications – Dermatology, Clinical Applications – Fetal Imaging, Clinical Applications – Lung, Clinical Applications – Musculoskeletal, Clinical Applications – Oncology, Clinical Applications – Ophthalmology, and Clinical Applications – Vascular
- Part VIII, LNCS Volume 14227: Clinical Applications – Neuroimaging and Microscopy
- Part IX, LNCS Volume 14228: Image-Guided Intervention, Surgical Planning, and Data Science
- Part X, LNCS Volume 14229: Image Reconstruction and Image Registration

The papers for the proceedings were selected after a rigorous double-blind peer-review process. The MICCAI 2023 Program Committee consisted of 133 area chairs and over 1600 reviewers, with representation from several countries across all major continents. It also maintained a gender balance with 31% of scientists who self-identified as women. With an increase in the number of area chairs and reviewers, the reviewer load on the experts was reduced this year, keeping to 16–18 papers per area chair and about 4–6 papers per reviewer. Based on the double-blinded reviews, area chairs' recommendations, and program chairs' global adjustments, 308 papers (14%) were provisionally accepted, 1196 papers (53%) were provisionally rejected, and 746 papers (33%) proceeded to the rebuttal stage. As in previous years, Microsoft's Conference Management Toolkit (CMT) was used for paper management and organizing the overall review process. Similarly, the Toronto paper matching system (TPMS) was employed to ensure knowledgeable experts were assigned to review appropriate papers. Area chairs and reviewers were selected following public calls to the community, and were vetted by the program chairs.

Among the new features this year was the emphasis on clinical translation, moving Medical Image Computing (MIC) and Computer-Assisted Interventions (CAI) research from theory to practice by featuring two clinical translational sessions reflecting the real-world impact of the field in the clinical workflows and clinical evaluations. For the first time, clinicians were appointed as Clinical Chairs to select papers for the clinical translational sessions. The philosophy behind the dedicated clinical translational sessions was to maintain the high scientific and technical standard of MICCAI papers in terms of methodology development, while at the same time showcasing the strong focus on clinical applications. This was an opportunity to expose the MICCAI community to the clinical challenges and for ideation of novel solutions to address these unmet needs. Consequently, during paper submission, in addition to MIC and CAI a new category of "Clinical Applications" was introduced for authors to self-declare.

MICCAI 2023 for the first time in its history also featured dual parallel tracks that allowed the conference to keep the same proportion of oral presentations as in previous years, despite the 30% increase in submitted and accepted papers.

We also introduced two new sessions this year focusing on young and emerging scientists through their Ph.D. thesis presentations, and another with experienced researchers commenting on the state of the field through a fireside chat format.

The organization of the final program by grouping the papers into topics and sessions was aided by the latest advancements in generative AI models. Specifically, Open AI's GPT-4 large language model was used to group the papers into initial topics which were then manually curated and organized. This resulted in fresh titles for sessions that are more reflective of the technical advancements of our field.

Although not reflected in the proceedings, the conference also benefited from keynote talks from experts in their respective fields including Turing Award winner Yann LeCun and leading experts Jocelyne Troccaz and Mihaela van der Schaar.

We extend our sincere gratitude to everyone who contributed to the success of MICCAI 2023 and the quality of its proceedings. In particular, we would like to express our profound thanks to the MICCAI Submission System Manager Kitty Wong whose meticulous support throughout the paper submission, review, program planning, and proceeding preparation process was invaluable. We are especially appreciative of the effort and dedication of our Satellite Events Chair, Bennett Landman, who tirelessly coordinated the organization of over 90 satellite events consisting of workshops, challenges and tutorials. Our workshop chairs Hongzhi Wang, Alistair Young, tutorial chairs Islem

Rekik, Guoyan Zheng, and challenge chairs, Lena Maier-Hein, Jayashree Kalpathy-Kramer, Alexander Seitel, worked hard to assemble a strong program for the satellite events. Special mention this year also goes to our first-time Clinical Chairs, Drs. Curtis Langlotz, Charles Kahn, and Masaru Ishii who helped us select papers for the clinical sessions and organized the clinical sessions.

We acknowledge the contributions of our Keynote Chairs, William Wells and Alejandro Frangi, who secured our keynote speakers. Our publication chairs, Kevin Zhou and Ron Summers, helped in our efforts to get the MICCAI papers indexed in PubMed. It was a challenging year for fundraising for the conference due to the recovery of the economy after the COVID pandemic. Despite this situation, our industrial sponsorship chairs, Mohammad Yaqub, Le Lu and Yanwu Xu, along with Dekon's Mehmet Eldegez, worked tirelessly to secure sponsors in innovative ways, for which we are grateful.

An active body of the MICCAI Student Board led by Camila Gonzalez and our 2023 student representatives Nathaniel Braman and Vaishnavi Subramanian helped put together student-run networking and social events including a novel Ph.D. thesis 3-minute madness event to spotlight new graduates for their careers. Similarly, Women in MICCAI chairs Xiaoxiao Li and Jayanthi Sivaswamy and RISE chairs, Islem Rekik, Pingkun Yan, and Andrea Lara further strengthened the quality of our technical program through their organized events. Local arrangements logistics including the recruiting of University of British Columbia students and invitation letters to attendees, was ably looked after by our local arrangement chairs Purang Abolmaesumi and Mehdi Moradi. They also helped coordinate the visits to the local sites in Vancouver both during the selection of the site and organization of our local activities during the conference. Our Young Investigator chairs Marius Linguraru, Archana Venkataraman, Antonio Porras Perez put forward the startup village and helped secure funding from NIH for early career scientist participation in the conference. Our communications chair, Ehsan Adeli, and Diana Cunningham were active in making the conference visible on social media platforms and circulating the newsletters. Niharika D'Souza was our cross-committee liaison providing note-taking support for all our meetings. We are grateful to all these organization committee members for their active contributions that made the conference successful.

We would like to thank the MICCAI society chair, Caroline Essert, and the MICCAI board for their approvals, support and feedback, which provided clarity on various aspects of running the conference. Behind the scenes, we acknowledge the contributions of the MICCAI secretariat personnel, Janette Wallace, and Johanne Langford, who kept a close eye on logistics and budgets, and Diana Cunningham and Anna Van Vliet for including our conference announcements in a timely manner in the MICCAI society newsletters. This year, when the existing virtual platform provider indicated that they would discontinue their service, a new virtual platform provider Conference Catalysts was chosen after due diligence by John Baxter. John also handled the setup and coordination with CMT and consultation with program chairs on features, for which we are very grateful. The physical organization of the conference at the site, budget financials, fund-raising, and the smooth running of events would not have been possible without our Professional Conference Organization team from Dekon Congress & Tourism led by Mehmet Eldegez. The model of having a PCO run the conference, which we used at

MICCAI, significantly reduces the work of general chairs for which we are particularly grateful.

Finally, we are especially grateful to all members of the Program Committee for their diligent work in the reviewer assignments and final paper selection, as well as the reviewers for their support during the entire process. Lastly, and most importantly, we thank all authors, co-authors, students/postdocs, and supervisors for submitting and presenting their high-quality work, which played a pivotal role in making MICCAI 2023 a resounding success.

With a successful MICCAI 2023, we now look forward to seeing you next year in Marrakesh, Morocco when MICCAI 2024 goes to the African continent for the first time.

October 2023

Tanveer Syeda-Mahmood
James Duncan
Russ Taylor
General Chairs

Hayit Greenspan
Anant Madabhushi
Parvin Mousavi
Septimiu Salcudean
Program Chairs

Organization

General Chairs

Tanveer Syeda-Mahmood	IBM Research, USA
James Duncan	Yale University, USA
Russ Taylor	Johns Hopkins University, USA

Program Committee Chairs

Hayit Greenspan	Tel-Aviv University, Israel and Icahn School of Medicine at Mount Sinai, USA
Anant Madabhushi	Emory University, USA
Parvin Mousavi	Queen's University, Canada
Septimiu Salcudean	University of British Columbia, Canada

Satellite Events Chair

Bennett Landman	Vanderbilt University, USA

Workshop Chairs

Hongzhi Wang	IBM Research, USA
Alistair Young	King's College, London, UK

Challenges Chairs

Jayashree Kalpathy-Kramer	Harvard University, USA
Alexander Seitel	German Cancer Research Center, Germany
Lena Maier-Hein	German Cancer Research Center, Germany

Tutorial Chairs

Islem Rekik   Imperial College London, UK
Guoyan Zheng   Shanghai Jiao Tong University, China

Clinical Chairs

Curtis Langlotz   Stanford University, USA
Charles Kahn   University of Pennsylvania, USA
Masaru Ishii   Johns Hopkins University, USA

Local Arrangements Chairs

Purang Abolmaesumi   University of British Columbia, Canada
Mehdi Moradi   McMaster University, Canada

Keynote Chairs

William Wells   Harvard University, USA
Alejandro Frangi   University of Manchester, UK

Industrial Sponsorship Chairs

Mohammad Yaqub   MBZ University of Artificial Intelligence,
                                   Abu Dhabi
Le Lu   DAMO Academy, Alibaba Group, USA
Yanwu Xu   Baidu, China

Communication Chair

Ehsan Adeli   Stanford University, USA

Publication Chairs

Ron Summers	National Institutes of Health, USA
Kevin Zhou	University of Science and Technology of China, China

Young Investigator Chairs

Marius Linguraru	Children's National Institute, USA
Archana Venkataraman	Boston University, USA
Antonio Porras	University of Colorado Anschutz Medical Campus, USA

Student Activities Chairs

Nathaniel Braman	Picture Health, USA
Vaishnavi Subramanian	EPFL, France

Women in MICCAI Chairs

Jayanthi Sivaswamy	IIIT, Hyderabad, India
Xiaoxiao Li	University of British Columbia, Canada

RISE Committee Chairs

Islem Rekik	Imperial College London, UK
Pingkun Yan	Rensselaer Polytechnic Institute, USA
Andrea Lara	Universidad Galileo, Guatemala

Submission Platform Manager

Kitty Wong	The MICCAI Society, Canada

Virtual Platform Manager

John Baxter	INSERM, Université de Rennes 1, France

Cross-Committee Liaison

Niharika D'Souza	IBM Research, USA

Program Committee

Sahar Ahmad	University of North Carolina at Chapel Hill, USA
Shadi Albarqouni	University of Bonn and Helmholtz Munich, Germany
Angelica Aviles-Rivero	University of Cambridge, UK
Shekoofeh Azizi	Google, Google Brain, USA
Ulas Bagci	Northwestern University, USA
Wenjia Bai	Imperial College London, UK
Sophia Bano	University College London, UK
Kayhan Batmanghelich	University of Pittsburgh and Boston University, USA
Ismail Ben Ayed	ETS Montreal, Canada
Katharina Breininger	Friedrich-Alexander-Universität Erlangen-Nürnberg, Germany
Weidong Cai	University of Sydney, Australia
Geng Chen	Northwestern Polytechnical University, China
Hao Chen	Hong Kong University of Science and Technology, China
Jun Cheng	Institute for Infocomm Research, A*STAR, Singapore
Li Cheng	University of Alberta, Canada
Albert C. S. Chung	University of Exeter, UK
Toby Collins	Ircad, France
Adrian Dalca	Massachusetts Institute of Technology and Harvard Medical School, USA
Jose Dolz	ETS Montreal, Canada
Qi Dou	Chinese University of Hong Kong, China
Nicha Dvornek	Yale University, USA
Shireen Elhabian	University of Utah, USA
Sandy Engelhardt	Heidelberg University Hospital, Germany
Ruogu Fang	University of Florida, USA

Jianming Liang	Arizona State University, USA
Jianfei Liu	National Institutes of Health Clinical Center, USA
Mingxia Liu	University of North Carolina at Chapel Hill, USA
Xiaofeng Liu	Harvard Medical School and MGH, USA
Herve Lombaert	École de technologie supérieure, Canada
Ismini Lourentzou	Virginia Tech, USA
Le Lu	Damo Academy USA, Alibaba Group, USA
Dwarikanath Mahapatra	Inception Institute of Artificial Intelligence, United Arab Emirates
Saad Nadeem	Memorial Sloan Kettering Cancer Center, USA
Dong Nie	Alibaba (US), USA
Yoshito Otake	Nara Institute of Science and Technology, Japan
Sang Hyun Park	Daegu Gyeongbuk Institute of Science and Technology, South Korea
Magdalini Paschali	Stanford University, USA
Tingying Peng	Helmholtz Munich, Germany
Caroline Petitjean	LITIS Université de Rouen Normandie, France
Esther Puyol Anton	King's College London, UK
Chen Qin	Imperial College London, UK
Daniel Racoceanu	Sorbonne Université, France
Hedyeh Rafii-Tari	Auris Health, USA
Hongliang Ren	Chinese University of Hong Kong, China and National University of Singapore, Singapore
Tammy Riklin Raviv	Ben-Gurion University, Israel
Hassan Rivaz	Concordia University, Canada
Mirabela Rusu	Stanford University, USA
Thomas Schultz	University of Bonn, Germany
Feng Shi	Shanghai United Imaging Intelligence, China
Yang Song	University of New South Wales, Australia
Aristeidis Sotiras	Washington University in St. Louis, USA
Rachel Sparks	King's College London, UK
Yao Sui	Peking University, China
Kenji Suzuki	Tokyo Institute of Technology, Japan
Qian Tao	Delft University of Technology, Netherlands
Mathias Unberath	Johns Hopkins University, USA
Martin Urschler	Medical University Graz, Austria
Maria Vakalopoulou	CentraleSupelec, University Paris Saclay, France
Erdem Varol	New York University, USA
Francisco Vasconcelos	University College London, UK
Harini Veeraraghavan	Memorial Sloan Kettering Cancer Center, USA
Satish Viswanath	Case Western Reserve University, USA
Christian Wachinger	Technical University of Munich, Germany

Hua Wang	Colorado School of Mines, USA
Qian Wang	ShanghaiTech University, China
Shanshan Wang	Paul C. Lauterbur Research Center, SIAT, China
Yalin Wang	Arizona State University, USA
Bryan Williams	Lancaster University, UK
Matthias Wilms	University of Calgary, Canada
Jelmer Wolterink	University of Twente, Netherlands
Ken C. L. Wong	IBM Research Almaden, USA
Jonghye Woo	Massachusetts General Hospital and Harvard Medical School, USA
Shandong Wu	University of Pittsburgh, USA
Yutong Xie	University of Adelaide, Australia
Fuyong Xing	University of Colorado, Denver, USA
Daguang Xu	NVIDIA, USA
Yan Xu	Beihang University, China
Yanwu Xu	Baidu, China
Pingkun Yan	Rensselaer Polytechnic Institute, USA
Guang Yang	Imperial College London, UK
Jianhua Yao	Tencent, China
Chuyang Ye	Beijing Institute of Technology, China
Lequan Yu	University of Hong Kong, China
Ghada Zamzmi	National Institutes of Health, USA
Liang Zhan	University of Pittsburgh, USA
Fan Zhang	Harvard Medical School, USA
Ling Zhang	Alibaba Group, China
Miaomiao Zhang	University of Virginia, USA
Shu Zhang	Northwestern Polytechnical University, China
Rongchang Zhao	Central South University, China
Yitian Zhao	Chinese Academy of Sciences, China
Tao Zhou	Nanjing University of Science and Technology, USA
Yuyin Zhou	UC Santa Cruz, USA
Dajiang Zhu	University of Texas at Arlington, USA
Lei Zhu	ROAS Thrust HKUST (GZ), and ECE HKUST, China
Xiahai Zhuang	Fudan University, China
Veronika Zimmer	Technical University of Munich, Germany

Reviewers

Alaa Eldin Abdelaal
John Abel
Kumar Abhishek
Shahira Abousamra
Mazdak Abulnaga
Burak Acar
Abdoljalil Addeh
Ehsan Adeli
Sukesh Adiga Vasudeva
Seyed-Ahmad Ahmadi
Euijoon Ahn
Faranak Akbarifar
Alireza Akhondi-asl
Saad Ullah Akram
Daniel Alexander
Hanan Alghamdi
Hassan Alhajj
Omar Al-Kadi
Max Allan
Andre Altmann
Pablo Alvarez
Charlems Alvarez-Jimenez
Jennifer Alvén
Lidia Al-Zogbi
Kimberly Amador
Tamaz Amiranashvili
Amine Amyar
Wangpeng An
Vincent Andrearczyk
Manon Ansart
Sameer Antani
Jacob Antunes
Michel Antunes
Guilherme Aresta
Mohammad Ali Armin
Kasra Arnavaz
Corey Arnold
Janan Arslan
Marius Arvinte
Muhammad Asad
John Ashburner
Md Ashikuzzaman
Shahab Aslani

Mehdi Astaraki
Angélica Atehortúa
Benjamin Aubert
Marc Aubreville
Paolo Avesani
Sana Ayromlou
Reza Azad
Mohammad Farid
    Azampour
Qinle Ba
Meritxell Bach Cuadra
Hyeon-Min Bae
Matheus Baffa
Cagla Bahadir
Fan Bai
Jun Bai
Long Bai
Pradeep Bajracharya
Shafa Balaram
Yaël Balbastre
Yutong Ban
Abhirup Banerjee
Soumyanil Banerjee
Sreya Banerjee
Shunxing Bao
Omri Bar
Adrian Barbu
Joao Barreto
Adrian Basarab
Berke Basaran
Michael Baumgartner
Siming Bayer
Roza Bayrak
Aicha BenTaieb
Guy Ben-Yosef
Sutanu Bera
Cosmin Bercea
Jorge Bernal
Jose Bernal
Gabriel Bernardino
Riddhish Bhalodia
Jignesh Bhatt
Indrani Bhattacharya

Binod Bhattarai
Lei Bi
Qi Bi
Cheng Bian
Gui-Bin Bian
Carlo Biffi
Alexander Bigalke
Benjamin Billot
Manuel Birlo
Ryoma Bise
Daniel Blezek
Stefano Blumberg
Sebastian Bodenstedt
Federico Bolelli
Bhushan Borotikar
Ilaria Boscolo Galazzo
Alexandre Bousse
Nicolas Boutry
Joseph Boyd
Behzad Bozorgtabar
Nadia Brancati
Clara Brémond Martin
Stéphanie Bricq
Christopher Bridge
Coleman Broaddus
Rupert Brooks
Tom Brosch
Mikael Brudfors
Ninon Burgos
Nikolay Burlutskiy
Michal Byra
Ryan Cabeen
Mariano Cabezas
Hongmin Cai
Tongan Cai
Zongyou Cai
Liane Canas
Bing Cao
Guogang Cao
Weiguo Cao
Xu Cao
Yankun Cao
Zhenjie Cao

Jaime Cardoso
M. Jorge Cardoso
Owen Carmichael
Jacob Carse
Adrià Casamitjana
Alessandro Casella
Angela Castillo
Kate Cevora
Krishna Chaitanya
Satrajit Chakrabarty
Yi Hao Chan
Shekhar Chandra
Ming-Ching Chang
Peng Chang
Qi Chang
Yuchou Chang
Hanqing Chao
Simon Chatelin
Soumick Chatterjee
Sudhanya Chatterjee
Muhammad Faizyab Ali
   Chaudhary
Antong Chen
Bingzhi Chen
Chen Chen
Cheng Chen
Chengkuan Chen
Eric Chen
Fang Chen
Haomin Chen
Jianan Chen
Jianxu Chen
Jiazhou Chen
Jie Chen
Jintai Chen
Jun Chen
Junxiang Chen
Junyu Chen
Li Chen
Liyun Chen
Nenglun Chen
Pingjun Chen
Pingyi Chen
Qi Chen
Qiang Chen

Runnan Chen
Shengcong Chen
Sihao Chen
Tingting Chen
Wenting Chen
Xi Chen
Xiang Chen
Xiaoran Chen
Xin Chen
Xiongchao Chen
Yanxi Chen
Yixiong Chen
Yixuan Chen
Yuanyuan Chen
Yuqian Chen
Zhaolin Chen
Zhen Chen
Zhenghao Chen
Zhennong Chen
Zhihao Chen
Zhineng Chen
Zhixiang Chen
Chang-Chieh Cheng
Jiale Cheng
Jianhong Cheng
Jun Cheng
Xuelian Cheng
Yupeng Cheng
Mark Chiew
Philip Chikontwe
Eleni Chiou
Jungchan Cho
Jang-Hwan Choi
Min-Kook Choi
Wookjin Choi
Jaegul Choo
Yu-Cheng Chou
Daan Christiaens
Argyrios Christodoulidis
Stergios Christodoulidis
Kai-Cheng Chuang
Hyungjin Chung
Matthew Clarkson
Michaël Clément
Dana Cobzas

Jaume Coll-Font
Olivier Colliot
Runmin Cong
Yulai Cong
Laura Connolly
William Consagra
Pierre-Henri Conze
Tim Cootes
Teresa Correia
Baris Coskunuzer
Alex Crimi
Can Cui
Hejie Cui
Hui Cui
Lei Cui
Wenhui Cui
Tolga Cukur
Tobias Czempiel
Javid Dadashkarimi
Haixing Dai
Tingting Dan
Kang Dang
Salman Ul Hassan Dar
Eleonora D'Arnese
Dhritiman Das
Neda Davoudi
Tareen Dawood
Sandro De Zanet
Farah Deeba
Charles Delahunt
Herve Delingette
Ugur Demir
Liang-Jian Deng
Ruining Deng
Wenlong Deng
Felix Denzinger
Adrien Depeursinge
Mohammad Mahdi
   Derakhshani
Hrishikesh Deshpande
Adrien Desjardins
Christian Desrosiers
Blake Dewey
Neel Dey
Rohan Dhamdhere

Maxime Di Folco
Songhui Diao
Alina Dima
Hao Ding
Li Ding
Ying Ding
Zhipeng Ding
Nicola Dinsdale
Konstantin Dmitriev
Ines Domingues
Bo Dong
Liang Dong
Nanqing Dong
Siyuan Dong
Reuben Dorent
Gianfranco Doretto
Sven Dorkenwald
Haoran Dou
Mitchell Doughty
Jason Dowling
Niharika D'Souza
Guodong Du
Jie Du
Shiyi Du
Hongyi Duanmu
Benoit Dufumier
James Duncan
Joshua Durso-Finley
Dmitry V. Dylov
Oleh Dzyubachyk
Mahdi (Elias) Ebnali
Philip Edwards
Jan Egger
Gudmundur Einarsson
Mostafa El Habib Daho
Ahmed Elazab
Idris El-Feghi
David Ellis
Mohammed Elmogy
Amr Elsawy
Okyaz Eminaga
Ertunc Erdil
Lauren Erdman
Marius Erdt
Maria Escobar

Hooman Esfandiari
Nazila Esmaeili
Ivan Ezhov
Alessio Fagioli
Deng-Ping Fan
Lei Fan
Xin Fan
Yubo Fan
Huihui Fang
Jiansheng Fang
Xi Fang
Zhenghan Fang
Mohammad Farazi
Azade Farshad
Mohsen Farzi
Hamid Fehri
Lina Felsner
Chaolu Feng
Chun-Mei Feng
Jianjiang Feng
Mengling Feng
Ruibin Feng
Zishun Feng
Alvaro Fernandez-Quilez
Ricardo Ferrari
Lucas Fidon
Lukas Fischer
Madalina Fiterau
Antonio
    Foncubierta-Rodríguez
Fahimeh Fooladgar
Germain Forestier
Nils Daniel Forkert
Jean-Rassaire Fouefack
Kevin François-Bouaou
Wolfgang Freysinger
Bianca Freytag
Guanghui Fu
Kexue Fu
Lan Fu
Yunguan Fu
Pedro Furtado
Ryo Furukawa
Jin Kyu Gahm
Mélanie Gaillochet

Francesca Galassi
Jiangzhang Gan
Yu Gan
Yulu Gan
Alireza Ganjdanesh
Chang Gao
Cong Gao
Linlin Gao
Zeyu Gao
Zhongpai Gao
Sara Garbarino
Alain Garcia
Beatriz Garcia Santa Cruz
Rongjun Ge
Shiv Gehlot
Manuela Geiss
Salah Ghamizi
Negin Ghamsarian
Ramtin Gharleghi
Ghazal Ghazaei
Florin Ghesu
Sayan Ghosal
Syed Zulqarnain Gilani
Mahdi Gilany
Yannik Glaser
Ben Glocker
Bharti Goel
Jacob Goldberger
Polina Golland
Alberto Gomez
Catalina Gomez
Estibaliz
    Gómez-de-Mariscal
Haifan Gong
Kuang Gong
Xun Gong
Ricardo Gonzales
Camila Gonzalez
German Gonzalez
Vanessa Gonzalez Duque
Sharath Gopal
Karthik Gopinath
Pietro Gori
Michael Götz
Shuiping Gou

Maged Goubran
Sobhan Goudarzi
Mark Graham
Alejandro Granados
Mara Graziani
Thomas Grenier
Radu Grosu
Michal Grzeszczyk
Feng Gu
Pengfei Gu
Qiangqiang Gu
Ran Gu
Shi Gu
Wenhao Gu
Xianfeng Gu
Yiwen Gu
Zaiwang Gu
Hao Guan
Jayavardhana Gubbi
Houssem-Eddine Gueziri
Dazhou Guo
Hengtao Guo
Jixiang Guo
Jun Guo
Pengfei Guo
Wenzhangzhi Guo
Xiaoqing Guo
Xueqi Guo
Yi Guo
Vikash Gupta
Praveen Gurunath Bharathi
Prashnna Gyawali
Sung Min Ha
Mohamad Habes
Ilker Hacihaliloglu
Stathis Hadjidemetriou
Fatemeh Haghighi
Justin Haldar
Noura Hamze
Liang Han
Luyi Han
Seungjae Han
Tianyu Han
Zhongyi Han
Jonny Hancox

Lasse Hansen
Degan Hao
Huaying Hao
Jinkui Hao
Nazim Haouchine
Michael Hardisty
Stefan Harrer
Jeffry Hartanto
Charles Hatt
Huiguang He
Kelei He
Qi He
Shenghua He
Xinwei He
Stefan Heldmann
Nicholas Heller
Edward Henderson
Alessa Hering
Monica Hernandez
Kilian Hett
Amogh Hiremath
David Ho
Malte Hoffmann
Matthew Holden
Qingqi Hong
Yoonmi Hong
Mohammad Reza
    Hosseinzadeh Taher
William Hsu
Chuanfei Hu
Dan Hu
Kai Hu
Rongyao Hu
Shishuai Hu
Xiaoling Hu
Xinrong Hu
Yan Hu
Yang Hu
Chaoqin Huang
Junzhou Huang
Ling Huang
Luojie Huang
Qinwen Huang
Sharon Xiaolei Huang
Weijian Huang

Xiaoyang Huang
Yi-Jie Huang
Yongsong Huang
Yongxiang Huang
Yuhao Huang
Zhe Huang
Zhi-An Huang
Ziyi Huang
Arnaud Huaulmé
Henkjan Huisman
Alex Hung
Jiayu Huo
Andreas Husch
Mohammad Arafat
    Hussain
Sarfaraz Hussein
Jana Hutter
Khoi Huynh
Ilknur Icke
Kay Igwe
Abdullah Al Zubaer Imran
Muhammad Imran
Samra Irshad
Nahid Ul Islam
Koichi Ito
Hayato Itoh
Yuji Iwahori
Krithika Iyer
Mohammad Jafari
Srikrishna Jaganathan
Hassan Jahanandish
Andras Jakab
Amir Jamaludin
Amoon Jamzad
Ananya Jana
Se-In Jang
Pierre Jannin
Vincent Jaouen
Uditha Jarayathne
Ronnachai Jaroensri
Guillaume Jaume
Syed Ashar Javed
Rachid Jennane
Debesh Jha
Ge-Peng Ji

Luping Ji
Zexuan Ji
Zhanghexuan Ji
Haozhe Jia
Hongchao Jiang
Jue Jiang
Meirui Jiang
Tingting Jiang
Xiajun Jiang
Zekun Jiang
Zhifan Jiang
Ziyu Jiang
Jianbo Jiao
Zhicheng Jiao
Chen Jin
Dakai Jin
Qiangguo Jin
Qiuye Jin
Weina Jin
Baoyu Jing
Bin Jing
Yaqub Jonmohamadi
Lie Ju
Yohan Jun
Dinkar Juyal
Manjunath K N
Ali Kafaei Zad Tehrani
John Kalafut
Niveditha Kalavakonda
Megha Kalia
Anil Kamat
Qingbo Kang
Po-Yu Kao
Anuradha Kar
Neerav Karani
Turkay Kart
Satyananda Kashyap
Alexander Katzmann
Lisa Kausch
Maxime Kayser
Salome Kazeminia
Wenchi Ke
Youngwook Kee
Matthias Keicher
Erwan Kerrien

Afifa Khaled
Nadieh Khalili
Farzad Khalvati
Bidur Khanal
Bishesh Khanal
Pulkit Khandelwal
Maksim Kholiavchenko
Ron Kikinis
Benjamin Killeen
Daeseung Kim
Heejong Kim
Jaeil Kim
Jinhee Kim
Jinman Kim
Junsik Kim
Minkyung Kim
Namkug Kim
Sangwook Kim
Tae Soo Kim
Younghoon Kim
Young-Min Kim
Andrew King
Miranda Kirby
Gabriel Kiss
Andreas Kist
Yoshiro Kitamura
Stefan Klein
Tobias Klinder
Kazuma Kobayashi
Lisa Koch
Satoshi Kondo
Fanwei Kong
Tomasz Konopczynski
Ender Konukoglu
Aishik Konwer
Thijs Kooi
Ivica Kopriva
Avinash Kori
Kivanc Kose
Suraj Kothawade
Anna Kreshuk
AnithaPriya Krishnan
Florian Kromp
Frithjof Kruggel
Thomas Kuestner

Levin Kuhlmann
Abhay Kumar
Kuldeep Kumar
Sayantan Kumar
Manuela Kunz
Holger Kunze
Tahsin Kurc
Anvar Kurmukov
Yoshihiro Kuroda
Yusuke Kurose
Hyuksool Kwon
Aymen Laadhari
Jorma Laaksonen
Dmitrii Lachinov
Alain Lalande
Rodney LaLonde
Bennett Landman
Daniel Lang
Carole Lartizien
Shlomi Laufer
Max-Heinrich Laves
William Le
Loic Le Folgoc
Christian Ledig
Eung-Joo Lee
Ho Hin Lee
Hyekyoung Lee
John Lee
Kisuk Lee
Kyungsu Lee
Soochahn Lee
Woonghee Lee
Étienne Léger
Wen Hui Lei
Yiming Lei
George Leifman
Rogers Jeffrey Leo John
Juan Leon
Bo Li
Caizi Li
Chao Li
Chen Li
Cheng Li
Chenxin Li
Chnegyin Li

Dawei Li
Fuhai Li
Gang Li
Guang Li
Hao Li
Haofeng Li
Haojia Li
Heng Li
Hongming Li
Hongwei Li
Huiqi Li
Jian Li
Jieyu Li
Kang Li
Lin Li
Mengzhang Li
Ming Li
Qing Li
Quanzheng Li
Shaohua Li
Shulong Li
Tengfei Li
Weijian Li
Wen Li
Xiaomeng Li
Xingyu Li
Xinhui Li
Xuelu Li
Xueshen Li
Yamin Li
Yang Li
Yi Li
Yuemeng Li
Yunxiang Li
Zeju Li
Zhaoshuo Li
Zhe Li
Zhen Li
Zhenqiang Li
Zhiyuan Li
Zhjin Li
Zi Li
Hao Liang
Libin Liang
Peixian Liang

Yuan Liang
Yudong Liang
Haofu Liao
Hongen Liao
Wei Liao
Zehui Liao
Gilbert Lim
Hongxiang Lin
Li Lin
Manxi Lin
Mingquan Lin
Tiancheng Lin
Yi Lin
Zudi Lin
Claudia Lindner
Simone Lionetti
Chi Liu
Chuanbin Liu
Daochang Liu
Dongnan Liu
Feihong Liu
Fenglin Liu
Han Liu
Huiye Liu
Jiang Liu
Jie Liu
Jinduo Liu
Jing Liu
Jingya Liu
Jundong Liu
Lihao Liu
Mengting Liu
Mingyuan Liu
Peirong Liu
Peng Liu
Qin Liu
Quan Liu
Rui Liu
Shengfeng Liu
Shuangjun Liu
Sidong Liu
Siyuan Liu
Weide Liu
Xiao Liu
Xiaoyu Liu

Xingtong Liu
Xinwen Liu
Xinyang Liu
Xinyu Liu
Yan Liu
Yi Liu
Yihao Liu
Yikang Liu
Yilin Liu
Yilong Liu
Yiqiao Liu
Yong Liu
Yuhang Liu
Zelong Liu
Zhe Liu
Zhiyuan Liu
Zuozhu Liu
Lisette Lockhart
Andrea Loddo
Nicolas Loménie
Yonghao Long
Daniel Lopes
Ange Lou
Brian Lovell
Nicolas Loy Rodas
Charles Lu
Chun-Shien Lu
Donghuan Lu
Guangming Lu
Huanxiang Lu
Jingpei Lu
Yao Lu
Oeslle Lucena
Jie Luo
Luyang Luo
Ma Luo
Mingyuan Luo
Wenhan Luo
Xiangde Luo
Xinzhe Luo
Jinxin Lv
Tianxu Lv
Fei Lyu
Ilwoo Lyu
Mengye Lyu

Qing Lyu
Yanjun Lyu
Yuanyuan Lyu
Benteng Ma
Chunwei Ma
Hehuan Ma
Jun Ma
Junbo Ma
Wenao Ma
Yuhui Ma
Pedro Macias Gordaliza
Anant Madabhushi
Derek Magee
S. Sara Mahdavi
Andreas Maier
Klaus H. Maier-Hein
Sokratis Makrogiannis
Danial Maleki
Michail Mamalakis
Zhehua Mao
Jan Margeta
Brett Marinelli
Zdravko Marinov
Viktoria Markova
Carsten Marr
Yassine Marrakchi
Anne Martel
Martin Maška
Tejas Sudharshan Mathai
Petr Matula
Dimitrios Mavroeidis
Evangelos Mazomenos
Amarachi Mbakwe
Adam McCarthy
Stephen McKenna
Raghav Mehta
Xueyan Mei
Felix Meissen
Felix Meister
Afaque Memon
Mingyuan Meng
Qingjie Meng
Xiangzhu Meng
Yanda Meng
Zhu Meng

Martin Menten
Odyssée Merveille
Mikhail Milchenko
Leo Milecki
Fausto Milletari
Hyun-Seok Min
Zhe Min
Song Ming
Duy Minh Ho Nguyen
Deepak Mishra
Suraj Mishra
Virendra Mishra
Tadashi Miyamoto
Sara Moccia
Marc Modat
Omid Mohareri
Tony C. W. Mok
Javier Montoya
Rodrigo Moreno
Stefano Moriconi
Lia Morra
Ana Mota
Lei Mou
Dana Moukheiber
Lama Moukheiber
Daniel Moyer
Pritam Mukherjee
Anirban Mukhopadhyay
Henning Müller
Ana Murillo
Gowtham Krishnan
    Murugesan
Ahmed Naglah
Karthik Nandakumar
Venkatesh
    Narasimhamurthy
Raja Narayan
Dominik Narnhofer
Vishwesh Nath
Rodrigo Nava
Abdullah Nazib
Ahmed Nebli
Peter Neher
Amin Nejatbakhsh
Trong-Thuan Nguyen

Truong Nguyen
Dong Ni
Haomiao Ni
Xiuyan Ni
Hannes Nickisch
Weizhi Nie
Aditya Nigam
Lipeng Ning
Xia Ning
Kazuya Nishimura
Chuang Niu
Sijie Niu
Vincent Noblet
Narges Norouzi
Alexey Novikov
Jorge Novo
Gilberto Ochoa-Ruiz
Masahiro Oda
Benjamin Odry
Hugo Oliveira
Sara Oliveira
Arnau Oliver
Jimena Olveres
John Onofrey
Marcos Ortega
Mauricio Alberto
    Ortega-Ruíz
Yusuf Osmanlioglu
Chubin Ou
Cheng Ouyang
Jiahong Ouyang
Xi Ouyang
Cristina Oyarzun Laura
Utku Ozbulak
Ece Ozkan
Ege Özsoy
Batu Ozturkler
Harshith Padigela
Johannes Paetzold
José Blas Pagador
    Carrasco
Daniel Pak
Sourabh Palande
Chengwei Pan
Jiazhen Pan

Jin Pan
Yongsheng Pan
Egor Panfilov
Jiaxuan Pang
Joao Papa
Constantin Pape
Bartlomiej Papiez
Nripesh Parajuli
Hyunjin Park
Akash Parvatikar
Tiziano Passerini
Diego Patiño Cortés
Mayank Patwari
Angshuman Paul
Rasmus Paulsen
Yuchen Pei
Yuru Pei
Tao Peng
Wei Peng
Yige Peng
Yunsong Peng
Matteo Pennisi
Antonio Pepe
Oscar Perdomo
Sérgio Pereira
Jose-Antonio
    Pérez-Carrasco
Mehran Pesteie
Terry Peters
Eike Petersen
Jens Petersen
Micha Pfeiffer
Dzung Pham
Hieu Pham
Ashish Phophalia
Tomasz Pieciak
Antonio Pinheiro
Pramod Pisharady
Theodoros Pissas
Szymon Płotka
Kilian Pohl
Sebastian Pölsterl
Alison Pouch
Tim Prangemeier
Prateek Prasanna

Raphael Prevost
Juan Prieto
Federica Proietto Salanitri
Sergi Pujades
Elodie Puybareau
Talha Qaiser
Buyue Qian
Mengyun Qiao
Yuchuan Qiao
Zhi Qiao
Chenchen Qin
Fangbo Qin
Wenjian Qin
Yulei Qin
Jie Qiu
Jielin Qiu
Peijie Qiu
Shi Qiu
Wu Qiu
Liangqiong Qu
Linhao Qu
Quan Quan
Tran Minh Quan
Sandro Queirós
Prashanth R
Febrian Rachmadi
Daniel Racoceanu
Mehdi Rahim
Jagath Rajapakse
Kashif Rajpoot
Keerthi Ram
Dhanesh Ramachandram
João Ramalhinho
Xuming Ran
Aneesh Rangnekar
Hatem Rashwan
Keerthi Sravan Ravi
Daniele Ravì
Sadhana Ravikumar
Harish Raviprakash
Surreerat Reaungamornrat
Samuel Remedios
Mengwei Ren
Sucheng Ren
Elton Rexhepaj

Mauricio Reyes
Constantino
    Reyes-Aldasoro
Abel Reyes-Angulo
Hadrien Reynaud
Razieh Rezaei
Anne-Marie Rickmann
Laurent Risser
Dominik Rivoir
Emma Robinson
Robert Robinson
Jessica Rodgers
Ranga Rodrigo
Rafael Rodrigues
Robert Rohling
Margherita Rosnati
Łukasz Roszkowiak
Holger Roth
José Rouco
Dan Ruan
Jiacheng Ruan
Daniel Rueckert
Danny Ruijters
Kanghyun Ryu
Ario Sadafi
Numan Saeed
Monjoy Saha
Pramit Saha
Farhang Sahba
Pranjal Sahu
Simone Saitta
Md Sirajus Salekin
Abbas Samani
Pedro Sanchez
Luis Sanchez Giraldo
Yudi Sang
Gerard Sanroma-Guell
Rodrigo Santa Cruz
Alice Santilli
Rachana Sathish
Olivier Saut
Mattia Savardi
Nico Scherf
Alexander Schlaefer
Jerome Schmid

Adam Schmidt
Julia Schnabel
Lawrence Schobs
Julian Schön
Peter Schueffler
Andreas Schuh
Christina
  Schwarz-Gsaxner
Michaël Sdika
Suman Sedai
Lalithkumar Seenivasan
Matthias Seibold
Sourya Sengupta
Lama Seoud
Ana Sequeira
Sharmishtaa Seshamani
Ahmed Shaffie
Jay Shah
Keyur Shah
Ahmed Shahin
Mohammad Abuzar
  Shaikh
S. Shailja
Hongming Shan
Wei Shao
Mostafa Sharifzadeh
Anuja Sharma
Gregory Sharp
Hailan Shen
Li Shen
Linlin Shen
Mali Shen
Mingren Shen
Yiqing Shen
Zhengyang Shen
Jun Shi
Xiaoshuang Shi
Yiyu Shi
Yonggang Shi
Hoo-Chang Shin
Jitae Shin
Keewon Shin
Boris Shirokikh
Suzanne Shontz
Yucheng Shu

Hanna Siebert
Alberto Signoroni
Wilson Silva
Julio Silva-Rodríguez
Margarida Silveira
Walter Simson
Praveer Singh
Vivek Singh
Nitin Singhal
Elena Sizikova
Gregory Slabaugh
Dane Smith
Kevin Smith
Tiffany So
Rajath Soans
Roger Soberanis-Mukul
Hessam Sokooti
Jingwei Song
Weinan Song
Xinhang Song
Xinrui Song
Mazen Soufi
Georgia Sovatzidi
Bella Specktor Fadida
William Speier
Ziga Spiclin
Dominik Spinczyk
Jon Sporring
Pradeeba Sridar
Chetan L. Srinidhi
Abhishek Srivastava
Lawrence Staib
Marc Stamminger
Justin Strait
Hai Su
Ruisheng Su
Zhe Su
Vaishnavi Subramanian
Gérard Subsol
Carole Sudre
Dong Sui
Heung-Il Suk
Shipra Suman
He Sun
Hongfu Sun

Jian Sun
Li Sun
Liyan Sun
Shanlin Sun
Kyung Sung
Yannick Suter
Swapna T. R.
Amir Tahmasebi
Pablo Tahoces
Sirine Taleb
Bingyao Tan
Chaowei Tan
Wenjun Tan
Hao Tang
Siyi Tang
Xiaoying Tang
Yucheng Tang
Zihao Tang
Michael Tanzer
Austin Tapp
Elias Tappeiner
Mickael Tardy
Giacomo Tarroni
Athena Taymourtash
Kaveri Thakoor
Elina Thibeau-Sutre
Paul Thienphrapa
Sarina Thomas
Stephen Thompson
Karl Thurnhofer-Hemsi
Cristiana Tiago
Lin Tian
Lixia Tian
Yapeng Tian
Yu Tian
Yun Tian
Aleksei Tiulpin
Hamid Tizhoosh
Minh Nguyen Nhat To
Matthew Toews
Maryam Toloubidokhti
Minh Tran
Quoc-Huy Trinh
Jocelyne Troccaz
Roger Trullo

Chialing Tsai
Apostolia Tsirikoglou
Puxun Tu
Samyakh Tukra
Sudhakar Tummala
Georgios Tziritas
Vladimír Ulman
Tamas Ungi
Régis Vaillant
Jeya Maria Jose Valanarasu
Vanya Valindria
Juan Miguel Valverde
Fons van der Sommen
Maureen van Eijnatten
Tom van Sonsbeek
Gijs van Tulder
Yogatheesan Varatharajah
Madhurima Vardhan
Thomas Varsavsky
Hooman Vaseli
Serge Vasylechko
S. Swaroop Vedula
Sanketh Vedula
Gonzalo Vegas
    Sanchez-Ferrero
Matthew Velazquez
Archana Venkataraman
Sulaiman Vesal
Mitko Veta
Barbara Villarini
Athanasios Vlontzos
Wolf-Dieter Vogl
Ingmar Voigt
Sandrine Voros
Vibashan VS
Trinh Thi Le Vuong
An Wang
Bo Wang
Ce Wang
Changmiao Wang
Ching-Wei Wang
Dadong Wang
Dong Wang
Fakai Wang
Guotai Wang

Haifeng Wang
Haoran Wang
Hong Wang
Hongxiao Wang
Hongyu Wang
Jiacheng Wang
Jing Wang
Jue Wang
Kang Wang
Ke Wang
Lei Wang
Li Wang
Liansheng Wang
Lin Wang
Ling Wang
Linwei Wang
Manning Wang
Mingliang Wang
Puyang Wang
Qiuli Wang
Renzhen Wang
Ruixuan Wang
Shaoyu Wang
Sheng Wang
Shujun Wang
Shuo Wang
Shuqiang Wang
Tao Wang
Tianchen Wang
Tianyu Wang
Wenzhe Wang
Xi Wang
Xiangdong Wang
Xiaoqing Wang
Xiaosong Wang
Yan Wang
Yangang Wang
Yaping Wang
Yi Wang
Yirui Wang
Yixin Wang
Zeyi Wang
Zhao Wang
Zichen Wang
Ziqin Wang

Ziyi Wang
Zuhui Wang
Dong Wei
Donglai Wei
Hao Wei
Jia Wei
Leihao Wei
Ruofeng Wei
Shuwen Wei
Martin Weigert
Wolfgang Wein
Michael Wels
Cédric Wemmert
Thomas Wendler
Markus Wenzel
Rhydian Windsor
Adam Wittek
Marek Wodzinski
Ivo Wolf
Julia Wolleb
Ka-Chun Wong
Jonghye Woo
Chongruo Wu
Chunpeng Wu
Fuping Wu
Huaqian Wu
Ji Wu
Jiangjie Wu
Jiong Wu
Junde Wu
Linshan Wu
Qing Wu
Weiwen Wu
Wenjun Wu
Xiyin Wu
Yawen Wu
Ye Wu
Yicheng Wu
Yongfei Wu
Zhengwang Wu
Pengcheng Xi
Chao Xia
Siyu Xia
Wenjun Xia
Lei Xiang

Tiange Xiang
Deqiang Xiao
Li Xiao
Xiaojiao Xiao
Yiming Xiao
Zeyu Xiao
Hongtao Xie
Huidong Xie
Jianyang Xie
Long Xie
Weidi Xie
Fangxu Xing
Shuwei Xing
Xiaodan Xing
Xiaohan Xing
Haoyi Xiong
Yujian Xiong
Di Xu
Feng Xu
Haozheng Xu
Hongming Xu
Jiangchang Xu
Jiaqi Xu
Junshen Xu
Kele Xu
Lijian Xu
Min Xu
Moucheng Xu
Rui Xu
Xiaowei Xu
Xuanang Xu
Yanwu Xu
Yanyu Xu
Yongchao Xu
Yunqiu Xu
Zhe Xu
Zhoubing Xu
Ziyue Xu
Kai Xuan
Cheng Xue
Jie Xue
Tengfei Xue
Wufeng Xue
Yuan Xue
Zhong Xue

Ts Faridah Yahya
Chaochao Yan
Jiangpeng Yan
Ming Yan
Qingsen Yan
Xiangyi Yan
Yuguang Yan
Zengqiang Yan
Baoyao Yang
Carl Yang
Changchun Yang
Chen Yang
Feng Yang
Fengting Yang
Ge Yang
Guanyu Yang
Heran Yang
Huijuan Yang
Jiancheng Yang
Jiewen Yang
Peng Yang
Qi Yang
Qiushi Yang
Wei Yang
Xin Yang
Xuan Yang
Yan Yang
Yanwu Yang
Yifan Yang
Yingyu Yang
Zhicheng Yang
Zhijian Yang
Jiangchao Yao
Jiawen Yao
Lanhong Yao
Linlin Yao
Qingsong Yao
Tianyuan Yao
Xiaohui Yao
Zhao Yao
Dong Hye Ye
Menglong Ye
Yousef Yeganeh
Jirong Yi
Xin Yi

Chong Yin
Pengshuai Yin
Yi Yin
Zhaozheng Yin
Chunwei Ying
Youngjin Yoo
Jihun Yoon
Chenyu You
Hanchao Yu
Heng Yu
Jinhua Yu
Jinze Yu
Ke Yu
Qi Yu
Qian Yu
Thomas Yu
Weimin Yu
Yang Yu
Chenxi Yuan
Kun Yuan
Wu Yuan
Yixuan Yuan
Paul Yushkevich
Fatemeh Zabihollahy
Samira Zare
Ramy Zeineldin
Dong Zeng
Qi Zeng
Tianyi Zeng
Wei Zeng
Kilian Zepf
Kun Zhan
Bokai Zhang
Daoqiang Zhang
Dong Zhang
Fa Zhang
Hang Zhang
Hanxiao Zhang
Hao Zhang
Haopeng Zhang
Haoyue Zhang
Hongrun Zhang
Jiadong Zhang
Jiajin Zhang
Jianpeng Zhang

Jiawei Zhang
Jingqing Zhang
Jingyang Zhang
Jinwei Zhang
Jiong Zhang
Jiping Zhang
Ke Zhang
Lefei Zhang
Lei Zhang
Li Zhang
Lichi Zhang
Lu Zhang
Minghui Zhang
Molin Zhang
Ning Zhang
Rongzhao Zhang
Ruipeng Zhang
Ruisi Zhang
Shichuan Zhang
Shihao Zhang
Shuai Zhang
Tuo Zhang
Wei Zhang
Weihang Zhang
Wen Zhang
Wenhua Zhang
Wenqiang Zhang
Xiaodan Zhang
Xiaoran Zhang
Xin Zhang
Xukun Zhang
Xuzhe Zhang
Ya Zhang
Yanbo Zhang
Yanfu Zhang
Yao Zhang
Yi Zhang
Yifan Zhang
Yixiao Zhang
Yongqin Zhang
You Zhang
Youshan Zhang

Yu Zhang
Yubo Zhang
Yue Zhang
Yuhan Zhang
Yulun Zhang
Yundong Zhang
Yunlong Zhang
Yuyao Zhang
Zheng Zhang
Zhenxi Zhang
Ziqi Zhang
Can Zhao
Chongyue Zhao
Fenqiang Zhao
Gangming Zhao
He Zhao
Jianfeng Zhao
Jun Zhao
Li Zhao
Liang Zhao
Lin Zhao
Mengliu Zhao
Mingbo Zhao
Qingyu Zhao
Shang Zhao
Shijie Zhao
Tengda Zhao
Tianyi Zhao
Wei Zhao
Yidong Zhao
Yiyuan Zhao
Yu Zhao
Zhihe Zhao
Ziyuan Zhao
Haiyong Zheng
Hao Zheng
Jiannan Zheng
Kang Zheng
Meng Zheng
Sisi Zheng
Tianshu Zheng
Yalin Zheng

Yefeng Zheng
Yinqiang Zheng
Yushan Zheng
Aoxiao Zhong
Jia-Xing Zhong
Tao Zhong
Zichun Zhong
Hong-Yu Zhou
Houliang Zhou
Huiyu Zhou
Kang Zhou
Qin Zhou
Ran Zhou
S. Kevin Zhou
Tianfei Zhou
Wei Zhou
Xiao-Hu Zhou
Xiao-Yun Zhou
Yi Zhou
Youjia Zhou
Yukun Zhou
Zongwei Zhou
Chenglu Zhu
Dongxiao Zhu
Heqin Zhu
Jiayi Zhu
Meilu Zhu
Wei Zhu
Wenhui Zhu
Xiaofeng Zhu
Xin Zhu
Yonghua Zhu
Yongpei Zhu
Yuemin Zhu
Yan Zhuang
David Zimmerer
Yongshuo Zong
Ke Zou
Yukai Zou
Lianrui Zuo
Gerald Zwettler

Outstanding Area Chairs

Mingxia Liu	University of North Carolina at Chapel Hill, USA
Matthias Wilms	University of Calgary, Canada
Veronika Zimmer	Technical University Munich, Germany

Outstanding Reviewers

Kimberly Amador	University of Calgary, Canada
Angela Castillo	Universidad de los Andes, Colombia
Chen Chen	Imperial College London, UK
Laura Connolly	Queen's University, Canada
Pierre-Henri Conze	IMT Atlantique, France
Niharika D'Souza	IBM Research, USA
Michael Götz	University Hospital Ulm, Germany
Meirui Jiang	Chinese University of Hong Kong, China
Manuela Kunz	National Research Council Canada, Canada
Zdravko Marinov	Karlsruhe Institute of Technology, Germany
Sérgio Pereira	Lunit, South Korea
Lalithkumar Seenivasan	National University of Singapore, Singapore

Honorable Mentions (Reviewers)

Kumar Abhishek	Simon Fraser University, Canada
Guilherme Aresta	Medical University of Vienna, Austria
Shahab Aslani	University College London, UK
Marc Aubreville	Technische Hochschule Ingolstadt, Germany
Yaël Balbastre	Massachusetts General Hospital, USA
Omri Bar	Theator, Israel
Aicha Ben Taieb	Simon Fraser University, Canada
Cosmin Bercea	Technical University Munich and Helmholtz AI and Helmholtz Center Munich, Germany
Benjamin Billot	Massachusetts Institute of Technology, USA
Michal Byra	RIKEN Center for Brain Science, Japan
Mariano Cabezas	University of Sydney, Australia
Alessandro Casella	Italian Institute of Technology and Politecnico di Milano, Italy
Junyu Chen	Johns Hopkins University, USA
Argyrios Christodoulidis	Pfizer, Greece
Olivier Colliot	CNRS, France

Lei Cui	Northwest University, China
Neel Dey	Massachusetts Institute of Technology, USA
Alessio Fagioli	Sapienza University, Italy
Yannik Glaser	University of Hawaii at Manoa, USA
Haifan Gong	Chinese University of Hong Kong, Shenzhen, China
Ricardo Gonzales	University of Oxford, UK
Sobhan Goudarzi	Sunnybrook Research Institute, Canada
Michal Grzeszczyk	Sano Centre for Computational Medicine, Poland
Fatemeh Haghighi	Arizona State University, USA
Edward Henderson	University of Manchester, UK
Qingqi Hong	Xiamen University, China
Mohammad R. H. Taher	Arizona State University, USA
Henkjan Huisman	Radboud University Medical Center, the Netherlands
Ronnachai Jaroensri	Google, USA
Qiangguo Jin	Northwestern Polytechnical University, China
Neerav Karani	Massachusetts Institute of Technology, USA
Benjamin Killeen	Johns Hopkins University, USA
Daniel Lang	Helmholtz Center Munich, Germany
Max-Heinrich Laves	Philips Research and ImFusion GmbH, Germany
Gilbert Lim	SingHealth, Singapore
Mingquan Lin	Weill Cornell Medicine, USA
Charles Lu	Massachusetts Institute of Technology, USA
Yuhui Ma	Chinese Academy of Sciences, China
Tejas Sudharshan Mathai	National Institutes of Health, USA
Felix Meissen	Technische Universität München, Germany
Mingyuan Meng	University of Sydney, Australia
Leo Milecki	CentraleSupelec, France
Marc Modat	King's College London, UK
Tiziano Passerini	Siemens Healthineers, USA
Tomasz Pieciak	Universidad de Valladolid, Spain
Daniel Rueckert	Imperial College London, UK
Julio Silva-Rodríguez	ETS Montreal, Canada
Bingyao Tan	Nanyang Technological University, Singapore
Elias Tappeiner	UMIT - Private University for Health Sciences, Medical Informatics and Technology, Austria
Jocelyne Troccaz	TIMC Lab, Grenoble Alpes University-CNRS, France
Chialing Tsai	Queens College, City University New York, USA
Juan Miguel Valverde	University of Eastern Finland, Finland
Sulaiman Vesal	Stanford University, USA

Wolf-Dieter Vogl RetInSight GmbH, Austria
Vibashan VS Johns Hopkins University, USA
Lin Wang Harbin Engineering University, China
Yan Wang Sichuan University, China
Rhydian Windsor University of Oxford, UK
Ivo Wolf University of Applied Sciences Mannheim, Germany
Linshan Wu Hunan University, China
Xin Yang Chinese University of Hong Kong, China

Contents – Part IV

Image Segmentation II

Category-Level Regularized Unlabeled-to-Labeled Learning for Semi-supervised Prostate Segmentation with Multi-site Unlabeled Data

Zhe Xu[1], Donghuan Lu[2(✉)], Jiangpeng Yan[3], Jinghan Sun[4], Jie Luo[5], Dong Wei[2], Sarah Frisken[5], Quanzheng Li[5], Yefeng Zheng[2], and Raymond Kai-yu Tong[1(✉)]

[1] Department of Biomedical Engineering, The Chinese University of Hong Kong, Hong Kong, China
jackxz@link.cuhk.edu.hk, kytong@cuhk.edu.hk
[2] Tencent Healthcare Co., Jarvis Lab, Shenzhen, China
caleblu@tencent.com
[3] Department of Automation, Tsinghua University, Beijing, China
[4] Xiamen University, Xiamen, China
[5] Harvard Medical School, Boston, MA, USA

Abstract. Segmenting prostate from MRI is crucial for diagnosis and treatment planning of prostate cancer. Given the scarcity of labeled data in medical imaging, semi-supervised learning (SSL) presents an attractive option as it can utilize both limited labeled data and abundant unlabeled data. However, if the local center has limited image collection capability, there may also not be enough unlabeled data for semi-supervised learning to be effective. To overcome this issue, other partner centers can be consulted to help enrich the pool of unlabeled images, but this can result in data heterogeneity, which could hinder SSL that functions under the assumption of consistent data distribution. Tailoring for this important yet under-explored scenario, this work presents a novel Category-level regularized Unlabeled-to-Labeled (CU2L) learning framework for semi-supervised prostate segmentation with multi-site unlabeled MRI data. Specifically, CU2L is built upon the teacher-student architecture with the following tailored learning processes: (i) local pseudo-label learning for reinforcing confirmation of the data distribution of the local center; (ii) category-level regularized non-parametric unlabeled-to-labeled learning for robustly mining shared information by using the limited expert labels to regularize the intra-class features across centers to be discriminative and generalized; (iii) stability learning under perturbations to further enhance robustness to heterogeneity. Our method is evaluated on prostate MRI data from six different clinical centers and shows superior performance compared to other semi-supervised methods.

Keywords: Prostate segmentation · Semi-supervised · Heterogeneity

© The Author(s), under exclusive license to Springer Nature Switzerland AG 2023
H. Greenspan et al. (Eds.): MICCAI 2023, LNCS 14223, pp. 3–13, 2023.
https://doi.org/10.1007/978-3-031-43901-8_1

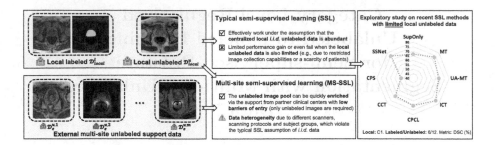

Fig. 1. Comparison between typical semi-supervised learning (SSL) and our focused multi-site semi-supervised learning (MS-SSL), and an exploratory validation on recent SSL methods with limited local unlabeled data.

Table 1. Details of the acquisition protocols and number of scans for the six different centers. Each center supplied T2-weighted MR images of the prostate.

Center	Source	#Scans	Field strength (T)	Resolution (in-plane/through-plane in mm)	Coil	Scanner
C1	RUNMC [1]	30	3	0.6–0.625/3.6–4	Surface	Siemens
C2	BMC [1]	30	1.5	0.4/3	Endorectal	Philips
C3	HCRUDB [4]	19	3	0.67–0.79/1.25	–	Siemens
C4	UCL [5]	13	1.5 and 3	0.325–0.625/3–3.6	–	Siemens
C5	BIDMC [5]	12	3	0.25/2.2–3	Endorectal	GE
C6	HK [5]	12	1.5	0.625/3.6	Endorectal	Siemens

1   Introduction

Prostate segmentation from magnetic resonance imaging (MRI) is a crucial step for diagnosis and treatment planning of prostate cancer. Recently, deep learning-based approaches have greatly improved the accuracy and efficiency of automatic prostate MRI segmentation [7,8]. Yet, their success usually requires a large amount of labeled medical data, which is expensive and expertise-demanding in practice. In this regard, semi-supervised learning (SSL) has emerged as an attractive option as it can leverage both limited labeled data and abundant unlabeled data [3,9–11,15,16,21–26,28]. Nevertheless, the effectiveness of SSL is heavily dependent on the *quantity* and *quality* of the unlabeled data.

Regarding *quantity*, the abundance of unlabeled data serves as a way to regularize the model and alleviate overfitting to the limited labeled data. Unfortunately, such "abundance" may be unobtainable in practice, i.e., the local unlabeled pool is also limited due to restricted image collection capabilities or scarce patient samples. As a specific case shown in Table 1, there are only limited prostate scans available per center. Taking C1 as a case study, if the amount of local unlabeled data is limited, existing SSL methods may still suffer from inferior performance when generalizing to unseen test data (Fig. 1). To efficiently enrich the unlabeled pool, seeking support from other centers is a viable solution, as illustrated in Fig. 1. Yet, due to differences in imaging protocols and variations in patient demographics, this solution usually introduces data heterogeneity, leading to a **quality** problem. Such heterogeneity may impede the performance of SSL which typically assumes that the distributions of labeled data and unlabeled data are independent and identically distributed (i.i.d.) [16]. Thus, proper mechanisms are called for this practical but challenging SSL scenario.

Here, we define this new SSL scenario as multi-site semi-supervised learning (MS-SSL), allowing to enrich the unlabeled pool with multi-site heterogeneous images. Being an under-explored scenario, few efforts have been made. To our best knowledge, the most relevant work is AHDC [2]. However, it only deals with additional unlabeled data from a specific source rather than multiple arbitrary sources. Thus, it intuitively utilizes image-level mapping to minimize dual-distribution discrepancy. Yet, their adversarial min-max optimization often leads to instability and it is difficult to align multiple external sources with the local source using a single image mapping network.

In this work, we propose a more generalized framework called Category-level regularized Unlabeled-to-Labeled (CU2L) learning, as depicted in Fig. 2, to achieve robust MS-SSL for prostate MRI segmentation. Specifically, CU2L is built upon the teacher-student architecture with customized learning strategies for local and external unlabeled data: (i) recognizing the importance of supervised learning in data distribution fitting (which leads to the failure of CPS [3] in MS-SSL as elaborated in Sec. 3), the local unlabeled data is involved into pseudo-label supervised-like learning to reinforce fitting of the local data distribution; (ii) considering that intra-class variance hinders effective MS-SSL, we introduce a non-parametric unlabeled-to-labeled learning scheme, which takes advantage of the scarce expert labels to explicitly constrain the prototype-propagated predictions, to help the model exploit discriminative and domain-insensitive features from heterogeneous multi-site data to support the local center. Yet, observing that such scheme is challenging when significant shifts and various distributions are present, we further propose category-level regularization, which advocates prototype alignment, to regularize the distribution of intra-class features from arbitrary external data to be closer to the local distribution; (iii) based on the fact that perturbations (e.g., Gaussian noises [15]) can be regarded as a simulation of heterogeneity, perturbed stability learning is incorporated to enhance the robustness of the model. Our method is evaluated on prostate MRI data from six different clinical centers and shows promising performance on tackling MS-SSL compared to other semi-supervised methods.

2 Methods

2.1 Problem Formulation and Basic Architecture

In our scenario of MS-SSL, we have access to a local target dataset $\mathcal{D}_{local}$ (consisted of a labeled sub-set $\mathcal{D}_{local}^l$ and an unlabeled sub-set $\mathcal{D}_{local}^u$) and the external unlabeled support datasets $\mathcal{D}_e^u = \bigcup_{j=1}^m \mathcal{D}_e^{u,j}$, where m is the number of support centers. Specifically, $\mathcal{D}_{local}^l = \{(X_{local(i)}^l, Y_{local(i)}^l)\}_{i=1}^{n_l}$ with n_l labeled scans and $\mathcal{D}_{local}^u = \{X_{local(i)}^u\}_{i=n_l+1}^{n_l+n_u}$ with n_u unlabeled scans.

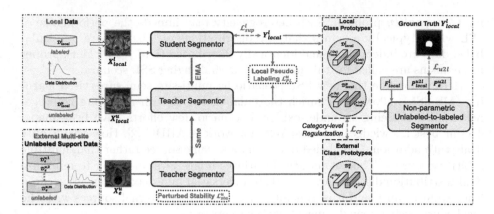

Fig. 2. Illustration of the proposed Category-level Regularized Unlabeled-to-labeled Learning (CU2L) framework. EMA: exponential moving average.

$X_{local(i)}^l, X_{local(i)}^u \in \mathbb{R}^{H \times W \times D}$ denote the scans with height H, width W and depth D, and $Y_{local(i)}^l \in \{0,1\}^{H \times W \times D}$ denotes the label of $X_{local(i)}^l$ (we focus on binary segmentation). Similarly, the j-th external unlabeled support dataset is denoted as $\mathcal{D}_e^{u,j} = \{X_{e(i)}^{u,j}\}_{i=1}^{n_j}$ with n_j unlabeled samples. Considering the large variance on slice thickness among different centers [7,8], our experiments are performed in 2D. Thus, we refer to pixels in the subsequent content. As shown in Fig. 2, our framework is built upon the popular teacher-student framework. Specifically, the student f_θ^s is an in-training model optimized by loss back-propagation as usual while the teacher model $f_{\tilde{\theta}}^t$ is slowly updated with a momentum term that averages previous weights with the current weights, where θ denotes the student's weights and $\tilde{\theta}$ the teacher's weights. $\tilde{\theta}$ is updated by $\tilde{\theta}_t = \alpha\tilde{\theta}_{t-1} + (1-\alpha)\theta_t$ at iteration t, where α is the exponential moving average (EMA) coefficient and empirically set to 0.99 [26]. Compared to the student, the teacher performs self-ensembling by nature which helps smooth out the noise and avoid sudden changes of predictions [15]. Thus, the teacher model is suitable for handling the heterogeneous external images and producing relatively stable pseudo labels (will be used later). As such, our task of MS-SSL can be formulated as optimizing the following loss:

$$\mathcal{L} = \mathcal{L}_{sup}^l\left(\theta, \mathcal{D}_{local}^l\right) + \lambda\mathcal{L}^u(\theta, \tilde{\theta}, \mathcal{D}_{local}^u, \mathcal{D}_e^u), \tag{1}$$

where $\mathcal{L}_{sup}^l$ is the supervised guidance from local labeled data and $\mathcal{L}^u$ denotes the additional guidance from the unlabeled data. λ is a trade-off weight scheduled by the time-dependent ramp-up Gaussian function [15] $\lambda(t) = w_{max} \cdot e^{-5(1-t/t_{max})^2}$, where w_{max} and t_{max} are the maximal weight and iteration, respectively. The key challenge of MS-SSL is the proper design of $\mathcal{L}^u$ for robustly exploiting multi-site unlabeled data $\{\mathcal{D}_{local}^u, \mathcal{D}_e^u\}$ to support the local center.

2.2   Pseudo Labeling for Local Distribution Fitting

As mentioned above, supervised-like learning is advocated for local unlabeled data to help the model fit local distribution better. Owning the self-ensembling property, the teacher model provides relatively stable pseudo labels for the student model. Given the predicted probability map $P^{u,t}_{local}$ of X^u_{local} from the teacher model, the pseudo label $\hat{Y}^{u,t}_{local}$ corresponds to the class with the maximal posterior probability. Yet, with limited local labeled data for training, it is difficult to generate high-quality pseudo labels. Thus, for each pixel, if $\max_c(p^{u,t}_{local}) \geq \delta$, where c denotes the c-th class and δ is a ramp-up threshold ranging from 0.75 to 0.9 as training goes, this pixel will be included in loss calculation. Considering that the cross-entropy loss has been found very sensitive to label noises [18], we adopt the partial Dice loss $\mathcal{L}_{\text{Dice}}$ [27] to perform pseudo label learning, formulated as: $\mathcal{L}^u_{PL} = \frac{1}{K}\sum^K_{k=1}\mathcal{L}_{\text{Dice}}\left(P^{u,s,k}_{local}, \hat{Y}^{u,t,k}_{local}\right)$, where $P^{u,s}_{local}$ denotes the prediction of X^u_{local} from the student model. The Dice loss is calculated for each of the K equally-sized regions of the image, and the final loss is obtained by taking their mean. Such a regional form [6] can help the model better perceive the local discrepancies for fine-grained learning.

2.3   Category-Level Regularized Unlabeled-to-Labeled Learning

Unlabeled-to-Labeled Learning. Inherently, the challenge of MS-SSL stems from intra-class variation, which results from different imaging protocols, disease progress and patient demographics. Inspired by prototypical networks [13,19,25] that compare class prototypes with pixel features to perform segmentation, here, we introduce a non-parametric unlabeled-to-labeled (U2L) learning scheme that utilizes expert labels to explicitly constrain the prototype-propagated predictions. Such design is based on two considerations: (i) a good prototype-propagated prediction requires both compact feature and discriminative prototypes, thus enhancing this prediction can encourage the model to learn in a variation-insensitive manner and focus on the most informative clues; (ii) using expert labels as final guidance can prevent error propagation from pseudo labels. Specifically, we denote the feature map of the external unlabeled image X^u_e before the penultimate convolution in the teacher model as $F^{u,t}_e$. Note that $F^{u,t}_e$ has been upsampled to the same size of X^u_e via bilinear interpolation but with L channels. With the argmax pseudo label $\hat{Y}^{u,t}_e$ and the predicted probability map $P^{u,t}_e$, the object prototype from the external unlabeled data can be computed via confidence-weighted masked average pooling: $c^{u(obj)}_e = \frac{\sum_v\left[\hat{Y}^{u,t,obj}_{e(v)}\cdot P^{u,t,obj}_{e(v)}\cdot F^{u,t}_{e(v)}\right]}{\sum_v\left[\hat{Y}^{u,t,obj}_{e(v)}\cdot P^{u,t,obj}_{e(v)}\right]}$.

Likewise, the background prototype $c^{u(bg)}_e$ can also be obtained. Considering the possible unbalanced sampling of prostate-containing slices, EMA strategy across training steps (with a decay rate of 0.9) is applied for prototype update. Then, as shown in Fig. 2, given the feature map F^l_{local} of the local labeled image X^l_{local} from the in-training student model, we can compare $\{c^{u(obj)}_e, c^{u(bg)}_e\}$ with the features F^l_{local} pixel-by-pixel and obtain the prototype-propagated prediction

P_e^{u2l} for X_{local}^l, formulated as: $P_e^{u2l} = \dfrac{\exp\left(\mathrm{sim}\left(F_{local}^l, c_e^{u(i)}\right)/T\right)}{\sum_{i \in \{obj, bg\}} \exp\left(\mathrm{sim}\left(F_{local}^l, c_e^{u(i)}\right)/T\right)}$, where we use cosine similarity for $\mathrm{sim}(\cdot, \cdot)$ and empirically set the temperature T to 0.05 [19]. Note that a similar procedure can also be applied to the local unlabeled data X_{local}^u, and thus we can obtain another prototype-propagated unlabeled-to-labeled prediction P_{local}^{u2l} for X_{local}^l. As such, given the accurate expert label Y_{local}^l, the unlabeled-to-labeled supervision can be computed as:

$$\mathcal{L}_{u2l} = \frac{1}{K} \left(\sum_{k=1}^{K} \mathcal{L}_{\mathrm{Dice}} \left(P_e^{u2l,k}, Y_{local}^{l,k} \right) + \sum_{k=1}^{K} \mathcal{L}_{\mathrm{Dice}} \left(P_{local}^{u2l,k}, Y_{local}^{l,k} \right) \right). \qquad (2)$$

Category-Level Regularization. Being a challenging scheme itself, the above U2L learning can only handle minor intra-class variation. Thus, proper mechanisms are needed to alleviate the negative impact of significant shift and multiple distributions. Specifically, we introduce category-level regularization, which advocates class prototype alignment between local and external data, to regularize the distribution of intra-class features from arbitrary external data to be closer to the local one, thus reducing the difficulty of U2L learning. In U2L, we have obtained prototypes from local unlabeled data $\{c_{local}^{u(obj)}, c_{local}^{u(bg)}\}$ and external unlabeled data $\{c_e^{u(obj)}, c_e^{u(bg)}\}$. Similarly, the prototypes of object $c_{local}^{l(obj)}$ and background $c_{local}^{l(bg)}$ of the local labeled data can be obtained but using expert labels and student's features. Then, the category-level regularization is formulated as:

$$\mathcal{L}_{cr} = \frac{3}{4} \left[d(c_{local}^{l(obj)}, c_e^{u(obj)}) + d(c_{local}^{u(obj)}, c_e^{u(obj)}) \right] + \frac{1}{4} \left[d(c_{local}^{l(bg)}, c_e^{u(bg)}) + d(c_{local}^{u(bg)}, c_e^{u(bg)}) \right],$$
$$(3)$$

where mean squared error is adopted as the distance function $d(\cdot, \cdot)$. The weight of background prototype alignment is smaller due to less relevant contexts.

Stability Under Perturbations. Although originally designed for typical SSL, encouraging stability under perturbations [26] can also benefit MS-SSL, considering that the perturbations can be regarded as a simulation of heterogeneity and enforcing such perturbed stability can regularize the model behavior for better generalizability. Specifically, for the same unlabeled input $X^u \in \{\mathcal{D}_{local}^u \cup \mathcal{D}_e^u\}$ with different perturbations ξ and ξ' (using the same Gaussian noises as in [26]), we encourage consistent pre-softmax predictions between the teacher and student models, formulated as $\mathcal{L}_{sta}^u = d\left(f_\theta^t(X^u + \xi), f_\theta^s(X^u + \xi') \right)$, where mean squared error is also adopted as the distance function $d(\cdot, \cdot)$.

Overall, the final loss for the multi-site unlabeled data is summarized as:

$$\mathcal{L}^u = \mathcal{L}_{PL}^u(\mathcal{D}_{local}^u) + [\mathcal{L}_{u2l}(\mathcal{D}_{local}, \mathcal{D}_e^u) + \mathcal{L}_{cr}(\mathcal{D}_{local}, \mathcal{D}_e^u)] + \mathcal{L}_{sta}^u(\mathcal{D}_{local}^u, \mathcal{D}_e^u). \quad (4)$$

3   Experiments and Results

Materials. We utilize prostate T2-weighted MR images from six different clinical centers (C1–6) [1,4,5] to perform a retrospective evaluation. Table 1 summa-

Table 2. Quantitative comparison. $*$ indicates $p \leq 0.05$ from the Wilcoxon signed rank test for pairwise comparison with our method. Standard deviations are shown in parentheses. The best mean results are shown in **bold**.

Method	# Scans Used			Local Site: C1		Local Site: C2	
	$\mathcal{D}_{local}^{l}$	$\mathcal{D}_{local}^{u}$	$\mathcal{D}_{e}^{u}$	DSC (%)	Jaccard (%)	DSC (%)	Jaccard (%)
Supervised	6	0	0	66.78 (23.26)*	54.24 (23.74)*	71.19 (16.01)*	57.33 (16.80)*
MT [15]	6	12	86	80.96 (11.15)*	70.14 (14.83)*	77.38 (10.24)*	64.21 (13.19)*
UA-MT [26]	6	12	86	81.86 (13.82)*	70.77 (17.31)*	78.31 (10.34)*	65.47 (13.26)*
ICT [17]	6	12	86	78.52 (16.82)*	67.25 (19.02)*	77.67 (9.22)*	64.38 (11.81)*
CCT [11]	6	12	86	79.95 (17.27)*	69.40 (19.55)*	73.20 (15.20)*	59.79 (17.28)*
FixMatch [14]	6	12	86	77.09 (18.45)*	65.69 (19.94)*	67.82 (14.80)*	53.16 (16.60)*
CPCL [25]	6	12	86	81.40 (14.42)*	71.27 (18.02)*	75.92 (13.59)*	62.96 (16.27)*
CPS [3]	6	12	86	65.02 (23.91)*	52.32 (23.65)*	43.41 (23.39)*	30.32 (17.49)*
SSNet [20]	6	12	86	80.37 (15.15)*	69.43 (17.88)*	75.62 (10.99)*	62.03 (14.06)*
AHDC [2]	6	12	86	79.52 (14.32)*	68.02 (15.34)*	74.65 (12.37)*	60.98 (14.33)*
CU2L (ours)	6	12	86	**85.93** (9.18)	**76.36** (12.87)	**80.29** (10.77)	**68.01** (13.92)
Supervised	8	0	0	69.04 (25.07)*	57.52 (25.34)*	75.86 (10.24)*	62.20 (13.18)*
MT [15]	8	10	86	80.18 (15.47)*	69.22 (17.85)*	77.90 (9.32)*	64.72 (11.99)*
UA-MT [26]	8	10	86	82.42 (12.45)*	71.75 (15.70)*	77.44 (8.94)*	64.02 (11.41)*
ICT [17]	8	10	86	79.88 (13.61)*	68.42 (16.96)*	76.80 (11.84)*	63.65 (13.75)*
CCT [11]	8	10	86	79.29 (14.21)*	67.71 (17.29)*	80.42 (7.60)*	67.31 (10.09)*
FixMatch [14]	8	10	86	80.46 (13.46)*	69.14 (16.29)*	66.17 (20.86)*	52.53 (20.05)*
CPCL [25]	8	10	86	81.34 (14.01)*	70.56 (17.13)*	77.49 (11.20)*	64.53 (13.93)*
CPS [3]	8	10	86	76.17 (15.58)*	63.74 (17.77)*	65.34 (13.53)*	50.04 (15.19)*
SSNet [20]	8	10	86	77.22 (15.09)*	65.19 (18.81)*	78.25 (9.90)*	65.27 (12.33)*
AHDC [2]	8	10	86	78.53 (16.23)*	66.01 (18.45)*	75.12 (11.23)*	61.97 (13.35)*
CU2L (ours)	8	10	86	**86.46** (6.72)	**76.74** (9.97)	**82.30** (9.93)	**70.71** (12.94)
Supervised (upper bound)	18	0	0	89.10 (4.33)	80.76 (6.71)	85.01 (4.35)	74.15 (6.44)

rizes the characteristics of the six data sources, following [7,8], where [7,8] also reveal the severity of inter-center heterogeneity here through extensive experiments. The heterogeneity comes from the differences in scanners, field strengths, coil types, disease and in-plane/through-plane resolution. Compared to C1 and C2, scans from C3 to C6 are taken from patients with prostate cancer, either for detection or staging purposes, which can cause inherent semantic differences in the prostate region to further aggravate heterogeneity. Following [7,8], we crop each scan to preserve the slices with the prostate region only and then resize and normalize it to 384×384 px in the axial plane with zero mean and unit variance. We take C1 or C2 as the local target center and randomly divide their 30 scans into 18, 3, and 9 samples as training, validation, and test sets, respectively.

Implementation and Evaluation Metrics. The framework is implemented on PyTorch using an NVIDIA GeForce RTX 3090 GPU. Considering the large variance in slice thickness among different centers, we adopt the 2D architecture. Specifically, 2D U-Net [12] is adopted as our backbone. The input patch size is set to 384×384, and the batch size is set to 36 including 12 labeled local slices, 12 unlabeled local slices and 12 unlabeled external slices. The supervised loss $\mathcal{L}_{sup}^{l}$

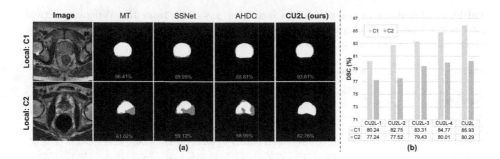

Fig. 3. (a) Exemplar results under the setting with 6 labeled scans. Grey color indicates the mismatch between the prediction and the ground truth. (b) Ablation results. (Color figure online)

consists of the cross-entropy loss and the K-regional Dice loss [6]. The maximum consistency weight w_{max} is set to 0.1 [20, 26]. $t_{\max}$ is set to 20,000. K is empirically set to 2. The network is trained using the SGD optimizer and the learning rate is initialized as 0.01 and decayed by multiplication with $(1.0 - t/t_{\max})^{0.9}$. Data augmentation is applied, including random flip and rotation. We adopt the Dice similarity coefficient (DSC) and Jaccard as the evaluation metrics and the results are the average over three runs with different seeds.

Comparison Study. Table 2 presents the quantitative results with either C1 or C2 as the local target center, wherein only 6 or 8 local scans are annotated. Besides the supervised-only baselines, we include recent top-performing SSL methods [2, 3, 11, 14, 15, 17, 20, 25, 26] for comparison. All methods are implemented with the same backbone and training protocols to ensure fairness. As observed, compared to the supervised-only baselines, our CU2L with {6, 8} local labeled scans achieves {19.15%, 17.42%} and {9.1%, 6.44%} DSC improvements in {C1, C2}, showing its effectiveness in leveraging multi-site unlabeled data. Despite the violation of the assumption of i.i.d. data, existing SSL methods can still benefit from the external unlabeled data to some extent compared to the results using local data only as shown in Fig. 1, revealing that the quantity of unlabeled data has a significant impact. However, due to the lack of proper mechanisms for learning from heterogeneous data, limited improvement can be achieved by them, especially for CPS [3] and FixMatch [14] in C2. Particularly, CPS relies on cross-modal pseudo labeling which exploits all the unlabeled data in a supervised-like fashion. We attribute its degradation to the fact that supervised learning is crucial for distribution fitting, which supports our motivation of performing pseudo-label learning on local unlabeled data only. As a result, its models struggle to determine which distribution to prioritize. Meanwhile, the most relevant AHDC [2] is mediocre in MS-SSL, mainly due to the instability of adversarial training and the difficulty of aligning multiple distributions to the local distribution via a single image-mapping network. In contrast, with specialized mechanisms for simultaneously learning informative representations from multi-site data and handling heterogeneity, our CU2L obtains the best performance over the recent SSL methods. Figure 3(a) further shows that the predictions of our method fit more accurately with the ground truth.

Ablation Study. To evaluate the effectiveness of each component, we conduct an ablation study under the setting with 6 local labeled scans, as shown in Fig. 2(b). Firstly, when we remove $\mathcal{L}^u_{PL}$ (CU2L-1), the performance drops by {5.69% (C1), 3.05%(C2)} in DSC, showing that reinforcing confirmation on local distribution is critical. CU2L-2 represents the removal of both $\mathcal{L}_{u2l}$ and $\mathcal{L}_{cr}$, and it can be observed that such an unlabeled-to-labeled learning approach combined with class-level regularization is crucial for exploring multi-site data. If we remove $\mathcal{L}_{cr}$ which accompanies with $\mathcal{L}_{u2l}$ (CU2L-3), the performance degrades, which justifies the necessity of this regularization to reduce the difficulty of unlabeled-to-labeled learning process. CU2L-4 denotes the removal of $\mathcal{L}^u_{sta}$. As observed, such a typical stability loss [15] can further improve the performance by introducing hand-crafted noises to enhance the robustness to real-world heterogeneity.

4 Conclusion

In this work, we presented a novel Category-level regularized Unlabeled-to-Labeled (CU2L) learning framework for semi-supervised prostate segmentation with multi-site unlabeled MRI data. CU2L robustly exploits multi-site unlabeled data via three tailored schemes: local pseudo-label learning for better local distribution fitting, category-level regularized unlabeled-to-labeled learning for exploiting the external data in a distribution-insensitive manner and stability learning for further enhancing robustness to heterogeneity. We evaluated our method on prostate MRI data from six different clinical centers and demonstrated its superior performance compared to other semi-supervised methods.

Acknowledgement. This research was done with Tencent Jarvis Lab and Tencent Healthcare (Shenzhen) Co., LTD and supported by General Research Fund from Research Grant Council of Hong Kong (No. 14205419) and the National Key R&D Program of China (No. 2020AAA0109500 and No. 2020AAA0109501).

References

1. Bloch, N., et al.: NCI-ISBI 2013 challenge: automated segmentation of prostate structures. The Cancer Imaging Arch. **370**(6), 5 (2015)
2. Chen, J., et al.: Adaptive hierarchical dual consistency for semi-supervised left atrium segmentation on cross-domain data. IEEE Trans. Med. Imaging **41**(2), 420–433 (2021)
3. Chen, X., Yuan, Y., Zeng, G., Wang, J.: Semi-supervised semantic segmentation with cross pseudo supervision. In: Proceedings of the IEEE/CVF Conference on Computer Vision and Pattern Recognition, pp. 2613–2622 (2021)

4. Lemaître, G., Martí, R., Freixenet, J., Vilanova, J.C., Walker, P.M., Meriaudeau, F.: Computer-aided detection and diagnosis for prostate cancer based on mono and multi-parametric MRI: a review. Comput. Biol. Med. **60**, 8–31 (2015)
5. Litjens, G., et al.: Evaluation of prostate segmentation algorithms for MRI: the PROMISE12 challenge. Med. Image Anal. **18**(2), 359–373 (2014)
6. Liu, J., Desrosiers, C., Zhou, Y.: Semi-supervised medical image segmentation using cross-model pseudo-supervision with shape awareness and local context constraints. In: Wang, L., Dou, Q., Fletcher, P.T., Speidel, S., Li, S. (eds.) MICCAI 2022. LNCS, vol. 13438, pp. 140–150. Springer, Cham (2022). https://doi.org/10.1007/978-3-031-16452-1_14
7. Liu, Q., Dou, Q., Heng, P.-A.: Shape-aware meta-learning for generalizing prostate MRI segmentation to unseen domains. In: Martel, A.L., et al. (eds.) MICCAI 2020. LNCS, vol. 12262, pp. 475–485. Springer, Cham (2020). https://doi.org/10.1007/978-3-030-59713-9_46
8. Liu, Q., Dou, Q., Yu, L., Heng, P.A.: MS-net: multi-site network for improving prostate segmentation with heterogeneous MRI data. IEEE Trans. Med. Imaging **39**(9), 2713–2724 (2020)
9. Luo, X., Chen, J., Song, T., Chen, Y., Wang, G., Zhang, S.: Semi-supervised medical image segmentation through dual-task consistency. In: AAAI Conference on Artificial Intelligence (2021)
10. Luo, X., et al.: Efficient semi-supervised gross target volume of nasopharyngeal carcinoma segmentation via uncertainty rectified pyramid consistency. In: de Bruijne, M., et al. (eds.) MICCAI 2021. LNCS, vol. 12902, pp. 318–329. Springer, Cham (2021). https://doi.org/10.1007/978-3-030-87196-3_30
11. Ouali, Y., Hudelot, C., Tami, M.: Semi-supervised semantic segmentation with cross-consistency training. In: Proceedings of the IEEE/CVF Conference on Computer Vision and Pattern Recognition, pp. 12674–12684 (2020)
12. Ronneberger, O., Fischer, P., Brox, T.: U-net: convolutional networks for biomedical image segmentation. In: Navab, N., Hornegger, J., Wells, W.M., Frangi, A.F. (eds.) MICCAI 2015. LNCS, vol. 9351, pp. 234–241. Springer, Cham (2015). https://doi.org/10.1007/978-3-319-24574-4_28
13. Snell, J., Swersky, K., Zemel, R.: Prototypical networks for few-shot learning. In: Advances in Neural Information Processing Systems, vol. 30, pp. 4080–4090 (2017)
14. Sohn, K., et al.: FixMatch: simplifying semi-supervised learning with consistency and confidence. arXiv preprint arXiv:2001.07685 (2020)
15. Tarvainen, A., Valpola, H.: Mean teachers are better role models: weight-averaged consistency targets improve semi-supervised deep learning results. In: Advances in Neural Information Processing Systems, pp. 1195–1204 (2017)
16. Van Engelen, J.E., Hoos, H.H.: A survey on semi-supervised learning. Mach. Learn. **109**(2), 373–440 (2020)
17. Verma, V., et al.: Interpolation consistency training for semi-supervised learning. Neural Netw. **145**, 90–106 (2022)
18. Wang, G., et al.: A noise-robust framework for automatic segmentation of COVID-19 pneumonia lesions from CT images. IEEE Trans. Med. Imaging **39**(8), 2653–2663 (2020)
19. Wang, K., Liew, J.H., Zou, Y., Zhou, D., Feng, J.: PANet: few-shot image semantic segmentation with prototype alignment. In: Proceedings of the IEEE/CVF International Conference on Computer Vision, pp. 9197–9206 (2019)

20. Wu, Y., Wu, Z., Wu, Q., Ge, Z., Cai, J.: Exploring smoothness and class-separation for semi-supervised medical image segmentation. In: Wang, L., Dou, Q., Fletcher, P.T., Speidel, S., Li, S. (eds.) MICCAI 2022. LNCS, vol. 13435, pp. 34–43. Springer, Cham (2022). https://doi.org/10.1007/978-3-031-16443-9_4
21. Wu, Y., Xu, M., Ge, Z., Cai, J., Zhang, L.: Semi-supervised left atrium segmentation with mutual consistency training. In: de Bruijne, M., et al. (eds.) MICCAI 2021. LNCS, vol. 12902, pp. 297–306. Springer, Cham (2021). https://doi.org/10.1007/978-3-030-87196-3_28
22. Xu, Z., et al.: Anti-interference from noisy labels: Mean-teacher-assisted confident learning for medical image segmentation. IEEE Trans. Med. Imaging **41**, 3062–3073 (2022)
23. Xu, Z., et al.: Noisy labels are treasure: mean-teacher-assisted confident learning for hepatic vessel segmentation. In: de Bruijne, M., et al. (eds.) MICCAI 2021. LNCS, vol. 12901, pp. 3–13. Springer, Cham (2021). https://doi.org/10.1007/978-3-030-87193-2_1
24. Xu, Z., et al.: Ambiguity-selective consistency regularization for mean-teacher semi-supervised medical image segmentation. Med. Image Anal. **88**, 102880 (2023)
25. Xu, Z., et al.: All-around real label supervision: cyclic prototype consistency learning for semi-supervised medical image segmentation. IEEE J. Biomed. Health Inform. **26**(7), 3174–3184 (2022)
26. Yu, L., Wang, S., Li, X., Fu, C.-W., Heng, P.-A.: Uncertainty-aware self-ensembling model for semi-supervised 3D left atrium segmentation. In: Shen, D., et al. (eds.) MICCAI 2019. LNCS, vol. 11765, pp. 605–613. Springer, Cham (2019). https://doi.org/10.1007/978-3-030-32245-8_67
27. Zhai, S., Wang, G., Luo, X., Yue, Q., Li, K., Zhang, S.: PA-seg: learning from point annotations for 3D medical image segmentation using contextual regularization and cross knowledge distillation. IEEE Trans. Med. Imaging **42**(8), 2235–2246 (2023). https://doi.org/10.1109/TMI.2023.3245068
28. Zhang, W., et al.: BoostMIS: boosting medical image semi-supervised learning with adaptive pseudo labeling and informative active annotation. In: Proceedings of the IEEE/CVF Conference on Computer Vision and Pattern Recognition, pp. 20666–20676 (2022)

Devil is in Channels: Contrastive Single Domain Generalization for Medical Image Segmentation

Shishuai Hu, Zehui Liao, and Yong Xia[✉]

National Engineering Laboratory for Integrated Aero-Space-Ground-Ocean Big Data Application Technology, School of Computer Science and Engineering, Northwestern Polytechnical University, Xi'an 710072, China
yxia@nwpu.edu.cn

Abstract. Deep learning-based medical image segmentation models suffer from performance degradation when deployed to a new healthcare center. To address this issue, unsupervised domain adaptation and multi-source domain generalization methods have been proposed, which, however, are less favorable for clinical practice due to the cost of acquiring target-domain data and the privacy concerns associated with redistributing the data from multiple source domains. In this paper, we propose a **C**hannel-level **C**ontrastive **S**ingle **D**omain **G**eneralization (C^2SDG) model for medical image segmentation. In C^2SDG, the shallower features of each image and its style-augmented counterpart are extracted and used for contrastive training, resulting in the disentangled style representations and structure representations. The segmentation is performed based solely on the structure representations. Our method is novel in the contrastive perspective that enables channel-wise feature disentanglement using a single source domain. We evaluated C^2SDG against six SDG methods on a multi-domain joint optic cup and optic disc segmentation benchmark. Our results suggest the effectiveness of each module in C^2SDG and also indicate that C^2SDG outperforms the baseline and all competing methods with a large margin. The code is available at https://github.com/ShishuaiHu/CCSDG.

Keywords: Single domain generalization · Medical image segmentation · Contrastive learning · Feature disentanglement

1 Introduction

It has been widely recognized that the success of supervised learning approaches, such as deep learning, relies on the i.i.d. assumption for both training and test samples [11]. This assumption, however, is less likely to be held on medical image segmentation tasks due to the imaging distribution discrepancy caused by

---

Supplementary Information The online version contains supplementary material available at https://doi.org/10.1007/978-3-031-43901-8_2.

H. Greenspan et al. (Eds.): MICCAI 2023, LNCS 14223, pp. 14–23, 2023.
https://doi.org/10.1007/978-3-031-43901-8_2

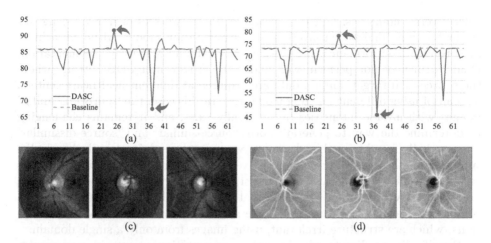

Fig. 1. Average OD (a) and OC (b) segmentation performance (Dice%) obtained on unseen target domain (BASE2) versus removed channel of shallow features. The Dice scores obtained before and after dropping a channel are denoted by 'Baseline' and 'DASC', respectively. The 24th channel (c) and 36th channel (d) obtained on three target-domain images are visualized.

non-uniform characteristics of the imaging equipment, inconsistent skills of the operators, and even compromise with factors such as patient radiation exposure and imaging time [14]. Therefore, the imaging distribution discrepancy across different healthcare centers renders a major hurdle that prevents deep learning-based medical image segmentation models from clinical deployment [7,18].

To address this issue, unsupervised domain adaptation (UDA) [8,17] and multi-source domain generalization (MSDG) [10,16] have been studied. UDA needs access to the data from source domain(s) and unlabeled target domain, while MSDG needs access to the data from multiple source domains. In clinical practice, both settings are difficult to achieve, considering the cost of acquiring target-domain data and the privacy concerns associated with redistributing the data from multiple source domains [9,22].

By contrast, single domain generalization (SDG) [2,13,15,19,22,23] is a more practical setting, under which only the labeled data from one source domain are used to train the segmentation model, which is thereafter applied to the unseen target-domain data. The difficulty of SDG is that, due to the existence of imaging distribution discrepancy, the trained segmentation model is prone to overfit the source-domain data but generalizes poorly on target-domain data. An intuitive solution is to increase the diversity of training data by performing data augmentation at the image-level [13,15,19,21]. This solution has recently been demonstrated to be less effective than a more comprehensive one, *i.e.*, conducting domain adaptation on both image- and feature-levels [2,8,12]. As a more comprehensive solution, Dual-Norm [23] first augments source-domain images into 'source-similar' images with similar intensities and 'source-dissimilar' images with inverse intensities, and then processes these two sets of images using different batch normalization layers in the segmentation model. Although achieving promising performance in cross-modality CT and MR image segmentation, Dual-Norm may not perform well under the cross-center SDG setting, where the source- and target-domain data are acquired at different healthcare centers, instead of using different imaging modalities. In this case, the 'source-dissimilar' images with inverse intensities do not really exist, and it remains challenging to determine the way to generate both 'source-similar' and 'source-dissimilar' images [1,4]. To address this challenge, we suggest resolving 'similar' and 'dissimilar' from the perspective of contrastive learning. Given a source image and its style-augmented counterpart, only the structure representations between them are 'similar', whereas their style representations should be 'dissimilar'. Based on contrastive learning, we can disentangle and then discard the style representations, which are structure-irrelevant, using images from only a single domain.

Specifically, to disentangle the style representations, we train a segmentation model, *i.e.*, the baseline, using single domain data and assess the impact of the features extracted by the first convolutional layer on the segmentation performance, since shallower features are believed to hold more style-sensitive information [8,18]. A typical example was given in Fig. 1(a) and (b), where the green line is the average Dice score obtained on the target domain (the BASE2 dataset) versus the index of the feature channel that has been dropped. It reveals that, in most cases, removing a feature does not affect the model performance, indicating that the removed feature is redundant. For instance, the performance even increases slightly after removing the 24th channel. This observation is consistent with the conclusion that there exists a sub-network that can achieve comparable performance [6]. On the contrary, it also shows that some features, such as the 36th channel, are extremely critical. Removing this feature results in a significant performance drop. We visualize the 24th and 36th channels obtained on three target-domain images in Fig. 1(c) and (d), respectively. It shows that the 36th channel is relatively 'clean' and most structures are visible on it, whereas the 24th channel contains a lot of 'shadows'. The poor quality of the 24th channel can be attributed to the fact that the styles of source- and target-domain images are different and the style representation ability learned on source-domain images cannot generalize well on target-domain images. Therefore, we suggest that the 24th channel is more style-sensitive, whereas the 36th channel contains more structure information. This phenomenon demonstrates that 'the devil is in channels'. Fortunately, contrastive learning provides us a promising way to identify and expel those style-sensitive 'devil' channels from the extracted image features.

In this paper, we incorporate contrastive feature disentanglement into a segmentation backbone and thus propose a novel SDG method called **C**hannel-level **C**ontrastive **S**ingle **D**omain **G**eneralization ($\mathbf{C^2SDG}$) for joint optic cup (OC) and optic disc (OD) segmentation on fundus images. In C^2SDG, the shallower features of each image and its style-augmented counterpart are extracted and used for contrastive training, resulting in the disentangled style representations

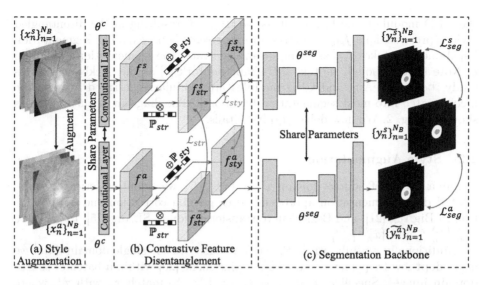

Fig. 2. Diagram of our C^2SDG. The rectangles in blue and green represent the convolutional layer and the segmentation backbone, respectively. The cubes represent different features. The projectors with parameters θ^p in (b) are omitted for simplicity. (Color figure online)

and structure representations. The segmentation is performed based solely on the structure representations. This method has been evaluated against other SDG methods on a public dataset and improved performance has been achieved. Our main contributions are three-fold: (1) we propose a novel contrastive perspective for SDG, enabling contrastive feature disentanglement using the data from only a single domain; (2) we disentangle the style representations and structure representations explicitly and channel-wisely, and then diminish the impact of style-sensitive 'devil' channels; and (3) our C^2SDG outperforms the baseline and six state-of-the-art SDG methods on the joint OC/OD segmentation benchmark.

2 Method

2.1 Problem Definition and Method Overview

Let the source domain be denoted by $\mathcal{D}^s = \{x_i^s, y_i^s\}_{i=1}^{N_s}$, where x_i^s is the i-th source domain image, and y_i^s is its segmentation mask. Our goal is to train a segmentation model $F_\theta : x \rightarrow y$ on $\mathcal{D}^s$, which can generalize well to an unseen target domain $\mathcal{D}^t = \{x_i^t\}_{i=1}^{N_t}$. The proposed C^2SDG mainly consists of a segmentation backbone, a style augmentation (StyleAug) module, and a contrastive feature disentanglement (CFD) module. For each image x^s, the StyleAug module generates its style-augmented counterpart x^a, which shares the same structure but different style to x^s. Then a convolutional layer extracts high-dimensional representations f^s and f^a from x^s and x^a. After that, f^s and f^a are fed to the CFD module to perform contrastive training, resulting in the disentangled style representations f_{sty} and structure representations f_{str}. The segmentation backbone only takes f_{str} as its input and generates the segmentation prediction $\tilde{y}$. Note that, although we take a U-shape network [5] as the backbone for this study, both StyleAug and CFD modules are modularly designed and can be incorporated into most segmentation backbones. The diagram of our C²SDG is shown in Fig. 2. We now delve into its details.

2.2 Style Augmentation

Given a batch of source domain data $\{x_n^s\}_{n=1}^{N_B}$, we adopt a series of style-related data augmentation approaches, i.e., gamma correction and noise addition in BigAug [21], and Bezier curve transformation in SLAug [15], to generate $\{x_n^{BA}\}_{n=1}^{N_B}$ and $\{x_n^{SL}\}_{n=1}^{N_B}$.

Additionally, to fully utilize the style diversity inside single domain data, we also adopt low-frequency components replacement [20] within a batch of source domain images. Specifically, We reverse $\{x_n^s\}_{n=1}^{N_B}$ to match x_n^s with x_r^s, where $r = N_B + 1 - n$ to ensure x_r^s provides a different reference style. Then we transform x_n^s and x_r^s to the frequency domain and exchange their low-frequency components $Low(Amp(x^s); \beta)$ in the amplitude map, where β is the cut-off ratio between low and high-frequency components and is randomly selected from (0.05, 0.15]. After that, we recover all low-frequency exchanged images to generate $\{x_n^{FR}\}_{n=1}^{N_B}$.

The style-augmented images batch $\{x_n^a\}_{n=1}^{N_B}$ is set to $\{x_n^{BA}\}_{n=1}^{N_B}$, $\{x_n^{SL}\}_{n=1}^{N_B}$, and $\{x_n^{FR}\}_{n=1}^{N_B}$ in turn to perform contrastive training and segmentation.

2.3 Contrastive Feature Disentanglement

Given x^s and x^a, we use a convolutional layer with parameter θ^c to generate their shallow features f^s and f^a, which are 64-channel feature maps for this study.

Then we use a channel mask prompt $\mathbb{P} \in \mathbb{R}^{2 \times 64}$ to disentangle each shallow feature map f into style representation f_{sty} and structure representation f_{str} explicitly channel-wisely

$$\begin{cases} f_{sty} = f \times \mathbb{P}_{sty} = f \times SM(\frac{\mathbb{P}}{\tau})_1 \\ f_{str} = f \times \mathbb{P}_{str} = f \times SM(\frac{\mathbb{P}}{\tau})_2, \end{cases} \tag{1}$$

where $SM(\cdot)$ is a softmax function, the subscript i denotes i-th channel, and $\tau = 0.1$ is a temperature factor that encourages $\mathbb{P}_{sty}$ and $\mathbb{P}_{str}$ to be binary-element vectors, i.e., approximately belonging to $\{0, 1\}^{64}$.

After channel-wise feature disentanglement, we have $\{f_{sty}^s, f_{str}^s\}$ from x^s and $\{f_{sty}^a, f_{str}^a\}$ from x^a. It is expected that (a) f_{sty}^s and f_{sty}^a are different since we want to identify them as the style-sensitive 'devil' channels, and (b) f_{str}^s and f_{str}^a are the same since we want to identify them as the style-irrelevant channels and x^s and x^a share the same structure. Therefore, we design two contrastive loss functions $\mathcal{L}_{sty}$ and $\mathcal{L}_{str}$

$$\begin{cases} \mathcal{L}_{str} = \sum |Proj(f_{str}^s) - Proj(f_{str}^a)| \\ \mathcal{L}_{sty} = -\sum |Proj(f_{sty}^s) - Proj(f_{sty}^a)|, \end{cases} \tag{2}$$

where the $Proj(\cdot)$ with parameters θ^p reduces the dimension of f_{str} and f_{sty}.

Only f_{str}^s and f_{str}^a are fed to the segmentation backbone with parameters θ^{seg} to generate the segmentation predictions $\widetilde{y^s}$ and $\widetilde{y^a}$.

2.4   Training and Inference

Training. For the segmentation task, we treat OC/OD segmentation as two binary segmentation tasks and adopt the binary cross-entropy loss as our objective

$$\mathcal{L}_{ce}(y, \widetilde{y}) = -(\widetilde{y} \log y + (1 - \widetilde{y}) \log (1 - y)) \tag{3}$$

where y represents the segmentation ground truth and $\widetilde{y}$ is the prediction. The total segmentation loss can be calculated as

$$\mathcal{L}_{seg} = \mathcal{L}_{seg}^s + \mathcal{L}_{seg}^a = \mathcal{L}_{ce}(y^s, \widetilde{y^s}) + \mathcal{L}_{ce}(y^s, \widetilde{y^a}). \tag{4}$$

During training, we alternately minimize $\mathcal{L}_{seg}$ to optimize $\{\theta^c, \mathbb{P}, \theta^{seg}\}$, and minimize $\mathcal{L}_{str} + \mathcal{L}_{sty}$ to optimize $\{\mathbb{P}, \theta^p\}$.

Inference. Given a test image x^t, its shallow feature map f^t can be extracted by the first convolutional layer. Based on f^t, the optimized channel mask prompt $\mathbb{P}$ can separate it into f_{sty}^t and f_{str}^t. Only f_{str}^t is fed to the segmentation backbone to generate the segmentation prediction $\widetilde{y^t}$.

3   Experiments and Results

Materials and Evaluation Metrics. The multi-domain joint OC/OD segmentation dataset RIGA+ [1,4,8] was used for this study. It contains annotated fundus images from five domains, including 195 images from BinRushed, 95 images from Magrabia, 173 images from BASE1, 148 images from BASE2, and 133 images from BASE3. Each image was annotated by six raters, and only the first rater's annotations were used in our experiments. We chose BinRushed and Magrabia, respectively, as the source domain to train the segmentation model, and evaluated the model on the other three (target) domains. We adopted the Dice Similarity Coefficient (D, %) to measure the segmentation performance.

Implementation Details. The images were center-cropped and normalized by subtracting the mean and dividing by the standard deviation. The input batch contains eight images of size 512×512. The U-shape segmentation network, whose encoder is a modified ResNet-34, was adopted as the segmentation backbone of our C^2SDG and all competing methods for a fair comparison. The

Table 1. Average performance of three trials of our C^2SDG and six competing methods in joint OC/OD segmentation using BinRushed (row 2–row 9) and Magrabia (row 10–row 17) as source domain, respectively. Their standard deviations are reported as subscripts. The performance of 'Intra-Domain' and 'w/o SDG' is displayed for reference. The best results except for 'Intra-Domain' are highlighted in blue.

Methods	BASE1		BASE2		BASE3		Average	
	D_{OD}	D_{OC}	D_{OD}	D_{OC}	D_{OD}	D_{OC}	D_{OD}	D_{OC}
Intra-Domain	$94.71_{0.07}$	$84.07_{0.35}$	$94.84_{0.18}$	$86.32_{0.14}$	$95.40_{0.05}$	$87.34_{0.11}$	94.98	85.91
w/o SDG	$91.82_{0.54}$	$77.71_{0.88}$	$79.78_{2.10}$	$65.18_{3.24}$	$88.83_{2.15}$	$75.29_{3.23}$	86.81	72.73
BigAug [23]	$94.01_{0.34}$	$81.51_{0.58}$	$85.81_{0.68}$	$71.12_{1.64}$	$92.19_{0.51}$	$79.75_{1.44}$	90.67	77.46
CISDG [13]	$93.56_{0.13}$	$81.00_{1.01}$	$94.38_{0.23}$	$83.79_{0.58}$	$93.87_{0.03}$	$83.75_{0.89}$	93.93	82.85
ADS [19]	$94.07_{0.29}$	$79.60_{5.06}$	$94.29_{0.38}$	$81.17_{3.72}$	$93.64_{0.28}$	$81.08_{4.97}$	94.00	80.62
MaxStyle [2]	$94.28_{0.14}$	$82.61_{0.67}$	$86.65_{0.76}$	$74.71_{2.07}$	$92.36_{0.39}$	$82.33_{1.24}$	91.09	79.88
SLAug [15]	$95.28_{0.12}$	$83.31_{1.10}$	$95.49_{0.16}$	$81.36_{2.51}$	$95.57_{0.06}$	$84.38_{1.39}$	95.45	83.02
Dual-Norm [23]	$94.57_{0.10}$	$81.81_{0.76}$	$93.67_{0.11}$	$79.16_{1.80}$	$94.82_{0.28}$	$83.67_{0.60}$	94.35	81.55
Ours	$95.73_{0.08}$	$86.13_{0.07}$	$95.73_{0.09}$	$86.82_{0.58}$	$95.45_{0.04}$	$86.77_{0.19}$	95.64	86.57
w/o SDG	$89.98_{0.54}$	$77.21_{1.15}$	$85.32_{1.79}$	$73.51_{0.67}$	$90.03_{0.27}$	$80.71_{0.63}$	88.44	77.15
BigAug [23]	$92.32_{0.13}$	$79.68_{0.38}$	$88.24_{0.82}$	$76.69_{0.37}$	$91.35_{0.14}$	$81.43_{0.78}$	90.64	79.27
CISDG [13]	$89.67_{0.76}$	$75.39_{3.22}$	$87.97_{1.04}$	$76.44_{3.48}$	$89.91_{0.64}$	$81.35_{2.81}$	89.18	77.73
ADS [19]	$90.75_{2.42}$	$77.78_{4.23}$	$90.37_{2.07}$	$79.60_{3.34}$	$90.34_{2.93}$	$79.99_{4.02}$	90.48	79.12
MaxStyle [2]	$91.63_{0.12}$	$78.74_{1.95}$	$90.61_{0.45}$	$80.12_{0.90}$	$91.22_{0.07}$	$81.90_{1.14}$	91.15	80.25
SLAug [15]	$93.08_{0.17}$	$80.70_{0.35}$	$92.70_{0.12}$	$80.15_{0.43}$	$92.23_{0.16}$	$80.89_{0.14}$	92.67	80.58
Dual-Norm [23]	$92.35_{0.37}$	$79.02_{0.39}$	$91.23_{0.29}$	$80.06_{0.26}$	$92.09_{0.28}$	$79.87_{0.25}$	91.89	79.65
Ours	$94.78_{0.03}$	$84.94_{0.36}$	$95.16_{0.09}$	$85.68_{0.28}$	$95.00_{0.09}$	$85.98_{0.29}$	94.98	85.53

projector in our CFD module contains a convolutional layer followed by a batch normalization layer, a max pooling layer, and a fully connected layer to convert f_{sty} and f_{str} to 1024-dimensional vectors. The SGD algorithm with a momentum of 0.99 was adopted as the optimizer. The initial learning rate was set to $lr_0 = 0.01$ and decayed according to $lr = lr_0 \times (1-e/E)^{0.9}$, where e is the current epoch and $E = 100$ is the maximum epoch. All experiments were implemented using the PyTorch framework and performed with one NVIDIA 2080Ti GPU.

Comparative Experiments. We compared our C^2SDG with two baselines, including 'Intra-Domain' (*i.e.*, training and testing on the data from the same target domain using 3-fold cross-validation) and 'w/o SDG' (*i.e.*, training on the source domain and testing on the target domain), and six SDG methods, including BigAug [21], CISDG [13], ADS [19], MaxStyle [2], SLAug [15], and Dual-Norm [23]. In each experiment, only one source domain is used for training, ensuring that only the data from a single source domain can be accessed during training. For a fair comparison, all competing methods are re-implemented using the same backbone as our C^2SDG based on their published code and paper. The results of C^2SDG and its competitors were given in Table 1. It shows that C^2SDG improves the performance of 'w/o SDG' with a large margin and outperforms all competing SDG methods. We also visualize the segmentation predictions generated by our C^2SDG and six competing methods in Fig. 3. It reveals that our C^2SDG can produce the most accurate segmentation map.

Ablation Analysis. To evaluate the effectiveness of low-frequency components replacement (FR) in StyleAug and CFD, we conducted ablation experiments using BinRushed and Magrabia as the source domain, respectively. The average performance is shown in Table 2. The performance of using both BigAug and SLAug is displayed as 'Baseline'. It reveals that both FR and CFD contribute to performance gains.

Analysis of CFD. Our CFD is modularly designed and can be incorporated into other SDG methods. We inserted our CFD to ADS [19] and SLAug [15], respectively. The performance of these two approaches and their variants, denoted as C^2-ADS and C^2-SLAug, was shown in Table 3. It reveals that our CFD module can boost their ability to disentangle structure representations and improve the segmentation performance on the target domain effectively. We also adopted 'Ours w/o CFD' as 'Baseline' and compared the channel-level contrastive feature disentanglement strategy with the adversarial training strategy and channel-level

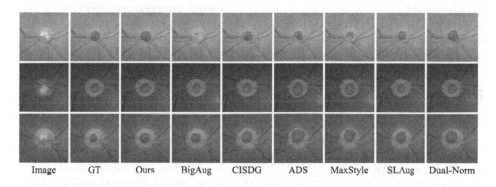

| Image | GT | Ours | BigAug | CISDG | ADS | MaxStyle | SLAug | Dual-Norm |

Fig. 3. Visualization of segmentation masks predicted by our C^2SDG and six competing methods, together with ground truth.

Table 2. Average performance of our C^2SDG and three variants.

Methods	Average	
	D_{OD}	D_{OC}
Baseline	94.13	81.62
w/o FR	95.07	84.90
w/o CFD	95.07	84.83
Ours	95.31	86.05

Table 3. Average performance of ADS, SLAug, and their two variants.

Methods	Average	
	D_{OD}	D_{OC}
ADS [19]	92.24	79.87
C^2-ADS	93.76	81.35
SLAug [15]	94.06	81.80
C^2-SLAug	94.24	83.68

Table 4. Average performance of using contrastive and other strategies.

Methods	Average	
	D_{OD}	D_{OC}
Baseline	95.07	84.83
Dropout	95.14	84.95
Adversarial	90.27	78.47
Ours	95.31	86.05

dropout (see Table 4). It shows that the adversarial training strategy fails to perform channel-level feature disentanglement, due to the limited training data [3] for SDG. Nonetheless, our channel-level contrastive learning strategy achieves the best performance compared to other strategies, further confirming the effectiveness of our CFD module.

4   Conclusion

In this paper, we propose a novel SDG method called C^2SDG for medical image segmentation. In C^2SDG, the StyleAug module generates style-augmented counterpart of each source domain image and enables contrastive learning, the CFD module performs channel-level style and structure representations disentanglement via optimizing a channel prompt $\mathbb{P}$, and the segmentation is performed based solely on structure representations. Our results on a multi-domain joint OC/OD segmentation benchmark indicate the effectiveness of StyleAug and CFD and also suggest that our C^2SDG outperforms the baselines and six completing SDG methods with a large margin.

Acknowledgement. This work was supported in part by the National Natural Science Foundation of China under Grant 62171377, in part by the Key Technologies Research and Development Program under Grant 2022YFC2009903/2022YFC2009900, in part by the Key Research and Development Program of Shaanxi Province, China, under Grant 2022GY-084, in part by the Innovation Foundation for Doctor Dissertation of Northwestern Polytechnical University under Grant CX2023016.

References

1. Almazroa, A., et al.: Retinal fundus images for glaucoma analysis: the RIGA dataset. In: Medical Imaging 2018: Imaging Informatics for Healthcare, Research, and Applications, vol. 10579, p. 105790B. International Society for Optics and Photonics (2018)
2. Chen, C., Li, Z., Ouyang, C., Sinclair, M., Bai, W., Rueckert, D.: MaxStyle: adversarial style composition for robust medical image segmentation. In: Wang, L., Dou, Q., Fletcher, P.T., Speidel, S., Li, S. (eds.) MICCAI 2022. LNCS, vol. 13435, pp. 151–161. Springer, Cham (2022). https://doi.org/10.1007/978-3-031-16443-9_15
3. Clarysse, J., Hörrmann, J., Yang, F.: Why adversarial training can hurt robust accuracy. In: International Conference on Learning Representations (ICLR) (2023). https://openreview.net/forum?id=-CA8yFkPc7O
4. Decencière, E., et al.: Feedback on a publicly distributed image database: the Messidor database. Image Anal. Stereol. **33**(3), 231–234 (2014)
5. Falk, T., et al.: U-net: deep learning for cell counting, detection, and morphometry. Nat. Methods **16**(1), 67–70 (2019)
6. Frankle, J., Carbin, M.: The lottery ticket hypothesis: finding sparse, trainable neural networks. In: International Conference on Learning Representations (ICLR) (2018)
7. Guan, H., Liu, M.: Domain adaptation for medical image analysis: a survey. IEEE Trans. Biomed. Eng. **69**(3), 1173–1185 (2021)

8. Hu, S., Liao, Z., Xia, Y.: Domain specific convolution and high frequency reconstruction based unsupervised domain adaptation for medical image segmentation. In: Wang, L., Dou, Q., Fletcher, P.T., Speidel, S., Li, S. (eds.) MICCAI 2022. LNCS, vol. 13437, pp. 650–659. Springer, Cham (2022). https://doi.org/10.1007/978-3-031-16449-1_62

9. Hu, S., Liao, Z., Xia, Y.: ProSFDA: prompt learning based source-free domain adaptation for medical image segmentation. arXiv preprint arXiv:2211.11514 (2022)

10. Hu, S., Liao, Z., Zhang, J., Xia, Y.: Domain and content adaptive convolution based multi-source domain generalization for medical image segmentation. IEEE Trans. Med. Imaging **42**(1), 233–244 (2022)

11. Litjens, G., et al.: A survey on deep learning in medical image analysis. Med. Image Anal. **42**, 60–88 (2017). https://doi.org/10.1016/j.media.2017.07.005

12. Ma, H., Lin, X., Yu, Y.: I2F: a unified image-to-feature approach for domain adaptive semantic segmentation. IEEE Trans. Pattern Anal. Mach. Intell. (2022)

13. Ouyang, C., et al.: Causality-inspired single-source domain generalization for medical image segmentation. IEEE Trans. Med. Imaging **42**, 1095–1106 (2022)

14. Sprawls, P.: Image characteristics and quality. In: Physical Principles of Medical Imaging, pp. 1–16. Aspen Gaithersburg (1993)

15. Su, Z., Yao, K., Yang, X., Wang, Q., Sun, J., Huang, K.: Rethinking data augmentation for single-source domain generalization in medical image segmentation. In: AAAI Conference on Artificial Intelligence (AAAI) (2023)

16. Wang, J., et al.: Generalizing to unseen domains: a survey on domain generalization. IEEE Trans. Knowl. Data Eng. **35**, 8052–8072 (2022)

17. Wilson, G., Cook, D.J.: A survey of unsupervised deep domain adaptation. ACM Trans. Intell. Syst. Technol. **11**(5), 1–46 (2020)

18. Xie, X., Niu, J., Liu, X., Chen, Z., Tang, S., Yu, S.: A survey on incorporating domain knowledge into deep learning for medical image analysis. Med. Image Anal. **69**, 101985 (2021). https://doi.org/10.1016/j.media.2021.101985

19. Xu, Y., Xie, S., Reynolds, M., Ragoza, M., Gong, M., Batmanghelich, K.: Adversarial consistency for single domain generalization in medical image segmentation. In: Wang, L., Dou, Q., Fletcher, P.T., Speidel, S., Li, S. (eds.) MICCAI 2022. LNCS, vol. 13437, pp. 671–681. Springer, Cham (2022). https://doi.org/10.1007/978-3-031-16449-1_64

20. Yang, Y., Soatto, S.: FDA: fourier domain adaptation for semantic segmentation. In: IEEE/CVF Conference on Computer Vision and Pattern Recognition (CVPR), pp. 4085–4095 (2020)

21. Zhang, L., et al.: Generalizing deep learning for medical image segmentation to unseen domains via deep stacked transformation. IEEE Trans. Med. Imaging **39**(7), 2531–2540 (2020)

22. Zhou, K., Liu, Z., Qiao, Y., Xiang, T., Loy, C.C.: Domain generalization: a survey. IEEE Trans. Pattern Anal. Mach. Intell. **45**, 4396–4415 (2022)

23. Zhou, Z., Qi, L., Yang, X., Ni, D., Shi, Y.: Generalizable cross-modality medical image segmentation via style augmentation and dual normalization. In: IEEE/CVF Conference on Computer Vision and Pattern Recognition (CVPR), pp. 20856–20865 (2022)

Transformer-Based Annotation Bias-Aware Medical Image Segmentation

Zehui Liao[1], Shishuai Hu[1], Yutong Xie[2(✉)], and Yong Xia[1(✉)]

[1] National Engineering Laboratory for Integrated Aero-Space-Ground-Ocean Big Data Application Technology, School of Computer Science and Engineering, Northwestern Polytechnical University, Xi'an 710072, China
yxia@nwpu.edu.cn
[2] Australian Institute for Machine Learning, The University of Adelaide, Adelaide, SA, Australia
yutong.xie678@gmail.com

Abstract. Manual medical image segmentation is subjective and suffers from annotator-related bias, which can be mimicked or amplified by deep learning methods. Recently, researchers have suggested that such bias is the combination of the annotator preference and stochastic error, which are modeled by convolution blocks located after decoder and pixelwise independent Gaussian distribution, respectively. It is unlikely that convolution blocks can effectively model the varying degrees of preference at the full resolution level. Additionally, the independent pixel-wise Gaussian distribution disregards pixel correlations, leading to a discontinuous boundary. This paper proposes a Transformer-based Annotation Bias-aware (TAB) medical image segmentation model, which tackles the annotator-related bias via modeling annotator preference and stochastic errors. TAB employs the Transformer with learnable queries to extract the different preference-focused features. This enables TAB to produce segmentation with various preferences simultaneously using a single segmentation head. Moreover, TAB takes the multivariant normal distribution assumption that models pixel correlations, and learns the annotation distribution to disentangle the stochastic error. We evaluated our TAB on an OD/OC segmentation benchmark annotated by six annotators. Our results suggest that TAB outperforms existing medical image segmentation models which take into account the annotator-related bias. The code is available at https://github.com/Merrical/TAB.

Keywords: Medical image segmentation · Multiple annotators · Transformer · Multivariate normal distribution

1 Introduction

Deep convolutional neural networks (DCNNs) have significantly advanced medical image segmentation [8,15,22]. However, their success relies heavily on accurately labeled training data [27], which are often unavailable for medical image

Supplementary Information The online version contains supplementary material available at https://doi.org/10.1007/978-3-031-43901-8_3.

segmentation tasks since manual annotation is highly subjective and requires the observer's perception, expertise, and concentration [4,10,12,20,24]. In a study of liver lesion segmentation using abdominal CT, three trained observers delineated the lesion twice over a one-week interval, resulting in the variation of delineated areas up to 10% per observer and more than 20% between observers [21].

To analyze the annotation process, it is assumed that a latent true segmentation, called meta segmentation for this study, exists as the consensus of annotators [14,29]. Annotators prefer to produce annotations that are different from the meta segmentation to facilitate their own diagnosis. Additionally, stochastic errors may arise during the annotation process. To predict accurate meta segmentation and annotator-specific segmentation, research efforts have been increasingly devoted to addressing the issue of annotator-related bias [9,14,29].

Existing methods can be categorized into three groups. *Annotator decision fusion* [9,18,25] methods model annotators individually and use a weighted combination of multiple predictions as the meta segmentation. Despite their advantages, these methods ignore the impact of stochastic errors on the modeling of annotator-specific segmentation [11]. *Annotator bias disentangling* [28,29] methods estimate the meta segmentation and confusion matrices and generate annotator-specific segmentation by multiplying them. Although confusion matrices characterize the annotator bias, their estimation is challenging due to the absence of ground truth. Thus, the less-accurate confusion matrices seriously affect the prediction of meta segmentation. *Preference-involved annotation distribution learning (PADL)* framework [14] disentangles annotator-related bias into annotator preference and stochastic errors and, consequently, outperforms previous methods in predicting both meta segmentation and annotator-specific segmentation. Although PADL has recently been simplified by replacing its multi-branch architecture with dynamic convolutions [6], it still has two limitations. First, PADL uses a stack of convolutional layers after the decoder to model the annotator preference, which may not be effective in modeling the variable degrees of preference at the full resolution level, and the structure of this block, such as the number of layers and kernel size, needs to be adjusted by trial and error. Second, PADL adopts the Gaussian assumption and learns the annotation distribution per pixel independently to disentangle stochastic errors, resulting in a discontinuous boundary.

To address these issues, we advocate extracting the features, on which different preferences focus. Recently, Transformer [16,17,23] has drawn increasing attention due to its ability to model the long-range dependency. Among its variants, DETR [2] has a remarkable ability to detect multiple targets at different locations in parallel, since it takes a set of different positional queries as conditions and focuses on features at different positions. Inspired by DETR, we suggest utilizing such queries to represent annotators' preferences, enabling Transformer to extract different preference-focused features. To further address the issue of missing pixel correlation, we suggest using a non-diagonal multivariate normal distribution [19] to replace the pixel-wise independent Gaussian distribution.

In this paper, we propose a **T**ransformer-based **A**nnotation **B**ias-aware (**TAB**) medical image segmentation model, which can characterize the annotator preference and stochastic errors and deliver accurate meta segmentation and annotator-specific segmentation. TAB consists of a CNN encoder, a Preference Feature Extraction (PFE) module, and a Stochastic Segmentation (SS) head. The CNN encoder performs image feature extraction. The PFE module takes the image feature as input and produces $R + 1$ preference-focused features in parallel for meta/annotator-specific segmentation under the conditions of $R + 1$ different queries. Each preference-focused feature is combined with the image feature and fed to the SS head. The SS head produces a multivariate normal distribution that models the segmentation and annotator-related error as the mean and variance respectively, resulting in more accurate segmentation. We conducted comparative experiments on a public dataset (two tasks) with multiple annotators. Our results demonstrate the superiority of the proposed TAB model as well as the effectiveness of each component.

The main contributions are three-fold. (1) TAB employs Transformer to extract preference-focused features under the conditions of various queries, based on which the meta/annotator-specific segmentation maps are produced simultaneously. (2) TAB uses the covariance matrix of a multivariate normal distribution, which considers the correlation among pixels, to characterize the stochastic errors, resulting in a more continuous boundary. (3) TAB outperforms existing methods in addressing the issue of annotator-related bias.

2 Method

2.1 Problem Formalization and Method Overview

Let a set of medical images annotated by R annotators be denoted by $D = \{x_i, y_{i1}, y_{i2}, \cdots, y_{iR}\}_{i=1}^N$, where $x_i \in \mathbb{R}^{C \times H \times W}$ represents the i-th image with C channels and a size of $H \times W$, and $y_{ir} \in \{0,1\}^{K \times H \times W}$ is the annotation with K classes given by the r-th annotator. We simplify the K-class segmentation problem as K binary segmentation problems. Our goal is to train a segmentation model on D so that the model can generate a meta segmentation map and R annotator-specific segmentation maps for each input image.

Our TAB model contains three main components: a CNN encoder for image feature extraction, a PFE module for preference-focused feature production, and a SS head for segmentation prediction (see Fig. 1). We now delve into the details of each component.

2.2 CNN Encoder

The ResNet34 [7] pre-trained on ImageNet is employed as the CNN encoder. We remove its average pooling layer and fully connected layer to adjust it to our tasks. Skip connections are built from the first convolutional block and the first three residual blocks in the CNN encoder to the corresponding locations of the

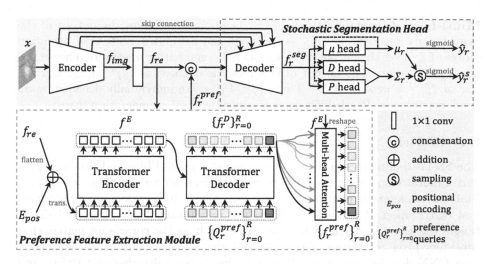

Fig. 1. Framework of TAB model, which consists of a CNN encoder, a PFE module, and a SS head. 'trans.' means the transpose operation.

decoder in the SS head. The CNN encoder takes an image $x \in \mathbb{R}^{C \times H \times W}$ as its input and generates a high-level low-resolution feature map $f_{img} \in \mathbb{R}^{C' \times H' \times W'}$, where $C' = 512$, $H' = \frac{H}{32}$, $W' = \frac{W}{32}$. Moreover, f_{img} is fed to a 1×1 convolutional layer for channel reduction, resulting in $f_{re} \in \mathbb{R}^{d \times H' \times W'}$, where $d = 256$.

2.3 PFE Module

The PFE module consists of an encoder-decoder Transformer and a multi-head attention block. Feeding the image feature f_{re} to the PFE module, we have $R+1$ enhanced feature maps $\{f_r^{pref}\}_{r=0}^R$, on which different preferences focus ($r = 0$ for meta segmentation and others for R annotator-specific segmentation). Note that meta segmentation is regarded as a special preference for simplicity.

The Transformer Encoder is used to enhance the image feature f_{re}. The Transformer encoder consists of a multi-head self-attention module and a feed-forward network. Since the encoder expects a sequence as input, we collapse the spatial dimensions of f_{re} and reshape it into the size of $d \times H'W'$. Next, f_{re} is added to the fixed positional encodings E_{pos} and fed to the encoder. The output of the Transformer encoder is denoted by f^E and its size is $d \times H'W'$.

The Transformer Decoder accepts f^E and $R+1$ learnable queries $\{Q_r^{pref}\}_{r=0}^R$ of size $d = 256$ as its input. We aim to extract different preference-focused features based on the conditions provided by $\{Q_r^{pref}\}_{r=0}^R$, which are called 'preference queries' accordingly. This decoder consists of a multi-head self-attention module for the intercommunication between queries, a multi-head attention module for feature extraction under the conditions of queries, and a feed-forward network. And it produces $R + 1$ features $\{f_r^D\}_{r=0}^R$ of size $d = 256$ in parallel.

Multi-head Attention Block has m heads in it. It takes the output of the Transformer decoder $\{f_r^D\}_{r=0}^R$ as its input and computes multi-head attention scores of f_r^D over the output of the encoder f^E, generating m attention heatmaps per segmentation. The output of this block is denoted as $\{f_r^{pref}\}_{r=0}^R$, and the size of f_r^{pref} is $m \times H' \times M'$. Then, $\{f_r^{pref}\}_{r=0}^R$ are individually decoded by SS head, resulting in $R+1$ different preference-involved segmentation maps.

2.4   SS Head

The SS head aims to disentangle the annotator-related bias and produce meta and annotator-specific segmentation maps. Given an input image, we assume that the annotation distribution over annotators follows a multivariant normal distribution. Thus, we can learn the annotation distribution and disentangle the annotator-related bias by modeling it as the covariance matrix that considers pixel correlation. First, we feed the concatenation of f_{re} and f_r^{pref} to a CNN decoder, which is followed by batch normalization and ReLU, and obtain the feature map $f_r^{seg} \in \mathbb{R}^{32 \times H \times W}$. Second, we establish the multivariant normal distribution $\mathcal{N}(\mu(x), \Sigma(x))$ via predicting the $\mu(x) \in \mathbb{R}^{K \times H \times W}$ and $\Sigma(x) \in \mathbb{R}^{(K \times H \times W)^2}$ based on f_r^{seg}. However, the size of the covariance matrix scales with $(K \times H \times W)^2$, making it computationally intractable. To reduce the complexity, we adopt the low-rank parameterization of the covariance matrix [19]

$$\Sigma = P \times P^T + D \tag{1}$$

where $P \in \mathbb{R}^{(K \times H \times W) \times \alpha}$ is the covariance factor, α defines the rank of the parameterization, D is a diagonal matrix with $K \times H \times W$ diagonal elements. We employ three convolutional layers with 1×1 kernel size to generate $\mu(x)$, P, and D, respectively. In addition, the concatenation of $\mu(x)$ and f_r^{seg} is fed to the P head and D head, respectively, to facilitate the learning of P and D. Finally, we get $R+1$ distributions. Among them, $\mathcal{N}(\mu_{MT}(x), \Sigma_{MT}(x))$ is for meta segmentation and $\mathcal{N}(\mu_r(x), \Sigma_r(x)), r = 1, 2, ..., R$ are for annotator-specific segmentation. The probabilistic meta/annotator-specific segmentation map $(\hat{y}_{MT}/\hat{y}_r)$ is calculated by applying the sigmoid function to the estimated μ_{MT}/μ_r. We can also produce the probabilistic annotator bias-involved segmentation maps $\hat{y}_{MT}^s/\hat{y}_r^s$ by applying the sigmoid function to the segmentation maps sampled from the established distribution $\mathcal{N}(\mu_{MT}(x), \Sigma_{MT}(x))/\mathcal{N}(\mu_r(x), \Sigma_r(x))$.

2.5   Loss and Inference

The loss of our TAB model contains two items: the meta segmentation loss L_{MT} and annotator-specific segmentation loss L_{AS}, shown as follows

$$L = L_{MT}(y^s, \hat{y}_{MT}^s) + \sum_{r=1}^R L_{AS}(y_r, \hat{y}_r^s) \tag{2}$$

where L_{MT} and L_{AS} are the binary cross-entropy loss, y^s is a randomly selected annotation per image, y_r is the delineation given by annotator A_r.

During inference, the estimated probabilistic meta segmentation map $\hat{y}_{MT}$ is evaluated against the mean voting annotation. The estimated probabilistic annotator-specific segmentation map $\hat{y}_r$ is evaluated against the annotation y_r given by the annotator A_r.

3   Experiments and Results

3.1   Dataset and Experimental Setup

Dataset. The RIGA dataset [1] is a public benchmark for optic disc and optic cup segmentation, which contains 750 color fundus images from three sources, including 460 images from MESSIDOR, 195 images from BinRushed, and 95 images from Magrabia. Six ophthalmologists from different centers labeled the optic disc/cup contours manually on each image. We use 655 samples from BinRushed and MESSIDOR for training and 95 samples from Magrabia for test [9,14,26]. The 20% of the training set that is randomly selected is used for validation.

Implementation Details. All images were normalized via subtracting the mean and dividing by the standard deviation. The mean and standard deviation were counted on training cases. We set the batch size to 8 and resized the input image to 256 × 256. The Adam optimizer [13] with default settings was adopted. The learning rate lr was set to $5e - 5$ and decayed according to the polynomial policy $lr = lr_0 \times (1 - t/T)^{0.9}$, where t is the epoch index and $T = 300$ is the maximum epoch. All results were reported over three random runs. Both mean and standard deviation are given.

Evaluation Metrics. We adopted Soft Dice (D^s) as the performance metric. At each threshold level, the Hard Dice is calculated between the segmentation and annotation maps. Soft Dice is calculated via averaging the hard Dice values obtained at multiple threshold levels, *i.e.*, (0.1, 0.3, 0.5, 0.7, 0.9) for this study. Based on the Soft Dice, there are two performance metrics, namely *Average* and *Mean Voting*. *Mean Voting* is the Soft Dice between the predicted meta segmentation and the mean voting annotation. A higher *Mean Voting* represents better performance on modeling the meta segmentation. The annotator-specific predictions are evaluated against each annotator's delineations, and the average Soft Dice of all annotators is denoted as *Average*. A higher *Average* represents better performance on mimicking all annotators.

3.2   Comparative Experiments

We compared our TAB to three baseline models and four recent segmentation models that consider the annotator-bias issue, including (1) the baseline 'Multi-Net' setting, under which R U-Nets (denoted by $M_r, r = 1, 2, ..., R$, $R = 6$ for RIGA) were trained and tested with the annotations provided by annotator A_r, respectively; (2) MH-UNet [5]: a U-Net variant with multiple heads, each

Table 1. Performance ($\mathcal{D}^s_{disc}$ (%), $\mathcal{D}^s_{cup}$ (%)) of our TAB, seven competing models, and two variants of TAB on the RIGA dataset. From left to right: Performance in mimicking the delineations of each annotator (A_r, $r = 1, 2, ..., 6$), *Average*, and *Mean Voting*. The standard deviation is shown as the subscript of the mean. Except for two variants of TAB and the 'Multi-Net' setting (M_r), the best results in *Average* and *Mean Voting* columns are highlighted in blue.

Methods	A_1	A_2	A_3	A_4
M_r	$96.20_{0.05},84.43_{0.16}$	$95.51_{0.02},84.81_{0.36}$	$96.56_{0.04},83.15_{0.11}$	$96.80_{0.03},87.89_{0.09}$
MH-UNet	$96.25_{0.29},83.31_{0.49}$	$95.27_{0.10},81.82_{0.24}$	$96.72_{0.28},77.32_{0.35}$	$97.00_{0.11},88.03_{0.29}$
MV-UNet	$95.11_{0.02},76.85_{0.34}$	$94.53_{0.05},78.45_{0.44}$	$95.57_{0.03},77.68_{0.22}$	$95.70_{0.02},76.27_{0.56}$
MR-Net	$95.33_{0.46},81.96_{0.47}$	$94.72_{0.43},81.13_{0.78}$	$95.65_{0.14},79.04_{0.81}$	$95.94_{0.03},84.13_{2.61}$
CM-Net	$96.26_{0.07},84.50_{0.14}$	$95.41_{0.08},81.46_{0.52}$	$96.55_{0.88},81.80_{0.41}$	$96.80_{1.17},87.50_{0.51}$
PADL	$96.43_{0.03},85.21_{0.26}$	$95.60_{0.05},85.13_{0.25}$	$96.67_{0.02},82.74_{0.37}$	$96.88_{0.11},88.80_{0.10}$
AVAP	$96.20_{0.13},85.79_{0.59}$	$95.44_{0.03},84.50_{0.51}$	$96.47_{0.03},81.65_{0.95}$	$96.82_{0.02},89.61_{0.20}$
Ours	$96.32_{0.05},86.13_{0.18}$	$96.21_{0.07},85.90_{0.17}$	$96.90_{0.08},84.64_{0.14}$	$97.01_{0.07},89.40_{0.12}$
Ours$_{w/o\ PFE}$	$95.67_{0.03},81.69_{0.13}$	$95.12_{0.02},80.16_{0.11}$	$96.08_{0.04},79.42_{0.16}$	$96.40_{0.02},78.53_{0.17}$
Ours$_{w/o\ SS}$	$96.39_{0.11},84.82_{0.46}$	$95.55_{0.02},83.68_{0.47}$	$96.41_{0.05},82.73_{0.29}$	$96.77_{0.07},88.21_{0.47}$

Methods	A_5	A_6	*Average*	*Mean Voting*
M_r	$96.70_{0.03},83.27_{0.15}$	$97.00_{0.00},80.45_{0.01}$	$96.46_{0.03},84.00_{0.16}$	/
MH-UNet	$97.09_{0.17},78.69_{0.90}$	$96.82_{0.41},75.89_{0.32}$	$96.54_{0.09},80.84_{0.41}$	$97.39_{0.06},85.27_{0.09}$
MV-UNet	$95.85_{0.04},78.64_{0.20}$	$95.62_{0.02},74.74_{0.15}$	$95.40_{0.02},77.11_{0.28}$	$97.45_{0.04},86.08_{0.12}$
MR-Net	$95.87_{0.20},79.00_{0.25}$	$95.71_{0.06},76.18_{0.27}$	$95.54_{0.19},80.24_{0.70}$	$97.50_{0.07},87.21_{0.19}$
CM-Net	$96.28_{1.06},82.43_{0.25}$	$96.85_{1.39},78.77_{0.33}$	$96.36_{0.06},82.74_{0.55}$	$96.41_{0.34},81.21_{0.12}$
PADL	$96.80_{0.08},83.42_{0.39}$	$96.89_{0.02},79.76_{0.36}$	$96.53_{0.01},84.32_{0.06}$	$97.77_{0.03},87.77_{0.08}$
AVAP	$96.29_{0.04},83.63_{0.35}$	$96.68_{0.02},80.06_{0.48}$	$96.32_{0.13},84.21_{0.46}$	$97.86_{0.02},87.60_{0.27}$
Ours	$96.80_{0.06},84.50_{0.13}$	$96.99_{0.04},82.55_{0.16}$	$96.70_{0.04},85.52_{0.06}$	$97.82_{0.04},88.22_{0.07}$
Ours$_{w/o\ PFE}$	$96.40_{0.04},79.84_{0.15}$	$96.25_{0.01},75.69_{0.16}$	$95.99_{0.05},79.22_{0.14}$	$97.71_{0.04},87.51_{0.14}$
Ours$_{w/o\ SS}$	$96.63_{0.09},82.88_{0.60}$	$96.81_{0.06},79.90_{0.52}$	$96.43_{0.03},83.71_{0.37}$	$97.18_{0.02},85.36_{0.67}$

accounting for imitating the annotations from a specific annotator; (3) MV-UNet [3]: a U-Net trained with the mean voting annotations; (4) MR-Net [9]: an annotator decision fusion method that uses an attention module to characterize the multi-rater agreement; (5) CM-Net [29]: an annotator bias disentangling method that uses a confusion matrix to model human errors; (6) PADL [14]: a multi-branch framework that models annotator preference and stochastic errors simultaneously; and (7) AVAP [6]: a method that uses dynamic convolutional layers to simplify the multi-branch architecture of PADL. The soft Dice of optic disc $\mathcal{D}^s_{disc}$ and optic cup $\mathcal{D}^s_{cup}$ obtained by our model and competing methods are listed in Table 1. It shows that our TAB achieves the second highest $\mathcal{D}^s_{disc}$ and highest $\mathcal{D}^s_{cup}$ on *Mean Voting*, and achieves the highest $\mathcal{D}^s_{disc}$ and $\mathcal{D}^s_{cup}$ on *Average*. We also visualize the segmentation maps predicted by TAB and other competing methods (see Fig. 2). It reveals that TAB can produce the most accurate segmentation map compared to the ground truth.

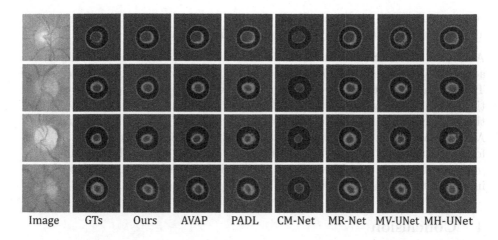

Fig. 2. Visualization of predicted meta segmentation maps obtained by applying six competing methods and our TAB to four cases from the RIGA dataset, together with ground truths (GTs, *i.e.*, mean voting annotations).

Table 2. Performance of the TAB with complete SS head and its three variants on the RIGA dataset. The *Average* and *Mean Voting* ($\mathcal{D}^s_{disc}$ (%), $\mathcal{D}^s_{cup}$ (%)) are used as the performance metrics. The standard deviation is shown as the subscript of the mean.

$\mathcal{N}(\mu,\sigma)$	$\mathcal{N}(\mu,\Sigma)$	μ prior	*Average*	*Mean Voting*
			$96.43_{0.03}$, $83.71_{0.37}$	$97.18_{0.02}$, $85.36_{0.67}$
$\checkmark$			$96.54_{0.02}$, $84.56_{0.09}$	$97.66_{0.02}$, $87.18_{0.12}$
	$\checkmark$		$96.67_{0.06}$, $85.24_{0.12}$	$97.76_{0.07}$, $87.87_{0.09}$
	$\checkmark$	$\checkmark$	$96.70_{0.04}$, $85.52_{0.06}$	$97.82_{0.04}$, $88.22_{0.07}$

3.3 Ablation Analysis

Both the PFE module and SS head play an essential role in the TAB model, characterizing the annotators' preferences and stochastic errors independently. To evaluate the contributions of them, we compared TAB with two of its variants, *i.e.*, 'Ours w/o PFE' and 'Ours w/o SS'. In 'Ours w/o SS', the SS head is replaced by the CNN decoder, which directly converts image features to a segmentation map. 'Ours w/o PFE' contains a CNN encoder and an SS head. This variant is, of course, not able to model the preference of each annotator. The performance of our TAB and its two variants was given in Table 1. It reveals that, when the PFE module was removed, the performance in reconstructing each annotator's delineations drops obviously. Meanwhile, without the SS head, *Mean Voting* score drops from 97.82% to 97.18% for optic disc segmentation and from 88.22% to 85.36% for optic cup segmentation. It indicates that the PFE module contributes substantially to the modeling of each annotator's preference, and the SS head can effectively diminish the impact of stochastic errors and produce accurate meta/annotator-specific segmentation.

Analysis of the SS Head. The effect of each component in the SS head was accessed using *Average* and *Mean Voting*. Table 2 gives the performance of the TAB with complete SS head and its three variants. We compare the stochastic errors modeling capacity of multivariate normal distribution $\mathcal{N}(\mu, \Sigma)$ and the pixel-wise independent Gaussian distribution $\mathcal{N}(\mu, \sigma)$ [14]. Though both $\mathcal{N}(\mu, \Sigma)$ and $\mathcal{N}(\mu, \sigma)$ can reduce the impact of stochastic errors, $\mathcal{N}(\mu, \Sigma)$ performs better than $\mathcal{N}(\mu, \sigma)$ on the test set. We also explored the influence of μ prior. It reveals that the μ prior can facilitate the learning of $\mathcal{N}(\mu, \Sigma)$ and improve the capacity of meta/annotator-specific segmentation.

4   Conclusion

In this paper, we propose the TAB model to address the issue of annotator-related bias that existed in medical image segmentation. TAB leverages the Transformer with multiple learnable queries on extracting preference-focused features in parallel and the multivariate normal distribution on modeling stochastic annotation errors. Extensive experimental results on the public RIGA dataset with annotator-related bias demonstrate that TAB achieves better performance than all competing methods.

Acknowledgement. This work was supported in part by the National Natural Science Foundation of China under Grant 62171377, in part by the Key Technologies Research and Development Program under Grant 2022YFC2009903 / 2022YFC2009900, in part by the Key Research and Development Program of Shaanxi Province, China, under Grant 2022GY-084, and in part by the Innovation Foundation for Doctor Dissertation of Northwestern Polytechnical University under Grant CX2022056.

References

1. Almazroa, A., et al.: Agreement among ophthalmologists in marking the optic disc and optic cup in fundus images. Int. Ophthalmol. **37**(3), 701–717 (2017)
2. Carion, N., Massa, F., Synnaeve, G., Usunier, N., Kirillov, A., Zagoruyko, S.: End-to-end object detection with transformers. In: Vedaldi, A., Bischof, H., Brox, T., Frahm, J.-M. (eds.) ECCV 2020. LNCS, vol. 12346, pp. 213–229. Springer, Cham (2020). https://doi.org/10.1007/978-3-030-58452-8_13
3. Falk, T., et al.: U-net: deep learning for cell counting, detection, and morphometry. Nat. Methods **16**(1), 67–70 (2019)
4. Fu, H., et al.: A retrospective comparison of deep learning to manual annotations for optic disc and optic cup segmentation in fundus photographs. Transl. Vision Sci. Technol. **9**(2), 33 (2020)
5. Guan, M., Gulshan, V., Dai, A., Hinton, G.: Who said what: Modeling individual labelers improves classification. In: Proceedings of the AAAI Conference on Artificial Intelligence, vol. 32 (2018)

6. Guo, X., et al.: Modeling annotator variation and annotator preference for multiple annotations medical image segmentation. In: 2022 IEEE International Conference on Bioinformatics and Biomedicine (BIBM), pp. 977–984. IEEE (2022)
7. He, K., Zhang, X., Ren, S., Sun, J.: Deep residual learning for image recognition. In: Proceedings of the IEEE Conference on Computer Vision and Pattern Recognition, pp. 770–778 (2016)
8. Hesamian, M.H., Jia, W., He, X., Kennedy, P.: Deep learning techniques for medical image segmentation: achievements and challenges. J. Digit. Imaging **32**, 582–596 (2019)
9. Ji, W., et al.: Learning calibrated medical image segmentation via multi-rater agreement modeling. In: Proceedings of the IEEE/CVF Conference on Computer Vision and Pattern Recognition, pp. 12341–12351 (2021)
10. Joskowicz, L., Cohen, D., Caplan, N., Sosna, J.: Inter-observer variability of manual contour delineation of structures in CT. Eur. Radiol. **29**, 1391–1399 (2019)
11. Jungo, A., et al.: On the effect of inter-observer variability for a reliable estimation of uncertainty of medical image segmentation. In: Frangi, A.F., Schnabel, J.A., Davatzikos, C., Alberola-López, C., Fichtinger, G. (eds.) MICCAI 2018. LNCS, vol. 11070, pp. 682–690. Springer, Cham (2018). https://doi.org/10.1007/978-3-030-00928-1_77
12. Karimi, D., Dou, H., Warfield, S.K., Gholipour, A.: Deep learning with noisy labels: exploring techniques and remedies in medical image analysis. Med. Image Anal. **65**, 101759 (2020)
13. Kingma, D.P., Ba, J.: Adam: a method for stochastic optimization. In: International Conference on Learning Representations (2015)
14. Liao, Z., Hu, S., Xie, Y., Xia, Y.: Modeling human preference and stochastic error for medical image segmentation with multiple annotators. arXiv preprint arXiv:2111.13410 (2021)
15. Liu, X., Song, L., Liu, S., Zhang, Y.: A review of deep-learning-based medical image segmentation methods. Sustainability **13**(3), 1224 (2021)
16. Liu, Z., et al.: Swin transformer V2: scaling up capacity and resolution. In: Proceedings of the IEEE/CVF Conference on Computer Vision and Pattern Recognition, pp. 12009–12019 (2022)
17. Liu, Z., et al.: Swin transformer: hierarchical vision transformer using shifted windows. In: Proceedings of the IEEE/CVF International Conference on Computer Vision, pp. 10012–10022 (2021)
18. Mirikharaji, Z., Yan, Y., Hamarneh, G.: Learning to segment skin lesions from noisy annotations. In: Wang, Q., et al. (eds.) DART/MIL3ID -2019. LNCS, vol. 11795, pp. 207–215. Springer, Cham (2019). https://doi.org/10.1007/978-3-030-33391-1_24
19. Monteiro, M., et al.: Stochastic segmentation networks: modelling spatially correlated aleatoric uncertainty. In: Advances in Neural Information Processing Systems, vol. 33, pp. 12756–12767 (2020)
20. Schaekermann, M., Beaton, G., Habib, M., Lim, A., Larson, K., Law, E.: Understanding expert disagreement in medical data analysis through structured adjudication. In: Proceedings of the ACM on Human-Computer Interaction, vol. 3, no. CSCW, pp. 1–23 (2019)
21. Suetens, P.: Fundamentals of Medical Imaging, 3rd edn. (2017)
22. Tajbakhsh, N., Jeyaseelan, L., Li, Q., Chiang, J.N., Wu, Z., Ding, X.: Embracing imperfect datasets: a review of deep learning solutions for medical image segmentation. Med. Image Anal. **63**, 101693 (2020)

23. Vaswani, A., et al.: Attention is all you need. In: Advances in Neural Information Processing Systems, vol. 30 (2017)
24. Wang, S., et al.: Annotation-efficient deep learning for automatic medical image segmentation. Nat. Commun. **12**(1), 5915 (2021)
25. Xiao, L., Li, Y., Qv, L., Tian, X., Peng, Y., Zhou, S.K.: Pathological image segmentation with noisy labels. arXiv preprint arXiv:2104.02602 (2021)
26. Yu, S., Xiao, D., Frost, S., Kanagasingam, Y.: Robust optic disc and cup segmentation with deep learning for glaucoma detection. Comput. Med. Imaging Graph. **74**, 61–71 (2019)
27. Zhang, C., Bengio, S., Hardt, M., Recht, B., Vinyals, O.: Understanding deep learning (still) requires rethinking generalization. Commun. ACM **64**(3), 107–115 (2021)
28. Zhang, L., et al.: Learning to segment when experts disagree. In: Martel, A.L., et al. (eds.) MICCAI 2020. LNCS, vol. 12261, pp. 179–190. Springer, Cham (2020). https://doi.org/10.1007/978-3-030-59710-8_18
29. Zhang, L., et al.: Disentangling human error from ground truth in segmentation of medical images. In: Advances in Neural Information Processing Systems, vol. 33, pp. 15750–15762 (2020)

Uncertainty-Informed Mutual Learning for Joint Medical Image Classification and Segmentation

Kai Ren[1,2], Ke Zou[1,2], Xianjie Liu[2], Yidi Chen[3], Xuedong Yuan[1,2(✉)],
Xiaojing Shen[1,4], Meng Wang[5], and Huazhu Fu[5(✉)]

[1] National Key Laboratory of Fundamental Science on Synthetic Vision,
Sichuan University, Chengdu, Sichuan, China
yxd@scu.edu.cn

[2] College of Computer Science, Sichuan University, Chengdu, Sichuan, China

[3] Department of Radiology, West China Hospital, Sichuan University, Chengdu,
Sichuan, China

[4] College of Mathematics, Sichuan University, Chengdu, Sichuan, China

[5] Institute of High Performance Computing, Agency for Science, Technology and
Research, Singapore, Singapore
hzfu@ieee.org

Abstract. Classification and segmentation are crucial in medical image analysis as they enable accurate diagnosis and disease monitoring. However, current methods often prioritize the mutual learning features and shared model parameters, while neglecting the reliability of features and performances. In this paper, we propose a novel Uncertainty-informed Mutual Learning (UML) framework for reliable and interpretable medical image analysis. Our UML introduces reliability to joint classification and segmentation tasks, leveraging mutual learning with uncertainty to improve performance. To achieve this, we first use evidential deep learning to provide image-level and pixel-wise confidences. Then, an uncertainty navigator is constructed for better using mutual features and generating segmentation results. Besides, an uncertainty instructor is proposed to screen reliable masks for classification. Overall, UML could produce confidence estimation in features and performance for each link (classification and segmentation). The experiments on the public datasets demonstrate that our UML outperforms existing methods in terms of both accuracy and robustness. Our UML has the potential to explore the development of more reliable and explainable medical image analysis models.

Keywords: Mutual learning · Medical image classification and segmentation · Uncertainty estimation

K. Ren and K. Zou—Denotes equal contribution.

Supplementary Information The online version contains supplementary material available at https://doi.org/10.1007/978-3-031-43901-8_4.

1   Introduction

Accurate and robust classification and segmentation of the medical image are powerful tools to inform diagnostic schemes. In clinical practice, the image-level classification and pixel-wise segmentation tasks are not independent [8,27]. Joint classification and segmentation can not only provide clinicians with results for both tasks simultaneously, but also extract valuable information and improve performance. However, improving the reliability and interpretability of medical image analysis is still reaching.

Considering the close correlation between the classification and segmentation, many researchers [6,8,20,22,24,27,28] proposed to collaboratively analyze the two tasks with the help of sharing model parameters or task interacting. Most of the methods are based on sharing model parameters, which improves the performance by fully utilizing the supervision from multiple tasks [8,27]. For example, Thomas et al. [20] combined whole image classification and segmentation of skin cancer using a shared encoder. Task interacting is also a widely used method [12,24,28] as it can introduce the high-level features and results produced by one task to benignly guide another. However, there has been relatively little research on introducing reliability into joint classification and segmentation. The reliability and interpretability of the model are particularly important for clinical tasks, a single result of the most likely hypothesis without any clues about how to make the decision might lead to misdiagnoses and sub-optimal treatment [10,22]. One potential way of improving reliability is to introduce uncertainty for the medical image analysis model.

The current uncertainty estimation method can roughly include the Dropout-based [11], ensemble-based [4,18,19], deterministic-based methods [21] and evidential deep learning [5,16,23,30,31]. All of these methods are widely utilized in classification and segmentation applications for medical image analysis. Abdar et al. [1] employed three uncertainty quantification methods (Monte Carlo dropout, Ensemble MC dropout, and Deep Ensemble) simultaneously to deal with uncertainty estimation during skin cancer image classification. Zou et al. [31] proposed TBraTS based on evidential deep learning to generate robust segmentation results for brain tumor and reliable uncertainty estimations. Unlike the aforementioned methods, which only focus on uncertainty in either medical image classification or segmentation. Furthermore, none of the existing methods have considered how pixel-wise and image-level uncertainty can help improve performance and reliability in mutual learning.

Based on the analysis presented above, we design a novel Uncertainty-informed Mutual Learning (UML) network for medical image analysis in this study. Our UML not only enhances the image-level and pixel-wise reliability of medical image classification and segmentation, but also leverages mutual learning under uncertainty to improve performance. Specifically, we adopt evidential deep learning [16,31] to simultaneously estimate the uncertainty of both to estimate image-level and pixel-wise uncertainty. We introduce an Uncertainty Navigator for segmentation (UN) to generate preliminary segmentation results, taking into account the uncertainty of mutual learning features. We also propose an Uncer-

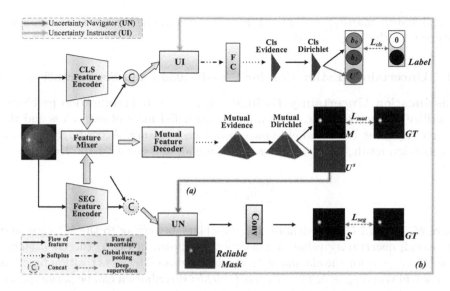

Fig. 1. The framework of Uncertainty-informed Mutual Learning network.

tainty Instructor for classification (UI) to screen reliable masks for classification based on the preliminary segmentation results. Our UML represents pioneering work in introducing reliability and interpretability to joint classification and segmentation, which has the potential to the development of more trusted medical analysis tools[1].

2 Method

The overall architecture of the proposed UML, which leverages mutual learning under uncertainty, is illustrated in Fig. 1. Firstly, Uncertainty Estimation for Classification and Segmentation adapts evidential deep learning to provide image-level and pixel-wise uncertainty. Then, Trusted Mutual Learning not only utilizes the proposed UN to fully exploit pixel-wise uncertainty as the guidance for segmentation but also introduces the UI to filter the feature flow between task interaction.

Given an input medical image $I, I \in \mathbb{R}^{H,W}$, where H, W are the height and width of the image, separately. To maximize the extraction of specific information required for two different tasks while adequately mingling the common feature which is helpful for both classification and segmentation, I is firstly fed into the dual backbone network that outputs the classification feature maps $f_i^c, i \in 1, ..., 4$ and segmentation feature maps $f_i^s, i \in 1, ..., 4$, where i denotes the ith layer of the backbone. Then following [29], we construct the Feature Mixer using Pairwise Channel Map Interaction to mix the original feature and get the

---

[1] Our code has been released in https://github.com/KarryRen/UML.

mutual feature maps $f_i^m, i \in 1, ..., 4$. Finally, we combine the last layer of mutual feature with the original feature.

2.1   Uncertainty Estimation for Classification and Segmentation

Classification Uncertainty Estimation. For the K classification problems, we utilize Subjective Logic [7] to produce the belief mass of each class and the uncertainty mass of the whole image based on *evidence*. Accordingly, given a classification result, its $K + 1$ mass values are all non-negative and their sum is one:

$$\sum_{k=1}^{K} b_k^c + U^c = 1, \tag{1}$$

where $b_k^c \geq 0$ and $U^c \geq 0$ denote the probability belonging to the kth class and the overall uncertainty value, respectively. As shown in Fig. 1, the *cls evidence* $e^c = [e_1^c, \ldots, e_K^c]$ for the classification result is acquired by an activation function layer softplus and $e_k^c \geq 0$. Then the *cls Dirichlet* distribution can be parameterize by $\alpha^c = [\alpha_1^c, \ldots, \alpha_K^c]$, which associated with the *cls evidence* e_k^c, i.e. $\alpha_k^c = e_k^c + 1$. In the end, the image-level belief mass and the uncertainty mass of the classification can be calculated by

$$b_k^c = \frac{e_k^c}{T^c} = \frac{\alpha_k^c - 1}{T^c}, \text{ and } U^c = \frac{K}{T^c}, \tag{2}$$

where $T^c = \sum_{k=1}^{K} \alpha_k^c = \sum_{k=1}^{K} (e_k^c + 1)$ represents the Dirichlet strength. Actually, Eq. 2 describes such a phenomenon that the higher the probability assigned to the kth class, the more evidence observed for k^{th} category should be.

Segmentation Uncertainty Estimation. Essentially, segmentation is the classification for each pixel of a medical image. Given a pixel-wise segmentation result, following [31] the *seg Dirichlet* distribution can be parameterized by $\alpha^{s(h,w)} = [\alpha_1^{s(h,w)}, \ldots, \alpha_Q^{s(h,w)}], (h, w) \in (H, W)$. We can compute the belief mass and uncertainty mass of the input image by

$$b_q^{s(h,w)} = \frac{e_q^{s(h,w)}}{T^{s(h,w)}} = \frac{\alpha_q^{s(h,w)} - 1}{T^{s(h,w)}}, \text{ and } u^{s(h,w)} = \frac{Q}{T^{s(h,w)}}, \tag{3}$$

where $b_q^{s(h,w)} \geq 0$ and $u^{s(h,w)} \geq 0$ denote the probability of the pixel at coordinate (h, w) for the qth class and the overall uncertainty value respectively. We also define $U^s = \{u^{s(h,w)}, (h, w) \in (H, W)\}$ as the pixel-wise uncertainty of the segmentation result.

2.2   Uncertainty-Informed Mutual Learning

Uncertainty Navigator for Segmentation. Actually, we have already obtained an initial segmentation mask $M = \alpha^s$, $M \in (Q, H, W)$ through estimating segmentation uncertainty, and achieved lots of valuable features such as

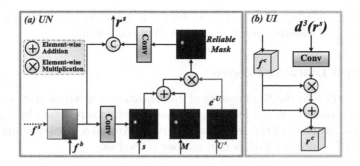

Fig. 2. Details of (a) Uncertainty Navigator (UN) and (b) Uncertainty Instructor (UI).

lesion location. In our method, appropriate uncertainty guided decoding on the feature list can obtain more reliable information and improve the performance of segmentation [3,9,26]. So we introduce Uncertainty Navigator for Segmentation(UN) as a feature decoder, which incorporates the pixel-wise uncertainty in U^s and lesion location information in M with the segmentation feature maps to generate the segmentation result and reliable features. Having a UNet-like architecture [15], UN computes segmentation $s_i, i \in 1, .., 4$ at each layer, as well as introduces the uncertainty in the bottom and top layer by the same way. Take the top layer as an example, as shown in Fig. 2(a), UN calculates the reliable mask M^r by:

$$M^r = (s_1 \oplus M) \otimes e^{-U^s}, \tag{4}$$

Then, the reliable segmentation feature r^s, which combines the trusted information in M^r with the original features, is generated by:

$$r^s = Cat(Conv(M^r), Cat(f_1^s, f_2^b)), \tag{5}$$

where f_1^s derives from jump connecting and f_2^b is the feature of the s_2 with one up-sample operation. $Conv(\cdot)$ represents the convolutional operation, $Cat(\cdot, \cdot)$ denotes the concatenation. Especially, the U^s is also used to guide the bottom feature with the dot product. The r^s is calculated from the segmentation result s_1 and contains uncertainty navigated information not found in s_1.

Uncertainty Instructor for Classification. In order to mine the complementary knowledge of segmentation as the instruction for the classification and eliminate intrusive features, we devise an Uncertainty Instructor for classification (UI) following [22]. Figure 2(b) shows the architecture of UI. It firstly generates reliable classification features r^c fusing the initial classification feature maps f_4^c and the rich information (e.g., lesion location and boundary characteristic) in r^s, which can be expressed by:

$$r^c = f_4^c \oplus (Conv(d^3(r^s)) \otimes f_4^c), \tag{6}$$

where $d^n(\cdot)$ denotes that the frequency of down-sampling operations is n. Then the produced features are transformed into a semantic feature vector by the global average pooling. The obtained vector is converted into the final result (belief values) of classification with uncertainty estimation.

2.3 Mutual Learning Process

In a word, to obtain the final results of classification and segmentation, we construct an end-to-end mutual learning process, which is supervised by a joint loss function. To obtain an initial segmentation result M and a pixel-wise uncertainty estimation U^s, following [31], a mutual loss is used as:

$$\mathcal{L}_m(\alpha^s, y^s) = \mathcal{L}_{ice}(\alpha^s, y^s) + \lambda_1^m \mathcal{L}_{KL}(\alpha^s) + \lambda_2^m \mathcal{L}_{Dice}(\alpha^s, y^s), \qquad (7)$$

where y^s is the Ground Truth (GT) of the segmentation. The hyperparameters λ_1^m and λ_2^m play a crucial role in controlling the Kullback-Leibler divergence (KL) and Dice score, as supported by [31]. Similarly, in order to estimate the image-level uncertainty and classification results. a classification loss is constructed following [5], as:

$$\mathcal{L}_c(\alpha^c, y^c) = \mathcal{L}_{ace}(\alpha^c, y^c) + \lambda^c \mathcal{L}_{KL}(\alpha^c), \qquad (8)$$

where y^c is the true class of the input image. The hyperparameter λ^c serves as a crucial hyperparameter governing the KL, aligning with previous work [5]. To obtain reliable segmentation results, we also adopt deep supervision for the final segmentation result $S = \{s_i, i = 1, ..., 4\}$, which can be denoted as:

$$\mathcal{L}_s = \frac{\sum_{i=1}^{4} \mathcal{L}_{Dice}(v^{i-1}(s_i), y^s)}{4}, \qquad (9)$$

where v^n indicates the number of up-sampling is 2^n. Thus, the overall loss function of our UML can be given as:

$$\mathcal{L}_{UML}(\alpha^s, \alpha^c, y^s, y^c) = w^m \mathcal{L}_m(\alpha^s, y^s) + w^c \mathcal{L}_c(\alpha^c, y^c) + w^s \mathcal{L}_s, \qquad (10)$$

where w^m, w^c, w^s denote the weights and are set $0.1, 0.5, 0.4$, separately.

3   Experiments

Dataset and Implementation. We evaluate the our UML network on two datasets REFUGE [14] and ISPY-1 [13]. REFUGE contains two tasks, classification of glaucoma and segmentation of optic disc/cup in fundus images. The overall 1200 images were equally divided for training, validation, and testing. All images are uniformly adjusted to 256×256 px. The tasks of ISPY-1 are the pCR prediction and the breast tumor segmentation. A total of 157 patients who suffer the breast cancer are considered - 43 achieve pCR and 114 non-pCR. For each case, we cut out the slices in the 3D image and totally got 1,570 2D images, which are randomly divided into the train, validation, and test datasets with 1,230, 170, and 170 slices, respectively.

Table 1. Evaluation of the classification and segmentation performance. The top-2 results are highlighted in bold and underlined ($p \leq 0.01$).

Method		REFUGE						ISPY-1			
		CLS		SEG				CLS		SEG	
		ACC	F1	DI_{disc}	$ASSD_{disc}$	DI_{cup}	$ASSD_{cup}$	ACC	F1	DI	ASSD
Single-task	EC	0.560	0.641	\	\	\	\	0.735	0.648	\	\
	TBraTS	\	\	0.776	1.801	0.787	1.798	\	\	0.784	4.075
	TransUNet	\	\	0.633	2.807	0.628	2.638	\	\	0.692	5.904
Multi-task	BCS	0.723	0.778	<u>0.802</u>	<u>1.692</u>	<u>0.831</u>	<u>1.532</u>	<u>0.758</u>	<u>0.692</u>	0.773	**3.804**
	DSI	<u>0.838</u>	<u>0.834</u>	0.793	2.030	0.811	1.684	0.741	0.673	0.760	4.165
	Ours	**0.853**	**0.875**	**0.855**	1.560	**0.858**	**1.251**	**0.771**	**0.713**	**0.785**	<u>3.927</u>

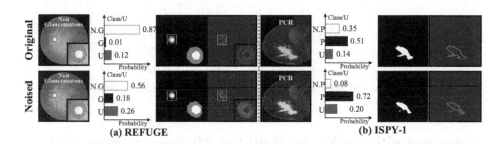

(a) REFUGE   **(b) ISPY-1**

Fig. 3. The visual result of segmentation and classification in REFUGE and ISPY-1. Top is the original image, and bottom is the input with Gaussian noise ($\sigma = 0.05$). From left to right, input (with GT), the result of classification (belief and image-level uncertainty), the result of segmentation, pixel-wise uncertainty.

We implement the proposed method via PyTorch and train it on NVIDIA GeForce RTX 2080Ti. The Adam optimizer is adopted to update the overall parameters with an initial learning rate 0.0001 for 100 epochs. The scale of the regularizer is set as 1×10^{-5}. We choose VGG-16 and Res2Net as the encoders for classification and segmentation, separately.

Compared Methods and Metrics. We compared our method with single-task methods and multi-task methods. (1) Single-task methods: (a) EC [17], (b) TBraTS [31] and (c) TransUNet [2]. Evidential deep learning for classification (EC) first proposed to parameterize classification probabilities as Dirichlet distributions to explain evidence. TBraTS then extended EC to medical image segmentation. Meriting both Transformers and U-Net, TransUNet is a strong model for medical image segmentation. (2) Multi-task methods: (d) BCS [25] and (e) DSI [28]. The baseline of the Joint Classification and Segmentation framework (BCS) is a simple but useful way to share model parameters, which utilize two different encoders and decoders for learning respectively. The Deep Synergistic Interaction Network (DSI) has demonstrated superior performance in joint task. We adopt overall Accuracy (ACC) and F1 score (F1) as the evaluation criteria for the classification task. Dice score (DI) and Average Symmetric Surface Distance (ASSD) are chosen for the segmentation task.

Table 2. The quantitative comparisons on the REFUGE dataset with vary NOISE levels.

REFUGE	0.030						0.050					
	CLS		SEG				CLS		SEG			
	ACC	$F1$	DI_{disc}	$ASSD_{disc}$	DI_{cup}	$ASSD_{cup}$	ACC	$F1$	DI_{disc}	$ASSD_{disc}$	DI_{cup}	$ASSD_{cup}$
BCS	0.620	0.694	<u>0.743</u>	<u>2.104</u>	<u>0.809</u>	<u>1.670</u>	0.430	0.510	<u>0.610</u>	<u>3.142</u>	<u>0.746</u>	<u>2.188</u>
DSI	<u>0.675</u>	<u>0.733</u>	0.563	9.196	0.544	9.705	<u>0.532</u>	<u>0.574</u>	0.409	8.481	0.364	9.794
Ours	**0.827**	**0.857**	**0.830**	**1.752**	**0.840**	**1.407**	**0.733**	**0.785**	**0.744**	**2.142**	**0.778**	**2.009**

Table 3. The result of Ablation Study.

MD	UN	UI	CLS		SEG			
			ACC	$F1$	DI_{disc}	$ASSD_{disc}$	DI_{cup}	$ASSD_{cup}$
✓			0.765	0.810	0.835	<u>1.454</u>	<u>0.841</u>	<u>1.423</u>
✓	✓		0.813	0.845	<u>0.836</u>	**1.333**	0.826	1.525
✓		✓	<u>0.828</u>	<u>0.856</u>	0.786	1.853	0.823	1.593
✓	✓	✓	**0.853**	**0.875**	**0.855**	1.560	**0.858**	**1.251**

Comparison with Single- and Multi-task Methods. As shown in Table 1, we report the performance on the two datasets of the proposed UML and other methods. By comparison, we can observe the fact that the accuracy of the model results is low if either classification or segmentation is done in isolation, the ACC has only just broken 0.5 in EC. But joint classification and segmentation changes this situation, the performance of BCS and DSI improves considerably, especially the ACC and the Dice score of optic cup. Excitingly, our UML not only achieves the best classification performance in ACC (85.3%) and F1 (0.875) with significant increments of 1.8%, 4.9%, but also obtains the superior segmentation performance with increments of 6.6% in DI_{disc} and 3.2% in DI_{cup}. A similar improvement can be observed in the experimental results in ISPY-1.

Comparison Under Noisy Data. To further valid the reliability of our model, we introduce Gaussian noise with various levels of standard deviations (σ) to the input medical images. The comparison results are shown in Table 2. As can be observed that, the accuracy of classification and segmentation significantly decreases after adding noise to the raw data. However, benefiting from the uncertainty-informed guiding, our UML consistently deliver impressive results. In Fig. 3, we show the output of our model under the noise. It is obvious that both the image-level uncertainty and the pixel-wise uncertainty respond reasonably well to noise. These experimental results can verify the reliability and interpre of the uncertainty guided interaction between the classification and segmentation in the proposed UML. The results of more qualitative comparisons can be found in the Supplementary Material.

Ablation Study. As illustrated in Table 3, both of the proposed UN and UI play important roles in trusted mutual learning. The baseline method is BCS.

MD represents the mutual feature decoder. It is clear that the performance of classification and segmentation is significantly improved when we introduce supervision of mutual features. As we thought, the introduction of UN and UI takes the reliability of the model to a higher level.

4 Conclusion

In this paper, we propose a novel deep learning approach, UML, for joint classification and segmentation of medical images. Our approach is designed to improve the reliability and interpretability of medical image classification and segmentation, by enhancing image-level and pixel-wise reliability estimated by evidential deep learning, and by leveraging mutual learning with the proposed UN and UI modules. Our extensive experiments demonstrate that UML outperforms baselines and introduces significant improvements in both classification and segmentation. Overall, our results highlight the potential of UML for enhancing the performance and interpretability of medical image analysis.

Acknowledgements. This work was supported by the National Research Foundation, Singapore under its AI Singapore Programme (AISG Award No: AISG2-TC-2021-003), A*STAR AME Programmatic Funding Scheme Under Project A20H4b0141, A*STAR Central Research Fund, the Science and Technology Department of Sichuan Province (Grant No. 2022YFS0071 & 2023YFG0273), and the China Scholarship Council (No. 202206240082).

References

1. Abdar, M., et al.: Uncertainty quantification in skin cancer classification using three-way decision-based Bayesian deep learning. Comput. Biol. Med. **135**, 104418 (2021)
2. Chen, J., et al.: TransUNet: transformers make strong encoders for medical image segmentation. arXiv preprint arXiv:2102.04306 (2021)
3. Cui, Y., Deng, W., Chen, H., Liu, L.: Uncertainty-aware distillation for semi-supervised few-shot class-incremental learning. arXiv preprint arXiv:2301.09964 (2023)
4. Gal, Y., Ghahramani, Z.: Dropout as a Bayesian approximation: representing model uncertainty in deep learning. In: International Conference on Machine Learning, pp. 1050–1059. PMLR (2016)
5. Han, Z., Zhang, C., Fu, H., Zhou, J.T.: Trusted multi-view classification. arXiv preprint arXiv:2102.02051 (2021)
6. Harouni, A., Karargyris, A., Negahdar, M., Beymer, D., Syeda-Mahmood, T.: Universal multi-modal deep network for classification and segmentation of medical images. In: 2018 IEEE 15th International Symposium on Biomedical Imaging (ISBI 2018), pp. 872–876. IEEE (2018)
7. Jsang, A.: Subjective Logic: A Formalism for Reasoning Under Uncertainty. Springer, Heidelberg (2016). https://doi.org/10.1007/978-3-319-42337-1
8. Kang, Q., et al.: Thyroid nodule segmentation and classification in ultrasound images through intra-and inter-task consistent learning. Med. Image Anal. **79**, 102443 (2022)

9. Kim, T., Lee, H., Kim, D.: UACANet: uncertainty augmented context attention for polyp segmentation. In: Proceedings of the 29th ACM International Conference on Multimedia, pp. 2167–2175 (2021)

10. Kohl, S., et al.: A probabilistic u-net for segmentation of ambiguous images. In: Advances in Neural Information Processing Systems, vol. 31 (2018)

11. Lakshminarayanan, B., Pritzel, A., Blundell, C.: Simple and scalable predictive uncertainty estimation using deep ensembles. In: Advances in Neural Information Processing Systems, vol. 30 (2017)

12. Mehta, S., Mercan, E., Bartlett, J., Weaver, D., Elmore, J.G., Shapiro, L.: Y-net: joint segmentation and classification for diagnosis of breast biopsy images. In: Frangi, A.F., Schnabel, J.A., Davatzikos, C., Alberola-López, C., Fichtinger, G. (eds.) MICCAI 2018, Part II. LNCS, vol. 11071, pp. 893–901. Springer, Cham (2018). https://doi.org/10.1007/978-3-030-00934-2_99

13. Newitt, D., Hylton, N., et al.: Multi-center breast DCE-MRI data and segmentations from patients in the I-SPY 1/ACRIN 6657 trials. Cancer Imaging Arch. **10**(7) (2016)

14. Orlando, J.I., et al.: Refuge challenge: a unified framework for evaluating automated methods for glaucoma assessment from fundus photographs. Med. Image Anal. **59**, 101570 (2020)

15. Ronneberger, O., Fischer, P., Brox, T.: U-net: convolutional networks for biomedical image segmentation. In: Navab, N., Hornegger, J., Wells, W.M., Frangi, A.F. (eds.) MICCAI 2015. LNCS, vol. 9351, pp. 234–241. Springer, Cham (2015). https://doi.org/10.1007/978-3-319-24574-4_28

16. Sensoy, M., Kaplan, L., Kandemir, M.: Evidential deep learning to quantify classification uncertainty. In: Advances in Neural Information Processing Systems, vol. 31 (2018)

17. Sensoy, M., Kaplan, L., Kandemir, M.: Evidential deep learning to quantify classification uncertainty. In: Proceedings of the 32nd International Conference on Neural Information Processing Systems, pp. 3183–3193 (2018)

18. Smith, L., Gal, Y.: Understanding measures of uncertainty for adversarial example detection. arXiv preprint arXiv:1803.08533 (2018)

19. Srivastava, N., Hinton, G., Krizhevsky, A., Sutskever, I., Salakhutdinov, R.: Dropout: a simple way to prevent neural networks from overfitting. J. Mach. Learn. Res. **15**(1), 1929–1958 (2014)

20. Thomas, S.M., Lefevre, J.G., Baxter, G., Hamilton, N.A.: Interpretable deep learning systems for multi-class segmentation and classification of non-melanoma skin cancer. Med. Image Anal. **68**, 101915 (2021)

21. Van Amersfoort, J., Smith, L., Teh, Y.W., Gal, Y.: Uncertainty estimation using a single deep deterministic neural network. In: International Conference on Machine Learning, pp. 9690–9700. PMLR (2020)

22. Wang, J., et al.: Information bottleneck-based interpretable multitask network for breast cancer classification and segmentation. Med. Image Anal. **83**, 102687 (2023)

23. Wang, M., et al.: Uncertainty-inspired open set learning for retinal anomaly identification. arXiv preprint arXiv:2304.03981 (2023)

24. Wang, X., et al.: Joint learning of 3D lesion segmentation and classification for explainable COVID-19 diagnosis. IEEE Trans. Med. Imaging **40**(9), 2463–2476 (2021)

25. Yang, X., Zeng, Z., Yeo, S.Y., Tan, C., Tey, H.L., Su, Y.: A novel multi-task deep learning model for skin lesion segmentation and classification. arXiv preprint arXiv:1703.01025 (2017)

26. Zhang, M., Xu, S., Piao, Y., Shi, D., Lin, S., Lu, H.: PreyNet: preying on cam-
    ouflaged objects. In: Proceedings of the 30th ACM International Conference on
    Multimedia, pp. 5323–5332 (2022)
27. Zhou, Y., et al.: Multi-task learning for segmentation and classification of tumors
    in 3d automated breast ultrasound images. Med. Image Anal. **70**, 101918 (2021)
28. Zhu, M., Chen, Z., Yuan, Y.: DSI-Net: deep synergistic interaction network for
    joint classification and segmentation with endoscope images. IEEE Trans. Med.
    Imaging **40**(12), 3315–3325 (2021)
29. Zou, K., Tao, T., Yuan, X., Shen, X., Lai, W., Long, H.: An interactive dual-branch
    network for hard palate segmentation of the oral cavity from CBCT images. Appl.
    Soft Comput. **129**, 109549 (2022)
30. Zou, K., et al.: EvidenceCap: towards trustworthy medical image segmentation via
    evidential identity cap. arXiv preprint arXiv:2301.00349 (2023)
31. Zou, K., Yuan, X., Shen, X., Wang, M., Fu, H.: TBraTS: Trusted brain tumor seg-
    mentation. In: Wang, L., Dou, Q., Fletcher, P.T., Speidel, S., Li, S. (eds.) MICCAI
    2022, Part VIII. LNCS, vol. 13438, pp. 503–513. Springer, Cham (2022). https://
    doi.org/10.1007/978-3-031-16452-1_48

A General Stitching Solution for Whole-Brain 3D Nuclei Instance Segmentation from Microscopy Images

Ziquan Wei[1,2], Tingting Dan[1], Jiaqi Ding[1,2], Mustafa Dere[1],
and Guorong Wu[1,2(✉)]

[1] Department of Psychiatry, University of North Carolina at Chapel Hill,
Chapel Hill, NC 27599, USA
guorong_wu@med.unc.edu
[2] Department of Computer Science, University of North Carolina at Chapel Hill,
Chapel Hill, NC 27599, USA

Abstract. High-throughput 3D nuclei instance segmentation (NIS) is critical to understanding the complex structure and function of individual cells and their interactions within the larger tissue environment in the brain. Despite the significant progress in achieving accurate NIS within small image stacks using cutting-edge machine learning techniques, there has been a lack of effort to extend this approach towards whole-brain NIS from light-sheet microscopy. This critical area of research has been largely overlooked, despite its importance in the neuroscience field. To address this challenge, we propose an efficient deep stitching neural network built upon a knowledge graph model characterizing 3D contextual relationships between nuclei. Our deep stitching model is designed to be agnostic, enabling existing limited methods (optimized for image stack only) to overcome the challenges of whole-brain NIS, particularly in addressing the issue of inter- and intra-slice gaps. We have evaluated the NIS accuracy on top of three state-of-the-art deep models with $128 \times 128 \times 64$ image stacks, and visualized results in both inter- and intra-slice gaps of whole brain. With resolved gap issues, our deep stitching model enables the whole-brain NIS (gigapixel-level) on entry-level GPU servers within 27 h.

Keywords: Image stitching · 3D microscopy image · Whole-brain nucleus instance segmentation · Graph neural network

1 Introduction

Light-sheet microscopy is a powerful imaging modality that allows for fast and high-resolution imaging of large samples, such as the whole brain of the

Supported by NIH R01NS110791, NIH R01MH121433, NIH P50HD103573, and Foundation of Hope.

Supplementary Information The online version contains supplementary material available at https://doi.org/10.1007/978-3-031-43901-8_5.

mouse [3,14]. Tissue-clearing techniques enable the removal of light-scattering molecules, thus improving the penetration of light through biological samples and allowing for better visualization of internal structures, including nuclei [2,16]. Together, light-sheet microscopy and tissue-clearing techniques have revolutionized the field of biomedical imaging and they have been widely used for studying the structure and function of tissues and organs at the cellular level.

Accurate 3D nuclei instance segmentation plays a crucial role in identifying and delineating individual nuclei within three-dimensional space, which is essential for understanding the complex structure and function of biological tissues in the brain. Previous [10] and [5] have applied graph-based approaches that model links of voxel and neuron region, respectively, for 3D neuron segmentation from electron microscopy image stacks. However, accurate segmentation of nuclei from light-sheet microscopy images of cleared tissue can be a challenging task due to the presence of complex tissue structures, cell shapes, and variations in nuclei size and shape [1]. Due to the high cost of 3D manual nuclei annotations and the complexity of learning, current end-to-end NIS models are typically limited to training and testing on small image stacks (e.g., $128 \times 128 \times 64$). Considering these limitations, one approach for achieving whole-brain NIS is dividing the whole stack into smaller stacks, so the existing NIS methods can handle each piece individually. In such a scenario, constructing the whole-brain nuclei instance segmentation in 3D from these smaller image stacks arises a new challenge. The gaps between these smaller stacks (intra-slice) and the slices (inter-slice) require a robust stitching method for accurate NIS. We show these gaps in Fig. 1. Note, the intra-slice gap, commonly referred to as the boundary gap, arises due to the existence of boundaries in the segmentation outcome of image stacks and poses a challenge in achieving smooth segmentation between neighboring image stacks. Current approaches may, however, undermine the overall quality of the whole-brain NIS results which leads to inaccurate reports of nuclei counts. Figure 1 (left) illustrates the typical examples of the boundary gap issues, where the red circle highlights the incidence of over-counting (both partial nuclei instances are recognized as a complete nucleus in the corresponding image stack), while the dashed blue box denotes the issue of under-counting (none of the partial nuclei instances has been detected in each image stack).

It is a common practice to use overlapped image stacks to stitch the intensity image (continuous values) by weighted averaging from multiple estimations [12,15]. However, when nuclei are in close proximity and represent the same entity, it becomes crucial to accurately match the indexes of nuclei instances, which refer to the segmentation labels. We call this the nuclei stitching issue. This issue presents a significant challenge in the pursuit of achieving whole-brain NIS. To address this non-trivial challenge, we formulate this problem as a knowledge graph (KP) task that is built to characterize the nuclei-to-nuclei relationships based on the feature presentation of partial image appearance. By doing so, the primary objective of this learning problem is to determine whether to merge two partial nuclei instances that exist across different slices or stacks. Drawing inspiration from recent research on object tracking using graph models,

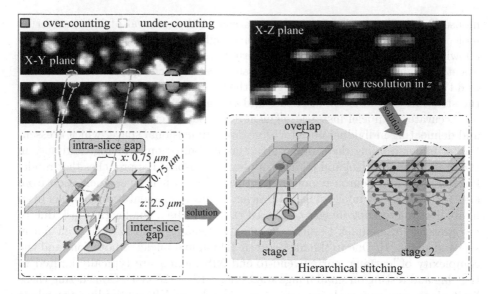

Fig. 1. Top left: Nuclei over-counting (in red) and under-counting (in dashed blue) issues due to the boundary gaps across image stacks. Bottom left: Undetected nuclei instances make stitching difficult due to the absence of partial nuclei instances (indicated by red crosses). Top right: The limited inter-slice resolution presents a challenge for NIS in each image stack, particularly for partial nuclei instances near the stack boundary. Bottom right: Our hierarchal stitching framework leverages overlapping strategy in X-Y plane to reduce the chance of absent nuclei instances in intra-slice stitching. (Color figure online)

we construct a graph contextual model to assemble 3D nuclei in a graph neural network (GNN). In the overlapping area, the complete 2D nucleus instance is represented as a graph node, and the links between these nodes correspond to the nuclei-to-nuclei relationship.

Conventional knowledge graph learning typically emphasizes the learning of the relationship function. In this context, it appears that the process of whole-brain NIS largely depends on the relationship function to link partial nuclei instances along the gap between slices. Nonetheless, a new challenge arises from the NIS backbone due to the anisotropic image resolution where inter-slice Z resolution (e.g., $2.5\,\mu m$) is often several times lower than the in-plane X-Y resolution ($0.75 \times 0.75\,\mu m^2$). That is, the (partial) nuclei instances located near the boundary across X-Y planes (along the inter-slice direction with poor image resolution) in the 3D image stack have a large chance of being misidentified, leading to the failure of NIS stitching due to the absence of nuclei instances that are represented as nodes in the contextual graph model. To alleviate this issue, we present a two-stage hierarchical whole-brain NIS framework that involves stitching 2D NIS results in the X-Y plane (stage 1) and then assembling these 2D instances into 3D nuclei using a graph contextual model (stage 2). The conjecture is that the high resolution in the X-Y plane minimizes the risk of missing

2D NIS instances, allowing the knowledge graph learning to be free of absent nodes in establishing correspondences between partial nuclei instances.

Stitching 2D nuclei instances in each image slice is considerably easier than stitching across image slices due to the finer image resolution in the X-Y plane. However, as shown in Fig. 1 (right), the poor resolution in the Z-axis makes it challenging to identify partial nuclei instances along this axis. Thus, pre-stitching in the X-Y plane of the first stage can reduce the probability of having 2D nuclei instances missing along the Z axis at the second stage. Among this, we train the graph contextual model to predict the nuclei-to-nuclei correspondence across image slices, where each node is the 2D nuclei instance without the intra-slice gap issue. Since stage 2 is on top of the existing NIS methods in stage 1, our stitching framework is agnostic and can support any state-of-the-art NIS methods to expand from small image stacks to the entire brain.

In the experiments, we have comprehensively evaluated the segmentation and stitching accuracy (correspondence matching between 2D instances) and whole-brain NIS results (both visual inspection and quantitative counting results). Compared to no stitching, Our deep stitching model has shown a significant improvement in the whole-brain NIS results with different state-of-the-art models, indicating its potential for practical applications in the field of neuroscience.

2 Methods

Current state-of-the-art NIS methods, such as Mask-RCNN [6,7], 3D Unet [4,8] and Cellpose [9,11] are designed to segment nuclei instances in a pre-defined small image stack only. To scale up to whole-brain NIS, we propose a graph-based contextual model to establish nuclei-to-nuclei correspondences across image stacks. On top of this backbone, we present a hierarchical whole-brain NIS stitching framework that is agnostic to existing NIS methods.

2.1 Graph Contextual Model

Problem Formulation. Our graph contextual model takes a set of partial nuclei instances, sliced by the inter-slice gap, as input. These nuclei instances can be obtained using a 2D instance segmentation method, such as Mask-RCNN. The output of our model is a collection of nuclei-to-nuclei correspondences, which enable us to stitch together the NIS results from different image slices (by running NIS separately). We formulate this correspondence matching problem as a knowledge graph learning task where the links between nodes in the graph contextual model represent the probability of them belonging to the same nuclei instance. In this regard, the key component of NIS stitching becomes seeking for a relationship function that estimates the likelihood of correspondence based on the node features, i.e., image appearance of to-be-stitched 2D nuclei instances.

Machine Learning Components in Graph Contextual Model. *First*, we construct an initial contextual graph $G = \{\mathbf{V}, \mathbf{E}\}$ for each 2D nucleus instance x (i.e., image appearance vector). The set of nodes $\mathbf{V} = \{x_i | \mathcal{D}(x, x_i) > \delta\}$ includes

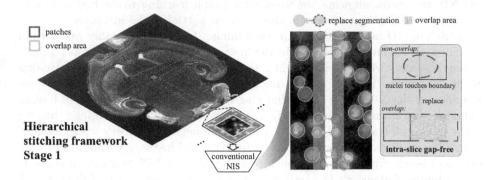

Fig. 2. Stage one of the proposed hierarchical stitching framework for whole-brain NIS. Resolve the intra-slice gap in the X-Y plane by overlap. (Color figure online)

all neighboring 2D nuclei instances, where the distance between the centers of two instances is denoted by $\mathcal{D}$, and δ is a predefined threshold. The matrix $\mathbf{E} \in \mathbb{R}^{N \times N}$ represents the edges between nodes, where N is the number of neighboring instances. Specifically, we compute the Intersection over Union (IoU) between the two instances and set the edge weight as $e_{ij} = IoU(x_i, x_j)$.

Second, we train the model on a set of contextual graphs G to recursively (1) find the mapping function γ to describe the local image appearance on each graph node and (2) learn the triplet similarity function ψ.

- *Graph feature representation learning.* For the kth iteration, we enable two connected nodes to exchange their feature representations constrained by the current relationship topology e_{ij}^k by the kth layer of the deep stitching model. In this context, we define the message-passing function as:

$$x_i^{(k+1)} = \gamma^{(k+1)} \left(x_i^{(k)}, \Sigma_{j \in \mathcal{N}(i)} \phi^{(k)}(x_i^{(k)}, x_j^{(k)}, e_{j,i}^{(k)}) \right). \tag{1}$$

Following the popular learning scheme in knowledge graphs [13], we employ Multilayer Perceptron (MLP) to act functions γ, ϕ.

- *Learning the link-wise similarity function to predict nuclei-to-nuclei correspondence.* Given the updated node feature representations $\{x_i^{(k+1)}\}$, we train another MLP to learn the similarity function ψ in a layer-by-layer manner. In the kth layer, we update each 2D-to-3D contextual correspondence $e_{j,i}^{(k+1)}$ for the next layer by

$$e_{j,i}^{(k+1)} = \psi^{(k+1)} \left(x_i^{(k+1)}, x_j^{(k+1)}, e_{j,i}^{(k)} \right). \tag{2}$$

2.2   Hierarchical Stitching Framework for Whole-Brain NIS

Our graph contextual model is able to stitch the NIS results across the intra-slice gap areas in X/Y-Z plane. As demonstrated in Fig. 1, the accuracy of no

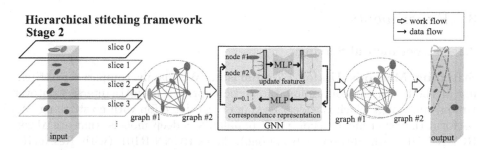

Fig. 3. Stage two of the proposed hierarchical stitching framework for whole-brain NIS. Graph contextual model for inter-slice gap.

stitching is limited by the absence of 2D partial nuclei instances due to the poor inter-slice resolution. With the graph model, as shown in Fig. 2 and 3, we propose a hierarchical stitching framework for whole-brain NIS in two stages.

1. *Resolve intra-slice gap in X-Y plane.* Suppose that each within-stack NIS result overlaps with its neighboring image stack in the X-Y plane. Then, we can resolve the intra-slice gap problem in X-Y plane in three steps: (i) identify the duplicated 2D nuclei instances from multiple overlapped image stacks, (ii) find the representative NIS result from the "gap-free" image stack, and (iii) unify multiple NIS estimations by using the "gap-free" NIS estimation as the appearance of the underlying 2D nuclei. The effect of this solution is shown in Fig. 2 right, where the gray areas are spatially overlapped. We use the arrows to indicate that the 2D nuclei instances (red dash circles) have been merged to the counterpart from the "gap-free" image stack (yellow circles).

2. *Inter-slice stitching using graph contextual model.* At each gap area along Z-axis, we deploy the graph contextual model to stitch the sliced nuclei instances. Specifically, we follow the partition of the whole-brain microscopy image in stage 1, that is a set of overlapped 3D image stacks, all the way from the top to the bottom as shown in the left of Fig. 3. It is worth noting that each 2D nuclei instance in the X-Y plane is complete, as indicated by the red-highlighted portion extending beyond the image stack. Next, we assign a stack-specific local index to each 2D nuclei instance. After that, we apply the (trained) graph contextual model to each 2D nuclei instance. By tracking the correspondences among local indexes, we remove the duplicated inter-slice correspondence and assign the global index to the 3D nuclei instance.

2.3   Implementation Details

We empirically use 18.75 μm for the overlap size between two neighboring image stacks in X-Y plane. The conventional NIS method is trained using 128×128 patches. For the graph contextual model, the MLPs consist of 12 fully-connected layers. Annotated imaging data has been split into training, validation, and testing sets in a ratio of 6:1:1. Adam is used with $lr = 5e - 4$ as the optimizer, $Dropout = 0.5$, and focal loss as the loss function.

3   Experiments

3.1   Experimental Settings

Stitching Methods Under Comparison. We mainly compare our stitching method with the conventional analytic approach by IoU (Intersection over Union) matching, which is widely used in object detection with no stitching. We perform the experiments using three popular NIS deep models, that are Mask-RCNN-R50 (with ResNet50 backbone), Mark-RCNN-R101 (with ResNet101 backbone) and CellPose, with two stitching methods, i.e., our hierarchical stitching framework and IoU-based matching scheme.

Data and Computing Environment. In the following experiments, we first train Mask-RCNN-R50, Mask-RCNN-R101, and CellPose on 16 image stacks ($128 \times 128 \times 64$), which include in total 6,847 manually labeled 3D nuclei. Then we make the methods comparison in two ways. For the stitching comparison based on 16 image stacks, We integrate each NIS model into two stitching methods respectively, which yields six NIS methods and corresponding results. For the stitching comparison based on whole-brain images with the size of $8,729 \times 9,097 \times 1,116 voxel^3$ and the resolution of $0.75\,\mu m \times 0.75\,\mu m \times 2.5\,\mu m$, we firstly partition the one whole-brain image into 89,424 ($69 \times 72 \times 18$) image stacks with the size of $128 \times 128 \times 64$, then employing the best NIS model to segment nuclei instances in each image stack in a parallel manner, finally we deploy our hierarchical stitching framework to stitch the image stacks together. All experiments are run on a Linux server with 4 GPUs (24 GB) and 48 CPUs with 125 GB RAM.

3.2   Evaluation Metrics

We use the common metrics precision, recall, and F1 score to evaluate the 3D NIS between annotated and predicted nuclei instances. Since the major challenge of NIS stitching is due to the large inter-slice gap, we also define the stitching accuracy for each 3D nuclei instance by counting the number of 2D NIS (in the X-Y plane) that both manual annotation and stitched nuclei share the same instance index.

3.3   Evaluating the Accuracy of NIS Stitching Results

Quantitative Evaluation. As shown in Fig. 4, there is a clear sign that NIS models with our hierarchical stitching method outperform IoU-based counterparts on NIS metrics, regardless of the NIS backbone models. In average, our hierarchical stitching method has improved 14.0%, 5.1%, 10.2%, and 3.4% in precision, recall, F1 score, and stitching accuracy, respectively compared with IoU-based results.

Visual Inspection. We also show the visual improvement of whole-brain NIS results before and after stitching in Fig. 5. Through the comparison, it is apparent that (1) the inconsistent NIS results along the intra-slice gap area have been

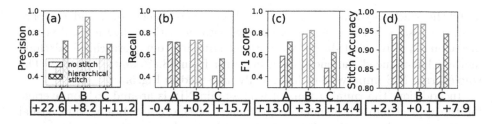

Fig. 4. The NIS precision (a), recall (b), F1 score (c), and stitching accuracy (d) by stitching or not, where the NIS backbones include Mask-RCNN-R50 (A, blue), Mask-RCNN-R101 (B, orange), CellPose (C, green). (Color figure online)

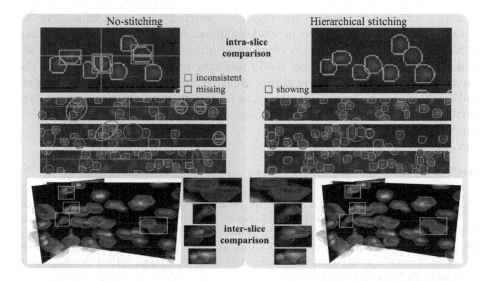

Fig. 5. Visual comparison before (left) and after (right) stitching. Red circles show the missing instances when there is no stitching, and Pink circles indicate inconsistency issues at the intra-slice gap areas. Inter-slice comparison is a zoom-in of 3D visualization (bottom, yellow-highlighted), where red instance denotes NIS results, and gray is GT. (Color figure online)

corrected after stitching (indicated by Pink circles), so there is no apparent boundary between the same nucleus. And (2) the graph contextual model on each nuclei instance alleviates the issue of missing nuclei (indicated by Red circles) by correspondence detection across the inter-slice gap areas.

3.4   Whole-Brain NIS in Neuroscience Applications

One of the important steps in neuroscience studies is the measurement of regional variations in terms of nuclei counts. In light of this, we evaluate the counting accuracy. Since we only have 16 image stacks with manual annotations, we simulate the stack-to-stack gap in the middle of each image stack and compare counting results between whole-brain NIS with or without hierarchical stitching. Compared to whole-brain NIS without stitching, our hierarchical stitching method has reduced the nuclei counting error from 48.8% down to 10.1%. The whole-brain NIS improvement has vividly appeared in Fig. 5.

In addition, the running time of whole-brain NIS is around 26 h on average for the typical light-sheet microscopy images of the mouse brain. More visual results of the whole-brain NIS can be found in supplementary materials.

4 Conclusion

In this work, we introduce a learning-based stitching approach to achieve 3D instance segmentation of nuclei in whole-brain microscopy images. Our stitching framework is flexible enough to incorporate existing NIS methods, which are typically trained on small image stacks and may not be able to scale up to the whole-brain level. Our method shows great improvement by addressing inter- and intra-slice gap issues. The promising results in simulated whole-brain NIS, particularly in terms of counting accuracy, also demonstrate the potential of our approach for neuroscience research.

References

1. Alahmari, S.S., Goldgof, D., Hall, L.O., Mouton, P.R.: A review of nuclei detection and segmentation on microscopy images using deep learning with applications to unbiased stereology counting. IEEE Trans. Neural Netw. Learn. Syst. (2022)
2. Banerjee, A., Poddar, R.: Enhanced visualization of tissue microstructures using swept-source optical coherence tomography and edible oil as optical clearing agent. Optik **267**, 169693 (2022)
3. Bennett, H.C., Kim, Y.: Advances in studying whole mouse brain vasculature using high-resolution 3D light microscopy imaging. Neurophotonics **9**(2), 021902 (2022)
4. Çiçek, Ö., Abdulkadir, A., Lienkamp, S.S., Brox, T., Ronneberger, O.: 3D U-net: learning dense volumetric segmentation from sparse annotation. In: Ourselin, S., Joskowicz, L., Sabuncu, M.R., Unal, G., Wells, W. (eds.) MICCAI 2016, Part II. LNCS, vol. 9901, pp. 424–432. Springer, Cham (2016). https://doi.org/10.1007/978-3-319-46723-8_49
5. Funke, J., Andres, B., Hamprecht, F.A., Cardona, A., Cook, M.: Efficient automatic 3D-reconstruction of branching neurons from EM data. In: 2012 IEEE Conference on Computer Vision and Pattern Recognition, pp. 1004–1011. IEEE (2012)
6. He, K., Gkioxari, G., Dollár, P., Girshick, R.: Mask R-CNN. In: Proceedings of the IEEE International Conference on Computer Vision, pp. 2961–2969 (2017)
7. Iqbal, A., Sheikh, A., Karayannis, T.: DeNerD: high-throughput detection of neurons for brain-wide analysis with deep learning. Sci. Rep. **9**(1), 13828 (2019)
8. Lin, Z., et al.: NucMM dataset: 3D neuronal nuclei instance segmentation at sub-cubic millimeter scale. In: de Bruijne, M., et al. (eds.) MICCAI 2021. LNCS, vol. 12901, pp. 164–174. Springer, Cham (2021). https://doi.org/10.1007/978-3-030-87193-2_16

9. Pachitariu, M., Stringer, C.: Cellpose 2.0: how to train your own model. Nat. Methods 1–8 (2022)
10. Pape, C., Beier, T., Li, P., Jain, V., Bock, D.D., Kreshuk, A.: Solving large multicut problems for connectomics via domain decomposition. In: Proceedings of the IEEE International Conference on Computer Vision Workshops, pp. 1–10 (2017)
11. Stringer, C., Wang, T., Michaelos, M., Pachitariu, M.: Cellpose: a generalist algorithm for cellular segmentation. Nat. Methods **18**(1), 100–106 (2021)
12. Vu, Q.D., Rajpoot, K., Raza, S.E.A., Rajpoot, N.: Handcrafted histological transformer (H2T): unsupervised representation of whole slide images. Med. Image Anal. **85**, 102743 (2023)
13. Wang, H., Ren, H., Leskovec, J.: Relational message passing for knowledge graph completion. In: Proceedings of the 27th ACM SIGKDD Conference on Knowledge Discovery & Data Mining, pp. 1697–1707 (2021)
14. Yang, B., et al.: DaXi-high-resolution, large imaging volume and multi-view single-objective light-sheet microscopy. Nat. Methods **19**(4), 461–469 (2022)
15. Yang, H., et al.: Deep learning-based six-type classifier for lung cancer and mimics from histopathological whole slide images: a retrospective study. BMC Med. **19**, 1–14 (2021)
16. You, S., et al.: High cell density and high-resolution 3D bioprinting for fabricating vascularized tissues. Sci. Adv. **9**(8), eade7923 (2023)

Adult-Like Phase and Multi-scale Assistance for Isointense Infant Brain Tissue Segmentation

Jiameng Liu[1], Feihong Liu[2,1], Kaicong Sun[1], Mianxin Liu[3], Yuhang Sun[1,4], Yuyan Ge[1,5], and Dinggang Shen[1,6,7(✉)]

[1] School of Biomedical Engineering, ShanghaiTech University, Shanghai, China
dgshen@shanghaitech.edu.cn
[2] School of Information Science and Technology, Northwest University, Xi'an, China
[3] Shanghai Artificial Intelligence Laboratory, Shanghai, China
[4] School of Biomedical Engineering, Southern Medical University, Guangzhou, China
[5] Institute of Artificial Intelligence and Robotics, Xi'an Jiaotong University, Xi'an, China
[6] Shanghai United Imaging Intelligence Co. Ltd., Shanghai, China
[7] Shanghai Clinical Research and Trial Center, Shanghai, China

Abstract. Precise brain tissue segmentation is crucial for infant development tracking and early brain disorder diagnosis. However, it remains challenging to automatically segment the brain tissues of a 6-month-old infant (isointense phase), even for manual labeling, due to inherent ongoing myelination during the first postnatal year. The intensity contrast between gray matter and white matter is extremely low in isointense MRI data. To resolve this problem, in this study, we propose a novel network with multi-phase data and multi-scale assistance to accurately segment the brain tissues of the isointense phase. Specifically, our framework consists of two main modules, *i.e.*, semantics-preserved generative adversarial network (SPGAN) and Transformer-based multi-scale segmentation network (TMSN). SPAGN bi-directionally transfers the brain appearance between the isointense phase and the adult-like phase. On the one hand, the synthesized isointense phase data augments the isointense dataset. On the other hand, the synthesized adult-like images provide prior knowledge to the ambiguous tissue boundaries in the paired isointense phase data. TMSN integrates features of multi-phase image pairs in a multi-scale manner, which exploits both the adult-like phase data, with much clearer boundaries as structural prior, and the surrounding tissues, with a larger receptive field to assist the isointense data tissue segmentation. Extensive experiments on the public dataset show that our proposed framework achieves significant improvement over the state-of-the-art methods quantitatively and qualitatively.

Keywords: Isointense infant tissue segmentation · Semantic-preserved GAN · Anatomical guidance segmentation

---

J. Liu and F. Liu—These authors contributed equally to this work.

H. Greenspan et al. (Eds.): MICCAI 2023, LNCS 14223, pp. 56–66, 2023.
https://doi.org/10.1007/978-3-031-43901-8_6

1   Introduction

Charting brain development during the first postnatal year is crucial for identifying typical and atypical changes of brain tissues, *i.e.*, grey matter (GM), white matter (WM), and cerebrospinal fluid (CSF), which can be utilized for diagnosis of autism and other brain disorders [16]. Early identification of autism is very important for effective intervention, and currently, an accurate diagnosis can be achieved as early as 12 months old [21]. Advancing diagnosis at an earlier stage could provide more time for greatly improving intervention [17]. Unfortunately, it remains challenging in obtaining a precise tissue map from the early infancy structural MRI (sMRI) data, such as T1 or (and) T2.

The acquisition of a mass of tissue maps relies on automatic segmentation techniques. However, the accuracy cannot be guaranteed, and the most difficult case is the segmentation of the isointense phase (6-9 months) data [18], as shown in Fig. 1, where the intensity distributions of GM and WM are highly overlapped compared with the adult-like phase ($\geq$ 9 months) [18] (clear boundaries between GM and WM). Such a peculiarity incurs two challenges to conventional segmentation methods: *i*) lack of well-annotated data for training automatic segmentation algorithms; and *ii*) GM and WM demonstrate a low contrast result in limited anatomical information for accurately distinguishing GM and WM, even with enough high-quality annotations as training data.

Conventional methods can mainly be classified into two categories: 1) the registration-based methods and 2) the learning-based methods, as introduced below. The registration-based methods usually utilize a single previous-defined atlas, for cross-sectional data, or a sequence of atlases, for longitudinal data-guided methods, to indirectly obtain the tissue maps [13,14,19,20]. Those methods require a substantial number of atlases and follow-up adult-like images to guide the segmentation process, which is known for low accuracy in segmenting infant tissues due to the rapid developmental changes in early life. The learning-based methods have gained significant prominence for individualized segmentation [8,22], which is also exploited to segment isointense brain tissues. For example, Nie et al. [9] used fully convolutional networks (FCNs) to directly segment tissue from MR images, and Wang et al. [23] employed an attention mechanism for better isointense infant brain segmentation. Also, Bui et al. proposed a 3D CycleGAN [2] to utilize the isointense data to synthesize adult-like data, which can be employed to make the segmentation of isointense data more accurate.

In this work, we propose a novel Transformer-based framework for isointense tissue segmentation via T1-weighted MR images, which is composed of two stages: *i.e.*, *i*) the semantics-preserved GAN (SPGAN), and *ii*) Transformer-based multi-scale segmentation network (TMSN). Specifically, SPGAN is a bidirectional synthesis model that enables both the isointense data synthesis using adult-like data and vice verse. The isointense structural MRI data is paired with the segmented tissue maps for extending the training dataset. Additionally, the synthesized adult-like data from isointense infant data are adopted to assist segmentation in TMSN. TMSN incorporates a Transformer-based cross-branch

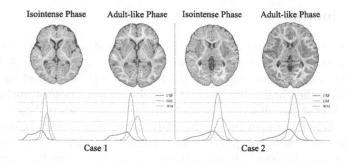

Fig. 1. T1-weighted images (the first row) and different tissue intensity distributions (the second row) of two typical infant brains scanned in isointense and adult-like phases, respectively

fusing (TCF) module which exploits supplementary tissue information from a patch with a larger receptive field to guide the local segmentation. Extensive experiments are conducted on the public dataset, National Database for Autism Research (NDAR) [5,12], and the results demonstrate that our proposed framework outperforms other state-of-the-art methods, particularly in accurately segmenting ambiguous tissue boundaries.

2 Methods

We propose a Transformer-based framework for isointense tissue segmentation, which includes two main modules: i) the semantic-preserved GAN (SPGAN), and ii) the Transformer-based multi-scale segmentation network (TMSN). An overview of the framework is provided in Fig. 2, SPGAN is designed to synthesize isointense or adult-like data to augment the training dataset and guide segmentation with anatomical information (Sect. 2.1). TMSN is designed for isointense tissue segmentation by utilizing the tissue information from the synthesized adult-like data (Sect. 2.2).

2.1 Semantics-Preserved Multi-phase Synthesis

Motivated by the effectiveness of classical unpaired image-to-image translation models [24,25], we propose a novel bi-directional generative model that synthesizes both isointense and adult-like data while introducing essential anatomical constraints to maintain structural consistency between the input images and the corresponding synthesized images. As shown in Fig. 2 (b.1), the SPGAN consists of two generators, G_{A2I} and G_{I2A}, along with their corresponding discriminators D_{A2I} and D_{I2A}. The adult-like infant images are aligned to the isointense infant images using affine registration methods [1], before training SPGAN.

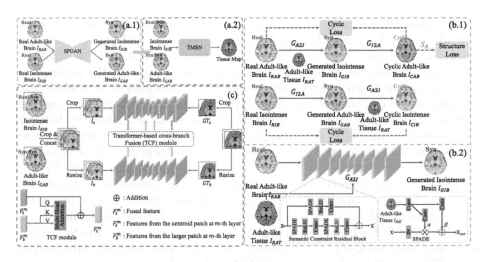

Fig. 2. (a) An overview of our proposed framework including: *i*) the semantics-preserved GAN (SPGAN), and *ii*) Transformer-based multi-scale segmentation (TMSN) network; (b) Detailed SPGAN architecture with the discriminator omitted for making the figure concise; (c) Detailed architecture of TMSN.

Isointense Phase Synthesis. The generator G_{A2I} is designed to synthesize isointense data from adult-like ones, which is constructed based on the 3D UNet [3] model, consisting of 4 encoder layers and 4 decoder layers. Inspired by GauGAN [10], we incorporate the spatial adaptive denormalization (SPADE) into G_{A2I} to maintain structural consistency between the synthesized isointense brain images and corresponding adult-like brain images, as shown in Fig. 2 (b.2). SPADE module takes the adult-like tissue maps to perform feature normalization during synthesis stage of isointense phase data. Thus, the generated isointense brain images can obtain consistent tissue structures with adult-like brain images. Specifically, X_{in} with a spatial resolution of $(H \times W \times D)$ represents the input feature map to each layer, consisting of a batch of N samples and C channels. For an arbitrary n-th sample at the c-th channel, the activation value associated with location (i,j,k) is denoted as $x_{n,c,i,j,k}$. The SPADE operator adopts two convolution blocks to learn the modulation maps $\alpha(M)$ and $\beta(M)$ for a given tissue map M, where the first convolution block makes the number of channels in the tissue map consistent with the number of channels in the feature map. The modulated value at (i,j,k) in each feature map is formulated as:

$$x'_{n,c,i,j,k} = \alpha(M)_{n,c,i,j,k} \frac{x_{n,c,i,j,k} - \mu_c}{\sigma_c} + \beta(M)_{n,c,i,j,k} \tag{1}$$

where μ_c and σ_c denote the mean and standard deviation, respectively. Additionally, we adopt PatchGAN [25] as the discriminator D_{A2I} to provide adversary loss to train the generator G_{A2I}.

Adult-Like Phase Synthesis. The generator G_{I2A} is designed to synthesize adult-like infant brain images from isointense infant brain images, which is employed to provide clear structural information for identifying the ambiguous tissue boundaries in isointense brain images. To ensure the synthesized adult-like infant brain images can provide tissue information as realistic and accurate as the real images, we utilize a pre-trained adult-like brain tissue segmentation network S_A (3D UNet) to preserve the structural similarity between the synthesized adult-like data and the real adult-like data and promote more reasonable anatomical structures in the synthesized images. To achieve that goal, during the training of SPGAN, we freeze the parameters of S_A and adopt the mean square error (MSE) to penalize the dissimilarity between the tissue probability maps of the real and synthesized brain images (extracted by the pre-trained segmentation model S_A). The MSE loss is formulated as $\mathcal{L}_{MSE} = MSE(S_A(I_{RAB}), S_A(I_{CAB}))$ where I_{RAB} and I_{CAB} denote the real adult-like brain images and synthesized adult-like brain images synthesized by G_{I2A}. The overall loss function of SPGAN is defined as:

$$\mathcal{L}_{SPGAN} = \gamma_1 \mathcal{L}_{G_{I2A}} + \gamma_2 \mathcal{L}_{G_{A2I}} + \gamma_3 \mathcal{L}_{MSE} + \gamma_4 \mathcal{L}_{cycle} \tag{2}$$

where $\mathcal{L}_{cycle}$ denotes the cycle consistency loss between two generators.

2.2   Transformer-Based Multi-scale Segmentation

The overview of our proposed segmentation network (TMSN) is shown in Fig. 2 (c). In order to guide the network with anatomical prior, TMSN takes the pair of isointense and the corresponding adult-like images as input. The isointense and adult-like images are concatenated and cropped to image patches and then, they are fed into the top branch of TMSN. Due to the fact that neighboring tissues can provide additional information and greatly improve segmentation performance, hence, we also employ a multi-scale strategy by taking an image pair of the larger receptive field as the input of the bottom branch. It is important to note that the image pairs of the top and bottom branches should have the same sizes.

Moreover, we fuse the features of the two branches at the same block using a novel Transformer-based cross-branch fusion (TCF) module, where the Transformer can better capture relationships across two different branches (with local and global features). Specifically, F_s^m and F_b^m denote the learned features of the two branches at the m-th stage, respectively. The TCF module treats the F_s^m and F_b^m as the input of the encoder and decoder in the multi-head Transformer to learn the relationships between the centroid tissues in the top branch and the corresponding surrounding tissues in the bottom branch and then fuses the bottom branch features with the ones of the top branch by utilizing the learned relationships. The proposed TCF module can be formulated as:

$$F_s^{m'} = F_s^m + [Q(F_s^m).(K(F_b^m))^\top] * V(F_b^m) \tag{3}$$

where Q, K, and V denote the query, key, and value in the Transformer.

Table 1. Ablation analysis of our proposed framework. DA: data augmentation; SE: structural enhancement; SegNet: single-branch segmentation network; SegNetMS: multi-branch segmentation network based on channel concatenation; SegNetMSAtt: proposed Transformer-based multi-branch segmentation network.

Method	DA	SE	Dice (%) ↑			HD (mm) ↓			ASD ×10 (mm) ↓		
			WM	GM	CSF	WM	GM	CSF	WM	GM	CSF
SegNet			93.68±0.55	92.45±0.62	93.31±0.76	5.44±1.56	11.19±1.86	8.69±1.19	1.44±0.17	1.17±0.10	0.88±0.09
SegNet	✓		94.31±0.54	93.35±0.60	94.50±0.64	5.34±1.27	11.26±2.01	10.60±2.40	1.30±0.16	1.01±0.09	0.70±0.07
SegNet		✓	94.01±0.61	93.28±0.61	94.92±0.58	7.40±2.11	8.19±2.62	10.15±2.29	1.43±0.19	1.00±0.10	0.65±0.06
SegNet	✓	✓	94.40±0.56	93.83±0.57	94.74±0.62	5.44±1.05	8.38±2.51	10.04±2.36	1.25±0.15	0.93±0.09	0.67±0.06
SegNetMS			95.02±0.41	94.50±0.46	95.44±0.48	6.51±1.94	8.29±2.36	7.95±1.16	1.12±0.13	0.84±0.09	0.57±0.05
SegNetMS	✓		95.38±0.37	94.99±0.44	95.79±0.51	4.96±1.17	8.87±2.32	8.20±1.75	1.04±0.11	0.77±0.08	0.53±0.05
SegNetMS		✓	95.88±0.41	95.40±0.42	96.18±0.43	4.86±0.94	8.11±2.79	7.87±1.72	0.94±0.12	0.70±0.08	0.48±0.04
SegNetMS	✓	✓	96.06±0.40	95.62±0.45	96.35±0.48	5.14±1.18	7.61±2.37	8.22±1.99	0.90±0.11	0.66±0.07	0.45±0.05
SegNetMSAtt			95.68±0.36	95.10±0.47	95.85±0.55	4.90±0.97	8.44±2.77	8.33±1.35	1.00±0.12	0.74±0.07	0.53±0.06
SegNetMSAtt	✓		95.82±0.36	95.40±0.46	96.23±0.48	4.82±0.78	**7.59±2.53**	8.32±1.63	0.95±0.12	0.68±0.07	0.47±0.05
SegNetMSAtt		✓	96.21±0.38	95.79±0.42	96.50±0.46	5.24±0.95	8.39±2.87	7.44±1.22	0.86±0.11	0.64±0.07	0.43±0.05
SegNetMSAtt	✓	✓	**96.31±0.36**	**95.89±0.41**	**96.60±0.42**	**4.71±0.99**	7.89±2.65	**7.30±1.24**	**0.84±0.11**	**0.62±0.07**	**0.42±0.04**

We take a hybrid Dice and focal loss to supervise the two segmentation branches as follows:

$$\mathcal{L}_{Seg} = \lambda_1 \mathcal{L}_s(I_s, GT_s) + (1 - \lambda_1)\mathcal{L}_b(I_b, GT_b) \tag{4}$$

where I_s and GT_s, respectively, denote the input and ground truth (GT) of the top branch, and I_b and GT_b represent the ones of the bottom branch. The final tissue segmentation results are obtained from the top branch.

3 Experiments and Results

3.1 Dataset and Evaluation Metrics

We evaluated our proposed framework for isointense infant brain tissue segmentation on the public dataset NDAR [5,12]. The NDAR dataset comprises T1-weighted brain images of 331 cases at the isointense phase and 368 cases at the adult-like phase (12-month-old), where 180 cases contain both time points. The GT annotations for GM, WM, and CSF were manually labeled by experienced radiologists. The data was aligned to MNI space and normalized to standard distribution using z-score normalization with dimensions of 182, 218, and 182 in the x, y, and z axis, respectively. The dataset was randomly divided into three subsets, *i.e.*, 70% for training, 10% for validation, and 20% for testing. For quantitative evaluation, we employed three assessment metrics including Dice score, Hausdorff distance (HD), and average surface distance (ASD) [15].

3.2 Implementation Details

Our proposed framework is implemented based on PyTorch 1.7.1 [11] and trained on a workstation equipped with two NVIDIA V100s GPUs. We employ the Adam optimizer [7] with momentum of 0.99 and weight decay of 0.001. The learning

Table 2. Quantitative comparison with the state-of-the-art (SOTA) methods.

Method	Dice (%) ↑			HD (mm) ↓			ASD ×10 (mm) ↓		
	WM	GM	CSF	WM	GM	CSF	WM	GM	CSF
3D UNet [3]	93.68±0.55	92.45±0.62	93.32±0.76	5.44±1.57	11.19±1.86	8.69±1.19	1.44±0.17	1.17±0.10	0.88±0.09
DualAtt [4]	95.21±0.34	94.61±0.41	95.51±0.47	5.24±1.22	10.96±1.89	7.94±1.42	1.07±0.11	0.85±0.08	0.56±0.04
SwinUNETR [6]	94.99±0.40	94.17±0.48	94.88±0.54	4.88±1.14	**7.60±2.58**	8.47±1.82	1.13±0.14	0.86±0.08	0.64±0.05
Ours	**96.31±0.36**	**95.89±0.41**	**96.60±0.42**	**4.71±0.99**	7.89±2.65	**7.30±1.23**	**0.84±0.11**	**0.62±0.07**	**0.42±0.04**

rate is set as 0.001 and reduced to 90% of the current learning rate every 50 epochs for both SPGAN and TMSN. For SPGAN the hyperparameters γ_1, γ_2, γ_2, γ_4 are set to 10, 10, 1 and 1 in our experiments, respectively. For TMSN, the patch sizes of the top and bottom branches are set as $128 \times 128 \times 128$ and $160 \times 160 \times 160$, respectively. The total parameter number of SPGAN and TMSN is 44.51M. The hyperparameter λ_1 in TMSN is set to 0.6. The inference for one single image took about 8.82 s. Code is available at this link.

3.3 Evaluation and Discussion

We conduct extensive experiments to evaluate the effectiveness of our infant brain tissue segmentation framework. Table 1 summarizes the results of the ablation study and demonstrates the contribution of each component in our framework. Particularly, SegNet represents the baseline model using a single scale, based on which, SegNetMS involves an additional branch that captures a larger receptive field and employs channel concatenation to fuse the features of these two branches. SegNetMSAtt replaces feature concatenation with the Transformer-based TCF module. Each of the above three configurations contains four different settings, as listed in Table 1. The results demonstrate the benefits of using both isointense phase data augmentation and adult-like phase structural enhancement.

Effectiveness of Multi-phase Synthesis. In Table 1, we demonstrate the effectiveness of the synthesis of isointense and adult-like phase images by the proposed synthesis network SPGAN. When comparing with the methods based on limited samples and augmented samples, we witness that data augmentation (DA) effectively improves segmentation performance. Furthermore, by comparing with the methods with or without structural enhancement (SE), we observe that the generated adult-like infant brain images provide richer and more reliable structural information to guide the identification of ambiguous tissue boundaries at the isointense phase. When combining DA and SE, an improvement of the segmentation performance on all evaluating metrics emerge, which demonstrates that the DA and SE can be effectively integrated to bring significant benefits for improving tissue segmentation performance.

Effectiveness of Multi-scale Assistance. In order to demonstrate the effectiveness of the additional branch in TMSN, which contains a larger receptive

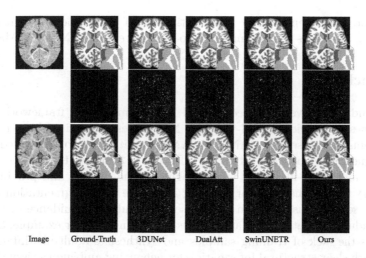

Image   Ground-Truth   3DUNet   DualAtt   SwinUNETR   Ours

Fig. 3. Qualitative comparison with SOTA methods: two representative cases with corresponding error maps.

field, we show the corresponding results as one part of the ablation study in Table 1. Comparing the multi-scale SegNetMS with the single-scale variant Seg-Net, we can see that SegNetMS improves the Dice score from 94.32% to 96.01% and achieves a decrease in HD from 7.95 mm to 6.99 mm on average. This indicates that the multi-scale strategy enables the segmentation model to make use of the intrinsic relations between the centroid tissues and the corresponding counterparts with a larger receptive field, which leads to great improvement in the segmentation performance. In our proposed model SegNetMSAtt, we replace the channel concatenation in SegNetMS with TCF. We can find that the TCF module can more effectively build up the correspondence between top and bottom branches such that the surrounding tissues can well guide the segmentation.

Comparison with State-of-the-Arts (SOTAs). We conduct a comparative evaluation of our framework against several SOTA learning-based tissue segmentation networks, including 1) the 3D UNet segmentation network [3], 2) the DualAtt network [4] utilizing both spatial and channel attention, and 3) the SwinUNETR network [6] employing a 3D Swin Transformer. From the results listed in Table 2, we can find that our framework achieves the highest average Dice score and lower HD and ASD compared with the SOTA methods.

To further illustrate the advanced performance of our framework, we provide a visual comparison of two typical cases in Fig. 3. It can be seen that the predicted tissue maps (the last column) obtained by our method are much closer to the GT (the second column), especially in the subcortical regions. We can draw the same conclusion from the error maps (the second and the fourth rows in Fig. 3) between the segmented tissue maps obtained by each method and GT. Both the quantitative improvement and the superiority from the qualitative results performance consistently show the effectiveness of our proposed segmentation framework, as well as the two main components, *i.e.*, SPGAN, and TMSN.

4 Conclusion

In this study, we have presented a novel Transformer-based framework to segment tissues from isointense T1-weighted MR images. We designed two modules, *i.e.*, *i*) semantic-preserved GAN (SPGAN), and *ii*) Transformer-based multi-scale segmentation network (TMSN). SPGAN is designed to synthesize both isointense and adult-like data, which augments the dataset and provides supplementary tissue constraint for assisting isointense tissue segmentation. TMSN is used to segment tissues from isointense data under the guidance of the synthesized adult-like data. Their advantages are distinctive. For example, SPGAN overcomes the lack of training samples and synthesizes adult-like infant brain images with clear structural information for enhancing ambiguous tissue boundaries. TMSN exploits the pair of isointense and adult-like phase images, as well as the multi-scale scheme, to provide more information for achieving accurate segmentation performance. Extensive experiments demonstrate that our proposed framework outperforms the SOTA approaches, which shows the potential in early brain development or abnormal brain development studies.

Acknowledgment. This work was supported in part by National Natural Science Foundation of China (No. 62131015 and 62203355), and Science and Technology Commission of Shanghai Municipality (STCSM) (No. 21010502600), and The Key R&D Program of Guangdong Province, China (No. 2021B0101420006).

References

1. Avants, B.B., Tustison, N., Song, G., et al.: Advanced normalization tools (ants). Insight J. **2**(365), 1–35 (2009)
2. Bui, T.D., Wang, L., Lin, W., Li, G., Shen, D.: 6-month infant brain MRI segmentation guided by 24-month data using cycle-consistent adversarial networks. In: 2020 IEEE 17th International Symposium on Biomedical Imaging (ISBI), pp. 359–362. IEEE (2020)
3. Çiçek, Ö., Abdulkadir, A., Lienkamp, S.S., Brox, T., Ronneberger, O.: 3D U-Net: learning dense volumetric segmentation from sparse annotation. In: Ourselin, S., Joskowicz, L., Sabuncu, M.R., Unal, G., Wells, W. (eds.) MICCAI 2016. LNCS, vol. 9901, pp. 424–432. Springer, Cham (2016). https://doi.org/10.1007/978-3-319-46723-8_49
4. Fu, J., et al.: Dual attention network for scene segmentation. In: Proceedings of the IEEE/CVF Conference on Computer Vision and Pattern Recognition, pp. 3146–3154 (2019)
5. Hall, D., Huerta, M.F., McAuliffe, M.J., Farber, G.K.: Sharing heterogeneous data: the national database for autism research. Neuroinformatics **10**, 331–339 (2012)

6. Hatamizadeh, A., Nath, V., Tang, Y., Yang, D., Roth, H.R., Xu, D.: Swin UNETR: swin transformers for semantic segmentation of brain tumors in MRI images. In: Crimi, A., Bakas, S. (eds.) Brainlesion: Glioma, Multiple Sclerosis, Stroke and Traumatic Brain Injuries. BrainLes 2021. LNCS, vol. 12962. Springer, Cham (2022). https://doi.org/10.1007/978-3-031-08999-2_22

7. Kingma, D.P., Ba, J.: Adam: a method for stochastic optimization. arXiv preprint arXiv:1412.6980 (2014)

8. Liu, J., et al.: Multi-scale segmentation network for Rib fracture classification from CT images. In: Lian, C., Cao, X., Rekik, I., Xu, X., Yan, P. (eds.) MLMI 2021. LNCS, vol. 12966, pp. 546–554. Springer, Cham (2021). https://doi.org/10.1007/978-3-030-87589-3_56

9. Nie, D., Wang, L., Gao, Y., Shen, D.: Fully convolutional networks for multi-modality isointense infant brain image segmentation. In: 2016 IEEE 13Th International Symposium on Biomedical Imaging (ISBI), pp. 1342–1345. IEEE (2016)

10. Park, T., Liu, M.Y., Wang, T.C., Zhu, J.Y.: GauGAN: semantic image synthesis with spatially adaptive normalization. In: ACM SIGGRAPH 2019 Real-Time Live!, p. 1 (2019)

11. Paszke, A., et al.: PyTorch: an imperative style, high-performance deep learning library. In: Advances in Neural Information Processing Systems 32 (2019)

12. Payakachat, N., Tilford, J.M., Ungar, W.J.: National Database for Autism Research (NDAR): big data opportunities for health services research and health technology assessment. Pharmacoeconomics 34(2), 127–138 (2016)

13. Prastawa, M., Gilmore, J.H., Lin, W., Gerig, G.: Automatic segmentation of MR images of the developing newborn brain. Med. Image Anal. 9(5), 457–466 (2005)

14. Shi, F., Yap, P.T., Fan, Y., Gilmore, J.H., Lin, W., Shen, D.: Construction of multi-region-multi-reference atlases for neonatal brain MRI segmentation. Neuroimage 51(2), 684–693 (2010)

15. Taha, A.A., Hanbury, A.: Metrics for evaluating 3D medical image segmentation: analysis, selection, and tool. BMC Med. Imaging 15(1), 1–28 (2015)

16. Tierney, A.L., Nelson, C.A., III.: Brain development and the role of experience in the early years. Zero Three 30(2), 9 (2009)

17. Wang, L., et al.: Anatomy-guided joint tissue segmentation and topological correction for 6-month infant brain MRI with risk of autism. Hum. Brain Mapp. 39(6), 2609–2623 (2018)

18. Wang, L., et al.: Benchmark on automatic six month old infant brain segmentation algorithms: the iSeg-2017 challenge. IEEE Trans. Med. Imaging 38(9), 2219–2230 (2019)

19. Wang, L., et al.: Integration of sparse multi-modality representation and anatomical constraint for isointense infant brain MR image segmentation. Neuroimage 89, 152–164 (2014)

20. Wang, L., et al.: Segmentation of neonatal brain MR images using patch-driven level sets. Neuroimage 84, 141–158 (2014)

21. Wang, L., Shi, F., Yap, P.T., Gilmore, J.H., Lin, W., Shen, D.: 4D multi-modality tissue segmentation of serial infant images (2012)

22. Wang, R., Lei, T., Cui, R., Zhang, B., Meng, H., Nandi, A.K.: Medical image segmentation using deep learning: a survey. IET Image Proc. 16(5), 1243–1267 (2022)

23. Wang, Z., Zou, N., Shen, D., Ji, S.: Non-local U-Nets for biomedical image segmentation. In: Proceedings of the AAAI Conference on Artificial Intelligence, vol. 34, pp. 6315–6322 (2020)

24. Yang, H., et al.: Unpaired brain MR-to-CT synthesis using a structure-constrained CycleGAN. In: Stoyanov, D., et al. (eds.) DLMIA/ML-CDS -2018. LNCS, vol. 11045, pp. 174–182. Springer, Cham (2018). https://doi.org/10.1007/978-3-030-00889-5_20
25. Zhu, J.Y., Park, T., Isola, P., Efros, A.A.: Unpaired image-to-image translation using cycle-consistent adversarial networks. In: Proceedings of the IEEE International Conference on Computer Vision, pp. 2223–2232 (2017)

Robust Segmentation via Topology Violation Detection and Feature Synthesis

Liu Li[1](✉), Qiang Ma[1], Cheng Ouyang[1], Zeju Li[1,2], Qingjie Meng[1],
Weitong Zhang[1], Mengyun Qiao[1], Vanessa Kyriakopoulou[3], Joseph V. Hajnal[3],
Daniel Rueckert[1,4], and Bernhard Kainz[1,5]

[1] Imperial College London, London, UK
liu.li20@imperial.ac.uk
[2] University of Oxford, Oxford, UK
[3] King's College London, London, UK
[4] Technical University of Munich, Munich, Germany
[5] FAU Erlangen-Nürnberg, Erlangen, Germany

Abstract. Despite recent progress of deep learning-based medical image segmentation techniques, fully automatic results often fail to meet clinically acceptable accuracy, especially when topological constraints should be observed, *e.g.*, closed surfaces. Although modern image segmentation methods show promising results when evaluated based on conventional metrics such as the Dice score or Intersection-over-Union, these metrics do not reflect the correctness of a segmentation in terms of a required topological genus. Existing approaches estimate and constrain the topological structure via persistent homology (PH). However, these methods are not computationally efficient as calculating PH is not differentiable. To overcome this problem, we propose a novel approach for topological constraints based on the multi-scale Euler Characteristic (EC). To mitigate computational complexity, we propose a fast formulation for the EC that can inform the learning process of arbitrary segmentation networks via topological violation maps. Topological performance is further facilitated through a corrective convolutional network block. Our experiments on two datasets show that our method can significantly improve topological correctness.

1 Introduction

Performance of automated medical image segmentation methods is widely evaluated by pixel-wise metrics, *e.g.*, the Dice score or the Intersection-over-Union [16]. Taking the average accuracy of each pixel/voxel into account, these evaluation methods are well suited to describe the overall segmentation performance. However, errors in some regions may be more important than others, *e.g.*, certain errors may lead to the misrepresented topology. In some cases, topological correctness is more important than pixel-wise classification correctness. For example, when segmentation is a fundamental step for surface reconstruction [4].

Supplementary Information The online version contains supplementary material available at https://doi.org/10.1007/978-3-031-43901-8_7.

H. Greenspan et al. (Eds.): MICCAI 2023, LNCS 14223, pp. 67–77, 2023.
https://doi.org/10.1007/978-3-031-43901-8_7

To address the above challenge, several methods explore the idea of topology constraints in image segmentation. Many utilize persistent homology (PH) [2, 3,9,24] by adding an additional topology loss term to the pixel-wise, *e.g.*, cross entropy, loss. These methods manipulate the prediction probability of critical points that are sensitive to topological changes. However, the detection of these critical points is neither computationally efficient nor accurate.

In this work, we propose a novel deep learning (DL)-based topology-aware segmentation method, utilizing the concept of the Euler Characteristic (EC) [7], which is an integer defined as the number of connected components (β_0) minus the number of holes (β_1) in a 2D image. We develop a DL-compatible extension to the classical EC calculation approach, *i.e.*, the Gray algorithm [7]. Given the difference in EC between prediction and ground truth (GT), which we refer to as *EC error*, we can visualize topology violations in the predictions, denoted as a *topology violation map*, by backpropagating the EC error with respect to the segmentations. With the resulting topology violation map, we further design an efficient correction network that can improve the connectedness of foreground elements by synthesising plausible alternatives that preserve learned topological priors (i.e. the EC).

Contributions. (1) This paper presents a novel DL-based method for fast EC calculation, which, to the best of our knowledge, is the first paper introducing DL-compatible EC computation. (2) The proposed EC calculation network enables the visualization of a topology violation map, by backpropagating the EC error between the predicted and GT segmentations. (3) Leveraging the topology violation map, we design a topology-aware feature synthesis network to correct regions with topological errors. It can be easily incorporated into any existing segmentation pipeline for topology correction. (4) We demonstrate the effectiveness of our method on two datasets with different foreground structures, and achieve superior performance compared to existing methods.

Related Work. Segmentation networks such as U-Net [21] are typically trained using loss functions such as binary cross-entropy or soft Dice loss. These metrics measure the pixel-wise overlap between the prediction and the GT. However, they do not guarantee topological correctness. Shape priors have been explored to mitigate this issue, *e.g.*, utilizing shape templates [14,19] with diffeomorphic deformation. These methods require the predictions to be very close to the shape priors, which is often unachievable in practice. Other methods inject shape information explicitly into the training process [20]. Implicit shape awareness has also been explored in [12,22] based on boundary and centerline information that can model connectivity. Unfortunately, neither of these methods is directly based on insights from topology theory, nor do they guarantee the corrections of topology.

So far, topology in segmentation has only been explored through the use of PH [1–3,5,9,15,23]. PH extracts topological features by tracking the birth and death of connected components, holes, and voids, as the filtration varies. In practice, these PH-based methods first detect critical points that are related to changes in topology and then manipulate the values of these critical points to encourage a predicted topology that matches the GT topology. This loss term is differentiable since it only changes the probability of critical points in the segmentation. Despite the available packages for computing PH [11,18], the polynomial computational complexity of the PH is a limitation that hinders its use in large-scale datasets. To address this unmet need, we propose an EC-based method that can serve as a guidance function for DL segmentation networks and can achieve real-time performance. Our code is publicly available here[1].

2   Method

Preliminaries. Algebraic topology utilizes algebraic techniques to study the properties of topological spaces. This field aims to identify and characterize topological invariants, such as the number of holes, the connectivity, or the homotopy groups of a space. For image segmentation, gray-scale images defined on grids $\mathbf{I} \in \mathbb{R}^{h \times w}$ can be modeled as cubical complexes $\mathcal{K}$, which is a set consisting of points, line segments, squares, and cubes. We show the modeling process in Fig. 1 in the appendix.

Euler Characteristic (EC), denoted as $\chi \in \mathbb{Z}$, is a topological invariant that describes the structure and properties of a given topological space. For a cubical complex $\mathcal{K}$ defined in a n-dimensional space, EC χ is defined as the alternating sum of the number of k-dimensional cells N_k, or alternating sum of the ranks of the k-dimensional homology groups, called Betti number β_k. Mathematically, EC can be formalized as:

$$\chi(\mathcal{K}) = \sum_{k=0}^{n}(-1)^k N_k = \sum_{k=0}^{n-1}(-1)^k \beta_k(\mathcal{K}). \tag{1}$$

Specifically, for 2-dimensional images, EC $\chi(\mathcal{K}) = N_0 - N_1 + N_2 = \beta_0 - \beta_1$, where N_0, N_1 and N_2 are the number of vertices, edges, and faces; and β_0 and β_1 are the number of connected components and holes.

Gray Algorithm for EC Calculation. The Gray algorithm is a conventional approach to calculate EC on binary images [7,26]. Specifically, instead of directly counting the number of vertices N_0, edges N_1, and faces N_2, the Gray algorithm calculates the EC by counting the number of occurrences of 10 different 2×2 patterns, called bit-quads, which are categorized into $\mathbf{K}_1$, $\mathbf{K}_2$ and $\mathbf{K}_3$. We show these patterns in Fig. 1. The number of occurrences of all the bit-quads from $\mathbf{K}_l$ is notated as M_l, $l \in \{1, 2, 3\}$. The EC χ can then be calculated by a linear transformation f:

$$\chi = f(M_1, M_2, M_3) = (M_1 - M_2 - 2 \times M_3)/4. \tag{2}$$

---

[1] https://github.com/smilell/Topology-aware-Segmentation-using-Euler-Characteristic.

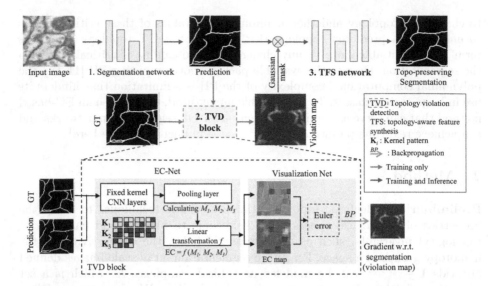

Fig. 1. Overview of our method in three steps: (1) We first generate an initial prediction from the segmentation network, where topological errors may occur. (2) Second, we propose a TVD block to detect topology-violating regions. (3) Based on this violation map, we use a topology-aware feature synthesis network to correct errors and improve the segmentation.

Overview. We illustrate our method in Fig. 1, which consists of three components: (1) a segmentation network, (2) a *topological violation detection* (TVD) block and (3) a *topology-aware feature synthesis* (TFS) network.

In (1), we utilize a U-Net to predict a segmentation probability map $\widehat{\mathbf{S}} \in [0,1]^{h \times w}$, with the supervision of GT segmentation $\mathbf{S} \in \{0,1\}^{h \times w}$. The predicted map may contain topological errors. (2) is the main contribution of our approach, consisting of two sub-nets: an *EC-Net* and a *Visualization Net*. The EC-Net takes the predicted $\widehat{\mathbf{S}}$ and GT segmentation $\mathbf{S}$ as inputs and predicts their corresponding EC maps $\widehat{\mathbf{X}} \in \mathbb{Z}^{h \times w}$ and $\mathbf{X} \in \mathbb{Z}^{h \times w}$. We then measure the Euler error $e \in \mathbb{R}$ by L1 distance as $e = \|\mathbf{X} - \widehat{\mathbf{X}}\|_1$. The Visualization Net takes e as input and produces a topology violation map $\mathbf{V} \in \mathbb{R}^{h \times w}$ that highlights the regions with topological errors. Finally, in (3), we design a TFS network, which learns to fill in the missing segmentation in the erroneous regions. This subnetwork takes the predicted segmentation $\widehat{\mathbf{S}}$ and violation map $\mathbf{V}$ as input and generates a topology-preserving segmentation , which is the final prediction.

During training, we use the TVD block (red arrows in Fig. 1) to generate the violation maps $\mathbf{V}$ to further guide the next feature synthesis network. During inference, we only run the first segmentation network and TFS network, as indicated by the upper blue arrows in Fig. 1 to produce the final topology-preserving segmentation results.

Topology-violation Detection. TVD consists of two parts: an EC-Net and a Visualization Net (Fig. 1 bottom). The Gray algorithm that calculates EC cannot be directly integrated into the gradient-based optimization process as is not differentiable. To overcome this problem, we propose a DL-compatible EC-Net as a CNN-based method that leverages the Gray algorithm to calculate the EC. The EC-Net serves as a function that maps the segmentation space to the Euler number space $g : \mathbf{S} \to \chi$. It is worth mentioning that in order to preserve spatial information, for each segmentation $\mathbf{S}$, EC-Net locally produces Euler numbers $\chi_{i,j}$ with the input of a segmentation patch $\mathbf{P}_{i,j,\Delta} = \mathbf{S}_{i:i+\Delta,j:j+\Delta}$, where Δ is the patch size. We can therefore obtain an Euler map $\mathbf{X}$ by combining all the local Euler numbers $\chi_{i,j}$.

EC-Net consists of three parts: 1) fixed kernel CNN layers, 2) an averaged pooling layer and 3) a linear transformation f. Following the Gray algorithm [7], we first utilize three CNN layers with fixed kernels to localize the bit-quads in the segmentation. The values of the kernels are the same as the bit-quads $\mathbf{K}_1 \in \{-1,1\}^{2\times2\times4}$, $\mathbf{K}_2 \in \{-1,1\}^{2\times2\times4}$, and $\mathbf{K}_3 \in \{-1,1\}^{2\times2\times2}$, as shown in Fig. 1. Note that we first binarize the prediction probability map $\widehat{\mathbf{S}}$ and further normalize it to $\{-1,1\}$. Therefore, if and only if the prediction has the same pattern as the bit-quads, it will be activated to 4 after convolution. Subsequently, we apply an average pooling layer to obtain the local number of bit-quads M_1, M_2, and M_3. The process can be summarized as:

$$M_{l,i,j,\Delta} = \Delta^2 \cdot \text{AvgPool}(\mathbb{1}(\mathbf{P}_{i,j,\Delta} * \mathbf{K}_l = 4)), \tag{3}$$

where $l \in \{1,2,3\}$, $*$ represents the convolutional operation, and $\mathbb{1}(\cdot)$ is an indicator function that equals 1 if and only if the input is true. Note that the patch size of average pooling is the same as the patch size of the segmentation Δ. Finally, following Eq. 2, a linear transformation is used to calculate the EC χ. During training, we separately take both $\mathbf{S}$ and $\widehat{\mathbf{S}}$ as the input of EC-Net and obtain their corresponding Euler maps $\mathbf{X}$ and $\widehat{\mathbf{X}}$.

In the second part of TVD, we measure the Euler error by the L1 distance as $e = \|\mathbf{X} - \widehat{\mathbf{X}}\|_1$. We can calculate the gradient of e with respect to the segmentation maps as the visualization of the EC error, called topology violation map $\mathbf{V} = \partial e / \partial \mathbf{S}$, which is the output of TVD.

Topology-Aware Feature Synthesis. In this module, we aim to improve the segmentation's topological correctness by utilizing the detected topology violation maps. We observe that topological errors are often caused by poor feature representation in the input image, *e.g.* blurry boundary regions. These errors are difficult to be corrected when trained from the image space. Therefore, we propose a TFS network that directly learns how to repair the topological structures from the segmentations.

During training, we mask out the topological error regions in the segmentation map and send it as the input of the feature synthesis network. We then use the GT segmentation to supervise this network to learn to repair these error regions. The input $\widetilde{\mathbf{S}}$ (with its element $\widetilde{S}_{i,j}$) of the feature synthesis network is

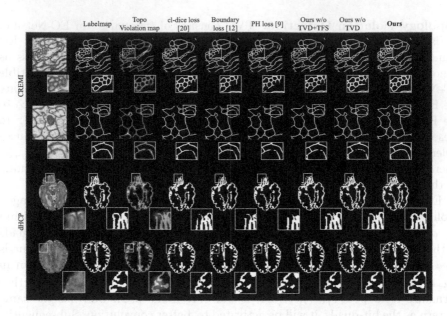

Fig. 2. Generated topology violation maps and the segmentation results. Our method can correct the disconnected structures compared with the existing method.

generated from the segmentation probability map $\widehat{\mathbf{S}}$ (with its element $\hat{S}_{i,j}$) and the topology violation map $\mathbf{V}$ (with its element $V_{i,j}$) as follows: We first filter out the coordinates $\{(i,j)\}$ with severe topological errors if $\mid V_{i,j} \mid \geq t$, where t is a filtration threshold. Then we replace the values of the segmentation at these coordinates by a random probability sampled from standard Gaussian distribution $\sigma \sim \mathcal{N}(0,1)$ and followed by a Sigmoid function to map it to $(0,1)$. The process of generating $\tilde{S}_{i,j}$ can be summarized as:

$$\tilde{S}_{i,j} = \begin{cases} \dfrac{1}{1+e^{-\sigma}}, & \mid V_{i,j} \mid \geq t, \\ \hat{S}_{i,j}, & \text{otherwise.} \end{cases} \qquad (4)$$

During inference, we feed the feature synthesis network with pure predictions $\widehat{\mathbf{S}}$. We show the effectiveness of our design in the next section.

3   Evaluation

Datasets: We conduct experiments on two datasets: CREMI for neuron boundary segmentation [6] and the developing human connectome project (dHCP[2]) dataset for fetal cortex segmentation. These datasets are employed to evaluate

---

[2] www.developingconnectome.org.

Table 1. Segmentation performance of our method on CREMI and dHCP data with two ablation experiments (Ours without TVD and TFS). e, e_0 and e_1 represent Betti error, Betti error 0 and Betti error 1. ASD is average surface distance. The unit for training time is seconds per batch of images. Best values are highlighted in **bold**.

Data	Method	Dice ↑	e ↓	e_0 ↓	e_1 ↓	ASD ↓	s/batch ↓
CREMI	cl-Dice loss [22]	83.48	10.403	7.445	2.958	1.612	0.047
	Boundary loss [12]	84.00	9.877	6.938	2.939	1.612	**0.041**
	PH loss [9]	84.07	9.102	6.213	2.889	1.690	8.772
	Ours w/o TVD+TFS	83.92	10.787	7.926	2.861	1.571	**0.041**
	Ours w/o TVD	85.17	5.183	2.968	2.215	1.245	0.083
	Ours	**85.25**	**4.970**	**2.793**	**2.177**	**1.241**	0.189
dHCP	cl-Dice loss [22]	88.21	2.573	1.081	1.492	0.232	0.022
	Boundary loss [12]	88.52	2.470	0.999	1.471	**0.210**	0.024
	PH loss [9]	88.17	2.546	1.075	1.471	0.232	3.300
	Ours w/o TVD+TFS	88.14	2.542	1.091	1.451	0.227	**0.021**
	Ours w/o TVD	88.45	2.220	0.824	1.395	0.228	0.042
	Ours	**88.56**	**2.032**	**0.737**	**1.295**	0.218	0.109

different topology challenges, where neuron boundaries in CREMI have a random and diverse number of holes β_1 and connected components β_0, while the cortex in the dHCP dataset is expected to have fixed β_1 and β_0.

CREMI consists of 3 subsets, each of which consists of 125 1250 × 1250 grayscale images and corresponding label maps. We randomly selected 100 samples from each subset (300 samples in total) as the training set and use the remaining 25 samples (75 samples in total) as validation and test set. We further divided each 1250 × 1250 image into 25 256 × 256 patches with an overlap of 8 pixels, in order to fit the GPU memory and enlarge the size of the training set. Thus, training and test sets consist of 7,500 and 1,875 samples, respectively.

The dHCP dataset has 242 fetal brain T2 Magnetic Resonance Imaging (MRI) scans with gestational ages ranging from 20.6 to 38.2 weeks. All MR images were motion corrected and reconstructed to 0.8 mm isotropic resolution for the fetal head region of interest (ROI) [10,13]. The images are affinely aligned to the MNI-152 coordinate space and clipped to the size of 144 × 192 × 192. We randomly split the data into 145 samples for training, 73 for testing, and 24 for validation. The GT cortical gray matter label is first generated by DrawEM method [17] and then refined manually to improve the segmentation accuracy.

Settings: The hyper-parameters in Eq. 3 and Eq. 4 are empirically chosen as $\Delta = 32$ for the CREMI dataset, $\Delta = 8$ for the dHCP dataset, and $t = 0.6$ for both datasets. We choose Δ to be higher for CREMI than for dHCP because: (1) the resolution of CREMI is higher and (2) the topology of the fetal cortex may change in smaller regions. We choose a default U-Net as the backbone for all methods for comparison, but our method can also be Incorporated into other segmentation frameworks. For the training process, we first use default cross-entropy loss to train the first segmentation network, then the TFS network. Note that the parameters in TVD are fixed.

Implementation: We use PyTorch 1.13.1 and calculate the Betti number with the GUDHI package [18]. The training time is evaluated on an NVIDIA RTX 3080 GPU with a batch size of 20.

Evaluation Metrics. Segmentation performance is evaluated by Dice score and averaged surface distance (ASD), and the performance of topology is evaluated by Betti errors, which is defined as: $e_i = | \beta_i^{\mathrm{pred}} - \beta_i^{\mathrm{gt}} |$, where $i \in \{0, 1\}$ indicates the dimension. We also report the mean Betti error as $e = e_0 + e_1$.

Quantitative Evaluation. We compare the segmentation performance of our method with three baselines which are proposed to preserve shape and topology: cl-Dice loss [22], boundary loss [12], warp loss [8] and PH loss [9]. For pixel-wise accuracy, our method achieves the best Dice score on both datasets, as demonstrated in Tab. 1. Despite the comparable ASD score in the dHCP dataset, we achieve significant improvements in terms of Betti errors. Compared to all baseline methods, our approach achieves an average Betti error of 4.970 on the CREMI dataset, representing an improvement of at least 4.132 (45.25%) in the mean average Betti error. We observe similar improvements for the dHCP dataset. We also conduct Wilcoxon signed-rank tests [25] comparing our method to all other baselines. The p-values are <0.05 across all metrics. Note that compared to the PH based loss [9], which is a similarly well grounded concept in algebraic topology, our method is 46.4 times faster at $0.189\,\mathrm{s/batch}$.

Qualitative Evaluation. As highlighted in Fig. 2, our method can effectively eliminate the topological errors in both the CREMI and dHCP datasets, outperforming all the other methods. For instance, in the first row of the CREMI dataset, none of the baseline methods could segment the tiny neuron structure when the boundary of the cells is blurry. Similarly, in the dHCP dataset, all the baseline methods fail to segment the fetal cortex as a closed surface, whereas our method can successfully resolve this issue. Second, we show the topology violation maps from the TVD block in the third column in Fig. 2, which indicate the topology error regions between the GT (second column) and the prediction from our first segmentation network (fourth column). For example, in the second raw of the CREMI evaluation, we observe that the topology violation map can highlight the disconnected boundary, therefore driving our method to correct these errors. We provide more visual examples in the supplementary materials.

Ablation Study. We first evaluate the effectiveness of our TVD design. We train our pipeline without the TVD block. Instead, we use the difference map between the prediction and GT as a substitute for the topology violation map. We also summarize the qualitative results in Fig. 2. Feature synthesis with the difference map can correct some of the false negative errors, however, the ring structure remains to be incorrect for the CREMI dataset. In contrast, our approach successfully predicts all the ring structures. Secondly, we further remove the TFS network. Quantitative results are provided in Tab. 1. The topology performance without TVD+TFS is significantly inferior to our method in terms of Betti errors, which illustrates the effectiveness of our design.

Discussion. This study sheds new light on improving and evaluation of topology-aware medical image segmentation. We observe that most existing methods either do not consider topological constraints, or are limited by their high computational complexity. As a computation-efficient block, our method can be easily integrated into existing segmentation methods to improve the topological structure. A limitation for topology-aware segmentation methods is that they are easy to be affected by noises, so they might be more suitable to datasets with clear topology structures.

4 Conclusion

We propose a novel EC-based method to include topology constraints in the segmentation network. Different from PH-based approaches, our method has a distinct advantage in computational efficiency while providing improved performance. In this paper, we generate a topology violation map from TVD and employ a post-processing feature synthesis network to correct topological errors. We believe this map is valuable and could be explored for other scenarios, such as serving as a spatial prior to regularize various loss functions.

Acknowledgements. This project is supported by Lee Family Scholarship from Imperial College London. HPC resources are provided by the Erlangen National High Performance Computing Center (NHR@FAU) of the Friedrich-Alexander-Universität Erlangen-Nürnberg (FAU) under the NHR project b143dc. NHR funding is provided by federal and Bavarian state authorities. NHR@FAU hardware is partially funded by the German Research Foundation (DFG) - 440719683. Support was also received by the ERC - project MIA-NORMAL 101083647 and DFG KA 5801/2-1, INST 90/1351-1.

References

1. Byrne, N., Clough, J.R., Valverde, I., Montana, G., King, A.P.: A persistent homology-based topological loss for CNN-based multiclass segmentation of CMR. IEEE Trans. Med. Imaging **42**(1), 3–14 (2022)
2. Clough, J.R., Byrne, N., Oksuz, I., Zimmer, V.A., Schnabel, J.A., King, A.P.: A topological loss function for deep-learning based image segmentation using persistent homology. IEEE Trans. Pattern Anal. Mach. Intell. **44**(12), 8766–8778 (2020)

3. Clough, J.R., Oksuz, I., Byrne, N., Schnabel, J.A., King, A.P.: Explicit topological priors for deep-learning based image segmentation using persistent homology. In: Chung, A.C.S., Gee, J.C., Yushkevich, P.A., Bao, S. (eds.) IPMI 2019. LNCS, vol. 11492, pp. 16–28. Springer, Cham (2019). https://doi.org/10.1007/978-3-030-20351-1_2

4. Dale, A.M., Fischl, B., Sereno, M.I.: Cortical surface-based analysis: I. segmentation and surface reconstruction. Neuroimage **9**(2), 179–194 (1999)

5. de Dumast, P., Kebiri, H., Atat, C., Dunet, V., Koob, M., Cuadra, M.B.: Segmentation of the cortical plate in fetal brain MRI with a topological loss. In: Sudre, C.H., et al. (eds.) UNSURE/PIPPI -2021. LNCS, vol. 12959, pp. 200–209. Springer, Cham (2021). https://doi.org/10.1007/978-3-030-87735-4_19

6. Funke, J., et al.: Large scale image segmentation with structured loss based deep learning for connectome reconstruction. IEEE Trans. Pattern Anal. Mach. Intell. **41**(7), 1669–1680 (2018)

7. Gray, S.B.: Local properties of binary images in two dimensions. IEEE Trans. Comput. **100**(5), 551–561 (1971)

8. Hu, X.: Structure-aware image segmentation with homotopy warping. Adv. Neural. Inf. Process. Syst. **35**, 24046–24059 (2022)

9. Hu, X., Li, F., Samaras, D., Chen, C.: Topology-preserving deep image segmentation. In: Advances in Neural Information Processing Systems 32 (2019)

10. Kainz, B., et al.: Fast volume reconstruction from motion corrupted stacks of 2D slices. IEEE Trans. Med. Imaging **34**(9), 1901–1913 (2015)

11. Kaji, S., Sudo, T., Ahara, K.: Cubical Ripser: software for computing persistent homology of image and volume data. arXiv:2005.12692 (2020)

12. Kervadec, H., Bouchtiba, J., Desrosiers, C., Granger, E., Dolz, J., Ayed, I.B.: Boundary loss for highly unbalanced segmentation. In: International Conference on Medical Imaging with Deep Learning, pp. 285–296. PMLR (2019)

13. Kuklisova-Murgasova, M., Quaghebeur, G., Rutherford, M.A., Hajnal, J.V., Schnabel, J.A.: Reconstruction of fetal brain MRI with intensity matching and complete outlier removal. Med. Image Anal. **16**(8), 1550–1564 (2012)

14. Lee, M.C.H., Petersen, K., Pawlowski, N., Glocker, B., Schaap, M.: Tetris: Template transformer networks for image segmentation with shape priors. IEEE Trans. Med. Imaging **38**(11), 2596–2606 (2019)

15. Li, L., et al.: Fetal cortex segmentation with topology and thickness loss constraints. In: Baxter, J.S.H., et al. (eds.) Ethical and Philosophical Issues in Medical Imaging, Multimodal Learning and Fusion Across Scales for Clinical Decision Support, and Topological Data Analysis for Biomedical Imaging. EPIMI ML-CDS TDA4BiomedicalImaging 2022. LNCS, vol. 13755. Springer, Cham (2022). https://doi.org/10.1007/978-3-031-23223-7_11

16. Maier-Hein, L., Menze, B., et al.: Metrics reloaded: pitfalls and recommendations for image analysis validation. arXiv. org (2206.01653) (2022)

17. Makropoulos, A.: The developing human connectome project: a minimal processing pipeline for neonatal cortical surface reconstruction. Neuroimage **173**, 88–112 (2018)

18. Maria, C., Boissonnat, J.-D., Glisse, M., Yvinec, M.: The Gudhi library: simplicial complexes and persistent homology. In: Hong, H., Yap, C. (eds.) ICMS 2014. LNCS, vol. 8592, pp. 167–174. Springer, Heidelberg (2014). https://doi.org/10.1007/978-3-662-44199-2_28

19. McInerney, T., Terzopoulos, D.: Deformable models in medical image analysis: a survey. Med. Image Anal. **1**(2), 91–108 (1996)

20. Oktay, O., et al.: Anatomically constrained neural networks (acnns): application to cardiac image enhancement and segmentation. IEEE Trans. Med. Imag. **37**(2), 384–395 (2017)
21. Ronneberger, O., Fischer, P., Brox, T.: U-Net: convolutional networks for biomedical image segmentation. In: Navab, N., Hornegger, J., Wells, W.M., Frangi, A.F. (eds.) MICCAI 2015. LNCS, vol. 9351, pp. 234–241. Springer, Cham (2015). https://doi.org/10.1007/978-3-319-24574-4_28
22. Shit, S., et al.: clDice-a novel topology-preserving loss function for tubular structure segmentation. In: Proceedings of the IEEE/CVF Conference on Computer Vision and Pattern Recognition, pp. 16560–16569 (2021)
23. Stucki, N., Paetzold, J.C., Shit, S., Menze, B., Bauer, U.: Topologically faithful image segmentation via induced matching of persistence barcodes. arXiv preprint arXiv:2211.15272 (2022)
24. Stucki, N., Paetzold, J.C., Shit, S., Menze, B., Bauer, U.: Topologically faithful image segmentation via induced matching of persistence barcodes. In: International Conference on Machine Learning, pp. 32698–32727. PMLR (2023)
25. Wilcoxon, F.: Individual comparisons by ranking methods. In: Kotz, S., Johnson, N.L. (eds.) Breakthroughs in Statistics. Springer Series in Statistics. Springer, NY (1992). https://doi.org/10.1007/978-1-4612-4380-9_16
26. Yao, B., He, L., Kang, S., Chao, Y., Zhao, X.: A novel bit-quad-based Euler number computing algorithm. Springerplus **4**, 1–16 (2015)

GL-Fusion: Global-Local Fusion Network for Multi-view Echocardiogram Video Segmentation

Ziyang Zheng[1], Jiewen Yang[1], Xinpeng Ding[1], Xiaowei Xu[2(✉)], and Xiaomeng Li[1(✉)]

[1] The Hong Kong University of Science and Technology, Hong Kong SAR, China
eexmli@ust.hk
[2] Guangdong Cardiovascular Institute, Guangdong Provincial People's Hospital (Guangdong Academy of Medical Sciences), Southern Medical University, Guangzhou, China
xiao.wei.xu@foxmail.com

Abstract. Cardiac structure segmentation from echocardiogram videos plays a crucial role in diagnosing heart disease. The combination of multi-view echocardiogram data is essential to enhance the accuracy and robustness of automated methods. However, due to the visual disparity of the data, deriving cross-view context information remains a challenging task, and unsophisticated fusion strategies can even lower performance. In this study, we propose a novel **G**obal-**L**ocal fusion (**GL-Fusion**) network to jointly utilize multi-view information globally and locally that improve the accuracy of echocardiogram analysis. Specifically, a **M**ulti-view **G**lobal-based **F**usion **M**odule (MGFM) is proposed to extract global context information and to explore the cyclic relationship of different heartbeat cycles in an echocardiogram video. Additionally, a **M**ulti-view **L**ocal-based **F**usion **M**odule (MLFM) is designed to extract correlations of cardiac structures from different views. Furthermore, we collect a multi-view echocardiogram video dataset (MvEVD) to evaluate our method. Our method achieves an 82.29% average dice score, which demonstrates a 7.83% improvement over the baseline method, and outperforms other existing state-of-the-art methods. To our knowledge, this is the first exploration of a multi-view method for echocardiogram video segmentation. Code available at: https://github.com/xmed-lab/GL-Fusion

Keywords: Multi-view fusion · Echocardiogram videos · Cardiac structure segmentation

1 Introduction

Accurate segmentation of the cardiac structure from echocardiogram videos is integral to several analysis tasks [11] and has a significant impact on clinical

---

Z. Zheng and J. Yang—Two authors contributed equally to this work.
Z. Zheng—Work completed during the internship at HKUST.

© The Author(s), under exclusive license to Springer Nature Switzerland AG 2023
H. Greenspan et al. (Eds.): MICCAI 2023, LNCS 14223, pp. 78–88, 2023.
https://doi.org/10.1007/978-3-031-43901-8_8

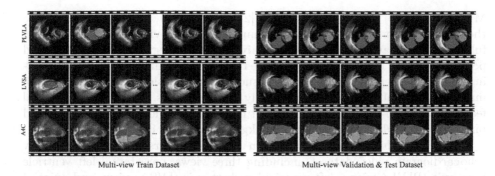

Multi-view Train Dataset   Multi-view Validation & Test Dataset

Fig. 1. Examples of multi-view echocardiogram dataset MvEVD, including PLVLA, LVSA, and A4C from top to bottom row. The colours red, green, blue, and cyan denote the LV, RV, LA, and RA cardiac structures. Our train set is sparsely annotated (5 frames per video), while the validation set and test set are fully annotated for each video frame. (Color figure online)

practice [26]. For example, segmentation of the left ventricle (LV) enables quantifiable functional analysis of the heart, facilitating the detection and diagnosis of heart diseases [3,20,21]. Compared with the single view segmentation, multi-view information is crucial to diagnose heart disease, *e.g.*, the diagnosis of congenital heart disease requires the analysis of four views: parasternal long-axis view (PSLAX), parasternal short-axis view (PSSAX), subxiphoid long-axis view (SXLAX), and suprasternal long-axis view (SSLAX) [22]. Consequently, to assist clinicians in diagnostic decision-making, there is a high demand for developing automated multi-view cardiac structure segmentation methods from echocardiogram videos in clinical practice. Existing echocardiogram segmentation approaches are primarily designed for single-view images or videos. For instance, Li et al. [26] proposed a dynamic neural network capable of segmenting the LV from a long-axis fetal echocardiogram. In comparison, Leclerc et al. [12] evaluated an encoder-decoder deep convolutional neural network that independently segments two and four-chamber images. However, these approaches have not addressed multi-view segmentation, where multi-view segmentation methods already exist in other medical domains, such as the CT-MRI [9,17,18], multi-view cardiac MRI [4,13,15,16], multi-view mammogram [2], and longitudinal multiple sclerosis [1]. Applying the proposed methods to multi-view echocardiogram segmentation presents several limitations: (1) Some methods are built for specific datasets and cannot adapt to our task. For instance, UMCT [25] designated supervised training in one view by generating pseudo segmentation labels from other views, but has limitations in our task due to the significant gaps between views. In contrast, InfoTrans [13] is designed for transmitting information between views instead of fusion them. While VCN [6] employs contrastive learning to predict volume but may not be suitable for our task since defining positive and negative pairs is challenging due to the significant gap between views and labels. (2) Methods such as JOIN [2], ROI-based fine-grained CNN [14],

MIMTP [1], MV U-Net [4], MV-CNN [23], and Type-I, II, III [9] concatenate the features or predicted probability maps of different views and then apply a fully-connected layer. However, these naive fusion strategies have shown limited performance and may even lead to worse results; see results in Table 1. (3) Existing multi-view segmentation methods such as TransFusion [15] and rDLA [16] mainly apply multi-view fusion with only global features. However, using global features for multi-view fusion may result in tangling the foreground/background pixels [10] or leads to high levels of background noise in echocardiograms.

To address this limitation, as shown in Fig. 1, we first collect a multi-view echocardiogram video dataset, including three views: parasternal left ventricle long axis (PLVLA view), left ventricular short axis (LVSA) view, and apical 4 chamber (A4C) view. Different views of echocardiograms contain annotations for different chambers, such as, the PLVLA view contains the left ventricle (LV) and right ventricle (RV), the LVSA view contains the LV and RV, and the A4C view contains the LV, left atrium (LA), right atrium (RA), and RV. Furthermore, we propose a novel global-local fusion (GL-Fusion) network for multi-view echocardiogram video segmentation, where GL-Fusion includes a multi-view local-global fusion module designed to aggregate information from different views and improve the representation of each view. The GL-Fusion comprises two components. First, a multi-view global fusion module (MGFM) interacts with the global semantics between different views and thus enhances the representation of each view. Second, since the global semantics may contain a significant amount of noisy information, a multi-view local fusion module (MLFM) is introduced to encourage the model to focus on foreground information.

In addition to capturing multi-view information, we propose a novel dense cycle loss designed to utilize unlabelled video data for improved representation learning. Our motivation is based on the idea that standard multi-view data is obtained from the same patient and under the same stable conditions, without abnormal behaviours such as suffocating or exercising, ensuring consistent cardiac cycles. Previous work [7] proposed an unsupervised method called cycle loss, which trains the model with unlabelled frames based on the heartbeat cycle's characteristics. Nevertheless, the proposed cycle loss only focuses on a pair in two different cycles but ignores possibly similar images that may appear simultaneously in a systolic or diastolic period, resulting in features from similar frames being considered distant. To address this issue, our dense cycle loss examines all possible pairings throughout the heartbeat cycle. In summary, our contributions are as follows:

- To the best of our knowledge, this is the first study to examine multi-view echocardiogram video segmentation.
- Our proposed GL-Fusion uses a multi-view local-global fusion module to combine information from different views and improve the representation of each view.
- We further design a dense cycle loss that utilizes unlabelled data to enforce feature similarity based on temporal cyclicality.

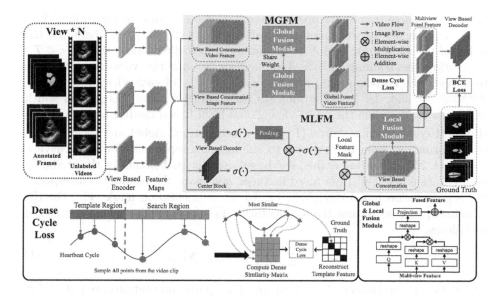

Fig. 2. The overview framework of GL-fusion. The Multi-view Global-based Fusion Module (MGFM) is proposed for global context information extraction and introduces dense cycle loss to devise the enforcement of the similarity of dense features between two heartbeat cycles from an echocardiogram video. The proposed Multi-view Local-based Fusion Module (MLFM) focuses on mining the correlation of local features of chambers in a different view.

- Extensive experiments demonstrate our method improved performance over existing methods, achieving an average dice score of 0.81. We plan to make our code publicly available upon paper acceptance.

2 Methodology

2.1 The Overall Framework

Figure 2 shows the overall pipeline of our proposed Multi-view Echocardiogram Global-Local Fusion Network (GL-Fusion), which consists of four main components: a view-based encoder, a multi-view global-local fusion module, a dense cycle loss module and a view-based decoder, where view-based indicate that parameters of the network of each independent view are non-shared. In our experiment, we use DeeplabV3 [5] as our view-based encoder and decoder.

Formally, we denoted the sample echocardiogram videos as $\mathbf{V} = \{\mathbf{X}^i\}_{i=1}^{V}$, where $\mathbf{X}^i \in \mathbb{R}^{C \times H \times W \times T}$ is the i-th view video and V is the number of views, and, C, H, W and T indicate the channels, height, width, and length of input images. Each video consists of T frames, $i.e.$, $\mathbf{X}^i = \{\mathbf{x}_t^i\}_{t=1}^{T}$, where T remain the same for different view and $\mathbf{x}_t^i \in \mathbb{R}^{C \times H \times W}$ indicate t-th frame of i-th view video.

Since only sparse frames are provided segmentation annotation for training in a video, thus we denote the annotation frame pair as $\{\mathbf{x}_{t_n}^i, \mathbf{y}_{t_n}^i\}_{n=1}^N$, where t_n is the index of the annotation and N is the number of labelled frames that $N \ll T$.

During the training, We feed the videos $\mathbf{V}$ into the view-based encoder to extract the corresponding feature maps $\{\mathbf{F}^i\}_{i=1}^V$ of each view, where $\mathbf{F}^i \in \mathbb{R}^{D \times h \times w \times T}$, and, D, h and w indicate the channel number, height and width of feature maps. Then the multi-view global-local fusion module aims to obtain the multi-view fused features $\{\overline{\mathbf{F}}^i\}_{i=1}^V$, which extract global and local semantics information from other views to enhance the representation of each view (See Sect. 2.2). Following is the view-based decoder that generates the predicted segmentation result $\mathbf{y}^i$ from fused features, and maps the results to corresponding segmentation annotation, i.e., $\hat{\mathbf{y}}_{t_n}^i$ to the segmentation masks $\mathbf{y}_{t_n}^i$. For the annotated frames, we use the segmentation loss to supervise them, formulated as follows:

$$\mathcal{L}_{seg} = \sum_{i=1}^V \sum_{t_n=1}^N \mathcal{L}_{bce}(\hat{\mathbf{y}}_{t_n}^i, \mathbf{y}_{t_n}^i), \tag{1}$$

where $\mathcal{L}_{bce}$ is the Binary Cross Entropy. The sparse annotations are only a few frames in the whole video; thus can not obtain a robust model. To leverage a large number of unlabelled frames, we design the dense cycle loss $\mathcal{L}_{cyc}$ to enforce temporal feature similarity of videos based on cyclicality; See Sect. 2.3. The overall loss function of our model is as follows:

$$\mathcal{L} = \mathcal{L}_{seg} + \alpha \mathcal{L}_{cyc}, \tag{2}$$

where α is the hyper-parameter to control the weight between two losses. In the following, we will illustrate the multi-view global-local fusion module and the dense cycle loss in detail.

2.2   Multi-view Global-Local Fusion Module

In this section, we describe the multi-view global-local fusion module that aggregates the information from different views to enhance their feature representation. To this end, we first concatenate extracted feature $\{\mathbf{F}^i\}_{i=1}^V$ from different views in a view-wise manner to obtain $\mathbf{F} = \{\mathbf{f}_t\}_{t=1}^T$, where $\mathbf{f}_t$ is the t-th feature vector in $\mathbf{F}$, and $\mathbf{f}_t \in \mathbb{R}^{D \times V \times w \times h}$. Then, we describe the multi-view global and local fusion with $\mathbf{F}_{global}$ and $\mathbf{F}_{local}$, respectively.

Multi-view Global Fusion. In order to enhance the representation of each view, we propose the global-based fusion module (MGFM) to interact with the global semantics between different views. To this end, we introduce a view-wise non-local block, which extracts the context information across views. Similarly to the previous research [8,24] that applied attention to fuse the information, we here introduce the view-wise attention module to aggregate the cross-view information (see Fig. 2). Then fused feature $\overline{\mathbf{F}}_{global}$ will be sent to both compute the dense cycle loss and cooperate with the local fused feature for segmentation prediction.

Multi-view Local Fusion. Since each view represents different morphological information of the heart and may contain the same cardiac structure as others, for example, the view PLVLA and LVSA both contain left ventricle(LA) and right ventricle(RV). Hence, extracting the local feature that represents the cardiac structure can contribute to feature fusion more efficiently. In this module, the extracted feature $\mathbf{F}_{local}$ will first pass to both the view-based decoder and a center block, where the decoder and center block has the same components with different output. The decoder provides the pseudo label $\{\hat{y}^i\}_{i=1}^V$ of different cardiac structures. A center block is introduced to acquire the weight $\{w^i\}_{i=1}^V$ of $\{\hat{y}^i\}_{i=1}^V$ and compute the local feature masks $\{\mathcal{M}^i\}_{i=1}^V$ as Eq. 3,

$$\mathcal{M}^i = \sigma(pooling(\sigma(\hat{y}^i)) \times \sigma(w^i)), \tag{3}$$

where weight w has the greatest volume in the central area of the segmented regions and attenuation with distance, σ denotes the sigmoid function and $\mathcal{M} \in \mathbb{R}^{1 \times H \times W \times T}$. These masks highlight features with a stronger intensity that are closer to the object center, while discarding background information that is farther away from the center. This selection is based on the understanding that morphological information should remain consistent closer to the center. In the final, similar to the process of MLFM, the view-wise local feature will be conducted view-wise concatenation operation and multiplied with local feature mask $\{\mathcal{M}^i\}_{i=1}^V$. Then sent to the view-wise attention module to acquire the local fused feature $\overline{\mathbf{F}}_{local}$.

2.3 Dense Cycle Loss

In echocardiogram videos, since only sparse annotation is available for the supervised training, involving the unlabelled data for our training and enhancing the performance is a challenge. The previous research [7] proposes an unsupervised method named cycle loss, which jointly trains the model with the unlabelled frames according to the characteristic of the heartbeat cycle. However, the proposed cycle loss considers only one clip in an iteration, which has the possibility to match frames that are morphologically identical but not in the same state, such as the search region being end-diastole while the template region is end-systole.

Thus, we propose the dense cycle loss, which considers all the possible matching across all template and search regions in each view independently. For the multi-view fused feature $\overline{\mathbf{F}}_{global}$ of each video will be separated to template region P^i and search region Q^i with a ratio in 2:3 according to total frame length T. Then we densely sample all feature intervals $\{p_1^i, ..., p_n^i\}$ from P^i and $\{q_1^i, ..., q_m^i\}$ from Q^i, respectively, both sampling use the same chunk size s and in our experiment, n and m is $\frac{2}{5} \times \frac{T}{s}$ and $\frac{3}{5} \times \frac{T}{s}$. Then we compute the similarity between candidate interval p_k^i and target intervals q_j^i of Q^i.

$$\alpha_j^i = \sum \mathcal{W}(\{p\}_k^i, \{q\}_j^i) \times \{q\}_j^i, \tag{4}$$

Table 1. The comparison with other methods. all results are reported in Dice Score.

	Method	PLVLA	LVSA	A4C	Average Dice (%)
Single-view	DeeplabV3 [5]	70.93	75.14	77.33	74.46
	U-Net [19]	73.35	77.57	76.60	75.84
	CSS [7]	79.09	79.70	77.71	78.83
Fusion-based	Early-fusion	79.78	77.07	77.58	78.14
	Mid-fusion	77.89	76.75	72.44	75.69
	Late-fusion	71.62	75.31	74.68	73.87
	TransFusion [15]	78.79	80.23	59.31	72.78
	Ours	**83.84**	**81.76**	**81.28**	**82.29**

where $\mathcal{W}(\cdot)$ is the computation of the similarity matrix. The similarity will be used as the weight to reconstruct the feature interval $\tilde{p}_k^i$. Then we back to template region P^i and compute the similarity between $\tilde{p}_k^i$ and all feature intervals$\{p_1^i, ..., p_n^i\}$ in P^i. Then we consider the index of p_k^i as one-hot label g of the most similar interval of $\tilde{p}_k^i$ and compute view-wise cycle loss $\mathcal{L}_{cyc}$ with label g as shown in the following equation:

$$\mathcal{L}_{cyc} = \sum_{i=1}^{V} \sum_{j \in P^i} \mathbb{1}_{j=g} log(\alpha_j^i) \tag{5}$$

3   Experiment

Datasets. We collect a large multi-view echocardiogram video dataset named **MvEVD** from one medical institution, with a total of 254 sparsely annotated videos and 10 fully annotated videos with 800×600 resolution across three cardiac views (PLVLA, LVSA and A4C view). Each video includes 5 annotated frames. The average length of each video is larger than 100 frames that are able to cover more than one cardiac cycle.

Implementation Details. We use the model DeeplabV3 [5] as our view-based encoder and decoder, and select Adam optimizer for the model training with initial learning rate as $3e^{-4}$ and weight decay of $1e^{-5}$. When training, we use all sparsely annotated videos. All annotated frames are selected to supervise training while randomly selecting 40 consecutive frames from videos for semi-supervised training. The training batch size of annotated images and unlabeled videos is 8 and 1, respectively. In the final, we use CosineAnnealing as a scheduler and set the total training epoch to 100. The framework is built with Pytorch with 4 NVIDIA RTX3090 GPUs for training. For the data augmentation in the training stage, we resize each frame in 144×144 size and then randomly crop them to 112×112.

Table 2. Effectiveness of MGFM and MLFM. This table shows the performance of the Global and Local fusion modules

	MGFM	MLFM	Avg. Dice(%)
Base	✗	✗	74.46
Base+MGFM	✓	✗	80.20
Base+MLFM	✗	✓	78.41
Ours	✓	✓	**82.29**

Table 3. Effectiveness of Cyc. and Dense Cyc.. This table shows the effectiveness of vanilla cycle loss [7] (Noted by Cyc.) and our proposed dense cycle (Noted by Dense Cyc.).

	Cyc.	Dense Cyc.	Avg. Dice(%)
Fusion-only	✗	✗	80.36
Fusion+Cyc.	✓	✗	79.33
GL-Fusion	✓	✓	**82.29**

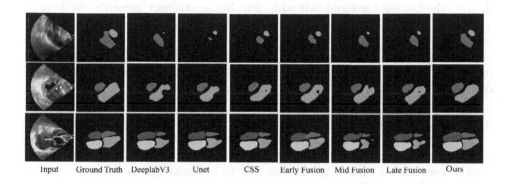

Input   Ground Truth   DeeplabV3   Unet   CSS   Early Fusion   Mid Fusion   Late Fusion   Ours

Fig. 3. Segmentation results from three views of echocardiogram videos, including PLVLA, LVSA, and A4C from top to bottom row. The red, green, blue, and cyan colours refer to LV, RV, LA, and RA cardiac structures, respectively. (Color figure online)

Validation and Testing. We use all fully annotated videos and split them into validation and testing with a ratio of 2:8. In this stage, we resize each frame in 144×144 size and conduct center cropping to them with the size of 112×112. Selecting the best model based on validation performance and report results in the testing set with Dice score.

3.1 Comparison with the State-of-the-Art Methods

To evaluate the performance of our method, we do the comparison with two types of methods: single-view methods and fusion-based methods in Table 1. To be specific, single-view methods independently train segmentation networks for each view without using any strategy across views or simply conducting semi-supervised approaches [7]. Fusion-based methods use feature-fusion modules to aggregate features and predict the segmentation masks. Our GL-Fusion method can reach 83.84%, 81.76% and 81.28% performance in Dice score across three different views, with 10.49%, 4.19% and 4.68% boosts when compared with the best single-view method [19], and 4.75%, 2.06%, 3.57% enhancement when compared to the best single-view with semi-supervised method CSS [7]. Also, compared with the different global fusion methods, our global and local fusion methods conduct significant improvements compared with the early-fusion approach. The visualization in Fig. 3 compares the segmentation quality with our GL-fusion method and others across three different views.

3.2 Ablation Study

In this section, we analyze the contribution to the performance of the proposed modules Multi-view Global Fusion Module (MGFM) and Multi-view Global Fusion Module (MGFM) of our framework. All results are illustrated in Table **2**. a-b, the baseline without adapting any fusion strategy presents the lowest average dice, while using only MGFM or MLFM module can boost the result to 80.20% and 78.41%, respectively. The combination of these two modules can reach 82.29% dice score with a 2.09% increase in Dice score. In contrast, using the fusion method and cycle loss will lead to worse performance, while our proposed dense cycle loss can boost the result from 80.36% to 82.29%.

4 Conclusion

In this paper, we propose a novel fusion framework called GL-Fusion, which jointly uses global and local information to enhance the segmentation performance of echocardiogram videos. Additionally, to ensure fair evaluation of the multi-view segmentation results, we introduce a multi-view echocardiogram video dataset called **MvEVD**, which provides full annotation for validating and testing performance. Our results demonstrate that the proposed GL-Fusion framework significantly outperforms other methods. In the future, we aim to further improve our method and make it more efficient.

Acknowledgements. This work was partially supported by the Beijing Institute of Collaborative Innovation (BICI) under Grant HCIC-004, in collaboration with HKUST; the Foshan HKUST Projects under Grants FSUST21-HKUST10E and FSUST21-HKUST11E; and the Hong Kong Innovation and Technology Fund under Project ITS/030/21.

References

1. Birenbaum, A., Greenspan, H.: Longitudinal multiple sclerosis lesion segmentation using multi-view convolutional neural networks. In: Carneiro, G., et al. (eds.) LABELS/DLMIA -2016. LNCS, vol. 10008, pp. 58–67. Springer, Cham (2016). https://doi.org/10.1007/978-3-319-46976-8_7
2. Carneiro, G., Nascimento, J., Bradley, A.P.: Automated analysis of unregistered multi-view mammograms with deep learning. IEEE Trans. Med. Imaging **36**(11), 2355–2365 (2017). https://doi.org/10.1109/TMI.2017.2751523
3. Carneiro, G., Nascimento, J.C., Freitas, A.: The segmentation of the left ventricle of the heart from ultrasound data using deep learning architectures and derivative-based search methods. IEEE Trans. Image Process. **21**(3), 968–982 (2012). https://doi.org/10.1109/TIP.2011.2169273

4. Chen, C., Biffi, C., Tarroni, G., Petersen, S., Bai, W., Rueckert, D.: Learning shape priors for robust cardiac MR segmentation from multi-view images. In: Shen, D., et al. (eds.) MICCAI 2019. LNCS, vol. 11765, pp. 523–531. Springer, Cham (2019). https://doi.org/10.1007/978-3-030-32245-8_58

5. Chen, L.C., Papandreou, G., Schroff, F., Adam, H.: Rethinking atrous convolution for semantic image segmentation. arXiv preprint arXiv:1706.05587 (2017)

6. Cheng, L.H., Sun, X., van der Geest, R.J.: Contrastive learning for echocardiographic view integration. In: Wang, L., Dou, Q., Fletcher, P.T., Speidel, S., Li, S. (eds.) Medical Image Computing and Computer Assisted Intervention – MICCAI 2022. MICCAI 2022. LNCS, vol. 13434. Springer, Cham (2022). https://doi.org/10.1007/978-3-031-16440-8_33

7. Dai, W., Li, X., Ding, X., Cheng, K.T.: Cyclical self-supervision for semi-supervised ejection fraction prediction from echocardiogram videos. IEEE Transactions on Medical Imaging (2022)

8. Ding, X., et al.: Support-set based cross-supervision for video grounding. In: Proceedings of the IEEE/CVF International Conference on Computer Vision, pp. 11573–11582 (2021)

9. Guo, Z., Li, X., Huang, H., Guo, N., Li, Q.: Deep learning-based image segmentation on multimodal medical imaging. IEEE Trans. Radiat. Plasma Med. Sci. 3(2), 162–169 (2019)

10. Hsu, C.-C., Tsai, Y.-H., Lin, Y.-Y., Yang, M.-H.: Every pixel matters: center-aware feature alignment for domain adaptive object detector. In: Vedaldi, A., Bischof, H., Brox, T., Frahm, J.-M. (eds.) ECCV 2020. LNCS, vol. 12354, pp. 733–748. Springer, Cham (2020). https://doi.org/10.1007/978-3-030-58545-7_42

11. Hu, Y., et al.: Fully automatic pediatric echocardiography segmentation using deep convolutional networks based on biSeNet. In: 2019 41st Annual International Conference of the IEEE Engineering in Medicine and Biology Society (EMBC), pp. 6561–6564. IEEE (2019)

12. Leclerc, S., et al.: Deep learning for segmentation using an open large-scale dataset in 2D echocardiography. IEEE Trans. Med. Imaging 38(9), 2198–2210 (2019)

13. Li, L., Ding, W., Huang, L., Zhuang, X.: Right ventricular segmentation from short- and long-axis MRIs via information transition. In: Puyol Antón, E., et al. (eds.) STACOM 2021. LNCS, vol. 13131, pp. 259–267. Springer, Cham (2022). https://doi.org/10.1007/978-3-030-93722-5_28

14. Liang, S., Thung, K.H., Nie, D., Zhang, Y., Shen, D.: Multi view spatial aggregation framework for joint localization and segmentation of organs at risk in head and neck CT images. IEEE Trans. Med. Imaging 39(9), 2794–2805 (2020). https://doi.org/10.1109/TMI.2020.2975853

15. Liu, D., et al.: TransFusion: multi-view divergent fusion for medical image segmentation with transformers. In: Wang, L., Dou, Q., Fletcher, P.T., Speidel, S., Li, S. (eds.) Medical Image Computing and Computer Assisted Intervention – MICCAI 2022. MICCAI 2022. LNCS, vol. 13435. Springer, Cham (2022). https://doi.org/10.1007/978-3-031-16443-9_47

16. Liu, D., Yan, Z., Chang, Q., Axel, L., Metaxas, D.N.: Refined deep layer aggregation for multi-disease, multi-view & multi-center cardiac MR segmentation. In: Puyol Antón, E., et al. (eds.) STACOM 2021. LNCS, vol. 13131, pp. 315–322. Springer, Cham (2022). https://doi.org/10.1007/978-3-030-93722-5_34

17. Patel, J.M., Parikh, M.C.: Medical image fusion based on multi-scaling (drt) and multi-resolution (dwt) technique. In: 2016 International Conference on Communication and Signal Processing (ICCSP), pp. 0654–0657. IEEE (2016)

18. Peiris, H., Chen, Z., Egan, G., Harandi, M.: Duo-SegNet: adversarial dual-views for semi-supervised medical image segmentation. In: de Bruijne, M., et al. (eds.) MICCAI 2021. LNCS, vol. 12902, pp. 428–438. Springer, Cham (2021). https://doi.org/10.1007/978-3-030-87196-3_40
19. Ronneberger, O., Fischer, P., Brox, T.: U-Net: convolutional networks for biomedical image segmentation. In: Navab, N., Hornegger, J., Wells, W.M., Frangi, A.F. (eds.) MICCAI 2015. LNCS, vol. 9351, pp. 234–241. Springer, Cham (2015). https://doi.org/10.1007/978-3-319-24574-4_28
20. Storve, S., Grue, J.F., Samstad, S., Dalen, H., Haugen, B.O., Torp, H.: Realtime automatic assessment of cardiac function in echocardiography. IEEE Trans. Ultrason. Ferroelectr. Freq. Control 63(3), 358–368 (2016)
21. Tobon-Gomez, C., et al.: Benchmarking framework for myocardial tracking and deformation algorithms: an open access database. Med. Image Anal. 17(6), 632–648 (2013)
22. Wang, J., et al.: Automated interpretation of congenital heart disease from multi-view echocardiograms. Med. Image Anal. 69, 101942 (2021)
23. Wang, S., et al.: A multi-view deep convolutional neural networks for lung nodule segmentation. In: 2017 39th Annual International Conference of the IEEE Engineering in Medicine and Biology Society (EMBC), pp. 1752–1755 (2017). https://doi.org/10.1109/EMBC.2017.8037182
24. Wang, X., Girshick, R., Gupta, A., He, K.: Non-local neural networks. In: Proceedings of the IEEE Conference on Computer Vision And Pattern Recognition, pp. 7794–7803 (2018)
25. Xia, Y., et al.: Uncertainty-aware multi-view co-training for semi-supervised medical image segmentation and domain adaptation. Med. Image Anal. 65, 101766 (2020)
26. Yu, L., Guo, Y., Wang, Y., Yu, J., Chen, P.: Segmentation of fetal left ventricle in echocardiographic sequences based on dynamic convolutional neural networks. IEEE Trans. Biomed. Eng. 64(8), 1886–1895 (2016)

Treasure in Distribution: A Domain Randomization Based Multi-source Domain Generalization for 2D Medical Image Segmentation

Ziyang Chen[1], Yongsheng Pan[2], Yiwen Ye[1], Hengfei Cui[1], and Yong Xia[1]($\boxtimes$)

[1] National Engineering Laboratory for Integrated Aero-Space-Ground-Ocean Big Data Application Technology, School of Computer Science and Engineering, Northwestern Polytechnical University, Xi'an 710072, China
yxia@nwpu.edu.cn

[2] School of Biomedical Engineering, ShanghaiTech University, Shanghai 201210, China

Abstract. Although recent years have witnessed the great success of convolutional neural networks (CNNs) in medical image segmentation, the domain shift issue caused by the highly variable image quality of medical images hinders the deployment of CNNs in real-world clinical applications. Domain generalization (DG) methods aim to address this issue by training a robust model on the source domain, which has a strong generalization ability. Previously, many DG methods based on feature-space domain randomization have been proposed, which, however, suffer from the limited and unordered search space of feature styles. In this paper, we propose a multi-source DG method called **Treasure in Distribution** (TriD), which constructs an unprecedented search space to obtain the model with strong robustness by randomly sampling from a uniform distribution. To learn the domain-invariant representations explicitly, we further devise a style-mixing strategy in our TriD, which mixes the feature styles by randomly mixing the augmented and original statistics along the channel wise and can be extended to other DG methods. Extensive experiments on two medical segmentation tasks with different modalities demonstrate that our TriD achieves superior generalization performance on unseen target-domain data. Code is available at https://github.com/Chen-Ziyang/TriD.

Keywords: Domain generalization · Domain randomization · Medical image segmentation · Deep learning

---

Z. Chen and Y. Pan—Contributed equally.

Supplementary Information The online version contains supplementary material available at https://doi.org/10.1007/978-3-031-43901-8_9.

H. Greenspan et al. (Eds.): MICCAI 2023, LNCS 14223, pp. 89–99, 2023.
https://doi.org/10.1007/978-3-031-43901-8_9

1   Introduction

Medical image segmentation is an essential task in computer-aided diagnosis. On this task, convolutional neural networks (CNNs) have demonstrated their effectiveness in an extensive literature [2,12]. However, these CNN models obtained on training (*i.e.*, source domain) data can hardly generalize well on the unseen test (*i.e.*, target domain) data. The poor generalization ability, which hinders CNNs to be used in real-world clinical applications, can be attributed to the fact that the quality of medical images varies greatly across healthcare centers with different scanners and imaging protocols, resulting in large distribution discrepancy (*a.k.a.*, domain shift). To improve the generalization ability, domain generalization (DG) methods have been proposed [20,24]. These methods can be trained on the data from one or multiple source domains. Considering the diversity of training data, we focus on multi-source DG in this study.

Most studies on DG attempt to alleviate the distribution discrepancy by standardizing the features [4,14–16] and/or adding extra structures [7,18] to the network. However, the former suffers from over-standardization and may hinder the network to preserve semantic contents, while the latter may introduce excess misjudgment risk when estimating the distance between source- and target-domain data.

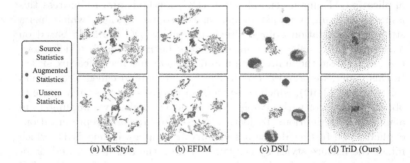

Fig. 1. Visualization of the statistics (1st row: standard deviation, 2nd row: mean) computed from the features of the first residual block of ResNet-34 trained on prostate dataset, which has six different domains. We take five as source domains and the left one as unseen target domain and visualize the augmented statistics produced by MixStyle, EFDM, DSU and our SR using 2D t-SNE. (a) MixStyle: Using a linear combination of different statistics. (b) EFDM: Using exact histogram matching and the statistics-fusion operation of MixStyle. (c) DSU: Sampling from a normal distribution constructed based on the original statistics. (d) TriD (Ours): Sampling from a uniform distribution.

A recent mainstream in DG research is to simulate the distributions of unseen target-domain data via domain randomization, *i.e.*, perturbing the styles of source-domain data. The perturbation function can be defined in the input space [11,17,21]. Thus, it is easy to evaluate the quality of perturbed images, but the definition generally requires domain knowledge and expertise [3]. By contrast, the perturbation can be performed in the feature space [9,22,25]. However, this may cause difficulties in monitoring the perturbation degree of semantic contents in the feature space due to the lack of visualization. Recently, two critical attributes of the feature space are revealed by IBN-Net [14] and AdaIN [8], respectively. First, most style-texture information resides in the low-level features extracted by shallow layers. Second, the content-preserving style transformation can be performed by changing the statistics (*e.g.*, mean and standard deviation) of the low-level features. Inspired by them, MixStyle [25] perturbs the feature styles using augmented statistics, which are generated by randomly mixing the statistics of the low-level features from two samples. Subsequently, more research efforts have been devoted to designing the search space that covers a larger area in the feature-style space [9,22]. Despite their improved performance, using the statistics of source-domain data for feature perturbation may limit the search space and can hardly explore in the feature-style space evenly (see Fig. 1(a), (b) and (c)). The points indicating the augmented statistics are scattered and do not completely cover the points from unseen target domain. Moreover, since all feature channels are perturbed, these methods lack a reference to the original feature, which prevents them from learning the domain-invariant representations explicitly.

To address these issues, in this paper, we propose a simple but effective multi-source DG method called **Treasure in Distribution** (TriD), which consists of two major steps: statistics randomization (SR) and style mixing (SM). SR aims to tap the potential of distribution by randomly sampling the augmented statistics from a uniform distribution to perturb the original intermediate features, which can expand the search space to cover more cases evenly (see Fig. 1(d)). It can be observed that the red points are distributed evenly and cover not only the unseen target domain, but also the source domains. This leads to the issue that the perturbed features may have unreal styles. We hypothetically extend the unreal styles to the feature space with the inspiration from [17], which demonstrated the effectiveness of unreal styles in the input space. SM is devised to mix the feature styles by randomly mixing the augmented and original statistics in the channel dimension, thus making it feasible to learn the domain-invariant representations explicitly. We have evaluated our proposed TriD on two medical segmentation tasks: (1) the prostate segmentation using magnetic resonance imaging (MRI) from six domains and (2) joint segmentation of optic disc (OD) and optic cup (OC) in fundus images from five domains. Extensive experiments demonstrate that our TriD achieves a superior generalization ability to the state-of-the-art DG methods on unseen target-domain data.

Our contributions are three-fold: (1) The proposed multi-source DG method called TriD can boost the robustness of model and alleviate the performance drop on the unseen target-domain data. (2) We focus on expanding the search space of feature styles and therefore devise the statistics-randomization strategy, which allows exploring in the feature-style space evenly. (3) Different from perturbing all feature channels, we introduce the original statistics to the augmented statistics to learn the domain-invariant representations explicitly.

2   Method

2.1   Preliminaries

Let $f \in R^{B \times C \times H \times W}$ be the intermediate features in a mini-batch, where B, C, H, and W respectively denote the mini-batch size, channel, height, and width. MixStyle [25] perturbs the features by randomly mixing different feature statistics, formulated as follows:

$$MixStyle(f_i) = \gamma_m \frac{f_i - \mu(f_i)}{\sigma(f_i)} + \beta_m, \tag{1}$$

$$\gamma_m = \lambda_m \sigma(f_i) + (1 - \lambda_m)\sigma(f_j), \ \beta_m = \lambda_m \mu(f_i) + (1 - \lambda_m)\mu(f_j), \tag{2}$$

where $\lambda_m \in R^B$ is a weight coefficient sampled from a Beta distribution [25], $f_i, f_{j(j \neq i)} \in R^{C \times H \times W}$ indicate the features from two different images in a mini-batch, and $\mu(*), \sigma(*) \in R^{B \times C}$ are the mean and standard deviation computed across the spatial dimension within each channel of each image.

2.2   Treasure in Distribution (TriD)

The TriD is designed to perturb the intermediate feature styles by randomly changing the feature statistics (*i.e.*, mean and standard deviation), as shown in Fig. 2. The feature statistics is substituted by the mixed statistics, which is generated by mixing the augmented and original statistics in channel dimension. It can be implemented as a plug-and-play module inserted into any CNN-based architecture. In this study, we use ResNet-34 [5] as the backbone to construct the segmentation network in a U-shape architecture [6]. The TriD is inserted behind the first and second residual blocks during training, and will be removed in the inference phase. We now delve into the details of our TriD.

Statistics Randomization (SR). Inspired by effectiveness of unreal styles in the input space [17], we hypothesis that unreal styles can also be extended to the feature space. To cover more cases evenly, we randomly sample the augmented statistics $\sigma_r, \mu_r \in R^{B \times C}$ from a uniform distribution which contains most feature statistics: $\sigma_r \sim U(0, 1)$, $\mu_r \sim U(0, 1)$.

Style Mixing (SM). To learn the domain-invariant representations explicitly, SM strategy is designed to randomly mix the augmented and original statistics along the channel wise. We first sample $P \in R^{B \times C}$ from the Beta distribution: $P \sim Beta(\alpha, \alpha)$, and use P as the probability to generate the Bernoulli distribution from which to sample $\lambda \in R^{B \times C}$: $\lambda \sim Bern(P)$, where α is set to 0.1 empirically [25]. Then the mixed statistics is calculated as:

$$\gamma_{mix} = \lambda \sigma(r) + (1 - \lambda)\sigma(f), \ \beta_{mix} = \lambda \mu(r) + (1 - \lambda)\mu(f), \tag{3}$$

where f denotes the intermediate features. Finally, the mixed feature statistics is applied to perturb the normalized f similar to Eq. (1),

$$TriD(f) = \gamma_{mix} \frac{f - \mu(f)}{\sigma(f)} + \beta_{mix}. \tag{4}$$

Different from MixStyle, we replace the batch-wise fusion with channel-wise mixing, which avoids the sampling preference and introduces original-feature reference, so as to learn the domain-invariant representations explicitly.

2.3   Training and Inference

Let $\mathcal{D}_s = \{(x_{di}, y_{di})_{i=1}^{N_d}\}_{d=1}^{K}$ be a set including K source domains, where x_{di} is the i-th image in the d-th source domain, and y_{di} is the corresponding segmentation mask of x_{di}. Our goal is to train a segmentation model that can generalize well to an unseen target domain $\mathcal{D}_t = (x_i)_{i=1}^{N_t}$.

Training. During training, we empirically set a probability of 0.5 to activate TriD in the forward pass [25]. The segmentation network is trained on the source domains $\mathcal{D}_s$ by using the combination of Dice loss ($\mathcal{L}_{Dice}$) and cross-entropy loss ($\mathcal{L}_{ce}$) as the objective: $\mathcal{L}_{seg} = \mathcal{L}_{Dice} + \mathcal{L}_{ce}$.

Inference. During inference, all the TriD modules are removed, and the segmentation network is tested on the unseen target domain $\mathcal{D}_t$.

3   Experiments and Results

3.1   Datasets and Evaluation Metrics

Two datasets are used for this study, whose details are summarized in Table 1.

The first dataset contains 116 MRI cases from six domains for prostate segmentation [10]. We preprocess these MRI cases same as a previous study [7] and only preserve the slices with the prostate region for consistent and objective segmentation evaluation. These slices are resized to 384×384 with same voxel spacing. On this dataset, we employ the Dice Similarity Coefficient (DSC) and Average Surface Distance (ASD) to evaluate the prostate segmentation. Note that we regard prostate segmentation as a 2D segmentation task, but calculate metrics on 3D volumes.

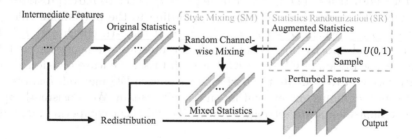

Fig. 2. Overview of our TriD for feature perturbation.

The second dataset is a collection of two large and three small public datasets used for joint segmentation of optic disc (OD) and optic cup (OC) [1,13,19,23],

Table 1. Details of two datasets used for this study. Number of cases with '/' denotes the data split (training/test cases).

Task	Modality	Number of Domains	Cases in Each Domain
Prostate Segmentation	MRI	6	30; 30; 19; 13; 12; 12
OD/OC Segmentation	Color Fundus Image	5	156/39; 76/19; 320/80; 500/150; 50/51

Table 2. Performance (OD, OC) of Intra-Domain, DeepAll, our TriD and seven DG methods in joint segmentation of OD and OC. The best results except for the ones of Intra-Domain are highlighted with **bold**.

Methods	Domain 1	Domain 2	Domain 3	Domain 4	Domain 5	Average
	DSC↑	DSC↑	DSC↑	DSC↑	DSC↑	DSC↑
Intra-Domain	(95.53, 82.53)	(94.92, 83.94)	(96.08, 86.30)	(92.37, 85.44)	(96.07, 86.44)	89.96
DeepAll	(92.87, 77.73)	(91.33, 77.41)	(91.45, 79.27)	(83.51, 73.84)	(90.82, 78.54)	83.68
SAN-SAW (CVPR 2022) [16]	(93.34, 76.31)	(92.88, 82.65)	(90.78, 81.18)	(88.07, 77.61)	(93.43, 83.97)	86.02
DCAC (TMI 2022) [7]	(94.34, 76.72)	(93.70, 79.21)	(91.05, 81.23)	(88.12, 77.87)	(95.71, 85.32)	86.33
RandConv (ICLR 2021) [21]	(93.11, 76.21)	(92.50, 81.33)	(89.01, 81.33)	(88.33, 76.56)	(95.32, 85.16)	85.89
MixStyle (ICLR 2021) [25]	(94.40, 79.11)	(92.02, 79.19)	(91.64, 80.79)	(86.41, 76.44)	(93.09, 83.35)	85.64
EFDM (CVPR 2022) [22]	(93.84, 77.04)	(91.00, 79.53)	(91.53, 81.59)	(86.31, 76.39)	(93.69, 81.31)	85.22
DSU (ICLR 2022) [9]	(93.71, 77.48)	(91.79, 81.65)	(92.11, 79.78)	(87.96, 76.76)	(93.58, 84.91)	85.97
MaxStyle (MICCAI 2022) [3]	(94.57, 77.59)	(93.67, 82.66)	(**92.40**, 79.34)	(88.81, 76.93)	(**96.02**, 84.39)	86.64
TriD (Ours)	(**94.72, 80.26**)	(**93.95, 82.70**)	(92.09, **81.92**)	(**90.37, 78.02**)	(95.64, **86.54**)	**87.62**

which can evaluate our TriD under different data-amount scenarios. Each of these five datasets has a training/test split, and in total, we have 1,102 cases for training and 339 cases for test. Each image is center-cropped and resized to 512×512 [6]. We employ DSC to evaluate the joint segmentation of OD and OC.

3.2 Implementation Details

We set the mini-batch size to 8 and adopt the SGD optimizer with a momentum of 0.99 for both tasks. The initial learning rate l_0 is set to 0.01 (prostate) and 0.001 (OD/OC) respectively and decays according to the polynomial rule $l_t = l_0 \times (1 - t/T)^{0.9}$, where l_t is the learning rate of the t-th epoch and T is the number of total epochs that is set to 200 for prostate segmentation and 100 for joint segmentation of OD and OC. For both tasks, the leave-one-domain-out strategy was used to evaluate the performance of each DG method, *i.e.*, training on K-1 source domains and evaluating on the left domain. We consistently apply the above implementation settings to our TriD and other competing methods.

3.3 Results

Comparing to Other DG Methods. We used the same segmentation network and loss function to compare our TriD with seven DG methods, including (1) DCAC: dynamic structure [7], (2) SAN-SAW: based on normalization

Table 3. Performance of Intra-Domain, DeepAll, our TriD and six DG methods in prostate segmentation. The best results except for the ones of Intra-Domain are highlighted with **bold**.

Methods	Domain 1		Domain 2		Domain 3		Domain 4		Domain 5		Domain 6		Average	
	DSC↑	ASD↓	DSC↑	ASD↓	DSC↑	ASD↓	DSC↑	ASD↓	DSC↑	ASD↓	DSC↑	ASD↓	DSC↑	ASD↓
Intra-Domain	93.24	0.59	91.85	0.59	90.52	1.57	89.69	0.81	88.19	1.29	91.09	0.69	90.76	0.93
DeepAll	90.72	1.04	88.53	0.77	85.10	3.30	88.04	0.91	85.84	1.98	89.01	0.81	87.87	1.47
DCAC (TMI 2022) [7]	90.51	0.98	88.18	1.34	84.35	4.05	88.32	0.83	87.01	2.73	89.95	0.64	88.05	1.76
RandConv (ICLR 2021) [21]	90.21	1.01	88.59	0.76	84.18	3.39	88.40	0.73	86.80	2.58	89.17	0.76	87.89	1.54
MixStyle (ICLR 2021) [25]	91.60	0.70	90.10	0.69	85.62	3.09	88.45	0.87	87.21	1.45	90.02	**0.61**	88.83	1.23
EFDM (CVPR 2022) [22]	91.57	0.72	90.18	0.70	85.34	3.30	89.25	0.71	86.82	1.71	89.52	0.71	88.78	1.31
DSU (ICLR 2022) [9]	90.92	0.79	88.19	0.82	84.57	3.86	88.68	0.73	86.25	1.80	89.13	0.70	87.96	1.45
MaxStyle (MICCAI 2022) [3]	90.33	0.79	89.17	0.74	85.34	2.91	88.72	**0.67**	87.46	1.38	88.15	0.72	88.19	1.20
TriD (Ours)	**91.63**	**0.66**	**90.71**	**0.64**	**86.91**	2.77	**89.42**	0.68	**88.67**	1.33	**90.11**	0.63	**89.57**	**1.12**

Table 4. Performance (OD, OC) of DeepAll, our TriD and its two variants in joint segmentation of OD and OC. The best results are highlighted with **bold**.

Methods	Domain 1	Domain 2	Domain 3	Domain 4	Domain 5	Average
	DSC↑	DSC↑	DSC↑	DSC↑	DSC↑	DSC↑
DeepAll	(92.87, 77.73)	(91.33, 77.41)	(91.45, 79.27)	(83.51, 73.84)	(90.82, 78.54)	83.68
DeepAll+SR	(94.30, 79.23)	(91.70, 81.15)	(91.30, 81.10)	(87.87, 76.96)	(95.08, 85.77)	86.45
DeepAll+SR+Mixup	(94.68, 78.62)	(92.08, 80.76)	(91.06, 80.81)	(87.18, 75.26)	(94.58, 84.15)	85.92
TriD (DeepAll+SR+SM)	**(94.72, 80.26)**	**(93.95, 82.70)**	**(92.09, 81.92)**	**(90.37, 78.02)**	**(95.64, 86.54)**	**87.62**

and whitening [16], (3) RandConv: input-space domain randomization [21], (4–6) MixStyle, EFDM, DSU: feature-space domain randomization [9,22,25] and (7) MaxStyle: adversarial noise [3]. Note that since SAN-SAW requires the data with at least two classes, we did not provide its results for prostate segmentation. Besides, we also compared our method with another two settings, including the 'Intra-Domain' and 'DeepAll'. Under the 'Intra-Domain' setting, training and test data are from the same domain, where three-fold cross-validation is used for prostate segmentation due to the lack of data split. Under the 'DeepAll' setting, the model is directly trained on the data aggregated from all source domains and tested on the unseen target domain. The results are shown in Table 2 and Table 3. It can be observed that the overall performance of our TriD is not only superior to the 'DeepAll' baseline but also better than other DG approaches. Furthermore, we found that the performance ranking of MixStyle, EFDM, DSU and our TriD is $TriD > MixStyle \approx EFDM > DSU$, which is consistent with the ranking of search scope in Fig. 1. It reveals that the unreal feature styles are indeed effective, and a larger search space is beneficial to boost the robustness.

Contribution of Each Component. To evaluate the contribution of statistics randomization (SR) and style mixing (SM), we chose the model trained in joint segmentation of OD and OC as an example and conducted a series of ablation experiments, as shown in Table 4. Note that the 'Mixup' denotes the fusion strategy proposed in MixStyle. It shows that (1) introducing SR to baseline can lead to huge performance gains; (2) adding the Mixup operation will degrade the

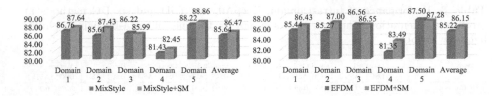

Fig. 3. Performance of two DG methods with/without SM in joint segmentation of OD and OC. The value of each domain indicate the mean DSC of OD and OC.

Table 5. Performance (OD, OC) of different locations of TriD in joint segmentation of OD and OC. The best results are highlighted with **bold**.

Methods	Domain 1	Domain 2	Domain 3	Domain 4	Domain 5	Average
	DSC↑	DSC↑	DSC↑	DSC↑	DSC↑	DSC↑
res1	$(94.52, 78.12)$	$(93.40, \mathbf{82.92})$	$(\mathbf{92.14}, 79.94)$	$(87.36, 77.22)$	$(94.84, 85.20)$	86.57
res2	$(93.65, 79.20)$	$(93.07, 81.16)$	$(92.12, 78.91)$	$(87.29, 77.12)$	$(93.47, 84.15)$	86.01
res12	$(\mathbf{94.72}, \mathbf{80.26})$	$(\mathbf{93.95}, 82.70)$	$(92.09, \mathbf{81.92})$	$(\mathbf{90.37}, \mathbf{78.02})$	$(\mathbf{95.64}, \mathbf{86.54})$	**87.62**
res123	$(92.52, 75.11)$	$(90.64, 80.40)$	$(91.24, 80.06)$	$(87.82, 77.42)$	$(95.14, 84.18)$	85.45
res1234	$(90.14, 75.27)$	$(90.35, 79.45)$	$(91.59, 80.76)$	$(86.77, 77.04)$	$(94.21, 84.17)$	84.97

Table 6. Performance (OD, OC) of using normal distribution and uniform distribution in joint segmentation of OD and OC. The best results are highlighted with **bold**.

Methods	Domain 1	Domain 2	Domain 3	Domain 4	Domain 5	Average
	DSC↑	DSC↑	DSC↑	DSC↑	DSC↑	DSC↑
Normal Distribution	$(92.93, 78.22)$	$(91.37, 81.06)$	$(91.64, 80.60)$	$(84.66, 72.20)$	$(93.55, 83.39)$	84.96
Uniform Distribution	$(\mathbf{94.72}, \mathbf{80.26})$	$(\mathbf{93.95}, \mathbf{82.70})$	$(\mathbf{92.09}, \mathbf{81.92})$	$(\mathbf{90.37}, \mathbf{78.02})$	$(\mathbf{95.64}, \mathbf{86.54})$	**87.62**

robustness of model due to the limited search space; (3) the best performance is achieved when SR and SM are jointly used (*i.e.*, our TriD).

Extendibility of SM. Under the same segmentation task, we further evaluate the extendibility of SM by combining it with other DG methods (*i.e.*, MixStyle and EFDM), and the results are shown in Fig. 3. It shows that each approach plus SM achieves better performance in most scenarios and has superior average DSC, proving that our SM strategy can be extended to other DG methods.

Location of TriD. To discuss where to apply our TriD, we repeated the experiments in joint segmentation of OD and OC and listed the results in Table 5. These four residual blocks of ResNet-34 are denoted as 'res1-4', and we trained different variants via applying TriD to different blocks. The results reveal that (1) the best performance is achieved by applying TriD to 'res12' that extract the low-level features with the most style-texture information; (2) the performance degrades when applying TriD to the third and last blocks that tend to capture semantic content rather than style texture [14].

Uniform Distribution vs. Normal Distribution. It is particularly critical to choose the distribution from which to randomly sample the augmented statistics. To verify the advantages of uniform distribution, we repeated the experiments in joint segmentation of OD and OC by replacing the uniform distribution with a normal distribution $N(0.5, 1)$ and compared the effectiveness of them in Table 6. It shows that the normal distribution indeed results in performance drop due to the limited search space.

4   Conclusion

We proposed the TriD, a domain-randomization based multi-source domain generalization method, for medical image segmentation. To solve the limitations existing in preview methods, TriD perturbs the intermediate features with two steps: (1) SR: randomly sampling the augmented statistics from a uniform distribution to expand the search space of feature styles; (2) SM: mixing the feature styles for explicit domain-invariant representation learning. Through extensive experiments on two medical segmentation tasks with different modalities, the proposed TriD is demonstrated to achieve superior performance over the baselines and other state-of-the-art DG methods.

Acknowledgment. This work was supported in part by the National Natural Science Foundation of China under Grant 62171377, in part by the Key Technologies Research and Development Program under Grant 2022YFC2009903/2022YFC2009900, in part by the Key Research and Development Program of Shaanxi Province, China, under Grant 2022GY-084, and in part by the China Postdoctoral Science Foundation 2021M703340/BX2021333.

References

1. Almazroa, A., et al.: Retinal fundus images for glaucoma analysis: the RIGA dataset. In: Medical Imaging 2018: Imaging Informatics for Healthcare, Research, and Applications, vol. 10579, pp. 55–62. SPIE (2018)
2. Asgari Taghanaki, S., Abhishek, K., Cohen, J.P., Cohen-Adad, J., Hamarneh, G.: Deep semantic segmentation of natural and medical images: a review. Artif. Intell. Rev. **54**, 137–178 (2021)
3. Chen, C., Li, Z., Ouyang, C., Sinclair, M., Bai, W., Rueckert, D.: MaxStyle: adversarial style composition for robust medical image segmentation. In: Wang, L., Dou, Q., Fletcher, P.T., Speidel, S., Li, S. (eds.) Medical Image Computing and Computer Assisted Intervention, MICCAI 2022. LNCS, vol. 13435. Springer, Cham (2022). https://doi.org/10.1007/978-3-031-16443-9_15
4. Choi, S., Jung, S., Yun, H., Kim, J.T., Kim, S., Choo, J.: RobustNet: improving domain generalization in urban-scene segmentation via instance selective whitening. In: Proceedings of the IEEE/CVF Conference on Computer Vision and Pattern Recognition, pp. 11580–11590 (2021)
5. He, K., Zhang, X., Ren, S., Sun, J.: Deep residual learning for image recognition. In: Proceedings of the IEEE Conference on Computer Vision and Pattern Recognition, pp. 770–778 (2016)

6. Hu, S., Liao, Z., Xia, Y.: Domain specific convolution and high frequency reconstruction based unsupervised domain adaptation for medical image segmentation. In: Wang, L., Dou, Q., Fletcher, P.T., Speidel, S., Li, S. (eds.) Medical Image Computing and Computer Assisted Intervention – MICCAI 2022, Part VII. LNCS, vol. 13437. Springer, Cham (2022). https://doi.org/10.1007/978-3-031-16449-1_62

7. Hu, S., Liao, Z., Zhang, J., Xia, Y.: Domain and content adaptive convolution based multi-source domain generalization for medical image segmentation. IEEE Trans. Med. Imaging **42**(1), 233–244 (2022)

8. Huang, X., Belongie, S.: Arbitrary style transfer in real-time with adaptive instance normalization. In: Proceedings of the IEEE International Conference on Computer Vision, pp. 1501–1510 (2017)

9. Li, X., Dai, Y., Ge, Y., Liu, J., Shan, Y., Duan, L.: Uncertainty modeling for out-of-distribution generalization. In: International Conference on Learning Representations (2022)

10. Liu, Q., Dou, Q., Heng, P.-A.: Shape-aware meta-learning for generalizing prostate MRI segmentation to unseen domains. In: Martel, A.L., et al. (eds.) MICCAI 2020, Part II. LNCS, vol. 12262, pp. 475–485. Springer, Cham (2020). https://doi.org/10.1007/978-3-030-59713-9_46

11. Liu, X.C., Yang, Y.L., Hall, P.: Geometric and textural augmentation for domain gap reduction. In: Proceedings of the IEEE/CVF Conference on Computer Vision and Pattern Recognition, pp. 14340–14350 (2022)

12. Olabarriaga, S.D., Smeulders, A.W.: Interaction in the segmentation of medical images: a survey. Med. Image Anal. **5**(2), 127–142 (2001)

13. Orlando, J.I., et al.: REFUGE challenge: a unified framework for evaluating automated methods for glaucoma assessment from fundus photographs. Med. Image Anal. **59**, 101570 (2020)

14. Pan, X., Luo, P., Shi, J., Tang, X.: Two at once: enhancing learning and generalization capacities via IBN-Net. In: Ferrari, V., Hebert, M., Sminchisescu, C., Weiss, Y. (eds.) ECCV 2018. LNCS, vol. 11208, pp. 484–500. Springer, Cham (2018). https://doi.org/10.1007/978-3-030-01225-0_29

15. Pan, X., Zhan, X., Shi, J., Tang, X., Luo, P.: Switchable whitening for deep representation learning. In: Proceedings of the IEEE/CVF International Conference on Computer Vision, pp. 1863–1871 (2019)

16. Peng, D., Lei, Y., Hayat, M., Guo, Y., Li, W.: Semantic-aware domain generalized segmentation. In: Proceedings of the IEEE/CVF Conference on Computer Vision and Pattern Recognition, pp. 2594–2605 (2022)

17. Peng, D., Lei, Y., Liu, L., Zhang, P., Liu, J.: Global and local texture randomization for synthetic-to-real semantic segmentation. IEEE Trans. Image Process. **30**, 6594–6608 (2021)

18. Segu, M., Tonioni, A., Tombari, F.: Batch normalization embeddings for deep domain generalization. Pattern Recogn. **135**, 109115 (2023)

19. Sivaswamy, J., Krishnadas, S., Joshi, G.D., Jain, M., Tabish, A.U.S.: Drishti-GS: retinal image dataset for optic nerve head (ONH) segmentation. In: 2014 IEEE 11th International Symposium on Biomedical Imaging (ISBI), pp. 53–56. IEEE (2014)

20. Wang, J., et al.: Generalizing to unseen domains: a survey on domain generalization. IEEE Trans. Knowl. Data Eng. **35**, 8052–8072 (2022)

21. Xu, Z., Liu, D., Yang, J., Raffel, C., Niethammer, M.: Robust and generalizable visual representation learning via random convolutions. In: International Conference on Learning Representations (2021)

22. Zhang, Y., Li, M., Li, R., Jia, K., Zhang, L.: Exact feature distribution matching for arbitrary style transfer and domain generalization. In: Proceedings of the IEEE/CVF Conference on Computer Vision and Pattern Recognition, pp. 8035–8045 (2022)

23. Zhang, Z., et al.: ORIGA-light: an online retinal fundus image database for glaucoma analysis and research. In: 2010 Annual International Conference of the IEEE Engineering in Medicine and Biology, pp. 3065–3068. IEEE (2010)

24. Zhou, K., Liu, Z., Qiao, Y., Xiang, T., Loy, C.C.: Domain generalization: a survey. IEEE Trans. Pattern Anal. Mach. Intell. **45**, 4396–4415 (2022)

25. Zhou, K., Yang, Y., Qiao, Y., Xiang, T.: Domain generalization with MixStyle. In: International Conference on Learning Representations (2021)

Diffusion Kinetic Model for Breast Cancer Segmentation in Incomplete DCE-MRI

Tianxu Lv[1], Yuan Liu[1], Kai Miao[3], Lihua Li[2], and Xiang Pan[1,3(✉)]

[1] School of Artificial Intelligence and Computer Science, Jiangnan University, Wuxi 214122, China
xiangpan@jiangnan.edu.cn
[2] Institute of Biomedical Engineering and Instrumentation, Hangzhou Dianzi University, Hangzhou, China
[3] Cancer Center, Faculty of Health Sciences, University of Macau, Macau SAR, China

Abstract. Recent researches on cancer segmentation in dynamic contrast enhanced magnetic resonance imaging (DCE-MRI) usually resort to the combination of temporal kinetic characteristics and deep learning to improve segmentation performance. However, the difficulty in accessing complete temporal sequences, especially post-contrast images, hinders segmentation performance, generalization ability and clinical application of existing methods. In this work, we propose a diffusion kinetic model (DKM) that implicitly exploits hemodynamic priors in DCE-MRI and effectively generates high-quality segmentation maps only requiring pre-contrast images. We specifically consider the underlying relation between hemodynamic response function (HRF) and denoising diffusion process (DDP), which displays remarkable results for realistic image generation. Our proposed DKM consists of a diffusion module (DM) and segmentation module (SM) so that DKM is able to learn cancer hemodynamic information and provide a latent kinetic code to facilitate segmentation performance. Once the DM is pretrained, the latent code estimated from the DM is simply incorporated into the SM, which enables DKM to automatically and accurately annotate cancers with pre-contrast images. To our best knowledge, this is the first work exploring the relationship between HRF and DDP for dynamic MRI segmentation. We evaluate the proposed method for tumor segmentation on public breast cancer DCE-MRI dataset. Compared to the existing state-of-the-art approaches with complete sequences, our method yields higher segmentation performance even with pre-contrast images. The source code will be available on https://github.com/Medical-AI-Lab-of-JNU/DKM.

Keywords: Deep learning · Kinetic representation · DCE-MRI · Cancer segmentation · Denoising Diffusion model

1 Introduction

Dynamic contrast-enhanced magnetic resonance imaging (DCE-MRI) revealing tumor hemodynamics information is often applied to early diagnosis and

H. Greenspan et al. (Eds.): MICCAI 2023, LNCS 14223, pp. 100–109, 2023.
https://doi.org/10.1007/978-3-031-43901-8_10

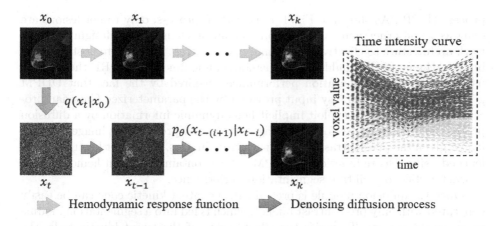

Fig. 1. Illustration of hemodynamic response function and denoising diffusion process as well as their underlying relation. Right is the time intensity curve (TIC). x_0 and x_k represent pre-contrast images and post-contrast images in DCE-MRI, respectively.

treatment of breast cancer [1]. In particular, automatically and accurately segmenting tumor regions in DCE-MRI is vital for computer-aided diagnosis (CAD) and various clinical tasks such as surgical planning. For the sake of promoting segmentation performance, recent methods utilize the dynamic MR sequence and exploit its temporal correlations to acquire powerful representations [2–4]. More recently, a handful of approaches take advantage of hemodynamic knowledge and time intensity curve (TIC) to improve segmentation accuracy [5,6]. However, the aforementioned methods require the complete DCE-MRI sequences and overlook the difficulty in assessing complete temporal sequences and the missing time point problem, especially post-contrast phase, due to the privacy protection and patient conditions. Hence, these breast cancer segmentation models cannot be deployed directly in clinical practice.

Recently, denoising diffusion probabilistic model (DDPM) [7,8] has produced a tremendous impact on image generation field due to its impressive performance. Diffusion model is composed of a forward diffusion process that add noise to images, along with a reverse generation process that generates realistic images from the noisy input [8]. Based on this, several methods investigate the potential of DDPM for natural image segmentation [9] and medical image segmentation [10–12]. Specifically, Baranchuk et al. [9] explores the intermediate activations from the networks that perform the markov step of the reverse diffusion process and find these activations can capture semantic information for segmentation. However, the applicability of DDPM to medical image segmentation are still limited. In addition, existing DDPM-based segmentation networks are generic and are not optimized for specific applications. In particular, a core question for DCE-MRI segmentation is how to optimally exploit hemodynamic priors.

Based on the above observations, we innovatively consider the underlying relation between hemodynamic response function (HRF) and denoising diffusion process (DDP). As shown in Fig. 1, during HRF process, only tumor lesions are enhanced and other non-tumor regions remain unchanged. By designing a network architecture to effectively transmute pre-contrast images into post-contrast images, the network should acquire hemodynamic inherent in HRF that can be used to improve segmentation performance. Inspired by the fact that DDPM generates images from noisy input provided by the parameterized Gaussian process, this work aims to exploit implicit hemodynamic information by a diffusion process that predict post-contrast images from noisy pre-contrast images. Specifically, given the pre-contrast and post-contrast images, the latent kinetic code is learned using a score function of DDPM, which contains sufficient hemodynamic characteristics to facilitate segmentation performance.

Once the diffusion module is pretrained, the latent kinetic code can be easily generated with only pre-contrast images, which is fed into a segmentation module to annotate cancers. To verify the effectiveness of the latent kinetic code, the SM adopts a simple U-Net-like structure, with an encoder to simultaneously conduct semantic feature encoding and kinetic code fusion, along with a decoder to obtain voxel-level classification. In this manner, our latent kinetic code can be interpreted to provide TIC information and hemodynamic characteristics for accurate cancer segmentation.

We verify the effectiveness of our proposed diffusion kinetic model (DKM) on DCE-MRI-based breast cancer segmentation using Breast-MRI-NACT-Pilot dataset [13]. Compared to the existing state-of-the-art approaches with complete sequences, our method yields higher segmentation performance even with pre-contrast images. In summary, the main contributions of this work are listed as follows:

- We propose a diffusion kinetic model that implicitly exploits hemodynamic priors in DCE-MRI and effectively generates high-quality segmentation maps only requiring pre-contrast images.
- We first consider the underlying relation between hemodynamic response function and denoising diffusion process and provide a DDPM-based solution to capture a latent kinetic code for hemodynamic knowledge.
- Compared to the existing approaches with complete sequences, the proposed method yields higher cancer segmentation performance even with pre-contrast images.

2   Methodology

The overall framework of the proposed diffusion kinetic model is illustrated in Fig. 2. It can be observed that the devised model consists of a diffusion module (DM) and a segmentation module (SM). Let $\{x_K, K = 0, 1, ..., k\}$ be a sequence of images representing the DCE-MRI protocol, in which x_0 represents the pre-contrast image and x_k represents the late post-contrast image. The DM takes a noisy pre-contrast image x_t as input and generates post-contrast images to estimate the latent kinetic code. Once the DM is trained, the learned kinetic code

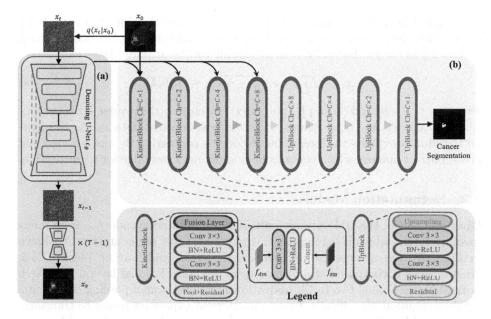

Fig. 2. Illustration of our method for implicitly exploiting hemodynamic information from pre-contrast images. The combination of learned kinetic code is an example.

is incorporated into the SM as hemodynamic priors to guide the segmentation process. Model details are shown as follows.

2.1 Diffusion Module

The diffusion module is following the denoising diffusion probabilistic model [8,14]. Based on the consideration from nonequilibrium thermodynamics, DDPM approximates the data distribution by learning a Markov chain process which originates from the Gaussian distribution. The forward diffusion process gradually adds Gaussian noise to the data x_0 according to a variance schedule $\beta_1, ..., \beta_T$ [8]:

$$q(x_t|x_{t-1}) := \mathcal{N}(x_t; \sqrt{1-\beta_t}x_{t-1}, \beta_t \mathbf{I}) \tag{1}$$

Particularly, a noisy image x_t can be directly obtained from the data x_0:

$$q(x_t|x_0) := \mathcal{N}(x_t; \sqrt{\bar{\alpha}_t}x_0, (1-\bar{\alpha}_t)\mathbf{I}) \tag{2}$$

where $\alpha_t := 1 - \beta_t$ and $\bar{\alpha}_t := \prod_{s=1}^{t} \alpha_s$. Afterwards, DDPM approximates the reverse diffusion process by the following parameterized Gaussian transitions:

$$p_\theta(x_{t-1}|x_t) := \mathcal{N}(x_{t-1}; \mu_\theta(x_t, t), \sum_\theta(x_t, t)) \tag{3}$$

where $\mu_\theta(x_t, t)$ is the learned posterior mean and $\sum_\theta(x_t; t)$ is a fixed set of scalar covariances. In particular, we employ a noise predictor network $(\epsilon_\theta(x_t, t))$ to predict the noise component at the step t (As shown in Fig. 2.(a)).

Inspired by the property of DDPM [8], we devise the diffusion module by considering the pre-contrast images x_0 as source and regarding the post-contrast images x_k as target. Formally, a noisy sample can be acquired by:

$$x_t = \sqrt{\bar{\alpha}_t}x_0 + \sqrt{1 - \bar{\alpha}_t}\epsilon, \epsilon \sim \mathcal{N}(0, \mathbf{I}) \tag{4}$$

where $\alpha_t := 1 - \beta_t$ and $\bar{\alpha}_t := \prod_{s=1}^t \alpha_s$. Next, we employ the reverse diffusion process to transform the noisy sample x_t to the post-contrast data x_k. As thus, the DM gradually exploits the latent kinetic code by comparing the pre-contrast and post-contrast images, which contains hemodynamic knowledge for segmentation.

2.2 Segmentation Module

Once pretrained, the DM outputs multi-scale latent kinetic code f_{dm} from intermediate layers, which is fed into the SM to guide cancer segmentation. As shown in Fig. 2(b), the SM consists of four kinetic blocks and four up blocks. Each kinetic block is composed of a fusion layer, two convolutional layers, two batch normalization layers, two ReLU activation functions, a max pooling layer and a residual addition. Specifically, to obtain sufficient expressive power to transform the learned kinetic code into higher-level features, at least one learnable linear transformation is required. To this end, a linear transformation, parametrized by a weight matrix W, is applied to the latent code f_{dm}, followed by a batch normalization, ReLU activation layer and concatenation, which can be represented as follows:

$$\hat{f} = \mathbb{C}(\phi(\text{BN}(W * f_{dm}); f_{sm}) \tag{5}$$

where $*$ represents 1×1 based convolution operation, W is the weight matrix, BN represents batch normalization, ϕ represents ReLU activation function and $\mathbb{C}$ is concatenation operation. In this way, the hemodynamic knowledge can be incorporated into the SM to capture more expressive representations to improve segmentation performance.

2.3 Model Training

To maintain training stability, the proposed DKM adopts a two-step training procedure for cancer annotation. In the first step, the DM is trained to transform pre-contrast images into post-contrast images for a latent space where hemodynamic priors are exploited. In particular, the diffusion loss for the reverse diffusion process can be formulated as follows:

$$\mathcal{L}_{\text{DM}} = \mathbb{E}_{t,\epsilon,x} ||\epsilon_\theta(x_t, t; x_0, x_k) - \epsilon||^2 \tag{6}$$

where ϵ_θ represents the denoising model that employs an U-Net structure, x_0 and x_k are the pre-contrast and post-contrast images, respectively, ϵ is Gaussian distribution data $\sim \mathcal{N}(0, \mathbf{I})$, and t is a timestep.

For a second step, we train the SM that integrates the previously learned latent kinetic code to provide tumor hemodynamic information for voxel-level prediction. Considering the varying sizes, shapes and appearances of tumors that results from intratumor heterogeneity and results in difficulties of accurate cancer annotation, we design the segmentation loss as follows:

$$\mathcal{L}_{SM} = \mathcal{L}_{seg} + \lambda \mathcal{L}_{SSIM}$$

$$= \mathcal{L}_{CE}(S,G) + \mathcal{L}_{Dice}(S,G) + \lambda(1 - \frac{(2\mu_S\mu_G + C_1)(2\varphi_{SG} + C_2)}{(\mu_S^2 + \mu_G^2 + C_1)(\varphi_S^2 + \varphi_G^2 + C_2)}) \quad (7)$$

where $\mathcal{L}_{SSIM}$ is used to evaluate tumor structural characteristics, S and G represents segmentation map and ground truth, respectively; μ_S is the mean of S and μ_G is the mean of G; φ_S represents the variance of S and φ_G represents the variance of G; C_1 and C_2 denote the constant to hold training stable [15], and φ_{SG} is the covariance between S and G. The λ is set as 0.5 empirically. Following [16], $C_1 = (k_1 L)^2$ and $C_2 = (k_2 L)^2$, where k_1 is set as 0.01, k_2 is set as 0.03 and L is set as the range of voxel values.

3   Experiments

Dataset: To demonstrate the effectiveness of our proposed DKM, we evaluate our method on 4D DCE-MRI breast cancer segmentation using the Breast-MRI-NACT-Pilot dataset [13], which contains a total of 64 patients with the contrast-enhanced MRI protocol: a pre-contrast scan, followed by 2 consecutive post-contrast time points (As shown in Fig. 3). Each MR volume consists of 60 slices and the size of each slice is 256×256. Regarding preprocessing, we conduct zero-mean unit-variance intensity normalization for the whole volume. We divided the original dataset into training (70%) and test set (30%) based on the scans. Ground truth segmentations of the data are provided in the dataset for tumor annotation. No data augmentation techniques are used to ensure fairness.

Competing Methods and Evaluation Metrics: To comprehensively evaluate the proposed method, We compare it with 3D segmentation methods, including Dual Attention Net (DANet) [17], MultiResUNet [18] and multi-task learning network (MTLN) [19], and 4D segmentation methods, including LNet [20], 3D patch U-Net [21], and HybridNet [5]. All approaches are evaluated using 1) Dice Similarity Coefficient (DSC) and 2) Jaccard Index (JI).

Implementation Details: We implement our proposed framework with PyTorch using two NVIDIA RTX 2080Ti GPUs to accelerate model training. Following DDPM [8], we set 128, 256, 256, 256 channels for each stage in the DM and set the noise level from 10^{-4} to 10^{-2} using a linear schedule with $T = 1000$. Once the DM is trained, we extract intermediate feature maps from four resolutions for further segmentation task. Similar to DM, the SM also consists of four resolution blocks. However, unlike channel settings of DM, we set 128, 256,

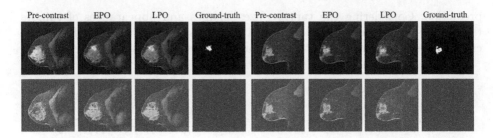

Fig. 3. Examples of breast DCE-MRI sequences. The bottom is heatmaps to observe the intensity change of cancers. EPO: early post-contrast, and LPO: late post-contrast.

Table 1. Cancer segmentation comparison between our method and previous models (Mean ± Std). The scans are employed for testing.

Method	Scans	Dice (%) ↑	JI (%) ↑
DANet [17]	post-contrast	52.3 ± 3.1	40.5 ± 3.1
MultiResUNet [18]	post-contrast	55.6 ± 2.8	43.2 ± 3.0
MTLN [19]	post-contrast	54.2 ± 2.5	41.6 ± 2.8
LNet [20]	complete sequence	52.3 ± 2.9	40.4 ± 3.2
3D patch U-Net [21]	complete sequence	53.8 ± 2.8	41.1 ± 3.0
HybridNet [5]	complete sequence	64.4 ± 2.4	51.5 ± 2.6
DKM (ours)	pre-contrast	**71.5 ± 2.5**	**58.5 ± 2.6**

512, 1024 channels for each stage in the SM to capture expressive and sufficient semantic information. The SM is optimized by Adam with a learning rate 2×10^{-5} and a weight decay 10^{-6}. The model is trained for 500 epochs with the batch size to 1. No data augmentation techniques are used to ensure fairness.

Comparison with SOTA Methods: The quantitative comparison of the proposed method to recent state-of-the-art methdos is reported in Table 1. Experimental results demonstrate that the proposed method comprehensively other models with less scans (i.e., pre-contrast) in testing. We attribute it to the ability of diffusion module to exploit hemodynamic priors to guide the segmentation task. Specifically, in comparison with 3D segmentation models (e.g. MTLN), our method yields higher segmentation scores. The possible reason is that our method is able to exploit the time intensity curve, which contains richer information compared to post-contrast scan. Besides, we can observe that our method achieves improvements when compared to 4D segmentation models using complete sequence. Our method outperform the HybridNet by 7.1% and 7.0% in DSC and JI, respectively. It probably due to two aspects: 1) The hemodynamic knowledge is implicitly exploited by diffusion module from pre-contrast images, which is useful for cancer segmentation. 2) The intermediate activations from

Pre-contrast	Post-contrast	Ground-truth	Baseline	Ours

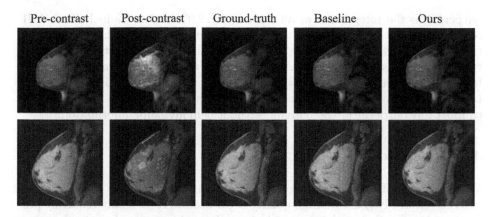

Fig. 4. Visual comparison of segmentation performance. The baseline is implemented without the incorporation of kinetic code.

Table 2. Ablation study for the incorporation of intermediate kinetic code.

f_1	f_2	f_3	f_4	Dice (%) ↑	JI (%) ↑
✓				67.9 ± 2.4	54.4 ± 2.4
	✓			68.6 ± 2.3	55.0 ± 2.2
		✓		68.3 ± 2.4	54.8 ± 2.5
			✓	69.3 ± 2.0	55.5 ± 2.1
	✓	✓		67.6 ± 2.3	53.2 ± 2.4
	✓		✓	70.1 ± 2.1	56.2 ± 2.3
		✓	✓	$\mathbf{71.5 \pm 2.5}$	$\mathbf{58.5 \pm 2.6}$
	✓	✓	✓	70.2 ± 2.3	56.4 ± 2.3
✓	✓	✓	✓	69.5 ± 2.1	55.9 ± 2.4

diffusion models effectively capture the semantic information and are excellent pixel-level representations for the segmentation problem [9]. Thus, combining the intermediate features can further promote the segmentation performance. In a word, the proposed framework can produce accurate prediction masks only requiring pre-contrast images. This is useful when post-contrast data is limited.

Ablation Study: To explore the effectiveness of the latent kinetic code, we first conduct ablation studies to select the optimal setting. We denote the intermediate features extracted from each stage in the DM as f_1, f_2, f_3, and f_4, respectively, where f_i represents the feature map of i-th stage. Table 2 reports the segmentation performance with different incorporations of intermediate kinetic codes. It can be observed that the latent kinetic code is able to guide the network training for better segmentation results. Specifically, we note that the incorporation of f_3 and f_4 achieves the highest scores among these combinations, and outperforms the integration of all features by 2.0% and 2.6% in DSC and JI, respectively. We attribute it to the denoising diffusion model that receives the noisy input, leading to the noise of shallow features. In contrast, the deep features capture essential characteristics to reveal the structural information and hemodynamic changes of tumors. Figure 4 shows visual comparison of segmentation performance. The above results reveal that incorporation of kinetic code comfortably outperform the baseline without hemodynamic information.

4 Conclusion

We propose a diffusion kinetic model by exploiting hemodynamic priors in DCE-MRI to effectively generate high-quality segmentation results only requiring pre-contrast images. Our models learns the hemodynamic response function based on the denoising diffusion process and estimates the latent kinetic code to guide the segmentation task. Experiments demonstrate that our proposed framework has the potential to be a promising tool in clinical applications to annotate cancers.

Acknowledgements. This work is supported in part by the National Key R&D Program of China under Grants 2021YFE0203700 and 2018YFA0701700, the Postgraduate Research & Practice Innovation Program of Jiangsu Province KYCX23_2524, and is supported by National Natural Science Foundation of China grants 61602007, U21A20521 and 61731008, Zhejiang Provincial Natural Science Foundation of China (LZ15F010001), Jiangsu Provincial Maternal and Child Health Research Project (F202034), Wuxi Health Commission Precision Medicine Project (J202106), Jiangsu Provincial Six Talent Peaks Project (YY-124), and the Science and Technology Development Fund, Macau SAR (File no. 0004/2019/AFJ and 0011/2019/AKP).

References

1. Fan, M., Xia, P., Clarke, R., Wang, Y., Li, L.: Radiogenomic signatures reveal multiscale intratumour heterogeneity associated with biological functions and survival in breast cancer. Nat. Commun. **11**(1), 4861 (2020)
2. Vidal, J., Vilanova, J.C., Martí, R., et al.: A U-net ensemble for breast lesion segmentation in DCE MRI. Comput. Biol. Med. **140**, 105093 (2022)
3. Qiao, M., et al.: Three-dimensional breast tumor segmentation on DCE-MRI with a multilabel attention-guided joint-phase-learning network. Comput. Med. Imaging Graph. **90**, 101909 (2021)
4. Nalepa, J., et al.: Fully-automated deep learning-powered system for DCE-MRI analysis of brain tumors. Artif. Intell. Med. **102**, 101769 (2020)
5. Lv, T., et al.: A hybrid hemodynamic knowledge-powered and feature reconstruction-guided scheme for breast cancer segmentation based on dce-mri. Med. Image Anal. **82**, 102572 (2022)
6. Wang, S., et al.: Breast tumor segmentation in DCE-MRI with tumor sensitive synthesis. IEEE Trans. Neural Netw. Learn. Syst. (2021)
7. Rombach, R., Blattmann, A., Lorenz, D., Esser, P., Ommer, B.: High-resolution image synthesis with latent diffusion models. In: Proceedings of the IEEE/CVF Conference on Computer Vision and Pattern Recognition, pp. 10684–10695 (2022)

8. Ho, J., Jain, A., Abbeel, P.: Denoising diffusion probabilistic models. Adv. Neural. Inf. Process. Syst. **33**, 6840–6851 (2020)
9. Baranchuk, D., Rubachev, I., Voynov, A., Khrulkov, V., Babenko, A.: Label-efficient semantic segmentation with diffusion models. In: International Conference on Learning Representation (ICLR) (2022)
10. Fernandez, V., et al.: Can segmentation models be trained with fully synthetically generated data? In: Zhao, C., Svoboda, D., Wolterink, J.M., Escobar, M. (eds.) SASHIMI 2022. LNCS, vol. 13570, pp. 79–90. Springer, Cham (2022). https://doi.org/10.1007/978-3-031-16980-9_8
11. Wolleb, J., Sandkühler, R., Bieder, F., Valmaggia, P., Cattin, P.C.: Diffusion models for implicit image segmentation ensembles. In: International Conference on Medical Imaging with Deep Learning, pp. 1336–1348. PMLR (2022)
12. Junde, W., et al.: Medsegdiff: medical image segmentation with diffusion probabilistic model (2023)
13. Newitt, D., Hylton, N.: Single site breast DCE-MRI data and segmentations from patients undergoing neoadjuvant chemotherapy. Cancer Imaging Arch. **2** (2016)
14. Kim, B., Ye, J.C.: Diffusion deformable model for 4D temporal medical image generation. In: Wang, L., Dou, Q., Fletcher, P.T., Speidel, S., Li, S. (eds.) MICCAI 2022. LNCS, vol. 13431, pp. 539–548. Springer, Cham (2022). https://doi.org/10.1007/978-3-031-16431-6_51
15. Wang, Z., Bovik, A.C., Sheikh, H.R., Simoncelli, E.P.: Image quality assessment: from error visibility to structural similarity. IEEE Trans. Image Process. **13**(4), 600–612 (2004)
16. Lin, J., Lin, H., Zhang, Z., Yiwen, X., Zhao, T.: SSIM-variation-based complexity optimization for versatile video coding. IEEE Signal Process. Lett. **29**, 2617–2621 (2022)
17. Fu, J., et al.: Dual attention network for scene segmentation. In: Proceedings of the IEEE/CVF Conference on Computer Vision and Pattern Recognition, pp. 3146–3154 (2019)
18. Ibtehaz, N., Rahman, M.S.: Multiresunet: rethinking the U-net architecture for multimodal biomedical image segmentation. Neural Netw. **121**, 74–87 (2020)
19. Zhou, Y., et al.: Multi-task learning for segmentation and classification of tumors in 3D automated breast ultrasound images. Med. Image Anal. **70**, 101918 (2021)
20. Denner, S., et al.: Spatio-temporal learning from longitudinal data for multiple sclerosis lesion segmentation. In: Crimi, A., Bakas, S. (eds.) BrainLes 2020. LNCS, vol. 12658, pp. 111–121. Springer, Cham (2021). https://doi.org/10.1007/978-3-030-72084-1_11
21. Khaled, R., Vidal, J., Martí, R.: Deep learning based segmentation of breast lesions in DCE-MRI. In: Del Bimbo, A., et al. (eds.) ICPR 2021. LNCS, vol. 12661, pp. 417–430. Springer, Cham (2021). https://doi.org/10.1007/978-3-030-68763-2_32

CAS-Net: Cross-View Aligned Segmentation by Graph Representation of Knees

Zixu Zhuang[1,3], Xin Wang[1], Sheng Wang[1,2,3], Zhenrong Shen[1],
Xiangyu Zhao[1], Mengjun Liu[1], Zhong Xue[3], Dinggang Shen[2,3],
Lichi Zhang[1(✉)], and Qian Wang[2(✉)]

[1] School of Biomedical Engineering, Shanghai Jiao Tong University, Shanghai, China
lichizhang@sjtu.edu.cn
[2] School of Biomedical Engineering, ShanghaiTech University, Shanghai, China
qianwang@shanghaitech.edu.cn
[3] Shanghai United Imaging Intelligence Co., Ltd., Shanghai, China

Abstract. Magnetic Resonance Imaging (MRI) has become an essential tool for clinical knee examinations. In clinical practice, knee scans are acquired from multiple views with stacked 2D slices, ensuring diagnosis accuracy while saving scanning time. However, obtaining fine 3D knee segmentation from multi-view 2D scans is challenging, which is yet necessary for morphological analysis. Moreover, radiologists need to annotate the knee segmentation in multiple 2D scans for medical studies, bringing additional labor. In this paper, we propose the Cross-view Aligned Segmentation Network (CAS-Net) to produce 3D knee segmentation from multi-view 2D MRI scans and annotations of sagittal views only. Specifically, a knee graph representation is firstly built in a 3D isotropic space after the super-resolution of multi-view 2D scans. Then, we utilize a graph-based network to segment individual multi-view patches along the knee surface, and piece together these patch segmentations into a complete knee segmentation with help of the knee graph. Experiments conducted on the Osteoarthritis Initiative (OAI) dataset demonstrate the validity of the CAS-Net to generate accurate 3D segmentation.

Keywords: Knee Segmentation · Multi-view MRI · Super Resolution · Graph Representation

1 Introduction

Magnetic resonance imaging (MRI) plays a pivotal role in knee clinical examinations. It provides a non-invasive and accurate way to visualize various internal knee structures and soft tissues [2]. With its excellent contrast and resolution, MRI can help radiologists detect and diagnose knee injuries or diseases. The choice of MRI sequences and acquisition techniques depends on the specific clinical scenario. While the acquisition of a 3D sequence can render the entire knee

---

Z. Zhuang, X. Wang and S. Wang—These authors contributed equally to this work.

H. Greenspan et al. (Eds.): MICCAI 2023, LNCS 14223, pp. 110–119, 2023.
https://doi.org/10.1007/978-3-031-43901-8_11

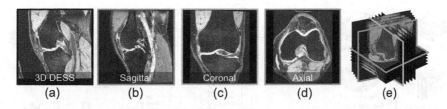

Fig. 1. Example of knee MRI scans in OAI dataset. (a) is a slice of 3D knee volume, which takes much longer scanning time; (b-d) are slices of multi-view 2D knee scans from sagittal, coronal, and axial views; (e) is the multi-view knee MRI scans aligned in the world coordinate system (WCS).

as shown in Fig. 1(a), it is not always feasible in clinical practice where scanner resource is limited [13]. In contrast, multi-view 2D scans can offer higher intra-slice resolution, better tissue contrast, and shorter scanning time [7,8], as displayed in Fig. 1(b)–(d). Therefore, multi-view 2D scans are widely used for diagnostics in clinical practice.

Although multi-view 2D MRI scanning can provide sufficient image quality for clinical diagnosis in most cases [7,16], it is challenging for estimating the corresponding 3D knee segmentation, which is essential for the functional and morphological analysis [15,17]. Previous studies mostly rely on 3D sequences to conduct image segmentation and morphology-based knee analysis [3,20], which is impractical for clinical data of only 2D scans.

The cross-view consistency of multi-view knee MR scans provides the basis for generating 3D segmentation from 2D scans, as shown in Fig. 1(e). Some studies have already utilized the cross-view consistency for medical image applications. For example, Liu et al. [10] proposed a Transformer-based method to fuse multi-view 2D scans for ventricular segmentation, by learning the cross-view consistency with the self-attention mechanism. Perslev et al. proposed to extract multi-view 2D slices from 3D images for 3D segmentation on knees [12] and brains [11]. Li et al. [9] transferred widely used sagittal segmentation to coronal and axial views based on cross-view consistency. Zhuang et al. [22] introduced a unified knee graph architecture to fuse the multi-view MRIs in a local manner for knee osteoarthritis diagnosis. These studies have demonstrated the value of cross-view consistency in medical analysis.

In this paper, we propose a novel framework, named Cross-view Aligned Segmentation Network (CAS-Net), to generate 3D knee segmentation from clinical multi-view 2D scans. Moreover, following the convention of radiologists, we require the supervising annotation for the sagittal segmentation only. We first align the multi-view knee MRI scans into an isotropic 3D volume by super-resolution. Then we sample multi-view patches and construct a knee graph to cover the whole knee joint. Next, we utilize a graph-based network to derive a fine 3D segmentation. We evaluate our proposed CAS-Net on the Osteoarthritis Initiative (OAI) dataset, demonstrating its effectiveness in cross-view 3D segmentation.

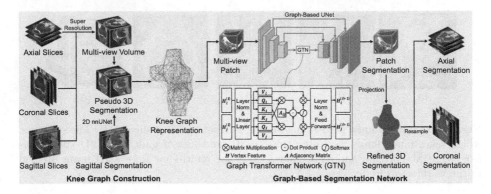

Fig. 2. The overall architecture of the proposed Cross-view Aligned Segmentation Network (CAS-Net). It is composed of two major parts: (1) Knee Graph Construction, which involves super-resolution of the multi-view 2D scans into a 3D volume and representing with a knee graph; (2) Graph-based Segmentation Network, which segments patches on the knee graph, resulting in 3D segmentation.

2   Method

The goal of CAS-Net is to generate 3D knee segmentation with multi-view 2D scans and sagittal segmentation annotation. The overall architecture is shown in Fig. 2, which consists of two major parts.

1 **Knee Graph Construction.** We conduct super-resolution on multi-view 2D scans and align them in a 3D volume. Then, we sample multi-view patches along bone-cartilage interfaces and construct a knee graph.
2 **Graph-based Segmentation Network.** We deal with the 3D multi-view patches on the knee graph by a 3D UNet [14] combined with Graph Transformer Network (GTN) [4,21]. We also randomly mask out views of patches to enforce cross-view consistency during network learning.

2.1   Knee Graph Construction

The 2D clinical multi-view scans are acquired from non-orthogonal angles and have large differences in inter- and intra-slice spacing (e.g. 3.5 mm v.s. 0.3 mm in clinical practice). Therefore, we use super-resolution to unify them into one 3D volume with a pseudo segmentation (with tolerable errors), and sample the volume into graph representation.

Isotropic Super-Resolution. We first apply super-resolution [19] to the multi-view 2D scans, which include sagittal, coronal and axial scans, for representing them in the same space and reducing the large inter-slice spacing. The process is shown in the left part of Fig. 2.

Specifically, we follow [19] by projecting multi-view 2D scans into the world coordinate system (WCS) and treating each scan as a continuous function of coordinates. Then, we take advantage of the continuity of coordinates to synthesize an isotropic 3D volume, by querying from the learned function for the coordinates at a equidistantly distributed interval. Since each 2D scan is reconstructed from the same set of WCS coordinates, the multi-view volumes after super-resolution are naturally aligned in the WCS space. Therefore, we concatenate these multi-view volumes as a multi-channel 3D image.

Moreover, we train a 2D nnUNet [6] on the sagittal 2D scan and apply it to the sagittal 3D volume slice-by-slice, to acquire a pseudo-3D segmentation. The pseudo-3D segmentation is the basis of knee graph construction. It also provides supervision for the graph-based segmentation network; yet it contains many errors caused by 2D segmentation, e.g., unsmooth segmentation boundaries of bones and cartilages.

Knee Graph Sampling. Errors in the pseudo-3D segmentation are distributed around bone surfaces. We sample multi-view patches from these areas and refine their segmentations in 3D. Previous studies have demonstrated the usefulness of organizing knee patches with graph-based representation [22]. Therefore, we follow this approach and construct the knee graph G.

We start extracting points on the bone surface by calculating the bone boundaries on the pseudo-3D segmentation. Then, we uniformly sample from these points until the average distance between them is about 0.8 times the patch size. These sampled points are collected as the vertices V in the graph. V stores the center-cropped multi-view patches P and their WCS positional coordinates C. The distribution of the sampled points ensures that multi-view patches center-cropped on V can fully cover the bone surfaces and bone-cartilage interfaces. Finally, we connect the vertices in V using edges E, which enables the vertex features to be propagated on the knee graph. The edges are established by connecting each vertex to its k ($k = 10$, following [22]) nearest neighbors in V.

2.2   Graph-Based Segmentation Network

Graph-Based UNet. The knee graph, which is constructed early, is input to a UNet combined with a 3-layer Graph Transformer Network (GTN) [4,21] to generate patch segmentation, as shown on the right part of Fig. 2.

To encode the multi-view patches P as vertex features H in the bottleneck of the UNet, we first apply average pooling to H and then add the linearly-projected WCS coordinate C to enable convolution of H along the graph. Specifically, we compute $H^{(0)} = \text{AvgPooling}(H) + \text{Linear}(C)$. Next, we pass the resulting vertex features $H^{(0)}$ through a graph convolutional network. Note that the knee graphs have varying numbers of vertices and edges, we utilize the Graph Transformer Networks (GTN) to adapt it.

Similar to Transformers [18], the GTN generates three attention embeddings Q, K, and V from $H^{(i)}$ by linear projection in the i-th layer. The difference is that GTN uses the adjacency matrix A to restrict the self-attention computation to between adjacent vertices instead of all of them: $\alpha = \text{Softmax}(A \odot \frac{QK^T}{\sqrt{d}})$, where α is the self-attention, $\odot$ denotes dot product, and d is the length of attention embeddings. Then, the local feature $H^{(i)}$ in the i-th layer can be computed as $H^{(i)} = \sigma(\alpha V)$, where σ is feedforward operation.

After the attention-based graph representation, $H^{(3)}$ is repeated to match the shape of H and then summed up, serving as the input to the UNet decoder. And the UNet decoder derives the 3D patch segmentation according to $H^{(3)}$ and skip connection. We further project the patch segmentations back into the pseudo-3D segmentation, thereby obtaining a rectified 3D knee segmentation.

Random View Masking. To reduce impacts of the errors in the initial pseudo-3D segmentation, we randomly masked out views (channels) in the multi-view (multi-channel) patches during training. The 3D volume, obtained from 2D scans by super-resolution, has a clearer tissue texture in the plane parallel to its original scanning plane. Thus, the random view masking helps to produce more accurate segmentation boundaries by forcing the network to learn from different combinations of patch views (channels). We set the probability of each view to be masked as 0.25. During inference, complete multi-view patches are used, reducing errors caused by the pseudo-3D segmentation.

3   Experimental Results

3.1   Data and Experimental Settings

We have tested the CAS-Net on the Osteoarthritis Initiative (OAI) dataset [5], which contains multi-view 2D scans and 3D volumes aligned in the same WCS space. The 3D DESS modality in the OAI dataset has segmentation annotation [1], which is used as the ground truth to evaluate the 3D segmentation performance in our work. The SAG_IW_TSE_LEFT, COR_MPR_LEFT, and AX_MPR_LEFT modalities in the OAI dataset are selected as multi-view 2D scans. For the 3D DESS is unsmooth when it is resampled to 2D scans, we recruit a radiologist to manually annotate the femur, tibia, their attached cartilages, and meniscus in 2D scans. The sagittal annotation is used in our method, and the coronal or axial annotation is set as the ground truth for evaluation. We extract 350 subjects for our experiments, with 210 for training, 70 for validation, and 70 for testing. We report the results of the testing set in the tables below. The 2D annotations are released in https://github.com/zixuzhuang/OAI_seg.

In the following, we refer to the femur bone and tibia bone as FB and TB for short, their attached cartilages as FC and TC, and meniscus as M. We use GTN and RM to denote the Graph Transformer Network and random masking strategies, respectively. SR and NN stand for super-resolution and nearest neighbor interpolation.

To ensure the fairness of the comparison, we use the same training and testing settings for all experiments: PyTorch1.9.0, RTX 3090, learning rate is 3e-4, 100 epochs, cosine learning rate decay, Adam optimizer.

3.2 Ablation Study

To find the optimal setting for the CAS-Net, we apply different hyper-parameters on it and test them on cross-view 2D segmentation. We evaluate the performance of the CAS-Net with different patch size, and then remove the GTN, random masking, and super-resolution to see the effect of each component. To ensure the multi-view slices can be aligned in the same space without super-resolution, we use trilinear interpolation as the alternative. The results are shown in Table 1.

Table 1. Results from a controlled ablation study of the proposed CAS-Net, examining the effects of patch size, GTN, randomly masking (RM), and super-resolution (SR).

Patch Size	Method	Coronal Dice Ratio (%)						Axial Dice Ratio (%)					
		FB	FC	TB	TC	M	Mean	FB	FC	TB	TC	M	Mean
64		96.1	79.3	95.3	80.7	79.1	86.1	96.9	79.9	96.9	78.6	76.9	85.8
128		**96.5**	**79.7**	**96.0**	**81.9**	**79.9**	**86.8**	**97.1**	**80.1**	**97.1**	**78.8**	**77.8**	**86.2**
128	w/o GTN	96.1	79.0	95.5	81.0	79.3	86.2	**97.1**	79.7	**97.1**	78.4	**77.8**	86.0
128	w/o RM	95.4	78.4	95.0	80.7	79.0	85.7	97.0	**80.1**	**97.1**	78.0	77.0	85.8
128	w/o SR	96.0	74.1	95.2	74.7	72.0	82.4	96.5	74.2	96.3	73.6	70.7	82.3

As shown in Table 1, the best performance appears when patch size is 128 (86.8% mean Dice ratio in coronal view), and the removing of GTN (86.2%), random masking (85.7%) or super-resolution (82.4%) always lead to performance drop. The patch size of 64 is too small to capture local appearance of the knee, and thus causes performance degradation. GTN enables the network to capture knee features globally and therefore brings performance improvement. Random masking encourages CAS-Net to learn knee segmentation with aligned views across different planes, avoiding errors caused by super-resolution and 3D pseudo segmentation, thus resulting in better performance. Super-resolution reduces the aliasing effect during resampling, resulting in better segmentation. Therefore, we select the patch size as 128, and keep GTN, RM and SR enabled in the next experiments.

3.3 Evaluation of 3D Segmentation Performance

We report the performance of the CAS-Net for 3D knee segmentation. To the best of our knowledge, CAS-Net is the first method that can perform knee segmentation from multi-view 2D scans and sagittal segmentation annotation. Therefore, we compare CAS-Net with the following settings: (1) the sagittal annotation that is resampled to 3D DESS space by nearest neighbor interpolation; (2) the pseudo-3D segmentation generated in Sect. 2.1; (3) the 3D prediction from 3D nnUNet based on 3D DESS images and annotations, which is considered as the upper bound for this task.

As can be seen from Table 2, the CAS-Net is capable of producing excellent 3D segmentation results with only sagittal annotation, as demonstrated in

Table 2. Results of the comparison of different methods in 3D segmentation.

Methods	Image View	Annotation View	3D Dice Ratio (%)					
			FB	FC	TB	TC	M	Mean
NN	-	Sagittal	96.1	76.4	95.2	72.5	79.7	83.9
Pseudo Seg	Sagittal	Sagittal	96.3	79.0	95.2	72.6	79.7	84.6
3D nnUNet	3D DESS	3D DESS	92.2	**97.8**	90.2	**90.2**	**89.8**	**92.0**
Ours	Multi-View	Sagittal	**96.7**	81.7	**95.8**	74.9	81.0	86.0

Fig. 3(f). In comparison to 3D nnUNet that utilizes 3D images and annotations, CAS-Net provides better performance on FB (96.7% vs 92.2% mean 3D Dice) and TB (95.8% vs 90.2%), since the 3D DESS sequence has a lower contrast of bone and muscle, resulting in more errors. Additionally, although there is roughly a 16% performance gap of cartilages compared to 3D nnUNet, CAS-Net demonstrates satisfactory outcomes in FC (81.7%) and M (81.0%) due to multi-view image localization errors like slight knee movements during scanning, resulting in misaligned segmentation generation from the multi-view scans and 3D annotation, as shown in Fig. 4. Furthermore, our method significantly improves 3D mean segmentation performance (86.0%) in comparison to the two techniques of using NN interpolation directly (83.9%) and employing super-resolution to generate pseudo sagittal segmentation (84.6%). These experimental findings indicate that our approach effectively leverages multi-view 2D scans and sagittal segment annotations to produce superior 3D segmentation.

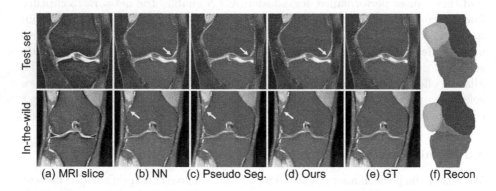

(a) MRI slice   (b) NN   (c) Pseudo Seg.   (d) Ours   (e) GT   (f) Recon

Fig. 3. We test our model on OAI data and in-the-wild clinical data. Here are two example cases with comparison of different methods.

3.4 Evaluation of Cross-View Projection of 3D Segmentation

The 3D segmentation results inferred by CAS-Net can be resampled onto unannotated 2D scans to achieve cross-view segmentation. We evaluate the cross-view segmentation performance on the coronal view, which is shown in Table 3. We compare with the three alternative methods: (1) the sagittal annotation that is projected to the coronal view by nearest neighbor interpolation, named NN in Table 3, which is considered as the lower bound for the cross-view segmentation task; (2) the pseudo-segmentation generated in Sect. 2.1 projected to the coronal view, named as Pseudo Seg. in Table 3; (3) the 2D prediction from 2D nnUNet based on coronal 2D scans and annotations, which is considered as the upper bound for the cross-view segmentation task. The visualized results are shown in Fig. 3(b)–(d).

Table 3. Results of the comparison of different methods in 2D segmentation.

Methods	Image View	Annotation View	Coronal Dice Ratio (%)					
			FB	FC	TB	TC	M	Mean
NN	-	Sagittal	96.1	76.4	95.2	71.5	79.7	83.8
Pseudo Seg	Sagittal	Sagittal	96.3	79.1	95.3	72.1	79.7	84.5
nnUNet	3D DESS	Coronal	98.9	92.2	98.8	92.8	92.9	95.1
Ours	Sagittal	Sagittal	96.5	79.7	96.0	81.9	79.9	86.8

CAS-Net not only eliminates some segmentation errors but also provides smooth segmentation boundaries in cross-view segmentation tasks. As seen in Table 3 (Row 1/2) and Fig. 3(b) and (c), while the super-resolution method has improved the mean Dice ratio by only 0.7%, it has resulted in noticeably smoother segmentation boundary of the knee tissue. The CAS-Net then further eliminates errors in Pseudo Segmentation, resulting in better segmentation outcomes and leading to additional improvements (86.8% mean coronal Dice).

Compared with nnUNet, there is still improvement room for CAS-Net in the case of small tissues, such as meniscus (79.9% vs. 92.9% Dice ratio). This is likely caused by subtle movements of subjects during the acquisition of multi-view 2D scans. As shown in Fig. 4, compared to sagittal MRI, FC is slightly shifted upwards in coronal and 3D MRIs. Thus the segmentation is aligned in sagittal view but not in coronal and 3D. We have examined subjects in the dataset

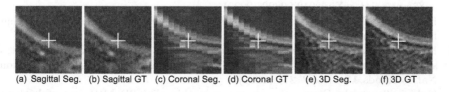

(a) Sagittal Seg.   (b) Sagittal GT   (c) Coronal Seg.   (d) Coronal GT   (e) 3D Seg.   (f) 3D GT

Fig. 4. The multi-view scans and 3D image are misaligned due to knee motion during scanning, though the segmentation have the correct shape as in sagittal view. The white cross indicates the same position in WCS space but cartilage is moved.

that have lower performance (around 50% mean Dice ratio) and confirmed this observation. On the contrary, given correctly aligned multi-view 2D scans, CAS-Net can achieve a mean Dice ratio of 95%.

4   Conclusion

We have proposed a novel framework named CAS-Net for generating cross-view consistent 3D knee segmentation via super-resolution and graph representation with clinical 2D multi-view scans and sagittal annotations. By creating a detailed 3D knee segmentation from clinical 2D multi-view MRI, our framework provides significant benefits to morphology-based knee analysis, with promising applications in knee disease analysis. We believe that this framework can be extended to other 3D segmentation tasks, and we intend to explore those possibilities. However, it should be noted that the performance of the CAS-Net is affected by misaligned multi-view images, as shown in Fig. 4. After removing the misaligned cases, the femur cartilage dice can be increased from 81.7% to 87.3% in 3D segmentation. We plan to address this issue by incorporating multimodal alignment in future work.

Acknowledgement. This work was supported by the National Natural Science Foundation of China (No. 62001292), and partially supported by the National Natural Science Foundation of China (U22A20283), and the Interdisciplinary Program of Shanghai Jiao Tong University (No. YG2023LC07).

References

1. Ambellan, F., Tack, A., Ehlke, M., Zachow, S.: Automated segmentation of knee bone and cartilage combining statistical shape knowledge and convolutional neural networks: data from the osteoarthritis initiative. Med. Image Anal. **52**, 109–118 (2019)
2. Calivà, F., Namiri, N.K., Dubreuil, M., Pedoia, V., Ozhinsky, E., Majumdar, S.: Studying osteoarthritis with artificial intelligence applied to magnetic resonance imaging. Nat. Rev. Rheumatol. **18**, 1–10 (2021)
3. Carballido-Gamio, J., et al.: Inter-subject comparison of MRI knee cartilage thickness. Med. Image Anal. **12**(2), 120–135 (2008)
4. Dwivedi, V.P., Bresson, X.: A generalization of transformer networks to graphs. arXiv preprint arXiv:2012.09699 (2020)
5. Eckstein, F., Wirth, W., Nevitt, M.C.: Recent advances in osteoarthritis imaging-the osteoarthritis initiative. Nat. Rev. Rheumatol. **8**(10), 622–630 (2012)
6. Isensee, F., Jaeger, P.F., Kohl, S.A., Petersen, J., Maier-Hein, K.H.: nnU-Net: a self-configuring method for deep learning-based biomedical image segmentation. Nat. Methods **18**(2), 203–211 (2021)
7. Kakigi, T., et al.: Diagnostic advantage of thin slice 2D MRI and multiplanar reconstruction of the knee joint using deep learning based denoising approach. Sci. Rep. **12**(1), 1–14 (2022)

8. Kijowski, R., Davis, K.W., Blankenbaker, D.G., Woods, M.A., Del Rio, A.M., De Smet, A.A.: Evaluation of the menisci of the knee joint using three-dimensional isotropic resolution fast spin-echo imaging: diagnostic performance in 250 patients with surgical correlation. Skeletal Radiol. **41**, 169–178 (2012)

9. Li, T., Xuan, K., Xue, Z., Chen, L., Zhang, L., Qian, D.: Cross-view label transfer in knee MR segmentation using iterative context learning. In: Albarqouni, S., et al. (eds.) DART/DCL -2020. LNCS, vol. 12444, pp. 96–105. Springer, Cham (2020). https://doi.org/10.1007/978-3-030-60548-3_10

10. Liu, D., et al.: Transfusion: multi-view divergent fusion for medical image segmentation with transformers. In: Wang, L., Dou, Q., Fletcher, P.T., Speidel, S., Li, S. (eds.) MICCAI 2022. LNCS, vol. 13435, pp. 485–495. Springer, Cham (2022). https://doi.org/10.1007/978-3-031-16443-9_47

11. Perslev, M., Dam, E.B., Pai, A., Igel, C.: One network to segment them all: a general, lightweight system for accurate 3D medical image segmentation. In: Shen, D., et al. (eds.) MICCAI 2019. LNCS, vol. 11765, pp. 30–38. Springer, Cham (2019). https://doi.org/10.1007/978-3-030-32245-8_4

12. Perslev, M., Pai, A., Runhaar, J., Igel, C., Dam, E.B.: Cross-cohort automatic knee MRI segmentation with multi-planar U-nets. J. Magn. Reson. Imaging **55**(6), 1650–1663 (2022)

13. Recht, M.P., et al.: Using deep learning to accelerate knee MRI at 3 T: results of an interchangeability study. AJR Am. J. Roentgenol. **215**(6), 1421 (2020)

14. Ronneberger, O., Fischer, P., Brox, T.: U-Net: convolutional networks for biomedical image segmentation. In: Navab, N., Hornegger, J., Wells, W.M., Frangi, A.F. (eds.) MICCAI 2015. LNCS, vol. 9351, pp. 234–241. Springer, Cham (2015). https://doi.org/10.1007/978-3-319-24574-4_28

15. Schmidt, A.M., et al.: Generalizability of deep learning segmentation algorithms for automated assessment of cartilage morphology and MRI relaxometry. J. Magn. Reson. Imaging **57**(4), 1029–1039 (2022)

16. Shakoor, D., et al.: Diagnosis of knee meniscal injuries by using three-dimensional MRI: a systematic review and meta-analysis of diagnostic performance. Radiology **290**(2), 435–445 (2019)

17. Vanwanseele, B., et al.: The relationship between knee adduction moment and cartilage and meniscus morphology in women with osteoarthritis. Osteoarthritis Cartilage **18**(7), 894–901 (2010)

18. Vaswani, A., et al.: Attention is all you need. In: Advances in Neural Information Processing Systems, vol. 30 (2017)

19. Wang, X., Xuan, K., Wang, S., Xiong, H., Zhang, L., Wang, Q.: Arbitrary reduction of MRI slice spacing based on local-aware implicit representation. arXiv preprint arXiv:2205.11346 (2022)

20. Wenger, A., et al.: Relationship of 3D meniscal morphology and position with knee pain in subjects with knee osteoarthritis: a pilot study. Eur. Radiol. **22**, 211–220 (2012)

21. Zhuang, Z., et al.: Knee cartilage defect assessment by graph representation and surface convolution. IEEE Trans. Med. Imaging **42**(2), 368–379 (2022)

22. Zhuang, Z., et al.: Local graph fusion of multi-view MR images for knee osteoarthritis diagnosis. In: Wang, L., Dou, Q., Fletcher, P.T., Speidel, S., Li, S. (eds.) MICCAI 2022. LNCS, vol. 13433, pp. 554–563. Springer, Cham (2022). https://doi.org/10.1007/978-3-031-16437-8_53

One-Shot Traumatic Brain Segmentation with Adversarial Training and Uncertainty Rectification

Xiangyu Zhao[1], Zhenrong Shen[1], Dongdong Chen[1], Sheng Wang[1,2],
Zixu Zhuang[1,2], Qian Wang[3], and Lichi Zhang[1(✉)]

[1] Shanghai Jiao Tong University, Shanghai, China
lichizhang@sjtu.edu.cn
[2] Shanghai United Imaging Intelligence Co., Ltd., Shanghai, China
[3] ShanghaiTech University, Shanghai, China

Abstract. Brain segmentation of patients with severe traumatic brain injuries (sTBI) is essential for clinical treatment, but fully-supervised segmentation is limited by the lack of annotated data. One-shot segmentation based on learned transformations (OSSLT) has emerged as a powerful tool to overcome the limitations of insufficient training samples, which involves learning spatial and appearance transformations to perform data augmentation, and learning segmentation with augmented images. However, current practices face challenges in the limited diversity of augmented samples and the potential label error introduced by learned transformations. In this paper, we propose a novel one-shot traumatic brain segmentation method that surpasses these limitations by adversarial training and uncertainty rectification. The proposed method challenges the segmentation by adversarial disturbance of augmented samples to improve both the diversity of augmented data and the robustness of segmentation. Furthermore, potential label error introduced by learned transformations is rectified according to the uncertainty in segmentation. We validate the proposed method by the one-shot segmentation of consciousness-related brain regions in traumatic brain MR scans. Experimental results demonstrate that our proposed method has surpassed state-of-the-art alternatives. Code is available at https://github.com/hsiangyuzhao/TBIOneShot.

Keywords: One-Shot Segmentation · Adversarial Training · Traumatic Brain Injury · Uncertainty Rectification

1 Introduction

Automatic brain ROI segmentation for magnetic resonance images (MRI) of severe traumatic brain injuries (sTBI) patients is crucial in brain damage assessment and brain network analysis [8,11], since manual labeling is time-consuming

Supplementary Information The online version contains supplementary material available at https://doi.org/10.1007/978-3-031-43901-8_12.

and labor-intensive. However, conventional brain segmentation pipelines, such as FSL [14] and FreeSurfer [4], suffer significant performance deteriorations due to skull deformation and lesion erosions in traumatic brains. Although automatic segmentation based on deep learning has shown promises in accurate segmentation [10,12], these methods are still constrained by the scarcity of annotated sTBI scans. Thus, researches on traumatic brain segmentation under insufficient annotations needs further exploration.

Recently, one-shot medical image segmentation based on learned transformations (OSSLT) has shown great potential [3,17] to deal with label scarcity. These methods typically utilize deformable image registration to learn spatial and appearance transformations and perform data augmentation on the single labeled image to train the segmentation, which is shown in Fig. 1(a). Given a labeled image as the atlas, two unlabeled images are provided as spatial and appearance references. Appearance transform and spatial transform learned by deformable registration are applied to the atlas image to generate a pseudo-labeled image to train the segmentation, and the label warped by spatial transform serves as the ground-truth. In this way, the data diversity is ensured by a large amount of unlabeled data, and the segmentation is learned by abundant pseudo images.

However, despite the previous success, the generalization ability of these methods is challenged by two issues in traumatic brain segmentation: 1) Limited diversity of generated data due to the amount of available unlabeled images. Although several studies [6,7] have proposed transformation sampling to introduce extra diversity for alleviating this issue, their strategies rely on a manual-designed distribution, which is not learnable and limits the capacity of data augmentation. 2) The assumption that appearance transforms in atlas augmentation do not affect semantic labels in the images [6,17]. However, this assumption neglects the presence of abnormalities in traumatic brains, such as brain edema, herniation, and erosions, which affect the appearance of brain tissues and introduce label errors.

To address the aforementioned issues, we propose a novel one-shot traumatic brain segmentation method that leverages adversarial training and uncertainty rectification. We introduce an adversarial training strategy that improves both the diversity of generated data and the robustness of segmentation, and incorporate an uncertainty rectification strategy that mitigates potential label errors in generated samples. We also quantify the segmentation difference of the same image with and without the appearance transform, which is used to estimate the uncertainty of segmentation and rectify the segmentation results accordingly. The main contributions of our method are summarized as follows: First, we develop an adversarial training strategy to enhance the capacity of data augmentation, which brings better data diversity and segmentation robustness. Second, we notice the potential label error introduced by appearance transform in current one-shot segmentation attempts, and introduce uncertainty rectification for compensation. Finally, we evaluate the proposed method on brain segmentation of sTBI patients, where our method outperforms current state-of-the-art methods.

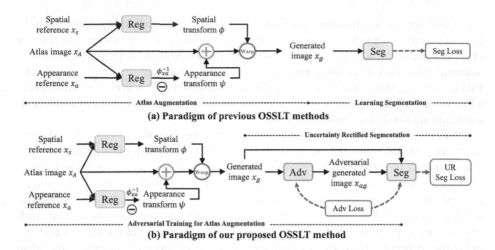

Fig. 1. An illustration of one-shot segmentation based on learned transformations (OSSLT). Compared with previous methods, we aim to use adversarial training and uncertainty rectification to address the current challenges in OSSLT.

2 Method

2.1 Overview of One-Shot Medical Image Segmentation

After training the unsupervised deformable registration, one-shot medical image segmentation based on learned transformations typically consists of two steps: 1) Data augmentation on the atlas image by learned transformations; 2) Training the segmentation using the augmented images. The basic workflow is shown in Fig. 1(a). Specifically, given a labeled image x_A as the atlas and its semantic labels y_A, two reference images, including the spatial reference x_s and appearance reference x_a, are provided to augment the atlas by spatial transform ϕ and appearance transform ψ that are calculated by the same pretrained registration network.

For spatial transform, given an atlas image x_A and a spatial reference x_s, the registration network performs the deformable registration between them and predicts a deformation field ϕ, which is used as the spatial transform to augment the atlas image spatially. For appearance transform, given an atlas image x_A and an appearance reference x_a, we warp x_a to x_A via an inverse registration ϕ_{xa}^{-1} and generates a inverse-warped $\tilde{x}_a = x_a \circ \phi_{xa}^{-1}$, and appearance transform $\psi = \tilde{x}_a - x_A$ is calculated by the residual of inverse-warped appearance reference $\tilde{x}_a$ and the atlas image x_A. It should be noted that the registration here is diffeomorphic to allow for inverse registration.

After acquiring both the spatial and appearance transform, the augmented atlas $x_g = (x_A + \psi) \circ \phi$ is generated by applying both transformations. The corresponding ground-truth $y_g = y_A \circ \phi$ is the atlas label warped by ϕ, as it is hypothesized that appearance transform does not alter the semantic labels in

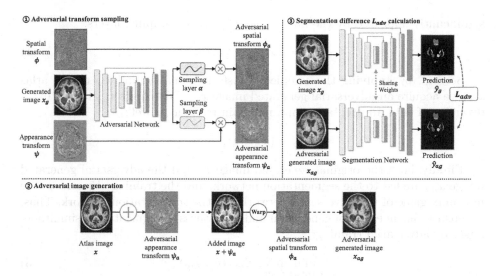

Fig. 2. Framework of adversarial training, which includes adversarial transform sampling, adversarial image generation and segmentation difference calculation.

the atlas image. During segmentation training, a large amount of spatial and appearance reference images are sampled to ensure the diversity of x_g, and the segmentation is trained with generated pairs of x_g and y_g.

In this work, we focus on both of the two steps in OSSLT by adversarial training and uncertainty rectification, which is shown in Fig. 1(b). Specifically, we generate an adversarial image x_{ag} along with x_g by adversarial training to learn better data augmentation for the atlas image, and uncertainty rectification is utilized during segmentation learning to bypass the potential label errors introduced by appearance transforms. We discuss our proposed adversarial training and uncertainty rectification in Sect. 2.2 and Sect. 2.3, respectively.

2.2 Adversarial Training

Although the diversity of generated pairs of x_g and y_g is ensured by the increased number of unlabeled images as references, such a setting requires a large amount of unlabeled data. Inspired by [2,9], we adopt the adversarial training strategy to increase both the data diversity and the segmentation robustness, which is shown in Fig. 2. Given a learned spatial transform ϕ, appearance transform ψ, and a generated image x_g augmented by ϕ and ψ, our adversarial training is decomposed into the following 3 steps:

First, we feed x_g into the adversarial network and generate two sampling layers α and β activated by Sigmoid function. The sampling layers α and β have the same spatial shape with ϕ and ψ respectively, and each location in the sampling layers represents the sampling amplitude of the original transform, ranging from 0 to 1. In this way, the diversity of spatial and appearance transforms is significantly improved by the infinite possibilities of sampling layers:

$$\phi_a = \phi \times \alpha, \psi_a = \psi \times \beta \tag{1}$$

Second, by applying the sampled transformations ϕ_a and ψ_a to the atlas x_A, we acquire an adversarial generated image x_{ag}. We expect x_{ag} to add extra diversity of data augmentation and maintain realistic as well:

$$x_{ag} = (x_A + \psi_a) \circ \phi_a \tag{2}$$

Finally, both the original generated image x_g and the adversarial generated image x_{ag} are fed to the segmentation network, and the training objective is the min-max game of the adversarial network and the segmentation network. Thus, we ensure the diversity of generation and robustness of segmentation simultaneously by adversarial training:

$$\min_{g(\cdot;\theta_h)} \max_{f(\cdot;\theta_g)} \mathcal{L}_{adv}(\hat{y}_g, \hat{y}_{ag}) \tag{3}$$

$$\mathcal{L}_{adv}(\hat{y}_g, \hat{y}_{ag}) = \frac{\hat{y}_g \cdot \hat{y}_{ag}}{\| \hat{y}_g \|_2 \cdot \| \hat{y}_{ag} \|_2} \tag{4}$$

where $f(\cdot; \theta_g)$ and $g(\cdot; \theta_h)$ denote the adversarial network and the segmentation network, respectively. $\hat{y}_g$ and $\hat{y}_{ag}$ are the segmentation predictions of x_g and x_{ag}. It should be noted that since the spatial transformation applied to x_g and x_{ag} is different, the loss calculation is performed in atlas space by inverse registration.

2.3   Uncertainty Rectification

Most of the current methods hypothesize that appearance transformation does not alter the label of the atlas. However, in brain scans with abnormalities such as sTBI, the appearance transformation may include edema, lesions, and *etc*, which may affect the actual semantic labels of the atlas and weaken the accuracy of segmentation. Inspired by [18], we introduce uncertainty rectification to bypass the potential label errors.

Specifically, given a segmentation network, fully augmented image $x_g = (x_A + \psi) \circ \phi$ and spatial-augmented image $x_{As} = x_A \circ \phi$ are fed to the network. The only difference between x_g and x_{As} is that the latter lacks the transformation on appearance. Thus, the two inputs x_g and x_{As} serve different purposes. Fully augmented image x_g is equipped with more diversity compared with x_{As}, as appearance transform has been applied to it, while the spatial augmented image x_{As} has more label authenticity and could guide a more accurate segmentation.

The overall supervised loss consists of two items. First, the segmentation loss $\mathcal{L}_{seg} = \mathcal{L}_{ce}(\hat{y}_{As}, y_g)$ of spatial-augmented image x_{As} guides the network to learn spatial variance only, where $\hat{y}_{As}$ is the prediction of x_{As}, and $\mathcal{L}_{ce}$ denotes cross-entropy loss. Second, the rectified segmentation loss $\mathcal{L}_{rseg}$ of x_g guides the network to learn segmentation under both spatial and appearance transformations. We adopt the KL-divergence D_{KL} of the segmentation results $\hat{y}_{As}$ and

$\hat{y}_g$ as the uncertainty in prediction [18]. Compared with Monte-Carlo dropout [5], KL-divergence for uncertainty estimation does not require multiple forward runs. A voxel with a greater uncertainty indicates a higher possibility of label error in the corresponding location of x_g, thus, the supervision signal of this location should be weakened to reduce the effect of label errors:

$$\mathcal{L}_{rseg} = \exp[-D_{KL}(\hat{y}_g, \hat{y}_{As})]\mathcal{L}_{ce}(\hat{y}_g, y_g) + D_{KL}(\hat{y}_g, \hat{y}_{As}) \qquad (5)$$

Thus, the overall supervised loss $\mathcal{L}_{sup} = \mathcal{L}_{seg} + \mathcal{L}_{rseg}$ is the segmentation loss $\mathcal{L}_{seg}$ of x_{As} and the rectified segmentation loss $\mathcal{L}_{rseg}$ of x_g. We apply the overall supervised loss $\mathcal{L}_{sup}$ on both x_g and x_{ag} in practice. During segmentation training, the linear summation of supervised segmentation loss $\mathcal{L}_{sup}$ and adversarial loss $\mathcal{L}_{adv}$ is minimized.

3   Experiments

3.1   Data

We have collected 165 MR T1-weighted scans with sTBI from 2014–2017, acquired on a 3T Siemens MR scanner from Huashan hospital. Among the 165 MR scans, 42 scans are labeled with the 17 consciousness-related brain regions (see appendix for details) while the remaining are left unlabeled, since the manual labeling requires senior-level expertise. Informed consent was obtained from all patients for the use of their information, medical records, and MRI data. All MR scans are linearly aligned to the MNI152 template using FSL [14]. For the atlas image, we randomly collect a normal brain scan at the same institute and label its 17 consciousness-related brain regions as well. During training, the labeled normal brain scan serves as the atlas image, and the 123 unlabeled sTBI scans are used as spatial or appearance references. For one-shot setting, the labeled sTBI scans are used for evaluation only and completely hidden during training. In order to validate the effectiveness of the proposed one-shot segmentation method, a U-Net [13] trained on the labeled scans by 5-fold cross validation is used as the reference of fully supervised segmentation.

3.2   Implementation Details

The framework is implemented with PyTorch 1.12.1 on a Debian Linux server with an NVIDIA RTX 3090 GPU. In practice, the registration network is based on VoxelMorph [1] and pretrained on the unlabeled sTBI scans. During adversarial training, the registration is fixed, while the adversarial network and segmentation network are trained alternately. Both the adversarial network and the segmentation network is based on U-Net [13] architectures and optimized by SGD optimizer with a momentum of 0.9 and weight decay of 1×10^{-4}. The initial learning rate is set to 1×10^{-2} and is slowly reduced with polynomial strategy. We have pretrained the registration network for 100 epochs, and trained the adversarial network and segmentation network for 100 epochs as well. The batch size is set to 1 during training.

Table 1. Ablation studies on different components. **Rand.** denotes uniform transform sampling following [7], **Adv.** denotes adversarial training, and **UR** denotes uncertainty rectification.

Experiments	Dice Coefficient (%)
(1) Baseline	51.4 ± 20.2
(2) Baseline + Rand	53.1 ± 20.6
(3) Baseline + Adv	54.3 ± 19.7
(4) Baseline + UR (wo/ $\mathcal{L}_{seg}$)	48.8 ± 23.5
(5) Baseline + UR (w/ $\mathcal{L}_{seg}$)	53.3 ± 19.0
(6) Baseline + Adv. + UR (w/ $\mathcal{L}_{seg}$)	$\mathbf{56.3 \pm 18.8}$
(Upper Bound) Fully-Supervised U-Net	61.5 ± 16.4

Table 2. Comparison with alternative segmentation methods (%).

	IR	IL	TR	TL	ICRA	ICRP
BrainStorm	46.2±21.9	43.2±21.3	66.7±18.1	57.1±7.3	46.8±11.7	34.5±21.0
LT-Net	45.4±24.6	**52.4±21.0**	59.5±17.9	**58.3±10.4**	47.0±19.7	**42.6±17.3**
DeepAtlas	**50.3±25.4**	44.2±27.0	56.4±16.1	57.5±8.7	**50.4±13.8**	38.8±18.7
Proposed	52.6±24.8	54.4±22.9	**62.0±16.2**	58.4±10.2	51.1±16.6	44.2±17.5
	ICLA	ICLP	CRA	CRP	CLA	CLP
BrainStorm	46.1±14.3	38.8±14.6	50.1±16.6	48.0±13.2	**50.9±11.1**	54.2±10.8
LT-Net	43.5±19.2	42.6±17.2	**52.2±18.4**	48.4±12.5	48.5±13.0	**56.5±11.9**
DeepAtlas	53.8±15.6	46.2±16.9	46.4±19.3	42.5±13.9	43.4±15.8	52.1±13.0
Proposed	**53.0±16.7**	**44.7±18.1**	56.1±15.0	53.2±12.5	51.7±12.9	60.8±11.1
	MCR	MCL	IPL	IPR	B	**Average**
BrainStorm	52.5±18.7	57.9±11.9	50.1±17.1	52.2±16.5	86.4±4.3	51.9±19.0
LT-Net	**56.1±20.1**	**59.2±12.3**	43.0±15.4	**53.5±18.2**	87.0±6.2	**52.7±19.6**
DeepAtlas	55.9±17.9	54.6±13.2	44.0±17.5	50.8±15.3	**88.9±4.0**	51.5±19.8
Proposed	61.0±16.9	61.1±11.1	**48.1±18.9**	53.7±17.0	90.1±4.2	56.3±18.8

3.3   Evaluation

We have evaluated our one-shot segmentation framework on the labeled sTBI MR scans in the one-shot setting, which means that only one labeled image is available to learn the segmentation. We explore the effectiveness of the proposed adversarial training and uncertainty rectification, and make the comparison with state-of-the-art alternatives, by reporting the Dice coefficients.

First, we conduct an ablation study to evaluate the impact of adversarial training and uncertainty rectification, which is shown in Table 1. In No. (1), the baseline OSSLT method yields an average Dice of 51.42%. In No. (3), adversarial training adds a performance gain of approximately 3% compared with No. (1), and is also superior to predefined uniform transform sampling in No. (2). The results indicate that adversarial training brings both extra diversity of

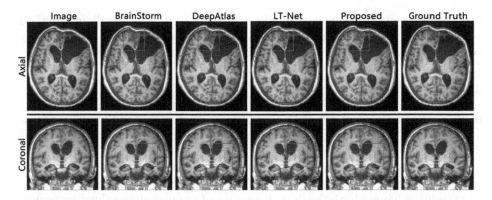

Fig. 3. Visual comparison of different segmentation methods. (Color figure online)

the generated samples and robustness of the segmentation network. The introduction of uncertainty rectification along with segmentation loss $\mathcal{L}_{seg}$ brings a performance gain of approximately 2%, which is shown in No. (5). But in No. (4), if we use rectified segmentation loss $\mathcal{L}_{rseg}$ only (without the segmentation loss $\mathcal{L}_{seg}$ of spatial-augmented image x_{As}), the segmentation performance drops significantly compared with No. (1). This is because the rectified segmentation loss $\mathcal{L}_{rseg}$ does not provide enough supervision signal in regions where the segmentation uncertainty is too high, and thus we need segmentation loss $\mathcal{L}_{seg}$ to compensate it. Finally, by applying both adversarial training and uncertainty rectification, the proposed method yields the best results with an improved Dice coefficient of approximately 5% and a lower standard deviation of segmentation performance, which is shown in No. (5). Experimental results demonstrate that the proposed adversarial training and uncertainty rectification can both contribute to the segmentation performance, compared with the baseline setting.

Then, we compare the proposed method with three cutting-edge alternatives in one-shot medical image segmentation, including BrainStorm [17], LT-Net [15], and DeepAtlas [16], which is shown in Table 2. The proposed method outperforms other segmentation methods with an average Dice score of 56.3%, higher than all of the previous state-of-the-art methods, and achieves the highest and second highest segmentation performance in all of the 17 brain regions. Also, it should be noted that the proposed method has a lower standard deviation in terms of segmentation performance, which also demonstrates the robustness of our method. However, despite the promising results of the proposed method, we have observed that the performance gain of proposed method in certain brain regions that are usually very small is not significant. The plausible reason is that the uncertainty of these small brain regions is too high and affects the segmentation.

For qualitative evaluation, we have visualized the segmentation results of the proposed method and the above-mentioned alternatives, which are shown in Fig. 3. The red bounding boxes indicate the regions where our method achieves better segmentation results compared with the alternatives. Overall, our method achieves more accurate segmentation, especially in the brain regions affected by ventriculomegaly, compared with BrainStorm, LT-Net, and DeepAtlas.

4   Conclusion

In this work, we present a novel one-shot segmentation method for severe traumatic brain segmentation, a difficult clinical scenario where limited annotated data is available. Our method addresses the critical issues in sTBI brain segmentation, namely, the need for diverse training data and mitigation of potential label errors introduced by appearance transforms. The introduction of adversarial training enhances both the data diversity and segmentation robustness, while uncertainty rectification is designed to compensate for the potential label errors. The experimental results on sTBI brains demonstrate the efficacy of our proposed method and its advantages over state-of-the-art alternatives, highlighting the potential of our method in enabling more accurate segmentation in severe traumatic brains, which may aid clinical pipelines.

Acknowledgements. This work was supported by the National Natural Science Foundation of China (No. 62001292).

References

1. Balakrishnan, G., Zhao, A., Sabuncu, M.R., Guttag, J., Dalca, A.V.: Voxelmorph: a learning framework for deformable medical image registration. IEEE Trans. Med. Imaging **38**(8), 1788–1800 (2019)
2. Chen, C., et al.: Enhancing MR image segmentation with realistic adversarial data augmentation. Med. Image Anal. **82**, 102597 (2022)
3. Ding, Y., Yu, X., Yang, Y.: Modeling the probabilistic distribution of unlabeled data for one-shot medical image segmentation. In: Proceedings of the AAAI Conference on Artificial Intelligence, vol. 35, pp. 1246–1254 (2021)
4. Fischl, B.: Freesurfer. Neuroimage **62**(2), 774–781 (2012)
5. Gal, Y., Ghahramani, Z.: Dropout as a Bayesian approximation: representing model uncertainty in deep learning. In: International Conference on Machine Learning, pp. 1050–1059. PMLR (2016)
6. He, Y., et al.: Learning better registration to learn better few-shot medical image segmentation: authenticity, diversity, and robustness. IEEE Trans. Neural Netw. Learn. Syst. (2022)
7. He, Y., et al.: Deep complementary joint model for complex scene registration and few-shot segmentation on medical images. In: Vedaldi, A., Bischof, H., Brox, T., Frahm, J.-M. (eds.) ECCV 2020. LNCS, vol. 12363, pp. 770–786. Springer, Cham (2020). https://doi.org/10.1007/978-3-030-58523-5_45
8. Huang, Z., et al.: The self and its resting state in consciousness: an investigation of the vegetative state. Hum. Brain Mapp. **35**(5), 1997–2008 (2014)
9. Olut, S., Shen, Z., Xu, Z., Gerber, S., Niethammer, M.: Adversarial data augmentation via deformation statistics. In: Vedaldi, A., Bischof, H., Brox, T., Frahm, J.-M. (eds.) ECCV 2020. LNCS, vol. 12374, pp. 643–659. Springer, Cham (2020). https://doi.org/10.1007/978-3-030-58526-6_38

10. Qiao, Y., Tao, H., Huo, J., Shen, W., Wang, Q., Zhang, L.: Robust hydrocephalus brain segmentation via globally and locally spatial guidance. In: Abdulkadir, A., et al. (eds.) MLCN 2021. LNCS, vol. 13001, pp. 92–100. Springer, Cham (2021). https://doi.org/10.1007/978-3-030-87586-2_10
11. Qin, P., et al.: How are different neural networks related to consciousness? Ann. Neurol. **78**(4), 594–605 (2015)
12. Ren, X., Huo, J., Xuan, K., Wei, D., Zhang, L., Wang, Q.: Robust brain magnetic resonance image segmentation for hydrocephalus patients: hard and soft attention. In: 2020 IEEE 17th International Symposium on Biomedical Imaging (ISBI), pp. 385–389. IEEE (2020)
13. Ronneberger, O., Fischer, P., Brox, T.: U-Net: convolutional networks for biomedical image segmentation. In: Navab, N., Hornegger, J., Wells, W.M., Frangi, A.F. (eds.) MICCAI 2015. LNCS, vol. 9351, pp. 234–241. Springer, Cham (2015). https://doi.org/10.1007/978-3-319-24574-4_28
14. Smith, S.M., et al.: Advances in functional and structural MR image analysis and implementation as FSL. Neuroimage **23**, S208–S219 (2004)
15. Wang, S., et al.: LT-Net: label transfer by learning reversible voxel-wise correspondence for one-shot medical image segmentation. In: Proceedings of the IEEE/CVF Conference on Computer Vision and Pattern Recognition, pp. 9162–9171 (2020)
16. Xu, Z., Niethammer, M.: DeepAtlas: joint semi-supervised learning of image registration and segmentation. In: Shen, D., et al. (eds.) MICCAI 2019. LNCS, vol. 11765, pp. 420–429. Springer, Cham (2019). https://doi.org/10.1007/978-3-030-32245-8_47
17. Zhao, A., Balakrishnan, G., Durand, F., Guttag, J.V., Dalca, A.V.: Data augmentation using learned transformations for one-shot medical image segmentation. In: Proceedings of the IEEE/CVF Conference on Computer Vision and Pattern Recognition, pp. 8543–8553 (2019)
18. Zheng, Z., Yang, Y.: Rectifying pseudo label learning via uncertainty estimation for domain adaptive semantic segmentation. Int. J. Comput. Vision **129**(4), 1106–1120 (2021)

MI-SegNet: Mutual Information-Based US Segmentation for Unseen Domain Generalization

Yuan Bi[1], Zhongliang Jiang[1(✉)], Ricarda Clarenbach[2], Reza Ghotbi[2], Angelos Karlas[3], and Nassir Navab[1]

[1] Chair for Computer-Aided Medical Procedures and Augmented Reality, Technical University of Munich, Munich, Germany
zl.jiang@tum.de
[2] Clinic for Vascular Surgery, Helios Klinikum München West, Munich, Germany
[3] Department for Vascular and Endovascular Surgery, rechts der Isar University Hospital, Technical University of Munich, Munich, Germany

Abstract. Generalization capabilities of learning-based medical image segmentation across domains are currently limited by the performance degradation caused by the domain shift, particularly for ultrasound (US) imaging. The quality of US images heavily relies on carefully tuned acoustic parameters, which vary across sonographers, machines, and settings. To improve the generalizability on US images across domains, we propose MI-SegNet, a novel mutual information (MI) based framework to explicitly disentangle the anatomical and domain feature representations; therefore, robust domain-independent segmentation can be expected. Two encoders are employed to extract the relevant features for the disentanglement. The segmentation only uses the anatomical feature map for its prediction. In order to force the encoders to learn meaningful feature representations a cross-reconstruction method is used during training. Transformations, specific to either domain or anatomy are applied to guide the encoders in their respective feature extraction task. Additionally, any MI present in both feature maps is punished to further promote separate feature spaces. We validate the generalizability of the proposed domain-independent segmentation approach on several datasets with varying parameters and machines. Furthermore, we demonstrate the effectiveness of the proposed MI-SegNet serving as a pre-trained model by comparing it with state-of-the-art networks (The code is available at: https://github.com/yuan-12138/MI-SegNet).

Keywords: Ultrasound segmentation · feature disentanglement · domain generalization

1 Introduction

Deep neural networks (DNNs) have achieved phenomenal success in image analysis and comparable human performance in many semantic segmentation tasks.

Supplementary Information The online version contains supplementary material available at https://doi.org/10.1007/978-3-031-43901-8_13.

However, based on the assumption of DNNs, the training and testing data of the network should come from the same probability distribution [23]. The generalization ability of DNNs on unseen domains is limited. The lack of generalizability hinders the further implementation of DNNs in real-world scenarios.

Ultrasound (US), as one of the most popular means of medical imaging, is widely used in daily medical practice to diagnose internal organs, such as vascular structures. Compared to other imaging methods, e.g., computed tomography (CT) and magnetic resonance imaging (MRI), US shows its advantages in terms of being radiation-free and portable. To accurately and robustly extract the vascular lumen for diagnosis, the Doppler signal [9] and artery pulsation signal [6] were employed to facilitate vessel segmentation. However, the US image quality is operator-dependent and sensitive to inter-machine and inter-patient variations. Therefore, the performance of the US segmentation is often decayed due to the domain shift caused by the inconsistency between the training and test data [8].

Data Augmentation. One of the most common ways of improving the generalization ability of DNNs is to increase the variability of the dataset [26]. However, in most clinical cases, the number of data is limited. Therefore, data augmentation is often used as a feasible method to increase diversity. Zhang *et al.* proposed BigAug [27], a deep stacked transformation method for 3D medical image augmentation. By applying a wide variety of augmentation methods to the single source training data, they showed the trained network is able to increase its performance on unseen domains. In order to take the physics of US into consideration, Tirindelli *et al.* proposed a physics-inspired augmentation method to generate realistic US images [21].

Image-Level Domain Adaptation. To make the network generalizable to target domains that are different from the source domain, the most intuitive way is to transfer the image style to the same domain. The work from Chen *et al.* achieved impressive segmentation results in MRI to CT adaptation by applying both image and feature level alignment [4]. To increase the robustness of segmentation networks for US images, Yang *et al.* utilized a rendering network to unify the image styles of training and test data so that the model is able to perform equally well on different domains [25]. Velikova *et al.* extended this idea by defining a common anatomical CT-US space so that the labeled CT data can be exploited to train a segmentation network for US images [24].

Feature Disentanglement. Instead of solving the domain adaptation problem directly at the image-level, many researchers focused on disentangling the features in latent space, forcing the network to learn the shared statistical shape model across different domains [2]. One way of realizing this is through adversarial learning [7,11,15,28]. However, adversarial learning optimization remains difficult and unstable in practice [12]. A promising solution for decoupling latent representations is to minimize a metric that can explicitly measure the shared information between different features. Mutual information (MI), which measures the amount of shared information between two random variables [10], suits this demand. Previous researches have exploited its usage in increasing the generalizability for classification networks when solving the vision recognition [3,13,16] and US image classification [14] problems. In this study, we investigate the effective way to integrate MI into a segmentation network in order to improve the adaptiveness on unseen images.

To solve the performance drop caused by the domain shift in segmentation networks, the aforementioned methods require a known target domain, e.g., CT [4], MRI [15], contrast enhanced US [28]. However, compared to MRI and CT, the image quality of US is more unstable and unpredictable. It is frequently observed that the performance of a segmentation network decreases dramatically for the US images acquired from a different machine or even with a different set of acquisition parameters. In such cases, it is impractical to define a so-called target US domain. Here we introduce MI-SegNet, an MI-based segmentation network, to address the domain shift problem in US image segmentation. Specifically, the proposed network extracts the disentangled domain (image style) and anatomical (shape) features from US images. The segmentation mask is generated based on the anatomical features, while the domain features are explicitly excluded. Thereby, the segmentation network is able to understand the statistical shape model of the target anatomy and generalize to different unseen scenarios. The ablation study shows that the proposed MI-SegNet is able to increase the generalization ability of the segmentation network in unseen domains.

2 Method

Our goal is to train a segmentation network that can generalize to unseen domains and serve as a good pre-trained model for downstream tasks, while the training dataset only contains images from a single domain. To this end, the training framework should be designed to focus on the shape of the segmentation target rather than the background or appearance of the images. Following this concept of design, we propose MI-SegNet. During the training phase, a parameterised data transformation procedure is undertaken for each training image (x). Two sets of parameters are generated for spatial (a_1, a_2) and domain (d_1, d_2) transformation respectively. For individual input, four transformed images $(x_{a_1 d_1}, x_{a_2 d_2}, x_{a_1 d_2}, x_{a_2 d_1})$ are created according to the four possible combinations of the spatial and domain configuration parameters. Two encoders (E_a, E_d) are applied to extract the anatomical features (f_{a_1}, f_{a_2}) and domain features (f_{d_1}, f_{d_2}) separately. The mutual information between the extracted anatomical features and the domain features from the same image is computed using mutual information neural estimator (MINE) [1] and minimized during training. Only the anatomical features are used to compute segmentation masks (m_1, m_2). The extracted anatomical and domain features are then combined and fed into the generator network (G) to reconstruct the images $(\hat{x}_{a_1 d_1}, \hat{x}_{a_1 d_2}, \hat{x}_{a_2 d_1}, \hat{x}_{a_2 d_2})$ accordingly. Since the images are transformed explicitly, it is possible to provide direct supervision to the reconstructed images.

Notably, only two of the transformed images $(x_{a_1d_1}, x_{a_2d_2})$ are fed into the network, while the other two $(x_{a_1d_2}, x_{a_2d_1})$ are used as ground truth for reconstructions (Fig. 1).

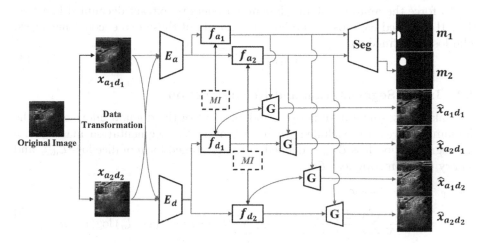

Fig. 1. Network structure of MI-SegNet. The green and blue arrows represent the data flow of the first $(x_{a_1d_1})$ and the second input image $(x_{a_2d_2})$, respectively. (Color figure online)

2.1 Mutual Information

In order to decouple the anatomical and domain features intuitively, a metric that can evaluate the dependencies between two variables is needed. Mutual information, by definition, is a metric that measures the amount of information obtained from a random variable by observing another random variable. The MI is defined as the Kullback-Leibler (KL) divergence between the joint distribution and the product of marginal distributions of random variables f_a and f_d:

$$\mathcal{MI}(f_a; f_d) = \mathcal{KL}(p(f_a, f_d) \| p(f_a) \otimes p(f_d)) \tag{1}$$

where $p(f_a, f_d)$ is the joint distribution and $p(f_a) \otimes p(f_d)$ is the product of the marginal distributions. Based on the Donsker-Varadhan representation [5], the lower bound of MI can be represented as:

$$\mathcal{MI}(f_a; f_d) \geq E_{p(f_a, f_d)}[\mathcal{T}(f_a, f_d)] - \log(E_{p(f_a) \otimes p(f_d)}[e^{\mathcal{T}(f_a, f_d)}]) \tag{2}$$

where $\mathcal{T}$ is any arbitrary given continuous function. By replacing $\mathcal{T}$ with a neural network $\mathcal{T}_{\theta_{MINE}}$ and applying Monte Carlo method [16], the lower bound can be calculated as:

$$\widehat{\mathcal{MI}(f_a; f_d)} = \frac{1}{N} \sum_{i=1}^{N} \mathcal{T}_{\theta_{MINE}}(f_a, f_d) - \log \frac{1}{N} \sum_{i=1}^{N} e^{\mathcal{T}_{\theta_{MINE}}(f_a', f_d')} \tag{3}$$

where (f_a, f_d) are drawn from the joint distribution and (f'_a, f'_d) are drawn from the product of marginal distributions. By updating the parameters θ_{MINE} to maximize the lower bound expression in Eq. 3, a loose estimation of MI is achieved, also known as MINE [1].

To force the anatomical and domain encoders to extract decoupled features, the MI is served as a loss to update the weights of these two encoder networks. The loss is defined as:

$$\mathcal{L}_{MI} = \widehat{\mathcal{MI}(f_a; f_d)} \tag{4}$$

2.2 Image Segmentation and Reconstruction

To make the segmentation network independent of the domain information, the domain features are excluded when generating the segmentation mask. Here, the segmentation loss $\mathcal{L}_{seg}$ is defined in the combined form of dice loss $\mathcal{L}_{dice}$ and binary cross-entropy loss $\mathcal{L}_{bce}$.

$$
\begin{aligned}
\mathcal{L}_{seg} &= \mathcal{L}_{dice} + \mathcal{L}_{bce} \\
&= 1 - \frac{1}{N}\sum_{n=1}^{N}\frac{2l_n m_n + s}{l_n + m_n + s} - \frac{1}{N}\sum_{n=1}^{N}(l_n \log m_n + (1 - l_n)\log(1 - m_n))
\end{aligned} \tag{5}
$$

where l is the ground truth label, m represents the predicted mask, s is added to ensure the numerical stability, and N is the mini batch size.

To ensure that the extracted anatomical and domain features can contain all the information of the input image, a generator network is used to reconstruct the image based on both features. The reconstruction loss is then defined as:

$$\mathcal{L}_{rec} = \frac{1}{N}\sum_{n=1}^{N}\frac{1}{wh}(x_n - \widehat{x}_n)^2 \tag{6}$$

where x_n is the ground truth image, $\widehat{x}_n$ is the reconstructed image, w and h are the width and height of the image in pixel accordingly.

2.3 Data Transformation

Since the training dataset only contains images from one single domain, it is necessary to enrich the diversity of the training data so that overfitting can be prevented and the generalization ability is increased. The transformation methods are divided into two categories, domain and spatial transformations. Each transformation (T) is controlled by two parameters, probability (p) and magnitude (λ).

Domain Transformations aim to transfer the single domain images to different domain styles. Five types of transformation methods are involved in this aspect, i.e., *blurriness, sharpness, noise level, brightness,* and *contrast.* The implementations are identical to [27], except the Gaussian noise is replaced by Rayleigh noise. The possibility of all the domain transformations are empirically set to 10%.

Spatial Transformations mainly consist of two parts, *crop* and *flip*. For cropping, a window with configurable sizes ([0.7, 0.9] of the original image size) is randomly masked on the original image. Then the cropped area is resized to the original size to introduce varying shapes of anatomy. Here λ controls the size and the position of the cropping window. Besides cropping, horizontal flipping is also involved. Unlike domain transformations, the labels are also transformed accordingly by the same spatial transformation. The probability (p) of flipping is 5%, while the p for cropping is 50% to introduce varying anatomy sizes. The images are then transformed in a stacked way:

$$x_{aug} = T^{(P[n],\Lambda[n])}(T^{(P[n-1],\Lambda[n-1])} \cdots (T^{(P[1],\Lambda[1])}(x))) \tag{7}$$

where $n = 7$ represents the seven different transformation methods involved in our work, $\Lambda = [\lambda_n, \lambda_{n-1}, \cdots, \lambda_1]$ represents the magnitude parameter, and $P = [p_n, p_{n-1}, \cdots, p_1]$ contains all the probability parameters for each transformations. In our setup, Λ and P can be further separated into $a = [\Lambda_a; P_a]$ and $d = [\Lambda_d; P_d]$ for spatial and domain transformations respectively.

2.4 Cross Reconstruction

According to experimental findings, the MI loss indeed forces the two representations to have minimal shared information. However, the minimization of MI between the anatomical and domain features cannot necessarily make both features contain the respective information. The network goes into local optimums frequently, where the domain features are kept constant, and all the information is stored in the anatomical features. Because there is no information in the domain features, the MI between two representations is thus approaching zero. However, this is not our original intention. As a result, cross reconstruction strategy is introduced to tackle this problem. The cross reconstruction loss will punish the behavior of summarizing all the information into one representation. Thus, it can force each encoder to extract informative features accordingly and prevent the whole network from going into the local optimums.

3   Experiments

3.1   Implementation Details

The training dataset consists of 2107 carotid US images of one adult acquired using Siemens Juniper US Machine (ACUSON Juniper, SIEMENS AG, Germany) with a system-predefined "Carotid" acquisition parameter. The test dataset consists of (1) ValS: 200 carotid US images which are left out from the training dataset, (2) TS1: 538 carotid US images of 15 adults from Ultrasonix device, (3) TS2: 433 US images of 2 adults and one child from Toshiba device, and (4) TS3: 540 US images of 6 adults from Cephasonics device (Cephasonics, California, USA). TS1 and TS2 are from a public database of carotid artery [17]. Notably, due to the absence of annotations, the publicly accessed images were also annotated by ourselves under the supervision of US experts. The acquisition was performed within the Institutional Review Board Approval by the Ethical Commission of the Technical University of Munich (reference number 244/19 S). All the images are resized to 256×256 for training and testing.

We use Adam optimizer with a learning rate of 1×10^{-4} to optimize all the parameters. The training is carried out on a single GPU (Nvidia TITAN Xp) with 12 GB memory.

3.2   Performance Comparison on Unseen Datasets

In this section, we compare the performance of the proposed MI-SegNet with other state-of-art segmentation networks. All the networks are trained on the same dataset described in Sect. 3.1 with 200 episodes.

Without Adaptation: The trained models are then tested directly on 4 different datasets described in Sect. 3.1 without further training or adaptation on the unseen domains. The dice score (DSC) is applied as the evaluation metrics. The results are shown in Table 1.

Table 1. Performance comparison of the proposed MI-SegNet with different segmentation networks on the US carotid artery datasets without adaptation.

Method	DSC			
	ValS	TS1	TS2	TS3
UNet [18]	0.920 ± 0.080	0.742 ± 0.283	0.572 ± 0.388	0.529 ± 0.378
GLFR [20]	0.927 ± 0.045	0.790 ± 0.175	0.676 ± 0.272	0.536 ± 0.347
Att-UNet [19]	$\mathbf{0.932 \pm 0.046}$	0.687 ± 0.254	0.602 ± 0.309	0.438 ± 0.359
MedT [22]	0.875 ± 0.056	0.674 ± 0.178	0.583 ± 0.303	0.285 ± 0.291
MI-SegNet w/o $\mathcal{L}_{MI}$	0.928 ± 0.057	0.768 ± 0.217	0.627 ± 0.346	0.620 ± 0.344
MI-SegNet w/o cross rec.	0.921 ± 0.050	0.790 ± 0.227	0.662 ± 0.309	0.599 ± 0.344
MI-SegNet	0.928 ± 0.046	$\mathbf{0.821 \pm 0.146}$	$\mathbf{0.725 \pm 0.215}$	$\mathbf{0.744 \pm 0.251}$

Compared to the performance on ValS, all networks demonstrate a performance degradation on unseen datasets (TS1, TS2, and TS3). In order to validate the effectiveness of the MI loss as well as the cross reconstruction design, two ablation networks (MI-SegNet w/o $\mathcal{L}_{MI}$ and MI-SegNet w/o cross rec.) are introduced here for comparison. The visual comparisons are shown in Fig. 2. The results on TS1 are the best among all three unseen datasets while the scores on TS3 are the worst for most networks, which indicates that the domain similarity between the source and target domain decreases accordingly from TS1 to TS3. The MI-SegNet performs the best among others on all three unseen datasets, which showcases the high generalization ability of the proposed framework.

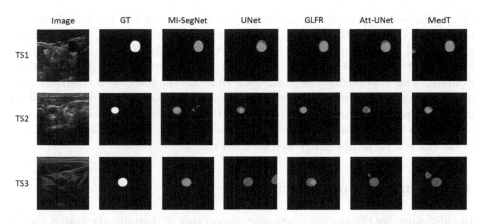

Fig. 2. Visual comparison between MI-SegNet and other segmentation networks on US carotid artery datasets without adaptation. For each row, we show the input US image, the ground truth (GT), and the output of each network. Red, pink and green regions represent the false negative, false positive and true positive, respectively. (Color figure online)

After Adaptation: Although the proposed network achieves the best scores when applied directly to unseen domains, performance decay still occurs. Using it directly to unseen dataset with degraded performance is not practical. As a result, adaptation on the target domain is needed. The trained models in Sect. 3.2 are further trained with 5% data of each unseen test dataset. The adapted models are then tested on the rest 95% of each dataset. Notably, for the MI-SegNet only the anatomical encoder and segmentor are involved in this adaptation process, which means the network is updated solely based on $\mathcal{L}_{seg}$.

Table 2. Performance comparison of the proposed MI-SegNet with different segmentation networks after adaptation when 5% data of each test dataset is used for adaptation.

Method	DSC		
	TS1	TS2	TS3
UNet [18]	0.890 ± 0.183	0.707 ± 0.328	0.862 ± 0.139
GLFR [20]	0.915 ± 0.154	0.875 ± 0.099	0.907 ± 0.049
Att-UNet [19]	0.916 ± 0.117	0.876 ± 0.145	0.893 ± 0.087
MedT [22]	0.870 ± 0.118	0.837 ± 0.137	0.795 ± 0.170
MI-SegNet	**0.919 ± 0.095**	**0.881 ± 0.111**	**0.916 ± 0.061**

The intention of this experiment is to validate whether the proposed network can serve as a good pre-trained model for the downstream task. A well-trained pre-trained model, which can achieve good results when only a limited amount of annotations is provided, has the potential to release the burden of manual labeling and adapts to different domains with few annotations. Table 2 shows that the MI-SegNet performs the best on all test datasets. However, the difference is not that significant as in Table 1 when no data is provided for the target domain. This is partially due to the fact that carotid artery is a relatively easy anatomy for segmentation. It is observed that when more data (10%) is involved in the adaptation process GLFR and Att-UNet tend to outperform the others and it can be therefore expected when the data size further increases all the networks will perform equally well on each test set.

4   Discussion and Conclusion

In this paper, we discuss the particular importance of domain adaptation for US images. Due to the low speed of sound compared to light and X-ray, the complexity of US imaging and its dependency on many parameters are more remarkable than optical imaging, X-ray, and CT. Therefore, the performance decay caused by the domain shift is a prevalent issue when applying DNNs in US images. To address this problem, a MI-based disentanglement method is applied to increase the generalization ability of the segmentation networks for US image segmentation. The ultimate goal of increasing the generalizability of the segmentation network is to apply the network to different unseen domains directly without any adaptation process. However, from the authors' point of view, training a good pre-trained model that can be adapted to an unseen dataset with minimal annotated data is still meaningful. As demonstrated in Sect. 3.2, the proposed model also shows the best performance in the downstream adaptation tasks. Currently, only the conventional image transformation methods are involved. In the future work, more realistic and US specific image transformations could be implemented to strengthen the feature disentanglement.

References

1. Belghazi, M.I., et al.: Mutual information neural estimation. In: International Conference on Machine Learning, pp. 531–540. PMLR (2018)
2. Bengio, Y., Courville, A., Vincent, P.: Representation learning: a review and new perspectives. IEEE Trans. Pattern Anal. Mach. Intell. **35**(8), 1798–1828 (2013)
3. Cha, J., Lee, K., Park, S., Chun, S.: Domain generalization by mutual-information regularization with pre-trained models. In: Avidan, S., Brostow, G., Cissé, M., Farinella, G.M., Hassner, T. (eds.) Computer Vision, ECCV 2022. LNCS, vol. 13683, pp. 440–457. Springer, Cham (2022). https://doi.org/10.1007/978-3-031-20050-2_26
4. Chen, C., Dou, Q., Chen, H., Qin, J., Heng, P.A.: Unsupervised bidirectional cross-modality adaptation via deeply synergistic image and feature alignment for medical image segmentation. IEEE Trans. Med. Imag. **39**(7), 2494–2505 (2020)
5. Donsker, M.D., Varadhan, S.R.S.: Asymptotic evaluation of certain Markov process expectations for large time. IV. Commun. Pure Appl. Math. **36**(2), 183–212 (1983)

6. Huang, D., Bi, Y., Navab, N., Jiang, Z.: Motion magnification in robotic sonography: enabling pulsation-aware artery segmentation. arXiv preprint arXiv:2307.03698 (2023)

7. Huang, X., Liu, M.-Y., Belongie, S., Kautz, J.: Multimodal unsupervised image-to-image translation. In: Ferrari, V., Hebert, M., Sminchisescu, C., Weiss, Y. (eds.) ECCV 2018. LNCS, vol. 11207, pp. 172–189. Springer, Cham (2018). https://doi.org/10.1007/978-3-030-01219-9_11

8. Huang, Y., et al.: Online Reflective learning for robust medical image segmentation. In: Wang, L., Dou, Q., Fletcher, P.T., Speidel, S., Li, S. (eds.) Medical Image Computing and Computer Assisted Intervention, MICCAI 2022. LNCS, vol. 13438, pp. 652–662. Springer, Cham (2022). https://doi.org/10.1007/978-3-031-16452-1_62

9. Jiang, Z., Duelmer, F., Navab, N.: DopUS-Net: quality-aware robotic ultrasound imaging based on doppler signal. IEEE Trans. Autom. Sci. Eng. (2023)

10. Kraskov, A., Stögbauer, H., Grassberger, P.: Estimating mutual information. Phys. Rev. E **69**(6), 066138 (2004)

11. Lee, H.-Y., Tseng, H.-Y., Huang, J.-B., Singh, M., Yang, M.-H.: Diverse image-to-image translation via disentangled representations. In: Ferrari, V., Hebert, M., Sminchisescu, C., Weiss, Y. (eds.) ECCV 2018. LNCS, vol. 11205, pp. 36 52. Springer, Cham (2018). https://doi.org/10.1007/978-3-030-01246-5_3

12. Lezama, J.: Overcoming the disentanglement vs reconstruction trade-off via Jacobian supervision. In: International Conference on Learning Representations (2018)

13. Liu, X., Yang, C., You, J., Kuo, C.C.J., Kumar, B.V.: Mutual information regularized feature-level Frankenstein for discriminative recognition. IEEE Trans. Pattern Anal. Mach. Intell. **44**(9), 5243–5260 (2021)

14. Meng, Q., et al.: Mutual information-based disentangled neural networks for classifying unseen categories in different domains: application to fetal ultrasound imaging. IEEE Trans. Med. Imag. **40**(2), 722–734 (2020)

15. Ning, M., et al.: A new bidirectional unsupervised domain adaptation segmentation framework. In: Feragen, A., Sommer, S., Schnabel, J., Nielsen, M. (eds.) IPMI 2021. LNCS, vol. 12729, pp. 492–503. Springer, Cham (2021). https://doi.org/10.1007/978-3-030-78191-0_38

16. Peng, X., Huang, Z., Sun, X., Saenko, K.: Domain agnostic learning with disentangled representations. In: International Conference on Machine Learning, pp. 5102–5112. PMLR (2019)

17. Říha, K., Mašek, J., Burget, R., Beneš, R., Závodná, E.: Novel method for localization of common carotid artery transverse section in ultrasound images using modified Viola-Jones detector. Ultrasound Med. Biol. **39**(10), 1887–1902 (2013)

18. Ronneberger, O., Fischer, P., Brox, T.: U-Net: convolutional networks for biomedical image segmentation. In: Navab, N., Hornegger, J., Wells, W.M., Frangi, A.F. (eds.) MICCAI 2015. LNCS, vol. 9351, pp. 234–241. Springer, Cham (2015). https://doi.org/10.1007/978-3-319-24574-4_28

19. Schlemper, J., et al.: Attention gated networks: learning to leverage salient regions in medical images. Med. Image Anal. **53**, 197–207 (2019)

20. Song, J., et al.: Global and local feature reconstruction for medical image segmentation. IEEE Trans. Med. Imag. **41**, 2273–2284 (2022)

21. Tirindelli, M., Eilers, C., Simson, W., Paschali, M., Azampour, M.F., Navab, N.: Rethinking ultrasound augmentation: a physics-inspired approach. In: de Bruijne, M., et al. (eds.) MICCAI 2021. LNCS, vol. 12908, pp. 690–700. Springer, Cham (2021). https://doi.org/10.1007/978-3-030-87237-3_66

22. Valanarasu, J.M.J., Oza, P., Hacihaliloglu, I., Patel, V.M.: Medical transformer: gated axial-attention for medical image segmentation. In: de Bruijne, M., et al. (eds.) MICCAI 2021. LNCS, vol. 12901, pp. 36–46. Springer, Cham (2021). https:// doi.org/10.1007/978-3-030-87193-2_4

23. Valiant, L.G.: A theory of the learnable. Commun. ACM **27**(11), 1134–1142 (1984)

24. Velikova, Y., Simson, W., Salehi, M., Azampour, M.F., Paprottka, P., Navab, N.: CACTUSS: common anatomical CT-US space for US examinations. In: Wang, L., Dou, Q., Fletcher, P.T., Speidel, S., Li, S. (eds.) Medical Image Computing and Computer Assisted Intervention, MICCAI 2022. LNCS, vol. 13433, pp. 492–501. Springer, Cham (2022). https://doi.org/10.1007/978-3-031-16437-8_47

25. Yang, X., et al.: Generalizing deep models for ultrasound image segmentation. In: Frangi, A.F., Schnabel, J.A., Davatzikos, C., Alberola-López, C., Fichtinger, G. (eds.) MICCAI 2018. LNCS, vol. 11073, pp. 497–505. Springer, Cham (2018). https://doi.org/10.1007/978-3-030-00937-3_57

26. Zhang, C., Bengio, S., Hardt, M., Recht, B., Vinyals, O.: Understanding deep learning (still) requires rethinking generalization. Commun. ACM **64**(3), 107–115 (2021)

27. Zhang, L., et al.: Generalizing deep learning for medical image segmentation to unseen domains via deep stacked transformation. IEEE Trans. Med. Imag. **39**(7), 2531–2540 (2020)

28. Zhao, Q., et al.: A multi-modality ovarian tumor ultrasound image dataset for unsupervised cross-domain semantic segmentation. arXiv preprint arXiv:2207.06799 (2022)

M-GenSeg: Domain Adaptation for Target Modality Tumor Segmentation with Annotation-Efficient Supervision

Malo Alefsen[1(✉)], Eugene Vorontsov[3], and Samuel Kadoury[1,2]

[1] Ecole Polytechnique de Montréal, Montréal, QC, Canada
`malo.alefsen-de-boisredon-dassier@polytechnique.edu`
[2] Centre de Recherche du CHUM, Montréal, QC, Canada
[3] Paige, Montréal, QC, Canada

Abstract. Automated medical image segmentation using deep neural networks typically requires substantial supervised training. However, these models fail to generalize well across different imaging modalities. This shortcoming, amplified by the limited availability of expert annotated data, has been hampering the deployment of such methods at a larger scale across modalities. To address these issues, we propose M-GenSeg, a new semi-supervised generative training strategy for cross-modality tumor segmentation on unpaired bi-modal datasets. With the addition of known healthy images, an unsupervised objective encourages the model to disentangling tumors from the background, which parallels the segmentation task. Then, by teaching the model to convert images across modalities, we leverage available pixel-level annotations from the source modality to enable segmentation in the unannotated target modality. We evaluated the performance on a brain tumor segmentation dataset composed of four different contrast sequences from the public BraTS 2020 challenge data. We report consistent improvement in Dice scores over state-of-the-art domain-adaptive baselines on the unannotated target modality. Unlike the prior art, M-GenSeg also introduces the ability to train with a partially annotated source modality.

Keywords: Image Segmentation · Semi-supervised Learning · Unpaired Image-to-image Translation

1 Introduction

Deep learning methods have demonstrated their tremendous potential when it comes to medical image segmentation. However, the success of most existing architectures relies on the availability of pixel-level annotations, which are

Supplementary Information The online version contains supplementary material available at https://doi.org/10.1007/978-3-031-43901-8_14.

difficult to produce [1]. Furthermore, these methods are known to be inadequately equipped for distribution shifts. Therefore, cross-modality generalization is needed when one imaging modality has insufficient training data. For instance, conditions such as Vestibular Schwannoma, where new hrT2 sequences are set to replace ceT1 for diagnosis to mitigate the use of contrast agents, is a sample use case [2]. Recently Billot et al. [3] proposed a domain randomisation strategy to segment images from a wide range of target contrasts without any fine-tuning. The method demonstrated great generalization capability for brain parcellation, but the model performance when exposed to tumors and pathologies was not quantified. This challenge could also be addressed through unsupervised domain-adaptive approaches, which transfer the knowledge available in the "source" modality S from pixel-level labels to the "target" imaging modality T lacking annotations [4].

Several generative models attempt to generalize to a target modality by performing unsupervised domain adaptation through image-to-image translation and image reconstruction. In [5], by learning to translate between CT and MR cardiac images, the proposed method jointly disentangles the domain specific and domain invariant features between each modality and trains a segmenter from the domain invariant features. Other methods [6–12] also integrate this translation approach, but the segmenter is trained in an end-to-end manner on the synthetic target images generated from the source modality using a Cycle-GAN [13] model. These methods perform well but do not explicitly use the unannotated target modality data to further improve the segmentation.

In this paper, we propose M-GenSeg, a novel training strategy for cross-modality domain adaptation, as illustrated in Fig. 1. This work leverages and extends GenSeg [14], a generative method that uses image-level "diseased" or "healthy" labels for semi-supervised segmentation. Given these labels, the model imposes an image-to-image translation objective between the image domain presenting tumor lesions and the domain corresponding to an absence of lesions. Therefore, like in low-rank atlas based methods [15–17] the model is taught to find and remove a lesion, which acts as a guide for the segmentation. We incorporate cross-modality image segmentation with an image-to-image translation objective between source and target modalities. We hypothesize both objectives are complementary since GenSeg helps localizing the tumors on unannotated target images, while modality translation enables fine-tuning the segmenter on the target modality by displaying annotated pseudo-target images. We evaluate M-GenSeg on a modified version of the BraTS 2020 dataset, in which each type of sequence (T1, T2, T1ce and FLAIR) is considered as a distinct modality. We demonstrate that our model can better generalize than other state-of-the-art methods to the target modality.

2   Methods

2.1   M-GenSeg: Semi-supervised Segmentation

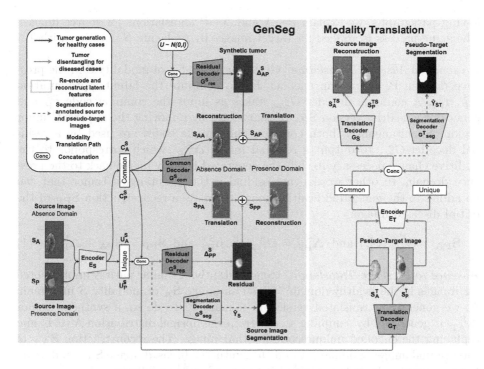

Fig. 1. M-GenSeg: Latent representations are shared for simultaneous cross-modality translation (green) and semi-supervised segmentation (blue). Source images are passed through the source GenSeg module and the S→T*→S modality translation cycle. Domain adaptation is achieved when training the segmentation on annotated pseudo-target T* images ($\mathbf{S_P^T}$). It is not shown but, symmetrically, target images are treated in an other branch to train the T→S→T cyclic translation, and the target GenSeg module to further close the domain gap. (Color figure online)

Healthy-Diseased Translation. We propose to integrate image-level supervision to the cross-modality segmentation task with GenSeg, a model that introduces translation between domains with a presence (P) or absence (A) of tumor lesions. Leveraging this framework has a two-fold advantage here. Indeed, *(i)* training a GenSeg module on the source modality makes the model aware of the tumor appearances in the source images even with limited source pixel-level annotations. This helps to preserve tumor structures during the generation of pseudo-target samples (see Sect. 2.1). Furthermore, *(ii)* training a second GenSeg module on the target modality allows to further close the domain gap by extending the segmentation objective to unannotated target data.

In order to disentangle the information common to A and P, and the information specific to P, we split the latent representation of each image into a common code **c** and a unique code **u**. Essentially, the common code contains information inherent to both domains, which represents organs and other structures, while the unique code stores features like tumor shapes and location. In the two following paragraphs, we explain P→A and A→P translations for source images. The same process is applied for target images by replacing S notation with T.

Presence to Absence Translation. Given an image $\mathbf{S_P}$ of modality S in the presence domain P, we use an encoder E_S to compute the latent representation $[\mathbf{c_P^S}, \mathbf{u_P^S}]$. A common decoder G_{com}^S takes as input the common code $\mathbf{c_P^S}$ and generates a healthy version $\mathbf{S_{PA}}$ of that image by removing the apparent tumor region. Simultaneously, both common and unique codes are used by a residual decoder G_{res}^S to output a residual image $\mathbf{\Delta_{PP}^S}$, which corresponds to the additive change necessary to shift the generated healthy image back to the presence domain. In other words, the residual is the disentangled tumor that can be added to the generated healthy image to create a reconstruction $\mathbf{S_{PP}}$ of the initial diseased image:

$$\mathbf{S_{PA}} = G_{com}^S(\mathbf{c_P^S}) \text{ and } \mathbf{\Delta_{PP}^S} = G_{res}^S(\mathbf{c_P^S}, \mathbf{u_P^S}) \text{ and } \mathbf{S_{PP}} = \mathbf{S_{PA}} + \mathbf{\Delta_{PP}^S} \quad (1)$$

Absence to Presence Translation. Concomitantly, a similar path is implemented for images in the healthy domain. Given an image $\mathbf{S_A}$ of modality S in domain A, we generate a translated version in domain P. To do so, a synthetic tumor $\mathbf{\Delta_{AP}^S}$ is generated by sampling a code from the normal distribution $\mathcal{N}(0, \mathbf{I})$ and replacing the encoded unique code for that image. The reconstruction $\mathbf{S_{AA}}$ of the original image in domain A and the synthetic diseased image $\mathbf{S_{AP}}$ in domain P are computed from the encoded features $[\mathbf{c_A^S}, \mathbf{u_A^S}]$ as follows:

$$\mathbf{S_{AA}} = G_{com}^S(\mathbf{c_A^S}) \quad \text{and} \quad \mathbf{S_{AP}} = \mathbf{S_{AA}} + G_{res}^S(\mathbf{c_A^S}, \mathbf{u} \sim \mathcal{N}(0, \mathbf{I})) \quad (2)$$

Like approaches in [18–20] we therefore generate diseased samples from healthy ones for data augmentation. However, M-GenSeg aims primarily at tackling cross-modality lesion segmentation tasks, which is not addressed in these studies. Furthermore, note that these methods are limited to data augmentation and do not incorporate any unannotated diseased samples when training the segmentation network, as achieved by our model with the P→A translation.

Modality Translation. Our objective is to learn to segment tumor lesions in a target modality by reusing potentially scarce image annotations in a source modality. Note that for each modality $m \in \{S, T\}$, M-GenSeg holds a segmentation decoder G_{seg}^m that shares most of its weights with the residual decoder G_{res}^m, but has its own set of normalization parameters and a supplementary classifying layer. Thus, through the Absence and Presence translations, these segmenters have already learned how to disentangle the tumor from the background. However, supervised training on a few example annotations is still required to learn how to transform the resulting residual representation into appropriate segmentation maps. While this is a fairly straightforward task for the source modality using pixel-level annotations, achieving this for the target modality is more complex, justifying the second unsupervised translation objective between source and target modalities. Based on the CycleGan [13] approach, modality translations are performed via two distinct generators that share their encoder with the GenSeg task. More precisely, combined with the encoder E_S a decoder G_T enables performing S→T modality translation, while the encoder E_T and a second decoder G_S perform the T→S modality translation. To maintain the anatomical information, we ensure cycle-consistency by reconstructing the initial images after mapping them back to their original modality. We note $\mathbf{S_d^T} = G_T \circ E_S(\mathbf{S_d})$ and $\mathbf{S_d^{TS}} = G_S \circ E_T(\mathbf{S_d^T})$, respectively the translation and reconstruction of $\mathbf{S_d}$ in the S→T→S translation loop, with domain $\mathbf{d} \in \{A, P\}$ and $\circ$ the composition operation. Similarly we have $\mathbf{T_d^S} = G_S \circ E_T(\mathbf{T_d})$ and $\mathbf{T_d^{ST}} = G_T \circ E_S(\mathbf{T_d^S})$ for the T→S→T cycle.

Note that to perform the domain adaptation, training the model to segment only the pseudo-target images generated by the S→T modality generator would suffice (in addition to the diseased/healthy target translation). However, training the segmentation on diseased source images also imposes additional constraints on encoder E_S, ensuring the preservation of tumor structures. This constraint proves beneficial for the translation decoder G_T as it generates pseudo-target tumoral samples that are more reliable. Segmentation is therefore trained on both diseased source images $\mathbf{S_P}$ and their corresponding synthetic target images $\mathbf{S_P^T}$, when provided with annotations $\mathbf{y_S}$. To such an extent, two segmentation masks are predicted $\hat{\mathbf{y}}_\mathbf{S} = G_{seg}^S \circ E_S(\mathbf{S_P})$ and $\hat{\mathbf{y}}_\mathbf{ST} = G_{seg}^T \circ E_T(\mathbf{S_P^T})$.

2.2   Loss Functions

Segmentation Loss. For the segmentation objective, we compute a soft Dice loss [21] on the predictions for both labelled source images and their translations:

$$\mathcal{L}_{seg} = Dice\,(\mathbf{y_S}, \hat{\mathbf{y}}_\mathbf{S}) + Dice\,(\mathbf{y_S}, \hat{\mathbf{y}}_\mathbf{ST}) \tag{3}$$

Reconstruction Losses. $\mathcal{L}_{cyc}^{mod}$ and $\mathcal{L}_{rec}^{Gen}$ respectively impose pixel-level image reconstruction constraints on modality translation and GenSeg tasks. Note that $\mathcal{L}_1$ refers to the standard L1 norm:

$$\begin{aligned} \mathcal{L}_{cyc}^{mod} &= \mathcal{L}_1\left(\mathbf{S_A^{TS}}, \mathbf{S_A}\right) + \mathcal{L}_1\left(\mathbf{T_A^{ST}}, \mathbf{T_A}\right) + \mathcal{L}_1\left(\mathbf{S_P^{TS}}, \mathbf{S_P}\right) + \mathcal{L}_1\left(\mathbf{T_P^{ST}}, \mathbf{T_P}\right) \\ \mathcal{L}_{rec}^{Gen} &= \mathcal{L}_1\left(\mathbf{S_{AA}}, \mathbf{S_A}\right) + \mathcal{L}_1\left(\mathbf{S_{PP}}, \mathbf{S_P}\right) + \mathcal{L}_1\left(\mathbf{T_{AA}}, \mathbf{T_A}\right) + \mathcal{L}_1\left(\mathbf{T_{PP}}, \mathbf{T_P}\right) \end{aligned} \tag{4}$$

Moreover, like in [14] we compute a loss $\mathcal{L}_{lat}^{Gen}$ that ensures that the translation task holds the information relative to the initial image, by reconstructing their latent codes with the L1 norm. It also enforces the distribution of unique codes to match the prior $\mathcal{N}(0, \mathbf{I})$ by making $\mathbf{u_{AP}}$ match $\mathbf{u}$, where $\mathbf{u_{AP}}$ is obtained by encoding the fake diseased sample $\mathbf{x_{AP}}$ produced with random sample $\mathbf{u}$.

Adversarial Loss. For the healthy-diseased translation adversarial objective, we compute a hinge loss $\mathcal{L}_{adv}^{Gen}$ as in GenSeg, learning to discriminate between pairs of real/synthetic images of the same output domain and always in the same imaging modality, e.g. $\mathbf{S_A}$ vs $\mathbf{S_{PA}}$. In the modality translation task, the $\mathcal{L}_{adv}^{mod}$ loss is computed between pairs of images of the same modality without distinction between domains A and P, e.g. $\{\mathbf{S_A}, \mathbf{S_P}\}$ vs $\{\mathbf{T_A^S}, \mathbf{T_P^S}\}$.

Overall Loss. The overall loss for M-GenSeg is a weighted sum of the afore-mentioned losses. These are tuned separately. All weights sum to 1. First, λ_{adv}^{Gen}, λ_{rec}^{Gen}, and λ_{lat}^{Gen} weights are tuned for successful translation between diseased and healthy images. Then, λ_{adv}^{mod} and λ_{cyc}^{mod} are tuned for successful modality translation. Finally, λ_{seg} is tuned for segmentation performance.

$$\mathcal{L}_{Total} = \lambda_{seg}\mathcal{L}_{seg} + \lambda_{adv}^{mod}\mathcal{L}_{adv}^{mod} + \lambda_{cyc}^{mod}\mathcal{L}_{cyc}^{mod}$$
$$+\lambda_{adv}^{Gen}\mathcal{L}_{adv}^{Gen} + \lambda_{rec}^{Gen}\mathcal{L}_{rec}^{Gen} + \lambda_{lat}^{Gen}\mathcal{L}_{lat}^{Gen} \tag{5}$$

2.3 Implementation Details

Training and Hyper-Parameters. All models are implemented using PyTorch and are trained on one NVIDIA A100 GPU with 40 GB memory. We used a batch size of 15, an AMSGrad optimizer ($\beta_1 = 0.5$ and $\beta_2 = 0.999$) and a learning rate of 10^{-4}. Our models were trained for 300 epochs and weights of the segmentation model with the highest validation Dice score were saved for evaluation. The same on-the-fly data augmentation as in [14] was applied for all runs. Each training experiment was repeated three times with a different random seed for weight initialization. The performance reported is the mean of all test Dice scores, with standard deviation, across the three runs. The following parameters yielded both great modality and absence/presence translations: $\lambda_{adv}^{mod} = 3$, $\lambda_{cyc}^{mod} = 20$, $\lambda_{adv}^{Gen} = 6$, $\lambda_{rec}^{Gen} = 20$ and $\lambda_{lat}^{Gen} = 2$. Note that optimal λ_{seg} varies depending on the fraction of pixel-level annotations provided to the network for training.

Architecture. One distinct encoder, common decoder, residual/segmentation decoder, and modality translation decoder are used for each modality. The architecture used for encoders, decoders and discriminators is the same as in [14]. However, in order to give insight on the model's behaviour and properly choose the semantic information relevant for each objective, we introduced attention gates [22] in the skip connections. Figure 2a shows the attention maps generated for each type of decoder. As expected, residual decoders focus towards tumor areas. More interestingly, in order not to disturb the process of healthy image generation, common decoders avoid lesion locations. Finally, modality translators tend to focus on salient details of the brain tissue, which facilitates contrast redefinition needed for accurate translation.

3 Experimental Results

3.1 Datasets

Experiments were performed on the BraTS 2020 challenge dataset [23–25], adapted for the cross-modality tumor segmentation problem where images are known to be diseased or healthy. Amongst the 369 brain volumes available in BraTS, 37 were allocated each for validation and test steps, while the 295 left were used for training. We split the 3D brain volumes into 2 hemispheres and

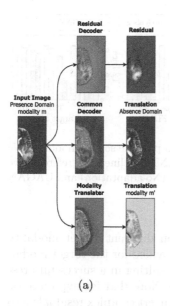
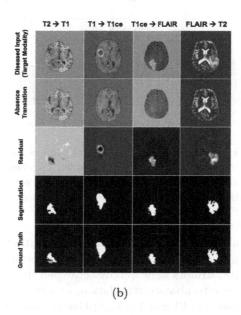

(a)   (b)

Fig. 2. (a) Attention maps for Presence → Absence and modality translations. Red indicates areas of focus while dark blue correspond to locations ignored by the network. (b) Examples of translations from Presence to Absence domains and resulting segmentation. Each column represents a domain adaptation scenario where target modality had no pixel-level annotations provided. (Color figure online)

extracted 2D axial slices. Any slices with at least 1% tumor by brain surface area were considered diseased. Those that didn't show any tumor lesion were labelled as healthy images. Datasets were then assembled from each distinct pair of the four MRI contrasts available (T1, T2, T1ce and FLAIR). To constitute unpaired training data, we used only one modality (source or target) per training volume. All the images are provided with healthy/diseased weak labels, distinct from the pixel-level annotations that we provide only to a subset of the data. Note that the interest for cross-sequence segmentation is limited if multi-parametric acquisitions are performed as is the case in BraTS. However, this modified version of the dataset provides an excellent study case for the evaluation of any modality adaptation method for tumor segmentation.

3.2 Model Evaluation

Domain Adaptation. We compared M-GenSeg with AccSegNet [10] and AttENT [6], two high performance models for domain-adaptative medical image segmentation. To that extent, we performed domain-adaptation experiments with source and target modalities drawn from T1, T2, FLAIR and T1ce. We used available GitHub code for the two baselines and performed fine-tuning on our data. For each possible source/target pair, pixel-level annotations were only retained for the source modality. We show in Fig. 2b several presence to

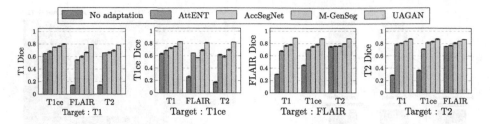

Fig. 3. Dice performance on the target modality for each possible source modality. We compare results for M-GenSeg with AccSegNet and AttENT baselines. For reference we also show Dice scores for source supervised segmentation (No adaptation) and UAGAN trained with all source and target annotations.

absence translations and segmentation examples on different target modality images. Although no pixel-level annotations were provided for the target modality, tumors were well disentangled from the brain, resulting in a successful presence to absence translation, as well as segmentation. Note that for hypo-intense lesions (T1 and T1ce), M-GenSeg still manages to convert complex residuals into consistent segmentation maps. We plot in Fig. 3 the Dice performance on the target modality for *(i)* supervised segmentation on source data without domain adaptation, *(ii)* domain adaptation methods and *(iii)* UAGAN [26], a model designed for unpaired multi-modal datasets, trained on all source and target data. Over all modality pairs our model shows an absolute Dice score increase of 0.04 and 0.08, respectively, compared to AccSegNet and AttENT.

Annotation Deficit. M-GenSeg introduces the ability to train with limited pixel-level annotations available in the source modality. We show in Fig. 4 the Dice scores for models trained when only 1%, 10%, 40%, or 70% of the source T1 modality and 0% of the T2 target modality annotations were available. While performance is severely dropping at 1% of annotations for the baselines, our model shows in comparison only a slight decrease. We thus claim that M-GenSeg can yield robust performance even when a small fraction of the source images is annotated.

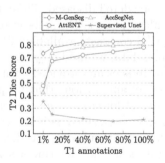

Fig. 4. T2 domain adaptation with T1 annotation deficit.

Reaching Supervised Performance. We report that, when the target modality is completely unannotated, M-GenSeg reaches 90% of UAGAN's performance (vs 81% and 85% for AttENT and AccSegNet). Further experiments showed that with a fully annotated source modality, it is sufficient to annotate 25% of the target modality to reach 99% of the performance of fully-supervised UAGAN (e.g.

M-GenSeg: 0.861 ± 0.004 vs UAGAN: 0.872 ± 0.003 for T1 $\rightarrow$ T2 experiment). Thus, the annotation burden could be reduced with M-GenSeg.

3.3   Ablation Experiments

We conducted ablation tests to validate our methodological choices. We report in Table 1 the relative loss in Dice scores on target modality as compared to the proposed model. We assessed the value of doing image-level supervision by setting all the λ^{Gen} loss weights to 0 ◯. Also, we showed that training modality translation only on diseased data is sufficient ●. However, doing it for healthy data as well provides additional training examples for this task. Likewise, performing translation from absence to presence domain is not necessary ● but makes more efficient use of the data. Finally, we evaluated M-GenSeg with separate latent spaces ◯ for the image-level supervision and modality translation, and we contend that M-GenSeg efficiently combines both tasks when the latent representations share model updates.

Table 1. Ablation studies: relative Dice change on target modality.

Ablation	Mean		Std	
◯ No image-level supervision	-8.22	$\pm$	2.71	%
● No healthy modality translation	-2.41	$\pm$	1.29	%
● No absence to presence translation	-3.84	$\pm$	1.71	%
◯ Unshared latent spaces	-4.39	$\pm$	1.91	%

4   Conclusion

We propose M-GenSeg, a new framework for unpaired cross-modality tumor segmentation. We show that M-GenSeg is an annotation-efficient framework that greatly reduces the performance gap due to domain shift in cross-modality tumor segmentation. We claim that healthy tissues, if adequately incorporated to the training process of neural networks like in M-GenSeg, can help to better delineate tumor lesions in segmentation tasks. However, top performing methods on BraTS are 3D models. Thus, future work will explore the use of full 3D images rather than 2D slices, along with more optimal architectures. Our code is available: https://github.com/MaloADBA/MGenSeg_2D.

References

1. Prevedello, L.M., et al.: Challenges related to artificial intelligence research in medical imaging and the importance of image analysis competitions. Radiol. Artif. Intell. **1**(1), e180031 (2019)

2. Dorent, R., et al.: CrossMoDA 2021 challenge: benchmark of cross-modality domain adaptation techniques for vestibular schwannoma and cochlea segmentation. Med. Image Anal. **83**, 102628 (2023)
3. Billot, B., et al.: SynthSeg: domain randomisation for segmentation of brain scans of any contrast and resolution (2021)
4. Guan, H., Liu, M.: Domain adaptation for medical image analysis: a survey. IEEE Trans. Biomed. Eng. **3**, 1173–1185 (2022)
5. Pei, C., Wu, F., Huang, L.: Disentangle domain features for cross-modality cardiac image segmentation. Med. Image Anal. **71**, 102078 (2021)
6. Li, C., Luo, X., Chen, W., He, Y. Wu, M., Tan, Y.: AttENT: domain-adaptive medical image segmentation via attention-aware translation and adversarial entropy minimization. In: 2021 IEEE International Conference on Bioinformatics and Biomedicine (BIBM), pp. 952–959. IEEE (2021)
7. Huo, Y., et al.: SynSeg-Net: synthetic segmentation without target modality ground truth. IEEE Trans. Med. Imaging **38**(4), 1016–1025 (2019)
8. Chen, C., Dou, Q., Chen, H., Qin, J., Heng, P.-A.: Synergistic image and feature adaptation: towards cross-modality domain adaptation for medical image segmentation, vol. 38, pp. 865–872 (2019)
9. Jiang, J., et al.: Self-derived organ attention for unpaired CT-MRI deep domain adaptation based MRI segmentation. Phys. Med. Biol. **65**(20), 205001 (2020)
10. Zhou, B., Liu, C., Duncan, J.S.: Anatomy-constrained contrastive learning for synthetic segmentation without ground-truth. In: de Bruijne, M., et al. (eds.) MICCAI 2021. LNCS, vol. 12901, pp. 47–56. Springer, Cham (2021). https://doi.org/10.1007/978-3-030-87193-2_5
11. Hoffman, J., et al.: CyCADA: cycle-consistent adversarial domain adaptation (2017)
12. Zhang, Y., Miao, S., Mansi, T., Liao, R.: Task driven generative modeling for unsupervised domain adaptation: application to X-ray image segmentation. In: Frangi, A.F., Schnabel, J.A., Davatzikos, C., Alberola-López, C., Fichtinger, G. (eds.) MICCAI 2018. LNCS, vol. 11071, pp. 599–607. Springer, Cham (2018). https://doi.org/10.1007/978-3-030-00934-2_67
13. Zhu, J.-Y., Park, T., Isola, P. Efros, A.A.: Unpaired image-to-image translation using cycle-consistent adversarial networks. In: 2017 IEEE International Conference on Computer Vision (ICCV), pp. 2242–51. IEEE (2017)
14. Vorontsov, E., Molchanov, P., Gazda, M., Beckham, C., Kautz, J., Kadoury, S.: Towards annotation-efficient segmentation via image-to-image translation. Med. Image Anal. **82**, 102624 (2022)
15. Liu, X., Niethammer, M., Kwitt, R., Singh, N., McCormick, M., Aylward, S.: Low-rank atlas image analyses in the presence of pathologies. IEEE Trans. Med. Imaging **34**(12), 2583–2591 (2015)
16. Lin, C., Wang, Y., Wang, T., Ni, D.: Low-rank based image analyses for pathological MR image segmentation and recovery. Front. Neurosci. **13**(13), 333 (2019)
17. Changfa, S., Min, X., Xiancheng, Z., Haotian, W., Heng-Da, C.: Multi-slice low-rank tensor decomposition based multi-atlas segmentation: application to automatic pathological liver CT segmentation. Med. Image Anal. **73**, 102152 (2021)
18. Shin, H.-C., et al.: Medical image synthesis for data augmentation and anonymization using generative adversarial networks. In: Gooya, A., Goksel, O., Oguz, I., Burgos, N. (eds.) SASHIMI 2018. LNCS, vol. 11037, pp. 1–11. Springer, Cham (2018). https://doi.org/10.1007/978-3-030-00536-8_1

19. Mok, T.C.W., Chung, A.C.S.: Learning data augmentation for brain tumor segmentation with coarse-to-fine generative adversarial networks. In: Crimi, A., Bakas, S., Kuijf, H., Keyvan, F., Reyes, M., van Walsum, T. (eds.) BrainLes 2018. LNCS, vol. 11383, pp. 70–80. Springer, Cham (2019). https://doi.org/10.1007/978-3-030-11723-8_7
20. Kim, S., Kim, B., Park, H.: Synthesis of brain tumor multicontrast MR images for improved data augmentation. Med. Phys. **48**(5), 2185–2198 (2021)
21. Drozdzal, M., Vorontsov, E., Chartrand, G., Kadoury, S., Pal, C.: The importance of skip connections in biomedical image segmentation. In: Carneiro, G., et al. (eds.) LABELS/DLMIA 2016. LNCS, vol. 10008, pp. 179–187. Springer, Cham (2016). https://doi.org/10.1007/978-3-319-46976-8_19
22. Oktay, O., et al.: Attention U-Net: learning where to look for the pancreas (2018)
23. Bakas, S., Reyes, M., Jakab, A., Bauer, S.: Identifying the best machine learning algorithms for brain tumor segmentation, progression assessment, and overall survival prediction in the BRATS challenge (2018)
24. Bakas, S., et al.: Advancing the cancer genome atlas glioma MRI collections with expert segmentation labels and radiomic features. Sci. Data **4**(1), 170117 (2017)
25. Menze, B.H., Jakab, A., Bauer, S., Kalpathy-Cramer, J.: The multimodal brain tumor image segmentation benchmark (BRATS). IEEE Trans. Med. Imaging **34**(10), 1993–2024 (2015)
26. Yuan, W., Wei, J., Wang, J., Ma, Q., Tasdizen, T.: Unified attentional generative adversarial network for brain tumor segmentation from multimodal unpaired images. In: Shen, D., et al. (eds.) MICCAI 2019. LNCS, vol. 11766, pp. 229–237. Springer, Cham (2019). https://doi.org/10.1007/978-3-030-32248-9_26

Factor Space and Spectrum for Medical Hyperspectral Image Segmentation

Boxiang Yun[1], Qingli Li[1], Lubov Mitrofanova[2], Chunhua Zhou[3],
and Yan Wang[1(✉)]

[1] Shanghai Key Laboratory of Multidimensional Information Processing, East China Normal University, Shanghai, China
52265904012@stu.ecnu.edu.cn, qlli@cs.ecnu.edu.cn, ywang@cee.ecnu.edu.cn
[2] Almazov National Medical Research Centre, St. Petersburg, Russia
[3] Rui Jin Hospital, School of Medicine, Shanghai Jiao Tong University, Shanghai, China

Abstract. Medical Hyperspectral Imaging (MHSI) brings opportunities for computational pathology and precision medicine. Since MHSI is a 3D hypercube, building a 3D segmentation network is the most intuitive way for MHSI segmentation. But, high spatiospectral dimensions make it difficult to perform efficient and effective segmentation. In this study, in light of information correlation in MHSIs, we present a computationally efficient, plug-and-play space and spectrum factorization strategy based on 2D architectures. Drawing inspiration from the low-rank prior of MHSIs, we propose spectral matrix decomposition and low-rank decomposition modules for removing redundant spatiospectral information. By plugging our dual-stream strategy into 2D backbones, we can achieve state-of-the-art MHSI segmentation performances with 3–13 times faster compared with existing 3D networks in terms of inference speed. Experiments show our strategy leads to remarkable performance gains in different 2D architectures, reporting an improvement up to 7.7% compared with its 2D counterpart in terms of DSC on a public Multi-Dimensional Choledoch dataset. Code is publicly available at https://github.com/boxiangyun/Dual-Stream-MHSI.

Keywords: Medical hyperspectral images · MHSI segmentation

1 Introduction

Medical Hyperspectral Imaging (MHSI) is an emerging imaging modality which acquires two-dimensional medical images across a wide range of electromagnetic spectrum. It brings opportunities for disease diagnosis, and computational pathology [16]. Typically, an MHSI is presented as a hypercube, with hundreds of narrow and contiguous spectral bands in spectral dimension, and thousands of pixels in spatial dimension (Fig. 1(a)).

Supplementary Information The online version contains supplementary material available at https://doi.org/10.1007/978-3-031-43901-8_15.

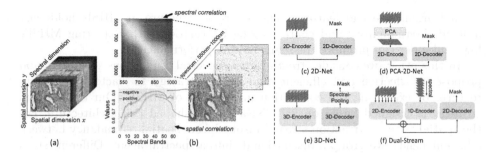

Fig. 1. (a): An example of MHSI. (b): Illustration of two types of correlation in MHSI. (c)–(f): HSI classification and segmentation backbones. 2D decoder can be changed into classification head for classification tasks.

Due to the success of 2-Dimensional (2D) deep neural network in natural images, the simplest way to classify/segment an MHSI is to treat its two spatial dimensions as input spatial dimension, and treat its spectral dimension as input channel dimension [25] (Fig. 1(c)). Dimensionality reduction [12] and recurrent approaches [1] are usually adopted to aggregate spectral information before feeding the HSI into 2D networks (Fig. 1(d)). These methods are not suitable for high spatial resolution MHSI, and they may bring noises in spatial features while reducing spectral dimension. The 2D networks are computationally efficient, usually much faster than 3D networks. But, they mix spectral information after the first convolutional layer, making the interband correlations of MHSIs underutilized. Building a 3D network usually suffers from high computational complexity, but it is the most straightforward way to learn interpixel and interband correlations of MHSIs [23] (Fig. 1(e)). Since spatiospectral orientations are not equally likely, there is no need to treat space and spectrum symmetrically, as is implicit in 3D networks. We might instead design a dual-stream strategy to "factor" the architecture. A few HSI classification backbones try to design dual-stream architectures that treat spatial structures and spectral intensities separately [2,20,30] (Fig. 1(f)). But, these methods simply adopt convolutional or MLP layers to extract spectral features. SpecTr [28], learning spectral and spatial features alternatively, utilizes Transformer to capture the global spectral feature. They overlook the low rankness in the spectral domain, which contains discriminative information for differentiating targets from the background.

High spatiospectral dimensions make it difficult to perform a thorough analysis of MHSI. In MHSIs, there exist two types of correlation. One is a spectral correlation in adjacent pixels. As shown in Fig. 1(b), the intensity values vs. spectral bands for the local positive (cancer) area and negative (normal) area are highly correlated. The other is spatial correlation between adjacent bands. Figure 1(b) plots the spatial similarity among all bands, and shows large cosine similarity scores among nearby bands (error band of line chart in the light color area) and small scores between bands in a long distance. The correlation implies spectral redundancy when representing spatial features, and spatial redundancy when learning spectral features. The low-rank structure in MHSIs holds significant discriminatory and characterizing information [11]. Exploring MHSI's low-rank prior can promote the segmentation performance.

In this paper, we consider treating spatiospectral dimensions separately and propose an effective and efficient dual-stream strategy to "factor" the architecture, by exploiting the correlation information of MHSIs. Our dual-stream strategy is designed based on 2D CNNs with U-shaped [16] architecture. For the spatial feature extraction stream, inspired from spatial redundancy between adjacent bands, we group adjacent bands into a spectral agent. Different spectral agents are fed into a 2D CNN backbone as a batch. For the spectral feature extraction stream, inspired by the low-rank prior on the spectral space, we propose a matrix factorization-based method to capture global spectral information. To remove the redundancy in the spatiospectral features and promote the capability of representing the low-rank prior of MHSI, we further design Low-rank Decomposition modules, and employ the Canonical-Polyadic decomposition method [9,32]. Our space and spectrum factorization strategy is plug-and-play. The effectiveness of the proposed strategy is compared and verified by plugging in different 2D architectures. We also show that with our proposed strategy, U-Net model using ResNet-34 can achieve state-of-the-art MHSI segmentation with 3–13 faster than other 3D architectures.

2   Methodology

Mathematically, let $\mathbf{Z} \in \mathbb{R}^{S \times H \times W}$ denote the 3D volume of a pathology MHSI, where $H \times W$ is the spatial resolution, and S is the number of spectral bands. The goal of MHSI segmentation is to predict the per-pixel annotation mask $\hat{\mathbf{Y}} \in \{0,1\}^{H \times W}$. Our training set is $\mathcal{D} = \{(\mathbf{Z}_i, \mathbf{Y}_i)\}_{i=1}^{N}$, where $\mathbf{Y}_i$ denotes the per-pixel groundtruth for MHSI $\mathbf{Z}_i$.

The overall architecture of our proposed method is shown in Fig. 2, where the 2D CNN in the figure is a proxy which may represent all widely-used 2D architectures. It represents a spatial stream, which focuses on extracting spatial features from spectral agents (Sect. 2.1). The lightweight spectral stream learns multi-granular spectral features, and it consists of three key modules: Depthwise Convolution (DwConv), Spectral Matrix Decomposition (SMD) and Feed Forward Network, where SMD module effectively leverages low-rank prior from spectral features (Sect. 2.1). Besides, the Low-rank Decomposition module (LD) represents high-level low-rank spatiospectral features (Sect. 2.2). The input MHSI $\mathbf{Z}$ is decomposed into a spatial input $\mathbf{Z}_{spa} \in \mathbb{R}^{G \times (S/G) \times H \times W}$ and a spectral input $\mathbf{Z}_{spe} \in \mathbb{R}^{S \times C_0^{spe} \times H \times W}$, where G indicates evenly dividing spectral bands into G groups, i.e., spectral agents. S/G and $C_0^{spe} = 1$ are the input feature dimensions for two streams respectively.

2.1   Dual-Stream Architecture with SpatioSpectral Representation

As mentioned above, for the spatial stream, we first reshape MHSI $\mathbf{Z} \in \mathbb{R}^{S \times H \times W}$ into $\mathbf{Z}_{spa} \in \mathbb{R}^{G \times (S/G) \times H \times W}$, which has G spectral agents. Each spectral agent is

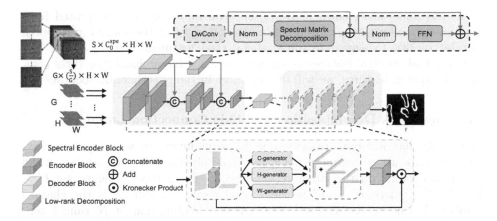

Fig. 2. The proposed dual-stream architecture. An MHSI goes through a spectral stream with the proposed spectral encoder block which consists of three key modules, *i.e*, DwConv, Spectral Matrix Decomposition (SMD) and Feed Forward Network (FFN), and it goes through a spatial stream after dividing into G spectral agents. Low-rank Decomposition (LD) is designed to characterize the low-rank prior.

treated as one sample. One sample contains highly correlated spectral bands, so that the spatial stream can focus on spatial feature extraction. For the spectral stream, to deal with problems of spatiospectral redundancy and the inefficiency of global spectral feature representation, we propose a novel and concise hierarchical structure shown in Fig. 2. We employ a basic transformer paradigm [21] but design it tailored for capturing global low-rank spectral features. Our spectral encoder block can be formulated by:

$$\mathbf{X} = DwConv(\mathbf{Z}_{in}), \ \mathbf{X}' = SMD(Norm(\mathbf{X})) + \mathbf{X}, \mathbf{Z}_{out} = FFN(Norm(\mathbf{X}')) + \mathbf{X}', \tag{1}$$

where $\mathbf{Z}_{in} \in \mathbb{R}^{S \times C_{spe} \times H \times W}$ indicates the input spectral token tensor, and C_{spe} is the spectral feature dimension. We introduce Depth-wise Conv (DwConv) for dynamically integrating redundant spatial information into spectral features to reduce spatial redundant noises, achieved by setting different strides of the convolutional kernel. Then, we represent long-distance dependencies among spectral inter-bands as a low-rank completion problem. $SMD(\cdot)$ indicates the spectral matrix decomposition operation. Concretely, we flatten feature map $\mathbf{X}$ to spectral sequence tokens $\mathbf{X}_{spe} \in \mathbb{R}^{H \cdot W \times S \times C_{spe}}$, which has S spectral tokens. We map $\mathbf{X}_{spe}$ to a feature space using a linear transform $\mathbf{W}_l \in \mathbb{R}^{C_{spe} \times C'_{spe}}$. We then apply a matrix decomposition method NMF (Non-negative Matrix Factorization) [10], denoted by $\mathcal{M}(\cdot)$, to identify and solve for a low-rank signal subspace and use iterative optimization algorithms backpropagate gradients [4]: $SMD(\mathbf{X}_{spe}) = \mathcal{M}(\mathbf{W}_l\mathbf{X}_{spe})$. Finally, to enhance the individual component of spectral tokens, we utilize a Feedforward Neural Network (FFN) in Transformer consisting of two linear layers and an activation layer.

In the framework shown in Fig. 2, spectral information is integrated from channel dimensions by performing concatenation, after the second and fourth encoder blocks, to aggregate the spatiospectral features. The reason for this design is that spectral features are simpler and lack hierarchical structures compared to spatial features, we will discuss more in the experimental section.

2.2   Low-Rank Decomposition and Skip Connection Ensemble

The MHSI has low-rank priority due to redundancy, so we propose a low-rank decomposition module using Canonical-Polyadic (CP) decomposition [9] to set constraints on the latent representation. For a three-order tensor $U \in \mathbb{R}^{C' \times H' \times W'}$, where $H' \times W'$ is the spatial resolution and C' is the channel number. It can be decomposed into a linear combination of N rank-1 tensors. The mathematical formulation of CP decomposition can be expressed as $U = \sum_{i=1}^{r} \lambda_i a_{ci} \otimes a_{hi} \otimes a_{wi}$. Where $\otimes$ denote Kronecker Product, $a_{ci} \in \mathbb{R}^{C' \times 1 \times 1}$, $a_{hi} \in \mathbb{R}^{C' \times 1 \times 1}$, $a_{wi} \in \mathbb{R}^{C' \times 1 \times 1}$, r is the tensor rank and λ_i is a scaling factor. Recent research [3,32] has proposed new methods based on DNNs to address this problem of representing MHSIs as low-rank tensors. As shown in Fig. 2, rank-1 generators are used to create rank-1 tensors in different directions, which are then aggregated by Kronecker Product to synthesize a sub-attention map A_1. The residual part between the input of features and the generated rank-1 tensor is used to generate second rank-1 tensors A_2. It can obtain r rank-1 tensors by repeating r times. Mathematically, this process can be expressed as:

$$
\begin{aligned}
A_1 &= \mathcal{G}_c(U) \otimes \mathcal{G}_h(U) \otimes \mathcal{G}_w(U), \\
A_2 &= \mathcal{G}_c(U - A_1) \otimes \mathcal{G}_h(U - A_1) \otimes \mathcal{G}_w(U - A_1), \\
A_r &= \mathcal{G}_c\left(U - \sum_{i=1}^{r-1} A_i\right) \otimes \mathcal{G}_h\left(U - \sum_{i=1}^{r-1} A_i\right) \otimes \mathcal{G}_w\left(U - \sum_{i=1}^{r-1} A_i\right),
\end{aligned}
\tag{2}
$$

where $\mathcal{G}_c(\cdot)$, $\mathcal{G}_h(\cdot)$ and $\mathcal{G}_w(\cdot)$ are the channel, height and width generators. Finally, we aggregate all rank-1 tensors (from A_1 to A_r) into the attention map along the channel dimension, followed by a linear layer used to reduce the feature dimension to obtain the low-rank feature U_{low}:

$$
U_{low} = U \odot Linear(Concate(A_1, A_2, ..., A_r)),
\tag{3}
$$

where $\odot$ is the element-wise product, and $U_{low} \in \mathbb{R}^{C' \times H' \times W'}$. We employ a straightforward non-parametric ensemble approach for grouping spectral agents. This approach involves multiple agents combining their features by averaging the vote. The encoders in the spatial stream produce 2D feature maps with G spectral agents, defined as $F_i \in \mathbb{R}^{G \times C_i \times H/2^i \times W/2^i}$ for the ith encoder, where G, C_i, $H/2^i$, and $W/2^i$ represent the spectral, channel, and two spatial dimensions, respectively. The ensemble is computed by $F_i^{out} = Mean(F_i^1, F_i^2, ..., F_i^G)$, where $F_i^G \in \mathbb{R}^{C_i \times H/2^i \times W/2^i}$ represents the 2D feature map of the Gth agent. The ensemble operation aggregates spectral agents to produce a 2D feature map

Table 1. Ablation study (in "mean (std)") on MDC dataset using RegNetX40 [26] as the backbone. SA denote the spectral agent. L1 to L4 represent the locations where output spectral features from the spectral flow module are inserted into the spatial flow. Tr and Conv mean we replace the SMD module in the spectral stream with self-attention and convolutional blocks. Best results are **highlighted**.

SA	Spectral Stream				LD	IoU ↑	DSC ↑	HD ↓
	L1	L2	L3	L4				
×	×	×	×	×	×	48.94 (20.53)	63.11 (19.13)	92.04 (34.14)
✓	×	×	×	×	×	51.64 (18.43)	66.05 (17.15)	86.04 (32.06)
✓	×	✓	×	×	×	52.05 (18.92)	66.27 (17.88)	83.05 (31.07)
✓	×	×	×	✓	×	53.87 (19.08)	67.95 (16.88)	88.75 (31.49)
✓	×	✓	×	✓	×	55.81 (18.41)	69.73 (16.31)	88.13 (31.47)
✓	✓	✓	✓	✓	×	55.01 (15.54)	69.58 (14.10)	86.25 (35.40)
✓	×	✓	×	✓	✓	**56.90 (17.38)**	**70.88 (15.05)**	**82.72 (31.77)**
✓	×	Conv	×	Conv	✓	55.19 (18.58)	69.15 (16.63)	83.13 (31.08)
✓	×	Tr	×	Tr	✓	55.72 (17.16)	69.89 (15.26)	84.48 (32.13)

with enhanced information interactions learned from the multi-spectral agents. The feature maps obtained from the ensemble can be decoded using lightweight 2D decoders to generate segmentation masks.

3   Experimental Results

3.1   Experimental Setup

We conducted experiments on the public Multi-Dimensional Choledoch (MDC) Dataset [31] with 538 scenes and Hyperspectral Gastric Carcinoma (HGC) Dataset [33] (data provided by the author) with 414 scenes, both with high-quality labels for binary MHSI segmentation tasks. These MHSIs are collected by hyperspectral system with an objective lens of 20x, and wavelengths from 550 nm to 1000 nm for MDC and 450 nm to 750 nm for HGC, resulting in 60 and 40 spectral bands for each scene. The size of a single band image in MDC and HGC are both resized to 256 × 320. Following [23, 27], we partition the datasets into training, validation, and test sets using a patient-centric hard split approach with a ratio of 3:1:1. Specifically, each patient's data is allocated entirely to one of the three sets, ensuring that the same patient's data do not appear in multiple sets.

We use data augmentation techniques such as rotation and flipping, and train with an Adam optimizer using a combination of dice loss and cross-entropy loss for 8 batch size and 100 epochs. The segmentation performance is evaluated using Dice-Sørensen coefficient (DSC), Intersection of Union (IoU), and Hausdorff Distance (HD) metrics, and Throughput (images/s) is reported for

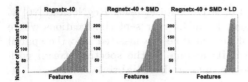

Method	DSC	Redundancy
RegNetX-40	66.06	0.4089
RegNetX-40+SMD	69.73	0.3924
RegNetX-40+SMD+LD	70.88	0.3699

Fig. 3. Feature redundancy of three methods on MDC dataset. Left figures plot each feature embedding in ascending order of the number of times they are dominant in the population (y-axis) and feature dimension (x-axis) on the statistical results of the test set. Right table shows the influence of SMD and LD modules in reducing redundancy.

comparison. Pytorch framework and four NVIDIA GeForce RTX 3090 are used for implementation.

3.2 Evaluation of the Proposed Dual-Stream Strategy

Ablation Study. Our dual-stream strategy is plug-and-play. We first conduct an ablation study to show the effectiveness of each component. We use a dual-stream strategy with RegNetX40 and U-Net architecture. As shown in Table 1, Our ablation study shows that spectral agent strategy improves segmentation performance by more than 2.5% (63.11 vs. 66.05). If we utilize spectral information from the spectral stream to assist in the spatial stream, we find that inserting spectral information at L2 and L4 yields a significant improvement of 3.7% (69.73 vs. 66.05), while inserting at L4 alone also results in a significant increase of 1.9% in DSC (67.95 vs. 66.05). A slight improvement is observed when inserting at L2, possibly due to the coarse features of shallow spectral information. Inserting spectral information at all spatial layers (*i.e.*, L1 to L4) and only at L2 and L4 produce similar results, indicating that spectral features do not possess complex multilevel characteristics relative to spatial features. Therefore, we adopt a simple and efficient two-layer spectral flow design. Replacing the spectral stream with transformer layers results in a 0.96% (70.88 vs. 69.89) lower DSC, possibly because transformers are difficult to optimize on small datasets. Our proposed LD module is crucial, resulting in a 1.12% performance drop in terms of DSC without it.

It is known that high feature redundancy limits the generalization of neural networks [29]. Here we show our low-rank representation effectively reduces the redundancy of features. Our quantitative and qualitative analysis demonstrated that the proposed MDC and LD modules effectively reduces the redundancy of output features. Following [8], we define the dominant features for the feature embedding of i-th MHSI $\mathbf{h}_i \in \mathbb{R}^{C_d}$ as $L_i = j : h_{ij} > \mu + \sigma$, where μ is mean of $\mathbf{h}_i$ and σ is stand deviation of $\mathbf{h}_i$. As shown in the left part of Fig. 3, our designed modules effectively reduce the number of dominant features and maintain sparsity in the entire spatiospectral feature space. Inspired by [24], we evaluate the degree of feature redundancy by computing the Pearson correlation coefficient between different feature channels. As shown in the right part of Fig. 3, the

Table 2. Performance comparison in "mean (std)" in MDC and HCG dataset. The best results of each comparison are **highlighted**.

Backbone	Method	MDC		HGC	
		DSC ↑	HD ↓	DSC ↑	HD ↓
ResNet50	U-Net	72.10 (14.17)	82.62 (29.70)	81.72 (15.79)	77.02 (44.22)
+Ours	U-Net	**74.12 (13.76)**	**78.32 (29.80)**	**85.08 (13.40)**	**69.78 (43.41)**
Convnext	U-Net	71.29 (13.68)	84.28 (30.50)	76.21 (15.15)	87.18 (43.85)
+Ours	U-Net	**72.82 (15.39)**	**82.12 (30.97)**	**77.51 (15.59)**	**79.73 (45.40)**
Swin-Transformer	U-Net	70.61 (14.42)	85.77 (31.76)	73.89 (18.05)	70.99 (36.10)
+Ours	U-Net	**72.10 (13.60)**	**82.94 (30.67)**	**78.07 (15.76)**	**63.35 (35.49)**
Efficinet-b2	U-Net	62.54 (19.28)	87.34 (32.72)	70.60 (17.41)	94.28 (46.95)
+Ours	U-Net	**68.72 (15.16)**	**84.91 (33.62)**	**77.44 (14.40)**	**81.24 (40.36)**
RegNetX40	U-Net	63.11 (19.13)	92.04 (34.14)	74.32 (18.39)	88.54 (45.34)
+Ours	U-Net	**70.88 (15.05)**	**82.72 (31.77)**	**79.86 (15.18)**	**80.35 (43.16)**
ResNet50	FPN	71.23 (16.01)	83.17 (36.84)	79.98 (15.30)	68.52 (43.32)
+Ours	FPN	**73.01 (13.84)**	**78.37 (30.26)**	**81.88 (14.75)**	**68.60 (45.63)**

Table 3. Performance comparison with SOTA methods with Throughput(images/s) on MDC dataset and HGC dataset. The best results are **highlighted**.

Method		MDC			HGC		
		DSC ↑	HD ↓	Throughput ↑	DSC ↑	HD ↓	Throughput ↑
2D	PCA-UNet [22]	70.83	80.70	12.37	78.27	78.06	**16.49**
	CGRU-UNet [1]	73.14	82.49	7.50	80.68	74.23	7.62
	Ours (Resnet34)	**75.44**	77.70	**13.84**	**85.80**	**64.04**	14.82
3D	3D-UNet [6]	72.55	79.87	4.04	83.37	79.97	4.34
	nnUNet [7]	74.12	79.87	1.92	85.36	75.83	2.83
	HyperNet [23]	72.47	83.75	0.99	84.00	67.05	1.67
	Swin-UNETR [19]	72.39	78.38	1.45	78.81	80.15	2.31
	SpecTr [28]	73.66	**76.92**	1.40	84.74	66.90	2.44

SMD and LD modules can reduce feature redundancy and lead to an increase in segmentation performance.

Comparisons Between w/ and w/o Dual-Stream Strategy on Different Backbones. To show the effectiveness of our dual-stream strategy in improving MHSI segmentation performance in various architectures, we plug it into different segmentation methods, *i.e.*, U-Net [17], FPN [13], with different spatial branch backbones *i.e.*, ResNet [5], Convnext [15], RegNetX40 [26], EfficientNet [18] and Swin-Transformer [14]. Results are summarized in Table 2. The results obtained with the proposed dual-stream strategy can consistently boost the segmentation performance by a large margin.

Comparisons with State-of-the-Art MHSI Segmentation Methods.
Table 3 shows comparisons on MDC and HGC datasets. We use a lightweight and efficient ResNet34 as the backbone of our dual-stream method. Experimental results show that 2D methods are generally faster than 3D methods in inference speed, but 3D methods have an advantage in segmentation performance (DSC & HD). However, our approach outperforms other methods in both inference speed and segmentation accuracy. It is also plug-and-play, with the potential to achieve better segmentation performance by selecting more powerful back-bones. The complete table (including IoU and variance) and qualitative results are shown in the supplementary material.

4   Conclusion

In this paper, we present to factor space and spectrum for accurate and fast medical hyperspectral image segmentation. Our dual-stream strategy, leveraging low-rank prior of MHSIs, is computationally efficient and plug-and-play, which can be easily plugged into any 2D architecture. We evaluate our approach on two MHSI datasets. Experiments show significant performance improvements on different evaluation metrics, *e.g.*, with our proposed strategy, we can obtain over 7.7% improvement in DSC compared with its 2D counterpart. After plugging our strategy into ResNet-34 backbone, we can achieve state-of-the-art MHSI segmentation accuracy with 3–13 times faster in terms of inference speed than existing 3D networks.

Acknowledgements. This work was supported by the National Natural Science Foundation of China (Grant No. 62101191), Shanghai Natural Science Foundation (Grant No. 21ZR1420800), and the Science and Technology Commission of Shanghai Municipality (Grant No. 22DZ2229004).

References

1. Bengs, M., et al.: Spectral-spatial recurrent-convolutional networks for *in-vivo* hyperspectral tumor type classification. In: Martel, A.L., et al. (eds.) MICCAI 2020. LNCS, vol. 12263, pp. 690–699. Springer, Cham (2020). https://doi.org/10.1007/978-3-030-59716-0_66
2. Chen, R., Li, G.: Spectral-spatial feature fusion via dual-stream deep architecture for hyperspectral image classification. Infrared Phys. Technol. **119**, 103935 (2021)
3. Chen, W., et al.: Tensor low-rank reconstruction for semantic segmentation. In: Vedaldi, A., Bischof, H., Brox, T., Frahm, J.-M. (eds.) ECCV 2020. LNCS, vol. 12362, pp. 52–69. Springer, Cham (2020). https://doi.org/10.1007/978-3-030-58520-4_4
4. Geng, Z., Guo, M.H., Chen, H., Li, X., Wei, K., Lin, Z.: Is attention better than matrix decomposition? In: International Conference on Learning Representations (2021)
5. He, K., Zhang, X., Ren, S., Sun, J.: Deep residual learning for image recognition. In: Computer Vision and Pattern Recognition (2016)

6. Çiçek, Ö., Abdulkadir, A., Lienkamp, S.S., Brox, T., Ronneberger, O.: 3D U-Net: learning dense volumetric segmentation from sparse annotation. In: Ourselin, S., Joskowicz, L., Sabuncu, M.R., Unal, G., Wells, W. (eds.) MICCAI 2016. LNCS, vol. 9901, pp. 424–432. Springer, Cham (2016). https://doi.org/10.1007/978-3-319-46723-8_49

7. Isensee, F., et al.: nnU-Net: self-adapting framework for u-net-based medical image segmentation. Nat. Methods (2021)

8. Kalibhat, N.M., Narang, K., Firooz, H., Sanjabi, M., Feizi, S.: Towards better understanding of self-supervised representations. In: Workshop on Spurious Correlations, Invariance and Stability, ICML 2022 (2022)

9. Kolda, T.G., Bader, B.W.: Tensor decompositions and applications. SIAM Rev. **51**(3), 455–500 (2009)

10. Lee, D.D., Seung, H.S.: Learning the parts of objects by non-negative matrix factorization. Nature **401**(6755), 788–791 (1999)

11. Li, L., Li, W., Du, Q., Tao, R.: Low-rank and sparse decomposition with mixture of gaussian for hyperspectral anomaly detection. IEEE Trans. Cybern. **51**(9), 4363–4372 (2020)

12. Li, X., Li, W., Xu, X., Hu, W.: Cell classification using convolutional neural networks in medical hyperspectral imagery. In: International Conference on Image, Vision and Computing (2017)

13. Lin, T.Y., Dollár, P., Girshick, R., He, K., Hariharan, B., Belongie, S.: Feature pyramid networks for object detection. In: Computer Vision and Pattern Recognition (2017)

14. Liu, Z., et al.: Swin transformer: hierarchical vision transformer using shifted windows. arXiv, Computer Vision and Pattern Recognition (2021)

15. Liu, Z., Mao, H., Wu, C.Y., Feichtenhofer, C., Darrell, T., Xie, S.: A convnet for the 2020s. In: 2022 IEEE/CVF Conference on Computer Vision and Pattern Recognition (CVPR), pp. 11966–11976 (2022). https://doi.org/10.1109/CVPR52688.2022.01167

16. Lu, G., Fei, B.: Medical hyperspectral imaging: a review. J. Biomed. Opt. **19**, 010901 (2014)

17. Ronneberger, O., Fischer, P., Brox, T.: U-Net: convolutional networks for biomedical image segmentation. In: Navab, N., Hornegger, J., Wells, W.M., Frangi, A.F. (eds.) MICCAI 2015. LNCS, vol. 9351, pp. 234–241. Springer, Cham (2015). https://doi.org/10.1007/978-3-319-24574-4_28

18. Tan, M., Le, Q.V.: EfficientNet: rethinking model scaling for convolutional neural networks. In: International Conference on Machine Learning (2019)

19. Tang, Y., et al.: Self-supervised pre-training of swin transformers for 3D medical image analysis. In: Proceedings of the IEEE/CVF Conference on Computer Vision and Pattern Recognition, pp. 20730–20740 (2022)

20. Trajanovski, S., Shan, C., Weijtmans, P.J., de Koning, S.G.B., Ruers, T.J.: Tongue tumor detection in hyperspectral images using deep learning semantic segmentation. IEEE Trans. Biomed. Eng. **68**(4), 1330–1340 (2020)

21. Vaswani, A., et al.: Attention is all you need. In: Advances in Neural Information Processing Systems, vol. 30 (2017)

22. Wang, J., et al.: PCA-U-Net based breast cancer nest segmentation from microarray hyperspectral images. Fundam. Res. **1**(5), 631–640 (2021)

23. Wang, Q., et al.: Identification of melanoma from hyperspectral pathology image using 3D convolutional networks. IEEE Trans. Med. Imaging **40**(1), 218–227 (2020)

24. Wang, Y., et al.: Revisiting the transferability of supervised pretraining: an MLP perspective. In: 2022 IEEE/CVF Conference on Computer Vision and Pattern Recognition (CVPR), pp. 9173–9183 (2022). https://doi.org/10.1109/CVPR52688.2022.00897
25. Wei, X., Li, W., Zhang, M., Li, Q.: Medical hyperspectral image classification based on end-to-end fusion deep neural network. IEEE Trans. Instrum. Measur. **68**, 4481–4492 (2019)
26. Xie, S., Girshick, R., Dollár, P., Tu, Z., He, K.: Aggregated residual transformations for deep neural networks. In: CVPR (2017)
27. Xie, X., Wang, Y., Li, Q.: S^{3}r: self-supervised spectral regression for hyperspectral histopathology image classification. In: Wang, L., Dou, Q., Fletcher, P.T., Speidel, S., Li, S. (eds.) MICCAI 2022. LNCS, vol. 13432, pp. 46–55. Springer, Cham (2022). https://doi.org/10.1007/978-3-031-16434-7_5
28. Yun, B., Wang, Y., Chen, J., Wang, H., Shen, W., Li, Q.: Spectr: spectral transformer for hyperspectral pathology image segmentation. arXiv, Image and Video Processing (2021)
29. Zbontar, J., Jing, L., Misra, I., LeCun, Y., Deny, S.: Barlow twins: self-supervised learning via redundancy reduction. In: ICML (2021)
30. Zhang, H., Li, Y., Zhang, Y., Shen, Q.: Spectral-spatial classification of hyperspectral imagery using a dual-channel convolutional neural network. Remote Sens. Lett. **8**, 438–447 (2017)
31. Zhang, Q., Li, Q., Yu, G., Sun, L., Zhou, M., Chu, J.: A multidimensional choledoch database and benchmarks for cholangiocarcinoma diagnosis. IEEE Access **7**, 149414–149421 (2019)
32. Zhang, S., Wang, L., Zhang, L., Huang, H.: Learning tensor low-rank prior for hyperspectral image reconstruction. In: Proceedings of the IEEE/CVF Conference on Computer Vision and Pattern Recognition, pp. 12006–12015 (2021)
33. Zhang, Y., Wang, Y., Zhang, B., Li, Q.: A hyperspectral dataset of precancerous lesions in gastric cancer and benchmarks for pathological diagnosis. J. Biophotonics e202200163 (2022)

RBGNet: Reliable Boundary-Guided Segmentation of Choroidal Neovascularization

Tao Chen[1,2], Yitian Zhao[1(✉)], Lei Mou[1], Dan Zhang[3], Xiayu Xu[4], Mengting Liu[5], Huazhu Fu[6], and Jiong Zhang[1(✉)]

[1] Institute of Biomedical Engineering, Ningbo Institute of Materials Technology and Engineering, Chinese Academy of Sciences, Ningbo, China
{yitian.zhao,zhangjiong}@nimte.ac.cn
[2] Cixi Biomedical Research Institute, Wenzhou Medical University, Ningbo, China
[3] School of Cyber Science and Engineering, Ningbo University of Technology, Ningbo, China
[4] School of Life Science and Technology, Xi'an Jiaotong University, Xi'an, China
[5] School of Biomedical Engineering, Sun Yat-sen University, Shenzhen, China
[6] Institute of High Performance Computing, A*STAR, Singapore, Singapore

Abstract. Choroidal neovascularization (CNV) is a leading cause of visual impairment in retinal diseases. Optical coherence tomography angiography (OCTA) enables non-invasive CNV visualization with micrometerscale resolution, aiding precise extraction and analysis. Nevertheless, the irregular shape patterns, variable scales, and blurred lesion boundaries of CNVs present challenges for their precise segmentation in OCTA images. In this study, we propose a **R**eliable **B**oundary-**G**uided choroidal neovascularization segmentation **Net**work (RBGNet) to address these issues. Specifically, our RBGNet comprises a dual-stream encoder and a multi-task decoder. The encoder consists of a convolutional neural network (CNN) stream and a transformer stream. The transformer captures global context and establishes long-range dependencies, compensating for the limitations of the CNN. The decoder is designed with multiple tasks to address specific challenges. Reliable boundary guidance is achieved by evaluating the uncertainty of each pixel label, By assigning it as a weight to regions with highly unstable boundaries, the network's ability to learn precise boundary locations can be improved, ultimately leading to more accurate segmentation results. The prediction results are also used to adaptively adjust the weighting factors between losses to guide the network's learning process. Our experimental results demonstrate that RBGNet outperforms existing methods, achieving a Dice score of 90.42% for CNV region segmentation and 90.25% for CNV vessel segmentation. https://github.com/iMED-Lab/RBGnet-Pytorch.git.

Keywords: CNV · OCTA · Transformer · Uncertainty · Multi-task

1 Introduction

Age-related macular degeneration (AMD) is a leading cause of blindness worldwide, primarily attributable to choroidal neovascularization (CNV) [1]. Optical

© The Author(s), under exclusive license to Springer Nature Switzerland AG 2023
H. Greenspan et al. (Eds.): MICCAI 2023, LNCS 14223, pp. 163–172, 2023.
https://doi.org/10.1007/978-3-031-43901-8_16

Fig. 1. Interference to CNV in OCTA images. Red arrows indicate artifacts and noise interference, respectively. (Color figure online)

coherence tomography angiography (OCTA), a non-invasive imaging technique, has gained popularity in recent years due to its ability to visualize blood flow in the retina and choroid with micrometer depth resolution [2]. Thus, automated CNV segmentation based on OCTA images can facilitate quantitative analysis and enhance the diagnosis performance of AMD [3]. However, the accurate segmentation of CNV from OCTA images poses a significant challenge due to the complex morphology of CNVs and the presence of imaging artifacts [4], as illustrated in Fig. 1. Hence, reliable CNV segmentation is promptly needed to assist ophthalmologists in making informed clinical decisions.

Several methods have been proposed to segment CNV regions from OCTA images, including handcraft feature descriptors [5,6] and deep learning-based techniques [7]. For example, a saliency-based method for automated segmentation of CNV regions in OCTA images was proposed by Liu *et al.* [5], which capitalizes on distinguishing features of CNV regions with higher intensity compared to background artifacts and noise. In [6], an unsupervised algorithm for CNV segmentation was proposed, which utilizes a density cell-like P system. However, their accuracy is restricted by weak saliency and ambiguous boundaries. With the recent advancements of deep learning, several methods have also been proposed for CNV segmentation in OCT images. U-shaped multiscale information fusion networks are proposed in [8] and [9] for segmenting CNVs with multiscale scenarios. Wang *et al.* [7] further proposed a two-stage CNN-based architecture based on OCTA images that is capable of extracting both CNV regions and vessel details. However, common issues including substantial scale variations of CNV regions and low-contrast microvascular boundaries were not fully deliberated in previous network designs. Thus, more dedicated modules with scale adaptivity and boundary refinement properties need to be explored to solve existing challenges.

Previously, Gal *et al.* [10] proposed a theory to effectively model uncertainty with dropout NNs. Afterward, Bragman *et al.* [11] applied this method to the field of medical image analysis. Nair *et al.* [12] showed the success of using dropout for the detection of three-dimensional multiple sclerosis lesions. Motivated by these findings, we also consider taking advantage of the pixel-level uncertainty estimation and making it adaptive to the segmentation of ambiguous CNV boundaries.

In this work, we propose a reliable boundary-guided network (RBGNet) to simultaneously segment both the CNV regions and vessels. Our proposed method is composed of a dual-branch encoder and a boundary uncertainty-guided multi-task decoder. The dual-branch encoder is designed to capture both of the global long-range dependencies and the local context of CNVs with significant scale variations, while the proposed uncertainty-guided multi-task decoder is designed to strengthen the model to segment ambiguous boundaries. The uncertainty is achieved by approximating a Bayesian network through Monte Carlo dropout.

The main contributions are summarized as follows:

(a) We propose a multi-task joint optimization method to interactively learn shape patterns and boundary contours for more accurate segmentation of CNV regions and vessels.

(b) We design a dual-stream encoder structure to take advantages of the CNN and transformer, which promote the network to effectively learn both local and global information.

(c) We propose an uncertainty estimation-guided weight optimization strategy to provide reliable guidance for multi-task network training.

2   Proposed Method

The proposed RBGNet comprises three primary components: a dual-stream encoder, a multi-task decoder, and an uncertainty-guided weight optimization strategy, as depicted in Fig. 2. The OCTA image is firstly processed using a dual-stream encoder that combines Convolutional Neural Networks (CNNs) and Vision Transformer (ViT) models [13] to produce high-dimensional semantic features. This approach enhances the local context and global dependencies of the image, thereby improving the CNV feature representation. These features are then fed into a multi-task decoder, which integrates information from multiple tasks to achieve better CNV segmentation. To further optimize the model's performance, we introduce a pixel-level uncertainty estimation approach that enhances the model's capacity to handle ambiguous region boundaries.

2.1   Dual-Stream Encoder

The CNVs in OCTA images are of various shapes and a wide range of scales, which pose challenges to the accurate segmentation of CNVs. To extract the long-range dependencies of cross-scale CNV information, we employ VIT [13] as an independent stream in the feature encoder of the proposed method. Moreover, the proposed dual-stream encoder is also embedded with a CNN-based multi-scale encoder to obtain the local context of CNV regions. Specifically, the VIT stream utilizes a stack of twelve transformer layers to extract features from flattened uniform non-overlapping patches, which seamlessly integrates with the CNN encoder stream through skip connections, similar to [14]. However, unlike [14], we divide the output representations of the transformer

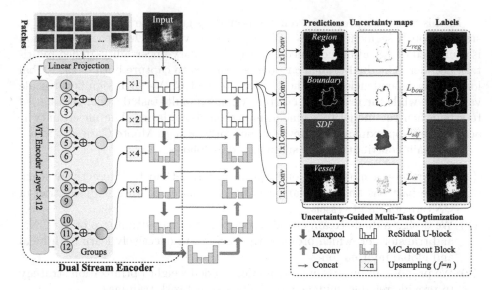

Fig. 2. Schematic diagram of the proposed RBGNet, which contains a dual-stream encoder and a multi-task decoder.

layers in VIT into four groups, each containing three feature representations. These groups correspond to the scales of the CNN stream, uniformly arranged from shallow to deep. Then, we perform element-wise summation for each group of representations, followed by reshaping them into non-overlapping patch sizes. The reshaped representation is further upsampled to the corresponding CNN feature resolution. The CNN branch is a U-shaped network [15] that extracts multi-scale features using ReSidual U-block (RSU) proposed by [16] to preserve high-resolution information locally. To integrate complementary information from the CNN and VIT features, we concatenate them in each of the first four feature extraction layers of the CNN branch. To enhance the features from the dual-branch encoder, we apply a bottleneck layer that consists of a feature extraction block, max-pooling, upsampling, and another feature extraction block.

2.2   Multi-task Decoder

The proposed multi-task decoder performs two main tasks: CNV region segmentation and vessel segmentation. The CNV region segmentation task contains two auxiliary subtasks including boundary prediction and shape regression. Each task is implemented at the end of the decoder using a 1×1 convolutional layer followed by a Sigmoid activation, as shown in Fig. 2.

Region Segmentation: Region segmentation of CNV allows accurate assessment of lesion size. This task aims to accurately segment the entire CNV regions in OCTA images via boundary prediction and shape regression. Region segmentation is typically accomplished by categorizing individual pixels as either belonging to the CNV region or the background region. The purpose of region boundary prediction is to explicitly enhance the model's focus on ambiguous boundaries, allowing for more accurate region segmentation. The process of shape regression for region segmentation involves the transformation of boundary regression into a task of signed distance field regression. This is achieved by assigning a signed distance to each pixel, representing its distance from the boundary, with negative values inside the boundary, positive values outside the boundary, and zero values on the boundary. By converting the ground truth into a signed distance map (SDM), the network can learn CNV shape patterns from the rich shape pattern information contained in the SDMs.

Vessel Segmentation: To improve the vessel segmentation performance, we propose to guide the model to focus on low-contrast vessel details by estimating their pixel uncertainty. Simultaneously, the complementary information from the CNV region segmentation task is further utilized to eliminate the interference of vessel pixels outside the region, thus better refining the vessel segmentation results. The proposed multi-task decoder improves the segmentation accuracy of regions and vessels by explicitly or implicitly using the information between individual tasks and optimizing each task itself.

2.3   Uncertainty-Guided Multi-Task Optimization

Uncertainty Estimation: In contrast to traditional deep learning models that produce deterministic predictions, Bayesian neural networks [17] can provide not only predictions but also uncertainty. It treats the network weights as random variables with a priori distribution and infers the posterior distribution of the weights. In this paper, we employ Monte Carlo dropout (MC-dropout) [10] to approximate Bayesian networks and capture the uncertainty of the model. Bayesian inference offers a rigorous method for making decisions in the presence of uncertainty. However, the computational complexity of computing the posterior distribution often renders it infeasible. This issue is usually solved by finding the best approximation in a finite space.

To learn the weight distribution of the network, we minimize the Kullback-Leibler (KL) scatter between the true posterior distribution and its approximation. The probability distribution of each pixel is obtained based on Dropout to sample the posterior weight distribution M times. Then, the mean P_i of each pixel is used to generate the prediction, while the variance V_i is used to quantify the uncertainty of the pixel. This process can be described as follows.

$$P_i = \frac{1}{M} \sum_{m=1}^{M} \hat{P}_m, \text{ and } V_i = \frac{1}{M} \sum_{m=1}^{M} \left(\hat{P}_m - P_i \right)^2. \tag{1}$$

Uncertainty-Weighted Loss: In CNV region segmentation, the importance of each pixel may vary, especially for ambiguous boundaries, while assigning equal weights to all samples may not be optimal. To address this issue, uncertainty maps are utilized to assign increased weights to pixels with higher levels of uncertainty. This, in turn, results in a more substantial impact on the update of the model parameters. Moreover, the incorporation of multiple tasks can generate different uncertainty weights for a single image, enabling a more comprehensive exploration of CNV boundary features via joint optimization. We employ a combination of loss functions, including binary cross-entropy (BCE) loss, mean squared error (MSE) loss, and Dice loss, to optimize the model parameters across all tasks. However, for the region shape regression task, we restricted the loss functions to only BCE and MSE. We incorporate uncertainty weights into the BCE loss by weighting each pixel to guide uncertainty on model training, i.e.,

$$\mathcal{L}_{UBCE} = -\frac{1}{N} \sum_{i=1}^{N} (1 + V_i) \cdot [y_i \log(\hat{y}_i) + (1 - y_i) \log(1 - \hat{y}_i)], \qquad (2)$$

where y_i and $\hat{y}_i$ are respective ground truth and prediction for pixel i. The total loss function can be expressed as $\mathcal{L} = \sum_t^T \lambda_t \mathcal{L}_t$, where $\mathcal{L}_t$ denotes the loss function for t^{th} task. λ_t denotes the loss weight, obtained by averaging the uncertainty map of the corresponding task and normalizing them to sum to 1.

3   Experimental Results

Dataset: The proposed RBGNet was evaluated on a dataset consisting of 74 OCTA images obtained using the Heidelberg OCT2 system (Heidelberg, Germany). All images were from AMD patients with CNV progression, captured in a 3×3 mm^2 area centered at the fovea. The enface projected OCTA images of the avascular complex were used for our experiments. All the images were resized into a resolution of 384×384 for experiments. The CNV areas and vessels were manually annotated by one senior ophthalmologist, and then reviewed and refined by another senior ophthalmologist. All images were acquired with regulatory approvals and patient consents as appropriate, following the Declaration of Helsinki.

Implementation Details: Our method is implemented based on the PyTorch framework with NVIDIA GeForce GTX 1080Ti. We train the model using an Adam optimizer with an initial learning rate of 0.0001 and a batch size of 4 for 300 epochs, without implementing a learning rate decay strategy. During training, the model inputs were subject to standard data augmentation pipelines, including random horizontal, vertical flips, random rotation, and random cropping. A 5-fold cross-validation approach is adopted to evaluate the performance.

Comparison with State-of-the-Arts: To benchmark our model's performance, we compared it with several state-of-the-art methods in the medical image segmentation field, including U-Net [15], CE-Net [18], CS-Net [19], TransUNet [20], and the backbone method U^{2}Net [16]. We use the Dice coefficient (Dice), intersection over union (IoU), false discovery rate (FDR), and area under the ROC curve (AUC) to evaluate the segmentation performance. The quantitative results are demonstrated in Table 1. Our results demonstrate that the

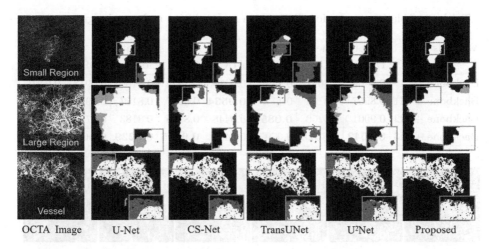

Fig. 3. Performance comparisons of different methods. Under-segmentation is shown in red, and over-segmentation is shown in green. (Color figure online)

Table 1. Performance comparisons for CNV segmentation.

Methods	Leison segmentation				Vessel segmentation			
	DICE	IoU	FDR	AUC	DICE	IoU	FDR	AUC
MF-Net [8]	0.8166	0.7142	0.1574	0.9084	0.7582	0.7191	0.1734	0.9062
U-Net [15]	0.8228	0.7290	0.1694	0.9148	0.8193	0.7280	0.1685	0.9175
CE-Net [18]	0.8690	0.7806	0.1173	0.9331	0.8372	0.7308	0.1459	0.9156
CS-Net [19]	0.8567	0.7689	0.1423	0.9365	0.8518	0.7622	0.1519	0.9378
TransUNet [20]	0.8414	0.7438	0.1254	0.9197	0.8301	0.7283	0.1423	0.9171
U^2Net [16]	0.8770	0.7930	0.1072	0.9383	0.8561	0.7662	0.0982	0.9214
Proposed	**0.9042**	**0.8328**	**0.0797**	**0.9464**	**0.9025**	**0.8229**	**0.0688**	**0.9455**

proposed method surpasses the existing state-of-the-art methods in both tasks. Specifically, our method achieved outstanding results on the test set for region segmentation, with a Dice of 90.42%, an IOU of 83.28%, and an AUC of 94.64%. The results in Table 1 indicate that superior vessel segmentation is positively associated with the model's ability to segment CNV regions.

The proposed method exhibits superior performance in precisely segmenting ambiguous boundaries of CNV regions, as demonstrated in the first two rows of Fig. 3. In contrast, existing state-of-the-art methods such as U-Net [15], and CS-Net [19] exhibit limitations in accurately segmenting complex and variable structures, leading to the misidentification of background structures as CNV regions. The illustrated quantitative results and performance comparisons serve as evidence of the proposed method's ability to simultaneously segment CNV regions and vessels with state-of-the-art performance. Accurate segmentation of

Table 2. Ablation results for CNV region segmentation and vessel segmentation.

Methods	Leison segmentation				Vessel segmentation			
	DICE	IoU	FDR	AUC	DICE	IoU	FDR	AUC
Backbone	0.8869	0.8051	0.1149	0.9473	0.8934	0.8160	0.0918	0.9401
Backbone + M1	0.8957	0.8194	0.1119	**0.9534**	0.8947	0.8177	0.0932	**0.9530**
Backbone + M2	0.9001	0.8260	**0.0856**	0.9448	0.8985	0.8183	0.0725	0.9322
Backbone + M3	**0.9042**	**0.8328**	0.0797	0.9464	**0.9025**	**0.8229**	**0.0688**	0.9455

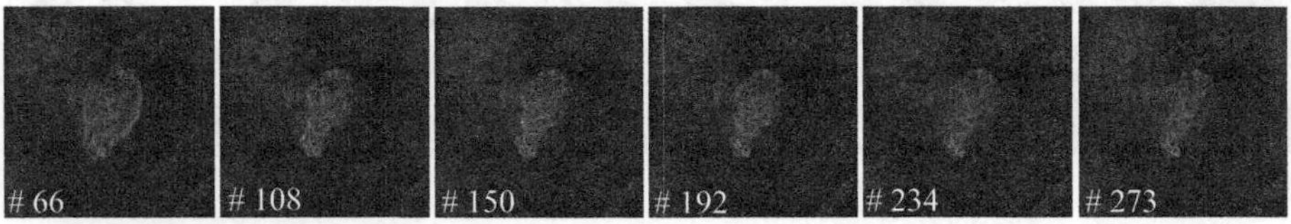

Fig. 4. Uncertainty maps during model training. #n represents the nth epoch.

CNV region boundaries leads to improved vessel segmentation accuracy, which in turn refines the boundary of CNV regions.

Ablation Study: To validate the effectiveness of multi-task processing in enhancing CNV segmentation, we systematically integrated individual tasks into the multi-task training framework, by taking CNV region and vessel segmentation as the backbone. We successively incorporate M1: boundary prediction, M2: M1 and shape regression, and M3: M2 and uncertainty prediction, into the Backbone. The results for ablation are summarized in Table 2. We also illustrate in Fig. 4 the uncertainty probability maps generated by the proposed method during training. Note that the notation #n indicates the nth training epoch, where the total number of training epochs is 300, and the model has specified to output the result in every 3 epochs. The best result is achieved at the $\#273^{th}$ epoch. Through Fig. 4, we can observe that the degree of uncertainty surrounding the segmented CNV regions diminishes gradually with the increased training epoch. These observations show the effectiveness of the uncertainty-weighted loss.

3.1 Conclusion

In summary, this study proposes a novel method to address the challenges of CNV segmentation in OCTA images. It incorporates a dual-branch encoder, multi-task optimization, and uncertainty-weighted loss to accurately segment CNV regions and vessels. The findings indicate that the utilization of cross-scale information, multi-task optimization, and uncertainty maps improve CNV segmentations. The proposed method exhibits superior performance compared to state-of-the-art methods, which suggests potential clinical implications for the diagnosis of CNV-related diseases. Nevertheless, further research is needed to validate the effectiveness of the proposed approach in large-scale clinical studies.

Acknowledgment. This work was supported in part by the National Science Foundation Program of China (62103398, 62272444), Zhejiang Provincial Natural Science Foundation of China (LZ23F010002, LR22F020008, LQ23F010002), in part by the Ningbo Natural Science Foundation (2022J143), and A*STAR AME Programmatic Fund (A20H4b0141) and Central Research Fund.

References

1. Friedman, D.S., et al.: Prevalence of age-related macular degeneration in the United States. Arch. Ophthalmol. **122**(4), 564–572 (2004)
2. Spaide, R.F., Klancnik, J.M., Cooney, M.J.: Retinal vascular layers imaged by fluorescein angiography and optical coherence tomography angiography. JAMA Ophthalmol. **133**(1), 45–50 (2015)
3. Jia, Y., et al.: Quantitative optical coherence tomography angiography of choroidal neovascularization in age-related macular degeneration. Ophthalmology **121**(7), 1435–1444 (2014)
4. Falavarjani, K.G., Al-Sheikh, M., Akil, H., Sadda, S.R.: Image artefacts in swept-source optical coherence tomography angiography. Br. J. Ophthalmol. **101**(5), 564–568 (2017)
5. Liu, L., Gao, S.S., Bailey, S.T., Huang, D., Li, D., Jia, Y.: Automated choroidal neovascularization detection algorithm for optical coherence tomography angiography. Biomed. Opt. Express **6**(9), 3564–3576 (2015)
6. Xue, J., Camino, A., Bailey, S.T., Liu, X., Li, D., Jia, Y.: Automatic quantification of choroidal neovascularization lesion area on OCT angiography based on density cell-like p systems with active membranes. Biomed. Opt. Express **9**(7), 3208–3219 (2018)
7. Wang, J., et al.: Automated diagnosis and segmentation of choroidal neovascularization in OCT angiography using deep learning. Biomed. Opt. Express **11**(2), 927–944 (2020)
8. Meng, Q., et al.: MF-Net: multi-scale information fusion network for CNV segmentation in retinal OCT images. Front. Neurosci. **15**, 743769 (2021)
9. Su, J., Chen, X., Ma, Y., Zhu, W., Shi, F.: Segmentation of choroid neovascularization in OCT images based on convolutional neural network with differential amplification blocks. In: Medical Imaging 2020: Image Processing, vol. 11313, pp. 491–497. SPIE (2020)
10. Gal, Y., Ghahramani, Z.: Dropout as a Bayesian approximation: representing model uncertainty in deep learning. In: International Conference on Machine Learning, pp. 1050–1059. PMLR (2016)
11. Bragman, F.J.S., et al.: Uncertainty in multitask learning: joint representations for probabilistic MR-only radiotherapy planning. In: Frangi, A.F., Schnabel, J.A., Davatzikos, C., Alberola-López, C., Fichtinger, G. (eds.) MICCAI 2018. LNCS, vol. 11073, pp. 3–11. Springer, Cham (2018). https://doi.org/10.1007/978-3-030-00937-3_1
12. Nair, T., Precup, D., Arnold, D.L., Arbel, T.: Exploring uncertainty measures in deep networks for multiple sclerosis lesion detection and segmentation. Med. Image Anal. **59**, 101557 (2020)
13. Dosovitskiy, A., et al.: An image is worth 16 × 16 words: transformers for image recognition at scale. arXiv preprint arXiv:2010.11929 (2020)

14. Hatamizadeh, A., et al.: UNETR: transformers for 3D medical image segmentation. In: Proceedings of the IEEE/CVF Winter Conference on Applications of Computer Vision, pp. 574–584 (2022)
15. Ronneberger, O., Fischer, P., Brox, T.: U-Net: convolutional networks for biomedical image segmentation. In: Navab, N., Hornegger, J., Wells, W.M., Frangi, A.F. (eds.) MICCAI 2015. LNCS, vol. 9351, pp. 234–241. Springer, Cham (2015). https://doi.org/10.1007/978-3-319-24574-4_28
16. Qin, X., Zhang, Z., Huang, C., Dehghan, M., Zaiane, O.R., Jagersand, M.: U2-Net: going deeper with nested U-structure for salient object detection. Pattern Recogn. **106**, 107404 (2020)
17. Sedai, S., et al.: Uncertainty guided semi-supervised segmentation of retinal layers in OCT images. In: Shen, D., et al. (eds.) MICCAI 2019. LNCS, vol. 11764, pp. 282–290. Springer, Cham (2019). https://doi.org/10.1007/978-3-030-32239-7_32
18. Gu, Z., et al.: CE-Net: context encoder network for 2D medical image segmentation. IEEE Trans. Med. Imaging **38**(10), 2281–2292 (2019)
19. Mou, L., et al.: CS-Net: channel and spatial attention network for curvilinear structure segmentation. In: Shen, D., et al. (eds.) MICCAI 2019. LNCS, vol. 11764, pp. 721–730. Springer, Cham (2019). https://doi.org/10.1007/978-3-030-32239-7_80
20. Chen, J., et al.: TransUNet: transformers make strong encoders for medical image segmentation. arXiv preprint arXiv:2102.04306 (2021)

QCResUNet: Joint Subject-Level and Voxel-Level Prediction of Segmentation Quality

Peijie Qiu[1(✉)], Satrajit Chakrabarty[2], Phuc Nguyen[3],
Soumyendu Sekhar Ghosh[2], and Aristeidis Sotiras[1,4]

[1] Mallinckrodt Institute of Radiology, Washington University School of Medicine,
St. Louis, MO, USA
`peijie.qiu@wustl.edu`
[2] Department of Electrical and Systems Engineering,
Washington University in St. Louis, St. Louis, MO, USA
[3] Department of Biomedical Engineering, University of Cincinnati,
Cincinnati, OH, USA
[4] Institute for Informatics, Data Science and Biostatistics,
Washington University School of Medicine, St. Louis, MO, USA

Abstract. Deep learning has achieved state-of-the-art performance in automated brain tumor segmentation from magnetic resonance imaging (MRI) scans. However, the unexpected occurrence of poor-quality outliers, especially in out-of-distribution samples, hinders their translation into patient-centered clinical practice. Therefore, it is important to develop automated tools for large-scale segmentation quality control (QC). However, most existing QC methods targeted cardiac MRI segmentation which involves a single modality and a single tissue type. Importantly, these methods only provide a subject-level segmentation-quality prediction, which cannot inform clinicians where the segmentation needs to be refined. To address this gap, we proposed a novel network architecture called QCResUNet that simultaneously produces segmentation-quality measures as well as voxel-level segmentation error maps for brain tumor segmentation QC. To train the proposed model, we created a wide variety of segmentation-quality results by using i) models that have been trained for a varying number of epochs with different modalities; and ii) a newly devised segmentation-generation method called SegGen. The proposed method was validated on a large public brain tumor dataset with segmentations generated by different methods, achieving high performance on the prediction of segmentation-quality metric as well as voxel-wise localization of segmentation errors. The implementation will be publicly available at https://github.com/peijie-chiu/QC-ResUNet.

Keywords: Automatic quality control · Brain tumor segmentation · Deep Learning

H. Greenspan et al. (Eds.): MICCAI 2023, LNCS 14223, pp. 173–182, 2023.
https://doi.org/10.1007/978-3-031-43901-8_17

1   Introduction

Gliomas are the most commonly seen central nervous system malignancies with aggressive growth and low survival rates [19]. Accurate multi-class segmentation of gliomas in multimodal magnetic resonance imaging (MRI) plays an indispensable role in quantitative analysis, treatment planning, and monitoring of progression and treatment. Although deep learning-based methods have achieved state-of-the-art performance in automated brain tumor segmentation [6–8,14], their performance often drops when tasked with segmenting out-of-distribution samples and poor-quality artifactual images. However, segmentations of desired quality are required to reliably drive treatment decisions and facilitate clinical management of gliomas. Therefore, tools for automated quality control (QC) are essential for the clinical translation of automated segmentation methods. Such tools can enable a streamlined clinical workflow by identifying catastrophic segmentation failures, informing clinical experts where the segmentations need to be refined, and providing a quantitative measure of quality that can be taken into account in downstream analyses.

Most previous studies of segmentation QC only provide subject-level quality assessment by either directly predicting segmentation-quality metrics or their surrogates. Specifically, Wang et al. [18] leveraged a variational autoencoder to learn the latent representation of good-quality image-segmentation pairs in the context of cardiac MRI segmentation. During the inference, an iterative search scheme was performed in the latent space to find a surrogate segmentation. This segmentation is assumed to be a good proxy of the (unknown) ground-truth segmentation of the query image, and can thus be compared to the at-hand predicted segmentation to estimate its quality. Another approach that takes advantage of the pairs of images and ground-truth segmentation is the reverse classification accuracy (RCA) framework [13,17]. In this framework, the test image is registered to a preselected reference dataset with known ground-truth segmentation. The quality of a query segmentation is assessed by warping the query image to the reference dataset. However, these methods primarily targeted QC of cardiac MRI segmentation, which involves a single imaging modality and a single tissue type with a welch-characterized location and appearance. In contrast, brain tumor segmentation involves the delineation of heterogeneous tumor regions, which are manifested through intensity changes relative to the surrounding healthy tissue across multiple modalities. Importantly, there is significant variability in brain tumor appearances, including multifocal masses and complex shapes with heterogeneous textures. Consequently, adapting approaches for automated QC of cardiac segmentation to brain tumor segmentation is challenging. Additionally, iterative search or registration during inference makes the existing methods computationally expensive and time-consuming, which limits their applicability in large-scale segmentation QC.

Multiple studies have also explored regression-based methods to directly predict segmentation-quality metrics, e.g., Dice Similarity Coefficient (DSC). For example, Kohlberger et al. [10] used Support Vector Machine (SVM) with hand-crafted features to detect cardiac MRI segmentation failures. Robinson et al. [12]

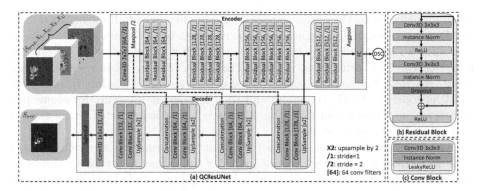

Fig. 1. (a) The proposed QCResUNet model adopts an encoder-decoder neural network architecture that takes four modalities and the segmentation to be evaluated. Given this input, QCResUNet predicts the DSC and segmentation error map. (b) The residual block in the encoder. (c) The convolutional block in the decoder.

proposed a convolutional neural network (CNN) to automatically extract features from segmentations generated by a series of Random Forest segmenters to predict DSC for cardiac MRI segmentation. Kofler et al. [9] proposed a CNN to predict holistic ratings of segmentations, which were annotated by neuroradiologists, with the goal of better emulating how human experts. Though these regression-based methods are advantageous for fast inference, they do not provide voxel-level localization of segmentation failures, which can be crucial for both auditing purposes and guiding manual refinements.

In summary, while numerous efforts have been devoted to segmentation QC, most works were in the context of cardiac MRI segmentation with few works tackling segmentation QC of brain tumors, which have more complex and heterogeneous appearances than the heart. Furthermore, most of the existing methods do not localize segmentation errors, which is meaningful for both auditing purposes and guiding manual refinement. To address these challenges, we propose a novel framework for joint subject-level and voxel-level prediction of segmentation quality from multimodal MRI. The contribution of this work is four-fold. First, we proposed a predictive model (QCResUNet) that simultaneously predicts DSC and localizes segmentation errors at the voxel level. Second, we devised a data-generation approach, called SegGen, that generates a wide range of segmentations of varying quality, ensuring unbiased model training and testing. Third, our end-to-end predictive model yields fast inference. Fourth, the proposed method achieved a good performance in predicting subject-level segmentation quality and identifying voxel-level segmentation failures.

2   Method

Given four imaging modalities denoted as $[X_1, X_2, X_3, X_4]$ and a predicted multi-class brain tumor segmentation mask (S_{pred}), the goal of our approach is to automatically assess the tumor segmentation quality by simultaneously predicting DSC and identifying segmentation errors as a binary mask (S_{err}). Toward this end, we proposed a 3D encoder-decoder architecture termed QCResUNet (see Fig. 1(a)) for simultaneously predicting DSC and localizing segmentation errors. QCResUNet has two parts trained in an end-to-end fashion: i) a ResNet-34 [4] encoder for DSC prediction; and ii) a decoder architecture for segmentation error map prediction (i.e., the difference between predicted segmentation and ground-truth segmentation).

The ResNet-34 encoder enables the extraction of semantically rich features that are useful for characterizing the quality of the segmentation. We maintained the main structure of the vanilla 2D ResNet-34 [4] but made the following modifications, which were necessary to account for the 3D nature of the input data (see Fig. 1(b)). First, all the 2D convolutional layers and pooling layers in the vanilla ResNet were changed to 3D. Second, the batch normalization [5] was replaced by instance normalization [16] to accommodate the small batch size in 3D model training. Third, spatial dropout [15] with a probability of 0.3 was added to each residual block to prevent overfitting.

The building block of the decoder consisted of an upsampling by a factor of two, which was implemented by a nearest neighbor interpolation in the feature map, followed by two convolutional blocks that halve the number of feature maps. Each convolutional block comprised a $3 \times 3 \times 3$ convolutional layer followed by an instance normalization layer and a leaky ReLU activation [11] (see Fig. 1(c)). The output of each decoder block was concatenated with features from the corresponding encoder level to facilitate information flow from the encoder to the decoder. Compared to the encoder, we used a shallower decoder with fewer parameters to prevent overfitting and reduce computational complexity.

The objective function for training QCResUNet consists of two parts. The first part corresponds to the DSC regression task. It consists of a mean absolute error (MAE) loss ($\mathcal{L}_{MAE}$) term that penalizes differences between ground truth (DSC_{gt}) and predicted DSC (DSC_{pred}):

$$\mathcal{L}_{MAE} = \frac{1}{N} \sum_{n=1}^{N} |DSC_{gt}^{(n)} - DSC_{pred}^{(n)}|_1, \tag{1}$$

where N denotes the number of samples in a batch. The second part of the objective function corresponds to the segmentation error prediction. It consists of a dice loss [3] and a binary cross-entropy loss, given by:

$$\mathcal{L}_{dice} = -\frac{2 \cdot \sum_{i=1}^{I} S_{err_{gt}}^{(i)} \cdot S_{err_{pred}}^{(i)}}{\sum_{i=1}^{I} S_{err_{gt}}^{(i)} + \sum_{i=1}^{I} S_{err_{pred}}^{(i)}}$$

$$\mathcal{L}_{BCE} = -\frac{1}{I} \sum_{i=1}^{I} S_{err_{gt}}^{(i)} \log S_{err_{pred}}^{(i)} + (1 - S_{err_{gt}}^{(i)}) \log(1 - S_{err_{pred}}^{(i)}), \tag{2}$$

where $S_{err_{gt}}$, $S_{err_{pred}}$ denote the binary ground-truth segmentation error map and the predicted error segmentation map from the sigmoid output of the

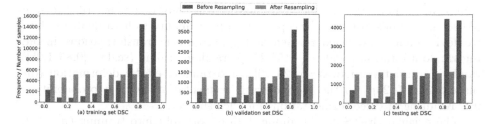

Fig. 2. DSC distribution of the generated dataset before and after resampling. (a), (b) and (c) are the DSC distribution for the training, validation, and testing set.

decoder, respectively. The dice loss and cross-entropy loss were averaged across the number of pixels I in a batch. The two parts are combined using a weight parameter λ to balance the different loss components:

$$\mathcal{L}_{total} = \mathcal{L}_{MAE} + \lambda \left(\mathcal{L}_{dice} + \mathcal{L}_{BCE} \right). \tag{3}$$

3 Experiments

For this study, pre-operative multimodal MRI scans of varying grades of glioma were obtained from the 2021 Brain Tumor Segmentation (BraTS) challenge [1] training dataset ($n = 1251$). For each subject, four modalities viz. pre-contrast T1-weighted (T1), T2-weighted (T2), post-contrast T1-weighted (T1c), and Fluid attenuated inversion recovery (FLAIR) are included in the dataset. It also included expert-annotated multi-class tumor segmentation masks comprising enhancing tumor (ET), necrotic tumor core (NCR), and edema (ED) classes. All data were already registered to a standard anatomical atlas and skull-stripped. The skull-stripped scans were then z-scored to zero mean and unit variance. All the data was first cropped to non-zero value regions, and then zero-padded to a size of $160 \times 192 \times 160$ to be fed into the network.

3.1 Data Generation

The initial dataset was expanded by producing segmentation results at different levels of quality to provide an unbiased estimation of segmentation quality. To this end, we adopted a three-step approach. First, a nnUNet framework [6] was adopted and trained five times separately using different modalities as input (i.e., T1-only, T1c-only, T2-only, FLAIR-only, and all four modalities). As only certain tissue-types are captured in each modality (e.g., enhancing tumor is captured well in T1c but not in FLAIR), this allowed us to generate segmentations of a wide range of qualities. nnUNet was selected for this purpose due to its wide success in brain tumor segmentation tasks. Second, to further enrich our dataset with segmentations of diverse quality, we sampled segmentations along the training routines at different iterations. A small learning rate (1×10^{-6}) was chosen in training all the models to slower their convergence in order to sample segmentations gradually sweeping from poor quality to high quality. Third, we devised a method called SegGen that applied image transformations, including random rotation (angle $= [-15°, 15°]$), random scaling (scale $= [0.85, 1.25]$), random translation (moves $= [-20, 20]$), and random elastic deformation (displacement $= [0, 20]$), to the ground-truth segmentations with a probability of 0.5, resulting in three segmentations for each subject.

The original BraTS 2021 training dataset was split into training ($n = 800$), validation ($n = 200$), and testing ($n = 251$) sets. After applying the three-step approach, it resulted in 48000, 12000, and 15060 samples for the three sets, respectively. However, this generated dataset suffered from imbalance (Fig. 2(a), (b), and (c)) because the CNN models could segment most of the cases correctly. Training using such an imbalanced dataset is prone to producing biased models that do not generalize well. To mitigate this issue, we proposed a resampling strategy during the training to make the DSC more uniformly distributed. Specifically, we used the Quantile transform to map the distribution of a variable to a target distribution by randomly smoothing out the samples unrelated to the target distribution. Using the Quantile transform, the data generator first transformed the distribution of the generated DSC to a uniform distribution. Next, the generated samples closest to the transformed uniform distribution in terms of Euclidean distance were chosen to form the resampled dataset. After applying our proposed resampling strategy, the DSC in the training and validation set approached a uniform distribution (Fig. 2(a), (b), and (c)). The total number of samples before and after resampling remained the same with repeating samples. We kept the resampling stochastic at each iteration during training to make all the generated samples seen by the model. The generated testing set was also resampled to perform an unbiased estimation of the quality at different levels resulting in 4895 samples.

In addition to the segmentations generated by the nnUNet framework and the SegGen method, we also generated out-of-distribution segmentation samples for the testing set to validate the generalizability of our proposed model. For this purpose, five models were trained on the training set using the DeepMedic framework [8] with different input modalities (i.e., T1-only, T1c-only, T2-only, FLAIR-only, and all four modalities). This resulted in $251 \times 5 = 1255$ out-of-distribution samples in the testing set.

3.2   Experimental Design

Baseline Methods: In this study, we compared the performance of the proposed model to three baseline models: (i) a UNet model [14], (ii) a ResNet-34 [4], and (iii) the ReNet-50 model used by Robinson et al. [12]. For a fair comparison, the residual blocks in the ResNet-34 and ResNet-50 were the same as that in the QCResUNet. We added an average pooling followed by a fully-connected layer to the last feature map of the UNet to predict a single DSC value. The evaluation was conducted on in-sample (nnUNet and SegGen) and out-of-sample segmentations generated by DeepMedic.

Table 1. The QC performance of three baseline methods and the proposed method was evaluated on in-sample (nnUNet and SegGen) and out-of-sample (DeepMedic) segmentations. The best metrics in each column are highlighted in bold. DSC_{err} denotes the median DSC between $S_{err_{gt}}$ and $S_{err_{pred}}$ across all samples.

Method	In-sample (nnUNet+SegGen)			Out-of-sample (DeepMedic)		
	r	MAE	DSC_{err}	r	MAE	DSC_{err}
UNet	0.895	0.1106 ± 0.085	–	0.923	0.0835 ± 0.063	–
ResNet34	0.942	0.0711 ± 0.063	–	0.942	0.0751 ± 0.065	–
ResNet50 [12]	0.944	0.0662 ± 0.065	–	0.943	0.0782 ± 0.067	–
QCResUNet(Ours)	**0.964**	**0.0570 ± 0.050**	**0.834**	**0.966**	**0.0606 ± 0.049**	**0.867**

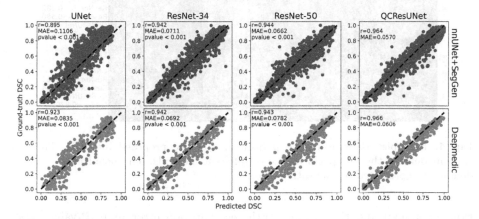

Fig. 3. Comparison of QC performance between three baseline methods (UNet, ResNet-34, ResNet-50) and the proposed method (QCResUNet) for segmentations generated using nnUNet and SegGen (top row) as well as DeepMedic (bottom row).

Training Procedure: All models were trained for 150 epochs using an Adam optimizer with a L_2 weight decay of 5×10^{-4}. The batch size was set to 4. Data augmentation, including random rotation, random scaling, random mirroring, random Gaussian noise, and Gamma intensity correction, was applied to prevent overfitting during training. We performed a random search [2] to determine the optimal hyperparameters (i.e., initial learning rate and loss weight balance parameter λ) on the training and validation set. The hyperparameters that yielded the best results were $\lambda = 1$ and an initial learning rate of 1×10^{-4}. The learning rate was exponentially decayed by a factor of 0.9 at each epoch until 1×10^{-6}. Model training was performed on four NVIDIA Tesla A100 and V100S GPUs. The proposed method was implemented in PyTorch v1.12.1.

Statistical Analysis: We assessed the performance of the subject-level segmentation quality prediction in terms of Pearson coefficient r and MAE between the predicted DSC and the ground-truth DSC. The performance of the segmentation error localization was assessed by the DSC_{err} between the predicted segmenta-

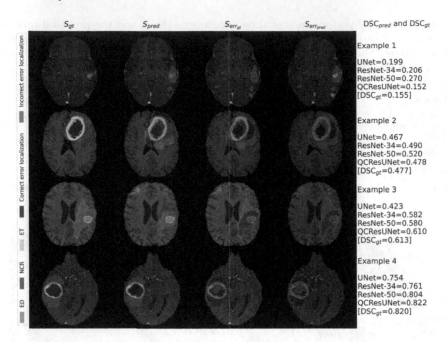

Fig. 4. Examples showcasing the performance of the proposed methods. The last column denotes the QC performance of different methods. The penultimate column denotes the predicted segmentation error.

tion error map and the ground-truth segmentation error map. P-values were computed using a paired t-test between DSC predicted by QCResUNet versus ones predicted by corresponding baselines.

4 Results

The proposed QCResUNet achieved good performance in predicting subject-level segmentation quality for in-sample (MAE $= 0.0570$, $r = 0.964$) and out-of-sample (MAE $= 0.0606$, $r = 0.966$) segmentations. The proposed method also showed statistically significant improvement against all three baselines (Table 1 and Fig. 3). We found that the DSC prediction error (MAE) of the proposed method was distributed more evenly across different levels of quality than all baselines (see Fig. 3) with a smaller standard deviation of 0.050 for in-sample segmentations and 0.049 for out-of-sample segmentations. A possible explanation is that the joint training of predicting subject-level and voxel-level quality enabled the QCResUNet to learn deep features that better characterize the segmentation quality. For the voxel-level segmentation error localization task, the model achieved a median DSC of 0.834 for in-sample segmentations and 0.867 for out-of-sample segmentations. This error localization is not provided by any of the baselines and enables QCResUnet to track segmentation failures at different levels of segmentation quality (Fig. 4).

5 Conclusion

In this work, we proposed a novel CNN architecture called QCResUNet to perform automatic brain tumor segmentation QC in multimodal MRI scans. QCResUNet simultaneously provides subject-level segmentation-quality prediction and localizes segmentation failures at the voxel level. It achieved superior DSC prediction performance compared to all baselines. In addition, the ability to localize segmentation errors has the potential to guide the refinement of predicted segmentations in a clinical setting. This can significantly expedite clinical workflows, thus improving the overall clinical management of gliomas.

Acknowledgements. All computations were supported by the Washington University Center for High Performance Computing, which was partially funded by NIH grants S10OD025200, 1S10RR022984-01A1, and 1S10OD018091-01.

References

1. Baid, U., et al.: The RSNA-ASNR-MICCAI BraTS 2021 benchmark on brain tumor segmentation and radiogenomic classification. arXiv preprint arXiv:2107.02314 (2021)
2. Bergstra, J., Bengio, Y.: Random search for hyper-parameter optimization. J. Mach. Learn. Res. **13**(2) (2012)
3. Drozdzal, M., Vorontsov, E., Chartrand, G., Kadoury, S., Pal, C.: The importance of skip connections in biomedical image segmentation. In: Carneiro, G., et al. (eds.) LABELS/DLMIA -2016. LNCS, vol. 10008, pp. 179–187. Springer, Cham (2016). https://doi.org/10.1007/978-3-319-46976-8_19
4. He, K., Zhang, X., Ren, S., Sun, J.: Deep residual learning for image recognition. In: Proceedings of the IEEE Conference on Computer Vision and Pattern Recognition, pp. 770–778 (2016)
5. Ioffe, S., Szegedy, C.: Batch normalization: accelerating deep network training by reducing internal covariate shift. In: International Conference on Machine Learning, pp. 448–456 (2015)
6. Isensee, F., Jaeger, P.F., Kohl, S.A., Petersen, J., Maier-Hein, K.H.: nnU-Net: a self-configuring method for deep learning-based biomedical image segmentation. Nat. Methods **18**(2), 203–211 (2021)
7. Isensee, F., Kickingereder, P., Wick, W., Bendszus, M., Maier-Hein, K.H.: Brain tumor segmentation and radiomics survival prediction: contribution to the BRATS 2017 challenge. In: Crimi, A., Bakas, S., Kuijf, H., Menze, B., Reyes, M. (eds.) BrainLes 2017. LNCS, vol. 10670, pp. 287–297. Springer, Cham (2018). https://doi.org/10.1007/978-3-319-75238-9_25
8. Kamnitsas, K., et al.: Efficient multi-scale 3D CNN with fully connected CRF for accurate brain lesion segmentation. Med. Image Anal. **36**, 61–78 (2017)
9. Kofler, F., et al.: Deep quality estimation: creating surrogate models for human quality ratings. arXiv preprint arXiv:2205.10355 (2022)
10. Kohlberger, T., Singh, V., Alvino, C., Bahlmann, C., Grady, L.: Evaluating segmentation error without ground truth. In: Ayache, N., Delingette, H., Golland, P., Mori, K. (eds.) MICCAI 2012. LNCS, vol. 7510, pp. 528–536. Springer, Heidelberg (2012). https://doi.org/10.1007/978-3-642-33415-3_65

11. Maas, A.L., Hannun, A.Y., Ng, A.Y., et al.: Rectifier nonlinearities improve neural network acoustic models. In: Proceedings of the ICML, Atlanta, Georgia, USA, vol. 30, p. 3 (2013)

12. Robinson, R., et al.: Real-time prediction of segmentation quality. In: Frangi, A.F., Schnabel, J.A., Davatzikos, C., Alberola-López, C., Fichtinger, G. (eds.) MICCAI 2018. LNCS, vol. 11073, pp. 578–585. Springer, Cham (2018). https://doi.org/10.1007/978-3-030-00937-3_66

13. Robinson, R., et al.: Automated quality control in image segmentation: application to the UK Biobank cardiovascular magnetic resonance imaging study. J. Cardiovasc. Magn. Reson. **21**(1), 1–14 (2019)

14. Ronneberger, O., Fischer, P., Brox, T.: U-Net: convolutional networks for biomedical image segmentation. In: Navab, N., Hornegger, J., Wells, W.M., Frangi, A.F. (eds.) MICCAI 2015. LNCS, vol. 9351, pp. 234–241. Springer, Cham (2015). https://doi.org/10.1007/978-3-319-24574-4_28

15. Tompson, J., Goroshin, R., Jain, A., LeCun, Y., Bregler, C.: Efficient object localization using convolutional networks. In: Proceedings of the IEEE Conference on Computer Vision and Pattern Recognition, pp. 648–656 (2015)

16. Ulyanov, D., Vedaldi, A., Lempitsky, V.: Instance normalization: the missing ingredient for fast stylization. arXiv preprint arXiv:1607.08022 (2016)

17. Valindria, V.V., et al.: Reverse classification accuracy: predicting segmentation performance in the absence of ground truth. IEEE Trans. Med. Imaging **36**(8), 1597–1606 (2017)

18. Wang, S., et al.: Deep generative model-based quality control for cardiac MRI segmentation. In: Martel, A.L., et al. (eds.) MICCAI 2020. LNCS, vol. 12264, pp. 88–97. Springer, Cham (2020). https://doi.org/10.1007/978-3-030-59719-1_9

19. Zhuge, Y., et al.: Brain tumor segmentation using holistically nested neural networks in MRI images. Med. Phys. **44**(10), 5234–5243 (2017)

Consistency-Guided Meta-learning for Bootstrapping Semi-supervised Medical Image Segmentation

Qingyue Wei[1], Lequan Yu[2], Xianhang Li[3], Wei Shao[4], Cihang Xie[3], Lei Xing[1], and Yuyin Zhou[3(✉)]

[1] Stanford University, Stanford, CA, USA
[2] The University of Hong Kong, Pok Fu Lam, Hong Kong, China
[3] University of California, Santa Cruz, CA, USA
yzhou284@ucsc.edu
[4] University of Florida, Gainesville, FL, USA

Abstract. Medical imaging has witnessed remarkable progress but usually requires a large amount of high-quality annotated data which is time-consuming and costly to obtain. To alleviate this burden, semi-supervised learning has garnered attention as a potential solution. In this paper, we present **Meta-Learning for Bootstrapping Medical Image Segmentation** (MLB-Seg), a novel method for tackling the challenge of semi-supervised medical image segmentation. Specifically, our approach first involves training a segmentation model on a small set of clean labeled images to generate initial labels for unlabeled data. To further optimize this bootstrapping process, we introduce a per-pixel weight mapping system that dynamically assigns weights to both the initialized labels and the model's own predictions. These weights are determined using a meta-process that prioritizes pixels with loss gradient directions closer to those of clean data, which is based on a small set of precisely annotated images. To facilitate the meta-learning process, we additionally introduce a consistency-based Pseudo Label Enhancement (PLE) scheme that improves the quality of the model's own predictions by ensembling predictions from various augmented versions of the same input. In order to improve the quality of the weight maps obtained through multiple augmentations of a single input, we introduce a mean teacher into the PLE scheme. This method helps to reduce noise in the weight maps and stabilize its generation process. Our extensive experimental results on public atrial and prostate segmentation datasets demonstrate that our proposed method achieves state-of-the-art results under semi-supervision. Our code is available at https://github.com/aijinrjinr/MLB-Seg.

Keywords: semi-supervised learning · meta-learning · medical image segmentation

---

Supplementary Information The online version contains supplementary material available at https://doi.org/10.1007/978-3-031-43901-8_18.

1   Introduction

Reliable and robust segmentation of medical images plays a significant role in clinical diagnosis [12]. In recent years, deep learning has led to significant progress in image segmentation tasks [6,26]. However, training these models [23] requires large-scale image data with precise pixel-wise annotations, which are usually time-consuming and costly to obtain. To address this challenge, recent studies have explored semi-supervised learning approaches for medical image segmentation, leveraging unlabeled data to enhance performance [1,21]. Semi-supervised learning (SSL) commonly uses two methods: pseudo labeling and consistency regularization. Pseudo labeling uses a model's predictions on unlabeled data to create "pseudo labels" and augment the original labeled data set, allowing for improved accuracy and reduced cost in training [2,7,10]. Despite its potential benefits, the pseudo labeling approach may pose a risk to the accuracy of the final model by introducing inaccuracies into the training data. Consistency regularization, on the other hand, aims to encourage the model's predictions to be consistent across different versions of the same input. UA-MT [22] and DTC [11] are examples of consistency regularization-based methods that use teacher-student models for segmentation and emphasize dual consistency between different representations, respectively.

Combining both methods can potentially lead to improved performance, as pseudo labeling leverages unlabeled data and consistency regularization improves the model's robustness [3,20]. However, pseudo labeling still inevitably introduces inaccuracies that can hinder the model's performance. To overcome this challenge, we propose a novel consistency-guided meta-learning framework called **M**eta-**L**earning for **B**ootstrapping Medical Image **Seg**mentation (MLB-Seg). Our approach uses pixel-wise weights to adjust the importance of each pixel in the initialized labels and pseudo labels during training. We learn these weights through a meta-process that prioritizes pixels with loss gradient direction closer to those of clean data, using a small set of clean labeled images. To further improve the quality of the pseudo labels, we introduce a consistency-based Pseudo Label Enhancement (PLE) scheme that ensembles predictions from augmented versions of the same input. To address the instability issue arising from using data augmentation, we incorporate a mean-teacher model to stabilize the weight map generation from the student meta-learning model, which leads to improved performance and network robustness. Our proposed approach has been extensively evaluated on two benchmark datasets, LA [4] and PROMISE12 [9], and has demonstrated superior performance compared to existing methods.

2   Method

We present MLB-Seg, a novel consistency-guided meta-learning framework for semi-supervised medical image segmentation. Assume that we are given a training dataset consisting of clean data $D_c = \{(x_i^c, y_i^c)\}_{i=1}^N$, and unlabeled data $D_u = \{(x_k^u)\}_{k=1}^K$ ($N \ll K$), where the input image x_i^c, x_k^u are of size $H \times W$ with the corresponding clean ground-truth mask y_i^c.

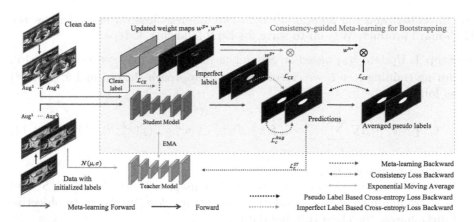

Fig. 1. Schematic of the proposed MLB-Seg. Weight maps w^{n*}, w^{p*} associated with the initialized labels and pseudo labels are meta-learned and optimization is further improved by enhancing the pseudo label estimation. A mean teacher model is used to provide guidance for stabilizing the weight meta-update in the student model.

2.1   Meta-learning for Bootstrapping

We first estimate labels for all unlabeled data using the baseline model which is trained on the clean data, denoted as follows

$$y_k^{\tilde{n}} = f_c(x_k^u; \theta_c), \tag{1}$$

where $f_c(:; \theta_c)$ denotes the trained model parameterized by θ_c and $k = 1, 2, ..., K$. We further denote them as $\widetilde{D_n} = \{(x_j^{\tilde{n}}, y_j^{\tilde{n}})\}_{j=1}^K$.

We then develop a novel meta-learning model for medical image segmentation, which learns from the clean set D_c to bootstrap itself up by leveraging the learner's own predictions (*i.e.*, pseudo labels), called Meta-Learning for Bootstrapping (MLB). As shown in Fig. 1, by adaptively adjusting the contribution between the initialized and pseudo labels commensurately in the loss function, our method effectively alleviates the negative effects from the erroneous pixels. Specifically, at training step t, given a training batch from $\widetilde{D_n}$ that $S^n = \{(x_j^{\tilde{n}}, y_j^{\tilde{n}}), 1 \leq j \leq b_n\}$ and a clean training batch $S^c = \{(x_i^c, y_i^c), 1 \leq i \leq b_c\}$ where b_n, b_c are the batch size respectively. Our objective is:

$$\theta^*(w^n, w^p) = \arg\min_\theta \sum_{j=1}^K w_j^n \circ \mathcal{L}(f(x_j^{\tilde{n}}; \theta), y_j^{\tilde{n}}) + w_j^p \circ \mathcal{L}(f(x_j^{\tilde{n}}; \theta), y_j^p), \tag{2}$$

$$y_j^p = \arg\max_{c=1,...,C} f(x_j^{\tilde{n}}; \theta), \tag{3}$$

where y_j^p is the pseudo label generated by $f(x_j^{\tilde{n}}; \theta)$, $\mathcal{L}(\cdot)$ is the cross-entropy loss function, C is the number of classes (we set $C = 2$ throughout this paper), $w_j^n, w_j^p \in \mathbb{R}^{H \times W}$ are the weight maps used for adjusting the contribution between the initialized and the pseudo labels in two different loss terms. $\circ$ denotes the Hadamard product. We aim to solve for Eq. 2 following 3 steps:

- **Step 1:** Update $\hat{\theta}_{t+1}$ based on S^n and current weight map set. Following [15], during training step t, we calculate $\hat{\theta}_{t+1}$ to approach the optimal $\theta^*(w^n, w^p)$ as follows:

$$\hat{\theta}_{t+1} = \theta_t - \alpha\nabla\left(\sum_{j=1}^{b_n} w_j^n \circ \mathcal{L}(f(x_j^{\tilde{n}}; \theta), y_j^{\tilde{n}}) + w_j^p \circ \mathcal{L}(f(x_j^{\tilde{n}}; \theta), y_j^p))\right)\Big|_{\theta=\theta_t}, \quad (4)$$

where α represents the step size.

- **Step 2:** Generate the meta-learned weight maps w^{n*}, w^{p*} based on S^c and $\hat{\theta}_{t+1}$ by minimizing the standard cross-entropy loss in the meta-objective function over the clean training data:

$$w^{n*}, w^{p*} = \underset{w^n, w^p \geq 0}{\arg\min} \frac{1}{N} \sum_{i=1}^{N} \mathcal{L}(f(x_i^c; \theta^*(w^n, w^p)), y_i^c). \quad (5)$$

Note that here we restrict every element in $w^{n/p}$ to be non-negative to prevent potentially unstable training [15]. Such a meta-learned process yields weight maps which can better balance the contribution of the initialized and the pseudo labels, thus reducing the negative effects brought by the erroneous pixels. Following [15,25], we only apply one-step gradient descent of $w_j^{n/p}$ based on a small clean-label data set S^c, to reduce the computational expense. To be specific, at training step t, $w_j^{n/p}$ is first initialized as $\mathbf{0}$, then we estimate w^{n*}, w^{p*} as:

$$(w_{j,t}^n, w_{j,t}^p) = -\beta\nabla\left(\frac{1}{b_c} \sum_{i=1}^{b_c} \mathcal{L}(f(x_i^c; \hat{\theta}_{t+1}), y_i^c))\right)\Big|_{w_j^n, w_j^p = 0}, \quad (6)$$

$$\tilde{w}_{j,t}^{n_r,s} = \max(w_{j,t}^{n_r,s}, 0), \quad \tilde{w}_{j,t}^{p_r,s} = \max(w_{j,t}^{p_r,s}, 0), \quad (7)$$

$$\tilde{w}_{j,t}^{n_r,s} = \frac{\tilde{w}_{j,t}^{n_r,s}}{\sum_{j=1}^{b_n}\sum_{r,s}\tilde{w}_{j,t}^{n_r,s} + \epsilon}, \tilde{w}_{j,t}^{p_r,s} = \frac{\tilde{w}_{j,t}^{p_r,s}}{\sum_{j=1}^{b_n}\sum_{r,s}\tilde{w}_{j,t}^{p_r,s} + \epsilon}, \quad (8)$$

where β stands for the step size and $w_{j,t}^{n/p_r,s}$ indicates the value at r^{th} row, s^{th} column of $w_j^{n/p}$ at time t. Equation 7 is used to enforce all weights to be strictly non-negative. Then Eq. 8 is introduced to normalize the weights in a single training batch so that they sum up to one. Here, we add a small number ϵ to keep the denominator greater than 0.

- **Step 3:** The meta-learned weight maps are used to spatially modulate the pixel-wise loss to update θ_{t+1}:

$$\theta_{t+1} = \theta_t - \alpha\nabla\left(\sum_{j=1}^{b_n} \tilde{w}_{j,t}^n \circ \mathcal{L}(f(x_j^{\tilde{n}}; \theta), y_j^{\tilde{n}}) + \tilde{w}_{j,t}^p \circ \mathcal{L}(f(x_j^{\tilde{n}}; \theta), y_j^p))\right)\Big|_{\theta=\theta_t}. \quad (9)$$

2.2   Consistency-Based Pseudo Label Enhancement

To generate more reliable pseudo labels, we propose Pseudo Label Enhancement (PLE) scheme based on consistency, which enforces consistency across augmented versions of the same input. Specifically, we perform Q augmentations on the same input image and enhance the pseudo label by averaging the outputs of the Q augmented versions and the original input:

$$\widehat{y}_j^p = \arg\max_{c=1,...,C} \frac{1}{Q+1}\left(\sum_{q=1}^{Q} \tau_q^{-1}(f(x_j^{\tilde{n}^q};\theta)) + f(x_j^{\tilde{n}^o};\theta)\right), \tag{10}$$

where $f(x_j^{\tilde{n}^q};\theta)$ is the output of q-th augmented sample, $f(x_{N+j}^{\tilde{n}^o};\theta)$ is the output of the original input, and τ_q^{-1} means the corresponding inverse transformation of the q-th augmented sample. Meanwhile, to further increase the output consistency among all the augmented samples and original input, we introduce an additional consistency loss $\mathcal{L}_c^{Aug}$ to the learning objective:

$$\mathcal{L}_c^{Aug}(x_j^{\tilde{n}}) = \frac{2}{(Q+1)Q} \frac{1}{HW} \sum_{q,v} \sum_{r,s} ||f(x_j^{\tilde{n}^q};\theta)_{r,s} - \tau_q(\tau_v^{-1}(f(x_j^{\tilde{n}^v};\theta)))_{r,s}||^2, \tag{11}$$

where (r,s) denotes the pixel index. τ_q is the corresponding transformation to generate the q-th augmented sample. (q,v) denotes the pairwise combination among all augmented samples and the original input. The final loss is the mean square distance among all $\frac{(Q+1)Q}{2}$ pairs of combinations.

2.3   Mean Teacher for Stabilizing Meta-learned Weights

Using PLE alone can result in performance degradation with increasing numbers of augmentations due to increased noise in weight maps. This instability can compound during subsequent training iterations. To address this issue during meta-learning, we propose using a mean teacher model [17] with consistency loss (Eq. 13). The teacher network guides the student meta-learning model, stabilizing weight updates and resulting in more reliable weight maps. Combining PLE with the mean teacher model improves output robustness. The student model is used for meta-learning, while the teacher model is used for weight ensemble with Exponential Moving Average (EMA) [17] applied to update it. The consistency loss maximizes the similarity between the teacher and student model outputs, adding reliability to the student model and stabilizing the teacher model. For each input $x_j^{\tilde{n}}$ in the batch S^n, we apply disturbance to the student model input to become the teacher model input. Then a consistency loss $\mathcal{L}_c^{ST}$ is used to maximize the similarity between the outputs from the teacher model and student model, further increasing the student model's reliability while stabilizing the teacher model. Specifically, for each input $x_j^{\tilde{n}}$ in the batch S^n, then corresponding input of teacher model is

$$x_j^T = x_j^{\tilde{n}} + \gamma \mathcal{N}(\mu,\sigma), \tag{12}$$

where $\mathcal{N}(\mu, \sigma) \in \mathbb{R}^{H \times W}$ denotes the Gaussian distribution with μ as mean and σ as standard deviation. And γ is used to control the noise level. The consistency loss is implemented based on pixel-wise mean squared error (MSE) loss:

$$\mathcal{L}_c^{ST}(x_j^{\tilde{n}}) = \frac{1}{HW} \sum_{r,s} ||f(x_j^{\tilde{n}}; \theta_t^S)_{r,s} - f(x_j^T; \theta_t^T)_{r,s}||^2. \tag{13}$$

3   Experiments

3.1   Experimental Setup

Datasets. We evaluate our proposed method on two different datasets including 1) the left atrial (LA) dataset from the 2018 Atrial Segmentation Challenge [4] as well as 2) the Prostate MR Image Segmentation 2012 (PROMISE2012) dataset [9]. Specifically, LA dataset is split into 80 scans for training and 20 scans for evaluation following [22]. From the training set, 8 scans are randomly selected as the meta set. All 2D input images were resized to 144×144. For PROMISE2012, we randomly split 40/4/6 cases for training and 10 for evaluation (4 for validation, 6 for test) following [13]. We randomly pick 3 cases from the training set as the meta set and resize all images to 144×144. We evaluate our segmentation performances using four metrics: the Dice coefficient, Jaccard Index (JI), Hausdorff Distance (HD), and Average Surface Distance (ASD).

Implementation Details. All of our experiments are based on 2D images. We adopt UNet++ as our baseline. Network parameters are optimized by SGD setting the learning rate at 0.005, momentum to be 0.9 and weight decay as 0.0005. The exponential moving average (EMA) decay rate is set as 0.99 following [17]. For the label generation process, we first train with all clean labeled data for 30 epochs with batch size set as 16. We then use the latest model to generate labels for unlabeled data. Next, we train our MLB-Seg for 100 epochs.

3.2   Results Under Semi-supervision

To illustrate the effectiveness of MLB-Seg under semi-supervision. We compare our method with the baseline (UNet++ [26]) and previous semi-supervised methods on LA dataset (Table 1) and PROMISE12 (Table 2), including an adversarial learning method [24], consistency based methods [5,11,18,20,22], an uncertainty-based strategy [1], and contrastive learning based methods [13,19]. For a fair comparison in Table 2, we also use UNet [16] as the backbone and resize images to 256×256, strictly following the settings in Self-Paced [13]. We split the evaluation set (10 cases) into 4 cases as the validation set, and 6 as the test set. We then select the best checkpoint based on the validation set and report the results on the test set. As shown in Table 1 and 2, our MLB-Seg outperforms recent state-of-the-art methods on both PROMISE12 (under different combinations of backbones and image sizes) and the LA dataset across almost all evaluation measures.

Table 1. Comparison with existing methods under semi-supervision on LA dataset.

Method	Dice (%)↑	JI (%)↑	HD (voxel)↓	ASD (voxel)↓
UNet++ [26]	81.33	70.87	14.79	4.62
DAP [24]	81.89	71.23	15.81	3.80
UA-MT [22]	84.25	73.48	13.84	3.36
SASSNet [8]	87.32	77.72	9.62	2.55
LG-ER-MT [5]	85.54	75.12	13.29	3.77
DUWM [18]	85.91	75.75	12.67	3.31
DTC [11]	86.57	76.55	14.47	3.74
MC-Net [20]	87.71	78.31	9.36	**2.18**
Uncertainty-Based [1]	86.58	76.34	11.82	–
MLB-Seg	**88.69**	**79.86**	**8.99**	2.61

Table 2. Comparison with existing methods under semi-supervision on PROMISE12. All methods included in the comparison have been re-implemented using our data split to ensure a fair evaluation, with the provided source codes.

Method	Dice (%)↑
UNet++ [26]	68.85
UA-MT [22]	65.05
DTC [11]	63.44
SASSNet [8]	73.43
MC-Net [20]	72.66
SS-Net [19]	73.19
Self-Paced [13] (UNet, 256)	74.02
MLB-Seg (UNet, 144)	76.41
MLB-Seg (UNet, 256)	76.15
MLB-Seg (UNet++, 144)	77.22
MLB-Seg (UNet++, 256)	**78.27**

3.3 Ablation Study

To explore how different components of our MLB-Seg contribute to the final result, we conduct the following experiments under semi-supervision on PROMISE12: 1) the bootstrapping method [14] (using fixed weights without applying meta-learning); 2) **MLB**, which only reweights the initialized labels and pseudo labels without applying PLE and mean teacher; 3) **MLB + mean teacher** which combines MLB with mean teacher scheme; 4) **MLB + PLE** which applies PLE strategy with MLB. When applying multiple data augmentations (*i.e.*, for $Q = 2, 4$), we find the best performing combinations are 2 × PLE (using one zoom in and one zoom out), 4 × PLE (using one zoom in and

Table 3. Ablation study on different components used in MLB-Seg based on PROMISE12.

MLB	mean teacher	2× PLE	4× PLE	Dice (%)↑	JI (%)↑	HD (voxel)↓	ASD (voxel)↓
				70.85	55.85	10.02	4.35
✓				73.97	59.71	8.49	3.49
✓	✓			73.76	59.37	8.04	3.00
✓		✓		75.01	60.92	**7.58**	2.70
✓			✓	73.10	58.45	9.42	3.57
✓	✓		✓	**76.68**	**63.14**	7.85	**2.64**

two zoom out and one flip for each input); 5) **MLB + PLE + mean teacher** which combines PLE, mean teacher with MLB simultaneously to help better understand how mean teacher will contribute to PLE. Our results are summarized in Table 3, which shows the effectiveness of our proposed components. The best results are achieved when all components are used.

To demonstrate how PLE combined with the mean teacher model help stabilize the meta-weight update, we compare the performance of MLB + PLE (w/mean teacher) with MLB + PLE + mean teacher under different augmentations (Q) on PROMISE12 dataset (See supplementary materials for details). We find out that for MLB + PLE (w/o mean teacher), performance improves from 74.34% to 74.99% when Q is increased from 1 to 2, but decreases significantly when $Q \geq 4$. Specifically, when Q reaches 4 and 6, the performance significant drops from 74.99% to 72.07% ($Q = 4$) and from 74.99% to 70.91% ($Q = 6$) respectively. We hypothesize that this is due to increased noise from initialized labels in some training samples, which can lead to instability in weight updates. To address this issue, we introduce the mean-teacher model [17] into PLE to stabilize weight map generation from the student meta-learning model. And MLB + PLE + mean teacher turns out to consistently improve the stability of meta-learning compared with its counterpart without using mean teacher, further validating the effectiveness of our method (see Supplementary for more examples). Specifically, for MLB + PLE + mean teacher, the performance reaches 76.63% (from 72.07%) when $Q = 4$, 75.84% (from 70.91%) when $Q = 6$.

Qualitative Analysis. To illustrate the benefits of MLB-Seg for medical image segmentation, we provide a set of qualitative examples in Fig. 2. In the visualization of weight maps of Fig. 2, the blue/purple represents for the initialized label in $y^{\tilde{n}}/y^p$, while the red indicates pixels in w^{p*} have higher values. We observe that MLB-Seg places greater emphasis on edge information. It is evident that higher weights are allotted to accurately predicted pseudo-labeled pixels that were initially mislabeled, which effectively alleviates the negative effects from erroneously initialized labels.

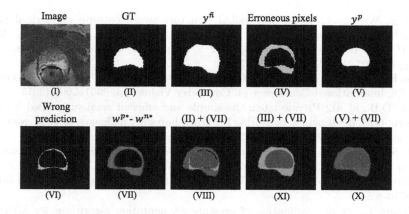

Fig. 2. Weight map visualization.

4 Conclusion

In this paper, we propose a novel meta-learning based segmentation method for medical image segmentation under semi-supervision. With few expert-level labels as guidance, our model bootstraps itself up by dynamically reweighting the contributions from initialized labels and its own outputs, thereby alleviating the negative effects of the erroneous voxels. In addition, we address an instability issue arising from the use of data augmentation by introducing a mean teacher model to stabilize the weights. Extensive experiments demonstrate the effectiveness and robustness of our method under semi-supervision. Notably, our approach achieves state-of-the-art results on both the LA and PROMISE12 benchmarks.

Acknowledgment. This work is supported by the Stanford 2022 HAI Seed Grant and National Institutes of Health 1R01CA256890 and 1R01CA275772.

References

1. Adiga Vasudeva, S., Dolz, J., Lombaert, H.: Leveraging labeling representations in uncertainty-based semi-supervised segmentation. In: Wang, L., Dou, Q., Fletcher, P.T., Speidel, S., Li, S. (eds.) MICCAI 2022. LNCS, vol. 13438, pp. 265–275. Springer, Cham (2022). https://doi.org/10.1007/978-3-031-16452-1_26
2. Bachman, P., Alsharif, O., Precup, D.: Learning with pseudo-ensembles. In: Advances in Neural Information Processing Systems, vol. 27 (2014)
3. Berthelot, D., Carlini, N., Goodfellow, I., Papernot, N., Oliver, A., Raffel, C.A.: MixMatch: a holistic approach to semi-supervised learning. In: Advances in Neural Information Processing Systems, vol. 32 (2019)
4. Chen, C., Bai, W., Rueckert, D.: Multi-task learning for left atrial segmentation on GE-MRI. In: Pop, M., et al. (eds.) STACOM 2018. LNCS, vol. 11395, pp. 292–301. Springer, Cham (2019). https://doi.org/10.1007/978-3-030-12029-0_32

5. Hang, W., et al.: Local and global structure-aware entropy regularized mean teacher model for 3D left atrium segmentation. In: Martel, A.L., et al. (eds.) MICCAI 2020. LNCS, vol. 12261, pp. 562–571. Springer, Cham (2020). https://doi.org/10.1007/978-3-030-59710-8_55

6. He, K., Gkioxari, G., Dollár, P., Girshick, R.: Mask R-CNN. In: Proceedings of the IEEE International Conference on Computer Vision, pp. 2961–2969 (2017)

7. Lee, D.H., et al.: Pseudo-label: the simple and efficient semi-supervised learning method for deep neural networks. In: Workshop on Challenges in Representation Learning, ICML, vol. 3, p. 896 (2013)

8. Li, S., Zhang, C., He, X.: Shape-aware semi-supervised 3D semantic segmentation for medical images. In: Martel, A.L., et al. (eds.) MICCAI 2020. LNCS, vol. 12261, pp. 552–561. Springer, Cham (2020). https://doi.org/10.1007/978-3-030-59710-8_54

9. Litjens, G., et al.: Evaluation of prostate segmentation algorithms for MRI: the PROMISE12 challenge. Med. Image Anal. **18**(2), 359–373 (2014)

10. Liu, F., Tian, Y., Chen, Y., Liu, Y., Belagiannis, V., Carneiro, G.: ACPL: anti-curriculum pseudo-labelling for semi-supervised medical image classification. In: Proceedings of the IEEE/CVF Conference on Computer Vision and Pattern Recognition, pp. 20697–20706 (2022)

11. Luo, X., Chen, J., Song, T., Wang, G.: Semi-supervised medical image segmentation through dual-task consistency. arXiv preprint arXiv:2009.04448 (2020)

12. Masood, S., Sharif, M., Masood, A., Yasmin, M., Raza, M.: A survey on medical image segmentation. Curr. Med. Imaging **11**(1), 3–14 (2015)

13. Peng, J., Wang, P., Desrosiers, C., Pedersoli, M.: Self-paced contrastive learning for semi-supervised medical image segmentation with meta-labels. In: Advances in Neural Information Processing Systems, vol. 34 (2021)

14. Reed, S., Lee, H., Anguelov, D., Szegedy, C., Erhan, D., Rabinovich, A.: Training deep neural networks on noisy labels with bootstrapping. arXiv preprint arXiv:1412.6596 (2014)

15. Ren, M., Zeng, W., Yang, B., Urtasun, R.: Learning to reweight examples for robust deep learning. In: International Conference on Machine Learning, pp. 4334–4343. PMLR (2018)

16. Ronneberger, O., Fischer, P., Brox, T.: U-Net: convolutional networks for biomedical image segmentation. In: Navab, N., Hornegger, J., Wells, W.M., Frangi, A.F. (eds.) MICCAI 2015. LNCS, vol. 9351, pp. 234–241. Springer, Cham (2015). https://doi.org/10.1007/978-3-319-24574-4_28

17. Tarvainen, A., Valpola, H.: Mean teachers are better role models: weight-averaged consistency targets improve semi-supervised deep learning results. In: Advances in Neural Information Processing Systems, vol. 30 (2017)

18. Wang, Y., et al.: Double-uncertainty weighted method for semi-supervised learning. In: Martel, A.L., et al. (eds.) MICCAI 2020. LNCS, vol. 12261, pp. 542–551. Springer, Cham (2020). https://doi.org/10.1007/978-3-030-59710-8_53

19. Wu, Y., Wu, Z., Wu, Q., Ge, Z., Cai, J.: Exploring smoothness and class-separation for semi-supervised medical image segmentation. In: Wang, L., Dou, Q., Fletcher, P.T., Speidel, S., Li, S. (eds.) MICCAI 2022. LNCS, vol. 13435, pp. 34–43. Springer, Cham (2022). https://doi.org/10.1007/978-3-031-16443-9_4

20. Wu, Y., Xu, M., Ge, Z., Cai, J., Zhang, L.: Semi-supervised left atrium segmentation with mutual consistency training. In: de Bruijne, M., et al. (eds.) MICCAI 2021. LNCS, vol. 12902, pp. 297–306. Springer, Cham (2021). https://doi.org/10.1007/978-3-030-87196-3_28

21. Xiang, J., Qiu, P., Yang, Y.: FUSSNet: fusing two sources of uncertainty for semi-supervised medical image segmentation. In: Wang, L., Dou, Q., Fletcher, P.T., Speidel, S., Li, S. (eds.) MICCAI 2022. LNCS, vol. 13438, pp. 481–491. Springer, Cham (2022). https://doi.org/10.1007/978-3-031-16452-1_46

22. Yu, L., Wang, S., Li, X., Fu, C.-W., Heng, P.-A.: Uncertainty-aware self-ensembling model for semi-supervised 3D left atrium segmentation. In: Shen, D., et al. (eds.) MICCAI 2019. LNCS, vol. 11765, pp. 605–613. Springer, Cham (2019). https://doi.org/10.1007/978-3-030-32245-8_67

23. Zhao, X., Wu, Y., Song, G., Li, Z., Zhang, Y., Fan, Y.: A deep learning model integrating FCNNs and CRFs for brain tumor segmentation. Med. Image Anal. **43**, 98–111 (2018)

24. Zheng, H., et al.: Semi-supervised segmentation of liver using adversarial learning with deep atlas prior. In: Shen, D., et al. (eds.) MICCAI 2019. LNCS, vol. 11769, pp. 148–156. Springer, Cham (2019). https://doi.org/10.1007/978-3-030-32226-7_17

25. Zhou, Y., et al.: Learning to bootstrap for combating label noise. arXiv preprint arXiv:2202.04291 (2022)

26. Zhou, Z., Rahman Siddiquee, M.M., Tajbakhsh, N., Liang, J.: UNct++: a nested U-Net architecture for medical image segmentation. In: Stoyanov, D., et al. (eds.) DLMIA/ML-CDS 2018. LNCS, vol. 11045, pp. 3–11. Springer, Cham (2018). https://doi.org/10.1007/978-3-030-00889-5_1

ACTION++: Improving Semi-supervised Medical Image Segmentation with Adaptive Anatomical Contrast

Chenyu You[1]([✉]), Weicheng Dai[2], Yifei Min[4], Lawrence Staib[1,2,3], Jas Sekhon[4,5], and James S. Duncan[1,2,3,4]

[1] Department of Electrical Engineering, Yale University, New Haven, USA
chenyu.you@yale.edu
[2] Department of Radiology and Biomedical Imaging,
Yale University, New Haven, USA
[3] Department of Biomedical Engineering, Yale University, New Haven, USA
[4] Department of Statistics and Data Science, Yale University, New Haven, USA
[5] Department of Political Science, Yale University, New Haven, USA

Abstract. Medical data often exhibits long-tail distributions with heavy class imbalance, which naturally leads to difficulty in classifying the minority classes (*i.e.*, boundary regions or rare objects). Recent work has significantly improved semi-supervised medical image segmentation in long-tailed scenarios by equipping them with unsupervised contrastive criteria. However, it remains unclear how well they will perform in the labeled portion of data where class distribution is also highly imbalanced. In this work, we present **ACTION++**, an improved contrastive learning framework with adaptive anatomical contrast for semi-supervised medical segmentation. Specifically, we propose an adaptive supervised contrastive loss, where we first compute the optimal locations of class centers uniformly distributed on the embedding space (*i.e.*, off-line), and then perform online contrastive matching training by encouraging different class features to adaptively match these distinct and uniformly distributed class centers. Moreover, we argue that blindly adopting a *constant* temperature τ in the contrastive loss on long-tailed medical data is not optimal, and propose to use a *dynamic* τ via a simple cosine schedule to yield better separation between majority and minority classes. Empirically, we evaluate ACTION++ on ACDC and LA benchmarks and show that it achieves state-of-the-art across two semi-supervised settings. Theoretically, we analyze the performance of adaptive anatomical contrast and confirm its superiority in label efficiency.

Keywords: Semi-Supervised Learning · Contrastive Learning · Imbalanced Learning · Long-tailed Medical Image Segmentation

Supplementary Information The online version contains supplementary material available at https://doi.org/10.1007/978-3-031-43901-8_19.

1   Introduction

With the recent development of semi-supervised learning (SSL) [3], rapid progress has been made in medical image segmentation, which typically learns rich anatomical representations from few labeled data and the vast amount of unlabeled data. Existing SSL approaches can be generally categorized into adversarial training [16,32,36], deep co-training [23,40], mean teacher schemes [7,13–15,27,34,38,39], multi-task learning [11,19,22], and contrastive learning [2,24,29,33,35,37].

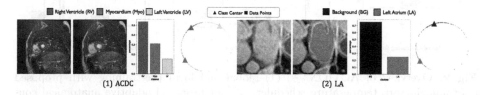

Fig. 1. Examples of two benchmarks (*i.e.*, ACDC and LA) with imbalanced class distribution. From left to right: input image, ground-truth segmentation map, class distribution chart, training data feature distribution for multiple classes.

Contrastive learning (CL) has become a remarkable approach to enhance semi-supervised medical image segmentation performance without significantly increasing the amount of parameters and annotation costs [2,29,35]. In real-world clinical scenarios, since the classes in medical images follow the Zipfian distribution [41], the medical datasets usually show a long-tailed, even heavy-tailed class distribution, *i.e.*, some minority (tail) classes involving significantly fewer pixel-level training instances than other majority (head) classes, as illustrated in Fig. 1. Such imbalanced scenarios are usually very challenging for CL methods to address, leading to noticeable performance drop [18].

To address long-tail medical segmentation, our motivations come from the following two perspectives in CL training schemes [2,35]: ❶ **Training objective** – the main focus of existing approaches is on designing proper unsupervised contrastive loss in learning high-quality representations for long-tail medical segmentation. While extensively explored in the unlabeled portion of long-tail medical data, supervised CL has rarely been studied from empirical and theoretical perspectives, which will be one of the focuses in this work; ❷ **Temperature scheduler** – the temperature parameter τ, which controls the strength of attraction and repulsion forces in the contrastive loss [4,5], has been shown to play a crucial role in learning useful representations. It is affirmed that a large τ emphasizes anatomically meaningful group-wise patterns by group-level discrimination, whereas a small τ ensures a higher degree of pixel-level (instance) discrimination [25,28]. On the other hand, as shown in [25], group-wise discrimination often results in reduced model's instance discrimination capabilities, where the model will be biased to "easy" features instead of "hard" features. It is thus unfavorable for long-tailed medical segmentation to blindly treat τ as a *constant* hyperparameter, and a dynamic temperature parameter for CL is worth investigating.

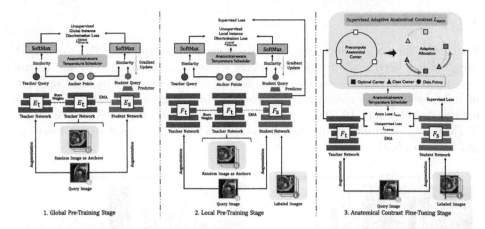

Fig. 2. Overview of ACTION++: (1) global and local pre-training with proposed anatomical-aware temperature scheduler, (2) our proposed adaptive anatomical contrast fine-tuning, which first pre-computes the optimal locations of class centers uniformly distributed on the embedding space (*i.e.*, off-line), and then performs online contrastive matching training by encouraging different class features to adaptively match these distinct and uniformly distributed class centers with respect to anatomical features.

In this paper, we introduce ACTION++, which further optimizes anatomically group-level and pixel-level representations for better head and tail class separations, on both labeled and unlabeled medical data. Specifically, we devise two strategies to improve overall segmentation quality by focusing on the two aforementioned perspectives: (1) we propose supervised adaptive anatomical contrastive learning (SAACL) for long-tail medical segmentation. To prevent the feature space from being biased toward the dominant head class, we first pre-compute the optimal locations of class centers uniformly distributed on the embedding space (*i.e.*, off-line), and then perform online contrastive matching training by encouraging different class features to adaptively match these distinct and uniformly distributed class centers; (2) we find that blindly adopting the *constant* temperature τ in the contrastive loss can negatively impact the segmentation performance. Inspired by an average distance maximization perspective, we leverage a *dynamic* τ via a simple cosine schedule, resulting in significant improvements in the learned representations. Both of these enable the model to learn a balanced feature space that has similar separability for both the majority (head) and minority (tail) classes, leading to better generalization in long-tail medical data. We evaluated our ACTION++ on the public ACDC and LA datasets [1,31]. Extensive experimental results show that our ACTION++ outperforms prior methods by a significant margin and sets the new state-of-the-art across two semi-supervised settings. We also theoretically show the superiority of our method in label efficiency (Appendix A). Code is released at here.

2 Method

2.1 Overview

Problem Statement. Given a medical image dataset $(\boldsymbol{X}, \boldsymbol{Y})$, our goal is to train a segmentation model $\boldsymbol{F}$ that can provide accurate predictions that assign each pixel to their corresponding K-class segmentation labels.

Setup. Figure 2 illustrates an overview of ACTION++. By default, we build this work upon ACTION pipeline [35], the state-of-the-art CL framework for semi-supervised medical image segmentation. The backbone model adopts the student-teacher framework that shares the same architecture, and the parameters of the teacher are the exponential moving average of the student's parameters. Hereinafter, we adopt their model as our backbone and briefly summarize its major components: (1) global contrastive distillation pre-training; (2) local contrastive distillation pre-training; and (3) anatomical contrast fine-tuning.

Global and Local Pre-training. [35] first creates two types of anatomical views as follows: (1) *augmented views* - $\mathbf{x}^1$ and $\mathbf{x}^2$ are augmented from the unlabeled input scan with two separate data augmentation operators; (2) *mined views* - n samples (*i.e.*, $\mathbf{x}^3$) are randomly sampled from the unlabeled portion with additional augmentation. The pairs $[\mathbf{x}^1, \mathbf{x}^2]$ are then processed by student-teacher networks $[F_s, F_t]$ that share the same architecture and weight, and similarly, $\mathbf{x}^3$ is encoded by F_t. Their global latent features after the encoder $\boldsymbol{E}$ (*i.e.*, $[\mathbf{h}^1, \mathbf{h}^2, \mathbf{h}^3]$) and local output features after decoder $\boldsymbol{D}$ (*i.e.*, $[\mathbf{f}^1, \mathbf{f}^2, \mathbf{f}^3]$) are encoded by the two-layer nonlinear projectors, generating global and local embeddings $\mathbf{v}_g$ and $\mathbf{v}_l$. $\mathbf{v}$ from F_s are separately encoded by the non-linear predictor, producing $\mathbf{w}$ in both global and local manners [1]. Third, the relational similarities between augmented and mined views are processed by SoftMax function as follows: $\mathbf{u}_s = \log \frac{\exp\left(\text{sim}\left(\mathbf{w}^1, \mathbf{v}^3\right)/\tau_s\right)}{\sum_{n=1}^{N} \exp\left(\text{sim}\left(\mathbf{w}^1, \mathbf{v}_n^3\right)/\tau_s\right)}$, $\mathbf{u}_t = \log \frac{\exp\left(\text{sim}\left(\mathbf{w}^2, \mathbf{v}^3\right)/\tau_t\right)}{\sum_{n=1}^{N} \exp\left(\text{sim}\left(\mathbf{w}^2, \mathbf{v}_n^3\right)/\tau_t\right)}$, where τ_s and τ_t are two temperature parameters. Finally, we minimize the unsupervised instance discrimination loss (*i.e.*, Kullback-Leibler divergence $\mathcal{KL}$) as:

$$\mathcal{L}_{\text{inst}} = \mathcal{KL}(\mathbf{u}_s \| \mathbf{u}_t). \tag{1}$$

We formally summarize the pretraining objective as the equal combination of the global and local $\mathcal{L}_{\text{inst}}$, and supervised segmentation loss $\mathcal{L}_{\text{sup}}$ (*i.e.*, equal combination of Dice loss and cross-entropy loss).

Anatomical Contrast Fine-Tuning. The underlying motivation for the fine-tuning stage is that it reduces the vulnerability of the pre-trained model to long-tailed unlabeled data. To mitigate the problem, [35] proposed to fine-tune the model by anatomical contrast. First, the additional representation head φ is used to provide dense representations with the same size as the input scans.

---

[1] For simplicity, we omit details of local instance discrimination in the following.

Then, [35] explore pulling queries $\mathbf{r}_q \in \mathcal{R}$ to be similar to the positive keys $\mathbf{r}_k^+ \in \mathcal{R}$, and push apart the negative keys $\mathbf{r}_k^- \in \mathcal{R}$. The AnCo loss is defined as follows:

$$\mathcal{L}_{\text{anco}} = \sum_{c \in \mathcal{C}} \sum_{\mathbf{r}_q \sim \mathcal{R}_q^c} - \log \frac{\exp(\mathbf{r}_q \cdot \mathbf{r}_k^{c,+} / \tau_{an})}{\exp(\mathbf{r}_q \cdot \mathbf{r}_k^{c,+} / \tau_{an}) + \sum_{\mathbf{r}_k^- \sim \mathcal{R}_k^c} \exp(\mathbf{r}_q \cdot \mathbf{r}_k^- / \tau_{an})}, \quad (2)$$

where $\mathcal{C}$ denotes a set of all available classes in the current mini-batch, and τ_{an} is a temperature hyperparameter. For class c, we select a query representation set $\mathcal{R}_q^c$, a negative key representation set $\mathcal{R}_k^c$ whose labels are not in class c, and the positive key $\mathbf{r}_k^{c,+}$ which is the c-class mean representation. Given $\mathcal{P}$ is a set including all pixel coordinates with the same size as R, these queries and keys can be defined as: $\mathcal{R}_q^c = \bigcup_{[i,j] \in \mathcal{A}} \mathbb{1}(\mathbf{y}_{[i,j]} = c)\, \mathbf{r}_{[i,j]}$, $\mathcal{R}_k^c = \bigcup_{[i,j] \in \mathcal{A}} \mathbb{1}(\mathbf{y}_{[i,j]} \neq c)\, \mathbf{r}_{[i,j]}$, $\mathbf{r}_k^{c,+} = \frac{1}{|\mathcal{R}_q^c|} \sum_{\mathbf{r}_q \in \mathcal{R}_q^c} \mathbf{r}_q$. We formally summarize the fine-tuning objective as the equal combination of unsupervised $\mathcal{L}_{\text{anco}}$, unsupervised cross-entropy loss $\mathcal{L}_{\text{unsup}}$, and supervised segmentation loss $\mathcal{L}_{\text{sup}}$. For more details, we refer the reader to [35].

2.2   Supervised Adaptive Anatomical Contrastive Learning

The general efficacy of anatomical contrast on long-tail unlabeled data has previously been demonstrated by the authors of [35]. However, taking a closer look, we observe that the well-trained F shows a downward trend in performance, which often fails to classify tail classes on labeled data, especially when the data shows long-tailed class distributions. This indicates that such well-trained F is required to improve the segmentation capabilities in long-tailed labeled data. To this end, inspired by [17] tailored for the image classification tasks, we introduce supervised adaptive anatomical contrastive learning (SAACL), a training framework for generating well-separated and uniformly distributed latent feature representations for both the head and tail classes. It consists of three main steps, which we describe in the following.

Anatomical Center Pre-computation. We first pre-compute the anatomical class centers in latent representation space. The optimal class centers are chosen as K positions from the unit sphere $\mathbb{S}^{d-1} = \{v \in \mathbb{R}^d : \|v\|_2 = 1\}$ in the d-dimensional space. To encourage good separability and uniformity, we compute the class centers $\{\psi_c\}_{c=1}^K$ by minimizing the following uniformity loss $\mathcal{L}_{\text{unif}}$:

$$\mathcal{L}_{\text{unif}}(\{\psi_c\}_{c=1}^K) = \sum_{c=1}^K \log \left(\sum_{c'=1}^K \exp(\psi_c \cdot \psi_{c'} / \tau) \right). \quad (3)$$

In our implementation, we use gradient descent to search for the optimal class centers constrained to the unit sphere $\mathbb{S}^{d-1}$, which are denoted by $\{\psi_c^\star\}_{c=1}^K$. Furthermore, the latent dimension d is a hyper-parameter, which we set such that $d \gg K$ to ensure the solution found by gradient descent indeed maximizes the minimum distance between any two class centers [6]. It is also known that any analytical minimizers of Eq. 3 form a perfectly regular K-vertex inscribed simplex of the sphere $\mathbb{S}^{d-1}$ [6]. We emphasize that this first step of pre-computation of class centers is completely off-line as it does not require any training data.

Adaptive Allocation. As the second step, we explore adaptively allocating these centers among classes. This is a combinatorial optimization problem and an exhaustive search of all choices would be computationally prohibited. Therefore, we draw intuition from the empirical mean in the K-means algorithm and adopt an adaptive allocation scheme to iteratively search for the optimal allocation during training. Specifically, consider a batch $\mathcal{B} = \{\mathcal{B}_1, \cdots, \mathcal{B}_K\}$ where $\mathcal{B}_c$ denotes a set of samples in a batch with class label c, for $c = 1, \cdots, K$. Define $\overline{\phi}_c(\mathcal{B}) = \sum_{i \in \mathcal{B}_c} \phi_i / \|\sum_{i \in \mathcal{B}_c} \phi_i\|_2$ be the empirical mean of class c in current batch, where ϕ_i is the feature embedding of sample i. We compute assignment π by minimizing the distance between pre-computed class centers and the empirical means:

$$\pi^\star = \arg\min_{\pi} \sum_{c=1}^{K} \|\psi^\star_{\pi(c)} - \overline{\phi}_c\|_2. \tag{4}$$

In implementation, the empirical mean is updated using moving average. That is, for iteration t, we first compute the empirical mean $\overline{\phi}_c(\mathcal{B})$ for batch $\mathcal{B}$ as described above, and then update by $\overline{\phi}_c \leftarrow (1 - \eta)\overline{\phi}_c + \eta\overline{\phi}_c(\mathcal{B})$.

Adaptive Anatomical Contrast. Finally, the allocated class centers are well-separated and should maintain the semantic relation between classes. To utilize these optimal class centers, we want to induce the feature representation of samples from each class to cluster around the corresponding pre-computed class center. To this end, we adopt a supervised contrastive loss for the label portion of the data. Specifically, given a batch of pixel-feature-label tuples $\{(\omega_i, \phi_i, y_i)\}_{i=1}^n$ where ω_i is the i-th pixel in the batch, ϕ_i is the feature of the pixel and y_i is its label, we define supervised <u>a</u>daptive <u>a</u>natomical <u>co</u>ntrastive loss for pixel i as:

$$\mathcal{L}_{\text{aaco}} = \frac{-1}{n} \sum_{i-1}^{n} \left(\sum_{\phi_i^+} \log \frac{\exp(\phi_i \cdot \phi_i^+/\tau_{sa})}{\sum_{\phi_j} \exp(\phi_i \cdot \phi_j/\tau_{sa})} + \lambda_a \log \frac{\exp(\phi_i \cdot \nu_i/\tau_{sa})}{\sum_{\phi_j} \exp(\phi_i \cdot \phi_j/\tau_{sa})} \right), \tag{5}$$

where $\nu_i = \psi^\star_{\pi^\star(y_i)}$ is the pre-computed center of class y_i. The first term in Eq. 5 is supervised contrastive loss, where the summation over ϕ_i^+ refers to the uniformly sampled positive examples from pixels in batch with label equal to y_i. The summation over ϕ_j refers to all features in the batch excluding ϕ_i. The second term is contrastive loss with the positive example being the pre-computed optimal class center.

2.3 Anatomical-Aware Temperature Scheduler (ATS)

Training with a varying τ induces a more isotropic representation space, wherein the model learns both group-wise and instance-specific features [12]. To this end, we are inspired to use an anatomical-aware temperature scheduler in both the supervised and the unsupervised contrastive losses, where the temperature parameter τ evolves within the range $[\tau^-, \tau^+]$ for $\tau^+ > \tau^-$. Specifically, for iteration $t = 1, \cdots, T$ with T being the total number of iterations, we set τ_t as:

$$\tau_t = \tau^- + 0.5(1 + \cos(2\pi t/T))(\tau^+ - \tau^-). \tag{6}$$

3   Experiments

Experimental Setup. We evaluate ACTION++ on two benchmark datasets: the LA dataset [31] and the ACDC dataset [1]. The LA dataset consists of 100 gadolinium-enhanced MRI scans, with the fixed split [29] using 80 and 20 scans for training and validation. The ACDC dataset consists of 200 cardiac cine MRI scans from 100 patients including three segmentation classes, *i.e.*, left ventricle (LV), myocardium (Myo), and right ventricle (RV), with the fixed split[2] using 70, 10, and 20 patients' scans for training, validation, and testing. For all our experiments, we follow the identical setting in [19,29,30,39], and perform evaluations under two label settings (*i.e.*, 5% and 10%) for both datasets.

Implementation Details. We use an SGD optimizer for all experiments with a learning rate of 1e-2, a momentum of 0.9, and a weight decay of 0.0001. Following [19,29,30,39] on both datasets, all inputs were normalized as zero mean and unit

Table 1. Quantitative comparison (DSC[%]/ASD[voxel]) for LA under two unlabeled settings (5% or 10%). All experiments are conducted as [16,19,20,29,30,35,39] in the identical setting for fair comparisons. The best results are indicated in **bold**. VNet-F (fully-supervised) and VNet-L (semi-supervised) are considered as the upper bound and the lower bound for the performance comparison.

Method	4 Labeled (5%)		8 Labeled (10%)	
	DSC[%]↑	ASD[voxel]↓	DSC[%]↑	ASD[voxel]↓
VNet-F [21]	91.5	1.51	91.5	1.51
VNet-L	52.6	9.87	82.7	3.26
UAMT [39]	82.3	3.82	87.8	2.12
SASSNet [16]	81.6	3.58	87.5	2.59
DTC [19]	81.3	2.70	87.5	2.36
URPC [20]	82.5	3.65	86.9	2.28
MC-Net [30]	83.6	2.70	87.6	1.82
SS-Net [29]	86.3	2.31	88.6	1.90
ACTION [35]	86.6	2.24	88.7	2.10
• ACTION++ (ours)	**87.8**	**2.09**	**89.9**	**1.74**

---

[2] https://github.com/HiLab-git/SSL4MIS/tree/master/data/ACDC.

variance. The data augmentations are rotation and flip operations. Our work is built on ACTION [35], thus we follow the identical model setting except for temperature parameters because they are of direct interest to us. For the sake of completeness, we refer the reader to [35] for more details. We set λ_a, d as 0.2, 128, and regarding all τ, we use $\tau^+ = 1.0$ and $\tau^- = 0.1$ if not stated otherwise. On ACDC, we use the U-Net model [26] as the backbone with a 2D patch size of 256×256 and batch size of 8. For pre-training, the networks are trained for 10K iterations; for fine-tuning, 20K iterations. On LA, we use the V-Net [21] as the backbone. For training, we randomly crop $112 \times 112 \times 80$ patches and the batch size is 2. For pre-training, the networks are trained for 5K iterations. For fine-tuning, the networks are for 15K iterations. For testing, we adopt a sliding window strategy with a fixed stride ($18 \times 18 \times 4$). All experiments are conducted in the same environments with fixed random seeds (Hardware: Single NVIDIA GeForce RTX 3090 GPU; Software: PyTorch 1.10.2+cu113, and Python 3.8.11).

Main Results. We compare our ACTION++ with current state-of-the-art SSL methods, including UAMT [39], SASSNet [16], DTC [19], URPC [20], MC-Net [30], SS-Net [29], and ACTION [35], and the supervised counterparts (UNet [26]/VNet [21]) trained with Full/Limited supervisions – using their released code. To evaluate 3D segmentation ability, we use Dice coefficient (DSC) and Average Surface Distance (ASD). Table 2 and Table 1 display the results on the public ACDC and LA datasets for the two labeled settings, respectively. We next discuss our main findings as follows. (1) **LA**: As shown in Table 1, our method generally presents better performance than the prior SSL methods under all settings. Figure 4 (Appendix) also shows that our model consistently outperforms all other competitors, especially in the boundary region; (2) **ACDC**: As Table 2 shows, ACTION++ achieves the best segmentation performance in terms of Dice and ASD, consistently outperforming the previous SSL methods across two labeled settings. In Fig. 3 (Appendix), we can observe that ACTION++ can yield the segmentation boundaries accurately, even for very challenging regions (*i.e.*, RV and Myo). This suggests that ACTION++ is inherently better at long-tailed learning, in addition to being a better segmentation model in general.

Table 2. Quantitative comparison (DSC[%]/ASD[voxel]) for ACDC under two unlabeled settings (5% or 10%). All experiments are conducted as [16,19,20,29,30,35,39] in the identical setting for fair comparisons. The best results are indicated in **bold**.

Method	Average	3 Labeled (5%) RV	Myo	LV	Average	7 Labeled (10%) RV	Myo	LV
UNet-F [26]	91.5/0.996	90.5/0.606	88.8/0.941	94.4/1.44	91.5/0.996	90.5/0.606	88.8/0.941	94.4/1.44
UNet-L	51.7/13.1	36.9/30.1	54.9/4.27	63.4/5.11	79.5/2.73	65.9/0.892	82.9/2.70	89.6/4.60
UAMT [39]	48.3/9.14	37.6/18.9	50.1/4.27	57.3/4.17	81.8/4.04	79.9/2.73	80.1/3.32	85.4/6.07
SASSNet [16]	57.8/6.36	47.9/11.7	59.7/4.51	65.8/2.87	84.7/1.83	81.8/0.769	82.9/1.73	89.4/2.99
URPC [20]	58.9/8.14	50.1/12.6	60.8/4.10	65.8/7.71	83.1/1.68	77.0/0.742	82.2/0.505	90.1/3.79
DTC [19]	56.9/7.59	35.1/9.17	62.9/6.01	72.7/7.59	84.3/4.04	83.8/3.72	83.5/4.63	85.6/3.77
MC-Net [30]	62.8/2.59	52.7/5.14	62.6/0.807	73.1/1.81	86.5/1.89	85.1/0.745	84.0/2.12	90.3/2.81
SS-Net [29]	65.8/2.28	57.5/3.91	65.7/2.02	74.2/0.896	86.8/1.40	85.4/1.19	84.3/1.44	90.6/1.57
ACTION [35]	87.5/1.12	85.4/0.915	85.8/0.784	91.2/1.66	89.7/0.736	89.8/0.589	86.7/0.813	92.7/0.804
• ACTION++ (ours)	**88.5/0.723**	**86.9/0.662**	**86.8/0.689**	**91.9/0.818**	**90.4/0.592**	**90.5/0.448**	**87.5/0.628**	**93.1/0.700**

Table 3. Ablation studies of Supervised Adaptive Anatomical Contrast (SAACL).

Method	DSC[%]↑	ASD[voxel]↓
KCL [9]	88.4	2.19
CB-KCL [10]	86.9	2.47
SAACL (Ours)	**89.9**	**1.74**
SAACL (random assign)	88.0	2.79
SAACL (adaptive allocation)	**89.9**	**1.74**

Table 4. Effect of cosine boundaries in with the largest difference between τ^- and τ^+.

τ^-	τ^+				
	0.2	0.3	0.4	0.5	1.0
0.07	84.1	85.0	86.9	87.9	89.7
0.1	84.5	85.9	87.1	88.3	**89.9**
0.2	84.2	84.4	85.8	87.1	87.6

Ablation Study. We first perform ablation studies on LA with 10% label ratio to evaluate the importance of different components. Table 3 shows the effectiveness of supervised adaptive anatomical contrastive learning (SAACL). Table 4 (Appendix) indicates that using anatomical-aware temperature scheduler (ATS) and SAACL yield better performance in both pre-training and fine-tuning stages. We then theoretically show the superiority of our method in Appendix A.

Finally, we conduct experiments to study the effects of cosine boundaries, cosine period, different methods of varying τ, and λ_a in Table 5, Table 6 (Appendix), respectively. Empirically, we find that using our settings (*i.e.*, $\tau^- = 0.1$, $\tau^+ = 1.0$, $T/\#$iterations=1.0, cosine scheduler, $\lambda_a = 0.2$) attains optimal performance (Table 4).

4   Conclusion

In this paper, we proposed ACTION++, an improved contrastive learning framework with adaptive anatomical contrast for semi-supervised medical segmentation. Our work is inspired by two intriguing observations that, besides the unlabeled data, the class imbalance issue exists in the labeled portion of medical data and the effectiveness of temperature schedules for contrastive learning on longtailed medical data. Extensive experiments and ablations demonstrated that our model consistently achieved superior performance compared to the prior semi-supervised medical image segmentation methods under different label ratios. Our theoretical analysis also revealed the robustness of our method in label efficiency. In future, we will validate CT/MRI datasets with more foreground labels and try t-SNE.

References

1. Bernard, O., et al.: Deep learning techniques for automatic MRI cardiac multi-structures segmentation and diagnosis: is the problem solved? IEEE Trans. Med. Imaging **37**, 2514–2525 (2018)
2. Chaitanya, K., Erdil, E., Karani, N., Konukoglu, E.: Contrastive learning of global and local features for medical image segmentation with limited annotations. In: NeurIPS (2020)

3. Chapelle, O., Scholkopf, B., Zien, A.: Semi-supervised learning (Chapelle, o., et al., eds.; 2006) [book reviews]. IEEE Trans. Neural Netw. **20**(3), 542-542 (2009)
4. Chen, T., Kornblith, S., Norouzi, M., Hinton, G.: A simple framework for contrastive learning of visual representations. In: ICML, pp. 1597–1607. PMLR (2020)
5. Chen, X., Fan, H., Girshick, R., He, K.: Improved baselines with momentum contrastive learning. arXiv preprint arXiv:2003.04297 (2020)
6. Graf, F., Hofer, C., Niethammer, M., Kwitt, R.: Dissecting supervised contrastive learning. In: ICML. PMLR (2021)
7. He, Y., Lin, F., Tzeng, N.F., et al.: Interpretable minority synthesis for imbalanced classification. In: IJCAI (2021)
8. Huang, W., Yi, M., Zhao, X.: Towards the generalization of contrastive self-supervised learning. arXiv preprint arXiv:2111.00743 (2021)
9. Kang, B., Li, Y., Xie, S., Yuan, Z., Feng, J.: Exploring balanced feature spaces for representation learning. In: ICLR (2021)
10. Kang, B., et al.: Decoupling representation and classifier for long-tailed recognition. arXiv preprint arXiv:1910.09217 (2019)
11. Kervadec, H., Dolz, J., Granger, É., Ben Ayed, I.: Curriculum semi-supervised segmentation. In: Shen, D., et al. (eds.) MICCAI 2019. LNCS, vol. 11765, pp. 568–576. Springer, Cham (2019). https://doi.org/10.1007/978-3-030-32245-8_63
12. Kukleva, A., Böhle, M., Schiele, B., Kuehne, H., Rupprecht, C.: Temperature schedules for self-supervised contrastive methods on long-tail data. In: ICLR (2023)
13. Lai, Z., Wang, C., Cheung, S.C., Chuah, C.N.: Sar: self-adaptive refinement on pseudo labels for multiclass-imbalanced semi-supervised learning. In: CVPR, pp. 4091–4100 (2022)
14. Lai, Z., Wang, C., Gunawan, H., Cheung, S.C.S., Chuah, C.N.: Smoothed adaptive weighting for imbalanced semi-supervised learning: Improve reliability against unknown distribution data. In: ICML, pp. 11828–11843 (2022)
15. Lai, Z., Wang, C., Oliveira, L.C., Dugger, B.N., Cheung, S.C., Chuah, C.N.: Joint semi-supervised and active learning for segmentation of gigapixel pathology images with cost-effective labeling. In: ICCV, pp. 591–600 (2021)
16. Li, S., Zhang, C., He, X.: Shape-aware semi-supervised 3D semantic segmentation for medical images. In: Martel, A.L., et al. (eds.) MICCAI 2020. LNCS, vol. 12261, pp. 552–561. Springer, Cham (2020). https://doi.org/10.1007/978-3-030-59710-8_54
17. Li, T., et al.: Targeted supervised contrastive learning for long-tailed recognition. In: CVPR (2022)
18. Li, Z., Kamnitsas, K., Glocker, B.: Analyzing overfitting under class imbalance in neural networks for image segmentation. IEEE Trans. Medi. Imaging **40**, 1065–1077 (2020)
19. Luo, X., Chen, J., Song, T., Wang, G.: Semi-supervised medical image segmentation through dual-task consistency. In: AAAI (2020)
20. Luo, X., et al.: Efficient semi-supervised gross target volume of nasopharyngeal carcinoma segmentation via uncertainty rectified pyramid consistency. In: de Bruijne, M., et al. (eds.) MICCAI 2021. LNCS, vol. 12902, pp. 318–329. Springer, Cham (2021). https://doi.org/10.1007/978-3-030-87196-3_30
21. Milletari, F., Navab, N., Ahmadi, S.A.: V-net: fully convolutional neural networks for volumetric medical image segmentation. In: 3DV, pp. 565–571. IEEE (2016)
22. Oliveira, L.C., Lai, Z., Siefkes, H.M., Chuah, C.N.: Generalizable semi-supervised learning strategies for multiple learning tasks using 1-d biomedical signals. In: NeurIPS Workshop on Learning from Time Series for Health (2022)

23. Qiao, S., Shen, W., Zhang, Z., Wang, B., Yuille, A.: Deep co-training for semi-supervised image recognition. In: ECCV (2018)

24. Quan, Q., Yao, Q., Li, J., Zhou, S.K.: Information-guided pixel augmentation for pixel-wise contrastive learning. arXiv preprint arXiv:2211.07118 (2022)

25. Robinson, J., Sun, L., Yu, K., Batmanghelich, K., Jegelka, S., Sra, S.: Can contrastive learning avoid shortcut solutions? In: NeurIPS (2021)

26. Ronneberger, O., Fischer, P., Brox, T.: U-Net: convolutional networks for biomedical image segmentation. In: Navab, N., Hornegger, J., Wells, W.M., Frangi, A.F. (eds.) MICCAI 2015. LNCS, vol. 9351, pp. 234–241. Springer, Cham (2015). https://doi.org/10.1007/978-3-319-24574-4_28

27. Tarvainen, A., Valpola, H.: Mean teachers are better role models: weight-averaged consistency targets improve semi-supervised deep learning results. In: NeurIPS. pp. 1195–1204 (2017)

28. Wang, F., Liu, H.: Understanding the behaviour of contrastive loss. In: CVPR (2021)

29. Wu, Y., Wu, Z., Wu, Q., Ge, Z., Cai, J.: Exploring smoothness and class-separation for semi-supervised medical image segmentation. In: Wang, L., Dou, Q., Fletcher, P.T., Speidel, S., Li, S. (eds.) Medical Image Computing and Computer Assisted Intervention - MICCAI 2022. LNCS, vol. 13435, pp. 34–43. Springer, Cham (2022). https://doi.org/10.1007/978-3-031-16443-9_4

30. Wu, Y., Xu, M., Ge, Z., Cai, J., Zhang, L.: Semi-supervised left atrium segmentation with mutual consistency training. In: de Bruijne, M., et al. (eds.) MICCAI 2021. LNCS, vol. 12902, pp. 297–306. Springer, Cham (2021). https://doi.org/10.1007/978-3-030-87196-3_28

31. Xiong, Z., et al.: A global benchmark of algorithms for segmenting the left atrium from late gadolinium-enhanced cardiac magnetic resonance imaging. Med. Image Anal. **67**, 101832 (2021)

32. Xue, Y., Xu, T., Zhang, H., Long, L.R., Huang, X.: SegAN: adversarial network with multi-scale l 1 loss for medical image segmentation. Neuroinformatics **16**, 383–392 (2018)

33. You, C., et al.: Mine your own anatomy: Revisiting medical image segmentation with extremely limited labels. arXiv preprint arXiv:2209.13476 (2022)

34. You, C., et al.: Rethinking semi-supervised medical image segmentation: a variance-reduction perspective. arXiv preprint arXiv:2302.01735 (2023)

35. You, C., Dai, W., Staib, L., Duncan, J.S.: Bootstrapping semi-supervised medical image segmentation with anatomical-aware contrastive distillation. In: Frangi, A., de Bruijne, M., Wassermann, D., Navab, N. (eds.) IPMI 2023. LNCS, vol. 13939, pp. 641–653. Springer, Cham (2023). https://doi.org/10.1007/978-3-031-34048-2_49

36. You, C., et al.: Class-aware adversarial transformers for medical image segmentation. In: NeurIPS (2022)

37. You, C., Zhao, R., Staib, L.H., Duncan, J.S.: Momentum contrastive voxel-wise representation learning for semi-supervised volumetric medical image segmentation. In: Wang, L., Dou, Q., Fletcher, P.T., Speidel, S., Li, S. (eds.) MICCAI 2022. LNCS, vol. 13434, pp. 639–652. Springer, Cham (2022). https://doi.org/10.1007/978-3-031-16440-8_61

38. You, C., Zhou, Y., Zhao, R., Staib, L., Duncan, J.S.: SimCVD: simple contrastive voxel-wise representation distillation for semi-supervised medical image segmentation. IEEE Trans. Med. Imaging **41**, 2228–2237 (2022)

39. Yu, L., Wang, S., Li, X., Fu, C.-W., Heng, P.-A.: Uncertainty-aware self-ensembling model for semi-supervised 3D left atrium segmentation. In: Shen, D., et al. (eds.) MICCAI 2019. LNCS, vol. 11765, pp. 605–613. Springer, Cham (2019). https://doi.org/10.1007/978-3-030-32245-8_67

40. Zhou, Y., et al.: Semi-supervised 3D abdominal multi-organ segmentation via deep multi-planar co-training. In: WACV. IEEE (2019)

41. Zipf, G.K.: The Psycho-Biology of Language: An Introduction to Dynamic Philology. Routledge, Milton Park (2013)

TransNuSeg: A Lightweight Multi-task Transformer for Nuclei Segmentation

Zhenqi He[1], Mathias Unberath[2], Jing Ke[3], and Yiqing Shen[2]($\boxtimes$)

[1] The University of Hong Kong, Pok Fu Lam, Hong Kong
[2] Johns Hopkins University, Baltimore, USA
yshen92@jhu.edu
[3] Shanghai Jiao Tong University, Shanghai, China

Abstract. Nuclei appear small in size, yet, in real clinical practice, the global spatial information and correlation of the color or brightness contrast between nuclei and background, have been considered a crucial component for accurate nuclei segmentation. However, the field of automatic nuclei segmentation is dominated by Convolutional Neural Networks (CNNs), meanwhile, the potential of the recently prevalent Transformers has not been fully explored, which is powerful in capturing local-global correlations. To this end, we make the first attempt at a pure Transformer framework for nuclei segmentation, called TransNuSeg. Different from prior work, we decouple the challenging nuclei segmentation task into an intrinsic multi-task learning task, where a tri-decoder structure is employed for nuclei instance, nuclei edge, and clustered edge segmentation respectively. To eliminate the divergent predictions from different branches in previous work, a novel self distillation loss is introduced to explicitly impose consistency regulation between branches. Moreover, to formulate the high correlation between branches and also reduce the number of parameters, an efficient attention sharing scheme is proposed by partially sharing the self-attention heads amongst the tri-decoders. Finally, a token MLP bottleneck replaces the over-parameterized Transformer bottleneck for a further reduction in model complexity. Experiments on two datasets of different modalities, including MoNuSeg have shown that our methods can outperform state-of-the-art counterparts such as CA$^{2.5}$-Net by 2–3% Dice with 30% fewer parameters. In conclusion, TransNuSeg confirms the strength of Transformer in the context of nuclei segmentation, which thus can serve as an efficient solution for real clinical practice. Code is available at https://github.com/zhenqi-he/transnuseg.

Keywords: Lightweight Multi-Task Framework · Shared Attention Heads · Nuclei · Edge and Clustered Edge Segmentation

1 Introduction

Accurate cancer diagnosis, grading, and treatment decisions from medical images heavily rely on the analysis of underlying complex nuclei structures [7]. Yet, due

H. Greenspan et al. (Eds.): MICCAI 2023, LNCS 14223, pp. 206–215, 2023.
https://doi.org/10.1007/978-3-031-43901-8_20

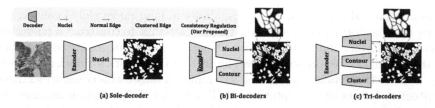

Fig. 1. Semantic illustrations of the nuclei segmentation networks with different numbers of decoders. (a) Sole-decoder to perform a single task of nuclei segmentation. (b) Bi-decoder to segment nuclei and locate nuclei edges simultaneously. (c) Tri-decoder with the third encoder path to specify the challenging clustered edge (ours), where the consistency regularization is designed across the predictions from the other two branches (dashed line).

to the numerous nuclei contained in a digitized whole-slide image (WSI), or even in an image patch of deep learning input, dense annotation of nuclei contouring is extremely time-consuming and labor-expensive [11]. Consequently, automated nuclei segmentation approaches have emerged to satisfy a broad range of computer-aided diagnostic systems, where the deep learning methods, particularly the convolutional neural networks [5,12,14,19,21] have received notable attention due to their simplicity and generalization ability.

In the literature work, the sole-decoder design in these UNet variants (Fig. 1(a)) is susceptible to failures in splitting densely clustered nuclei when precise edge information is absent. Hence, deep contour-aware neural network (DCAN) [3] with bi-decoder structure achieves improved instance segmentation performance by adopting multi-task learning, in which one decoder learns to segment the nuclei and the other recognizes edges as described in Fig. 1(b). Similarly, CIA-Net [20] extends DCAN with an extra information aggregator to fuse the features from two decoders for more precise segmentation. Much recently, $CA^{2.5}$-Net [6] shows identifying the clustered edges in a multiple-task learning manner can achieve higher performance, and thereby proposes an extra output path to learn the segmentation of clustered edges explicitly. A significant drawback of the aforementioned multi-decoder networks is the ignorance of the prediction consistency between branches, resulting in sub-optimal performance and missing correlations between the learned branches. Specifically, a prediction mismatch between the nuclei and edge branches is observed in previous work [8], implying a direction for performance improvement. To narrow this gap, we propose a consistency distillation between the branches, as shown by the dashed line in Fig. 1(c). Furthermore, to resolve the cost of involving more decoders, we propose an attention sharing scheme, along with an efficient token MLP bottleneck [16], which can both reduce the number of parameters.

Additionally, existing methods are CNN-based, and their intrinsic convolution operation fails to capture global spatial information or the correlation amongst nuclei [18], which domain experts rely heavily on for accurate nuclei allocation. It suggests the presence of long-range correlation in practical nuclei segmentation tasks. Inspired by the capability in long-range global context capturing by Transformers [17], we make the first attempt to construct a tri-decoder based Transformer model to segment nuclei. In short, our major contributions are three-fold: (1) We propose a novel multi-task framework for nuclei segmentation, namely `TransNuSeg`, as the first attempt at a fully Swin-Transformer driven architecture for nuclei segmentation. (2) To alleviate the prediction inconsistency between branches, we propose a novel self distillation loss that regulates the consistency between the nuclei decoder and normal edge decoder. (3) We propose an innovative attention sharing scheme that shares attention heads amongst all decoders. By leveraging the high correlation between tasks, it can communicate the learned features efficiently across decoders and sharply reduce the number of parameters. Furthermore, the incorporation of a light-weighted MLP bottleneck leads to a sharp reduction of parameters at no cost of performance decline.

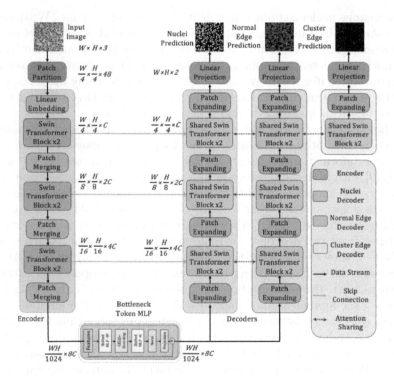

Fig. 2. The overall framework of the proposed `TransNuSeg` of three output branches to separate the nuclei, normal edges, and cluster edges, respectively. In the novel design, a pre-defined proportion of the attention heads are shared between the decoders via the proposed sharing scheme, which considerably reduces the number of parameters and enables more efficient information communication.

2   Methodology

Network Architecture Overview. Figure 2 illustrates the overall architecture of the proposed multi-task tri-decoder Transformer network, named TransNuSeg. Both the encoder and decoders utilize the Swin Transformer [13] as the building blocks to capture the long-range feature correlations in the nuclei segmentation context. Our network consists of three individual output decoder paths for nuclei segmentation, normal edges segmentation, and clustered edges segmentation. Given the high dependency between edge and clustered edge, we are inspired to propose a novel attention sharing scheme, which can communicate the information and share learned features across decoders while also reducing the number of parameters. Additionally, a token MLP bottleneck is incorporated to further increase the model efficiency.

Attention Sharing Scheme. To capture the strong correlation between nuclei segmentation and contour segmentation between multiple decoders [15], we introduce a novel attention sharing scheme that is designed as an enhancement to the multi-headed self-attention (MSA) module in the plain Transformer [17]. Based on the attention sharing scheme, we design a shared MSA module, which is similar in structure to vanilla MSA. Specifically, it consists of a `LayerNorm` layer [1], residual connection, and feed-forward layer. Innovatively, it differs from the vanilla MSA by sharing a proportion of globally-shared self-attention (SA) heads amongst all the parallel Transformer blocks in decoders, while keeping the remaining SA heads unshared *i. e.* learn the weights separately. A schematic illustration of the shared MSA module in the Swin Transformer block is demonstrated in Fig. 3, as is formally formulated as follows:

$$\text{Shared-MSA}(\mathbf{z}) = \Big[\text{SA}_1^s(\mathbf{z}), \cdots, \text{SA}_m^s(\mathbf{z}), \text{SA}_1^u(\mathbf{z}), \cdots, \text{SA}_n^u(\mathbf{z})\Big]\mathbf{U}_{\text{MSA}}^u, \quad (1)$$

$[\cdot]$ writes for the concatenation, $\text{SA}(\cdot)$ denotes the self-attention head whose output dimension is D_h, and $\mathbf{U}_{\text{MSA}}^u \in \mathbb{R}^{(m+n)\cdot D_h \times D}$ is a learnable matrix. The superscript s and u refer to the globally-shared and unshared weights across all decoders, respectively.

Token MLP Bottleneck. To reduce the complexity of the model, we leverage a token MLP bottleneck as a light-weight alternative for the Swin Transformer bottleneck. Specifically, this approach involves shifting the latent features extracted by the encoder via two MLP blocks across the width and height channels, respectively [16]. The objective of this process is to attend to specific areas, which mimics the shifted window attention mechanism in Swin Transformer [13]. The shifted features are then projected by a learnable MLP and normalized through a `LayerNorm` [1] before being fed to a reprojection MLP layer.

Consistency Self Distillation. To alleviate the inconsistency between the contour generated from the nuclei segmentation prediction and the predicted edge, we propose a novel consistency self distillation loss, denoted as $\mathcal{L}_{SD}$. Formally, this regularization is defined as the dice loss between the contour generated from the nuclei branch prediction (y_n) using the Sobel operation $(\texttt{sobel}(y_n))$ and the predicted edges y_e from the normal edge decoder. Specifically, the self distillation loss $\mathcal{L}_D$ is formulated by $\mathcal{L}_{sd} = \texttt{Dice}(\texttt{sobel}(y_n), y_e)$.

Multi-task Learning Objective. We employ a multi-task learning paradigm to train the tri-decoder network, aiming to improve model performance by leveraging the additional supervision signal from edges. Particularly, the nuclei semantic segmentation is considered the primary task, while the normal edge and clustered edge semantic segmentation are viewed as auxiliary tasks. All decoder branches follow a uniform scheme that combines the cross-entropy loss and the dice loss, with the balancing coefficients set to 0.60 and 0.40 respectively, as previous work [6]. Subsequently, the overall loss $\mathcal{L}$ is calculated as a weighted summation of semantic nuclei mask loss $(\mathcal{L}_n)$, normal edge loss $(\mathcal{L}_e)$, and clustered edge loss $(\mathcal{L}_c)$, and the self distillation loss $(\mathcal{L}_{SD})$ i. e.

$\mathcal{L} = \gamma_n \cdot \mathcal{L}_n + \gamma_e \cdot \mathcal{L}_e + \gamma_c \cdot \mathcal{L}_c + \gamma_{sd} \cdot \mathcal{L}_{sd}$, where coefficients γ_n, γ_e and γ_c are set to 0.30, 0.35, 0.35 respectively, and γ_{sd} is initially set to 1 with a 0.3 decrease for every 10 epochs until it reaches 0.4.

3   Experiments

Dataset. We evaluated the applicability of our approach across multiple modalities by conducting evaluations on microscopy and histology datasets. (1) *Fluorescence Microscopy Image Dataset*: This set combines three different data sources to simulate the heterogeneous nature of medical images [9]. It consists of 524 fluorescence images, each with a resolution of 512×512 pixels. (2) *Histology Image Dataset*: This set is the combination of the open dataset MoNuSeg [10] and another private histology dataset [8] of 462 images. We crop each image in the MoNuSeg dataset into four partially overlapping 512×512 images.

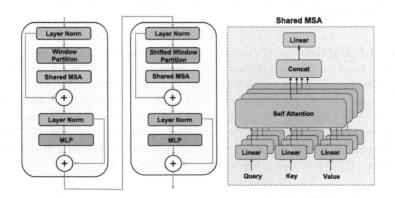

Fig. 3. A schematic illustration of the proposed Attention Sharing scheme.

Table 1. Quantitative comparisons with counterparts. The best performance with respect to each metric is highlighted in **boldface**.

Dataset	Methods	DSC (%)	F1 (%)	Acc (%)	IoU (%)	ErCnt (%)
Microscopy	UNet	85.51 ± 0.35	91.05 ± 0.13	92.19 ± 0.20	85.44 ± 0.29	55.2 ± 2.7
	UNet++	94.14 ± 0.58	92.34 ± 0.63	93.87 ± 0.61	86.20 ± 1.02	69.3 ± 1.4
	TransUNet	94.14 ± 0.47	92.31 ± 0.34	93.76 ± 0.50	86.16 ± 0.56	51.9 ± 1.0
	SwinUNet	96.05 ± 0.27	95.02 ± 0.23	96.08 ± 0.23	91.06 ± 0.43	31.2 ± 0.6
	CA$^{2.5}$-Net	91.08 ± 0.49	90.05 ± 0.27	93.40 ± 0.14	86.89 ± 0.87	18.6 ± 1.3
	Ours	$\mathbf{97.01 \pm 0.74}$	$\mathbf{96.67 \pm 0.60}$	$\mathbf{97.11 \pm 1.02}$	$\mathbf{92.97 \pm 0.41}$	$\mathbf{9.78 \pm 2.1}$
Histology	UNet	80.97 ± 0.75	72.17 ± 0.49	90.14 ± 0.24	61.63 ± 0.36	45.7 ± 1.6
	UNet++	87.10 ± 0.16	75.20 ± 0.19	91.34 ± 0.14	62.89 ± 0.27	38.0 ± 2.4
	TransUNet	85.80 ± 0.20	72.87 ± 0.49	90.53 ± 0.27	60.21 ± 0.46	35.2 ± 0.8
	SwinUNet	88.73 ± 0.90	78.11 ± 1.88	91.23 ± 0.73	64.41 ± 0.15	27.6 ± 2.3
	CA$^{2.5}$-Net	86.74 ± 0.18	77.42 ± 0.30	91.52 ± 0.78	66.79 ± 0.34	23.7 ± 0.7
	Ours	$\mathbf{90.81 \pm 0.22}$	$\mathbf{81.52 \pm 0.44}$	$\mathbf{92.77 \pm 0.64}$	$\mathbf{69.49 \pm 0.17}$	$\mathbf{11.4 \pm 1.1}$

The private dataset contains 300 images sized at 512×512 tessellated from 50 WSIs scanned at $20\times$, and meticulously labeled by five pathologists according to the labeling guidelines of the MoNuSeg [10]. For both datasets, we randomly split 80% of the samples on the patient level as the training set and the remaining 20% as the test set.

Table 2. Comparison of the model complexity in terms of the number of parameters, FLOPs, as well as the training cost in the form of the averaged training time per epoch. The average training time is computed using the same batch size for both datasets, with the first number indicating the averaged time on the Fluorescence Microscopy Image Dataset and the second on the Histology Image Dataset. The token MLP bottleneck and attention sharing scheme are denoted as 'MLP', and 'AS', respectively.

Methods	#Params ($\times 10^6$)	FLOPs ($\times 10^9$)	Training (s)
UNet [14]	31.04	219.03	43.4/27.7
UNet++ [21]	9.05	135.72	41.8/31.7
TransUNet [4]	67.87	129.97	37.1/34.5
SwinUNet [2]	27.18	30.67	37.8/35.2
CA$^{2.5}$-Net [6]	24.27	460.70	73.8/70.2
Ours (w/o MLP & w/o AS)	34.33	93.98	76.1/74.3
Ours (w/o MLP)	30.82	123.60	62.6/61.2
Ours (w/o AS)	21.33	116.95	53.1/51.2
Ours (full settings)	17.82	165.95	51.5/50.8

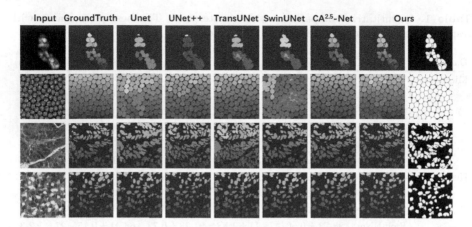

Fig. 4. Exemplary samples and their segmentation results using different methods. TransNuSeg demonstrates superior segmentation performance compared to its counterparts, which can successfully distinguish severely clustered nuclei from normal edges.

Implementations. All experiments are performed on one NVIDIA RTX 3090 GPU with 24 GB memory. We use Adam optimizer with an initial learning rate of 1×10^{-4}. We compare TransNuSeg with UNet [14], UNet++ [21], TransUNet [4], SwinUNet [2], and $CA^{2.5}$-Net [6]. We evaluate the results by using Dice Score (DSC), Intersection over Union (IoU), pixel-level accuracy (Acc), and F1-score(F1) as metrics, and ErCnt [8]. To ensure statistical significance, we run all methods five times with different fixed seeds and report the results as mean ± standard deviation.

Results. Table 1 shows the quantitative comparisons for the nuclei segmentation. The large margin between the SwinUNet and the other CNN-based or hybrid networks also confirms the superiority of the Transformer in fine-grained nuclei segmentation. More importantly, our method can outperform SwinUNet and the previous methods on both datasets. For example, in the histology image dataset, TransNuSeg improves the dice score, F1 score, accuracy, and IoU by 2.08%, 3.41%, 1.25%, and 2.70% respectively, over the second-best models. Similarly, in the fluorescence microscopy image dataset, our proposed model improves DSC by 0.96%, while also leading to 1.65%, 1.03% and 1.91% increment in F1 score, accuracy, and IoU to the second-best performance. For better visualization, representative samples and their segmentation results using different methods are demonstrated in Fig. 4. Furthermore, Table 2 compares the model complexity in terms of the number of parameters, floating point operations per second (FLOPs), and the training computational cost, where our approach can significantly reduce around 28% of the training time compared to the state-of-the-art CNN multi-task method $CA^{2.5}$-Net, while also boosting performance.

Table 3. The ablation on each functional block, where 'MLP', 'AS', and 'SD' represent the token MLP bottleneck, attention sharing scheme, and the self distillation.

MLP	AS	SD	Microscopy				Histology			
			DSC (%)	F1 (%)	Acc (%)	IoU (%)	DSC (%)	F1 (%)	Acc (%)	IoU (%)
×	×	×	95.31	94.05	96.06	90.05	88.76	78.20	90.96	64.48
•	×	×	95.49	94.48	95.95	89.97	89.41	77.94	91.02	65.17
×	•	×	95.88	93.51	96.11	90.55	90.23	80.46	92.03	67.84
•	•	×	96.95	95.72	96.92	91.98	90.27	81.04	92.01	67.56
×	×	•	96.99	95.74	97.02	92.22	90.25	80.81	92.45	68.14
•	×	•	96.58	95.65	97.03	92.07	90.17	80.62	92.35	67.88
×	•	•	96.89	95.78	97.12	92.08	90.34	80.88	92.49	68.05
•	•	•	**97.01**	**96.67**	**97.11**	**92.97**	**90.81**	**81.52**	**92.77**	**69.49**

Ablation. Our ablation study yields that token MLP bottleneck and attention sharing schemes can complementarily reduce the training cost while increasing efficiency, as shown in Table 2 (the last 4 rows). To further show the effectiveness of these schemes, as well as consistency self distillation, we conduct a comprehensive ablation study on both datasets. As described in Table 3, each component proportionally contributes to the improvement to reach the overall performance

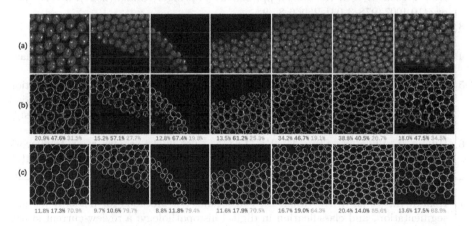

Fig. 5. The impact of self distillation regularization on mismatch reduction across three decoders. (a) Raw input image. Segmentation results by `TransNuSeg` trained (b) w/o self distillation, and (c) w/ self distillation. The predicted normal edges from the normal edge decoder are shown in green; while the edges generated from the nuclei decoder and processed with the Sobel operation are in red. The yellow color indicates the overlap between both. Accordingly, the numbers below images indicate the proportion of the pixels belonging to the three parts. Compared to the results without self distillation, the outputs with self distillation exhibit reduced mismatches, resulting in improved segmentation performance. (Color figure online)

boost. Moreover, self distillation can enhance the intrinsic consistency between two branches, as visualized in Fig. 5.

4   Conclusion

In this paper, we make the first attempt at an efficient but effective multi-task Transformer framework for modality-agnostic nuclei segmentation. Specifically, our tri-decoder framework `TransNuSeg` leverages an innovative self distillation regularization to impose consistency between the different branches. Experimental results on two datasets demonstrate the excellence of our `TransNuSeg` against state-of-the-art counterparts for potential real-world clinical deployment. Additionally, our work opens a new architecture to perform nuclei segmentation tasks with Swin Transformer, where further investigations can be performed to explore the generalizability to the top of our methods with different modalities.

References

1. Ba, J.L., Kiros, J.R., Hinton, G.E.: Layer normalization. arXiv preprint arXiv:1607.06450 (2016)
2. Cao, H., Wang, Y., Chen, J., et al.: Swin-unet: Unet-like pure transformer for medical image segmentation. In: Karlinsky, L., Michaeli, T., Nishino, K. (eds.) ECCV 2022. LNCS, vol. 13803, pp. 205–218. Springer, Cham (2021). https://doi.org/10.1007/978-3-031-25066-8_9
3. Chen, H., Qi, X., Yu, L., Heng, P.A.: DCAN: deep contour-aware networks for accurate gland segmentation. CoRR, abs/1604.02677 (2016)
4. Chen, J., et al.: TransuNet: transformers make strong encoders for medical image segmentation. CoRR, abs/2102.04306 (2021)
5. Guo, R., Pagnucco, M., Song, Y.: Learning with noise: mask-guided attention model for weakly supervised nuclei segmentation. In: de Bruijne, M., et al. (eds.) MICCAI 2021. LNCS, vol. 12902, pp. 461–470. Springer, Cham (2021). https://doi.org/10.1007/978-3-030-87196-3_43
6. Huang, J., Shen, Y., Shen, D., Ke, J.: CA$^{2.5}$-net nuclei segmentation framework with a microscopy cell benchmark collection. In: de Bruijne, M., et al. (eds.) MICCAI 2021. LNCS, vol. 12908, pp. 445–454. Springer, Cham (2021). https://doi.org/10.1007/978-3-030-87237-3_43
7. Irshad, H., Veillard, A., Roux, L., Racoceanu, D.: Methods for nuclei detection, segmentation, and classification in digital histopathology: a review-current status and future potential. IEEE Rev. Biomed. Eng. 7, 97–114 (2014)
8. Ke, J., et al.: ClusterSeg: a crowd cluster pinpointed nucleus segmentation framework with cross-modality datasets. Med. Image Anal. 85, 102758 (2023)
9. Kromp, F., et al.: An annotated fluorescence image dataset for training nuclear segmentation methods. Sci. Data 7(1), 262 (2020)
10. Kumar, N., Verma, R., Anand, D., et al.: A multi-organ nucleus segmentation challenge. IEEE Trans. Med. Imaging 39(5), 1380–1391 (2020)
11. Lagree, A., et al.: A review and comparison of breast tumor cell nuclei segmentation performances using deep convolutional neural networks. Sci. Rep. 11(1), 8025 (2021)

12. Lal, S., Das, D., Alabhya, K., Kanfade, A., Kumar, A., Kini, J.: Nucleiseg-net: robust deep learning architecture for the nuclei segmentation of liver cancer histopathology images. Comput. Biol. Med. **128**, 01 (2021)
13. Liu, Z., et al.: Swin transformer: hierarchical vision transformer using shifted windows. CoRR, abs/2103.14030 (2021)
14. Ronneberger, O., Fischer, P., Brox, T.: U-net: convolutional networks for biomedical image segmentation. CoRR, abs/1505.04597 (2015)
15. Shen, Y.: Federated learning for chronic obstructive pulmonary disease classification with partial personalized attention mechanism. arXiv preprint arXiv:2210.16142 (2022)
16. Valanarasu, J.M.J., Patel, V.M.: UneXt: MLP-based rapid medical image segmentation network. In: Wang, L., Dou, Q., Fletcher, P.T., Speidel, S., Li, S. (eds.) MICCAI 2022. LNCS, vol. 13435, pp. 23–33. Springer, Cham (2022). https://doi.org/10.1007/978-3-031-16443-9_3
17. Vaswani, A., et al.: Attention is all you need. In: Guyon, I., Von Luxburg, U., et al. (eds.) Advances in Neural Information Processing Systems, vol. 30. Curran Associates Inc, (2017)
18. Wang, C., Xu, R., Xu, S., Meng, W., Zhang, X.: DA-Net: dual branch transformer and adaptive strip upsampling for retinal vessels segmentation. In: Wang, L., Dou, Q., Fletcher, P.T., Speidel, S., Li, S. (eds.) MICCAI 2022. LNCS, vol. 13432, pp. 528–538. Springer, Cham (2022). https://doi.org/10.1007/978-3-031-16434-7_51
19. Wazir, S., Fraz, M.M.: HistoSeg: quick attention with multi-loss function for multi-structure segmentation in digital histology images. In: 2022 12th International Conference on Pattern Recognition Systems (ICPRS). IEEE (2022)
20. Zhou, Y., Onder, O.F., Dou, Q., Tsougenis, E., Chen, H., Heng, P.A.: Cia-net: robust nuclei instance segmentation with contour-aware information aggregation. CoRR, abs/1903.05358 (2019)
21. Zhou, Z., Rahman Siddiquee, M.M., Tajbakhsh, N., Liang, J.: UNet++: a nested U-Net architecture for medical image segmentation. In: Stoyanov, D., et al. (eds.) DLMIA/ML-CDS -2018. LNCS, vol. 11045, pp. 3–11. Springer, Cham (2018). https://doi.org/10.1007/978-3-030-00889-5_1

Learnable Cross-modal Knowledge Distillation for Multi-modal Learning with Missing Modality

Hu Wang[1][✉], Congbo Ma[1], Jianpeng Zhang[2], Yuan Zhang[1], Jodie Avery[1], Louise Hull[1], and Gustavo Carneiro[3]

[1] The University of Adelaide, Adelaide, Australia
hu.wang@adelaide.edu.au
[2] Alibaba DAMO Academy, Hangzhou, China
[3] Centre for Vision, Speech and Signal Processing, University of Surrey, Guildford, UK

Abstract. The problem of missing modalities is both critical and non-trivial to be handled in multi-modal models. It is common for multi-modal tasks that certain modalities contribute more compared to other modalities, and if those important modalities are missing, the model performance drops significantly. Such fact remains unexplored by current multi-modal approaches that recover the representation from missing modalities by feature reconstruction or blind feature aggregation from other modalities, instead of extracting useful information from the best performing modalities. In this paper, we propose a Learnable Cross-modal Knowledge Distillation (LCKD) model to adaptively identify important modalities and distil knowledge from them to help other modalities from the cross-modal perspective for solving the missing modality issue. Our approach introduces a teacher election procedure to select the most "qualified" teachers based on their single modality performance on certain tasks. Then, cross-modal knowledge distillation is performed between teacher and student modalities for each task to push the model parameters to a point that is beneficial for all tasks. Hence, even if the teacher modalities for certain tasks are missing during testing, the available student modalities can accomplish the task well enough based on the learned knowledge from their automatically elected teacher modalities. Experiments on the Brain Tumour Segmentation Dataset 2018 (BraTS2018) shows that LCKD outperforms other methods by a considerable margin, improving the state-of-the-art performance by 3.61% for enhancing tumour, 5.99% for tumour core, and 3.76% for whole tumour in terms of segmentation Dice score.

Keywords: Missing modality issue · Multi-modal learning · Learnable cross-modal knowledge distillation

---

Supplementary Information The online version contains supplementary material available at https://doi.org/10.1007/978-3-031-43901-8_21.

1 Introduction

Multi-modal learning has become a popular research area in computer vision and medical image analysis, with modalities spanning across various media types, including texts, audio, images, videos and multiple sensor data. This approach has been utilised in Robot Control [15,17], Visual Question Answering [12] and Audio-Visual Speech Recognition [10], as well as in the medical field to improve diagnostic system performance [7,18]. For instance, Magnetic Resonance Imaging (MRI) is a common tool for brain tumour detection that relies on multiple modalities (Flair, T1, T1 contrast-enhanced known as T1c, and T2) rather than a single type of MRI images. However, most existing multi-modal methods require complete modalities during training and testing, which limits their applicability in real-world scenarios, where subsets of modalities may be missing during training and testing.

The missing modality issue is a significant challenge in the multi-modal domain, and it has motivated the community to develop approaches that attempt to address this problem. Havaei et al. [8] developed HeMIS, a model that handles missing modalities using statistical features as embeddings for the model decoding process. Taking one step ahead, Dorent et al. [6] proposed an extension to HeMIS via a multi-modal variational auto-encoder (MVAE) to make predictions based on learned statistical features. In fact, variational auto-encoder (VAE) has been adopted to generate data from other modalities in the image or feature domains [3,11]. Yin et al. [20] aimed to learn a unified subspace for incomplete and unlabelled multi-view data. Chen et al. [4] proposed a feature disentanglement and gated fusion framework to separate modality-robust and modality-sensitive features. Ding et al. [5] proposed an RFM module to fuse the modal features based on the sensitivity of each modality to different tumor regions and a segmentation-based regularizer to address the imbalanced training problem. Zhang et al. [22] proposed an MA module to ensure that modality-specific models are interconnected and calibrated with attention weights for adaptive information exchange. Recently, Zhang et al. [21] introduced a vision transformer architecture, MMFormer, that fuses features from all modalities into a set of comprehensive features. There are several existing works [9,16,19] proposed to approximate the features from full modalities when one or more modalities are absent. But none work performs cross-modal knowledge distillation. From another point of view, Wang et al. [19] introduced a dedicated training strategy that separately trains a series of models specifically for each missing situation, which requires significantly more computation resources compared with a non-dedicated training strategy. An interesting fact about multi-modal problems is that there is always one modality that contributes much more than other modalities for a certain task. For instance, for brain tumour segmentation, it is known from domain knowledge that T1c scans clearly display the enhanced tumour, but not edema [4]. If the knowledge of these modalities can be successfully preserved, the model can produce promising results even when these best performing modalities are not available. However, the aforementioned methods neglect the contribution biases of different modalities and failed to consider keeping that knowledge.

Aiming at this issue, we propose the non-dedicated training model[1] **L**earnable **C**ross-modal **K**nowledge **D**istillation (LCKD) for tackling the missing modality issue. LCKD is able to handle missing modalities in both training and testing by automatically identifying important modalities and distilling knowledge from them to learn the parameters that are beneficial for all tasks while training for other modalities (e.g., there are four modalities and three tasks for the three types of tumours in BraTS2018). Our main contributions are:

- We propose the Learnable Cross-modal Knowledge Distillation (LCKD) model to address missing modality problem in multi-modal learning. It is a simple yet effective model designed from the viewpoint of distilling cross-modal knowledge to maximise the performance for all tasks;
- The LCKD approach is designed to automatically identify the important modalities per task, which helps the cross-modal knowledge distillation process. It also can handle missing modality during both training and testing.

The experiments are conducted on the Brain Tumour Segmentation benchmark BraTS2018 [1,14], showing that our LCKD model achieves state-of-the-art performance. In comparison to recently proposed competing methods on BraTS2018, our model demonstrates better performance in segmentation Dice score by 3.61% for enhancing tumour, 5.99% for tumour core, and 3.76% for whole tumour, on average.

2 Methodology

2.1 Overall Architecture

Let us represent the N-modality data with $\mathcal{M}_l = \{\mathbf{x}_l^{(i)}\}_{i=1}^N$, where $\mathbf{x}_l^{(i)} \in \mathcal{X}$ denotes the lth data sample and the superscript $^{(i)}$ indexes the modality. To simplify the notation, we omit the subscript l when that information is clear from the context. The label for each set $\mathcal{M}$ is represented by $\mathbf{y} \in \mathcal{Y}$, where $\mathcal{Y}$ represents the ground-truth annotation space. The framework of LCKD is shown in Fig. 1.

Multi-modal segmentation is composed not only of multiple modalities, but also of multiple tasks, such as the three types of tumours in BraTS2018 dataset that represent the three tasks. Take one of the tasks for example. Our model undergoes an external Teacher Election Procedure prior to processing all modalities $\{\mathbf{x}^{(i)}\}_{i=1}^N \in \mathcal{M}$ in order to select the modalities that exhibit promising performance as teachers. This is illustrated in Fig. 1, where one of the modalities, $\mathbf{x}^{(2)}$, is selected as a teacher, $\{\mathbf{x}^{(1)}, \mathbf{x}^{(3)}, ..., \mathbf{x}^{(N)}\}$ are the students, and $\mathbf{x}^{(n)}$ (with $n \neq 2$) is assumed to be absent. Subsequently, the modalities are encoded to output features $\{\mathbf{f}^{(i)}\}_{i=1}^N$, individually. For the modalities that are available, namely $\mathbf{x}^{(1)}, ..., \mathbf{x}^{(n-1)}, \mathbf{x}^{(n+1)}, ..., \mathbf{x}^{(N)}$, knowledge distillation is carried out between each pair of teacher and student modalities. However, for the absent

---

[1] We train one model to handle all of the different missing modality situations.

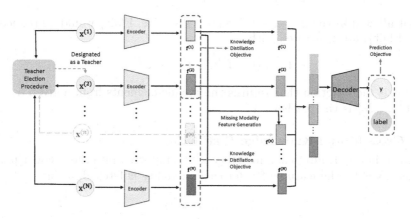

Fig. 1. LCKD model framework for training and testing. The N modalities $\{x^{(i)}{}_{i=1}^{N}\}$ are processed by the encoder to produce the features $\{f^{(i)}{}_{i=1}^{N}\}$, which are concatenated and used by the decoder to produce the segmentation. The teacher is elected using a validation process that selects the top-performing modalities as teachers. Cross-modal distillation is performed by approximating the available students' features to the available teachers' features. Features from missing modalities are generated by averaging the other modalities' features.

modality $\mathbf{x}^{(n)}$, its features $\mathbf{f}^{(n)}$ are produced through a missing modality feature generation process from the available features $\mathbf{f}^{(1)}, ..., \mathbf{f}^{(n-1)}, \mathbf{f}^{(n+1)}, ..., \mathbf{f}^{(N)}$.

In the next sections, we explain each module of the proposed Learnable Cross-modal Knowledge Distillation model training and testing with full and missing modalities.

2.2 Teacher Election Procedure

Usually, one of the modalities is more useful than others for a certain task, e.g. for brain tumour segmentation, T1c scan clearly displays the enhanced tumour, but it does not clearly show edema [4]. Following knowledge distillation (KD) [9], we propose to transfer the knowledge from modalities with promising performance (known as teachers) to other modalities (known as students). The teacher election procedure is further introduced to automatically elect proper teachers for different tasks.

More specifically, in the teacher election procedure, a validation process is applied: for each task k (for $k \in \{1, ..., K\}$), the modality with the best performance is selected as the teacher $\mathbf{t}^{(k)}$. Formally, we have:

$$\mathbf{t}^{(k)} = \arg\max_{i \in \{1,...,N\}} \sum_{l=1}^{L} d(F(\mathbf{x}_l^{(i)}; \boldsymbol{\Theta}), \mathbf{y}_l), \tag{1}$$

where i indexes different modalities, $F(\cdot; \boldsymbol{\Theta})$ is the LCKD segmentation model parameterised by $\boldsymbol{\Theta}$, including the encoder and decoder parameters $\{\theta^{enc}, \theta^{dec}\} \in \boldsymbol{\Theta}$, and $d(\cdot, \cdot)$ is the function to calculate the Dice score. Based on the elected teachers for different tasks, a list of unique teachers (i.e., repetitions are not allowed in the list, so for BraTS, {T1c, T1c, Flair} would be reduced to {T1c, Flair}) are generated with:

$$\mathbf{T} = \phi(\mathbf{t}^{(1)}, \mathbf{t}^{(2)}, ..., \mathbf{t}^{(k)}, ..., \mathbf{t}^{(K)}), \tag{2}$$

where ϕ is the function that returns the unique elements from a given list, and $\mathbf{T} \subseteq \{1, ..., N\}$ is the teacher set.

2.3   Cross-Modal Knowledge Distillation

As shown in Fig. 1, after each modality $\mathbf{x}^{(i)}$ is inputted into the encoder parameterised by θ^{enc}, the features $\mathbf{f}^{(i)}$ for each modality is fetched, as in:

$$\mathbf{f}^{(i)} = f_{\theta^{enc}}(\mathbf{x}^{(i)}). \tag{3}$$

The cross-modal knowledge distillation (CKD) is defined by a loss function that approximates all available modalities' features to the available teacher modalities in a pairwise manner for all tasks, as follows:

$$\ell_{ckd}(\mathcal{D}; \theta^{enc}) = \sum_{i \in \mathbf{T}; i, j \notin \mathbf{m}}^{N} \|\mathbf{f}^{(i)} - \mathbf{f}^{(j)}\|_p, \tag{4}$$

where $\| \cdot \|_p$ presents the p-norm operation, and here we expended the notation of missing modalities to make it more general by assuming a set of modalities $\mathbf{m}$ is missing. The minimisation of this loss pushes the model parameter values to a point in the parameter space that can maximise the performance of all tasks for all modalities.

2.4   Missing Modality Feature Generation

Because of the knowledge distillation between each pair of teachers and students, the features of modalities in the feature space ought to be close to the "genuine" features that can uniformly perform well for different tasks. Still assuming that modality set $\mathbf{m}$ is missing, the missing features $\mathbf{f}^{(n)}$ can thus be generated from the available features:

$$\mathbf{f}^{(n)} = \frac{1}{N - |\mathbf{m}|} \sum_{i=1; i \notin \mathbf{m}}^{N} \mathbf{f}^{(i)}, \tag{5}$$

where $|\mathbf{m}|$ denotes the number of missing modalities.

2.5   Training and Testing

All features encoded from Eq. 3 or generated from Eq. 5 are then concatenated to be fed into the decoder parameterised by θ^{dec} for predicting

$$\tilde{\mathbf{y}} = f_{\theta^{dec}}(\mathbf{f}^{(1)}, ..., \mathbf{f}^{(N)}), \tag{6}$$

where $\tilde{\mathbf{y}} \in \mathcal{Y}$ is the prediction of the task.

The training of the whole model is achieved by minimising the following objective function:

$$\ell_{tot}(\mathcal{D}, \boldsymbol{\Theta}) = \ell_{task}(\mathcal{D}, \theta^{enc}, \theta^{dec}) + \alpha \ell_{ckd}(\mathcal{D}; \theta^{enc}), \qquad (7)$$

where $\ell_{task}(\mathcal{D}, \theta^{enc}, \theta^{dec})$ is the objective function for the whole task (e.g., Cross-Entropy and Dice losses are adopted for brain tumour segmentation), and α is the trade-off factor between the task objective and cross-modal KD objective.

Testing is based on taking all image modalities available in the input to produce the features from Eq. 3, and generating the features from the missing modalities with Eq. 5, which are then provided to the decoder to predict the segmentation with Eq. 6.

3   Experiments

3.1   Data and Implementation Details

Our model and competing methods are evaluated on the BraTS2018 Segmentation Challenge dataset [1,14]. The task involves segmentation of three sub-regions of brain tumours, namely enhancing tumour (ET), tumour core (TC), and whole tumour (WT). The dataset consists of 3D multi-modal brain MRIs, including Flair, T1, T1 contrast-enhanced (T1c), and T2, with ground-truth annotations. The dataset comprises 285 cases for training, and 66 cases for evaluation. The ground-truth annotations for the training set are publicly available, while the validation set annotations are hidden[2].

3D UNet architecture (with 3D convolution and normalisation) is adopted as our backbone network, where the CKD process occurs at the bottom stage of the UNet structure. To optimise our model, we adopt a stochastic gradient descent optimiser with Nesterov momentum [2] set to 0.99. L1 loss is adopted for $\ell_{ckd}(.)$ in Eq. 4. Batch-size is set to 2. The learning rate is initially set to 10^{-2} and gradually decreased via the cosine annealing [13] strategy. We trained the LCKD model for 115,000 iterations and use 20% of the training data as the validation task for teacher election. To simulate modality-missing situations with non-dedicated training of models, we randomly dropped 0 to 3 modalities for each iteration. Our training time is 70.12 h and testing time is 6.43 s per case on one Nvidia 3090 GPU. 19795 MiB GPU memory is used for model training with batch-size 2 and 3789 MiB GPU memory is consumed for model testing with batch-size 1.

3.2   Overall Performance

Table 1 shows the overall performance on all 15 possible combinations of missing modalities for three sub-regions of brain tumours. Our models are compared with several strong baseline models: U-HeMIS (abbreviated as HMIS in the figure) [8], U-HVED (HVED) [6], Robust-MSeg (RSeg) [4] and mmFormer (mmFm) [21]. We can clearly observe that with T1c, the model performs considerably better

Table 1. Model performance comparison of segmentation Dice score (normalised to 100%) on BraTS2018. The best results for each row within tumour types are bolded. "•" and "○" indicate the availability and absence of the modality for testing, respectively. Last row shows p-value from one-tailed paired t-test.

Modalities				Enhancing Tumour					Tumour Core					Whole Tumour				
Fl	T1	T1c	T2	HMIS	HVED	RSeg	mmFm	LCKD	HMIS	HVED	RSeg	mmFm	LCKD	HMIS	HVED	RSeg	mmFm	LCKD
•	○	○	○	11.78	23.80	25.69	39.33	**45.48**	26.06	57.90	53.57	61.21	**72.01**	52.48	84.39	85.69	86.10	**89.45**
○	•	○	○	10.16	8.60	17.29	32.53	**43.22**	37.39	33.90	47.90	56.55	**66.58**	57.62	49.51	70.11	67.52	**76.48**
○	○	•	○	62.02	57.64	67.07	72.60	**75.65**	65.29	59.59	76.83	75.41	**83.02**	61.53	53.62	73.31	72.22	**77.23**
○	○	○	•	25.63	22.82	28.97	43.05	**47.19**	57.20	54.67	57.49	64.20	**70.17**	80.96	79.83	82.24	81.15	**84.37**
•	•	○	○	10.71	27.96	32.13	42.96	**48.30**	41.12	61.14	60.68	65.91	**74.58**	64.62	85.71	88.24	87.06	**89.97**
•	○	•	○	66.10	68.36	70.30	75.07	**78.75**	71.49	75.07	80.62	77.88	**85.67**	68.99	85.93	88.51	87.30	**90.47**
•	○	○	•	30.22	32.31	33.84	47.52	**49.01**	57.68	62.70	61.16	69.75	**75.41**	82.95	87.58	88.28	87.59	**90.39**
○	•	•	○	66.22	61.11	69.06	74.04	**76.09**	72.46	67.55	78.72	78.59	**82.49**	68.47	64.22	77.18	74.42	**80.10**
○	•	○	•	32.39	24.29	32.01	44.99	**50.09**	60.92	56.26	62.19	69.42	**72.75**	82.41	81.56	84.78	82.20	**86.05**
○	○	•	•	67.83	67.83	69.71	74.51	**76.01**	76.64	73.92	80.20	78.61	**84.85**	82.48	81.32	85.19	82.99	**86.49**
•	•	•	○	68.54	68.60	70.78	75.47	**77.78**	76.01	77.05	81.06	79.80	**85.24**	72.31	86.72	88.73	87.33	**90.50**
•	•	○	•	31.07	32.34	36.41	47.70	**49.96**	60.32	63.14	64.38	71.52	**76.68**	83.43	88.07	88.81	87.75	**90.46**
•	○	•	•	68.72	68.93	70.88	75.67	**77.48**	77.53	76.75	80.72	79.55	**85.56**	83.85	88.09	89.27	88.14	**90.90**
○	•	•	•	69.92	67.75	70.10	74.75	**77.60**	78.96	75.28	80.33	80.39	**84.02**	83.94	82.32	86.01	82.71	**86.73**
•	•	•	•	70.24	69.03	71.13	77.61	**79.33**	79.48	77.71	80.86	**85.78**	85.31	84.74	88.46	89.45	89.64	**90.84**
Average				46.10	46.76	51.02	59.85	**63.46**	62.57	64.84	69.78	72.97	**78.96**	74.05	79.16	84.39	82.94	**86.70**
p-value				5.3e-6	3.8e-7	8.6e-7	2.8e-5	-	3.7e-5	4.1e-7	7.2e-6	5.3e-7	-	1.1e-4	1.2e-3	1.1e-5	5.1e-7	-

than other modalities for ET. Similarly, T1c for TC and Flair for WT contribute the most, which confirm our motivation.

The LCKD model significantly outperforms (as shown by the one-tailed paired t-test for each task between models in the last row of Table 1) U-HeMIS, U-HVED, Robust-MSeg and mmFormer in terms of the segmentation Dice for enhancing tumour and whole tumour on all 15 combinations and the tumour core on 14 out of 15. It is observed that, on average, the proposed LCKD model improves the state-of-the-art performance by 3.61% for enhancing tumour, 5.99% for tumour core, and 3.76% for whole tumour in terms of the segmentation Dice score. Especially in some combinations without the best modality, e.g. ET/TC without T1c and WT without Flair, LCKD has a 6.15% improvement with only Flair and 10.69% with only T1 over the second best model for ET segmentation; 10.8% and 10.03% improvement with only Flair and T1 for TC; 8.96% and 5.01% improvement with only T1 and T1c for WT, respectively. These results demonstrate that useful knowledge of the best modality has been successfully distilled into the model by LCKD for multimodal learning with missing modalities.

3.3   Analyses

Single Teacher vs. Multi-teacher. To analyse the effectiveness of knowledge distillation from multiple teachers of all tasks in the proposed LCKD model, we perform a study to compare the model performance of adopting single teacher and multi-teachers for knowledge distillation. We enable multi-teachers for LCKD by default to encourage the model parameters to move to a point that can perform well for all tasks. However, for single teacher, we modify the

---

[2] Online evaluation is required at https://ipp.cbica.upenn.edu/.

Table 2. Different LCKD variants Dice score. LCKD-s and LCKD-m represent LCKD with single teacher and multi-teacher, respectively.

Modalities				Enhancing Tumour		Tumour Core		Whole Tumour	
Fl	T1	T1c	T2	LCKD-s	LCKD-m	LCKD-s	LCKD-m	LCKD-s	LCKD-m
●	○	○	○	**46.19**	45.48	**72.51**	72.01	89.38	**89.45**
○	●	○	○	43.05	**43.22**	65.79	**66.58**	75.86	**76.48**
○	○	●	○	74.26	**75.65**	81.93	**83.02**	77.14	**77.23**
○	○	○	●	**48.59**	47.19	**70.64**	70.17	84.25	**84.37**
●	●	○	○	**49.65**	48.30	**74.98**	74.58	**90.12**	89.97
●	○	●	○	77.72	**78.75**	85.37	**85.67**	90.33	**90.47**
●	○	○	●	**49.86**	49.01	**75.65**	75.41	90.28	**90.39**
○	●	●	○	75.32	**76.09**	81.77	**82.49**	79.96	**80.10**
○	●	○	●	**51.65**	50.09	**73.95**	72.75	**86.39**	86.05
○	○	●	●	75.34	**76.01**	84.21	**84.85**	86.05	**86.49**
●	●	●	○	77.42	**77.78**	84.79	**85.24**	**90.50**	**90.50**
●	●	○	●	**51.05**	49.96	**76.84**	76.68	90.39	**90.46**
●	○	●	●	76.97	**77.48**	84.93	**85.56**	90.83	**90.90**
○	●	●	●	77.53	**77.60**	83.95	**84.02**	86.71	**86.73**
●	●	●	●	78.39	**79.33**	85.26	**85.31**	90.74	**90.84**
Average				**63.53**	63.46	78.84	**78.96**	86.60	**86.70**
Best Teacher				T1c		T1c		Fl	

function $\phi(.)$ in Eq. 2 to pick the modality with max appearance time (e.g. if we have {T1c, T1c, Flair} then $\phi(.)$ returns {T1c}), while keeping other settings the same.

From Table 2, compared with multi-teacher model LCKD-m, we found that the single teacher model LCKD-s receives comparable results for ET and TC segmentation (it even has better average performance on ET), but it cannot outperform LCKD-m on WT. This phenomenon, also shown in Fig. 2, demonstrates that LCKD-m has better overall segmentation performance. This resonates with our expectations because there are 3 tasks in BraTS, and the best teachers for ET and TC are the same, which is T1c, but for WT, Flair is the best one. Therefore, for LCKD-s, the knowledge of the best teacher for ET and TC can be distilled into the model, but not for WT. The LCKD-m model can overcome this issue since it attempts to find a point in the parameter space that is beneficial for all tasks. Empirically, we observed that both models found the correct teacher(s) quickly: the best teacher of the single teacher model alternated between T1c and Flair for a few validation rounds and stabilised at T1c; while the multi-teacher model found the best teachers (T1c and Flair) from the first validation round.

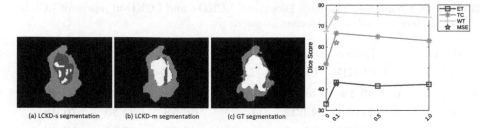

(a) LCKD-s segmentation   (b) LCKD-m segmentation   (c) GT segmentation

Fig. 2. Segmentation Visualisation with only T2 input available. Light grey, dark grey and white represent different tumour sub-regions. (Color figure online)

Fig. 3. T1 Dice score as function of α (L1 loss for $\ell_{ckd}(.)$ in (4)). Star markers show the Dice score for L2 loss for $\ell_{ckd}(.)$ ($\alpha = .1$). Colors denote different tumors.

Role of α and CKD Loss Function. As shown in Fig. 3, we set α in Eq. 7 to $\{0, 0.1, 0.5, 1\}$ using T1 input only and L1 loss for $\ell_{ckd}(.)$ in (4). If $\alpha = 0$, the model performance drops greatly, but when $\alpha > 0$, results improve, where $\alpha = 0.1$ produces the best result. This shows the importance of the cross-modal knowledge distillation loss in (7). To study the effect of a different CKD loss, we show Dice score with L2 loss for $\ell_{ckd}(.)$ in (4), with $\alpha = 0.1$. Compared with the L1 loss, we note that Dice decreases slightly with the L2 loss, especially for TC and WT.

4   Conclusion

In this paper we introduced the Learnable Cross-modal Knowledge Distillation (LCKD), which is the first method that can handle missing modality during training and testing by distilling knowledge from automatically selected important modalities for all training tasks to train other modalities. Experiments on BraTS2018 [1,14] show that LCKD reaches state-of-the-art performance in missing modality segmentation problems. We believe that our proposed LCKD has the potential to allow the use of multimodal data for training and missing-modality data per testing. One point to improve about LCKD is the greedy teacher selection per task. We plan to improve this point by transforming this problem into a meta-learning strategy, where the meta parameter is the weight for each modality, which will be optimised per task.

References

1. Bakas, S., et al.: Identifying the best machine learning algorithms for brain tumor segmentation, progression assessment, and overall survival prediction in the brats challenge. arXiv preprint arXiv:1811.02629 (2018)
2. Botev, A., Lever, G., Barber, D.: Nesterov's accelerated gradient and momentum as approximations to regularised update descent. In: 2017 International Joint Conference on Neural Networks (IJCNN), pp. 1899–1903. IEEE (2017)

3. Chartsias, A., Joyce, T., Giuffrida, M.V., Tsaftaris, S.A.: Multimodal MR synthesis via modality-invariant latent representation. IEEE Trans. Med. Imaging **37**(3), 803–814 (2017)
4. Chen, C., Dou, Q., Jin, Y., Chen, H., Qin, J., Heng, P.-A.: Robust multimodal brain tumor segmentation via feature disentanglement and gated fusion. In: Shen, D., et al. (eds.) MICCAI 2019. LNCS, vol. 11766, pp. 447–456. Springer, Cham (2019). https://doi.org/10.1007/978-3-030-32248-9_50
5. Ding, Y., Yu, X., Yang, Y.: RFNet: region-aware fusion network for incomplete multi-modal brain tumor segmentation. In: Proceedings of the IEEE/CVF International Conference on Computer Vision, pp. 3975–3984 (2021)
6. Dorent, R., Joutard, S., Modat, M., Ourselin, S., Vercauteren, T.: Hetero-modal variational encoder-decoder for joint modality completion and segmentation. In: Shen, D., et al. (eds.) MICCAI 2019. LNCS, vol. 11765, pp. 74–82. Springer, Cham (2019). https://doi.org/10.1007/978-3-030-32245-8_9
7. Dou, Q., Liu, Q., Heng, P.A., Glocker, B.: Unpaired multi-modal segmentation via knowledge distillation. IEEE Trans. Med. Imaging **39**, 2415–2425 (2020)
8. Havaei, M., Guizard, N., Chapados, N., Bengio, Y.: HeMIS: hetero-modal image segmentation. In: Ourselin, S., Joskowicz, L., Sabuncu, M.R., Unal, G., Wells, W. (eds.) MICCAI 2016. LNCS, vol. 9901, pp. 469–477. Springer, Cham (2016). https://doi.org/10.1007/978-3-319-46723-8_54
9. Hu, M., et al.: Knowledge Distillation from Multi-modal to Mono-modal Segmentation Networks. In: Martel, A.L., et al. (eds.) MICCAI 2020. LNCS, vol. 12261, pp. 772–781. Springer, Cham (2020). https://doi.org/10.1007/978-3-030-59710-8_75
10. Huang, J., Kingsbury, B.: Audio-visual deep learning for noise robust speech recognition. In: 2013 IEEE International Conference on Acoustics, Speech and Signal Processing, pp. 7596–7599. IEEE (2013)
11. Jing, M., Li, J., Zhu, L., Lu, K., Yang, Y., Huang, Z.: Incomplete cross-modal retrieval with dual-aligned variational autoencoders. In: Proceedings of the 28th ACM International Conference on Multimedia, pp. 3283–3291 (2020)
12. Kazemi, V., Elqursh, A.: Show, ask, attend, and answer: a strong baseline for visual question answering. arXiv preprint arXiv:1704.03162 (2017)
13. Loshchilov, I., Hutter, F.: SGDR: stochastic gradient descent with warm restarts. arXiv preprint arXiv:1608.03983 (2016)
14. Menze, B.H., et al.: The multimodal brain tumor image segmentation benchmark (brats). IEEE Trans. Med. Imaging **34**(10), 1993–2024 (2014)
15. Noda, K., Arie, H., Suga, Y., Ogata, T.: Multimodal integration learning of robot behavior using deep neural networks. Robot. Auton. Syst. **62**(6), 721–736 (2014)
16. Shen, Y., Gao, M.: Brain tumor segmentation on MRI with missing modalities. In: Chung, A.C.S., Gee, J.C., Yushkevich, P.A., Bao, S. (eds.) IPMI 2019. LNCS, vol. 11492, pp. 417–428. Springer, Cham (2019). https://doi.org/10.1007/978-3-030-20351-1_32
17. Wang, H., Wu, Q., Shen, C.: Soft expert reward learning for vision-and-language navigation. In: Vedaldi, A., Bischof, H., Brox, T., Frahm, J.-M. (eds.) ECCV 2020. LNCS, vol. 12354, pp. 126–141. Springer, Cham (2020). https://doi.org/10.1007/978-3-030-58545-7_8
18. Wang, H., et al.: Uncertainty-aware multi-modal learning via cross-modal random network prediction. arXiv preprint arXiv:2207.10851 (2022)
19. Wang, Y., Zhang, Y., Liu, Y., Lin, Z., Tian, J., Zhong, C., Shi, Z., Fan, J., He, Z.: ACN: adversarial co-training network for brain tumor segmentation with missing modalities. In: de Bruijne, M., et al. (eds.) MICCAI 2021. LNCS, vol. 12907, pp. 410–420. Springer, Cham (2021). https://doi.org/10.1007/978-3-030-87234-2_39

20. Yin, Q., Wu, S., Wang, L.: Unified subspace learning for incomplete and unlabeled multi-view data. Pattern Recogn. **67**, 313–327 (2017)
21. Zhang, Y., et al.: mmFormer: multimodal medical transformer for incomplete multimodal learning of brain tumor segmentation. arXiv preprint arXiv:2206.02425 (2022)
22. Zhang, Y., et al.: Modality-aware mutual learning for multi-modal medical image segmentation. In: de Bruijne, M., et al. (eds.) MICCAI 2021. LNCS, vol. 12901, pp. 589–599. Springer, Cham (2021). https://doi.org/10.1007/978-3-030-87193-2_56

EchoGLAD: Hierarchical Graph Neural Networks for Left Ventricle Landmark Detection on Echocardiograms

Masoud Mokhtari[1] , Mobina Mahdavi[1], Hooman Vaseli[1] , Christina Luong[2],
Purang Abolmaesumi[1(✉)], Teresa S. M. Tsang[2], and Renjie Liao[1(✉)]

[1] Electrical and Computer Engineering, University of British Columbia, Vancouver, BC, Canada
{masoud,mobina,hoomanv,purang,rjliao}@ece.ubc.ca
[2] Vancouver General Hospital, Vancouver, BC, Canada
{christina.luong,t.tsang}@ubc.ca

Abstract. The functional assessment of the left ventricle chamber of the heart requires detecting four landmark locations and measuring the internal dimension of the left ventricle and the approximate mass of the surrounding muscle. The key challenge of automating this task with machine learning is the sparsity of clinical labels, i.e., only a few landmark pixels in a high-dimensional image are annotated, leading many prior works to heavily rely on isotropic label smoothing. However, such a label smoothing strategy ignores the anatomical information of the image and induces some bias. To address this challenge, we introduce an **echo**cardiogram-based, hierarchical **g**raph neural network (GNN) for **l**eft ventricle l**a**ndmark **d**etection (EchoGLAD). Our main contributions are: 1) a hierarchical graph representation learning framework for multi-resolution landmark detection via GNNs; 2) induced hierarchical supervision at different levels of granularity using a multi-level loss. We evaluate our model on a public and a private dataset under the in-distribution (ID) and out-of-distribution (OOD) settings. For the ID setting, we achieve the state-of-the-art mean absolute errors (MAEs) of 1.46 mm and 1.86 mm on the two datasets. Our model also shows better OOD generalization than prior works with a testing MAE of 4.3 mm.

Keywords: Graph Neural Networks · Landmark Detection · Ultrasound

1 Introduction

Left Ventricular Hypertrophy (LVH), one of the leading predictors of adverse cardiovascular outcomes, is the condition where heart's mass abnormally increases

Supplementary Information The online version contains supplementary material available at https://doi.org/10.1007/978-3-031-43901-8_22.

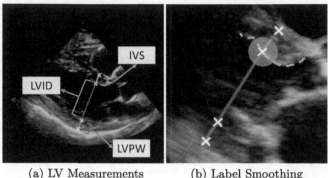

(a) LV Measurements   (b) Label Smoothing

Fig. 1. (a) IVS, LVID and LVPW measurements visualized on a PLAX echo frame. (b) If the wall landmark labels are smoothed by an isotropic Gaussian distribution, points along the visualized wall and ones perpendicular are penalized equally. Ideally, points along the walls must be penalized less.

secondary to anatomical changes in the Left Ventricle (LV) [10]. These anatomical changes include an increase in the septal and LV wall thickness, and the enlargement of the LV chamber. More specifically, Inter-Ventricular Septal (IVS), LV Posterior Wall (LVPW) and LV Internal Diameter (LVID) are assessed to investigate LVH and the risk of heart failure [21]. As shown in Fig. 1(a), four landmarks on a parasternal long axis (PLAX) echo frame can characterize IVS, LVPW and LVID, and allow cardiac function assessment. To automate this, machine learning-based (ML) landmark detection methods have gained traction.

It is difficult for such ML models to achieve high accuracy due to the sparsity of positive training signals (four or six) pertaining to the correct pixel locations. In an attempt to address this, previous works use 2D Gaussian distributions to smooth the ground truth landmarks of the LV [9,13,18]. However, as shown in Fig. 1(b), for LV landmark detection where landmarks are located at the wall boundaries (as illustrated by the dashed line), we argue that an isotropic Gaussian label smoothing approach confuses the model by being agnostic to the structural information of the echo frame and penalizing the model similarly whether the predictions are perpendicular or along the LV walls.

In this work, to address the challenge brought by sparse annotations and label smoothing, we propose a hierarchical framework based on Graph Neural Networks (GNNs) [25] to detect LV landmarks in ultrasound images. As shown in Fig. 2, our framework learns useful representations on a hierarchical grid graph built from the input echo image and performs multi-level prediction tasks.

Our contributions are summarized below.

- We propose a novel GNN framework for LV landmark detection, performing message passing over hierarchical graphs constructed from an input echo;
- We introduce a hierarchical supervision that is automatically induced from sparse annotations to alleviate the issue of label smoothing;

- We evaluate our model on two LV landmark datasets and show that it not only achieves state-of-the-art mean absolute errors (MAEs) (1.46 mm and 1.86 mm across three LV measurements) but also outperforms other methods in out-of-distribution (OOD) testing (achieving 4.3 mm).

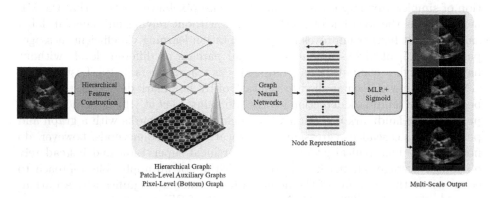

Fig. 2. Overview of our proposed model architecture. **Hierarchical Feature Construction** provides node features for the hierarchical graph representation of each echo frame where the nodes in the main graph correspond to pixels in the image, and nodes in the auxiliary graphs correspond to patches of different granularity in the image. **Graph Neural Networks** are used to process the hierarchical graph representation and produce node embeddings for the auxiliary graphs and the main graph. **Multi-Layer Perceptrons (MLPs)** are followed by a Sigmoid output function to map the node embeddings into landmark heatmaps of different granularity over the input echo frame.

2   Related Work

Various convolution-based LV landmark detection works have been proposed. Sofka *et al.* [26] use Fully Convolutional Networks to generate prediction heatmaps followed by a center of mass layer to produce the coordinates of the landmark locations. Another work [18] uses a modified U-Net [24] model to produce a segmentation map followed by a focal loss to penalize pixel predictions in close proximity of the ground truth landmark locations modulated by a Gaussian distribution. Jafari *et al.* [13] use a similar U-Net model with Bayesian neural networks [8] to estimate the uncertainty in model predictions and reject samples that exhibit high uncertainties. Gilbert *et al.* [6] smooth ground truth labels by placing 2D Gaussian heatmaps around landmark locations at angles that are statistically obtained from training data. Lastly, Duffy *et al.* [4] use atrous convolutions [1] to make predictions for LVID, IVS and LVPW measurements.

Other related works focus on the detection of cephalometric landmarks from X-ray images. These works are highly transferable to the task of LV landmark detection as they must also detect a sparse number of landmarks. McCouat *et al.* [20] is one of these works that abstains from using Gaussian label smoothing, but still relies on one-hot labels and treats landmark detection as a pixel-wise classification task. Chen *et al.* [2] is another cephalometric landmark detection work that creates a feature pyramid from the intermediate layers of a ResNet [11].

Our approach is different from prior works in that it aims to avoid the issue shown in Fig. 1(b) and the sparse annotations problem by the introduction of simpler auxiliary tasks to guide the main pixel-level task, so that the ML model learns the location of the landmarks without relying on Gaussian label smoothing. It further improves the representation learning via efficient message-passing [7,25] of GNNs among pixels and patches at different levels without having as high a computational complexity as transformers [3,19]. Lastly, while GNNs have never been applied to the task of LV landmark detection, they have been used for landmark detection in other domains. Li *et al.* [16] and Lin *et al.* [17] perform face landmark detection via modeling the landmarks with a graph and performing a cascaded regression of the locations. These methods, however, do not leverage hierarchical graphs and hierarchical supervision and instead rely on initial average landmark locations, which is not an applicable approach to echo, where the anatomy of the depicted heart can vary significantly. Additionally, Mokhtari *et al.* [22] use GNNs for the task of EF prediction from echo cine series. However, their work focuses on regression tasks.

3   Method

3.1   Problem Setup

We consider the following supervised setting for LV wall landmark detection. We have a dataset $D = \{X, Y\}$, where $|D| = n$ is the number of $\{x^i, y^i\}$ pairs such that $x^i \in X$, $y^i \in Y$, and $i \in [1, n]$. Each $x^i \in \mathbb{R}^{H \times W}$ is an echo image of the heart, where H and W are height and width of the image, respectively, and each y^i is the set of four point coordinates $[(h_1^i, w_1^i), (h_2^i, w_2^i), (h_3^i, w_3^i), (h_4^i, w_4^i)]$ indicating the landmark locations in x^i. Our goal is to learn a function $f :$ $\mathbb{R}^{H \times W} \mapsto \mathbb{R}^{4 \times 2}$ that predicts the four landmark coordinates for each input image. *A figure in the supp. material further clarifies how the model generates landmark location heatmaps on different scales (Fig. 2).*

3.2   Model Overview

As shown in Fig. 2, each input echo frame is represented by a hierarchical grid graph where each sub-graph corresponds to the input echo frame at a different resolution. The model produces heatmaps over both the main pixel-level task as well as the coarse auxiliary tasks. While the pixel-level heatmap prediction is of main interest, we use a hierarchical multi-level loss approach where the model's prediction over auxiliary tasks is used during training to optimize the model through comparisons to coarser versions of the ground truth. The intuition behind such an approach is that the model learns nuances in the data by performing landmark detection on the easier auxiliary tasks and uses this established reasoning when performing the difficult pixel-level task.

3.3   Hierarchical Graph Construction

To learn representations that better capture the dependencies among pixels and patches, we introduce a hierarchical grid graph along with multi-level prediction tasks. As an example, the simplest task consists of a grid graph with only four nodes, where each node corresponds to four equally-sized patches in the original echo image. In the main task (the one that is at the bottom in Fig. 2 and is the most difficult), the number of nodes is equal to the total number of pixels.

More formally, let us denote a graph as $G = (V, E)$, where V is the set of nodes, and E is the set of edges in the graph such that if $v_i, v_j \in V$ and there is an edge from v_i to v_j, then $e_{i,j} \in E$. To build hierarchical task representations, for each image $x \in X$ and the ground truth $y \in Y$, K different auxiliary graphs $G_k(V_k, E_k)$ are constructed using the following steps for each $k \in [1, K]$:

1. $2^k \times 2^k = 4^k$ nodes are added to V_k to represent each patch in the image. Note that the larger values of k correspond to graphs of finer resolution, while the smaller values of k correspond to coarser graphs.
2. Grid-like, undirected edges are added such that $e_{m-1,q}, e_{m+1,q}, e_{m,q-1}, e_{m,q+1} \in E_k$ for each $m, q \in [1 \ldots 2^k]$ if these neighbouring nodes exist in the graph (border nodes will not have four neighbouring nodes).
3. A patch feature embedding z_j^k, where $j \in [1 \ldots 4^k]$ is generated and associated with that patch (node) $v_j \in V_k$. The patch feature construction technique is described in Sect. 3.4.
4. Binary node labels $\hat{y}_k \in \{0,1\}^{4^k \times 4}$ are generated such that $\hat{y}_{kj} = 1$ if at least one of the ground truth landmarks in y is contained in the patch associated with node $v_j \in V_k$. Note that for each auxiliary graph, four different one-hot labels are predicted, which correspond to each of the four landmarks required to characterize LV measurements.

The main graph, G_{main}, has a grid structure and contains $H \times W$ nodes regardless of the value of K, where each node corresponds to a pixel in the image. Additionally, to allow the model to propagate information across levels, we add inter-graph edges such that each node in a graph is connected to four nodes in the corresponding region in the next finer graph as depicted in Fig. 2.

3.4   Node Feature Construction

The graph representation described in Sect. 3.3 is not complete without proper node features, denoted by $z \in \mathbb{R}^{|V| \times d}$, characterizing patches or pixels of the image. To achieve this, the grey-scale image is initially expanded in the channel dimension using a CNN. The features are then fed into a U-Net where the decoder part is used to obtain node features such that deeper layer embeddings correspond to the node features for the finer graphs. This means that the main pixel-level graph would have the features of the last layer of the network. *A figure clarifying node feature construction is provided in the supp. material (Fig. 1).*

3.5 Hierarchical Message Passing

We now introduce how we perform message passing on our constructed hierarchical graph using GNNs to learn node representations for predicting landmarks.

The whole hierarchical graph created for each sample, *i.e.*, the main graph, auxiliary graphs, and cross-level edges, are collectively denoted as G^i, where $i \in [1, \ldots, n]$. Each G^i is fed into GNN layers followed by an MLP:

$$h_{\text{nodes}}^{l+1} = \text{ReLU}(\text{GNN}_l(G^i), h_{\text{nodes}}^l), \quad l \in [0, \ldots, L] \tag{1}$$

$$h_{\text{out}} = \sigma(\text{MLP}(h_{\text{nodes}^{L+1}})), \tag{2}$$

where σ is the Sigmoid function, $h_{\text{nodes}}^l \in \mathbb{R}^{|V_{G^i}| \times d}$ is the set of d-dimensional embeddings for all nodes in the graph at layer l, and $h_{\text{out}} \in [0, 1]^{|V_{G^i}| \times 4}$ is the four-channel prediction for each node with each channel corresponding to a heatmap for each of the pixel landmarks. The initial node features h_{nodes}^1 are set to the features z described in Sects. 3.3 and 3.4. The coordinates $(x_{\text{out}}^p, y_{\text{out}}^p)$ for each landmark location $p \in [1, 2, 3, 4]$ are obtained by taking the expected value of individual heatmaps h_{out}^p along the x and y directions such that:

$$x_{\text{out}}^p = \sum_{s=1}^{|V_{G^i}|} \text{softmax}(h_{\text{out}}^p)_s * \text{loc}_x(s), \tag{3}$$

where similar operations are performed in the y direction for y_{out}^p. Here, we vectorize the 2D heatmap into a single vector and then feed it to the softmax. loc_x and loc_y return the x and y positions of a node in the image. It must be noted that unlike some prior works such as Duffy *et al.* [4] that use post-processing steps such as imposing thresholds on the heatmap values, our work directly uses the output heatmaps to find the final predictions.

3.6 Training and Objective Functions

To train the network, we leverage two types of objective functions. 1) *Weighted Binary Cross Entropy (BCE):* Since the number of landmark locations is much smaller than non-landmark locations, we use a weighted BCE loss; 2) *L2 regression of landmark coordinates:* We add a regression objective which is the L2 loss between the predicted coordinates and the ground truth labels.

4 Experiments

4.1 Datasets

Internal Dataset: Our private dataset contains 29,867 PLAX echo frames, split in a patient-exclusive manner with 23824, 3004, and 3039 frames for training, validation, and testing, respectively. **External Dataset:** The public Unity

Imaging Collaborative (UIC) [12] LV landmark dataset consists of a combination of 3822 end-systolic and end-diastolic PLAX echo frames acquired from seven British echocardiography labs. The provided splits contain 1613, 298, and 1911 training, validation, and testing samples, respectively. For both datasets, we down-sample the frames to a fixed size of 224 × 224.

4.2 Implementation Details

Our model creates $K = 7$ auxiliary graphs. For the node features, the initial single-layer CNN uses a kernel size of 3 and zero-padding to output features with a dimension of $224 \times 224 \times 4$ ($C = 4$). The U-Net's encoder contains 7 layers with $128 \times 128, 64 \times 64, 32 \times 32, 16 \times 16, 8 \times 8, 4 \times 4$, and 2×2 spatial dimensions, and 8, 16, 32, 64, 128, 256, and 512 number of channels, respectively. Three Graph Convolutional Network (GCN) [15] layers ($L = 3$) with a hidden node dimension of 128 are used. To optimize the model, we use the Adam optimizer [14] with an initial learning rate of 0.001, β of (0.9, 0.999) and a weight decay of 0.0001, and for the weighted BCE loss, we use a weight of 9000. The model is implemented using PyTorch [23] and Pytorch Geometric [5] and is trained on two 32-GB Nvidia Titan GPUs. Our code-base is publicly available at https://github.com/MasoudMo/echoglad.

4.3 Results

We evaluate models using Mean Absolute Error (MAE) in mm, and Mean Percent Error (MPE) in percents, which is formulated as $\text{MPE} = 100 \times \frac{|L_{\text{pred}} - L_{\text{true}}|}{L_{\text{true}}}$, where L_{pred} and L_{true} are the prediction and ground truth values for every measurement. We also report the Success Detection Rate (SDR) for LVID for 2 and 6 mm thresholds. This rate shows the percentage of samples where the absolute error between ground truth and LVID predictions is below the specific threshold. These thresholds are chosen based on the healthy ranges for IVS (0.6–1.1 cm), LVID (2.0–5.6 cm), and LVPW (0.6–0.1 cm). Hence, the 2 mm threshold provides a stringent evaluation of the models, while the 6 mm threshold facilitates the assessment of out-of-distribution performance.

In-Distribution (ID) Quantitative Results. In Table 1, we compare the performance of our model with previous works in the ID setting where the training and test sets come from the same distribution (e.g., the same clinical setting), we separately train and test the models on the private and the public dataset. *The results for the public dataset are provided in the supp. material (Table 1).*

Out-of-Distribution (OOD) Quantitative Results. To investigate the generalization ability of our model compared to previous works, we train all models on the private dataset (which consists of a larger number of samples compared to UIC), and test the trained models on the public UIC dataset as shown in Table 2. Based on our visual assessment, the UIC dataset looks very different compared to the private dataset, thus serving as an OOD test-bed.

Table 1. Quantitative results on the private test set for models trained on the private training set. We see that our model has the best average performance over the three measurements, which shows the superiority of our model in the in-distribution setting for high-data regime.

Model	MAE [mm] ↓			MPE [%] ↓			SDR[%] of LVID <↑	
	LVID	IVS	LVPW	LVID	IVS	LVPW	2.0 mm	6.0 mm
Gilbert et al. [6]	2.9	1.4	1.4	6.5	14.5	15.2	48.1	88.9
Lin et al. [18]	9.4	11.2	9.0	21.2	116.5	92.9	26.0	49.1
McCouat et al. [20]	**2.2**	1.3	1.4	**4.8**	13.5	15.1	58.3	93.9
Chen et al. [2]	2.3	1.2	1.2	5.2	12.6	13.8	60.4	92.6
Duffy et al. [4]	2.5	1.2	1.2	5.4	13.2	13.5	52.1	93.0
Ours	**2.2**	**1.1**	**1.1**	**4.8**	**11.2**	**12.2**	**62.4**	**94.4**

Table 2. Quantitative results on the public UIC test set for models trained on the private training set. This table shows the out-of-distribution performance of the models when trained on a larger dataset and tested on a smaller external dataset. We can see that in this case, our model outperforms previous works by a large margin, which attests to the generalizability of our framework.

Model	MAE [mm] ↓			MPE [%] ↓			SDR[%] of LVID < ↑	
	LVID	IVS	LVPW	LVID	IVS	LVPW	2.0 mm	6.0 mm
Gilbert et al. [6]	9.5	4.8	4.1	23.5	32.3	26.8	22.5	52.2
Lin et al. [18]	51.5	51.7	41.3	121.0	375.8	298.0	11.3	24.6
McCouat et al. [20]	5.9	3.6	4.4	18.5	30.5	36.4	34.6	72.3
Chen et al. [2]	7.4	5.3	6.9	22.5	49.4	62.4	28.9	65.3
Duffy et al. [4]	13.7	4.1	5.5	36.8	36.4	45.4	6.2	20.6
Ours	**5.8**	**2.8**	**4.3**	**18.4**	**23.8**	**34.6**	**35.8**	**74.9**

Table 3. Ablation results on the validation set of our private dataset. Vanilla U-Net uses a simple U-Net model, while U-Net Main Graph only uses the pixel-level graph (no aux. graphs). Main Model is our proposed approach. Lastly, Single-Scale Loss has the same framework as the Main Model but only computes the loss for the model's predictions on the main graph (no multi-scale loss).

Model	MPE [%]		
	LVID	IVS	LVPW
Vanilla U-Net	5.31	13.17	13.47
U-Net Main Graph	4.98	11.67	12.78
Single-Scale Loss	5.41	12.37	12.8
Main Model	**4.91**	**11.45**	**12.36**

Qualitative Results. *Failure cases are shown in supp. material (Fig. 3).*

Ablation Studies. In Table 3, we show the benefits of a hierarchical graph representation with a multi-scale objective for the task of LV landmark detection. *We provide a qualitative view of the ablation study in supp. material (Fig. 4).*

5   Conclusion and Future Work

In this work, we introduce a novel hierarchical GNN for LV landmark detection. The model performs better than the state-of-the-art on most measurements without relying on label smoothing. We attribute this gain in performance to two main contributions. First, our choice of representing each frame with a hierarchical graph has facilitated direct interaction between pixels at differing scales. This approach is effective in capturing the nuanced dependencies amongst the landmarks, bolstering the model's performance. Secondly, the implementation of a multi-scale objective function as a supervisory mechanism has enabled the model to construct a superior inductive bias. This approach allows the model to leverage simpler tasks to optimize its performance in the more challenging pixel-level landmark detection task.

For future work, we believe that the scalability of the framework for higher-resolution images must be studied. Additionally, extension of the model to video data can be considered since the concept of intra-scale and inter-scale edges connecting nodes could be extrapolated to include temporal edges linking similar spatial locations across frames. Such an approach could greatly enhance the model's performance in unlabeled frames, mainly through the enforcement of consistency in predictions from frame to frame.

References

1. Chen, L.C., Papandreou, G., Kokkinos, I., Murphy, K., Yuille, A.L.: DeepLab: semantic image segmentation with deep convolutional nets, atrous convolution, and fully connected CRFs. IEEE Trans. Pattern Anal. Mach. Intell. **40**(4), 834–848 (2018)
2. Chen, R., Ma, Y., Chen, N., Lee, D., Wang, W.: Cephalometric landmark detection by attentive feature pyramid fusion and regression-voting. In: Shen, D., et al. (eds.) MICCAI 2019. LNCS, vol. 11766, pp. 873–881. Springer, Cham (2019). https://doi.org/10.1007/978-3-030-32248-9_97
3. Dosovitskiy, A., et al.: An image is worth 16x16 words: transformers for image recognition at scale. In: International Conference on Learning Representations (2021)
4. Duffy, G., et al.: High-throughput precision phenotyping of left ventricular hypertrophy with cardiovascular deep learning. JAMA Cardiol. **7**(4), 386–395 (2022)
5. Fey, M., Lenssen, J.E.: Fast graph representation learning with PyTorch geometric. In: ICLR Workshop on Representation Learning on Graphs and Manifolds (2019)

6. Gilbert, A., Holden, M., Eikvil, L., Aase, S.A., Samset, E., McLeod, K.: Automated left ventricle dimension measurement in 2D cardiac ultrasound via an anatomically meaningful CNN approach. In: Wang, Q., et al. (eds.) PIPPI/SUSI -2019. LNCS, vol. 11798, pp. 29–37. Springer, Cham (2019). https://doi.org/10.1007/978-3-030-32875-7_4
7. Gilmer, J., Schoenholz, S.S., Riley, P.F., Vinyals, O., Dahl, G.E.: Neural message passing for quantum chemistry. In: Proceedings of the 34th International Conference on Machine Learning, vol. 70, pp. 1263–1272. JMLR.org (2017)
8. Goan, E., Fookes, C.: Bayesian neural networks: an introduction and survey. In: Mengersen, K.L., Pudlo, P., Robert, C.P. (eds.) Case Studies in Applied Bayesian Data Science. LNM, vol. 2259, pp. 45–87. Springer, Cham (2020). https://doi.org/10.1007/978-3-030-42553-1_3
9. Goco, J.A.D., Jafari, M.H., Luong, C., Tsang, T., Abolmaesumi, P.: An efficient deep landmark detection network for PLAX EF estimation using sparse annotations. In: Linte, C.A., Siewerdsen, J.H. (eds.) Medical Imaging 2022: Image-Guided Procedures, Robotic Interventions, and Modeling, vol. 12034, p. 120340N. International Society for Optics and Photonics, SPIE (2022)
10. Gradman, A.H., Alfayoumi, F.: From left ventricular hypertrophy to congestive heart failure: Management of hypertensive heart disease. Progress Cardiovas. Dis. 48(5), 326–341 (2006). hypertension 2006 Update
11. He, K., Zhang, X., Ren, S., Sun, J.: Deep residual learning for image recognition. In: 2016 IEEE Conference on Computer Vision and Pattern Recognition, pp. 770–778 (2016)
12. Howard, J.P., et al.: Automated left ventricular dimension assessment using artificial intelligence developed and validated by a UK-wide collaborative. Circ. Cardiovasc. Imaging 14, e011951–e011951 (2021)
13. Jafari, M.H., et al.: U-land: uncertainty-driven video landmark detection. IEEE Trans. Med. Imaging 41(4), 793–804 (2022)
14. Kingma, D., Ba, J.: Adam: a method for stochastic optimization. In: International Conference on Learning Representations (2014)
15. Kipf, T.N., Welling, M.: Semi-supervised classification with graph convolutional networks. In: International Conference on Learning Representations (2017)
16. Li, W., et al.: Structured landmark detection via topology-adapting deep graph learning. In: Vedaldi, A., Bischof, H., Brox, T., Frahm, J.-M. (eds.) ECCV 2020. LNCS, vol. 12354, pp. 266–283. Springer, Cham (2020). https://doi.org/10.1007/978-3-030-58545-7_16
17. Lin, C., et al.: Structure-coherent deep feature learning for robust face alignment. IEEE Trans. Image Process. 30, 5313–5326 (2021)
18. Lin, J., et al.: Reciprocal landmark detection and tracking with extremely few annotations. In: IEEE/CVF Conference on Computer Vision and Pattern Recognition, pp. 15165–15174. IEEE Computer Society (2021)
19. Liu, Z., et al.: Swin transformer: hierarchical vision transformer using shifted windows. In: 2021 IEEE/CVF International Conference on Computer Vision, pp. 9992–10002. IEEE Computer Society (oct 2021)
20. McCouat, J., Voiculescu, I.: Contour-hugging heatmaps for landmark detection. In: 2022 IEEE/CVF Conference on Computer Vision and Pattern Recognition, pp. 20565–20573 (2022)
21. McFarland, T.M., Alam, M., Goldstein, S., Pickard, S.D., Stein, P.D.: Echocardiographic diagnosis of left ventricular hypertrophy. Circulation 50 (1978)

22. Mokhtari, M., Tsang, T., Abolmaesumi, P., Liao, R.: Echognn: explainable ejection fraction estimation with graph neural networks. In: Wang, L., Dou, Q., Fletcher, P.T., Speidel, S., Li, S. (eds.) MICCAI. LNCS, vol. 13434, pp. 360–369. Springer, Cham (2022)

23. Paszke, A., et al.: Pytorch: an imperative style, high-performance deep learning library. In: Wallach, H., Larochelle, H., Beygelzimer, A., d' Alché-Buc, F., Fox, E., Garnett, R. (eds.) Advances in Neural Information Processing Systems, vol. 32. Curran Associates, Inc. (2019)

24. Ronneberger, O., Fischer, P., Brox, T.: U-Net: convolutional networks for biomedical image segmentation. In: Navab, N., Hornegger, J., Wells, W.M., Frangi, A.F. (eds.) MICCAI 2015. LNCS, vol. 9351, pp. 234–241. Springer, Cham (2015). https://doi.org/10.1007/978-3-319-24574-4_28

25. Scarselli, F., Gori, M., Tsoi, A.C., Hagenbuchner, M., Monfardini, G.: The graph neural network model. IEEE Trans. Neural Networks 20(1), 61–80 (2009)

26. Sofka, M., Milletari, F., Jia, J., Rothberg, A.: Fully convolutional regression network for accurate detection of measurement points. In: Cardoso, M.J., et al. (eds.) DLMIA/ML-CDS -2017. LNCS, vol. 10553, pp. 258–266. Springer, Cham (2017). https://doi.org/10.1007/978-3-319-67558-9_30

Class-Aware Feature Alignment
for Domain Adaptive Mitochondria
Segmentation

Dan Yin[1,2], Wei Huang[1,2], Zhiwei Xiong[1,2], and Xuejin Chen[1,2(✉)]

[1] National Engineering Laboratory for Brain-inspired Intelligence Technology and
Application, University of Science and Technology of China, Hefei 230027, China
{ydaugust,weih527}@email.ustc.edu.cn, {zwxiong,xjchen99}@ustc.edu.cn
[2] Institute of Artificial Intelligence, Hefei Comprehensive National Science Center,
Hefei 230088, China

Abstract. Unsupervised domain adaptation (UDA) has gained great
popularity in mitochondria segmentation, aiming to improve the adaptability of models from the labeled source domain to the unlabeled target
domain via domain alignment. However, existing UDA methods only
focus on aligning domains on the prediction level, while ignoring the feature space containing more adequate information than the predictions. In
this paper, we propose a class-aware domain adaptation method for mitochondria segmentation on the feature level, which relies on the prototype
representation to achieve more fine-grained alignment. In particular, we
first extract the feature centroids of classes from the source domain as
prototypes. Leveraging the extracted prototypes as a bridge, we constrain
that features belonging to the same class but from different domains are
pulled closer to each other, achieving the class-aware alignment. Meanwhile, we derive a segmentation prediction directly from feature space
based on the distance between target features and source prototypes.
By incorporating a pseudo label to supervise the learning of this prediction, the feature distribution gap across domains is further reduced.
Furthermore, to take full advantage of the potential of target domain,
we propose an intra-domain consistency constraint to maintain consistent predictions of samples perturbed differently from the target image.
Extensive experiments on different datasets demonstrate the superiority
of our proposed method over existing UDA methods. Code is available
at https://github.com/Danyin813/CAFA.

Keywords: Unsupervised domain adaptation · Class-aware
alignment · Mitochondria segmentation

1 Introduction

Mitochondria segmentation from electron microscopy (EM) images is pivotal to
mitochondria morphological analysis [15]. With pixel-wise annotations, existing supervised mitochondria segmentation methods [3,9,10,14] have achieved

---

Supplementary Information The online version contains supplementary material
available at https://doi.org/10.1007/978-3-031-43901-8_23.

extraordinary advances in the test data, when the training data and test data come from the same distribution. However, the data distribution of EM images varies in real scenarios due to the diversity in imaging devices, collected organisms and tissues. The different distributions between training and test data, *i.e.*, *domain shift* [13], lead to drastic performance drops on the test data. Manual annotations and model finetuning can ameliorate this problem but at a huge cost. Instead, unsupervised domain adaptive (UDA) mitochondria segmentation methods [2,4,7,11,16,17,21], aiming to transfer the knowledge learned in the labeled dataset (source domain) to unlabeled data (target domain) without any annotations, have gained great popularity in the community.

The mainstream of previous works focus on aligning the distributions of the source and target domains with supervision directly *on the output segmentation maps*. One line of these works use the model pretrained on source domain to obtain **pseudo labels** for target domain [1,21], the performance of which highly relies on the quality of the pseudo labels. Other methods are mainly based on GAN [7,16], where an additional **domain adversarial learning** task, designed to reduce the domain gap between source and target domains, is jointly optimized with segmentation task. The feature space, having higher dimension than the predictions, can express more adequate class-aware knowledge. However, these methods perform domain alignment on the output space, which has insufficient information compared with the feature space, hindering effective alignment.

In this work, to take full advantage of the sufficient class related information in the feature space, we propose a **class-aware domain alignment in the feature space** for mitochondria segmentation, which relies on the *prototype representation* [8,22] to achieve fine-grained feature alignment. Specifically, 1) we first extract the source feature centroid of each class as prototype. To make the prototypes represent source class knowledge better, we minimize the distance between prototype and its within-class source features, as well as push different prototypes away from each other. Also, we select *partial* target features close enough to source prototypes and minimize their distance to align domains at class level. 2) To further supervise *all* target features, we derive the closest prototype as predicted result for each target feature vector based on its distance to prototypes, resulting in a segmentation result directly from the feature space without the segmentation head. A pseudo label is utilized to supervise these predictions for further cross-domain alignment with class knowledge. 3) Though cross-domain alignment can introduce knowledge from source domain to the target domain, there still exists *additional* potential information useful to segmentation in the target domain [18]. Taking this into consideration, we further propose an intra-domain consistency constraint for target samples, where two input images perturbed differently from the same image are enforced to generate the same features and predictions.

Our contributions can be summarized as follows: 1) We propose a class-aware feature alignment method for domain adaptive mitochondria segmentation. To our best knowledge, it is the first attempt to align source and target domains on the feature level in UDA for EM mitochondria segmentation. 2) Our class-aware

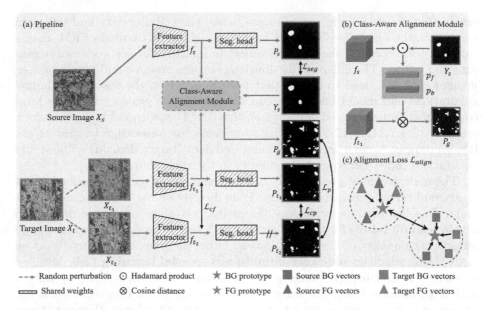

Fig. 1. The overall framework of our proposed method. (a) Images are first fed into the feature extractor to extract image features, which are then used to realize feature alignment by the class-aware alignment module and obtain the segmentation predictions by the segmentation (Seg.) head. We randomly perturb the target image X_t to obtain two augmented counterparts X_{t_1} and X_{t_2} with different augmentations. An intra-domain consistency constraint is incorporated in the feature level ($\mathcal{L}_{cf}$) and the prediction level ($\mathcal{L}_{cp}$). (b) In the Class-Aware Alignment Module, the source feature f_s and its corresponding groundtruth label are used to extract the centroids/prototypes (*i.e.*, p_f and p_b) of each class. Based on the distance between target features f_{t_1} and the prototypes, we can obtain the distance map P_g, representing the segmentation prediction directly from the features, which is further supervised by the pseudo label P_{t_2} ($\mathcal{L}_p$). (c) The illustration of the class-aware alignment loss $\mathcal{L}_{align}$, where the features within the same class are pulled together and the features belonging to different classes are pushed away from each other.

feature alignment relies on the source prototypes, which represent class knowledge from the feature space. With these prototypes, an innovative distance-based alignment and pseudo-labeling are incorporated to achieve class-aware feature alignment. 3) We propose an intra-domain consistency constraint in the target domain to tap into the potential target domain information. 4) We conduct thorough experiments on various EM dataset benchmarks and our proposed method achieves state-of-the-art performance for mitochondria segmentation.

2   Class-Aware Feature Alignment

Problem Formulation. Unsupervised domain adaptation (UDA) aims to transfer the knowledge learned from the labeled source domain to the unlabeled target domain. In our work, we denote the source domain as $\mathcal{D}_S = \{X^S, Y^S\} = \{(x_i^s, y_i^s)\}_{i=1}^M$ with M samples, where y_i^s is the groundtruth binary segmentation map of the input image x_i^s. The unlabeled target domain is denoted as $\mathcal{D}_T = \{X^t\} = \{x_i^t\}_{i=1}^N$ with N samples. The overall framework of our proposed method is shown in Fig. 1.

Prototype Extraction. Considering there exists more plentiful class-aware information in the feature space than the predictions, we propose the class-aware alignment for better adaptation in the feature space. To achieve class-aware alignment, we first derive the class-aware source prototypes from the source features with the corresponding labels. The prototypes can be calculated as the centroid of each class in the feature space:

$$p_c^s = \frac{\sum_{b=1}^{B_s} \sum_{h=1}^{H_s} \sum_{w=1}^{W_s} f_{b,h,w}^s \mathbb{1}[Y_{b,h,w}^s = c]}{\sum_{b=1}^{B_s} \sum_{h=1}^{H_s} \sum_{w=1}^{W_s} \mathbb{1}[Y_{b,h,w}^s = c]}, \tag{1}$$

where $f_{b,h,w}^s \in \mathbb{R}$ is the source feature vectors, B_s is the batch size, and H_s, W_s is the height and width of the features. c is the index of class number C. The maximum of C is 1. The prototypes can represent the class knowledge in the source domain.

Inter- and Intra-class Constraints. To make the source prototypes represent the class-discriminative source knowledge more accurately, we incorporate inter- and intra-class constraints on the prototypes, which can further help better class-aware alignment across domains. The inter-class loss $\mathcal{L}_{inter}^s$ intends to push the prototypes of different classes far away from each other, which can be implemented by minimizing the average cosine distance of different prototype pairs:

$$\mathcal{L}_{inter}^s = \sum_{i=0}^{C} \sum_{j>i}^{C} \frac{p_i^s p_j^s}{|p_i^s||p_j^s|}. \tag{2}$$

In contrast, the intra-class loss $\mathcal{L}_{intra}^s$ is designed to pull the feature instance point closer to its corresponding prototype, *i.e.*, making the feature distribution of the same class more concentrated/compact. The intra-class loss can be formulated as maximizing the average cosine distance between the prototype and the features belonging to the same class:

$$\mathcal{L}_{intra}^s = 1 - \sum_{c=0}^{C} \frac{f_c^s p_c^s}{|f_c^s||p_c^s|}, \tag{3}$$

It is not straightforward to align target domain to source domain in the class level, considering the lack of groundtruth labels in target domain. To achieve more reliable class-aware alignment for target samples, we only perform alignment on the instances with higher confidence. Specifically, we first calculate the cosine distance between each target feature and all the source prototypes, and only select instances $\{\tilde{f}^t\}$ the distance of which is closer than a preset threshold τ. The intra-class alignment loss enforces $\tilde{f}_c^t$ to be closer to its corresponding prototype p_c^t:

$$\mathcal{L}_{intra}^t = 1 - \sum_{c=0}^{C} \frac{\tilde{f}_c^t p_c^s}{|\tilde{f}_c^t||p_c^s|}, \tag{4}$$

The class-aware alignment loss $\mathcal{L}_{align}$ is the combination of these three losses, i.e., $\mathcal{L}_{align} = \mathcal{L}_{intra}^s + \mathcal{L}_{inter}^s + \mathcal{L}_{intra}^t$. It is noteworthy that the alignment loss is optimized directly on the feature space instead of the final output predictions, considering there is more abundant information in the feature space.

Pseudo Supervision. The above mentioned alignment loss $\mathcal{L}_{align}$ only affects partial target features with higher confidence. To further force the alignment across domains in the feature space, we incorporate a pseudo supervision on the feature space. Specifically, based on the cosine distance between the feature of each location and the source prototypes, we can attain a distance map P_g, which can be regarded as a segmentation prediction directly from feature space instead of the prediction head. We utilize a pseudo label map P_{t2} as groundtruth to supervise the learning of P_g, leading to alignment directly on feature space. The formulation of P_{t2} will be discussed in the later section. The supervision is the standard cross entropy loss:

$$\mathcal{L}_p = CE(P_g, P_{t_2}). \tag{5}$$

Intra-domain Consistency. The alignment cross domains will borrow the knowledge from source domain to target domain. However, there exists abundant knowledge and information in the target domain itself [18]. To further exploit the sufficient knowledge existed in target domain, we propose an intra-domain consistency constraint in target domain. Specifically, for each target input image I_t, we first augment it with two different random augmentation strategies, resulting in I_{t1} and I_{t2}, which are then fed into the network for segmentation prediction. We incorporate two consistency losses on the feature level $\mathcal{L}_{cf}$ and the final prediction level $\mathcal{L}_{cp}$, respectively:

$$\mathcal{L}_{cf} = MSE(f_{t_1}, f_{t_2}), \quad \mathcal{L}_{cp} = CE(P_{t_1}, P_{t_2}), \tag{6}$$

where MSE denotes the standard mean squared error loss.

Training and Inference. During the training phase, the total training objective $\mathcal{L}_{total}$ is formulated as :

$$\mathcal{L}_{total} = \mathcal{L}_{seg}^s + \lambda_{align}\mathcal{L}_{align} + \lambda_p\mathcal{L}_p + \lambda_{cf}\mathcal{L}_{cf} + \lambda_{cp}\mathcal{L}_{cp}, \tag{7}$$

where $\mathcal{L}_{seg}^s$ denotes the supervised segmentation loss with the cross-entropy loss and $\lambda_{\{align,p,cf,cp\}}$ are the hyperparameters for balancing different terms. Note

Table 1. Quantitative comparisons on the Lucchi and MitoEM datasets. *Oracle* denotes the model is trained on target with groundtruth labels, while *NoAdapt* represents the model pretrained on source is directly applied in target for inference without any adaptation strategy. The results of Oracle, NoAdapt, UALR, DAMT-Net, DA-VSN and DA-ISC are adopted from [7].

Methods	VNC III → Lucchi (Subset1)				VNC III → Lucchi (Subset2)			
	mAP(%)	F1(%)	MCC(%)	IoU(%)	mAP(%)	F1(%)	MCC(%)	IoU(%)
Oracle	-	92.7	-	86.5	-	93.9	-	88.6
NoAdapt	-	57.3	-	40.3	-	61.3	-	44.3
Advent [19]	78.9	74.8	73.3	59.7	90.5	82.8	81.8	70.7
UALR [21]	80.2	72.5	71.2	57.0	87.2	78.8	77.7	65.2
DAMT-Net [16]	-	74.7	-	60.0	-	81.3	-	68.7
DA-VSN [6]	82.8	75.2	73.9	60.3	91.3	83.1	82.2	71.1
DA-ISC [7]	89.5	81.3	80.5	68.7	92.4	85.2	84.5	74.3
Ours	**91.1**	**83.4**	**82.8**	**71.8**	**94.8**	**85.8**	**85.4**	**75.4**
Methods	MitoEM-R → MitoEM-H				MitoEM-H → MitoEM-R			
	mAP(%)	F1(%)	MCC(%)	IoU(%)	mAP(%)	F1(%)	MCC(%)	IoU(%)
Oracle	97.0	91.6	91.2	84.5	98.2	93.2	92.9	87.3
NoAdapt	74.6	56.8	59.2	39.6	88.5	76.5	76.8	61.9
Advent [19]	89.7	82.0	81.3	69.6	93.5	85.4	84.8	74.6
UALR [21]	90.7	83.8	83.2	72.2	92.6	86.3	85.5	75.9
DAMT-Net [16]	92.1	84.4	83.7	73.0	94.8	86.0	85.7	75.4
DA-VSN [6]	91.6	83.3	82.6	71.4	94.5	86.7	86.3	76.5
DA-ISC [7]	92.6	85.6	84.9	74.8	**96.8**	88.5	88.3	79.4
Ours	**92.8**	**86.6**	**86.0**	**76.3**	**96.8**	**89.2**	**88.9**	**80.6**

that the feature extractor and the segmentation head are shared weights in the training phase. Their detailed structures can be found in the supplementary material. During the inference phase, we only adopt the trained feature extractor and segmentation head to predict the target images.

3   Experiments

Datasets. Following the previous work [7], our experiments involve four challenging EM datasets for domain adaptive mitochondria segmentation tasks, *i.e.*, VNC III [5]→ Lucchi (Subset1) [12], VNC III→ Lucchi (Subset2) [12], MitoEM-H [20] → MitoEM-R [20] and MitoEM-R → MitoEM-H. **VNC III** [5] is imaged from the Drosophila ventral nerve cord by ssTEM. The physical resolution of the pixel is $50 \times 5 \times 5$ nm^3. The dataset consists of 20 images, and their resolution is 1024×1024. **Lucchi** [12] is imaged from the hippocampus of mice collected by FIB-SEM. The physical resolution of the pixel is $5 \times 5 \times 5$ nm^3, the training (Subset1) and test (Subset2) sets both have 165 images with 1024×768 resolution. **MitoEM** [20] contains two image volumes imaged by mbSEM, one is from the

Table 2. Quantitative comparisons for the 3D instance segmentation results on the MitoEM dataset.

Methods	MitoEM-R → MitoEM-H		MitoEM-H → MitoEM-R	
	AP^{50}(%)	AP^{75}(%)	AP^{50}(%)	AP^{75}(%)
Advent [19]	43.6	17.3	56.4	27.0
UALR [21]	56.4	29.1	55.4	33.6
DAMT-Net [16]	55.2	29.5	54.6	25.5
DA-VSN [6]	53.1	24.6	60.2	29.3
DA-ISC [7]	60.0	37.4	63.0	34.7
Ours	**65.6**	**46.3**	**68.5**	**42.4**

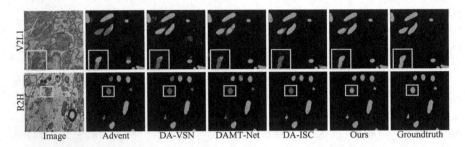

Fig. 2. Visual comparisons of different domain adaptive mitochondria segmentation methods, from VNC III to Lucchi (Subset1), *i.e.*, V2L1, and MitoEM-R to MitoEM-H, *i.e.*, R2H. The pixels in green, red and blue denote the true-positive, false-negative and false-positive segmentation results respectively. More 3D instance segmentation visualizations can be found in the supplementary material. (Color figure online)

rat cortex, named as MitoEM-R, and the other one is from the human cortex, named as MitoEM-H. Each volume contains 1000 images, where the instance-level annotations of the first 500 images are available. The physical resolution of the pixel is $30 \times 8 \times 8\,nm^3$, and the image size is 4096×4096.

Evaluation Metrics. To thoroughly evaluate the performance of models, we conduct comparisons both on semantic-level and instance-level predictions. 1) Following [7], we compare different methods with mAP, F1, MCC and IoU scores on the 2D binary segmentation. 2) Considering that the quantity, size and morphology of mitochondria are pivotal to related studies, we further evaluate on the 3D instance segmentation task. Following [20], we take AP^{50} and AP^{75} as the metrics to quantitatively compare the performance of different methods.

Implementation Details. Our network architecture is following [7]. We crop each image into 512×512 as patch to input feature extractor. All models are trained using the Adam optimizer with $\beta_1 = 0.9$ and $\beta_2 = 0.999$. The learning

Table 3. Ablation results for the effectiveness of each loss term.

MitoEM-R → MitoEM-H

Settings	$\mathcal{L}_{seg}^{s}$	$\mathcal{L}_{align}$	$\mathcal{L}_p$	$\mathcal{L}_{cf}$	$\mathcal{L}_{cp}$	mAP(%)	F1(%)	MCC(%)	IoU(%)
①	✓					87.2	78.7	78.1	64.9
②	✓	✓				88.8	83.3	82.8	71.5
③	✓	✓	✓			91.1	84.2	83.6	72.7
④	✓	✓	✓	✓		91.5	84.2	85.4	75.4
⑤	✓	✓	✓	✓	✓	**92.8**	**86.6**	**86.0**	**76.3**

rate is set at $1e-4$ and is polynomially decayed with a power of 0.9. We train models for 200k iterations in total. The balancing weights λ_{align}, λ_{proto}, λ_{cf}, and λ_{cp} in Eq. 7 are set as 0.1, 0.1, 0.1, and 0.1, respectively. The preset threshold τ is set as 0.7. To obtain 3D instance segmentation results, we adopt the marker-controlled watershed algorithm [20] on the predicted binary predictions.

Comparisons with Baselines. The binary segmentation result comparisons of our proposed method with previous works on the Lucchi and MitoEM datasets are shown in Table 1. The competitors include UALR [21], DAMT-Net [16], Advent [19], DA-VSN [6], and DA-ISC [7]. Our method achieves the new state-of-the-art results in all cases, which corroborates the superiority of the proposed class-aware alignment in the feature space. Especially, compared with the previous state-of-the-art DA-ISC [7], our method surpasses it by a large margin on the benchmarks of VNC III→ Lucchi (Subset1) (3.1% IoU). The mitochondria in MitoEM-H distribute more densely and are more complex than those in MitoEM-R, leading to the result of MitoEM-R → MitoEM-H is lower than that of MitoEM-H → MitoEM-R. However, our result on MitoEM-R → MitoEM-H has remarkable improvement, owing to that our method not only aligns domain in a fine-grained way but also explore the full potential of target.

As shown in Table 2, we also evaluate the effectiveness of our proposed method on 3D instance segmentation results. We only conduct experiments on the MitoEM dataset due to the lack of groundtruth for 3D instance segmentation in Lucchi. Our method not only deals with the domain adaptation for binary segmentation but also behaves well for the harder 3D instance segmentation, where the latter has rarely been studied in the literature. Furthermore, to further evaluate the effectiveness of our method, we visualize the predicted segmentation of our method and baselines in Fig. 2. Credited to the proposed class-aware feature alignment, our method estimates more fine-grained results on the target domain, substantiating that our method can alleviate the domain shift between source and target domains. In Fig. 2, the yellow and orange boxes represent mitochondria and background, respectively. In R2H, only our method segments the mitochondria in the yellow box. This is because our method is able to extract more representative features of mitochondria and background, and can separate these two types of features more effectively.

Ablation Study for Loss Functions. We conduct thorough ablation experiments to validate the contribution of each loss term in Eq. 7, where the results are shown in Table 3. The experiment ① with only supervised segmentation loss $\mathcal{L}^s_{seg}$ is the *Source-Only* method. All other variants with our proposed losses have superior performance than ①. Specifically, with the proposed class-aware feature alignment ($\mathcal{L}_{align}$ and $\mathcal{L}_p$), ② improves ① by 6.6% IoU and ③ further enlarges the gain to 7.8%. To take advantage of the potential information in target domain with the *intra-domain* losses $\mathcal{L}_{cf}$ and $\mathcal{L}_{cp}$, the final ⑤ improves by 3.6% IoU, leading to the 11.4% IoU improvement in total. To further explore the impact of different components in $\mathcal{L}_{align}$, *i.e.*, $\mathcal{L}^s_{intra}$, $\mathcal{L}^s_{inter}$ and $\mathcal{L}^t_{intra}$, we conduct experiments and the results are shown in the supplementary material. We find that the cross-domain alignment term $\mathcal{L}^t_{intra}$ plays the core role. It is noteworthy that $\mathcal{L}^s_{intra}$ and $\mathcal{L}^s_{inter}$ help to learn better class-aware prototypes.

4   Conclusion

In this paper, for the first time, we propose the class-aware alignment for domain adaptation on mitochondria segmentation in the feature space. Based on the extracted source prototypes representing class knowledge, we design intra-domain and inter-domain alignment constraint for fine-grained alignment cross domains. Furthermore, we incorporate an intra-domain consistency loss to take full advantage of the potential information existed in target domain. Comprehensive experiments demonstrate the effectiveness of our proposed method.

Acknowledgement. This work was supported by the National Natural Science Foundation of China under Grant 62076230.

References

1. Bermúdez-Chacón, R., Altingövde, O., Becker, C., Salzmann, M., Fua, P.: Visual correspondences for unsupervised domain adaptation on electron microscopy images. IEEE Trans. Med. Imaging **39**(4), 1256–1267 (2019)
2. Bermúdez-Chacón, R., Márquez-Neila, P., Salzmann, M., Fua, P.: A domain-adaptive two-stream u-net for electron microscopy image segmentation. In: ISBI (2018)
3. Franco-Barranco, D., Muñoz-Barrutia, A., Arganda-Carreras, I.: Stable deep neural network architectures for mitochondria segmentation on electron microscopy volumes. Neuroinformatics **20**(2), 437–450 (2022)

4. Franco-Barranco, D., Pastor-Tronch, J., González-Marfil, A., Muñoz-Barrutia, A., Arganda-Carreras, I.: Deep learning based domain adaptation for mitochondria segmentation on EM volumes. Comput. Methods Programs Biomed. **222**, 106949 (2022)
5. Gerhard, S., Funke, J., Martel, J., Cardona, A., Fetter, R.: Segmented anisotropic ssTEM dataset of neural tissue. Figshare (2013)
6. Guan, D., Huang, J., Xiao, A., Lu, S.: Domain adaptive video segmentation via temporal consistency regularization. In: ICCV (2021)
7. Huang, W., Liu, X., Cheng, Z., Zhang, Y., Xiong, Z.: Domain adaptive mitochondria segmentation via enforcing inter-section consistency. In: Wang, L., Dou, Q., Fletcher, P.T., Speidel, S., Li, S. (eds.) MICCAI 2022. LNCS, vol. 13434, pp. 89–98. Springer, Cham (2022)
8. Jiang, Z., et al.: Prototypical contrast adaptation for domain adaptive semantic segmentation. In: Avidan, S., Brostow, G., Cissé, M., Farinella, G.M., Hassner, T. (eds.) ECCV 2022. LNCS, vol. 13694, pp. 36–54. Springer, Cham (2022). https://doi.org/10.1007/978-3-031-19830-4_3
9. Li, M., Chen, C., Liu, X., Huang, W., Zhang, Y., Xiong, Z.: Advanced deep networks for 3d mitochondria instance segmentation. In: 2022 IEEE 19th International Symposium on Biomedical Imaging (ISBI), pp. 1–5. IEEE (2022)
10. Li, Z., Chen, X., Zhao, J., Xiong, Z.: Contrastive learning for mitochondria segmentation. In: EMBC (2021)
11. Liu, D., et al.: PDAM: a panoptic-level feature alignment framework for unsupervised domain adaptive instance segmentation in microscopy images. IEEE Trans. Med. Imaging **40**(1), 154–165 (2020)
12. Lucchi, A., Li, Y., Fua, P.: Learning for structured prediction using approximate subgradient descent with working sets. In: CVPR (2013)
13. Luo, Y., Zheng, L., Guan, T., Yu, J., Yang, Y.: Taking a closer look at domain shift: category-level adversaries for semantics consistent domain adaptation. In: Proceedings of the IEEE/CVF Conference on Computer Vision and Pattern Recognition, pp. 2507–2516 (2019)
14. Mekuč, M.Ž, Bohak, C., Hudoklin, S., Kim, B.H., Kim, M.Y., Marolt, M., et al.: Automatic segmentation of mitochondria and endolysosomes in volumetric electron microscopy data. Comput. Biol. Med. **119**, 103693 (2020)
15. Mumcuoglu, E., Hassanpour, R., Tasel, S., Perkins, G., Martone, M., Gurcan, M.: Computerized detection and segmentation of mitochondria on electron microscope images. J. Microsc. **246**(3), 248–265 (2012)
16. Peng, J., Yi, J., Yuan, Z.: Unsupervised mitochondria segmentation in EM images via domain adaptive multi-task learning. IEEE J. Sel. Top. Signal Process. **14**(6), 1199–1209 (2020)
17. Roels, J., Hennies, J., Saeys, Y., Philips, W., Kreshuk, A.: Domain adaptive segmentation in volume electron microscopy imaging. In: ISBI (2019)
18. Sohn, K., et al.: Fixmatch: simplifying semi-supervised learning with consistency and confidence. Adv. Neural. Inf. Process. Syst. **33**, 596–608 (2020)
19. Vu, T.H., Jain, H., Bucher, M., Cord, M., Pérez, P.: Advent: adversarial entropy minimization for domain adaptation in semantic segmentation. In: Proceedings of the IEEE/CVF Conference on Computer Vision and Pattern Recognition, pp. 2517–2526 (2019)
20. Wei, D., et al.: MitoEM dataset: large-scale 3D mitochondria instance segmentation from EM images. In: Martel, A.L., et al. (eds.) MICCAI 2020. LNCS, vol. 12265, pp. 66–76. Springer, Cham (2020). https://doi.org/10.1007/978-3-030-59722-1_7

21. Wu, S., Chen, C., Xiong, Z., Chen, X., Sun, X.: Uncertainty-aware label rectification for domain adaptive mitochondria segmentation. In: de Bruijne, M., Cattin, P.C., Cotin, S., Padoy, N., Speidel, S., Zheng, Y., Essert, C. (eds.) MICCAI 2021. LNCS, vol. 12903, pp. 191–200. Springer, Cham (2021). https://doi.org/10.1007/978-3-030-87199-4_18
22. Zhang, P., Zhang, B., Zhang, T., Chen, D., Wang, Y., Wen, F.: Prototypical pseudo label denoising and target structure learning for domain adaptive semantic segmentation. In: Proceedings of the IEEE/CVF Conference on Computer Vision and Pattern Recognition, pp. 12414–12424 (2021)

A Sheaf Theoretic Perspective for Robust Prostate Segmentation

Ainkaran Santhirasekaram[1]([envelope]), Karen Pinto[2], Mathias Winkler[2], Andrea Rockall[2], and Ben Glocker[1]

[1] Department of Computing, Imperial College London, London, UK
`a.santhirasekaram19@ic.ac.uk`
[2] Department of Surgery and Cancer, Imperial College London, London, UK

Abstract. Deep learning based methods have become the most popular approach for prostate segmentation in MRI. However, domain variations due to the complex acquisition process result in textural differences as well as imaging artefacts which significantly affects the robustness of deep learning models for prostate segmentation across multiple sites. We tackle this problem by using multiple MRI sequences to learn a set of low dimensional shape components whose combinatorially large learnt composition is capable of accounting for the entire distribution of segmentation outputs. We draw on the language of cellular sheaf theory to model compositionality driven by local and global topological correctness. In our experiments, our method significantly improves the domain generalisability of anatomical and tumour segmentation of the prostate. Code is available at https://github.com/AinkaranSanthi/A-Sheaf-Theoretic-Perspective-for-Robust-Segmentation.git.

1 Introduction

Segmenting the prostate anatomy and detecting tumors is essential for both diagnostic and treatment planning purposes. Hence, the task of developing domain generalisable prostate MRI segmentation models is essential for the safe translation of these models into clinical practice. Deep learning models are susceptible to textural shifts and artefacts which is often seen in MRI due to variations in the complex acquisition protocols across multiple sites [12].

The most common approach to tackle domain shifts is with data augmentation [16,33,35] and adversarial training [11,30]. However, this increases training time and we propose to tackle the problem head on by learning shape only embedding features to build a shape dictionary using vector quantisation [31] which can be sampled to compose the segmentation output. We therefore hypothesise by limiting the search space to a set of shape components, we can improve generalisability of a segmentation model. We also propose to correctly sample

---

Supplementary Information The online version contains supplementary material available at https://doi.org/10.1007/978-3-031-43901-8_24.

H. Greenspan et al. (Eds.): MICCAI 2023, LNCS 14223, pp. 249–259, 2023.
https://doi.org/10.1007/978-3-031-43901-8_24

and compose shape components with local and global topological constraints by tracking topological features as we compose the shape components in an ordered manner. This is achieved using a branch of algebraic topology called cellular sheaf theory [8,19]. We hypothesise this approach will produce more anatomically meaningful segmentation maps and improve tumour localisation.

The contributions of this paper are summarized as follows: 1. This work considers shape compositionality to enhance the generalisability of deep learning models to segment the prostate on MRI. 2. We use cellular sheaves to aid compositionality for segmentation as well as improve tumour localisation.

2   Preliminaries

2.1   Persistent Homology

Topological data analysis is a field which extracts topological features from complex data structures embedded in a topological space. One can describe a topological space through its connectivity which can be captured in many forms. One such form is the cubical complex. The *cubical complex* $\mathcal{C}$ is naturally equipped to deal with topological spaces represented as volumetric grid structured data such as images [32]. In a 3D image, a cubical complex consists of individual voxels serving as vertices, with information regarding their connections to neighboring voxels captured through edges, squares, and cubes. Matrix reduction algorithms enable us to represent the connectivity of $\mathcal{C}$ in terms of a series of mathematical groups, known as the homology groups. Each homology group encompasses a specific dimension, d of topological features, such as connected components ($d = 0$), holes ($d = 1$), and voids ($d = 2$). The number of topological features present in each group is quantified by the corresponding Betti number. Betti numbers provide useful topological descriptors of the binary label maps as it is a single scale topological descriptor. However, the output, $\mathcal{Y}$ from a segmentation model is continuous. Thus, the Betti number for a cubical complex where vertices are continuous will be a noisy topological descriptor. We therefore use *persistent homology* which tracks changes in the topological features at multiple scales [17]. A cubical complex can be constructed at some threshold, τ over the output defined as: $\mathcal{C}^\tau = \{y \in \mathcal{Y} | \mathcal{Y} \geq \tau\}$. We can now create q cubical complexes over q ordered thresholds. This leads to a sequence of nested cubical complexes shown in Eq. 1 known as a sublevel set filtration. The persistent homology defines d dimensional topological features such as connected components which are born at τ^i and dies at τ^j where $\tau^j > \tau^i$. This creates tuples (τ^i, τ^j) which are stored as points in a persistence diagram (Fig. 2b).

$$\emptyset = \mathcal{C}^{\tau^1} \subseteq \mathcal{C}^{\tau^2} ... \subseteq \mathcal{C}^{\tau^q} = \mathcal{Y} \tag{1}$$

2.2   Cellular Sheaves

The homology of segmentation maps provides a useful tool for analysing global topology but does not describe how local topology is related to construct global topological features. Sheaf theory provides a way of composing or 'gluing' local data together to build a global object (new data) that is consistent with the local information [8]. This lends well to modelling compositionality. Formally, a sheaf is a mathematical object which attaches to each open subset or subspace, U in a topological space, Y an algebraic object like a vector space or set (local data) such that it is well-behaved under restriction to smaller open sets [8].

We can consider a topological space, Y such as a segmentation output divided into a finite number of subspaces, $\{\emptyset, Y_1, Y_2...Y_n\}$ which are the base spaces for Y or equivalently the patches in a segmentation map. If we sequentially glue base spaces together in a certain order to form increasingly larger subspaces of Y starting with the $\emptyset$, one can construct a filtration of Y such that; $\emptyset \subseteq Y_1 \subseteq Y_1 \cup Y_2... \subseteq Y_1 \cup Y_2... \cup Y_n \subseteq Y$. We neatly formalise the subspaces and how subspaces are glued together with a poset. A *poset (P)* is a partially ordered set defined by a relation, $\leq$ between elements in P which is reflexive, anti-symmetric, and transitive [19]. In our work, we define a poset by the inclusion relation; $p_i \leq p_j$ implies $p_i \subseteq p_j$ for $p_i, p_j \in P$. Hence, we can map each element in P with a subspace in X which satisfies the inclusion relations in P like in X.

A *cellular sheaf*, $\mathcal{F}$ over a poset is constructed by mapping, each element, $p \in P$ to a vector space $\mathcal{F}(p)$ over a fixed field which preserves the ordering in P by linear transformations, $\rho_{.,.}$ which are inclusion maps in this case [19]. In our work each element in P maps to the vector space, $\mathbb{R}^2$ which preserves the inclusion relations in P. Specifically, we compute a persistence diagram, $\mathcal{D}$ for the subspace in X associated (homeomorphic) with $p \in P$ whereby (τ^i, τ^j) in the persistence diagram are a set of vectors in the vector space, $\mathbb{R}^2$. A cellular sheaf naturally arises in modelling the connectivity of a segmentation map and provides a mathematically precise justification for using cellular sheaves in our method. We show by approaching the composition of segmentation maps through this lens, one can significantly improved the robustness of segmentation models.

3   Related Work

There have been various deep learning based architectures developed for prostate tumour segmentation [3,15,18]. There is however no work looking at developing models which generalise well to target domains after training on one source domain known as single domain generalisation (SDG). Effective data augmentation techniques, such as CutOut [16], MixUp [34] and BigAug [35] offer a straightforward approach to enhance the generalisability of segmentation models across different domains. Recent methods have utilized adversarial techniques, such as AdvBias [11], which trains the model to generate bias field deformations and enhance its robustness.

RandConv [33] incorporates a randomized convolution layer to learn textural invariant features. Self-supervised strategies such as JiGen [9] can also improve generalisability. The principle of compositionality has been integrated into neural networks for tasks such as image classification [23], generation [2] and more recently, segmentation [26,28] to improve generalisability.

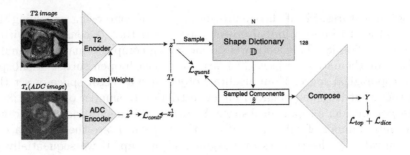

Fig. 1. Three part overview: 1. Use ADC maps to help learn shape equivariant features from T2 weighted images. 2. Construct a shape dictionary, $\mathbb{D}$ from the shape equivariant features. 3. Model the composition of the sampled shaped components to form the segmentation output using a cellular sheaf.

The utilization of persistent homology in deep learning-based segmentation is restricted to either generating topologically accurate segmentations in the output space [21] or as a subsequent processing step [14]. The novel approach of topological auto-encoders [27] marks the first instance of incorporating persistent homology to maintain the topological structure of the data manifold within the latent representation. Cellular sheaves were used to provide a topological insight into the poor performance of graph neural networks in the heterophilic setting [7]. Recently, cellular sheaves were used as a method of detecting patch based merging relations in binary images [20]. Finally, [4] recently proposed using sheaf theory to construct a shape space which allows one to precisely define how to glue shapes together in this shape space.

4   Methods

4.1   Shape Equivariant Learning

Given spatial, T_s and textural, T_i transformations of the input space, $\mathcal{X}$, the goal is to learn an encoder, Φ_e to map $\mathcal{X}$ to lower dimensional embedding features, $\mathcal{E}$ which are shape equivariant and texture invariant as shown in Eq. 2.

$$\Phi_e(T_s(T_i(\mathcal{X}))) = T_s(\mathcal{E}) \tag{2}$$

We assume T2 and ADC MRI images share the same spatial information and only have textural differences. We exploit this idea in Fig. 1, where firstly an ADC image under spatial transformation, T_s is mapped with an encoder, Φ_e to z^2 and the T2 image is mapped with the same encoder to z^1. Shape equivariance and texture invariance is enforced with the contrastive loss, $\mathcal{L}_{contr} = \|T_s(z^1) - z^2\|_2^2$. Specifically, we apply transformations from the dihedral group (D4) which consists of 90° rotations in the z plane and 180° rotations in the y plane. Note, a contrastive only learns equivariance as opposed to constraining the convolutional kernels to be equivariant. z^1 containing 128 channels is spatially quantised before

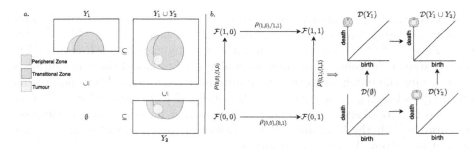

Fig. 2. Figure 2a shows 2 patches (Y_1, Y_2) glued together in an ordered manner defined by the poset, P, via inclusions maps to form the prostate segmentation output $Y = Y_1 \cup Y_2$. Figure 2b shows a cellular sheaf $\mathcal{F}$ over the poset P.

passing into the composer. In the test phase, the ADC image is not required and only the T2 image is used as input. T2 segmentations are used as the label.

4.2 Shape Component Learning

We posit that there is limited shape variation in the low dimensional embedding space across subjects which can be fully captured in N discrete shapes. N discrete shapes form a shape dictionary, $\mathbb{D}$ shown in Fig. 1 which is learnt with vector quantisation [31]. Given we enforce a texture invariant continuous embedding space and hence only contains shape information, quantisation converts this continuous embedding space to discrete shape features, $\hat{z}$. The quantisation process involves minimising the Euclidean distance between the embedding space, z^1 divided into m features, z_i^1 and its nearest shape component, $e_k \in \mathbb{D}$ shown in Eq. 3 where $k = argmin_j \|z_i^1 - e_j\|_2$ and $m = 3048$. Next, sampling $\mathbb{D}$ such that z_i^1 is replaced by e_k produces the spatially quantized embedding space $\hat{z}$. Straight-through gradient approximation is applied for backpropagation through the sampling process to update z^1 and $\mathbb{D}$ [31]. Gradient updates are applied to only the appropriate operands using stop gradients (sg) during optimization.

$$\mathcal{L}_{Quant} = \frac{1}{m} \sum_{i=0}^{i=m-1} \|sg(z_i^1) - e_k\|_2 + \beta \|z_i^1 - sg(e_k)\|_2 \tag{3}$$

4.3 Cellular Sheaves for Shape Composition

Shapes in $\mathbb{D}$ sampled with a uniform prior can lead to anatomically implausible segmentations after composition which we tackle through the language of cellular sheaves to model the connectivity of patches in an image which provides a connectivity-based loss function.

Composition: The quantised embedding space, $\hat{z}$, is split into c groups. The composition of each group in $\hat{z}$ to form each class segmentation, Y_c in the output,

Y involves two steps. Initially, a decoder with grouped convolutions equal to the number of classes followed by the softmax function maps, $\hat{z} \in \mathbb{R}^{128 \times 16 \times 16 \times 12}$ to $C \in \mathbb{R}^{p \times c \times 256 \times 256 \times 24}$ where p is the number of patches for each class c. The second step of the composition uses a cellular sheaf to model the composition of Y_c by gluing the patches together in an ordered manner defined by a poset while tracking its topology using persistent homology. This in turn enforces $\mathbb{D}$ to be sampled in a topological preserving manner as input into the decoder/composer to improve both the local and global topological correctness of each class segmentation output, Y_c after composition.

Illustration: We illustrate our methodology of using cellular sheaves with a simple example in Fig. 2. Here, we show Y as perfectly matching the ground truth label (not one-hot encoded) divided into 2 patches. Y is a topological space with the subspaces, $\mathcal{V} = \{\emptyset, Y_1, Y_2, Y_1 \cup Y_2\}$. A 4-element poset, $P = \{(0,0), (1,0), (0,1), (1,1)\}$ is constructed where given $(x_1, x_2), (y_1, y_2) \in P$ then $(x_1, x_2) \subseteq (y_1, y_2)$ only if $x_1 \leq y_1 \wedge x_2 \leq y_2$. Each element in P is associated with a subspace in $\mathcal{V}$ such that the inclusion relationship is satisfied. Therefore, in Fig. 2a, P defines that Y_1 and Y_2 associated with $(1,0)$ and $(0,1)$ respectively are glued together to form $Y = Y_1 \cup Y_2$ which maps with $(1,1)$. A cellular sheaf $\mathcal{F}$ over P is created by assigning a vector space to $p \in P$ by deriving a persistence diagram, $\mathcal{D}$ for each element in $\mathcal{V}$ associated with p as shown in Fig. 2b. The arrows in Fig. 2b are inclusion maps defined as $\rho_{\cdot,\cdot}$. Persistence diagrams are computed from the sequence of nested cubical complexes of each subspace in $\mathcal{V}$. The persistence diagrams in Fig. 2b are formed by overlapping the persistence diagrams for each class segmentation. Note, persistence diagrams contain infinite points in the form (τ, τ) (diagonal line in persistence diagrams) which always allows a bijection between two persistence diagrams. The main advantage of our approach is that in addition to ensuring correct local topology (patch level) and global topology (image level), we also force our network to produce topologically accurate patches correctly merged together in a topology preserving manner which matches the ground truth. For example in Fig. 2b, Y_2 contains 3 connected components glued onto Y_1 containing 2 connected components to form Y, which also has 3 connected components. This means an extra connected component is added by Y_2 due to tumour which therefore improves patch-wise tumour localisation. It also indicates the other 2 connected components in Y_2 are merged into the 2 connected components in Y_1 to form 2 larger connected components (peripheral and transitional zone) in Y. Hence, the same 2 vectors present in both $\mathcal{F}(1,0)$ and $\mathcal{F}(0,1)$ representing the peripheral and transitional zone are also in $\mathcal{F}(1,1)$. This is also known as a local section in $\mathcal{F}$.

Implementation: In practise during training, a cellular sheaf is built for each class in the output Y and label $\hat{Y}$, denoted Y^c and $\hat{Y}^c$ respectively. Y^c and $\hat{Y}^c$ are divided into $2 \times 2 \times 1$ patches. Starting with the top left patch, we sequentially glue on each patch in a zigzag manner until the entire image is formed which is formally defined by posets, P^c for Y^c and $\hat{P}^c$ for $\hat{Y}^c$, each containing 7 elements this time. Each element in the poset is associated with a patch i.e. Y_i^c or a sequences of patches glued together. For example, P^c is

Table 1. Mean dice, Hausdorff distance (HD) Betti error±standard deviation using our method, several SDG methods and the nnUNet for zonal segmentation.

	$RUNMC \rightarrow BMC$			$Internal \rightarrow RUNMC$		
	Dice	HD	Betti Error	Dice	HD	Betti Error
Baseline	0.51±0.13	0.40±0.11	2.98±0.91	0.67±0.17	0.35±0.09	2.01±0.72
nnUNet [16]	0.57±0.15	0.32±0.10	1.90±0.82	0.72±0.15	0.30±0.12	1.10±0.44
AdvBias [11]	0.56±0.13	0.33±0.15	1.92±0.16	0.73±0.19	0.29±0.13	1.09±0.22
RandConv [33]	0.59±0.15	0.29±0.08	1.54±0.24	0.73±.017	0.23±0.11	0.99±0.20
BigAug [35]	0.63±0.15	0.25±0.12	1.39±0.49	0.75±0.18	0.21±0.07	0.86±0.38
Jigen [9]	0.54±0.25	0.38±0.17	2.72±1.17	0.68±0.13	0.33±0.15	1.89±0.93
vMFNet [26]	0.61±0.15	0.28±0.14	1.48±0.39	0.72±0.16	0.24±0.09	0.99±0.28
Ours	**0.65±0.10**	**0.20±0.10**	**0.93±0.27**	**0.77±0.14**	**0.18±0.07**	**0.69±0.20**

Table 2. The average dice score, Betti error, Hausdorff distance(HD), sensitivity, specificity and positive predictive value (PPV) ± standard deviations using our method, several SDG methods and the nnUNet for tumour segmentation.

	Dice	HD	Betti Error	Sensitivity	Specificity	PPV
Baseline	0.38±0.17	1.03±0.32	5.11±2.90	0.37±0.11	0.60±0.20	0.29±0.11
nnUNet [16]	0.45±0.15	0.81±0.22	4.43±2.32	0.46±0.11	0.70±0.14	0.37±0.14
AdvBias [11]	0.42±0.10	0.90±0.19	4.41±2.16	0.45±0.13	0.66±0.21	0.35±0.17
RandConv [33]	0.43±0.18	0.80±0.27	4.19±2.01	0.47±0.16	0.65±0.17	0.35±0.19
BigAug [35]	0.47±0.12	0.68±0.19	4.03±1.89	0.48±0.18	0.73±0.22	0.40±0.19
Jigen [9]	0.42±0.11	0.88±0.21	4.51±2.43	0.42±0.18	0.65±0.13	0.33±0.09
vMFNet [26]	0.46±0.12	0.80 ±0.20	3.33±1.18	0.47±0.15	0.66±0.21	0.38±0.11
Ours	**0.51±0.13**	**0.57±0.16**	**2.99±0.97**	**0.50±0.18**	**0.79±0.20**	**0.45±0.18**

bijective with $\{Y_1^c, Y_2^c, Y_3^c, Y_4^c, Y_1^c \cup Y_2^c, Y_1^c \cup Y_2^c \cup Y_3^c, Y^c\}$. We construct cellular sheaves, $\mathcal{F}$ over P^c and $\hat{P}^c$ and minimise the distance between these cellular sheaves.

We firstly plot persistence diagrams, $\mathcal{D}$ from the set of vectors (τ^i, τ^j) in $\mathcal{F}(P_i^c)$ and $\mathcal{F}(\hat{P}_i^c)$. Next, we minimise the total p^{th} Wasserstein distance (topological loss) between the persistence diagrams $\mathcal{D}(\mathcal{F}(P_i^c))$ and $\mathcal{D}(\mathcal{F}(\hat{P}_i^c))$ shown in Eq. 4 where $\eta : \mathcal{D}(\mathcal{F}(P_i^c)) \rightarrow \mathcal{D}(\mathcal{F}(\hat{P}_i^c))$ is a bijection between the persistence diagrams [27] and $p = 2$. This loss function is proven to be stable to noise [29] and differentiable [10]. We add a dice loss between Y and $\hat{Y}$. The total loss to train our entire framework is: $\mathcal{L}_{total} = \mathcal{L}_{dice}(Y, \hat{Y}) + \mathcal{L}_{contr} + \mathcal{L}_{quant} + \mathcal{L}_{top}$.

$$\mathcal{L}_{top} = \sum_{c=1}^{c=4} \sum_{i=1}^{i=7} \left(inf_{\eta:\mathcal{D}(\mathcal{F}(P_i^c)) \rightarrow \mathcal{D}(\mathcal{F}(\hat{P}_i^c))} \sum_{x \in \mathcal{D}(\mathcal{F}(P_i^c))} \|x - \eta(x)\|_\infty^p \right)^{\frac{1}{p}} \quad (4)$$

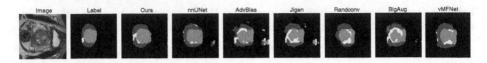

Fig. 3. Tumour and zonal Segmentation of a prostate slice from the RUNMC datasets when training on the internal dataset.

5 Experiments

5.1 Experimental Setup

Datasets: We firstly assess single domain generalisability of prostate zonal segmentation on 2 external datasets which has 2 labels; the transitional and peripheral zone. The training set contains the Prostate dataset from the Medical Segmentation Decathlon which has 32 T2 weighted and ADC images acquired from the Radboud University Nijmegen Medical Centre (RUNMC) [1]. The test set consists of 30 T2 weighted scans acquired from the Boston Medical Centre (BMC) as part of the NCI-ISBI13 Challenge [5,6,13]. The second experiment is a 3 class segmentation problem assessing single domain generalisability of prostate tumour and zonal segmentation. We compose an internal dataset for training which contains 168 T2 weighted and ADC images obtained from a single site across 6 years. The test set has 68 T2 weighted images from the PROSTATEx-2 challenge [24,25] acquired from RUNMC. All experiments were validated on 10 samples. See supplementary material for more details on the datasets.

Pre-processing: All images are resampled to $0.5 \times 0.5 \times 3\,\mathrm{mm}$, centre cropped to $256 \times 256 \times 24$ and normalised between 0 and 1.

Model: In order to address the anisotropic characteristics of Prostate MRI images, we have chosen a hybrid 2D/3D UNet as our baseline model. We use the same encoder and decoder architecture as the baseline model in our method. See supplementary material for further details.

Comparison: We compare our method with the nnUNet [22] and several approaches to tackle SDG segmentation namely, RandConv [33], AdvBias [11], Jigen [9] and BigAug [35] applied to the baseline model. We also compare to a compositionality driven segmentation method called the vMFNet [26].

Training: In all our experiments, the models were trained using Adam optimization with a learning rate of 0.0001 and weight decay of 0.05. Training was run for up to 500 epochs on three NVIDIA RTX 2080 GPUs. The performance of the models was evaluated using the Dice score, Betti error [21] and Hausdorff distance. We evaluate tumour localisation by determining a true positive if the tumour segmentation overlaps by a minimum of one pixel with the ground truth.

In our ablation studies, the minimum number of shape components required in $\mathbb{D}$ for the zonal and zonal + tumour segmentation experiments was 64 and 192 respectively before segmentation performance dropped. See supplementary material for ablation experiments analysing each component of our framework.

5.2 Results

In the task of anatomical segmentation, the first two columns of Table 1 show the results for the domain shift from RUNMC in the decathlon dataset to BMC. Here, we demonstrate that our method improves segmentation performance in all evaluation metrics compared to the baseline, nn-UNet and the other SDG methods. Similar findings are noted for the domain shift from the internal dataset to the RUNMC data in the ProstateX2 dataset (second two columns of Table 1).

In Table 2, we note our method significantly improves tumour segmentation and localisation performance. We visualise our findings with an example in Fig. 3, where there is improved localisation of the tumour and the correct number of tumour components enforced by our topological loss. This significantly reduces the false positive rate highlighted in Table 2. Also, note the more anatomically plausible zonal segmentations. However, our method is restricted by the number of low dimensional shape components in the shape dictionary used to compose the high dimensional segmentation output. Therefore, our approach can fail to segment the finer details of prostate tumours due to its high shape variability which leads to coarser but better localised tumour segmentations.

6 Conclusion

In conclusion, we propose shape compositionality as a way to improve the generalisability of segmentation models for prostate MRI. We devise a method to learn texture invariant and shape equivariant features used to create a dictionary of shape components. We use cellular sheaf theory to help model the composition of sampled shape components from this dictionary in order to produce more anatomically meaningful segmentations and improve tumour localisation.

Acknowledgements. This work was supported and funded by Cancer Research UK (CRUK) (C309/A28804).

References

1. Antonelli, M., et al.: The medical segmentation decathlon. Nat. Commun. **13**(1), 4128 (2022)
2. Arad Hudson, D., Zitnick, L.: Compositional transformers for scene generation. Adv. Neural. Inf. Process. Syst. **34**, 9506–9520 (2021)
3. Arif, M., et al.: Clinically significant prostate cancer detection and segmentation in low-risk patients using a convolutional neural network on multi-parametric MRI. Eur. Radiol. **30**, 6582–6592 (2020)
4. Arya, S., Curry, J., Mukherjee, S.: A sheaf-theoretic construction of shape space. arXiv preprint arXiv:2204.09020 (2022)
5. Bloch, N., et al.: Cancer imaging archive Wiki. http://doi.org/10.7937/K9/TCIA.2015.zF0vlOPv (2015)
6. Bloch, N., et al.: NCI-ISBI 2013 challenge: automated segmentation of prostate structures. Cancer Imaging Arch. **370**, 6 (2015)

7. Bodnar, C., Di Giovanni, F., Chamberlain, B.P., Liò, P., Bronstein, M.M.: Neural sheaf diffusion: a topological perspective on heterophily and oversmoothing in GNNs. arXiv preprint arXiv:2202.04579 (2022)

8. Bredon, G.E.: Sheaf Theory, vol. 170. Springer, Heidelberg (2012)

9. Carlucci, F.M., D'Innocente, A., Bucci, S., Caputo, B., Tommasi, T.: Domain generalization by solving jigsaw puzzles. In: Proceedings of the IEEE/CVF Conference on Computer Vision and Pattern Recognition, pp. 2229–2238 (2019)

10. Carriere, M., Chazal, F., Glisse, M., Ike, Y., Kannan, H., Umeda, Y.: Optimizing persistent homology based functions. In: International Conference on Machine Learning, pp. 1294–1303. PMLR (2021)

11. Chen, C., et al.: Realistic adversarial data augmentation for MR image segmentation. In: Martel, A.L., et al. (eds.) MICCAI 2020. LNCS, vol. 12261, pp. 667–677. Springer, Cham (2020). https://doi.org/10.1007/978-3-030-59710-8_65

12. Chen, Y.: Towards to robust and generalized medical image segmentation framework. arXiv preprint arXiv:2108.03823 (2021)

13. Clark, K., et al.: The cancer imaging archive (TCIA): maintaining and operating a public information repository. J. Digit. Imaging **26**, 1045–1057 (2013)

14. Clough, J.R., Byrne, N., Oksuz, I., Zimmer, V.A., Schnabel, J.A., King, A.P.: A topological loss function for deep-learning based image segmentation using persistent homology. arXiv preprint arXiv:1910.01877 (2019)

15. De Vente, C., Vos, P., Hosseinzadeh, M., Pluim, J., Veta, M.: Deep learning regression for prostate cancer detection and grading in bi-parametric MRI. IEEE Trans. Biomed. Eng. **68**(2), 374–383 (2020)

16. DeVries, T., Taylor, G.W.: Improved regularization of convolutional neural networks with cutout. arxiv 2017. arXiv preprint arXiv:1708.04552 (2017)

17. Edelsbrunner, H., Harer, J., et al.: Persistent homology-a survey. Contemp. Math. **453**(26), 257–282 (2008)

18. Gunashekar, D.D., et al.: Explainable AI for CNN-based prostate tumor segmentation in multi-parametric MRI correlated to whole mount histopathology. Radiat. Oncol. **17**(1), 1–10 (2022)

19. Hu, C.S.: A brief note for sheaf structures on posets. arXiv preprint arXiv:2010.09651 (2020)

20. Hu, C.S., Chung, Y.M.: A sheaf and topology approach to detecting local merging relations in digital images. In: Proceedings of the IEEE/CVF Conference on Computer Vision and Pattern Recognition, pp. 4396–4405 (2021)

21. Hu, X., Li, F., Samaras, D., Chen, C.: Topology-preserving deep image segmentation. In: Advances in Neural Information Processing Systems, vol. 32 (2019)

22. Isensee, F., Jaeger, P.F., Kohl, S.A., Petersen, J., Maier-Hein, K.H.: nnU-Net: a self-configuring method for deep learning-based biomedical image segmentation. Nat. Methods **18**(2), 203–211 (2021)

23. Kortylewski, A., He, J., Liu, Q., Yuille, A.L.: Compositional convolutional neural networks: a deep architecture with innate robustness to partial occlusion. In: Proceedings of the IEEE/CVF Conference on Computer Vision and Pattern Recognition, pp. 8940–8949 (2020)

24. Litjens, G., Debats, O., Barentsz, J., Karssemeijer, N., Huisman, H.: Cancer imaging archive Wiki. URL https://doi.org/10.7937/K9TCIA (2017)

25. Litjens, G., et al.: A survey on deep learning in medical image analysis. Med. Image Anal. **42**, 60–88 (2017)

26. Liu, X., Thermos, S., Sanchez, P., O'Neil, A.Q., Tsaftaris, S.A.: vMFNet: compositionality meets domain-generalised segmentation. In: Wang, L., Dou, Q., Fletcher, P.T., Speidel, S., Li, S. (eds.) Medical Image Computing and Computer Assisted Intervention-MICCAI 2022: 25th International Conference, Singapore, 18–22 September 2022, Proceedings, Part VII, pp. 704–714. Springer, Cham (2022). https://doi.org/10.1007/978-3-031-16449-1_67

27. Moor, M., Horn, M., Rieck, B., Borgwardt, K.: Topological autoencoders. In: International Conference on Machine Learning, pp. 7045–7054. PMLR (2020)

28. Santhirasekaram, A., Kori, A., Winkler, M., Rockall, A., Glocker, B.: Vector quantisation for robust segmentation. In: Wang, L., Dou, Q., Fletcher, P.T., Speidel, S., Li, S. (eds.) International Conference on Medical Image Computing and Computer-Assisted Intervention, pp. 663–672. Springer, Cham (2022). https://doi.org/10.1007/978-3-031-16440-8_63

29. Skraba, P., Turner, K.: Wasserstein stability for persistence diagrams. arXiv preprint arXiv:2006.16824 (2020)

30. Tramer, F., Boneh, D.: Adversarial training and robustness for multiple perturbations. In: Advances in Neural Information Processing Systems, vol. 32 (2019)

31. Van Den Oord, A., Vinyals, O., et al.: Neural discrete representation learning. In: Advances in Neural Information Processing Systems, vol. 30 (2017)

32. Wagner, H., Chen, C., Vuçini, E.: Efficient computation of persistent homology for cubical data. In: Peikert, R., Hauser, H., Carr, H., Fuchs, R. (eds.) Topological Methods in Data Analysis and Visualization II: Theory, Algorithms, and Applications, pp. 91–106. Springer, Cham (2011). https://doi.org/10.1007/978-3-642-23175-9_7

33. Xu, Z., Liu, D., Yang, J., Raffel, C., Niethammer, M.: Robust and generalizable visual representation learning via random convolutions. arXiv preprint arXiv:2007.13003 (2020)

34. Zhang, H., Cisse, M., Dauphin, Y.N., Lopez-Paz, D.: mixup: beyond empirical risk minimization. arXiv preprint arXiv:1710.09412 (2017)

35. Zhang, L., et al.: Generalizing deep learning for medical image segmentation to unseen domains via deep stacked transformation. IEEE Trans. Med. Imaging **39**(7), 2531–2540 (2020)

Semi-supervised Domain Adaptive Medical Image Segmentation Through Consistency Regularized Disentangled Contrastive Learning

Hritam Basak[✉] and Zhaozheng Yin

Department of Computer Science, Stony Brook University, Stony Brook, NY, USA
hbasak@cs.stonybrook.edu

Abstract. Although unsupervised domain adaptation (UDA) is a promising direction to alleviate domain shift, they fall short of their supervised counterparts. In this work, we investigate relatively less explored semi-supervised domain adaptation (SSDA) for medical image segmentation, where access to a few labeled target samples can improve the adaptation performance substantially. Specifically, we propose a two-stage training process. First, an encoder is pre-trained in a self-learning paradigm using a novel domain-content disentangled contrastive learning (CL) along with a pixel-level feature consistency constraint. The proposed CL enforces the encoder to learn discriminative content-specific but domain-invariant semantics on a global scale from the source and target images, whereas consistency regularization enforces the mining of local pixel-level information by maintaining spatial sensitivity. This pre-trained encoder, along with a decoder, is further fine-tuned for the downstream task, (i.e. pixel-level segmentation) using a semi-supervised setting. Furthermore, we experimentally validate that our proposed method can easily be extended for UDA settings, adding to the superiority of the proposed strategy. Upon evaluation on two domain adaptive image segmentation tasks, our proposed method outperforms the SoTA methods, both in SSDA and UDA settings. Code is available at GitHub.

Keywords: Contrastive Learning · Style-content disentanglement · Consistency Regularization · Domain Adaptation · Segmentation

1 Introduction

Despite their remarkable success in numerous tasks, deep learning models trained on a source domain face the challenges to generalize to a new target domain, especially for segmentation which requires dense pixel-level prediction. This is

Supplementary Information The online version contains supplementary material available at https://doi.org/10.1007/978-3-031-43901-8_25.

attributed to a large semantic gap between these two domains. Unsupervised Domain Adaptation (UDA) has lately been investigated to bridge this semantic gap between labeled source domain, and unlabeled target domain [27], including adversarial learning for aligning latent representations [23], image translation networks [24], etc. However, these methods produce subpar performance because of the lack of supervision from the target domain and a large semantic gap in style and content information between the source and target domains. Moreover, when an image's content-specific information is entangled with its domain-specific style information, traditional UDA approaches fail to learn the correct representation of the domain-agnostic content while being distracted by the domain-specific styles. So, they cannot be generalized for multi-domain segmentation tasks [4].

Compared to UDA, obtaining annotation for a few target samples is worthwhile if it can substantially improve the performance by providing crucial target domain knowledge. Driven by this speculation, and the recent success of semisupervised learning (SemiSL), we investigate semi-supervised domain adaptation (SSDA) as a potential solution. Recently, Liu et al. [14] proposed an asymmetric co-training strategy between a SemiSL and UDA task, that complements each other for cross-domain knowledge distillation. Xia et al. [22] proposed a co-training strategy through pseudo-label refinement. Gu et al. [7] proposed a new SSDA paradigm using cross-domain contrastive learning (CL) and self-ensembling mean-teacher. However, these methods force the model to learn the low-level nuisance variability, which is insignificant to the task at hand, hence failing to generalize if similar variational semantics are absent in the training set. Fourier Domain Adaptation (FDA) [26] was proposed to address these challenges by an effective spectral transfer method. Following [26], we design a new Gaussian FDA to handle this cross-domain variability, without feature alignment.

Contrastive learning (CL) is another prospective direction where we enforce models to learn discriminative information from (dis)similarity learning in a latent subspace [2,10]. Liu et al. [15] proposed a margin-preserving constraint along with a self-paced CL framework, gradually increasing the training data difficulty. Gomariz et al. [6] proposed a CL framework with an unconventional channel-wise aggregated projection head for inter slice representation learning. However, traditional CL utilized for DA on images with entangled style and content leads to mixed representation learning, whereas ideally, it should learn discriminative content features invariant to style representation. Besides, the instance-level feature alignment of CL is subpar for segmentation, where dense pixel-wise predictions are indispensable [1].

To alleviate these three underlined shortcomings, we propose a novel contrastive learning with pixel-level consistency constraint via disentangling the style and content information from the joint distribution of source and target domain. Precisely, our contributions are as follows: (1) We propose to disentangle the style and content information in their compact embedding space using a joint-learning framework; (2) We propose encoder pre-training with two CL strategies: *Style CL* and *Content CL* that learns the style and content information respectively from the embedding space; (3) The proposed CL is complemented

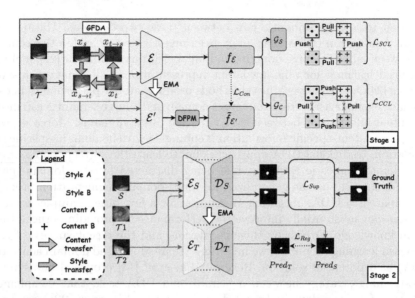

Fig. 1. Overall workflow of our proposed method. **Stage 1**: Encoder pre-training by GFDA and CL on disentangled style and content branches, and pixel-wise feature consistency module DFPM; **Stage 2**: Fine-tuning the encoder in a semi-supervised student-teacher setting.

with a pixel-level consistency constraint with dense feature propagation module, where the former provides better categorization competence whereas the later enforces effective spatial sensitivity; **(4)** We experimentally validate that our SSDA method can be extended in the UDA setting easily, achieving superior performance as compared to the SoTA methods on two widely-used domain adaptive segmentation tasks, both in SSDA and UDA settings.

2   Proposed Method

Given the source domain image-label pairs $\{(x_s^i, y_s^i)_{i=1}^{N_s} \in \mathcal{S}\}$, a few image-label pairs from target domain $\{(x_{t1}^i, y_{t1}^i)_{i=1}^{N_{t1}} \in \mathcal{T}1\}$, and a large number of unlabeled target images $\{(x_{t2}^i)_{i=1}^{N_{t2}} \in \mathcal{T}2\}$, our proposed pre-training stage learns from images in $\{\mathcal{S} \cup \mathcal{T}; \mathcal{T} = \mathcal{T}1 \cup \mathcal{T}2\}$ in a self-supervised way, without requiring any labels. The following fine-tuning in SSDA considers image-label pairs in $\{\mathcal{S} \cup \mathcal{T}1\}$ for supervised learning alongside unlabeled images $\mathcal{T}2$ in the target domain for unsupervised prediction consistency. Our workflow is shown in Fig. 1.

2.1   Gaussian Fourier Domain Adaptation (GFDA)

Manipulating the low-level amplitude spectrum of the frequency domain is the easiest way for style transfer between domains [26], without notable alteration in the visuals of high-level semantics. However, as observed in [26], the generated images consist of incoherent dark patches, caused by abrupt changes in amplitude around the rectangular mask. Instead, we propose a Gaussian mask for a smoother transition in frequency. Let, $\mathcal{F}_A(\cdot)$ and $\mathcal{F}_P(\cdot)$ be the amplitude and phase spectrum in frequency space of an RGB image, and $\mathcal{F}^{-1}$ indicates inverse Fourier transform. We define a 2D Gaussian mask g_σ of the same size as $\mathcal{F}_A$, with σ being the standard deviation. Given two randomly sampled images $x_s \sim \mathcal{S}$ and $x_t \sim \mathcal{T}$, our proposed GFDA can be formulated as:

$$x_{s \to t} = \mathcal{F}^{-1}[\mathcal{F}_P(x_s), \mathcal{F}_A(x_t) \odot g_\sigma + \mathcal{F}_A(x_s) \odot (1 - g_\sigma)], \tag{1}$$

where $\odot$ indicates element-wise multiplication. It generates an image preserving the semantic content from $\mathcal{S}$ but preserving the style from $\mathcal{T}$. Reciprocal pair $x_{t \to s}$ is also formulated using the same drill. The source and target images, and the style-transferred versions $\{x_s, x_{s \to t}, x_t, x_{t \to s}\}$ are then used for contrastive pre-training below. Visualization of GFDA is shown in the supplementary file.

2.2 CL on Disentangled Domain and Content

We aim to learn discriminative content-specific features that are invariant of the style of the source or target domain, for a better pre-training of the network for the task at hand. Hence, we propose to disentangle the style and content information from the images and learn them jointly in a novel disentangled CL paradigm: Style CL (*SCL*) and Content CL (*CCL*). The proposed *SCL* imposes learning of domain-specific attributes, whereas *CCL* enforces the model to identify the ROI, irrespective of the spatial semantics and appearance. In joint learning, they complement each other to render the model to learn domain-agnostic and content-specific information, thereby mitigating the domain dilemma. The set of images $\{x_s, x_{s \to t}, x_t, x_{t \to s}\}$, along with their augmented versions are passed through encoder $\mathcal{E}$, followed by two parallel projection heads, namely style head $(\mathcal{G}_S)$ and content head $(\mathcal{G}_C)$ to obtain the corresponding embeddings. Two different losses: style contrastive loss $\mathcal{L}_{SCL}$ and content contrastive loss $\mathcal{L}_{CCL}$, are derived below.

Assuming $\{x_s, x_{t \to s}\}$ (along with their augmentations) having source-style representation (style A), and $\{x_t, x_{s \to t}\}$ (and their augmentations) having target-style representation (style B), in style CL, embeddings from the same domain (style) are grouped together whereas embeddings from different domains are pushed apart in the latent space. Considering the ith anchor point $x_t^i \in \mathcal{T}$ in a minibatch and its corresponding style embedding $s_t^i \leftarrow \mathcal{G}_S(\mathcal{E}(x_t^i))$ (with style B), we define the *positive* set consisting of the same target domain representations as $\Lambda^+ = \{s_t^{j+}, s_{s \to t}^{j+}\} \leftarrow \mathcal{G}_S(\mathcal{E}(\{x_t^j, x_{s \to t}^j\})), \forall j \in$ minibatch, and *negative* set having unalike source domain representation as $\Lambda^- = \{s_s^{j-}, s_{t \to s}^{j-}\} \leftarrow \mathcal{G}_S(\mathcal{E}(\{x_s^j, x_{t \to s}^j\})), \forall j \in$ minibatch. Following SimCLR [5] our style contrastive loss can be formulated as:

$$\mathcal{L}_{SCL} = \sum_{i,j} -\log \frac{\exp(sim(s^i, s^{j+})/\tau)}{\exp(sim(s^i, s^{j+})/\tau) + \sum_{j \in \Lambda^-} \exp(sim(s^i, s^{j-})/\tau)}, \tag{2}$$

where $\{s^i, s^{j+}\} \in$ style B; $s^{j-} \in$ style A, $sim(\cdot, \cdot)$ defines cosine similarity, τ is the temperature parameter [5]. Similarly, we define $\mathcal{L}_{CCL}$ for content head as:

$$\mathcal{L}_{CCL} = \sum_{i,j} -\log \frac{\exp(sim(c^i, c^{j+})/\tau)}{\exp(sim(c^i, c^{j+})/\tau) + \sum_{j\in\Lambda^-} \exp(sim(c^i, c^{j-})/\tau)}, \quad (3)$$

where $\{c^i, c^j\} \leftarrow \mathcal{G}_C(\mathcal{E}(\{x^i, x^j\}))$. These contrastive losses, along with the consistency constraint below enforce the encoder to extract domain-invariant and content-specific feature embeddings.

2.3   Consistency Constraint

The disentangled CL aims to learn global image-level representation, which is useful for instance discrimination tasks. However, segmentation is attributed to learning dense pixel-level representations. Hence, we propose an additional Dense Feature Propagation Module (DFPM) along with a momentum encoder $\mathcal{E}'$ with exponential moving average (EMA) of parameters from $\mathcal{E}$. Given any pixel m of an image x, we transform its feature $f_{\mathcal{E}'}^m$ obtained from $\mathcal{E}'$ by propagating other pixel features from the same image:

$$\tilde{f}_{\mathcal{E}'}^m = \sum_{\forall n \in x} \mathcal{K}(f_{\mathcal{E}'}^m) \otimes \cos(f_{\mathcal{E}'}^m, f_{\mathcal{E}'}^n) \quad (4)$$

where $\mathcal{K}$ is a linear transformation layer, $\otimes$ denotes *matmul* operation. This spatial smoothing of learned representation is useful for structural sensitivity, which is fundamental for dense segmentation tasks. We enforce consistency between this smoothed feature $\tilde{f}_{\mathcal{E}'}$ from $\mathcal{E}'$ and the regular feature $f_{\mathcal{E}}$ from $\mathcal{E}$ as:

$$\mathcal{L}_{Con} = \sum_{[d(m,n)<Th]} -\left[\cos(\tilde{f}_{\mathcal{E}'}^m, f_{\mathcal{E}}^n) + \cos(f_{\mathcal{E}}^m, \tilde{f}_{\mathcal{E}'}^n)\right] \quad (5)$$

where $d(\cdot, \cdot)$ indicates the spatial distance, Th is a threshold. The overall pre-training objective can be summarized as:

$$\mathcal{L}_{Pre} = \lambda_1 \mathcal{L}_{SCL} + \lambda_2 \mathcal{L}_{CCL} + \mathcal{L}_{Con} \quad (6)$$

2.4   Semi-supervised Fine-Tuning

The pre-training stage is followed by semi-supervised fine-tuning using a student-teacher framework [18]. The pre-trained encoder $\mathcal{E}$, along with a decoder $\mathcal{D}$ are used as a student branch, whereas an identical encoder-decoder network (but differently initialized) is used as a teacher network. We compute a supervised loss on the labeled set $\{\mathcal{S} \cup \mathcal{T}1\}$ along with a regularization loss between the prediction of the student and teacher branches on the unlabeled set $\{\mathcal{T}2\}$ as:

$$\mathcal{L}_{Sup} = \frac{1}{\mathbb{N}_s + \mathbb{N}_{t1}} \sum_{x^i \in \{\mathcal{S}\cup\mathcal{T}1\}} CE\left[\mathcal{D}_S\left(\mathcal{E}_S(x^i)\right), y^i\right] \quad (7)$$

$$\mathcal{L}_{Reg} = \frac{1}{\mathbb{N}_{t2}} \sum_{x^i \in \{T2\}} CE \left[\mathcal{D}_S \left(\mathcal{E}_S(x^i) \right), \mathcal{D}_T \left(\mathcal{E}_T(x^i) \right) \right] \qquad (8)$$

where CE indicates cross-entropy loss, $\mathcal{E}_S, \mathcal{D}_S, \mathcal{E}_T, \mathcal{D}_T$ indicate the student and teacher encoder and decoder networks. The student branch is updated using a consolidated loss $\mathcal{L} = \mathcal{L}_{Sup} + \lambda_3 \mathcal{L}_{Reg}$, whereas the teacher parameters (θ_T) are updated using EMA from the student parameters (θ_S):

$$\theta_T(t) = \alpha \theta_T(t-1) + (1-\alpha)\theta_S(t) \qquad (9)$$

where t tracks the step number, and α is the momentum coefficient [9].

In summary, the overall SSDA training process contains pre-training (Subsect. 2.1–Subsect. 2.3) and fine-tuning (Subsect. 2.4), whereas, we only use the student branch $(\mathcal{E}_S, \mathcal{D}_S)$ for inference.

3   Experiments and Results

Datasets: We evaluate our work on two different DA tasks to evaluate its generalizability: (**1**) Polyp segmentation from colonoscopy images in Kvasir-SEG [11] and CVC-EndoScene Still [20], and (**2**) Brain tumor segmentation in MRI images from BraTS2018 [16]. Kvasir and CVC contain 1000 and 912 images respectively and were split into 4 : 1 training-testing sets following [10]. BraTS consists of brain MRIs from 285 patients with T1, T2, T1CE, and FLAIR scans. The data was split into 4 : 1 train-test ratio, following [14]. **Source→Target**: We perform experiments on $CVC \rightarrow Kvasir$ and $Kvasir \rightarrow CVC$ for polyp segmentation, and $T2 \rightarrow \{T1, T1CE, FLAIR\}$ for tumor segmentation. The SSDA accesses $10 - 50\%$ and $1 - 5$ labels from the target domain for the two tasks, respectively. For UDA, only $\mathcal{S}$ is used for $\mathcal{L}_{Sup}$, whereas $T1 \cup T2$ is used for $\mathcal{L}_{Reg}$. **Implementation details**: Implementation is done in a PyTorch environment using a Tesla V100 GPU with 32GB RAM. We use U-Net [17] backbone for the encoder-decoder structure, and the projection heads $\mathcal{G}_S$ and $\mathcal{G}_C$ are shallow FC layers. The model is trained for 300 epochs for pre-training and 500 epochs for fine-tuning using an ADAM optimizer with a batch size of 4 and a learning rate of $1e - 4$. $\lambda_1, \lambda_2, \lambda_3$, and Th are set to $0.75, 0.75, 0.5, 0.6$, respectively by validation, τ, α are set to $0.07, 0.999$ following [9]. Augmentations include random rotation and translation. **Metrics**: Segmentation performance is evaluated using Dice Similarity Score (DSC) and Hausdorff Distance (HD).

3.1   Performance on SSDA

Quantitative comparison of our proposed method with different SSDA methods [4,14,21,24] for both tasks are shown in Table 1 and Table 2. ACT [14] simply ignores the domain gap and only learns content semantics, resulting in substandard performance on the BraTS dataset that has a significant domain gap. FSM [24], on the other hand, is adaptable to learning explicit domain information, but

Table 1. Comparison with state-of-the-art UDA and SSDA methods for polyp segmentation on KVASIR and CVC. SSDA results are shown for 10%-labeled (10%L) and 50%-labeled (50%L) data in the target domain. The results of cited methods are directly reported from the corresponding papers. **No DA**: the encoder-decoder model trained only using labeled data from the source domain is applied to the target domain without adaptation. **Supervised**: model is trained using all labeled data from source and target domains. The best and second-best results are highlighted in RED and BLUE, respectively.

Task	Method	Target label	CVC → Kvasir		Kvasir → CVC	
			DSC↑	HD↓	DSC↑	HD↓
No DA	Source only	0%L	62.2	5.6	53.9	6.2
UDA	PCEDA [25]	0%L	73.6	4.4	70.1	4.7
	ASN [19]	0%L	80.1	3.6	83.7	3.7
	BDL [12]	0%L	77.8	4.0	81.7	4.1
	CoFo [10]	0%L	82.8	3.6	81.1	3.5
	FDA [26]	0%L	80.4	3.9	75.1	4.2
	Ours	**0%L**	83.8	3.4	84.5	3.1
SSDA	DLD [21]	10%L	84.2	3.2	85.1	3.1
	ACT [14]	10%L	86.9	3.0	87.3	2.9
	SLA [4]	10%L	85.5	3.1	86.2	3.3
	FSM [24]	10%L	85.8	3.4	86.2	3.1
	Ours	**10%L**	87.7	2.9	**86.9**	2.7
	DLD [21]	50%L	87.6	2.8	87.9	2.6
	ACT [14]	50%L	89.4	2.6	90.3	2.4
	SLA [4]	50%L	88.6	2.7	89.3	2.8
	FSM [24]	50%L	89.1	2.6	89.8	2.5
	Ours	**50%L**	90.6	2.4	90.8	2.2
Supervised	Source+Target	100%L	92.1	2.1	93.8	2.0

lacks strong pixel-level regularization on its prediction, resulting in subpar performance. We address both of these shortcomings in our work, resulting in superior performance on both tasks. Other methods like [4,21], which are originally designed for natural images, lack critical refining abilities even after fine-tuning for medical image segmentation and hence are far behind our performance in both tasks. The margins are even higher for less labeled data (1L) on the BraTS dataset, which is promising considering the difficulty of the task. Moreover, our method produces performance close to its fully-supervised counterpart (last row in Table 1 and Table 2), using only a few target labels.

3.2 Performance on UDA

Unlike SSDA methods, UDA fully relies on unlabeled data for domain-invariant representation learning. To analyze the effectiveness of DA, we extend our model to the UDA setting (explained in Sect. 3 [**Source → Target**]) and compare it with SoTA methods [3,8,10,12,13,25,26,28] in Table 1 and Table 2. Methods like [10,19] rely on adversarial learning for aligning multi-level feature space, which is not effective for small-sized medical data. Other methods [12,25] rely on an image-translation network but fail in effective style adaptation, resulting

Table 2. Comparison with state-of-the-art UDA and SSDA methods for whole tumor segmentation on BraTS2018, where source domain is T2. SSDA results are demonstrated for 1-labeled (1L) and 5-labeled (5L) data in the target domain.

Task	Method	Target Label	DSC↑			HD↓		
			T1	T1CE	FLAIR	T1	T1CE	FLAIR
No DA	Source only	0L	3.9	6.0	64.4	56.9	50.8	30.4
UDA	SSCA [13]	0L	59.3	63.5	82.9	12.5	11.2	7.9
	SIFA [3]	0L	51.7	58.2	68.0	19.6	15.0	16.9
	DSA [8]	0L	57.7	62.0	81.8	14.2	13.7	8.6
	DSFN [28]	0L	57.3	62.2	78.9	17.5	15.5	13.8
	Ours	**0L**	60.7	64.4	83.3	11.1	10.9	7.3
SSDA	DLD [21]	1L	65.8	66.5	81.5	12.0	10.3	7.1
	ACT [14]	1L	69.7	69.7	84.5	10.5	10.0	5.8
	ACT-EMD [14]	1L	67.4	69.0	83.9	10.9	10.3	6.4
	SLA [4]	1L	64.7	66.1	82.3	12.2	10.5	7.1
	Ours	**1L**	72.2	71.9	85.8	10.0	9.5	5.2
	DLD [21]	5L	67.8	68.3	83.3	11.2	9.9	6.6
	ACT [14]	5L	71.3	70.8	85.0	10.0	9.8	5.2
	ACT-EMD [14]	5L	70.3	69.8	84.4	10.4	10.2	5.7
	SLA [4]	5L	67.2	71.2	83.1	11.7	10.1	6.8
	Ours	**5L**	73.1	72.4	86.1	9.7	9.3	4.8
Supervised	Source+Target	all labeled	73.6	72.9	86.6	9.5	9.1	4.6

Table 3. Ablation experiment for polyp segmentation in SSDA(50%L) setting to identify the contribution of individual components. TCL: traditional CL [2], SCL: proposed style CL, CCL: proposed content CL. The last row, highlighted in RED, indicates our results.

Experiment#	Stage 1				Stage 2	CVC → Kvasir		Kvasir → CVC	
	TCL	SCL	CCL	DFPM	SemiSL	DSC↑	HD↓	DSC↑	HD↓
(a)	✓	×	×	×	✓	81.7	4.4	82.1	4.2
(b)	×	✓	×	×	✓	83.2	3.9	84.7	3.5
(c)	×	×	✓	×	✓	84.5	3.8	85.4	3.1
(d)	×	✓	✓	×	✓	89.5	2.8	89.1	2.4
(e)	×	✓	✓	✓	✓	90.6	2.4	90.8	2.2

in source domain-biased subpar performance. Our method, although relies on FDA [26], outperforms it with a large margin of upto 12.5% DSC for polyp segmentation, owing to its superior learning ability of disentangled style and content semantics. Similar results are observed for the BraTS dataset in Table 2, where our work achieved a margin of upto 2.4% DSC than its closest performer.

3.3 Ablation Experiments

We perform a detailed ablation experiment, as shown in Table 3. The effectiveness of disentangling and joint-learning of style and content information is evident from the experiment (b)&(c) as compared to (a), where the introduction of *SCL* and *CCL* boosts overall performance significantly. Moreover, when combined together (experiment (d)), they provide a massive 9.54% and

8.52% DSC gain over traditional CL (experiment (a)) for $CVC \rightarrow Kvasir$ and $Kvasir \rightarrow CVC$, respectively. This also points out a potential shortfall of traditional CL: its inability to adapt to a complex domain in DA. The proposed DFPM (experiment (e)) provides local pixel-level regularization, complementary to the global disentangled CL, resulting in a further boost in performance ($\sim 1.5\%$). We have similar ablation study observations on the BraTS2018 dataset, which is provided in the supplementary file, along with some qualitative examples along with available ground truth.

4   Conclusion

We propose a novel style-content disentangled contrastive learning, guided by a pixel-level feature consistency constraint for semi-supervised domain adaptive medical image segmentation. To the best of our knowledge, this is the first attempt for SSDA in medical image segmentation using CL, which is further extended to the UDA setting. Our proposed work, upon evaluation on two different domain adaptive segmentation tasks in SSDA and UDA settings, outperforms the existing SoTA methods, justifying its effectiveness and generalizability.

References

1. Basak, H., Chattopadhyay, S., Kundu, R., Nag, S., Mallipeddi, R.: Ideal: improved dense local contrastive learning for semi-supervised medical image segmentation. In: IEEE International Conference on Acoustics, Speech and Signal Processing (ICASSP) (2023)
2. Chaitanya, K., Erdil, E., Karani, N., Konukoglu, E.: Contrastive learning of global and local features for medical image segmentation with limited annotations. In: Advances in Neural Information Processing Systems, vol. 33, pp. 12546–12558 (2020)
3. Chen, C., Dou, Q., Chen, H., Qin, J., Heng, P.A.: Synergistic image and feature adaptation: towards cross-modality domain adaptation for medical image segmentation. In: Proceedings of the AAAI Conference on Artificial Intelligence, vol. 33, pp. 865–872 (2019)
4. Chen, S., Jia, X., He, J., Shi, Y., Liu, J.: Semi-supervised domain adaptation based on dual-level domain mixing for semantic segmentation. In: Proceedings of the IEEE/CVF Conference on Computer Vision and Pattern Recognition, pp. 11018–11027 (2021)
5. Chen, T., Kornblith, S., Norouzi, M., Hinton, G.: A simple framework for contrastive learning of visual representations. In: International Conference on Machine Learning, pp. 1597–1607. PMLR (2020)
6. Gomariz, A., et al.: Unsupervised domain adaptation with contrastive learning for OCT segmentation. In: Wang, L., Dou, Q., Fletcher, P.T., Speidel, S., Li, S. (eds.) Medical Image Computing and Computer Assisted Intervention-MICCAI 2022: 25th International Conference, Part VIII, pp. 351–361. Springer, Cham (2022). https://doi.org/10.1007/978-3-031-16452-1_34
7. Gu, R., et al.: Contrastive semi-supervised learning for domain adaptive segmentation across similar anatomical structures. IEEE Trans. Med. Imaging **42**(1), 245–256 (2022)

8. Han, X., et al.: Deep symmetric adaptation network for cross-modality medical image segmentation. IEEE Trans. Med. Imaging **41**(1), 121–132 (2022)
9. He, K., Fan, H., Wu, Y., Xie, S., Girshick, R.: Momentum contrast for unsupervised visual representation learning. In: Proceedings of the IEEE/CVF Conference on Computer Vision and Pattern Recognition, pp. 9729–9738 (2020)
10. Huy, T.D., Huyen, H.C., Nguyen, C.D., Duong, S.T., Bui, T., Truong, S.Q.: Adversarial contrastive Fourier domain adaptation for polyp segmentation. In: 2022 IEEE 19th International Symposium on Biomedical Imaging (ISBI), pp. 1–5. IEEE (2022)
11. Jha, D., et al.: Kvasir-SEG: a segmented polyp dataset. In: Ro, Y.M., et al. (eds.) MMM 2020. LNCS, vol. 11962, pp. 451–462. Springer, Cham (2020). https://doi.org/10.1007/978-3-030-37734-2_37
12. Li, Y., Yuan, L., Vasconcelos, N.: Bidirectional learning for domain adaptation of semantic segmentation. In: Proceedings of the IEEE/CVF Conference on Computer Vision and Pattern Recognition, pp. 6936–6945 (2020)
13. Liu, X., Xing, F., El Fakhri, G., Woo, J.: Self-semantic contour adaptation for cross modality brain tumor segmentation. In: 2022 IEEE 19th International Symposium on Biomedical Imaging (ISBI), pp. 1–5. IEEE (2022)
14. Liu, X., et al.: ACT: semi-supervised domain-adaptive medical image segmentation with asymmetric co-training. In: Wang, L., Dou, Q., Fletcher, P.T., Speidel, S., Li, S. (eds.) Medical Image Computing and Computer Assisted Intervention-MICCAI 2022: 25th International Conference, Proceedings, Part V, pp. 66–76. Springer, Cham (2022). https://doi.org/10.1007/978-3-031-16443-9_7
15. Liu, Z., Zhu, Z., Zheng, S., Liu, Y., Zhou, J., Zhao, Y.: Margin preserving self-paced contrastive learning towards domain adaptation for medical image segmentation. IEEE J. Biomed. Health Inform. **26**(2), 638–647 (2022)
16. Menze, B.H., Jakab, A., Bauer, S., Kalpathy-Cramer, J., Farahani, K., et al.: The multimodal brain tumor image segmentation benchmark (BRATS). IEEE Trans. Med. Imaging **34**(10), 1993–2024 (2014)
17. Ronneberger, O., Fischer, P., Brox, T.: U-Net: convolutional networks for biomedical image segmentation. In: Navab, N., Hornegger, J., Wells, W.M., Frangi, A.F. (eds.) MICCAI 2015. LNCS, vol. 9351, pp. 234–241. Springer, Cham (2015). https://doi.org/10.1007/978-3-319-24574-4_28
18. Tarvainen, A., Valpola, H.: Mean teachers are better role models: weight-averaged consistency targets improve semi-supervised deep learning results. In: Advances in Neural Information Processing Systems, vol. 30 (2017)
19. Tsai, Y.H., Hung, W.C., Schulter, S., Sohn, K., Yang, M.H., Chandraker, M.: Learning to adapt structured output space for semantic segmentation. In: Proceedings of the IEEE Conference on Computer Vision and Pattern Recognition, pp. 7472–7481 (2019)
20. Vázquez, D., et al.: A benchmark for endoluminal scene segmentation of colonoscopy images. J. Healthcare Eng. **2017** (2017)
21. Wang, Z., et al.: Alleviating semantic-level shift: a semi-supervised domain adaptation method for semantic segmentation. In: Proceedings of the IEEE/CVF Conference on Computer Vision and Pattern Recognition Workshops, pp. 936–937 (2020)
22. Xia, Y., et al.: Uncertainty-aware multi-view co-training for semi-supervised medical image segmentation and domain adaptation. Med. Image Anal. **65**, 101766 (2020)

23. Xing, F., Cornish, T.C.: Low-resource adversarial domain adaptation for cross-modality nucleus detection. In: Wang, L., Dou, Q., Fletcher, P.T., Speidel, S., Li, S. (eds.) Medical Image Computing and Computer Assisted Intervention-MICCAI 2022: 25th International Conference, Proceedings, Part VII, pp. 639–649. Springer, Cham (2022). https://doi.org/10.1007/978-3-031-16449-1_61
24. Yang, C., Guo, X., Chen, Z., Yuan, Y.: Source free domain adaptation for medical image segmentation with Fourier style mining. Med. Image Anal. **79**, 102457 (2022)
25. Yang, Y., Lao, D., Sundaramoorthi, G., Soatto, S.: Phase consistent ecological domain adaptation. In: Proceedings of the IEEE/CVF Conference on Computer Vision and Pattern Recognition, pp. 9011–9020 (2020)
26. Yang, Y., Soatto, S.: FDA: Fourier domain adaptation for semantic segmentation. In: Proceedings of the IEEE/CVF Conference on Computer Vision and Pattern Recognition, pp. 4085–4095 (2020)
27. Yao, K., et al.: A novel 3D unsupervised domain adaptation framework for cross-modality medical image segmentation. IEEE J. Biomed. Health Inform. **26**(10), 4976–4986 (2022)
28. Zou, D., Zhu, Q., Yan, P.: Unsupervised domain adaptation with dual-scheme fusion network for medical image segmentation. In: IJCAI, pp. 3291–3298 (2022)

Few-Shot Medical Image Segmentation via a Region-Enhanced Prototypical Transformer

Yazhou Zhu[1], Shidong Wang[2], Tong Xin[3], and Haofeng Zhang[1(✉)]

[1] School of Computer Science and Engineering, Nanjing University of Science
and Technology, Nanjing 210094, China
{zyz_nj,zhanghf}@njust.edu.cn
[2] School of Engineering, Newcastle University, Newcastle upon Tyne NE17RU, UK
shidong.wang@newcastle.ac.uk
[3] School of Computing, Newcastle University, Newcastle upon Tyne NE17RU, UK
tong.xin@newcastle.ac.uk

Abstract. Automated segmentation of large volumes of medical images
is often plagued by the limited availability of fully annotated data and
the diversity of organ surface properties resulting from the use of different
acquisition protocols for different patients. In this paper, we introduce a
more promising few-shot learning-based method named Region-enhanced
Prototypical Transformer (RPT) to mitigate the effects of large intra-
class diversity/bias. First, a subdivision strategy is introduced to produce
a collection of regional prototypes from the foreground of the support
prototype. Second, a self-selection mechanism is proposed to incorporate
into the Bias-alleviated Transformer (BaT) block to suppress or remove
interferences present in the query prototype and regional support prototypes. By stacking BaT blocks, the proposed RPT can iteratively optimize the generated regional prototypes and finally produce rectified and
more accurate global prototypes for Few-Shot Medical Image Segmentation (FSMS). Extensive experiments are conducted on three publicly
available medical image datasets, and the obtained results show consistent improvements compared to state-of-the-art FSMS methods. The
source code is available at: https://github.com/YazhouZhu19/RPT.

Keywords: Few-Shot Learning · Medical Image Segmentation · Bias
Alleviation · Transformer

1 Introduction

Automatic medical image segmentation is the implementation of data-driven
image segmentation concepts to identify a specific anatomical structure's surface or volume in a medical image ranging from X-ray and ultrasonography to
CT and MRI scans. Deep learning algorithms are exquisitely suited for this task

Supplementary Information The online version contains supplementary material
available at https://doi.org/10.1007/978-3-031-43901-8_26.

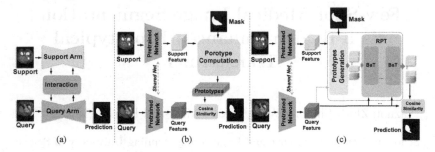

Fig. 1. Comparison between previous FSMS models and our model. (a) Interactive model. (b) Prototypical network based model. (c) Our proposed model.

because they can generate measurements and segmentations from medical images without the time-consuming manual work as in traditional methods. However, the performance of deep learning algorithms depends heavily on the availability of large-scale, high-quality, fully pixel-wise annotations, which are often expensive to acquire. To this end, few-shot learning is considered as a more promising approach and introduced into the medical image segmentation by [13].

Through revisiting existing FSMS algorithms [3–5,16,17,19], they can be grouped into two folders, including the interactive method originated from SENet [15] (shown in Fig. 1(a)) and the prototype networks [18,20] (demonstrated in Fig. 1(b)). For the interaction-based approach, the ideas of *attention* [19], and *contrastive learning* [22] are introduced to work interactively between parallel support and query arms. In contrast, prototype network-based approach almost dominates the FSMS research, such as SSL-ALPNet [13], ADNet [5] and SR&CL [21], whose core idea is to obtain semantic-level prototypes by compressing support features, and then make predictions by matching with query features. However, the problem of how to obtain an accurate and representative prototype remains.

The main reason affecting the representativeness of the prototype is the significant discrepancy between support and query. Specifically, in general, different protocols are taken for different patients, which results in a variety of superficial organ appearances, including the *size*, *shape*, and *contour* of features. In this case, the prototype generated from the support features may not accurately represent the key attributes of the target organ in the query image. In addition, it is also challenging to extract useful information (prototypes of novel classes) from the cluttered background due to the extremely heterogeneous texture between the target and its surroundings, which may contain information belonging to some novel classes or redundant information issue [19].

To mitigate the impact of intra-class diversity, it considers subdividing the foreground of the supporting prototypes to produce some regional prototypes, which are then rectified to suppress or exclude areas inconsistent with the query targets, as illustrated in Fig. 1(c). Concretely, in the prototype learning stage, multiple subdivided regional prototypes are enhanced with a more accurate class center, which can be derived from the newly designed Regional Prototype Generation (RPG) and Query Prototype Generation (QPG) modules. Then, a designed Region-enhanced Prototypical Transformer (RPT) that is mainly composed of a number of stacked Bias-alleviated Transformer (BaT) blocks, each of which contains the core debiasing function-Search and Filter (S&F) modules, to filter out undesirable prototypes. As shown in Fig. 2, Our contributions are summarized as follows:

- A Region-enhanced Prototypical Transformer (RPT) consisting of stacked Bias-alleviated Transformer (BaT) blocks is proposed to mitigate the effects of large intra-class variations present in FSMS through Search and Filter (S&F) modules devised based on the self-selection mechanism.
- A subdivision strategy is proposed to perform in the foreground of the support prototype to generate multiple regional prototypes, which can be further iteratively optimized by the RPT to produce the optimal prototype.
- The proposed method can achieve state-of-the-art performance on three experimental datasets commonly used in medical image segmentation.

2 Methodology

2.1 Overall Architecture

Before introducing the overall architecture, it is necessary to briefly explain how data is processed. Specifically, the 3D supervoxel clustering method [5] is employed to generate pseudo-masks as supervision, which is learned in a self-supervised learning manner without any manual annotations. Meta-learning-based episodic tasks can then be constructed using the generated pseudo-masks. Notably, the pseudo-masks obtained by the 3D clustering method is more consistent with the volumetric properties of medical images than the 2D superpixel clustering method adopted in [13].

As depicted in Fig. 2, the overall architecture includes three main components: the Regional Prototype Generation (RPG) module, the Query Prototype Generation (QPG) module and the Region-enhanced Prototypical Transformer (RPT) consisting of three Bias-alleviated Transformer (BaT) blocks. The pipeline first extracts features from support and query images using a weight-shared ResNet-101 [6] as a backbone, which has been pretrained on the MS-COCO dataset [10]. We employ the ResNet101 pretrained on MS-COCO for optimal performance, and the comparison with ResNet50 pretrained on ImageNet dataset [2] is also included in the appendix. The extracted features are then taken as the input of the RPG and QPG modules to generate multiple region prototypes, which will be rectified by the following RPT to produce the optimal prototype.

2.2 Regional Prototype Generation

The core problem considered in this paper is what causes prototype bias. By examining the input data, it can be observed that images of healthy and diseased organs have a chance to be considered as support or query. This means that

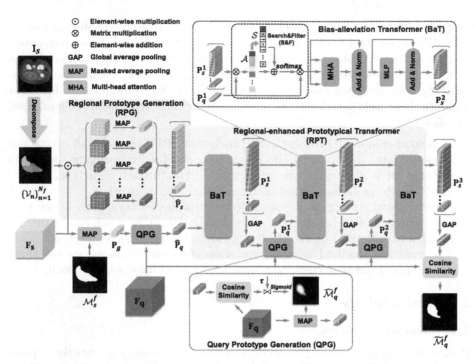

Fig. 2. Overview of the proposed Region-enhanced Prototypical Transformer.

if there are lesioned or edematous regions in some areas of the support images, they will be regarded as biased information which in reality cannot be accurately transferred for the query images containing only healthy organs. When these prototypes that contain the natural heterogeneity of the input images are processed by the Masked Average Pooling (MAP) operation, they inevitably lead to significant intra-class biases.

To cope with the above problems, we propose a Region Prototype Generation (RPG) module to generate multi-region prototypes by performing subdivisions in the foreground of the support images. Given an input support image $\mathbf{I}_s$ and the corresponding foreground mask $\mathcal{M}^f$, the foreground of this image can be obtained by calculating their product. The foreground image then can be partitioned into N_f regions, where N_f is set to 10 by default. By using the Voronoi-based partition method [1,23], a set of regional masks $\{\mathcal{V}_n\}_{n=1}^{N_f}$ can be derived for subsequent use of Masked Average Pooling (MAP) to generate a set of coarse regional prototypes $\hat{\mathcal{P}}_s = \{\hat{p}_n\}_{n=1}^{N_f}, \hat{p}_n \in \mathbb{R}^C$. Formally,

$$\hat{p}_n = \mathrm{MAP}(\mathbf{F}_s, \mathcal{V}_n) = \frac{1}{|\mathcal{V}_n|} \sum_{i=1}^{HW} \mathbf{F}_{s,i} \mathcal{V}_{n,i}, \tag{1}$$

where $\mathbf{F}_s \in \mathbb{R}^{C \times H \times W}$ is the feature extracted from the support images and $\mathcal{V}_n$ denotes the regional masks.

2.3   Query Prototype Generation

Once a set of coarse regional prototypes $\hat{\mathcal{P}}_s$ have been generated for the support images, we can employ the method introduced in [11] to learn the coarse query prototype $\hat{\mathbf{P}}_q \in \mathbb{R}^{1 \times C}$. Concretely, it first uses the MAP$(\cdot)$ operator as introduced in Eq. (1) to learn a global support prototype $\mathbf{P}_g = \text{MAP}(\mathbf{F}_s, \mathcal{M}_s)$ with $\mathbf{P}_g \in \mathbb{R}^{1 \times C}$, whose output can then be used to calculate the coarse query foreground mask $\hat{\mathcal{M}}_q^f$. Considering that the empirically designed threshold described in [11] may affect the quality of the $\hat{\mathcal{M}}_q^f$, we hereby introduce a learnable threshold τ. This process can be denoted as

$$\hat{\mathcal{M}}_q^f = 1 - \sigma(S(\mathbf{F}_q, \mathbf{P}_g) - \tau), \tag{2}$$

where $\mathbf{F}_q \in \mathbb{R}^{C \times H \times W}$ is feature extracted from query images, $S(a, b) = -\alpha cos(a, b)$ is the negative cosine similarity with a fixed scaling factor $\alpha = 20$, σ denotes the Sigmoid activation, and τ is obtained by applying one average-pooling and two fully-connected layers (FC) to the query feature, expressed as $\tau = \text{FC}(\mathbf{F}_q)$. After this, the coarse query foreground prototype can be achieved by using $\hat{\mathbf{P}}_q = \text{MAP}(\mathbf{F}_{q,i}, \hat{\mathcal{M}}_{q,i}^f)$.

2.4   Region-Enhanced Prototypical Transformer

The above received prototypes $\hat{\mathcal{P}}_s$ and $\hat{\mathbf{P}}_q$ are taken as input to the proposed Region-enhanced Prototypical Transformer (RPT) to rectify and regenerate the optimal global prototype $\mathbf{P}_s$. As shown in Fig. 2, our RPT mainly consists of L stacked Bias-alleviated Transformer (BaT) blocks each of which contains a Search and Filter (S&F) module, and QPG modules that maintain the query prototypes continuously updated. Taking the first BaT block as an example, it calculates an affinity map $\mathcal{A} = \hat{\mathbf{P}}_s \hat{\mathbf{P}}_q^\top \in \mathbb{R}^{N_f \times 1}$ to reveal the correspondence between the query and N_f support regional prototypes by taking an input containing the query prototype $\hat{\mathbf{P}}_q$ and the support prototype $\hat{\mathbf{P}}_s \in \mathbb{R}^{N_f \times C}$ obtained by concatenating all elements in $\hat{\mathcal{P}}_s$. Then, a selective map $\mathcal{S} \in \mathbb{R}^{N_f \times 1}$ can be derived from the proposed self-selection based S&F module by

$$\mathcal{S}_i(\mathcal{A}_i) = \begin{cases} 0 & \text{if } \mathcal{A}_i >= \xi \\ -\infty & otherwise \end{cases}, i \in \{0, 1, ..., N_f\}, \tag{3}$$

where ξ is the selection threshold achieved by $\xi = (min(\mathcal{A}) + mean(\mathcal{A}))/2$, $\mathcal{S}$ indicates the chosen regions from the support image that performs compatible with the query at the prototypical level. Then, the heterogeneous or disturbing regions of support foreground will be weeded out with softmax$(\cdot)$ function. The preliminary rectified prototypes $\hat{\mathbf{P}}_s^o \in \mathbb{R}^{N_f \times C}$ is aggregated as:

$$\hat{\mathbf{P}}_s^o = \text{softmax}(\hat{\mathbf{P}}_s \hat{\mathbf{P}}_q^\top + \mathcal{S})\hat{\mathbf{P}}_q. \tag{4}$$

The refined $\hat{\mathbf{P}}_s^o$ will be fed into the following components designed based on the multi-head attention mechanism to produce the output $\mathbf{P}_s^1 \in \mathbb{R}^{N_f \times C}$. Formally,

$$\hat{\mathbf{P}}_s^{o+1} = \mathrm{LN}(\mathrm{MHA}(\hat{\mathbf{P}}_s^o) + \hat{\mathbf{P}}_s^o), \qquad \mathbf{P}_s^1 = \mathrm{LN}(\mathrm{MLP}(\hat{\mathbf{P}}_s^{o+1}) + \hat{\mathbf{P}}_s^{o+1}), \qquad (5)$$

where $\hat{\mathbf{P}}_s^{o+1} \in \mathbb{R}^{N_f \times C}$ is the intermediate generated prototype, $\mathrm{LN}(\cdot)$ denotes the layer normalization, $\mathrm{MHA}(\cdot)$ represents the standard multi-head attention module and $\mathrm{MLP}(\cdot)$ is the multilayer perception.

By stacking multiple BaT blocks, our RPT can iteratively rectify and update all coarse support and the query prototype. Given the prototypes $\mathbf{P}_s^{l-1}$ and $\mathbf{P}_q^{l-1}$ from the previous BaT block, the updates for the current BaT block are computed by:

$$\mathbf{P}_s^l = \mathrm{BaT}(\mathbf{P}_s^{l-1}, \mathbf{P}_q^{l-1}), \qquad \mathbf{P}_q^l = \mathrm{QPG}(\mathrm{GAP}(\mathbf{P}_s^l), \mathbf{F}_q), \qquad (6)$$

where $\mathbf{P}_s^l \in \mathbb{R}^{N_f \times C}$ and $\mathbf{P}_q^l \in \mathbb{R}^{1 \times C}$ ($l = 1, 2, ..., L$) are updated prototypes, $\mathrm{GAP}(\cdot)$ denotes the global average pooling operation. The final output prototypes $\mathbf{P}_s$ optimized by the RPT can be used to predict the foreground of the query image by using Eq. (2: $\tilde{\mathcal{M}}_q^f = 1 - \sigma(S(\mathbf{F}_q, \mathrm{GAP}(\mathbf{P}_s^3)) - \tau)$, while its background can be obtained by $\tilde{\mathcal{M}}_q^b = 1 - \tilde{\mathcal{M}}_q^f$ accordingly.

2.5   Objective Function

The binary cross-entropy loss $\mathcal{L}_{ce}$ is adopted to determine the error between the predict masks $\tilde{\mathcal{M}}_q$ and the given ground-truth $\mathcal{M}_q$. Formally,

$$\mathcal{L}_{ce} = -\frac{1}{HW} \sum_h^H \sum_w^W \mathcal{M}_q^f(x,y) log(\tilde{\mathcal{M}}_q^f(x,y)) + \mathcal{M}_q^b(x,y) log(\tilde{\mathcal{M}}_q^b(x,y)). \quad (7)$$

Considering the prevalent class imbalance problem in medical image segmentation, the boundary loss [8] $\mathcal{L}_B$ is also adopted and it is written as

$$\mathcal{L}_B(\theta) = \int_\Omega \phi G(q) s_\theta(q) dq, \qquad (8)$$

where θ denotes the network parameters, Ω denotes the spatial domain, $\phi G : \Omega \rightarrow \mathbb{R}$ denotes the *level set* representation of the ground-truth boundary, $\phi G(q) = -D_G(q)$ if $q \in G$ and $\phi G(q) = D_G(q)$ otherwise, D_G is distance map between the boundary of prediction and ground-truth, and $s_\theta(q) : \Omega \rightarrow [0, 1]$ denotes softmax$(\cdot)$ function.

Overall, the loss used for training our RPT is defined as $\mathcal{L} = \mathcal{L}_{ce} + \eta \mathcal{L}_{dice} + (1 - \eta)\mathcal{L}_B$, where $\mathcal{L}_{dice}$ is the Dice loss [12], η is initially set to 1 and decreased by 0.01 every epoch.

Table 1. Quantitative Comparison (in Dice score %) of different methods on abdominal datasets under *Setting 1* and *Setting 2*.

Setting	Method	Reference	Abd-MRI					Abd-CT				
			Lower		Upper		Mean	Lower		Upper		Mean
			LK	RK	Spleen	Liver		LK	RK	Spleen	Liver	
1	ADNet [5]	MIA'22	73.86	85.80	72.29	82.11	78.51	72.13	79.06	63.48	77.24	72.97
	AAS-DCL [22]	ECCV'22	80.37	86.11	76.24	72.33	78.76	74.58	73.19	72.30	78.04	74.52
	SR&CL [21]	MICCAI'22	79.34	87.42	76.01	80.23	80.77	73.45	71.22	**73.41**	76.06	73.53
	CRAPNet [3]	WACV'23	**81.95**	86.42	74.32	76.46	79.79	74.69	74.18	70.37	75.41	73.66
	Ours (RPT)	—	80.72	**89.82**	**76.37**	**82.86**	**82.44**	**77.05**	**79.13**	72.58	**82.57**	**77.83**
2	ADNet [5]	MIA'22	59.64	56.68	59.44	**77.03**	63.20	48.41	40.52	50.97	70.63	52.63
	AAS-DCL [22]	ECCV'22	76.90	83.75	74.86	69.94	76.36	64.71	**69.95**	66.36	71.61	68.16
	SR&CL [21]	MICCAI'22	77.07	84.24	73.73	75.55	77.65	67.39	63.37	67.36	73.63	67.94
	CRAPNet [3]	WACV'23	74.66	82.77	70.82	73.82	75.52	70.91	67.33	70.17	70.45	69.72
	Ours (RPT)	—	**78.33**	**86.01**	**75.46**	76.37	**79.04**	**72.99**	67.73	70.80	**75.24**	**71.69**

3   Experiments

Experimental Datasets: The proposed method is comprehensively evaluated on three publicly available datasets, including **Abd-MRI**, **Abd-CT** and **Card-MRI**. Concretely, **Abd-MRI** [7] is an abdominal MRI dataset used in the ISBI 2019 Combined Healthy Abdominal Organ Segmentation Challenge. **Abd-CT** [9] is an abdominal CT dataset from MICCAI 2015 Multi-Atlas Abdomen Labeling Challenge. **Card-MRI** [24] is a cardiac MRI dataset from MICCAI 2019 Multi-Sequence Cardiac MRI Segmentation Challenge. All 3D scans are reformatted into 2D axial and 2D short-axis slices. The abdominal datasets **Abd-MRI** and **Abd-CT** share the same categories of labels which includes the liver, spleen, left kidney (LK) and right kidney (RK). The labels for **Card-MRI** include left ventricular myocardium (LV-MYO), right ventricular myocardium (RV), and blood pool (LV-BP).

Experiment Setup: The model is trained for 30k iterations with batch size set to 1. During training, the initial learning rate is set to 1×10^{-3} with a step decay of 0.8 every 1000 iterations. The values of N_f and iterations L are set to 10 and 3, respectively. To simulate the scarcity of labeled data in medical scenarios, all experiments embrace a 1-way 1-shot setting, and 5-fold cross-validation is also carried out in the experiments, where we only record the mean value.

Evaluation: For a fair comparison, the metric used to evaluate the performance of 2D slices on 3D volumetric ground-truth is the Dice score used in [13]. Furthermore, two different supervision settings are used to evaluate the generalization ability of the proposed method: in Setting 1, the test classes may appear in the background of the training slices, while in Setting 2, the training slices containing the test classes are removed from the dataset to ensure that the test classes are unseen. Note that Setting 2 is impractical for Card-MRI scans, since all classes typically co-occur on one 2D slice, making label exclusion impossible. In addition, as in [13], abdominal organs are categorized into *upper* abdomen

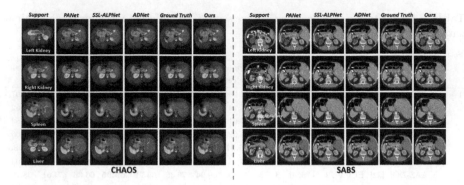

CHAOS   SABS

Fig. 3. Qualitative results of our model on Abd-MRI and Abd-CT.

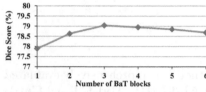

Fig. 4. Analysis of the number of BaT blocks.

Table 2. Ablation study of the three loss functions.

$\mathcal{L}_{ce}$	$\mathcal{L}_B$	$\mathcal{L}_{dice}$	Dice score
✓			78.43
✓	✓		78.81
✓	✓	✓	**79.04**

(liver, spleen) and *lower* abdomen (left, right kidney) to demonstrate whether the learned representations can encode spatial concepts.

3.1 Quantitative and Qualitative Results

Table 1 shows the performance comparison of the proposed method with state-of-the-art methods, including the vanilla PA-Net [20], SE-Net [15], ADNet [5], CRAPNet [3], SSL-ALPNet [13,14], AAS-DCL [22] and SR&CL [21] under two experimental settings. From Table 1, it can be seen that the proposed method outperforms all listed methods in terms of the Mean values obtained under two different settings. Especially, the Mean value on Abd-CT dataset under Setting 1 reaches 77.83, which is 3.31 higher than the best result achieved by AAS-DCL. Consistent improvements are also indicated for Card-MRI dataset and can be found in the Appendix. In addition to the quantitative comparisons, qualitative results of our model and the other model on Abd-MRI and Abd-CT are shown in Fig. 3 (See Appendix for CMR dataset). It is not difficult to see that our model shows considerable bound-preserving and generalization capabilities.

3.2 Ablation Studies

The ablation studies were conducted on Abd-MRI dataset under Setting 2. As can be seen from Fig. 4, the use of three stacked BaT blocks is suggested to obtain the best Dice score. From Table 2, using a combination of boundary and dice loss gives a 0.61 increase in terms of the dice score compared to using only the cross-entropy loss. More ablation study results can be found in Appendix.

4   Conclusion

In this paper, we introduced a Region-enhanced Prototypical Transformer (RPT) to mitigate the impact of large intra-class variations present in medical image segmentation. The model is mainly beneficial from a subdivision-based strategy used for generating a set of regional support prototypes and a self-selection mechanism introduced to the Bias-alleviated Transformer (BaT) blocks. The proposed RPT can iteratively optimize the generated regional prototypes and output a more precise global prototype for predictions. The results of extensive experiments and ablation studies can demonstrate the advancement and effectiveness of the proposed method.

References

1. Aurenhammer, F.: Voronoi diagrams: a survey of a fundamental geometric data structure. ACM Comput. Surv. (CSUR) **23**(3), 345–405 (1991)
2. Deng, J., Dong, W., Socher, R., Li, L.J., Li, K., Fei-Fei, L.: ImageNet: a large-scale hierarchical image database. In: 2009 IEEE Conference on Computer Vision and Pattern Recognition, pp. 248–255. IEEE (2009)
3. Ding, H., Sun, C., Tang, H., Cai, D., Yan, Y.: Few-shot medical image segmentation with cycle-resemblance attention. In: Proceedings of the IEEE/CVF Winter Conference on Applications of Computer Vision, pp. 2488–2497 (2023)
4. Feng, R., et al.: Interactive few-shot learning: limited supervision, better medical image segmentation. IEEE Trans. Med. Imaging **40**(10), 2575–2588 (2021)
5. Hansen, S., Gautam, S., Jenssen, R., Kampffmeyer, M.: Anomaly detection-inspired few-shot medical image segmentation through self-supervision with super-voxels. Med. Image Anal. **78**, 102385 (2022)
6. He, K., Zhang, X., Ren, S., Sun, J.: Deep residual learning for image recognition. In: Proceedings of the IEEE Conference on Computer Vision and Pattern Recognition, pp. 770–778 (2016)
7. Kavur, A.E., et al.: Chaos challenge-combined (CT-MR) healthy abdominal organ segmentation. Med. Image Anal. **69**, 101950 (2021)
8. Kervadec, H., Bouchtiba, J., Desrosiers, C., Granger, E., Dolz, J., Ayed, I.B.: Boundary loss for highly unbalanced segmentation. In: International Conference on Medical Imaging with Deep Learning, pp. 285–296 (2019)
9. Landman, B., Xu, Z., Igelsias, J., Styner, M., Langerak, T., Klein, A.: MICCAI multi-atlas labeling beyond the cranial vault-workshop and challenge. In: Proceedings of MICCAI Multi-Atlas Labeling Beyond Cranial Vault-Workshop Challenge, vol. 5, p. 12 (2015)
10. Lin, T.-Y., et al.: Microsoft COCO: common objects in context. In: Fleet, D., Pajdla, T., Schiele, B., Tuytelaars, T. (eds.) ECCV 2014. LNCS, vol. 8693, pp. 740–755. Springer, Cham (2014). https://doi.org/10.1007/978-3-319-10602-1_48
11. Liu, J., Qin, Y.: Prototype refinement network for few-shot segmentation. arXiv preprint arXiv:2002.03579 (2020)

12. Ma, J., et al.: Loss odyssey in medical image segmentation. Med. Image Anal. **71**, 102035 (2021)
13. Ouyang, C., Biffi, C., Chen, C., Kart, T., Qiu, H., Rueckert, D.: Self-supervision with superpixels: training few-shot medical image segmentation without annotation. In: Vedaldi, A., Bischof, H., Brox, T., Frahm, J.-M. (eds.) ECCV 2020. LNCS, vol. 12374, pp. 762–780. Springer, Cham (2020). https://doi.org/10.1007/978-3-030-58526-6_45
14. Ouyang, C., Biffi, C., Chen, C., Kart, T., Qiu, H., Rueckert, D.: Self-supervised learning for few-shot medical image segmentation. IEEE Trans. Med. Imaging **41**(7), 1837–1848 (2022)
15. Roy, A.G., Siddiqui, S., Pölsterl, S., Navab, N., Wachinger, C.: 'squeeze & excite' guided few-shot segmentation of volumetric images. Med. Image Anal. **59**, 101587 (2020)
16. Shen, Q., Li, Y., Jin, J., Liu, B.: Q-Net: query-informed few-shot medical image segmentation. arXiv preprint arXiv:2208.11451 (2022)
17. Shen, X., Zhang, G., Lai, H., Luo, J., Lu, J., Luo, Y.: PoissonSeg: semi-supervised few-shot medical image segmentation via poisson learning. In: IEEE International Conference on Bioinformatics and Biomedicine, pp. 1513–1518 (2021)
18. Snell, J., Swersky, K., Zemel, R.: Prototypical networks for few-shot learning. In: Advances in Neural Information Processing Systems, vol. 30 (2017)
19. Sun, L., et al.: Few-shot medical image segmentation using a global correlation network with discriminative embedding. Comput. Biol. Med. **140**, 105067 (2022)
20. Wang, K., Liew, J.H., Zou, Y., Zhou, D., Feng, J.: PANet: few-shot image semantic segmentation with prototype alignment. In: Proceedings of the IEEE/CVF International Conference on Computer Vision, pp. 9197–9206 (2019)
21. Wang, R., Zhou, Q., Zheng, G.: Few-shot medical image segmentation regularized with self-reference and contrastive learning. In: International Conference on Medical Image Computing and Computer-Assisted Intervention, pp. 514–523 (2022)
22. Wu, H., Xiao, F., Liang, C.: Dual contrastive learning with anatomical auxiliary supervision for few-shot medical image segmentation. In: Avidan, S., Brostow, G., Cissé, M., Farinella, G.M., Hassner, T. (eds.) European Conference on Computer Vision, pp. 417–434. Springer, Cham (2022). https://doi.org/10.1007/978-3-031-20044-1_24
23. Zhang, J.W., Sun, Y., Yang, Y., Chen, W.: Feature-proxy transformer for few-shot segmentation. In: Advance in Neural Information Processing Systems (2022)
24. Zhuang, X.: Multivariate mixture model for myocardial segmentation combining multi-source images. IEEE Trans. Pattern Anal. Mach. Intell. **41**(12), 2933–2946 (2018)

Morphology-Inspired Unsupervised Gland Segmentation via Selective Semantic Grouping

Qixiang Zhang[1], Yi Li[1], Cheng Xue[2], and Xiaomeng Li[1(✉)]

[1] Department of Electronic and Computer Engineering, The Hong Kong University of Science and Technology, Hong Kong, China
eexmli@ust.hk
[2] School of Computer Science and Engineering, Southeast University, Nanjing, China

Abstract. Designing deep learning algorithms for gland segmentation is crucial for automatic cancer diagnosis and prognosis. However, the expensive annotation cost hinders the development and application of this technology. In this paper, we make a first attempt to explore a deep learning method for unsupervised gland segmentation, where *no manual annotations are required*. Existing unsupervised semantic segmentation methods encounter a huge challenge on gland images. **They either over-segment a gland into many fractions or under-segment the gland regions by confusing many of them with the background.** To overcome this challenge, our key insight is to introduce an empirical cue about gland morphology as extra knowledge to guide the segmentation process. To this end, we propose a novel Morphology-inspired method via Selective Semantic Grouping. We first leverage the empirical cue to selectively mine out proposals for gland sub-regions with variant appearances. Then, a Morphology-aware Semantic Grouping module is employed to summarize the overall information about glands by explicitly grouping the semantics of their sub-region proposals. In this way, the final segmentation network could learn comprehensive knowledge about glands and produce well-delineated and complete predictions. We conduct experiments on the GlaS dataset and the CRAG dataset. Our method exceeds the second-best counterpart by over **10.56%** at mIOU.

Keywords: Whole Slide Image · Unsupervised Gland Segmentation · Morphology-inspired Learning · Semantic Grouping

1 Introduction

Accurate gland segmentation from whole slide images (WSIs) plays a crucial role in the diagnosis and prognosis of cancer, as the morphological features of glands can provide valuable information regarding tumor aggressiveness [11]. With the

---

Supplementary Information The online version contains supplementary material available at https://doi.org/10.1007/978-3-031-43901-8_27.

H. Greenspan et al. (Eds.): MICCAI 2023, LNCS 14223, pp. 281–291, 2023.
https://doi.org/10.1007/978-3-031-43901-8_27

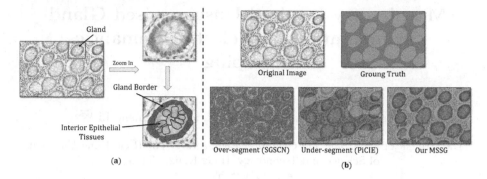

Fig. 1. (a): Example of a gland and its gland border and interior epithelial tissues. (b) Prior USS methods in medical image research [2] and natural image research [6] vs. Our MSSG. Green and orange regions denote the glands and the background respectively. (Color figure online)

emergence of deep learning (DL), there has been a growing interest in developing DL-based methods for semantic-level [9,12,36] and instance-level [5,13,25,27, 32,35] gland segmentation. However, such methods typically rely on large-scale annotated image datasets, which usually require significant effort and expertise from pathologists and can be prohibitively expensive [28].

To reduce the annotation cost, developing annotation-efficient methods for semantic-level gland segmentation has attracted much attention [10,18,23,37]. Recently, some researchers have explored weakly supervised semantic segmentation methods which use weak annotations (e.g., Bound Box [37] and Patch Tag [18]) instead of pixel-level annotations to train a gland segmentation network. However, these weak annotations are still laborious and require expert knowledge [37]. To address this issue, previous works have exploited conventional clustering [8,22,23] and metric learning [10,29] to design annotation-free methods for gland segmentation. However, the performance of these methods can vary widely, especially in cases of malignancy. This paper focuses on unsupervised gland segmentation, where **no annotations are required during training and inference.**

One potential solution is to adopt unsupervised semantic segmentation (USS) methods which have been successfully applied to medical image research and natural image research. On the one hand, existing USS methods have shown promising results in various medical modalities, e.g., magnetic resonance images [19], x-ray images [1,15] and dermoscopic images [2]. However, directly utilizing these methods to segment glands could lead to over-segment results where a gland is segmented into many fractions rather than being considered as one target (see Fig. 1(b)). This is because these methods are usually designed to be extremely sensitive to color [2], while gland images present a unique challenge due to their highly dense and complex tissues with intricate color distribution [18]. On the other hand, prior USS methods for natural images can be broadly categorized into coarse-to-fine-grained [4,14,16,21,31] and end-to-end (E2E) clustering [3,6,17]. The former ones typically rely on pre-generated coarse masks (e.g., super-pixel proposals [16], salience masks [31], and self-attention maps [4,14,21]) as prior, which is not always feasible on gland images. The E2E clustering methods, however, produce under-segment results on gland images by confusing many gland regions with the background; see Fig. 1(b). This is due to the fact that E2E clustering relies on the inherent connections between pixels of the same class [33], and essentially, grouping similar pixels and separate dissimilar ones. Nevertheless, the glands are composed of different parts (gland border and interior epithelial tissues, see Fig. 1(a)) with significant variations in appearance. Gland borders typically consist of dark-colored cells, whereas the interior epithelial tissues contain cells with various color distributions that may closely resemble those non-glandular tissues in the background. As such, the E2E clustering methods tend to blindly cluster pixels with similar properties and confuse many gland regions with the background, leading to under-segment results.

To tackle the above challenges, **our solution is to incorporate an empirical cue about gland morphology as additional knowledge to guide gland segmentation.** The cue can be described as: *Each gland is comprised of a border region with high gray levels that surrounds the interior epithelial tissues.* To this end, we propose a novel Morphology-inspired method via Selective Semantic Grouping, abbreviated as MSSG. To begin, we leverage the empirical cue to selectively mine out proposals for the two gland sub-regions with variant appearances. Then, considering that our segmentation target is the gland, we employ a Morphology-aware Semantic Grouping module to summarize the semantic information about glands by explicitly grouping the semantics of the sub-region proposals. In this way, we not only prioritize and dedicate extra attention to the target gland regions, thus avoiding under-segmentation; but also exploit the valuable morphology information hidden in the empirical cue, and force the segmentation network to recognize entire glands despite the excessive variance among the sub-regions, thus preventing over-segmentation. Ultimately, our method produces well-delineated and complete predictions; see Fig. 1(b).

Our contributions are as follows: (1) We identify the major challenge encountered by prior unsupervised semantic segmentation (USS) methods when dealing with gland images, and propose a novel MSSG for unsupervised gland segmentation. (2) We propose to leverage an empirical cue to select gland sub-regions and explicitly group their semantics into a complete gland region, thus avoiding over-segmentation and under-segmentation in the segmentation results. (3) We validate the efficacy of our MSSG on two public glandular datasets (i.e., the GlaS dataset [27] and the CRAG dataset [13]), and the experiment results demonstrate the effectiveness of our MSSG in unsupervised gland segmentation.

2   Methodology

The overall pipeline of MSSG is illustrated in Fig. 2. The proposed method begins with a Selective Proposal Mining (SPM) module which generates a proposal map that highlights the gland sub-regions. The proposal map is then used to train

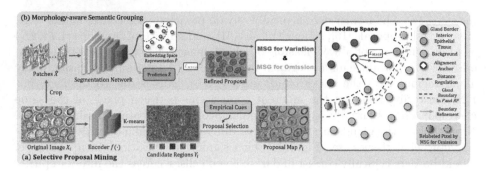

Fig. 2. Overview of our Morphology-inspired Unsupervised Gland Segmentation via Selective Semantic Grouping. (a) Selective Proposal Mining pipeline. We leverage an empirical cue to select proposals for gland sub-regions from the prediction of a shallow encoder $f(\cdot)$ which emphasizes low-level appearance features rather than high-level semantic features. (b) Morphology-aware Semantic Grouping (MSG) pipeline. We deploy a MSG for Variation module to group the two gland sub-regions in the embedding space with L_{MSGV}, and a MSG for Omission module to dynamically refine the proposal map generated by the proposal mining frame (see Gland boundary in P and RP).

a segmentation network. Meantime, a Morphology-aware Semantic Grouping (MSG) module is used to summarize the overall information about glands from their sub-region proposals. More details follow in the subsequent sections.

2.1   Selective Proposal Mining

Instead of generating pseudo-labels for the gland region directly from all the pixels of the gland images as previous works typically do, which could lead to over-segmentation and under-segmentation results, we propose using the empirical cue as extra hints to guide the proposal generation process.

Specifically, let the ith input image be denoted as $X_i \in \mathbb{R}^{C \times H \times W}$, where H, W, and C refer to the height, width, and number of channels respectively. We first obtain a normalized feature map F_i for X_i from a shallow encoder f with 3 convolutional layers, which can be expressed as $F_i = \|f(X_i)\|_2$. We train the encoder in a self-supervised manner, and the loss function $\mathcal{L}$ consists of a typical self-supervised loss $\mathcal{L}_{SS}$, which is the cross-entropy loss between the feature map F_i and the one-hot cluster label $C_i = \arg\max(F_i)$, and a spatial continuity loss $\mathcal{L}_{SC}$, which regularizes the vertical and horizontal variance among pixels within a certain area S to assure the continuity and completeness of the gland border regions (**see Fig. 1 in the Supplementary Material**). The expressions for $\mathcal{L}_{SS}$ and $\mathcal{L}_{SC}$ are given below:

$$\mathcal{L}_{SS}(F_i[:,h,w], C_i[:,h,w]) = -\sum_d^D C_i[d,h,w] \cdot \ln F_i[d,h,w] \tag{1}$$

$$\mathcal{L}_{SC}(F_i) = \sum_{s,h,w}^{S,H-s,W-s} (F_i[:,h+s,w] - F_i[:,h,w])^2 \tag{2}$$
$$+ (F_i[:,h,w+s] - F_i[:,h,w])^2.$$

Then we employ K-means [24] to cluster the feature map F_i into 5 candidate regions, denoted as $Y_i = \{y_{i,1} \in \mathbb{R}^{D \times n_0}, y_{i,2} \in \mathbb{R}^{D \times n_2}, ..., y_{i,5} \in \mathbb{R}^{D \times n_5}\}$, where $n_1 + n_2 + ... + n_5$ equals the total number of pixels in the input image $(H \times W)$.
Sub-region Proposal Selection via the Empirical Cue. The aforementioned empirical cue is used to select proposals for the gland border and interior epithelial tissues from the candidate regions Y_i. Particularly, we select the region with the highest average gray level as the proposal for the *gland border*. Then, we fill the areas surrounded by the gland border proposal and consider them as the proposal for the *interior epithelial tissues*, while the rest areas of the gland image are regarded as the background (i.e., non-glandular region). Finally, we obtain the proposal map $P_i \in \mathbb{R}^{3 \times H \times W}$, which contains the two proposals for two gland sub-regions and one background proposal.

2.2 Morphology-Aware Semantic Grouping

A direct merge of the two sub-region proposals to train a fully-supervised segmentation network may not be optimal for our case. Firstly, the two gland sub-regions exhibit significant **variation** in appearance, which can impede the segmentation network's ability to recognize them as integral parts of the same object. Secondly, the SPM module may produce proposals with inadequate highlighting of many gland regions, particularly the interior epithelial tissues, as shown in Fig. 2 where regions marked with $\times$ are **omitted**. Consequently, applying pixel-level cross-entropy loss between the gland image and the merged proposal map could introduce undesired noise into the segmentation network, thus leading to under-segment predictions with confusion between the glands and the background. As such, we propose two types of Morphology-aware Semantic Grouping (MSG) modules (i.e., MSG for Variation and MSG for Omission) to respectively reduce the confusion caused by the two challenges mentioned above and improve the overall accuracy and comprehensiveness of the segmentation results. The details of the two MSG modules are described as follows.

Here, we first slice the gland image and its proposal map into patches as inputs. Let the input patch and its corresponding sliced proposal map be denoted as $\hat{X} \in \mathbb{R}^{C \times \hat{H} \times \hat{W}}$ and $\hat{P} \in \mathbb{R}^{3 \times \hat{H} \times \hat{W}}$. We can obtain the feature embedding map $\hat{F}$ which is derived as $\hat{F} = f_{feat}(\hat{X})$ and the prediction map $\tilde{X}$ as $\tilde{X} = f_{cls}(\hat{F})$, where f_{feat} and f_{cls} refers to the feature extractor and pixel-wise classifier of the segmentation network respectively.

MSG for Variation is designed to mitigate the adverse impact of appearance variation between the gland sub-regions. It regulates the pixel-level feature embeddings of the two sub-regions by explicitly reducing the distance between them in the embedding space. Specifically, according to the proposal map $\hat{P}$, we divide the pixel embeddings in $\hat{F} \in \mathbb{R}^{D \times \hat{H} \times \hat{W}}$ into gland border set $G = \{g_0, g_1, ..., g_{k_g}\}$, interior epithelial tissue set $I = \{i_0, i_1, ..., i_{k_i}\}$ and non-glandular (i.e., background) set $N = \{n_0, n_1, ..., n_{k_n}\}$, where $k_g + k_i + k_n = \hat{H} \times \hat{W}$. Then, we use the average of the pixel embeddings in gland border set G as the alignment anchor and pull all pixels of I towards the anchor:

$$\mathcal{L}_{MSGV} = \frac{1}{I} \sum_{i \in I} \left(i - \frac{1}{G} \sum_{g \in G} g \right)^2. \tag{3}$$

MSG for Omission is designed to overcome the problem of partial omission in the proposals. It identifies and relabels the overlooked gland regions in the proposal map and groups them back into the gland semantic category. To achieve this, for each pixel n in the non-glandular (i.e., background) set N, two similarities are computed with the gland sub-regions G and I respectively:

$$S_n^G = \frac{1}{G} \sum_{g \in G} \frac{g}{\|g\|_2} \cdot \frac{n}{\|n\|_2}, S_n^I = \frac{1}{I} \sum_{i \in I} \frac{i}{\|i\|_2} \cdot \frac{n}{\|n\|_2}. \tag{4}$$

S_n^G (or S_n^I) represents the similarity between the background pixel n and gland borders (or interior epithelial tissues). If either of them is higher than a preset threshold β (set to 0.7), we consider n as an overlooked pixel of gland borders (or interior epithelial tissues), and relabel n to G (or I). In this way, we could obtain a refined proposal map RP. Finally, we impose a pixel-level cross-entropy loss on the prediction and refined proposal RP to train the segmentation network:

$$\mathcal{L}_{MSGO} = - \sum_{\hat{h}, \hat{w}}^{\hat{H}, \hat{W}} RP[:, \hat{h}, \hat{w}] \cdot \ln \widetilde{X}[:, \hat{h}, \hat{w}], \tag{5}$$

The total objective function $\mathcal{L}$ for training the segmentation network can be summarized as follows:

$$\mathcal{L} = \mathcal{L}_{MSGO} + \lambda_v \mathcal{L}_{MSGV}, \tag{6}$$

where λ_v (set to 1) is the coefficient.

3   Experiments

3.1   Datasets

We evaluate our MSSG on The Gland Segmentation Challenge (GlaS) dataset [27] and The Colorectal Adenocarcinoma Gland (CRAG) dataset [13]. The GlaS dataset contains 165 H&E-stained histopathology patches extracted from 16 WSIs. The CRAG dataset owns 213 H&E-stained histopathology patches extracted from 38 WSIs. The CRAG dataset has more irregular malignant glands, which makes it more difficult than GlaS, and we would like to emphasize that the results on CRAG are from the model trained on GlaS without retraining.

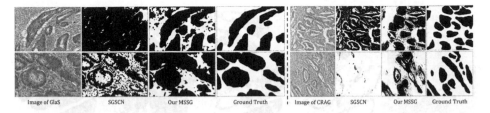

Fig. 3. Visualization of predictions on GlaS (left) and CRAG dataset(right). Black denotes glandular tissues and white denotes non-glandular tissues (**More in the Supplementary Material**).

Table 1. Comparison results on GlaS and CRAG dataset. **Bold** and underline denote best and second-best results of the *unsupervised methods*.

Dataset	Method	Backbone	Supervision	F1	DICE	mIOU
GlaS Dataset	Unet [26]	U-Net	Fully	77.78%	79.04%	65.34%
	ResUNet [34]	U-Net	Fully	78.83%	79.48%	65.95%
	MedT [30]	Transformer	Fully	81.02%	82.08%	69.61%
	Randomly Initial	PSPNet	None	49.72%	48.63%	32.13%
	DeepCluster* [3]	PSPNet	None	57.03%	57.32%	40.17%
	PiCIE* [6]	PSPNet	None	64.98%	65.61%	48.77%
	DINO [4]	PSPNet	None	56.93%	57.38%	40.23%
	DSM [21]	PSPNet	None	68.18%	66.92%	49.92%
	SGSCN* [2]	PSPNet	None	67.62%	68.72%	52.16%
	MSSG	PSPNet	None	**78.26%**	**77.09%**	**62.72%**
CRAG Dataset	Unet [26]	U-Net	Fully	82.70%	84.40%	70.21%
	VF-CNN [20]	RotEqNet	Fully	71.10%	72.10%	57.24%
	MILDNet [13]	MILD-Net	Fully	86.90%	88.30%	76.95%
	PiCIE* [6]	PSPNet	None	67.04%	64.33%	52.06%
	DSM [21]	PSPNet	None	67.22%	66.07%	52.28%
	SGSCN* [2]	PSPNet	None	69.29%	67.88%	55.31%
	MSSG	PSPNet	None	**77.43%**	**77.26%**	**65.89%**

3.2 Implementation Details

The experiments are conducted on four RTX 3090 GPUs. For the SPM, a 3-layer encoder is trained for each training sample. Each convolutional layer uses a 3 × 3 convolution with a stride of 1 and a padding size of 1. The encoder is trained for 50 iterations using an SGD optimizer with a polynomial decay policy and an initial learning rate of 1e−2. For the MSG, MMSegmentation [7] is used to construct a PSPNet [38] with a ResNet-50 backbone as the segmentation network. The network is trained for 20 epochs with an SGD optimizer, a learning rate of 5e−3, and a batch size of 16. For a fair comparison, the results of all

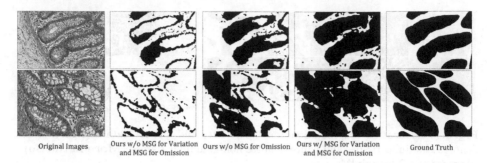

Original Images | Ours w/o MSG for Variation and MSG for Omission | Ours w/o MSG for Omission | Ours w/ MSG for Variation and MSG for Omission | Ground Truth

Fig. 4. Ablation Study on MSG modules. Without MSG, the performance is not good enough, due to *significant sub-region variation* and *gland omission*. With MSG modules, the performance of the network is progressively improved (**More in the Supplementary Material**).

Table 2. Performance gains with MSG modules. The segmentation performance is progressively improved as the involvement of MSG for Variation & MSG for Omission.

MSG for Variation	MSG for Omission	mIOU	Improvement(Δ)
✗	✗	48.42%	-
✓	✗	56.12%	+7.70%
✗	✓	50.18%	+1.64%
✓	✓	62.72%	+14.30%

unsupervised methods in Table 1 are obtained using the same backbone trained with the corresponding pseudo-labels. The code is available at https://github.com/xmed-lab/MSSG.

3.3   Comparison with State-of-the-art Methods

We compare our MSSG with multiple approaches with different supervision settings in Table 1. On the GlaS dataset, the end-to-end clustering methods (denoted by "*") end up with limited improvement over a randomly initialized network. Our MSSG, on the contrary, achieves significant advances. Moreover, MSSG surpasses all other unsupervised counterparts, with a huge margin of 10.56% at mIOU, compared with the second-best unsupervised counterpart. On CRAG dataset, even in the absence of any hints, MSSG still outperforms all unsupervised methods and even some of the fully-supervised methods. Additionally, we visualize the segmentation results of MSSG and its counterpart (i.e., SGSCN [2]) in Fig. 3. On both datasets, MSSG obtains more accurate and complete results.

3.4 Ablation Study

Table 2 presents the ablation test results of the two MSG modules. It can be observed that the segmentation performance without the MSG modules is not satisfactory due to the *significant sub-region variation* and *gland omission*. With the gradual inclusion of the MSG for Variation and Omission, the mIOU is improved by 6.42% and 2.57%, respectively. Moreover, with both MSG modules incorporated, the performance significantly improves to 62.72% (+14.30%). we also visualize the results with and without MSG modules in Fig. 4. It is apparent that the model without MSG ignores most of the interior epithelial tissues. With the incorporation of MSG for Variation, the latent distance between gland borders and interior epithelial tissues is becoming closer, while both of these two sub-regions are further away from the background. As a result, the model can highlight most of the gland borders and interior epithelial tissues. Finally, with both MSG modules, the model presents the most accurate and similar result to the ground truth. **More ablation tests on the SPM (Tab. 1 & 2) and hyper-parameters (Tab. 3) are in the Supplementary Material.**

4 Conclusion

This paper explores a DL method for unsupervised gland segmentation, which aims to address the issues of over/under segmentation commonly observed in previous USS methods. The proposed method, termed MSSG, takes advantage of an empirical cue to select gland sub-region proposals with varying appearances. Then, a Morphology-aware Semantic Grouping is deployed to integrate the gland information by explicitly grouping the semantics of the selected proposals. By doing so, the final network is able to obtain comprehensive knowledge about glands and produce well-delineated and complete predictions. Experimental results prove the superiority of our method qualitatively and quantitatively.

Acknowledgement. This work was supported in part by the Hong Kong Innovation and Technology Fund under Project ITS/030/21 and in part by a grant from Foshan HKUST Projects under FSUST21-HKUST11E.

References

1. Aganj, I., Harisinghani, M.G., Weissleder, R., Fischl, B.R.: Unsupervised medical image segmentation based on the local center of mass. Sci. Rep. **8**, 13012 (2018)
2. Ahn, E., Feng, D., Kim, J.: A spatial guided self-supervised clustering network for medical image segmentation. In: de Bruijne, M., et al. (eds.) MICCAI 2021. LNCS, vol. 12901, pp. 379–388. Springer, Cham (2021). https://doi.org/10.1007/978-3-030-87193-2_36
3. Caron, M., Bojanowski, P., Joulin, A., Douze, M.: Deep clustering for unsupervised learning of visual features. In: Proceedings of the ECCV (2018)
4. Caron, M., et al.: Emerging properties in self-supervised vision transformers. In: Proceedings of the ICCV, pp. 9630–9640 (2021)

5. Chen, H., Qi, X., Yu, L., Heng, P.A.: DCAN: deep contour-aware networks for accurate gland segmentation. In: Proceedings of the IEEE CVPR, pp. 2487–2496 (2016)
6. Cho, J.H., Mall, U., Bala, K., Hariharan, B.: PiCIE: Unsupervised semantic segmentation using invariance and equivariance in clustering. In: Proceedings of the CVPR, pp. 16794–16804 (2021)
7. Contributors, M.: MMSegmentation: Openmmlab semantic segmentation toolbox and benchmark (2020). https://github.com/open-mmlab/mmsegmentation
8. Datar, M., Padfield, D., Cline, H.: Color and texture based segmentation of molecular pathology images using HSOMs. In: 2008 5th IEEE International Symposium on Biomedical Imaging: From Nano to Macro, pp. 292–295 (2008)
9. Ding, H., Pan, Z., Cen, Q., Li, Y., Chen, S.: Multi-scale fully convolutional network for gland segmentation using three-class classification. Neurocomputing **380**, 150–161 (2020)
10. Egger, J.: PCG-cut: graph driven segmentation of the prostate central gland. PLoS ONE **8**(10), e76645 (2013)
11. Fleming, M., Ravula, S., Tatishchev, S.F., Wang, H.L.: Colorectal carcinoma: pathologic aspects. J. Gastrointest. Oncol. **3**(3), 153 (2012)
12. Gao, E., et al.: Automatic multi-tissue segmentation in pancreatic pathological images with selected multi-scale attention network. CBM **151**, 106228 (2022)
13. Graham, S., et al.: MILD-Net: minimal information loss dilated network for gland instance segmentation in colon histology images. Med. Image Anal. **52**, 199–211 (2019)
14. Hamilton, M., Zhang, Z., Hariharan, B., Snavely, N., Freeman, W.T.: Unsupervised semantic segmentation by distilling feature correspondences. In: Proceedings of the ICLR (2021)
15. Huang, Q., et al.: A Chan-vese model based on the Markov chain for unsupervised medical image segmentation. Tsinghua Sci. Technol. **26**, 833–844 (2021)
16. Hwang, J.J., et al.: SegSort: segmentation by discriminative sorting of segments. In: Proceedings of the ICCV, pp. 7333–7343 (2019)
17. Ji, X., Henriques, J.F., Vedaldi, A.: Invariant information clustering for unsupervised image classification and segmentation. In: Proceedings of the ICCV, pp. 9865–9874 (2019)
18. Li, Y., Yu, Y., Zou, Y., Xiang, T., Li, X.: Online easy example mining for weakly-supervised gland segmentation from histology images. In: Wang, L., Dou, Q., Fletcher, P.T., Speidel, S., Li, S. (eds.) MICCAI 2022. Lecture Notes in Computer Science, vol. 13434, pp. 578–587. Springer, Cham (2022). https://doi.org/10.1007/978-3-031-16440-8_55
19. Liu, L., Aviles-Rivero, A.I., Schönlieb, C.B.: Contrastive registration for unsupervised medical image segmentation. arXiv preprint arXiv:2011.08894 (2020)
20. Marcos, D., Volpi, M., Komodakis, N., Tuia, D.: Rotation equivariant vector field networks. In: Proceedings of the ICCV, pp. 5058–5067 (2017)
21. Melas-Kyriazi, L., Rupprecht, C., Laina, I., Vedaldi, A.: Deep spectral methods: a surprisingly strong baseline for unsupervised semantic segmentation and localization. In: Proceedings of the CVPR, pp. 8364–8375 (2022)
22. Nguyen, K., Jain, A.K., Allen, R.L.: Automated gland segmentation and classification for Gleason grading of prostate tissue images. In: ICPR, pp. 1497–1500. IEEE (2010)
23. Paul, A., Mukherjee, D.P.: Gland segmentation from histology images using informative morphological scale space. In: IEEE International Conference on Image Processing, pp. 4121–4125. IEEE (2016)

24. Pedregosa, F., et al.: Scikit-learn: machine learning in Python. J. Mach. Learn. Res. **12**, 2825–2830 (2011)
25. Qu, H., Yan, Z., Riedlinger, G.M., De, S., Metaxas, D.N.: Improving nuclei/gland instance segmentation in histopathology images by full resolution neural network and spatial constrained loss. In: Shen, D., et al. (eds.) MICCAI 2019. LNCS, vol. 11764, pp. 378–386. Springer, Cham (2019). https://doi.org/10.1007/978-3-030-32239-7_42
26. Ronneberger, O., Fischer, P., Brox, T.: U-Net: convolutional networks for biomedical image segmentation. In: Navab, N., Hornegger, J., Wells, W.M., Frangi, A.F. (eds.) MICCAI 2015. LNCS, vol. 9351, pp. 234–241. Springer, Cham (2015). https://doi.org/10.1007/978-3-319-24574-4_28
27. Sirinukunwattana, K., et al.: Gland segmentation in colon histology images: the GlaS challenge contest. Med. Image Anal. **35**, 489–502 (2017)
28. Srinidhi, C.L., Ciga, O., Martel, A.L.: Deep neural network models for computational histopathology: a survey. Med. Image Anal. **67**, 101813 (2021)
29. Tosun, A.B., Gunduz-Demir, C.: Graph run-length matrices for histopathological image segmentation. IEEE Trans. Med. Imaging **30**(3), 721–732 (2010)
30. Valanarasu, J.M.J., Oza, P., Hacihaliloglu, I., Patel, V.M.: Medical transformer: gated axial-attention for medical image segmentation. In: de Bruijne, M., et al. (eds.) MICCAI 2021. LNCS, vol. 12901, pp. 36–46. Springer, Cham (2021). https://doi.org/10.1007/978-3-030-87193-2_4
31. Van Gansbeke, W., Vandenhende, S., Georgoulis, S., Van Gool, L.: Unsupervised semantic segmentation by contrasting object mask proposals. In: Proceedings of the ICCV, pp. 10052–10062 (2021)
32. Wang, H., Xian, M., Vakanski, A.: Ta-net: topology-aware network for gland segmentation. In: Proceedings of the IEEE WACV, pp. 1556–1564 (2022)
33. Wu, Z., Xiong, Y., Yu, S.X., Lin, D.: Unsupervised feature learning via non-parametric instance discrimination. In: Proceedings of the CVPR, pp. 3733–3742 (2018)
34. Xiao, X., Lian, S., Luo, Z., Li, S.: Weighted Res-Unet for high-quality retina vessel segmentation. In: International Conference on Information Technology in Medicine and Education, pp. 327–331. IEEE (2018)
35. Xu, Y., et al.: Gland instance segmentation using deep multichannel neural networks. IEEE Trans. Biomed. Eng. **64**(12), 2901–2912 (2017)
36. Yang, J., Chen, H., Liang, Y., Huang, J., He, L., Yao, J.: ConCL: concept contrastive learning for dense prediction pre-training in pathology images. In: Avidan, S., Brostow, G., Cissé, M., Farinella, G.M., Hassner, T. (eds.) Computer Vision - ECCV 2022, ECCV 2022. Lecture Notes in Computer Science, vol. 13681, pp. 523–539. Springer, Cham (2022). https://doi.org/10.1007/978-3-031-19803-8_31
37. Yang, L., et al.: BoxNet: deep learning based biomedical image segmentation using boxes only annotation. arXiv (2018)
38. Zhao, H., Shi, J., Qi, X., Wang, X., Jia, J.: Pyramid scene parsing network. In: Proceedings of the CVPR, pp. 2881–2890 (2017)

Boundary Difference over Union Loss for Medical Image Segmentation

Fan Sun, Zhiming Luo[✉], and Shaozi Li

Department of Artificial Intelligence, Xiamen University, Xiamen, Fujian, China
zhiming.luo@xmu.edu.cn

Abstract. Medical image segmentation is crucial for clinical diagnosis. However, current losses for medical image segmentation mainly focus on overall segmentation results, with fewer losses proposed to guide boundary segmentation. Those that do exist often need to be used in combination with other losses and produce ineffective results. To address this issue, we have developed a simple and effective loss called the Boundary Difference over Union Loss (Boundary DoU Loss) to guide boundary region segmentation. It is obtained by calculating the ratio of the difference set of prediction and ground truth to the union of the difference set and the partial intersection set. Our loss only relies on region calculation, making it easy to implement and training stable without needing any additional losses. Additionally, we use the target size to adaptively adjust attention applied to the boundary regions. Experimental results using UNet, TransUNet, and Swin-UNet on two datasets (ACDC and Synapse) demonstrate the effectiveness of our proposed loss function. Code is available at https://github.com/sunfan-bvb/BoundaryDoULoss.

Keywords: Medical image segmentation · Boundary loss

1 Introduction

Medical image segmentation is a vital branch of image segmentation [4,5,9, 15,16,23], and can be used clinically for segmenting human organs, tissues, and lesions. Deep learning-based methods have made great progress in medical image segmentation tasks and achieved good performance, including early CNN-based methods [10,11,18,25], as well as more recent approaches utilizing Transformers [7,8,21,22,24].

From CNN to Transformer, many different model architectures have been proposed, as well as a number of training loss functions. These losses can be mainly divided into three categories. The first class is represented by the Cross-Entropy Loss, which calculates the difference between the predicted probability distribution and the ground truth. Focal Loss [14] is proposed for addressing hard-to-learn samples. The second category is Dice Loss and other improvements. Dice Loss [17] is based on the intersection and union between the prediction and ground truth. Tversky Loss [19] improves the Dice Loss by balancing precision and recall. The Generalized Dice Loss [20] extends the Dice Loss to

H. Greenspan et al. (Eds.): MICCAI 2023, LNCS 14223, pp. 292–301, 2023.
https://doi.org/10.1007/978-3-031-43901-8_28

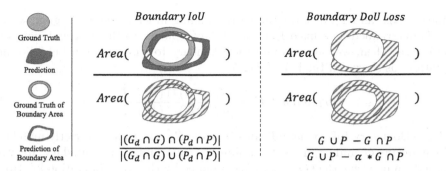

Fig. 1. The illustration of Boundary IoU (left) and Boundary DoU Loss (right), the shaded area in the figure will be calculated. G and P indicate ground truth and prediction, and G_d and P_d denote their corresponding boundary areas. α is a hyperparameter.

multi-category segmentation. The third category focuses on boundary segmentation. Hausdorff Distance Loss [12] is proposed to optimize the Hausdorff distance, and the Boundary Loss [13] calculates the distance between each point in the prediction and the corresponding ground truth point on the contour as the weight to sum the predicted probability of each point. However, the current loss for optimizing segmented boundaries dependent on combining different losses or training instability. To address these issues, we propose a simple boundary loss inspired by the Boundary IoU metrics [6], *i.e.*, Boundary DoU Loss.

Our proposed Boundary DoU Loss improves the focus on regions close to the boundary through a region-like calculation similar to Dice Loss. The error region near the boundary is obtained by calculating the difference set of ground truth and prediction, which is then reduced by decreasing its ratio to the union of the difference set and a partial intersection set. To evaluate the performance of our proposed Boundary DoU loss, we conduct experiments on the ACDC [1] and Synapse datasets by using the UNet [18], TransUNet [3] and Swin-UNet [2] models. Experimental results show the superior performance of our loss when compared with others.

2 Method

This section first revisit the Boundary IoU metric [6]. Then, we describe the details of our Boundary DoU loss function and adaptive size strategy. Next, we discuss the connection between our Boundary DoU loss with the Dice loss.

2.1 Boundary IoU Metric

The Boundary IoU is a segmentation evaluation metric which mainly focused on boundary quality. Given the ground truth binary mask G, the G_d denotes the inner boundary region within the pixel width of d. The P is the predicted binary mask, and P_d denotes the corresponding inner boundary region, whose size is determined as a fixed fraction of 0.5% relative to the diagonal length of the image. Then, we can compute the Boundary IoU metric by using following equation, as shown in the left of Fig. 1,

$$Boundary\ IoU = \frac{|(G_d \cap G) \cap (P_d \cap P)|}{|(G_d \cap G) \cup (P_d \cap P)|}. \tag{1}$$

A large Boundary IoU value indicates that the G_d and P_d are perfectly matched, which means G and P are with a similar shape and their boundary are well aligned. In practice, the G_d and P_d is computed by the erode operation [6]. However, the erode operation is non-differentiable, and we can not directly leverage the Boundary IoU as a loss function for training for increasing the consistency between two boundary areas.

2.2 Boundary DoU Loss

As shown in the left of Fig. 1, we can find that the union of the two boundaries $|(G_d \cap G) \cup (P_d \cap P)|$ actually are highly correlated to the difference set between G and P. The intersection $|(G_d \cap G) \cap (P_d \cap P)|$ is correlated to the inner boundary of the intersection of G and P. If the difference set $G \cup P - G \cap P$ decreases and the $G \cap P$ increases, the corresponding Boundary IoU will increase.

Based on the above analysis, we design a Boundary DoU loss based on the difference region to facilitate computation and backpropagation. First, we directly treat the difference set as the miss-matched boundary between G and P. Besides, we consider removing the middle part of the intersection area as the inner boundary, which is computed by $\alpha * G \cap P$ $(\alpha < 1)$ for simplicity. Then, we joint compute the $G \cup P - \alpha * G \cap P$ as the partial union. Finally, as shown in the right of Fig. 1, our Boundary DoU Loss can be computed by,

$$L_{DoU} = \frac{G \cup P - G \cap P}{G \cup P - \alpha * G \cap P}, \tag{2}$$

where α is a hyper-parameter controlling the influence of the partial union area.

Adaptive Adjusting α Based-on Target Size: On the other aspect, the proportion of the boundary area relative to the whole target varies for different sizes. When the target is large, the boundary area only accounts for a small proportion, and the internal regions can be easily segmented, so we are encouraged to focus more on the boundary area. In such a case, using a large α is preferred. However, when the target is small, neither the interior nor the boundary areas are easily distinguishable, so we need to focus simultaneously on the interior and boundary, and a small α is preferred. To achieve this goal, we future adaptively compute α based on the proportion,

$$\alpha = 1 - 2 \times \frac{C}{S}, \alpha \in [0, 1), \tag{3}$$

where C denotes the boundary length of the target, and S denotes its size.

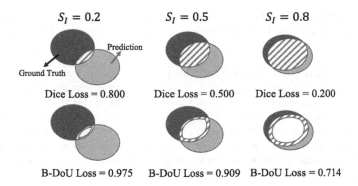

Fig. 2. Comparison of Boundary DoU Loss (B-DoU Loss) and Dice Loss. The figure shows the values of the two losses calculated at 20%, 50%, and 80% of the intersection of Ground Truth and Prediction, respectively. We assume that both Ground Truth and Prediction have an area of 1 and α is 0.8.

2.3 Discussion

In this part, we compare Boundary DoU Loss with Dice Loss. Firstly, we can re-write our Boundary DoU Loss as:

$$L_{DoU} = \frac{S_D}{S_D + S_I - \alpha S_I} = 1 - \frac{\alpha' * S_I}{S_D + \alpha' * S_I}, \tag{4}$$

where S_D denotes the area of the difference set between ground truth and prediction, S_I denotes the intersection area of them, and $\alpha' = 1 - \alpha$. Meanwhile, the Dice Loss can be expressed by the following:

$$L_{Dice} = 1 - \frac{2 * TP}{2 * TP + FP + FN} = 1 - \frac{2 * S_I}{2 * S_I + S_D}, \tag{5}$$

where TP, FP and FN denote True Positive, False Positive, and False Negative, respectively. It can be seen that Boundary DoU Loss and Dice loss differ only in the proportion of the intersection area. Dice is concerned with the whole intersection area, while Boundary DoU Loss is concerned with the boundary since $\alpha < 1$. Similar to the Dice loss function, minimizing the L_{DoU} will encourage an increase of the intersection area ($S_I \uparrow$) and a decrease of the different set ($S_D \downarrow$). Meanwhile, the L_{DoU} will penalize more over the ratio of S_D/S_I. To corroborate its effectiveness more clearly, we compare the values of L_{Dice} and L_{DoU} in different cases in Fig. 2. The L_{Dice} decreases linearly with the difference set, whereas L_{DoU} will decrease faster when S_I is higher enough.

Table 1. Experimental results on the Synapse dataset with the three models (average Dice score % and average Hausdorff Distance in mm, and average Boundary IoU %).

Model	UNet			TransUNet			Swin-UNet		
Loss	DSC↑	HD↓	B-IoU↑	DSC↑	HD↓	B-IoU↑	DSC↑	HD↓	B-IoU↑
Dice	76.38	31.45 ± 9.31	86.26	78.52	28.84 ± 2.47	87.34	77.98	25.95 ± 9.07	86.19
CE	65.95	40.31 ± 50.3	82.69	72.98	35.05 ± 14.1	84.84	71.77	33.20 ± 4.02	84.04
Dice + CE	76.77	30.20 ± 5.10	86.21	78.19	29.30 ± 11.8	87.18	78.30	24.71 ± 2.84	86.72
Tversky	63.61	65.12 ± 4.38	75.65	63.90	70.89 ± 218	70.77	68.22	41.22 ± 419	79.53
Boundary	76.23	34.54 ± 12.5	85.75	76.82	31.88 ± 3.01	86.66	76.00	26.74 ± 2.18	84.98
Ours	**78.68**	$\mathbf{26.29 \pm 2.35}$	**87.08**	**79.53**	$\mathbf{27.28 \pm 0.51}$	**88.11**	**79.87**	$\mathbf{19.80 \pm 2.34}$	**87.78**

Table 2. Experimental results on the ACDC dataset with the three models (average Dice score % and average Hausdorff Distance in mm, and average Boundary DoU %).

Model	UNet			TransUNet			Swin-UNet		
Loss	DSC↑	HD↓	B-IoU↑	DSC↑	HD↓	B-IoU↑	DSC↑	HD↓	B-IoU↑
Dice	90.17	1.34 ± 0.10	75.20	90.69	2.03 ± 0.00	76.66	90.17	1.34 ± 0.01	75.20
CE	88.08	1.48 ± 0.28	71.25	89.22	2.07 ± 0.05	73.78	88.08	1.48 ± 0.03	71.25
Dice + CE	89.94	$\mathbf{1.28 \pm 1.29}$	74.80	90.48	1.94 ± 0.00	76.27	89.94	$\mathbf{1.28 \pm 0.00}$	74.80
Tversky	83.60	9.88 ± 226	69.36	90.37	1.95 ± 0.02	76.20	89.55	1.48 ± 0.04	74.37
Boundary	89.25	2.28 ± 0.24	73.08	90.48	$\mathbf{1.84 \pm 0.08}$	76.31	88.95	1.53 ± 0.03	72.73
Ours	**90.84**	1.54 ± 0.33	**76.44**	**91.29**	2.16 ± 0.02	**78.45**	**91.02**	1.28 ± 0.00	**77.00**

3 Experiments

3.1 Datasets and Evaluation Metrics

Synapse:[1] The Synapse dataset contains 30 abdominal 3D CT scans from the MICCAI 2015 Multi-Atlas Abdomen Labeling Challenge. Each CT volume contains $86 \sim 198$ slices of 512×512 pixels. The slice thicknesses range from 2.5 mm to 5.0 mm, and in-plane resolutions vary from 0.54×0.54 mm^2 to 0.98×0.98 mm^2. Following the settings in TransUNet [3], we randomly select 18 scans for training and the remaining 12 cases for testing.

ACDC:[2] The ACDC dataset is a 3D MRI dataset from the Automated Cardiac Diagnosis Challenge 2017 and contains cardiac data from 150 patients in five categories. Cine MR images were acquired under breath-holding conditions, with slices 5–8 mm thick, covering the LV from basal to apical, with spatial resolutions from 1.37 to 1.68 mm/pixel and 28 to 40 images fully or partially covering the cardiac cycle. Following the TransUNet, we split the original training set with 100 scans into the training, validation, and testing sets with a ratio of 7:1:2.

---

[1] https://www.synapse.org/#!Synapse:syn3193805/wiki/217789.
[2] https://www.creatis.insa-lyon.fr/Challenge/acdc/.

Table 3. Experimental results for different target sizes on the ACDC and Synapse datasets (average Dice score%).

	ACDC						Synapse					
Model	UNet		TransUNet		Swin-UNet		UNet		TransUNet		Swin-UNet	
Loss	large	small	large	small	large	small	large	small	large	small	large	small
Dice	92.60	84.11	93.63	85.95	92.40	86.43	78.59	36.22	80.84	36.16	79.97	41.67
CE	91.40	81.72	92.55	83.86	91.49	82.58	68.89	0.12	75.21	32.32	74.27	26.25
Dice + CE	92.91	84.36	93.45	85.70	92.83	85.30	78.89	38.27	80.47	36.82	80.29	42.22
Tversky	91.82	70.37	93.28	85.86	92.67	84.53	65.76	24.60	66.00	25.74	70.08	34.44
Boundary	92.53	83.97	93.53	85.57	92.25	83.63	78.37	37.30	79.00	37.14	78.02	39.23
Ours	**93.73**	**86.04**	**94.01**	**86.93**	**93.63**	**86.83**	**80.88**	**38.68**	**81.85**	**37.32**	**81.83**	**44.08**

Evaluation Metrics: We use the most widely used Dice Similarity Coefficient (DSC) and Hausdorff Distances (HD) as evaluation metrics. Besides, the Boundary IoU [6] (B-IoU) is adopted as another evaluation metric for the boundary.

3.2 Implementation Details

We conduct experiments on three advanced models to evaluate the performance of our proposed Boundary DoU Loss, i.e., UNet, TransUNet, and Swin-UNet. The models are implemented with the PyTorch toolbox and run on an NVIDIA GTX A4000 GPU. The input resolution is set as 224×224 for both datasets. For the Swin-UNet [2] and TransUNet [3], we used the same training and testing parameters provided by the source code, i.e., the learning rate is set to 0.01, with a weight decay of 0.0001. The batch size is 24, and the optimizer uses SGD with a momentum of 0.9. For the UNet, we choose ResNet50 as the backbone and initialize the encoder with the ImageNet pre-trained weights following the setting in TransUNet [3]. The other configurations are the same as TransUNet. We train all models by 150 epochs on both Synapse and ACDC datasets.

We further train the three models by different loss functions for comparison, including Dice Loss, Cross-Entropy Loss (CE), Dice+CE, Tversky Loss [19], and Boundary Loss [13]. The training settings of different loss functions are as follows. For the $\lambda_1 \text{Dice} + \lambda_2 \text{CE}$, we set (λ_1, λ_2) as (0.5, 0.5) for the UNet and TransUNet, and (0.6, 0.4) for Swin-UNet. For the Tversky Loss, we set $\alpha = 0.7$ and $\beta = 0.3$ by referring to the best performance in [19]. Following the Boundary Loss [13], we use $L = \alpha * (\text{Dice} + \text{CE}) + (1 - \alpha) * \text{Boundary}$ for training . The α is initially set to 1 and decreases by 0.01 at each epoch until it equals 0.01.

3.3 Results

Quantitative Results: Table 1 shows the results of different losses on the Synapse dataset. From the table, we can have the following findings: 1) The original Dice Loss achieves the overall best performance among other losses. The CE loss function obtains a significantly lower performance than the Dice.

aorta   gallbladder   left kidney   right kidney   liver   pancreas   spleen   stomach

Ground Truth   Ours   Dice   Dice + CE   CE   Boundary

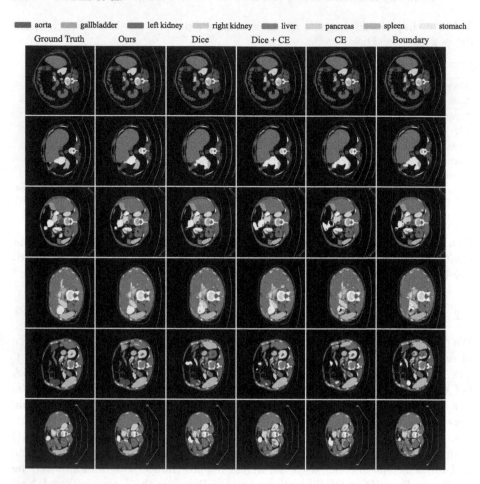

Fig. 3. The qualitative comparison of the segmentation on the Synapse dataset. (Row 1&2: Swin-UNet, Row 3&4: TransUNet, and Row 5&6: UNet)

Besides, the Dice+CE, Tversky, and Boundary do not perform better than Dice. 2) Compared with the Dice Loss, our Loss improves the DSC by 2.30%, 1.20%, and 1.89% on UNet, TransUNet, and Swin-UNet models, respectively. The Hausdorff Distance also shows a significant decrease. Meanwhile, we achieved the best performance on the Boundary IoU, which verified that our loss could improve the segmentation performance of the boundary regions.

Table 2 reports the results on the ACDC dataset. We can find that our Boundary DOU Loss effectively improves DSC on all three models. Compared with Dice Loss, the DSC of UNet, TransUNet, and Swin-UNet improved by 0.62%, 0.6%, and 0.85%, respectively. Although our Loss did not get all optimal performance for the Hausdorff Distance, we substantially outperformed all other losses on the Boundary IoU. These results indicate that our method can better segment the boundary regions. This capability can assist doctors in better identifying challenging object boundaries in clinical settings.

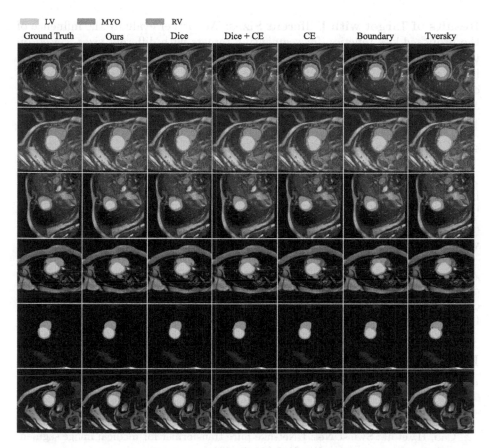

Fig. 4. The qualitative comparison of segmentation results on the ACDC dataset. (Row 1&2: Swin-UNet, Row 3&4: TransUNet, and Row 5&6: UNet)

Qualitative Results: Figures 3 and 4 show the qualitative visualization results of our loss and other losses. Overall, our method has a clear advantage for segmenting the boundary regions. In the Synapse dataset (Fig. 3), we can achieve more accurate localization and segmentation for complicated organs such as the stomach and pancreas. Our results from the 3rd and 5th rows substantially outperform the other losses when the target is small. Based on the 2nd and last rows, we can obtain more stable segmentation on the hard-to-segment objects. As for the ACDC dataset (Fig. 4), due to the large variation in the shape of the RV region as shown in Row 1, 3, 4 and 6, it is easy to cause under- or missegmentation. Our Loss resolves this problem better compared with other Losses. Whereas the MYO is annular and the finer regions are difficult to segment, as shown in the 2nd and 5th row, the other losses all result in different degrees of under-segmentation, while our loss ensures its completeness. Reducing the mis- and under-classification will allow for better clinical guidance.

Results of Target with Different Sizes: We further evaluate the influence of the proposed loss function for segmenting targets with different sizes. Based on the observation of C/S values for different targets, we consider a target to be a large one when $C/S < 0.2$ and otherwise as a small target. As shown in Table 3, our Boundary DoU Loss function can improve the performance for both large and small targets.

4   Conclusion

In this study, we propose a simple and effective loss (Boundary DoU) for medical image segmentation. It adaptively adjusts the penalty to regions close to the boundary based on the size of the different targets, thus allowing for better optimization of the targets. Experimental results on ACDC and Synapse datasets validate the effectiveness of our proposed loss function.

Acknowledgement. This work is supported by the National Natural Science Foundation of China (No. 62276221), the Natural Science Foundation of Fujian Province of China (No. 2022J01002), and the Science and Technology Plan Project of Xiamen (No. 3502Z20221025).

References

1. Bernard, O., et al.: Deep learning techniques for automatic MRI cardiac multi-structures segmentation and diagnosis: is the problem solved? IEEE Trans. Med. Imaging **37**(11), 2514–2525 (2018)
2. Cao, H., et al.: Swin-UNet: UNet-like pure transformer for medical image segmentation. arXiv preprint arXiv:2105.05537 (2021)
3. Chen, J., et al.: TransUNet: transformers make strong encoders for medical image segmentation. arXiv preprint arXiv:2102.04306 (2021)
4. Chen, L.C., Papandreou, G., Schroff, F., Adam, H.: Rethinking atrous convolution for semantic image segmentation. arXiv preprint arXiv:1706.05587 (2017)
5. Chen, L.C., Zhu, Y., Papandreou, G., Schroff, F., Adam, H.: Encoder-decoder with atrous separable convolution for semantic image segmentation. In: Proceedings of the European conference on computer vision (ECCV), pp. 801–818 (2018)
6. Cheng, B., Girshick, R., Dollár, P., Berg, A.C., Kirillov, A.: Boundary IoU: improving object-centric image segmentation evaluation. In: Proceedings of the IEEE/CVF Conference on Computer Vision and Pattern Recognition, pp. 15334–15342 (2021)
7. Gao, Y., Zhou, M., Metaxas, D.N.: UTNet: a hybrid transformer architecture for medical image segmentation. In: de Bruijne, M., et al. (eds.) MICCAI 2021. LNCS, vol. 12903, pp. 61–71. Springer, Cham (2021). https://doi.org/10.1007/978-3-030-87199-4_6
8. Hatamizadeh, A., et al.: Unetr: transformers for 3D medical image segmentation. In: Proceedings of the IEEE/CVF Winter Conference on Applications of Computer Vision, pp. 574–584 (2022)
9. He, K., Gkioxari, G., Dollár, P., Girshick, R.: Mask R-CNN. In: Proceedings of the IEEE International Conference on Computer Vision, pp. 2961–2969 (2017)

10. Huang, H., et al.: Unet 3+: a full-scale connected UNet for medical image segmentation. In: ICASSP 2020–2020 IEEE International Conference on Acoustics, Speech and Signal Processing (ICASSP), pp. 1055–1059. IEEE (2020)

11. Isensee, F., et al.: NNU-Net: self-adapting framework for u-net-based medical image segmentation. arXiv preprint arXiv:1809.10486 (2018)

12. Karimi, D., Salcudean, S.E.: Reducing the hausdorff distance in medical image segmentation with convolutional neural networks. IEEE Trans. Med. Imaging **39**(2), 499–513 (2019)

13. Kervadec, H., Bouchtiba, J., Desrosiers, C., Granger, E., Dolz, J., Ayed, I.B.: Boundary loss for highly unbalanced segmentation. In: International Conference on Medical Imaging with Deep Learning, pp. 285–296. PMLR (2019)

14. Lin, T.Y., Goyal, P., Girshick, R., He, K., Dollár, P.: Focal loss for dense object detection. In: Proceedings of the IEEE International Conference on Computer Vision, pp. 2980–2988 (2017)

15. Liu, S., Qi, L., Qin, H., Shi, J., Jia, J.: Path aggregation network for instance segmentation. In: Proceedings of the IEEE Conference on Computer Vision and Pattern Recognition, pp. 8759–8768 (2018)

16. Long, J., Shelhamer, E., Darrell, T.: Fully convolutional networks for semantic segmentation. In: Proceedings of the IEEE Conference on Computer Vision and Pattern Recognition, pp. 3431–3440 (2015)

17. Milletari, F., Navab, N., Ahmadi, S.A.: V-net: fully convolutional neural networks for volumetric medical image segmentation. In: The Fourth International Conference on 3D Vision (3DV), pp. 565–571 (2016)

18. Ronneberger, O., Fischer, P., Brox, T.: U-Net: convolutional networks for biomedical image segmentation. In: Navab, N., Hornegger, J., Wells, W.M., Frangi, A.F. (eds.) MICCAI 2015. LNCS, vol. 9351, pp. 234–241. Springer, Cham (2015). https://doi.org/10.1007/978-3-319-24574-4_28

19. Salehi, S.S.M., Erdogmus, D., Gholipour, A.: Tversky Loss function for image segmentation using 3D fully convolutional deep networks. In: Wang, Q., Shi, Y., Suk, H.-I., Suzuki, K. (eds.) MLMI 2017. LNCS, vol. 10541, pp. 379–387. Springer, Cham (2017). https://doi.org/10.1007/978-3-319-67389-9_44

20. Sudre, C.H., Li, W., Vercauteren, T., Ourselin, S., Jorge Cardoso, M.: Generalised dice overlap as a deep learning loss function for highly unbalanced segmentations. In: Cardoso, M.J., et al. (eds.) DLMIA/ML-CDS -2017. LNCS, vol. 10553, pp. 240–248. Springer, Cham (2017). https://doi.org/10.1007/978-3-319-67558-9_28

21. Valanarasu, J.M.J., Oza, P., Hacihaliloglu, I., Patel, V.M.: Medical transformer: gated axial-attention for medical image segmentation. In: de Bruijne, M., et al. (eds.) MICCAI 2021. LNCS, vol. 12901, pp. 36–46. Springer, Cham (2021). https://doi.org/10.1007/978-3-030-87193-2_4

22. Xu, G., Wu, X., Zhang, X., He, X.: LeViT-UNet: make faster encoders with transformer for medical image segmentation. arXiv preprint arXiv:2107.08623 (2021)

23. Zhao, H., Shi, J., Qi, X., Wang, X., Jia, J.: Pyramid scene parsing network. In: Proceedings of the IEEE Conference on Computer Vision and Pattern Recognition, pp. 2881–2890 (2017)

24. Zhou, H.Y., Guo, J., Zhang, Y., Yu, L., Wang, L., Yu, Y.: nnFormer: interleaved transformer for volumetric segmentation. arXiv preprint arXiv:2109.03201 (2021)

25. Zhou, Z., Rahman Siddiquee, M.M., Tajbakhsh, N., Liang, J.: UNet++: a nested U-Net architecture for medical image segmentation. In: Stoyanov, D., et al. (eds.) DLMIA/ML-CDS -2018. LNCS, vol. 11045, pp. 3–11. Springer, Cham (2018). https://doi.org/10.1007/978-3-030-00889-5_1

Do We Really Need
that Skip-Connection? Understanding Its Interplay with Task Complexity

Amith Kamath[1], Jonas Willmann[2,3], Nicolaus Andratschke[2],
and Mauricio Reyes[1,4(✉)]

[1] ARTORG Center for Biomedical Engineering Research, University of Bern, Bern,
Switzerland
mauricio.reyes@unibe.ch
[2] Department of Radiation Oncology, University Hospital Zurich, University of
Zurich, Zurich, Switzerland
[3] Center for Proton Therapy, Paul Scherrer Institute, Villigen, Switzerland
[4] Department of Radiation Oncology, Inselspital, Bern University Hospital and
University of Bern, Bern, Switzerland
https://github.com/amithjkamath/to_skip_or_not

Abstract. The U-Net architecture has become the preferred model used
for medical image segmentation tasks. Since its inception, several variants have been proposed. An important component of the U-Net architecture is the use of skip-connections, said to carry over image details
on its decoder branch at different scales. However, beyond this intuition,
not much is known as to what extent skip-connections of the U-Net are
necessary, nor what their interplay is in terms of model robustness when
they are subjected to different levels of task complexity. In this study,
we analyzed these questions using three variants of the U-Net architecture (the standard U-Net, a "No-Skip" U-Net, and an Attention-Gated
U-Net) using controlled experiments on varying synthetic texture images
and evaluated these findings on three medical image data sets. We measured task complexity as a function of texture-based similarities between
foreground and background distributions. Using this scheme, our findings suggest that the benefit of employing skip-connections is small for
low-to-medium complexity tasks, and its benefit appears only when the
task complexity becomes large. We report that such incremental benefit
is non-linear, with the Attention-Gated U-Net yielding larger improvements. Furthermore, we find that these benefits also bring along robustness degradations on clinical data sets, particularly in out-of-domain scenarios. These results suggest a dependency between task complexity and
the choice/design of noise-resilient skip-connections, indicating the need
for careful consideration while using these skip-connections.

Keywords: Image segmentation · U-Net · robustness

Supplementary Information The online version contains supplementary material
available at https://doi.org/10.1007/978-3-031-43901-8_29.

1 Introduction

Due to the broad success of U-Nets [17] for image segmentation, it has become the go-to architecture in the medical image computing community. Since its creation in 2015, much research has been dedicated to exploring variants and improvements over the standard base model [3]. However, Isensee et al. [12] showed with their not-new-U-Net (nnU-Net) that the success of the U-Net relies on a well-prepared data pipeline incorporating appropriate data normalization, class balancing checks, and preprocessing, rather than on architecture changes. Arguably the two most important challenges at present for medical image segmentation are generalization and robustness. A lack of generalization decreases the performance levels of a model on data sets not well characterized by the training data set, while poor robustness appears when models under-perform on data sets presenting noise or other corruptions [13]. Modern neural networks have been shown to be highly susceptible to distribution shifts and corruptions that are modality-specific [6]. While the average accuracy of U-Net-based models has increased over the years, it is evident from the literature that their robustness level has not improved at the same rate [4,5,9].

One of the key elements of the U-Net are the skip-connections, which propagate information directly (i.e., without further processing) from the encoding to the decoding branch at different scales. Azad et al. [3] mention that this novel design propagates essential high-resolution contextual information along the network, which encourages the network to re-use the low-level representation along with the high-context representation for accurate localization. Nonetheless, there is no clear evidence supporting this intuition and moreover, there is limited knowledge in the literature describing to what extent skip-connections of the U-Net are necessary, and what their interplay is in terms of model robustness when they are subjected to different levels of task complexity.

Currently, the U-Net is used more as a "Swiss-army knife" architecture across different image modalities and image quality ranges. In this paper, we describe the interplay between skip-connections and their effective role of "transferring information" into the decoding branch of the U-Net for different degrees of task complexity, based on controlled experiments conducted on synthetic images of varying textures as well as on clinical data comprising Ultrasound (US), Computed tomography (CT), and Magnetic Resonance Imaging (MRI). In this regard, the work of [10] showed that neural networks are biased toward texture information. Recently, [19,20] similarly showed the impact of texture modifications on the performance and robustness of trained U-Net models. Contrary to these prior works analyzing the impact of data perturbation to model performance (e.g. [6,13]), in this study we focus on analyzing the role of skip-connections to model performance and its robustness. We hypothesize therefore that skip-connections may not always lead to beneficial effects across varying task complexities as measured with texture modifications. Our major contributions through this paper are:

(i) We describe a novel analysis pipeline to evaluate the robustness of image segmentation models as a function of the difference in texture between foreground and background.

(ii) We confirm the hypothesis that severing these skip-connections could lead to more robust models, especially in the case of out-of-domain (OOD) test data. Furthermore, we show that severing skip-connections could work better than filtering feature maps from the encoder with attention-gating.

(iii) Finally, we also demonstrate failure modes of using skip-connections, where robustness across texture variations appear to be sacrificed in the pursuit of improvements within domain.

2   Materials and Methods

2.1   Experiment Design

Figure 1 describes our experimental setup to assess the impact of skip-connections in U-Net-like architectures under varying levels of task complexity.

Given a set of N pairs of labeled training images $\{(I, S)_i : 1 \leq i \leq N\}, I \in \mathbb{R}^{H \times W}$ and $S \in \mathbb{Z} : \{0,1\}^{H \times W}$, corresponding ground-truth segmentation, a deep learning segmentation model $M(I) \mapsto S$ is commonly updated by minimizing a standard loss term, such as the binary cross entropy or dice loss. To

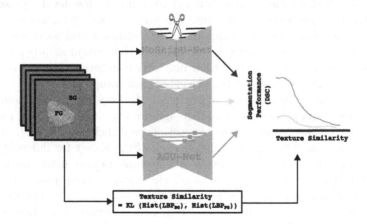

Fig. 1. Experimental design to evaluate the role of U-Net's skip-connections under different levels of task complexity. Given training images with controllable background (BG) and foreground (FG) textures, three variants of the U-Net were trained featuring no skip-connections (NoSkipU-Net), standard U-Net (U-Net) [17], and Attention-Gated U-Net (AGU-Net) [18], each characterizing a different strategy (zeroing information through skips, identity transform and filtering information through skips, respectively). Each model was trained with different levels of texture similarity between background and foreground, based on the Kullback-Leibler divergence of Local Binary Pattern (LBP) histograms for foreground and background regions. For each level of foreground-to-background texture similarity, the performance for each model was recorded in-domain, and robustness was measured with out-of-domain texture similarities.

evaluate how the model behaves at varying task complexities, we construct training data sets where each training sample is subjected to a linear transformation where its foreground is blended with the background: $I(x \mid Z(x) = 1) = \alpha I(x \mid Z(x) = 1) + (1 - \alpha)I(x \mid Z(x) = 0)$.

By increasing α from zero to one, more of the foreground texture is added in the foreground mask, which otherwise is made up of the background texture (See Fig. 2), while the background itself is unimpacted. We then quantify the similarity between foreground and background regions by measuring the Kullback-Leibler divergence between their local-binary-pattern (LBP) [16] histograms. We selected LBP since it is a commonly used and benchmarked texture descriptor in machine learning applications [8,15].

$$TS = KL(\mathcal{H}(\mathcal{L}(I)_{BG})||\mathcal{H}(\mathcal{L}(I)_{FG}) \tag{1}$$

$$\mathcal{L}(I)_{BG} = LBP(I(x \mid Z(x) = 0) \tag{2}$$

$$\mathcal{L}(I)_{FG} = LBP(I(x \mid Z(x) = 1) \tag{3}$$

where TS refers to the level of texture similarity, $\mathcal{H}()$ corresponds to histogram, and $\mathcal{L}(\mathcal{I})_{\{BG,FG\}}$ refers to LBP calculated for BG or FG. The LBP histogram was computed using a 3×3 neighbourhood with 8 points around each pixel in the image. Three U-Net models were trained featuring three different skip-connection strategies: NoSkipU-Net, U-Net, and AGU-Net, representing the absence of skip-connections, the use of an identity transform (i.e., information through skips is kept as is), and filtering information via attention through skip-connections, respectively. Models were trained at different levels of TS between the foreground and background regions, determined based on the Kullback-Leibler divergence of Local Binary Pattern (LBP) histograms, Eq. 1. For each level of α used to create a training set, we trained a model to be evaluated on a synthetic test set using the same α to measure within-domain performance and across a range of α, to measure their out-of-domain robustness.

Next, using Eq. 1 and ground truth labels, we computed the TS of images from the test set of the medical data sets and applied corruptions by way of noise or blurring in order to increase and decrease TS depending on the imaging modality being analyzed. Then we evaluated the robustness of these models to texture changes in these data sets. We did this at two levels of task complexity (easier - where TS is higher, and harder, where TS is lower) and different from the original TS. We report all model performances using dice scores.

2.2   Description of Data

Synthetic Textures: We took two representative grayscale textures from the synthetic toy data set described in [11] and used them as the background and foreground patterns. These patterns were chosen such that the TS values matched the range of medical data sets described next. We also generated synthetic segmentation masks using bezier curves setup such that the curvature and size of the foreground simulate clinical image segmentation problems. Examples of such images are shown in Fig. 2. We generated 100 such image-mask

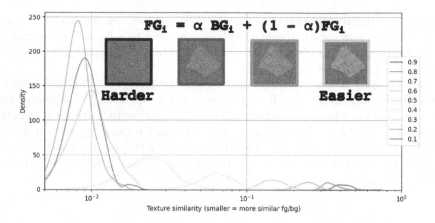

Fig. 2. Generation of synthetic data samples as a function of blending foreground texture into the background. Numbers in the legend indicate the proportion of foreground blended within the foreground mask.

pairs at 9 levels (for $\alpha \in \{0.1, 0.2, ..., 0.9\}$), so that we create training data sets at various task complexities. These images are generated by randomly cropping the grayscale textures to 256×256 pixels. 70 of these were used as the training set, 10 were reserved for validation, and the rest of the 20 formed the test set, identically split for all task complexities. Figure 2 show kernel density estimates of each of these 9 data sets along the texture similarity axis. The curve in orange ($\alpha = 0.1$) indicates that the foreground mask in this set contains only 10% of the actual foreground texture and 90% of the background texture blended together. This represents a situation where it is texturally hard for humans as well as for segmentation models. The data set in green ($\alpha = 0.9$) shows the reverse ratio - the foreground region now contains 90% of the foreground texture, thereby making it an easier task to segment.

Medical Data Sets: We tested the three variants of the U-Net architecture on three medical binary segmentation data sets: a Breast Ultrasound [1], a spleen CT and a heart MRI data set [2]. The breast ultrasound data set contained 647 images, 400 of which were used as training, 100 as validation and 147 as the test set. We used the benign and malignant categories in the breast ultrasound data and excluded images with no foreground to segment (i.e. the "normal" category). The spleen data set contained 899 images, 601 of which were used as training, 82 as validation and 216 as test set images. The heart data set contained 829 images, 563 of which were used as training, 146 as validation, and 120 as test set images. We selected 2D axial slices from the spleen, and sagittal slices from the heart data sets, both of which were originally 3D volumes, such that there is at least one pixel corresponding to the foreground. Care was taken to ensure that 2D slices were selected from 3D volumes and split at the patient level to avoid cross-contamination of images across training/test splits.

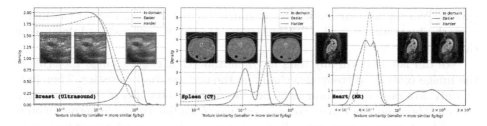

Fig. 3. Medical data test sets on the texture similarity ($\mathcal{TS}$) axis with in-domain (dashed gray), easier task (green, low similarity) and harder task (red, high similarity) distributions. Three modalities tested include US, CT, and MR, whose $\mathcal{TS}$ are in the same range as synthetic data in Fig. 2.

To vary $\mathcal{TS}$ of images in the test set, and to evaluate the robustness of the U-Net variants, speckle noise with variance 0.1 was added to both the foreground and background. This made the textures more similar, hence lowered $\mathcal{TS}$, and essentially rendered them harder to segment. This is shown in the red boxes in Fig. 3. We also created another test set with textures that are less similar by blurring the background using a Gaussian kernel of variance 3.0 while not blurring the foreground pixels. These are shown in the green boxes in Fig. 3, where it can be seen they are easier to segment.

2.3 Model Architecture and Training Settings

The network architectures were created using MONAI [7] v1.1 and were all trained with random weight initialization. The U-Net was implemented with one input and output channel, with input image size set to 256×256 pixels across all experiments. The model had five levels with $16, 32, 64, 128, 256$ channels each for synthetic experiments and six levels (an additional level with 512 channels) for medical image experiments, all intermediate channels with a stride of 2. The ReLU activation was used, and no residual units were included. To reduce stochasticity, no dropout was used in any variant.

The NoSkipU-Net was identical to the U-Net except for severed skip-connections. This led to the number of channels in the decoder to be smaller as there is no concatenation from the corresponding encoder level. The AGU-Net was setup to be the same as the U-Net, except with attention gating through the skip-connections.

The training parameters were kept constant across compared models for fair comparison. Our experiments[1] were implemented in Python 3.10.4 using the PyTorch implementation of the adam [14] optimizer. We set the learning rate to be $1e^{-3}$ for synthetic experiments (and $1e^{-2}$ for medical image experiments), maintaining it constant without using a learning rate scheduler. No early stopping criteria were used while training, and all models were allowed to train to 100

---

[1] Code to reproduce this is at https://github.com/amithjkamath/to_skip_or_not..

epochs. We trained our models to optimize the dice loss, and saved the model with the best validation dice (evaluated once every two epochs) for inference on the test set.

We did not perform any data augmentation that could change the scale of the image content, thereby also changing the texture characteristics. Therefore, we only do a random rotation by 90 degrees with a probability of 0.5 for training, and no other augmentations. We also refrained from fine-tuning hyperparameters and did not perform any ensembling as our study design is not meant to achieve the best possible performance metric as much as it attempts to reliably compare performance across architecture variants while keeping confounding factors to a minimum. We therefore trained each model using the same random seeds (three times) and report the dice score statistics. Training and testing were performed on an NVIDIA A5000 GPU with 24 GB RAM and CUDA version 11.4.

3   Results

3.1   On Synthetic Texture Variants

In-domain (Performance): Figure 4 (left) indicates the relative improvement in dice scores between the three U-Net variants using the NoSkipU-Net as the baseline. To make the interpretation easier, the α value is used as a proxy for TS on the horizontal axis. It is worth noting that for α values > 0.3, there is negligible difference between the dice score performances of all the U-Net variants, indicating their ability with or without the skip-connections to learn the distributions of the foreground and background textures at that level of task complexity. Below α values of 0.3, the benefits of using attention gating in the skip-connections start to appear. This indicates that the benefit of attention-gating as a function of complexity is non-linear: models do not benefit from skip-connections at lower ranges of task complexity, but at larger ones, filtering the information flowing through the skip connections is important. What is interesting is also how the standard U-Net performance is noisy compared to NoSkipU-Net, indicating that passing through the entire encoder feature map to be concatenated with the decoder feature maps may not always be beneficial.

Out-of-Domain (Robustness): Rows in Fig. 4 (right) includes a heatmap to represent the α values that the model was trained on, and columns correspond to the α value it was tested on. The entries in the matrix are normalized dice score differences between AGU-Net and NoSkipU-Net (comparisons between standard U-Net and NoSkipU-Net show similar trends). The diagonal entries here correspond to the AGU-Net plot in Fig. 4 (left). For α values 0.3 and 0.4 in training and 0.9 on testing (corresponding to an out-of-domain testing scenario), the NoSkipU-Net performs better than the AGU-Net, indicating that there indeed are situations where skip-connections cause more harm than benefit.

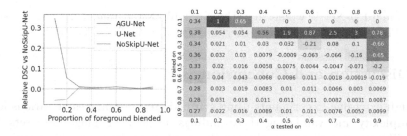

Fig. 4. Relative performance in-domain (left) across U-Net variants, and out-of-domain robustness metrics (right) for AGU-Net versus NoSkipU-Net.

3.2 On Medical Image Textures

In-Domain (performance): Looking at the "In-domain" rows in Table 1, on all three data sets, the AGU-Net outperforms both the other variants. However, the relative improvements in performance vary across modalities, with the performance differences on CT being the most stark. On the Ultrasound data set, the NoSkipU-Net performs as well as the standard U-Net, supporting our hypothesis that skip-connections may not always be beneficial.

Out-of-Domain (Robustness): Focusing on the rows "Harder" and "Easier" in Table 1, we observe for the Ultrasound data set that the AGU-Net improves in the easier task, but declines in performance in the harder one. The drop in performance is most pronounced for the U-Net, but moderate for the NoSkipU-Net. For the spleen data set, both the AGU-Net and the standard U-Net demonstrate severe drop in performance in the harder case. However, AGU-Net is better and

Table 1. Mean (standard deviation) of Dice scores for each of hard, in-domain and easy textures on the Breast (Ultrasound), Spleen (CT) and Heart (MR) data sets. Best performing model at each texture level is highlighted in bold.

Data set	Texture level	AGU-Net	U-Net	NoSkipU-Net
Breast (Ultrasound)	Harder	0.645 (0.291)	0.723 (0.281)	**0.735** (0.268)
	In-domain	**0.795** (0.206)	0.762 (0.261)	0.761 (0.258)
	Easier	**0.799** (0.200)	0.735 (0.244)	0.748 (0.243)
Spleen (CT)	Harder	0.310 (0.226)	0.074 (0.152)	**0.558** (0.265)
	In-domain	**0.927** (0.092)	0.745 (0.275)	0.606 (0.265)
	Easier	**0.809** (0.201)	0.394 (0.354)	0.486 (0.292)
Heart (MRI)	Harder	0.139 (0.242)	0.500 (0.316)	**0.815** (0.126)
	In-domain	**0.929** (0.055)	0.900 (0.080)	0.833 (0.111)
	Easier	**0.889** (0.073)	0.805 (0.129)	0.823 (0.103)

the standard U-Net is worse than the NoSkipU-Net in the easier texture situations. The heart data set shows the same trend as in the spleen data set.

4 Discussion and Conclusion

Through extensive experiments using synthetic texture images at various levels of complexity and validating these findings on medical image data sets from three different modalities, we show in this paper that the use of skip-connections can both be beneficial as well as harmful depending on what can be traded off: robustness or performance. A limitation of our work is that we vary only the foreground in synthetic experiments but background variations could demonstrate unexpected asymmetric behavior. We envision the proposed analysis pipeline to be useful in quality assurance frameworks where U-Net variants could be compared to analyse potential failure modes.

References

1. Al-Dhabyani, W., Gomaa, M., Khaled, H., Fahmy, A.: Dataset of breast ultrasound images. Data Brief **28**, 104863 (2020)
2. Antonelli, M., et al.: The medical segmentation decathlon. Nature Commun. **13**(1), 4128 (2022)
3. Azad, R., et al.: Medical image segmentation review: the success of u-net. arXiv preprint arXiv:2211.14830 (2022)
4. Bakas, S., et al.: Identifying the best machine learning algorithms for brain tumor segmentation, progression assessment, and overall survival prediction in the brats challenge. arXiv preprint arXiv:1811.02629 (2018)
5. Bernard, O., et al.: Deep learning techniques for automatic MRI cardiac multi-structures segmentation and diagnosis: is the problem solved? IEEE Trans. Med. Imaging **37**(11), 2514–2525 (2018)
6. Boone, L., et al.: Rood-MRI: Benchmarking the robustness of deep learning segmentation models to out-of-distribution and corrupted data in MRI. arXiv preprint arXiv:2203.06060 (2022)
7. Cardoso, M.J., et al.: Monai: An open-source framework for deep learning in healthcare. arXiv preprint arXiv:2211.02701 (2022)
8. Doshi, N.P., Schaefer, G.: A comprehensive benchmark of local binary pattern algorithms for texture retrieval. In: Proceedings of the 21st International Conference on Pattern Recognition (ICPR2012), pp. 2760–2763. IEEE (2012)
9. Galati, F., Ourselin, S., Zuluaga, M.A.: From accuracy to reliability and robustness in cardiac magnetic resonance image segmentation: a review. Appl. Sci. **12**(8), 3936 (2022)
10. Geirhos, R., Rubisch, P., Michaelis, C., Bethge, M., Wichmann, F.A., Brendel, W.: Imagenet-trained cnns are biased towards texture; increasing shape bias improves accuracy and robustness. arXiv preprint arXiv:1811.12231 (2018)
11. Hoyer, L., Munoz, M., Katiyar, P., Khoreva, A., Fischer, V.: Grid saliency for context explanations of semantic segmentation. Adv. Neural Inform. Process. Syst. **32** (2019)

12. Isensee, F., Jaeger, P.F., Kohl, S.A., Petersen, J., Maier-Hein, K.H.: NNU-net: a self-configuring method for deep learning-based biomedical image segmentation. Nat. Methods **18**(2), 203–211 (2021)
13. Kamann, C., Rother, C.: Benchmarking the robustness of semantic segmentation models with respect to common corruptions. Int. J. Comput. Vision **129**(2), 462–483 (2021)
14. Kingma, D.P., Ba, J.: Adam: A method for stochastic optimization. arXiv preprint arXiv:1412.6980 (2014)
15. Liu, L., Fieguth, P., Wang, X., Pietikäinen, M., Hu, D.: Evaluation of LBP and deep texture descriptors with a new robustness benchmark. In: Leibe, B., Matas, J., Sebe, N., Welling, M. (eds.) ECCV 2016. LNCS, vol. 9907, pp. 69–86. Springer, Cham (2016). https://doi.org/10.1007/978-3-319-46487-9_5
16. Ojala, T., Pietikäinen, M., Harwood, D.: A comparative study of texture measures with classification based on featured distributions. Pattern Recogn. **29**(1), 51–59 (1996)
17. Ronneberger, O., Fischer, P., Brox, T.: U-net: Convolutional networks for biomedical image segmentation. In: Medical Image Computing and Computer-Assisted Intervention–MICCAI 2015: 18th International Conference, Munich, Germany, October 5-9, 2015, Proceedings, Part III 18, pp. 234–241. Springer (2015)
18. Schlemper, J., et al.: Attention gated networks: learning to leverage salient regions in medical images. Med. Image Anal. **53**, 197–207 (2019)
19. Sheikh, R., Schultz, T.: Feature preserving smoothing provides simple and effective data augmentation for medical image segmentation. In: Martel, A.L., et al. (eds.) Medical Image Computing and Computer Assisted Intervention – MICCAI 2020: 23rd International Conference, Lima, Peru, October 4–8, 2020, Proceedings, Part I, pp. 116–126. Springer International Publishing, Cham (2020). https://doi.org/10.1007/978-3-030-59710-8_12
20. You, S., Reyes, M.: Influence of contrast and texture based image modifications on the performance and attention shift of u-net models for brain tissue segmentation. Front. Neuroimag. **1**, 1012639 (2022)

Conditional Temporal Attention Networks for Neonatal Cortical Surface Reconstruction

Qiang Ma[1(✉)], Liu Li[1], Vanessa Kyriakopoulou[2], Joseph V. Hajnal[2],
Emma C. Robinson[2], Bernhard Kainz[1,2,3], and Daniel Rueckert[1,4]

[1] BioMedIA, Department of Computing, Imperial College London, London, UK
q.ma20@imperial.ac.uk
[2] King's College London, London, UK
[3] FAU Erlangen-Nürnberg, Erlangen, Germany
[4] Klinikum rechts der Isar, Technical University of Munich, Munich, Germany

Abstract. Cortical surface reconstruction plays a fundamental role in modeling the rapid brain development during the perinatal period. In this work, we propose Conditional Temporal Attention Network (CoTAN), a fast end-to-end framework for diffeomorphic neonatal cortical surface reconstruction. CoTAN predicts multi-resolution stationary velocity fields (SVF) from neonatal brain magnetic resonance images (MRI). Instead of integrating multiple SVFs, CoTAN introduces attention mechanisms to learn a conditional time-varying velocity field (CTVF) by computing the weighted sum of all SVFs at each integration step. The importance of each SVF, which is estimated by learned attention maps, is conditioned on the age of the neonates and varies with the time step of integration. The proposed CTVF defines a diffeomorphic surface deformation, which reduces mesh self-intersection errors effectively. It only requires 0.21 s to deform an initial template mesh to cortical white matter and pial surfaces for each brain hemisphere. CoTAN is validated on the Developing Human Connectome Project (dHCP) dataset with 877 3D brain MR images acquired from preterm and term born neonates. Compared to state-of-the-art baselines, CoTAN achieves superior performance with only 0.12 ± 0.03 mm geometric error and $0.07 \pm 0.03\%$ self-intersecting faces. The visualization of our attention maps illustrates that CoTAN indeed learns coarse-to-fine surface deformations automatically without intermediate supervision.

1 Introduction

Cortical surface reconstruction aims to extract 3D meshes of inner (white matter) and outer (pial) surfaces of the cerebral cortex from brain magnetic resonance images (MRI). These surfaces provide both 3D visualization and estimation of morphological features for the cortex [7,11,12]. In addition to accurately representing the highly folded cortex, the cortical surfaces of each hemisphere are

---

Supplementary Information The online version contains supplementary material available at https://doi.org/10.1007/978-3-031-43901-8_30.

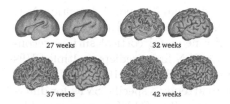

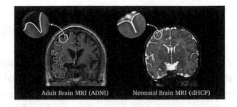

Fig. 1. Neonatal cortical surfaces at different post-menstrual ages.

Fig. 2. Left: adult brain MRI from the ADNI dataset [20]. Right: neonatal brain MRI from the dHCP dataset [9].

required to be closed manifolds and topologically homeomorphic to a sphere [7]. Traditional neuroimage analysis pipelines [10,13,24,29] such as FreeSurfer [10] comprise a series of processing steps to extract cortical surfaces from brain MRI. These pipelines provide an invaluable service to the research community. However, the current implementations have limited accuracy and require several hours to process a single MRI scan.

With the recent advances of geometric deep leaning [32,33], a growing number of fast learning-based approaches have been proposed to learn either implicit surface representation [6,15] or explicit mesh deformation [4,17,21–23,28] for cortical surface reconstruction. These approaches enhance the accuracy and reduce the processing time to a few seconds for a single subject. Recent studies focus on generating manifold cortical meshes [21,22,28] and preventing mesh self-intersections by learning diffeomorphic deformations, which have been widely adopted in medical image registration [1–3]. The basic idea is to learn stationary velocity fields (SVF) to deform an initial mesh template to target surfaces. Since a single SVF has limited representation capacity, several approaches [16,21,28,30] have proposed to train multiple neural networks to predict a sequence of SVFs for coarse-to-fine surface deformation. This improves the geometric accuracy but increases the computational burden for both training and inference.

Cortical surface reconstruction plays an essential role in modeling and quantifying the brain development in fetal and neonatal neuroimaging studies such as the Developing Human Connectome Project (dHCP) [9,24]. However, most learning-based approaches so far rely on adult MRI as training data [13,20,25]. Compared to adult data, neonatal brain MR images have lower resolution and contrast due to a smaller region of interest and the use of, *e.g.*, fast imaging sequences with sparse acquisition to minimize head motion artifacts of unsedated infants [5,19]. Besides, the rapid growth and continuously increasing complexity of the cortex during the perinatal period lead to considerable variation in shape and scale between neonatal cortical surfaces at different post-menstrual ages (PMA) (see Fig. 1). Moreover, since the neonatal head is still small, the cortical sulci of term-born neonates are much narrower than adults as shown in Fig. 2. Hence, the neonatal pial surfaces are prone to be affected by partial volume effects and more likely to produce surface self-intersections.

Contribution. In this work, we present Conditional Temporal Attention Network (CoTAN). CoTAN adopts attention mechanism [18,27] to learn a conditional time-varying velocity field (CTVF) for neonatal cortical surface reconstruction. Given an input brain MR image, CoTAN first predicts multiple SVFs at different resolutions. Rather than integrating all SVFs as [16,21,28], CoTAN learns conditional temporal attention maps to attend to specific SVFs for different time steps of integration and PMA of neonates. The CTVF is represented by the weighted sum of learned SVFs, and thus a single CoTAN model is sufficient to model the large deformation and variation of neonatal cortical surfaces. The evaluation on the dHCP neonatal dataset [9] verifies that CoTAN performs better in geometric accuracy, mesh quality and computational efficiency than state-of-the-art methods. The visualization of attention maps indicates that CoTAN learns coarse-to-fine deformations automatically without intermediate constraints. Our source code is released publicly at https://github.com/m-qiang/CoTAN.

2   Method

Diffeomorphic Surface Deformation. We define the diffeomorphic surface deformation $\phi_t : \mathbb{R}^3 \times \mathbb{R} \to \mathbb{R}^3$ as a flow ordinary differential equation (ODE) following previous work [1–3,16,21,22,28]:

$$\frac{\partial}{\partial t}\phi_t = v_t(\phi_t), \;\; \phi_0 = Id, \; t \in [0, T], \tag{1}$$

where v_t is a time-varying velocity field (TVF) and Id is the identity mapping. Given an initial surface $S_0 \subset \mathbb{R}^3$ with points $x_0 \in S_0$, we define $x_t := \phi_t(x_0)$ as the trajectories of the points on the deformable surface $S_t = \phi_t(S_0)$ for $t \in [0, T]$. Then the flow equation (1) can be rewritten as $\frac{d}{dt}x_t = v_t(x_t)$ with initial value x_0. By the existence and uniqueness theorem for ODE solutions [16], if $v_t(x)$ is Lipschitz continuous with respect to x, the trajectories x_t will not intersect with each other, so that the surface self-intersections can be prevented effectively. By integrating the ODE (1), we obtain a diffeomorphism ϕ_T that deforms S_0 to a manifold surface S_T, on which the points are $x_T = \phi_T(x_0) = x_0 + \int_0^T v_t(x_t)dt$.

Conditional Temporal Attention Network (CoTAN). An overview of the CoTAN architecture is shown in Fig. 3. CoTAN first predicts multiple SVFs given a 3D brain MRI volume. A 3D U-Net [26] is used to extract feature maps with R resolution levels, each of which scales the input size by the factor of 2^{r-R} for $r = 1, ..., R$. Then we upsample the multi-scale feature maps and learn M volumetric SVFs for each resolution. Let $\mathbf{V}$ denote all $R \times M$ discrete SVFs. The continuous multi-resolution SVFs $\mathbf{v} : \mathbb{R}^3 \to \mathbb{R}^{R \times M \times 3}$ can be obtained by $\mathbf{v}(x) = \mathrm{Lerp}(x, \mathbf{V})$, where $\mathrm{Lerp}(\cdot)$ is the trilinear interpolation function. Each element $\mathbf{v}^{r,m} : \mathbb{R}^3 \to \mathbb{R}^3$ is an SVF for $r = 1, ..., R$ and $m = 1, ..., M$. Note that $\mathbf{v}(x)$ is Lipschitz continuous since $\mathrm{Lerp}(\cdot)$ is continuous and piece-wise linear.

CoTAN adopts a channel-wise attention mechanism [18,27] to focus on specific SVFs since it is time-consuming to integrate all $R \times M$ SVFs [16,21,28].

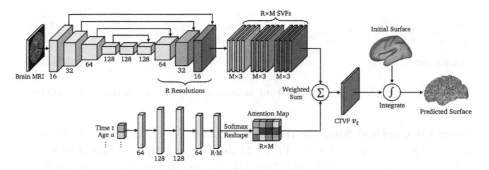

Fig. 3. The architecture of the proposed CoTAN framework. Given an input 3D brain MRI, CoTAN uses a U-Net to predict M SVFs for each resolution level R. An attention map is learned to focus on specific SVFs varying with the input integration time t and conditioned on the age a of the neonatal subjects. Further conditioning could be achieved with minimal effort, e.g., biological sex, diagnoses, etc. For each time step t, the CTVF v_t is represented by the weighted sum of all $R \times M$ SVFs. By integrating the CTVF, CoTAN deforms an input initial surface to the predicted cortical surface.

The attention is conditioned on both integration time $t \in [0, T]$ and information about the subject. To model the high variation between infant brains, we consider the post-menstrual ages (PMA) $a \in \mathbb{R}$ of the neonates at scan time as the conditioning variable in this work. Note that we do not use a self-attention module [8,31] to learn key and query pairs. Instead, we learn a probability attention map to measure the importance of each SVF. More precisely, as shown in Fig. 3, we use a fully connected network (FCN) to encode the input time t and PMA a into a $(R \cdot M) \times 1$ feature vector. After reshaping and softmax activation, the FCN learns conditional temporal attention maps $\mathbf{p}(t, a) \in \mathbb{R}^{R \times M}$ which satisfy $\sum_{r=1}^{R} \sum_{m=1}^{M} \mathbf{p}^{r,m}(t, a) = 1$ for any t and a. Then a *conditional time-varying velocity field* (CTVF) is predicted by computing the weighted sum of all SVFs:

$$v_t(x; a) = \sum_{r=1}^{R} \sum_{m=1}^{M} \mathbf{p}^{r,m}(t, a) \cdot \mathbf{v}^{r,m}(x). \tag{2}$$

The CTVF is adaptive to the integration time and the age of subjects, which can handle the large deformation and variation for neonatal cortical surfaces. Such an attention mechanism encourages CoTAN to learn coarse-to-fine surface deformation by attending to SVFs at different resolutions.

To deform the initial surface S_0 to the target surface, we integrate the flow ODE (1) with the CTVF through the forward Euler method. For $k = 0, ..., K-1$, the surface points are updated by $x_{k+1} = x_k + h v_k(x_k; a)$, where K is the total integration steps and $h = T/K$ is the step size with $T = 1$. For each step k, we only need to recompute the attention maps $\mathbf{p}(hk, a)$ and update the CTVF $v_k(x_k; a)$ accordingly by Eq. (2). CoTAN only integrates a single CTVF which saves considerable runtime compared to integrating multiple SVFs directly as [16,21,28].

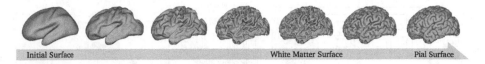

Initial Surface   White Matter Surface   Pial Surface

Fig. 4. Diffeomorphic deformation from an initial template mesh to cortical surfaces.

Neonatal Cortical Surface Reconstruction. We train two CoTAN models on the dHCP neonatal dataset [9] to 1) deform an initial surface into a white matter surface and to 2) expand the white matter surface into a pial surface as shown in Fig. 4. We use the same initial surface (leftmost in Fig. 4) for all subjects, which is created by iteratively applying Laplacian smoothing to a Conte-69 surface atlas [14]. We generate pseudo ground truth (GT) surfaces by the dHCP structural neonatal pipeline [24], which has been fully validated through quality control performed by clinical experts.

For white matter surface reconstruction, we consider loss functions that have been widely used in previous work [4,32,33]: the Chamfer distance loss $\mathcal{L}_{cd}$ computes the distance between two point clouds, the mesh Laplacian loss $\mathcal{L}_{lap}$ regularizes the smoothness of the mesh, and the normal consistency loss $\mathcal{L}_{nc}$ constrains the cosine similarity between the normals of two adjacent faces. The final loss is weighted by $\mathcal{L} = \mathcal{L}_{cd} + \lambda_{lap}\mathcal{L}_{lap} + \lambda_{nc}\mathcal{L}_{nc}$. We train CoTAN in two steps for white matter surface extraction. First, we pre-train the model using relatively large weights λ_{lap} and λ_{nc} for regularization. The Chamfer distance is computed between the vertices of the predicted and pseudo-GT surfaces. These ensure that the initial surface can be deformed robustly during training. Then, we fine-tune CoTAN using small weights to increase geometric accuracy. The distances are computed between 150k uniformly sampled points on the surfaces.

For pial surface reconstruction, we follow [22,23] and use the pseudo-GT white matter surfaces as the input for training. Then the MSE loss $\mathcal{L} = \sum_i \|\hat{x}_i - x_i^*\|^2$ can be computed between the vertices of predicted and pseudo-GT pial meshes. No point matching is required since the pseudo-GT white matter and pial surfaces have the same mesh connectivity. Therefore, the MSE loss provides stronger supervision than the Chamfer distance, while the latter is prone to mismatching the points in narrow sulci, resulting in mesh self-intersections.

3   Experiments

Implementation Details. We evaluate CoTAN on the third release of dHCP neonatal dataset [9] (https://biomedia.github.io/dHCP-release-notes/), which includes 877 T2-weighted (T2w) brain MRI scanned from newborn infants at PMA between 27 to 45 weeks. The MRI images are affinely aligned to the MNI152 space and clipped to the size of $112 \times 224 \times 160$ for each brain hemisphere. The dataset is split into 60/10/30% for training/validation/testing.

For the CoTAN model, we set the resolution $R=3$ and the number of SVFs $M=4$ for each resolution. For integration, we set the total number of steps to

Table 1. Comparative results of neonatal cortical surface reconstruction on the dHCP dataset. The geometric accuracy (ASSD and HD90) and mesh quality (the ratio of SIFs) are reported for white matter and pial surfaces. Smaller values mean better results. *CoTAN (ours) shows significant improvement ($p < 0.05$) compared to baselines.

Method	White Matter Surface			Pial Surface		
	ASSD (mm)	HD90 (mm)	SIF (%)	ASSD (mm)	HD90 (mm)	SIF (%)
CoTAN	**0.107 ± 0.026**	**0.217 ± 0.076**	**0.001 ± 0.004**	**0.121 ± 0.029**	**0.259 ± 0.075**	**0.071 ± 0.034**
CortexODE	0.109 ± 0.052	0.231 ± 0.326	**0.001 ± 0.002**	0.134 ± 0.052*	0.306 ± 0.358*	0.221 ± 0.114*
CFPP	0.118 ± 0.028*	0.241 ± 0.085*	0.075 ± 0.057*	0.124 ± 0.031*	0.273 ± 0.086*	2.457 ± 1.003*
CorticalFlow	0.122 ± 0.029*	0.247 ± 0.080*	0.048 ± 0.032*	0.157 ± 0.031*	0.331 ± 0.089*	9.798 ± 2.902*
Vox2Cortex	0.115 ± 0.035*	0.233 ± 0.110*	0.253 ± 0.169*	0.130 ± 0.039*	0.291 ± 0.141*	14.366 ± 2.262*
DeepCSR	0.129 ± 0.047*	0.276 ± 0.211*	—	0.299 ± 0.070*	1.214 ± 0.337*	—

Table 2. Comparative results of runtime, GPU memory cost, and the number of model parameters for both training and testing.

Method	Runtime		GPU (GB)		Model
	Train	Test	Train	Test	#Param
CoTAN (Ours)	57.9h	**0.21s**	8.71	4.05	2.47 M
CortexODE	51.6h	1.88 s	9.24	4.61	1.99 M
CFPP	131.5h	0.57 s	9.86	3.56	1.03 M
CorticalFlow	105.7h	0.51 s	9.12	3.56	1.03 M
Vox2Cortex	63.7h	1.48 s	6.96	6.26	6.54 M
DeepCSR	**15.1 h**	10.69 s	**5.26**	**2.33**	4.65 M

Table 3. The results of ablation experiments for CoTAN on white matter surface reconstruction.

Method	ASSD (mm)	HD90 (mm)
CoTAN (Ours)	**0.107 ± 0.026**	**0.217 ± 0.076**
Pre-train	0.116 ± 0.029	0.238 ± 0.095
$R=1$	0.112 ± 0.027	0.232 ± 0.081
$M=1$	0.111 ± 0.030	0.231 ± 0.097
SVF ($a=t=0$)	0.138 ± 0.037	0.291 ± 0.109
CVF ($t=0$)	0.135 ± 0.036	0.285 ± 0.111
TVF ($a=0$)	0.108 ± 0.028	0.222 ± 0.090

$K=50$ with step size $h=0.02$. We re-mesh the initial mesh to 140k vertices, of which the coordinates are normalized to $[-1, 1]$. We use the Adam optimizer for training. For the white matter surface, we first pre-train CoTAN for 100 epochs using a learning rate of $\gamma=10^{-4}$ and weights $\lambda_{lap}=0.5$, $\lambda_{nc}=5\times10^{-4}$ for the loss function. Then we fine-tune for 100 epochs using smaller weights $\lambda_{lap}=0.1$ and $\lambda_{nc}=10^{-4}$ with $\gamma=2\times10^{-5}$. For the pial surface, we set the maximum channel size of CoTAN as 32 to avoid overfitting and train for 200 epochs with $\gamma=10^{-4}$. We only consider left brain hemisphere in the experiments. All experiments are conducted on a Nvidia RTX3080 GPU with 12GB memory.

Comparative Results. We compare the performance of CoTAN with existing learning-based cortical surface extraction approaches including Cortex-ODE [22], CorticalFlow++ (CFPP) [28], CorticalFlow [21], Vox2Cortex [4] and DeepCSR [6]. We employ the fast topology correction [22] for DeepCSR. CoTAN can guarantee spherical topology and the Euler number is 2 for all predicted surfaces.

—*Geometric Accuracy*: We measure the geometric accuracy by commonly used metrics [4, 6, 22]: average symmetric surface distance (ASSD) and 90th percentile of Hausdorff distance (HD90). The distances are computed between uniformly sampled 100k points on the predicted and pseudo-GT surfaces. For fair com-

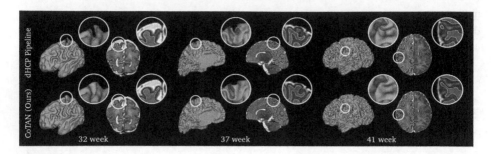

Fig. 5. Visualization of neonatal cortical surfaces generated by the dHCP structural pipeline [24] and CoTAN for different ages. CoTAN shows better anatomical accuracy as highlighted in both surface meshes and corresponding brain MRI images.

parison, the predicted cortical meshes have around 140k vertices for all baseline approaches. Note that CoTAN and [6,21,28] can generalize on high-resolution meshes with up to 600k vertices (see Appendix). We conduct a paired t-test to examine the statistical significance. As reported in Table 1, CoTAN achieves significantly superior geometric accuracy (p-value<0.05) compared to all state-of-the-art baselines, except for the white matter surfaces of CortexODE.

—*Mesh Quality*: We evaluate the mesh quality by the ratio of self-intersecting faces (SIF) as shown in Table 1. Note that due to the narrower sulcal gaps of neonatal cortex compared to adult [20,25] (see Fig. 2), all baseline methods produce more SIFs than reported in their original papers. Except that DeepCSR produces no SIFs since it uses the Marching Cubes algorithm, CoTAN achieves the best mesh quality with only 0.001% SIFs in white matter surfaces and 0.071% SIFs in pial surfaces. These remaining SIFs are likely introduced by the discretization of triangular mesh representation and ODE integration.

CoTAN produces fewer SIFs for three reasons. Firstly, CoTAN employs diffeomorphic deformation, while the non-diffeomorphic Vox2Cortex creates 14% SIFs. We further set the integration steps K=5 for CoTAN so that the deformation is no longer diffeomorphic. The SIFs of pial surfaces are increased to $2.99 \pm 1.19\%$. Secondly, CoTAN reconstructs the pial surface by expanding the input white matter surface. It is difficult to avoid collisions during the deformation from a smooth template into deep sulci, *e.g.*, in Vox2Cortex and CorticalFlow. Lastly, CoTAN uses MSE loss rather than Chamfer distance for pial surface extraction, which alleviates the mismatch between points in the narrow sulci.

—*Computational Efficiency*: We report the runtime and GPU memory cost for both training and testing, as well as the number of learnable parameters for CoTAN and all baseline approaches in Table 2. The runtime includes both model inference and post-processing. CoTAN only requires 0.21 s to extract cortical surfaces for each hemisphere, which is 2× faster than the best baseline. CoTAN can be trained end-to-end and reduces training time by 46% compared to CorticalFlow and CFPP, which have to train three U-Nets consecutively to parameterize three SVFs for a single surface. Although CoTAN uses relatively

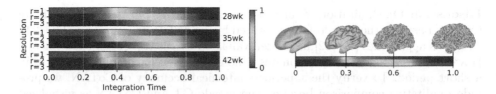

Fig. 6. Visualization of attention maps predicted by CoTAN for white matter surface reconstruction. Left: conditional temporal attention maps $\mathbf{p}^r(t, a)$ at different time and ages. Right: coarse-to-fine white matter surface deformations learned by CoTAN.

more parameters to learn surface deformations, it is still memory efficient which only costs 8.7GB and 4GB GPU memory for training and testing.

Ablation Study. We conduct ablation studies on CoTAN and evaluate the geometric accuracy on the white matter surface reconstruction. First, without fine-tuning, the geometric errors increase by 8% as reported in Table 3. Then we consider CoTAN with single resolution (R=1) or only predict a single SVF (M=1) for each resolution. The geometric distances increase in both cases.

Next, we examine the effectiveness of the CTVF $v_t(x; a)$ by fixing the input time t=0 or age a=0. We train CoTAN models to predict the TVF (a=0), CVF (t=0) and SVF (a=t=0), which are degraded from the CTVF. Table 3 shows that the SVF increases geometric errors by 30% due to its limited representation capacity. The CVF increases accuracy slightly by learning conditional deformations adaptive to the age of neonates. The TVF exploits temporal information to model a wider range of deformations and effectively improves the performance.

Lastly, we train a U-Net to predict $R \times M$ SVFs and integrate them directly without attention. Since the gradients are backpropagated through all SVFs, it requires >140 h training time which is 2.4× slower than our attention-based CoTAN. The model is also sensitive to small updates, which can affect all SVFs. This results in exploding gradients which we have observed in the training, whereas CoTAN can be trained robustly by integrating a single CTVF.

Attention Maps. We explore the attention maps $\mathbf{p}^{r,m}(t, a)$ learned by CoTAN. We define $\mathbf{p}^r = \sum_{m=1}^{M} \mathbf{p}^{r,m}$ to reflect the importance of the SVFs at each resolution level r=1, ..., R, where $R = 3$ and larger r means higher resolution. Figure 6 visualizes the attention maps $\mathbf{p}^r(t, a)$ for white matter surface reconstruction for integration time $t \in [0, 1]$ and age $a \in \{28, 35, 42\}$. It shows that the attention maps focus on low resolution (r=1) at the beginning of the integration, and then attends to high resolution (r=3) when $t \to 1$. Furthermore, Fig. 6 (Right) shows that an initial white matter surface deforms into a coarse shape for $t \leq 0.3$ and learns fine details for $t \geq 0.6$. This matches the attention maps and demonstrates that CoTAN learns coarse-to-fine deformations automatically without any supervision on the intermediate deformations. Additionally, Fig. 6 shows that CoTAN pays more attention to low resolution for younger subjects (28 week) whose brains have not fully developed yet. More deformations at higher resolutions ($r \geq 2$) are required for older neonates (≥ 35 week) with highly folded cortex.

Discussion. One limitation of our experiments is that we only train and evaluate CoTAN based on the pseudo-GT generated by the dHCP structural pipeline [24]. Previous approaches [4,6,22] have been validated by the test-retest experiments. However, this is infeasible for neonates, whose brain develops rapidly even within a short period. To verify the superior anatomical accuracy of CoTAN, we provide qualitative comparison between the pseudo-GT and CoTAN as visualized in Fig. 5. It shows that CoTAN can effectively mitigate corruptions introduced by the dHCP pipeline for neonatal subjects at different ages. In addition, the dHCP pipeline requires 4 h to process a single subject [24], while CoTAN extracts cortical surfaces in only 0.21 s for each brain hemisphere.

4 Conclusion

In this work, we propose CoTAN for diffeomorphic neonatal cortical surface reconstruction. CoTAN employs an attention mechanism to combine multiple SVFs to a CTVF, which outperforms existing baselines in geometric accuracy and mesh quality. CoTAN can also be extended and applied to extract adult cortical surfaces conditioned on the age, gender or pathological information of the subjects. Our future work will integrate CoTAN into a learning-based pipeline for universal cortical surface analysis across all age groups.

Acknowledgements. This work was supported by the President's PhD Scholarship at Imperial College London. Support was also received from the ERC project MIA-NORMAL 101083647 and ERC project Deep4MI 884622. Data were provided by the developing Human Connectome Project, KCL-Imperial-Oxford Consortium funded by the ERC under the European Union Seventh Framework Programme (FP/2007-2013) / ERC Grant Agreement no. [319456]. We are grateful to the families who generously supported this trial.

References

1. Ashburner, J.: A fast diffeomorphic image registration algorithm. Neuroimage **38**(1), 95–113 (2007)
2. Balakrishnan, G., Zhao, A., Sabuncu, M.R., Guttag, J., Dalca, A.V.: VoxelMorph: a learning framework for deformable medical image registration. IEEE Trans. Med. Imaging **38**(8), 1788–1800 (2019)
3. Beg, M.F., Miller, M.I., Trouvé, A., Younes, L.: Computing large deformation metric mappings via geodesic flows of diffeomorphisms. Int. J. Comput. Vision **61**, 139–157 (2005)
4. Bongratz, F., Rickmann, A.M., Pölsterl, S., Wachinger, C.: Vox2cortex: fast explicit reconstruction of cortical surfaces from 3D MRI scans with geometric deep neural networks. In: Proceedings of the IEEE/CVF Conference on Computer Vision and Pattern Recognition, pp. 20773–20783 (2022)
5. Cordero-Grande, L., Hughes, E.J., Hutter, J., Price, A.N., Hajnal, J.V.: Three-dimensional motion corrected sensitivity encoding reconstruction for multi-shot multi-slice MRI: application to neonatal brain imaging. Magn. Reson. Med. **79**(3), 1365–1376 (2018)

6. Cruz, R.S., Lebrat, L., Bourgeat, P., Fookes, C., Fripp, J., Salvado, O.: DeepCSR: a 3D deep learning approach for cortical surface reconstruction. In: Proceedings of the IEEE/CVF Winter Conference on Applications of Computer Vision, pp. 806–815 (2021)
7. Dale, A.M., Fischl, B., Sereno, M.I.: Cortical surface-based analysis: I. segmentation and surface reconstruction. Neuroimage **9**(2), 179–194 (1999)
8. Dosovitskiy, A., et al.: An image is worth 16×16 words: transformers for image recognition at scale. In: International Conference on Learning Representations (2021)
9. Edwards, A.D., et al.: The developing human connectome project neonatal data release. Front. Neurosci. **16**, 886772 (2022)
10. Fischl, B.: FreeSurfer. Neuroimage **62**(2), 774–781 (2012)
11. Fischl, B., Dale, A.M.: Measuring the thickness of the human cerebral cortex from magnetic resonance images. Proc. Natl. Acad. Sci. **97**(20), 11050–11055 (2000)
12. Fischl, B., Sereno, M.I., Dale, A.M.: Cortical surface-based analysis: II: inflation, flattening, and a surface-based coordinate system. Neuroimage **9**(2), 195–207 (1999)
13. Glasser, M.F., Sotiropoulos, S.N., Wilson, J.A., et al.: The minimal preprocessing pipelines for the human connectome project. Neuroimage **80**, 105–124 (2013)
14. Glasser, M.F., Van Essen, D.C.: Mapping human cortical areas in vivo based on myelin content as revealed by T1- and T2-weighted MRI. J. Neurosci. **31**(32), 11597–11616 (2011)
15. Gopinath, K., Desrosiers, C., Lombaert, H.: SEGRECON: learning joint brain surface reconstruction and segmentation from images. In: de Bruijne, M., et al. (eds.) MICCAI 2021. LNCS, vol. 12907, pp. 650–659. Springer, Cham (2021). https://doi.org/10.1007/978-3-030-87234-2_61
16. Gupta, K., Chandraker, M.: Neural mesh flow: 3D manifold mesh generation via diffeomorphic flows. In: Proceedings of the 34th International Conference on Neural Information Processing Systems, pp. 1747–1758 (2020)
17. Hoopes, A., Iglesias, J.E., Fischl, B., Greve, D., Dalca, A.V.: TopoFit: rapid reconstruction of topologically-correct cortical surfaces. In: Medical Imaging with Deep Learning (2021)
18. Hu, J., Shen, L., Sun, G.: Squeeze-and-excitation networks. In: Proceedings of the IEEE Conference on Computer Vision and Pattern Recognition, pp. 7132–7141 (2018)
19. Hughes, E.J., Winchman, T., Padormo, F., Teixeira, R., et al.: A dedicated neonatal brain imaging system. Magn. Reson. Med. **78**(2), 794–804 (2017)
20. Jack, C.R., Jr., et al.: The Alzheimer's disease neuroimaging initiative (ADNI): MRI methods. J. Magn. Reson. Imaging **27**(4), 685–691 (2008)
21. Lebrat, L., et al.: CorticalFlow: a diffeomorphic mesh transformer network for cortical surface reconstruction. Adv. Neural. Inf. Process. Syst. **34**, 29491–29505 (2021)
22. Ma, Q., Li, L., Robinson, E.C., Kainz, B., Rueckert, D., Alansary, A.: Cortex-ODE: learning cortical surface reconstruction by neural ODEs. IEEE Trans. Med. Imaging **42**, 430–443 (2022)
23. Ma, Q., Robinson, E.C., Kainz, B., Rueckert, D., Alansary, A.: PialNN: a fast deep learning framework for cortical pial surface reconstruction. In: Abdulkadir, A., et al. (eds.) MLCN 2021. LNCS, vol. 13001, pp. 73–81. Springer, Cham (2021). https://doi.org/10.1007/978-3-030-87586-2_8

24. Makropoulos, A., et al.: The developing human connectome project: a minimal processing pipeline for neonatal cortical surface reconstruction. Neuroimage **173**, 88–112 (2018)

25. Marcus, D.S., Wang, T.H., Parker, J., Csernansky, J.G., Morris, J.C., Buckner, R.L.: Open access series of imaging studies (OASIS): cross-sectional MRI data in young, middle aged, nondemented, and demented older adults. J. Cogn. Neurosci. **19**(9), 1498–1507 (2007)

26. Ronneberger, O., Fischer, P., Brox, T.: U-Net: convolutional networks for biomedical image segmentation. In: Navab, N., Hornegger, J., Wells, W.M., Frangi, A.F. (eds.) MICCAI 2015. LNCS, vol. 9351, pp. 234–241. Springer, Cham (2015). https://doi.org/10.1007/978-3-319-24574-4_28

27. Roy, A.G., Navab, N., Wachinger, C.: Concurrent spatial and channel 'squeeze & excitation' in fully convolutional networks. In: Frangi, A.F., Schnabel, J.A., Davatzikos, C., Alberola-López, C., Fichtinger, G. (eds.) MICCAI 2018. LNCS, vol. 11070, pp. 421–429. Springer, Cham (2018). https://doi.org/10.1007/978-3-030-00928-1_48

28. Santa Cruz, R., et al.: CorticalFlow++: boosting cortical surface reconstruction accuracy, regularity, and interoperability. In: Wang, L., Dou, Q., Fletcher, P.T., Speidel, S., Li, S. (eds.) MICCAI 2022. LNCS, vol. 13435, pp. 496–505. Springer, Cham (2022). https://doi.org/10.1007/978-3-031-16443-9_48

29. Shattuck, D.W., Leahy, R.M.: BrainSuite: an automated cortical surface identification tool. Med. Image Anal. **6**(2), 129–142 (2002)

30. Sun, S., Han, K., Kong, D., Tang, H., Yan, X., Xie, X.: Topology-preserving shape reconstruction and registration via neural diffeomorphic flow. In: Proceedings of the IEEE/CVF Conference on Computer Vision and Pattern Recognition, pp. 20845–20855 (2022)

31. Vaswani, A., et al.: Attention is all you need. In: Advances in Neural Information Processing Systems, vol. 30 (2017)

32. Wang, N., Zhang, Y., Li, Z., Fu, Y., Liu, W., Jiang, Y.G.: Pixel2Mesh: generating 3D mesh models from single RGB images. In: Proceedings of the European Conference on Computer Vision (ECCV), pp. 52–67 (2018)

33. Wickramasinghe, U., Remelli, E., Knott, G., Fua, P.: Voxel2Mesh: 3D mesh model generation from volumetric data. In: Martel, A.L., et al. (eds.) MICCAI 2020. LNCS, vol. 12264, pp. 299–308. Springer, Cham (2020). https://doi.org/10.1007/978-3-030-59719-1_30

Elongated Physiological Structure Segmentation via Spatial and Scale Uncertainty-Aware Network

Yinglin Zhang[1,2], Ruiling Xi[2], Huazhu Fu[4], Dave Towey[1], RuiBin Bai[1], Risa Higashita[2,3(✉)], and Jiang Liu[1,2,3(✉)]

[1] School of Computer Science, University of Nottingham Ningbo China, Ningbo 315100, China
[2] Research Institute of Trustworthy Autonomous Systems and Department of Computer Science and Engineering, Southern University of Science and Technology, Shenzhen 518055, China
rise@mail.sustech.edu.cn, liuj@sustech.edu.cn
[3] Tomey Corporation, Nagoya 451-0051, Japan
[4] Institute of High Performance Computing (IHPC), Agency for Science, Technology and Research (A*STAR), Singapore, Singapore

Abstract. Robust and accurate segmentation for elongated physiological structures is challenging, especially in the ambiguous region, such as the corneal endothelium microscope image with uneven illumination or the fundus image with disease interference. In this paper, we present a spatial and scale uncertainty-aware network (SSU-Net) that fully uses both spatial and scale uncertainty to highlight ambiguous regions and integrate hierarchical structure contexts. First, we estimate epistemic and aleatoric spatial uncertainty maps using Monte Carlo dropout to approximate Bayesian networks. Based on these spatial uncertainty maps, we propose the gated soft uncertainty-aware (GSUA) module to guide the model to focus on ambiguous regions. Second, we extract the uncertainty under different scales and propose the multiscale uncertainty-aware (MSUA) fusion module to integrate structure contexts from hierarchical predictions, strengthening the final prediction. Finally, we visualize the uncertainty map of final prediction, providing interpretability for segmentation results. Experiment results show that the SSU-Net performs best on cornea endothelial cell and retinal vessel segmentation tasks. Moreover, compared with counterpart uncertainty-based methods, SSU-Net is more accurate and robust.

Keywords: Uncertainty · Medical Image Segmentation · Elongated Physiological Structure · Deep Learning

H. Greenspan et al. (Eds.): MICCAI 2023, LNCS 14223, pp. 323–332, 2023.
https://doi.org/10.1007/978-3-031-43901-8_31

1 Introduction

Robust and accurate elongated physiological structure segmentation is crucial for computer-aided diagnosis and quantification of clinical parameters [26,27]. Manual delineation is tedious and laborious. Recently, deep learning-based methods [14,16,19] have been proposed to delineate targets automatically. However, they are not able to outline correctly in ambiguous regions where exist uneven illumination, artifacts, or interference from the disease.

Many researchers have tried to use uncertainty information to concentrate on the ambiguous region, and to evaluate the reliability of model's prediction. According to the source of prediction errors [3], uncertainty is categorized into two types: epistemic and aleatoric. The main methods for uncertainty estimation are as follows. Bayesian neural networks [15] place a probability distribution over model weights, but are hard to optimize. Monte Carlo dropout [10] approximates the Gaussian process by embedding the dropout operation into the neural network layers and calculating the variance of several times inference. Deep Ensembles [8] combine the outputs from a group of independent models to estimate uncertainty. Softmax uncertainty [13,17,18] performs well in distinguishing examples that are easy or fallible to classify. Once the uncertainty information has been estimated, we are able to pay more attention to the ambiguous region. Xie et al. [24] used the cross-attention module to extract influential features for ambiguous regions based on pixel-level uncertainty. Yang et al. [25] achieved uncertainty awareness by training with a multi-confidence mask, and further used self-attention block with feature aware filter together to highlight uncertain areas. Wang et al. [23] annotated alpha matte for medical images and used it as a soft label to intuitively promote the network to focus on uncertain areas. Kohl et al. [7] proposed a generative model to produce multiple reasonable hypotheses for clinical experts to select from, which improved the diagnosis reliability. However, existing works applied the 'hard' attention to utilize uncertainty, which lacks the ability of adaptive adjustment and ignores neighboring uncertain regions. In addition, features at different scales contain rich structural and semantic contexts, which are essential for elongated physiological structure segmentation, such as cobweb corneal endothelial cells and retinal vessels.

This paper proposes a spatial and scale uncertainty-aware network (SSU-Net) for elongated physiological structure segmentation, which fully uses both spatial and scale uncertainty to highlight ambiguous regions and integrate hierarchical structure contexts. First, we use a gated soft uncertainty-aware (GSUA) module to adaptively highlight ambiguous areas based on spatial uncertainty maps. Second, we extract the uncertainty under different scales and propose the multi-scale uncertainty-aware (MSUA) fusion module to integrate hierarchical predictions for enhancing the final segmentation. Experiment results on segmentation tasks of the cornea endothelium and retinal vessel show the effectiveness of SSU-Net.

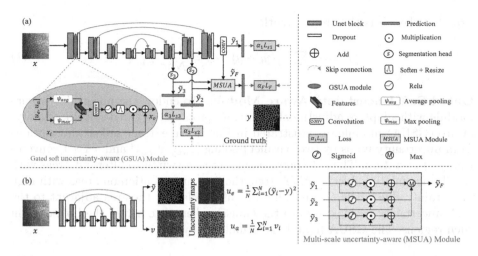

Fig. 1. The pipeline of the proposed algorithm. (a) The framework of spatial and scale uncertainty-aware network, SSU-Net. (b) We estimate the spatial uncertainty maps with Bayesian approximate network.

2   Method

Figure 1 (a) illustrates the framework of the proposed spatial and scale uncertainty-aware network, SSU-Net. The gated soft uncertainty-aware (GSUA) module enables the network to focus on the ambiguous region indicated by the spatial uncertainty maps $[u_e, u_a]$. Specifically, we construct a Bayesian approximate network to generate spatial uncertainty maps by introducing Monte Carlo dropout [2] into U-Net, as shown in Fig. 1 (b). The Bayesian approximate network has two outputs: segmentation prediction $\hat{y}$ and the estimation of aleatoric uncertainty v. We can calculate the epistemic and aleatoric uncertainty maps, u_e and u_a, after multiple inferences. Furthermore, we consider the sigmoid probabilities of predictions under different scales as the second uncertainty source, and fuse the predictions $\{\hat{y}_1, \hat{y}_2, \hat{y}_3\}$ from multiple scales using the multi-scale uncertainty-aware (MSUA) module. $\hat{y}_F$ is the final target output.

2.1   Spatial Uncertainty and Gated Soft Uncertainty-Aware Module

Spatial Uncertainty. Since the epistemic and aleatoric uncertainty maps are used to find the hard-to-classify spatial areas in this work, we regard them as spatial uncertainty. Referring to [6], we add dropout after each UNet block to approximate the Bayesian network, which learns the segmentation $\tilde{y}$ and aleatoric uncertainty v simultaneously. During inference, we sample a group of predictions $\{\tilde{y}_i\}_{i=1}^N$ and $\{v_i\}_{i=1}^N$ by N stochastic forward pass. In this work, we set $N = 16$. The epistemic u_e and aleatoric u_a uncertainty are formulated by

Eq. (1), where y is the ground truth.

$$u_e = \frac{1}{N}\sum_{i=1}^{N}(\tilde{y}_i - y)^2, u_a = \frac{1}{N}\sum_{i=1}^{N} v_i \tag{1}$$

Gated Soft Uncertainty-Aware Module. To endow the uncertainty-aware module with adaptive adjustment ability, we propose the gated soft uncertainty-aware (GSUA) module, as illustrated in Fig. 1. We extract salient descriptions from uncertainty maps by two parallel pooling $[\psi_{avg}, \psi_{max}]$ and a 1×1 convolution $f(\cdot)$ operation. The $relu$ operation is set as a switch to filter out areas with small uncertainty values, further strengthening our attention on areas with high uncertainty. Since it is also usually difficult to classify the area adjacent to the high-uncertainty regions, we use the Gaussian kernel to soften the boundary in such regions. The GSUA module is formulated by:

$$x_o = x_i \odot g_s(\sigma(relu(f([\psi_{avg}(\mathbf{u}), \psi_{max}(\mathbf{u})])))) + x_i \tag{2}$$

where $x_i, x_o \in R^{N \times c \times h \times w}$ are the input and output features respectively; $\mathbf{u} = [u_a, u_e] \in R^{N \times 2 \times H \times W}$ is a tensor of uncertainty maps; ψ_{avg} and ψ_{max} represent average and max pooling; σ is the sigmoid function; g_s denotes a convolution operation with Gaussian kernel and resizes the attention maps to the size of input features; $\odot$ is element-wise multiplication.

2.2 Scale Uncertainty and Multi-scale Uncertainty-Aware Module

Scale Uncertainty. To integrate the predictions from hierarchical layers during model training, we capture the uncertainty under multiple scales. The sigmoid function is a simple and effective way to estimate uncertainty for the binary classification task. We extract the multi-scale uncertainty by Equation (3), where u_s is the uncertainty map of prediction $\tilde{y}_s$ under scale $s \in \{1, 2, 3\}$.

$$u_s = \frac{1}{1 + e^{-\tilde{y}_s}} \tag{3}$$

Multi-scale Uncertainty-Aware Module. With the uncertainty maps from different scales, all the hierarchical predictions $\{\tilde{y}_1, \tilde{y}_2, \tilde{y}_3\}$ are fused by the MSUA module to generate the enhanced prediction $\tilde{y}_F$, as illustrate in Fig. 1. The uncertainty map u_s provides the classification confidence for each pixel. Therefore, we use u_s to highlight the confident region of $\tilde{y}_s$ and further extract the max value across the different scales. The process is formulated by:

$$\tilde{y}_F(i, j) = \max_{s \in \{1,2,3\}}(y_s(i, j) \odot \sigma(y_s(i, j)) + y_s(i, j)) \tag{4}$$

where $\tilde{y}_F(i, j)$ denotes the pixel value at location (i, j) of enhanced prediction; $y_s(i, j)$ is the value of prediction under scale $s \in \{1, 2, 3\}$; and σ denotes the sigmoid operation of Eq. (3).

2.3 Objective Function

As shown in Fig. 1, we optimize the model with supervision on four segmentation branches simultaneously, including supervision for predicting three scales and the final enhanced output. The loss function is summarized as follows:

$$\mathcal{L}_{total} = \alpha_1 \mathcal{L}_{s1} + \alpha_2 \mathcal{L}_{s2} + \alpha_3 \mathcal{L}_{s3} + \alpha_F \mathcal{L}_F \tag{5}$$

where $\alpha_1, \alpha_2, \alpha_3$, and α_F are the weight parameters for sub loss $\mathcal{L}_{s1}, \mathcal{L}_{s2}, \mathcal{L}_{s3}$, and $\mathcal{L}_F$. In this experiment, we set all the weight parameters as 1. For these sub-losses, we adopt binary cross-entropy loss, as shown in Eq. (6).

$$\mathcal{L} = -[y log \tilde{y} + (1 - y) log(1 - \tilde{y})] \tag{6}$$

where y is the ground truth; the positive class value of each pixel is 1; the negative class value is 0; and $\tilde{y} \in (0, 1)$ is the predicted probability value.

3 Experiment

3.1 Dataset

Two cornea endothelium microscope image datasets, TM-EM3000 and Rodrep, and one retinal fundus image dataset, FIVES, are used in this work. The private dataset **TM-EM3000** contains 183 images measured by EM3000 specular microscope (Tomey Corporation, Japan). Following Ruggeri et al. [20], we cropped a 192 × 192 pixels sub-region from its 260 × 480 pixels whole image. We used 155 images for model training, ten images for validation, and 18 images for testing. **Rodrep** [21] contains 52 in-vivo confocal corneal microscope images, from 23 Fuchs patients with endothelial corneal dystrophy. We used 40 for training, five images for validation, and seven for testing. **FIVES** [5] is the largest known high-resolution fundus image dataset: It covers normal eyes and three different eye diseases with a balanced distribution. There are 800 high-resolution images and the corresponding manual annotations, with 550 for training, 50 for validation, and 200 for testing.

3.2 Evaluation Metrics and Implementation Details

There is a class imbalance between foreground and background pixels. To better evaluate the segmentation performance, we choose $Dice$ score [22], $mIoU$, and $mAcc$ as evaluation metrics. We optimized the models using the RMSprop strategy with momentum = 0.9 and weight decay = 1e-8 for 100 epochs. The initial learning rate was 2e-4, and the input size of all networks was uniformly set to 256 × 256. Random shift and rescaling within a range of $[-0.3, +0.3]$ were used for data augmentation. We set the batch size to 1 based on our empirical observations. For uncertainty-based models, we set the dropout rate as 0.5 and no data augmentation. During testing, we inferred $N = 16$ times and obtained the final prediction $\bar{y} = \frac{1}{N} \sum_{i=1}^{N} \tilde{y}_F^i$ and the epistemic uncertainty $u'_e = \frac{1}{N} \sum_{i=1}^{N} (\tilde{y}_F^i - \bar{y})^2$.

3.3 Ablation Study

We investigated the influence of GSUA and MSUA modules on TM-EM3000, as shown in Table 1. The MSUA increased performance by 0.69% on the Dice score, and GSUA increased by 0.19%. The MSUA module brought more improvement than GSUA, which indicated that multi-scale context is crucial for cornea endothelium cell segmentation. When using both GSUA and MSUA modules simultaneously, we achieved the best performance.

Table 1. Ablation study on TM-EM3000. GSUA denotes the gated soft uncertainty-aware module. MSUA means the multi-scale uncertainty-aware fusion module.

Models	GSUA	MSUA	Dice(%) ↑	mIoU(%) ↑	mAcc(%) ↑
Variant1	✗	✗	76.04	76.32	85.52
Variant2	✗	✓	76.73	76.68	**87.38**
Variant3	✓	✗	76.23	76.42	86.08
Variant4	✓	✓	**77.16**	**77.02**	87.34

3.4 Comparison with State-of-the-Art Methods

To study the effectiveness of the proposed SSU-Net, we compared it with a series of state-of-the-art methods. On TM-EM3000 and Rodrep, we implemented several popular networks for comparison: UNet [19], D-LinkNet [28], AttentionUNet [16], TransUNet [1], and uncertainty-based counterparts, Monte Carlo (MC) BayesianNet [2], Lee's method [9]. On the fundus image dataset FIVES, we additionally implemented several recent retinal vessel segmentation algorithms: FR-UNet [12], SA-UNet [4], and IterNet [11].

As shown in Table 2, the proposed SSU-Net achieved the best performance. On the TM-EM3000 dataset, the uncertainty-based methods outperformed the typical convolution and attention methods, which proves that introducing uncertainty is beneficial. On the Rodrep dataset, SSU-Net performed considerably better than Lee's uncertainty method, improving the Dice score by 4.98%, mIoU by 3.48%, and mAcc by 3.14%. The results further suggest that the multi-scale predictions fusion module is crucial to elevate the robustness. According to the indication of uncertainty map u'_e, we cropped and zoomed in two ambiguous regions of each image, as shown in Fig. 2. The visualization results suggested that the proposed SSU-Net effectively improved the segmentation performance in ambiguous regions. On the FIVES dataset, the performance of the specialized network for retinal blood vessel segmentation in the fundus was similar to that of UNet, TransUNet, and AttettionUNet. The uncertainty-based methods are uniformly significantly superior to the above methods. The proposed SSU-Net achieved the best performance, increasing the Dice score by 10.08%, mIoU by 7.29%, and mAcc by 7.51% compared with UNet. Qualitative analysis is shown in Fig. 3, further supporting the conclusions of quantitative analysis.

Table 2. Comparison of SSU-Net with some SOTA methods on cornea endothelial cell and retinal vessel segmentation tasks.

Dataset	Models	Dice (%) ↑	mIoU (%) ↑	mAcc (%) ↑
TM-EM3000	UNet	71.28	72.67	81.34
	D-LinkNet	72.71	73.60	82.67
	AttentionUNet	72.29	73.46	82.06
	TransUNet	71.65	72.78	82.20
	BayesianNet	75.42	75.71	85.66
	Lee's	76.40	76.53	85.80
	SSU-Net	**77.16**	**77.02**	**87.34**
Rodrep	UNet	64.89	67.89	78.50
	D-LinkNet	65.40	68.22	79.21
	AttentionUNet	60.55	65.68	74.45
	TransUNet	66.56	68.87	80.43
	BayesianNet	65.62	68.15	80.24
	Lee's	63.51	66.62	79.44
	SSU-Net	**68.49**	**70.10**	**82.58**
FIVES	UNet	78.99	82.59	86.05
	D-LinkNet	73.45	78.03	82.89
	AttentionUNet	78.09	81.82	85.26
	TransUNet	80.81	83.19	88.18
	FR-UNet	78.47	82.09	85.51
	SA-UNet	79.15	82.05	88.27
	IterNet	79.32	82.54	85.56
	BayesianNet	88.70	89.65	93.01
	Lee's	88.79	89.70	93.37
	SSU-Net	**89.07**	**89.88**	**93.56**

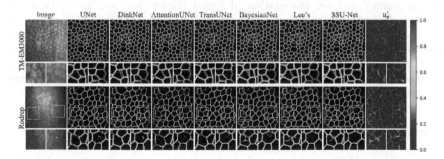

Fig. 2. Visualization for cornea endothelial cell segmentation. Red, green, and yellow line presents manual label, predicted segmentation result, and their overlap region. Indicated by the uncertainty map u'_e of the SSU-Net, we zoomed in two ambiguous local regions for clear observation. (Color figure online)

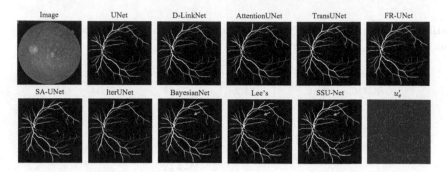

Fig. 3. Visualization results for retinal vessel segmentation on FIVES. Red, green, and yellow line presents manual label, predicted segmentation result, and their overlap region. The cyan arrow points to a sub-region with differences among uncertainty-based methods. We also visualize the epistemic uncertainty map u'_e of our SSU-Net. (Color figure online)

4   Conclusion

This paper proposes a spatial and scale uncertainty-aware network (SSU-Net) for elongated physiological structure segmentation. The ablation study shows the effectiveness of core components: the soft gated uncertainty-aware (GSUA) and the multi-scale uncertainty-aware (MSUA) fusion modules. Compared with some SOTA methods on cornea endothelial cell and retinal vessel image segmentation tasks, the proposed SSU-Net achieved the best segmentation performance and is more robust than other uncertainty-based methods. It is noteworthy that the SSU-Net performed considerably better than specialized retinal vessel segmentation networks. In the future, we plan to conduct experiments on various challenging situations to further explore the characteristics of SSU-Net.

Acknowledgements. This work was supported in part by General Program of National Natural Science Foundation of China (Grant No. 82272086), Guangdong Provincial Department of Education (Grant No. 2020ZDZX3043), Shenzhen Natural Science Fund (JCYJ20200109140820699), the Stable Support Plan Program (20200925174052004), the National Research Foundation, Singapore under its AI Singapore Programme (AISG Award No: AISG2-TC-2021-003), and A*STAR Advanced Manufacturing and Engineering (AME) Programmatic Fund (A20H4b0141) Central Research Fund (CRF).

References

1. Chen, J., et al.: Transunet: transformers make strong encoders for medical image segmentation. arXiv preprint arXiv:2102.04306 (2021)
2. Gal, Y., Ghahramani, Z.: Dropout as a bayesian approximation: Representing model uncertainty in deep learning. In: Balcan, M.F., Weinberger, K.Q. (eds.) Proceedings of ICML. Proceedings of Machine Learning Research, vol. 48, pp. 1050–1059. PMLR, New York, New York, USA (20–22 Jun 2016)

3. Gawlikowski, J., et al.: A survey of uncertainty in deep neural networks. arXiv preprint arXiv:2107.03342 (2021)
4. Guo, Changlu, et al.: Sa-unet: spatial attention u-net for retinal vessel segmentation. In: Proceedings of ICPR, pp. 1236–1242 (2021)
5. Jin, K., et al.: Fives: a fundus image dataset for artificial intelligence based vessel segmentation. Sci. Data 9(1), 475 (2022)
6. Kendall, A., Gal, Y.: What uncertainties do we need in Bayesian deep learning for computer vision? Proc. NeurIPS 30 (2017)
7. Kohl, S., et al.: A probabilistic u-net for segmentation of ambiguous images. Proc. of NeurIPS 31 (2018)
8. Lakshminarayanan, B., et al.: Simple and scalable predictive uncertainty estimation using deep ensembles. Proc. of NeurIPS 30 (2017)
9. Lee, J., et al.: Method to minimize the errors of AI: quantifying and exploiting uncertainty of deep learning in brain tumor segmentation. Sensors 22(6), 2406 (2022)
10. Leibig, C., et al.: Leveraging uncertainty information from deep neural networks for disease detection. Sci. Rep. 7(1), 1–14 (2017)
11. Li, L., et al.: Iternet: Retinal image segmentation utilizing structural redundancy in vessel networks. In: Proceedings of the IEEE/CVF Winter Conference on Applications of Computer Vision, pp. 3656–3665 (2020)
12. Liu, W., et al.: Full-resolution network and dual-threshold iteration for retinal vessel and coronary angiograph segmentation. IEEE J. Biomed. Health Inform. 26(9), 4623–4634 (2022)
13. Mehrtash, A., et al.: Confidence calibration and predictive uncertainty estimation for deep medical image segmentation. IEEE Trans. Med. Imaging 39(12), 3868–3878 (2020)
14. Mou, L., et al.: Cs2-net: deep learning segmentation of curvilinear structures in medical imaging. Med. Image Anal. 67, 101874 (2021)
15. Neal, R.M.: Bayesian learning for neural networks. IEEE Trans. Neural Netw. (1994)
16. Oktay, O., et al.: Attention u-net: learning where to look for the pancreas. arXiv preprint arXiv:1804.03999 (2018)
17. Pearce, T., et al.: Understanding softmax confidence and uncertainty. arXiv preprint arXiv:2106.04972 (2021)
18. Pidaparthy, H., et al.: Automatic play segmentation of hockey videos. In: Proceedings of CVPR, pp. 4585–4593 (2021)
19. Ronneberger, O., Fischer, P., Brox, T.: U-Net: convolutional networks for biomedical image segmentation. In: Navab, N., Hornegger, J., Wells, W.M., Frangi, A.F. (eds.) MICCAI 2015. LNCS, vol. 9351, pp. 234–241. Springer, Cham (2015). https://doi.org/10.1007/978-3-319-24574-4_28
20. Ruggeri, A., et al.: A system for the automatic estimation of morphometric parameters of corneal endothelium in Alizarine red-stained images. Br. J. Ophthalmol. 94(5), 643–647 (2010)
21. Selig, B., et al.: Fully automatic evaluation of the corneal endothelium from in vivo confocal microscopy. BMC Med. Imaging 15(1), 1–15 (2015)
22. Taha, A.A., Hanbury, A.: Metrics for evaluating 3D medical image segmentation: analysis, selection, and tool. BMC Med. Imaging 15(1), 1–28 (2015)
23. Wang, L., et al.: Medical matting: a new perspective on medical segmentation with uncertainty. In: Proceedings of MICCAI, pp. 573–583 (2021)
24. Xie, Y., et al.: Uncertainty-aware cascade network for ultrasound image segmentation with ambiguous boundary. In: Proceedings of MICCAI, pp. 268–278 (2022)

25. Yang, H., et al.: Uncertainty-guided lung nodule segmentation with feature-aware attention. In: Proceedings of MICCAI, pp. 44–54 (2022)
26. Zhang, Y., et al.: A multi-branch hybrid transformer network for corneal endothelial cell segmentation. In: Proceedings of MICCAI, pp. 99–108 (2021)
27. Zhao, Y., et al.: Automated tortuosity analysis of nerve fibers in corneal confocal microscopy. IEEE Trans. Med. Imaging **39**(9), 2725–2737 (2020)
28. Zhou, L., et al.: D-linknet: linknet with pretrained encoder and dilated convolution for high resolution satellite imagery road extraction. In: Proceedings of CVPR, pp. 182–186 (June 2018)

EoFormer: Edge-Oriented Transformer for Brain Tumor Segmentation

Dong She[1,2], Yueyi Zhang[1,2(✉)], Zheyu Zhang[1], Hebei Li[1], Zihan Yan[3], and Xiaoyan Sun[1,2(✉)]

[1] University of Science and Technology of China, Hefei 230026, China
{zhyuey,sunxiaoyan}@ustc.edu.cn
[2] Hefei Comprehensive National Science Center, Institute of Artificial Intelligence, Hefei 230088, China
[3] Beijing Tiantan Hospital, Capital Medical University, Beijing 100050, China

Abstract. Accurate segmentation of brain tumors in MRI images requires precise detection of the edges. However, this crucial information has been overlooked by existing methods. In this paper, we introduce the **Edge-oriented Transformer** (EoFormer) which specifically captures and enhances edge information for brain tumor segmentation. Our approach incorporates a CNN-Transformer encoder to comprehensively improve the feature representation capability. The CNN structure captures low-level local features in the image, while the Transformer structure establishes long-range dependencies between features to generate high-level global features. Additionally, the decoder of our approach utilizes two edge sharpening modules, the Edge-oriented Sobel and Laplacian modules, which enhance the edge information. We also introduce efficient attention and re-parameterization techniques that make EoFormer computationally efficient. Experimental results on the BraTS 2020 dataset and a private medulloblastoma dataset demonstrate the superiority of our approach compared with existing state-of-the-art methods. Moreover, our method achieves this with limtied model parameters and lower FLOPs, making it a promising approach for future research. The code is available at https://github.com/sd0809/EoFormer.

Keywords: Brain tumor segmentation · Edge-oriented module · Transformer

1 Introduction

Accurate segmentation of brain tumors from MRI images is of great significance as it enables more accurate assessment of tumor morphology, size, location, and distribution range, thereby providing clinicians with a reliable basis for diagnosis and treatment [16]. Physicians manually delineate the tumor regions based on the varying signal intensities between diseased and normal tissues. This signal

---

Supplementary Information The online version contains supplementary material available at https://doi.org/10.1007/978-3-031-43901-8_32.

H. Greenspan et al. (Eds.): MICCAI 2023, LNCS 14223, pp. 333–343, 2023.
https://doi.org/10.1007/978-3-031-43901-8_32

disparity constitutes the edge information in the images, making it essential for accurate tumor segmentation.

CNN-based networks, such as UNet [2], SegResNet [15], and nnUNet [8], have made significant progress in the field of medical image segmentation, including brain tumor segmentation. With the emergence of Transformer [19], which is capable of modeling long-range dependencies that CNNs struggle with, a number of CNN-Transformer hybrid networks have been proposed, such as TransBTS [21], UNETR [7], Swin-UNETR [6] and NestedFormer [23], leading to further improvements in brain tumor segmentation. However, the performance of existing brain tumor segmentation methods are still unsatisfactory, especially for the segmentation of edges between tumor lesion and normal tissues.

Considerable advancement has been achieved in the field of natural image segmentation by focusing on the edge information [3,11,18,25], and this idea is also being applied to medical image segmentation. Some methods utilize the distance-dependent objective functions to generate more accurate edge predictions. Karimi et al. [9] design a Hausdorff-based metric loss function to minimize Hausdorff distance (HD), which is used to measure the edge distance between two point sets. Other methods [1,12,20,22] involve post-processing uncertain regions to more accurately segment pixels near edges. For example, BAT [20] considers global context to coarsely locate lesion area and paying special attention to the ambiguous area to specify the exact edges of the skin cancers. Similarly, Xie et al. [22] use the confidence map to evaluate the uncertainty of each pixel to enhance the segmentation of the ambiguous edges of ultrasound images. However, the methods mentioned above are not suitable for brain tumor segmentation for two main reasons. (1) Efficiency. For instance, Karimi et al. [9] require the calculation of the HD at each iteration, which is both time-consuming and computationally demanding. Moreover, processing every slice of large volumes of MRI images at the pixel-level is impractical. (2) Task Complexity. Unlike many other medical image segmentation tasks that involve the segmentation of a single ROI, brain tumor segmentation requires the simultaneous segmentation of three regions: the whole tumor (WT), the tumor core (TC), and the enhancing tumor (ET) regions. Therefore, in addition to focusing on the edge between the tumor lesion and normal tissue to segment the WT, it is also necessary to consider the edges within the tumor in order to segment the TC and ET regions.

In this paper, we propose an **E**dge-oriented trans**Former** (EoFormer), for efficient and accurate brain tumor segmentation. We design a CNN-Transformer based encoder for more effective feature representation, called Efficient Hybrid Encoder (EHE). Specifically, the input image is first processed by the CNN blocks to extract low-level local features. Then, the extracted features are fed into the transformer blocks to create long-range dependencies, resulting in the formation of high-level semantic features. In addition, to provide more accurate edge predictions, we design two edge sharpening modules in the decoder, called Edge-oriented Sobel (EoS) and Laplacian (EoL) modules. By implicitly embedding Sobel and Laplacian filters into the convolution layers to extract 1st-order and 2nd-order differential features, the two modules could enhance the edge information contained in the feature maps. In order to reduce the computational and

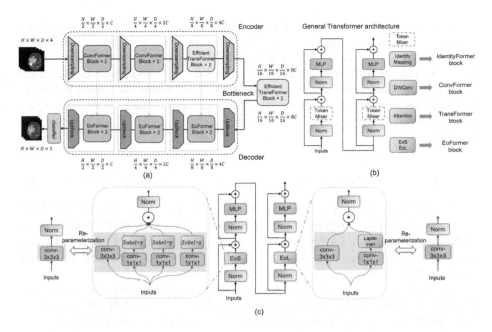

Fig. 1. (a) Overview of the proposed EoFormer. (b) The general transformer architecture and its variants. (c) The EoFormer block, the Edge-oriented Sobel Module (EoS) and the Edge-oriented Laplacian Module (EoL).

memory complexity of the model, we replace the vanilla attention module with our extended efficient attention module [17]. To simplify the model architecture and reduce inference time, we also introduce the re-parameterization technique [4,5]. Our model has been evaluated on both the publicly BraTS 2020 dataset and a private medulloblastoma segmentation dataset. The results demonstrate that EoFormer clearly outperforms the state-of-the-art methods with limited model parameters and lower FLOPs (see more in supplementary material).

2 Method

Figure 1(a) presents an overview of the proposed EoFormer architecture, which comprises two components: (1) an EHE encoder and bottleneck which are used to capture low-level local features and learn a comprehensive feature representation. (2) a decoder which incorporates edge-oriented modules to enhance the edge information in features.

2.1 Efficient Hybrid Encoder

The EHE, shown in Fig. 1(a), comprises four stages, each of which consists of a feature extraction module and a downsampling module. All four feature extraction modules follow the same paradigm of the general transformer architecture (see Fig. 1(b)), which regards the attention module in the transformer as a token mixer [24]. In the first two stages of EHE, we use depth-wise convolution (DWConv) to instantiate the token mixer, called the ConvFormer block. In the third stage and bottleneck, we use the multi-head self-attention (MSA) to instantiate the token mixer, which is the typical transformer block. For each stage i, given an input feature map X, the output of the ith block X'' is computed as follows:

$$X' = \text{TokenMixer}_i(\text{Norm}(X)) + X, \quad X'' = \text{MLP}(\text{Norm}(X')) + X', \quad (1)$$

where the $\text{TokenMixer}_i(\cdot)$ corresponds to DWConv ($i \in \{0,1\}$) and MSA ($i \in \{2,3\}$), Norm($\cdot$) represents layer normalization, and MLP($\cdot$) denotes the Multilayer Perceptron. Our approach combines the strengths of CNN and transformer to create a more powerful encoder that can extract both local and global information from input data.

We address the computational and memory complexity issues that arise from 3D input by replacing the vanilla attention with our extended 3D efficient attention. Assuming the size of the input feature is n and the dimensionality is d, the input feature $X \in \mathbb{R}^{n \times d}$ pass through three linear layers to generate the queries $Q \in \mathbb{R}^{n \times d_k}$, keys $K \in \mathbb{R}^{n \times d_k}$ and values $V \in \mathbb{R}^{n \times d_v}$. The vanilla attention $D(\cdot)$ and the efficient attention $E(\cdot)$ are computed as follows:

$$D(Q,K,V) = \rho(QK^T)V, \quad E(Q,K,V) = \rho(Q)(\rho(K)^T V), \quad (2)$$

where $\rho(\cdot)$ is the softmax activation function, T represents the matrix transpose operator. The efficient attention reduces the memory complexity and computational complexity of vanilla attention from $\mathcal{O}(n^2)$ and $\mathcal{O}(dn^2)$ to $\mathcal{O}(dn + d^2)$ and $\mathcal{O}(nd^2)$, where $d = d_v = 2d_k$.

2.2   Edge-Oriented Transformer Decoder

We design the EoFormer block (see Fig. 1(c)) in the decoder, which instantiates the token mixer with our proposed Edge-oriented Sobel module (EoS) and Edge-oriented Laplacian module (EoL). Each edge-oriented module includes a normal $3 \times 3 \times 3$ convolution and an edge detection path to extract the 1st-order or the 2nd-order spatial derivatives from intermediate features. This design allows the edge-oriented module to efficiently extract the edges and textures of the features. Moreover, to boost the segmentation performance without sacrificing efficiency, we incorporate the re-parameterization technique in the decoder.

Edge-Oriented Sobel Module. We use a dual-branch structure, where the input feature X is simultaneously processed by two different branches. The first branch contains a $3 \times 3 \times 3$ convolution that extracts basic features from the input. The second branch, which is responsible for edge extraction, first uses a $C \times C \times 1 \times 1 \times 1$ convolution to enhance the interaction between channel features of X, then utilizes a learnable scaled Sobel filter to extract the 1st-order differentiation edge information from X. This filter is capable of detecting edges in three directions (i.e. horizontal, vertical, and orthogonal directions), so it comprises three filters M_x, M_y, and M_z, each of which is represented by a 3 × 3 × 3 array. Take M_x as an example, which is described as:

$$M_x[0,:,:] = \begin{bmatrix} -1 & 0 & +1 \\ -2 & 0 & +2 \\ -1 & 0 & +1 \end{bmatrix}, M_x[1,:,:] = \begin{bmatrix} -2 & 0 & +2 \\ -4 & 0 & +4 \\ -2 & 0 & +2 \end{bmatrix}, M_x[2,:,:] = \begin{bmatrix} -1 & 0 & +1 \\ -2 & 0 & +2 \\ -1 & 0 & +1 \end{bmatrix}.$$

We then apply a learnable scaling matrix $S \in \mathbb{R}^{C \times 1 \times 1 \times 1}$ to M_x, which allows for dynamic adjustment of the scaling factor in each channel. The resulting feature extracted from the scaled Sobel-x filter is denoted as:

$$F_x = \text{DWConv}_{S \cdot M_x}(\text{Conv}_{1 \times 1 \times 1}(X)), \tag{3}$$

where the '·' denotes channel-wise multiplication; the $\text{DWConv}_{S \cdot M_x}$ indicates that $\text{DWConv}(\cdot)$ applies a $S \cdot M_x$ learnable scaled filter as its kernel weight. Similarly, F_y and F_z are processed in the same way. The final output of the EoS module, denoted as F_{sob}, is given by:

$$F_{sob} = \text{Norm}(\text{Conv}_{3 \times 3 \times 3}(X) + F_x + F_y + F_z). \tag{4}$$

Edge-Oriented Laplacian Module. Different from the Sobel filter that only extracts edges in the horizontal, vertical, and orthogonal directions, the Laplacian filter can extract edges in all directions. After extracting the 1st-order differentiation edge information, the intermediate features are then fed into the EoL module for extracting the 2nd-order differentiation edge information. Similarly, the feature F, obtained from the learnable scaling Laplacian filter, and the final output of the EoL module, denoted as F_{lap}, are defined as:

$$F = \text{DWConv}_{S \cdot M_{lap}}(\text{Conv}_{1 \times 1 \times 1}(X)),$$
$$F_{lap} = \text{Norm}(\text{Conv}_{3 \times 3 \times 3}(X) + F). \tag{5}$$

Re-parameterization of the Edge-Oriented Modules. We introduce the re-parameterization [4,5] into the edge-oriented modules to boost the segmentation performance while maintaining high efficiency. Specifically, we explain the re-parameterization of the EoL module as follows:

$$W_{\text{rep}} = W_{\text{conv}_{3 \times 3 \times 3}} + W_{\text{conv}_{1 \times 1 \times 1}} * W_{\text{conv}_{lap}} \tag{6}$$

$$B_{\text{rep}} = B_{\text{conv}_{3 \times 3 \times 3}} + W_{\text{conv}_{lap}} * \text{up}(B_{\text{conv}_{1 \times 1 \times 1}}) + B_{\text{conv}_{lap}} \tag{7}$$

where '*' represents the convolution operation, W_{conv} means the weights of the convolution and B_{conv} denotes the bias, and up$(\cdot)$ is the spatial broadcasting operation, which upgrades the bias $B \in \mathbb{R}^{1 \times C \times 1 \times 1 \times 1}$ into up$(B) \in \mathbb{R}^{1 \times C \times 3 \times 3 \times 3}$. In the inference stage, the output feature F is produced by a normal 3 × 3 × 3 convolution as follows:

$$F = W_{\text{rep}} * X + B_{\text{rep}}. \tag{8}$$

Table 1. Quantitative comparison on BraTS 2020 dataset. The '$*$' means the FLOPs we recalculate.

Method	Param(M)	FLOPs(G)	Dice (%) ↑				HD95 (mm) ↓			
			WT	TC	ET	Ave	WT	TC	ET	Ave
3D-UNet [2]	5.75	1449.59	90.01	80.68	79.18	83.29	8.591	10.91	5.932	8.477
SegResNet [15]	18.79	185.23	87.59	85.50	81.20	84.77	4.941	4.653	3.822	4.472
nnUNet [8]	5.75	1449.59	91.15	86.12	81.67	86.32	**3.532**	4.901	3.561	3.998
TransBTS [21]	32.99	333.00	90.54	84.93	79.91	85.12	3.916	4.843	4.501	4.420
UNETR [7]	102.06	203.32	90.77	84.11	79.12	84.67	4.917	5.054	3.943	4.638
Swin-UNETR [6]	61.98	793.92	**91.50**	84.06	80.98	85.51	3.386	5.080	3.640	4.035
NestedFormer [23]	10.48	209.58*	91.07	85.30	80.10	85.49	3.583	4.735	4.391	4.236
EoFormer	6.28	91.81	90.84	**86.38**	**83.22**	**86.81**	3.974	**4.500**	**3.432**	**3.968**

Table 2. Quantitative comparison on MedSeg dataset.

Method	Dice (%) ↑			HD95 (mm) ↓		
	WT	ET	Ave	WT	ET	Ave
3D-UNet [2]	61.52	50.71	56.11	17.43	14.62	16.03
SegResNet [15]	76.76	55.60	66.18	7.810	9.411	8.611
TransBTS [21]	72.35	55.56	63.96	11.09	11.19	11.14
UNETR [7]	73.38	56.02	64.70	9.112	12.70	10.90
Swin-UNETR [6]	70.10	**60.79**	65.44	9.766	9.339	9.552
NestedFormer [23]	**79.89**	55.76	67.83	7.099	12.08	9.587
EoFormer	79.74	59.10	**69.42**	**6.978**	**7.104**	**7.041**

3 Experiment

3.1 Dataset and Evaluation Metric

In order to validate the performance of EoFormer, we conduct extensive experiments on both the publicly available BraTS 2020 dataset and a private medulloblastoma segmentation dataset (MedSeg).

The BraTS 2020 dataset [14] consists of MRI image data from 369 patients, with each patient having four modalities (T1, T1ce, T2 and T2-FLAIR) of skull-striped MRI, which are aligned to a standard brain template. The training/validation/test split follows 315/16/37 according to recent works [10,23].

The MedSeg dataset includes MRI images of T1, T1ce, T2, and T2 FLAIR modalities from 255 patients with medulloblastoma. The dataset includes manual annotations of the WT and ET regions. These annotated masks are reviewed by two experienced physicians to ensure the accuracy of the annotated results. The images are registered to the size of $24 \times 256 \times 256$. The training/validation/test split ratio is 3:1:1. Four-fold cross-validation is performed on this dataset.

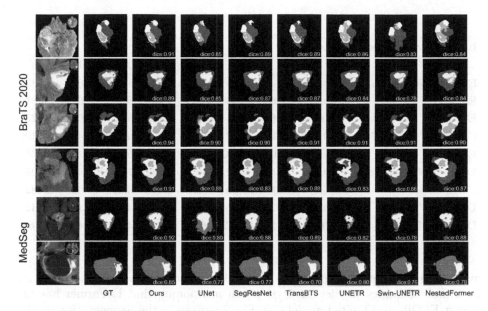

Fig. 2. Qualitative Comparison of segmentation results on BraTS and MedSeg. The red region represents WT, the yellow means TC and the white denotes ET (Color figure online).

3.2 Implementation Details

We implement EoFormer in Pytorch 1.11. Our model is trained from scratch for 300 epochs using two NVIDIA GTX 3090 GPUs. We select a combination of soft dice loss and cross-entropy as the loss function and utilize the AdamW optimizer [13] with a weight decay of 1×10^{-5}. The initial learning rate is 2×10^{-4}. For data augmentation, we apply image croping, flipping, identity scaling and shifting.

3.3 Results

We compare EoFormer with seven methods, including CNN-based methods (3D-UNet [2], SegResNet [15] and nnUNet [8]) and Transformer-based methods (TransBTS [21], UNETR [7], Swin-UNETR [6], NestedFormer [23]). The results are reproduced on our data split.

Table 1 displays the performance comparison of EoFormer against other methods on the BraTS 2020 dataset. EoFormer achieves the highest Dice scores on TC, ET, and the average. In addition, EoFormer attains the best HD95 scores on TC, ET, and the average. HD95 measures the edge distance between prediction and annotation, which is more sensitive to boundaries. Table 2 illustrates the performance of EoFormer and other methods on MedSeg. EoFormer outperforms the second-ranked NestedFormer by an average of 1.59% on Dice and achieves the top performance for both WT and ET on HD95. Furthermore, compared to the second-ranked SegResNet, EoFormer demonstrates an

Table 3. Ablation study on the encoder and decoder design.

Index	Encoder/Decoder	Dice (%) ↑				HD95 (mm) ↓			
		WT	TC	ET	Ave	WT	TC	ET	Ave
Enc1	UNet encoder	83.29	83.43	79.01	84.00	6.232	8.018	6.697	6.983
Enc2	CF×4	88.51	83.42	82.56	84.83	7.131	8.772	5.641	7.181
Enc3	CF×2+TF×2	**90.92**	**86.63**	81.05	86.20	**3.906**	5.227	5.637	4.923
Enc4	CF×2+ETF×2	90.84	86.38	83.22	**86.81**	3.974	**4.500**	3.432	**3.968**
Enc5	ETF×4	90.24	84.15	**85.40**	86.59	3.951	5.651	**3.405**	4.336
Dec1	UNet decoder	89.75	84.09	80.12	84.66	7.562	7.332	6.427	7.107
Dec2	IF×3	90.27	86.07	80.09	85.48	5.448	6.069	4.929	5.482
Dec3	CF×3	90.63	85.63	81.61	85.96	4.098	**4.467**	**3.023**	4.842
Dec4	EoF×3	**90.84**	**86.38**	**83.22**	**86.81**	**3.974**	4.500	3.432	**3.968**

average HD95 improvement of 1.57 mm, highlighting its superior performance in tumor boundary prediction. It is worth mentioning that EoFormer has the lowest FLOPs and limited model size. Fig. 2 represents the segmentation results on the BraTS 2020 and MedSeg datasets. The visualisation demonstrates that the EoFormer achieves the closest segmentation results to the ground truth. Specifically, EoFormer accurately segments both TC and ET region boundaries.

3.4 Ablation

We evaluate the effectiveness of our proposed EoFormer framework by conducting ablation experiments on the BraTS 2020. In Table 3, the abbreviations IF, CF, TF, ETF, and EoF represent IdentityFormer blocks (see Fig. 1(b)), ConvFormer blocks, TransFormer blocks, Efficient TransFormer blocks, and EoFormer blocks, respectively.

Encoder and Bottleneck Design. We compare our proposed EHE with different encoders and bottleneck in Table 3. Enc1 utilizes UNet encoder. Enc2 - 5 have the same encoder and bottleneck as EHE but with different configurations. Our results show that EHE outperforms other methods, with high average Dice and low average HD95. This is because a full CNN encoder (Enc2) is not good at capturing global dependencies, while a full Transformer encoder (Enc5) is inadequate at capturing low-level features. Our proposed EHE balances the strengths of both and achieves the best segmentation performance. Additionally, our extended efficient attention achieves better performance compared with Enc3 because it has a better ability to capture the periphery of objects [17].

Decoder Design. We compare the performance of various decoders in Table 3. Dec1 - 4 share the same EHE encoder, but employ different decoders: Dec1 uses the UNet decoder, Dec2 has three IdentityFormer blocks, Dec3 replaces the IdentityFormer blocks with ConvFormer blocks, and Dec4 is our proposed EoFormer decoder. Our results show that the EoFormer decoder achieves the highest Dice scores, and achieves the lowest average HD95 score due to the incorporation of the EoS and EoL modules within the EoFormer block.

4 Conclusion

In this paper, we propose the EoFormer, a novel approach for brain tumor segmentation. Our method comprises the Efficient Hybrid Encoder and the Edge-oriented Transformer Decoder. The encoder effectively extracts features from images by striking a balance between CNN and Transformer architectures. The decoder integrates the Sobel and Laplacian edge detection filters into our edge-oriented modules that enhance the extraction capability of edge and texture information. Besides, we introduce the efficient attention mechanism and the re-parameterization technology to improve the model efficiency. Our EoFormer outperforms other state-of-the-art methods on both BraTS 2020 and MedSeg. Our model is computationally efficient and can be readily applied to other 3D medical image segmentation tasks.

References

1. Chen, S., Ding, C., Tao, D.: Boundary-assisted region proposal networks for nucleus segmentation. In: Martel, A.L., et al. (eds.) MICCAI 2020. LNCS, vol. 12265, pp. 279–288. Springer, Cham (2020). https://doi.org/10.1007/978-3-030-59722-1_27
2. Çiçek, Ö., Abdulkadir, A., Lienkamp, S.S., Brox, T., Ronneberger, O.: 3D U-Net: learning dense volumetric segmentation from sparse annotation. In: Ourselin, S., Joskowicz, L., Sabuncu, M.R., Unal, G., Wells, W. (eds.) MICCAI 2016. LNCS, vol. 9901, pp. 424–432. Springer, Cham (2016). https://doi.org/10.1007/978-3-319-46723-8_49
3. Ding, H., Jiang, X., Liu, A.Q., Thalmann, N.M., Wang, G.: Boundary-aware feature propagation for scene segmentation. In: Proceedings of the IEEE/CVF International Conference on Computer Vision, pp. 6819–6829 (2019)
4. Ding, X., Guo, Y., Ding, G., Han, J.: ACNet: strengthening the kernel skeletons for powerful CNN via asymmetric convolution blocks. In: Proceedings of the IEEE/CVF International Conference on Computer Vision, pp. 1911–1920 (2019)
5. Ding, X., Zhang, X., Ma, N., Han, J., Ding, G., Sun, J.: RepVGG: making VGG-style convnets great again. In: Proceedings of the IEEE/CVF Conference on Computer Vision and Pattern Recognition, pp. 13733–13742 (2021)
6. Hatamizadeh, A., Nath, V., Tang, Y., Yang, D., Roth, H.R., Xu, D.: Swin UNETR: swin transformers for semantic segmentation of brain tumors in MRI images. In: Crimi, A., Bakas, S. (eds.) Brainlesion: Glioma, Multiple Sclerosis, Stroke and Traumatic Brain Injuries: 7th International Workshop, BrainLes 2021, Held in Conjunction with MICCAI 2021, Virtual Event, 27 September 2021, Revised Selected Papers, Part I, pp. 272–284. Springer, Cham (2022). https://doi.org/10.1007/978-3-031-08999-2_22
7. Hatamizadeh, A., et al.: UNETR: transformers for 3D medical image segmentation. In: Proceedings of the IEEE/CVF winter Conference on Applications of Computer Vision, pp. 574–584 (2022)

8. Isensee, F., Jaeger, P.F., Kohl, S.A., Petersen, J., Maier-Hein, K.H.: nnU-Net: a self-configuring method for deep learning-based biomedical image segmentation. Nat. Methods **18**(2), 203–211 (2021)
9. Karimi, D., Salcudean, S.E.: Reducing the hausdorff distance in medical image segmentation with convolutional neural networks. IEEE Trans. Med. Imaging **39**(2), 499–513 (2019)
10. Larrazabal, A.J., Martínez, C., Dolz, J., Ferrante, E.: Orthogonal ensemble networks for biomedical image segmentation. In: de Bruijne, M., et al. (eds.) MICCAI 2021. LNCS, vol. 12903, pp. 594–603. Springer, Cham (2021). https://doi.org/10.1007/978-3-030-87199-4_56
11. Li, X., et al.: Improving semantic segmentation via decoupled body and edge supervision. In: Vedaldi, A., Bischof, H., Brox, T., Frahm, J.-M. (eds.) ECCV 2020. LNCS, vol. 12362, pp. 435–452. Springer, Cham (2020). https://doi.org/10.1007/978-3-030-58520-4_26
12. Lin, L., et al.: BSDA-Net: a boundary shape and distance aware joint learning framework for segmenting and classifying OCTA images. In: de Bruijne, M., et al. (eds.) MICCAI 2021. LNCS, vol. 12908, pp. 65–75. Springer, Cham (2021). https://doi.org/10.1007/978-3-030-87237-3_7
13. Loshchilov, I., Hutter, F.: Decoupled weight decay regularization. arXiv preprint arXiv:1711.05101 (2017)
14. Menze, B.H., et al.: The multimodal brain tumor image segmentation benchmark (BRATS). IEEE Trans. Med. Imaging **34**(10), 1993–2024 (2014)
15. Myronenko, A.: 3D MRI brain tumor segmentation using autoencoder regularization. In: Crimi, A., Bakas, S., Kuijf, H., Keyvan, F., Reyes, M., van Walsum, T. (eds.) BrainLes 2018, Part II. LNCS, vol. 11384, pp. 311–320. Springer, Cham (2019). https://doi.org/10.1007/978-3-030-11726-9_28
16. Sharma, N., Aggarwal, L.M., et al.: Automated medical image segmentation techniques. J. Med. Phys. **35**(1), 3 (2010)
17. Shen, Z., Zhang, M., Zhao, H., Yi, S., Li, H.: Efficient attention: attention with linear complexities. In: Proceedings of the IEEE/CVF Winter Conference on Applications of Computer Vision, pp. 3531–3539 (2021)
18. Tang, C., Chen, H., Li, X., Li, J., Zhang, Z., Hu, X.: Look closer to segment better: boundary patch refinement for instance segmentation. In: Proceedings of the IEEE/CVF Conference on Computer Vision and Pattern Recognition, pp. 13926–13935 (2021)
19. Vaswani, A., et al.: Attention is all you need. In: Advances in Neural Information Processing Systems, vol. 30 (2017)
20. Wang, J., Wei, L., Wang, L., Zhou, Q., Zhu, L., Qin, J.: Boundary-aware transformers for skin lesion segmentation. In: de Bruijne, M., et al. (eds.) MICCAI 2021. LNCS, vol. 12901, pp. 206–216. Springer, Cham (2021). https://doi.org/10.1007/978-3-030-87193-2_20
21. Wang, W., Chen, C., Ding, M., Yu, H., Zha, S., Li, J.: TransBTS: multimodal brain tumor segmentation using transformer. In: de Bruijne, M., et al. (eds.) MICCAI 2021. LNCS, vol. 12901, pp. 109–119. Springer, Cham (2021). https://doi.org/10.1007/978-3-030-87193-2_11

22. Xie, Y., Liao, H., Zhang, D., Chen, F.: Uncertainty-aware cascade network for ultrasound image segmentation with ambiguous boundary. In: Wang, L., Dou, Q., Fletcher, P.T., Speidel, S., Li, S. (eds.) Medical Image Computing and Computer Assisted Intervention-MICCAI 2022: 25th International Conference, Singapore, 18–22 September 2022, Proceedings, Part IV, pp. 268–278. Springer, Cham (2022). https://doi.org/10.1007/978-3-031-16440-8_26

23. Xing, Z., Yu, L., Wan, L., Han, T., Zhu, L.: NestedFormer: nested modality-aware transformer for brain tumor segmentation. In: Wang, L., Dou, Q., Fletcher, P.T., Speidel, S., Li, S. (eds.) Medical Image Computing and Computer Assisted Intervention-MICCAI 2022: 25th International Conference, Singapore, 18–22 September 2022, Proceedings, Part V, pp. 140–150. Springer, Cham (2022). https://doi.org/10.1007/978-3-031-16443-9_14

24. Yu, W., et al.: MetaFormer is actually what you need for vision. In: Proceedings of the IEEE/CVF Conference on Computer Vision and Pattern Recognition, pp. 10819–10829 (2022)

25. Zou, N., Xiang, Z., Chen, Y., Chen, S., Qiao, C.: Boundary-aware CNN for semantic segmentation. IEEE Access **7**, 114520–114528 (2019)

Breast Ultrasound Tumor Classification Using a Hybrid Multitask CNN-Transformer Network

Bryar Shareef, Min Xian(✉), Aleksandar Vakanski, and Haotian Wang

Department of Computer Science, University of Idaho, Idaho Falls, ID 83402, USA
mxain@uidaho.edu

Abstract. Capturing global contextual information plays a critical role in breast ultrasound (BUS) image classification. Although convolutional neural networks (CNNs) have demonstrated reliable performance in tumor classification, they have inherent limitations for modeling global and long-range dependencies due to the localized nature of convolution operations. Vision Transformers have an improved capability of capturing global contextual information but may distort the local image patterns due to the tokenization operations. In this study, we proposed a hybrid multitask deep neural network called Hybrid-MT-ESTAN, designed to perform BUS tumor classification and segmentation using a hybrid architecture composed of CNNs and Swin Transformer components. The proposed approach was compared to nine BUS classification methods and evaluated using seven quantitative metrics on a dataset of 3,320 BUS images. The results indicate that Hybrid-MT-ESTAN achieved the highest accuracy, sensitivity, and F1 score of 82.7%, 86.4%, and 86.0%, respectively.

Keywords: Breast Ultrasound · Classification · Multitask Learning · Hybrid CNN-Transformer

1 Introduction

Breast cancer is the leading cause of cancer-related fatalities among women. Currently, it holds the highest incidence rate of cancer among women in the U.S., and in 2022 it accounted for 31% of all newly diagnosed cancer cases [1]. Due to the high incidence rate, early breast cancer detection is essential for reducing mortality rates and expanding treatment options. BUS imaging is an effective screening option because it is cost-effective, nonradioactive, and noninvasive. However, BUS image analysis is also challenging due to the large variations in tumor shape and appearance, speckle noise, low contrast, weak boundaries, and occurrence of artifacts.

In the past decade, deep learning-based approaches achieved remarkable advancements in BUS tumor classification [2,3]. The progress has been driven by the capability of CNN-based models to learn hierarchies of structured image representations as semantics. To extract deep context features, CNNs apply a series of convolutional and downsampling layers, frequently organized into blocks

© The Author(s), under exclusive license to Springer Nature Switzerland AG 2023
H. Greenspan et al. (Eds.): MICCAI 2023, LNCS 14223, pp. 344–353, 2023.
https://doi.org/10.1007/978-3-031-43901-8_33

with residual connections. Nevertheless, one disadvantage of such architectural choice is that the feature representations in the deeper layers become increasingly abstract, leading to a loss of spatial and contextual information. The intrinsic locality of convolutional operations hinders the ability of CNNs to model long-range dependencies while preserving spatial information in images effectively.

Vision Transformer (ViT) [5] and its variants recently demonstrated superior performance in image classification tasks. These models convert input images into smaller patches and utilize the self-attention mechanism to model the relationships between the patches. Self-attention enables ViTs to capture long-range dependencies and model complex relationships between different regions of the image. However, the effectiveness of ViT-based approaches heavily relies on access to large datasets for learning meaningful representations of input images. This is primarily because the architectural design of ViTs does not rely on the same inductive biases in feature extraction which allow CNNs to learn spatially invariant features.

Accordingly, numerous prior studies introduced modifications to the original ViT network specifically designed for BUS image classification [13,14,23]. In addition, several works proposed network architectures that combined Transformers and CNNs [4,15,16]. For instance, Mo et al. [15] proposed a hybrid CNN-Transformer incorporating BUS anatomical priors. Qu et al. [16] employed squeeze and excitation blocks to enhance the feature extraction capacity in a hybrid CNN-based VGG16 network and ViT. Similarly, Iqbal et al. [4] designed two hybrid CNN-Transformer networks intended either for classification or segmentation of multi-modal breast cancer images. Despite the promising results of such hybrid approaches, effectively capturing the local patterns and global long-range dependencies in BUS images remains challenging [4,5,24].

Multitask learning leverages shared information across related tasks by jointly training the model. It constrains models to learn representations that are relevant to all tasks rather than learning task-specific details. Moreover, multitask learning acts as a regularizer by introducing inductive bias and prevents overfitting [25] (particularly with ViTs), and with that, can mitigate the challenges posed by small BUS dataset sizes. In [3], the authors demonstrated that multitask learning outperforms single-task learning approaches for BUS classification.

In this study, we introduce a hybrid multitask approach, Hybrid-MT-ESTAN, which encompasses tumor classification as a primary task and tumor segmentation as a secondary task. Hybrid-MT-ESTAN combines the advantages of CNNs and Transformers in a framework incorporating anatomical tissue information in BUS images. Specifically, we designed a novel attention block named Anatomy-Aware Attention (AAA), which modifies the attention block of Swin Transformer by considering the breast anatomy. The anatomy of the human breast is categorized into four primary layers: the skin, premammary (subcutaneous fat), mammary, and retromammary layers, where each layer has a distinct texture and generates different echo patterns. The primary layers in BUS images are arranged in a vertical stack, with similar echo patterns appearing horizontally across the images. The kernels in the introduced AAA attention blocks are organized in rows and columns to capture the anatomical structure of the breast tissue. In the published literature, the closest approach to ours is the work by Iqbal et al. [4], in which the authors used hybrid single-task CNN-Transformer networks for either classification or segmentation of BUS images. Conversely, Hybrid-MT-ESTAN employs a multitask approach and introduces novel architectural design. The main contributions of this work are summarized as:

- The proposed architecture effectively integrates the advantages of CNNs for extracting hierarchical and local patterns in BUS images and Swin Transformers for leveraging long-range dependencies.
- The designed Anatomy-Aware Attention (AAA) block improves the learning of contextual information based on the anatomy of the breast.
- The multitask learning approach leverages the shared representations across the classification and segmentation tasks to improve the model performance.

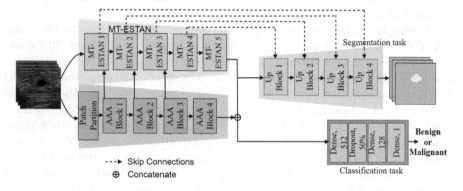

Fig. 1. Hybrid-MT-ESTAN consists of MT-ESTAN and AAA encoders, a segmentation branch, and a classification branch.

2   Proposed Method

2.1   Hybrid-MT-ESTAN

The architecture of Hybrid-MT-ESTAN is shown in Fig. 1, and consists of: (1) the MT-ESTAN encoder [3], and a Swin Transformer-based encoder with Anatomy-Aware Attention (AAA) blocks, (2) a decoder branch for the segmentation task, and (3) a branch with fully-connected layers for the classification task. MT-ESTAN [3] is a CNN-based multitask learning network that simultaneously performs BUS classification and segmentation. The encoder sub-network of MT-ESTAN is ESTAN [17], which employs row-column-wise kernels to learn and fuse context information in BUS images at different context scales (see Fig. 2). Specifically, each MT-ESTAN block is composed of two parallel branches consisting of four square convolutional kernels and two consecutive row-column-wise kernels. These specialized convolutional kernels effectively extract contextual information of small tumors in BUS images. Refer to [17,22], and [3] for the implementation details of ESTAN and MT-ESTAN. The source codes of these works are available at http://busbench.midalab.net.

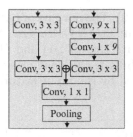

Fig. 2. MT-ESTAN blocks include parallel convolutional branches with different kernel size, followed by 1×1 convolution and a pooling layer.

2.2 Anatomy-Aware Attention (AAA) Block

Swin Transformer [18] is a hierarchical transformer-based approach that uses shifted windows to model global context information. Swin Transformer partitions an input image into non-overlapping patches of size 4×4, where each patch is treated as a "token". A linear layer receives the patches and projects them into an arbitrary dimension. Each Swin Transformer block consists of a LayerNorm layer (LN) layer, a multi-head self-attention module (MSA), and a multi-layer perceptron (MLP) with GELU activation. To model long-range dependencies, the original Swin Transformer relies on shifted windows, where the window-based multi-head self-attention (W-MSA) and shifted window-based multi-head self-attention (SW-MSA) modules are employed in each consecutive Swin block. The Swin block is formulated as follows.

$$\hat{f}^l = \text{W-MSA}(\text{LN}(f^{l-1})) + f^{l-1} \tag{1}$$

$$f^l = \text{MLP}(\text{LN}(\hat{f}^l)) + \hat{f}^l \tag{2}$$

$$\hat{f}^{l+1} = \text{SW-MSA}(\text{LN}(f^l)) + f^l \tag{3}$$

$$f^{l+1} = \text{MLP}(\text{LN}(\hat{f}^{l+1})) + \hat{f}^{l+1} \tag{4}$$

where f^l and $\hat{f}^l$ are the output features of the MLP module and the (S)W-MSA module for block l, respectively; in the proposed Anatomy-Aware Attention (AAA) block, we redesigned the Swin blocks to enhance their ability to model both global and local features by adding an attention block based on the breast anatomy (see Fig. 3). The additional layers are defined by

$$y^i = M(f^{l+1}) \tag{5}$$

$$B^i = U(\text{MAX-P}(y^i) + \text{AVG-P}(y^i)) \tag{6}$$

$$O^i = y^i \cdot (\sigma(A(B))) \tag{7}$$

Concretely, we first reconstruct the i-th feature map (y^i) by merging (M) all patches, and afterward, we applied average pooling (AVG-P) and max pooling

(MAX-P) layers with size (2, 2). The outputs of (AVG-P) and (MAX-P) layers are concatenated and UP-SAMPLED (U) with size (2, 2) and stride (2, 2). ROW-COLUMN-WISE kernels (A) with size (9 , 1) and (1 , 9) are then employed to adapt to the anatomy of the breast, and finally a sigmoid function (σ) is applied to the output of (A) multiplied by the input feature map (y^i).

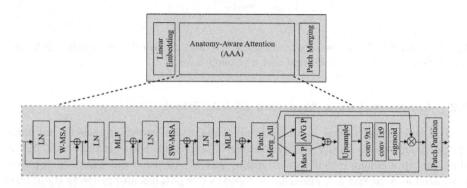

Fig. 3. Anatomy-Aware Attention (AAA) block.

2.3   Segmentation and Classification Branches/Tasks

The segmentation branch in Fig. 1 outputs dense mask predictions of BUS tumors. It consists of four Up Blocks, each with three convolutional layers and one upsampling layer (with size (2, 2) and stride (2, 2)). The settings of the convolutional layers are adopted from [3]. In addition, the blocks receive four skip connections from the MT-ESTAN encoder, i.e., there is a skip connection from each MT-ESTAN block 1 to 4. The classification branch consists of three dense layers, a dropout layer (50%), and the final dense layer that predicts the tumor class into benign or malignant.

2.4   Loss Function

We applied a multitask loss function (L_{mt}) that aggregates two terms: a focal loss L_{Focal} for the classification task and dice loss L_{Dice} for the segmentation task. Therefore, the composite loss function is $L_{mt} = w_1 \cdot L_{Focal} + L_{Dice}$, where the weight coefficient w_1 is set to apply greater importance to the classification task as the primary task. Since in medical image diagnosis achieving high sensitivity places emphasis on the detection of malignant lesions, we employed the focal loss for the classification task to trade off between sensitivity and specificity. Because malignant tumors are more challenging to detect due to greater differences in margin, shape, and appearance in BUS images, focal loss forces the model to focus more on difficult predictions. Specifically, focal loss adds a factor $(1 - p_i)^\gamma$ to the cross-entropy loss where γ is a focusing parameter, resulting in $L_{Focal} = -1/N \sum_{i=1}^{N}[(\alpha \cdot t_i \cdot (1-p_i)^\gamma \cdot log(p_i) + (1-\alpha) \cdot p_i \cdot log(1-p_i)]$. In the formulation, α

is a weighting coefficient, N denotes the number of image samples, t_i is the target label of the ith training sample, and p_i denotes the prediction. The segmentation loss is calculated using the commonly-employed Dice loss (L_{Dice}) function.

3   Experimental Results

3.1   Datasets

We evaluated the performance of Hybrid-MT-ESTAN using four public datasets, HMSS [9], BUSI [10], BUSIS [20], and Dataset B [6]. We combined all four datasets to build a large and diverse dataset with a total of 3,320 B-mode BUS images, of which 1,664 contain benign tumors and 1,656 have malignant tumors. Table 1 shows the detailed information for each dataset. HMSS dataset does not provide the segmentation ground-truth masks, and for this study we arranged with a group of experienced radiologists to prepare the masks for HMSS. Refer to the original publications of the datasets for more details.

Table 1. Breast ultrasound (BUS) datasets. 'b' denotes benign tumor and 'm' is malignant tumor.

BUS dataset	No. of images	Distribution	Source
HMSS	1,948	b:812, m:1136	Netherlands
BUSI	647	b:437, m:210	Egypt
BUSIS	562	b:306, m:256	China
Dataset B	163	b:109, m:54	Spain
Total	3,320	b: 1,664, m: 1,656	

3.2   Evaluation Metrics

For performance evaluation of the classification task, we used the following metrics: accuracy (Acc), sensitivity (Sens), specificity (Spec), F1 score, Area Under the Curve of Receiver Operating Characteristic (AUC), false positive rate (FPR), and false negative rate (FNR). To evaluate the segmentation performance, we used dice similarity coefficient (DSC) and Jaccard index (JI).

3.3   Implementation Details

The proposed approach was implemented with Keras and TensorFlow libraries. All experiments were performed on a machine with NVIDIA Quadro RTX 8000 GPUs and two Intel Xeon Silver 4210R CPUs (2.40GHz) with 512 GB of RAM. All BUS images in the dataset were zero-padded and reshaped to form square images. To avoid data leakage and bias, we selected the train, test, and validation sets based on the cases, i.e., the images from one case (patient) were

Table 2. Performance metrics of the compared methods for BUS image classification and segmentation.

Methods	Classification							Segmentation	
	Acc↑	Sens.↑	Spec.↑	F1↑	Auc↑	FNR↓	FPR↓	DSC↑	JI↑
SHA-MTL [8]	69.6	48.1	**90.8**	0.58	69.5	51.9	**9.2**	72.2	60.7
MobileNet [19]	71.0	82.0	61.0	0.74	71.5	18.0	39.0	-	-
VGGA-ViT [16]	73.6	61.8	79.8	0.61	70.8	38.2	20.2	74.9	64.9
DenseNet121 [7]	73.0	74.0	71.0	0.73	72.5	26.0	29.0	-	-
EMT-Net [12]	74.1	79.4	69.1	0.75	74.3	20.6	30.9	76.7	67.0
ViT [5]	72.1	74.1	69.3	0.73	71.7	25.9	30.7	-	-
Chowdery [10]	77.4	77.3	77.3	0.77	77.3	22.7	22.7	77.0	67.9
Swin Transformer	77.4	72.6	82.5	0.74	77.6	27.4	17.5	-	-
MT-ESTAN	78.6	83.7	72.6	0.83	78.2	16.3	27.4	78.2	69.3
Ours	**82.8**	**86.4**	79.2	**0.86**	**82.8**	**13.6**	20.8	**84.1**	**75.7**

Note: A dash '-' in the Segmentation column indicates that the model uses single-task learning.

assigned to only one of the training, validation, and test sets. Furthermore, we employed horizontal flip, height shift (20%), width shift (20%), and rotation (20°C) for data augmentation. The proposed approach utilizes the building blocks of ResNet50 and Swin-Transformer-V2, pretrained on ImageNet dataset. Namely, MT-ESTAN uses pretrained ResNet50 as a base model for the five encoder blocks (the implementation details of MT-ESTAN can be found in [3]). The encoder with AAA blocks uses the SwinTransformer_V2_Base_256 pretrained model as a backbone. For the composite loss function, we adopted a weight coefficient $w_1 = 3$, and in the focal loss $\alpha = 0.5$ and $\gamma = 2$. For model training we utilized Adam optimizer with a learning rate of 10^{-5} and mini batch size of 4 images.

3.4   Performance Evaluation and Comparative Analysis

We compared the performance of Hybrid-MT-ESTAN for BUS classification to nine deep learning approaches commonly used for medical image analysis. The compared models include CNN-based, ViT-based, and hybrid approaches. CNN-based networks are SHA-MTL [8], MobileNet [19], DenseNet121 [7], and EMT-Net [12]. ViT-based approaches include the original ViT [5], Chowdery [10], and Swin Transformer [18]. VGGA-ViT [16] is a hybrid CNN-Transformer network. The values of the performance metrics are shown in Table 2, indicating that the proposed Hybrid-MT-ESTAN outperformed all nine approaches by achieving the best accuracy, sensitivity, F1 score, and AUC with 82.8%, 86.4%, 86.0%, and 82.8%, respectively. Although SHA-MTL [8] obtained the highest specificity of 90.8% and FNR of 9.2%, the trade-off between sensitivity and specificity should be taken into consideration, as that approach had sensitivity of 48.1%.

Table 3. Ablation study for evaluating the components of Hybrid-MT-ESTAN.

Methods	Classification							Segmentation	
	Acc↑	Sens.↑	Spec.↑	F1↑	Auc↑	FNR↓	FPR↓	DSC↑	JI↑
MT-ESTAN [10]	78.6	83.7	72.6	0.83	78.2	16.3	27.4	78.2	69.3
Swin Trans.	77.4	72.6	**82.5**	0.74	77.6	27.4	17.5	-	-
MT-ESTAN + Swin Trans.	80.3	84.2	76.3	0.83	80.2	15.8	23.7	82.3	73.6
Ours	**82.8**	**86.4**	79.2	**0.86**	**82.8**	**13.6**	20.8	**84.1**	**75.7**

The preferred trade-off in medical image analysis typically is high sensitivity without significant degradation in specificity.

We evaluated the segmentation performance of Hybrid MT-ESTAN and compared the results to five multitask approaches, including SHA-MTL [8], EMT-Net [12], Chowdery [10], MT-ESTAN [3], and VGGA-ViT [16]. As shown in Table 2,the proposed Hybrid MT-ESTAN achieved the highest performance and increased DSC and JI by 5.9% and 6.4%, respectively compared to MT-ESTAN. Note that results of single-task models in Table 2 are not provided.

3.5 Effectiveness of the Anatomy-Aware Attention (AAA) Block

To verify the effectiveness of the Anatomy-Aware Attention (AAA) block, we conducted an ablation study that quantified the impact of the different components in Hybrid-MT-ESTAN on the classification and segmentation performance. Table 3 presents the values of the performance metrics for MT-ESTAN (pure CNN-based approach), Swin Transformer (pure Transformer network), a hybrid architecture of MT-ESTAN and Swin Transformer, and our proposed Hybrid-MT-ESTAN with AAA block. According to the results in Table 3, MT-ESTAN achieved better sensitivity and F1 score than Swin Transformer, with 83.7% and 83%, respectively. The hybrid architectures of MT-ESTAN with Swin Transformer improved the classification performance and has higher accuracy, sensitivity, F1 score, and AUC with 80.3%, 84.2%, 83%, and 80.2%, compared to MT-ESTAN and Swin Transformer individually. The proposed approach, Hybrid-MT-ESTAN with AAA block, further improved accuracy, sensitivity, F1 score, and AUC by 2.5%, 2.2%, 3%, and 2.6%, respectively, relative to the hybrid model without the AAA block.

To evaluate the segmentation performance, we compared the proposed approach with and without the AAA block and Swin Transformer. As shown in Table 3, MT-ESTAN combined with Swin Transformer improved DSC and JI by 4.1% and 4.3%, respectively compared to MT-ESTAN. Employing the proposed AAA block further improved DSC and JI by 1.8% and 2.1%, respectively.

4   Conclusion

In this paper, we introduced the Hybrid-MT-ESTAN, a multitask learning approach for BUS image analysis that alleviates the lack of global contextual information in the low-level layers of CNN-based approaches. Hybrid-MT-ESTAN concurrently performs BUS tumor classification and segmentation, with a hybrid architecture that employs CNN-based and Swin Transformer layers. The proposed approach exploits multi-scale local patterns and global long-range dependencies provided by MT-ESTA and AAA Transformer blocks for learning feature representations, resulting in improved generalization. Experimental validation demonstrated significant performance improvement by Hybrid-MT-ESTAN in comparison to current state-of-the-art models for BUS classification.

Acknowledgement. Research reported in this publication was supported by the National Institute Of General Medical Sciences of the National Institutes of Health under Award Number P20GM104420. The content is solely the responsibility of the authors and does not necessarily represent the official views of the National Institutes of Health.

References

1. American Cancer Society, Cancer Facts & Figures, https://www.cancer.org, (2022)
2. Zhuang, Z., Yang, Z., Raj, A., Noel, J., Wei, C.: Breast ultrasound tumor image classification using image decomposition and fusion based on adaptive multi-model spatial feature fusion. Comput. Methods Programs Biomed. **208**, 106221 (2021)
3. Shareef, B., et al.: A benchmark for breast ultrasound image classification. Available at SSRN (2023). https://ssrn.com/abstract=4339660
4. Iqbal, A., Sharif, M., BTS-ST: Swin transformer network for segmentation and classification of multimodality breast cancer images. Knowl. Based Syst. **262**, 110393 (2023)
5. Dosovitskiy, A., et al.: An image is worth 16x16 words: transformers for image recognition at scale, preprint arXiv:2010.11929 (2020)
6. Yap, M., et al.: Automated breast ultrasound lesions detection using convolutional neural networks. IEEE J. Biomed. Health Inf. **22**(4), 1218–1226 (2017)
7. Huang, G., Liu, Z., Mateen, L., Weinberger, K.: Densely connected convolutional networks. In: 2017 IEEE Conference on Computer Vision and Pattern Recognition (CVPR), pp. 4700–4708 (2017)
8. Zhang, G., Zhao, K., Hong, Y., Qiu, X., Zhang, K., Wei, B.: SHA-MTL: soft and hard attention multi-task learning for automated breast cancer ultrasound image segmentation and classification. Int. J. Comput. Assisted Radiol. Surg. **16**(10), 1719–1725 (2021). https://doi.org/10.1007/s11548-021-02445-7
9. Geertsma, T., Fujifilm.: Ultrasound cases, https://www.ultrasoundcases.info/ (2014)
10. Chowdary, J., Yogarajah, P., Chaurasia, P., Guruviah, V.: A multi-task learning framework for automated segmentation and classification of breast tumors from ultrasound images. Ultrason. Imaging **44**(1), 3–12 (2022)
11. Vakanski, A., Xian, M.: Evaluation of complexity measures for deep learning generalization in medical image analysis. In: 2021 IEEE 31st Int. Workshop on MLSP, pp. 1–6 (2021)
12. Shi, J., Vakanski, A., Xian, M., Ding, J., Ning, C.: EMT-NET: efficient multi-task network for computer-aided diagnosis of breast cancer. In: 2022 IEEE 19th International Symposium on Biomedical Imaging (ISBI), pp. 1–5 (2022)

13. Gheflati, B., Rivaz, H.: Vision transformers for classification of breast ultrasound images, In: 2022 44th Annual International Conference of the IEEE Engineering in Medicine & Biology Society (EMBC), pp. 480–483 (2022)
14. Hassanien, A., Singh, K., Puig, D., Abdel-Nasser, M.: Predicting breast tumor malignancy using deep ConvNeXt radiomics and quality-based score pooling in ultrasound sequences. Diagnostics **12**(5), 1053 (2022)
15. Mo, Y., et al.: Hover-trans: anatomy-aware hover-transformer for ROI-Free breast cancer diagnosis in ultrasound images IEEE Trans. Med. Imaging **42**, 1696–1706 (2023)
16. Qu, X., et al.: A VGG attention vision transformer network for benign and malignant classification of breast ultrasound images. Med. Phys. **49**(9), 5787–5798 (2022)
17. Shareef, B., Vakanski, A., Freer, P., Min Xian: ESTAN: enhanced small tumor-aware network for breast ultrasound image segmentation. Healthcare **10**(11), 2262 (2022)
18. Liu, Z., et al.: Swin transformer: hierarchical vision transformer using shifted windows. In: Proceedings of the IEEE/CVF International Conference on Computer Vision, pp. 10012–10022 (2021)
19. Howard, A., Zhu, M., Chen, B., Kalenichenko, D., Wang, W., Weyand, T.: MobileNets: efficient convolutional neural networks for mobile vision applications. preprint arXiv:1704.04861 (2017)
20. Zhang, Y., et al.: BUSIS: a benchmark for breast ultrasound image segmentation. Healthcare **10**(4), 729 (2022)
21. Yap, M., et al.: Breast ultrasound lesions recognition: end-to-end deep learning approaches. J. Med. Imaging **6**(1), 011007. SPIE (2018)
22. Shareef, B., Xian, M., Vakanski, A.: Stan: small tumor-aware network for breast ultrasound image segmentation. In: 2020 IEEE 17th International Symposium on Biomedical Imaging (ISBI), pp. 1–5 (2020)
23. Ayana, G., Choe, S.: BUViTNet: breast ultrasound detection via vision transformers, Diagnostics **12**(11), 2654 (2022). https://doi.org/10.3390/diagnostics12112654
24. Tang, S., et al.: Transformer-based multi-task learning for classification and segmentation of gastrointestinal tract endoscopic images. Comput. Bio. Med. **157**, 106723 (2023)
25. Sebastian, R.: An overview of multi-task learning in deep neural networks. arXiv preprint arXiv:1706.05098 (2017)

Shape-Aware 3D Small Vessel Segmentation with Local Contrast Guided Attention

Zhiwei Deng[1,2], Songnan Xu[1,2], Jianwei Zhang[1,2], Jiong Zhang[3], Danny J. Wang[1], Lirong Yan[4], and Yonggang Shi[1,2(✉)]

[1] Stevens Neuroimaging and Informatics Institute, Keck School of Medicine, University of Southern California (USC), Los Angeles, CA 90033, USA
[2] Ming Hsieh Department of Electrical and Computer Engineering, Viterbi School of Engineering, University of Southern California (USC), Los Angeles, CA 90089, USA
yonggans@usc.edu
[3] Cixi Institute of Biomedical Engineering, Ningbo Institute of Materials Technology and Engineering, Chinese Academy of Sciences, Ningbo 315300, China
[4] Department of Radiology, Feinberg School of Medicine, Northwestern University, Chicago, IL, USA

Abstract. The automated segmentation and analysis of small vessels from *in vivo* imaging data is an important task for many clinical applications. While current filtering and learning methods have achieved good performance on the segmentation of large vessels, they are sub-optimal for small vessel detection due to their apparent geometric irregularity and weak contrast given the relatively limited resolution of existing imaging techniques. In addition, for supervised learning approaches, the acquisition of accurate pixel-wise annotations in these small vascular regions heavily relies on skilled experts. In this work, we propose a novel self-supervised network to tackle these challenges and improve the detection of small vessels from 3D imaging data. First, our network maximizes a novel shape-aware flux-based measure to enhance the estimation of small vasculature with non-circular and irregular appearances. Then, we develop novel local contrast guided attention(LCA) and enhancement(LCE) modules to boost the vesselness responses of vascular regions of low contrast. In our experiments, we compare with four filtering-based methods and a state-of-the-art self-supervised deep learning method in multiple 3D datasets to demonstrate that our method achieves significant improvement in all datasets. Further analysis and ablation studies have also been performed to assess the contributions of various modules to the improved performance in 3D small vessel segmentation. Our code is available at https://github.com/dengchihwei/LCNetVesselSeg.

Keywords: Small vessel · Shape-aware flux · Local contrast

This work is supported by the National Institute of Health (NIH) under grants R01EB022744, RF1AG077578, RF1AG056573, RF1AG064584, RF1AG072490, R21AG064776, P41EB015922, U19AG078109.

1   Introduction

The automated detection and analysis of small vessels from non-invasive imaging data is critical for many clinical studies such as the research on cerebral small vessel disease(CSVD) [3], which is the most common vascular cause of dementia in Alzheimer's disease and related dementia (ADRD) [5]. According to [12], a brain vessel with a diameter less than 0.5mm is considered as a small vessel by most pathologists. Fortunately, with the recent advances in MRA at 7-Tesla [7] and black-blood MRI at 3- and 7-Tesla [10], it is now possible to detect the small cerebral vessels directly. Although the segmentation of large vascular structures has been well studied for many years [4,6,8,11,15], accurate and reliable small vessel segmentation remains a challenging task. In contrast to large vessels, small vessels usually exhibit the following two main characteristics (Fig. 1). (1) Its cross-section only occupies a few image voxels due to the limited resolution of imaging techniques such as the magnetic resonance angiography(MRA). This makes the regular assumption of tube-like shape for vessels often does not hold. (2) Small vessels often have weak intensities and low contrasts, which would be easily affected by noise or surrounding backgrounds. These characteristics are generally not well modeled by existing methods for vessel detection [4,6,8,11,15].

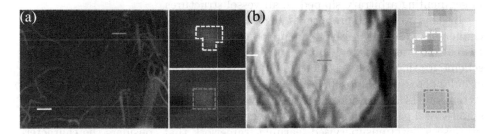

Fig. 1. Challenges in small vessel detection. (a) A high-resolution MRA image patch ($0.2 \times 0.2 \times 0.4$ mm) acquired by a 7T Simens Terra scanner. **(b)** A black-blood MRI image patch ($0.5 \times 0.5 \times 0.5$ mm) acquired by a Siemens 3T Prisma scanner. In each case, a maximal(minimal) intensity projection (left) and two cross sections of small vessels (right) were plotted. In (a) and (b), each cross-section corresponds to the line of the same color overlaid on the left panel (green: irregular appearances of small vessel cross-sections , red: low contrasts of small vessel cross-sections).

Traditional vesselness filters typically characterize blood vessels based on hand-crafted features [4,8,15]. The inherent assumption about the regularity of tube-like vessel geometry, however, makes them sub-optimal for small vessel segmentation. In addition, most Hessian-based filters rely on complicated pre-processing including smoothing for the calculation of second derivatives, which can further weaken or even eliminate the contrasts of small vascular structures. While deep learning methods have been successfully applied in various segmentation tasks, large-scale annotated labels are hard to obtain to train supervised networks for the segmentation of highly variable small vessels. To overcome this

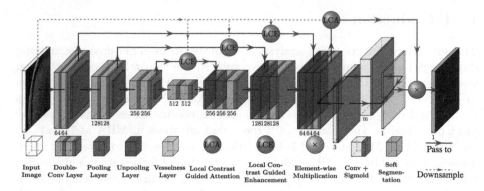

Fig. 2. Network overview. Our model jointly estimate the vessel directions and shapes using a vector field P representing the principal direction of tubular structures and a collection of scalar fields R denoting the radius of the vessel at different directions.

challenge, a self-supervised deep learning approach was recently proposed that combines geometric models and deep neural networks to learn vessel flow directions [6], but it also assumes a circular tube-like vessel shape and is thus limited in segmenting arbitrary-shaped small vessel structures.

In this work, we propose a novel self-supervised network that focus on challenges in small vessel segmentation with the following contributions: (1) Instead of assuming ideal tube-shaped vessels like [6,8], we propose an adaptive scheme for shape-aware estimation of oriented fluxes to model the irregular (non-circular) cross-section profiles of small vessels. (2) We propose the Local Contrast Attention (LCA) module based on a novel local contrast measure to enhance the small vessel pattern and suppress the background clutter simultaneously. (3) We propose a novel unsupervised learning framework that considers the characteristics of small vessel structures in relatively limited resolution for the first time. Comprehensive experiments show promising improvements in 3D datasets of multiple modalities compared with previous unsupervised approaches.

2 Method

Our proposed framework for small vessel detection is shown in Fig. 2, a U-shaped network augmented with multiple novel LCA and LCE modules to learn a general representation of irregular-shaped small vessels by optimizing a self-supervised loss of shape-aware flux. Formally, let $\Omega \subseteq \mathbb{R}^3$ denote the image domain. To characterize the irregular shape of the small vessel at each point $x \in \Omega$, we estimate a principal vessel direction $\vec{\rho_x}$ and a set of radii values $R(x) = \{r_i(x)|i = 1, 2, ..., m\}$ that represent the vessel radius along m sampling directions on the unit sphere S^2. Based on the estimated $R(x)$ and $\vec{\rho_x}$, we can compute the vesselness score f_{vs} at x, which represents the likelihood of x being a vascular structure. The proposed network is trained to maximize the vessel

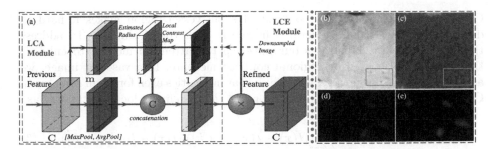

Fig. 3. Local contrast guided attention. (a) shows the proposed LCA and LCE module framework; **(b)** is an axial slice of a black-blood MRI image that contains multiple low-contrast small vessels (pink box); **(c)** shows the learnt attention map for (b); **(d-e)** show the vesselness maps generated by our model without and with the proposed modules for small vascular regions, respectively.

score across Ω. Next we describe in detail the shape-aware flux and the local contrast guided attention modules to enhance the detection of small vessels.

Shape-Aware Flux. Given an input image I, our proposed network will generate three outputs: P, $R = \{r_1, r_2, ..., r_m\}$ and the reconstructed image $\hat{I}$. We denote P as a vector field that represents the principal direction for every point $x \in \Omega$ and R as a collection of scalar fields, where $r_i \in R$ is a scalar field over Ω that estimates the vessel radius along $d_i \in S^2 (i = 1, 2, \cdots, m)$. A projected flux response along a direction $\overrightarrow{\rho}$, which generalizes conventional flux measures for circular shaped tubes [8], can be defined at x as follows:

$$f(x, R(x), \overrightarrow{\rho}) = -\frac{1}{m} \sum_i ((v(x + \overrightarrow{h_i}) \cdot \overrightarrow{\rho})\overrightarrow{\rho}) \cdot \overrightarrow{d_i} \tag{1}$$

after discretization and normalization, where $v(\cdot)$ is the image gradient, $\overrightarrow{h_i} = r_i(x)\overrightarrow{d_i}$ and $r_i(x) \in R(x)$. The vesselness score of x can then be computed as

$$f_{vs}(x, R(x), \overrightarrow{\rho_1}) = -f(x, R(x), \overrightarrow{\rho_2}) - f(x, R(x), \overrightarrow{\rho_3}) \tag{2}$$

where $\overrightarrow{\rho_1} = P(x)$ is the estimated principal direction at x and $\overrightarrow{\rho_2}, \overrightarrow{\rho_3}$ are two orthogonal vectors in the cross-sectional plane of the vascular structure. In contrast to conventional flux measures for vessel segmentation, where only isotropic radius value is estimated, radius in different sampling directions d_i are estimated adaptively in this work to fit the irregular-shaped vascular boundaries as f_{vs} is maximized only if the sampling vectors $\overrightarrow{h_i}$ fit the vessel edges. Since the radius values can be different from each other, our proposed network is designed to handle the small vessels with irregular and non-circular cross-sections.

Local Contrast Guided Attention. As shown in Fig. 1, geometric features of large vessels are well-presented due to their high signal-to-noise ratio(SNR).

On the other hand, the low contrast of small vessel region makes it hard to distinguish the vascular structures from the background clutters. To address this issue, Fig. 3(a) illustrates our novel spatial attention module to enhance the small vessels based on regional contrast measure. In a vascular image, we assume the pixels inside a vessel have similar intensities even for small structures. Consider the estimated radius $R(x)$ for $x \in \Omega$, we define a local contrast D as

$$D(I, x, R(x), s) = \frac{1}{2} \sum_{i=1}^{m} |I(x) - I(x + s \times \overrightarrow{h_i})||I(x) - I(x + s \times \overrightarrow{h_j})| \quad (3)$$

where $\overrightarrow{h_j} = r_j(x)\overrightarrow{d_j}$ and $\overrightarrow{d_j} = -\overrightarrow{d_i}$. With $s = 1$, D measures the intensity differences between x and $x + \overrightarrow{h_i}$ which locates on vascular boundaries for every sampling direction d_i. We can contract or expand the vascular regions along its edges with different s. To measure the contrast inside and outside the vascular regions, we design two measures D_{in} and D_{out} as follows:

$$D_{in}(I, x) = \int_0^1 D(I, x, R(x), s)ds, \quad D_{out}(I, x) = \int_1^2 D(I, x, R(x), s)ds \quad (4)$$

With $s \in [0, 1]$, D_{in} computes the intensity differences inside vessels. D_{out} computes outside intensity differences similarly with $s \in [1, 2]$. As for the relations between D_{in} and D_{out}, we consider 3 situations for different $x \in \Omega$. (1) For x inside vascular structures, $D_{in} << D_{out}$ should stand since $|I(x) - I(x + s \times \overrightarrow{h_i})| \approx 0$ when $s < 1$; (2) D_{in} and D_{out} would give similar measures if x is in the backgrounds, where we consider the local regional intensities are similar. (3) Since equation (3) is evaluated on two opposite directions, D_{in} and D_{out} will both give low measures if x is an one-sided edge. Based on these three scenarios, we design a novel local contrast measure as

$$D_{LC}(I, x) = Sigmoid(\frac{D_{out}(I, x)}{D_{in}(I, x) + \epsilon} - 1) \quad (5)$$

where ϵ is a small constant to prevent the numerical explosion. Since the ratio of D_{out} and D_{in} gives large value, $D_{LC} \approx 1$ for x inside a vascular structure, which means the D_{LC} measure of small vessels is on a similar scale to large vessels. For the latter two situations, $D_{LC} \approx \sigma(0) = 0.5$ will stand since $D_{in} \approx D_{out}$. Therefore, the proposed D_{LC} measure can enhance the small vascular structures and suppresses backgrounds and edges simultaneously.

Guided by the local contrast map of images, we propose to incorporate a novel spatial attention module in our DL network. Following CBAM [16], we use two pooling layers to abstract the previous features as two feature maps. A local contrast map is computed based on a coarsely estimated $\hat{R}$, which is generated by previous features. Then, all three maps are sent to a convolutional block to compute the final spatial attention map. By simply extending the LCA module, LCE module scales the previous features with the attention map to refine it. To include low-level contrast information, we insert our LCA and LCE module in the skip connections between encoders and decoders as shown in Fig. 2. As shown in Fig. 3(b-d), with the local contrast attention modules, our model can enhance the vesselness measure of the small vessels with low contrasts.

Self-supervised Losses. Based on (2), we propose our flux-based loss L_{flux} as the negative average vesselness score over the whole image spatial space. This cost function takes advantage of the clear edge separation of flux-based filter and robustness to irregular-shaped vessels due to our adaptive shape-aware flux computation. To ensure the vessel structure's continuity, we adopt the path continuity loss L_{path} from [6]. Let us denote $P(x)$ as the vascular principal direction at x, and $P(x+t)$ as the principal direction at the location $x+t$, which we obtain by *walking* for t from x along $P(x)$. L_{path} is designed to maximize the inner product of these two vectors to encourages P to have a consistent and smooth direction along the vessels. Formally, L_{flux} and L_{path} are computed as

$$L_{flux}(P, R) = -\frac{1}{N} \sum_{x \in \Omega} f_{vs}(x, R(x), \overrightarrow{\rho_x})$$

$$L_{path}(P, R) = -\int_{\Omega} \int_0^{2\overline{R(x)}} (P(x) \cdot P(x + t)) dt dx \qquad (6)$$

where N is the total number of voxels and $\overline{R(x)}$ is the average radius of $R(x)$ so that the length of the *walking* path is relative to the vessel size at x. Mean square error loss is applied between the reconstructed image $\hat{I}$ and original image I to make sure the network learns the semantic meaningful features based on the previous success of reconstruction-based self-supervised learning [9]. So, the overall objective of our network is expressed as

$$L = \lambda_1 L_{flux} + \lambda_2 L_{path} + \lambda_3 MSE(\hat{I}, I) \qquad (7)$$

where λ_1, λ_2 and λ_3 correspond to the coefficients of each objective.

3   Experimental Results

Datasets. We used two public datasets and two in-house datasets in our experiments for 3D vessel segmentation. The VESSEL12 [14] contains 23 CT lung images and 3 of them are provided with sparsely annotated vessel and non-vessel locations along 3 axial slices, which we used as a test set. The TubeTk [2] consists of 109 3D brain MRA images. We used the 42 images with ground truths as the test set and the rest of the dataset for training. To better demonstrate the small vessel detection ability of our method, we collected a new brain MRA dataset on a 7T Simens Terra scanner from 31 subjects with voxel size of $0.2 \times 0.2 \times 0.4$mm. For this dataset, we sparsely annotated 2200 small vessels(within 3×3 grids) locations on axial slices of 7 images according to the definition of the small vessel in [12] (Fig. 4(i-j)). The remaining 24 images constitute the training set. In addition, a black-blood MRI dataset was collected using a Siemens 3T

Table 1. Vessel Segmentation Performance Comparison on Different Datasets.

Method	VESSEL12				7T MRA			
	Acc	Sens	Spec	AUC	Acc	Sens	Spec	AUC
Sato	0.8299	0.7580	0.8636	0.9102	0.8716	0.9705	0.7347	0.8933
Meijering	0.9184	0.8932	0.9301	0.9720	0.7060	0.9377	0.3854	0.6547
Frangi	0.9672	0.9584	0.9669	0.9738	0.8130	0.7008	0.9683	0.8104
OOF	0.9331	0.9324	0.9334	0.9659	0.8416	0.8074	0.8889	0.8623
Flow-Based	0.9553	**0.9856**	0.9213	0.9871	0.9247	0.9432	0.9261	0.9485
Ours	**0.9761**	0.9809	**0.9793**	**0.9937**	**0.9824**	**0.9787**	**0.9875**	**0.9972**

Method	TubeTK				Black-blood LSA			
	Dice	Sens	Spec	AUC	Dice	Sens	Spec	AUC
Sato	0.3166	0.3933	0.9964	0.9262	0.0596	**0.7111**	0.92251	0.9432
Meijering	0.1573	0.1604	0.9970	0.9023	0.1669	0.6512	0.9797	0.9459
Frangi	0.3569	0.3641	0.9977	0.9319	0.2667	0.6221	0.9905	0.9281
OOF	0.3877	0.3874	0.9980	0.9426	0.3368	0.6454	0.9886	0.9584
Flow-Based	0.4003	0.4211	0.9978	0.9693	0.4608	0.6673	0.9927	**0.9732**
Ours	**0.5487**	**0.5061**	**0.9987**	**0.9878**	**0.5121**	0.6979	**0.9983**	0.9624

Prisma scanner with voxel size of 0.5 × 0.5 × 0.5mm and separated by left and right hemispheres for a total of 56 image volumes. Dense manual segmentation of the lenticulostriate arteries(LSAs) was carefully performed by two experts. The dataset was divided into a training set with 21 subjects (42 volumes) and a test set with 7 subjects (14 volumes).

Implementation Details. We performed the vessel segmentation tasks for each dataset to evaluate our model's performance. For comparison, we selected 4 conventional filters, including the Frangi, Sato, Meijering and OOF filter and a flow-based DL method [6]. For fair comparison, the same pre-processing procedures were applied for all methods using FreeSurfer 6.0 [13] and the SimpleITK toolbox [1]. In addition, for the black-blood LSA dataset, the image intensities were inversed to make sure the vascular voxels are brighter than backgrounds.

To tune the network's hyper-parameters, we used 15% of the training data as the validation set. Finally, we set $m = 128$, $\lambda_1 = 5$ and $\lambda_2 = \lambda_3 = 1$ for our model. For both DL models, patch size is set to 64 × 64 × 64 and the networks are trained for 100 epochs. All experiments were conducted with Pytorch on one NVIDIA A5000 GPU with the Adam optimizer and a learning rate of 0.001.

For all methods, the output is a vessel enhanced image. The enhanced image is binarized through a hard threshold. We thus found the best threshold by optimizing the metrics based on the validation sets and then applied the final threshold to test sets. We reported five metrics in Table 1, namely, the area-under-curve(AUC), accuracy, sensitivity, specificity and dice score. For VES-

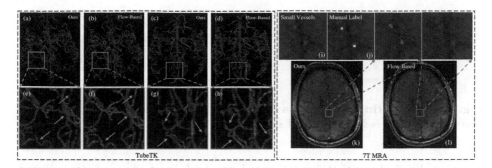

Fig. 4. Qualitative comparison. (a–h) is a comparison for TubeTK dataset. Green: true positives; red: false negatives. **(i–j)** is an example of our manual label for a 7T MRA image patch; **(k–l)** is small vessel detection comparison for the patch in (i–j).

SEL12 and 7T MRA dataset, we treated the segmentation tasks as classification problems and dropped the dice score metric since they do not have dense labels.

Results and Discussion. We can clearly observe from Table 1 that our model outperforms all other methods by a significant margin for all datasets. Since the flow-based DL outperformed the conventional filters, we will use it for visual comparisons with our method. From Fig. 4(a–d), we can observe that our model generates more accurate segmentation with more true positives and less false negatives as compared to the flow-based DL method [6]. These results show that our model can better detect general vessel structures including large vessels.

Small Vessel Segmentation. From Fig. 4(e-h), we can see the main improvement of our model over the flow-based DL method occurs often at areas of small vessels, where our results have much lower false negatives for small vessels missed by the other. The results on the 7T MRA data in Table 1 further demonstrate that our model achieved better performance on the small vessel detection task. Figure 4(k–l) provides a visual comparison of the performance of small vessel detection by our method and the flow-based DL, which shows that our model can more successfully segment the vessels with very small radius and low contrasts. In

Table 2. Ablation Analysis Results with the 7T high-resolution MRA Dataset.

Methods	Acc	Sens	Spec	AUC	# of Params
Flow-based [6]	0.9247	0.9432	0.9261	0.9485	32.63M
Ours (CS only)	0.9344	0.9478	0.9329	0.9554	32.63M
Ours (AS only)	0.9578	0.9669	0.9333	0.9662	32.64M
Ours (LCA only)	0.9747	0.9699	0.9732	0.9865	32.65M
Ours (AS + LCA)	0.9824	0.9787	0.9875	0.9972	32.66M

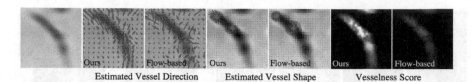

Fig. 5. Vessel Estimation Comparison. Comparison between our model and flow-based model on vessel direction and shape estimations.

addition, Fig. 5 shows that our estimated principal directions are better aligned with the vessel directions over the flow-based DL. Furthermore, our estimated radius can better capture the vascular shapes by fitting the asymmetric vessel boundaries. Thus, comparing with the flow-based DL, our model can produce sharper and clearer vesselness maps(Fig. 5).

Ablation Study. To investigate the contributions of each proposed module in our approach, we performed ablation studies by training the network with each proposed module using the 7T MRA dataset. The results are shown in Table 2, where we used the flow-based DL method as the baseline for comparison. Note that *CS* stands for circular sampling and *AS* stands for the proposed shape-aware adaptive sampling. As can be seen, with all the proposed modules, we increase the AUC metric of the baseline by 4.87% on the test set, showing significant improvements with only minimal extra computational cost. Furthermore, we observed that all the proposed modules improve both the sensitivity and specificity since complicated small vessels can be better modeled by these components.

4   Conclusions

In this paper, we proposed a self-supervised network for the detection of small vessels from 3D imaging data. Our method is designed to address existing challenges in small vessel detection arising from their irregular appearance in relatively limited resolution and low-contrast conditions. In comparison to previous methods, we demonstrated that our method is able to achieve superior performance on small vessel detection. For future work, we will also apply it to various clinical datasets to examine its power for CSVD detection in brain images.

References

1. Beare, R., Lowekamp, B., Yaniv, Z.: Image segmentation, registration and characterization in r with simpleITK. J. Stat. Softw. **86** (2018)
2. Bullitt, E., et al.: Vessel tortuosity and brain tumor malignancy: a blinded study1. Acad. Radiol. **12**(10), 1232–1240 (2005)

3. Cuadrado-Godia, E., et al.: Cerebral small vessel disease: a review focusing on pathophysiology, biomarkers, and machine learning strategies. J. Stroke **20**(3), 302 (2018)
4. Frangi, A.F., Niessen, W.J., Vincken, K.L., Viergever, M.A.: Multiscale vessel enhancement filtering. In: Wells, W.M., Colchester, A., Delp, S. (eds.) MICCAI 1998. LNCS, vol. 1496, pp. 130–137. Springer, Heidelberg (1998). https://doi.org/10.1007/BFb0056195
5. Gorelick, P.B., et al.: Vascular contributions to cognitive impairment and dementia. Stroke **42**(9), 2672–2713 (2011)
6. Jena, R., Singla, S., Batmanghelich, K.: Self-supervised vessel enhancement using flow-based consistencies. In: de Bruijne, M., et al. (eds.) Medical Image Computing and Computer Assisted Intervention – MICCAI 2021: 24th International Conference, Strasbourg, France, September 27–October 1, 2021, Proceedings, Part II, pp. 242–251. Springer, Cham (2021). https://doi.org/10.1007/978-3-030-87196-3_23
7. Kraff, O., Quick, H.H.: 7t: physics, safety, and potential clinical applications. J. Magn. Reson. Imaging **46**(6), 1573–1589 (2017)
8. Law, M.W.K., Chung, A.C.S.: Three dimensional curvilinear structure detection using optimally oriented flux. In: Forsyth, D., Torr, P., Zisserman, A. (eds.) Computer Vision – ECCV 2008: 10th European Conference on Computer Vision, Marseille, France, October 12-18, 2008, Proceedings, Part IV, pp. 368–382. Springer, Berlin, Heidelberg (2008). https://doi.org/10.1007/978-3-540-88693-8_27
9. Li, X., et al.: Self-supervised single-view 3D reconstruction via semantic consistency. In: Vedaldi, A., Bischof, H., Brox, T., Frahm, J.-M. (eds.) Computer Vision – ECCV 2020: 16th European Conference, Glasgow, UK, August 23–28, 2020, Proceedings, Part XIV, pp. 677–693. Springer, Cham (2020). https://doi.org/10.1007/978-3-030-58568-6_40
10. Ma, S.J., et al.: Characterization of lenticulostriate arteries with high resolution black-blood t1-weighted turbo spin echo with variable flip angles at 3 and 7 tesla. Neuroimage **199**, 184–193 (2019)
11. Meijering, E., Jacob, M., Sarria, J.C., Steiner, P., Hirling, H., Unser, E.M.: Design and validation of a tool for neurite tracing and analysis in fluorescence microscopy images. Cytometry Part A: J. Int. Soc. Analy. Cytol. **58**(2), 167–176 (2004)
12. Postmortem examination of vascular lesions in cognitive impairment: a survey among neuropathological services. Stroke **37**(4), 1005–1009 (2006)
13. Reuter, M., Rosas, H.D., Fischl, B.: Highly accurate inverse consistent registration: a robust approach. NeuroImage **53**(4), 1181–1196 (2010)
14. Rudyanto, R.D., et al.: Comparing algorithms for automated vessel segmentation in computed tomography scans of the lung: the vessel12 study. Medical image analysis **18**(7), 1217–1232 (2014)
15. Sato, Y., et al.: Three-dimensional multi-scale line filter for segmentation and visualization of curvilinear structures in medical images. Med. Image Anal. **2**(2), 143–168 (1998)
16. Woo, S., Park, J., Lee, J.Y., Kweon, I.S.: CBAM: convolutional block attention module. In: Proceedings of the European Conference on Computer Vision (ECCV), pp. 3–19 (2018)

HartleyMHA: Self-attention in Frequency Domain for Resolution-Robust and Parameter-Efficient 3D Image Segmentation

Ken C. L. Wong[✉], Hongzhi Wang, and Tanveer Syeda-Mahmood

IBM Research – Almaden Research Center, San Jose, CA, USA
{clwong,hongzhiw,stf}@us.ibm.com

Abstract. With the introduction of Transformers, different attention-based models have been proposed for image segmentation with promising results. Although self-attention allows capturing of long-range dependencies, it suffers from a quadratic complexity in the image size especially in 3D. To avoid the out-of-memory error during training, input size reduction is usually required for 3D segmentation, but the accuracy can be suboptimal when the trained models are applied on the original image size. To address this limitation, inspired by the Fourier neural operator (FNO), we introduce the HartleyMHA model which is robust to training image resolution with efficient self-attention. FNO is a deep learning framework for learning mappings between functions in partial differential equations, which has the appealing properties of zero-shot super-resolution and global receptive field. We modify the FNO by using the Hartley transform with shared parameters to reduce the model size by orders of magnitude, and this allows us to further apply self-attention in the frequency domain for more expressive high-order feature combination with improved efficiency. When tested on the BraTS'19 dataset, it achieved superior robustness to training image resolution than other tested models with less than 1% of their model parameters.

Keywords: Image segmentation · Transformer · Fourier neural operator · Hartley transform · Resolution-robust

1 Introduction

Convolutional neural networks (CNNs) have been widely used for medical image segmentation because of their speed and accuracy [11,18]. Nevertheless, given the local receptive fields of convolutional layers, long-range spatial correlations are mainly captured through consecutive convolutions and pooling. For computationally demanding 3D segmentation, the receptive fields and abstract levels can be more limited than in 2D as fewer layers can be used. To balance between computational complexity and network capability, input size reductions by image downsampling and patch-wise training are common approaches. However, CNNs trained with downsampled images can be suboptimal when applied on the original resolution, and the receptive field of patch-wise training can be largely reduced depending on the patch size.

With the introduction of Transformers [24] and their vision alternatives [6,19], the self-attention mechanism for long-range dependencies has been adopted to medical image segmentation with promising results [5,7,9,27]. These approaches form a sequence of samples by either using the pixel values of low-resolution features [7,27] or by dividing an image into smaller patches [5,9], and the multi-head attention is used to learn the dependencies among samples. Although these self-attention approaches allow capturing of long-range dependencies, as the computational requirements are proportional to sequence lengths and patch sizes which are proportional to image sizes, size-reduction approaches are needed for large images especially in 3D.

As image size reduction is usually required for large images, it is desirable to have a model that is robust to training image resolution so that the trained model can be applied to higher-resolution images with decent accuracy. Furthermore, as self-attention of Transformers allows better expressiveness through high-order channel and sample mixing [15,23], incorporating self-attention in an efficient way can be beneficial. To gain these advantages, here we propose the HartleyMHA model which is a resolution-robust and parameter-efficient network architecture with frequency-domain self-attention for 3D image segmentation. This model is based on the Fourier neural operator (FNO) [17], which is a deep learning model that learns mappings between functions in partial differential equations (PDEs) and has the appealing properties of zero-shot super-resolution and global receptive field. Our contributions include:

1. To utilize the FNO for computationally expensive 3D segmentation, we modify it by using the Hartley transform with shared model parameters in the frequency domain. Residual connections [10] and deep supervision [14] are also introduced. These reduce the number of model parameters by orders of magnitude and improve accuracy. We call it the HNOSeg model.
2. As only low-frequency components are required for decent segmentation results, multi-head self-attention can be efficiently applied in the frequency domain. This allows high-order combination of features to improve the expressiveness of the model. We call it the HartleyMHA model.
3. We compare our proposed models with other models on different training image resolutions to study their robustness. This provides useful insights that are usually unavailable in other studies.

Experimental results on the BraTS'19 dataset [2,3,22] show that the proposed models have superior robustness to training image resolution than other tested models with less than 1% of their model parameters.

2   Methodology

2.1   Fourier Neural Operator

FNO is a deep learning model for learning mappings between functions in PDEs without the PDEs provided [17]. By formulating the solution in the continuous space based on the Green's function [16], FNO can learn a single set of model parameters for multiple resolutions. For computationally expensive 3D segmentation, such zero-shot super-resolution capability is advantageous as a model trained with lower-resolution images can be applied on higher-resolution images with decent accuracy. The neural operator is formulated as iterative updates:

$$u_{t+1}(x) := \sigma \left(W u_t(x) + (\mathcal{K} u_t)(x) \right)$$
$$\text{with} \quad (\mathcal{K} u_t)(x) := \int_D \kappa(x - y) u_t(y) \, dy, \quad \forall x \in D \tag{1}$$

where $u_t(x) \in \mathbb{R}^{d_{u_t}}$ is a function of x. $W \in \mathbb{R}^{d_{u_{t+1}} \times d_{u_t}}$ is a learnable linear transformation and σ accounts for normalization and activation. In our work, $D \subset \mathbb{R}^3$ represents the 3D imaging space, and $u_t(x)$ are the outputs of hidden layers with d_{u_t} channels. $\mathcal{K}$ is the kernel integral operator with $\kappa \in \mathbb{R}^{d_{u_{t+1}} \times d_{u_t}}$ a learnable kernel function. As $(\mathcal{K} u_t)$ is a convolution, it can be efficiently solved by the convolution theorem which states that the Fourier transform $(\mathcal{F})$ of a convolution of two functions is the pointwise product of their Fourier transforms:

$$(\mathcal{K} u_t)(x) = \mathcal{F}^{-1} \left(\mathcal{F}(\kappa) \mathcal{F}(u_t) \right)(x) = \mathcal{F}^{-1} \left(R U_t \right)(x), \quad \forall x \in D \tag{2}$$

$R(k) = (\mathcal{F}\kappa)(k) \in \mathbb{C}^{d_{u_{t+1}} \times d_{u_t}}$ is a learnable function in the frequency domain and $U_t(k) = (\mathcal{F} u_t)(k) \in \mathbb{C}^{d_{u_t}}$. Therefore, each pointwise product at k is realized as a matrix multiplication. When the fast Fourier transform is used in implementation, $k \in \mathbb{N}^3$ are non-negative integer coordinates, and each k has a learnable $R(k)$. As mainly low-frequency components are required for image segmentation, only $k_i \leq k_{\max,i}$ corresponding to the lower frequencies in each dimension i are used to reduce model parameters and computation time.

2.2 Hartley Neural Operator (HNO)

As the FNO requires complex number operations in the frequency domain, the computational requirements such as memory and floating point operations are higher than with real numbers. Therefore, we use the Hartley transform instead, which is an integral transform alternative to the Fourier transform [8]. The Hartley transform $(\mathcal{H})$ converts real-valued functions to real-valued functions, which is related to the Fourier transform as $(\mathcal{H}f) = \text{Real}(\mathcal{F}f) - \text{Imag}(\mathcal{F}f)$. The convolution theorem of discrete Hartley transform is more complicated [4], and the kernel integration in (1) becomes:

$$\mathcal{H}(\mathcal{K} u_t)(k) = \frac{\hat{R}(k) \left(\hat{U}_t(k) + \hat{U}_t(N - k) \right) + \hat{R}(N - k) \left(\hat{U}_t(k) - \hat{U}_t(N - k) \right)}{2}$$
$$\tag{3}$$

with $\hat{R}(k) = (\mathcal{H}\kappa)(k) \in \mathbb{R}^{d_{u_{t+1}} \times d_{u_t}}$ and $\hat{U}_t(k) = (\mathcal{H} u_t)(k) \in \mathbb{R}^{d_{u_t}}$. $N \in \mathbb{N}^3$ is the size of the frequency domain. $\hat{R}$ and $\hat{U}$ are N-periodic in each dimension[1].

---

[1] In Python, this means $\hat{U}[N_x, :, :] = \hat{U}[0, :, :]$, etc.

Similar to using (2), the models built using (3) have tens of million parameters even with small $k_{\max}$ (e.g., (14, 14, 10)). Therefore, instead of using a different $\hat{R}(k)$ at each k, we use the same (shared) $\hat{R}$ for all k and (3) becomes:

$$\mathcal{H}\left(\mathcal{K}u_t\right)(k) = \hat{R}\hat{U}_t(k) \qquad (4)$$

This is equivalent to applying a convolution layer with the kernel size of one in the frequency domain. We find that using (4) simplifies the computation and largely reduces the number of parameters without affecting the accuracy.

2.3 Hartley Multi-head Attention (MHA)

As real instead of complex numbers are used in (4), multi-head attention in [24] can be applied in the frequency domain for high-order feature combination. As $k_{\max}$ can be much smaller than the image size for image segmentation, the sequence length (number of voxels) can be largely reduced. With (4), the query, key, and value matrices (Q, K, V) of self-attention can be computed as:

$$Q = \bar{U}_t \hat{R}_Q^{\mathrm{T}}, \ K = \bar{U}_t \hat{R}_K^{\mathrm{T}}, \ V = \bar{U}_t \hat{R}_V^{\mathrm{T}} \ \in \mathbb{R}^{N_f \times d_{u_{t+1}}} \qquad (5)$$

where $\bar{U}_t \in \mathbb{R}^{N_f \times d_{u_t}}$, with $N_f = 8k_{\max,x}k_{\max,y}k_{\max,z}$, is a 2D matrix formed by stacking $\hat{U}_t(k)^2$. Although N_f can be relatively small, the computation and memory requirements of computing QK^{T} can still be demanding. For example, $k_{\max} = (14, 14, 10)$ corresponds to an attention matrix with around 246M elements. To remedy this, for each Q, K, and V, we group the feature vectors with a patch size of $2 \times 2 \times 2$ voxels in the frequency domain and their matrix sizes become $\frac{N_f}{8} \times 8d_{u_{t+1}}$. This reduces the number of elements in QK^{T} by 64 times. The self-attention can then be computed as:

$$\mathrm{Attention}(Q, K, V) = \mathrm{SELU}\left(QK^{\mathrm{T}}/\sqrt{8d_{u_{t+1}}}\right) V \in \mathbb{R}^{\frac{N_f}{8} \times 8d_{u_{t+1}}} \qquad (6)$$

where SELU represents the scaled exponential linear unit [13]. Similar to [21], we find that using softmax in self-attention results in suboptimal segmentations, thus the SELU was chosen after testing with multiple activations. Furthermore, we find that position encoding is unnecessary. The result of (6) can be rearranged back to the original shape in the frequency domain so that the inverse Hartley transform can be applied. The multi-head attention can be used with (6).

2.4 Network Architectures – HNOSeg and HartleyMHA

Figure 1 shows the network architecture. We call it HNOSeg with the HNO blocks and HartleyMHA with the Hartley MHA blocks. Different from the FNO in [17], residual connections [10] and deep supervision [14] are used to improve the

---

[2] Note that $k_{\max} = (k_{\max,x}, k_{\max,y}, k_{\max,z})$ corresponds to a frequency domain of size $2k_{\max,x} \times 2k_{\max,y} \times 2k_{\max,z}$ to cover both positive and negative frequency terms.

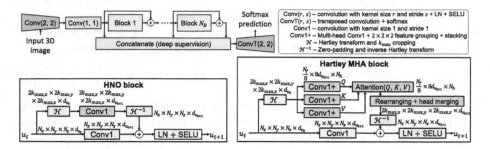

Fig. 1. Network architecture. The blocks are (1) with the kernel integral operator implemented by the Hartley transform (HNO block) or the Hartley multi-head attention (Hartley MHA block). $N_h = 4$ is the number of heads. We use $d_{u_{t+1}} = d_{u_t} = 12$, $k_{\max} = (14, 14, 10)$, and $N_B = 32$ with the HNO block and $N_B = 16$ with the Hartley MHA block. The red blocks are for learnable resampling.(Color figure online)

training stability, convergence, and accuracy. As the batch size is usually small for memory demanding 3D segmentation, layer normalization (LN) is used [1]. The SELU [13] is used as the activation function, and the softmax function is used to produce the final prediction scores. Similar to the Fourier transform, the Hartley transform provides a global receptive field as all voxels are used to compute the value at each k, thus pooling is not required. As using the original image resolution usually results in out-of-memory errors in 3D segmentation, downsampling the inputs and then upsampling the predictions may be required. Instead of using traditional image resampling methods, we use a convolutional layer with the kernel size and stride of two right after the input layer, and replace the output convolutional layer by a transposed convolutional layer with the kernel size and stride of two (red blocks in Fig. 1). In this way, the model can learn the optimal resampling approach. In the experiments, $k_{\max} = (14, 14, 10)$ so that it can be used with the lowest tested training resolution of $60 \times 60 \times 39$. Other hyperparameters such as $d_{u_{t+1}}$, N_h, and N_B were obtained empirically for decent segmentations when training with the original image resolution. As each HNO block and Hartley MHA block can be implemented as a deep-learning layer in commonly used libraries, they can be easily adopted by other architectures.

2.5 Training Strategy

The images of different modalities are stacked along the channel axis to provide a multi-channel input. As the intensity ranges across modalities can be quite different, intensity normalization is performed on each image of each modality. Image augmentation with rotation (axial, $\pm30°$), shifting ($\pm20\%$), and scaling ($[0.8, 1.2]$) is used and each image has an 80% chance to be transformed. The Adamax optimizer [12] is used with the cosine annealing learning rate scheduler [20], with the maximum and minimum learning rates as 10^{-2} and 10^{-3}, respectively. The Pearson's correlation coefficient loss is used as it is robust to learning rate for image segmentation [25], and it consistently outperformed the Dice loss

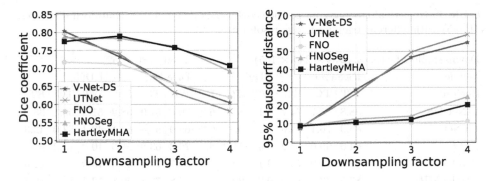

Fig. 2. Comparisons of robustness to training image resolution. Each point represents the average value from WT, TC, and ET of the 125 official validation cases of BraTS'19. The training images were downsampled by different factors while the trained models were tested with the original resolution.

and weighted cross-entropy in our experiments. An NVIDIA Tesla P100 GPU with 16 GB memory is used with a batch size of one and 100 epochs, and Keras in TensorFlow 2.6.2 is used for implementation. Note that small batch sizes are common in 3D segmentation given the high memory requirement.

3 Experiments

3.1 Data and Experimental Setups

The dataset of BraTS'19 with 335 cases of gliomas was used, each with four modalities of T1, post-contrast T1, T2, and T2-FLAIR images with $240 \times 240 \times 155$ voxels [3]. There is also an official validation dataset of 125 cases in the same format without given annotations. Models were trained with images downsampled by different factors (1, 2, 3, and 4) to study the robustness to image resolution. In training, we split the training dataset (335 cases) into 90% for training and 10% for validation. In testing, each model was tested on the official validation dataset (125 cases) with $240 \times 240 \times 155$ voxels regardless of the downsampling factor. The predictions were uploaded to the CBICA Image Processing Portal[3] for the results statistics of the "whole tumor" (WT), "tumor core" (TC), and "enhancing tumor" (ET) regions [3]. We compare our proposed HNOSeg and HartleyMHA models with three other models:

1. **V-Net-DS** [26]: a V-Net with deep supervision representing the commonly-used encoding-decoding architectures.
2. **UTNet** [7]: a U-Net enhanced by the Transformer's attention mechanism.
3. **FNO** [17]: original FNO without shared parameters, residual connections, and deep supervision. The same hyperparameters as HNOSeg were used.

---

[3] https://ipp.cbica.upenn.edu/.

Table 1. Numerical comparisons of Dice coefficients (%) and 95% Hausdorff distances (HD95) with different training image resolutions.

Downsampling factor	1 (240 × 240 × 155)						3 (80 × 80 × 52)					
Metric	Dice (%)			HD95			Dice (%)			HD95		
Region	WT	TC	ET	WT	TC	ET	WT	TC	ET	WT	TC	ET
V-Net-DS	88.8	77.7	74.7	7.1	8.9	6.2	70.6	57.9	68.2	51.9	53.1	35.2
UTNet	86.9	76.1	74.0	7.5	9.4	6.6	69.9	56.8	63.1	49.4	57.0	42.3
FNO	84.0	69.0	62.2	9.4	11.2	8.1	79.3	63.8	54.0	10.0	11.2	9.8
HNOSeg	87.7	75.0	73.2	9.1	9.6	6.5	86.7	71.9	69.8	15.0	14.4	12.6
HartleyMHA	86.9	73.1	72.5	9.2	9.8	7.3	84.8	72.8	69.8	12.5	13.1	10.7

Table 2. Number of parameters, inference time per image in seconds averaged from images of size 240 × 240 × 155, and memory in GB with a batch size of 1.

V-Net-DS			UTNet			FNO			HNOSeg			HartleyMHA		
Param	Time	Mem	Param	Time	Mem	Param	Time	Mem	Param	Time	Mem	Param	Time	Mem
5.7M	0.33	9.0	7.1M	0.41	4.9	144.5M	0.61	4.8	24.8k	0.71	8.9	47.7k	0.57	4.8

The learnable resampling approach in Sect. 2.4 was applied to all models. Note that our goal is not competing for the best accuracy but studying the robustness to image resolution. Although only the results of a dataset are shown because of the page limit, the characteristics of the proposed models can be demonstrated through this challenging multi-modal brain tumor segmentation problem.

3.2   Results and Discussion

Figure 2 and Table 1 show comparisons of resolution robustness among tested models, and Table 2 shows the computational costs during inference. At the original resolution, V-Net-DS and UTNet outperformed HNOSeg and HartleyMHA by less than 3% in the Dice coefficient on average, but HNOSeg and HartleyMHA only had less than 50k model parameters which were less than 1% of V-Net-DS and UTNet. FNO performed worst with the most parameters (144.5M). As the resolution decreased, the accuracies of V-Net-DS and UTNet decreased almost linearly with the downsampling factor, while HNOSeg and HartleyMHA were more robust. When the downsampling factor changed from 1 to 3, the average Dice coefficients of V-Net-DS and UTNet decreased by more than 14.8%, while those of HNOSeg and HartleyMHA only decreased by less than 2.5%. Similar trends can be observed for the 95% Hausdorff distance, except that FNO performed surprisingly well in this aspect. HartleyMHA performed better overall than HNOSeg. Note that we fixed $k_{\max}$ for the consistency among models in the experiments, which can be adjusted for better results in other situations.

For computation cost, Table 2 shows that V-Net-DS and UTNet had shorter inference times than HNOSeg and HartleyMHA, though all models used less

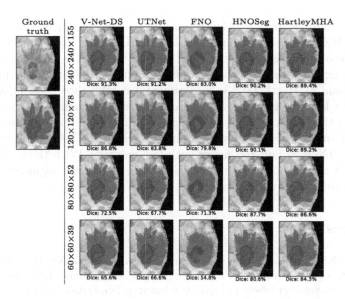

Fig. 3. Visual comparisons among models trained with different image resolutions, tested on an unseen case of size 240 × 240 × 155. The Dice coefficients were averaged from the WT, TC, and ET regions.

than 0.8 s per image of size 240 × 240 × 155. HartleyMHA ran faster than HNOSeg and used less memory, though HartleyMHA had more parameters.

Figure 3 shows the visual comparisons of the segmentation results on an unseen case. Consistent to Fig. 2, except FNO, the accuracies of the models were similar with the original training image resolution. As the training image resolution reduced, HNOSeg and HartleyMHA gradually outperformed V-Net-DS and UTNet. When the training images were downsampled from 240 × 240 × 155 to 60 × 60 × 39 (downsampling factor = 4), the average Dice coefficients of V-Net-DS and UTNet decreased by more than 24%, and HartleyMHA had the least reduction of 5.1%.

With such superior robustness to image resolution, HNOSeg and HartleyMHA can be trained with lower-resolution images using fewer computational resources to provide decent segmentation results on the original resolution during inference. While HartleyMHA performed better than HNOSeg in general, their similar performance is consistent with the findings in [15,23] that self-attention is sufficient for good performance but is not crucial. On the other hand, as the use of efficient self-attention improves the expressiveness of the Hartley MHA block, fewer layers can be used to reduce the overall computational costs.

4   Conclusion

In this paper, based on the idea of FNO which has the properties of zero-shot super-resolution and global receptive field, we propose the HNOSeg and

HartleyMHA models for resolution-robust and parameter-efficient 3D image segmentation. HNOSeg is FNO improved by the Hartley transform, residual connections, deep supervision, and shared parameters in the frequency domain. We further extend this concept for efficient multi-head attention in the frequency domain as HartleyMHA. Experimental results show that HNOSeg and HartleyMHA had similar accuracies as other tested segmentation models when trained with the original image resolution, but had superior performance when trained with images of much lower resolutions. HartleyMHA performed slightly better than HNOSeg and ran faster with less memory. With these advantages, HartleyMHA can be a promising alternative for 3D image segmentation especially when computational resources are limited.

References

1. Ba, J.L., Kiros, J.R., Hinton, G.E.: Layer normalization. arXiv:1607.06450 (2016)
2. Bakas, S., et al.: Advancing the cancer genome atlas glioma MRI collections with expert segmentation labels and radiomic features. Sci. Data **4**(170117), 1–13 (2017)
3. Bakas, S., et al.: Identifying the best machine learning algorithms for brain tumor segmentation, progression assessment, and overall survival prediction in the BRATS challenge. arXiv:1811.02629 (2018)
4. Bracewell, R.N.: Discrete Hartley transform. J. Opt. Soc. Am. **73**(12), 1832–1835 (1983)
5. Cao, H., et al.: Swin-Unet: Unet-like pure transformer for medical image segmentation. In: Karlinsky, L., Michaeli, T., Nishino, K. (eds.) ECCV 2022. LNCS, vol. 13803, pp. 205–218. Springer, Cham (2023). https://doi.org/10.1007/978-3-031-25066-8_9
6. Dosovitskiy, A., et al.: An image is worth 16x16 words: transformers for image recognition at scale. In: International Conference on Learning Representations (2021)
7. Gao, Y., Zhou, M., Metaxas, D.N.: UTNet: a hybrid transformer architecture for medical image segmentation. In: International Conference on Medical Image Computing and Computer-Assisted Intervention, pp. 61–71 (2021)
8. Hartley, R.V.L.: A more symmetrical Fourier analysis applied to transmission problems. Proc. IRE **30**(3), 144–150 (1942)
9. Hatamizadeh, A., et al.: UNETR: transformers for 3D medical image segmentation. In: Proceedings of the IEEE/CVF Winter Conference on Applications of Computer Vision, pp. 574–584 (2022)
10. He, K., Zhang, X., Ren, S., Sun, J.: Deep residual learning for image recognition. In: IEEE Conference on Computer Vision and Pattern Recognition, pp. 770–778 (2016)
11. Hesamian, M.H., Jia, W., He, X., Kennedy, P.: Deep learning techniques for medical image segmentation: achievements and challenges. J. Digital Imaging **32**(4), 582–596 (2019)
12. Kingma, D.P., Ba, J.L.: Adam: A method for stochastic optimization. arXiv:1412.6980 (2014)
13. Klambauer, G., Unterthiner, T., Mayr, A., Hochreiter, S.: Self-normalizing neural networks. In: Advances in Neural Information Processing Systems, pp. 972–981 (2017)

14. Lee, C.Y., Xie, S., Gallagher, P.W., Zhang, Z., Tu, Z.: Deeply-supervised nets. In: International Conference on Artificial Intelligence and Statistics, pp. 562–570 (2015)
15. Lee-Thorp, J., Ainslie, J., Eckstein, I., Ontanon, S.: FNet: Mixing tokens with Fourier transforms. arXiv:2105.03824 (2021)
16. Li, Z., et al.: Neural operator: Graph kernel network for partial differential equations. arXiv:2003.03485 (2020)
17. Li, Z., et al.: Fourier neural operator for parametric partial differential equations. In: International Conference on Learning Representations (2021)
18. Liu, X., Song, L., Liu, S., Zhang, Y.: A review of deep-learning-based medical image segmentation methods. Sustainability **13**(3), 1224 (2021)
19. Liu, Z., et al.: Swin transformer: hierarchical vision transformer using shifted windows. In: Proceedings of the IEEE/CVF International Conference on Computer Vision (ICCV), pp. 10012–10022 (2021)
20. Loshchilov, I., Hutter, F.: SGDR: stochastic gradient descent with warm restarts. In: International Conference on Learning Representations (2017)
21. Lu, J., et al.: SOFT: softmax-free transformer with linear complexity. In: Advances in Neural Information Processing Systems, vol. 34, pp. 21297–21309 (2021)
22. Menze, B.H., et al.: The multimodal brain tumor image segmentation benchmark (BRATS). IEEE Trans. Med. Imaging **34**(10), 1993–2024 (2015)
23. Tolstikhin, I.O., et al.: MLP-Mixer: an all-MLP architecture for vision. In: Advances in Neural Information Processing Systems, vol. 34, pp. 24261–24272 (2021)
24. Vaswani, A., et al.: Attention is all you need. In: Advances in Neural Information Processing Systems (2017)
25. Wong, K.C.L., Moradi, M.: 3D segmentation with fully trainable Gabor kernels and Pearson's correlation coefficient. In: Machine Learning in Medical Imaging, pp. 53–61 (2022)
26. Wong, K.C.L., Moradi, M., Tang, H., Syeda-Mahmood, T.: 3D segmentation with exponential logarithmic loss for highly unbalanced object sizes. In: Frangi, A.F., Schnabel, J.A., Davatzikos, C., Alberola-López, C., Fichtinger, G. (eds.) MICCAI 2018. LNCS, vol. 11072, pp. 612–619. Springer, Cham (2018). https://doi.org/10.1007/978-3-030-00931-1_70
27. Xie, Y., Zhang, J., Shen, C., Xia, Y.: CoTr: efficiently bridging CNN and transformer for 3D medical image segmentation. In: International Conference on Medical Image Computing and Computer-Assisted Intervention, pp. 171–180 (2021)

EdgeMixup: Embarrassingly Simple Data Alteration to Improve Lyme Disease Lesion Segmentation and Diagnosis Fairness

Haolin Yuan[1], John Aucott[3], Armin Hadzic[2], William Paul[2],
Marcia Villegas de Flores[1], Philip Mathew[1,2], Philippe Burlina[2],
and Yinzhi Cao[1(✉)]

[1] Johns Hopkins University, Baltimore, USA
{hyuan4,mvilleg5,yinzhi.cao}@jhu.edu
[2] Johns Hopkins Applied Physics Laboratory, Laurel, USA
{william.paul,philip.mathew,Philippe.Burlina}@jhuapl.edu
[3] Johns Hopkins University School of Medicine, Baltimore, USA
jaucott2@jhmi.edu

Abstract. Lyme disease is a severe skin disease caused by tick bites, which affects hundreds of thousands of people. One task in diagnosing Lyme disease is lesion segmentation, i.e., separating benign skin from lesions, which can not only help clinicians to focus on lesions but also improve downstream tasks such as disease classification. However, it is challenging to segment Lyme disease lesions due to the lack of well-segmented, labeled Lyme datasets and the nature of Lyme, e.g., the typical bull's eye lesion and its closeness to normal skin. In this paper, we design a simple yet novel data preprocessing and alteration method, called EDGEMIXUP, to help segment Lyme lesions on imbalanced training datasets. The key insight is to deploy a linear combination of lesion edge, either detected or computed, and the source image highlights the affected lesion area so that a learning model focuses more on the preserved lesion structure instead of skin tone, thus iteratively improving segmentation performance. Additionally, the improved edge from lesion segmentation can be further used for Lyme disease classification—e.g., in differentiating Lyme from other similar lesions including tinea corporis and herpes zoster—with improved model fairness on different subpopulations.

1 Introduction

Medical Image Analysis has greatly benefited from advances in AI [1] yet some improvements still remain to be addressed, importantly in areas that allow both algorithmic performance and fairness [2], and in certain medical applications that promise to significantly lessen morbidity and mortality. Early detection of skin lesions is such an endeavor as it can aid in identifying infectious diseases with

---

Supplementary Information The online version contains supplementary material available at https://doi.org/10.1007/978-3-031-43901-8_36.

cutaneous manifestations. Lyme disease is an example of that with a potentially diagnostic skin lesion [3]—which is caused by the bacterium *Borrelia burgdorferi* and leads to nearly 476,000 cases per annum during 2010–2018 [4]. The earliest and most treatable phase of Lyme disease is manifested via a red concentric lesion at the site of a tick bite, called erythema migrans (EM) [5]. While the EM pattern may appear simple to recognize, its diagnosis can be challenging for those with or without a medical background alike, as only 20% of United States patients have the stereotypical bull's eye lesion [6]. When skin lesions are atypical they can be mistaken for other diseases such as tinea corporis (TC) or herpes zoster (HZ), two other diseases acting as confusers for Lyme, considered herein. This has increased interest in medical applications of deep learning (DL), and using deep convolutional neural networks (CNNs), to assist clinicians in timely and accurate diagnosis of conditions including Lyme disease, TC and HZ [7–9].

One important diagnosis task is to segment Lyme lesion, particularly the EM pattern, from benign skins. Such DL-assisted segmentation not only helps clinicians in pre-screening patients but also improves downstream tasks such as lesion classification. However, while Lyme disease lesion segmentation is intuitively simple, it is challenging due to the following reasons. First, there lacks of a well-segmented dataset with manual labels on Lyme disease. On one hand, some datasets—such as HAM10000 [10] and ISBI Challenges [11]—have manual annotated segmentations for diseases like melanoma, but they do not have Lyme disease lesions. On the other hand, some datasets—such as Groh et al. [12]—have Lyme disease and skin tone and classification labels, but not segmentation.

Second, the segmentation of Lyme lesion is itself challenging due to the nature of EM pattern. Specifically, a typical Lyme lesion exhibits a bull's eye pattern with one central redness and one outer circle, which is different from darkness lesion in cancer-related skin disease like melanoma. Furthermore, clinical data collected for training is usually imbalanced in some properties, e.g., more samples with light skins compared with dark skins. Therefore, existing skin disease segmentation [13] as well as existing general segmentation works, such as U-Net [14], polar training [15], ViT-Adapter [16], and MFSNet [17], usually suffer from relatively low performance and reduced fairness [2, 18, 19].

In this paper, we present the first Lyme disease dataset that contains labeled segmentation and skin tones. Our Lyme disease dataset contains two parts: (i) a classification dataset, composed of more than 3,000 diseased skin images that are either obtained from public resources or clinicians with patient-informed consent, and (ii) a segmentation dataset containing 185 samples that are manually annotated for three regions—i.e., background, skin (light vs. dark), and lesion—conducted under clinician supervision and Institutional Review Boards (IRB) approval. Our dataset with manual labels is available at this URL [20].

Secondly, we design a simple yet novel data preprocessing and alternation method, called EDGEMIXUP, to improve Lyme disease segmentation and diagnosis fairness on samples with different skin-tones. The key insight is to alter a skin image with a linear combination of the source image and a detected lesion boundary so that the lesion structure is preserved while minimizing skin tone information. Such an improvement is an *iterative* process that gradually improves lesion edge detection and segmentation fairness until convergence. Then, the detected,

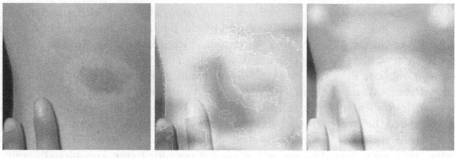

(a) Original image   (b) Heat map for EDGEMIXUP   (c) Heat map for legacy analysis

Fig. 1. A motivating example to illustrate why EDGEMIXUP improves model performance and reduces biases via mixing up lesion boundary with original image (Heatmap is generated via Grad-CAM). (Color figure online)

converged edge in the first step also helps classification of Lyme diseases via mixup with improved fairness. Our source code is available at this URL [20].

We evaluate EDGEMIXUP for skin disease segmentation and classification tasks. Our results show that EDGEMIXUP is able to increase segmentation utility and improve fairness. We also show that the improved segmentation further improves classification fairness as well as joint fairness-utility metrics compared to existing debiasing methods, e.g., AD [21] and ST-Debias [22].

2   Motivation

In this section, we motivate the design of EDGEMIXUP by showing that added lesion boundary helps a DL model focus more on the lesion part instead of other features such as skin or background. Note that not all skin disease datasets are carefully processed either due to the large amount of work required or the scarcity of data samples collected, e.g., SD-198 [23] contains samples that are taken under variant environments. Specifically, we train two ResNet-34 models using the same dataset with and without EDGEMIXUP for a classification task of skin disease. We keep all hyper-parameters exactly the same for two models, and only augment the same image with and without mixing lesion boundary up with the original image. We generate initial lesion edges using EdgeMixup, which we will elaborate in following sections. Figure 1 shows the original image (Fig. 1a) as well as two models' attention as heat-maps where red color represents the highest attention, yellow a higher attention, and purple the least attention. EDGEMIXUP helps the model to focus more on the lesion area comparing Fig. 1b and 1c. The reason is that a legacy diagnosis has no information about lesion and does not know where to locate its focus, thus easily gets distracted by fingers instead of the lesion pattern.

3   Method

In this section, we first give the definition for model fairness, and we then describe the design of EDGEMIXUP for the purpose of de-biasing in Fig. 2 and

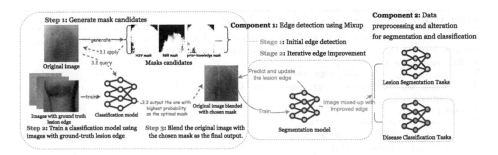

Fig. 2. EdgeMixup Process

Algorithm 1. We consider any model f, either a classification model f_{class} or a segmentation model f_{seg}, to be biased against certain skin-tone st_2 if given metrics M and samples x_{st_1} and x_{st_2} from class y, where st_1 and st_2 are different skin-tones according to their ITA scores, $M(f(x_{st_1}), y) > M(f(x_{st_2}), y)$. If there exists a model f such that $M(f(x_{st_1}), y) = M(f(x_{st_2}, y))$, we consider it perfectly fair for st_1 and st_2 skin-tone samples.

EDGEMIXUP improves model fairness on light and dark skin samples in both segmentation and classification tasks, and it has two major components: (i) edge detection using mixup, and (ii) data preprocessing and alteration for downstream tasks. More specifically, our proposed edge detection has two parts: initial edge detection and iterative improvement.

Initial Edge Detection: The purpose of initial detection, which is documented in the `Initial_edge_detection` function of Algorithm 1, is to provide a starting point, i.e., a rough boundary, for the next step of iterative improvement. The high-level idea is that EDGEMIXUP detects several edge candidates using the color range of ground-truth lesions in both Red-Green-Blue (RGB) and Hue-Saturation-Value (HSV) color space and then selects the target edge using a learning model based on the output confidence score. First, EDGEMIXUP trains a classification model based on a mixup of the ground-truth segmentation under clinician supervision and the original image (Line 7). Second, EDGEMIXUP generates many edge candidates. For example, EDGEMIXUP collects the mean range of lesion color from the training set and use the range as threshold to filter out any given sample for a candidate mask (Line 9). Lastly, EDGEMIXUP selects an edge candidate with the highest confidence score output by the learning model (Line 11) and returns it as the edge for this given sample. Note that the initial edge detection is irrelevant to the sample size of a particular subpopulation, thus improving the fairness. That is, even if the original dataset is imbalanced, as long as one sample from a subpopulation exists, the color range of the sample's lesion is considered in the initial detection.

Iterative Edge Improvement: EDGEMIXUP includes iterative edge improvement in the training phase of our segmentation model to further improve model utility. The intuitive reason of utilizing such algorithm is that by applying the

---

Algorithm 1. Pseudo-code of EDGEMIXUP
___

Require: A labelled sample $(x, y) \in D$, mixup weights α, ground-truth edged training set $D_{edge\_gt}^{train}$
Ensure: dataset $D_{\texttt{final\_edge}}$ in which each sample has it lesion edge highlighted $(x_{\texttt{edge}}, y)$
1: **function** main( )
2:     $D_{\texttt{initial_edge}} = $ Initial_edge_detection(D, α)
3:     $D_{\texttt{final_edge}} = $ Iterative_edge_improvement$(D_{\texttt{initial\_edge}}, \alpha)$
4:     **return** $D_{\texttt{final\_edge}}$
5: **end function**
6: **function** Initial_edge_detection(D, α)
7:     Train classification model $m_{\texttt{class}}$ using $D_{\texttt{edge\_gt}}^{\texttt{train}}$
8:     **for** each sample $x \in D$ **do**
9:         Get all edge candidates $\{\texttt{edge}_1, \texttt{edge}_2, .., \texttt{edge}_n\}$ for each sample x
10:         Mixup each edge candidate with x
11:         Query $m_{\texttt{class}}$ using all mixed-up $\{x_{\texttt{edge}_1}, ...x_{\texttt{edge}_n}\}$ and choose the optimal edge $\texttt{edge}_{\texttt{opt}}$
12:         Generate edged sample $x_{\texttt{edge}} = \texttt{Mixup}(x, \texttt{edge}_{\texttt{opt}}, \alpha)$
13:     **end for**
14:     **return** $D_{\texttt{edge}}$
15: **end function**
16: **function** Iterative_edge_improvement$(D_{\texttt{edge}}, \alpha)$
17:     Train the first model $m_{\texttt{iter}}$ using edged dataset $D_{\texttt{edge}}^{\texttt{train}}$
18:     Evaluate $m_{\texttt{iter}}$ using $D_{\texttt{edge}}^{\texttt{test}}$ and get current_Jaccard
19:     best_Jaccard $= 0$
20:     iter $= 1$
21:     **while** current_Jaccard $>$ best_Jaccard **do**
22:         best_Jaccard $=$ current_Jaccard
23:         Predict lesion masks using $m_{\texttt{iter}}$, convert them to lesion edge $edge$
24:         Generate new training set for next model $\texttt{Mixup}(D_{\texttt{train}}, \texttt{edge}, \alpha)$
25:         Train a model for next iteration $m_{\texttt{iter}+1}$
26:         Evaluate $m_{\texttt{iter}+1}$ using edged $D_{\texttt{edge}}^{\texttt{test}}$ and get current_Jaccard
27:         iter $+= 1$
28:     **end while**
29: **end function**
30: **function** Mixup$(x, edge, \alpha)$
31:     **return** $(\alpha \cdot x + (1 - \alpha) \cdot edge)$
32: **end function**

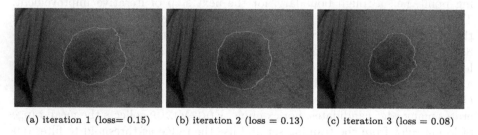

(a) iteration 1 (loss= 0.15)   (b) iteration 2 (loss = 0.13)   (c) iteration 3 (loss = 0.08)

Fig. 3. Illustration of iterative edge improvement on different iterations with train loss

mixup of detected edge and original image, given the lesion boundary feature detected in the previous iteration, the next-iteration segmentation model can converge better and the lesion boundary predicted by it is fine-grained. Specifically, EDGEMIXUP iteratively trains segmentation models from scratch, and we let the model trained in the previous iteration to predict lesion edge, which is then mixed-up with original training samples as the new training set for next-iteration model. The high-level idea is that when the lesion is restricted in a small

Table 1. Annotated segmentation and classification dataset characteristics, broken down by ITA-based skin tones (light skin/ dark skin) and disease types.

Split	Skin					SD-sub					
	NO	EM	HZ	TC	Total	DF	KA	PG	TC	TF	Total
seg	–	62	62	61	185	30	30	30	30	30	150
	–	47/15	46/16	40/21	133/52	23/7	27/3	27/3	24/6	29/1	130/20
class	885	740	698	704	3027	40	40	40	40	40	200
	822/63	682/58	608/90	609/95	2721/306	36/4	36/4	29/1	33/7	30/0	164/16

affected area, further detection will refine and constrain the detected boundary. Besides, EDGEMIXUP calculates a linear combination of original image and lesion boundary, i.e., by assigning the weight of original image as α and lesion boundary as $1 - \alpha$. Figure 3 shows the edge-mixed-up images for different iterations. EDGEMIXUP removes more skin areas after each iteration and gradually gets close to the real lesion at the third iteration.

4 Datasets

We present two datasets: (i) a dataset collected and annotated by us (called Skin), and (ii) a subset of SD-198 [23] with our annotation (called SD-sub). First, We collect and annotate a dataset with 3,027 images containing three types of disease/lesions, i.e., Tinea Corporis (TC), Herpes Zoster (HZ), and Erythema Migrans (EM). All skin images are either collected from publicly available sources or from clinicians with patient informed consent. Then, a medical technician and a clinician in our team manually annotate each image. For the segmentation task, we annotate skin images into three classes: background, skin, and lesion; then, for the classification task, we annotate skin images by classifying them into four classes: No Disease (NO), TC, HZ, and EM. We name it as Skin-class for later reference. Second, we select five classes from SD-198 [23], a benchmark dataset for skin disease classification, as another dataset for both segmentation and classification tasks. Note that due to the amount of manual work involved in annotation, we select those classes based on the number of samples in each class. The selected classes are Dermatofibroma (DF), Keratoacanthoma (KA), Pyogenic Granuloma (PG), Tinea Corporis (TC), and Tinea Faciale (TF). We choose 30 samples in each class for segmentation task, and we split them into 0.7, 0.1, and 0.2 ratio for training, validation, and testing, respectively.

Table 1 show the characteristics of these two datasets for both classification and segmentation tasks broken down by the disease type and skin tone, as calculated by the Individual Typology Angle (ITA) [24]. Specifically, we consider tan2, tan1, and dark as dark skin (ds) and others as light skin (ls). Compared to other skin tone classification schemas such as Fitzpartick scale [25], we divide ITA scores into more detailed categories (eight). One prominent observation is that ls images are more abundant than ds images due to a disparity in the availability of ds imagery found from either public sources or from clinicians with patient consent.

Table 2. Segmentation: Performance and Fairness (margin of error reported in parenthesis)

	Method	Unet	Polar	MFSNet	ViT-Adapter	**EDGEMIXUP**
Skin	Jaccard	0.7053(0.0035)	0.7126(0.0033)	0.5877(0.0080)	0.7027(0.0057)	**0.7807(0.0031)**
	J_{gap}	0.0809(0.0001)	0.0813(0.0001)	0.1291(0.0076)	0.2346(0.0035)	**0.0379(0.0001)**
SD-seg	Jaccard	0.7134(0.0031)	0.6527(0.0036)	0.6170(0.0052)	0.5088(0.0042)	**0.7799(0.0031)**
	J_{gap}	0.0753(0.0001)	0.1210(0.0003)	0.0636(0.0033)	0.2530(0.0021)	**0.0528(0.0001)**

5   Evaluation

We implement EDGEMIXUP using python 3.8 and Pytorch, and all experiments are performed using one GeForce RTX 3090 graphics card (NVIDIA).

Segmentation Evaluation. Our segmentation evaluation adopts four baselines, (i) a U-Net trained to segment skin lesions, (ii) a polar training [15] transforming images from Cartesian coordinates to polar coordinates, (iii) ViT-Adapter [16], a state-of-the-art semantic segmentation using a fine-tuned ViT model, (iv) MFSNet [17], a segmentation model with differently scaled feature maps to compute the final segmentation mask. We follow the default setting from each paper for evaluation. Our evaluation metrics include (i) Jaccard index (IoU score), which measures the similarity between a predicted mask and the manually annotated ground truth, and (ii) the gap between Jaccard values (J_{gap}) to measure fairness.

Table 2 shows the performance and fairness of EDGEMIXUP and different baselines. We compare predicted masks with the manually-annotated ground truth by calculating the Jaccard index, and computing the gap for subpopulations with ls and ds (based on ITA). EDGEMIXUP, a data preprocessing method, improves the utility of lesion segmentation in terms of Jaccard index compared with all existing baselines. One reason is that EDGEMIXUP preserves skin lesion information, thus improving the segmentation quality, while attenuating markers for protected factors. Note that EDGEMIXUP iteratively improves the segmentation results. Take our Skin-seg dataset for example. We trained our baseline Unet model for three iterations, and the model utility is increased by 0.0468 on Jaccard index while the J_{gap} between subpopulations is reduced by 0.0193.

Classification Evaluation. Our classification evaluation involves: (i) Adversarial Debiasing (AD) [21], (ii) DexiNed-avg, the average version of DexiNed [26] as an boundary detector used by EDGEMIXUP, and (iii) ST-Debias [22], a debiasing method augmenting data with conflicting shape and texture information. Our evaluation metrics include accuracy gap, the (Rawlsian) minimum accuracy across subpopulations, area under the receiver operating characteristic curve (AUC), and joint metrics (CAI_{α} and $CAUCI_{\alpha}$).

Table 3 shows utility performance (acc and AUC) and fairness results (gaps of acc and AUC between ls and ds subpopulations). We here list two variants of EDGEMIXUP, and one of which, "Unet", uses the lesion edge generated by

Table 3. Skin disease classification and associated bias. Samples contain skin tones as a protected factor. (margin of error reported in parentheses, subpopulation reported in brackets)

	Metrics	ResNet34	Baselines			EdgeMixup (ours)	
			AD	DexiNed-avg	ST-Debias	U-Net	Mask-based
Skin	acc	**88.08(3.66)**	81.79(4.35)	69.87(5.17)	76.52(5.23)	86.75(3.82)	86.09(3.90)
	acc_{gap}	16.38(12.21)	5.33(11.69)	19.79(13.52)	2.64(8.05)	8.280(9.66)	**1.923(8.49)**
	acc_{min}	73.33[ds]	76.92[ds]	51.85[ds]	71.12[ds]	79.41[ds]	**84.38[ds]**
	$CAI_{0.5}$	–	2.380	−10.81	1.090	3.385	**6.233**
	$CAI_{0.75}$	–	6.715	−7.110	7.415	5.743	**10.35**
	AUC	**0.977(0.02)**	0.956(0.02)	0.889(0.04)	0.933(0.03)	0.974(0.02)	0.973(0.02)
	AUC_{gap}	0.039(0.07)	0.009(0.05)	0.090(0.11)	0.035(0.04)	0.011(0.02)	**0.01 (0.05)**
	AUC_{min}	0.942[ds]	0.955[ds]	0.807[ds]	0.910[ds]	0.973[ds]	**0.964[ds]**
	$CAUCI_{0.5}$	–	0.004	−0.069	−0.024	0.012	**0.013**
	$CAUCI_{0.75}$	–	0.017	−0.060	−0.014	0.020	**0.022**
SD-sub	acc	75.60(14.26)	73.53(14.83)	63.13(16.08)	71.73(13.01)	74.17(13.30)	**76.47(11.26)**
	acc_{gap}	28.12(54.98)	25.00(54.30)	25.21(51.20)	18.66(13.21)	18.51(11.50)	**15.00(9.62)**
	acc_{min}	50.00[ds]	50.00[ds]	43.75 [ds]	70.59[ls]	72.11[ls]	**75.00[ls]**
	$CAI_{0.5}$	–	0.525	−4.780	2.795	4.090	**6.995**
	$CAI_{0.75}$	–	1.822	−0.934	6.127	6.850	**10.06**
	AUC	0.922(0.10)	0.962(0.06)	0.824(0.13)	0.941(0.08)	0.953(0.11)	**0.970(0.06)**
	AUC_{gap}	0.429(0.35)	0.319(0.36)	0.175(0.29)	0.255(0.31)	0.178(0.29)	**0.170(0.29)**
	AUC_{min}	0.500[ds]	0.650[ds]	0.650[ds]	0.711[ds]	0.784[ds]	**0.800[ds]**
	$CAUCI_{0.5}$	–	0.075	0.078	0.097	0.140	**0.153**
	$CAUCI_{0.75}$	–	0.092	0.166	0.135	0.196	**0.206**

the baseline Unet model while "mask-based" implements deep-learning model involved methodology introduced in Sect. 3. By adding the "Unet" variant, we demonstrate here that simply applying lesion edge predicetd by the baseline Unet model, while not optimal, efficiently reduces model bias on different skin-tone samples. EDGEMIXUP outperforms SOTA approaches in balancing the model's performance and fairness, i.e., the CAI_α and $CAUCI_\alpha$ values of EDGEMIXUP are the highest compared with the vanilla ResNet34 and other baselines.

6 Related Work

Skin Disease Classification and Segmentation: Previous researches mainly work on improving model utility for both medical image [27] and skin lesion [28] classification. As for skin lesion segmentation tasks, few works has been proposed due to the lack of datasets with ground-truth segmentation masks. International Skin Imaging Collaboration (ISIC) hosts challenges of International Symposium on Biomedical Imaging (ISBI) [11] to encourage researches studying lesion segmentation, feature detection, and image classification. However, official datasets released, e.g., HAM10000 [10] only contains melanoma samples and all of the samples are with light skins according to our inspection using ITA scores.

Bias Mitigation: Researchers have addressed bias and heterogeneity in deep learning models [18,29]. First, masking sensitive factors in imagery is shown to improve fairness in object detection and action recognition [30]. Second, adversarial debiasing operates on the principle of simultaneously training two networks with different objectives [31]. The competing two-player optimization paradigm is applied to maximizing equality of opportunity [32]. As a comparison, EDGEMIXUP is an effective preprocessing approach to debiasing when applied to skin disease particularly for Lyme-focused classification and segmentation tasks.

7   Conclusion

We present a simple yet novel approach to segment Lyme disease lesion, which can be further used for disease classification. The key insight is a novel data pre-processing method that utilizes edge detection and mixup to isolate and highlight skin lesions and reduce bias. EDGEMIXUP outperforms SOTAs in terms of Jaccord index for segmentation and CAI_α and $CAUCI_\alpha$ for disease classification.

Acknowledgement. This work was supported in part by Johns Hopkins University Institute for Assured Autonomy (IAA) with grants 80052272 and 80052273, and National Science Foundation (NSF) under grants CNS18-54000. The views and conclusions contained herein are those of the authors and should not be interpreted as necessarily representing the official policies or endorsements, either expressed or implied, of NSF or JHU-IAA.

References

1. Ting, D.S., Liu, Y., Burlina, P., Xu, X., Bressler, N.M., Wong, T.Y.: AI for medical imaging goes deep. Nat. Med. **24**(5), 539–540 (2018)
2. Burlina, P., Joshi, N., Paul, W., Pacheco, K.D., Bressler, N.M.: Addressing artificial intelligence bias in retinal disease diagnostics. Transl. Vis. Sci. Technol. (2020)
3. Hinckley, A.F., et al.: Lyme disease testing by large commercial laboratories in the United States. Clin. Infect. Diseases **59**, 676–681 (2014)
4. Kugeler, K.J., Schwartz, A.M., Delorey, M.J., Mead, P.S., Hinckley, A.F.: Estimating the frequency of Lyme disease diagnoses, United States, 2010–2018. Emerg. Infect. Diseases **27**, 616 (2021)
5. Nadelman, R.B.: Erythema migrans. Infectious Disease Clinics of North America (2015)
6. Tibbles, C.D., Edlow, J.A.: Does this patient have erythema migrans? JAMA **297**, 2617–2627 (2007)
7. Burlina, P.M., Joshi, N.J., Ng, E., Billings, S.D., Rebman, A.W., Aucott, J.N.: Automated detection of erythema migrans and other confounding skin lesions via deep learning. Comput. Biol. Med. **105**, 151–156 (2019)
8. Gu, Y., Ge, Z., Bonnington, C.P., Zhou, J.: Progressive transfer learning and adversarial domain adaptation for cross-domain skin disease classification. IEEE J. Biomed. Health Inf. **24**(5), 1379–1393 (2019)

9. Burlina, P.M., Joshi, N.J., Mathew, P.A., Paul, W., Rebman, A.W., Aucott, J.N.: AI-based detection of erythema migrans and disambiguation against other skin lesions. Comput. Biol. Med. **125**, 103977 (2020)

10. Tschandl, P., Rosendahl, C., Kittler, H.: The HAM10000 dataset, a large collection of multi-source dermatoscopic images of common pigmented skin lesions. Sci. Data **5**(1), 1–9 (2018)

11. Codella, N., et al.: Skin lesion analysis toward melanoma detection 2018: A challenge hosted by the international skin imaging collaboration (ISIC) (2019)

12. Groh, M., et al.: Evaluating deep neural networks trained on clinical images in dermatology with the fitzpatrick 17k dataset. In: Proceedings of the IEEE/CVF CVPR (2021)

13. Khan, M.A., Sharif, M., Akram, T., Damaševičius, R., Maskeliūnas, R.: Skin lesion segmentation and multiclass classification using deep learning features and improved moth flame optimization. Diagnostics **11**(5), 811 (2021)

14. Ronneberger, O., Fischer, P., Brox, T.: U-Net: convolutional networks for biomedical image segmentation. In: Navab, N., Hornegger, J., Wells, W., Frangi, A. (eds.) MICCAI 2015. LNCS, vol. 9351, pp. 234–241. Springer, Cham (2015). https://doi.org/10.1007/978-3-319-24574-4_28

15. Benčević, M., Galić, I., Habijan, M., Babin, D.: Training on polar image transformations improves biomedical image segmentation. IEEE Access **9**, 133365–133375 (2021)

16. Chen, Z., et al.: Vision transformer adapter for dense predictions. In: ICLR (2023)

17. Basak, H., Kundu, R., Sarkar, R.: MFSNet: a multi focus segmentation network for skin lesion segmentation. Pattern Recognit. **128**, 108673 (2022)

18. Caton, S., Haas, C.: Fairness in machine learning: A survey. arXiv preprint arXiv:2010.04053 (2020)

19. Burlina, P., Paul, W., Mathew, P., Joshi, N., Pacheco, K.D., Bressler, N.M.: Low-shot deep learning of diabetic retinopathy with potential applications to address artificial intelligence bias in retinal diagnostics and rare ophthalmic diseases. JAMA Ophthalmol. **138**, 1070–1077 (2020)

20. Edgemixup repository. https://github.com/Haolin-Yuan/EdgeMixup

21. Zhang, B.H., Lemoine, B., Mitchell, M.: Mitigating unwanted biases with adversarial learning. In: Proceedings of the 2018 AAAI/ACM Conference on AI, Ethics, and Society (2018)

22. Li, Y., et al.: Shape-texture debiased neural network training. In: International Conference on Learning Representations (2021)

23. Sun, X., Yang, J., Sun, M., Wang, K.: A benchmark for automatic visual classification of clinical skin disease images. In: Leibe, B., Matas, J., Sebe, N., Welling, M. (eds.) ECCV 2016. LNCS, vol. 9910, pp. 206–222. Springer, Cham (2016). https://doi.org/10.1007/978-3-319-46466-4_13

24. Wilkes, M., Wright, C.Y., du Plessis, J.L., Reeder, A.: Fitzpatrick skin type, individual typology angle, and melanin index in an African population: steps toward universally applicable skin photosensitivity assessments. JAMA Dermatol. **151**(8), 902–903 (2015)

25. Fitzpatrick, T.B.: Soleil et peau. J. Médecine Esthétique (in French) (1975)

26. Poma, X.S., Riba, E., Sappa, A.: Dense extreme inception network: towards a robust CNN model for edge detection. In: WACV (2020)

27. Yuan, Z., Yan, Y., Sonka, M., Yang, T.: Large-scale robust deep auc maximization: a new surrogate loss and empirical studies on medical image classification. In: ICCV (2021)

28. Shetty, B., Fernandes, R., Rodrigues, A.P., Chengoden, R., Bhattacharya, S., Lakshmanna, K.: Skin lesion classification of dermoscopic images using machine learning and convolutional neural network. Sci. Rep. **12**(1), 18134 (2022)
29. Yuan, H., Hui, B., Yang, Y., Burlina, P., Gong, N.Z., Cao, Y.: Addressing heterogeneity in federated learning via distributional transformation. In: Avidan, S., Brostow, G., Cisse, M., Farinella, G.M., Hassner, T. (eds.) ECCV 2022. LNCS, vol. 13698, pp. 179–195. Springer, Cham (2022). https://doi.org/10.1007/978-3-031-19839-7_11
30. Wang, T., Zhao, J., Yatskar, M., Chang, K.W., Ordonez, V.: Balanced datasets are not enough: estimating and mitigating gender bias in deep image representations. In: ICCV (2019)
31. Shafahi, A., et al.: Adversarial training for free!. In: Advances in Neural Information Processing Systems (2019)
32. Beutel, A., Chen, J., Zhao, Z., Chi, E.H.: Data decisions and theoretical implications when adversarially learning fair representations. arXiv preprint arXiv:1707.00075 (2017)

Learning Ontology-Based Hierarchical Structural Relationship for Whole Brain Segmentation

Junyan Lyu[1,2], Pengxiao Xu[1], Fatima Nasrallah[2], and Xiaoying Tang[1(✉)]

[1] Department of Electronic and Electrical Engineering, Southern University of Science and Technology, Shenzhen, China
tangxy@sustech.edu.cn
[2] Queensland Brain Institute, The University of Queensland, Brisbane, Australia

Abstract. Whole brain segmentation is vital for a variety of anatomical investigations in brain development, aging, and degradation. It is nevertheless challenging to accurately segment fine-grained brain structures due to the low soft-tissue contrast. In this work, we propose and validate a novel method for whole brain segmentation. By learning ontology-based hierarchical structural knowledge with a triplet loss enhanced by graph-based dynamic violate margin, our method can mimic experts' hierarchical perception of the brain anatomy and capture the relationship across different structures. We evaluate the whole brain segmentation performance of our method on two publicly-accessible datasets, namely JHU Adult Atlas and CANDI, respectively possessing fine-grained (282) and coarse-grained (32) manual labels. Our method achieves mean Dice similarity coefficients of 83.67% and 88.23% on the two datasets. Quantitative and qualitative results identify the superiority of the proposed method over representative state-of-the-art whole brain segmentation approaches. The code is available at https://github.com/CRazorback/OHSR.

Keywords: Whole brain segmentation · Hierarchical ontology · U-Net · Triplet loss · Structural MRI

1 Introduction

Quantitative analysis of structural MRI of the human brain is essential in various anatomical investigations in brain development, aging, and degradation. Accurate segmentation of brain structures is a prerequisite for quantitative and particularly morphometric analysis [5,8,23]. However, whole brain segmentation is challenging due to the low soft-tissue contrast, the high anatomical variability, and the limited labeled data, especially for fine-grained structures.

During the past decade, MRI based whole brain segmentation approaches have been explored. Multi-atlas based methods [19,24] are shown to be simple

Supplementary Information The online version contains supplementary material available at https://doi.org/10.1007/978-3-031-43901-8_37.

Coarse granularity   Fine granularity

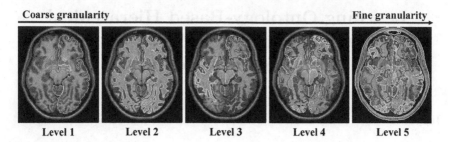

Level 1   Level 2   Level 3   Level 4   Level 5

Fig. 1. Illustration of the five-level ontology-based hierarchical structural relationship. The labels are created in a corase-to-fine manner.

yet effective and serve as the de facto standard whole brain segmentation methods. With a limited number of labeled data, multi-atlas based methods propagate labels from the atlas images to the target images through registration (mostly non-linear or even diffeomorphic registration), joint label fusion, and possibly corrective learning. With diffeomorphic registration, multi-atlas based methods are empowered of enhanced segmentation accuracy and topology-preserving capabilities that well accommodate the potential need of incorporating shape prior [4]. Nevertheless, multi-atlas based methods suffer from low computational efficiency and fail to leverage sophisticated contextual information when the tissue contrast in the structural MRI data is low.

Recently, deep learning has demonstrated state-of-the-art (SOTA) performance on various medical image segmentation tasks [6,12,16], also serving as an alternative solution to the whole brain segmentation task. QuickNAT [21] trains three 2D U-Nets respectively on the axial, coronal and sagittal views, and then aggregates them to infer the final segmentation. Pre-training with auxiliary labels derived from Freesurfer is conducted to alleviate QuickNAT's reliance on manually annotated data. SLANT [11] applies multiple independent 3D U-Nets to segment brain structures in overlapped MNI subspace, followed by label fusion. Wu et al. [26] propose a multi-atlas and diffeomorphism based encoding block to determine the most similar atlas patches to a target patch and propagate them into 3D fully convolutional networks. These deep learning methods improve over conventional multi-atlas based methods by considerable margins in terms of both computational efficiency and segmentation accuracy.

However, existing methods tend to ignore the ontology-based hierarchical structural relationship (OHSR) of the human brain's anatomy. Most of them assume all brain structures are disjoint and use multiple U-Nets to separately perform voxel-wise predictions for each of multiple structures of interest. It has been suggested that neuroanatomical experts recognize and delineate brain anatomy in a coarse-to-fine manner (Fig. 1) [17,18]. Concretely, at the highest level of the most coarse granularity (Level 1 in Fig. 1), the brain can be simply decomposed into telencephalon, diencephalon, mesencephalon, metencephalon, and myelencephalon. At a lower level, the cerebral cortex, cerebral nuclei and white matter can be identified within the telencephalon. At the lowest level of the most fine granularity, namely Level 5 in Fig. 1, the most fine-grained structures such as the hippocampus and the amygdala in the limbic system are characterized. It shall be most desirable if a neural network can learn brain anatomy in a similar fashion. Inspired by OHSR of brain anatomy, in the manifold space of a neural network, the distance between the feature vectors of the hippocampus and the amygdala should be smaller than that between the feature vectors of the hippocampus and any other structure that does not also belong to the limbic system. In other words, embeddings of fine-grained structures labeled as a same class at a higher level are supposed to be more similar than those of fine-grained structures labeled as different classes at the same higher level. Such prior knowledge on brain anatomy has been ignored or only implicitly learned in both multi-atlas based and deep learning based whole brain segmentation methods. Moreover, image contrast is not the only anatomical clue to discriminate structure boundary, especially for fine-grained structures. For instance, the anterior limb and posterior limb of the internal capsule are both part of white matter, which cannot be separated based on intensities and contrasts but can be differentiated by the sharp bend and feature-rich neighboring gray matter anatomy. This further suggests the importance of exploring and capturing OHSR.

To mimic experts' hierarchical perception of the brain anatomy, we here propose a novel approach to learn brain's hierarchy based on ontology, for a purpose of whole brain segmentation. Specifically, we encode the multi-level ontology knowledge into a voxel-wise embedding space. Deep metric learning is conducted to cluster contextually similar voxels and separate contextually dissimilar ones using a triplet loss with dynamic violate margin. By formatting the brain hierarchy into a directed acyclic graph, the violate margin can be easily induced by the height of the tree rooted at triplet's *least common subsumer*. As a result, the network is able to exploit the hierarchical relationship across different brain structures. The feature prototypes in the latent space are hierarchically organized following the brain hierarchy. To the best of our knowledge, this is the first work to incorporate ontology-based brain hierarchy into deep learning segmentation models. We evaluate our method on two whole brain segmentation datasets with different granularity and successfully establish SOTA performance.

2 Methodology

The proposed approach builds upon standard 3D U-Net and makes use of multi-level ontology knowledge from the brain hierarchy to enhance the whole brain segmentation performance. In subsequent subsections, we first revisit the standard triplet loss [9,14] and its dilemma in learning brain hierarchy. Then we describe how we construct the brain hierarchy graph and how we measure the semantic similarity between brain structures based on the constructed graph. After that, we introduce dynamic violate margin to the triplet loss.

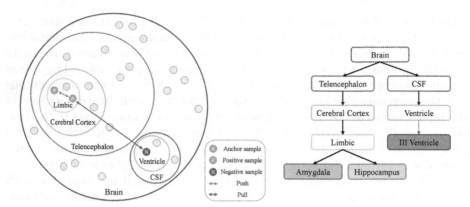

Fig. 2. An intuitive illustration of hierarchy-based triplet loss and its corresponding toy example graph of brain hierarchy. The loss tends to group structures labeled as a same class at a higher ontology level.

2.1 Triplet Loss

The goal of the triplet loss is to learn a feature embedding space wherein distances between features correspond to semantic dissimilarities between objects. Given a triplet $\mathcal{T}_i = \{f_i, f_i^+, f_i^-\}$ comprising an anchor voxel-wise feature vector f_i, a positive voxel-wise feature vector f_i^+ which is semantically similar to the anchor vector, and a negative voxel-wise feature vector f_i^- which is semantically dissimilar to the anchor vector, the triplet loss is formulated as

$$l_{\text{triplet}}(\mathcal{T}_i) = [\langle f_i, f_i^+ \rangle - \langle f_i, f_i^- \rangle + \mathcal{M}]_+, \tag{1}$$

where $\langle \cdot, \cdot \rangle$ is a distance function to evaluate the semantic dissimilarity between two feature vectors. The violate margin $\mathcal{M}$ is a hyperparameter that defines the minimum distance between positive and negative samples. It forces the gap between $\langle f, f^+ \rangle$ and $\langle f, f^- \rangle$ to be larger than $\mathcal{M}$ and ensures the model does not learn trivial solutions. $[\cdot]_+ = max\{0, \cdot\}$ is the hinge loss, which prevents the model from being updated when the triplet is already fulfilled. During training, the overall objective of the voxel-wise triplet loss is to minimize the sum of the loss over all triplets in a mini-batch, namely

$$\mathcal{L}_{\text{triplet}} = \frac{1}{N} \sum_{\mathcal{T}_i \in \mathcal{T}} l_{\text{triplet}}(\mathcal{T}_i), \tag{2}$$

where N is the total number of triplets in a mini-batch. Note that a triplet is allowed to consist of voxel-wise feature vectors from different subjects when the batch size is larger than 1. This strategy of sampling an anchor's neighbor enables the model to learn the global context in the brain instead of the local context in a subspace of the brain, since it is infeasible to train a 3D U-Net with whole brain MRI data.

However, it is challenging to apply the standard triplet loss to learn brain hierarchy: postive or negative is ill-defined with a fixed violate margin. For instance, the violate margin between $\langle f_{hippo}, f_{amyg} \rangle$ and $\langle f_{hippo}, f_{fimb} \rangle$ is certainly different from that between $\langle f_{hippo}, f_{amyg} \rangle$ and $\langle f_{hippo}, f_{IIIvent} \rangle$: the hippocampus, the amygdala, and the fimbria all belong to the limbic system while the third ventricle belongs to cerebrospinal fluid (CSF). As such, a distance function $d_{\mathcal{G}}(\cdot, \cdot)$ is required to measure the semantic dissimilarity between two brain structures, and can be then used to determine the corresponding violate margin.

2.2 Measuring Semantic Dissimilarity

Let $\mathcal{G} = (V, E)$ be a directed acyclic graph with vertices V and edges $E \subseteq V^2$. It specifies the hyponymy relationship between structures at different ontology levels. An edge $(u, v) \in E$ indicates u is an ancestor vertex of v. Specifically, v belongs to u at a higher ontology level. The structures of interest $S = \{s_1, ..., s_n\} \subseteq V$ are of the lowest ontology level. An example is shown in Fig. 2.

A common measure for the dissimilarity $d_{\mathcal{G}} : S^2 \to \mathbb{R}$ between two structures is the height of the tree rooted at the *least common subsumer* (LCS) divided by the height of the whole brain hierarchy tree, namely

$$d_{\mathcal{G}}(u, v) = \frac{h(\text{lcs}(u, v))}{h(\mathcal{G})}, \tag{3}$$

where the height of a tree $h(\cdot) = \max_{v \in V} \psi(\cdot, v)$ is defined as the length of the longest path from the root to a leaf. $\psi(\cdot, \cdot)$ is defined as the number of edges in the shortest path between two vertices. $\text{lcs}(\cdot, \cdot)$ refers to the ancestor shared by two vertices that do not have any child also being an ancestor of the same two vertices. With respect to the example hierarchy in Fig. 2, the LCS of the hippocampus and the amygdala is the limbic and the LCS of the hippocampus and the third ventricle is the brain. Given the height of the example hierarchy is 4, we can easily derive that $d_{\mathcal{G}}(\text{hippo}, \text{amyg}) = \frac{1}{4}$ and $d_{\mathcal{G}}(\text{hippo}, \text{IIIvent}) = 1$. Non-negativity, symmetry, identity of indiscernibles, and triangle inequality always hold for $d_{\mathcal{G}}(\cdot, \cdot)$ since the brain hierarchy is a tree and all structures of interest are leaf vertices, thus being a proper metric [2].

2.3 Dynamic Violate Margin

With $d_{\mathcal{G}}(\cdot, \cdot)$, we can define positive and negative samples and their violate margin in the triplet loss. We sample triplet $\mathcal{T}_i = \{f_i, f_i^+, f_i^-\}$ satisfying $d_{\mathcal{G}}(v_i, v_i^-) > d_{\mathcal{G}}(v_i, v_i^+)$, where f, f_i^+, f_i^- are the feature vectors of the voxels respectively labeled as v, v_i^+, v_i^-. Then the violate margin $\tilde{\mathcal{M}}$ can be determined dynamically

$$\tilde{\mathcal{M}} = 0.5(\mathcal{M}_\tau + \mathcal{M}_\epsilon),$$
$$\mathcal{M}_\tau = d_{\mathcal{G}}(v_i, v_i^-) - d_{\mathcal{G}}(v_i, v_i^+),$$
$$\mathcal{M}_\epsilon = \frac{1}{N_v(N_v - 1)} \sum_{a,b \in v} \langle f_a, f_b \rangle. \tag{4}$$

$\mathcal{M}_\tau \in (0,1]$ is the hierarchy-induced margin required between negative pairs and positive pairs in terms of $d_{\mathcal{G}}(\cdot,\cdot)$. $\mathcal{M}_\epsilon$ is the tolerance of the intra-class variance, which is computed as the average distance between samples in v. In this work, we adopt the cosine distance as our distance function in latent space: $\langle f_a, f_b \rangle = \frac{1}{2}(1 - \frac{f_a \cdot f_b}{\|f_a\|\|f_b\|}) \in (0,1]$. The triplet loss can thus be reformulated as

$$l_{\text{triplet}}(\mathcal{T}_i) = [\langle f_i, f_i^+ \rangle - \langle f_i, f_i^- \rangle + \tilde{\mathcal{M}}]_+. \tag{5}$$

Collectively, the overall training objective to learn OHSR is

$$\mathcal{L} = \mathcal{L}_{\text{seg}} + \lambda \mathcal{L}_{\text{triplet}}, \tag{6}$$

where λ is a hyperparameter.

3 Experiment

3.1 Datasets and Implementation

We evaluate our method on two public-accessible datasets with manually labeled fine-grained or coarse-grained brain structures. The first one is JHU Adult Atlas [25], containing 18 subjects with an age range of 27–55 years. T1-weighted MRI images are acquired in the MPRAGE sequence at Johns Hopkins University using 3T Philips scanners. All images are normalized to the MNI152 1mm space. The images are initially segmented into 289 structures using a single-subject atlas followed by substantial manual corrections. The optic tract, skull and bone marrow are excluded from our experiment, ending up with 282 structures. Two types of five-level OHSR are provided. Type I is based on classical definitions of the brain ontology and Type II is more commonly used in clinical descriptions [1,7,10]. The second one is Child and Adolescent Neuro Development Initiative (CANDI) [13], consisting of 103 1.5T T1-weighted MRI scans with 32 labeled structures. The subjects are aged 4–17 and come from both healthy and neurological disorder groups. A two-level hierarchical relationship is built by grouping the 32 structures into white matter, gray matter or CSF.

For JHU Adult Atlas, 8, 2 and 8 images are randomly selected and used for training, validation and testing. For CANDI, we split the dataset into training (60%), validation (10%), and testing (30%) sets following a previous study [15]. Foreground image patches of size $96 \times 112 \times 96$ are randomly cropped as the input of our model. To enlarge the training set, random scaling, gamma correction and rotation are applied. Image intensities are normalized using Z-score normalization. AdamW with a batch size of 4 is used to optimize the training objective, wherein $\mathcal{L}_{\text{seg}} = \mathcal{L}_{\text{dice}} + \mathcal{L}_{\text{ce}}$. $\mathcal{L}_{\text{dice}}$ and $\mathcal{L}_{\text{ce}}$ are respectively the Dice loss and the cross entropy loss. The hyperparameter $\lambda \in [0, 0.5]$ is scheduled following a cosine annealing policy. The initial learning rate is 5×10^{-4} and decays following a polynomial function. The model is trained for 30,000 steps, with the best model saved based on the validation Dice. All experiments are conducted using PyTorch 1.13.1 with NVIDIA Tesla V100 GPUs. More dataset and implementation details are provided in the supplementary material.

Table 1. Comparisons with SOTA for segmenting 282 fine-grained brain structures on JHU Adult Atlas, in terms of DSC. "S" denotes small structures with a size smaller than 1000 mm³. "M" denotes medium structures with a size between 1000 mm³ and 5000 mm³. "L" denotes large structures with a size larger than 5000 mm³. "A" indicates all structures. The best results are highlighted in **bold**. Please note all compared results are based on self-reimplementation.

Method	DSC(%)			
	S	M	L	A
U-Net [3]	66.56 ± 15.31	73.37 ± 16.22	78.58 ± 6.79	73.82 ± 14.07
QuickNAT [21]	69.30 ± 19.55	72.09 ± 20.76	78.97 ± 9.15	74.25 ± 17.25
nnU-Net [12]	80.04 ± 14.60	82.85 ± 13.06	84.79 ± 5.84	82.97 ± 11.56
Proposed	**80.76 ± 13.69**	**83.51 ± 13.13**	**85.52 ± 5.38**	**83.67 ± 11.26**

Table 2. Comparisons with SOTA for segmenting 32 coarse-grained brain structures on CANDI, in terms of DSC. The best result is highlighted in **bold**.

Method	DSC(%)
U-Net [20]	85.70 ± 9.70
QuickNAT V2 [22]	86.20 ± 9.50
nnU-Net [12]	87.97 ± 8.46
ACEnet [15]	88.10 ± 7.40
Proposed	**88.23 ± 6.32**

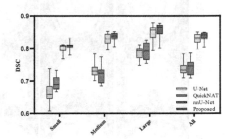

Fig. 3. DSC boxplots of small, medium, large and all structures from different methods on JHU Adult Atlas.

3.2 Evaluation Results

We now report quantitative and qualitative evaluation results. The Dice similarity coefficient (DSC) is used to quantitatively measure the segmentation accuracy. We compare our proposed method with SOTA methods and conduct several ablation studies to demonstrate the effectiveness of our method.

Comparisons with SOTA. To fairly compare with SOTA methods, we use the identical 3D U-Net and data augmentation hyperparameters as the "3d_fullres nnUNetTrainerV2_noMirroring" configuration in nnU-Net. As summarized in Table 1, our method obtains the highest overall mean DSC of 83.67% on JHU Adult Atlas, with an improvement of 0.70% over previous best method, i.e., nnU-Net. The improvements are consistent over small, medium and large brain structures, demonstrating the robustness of our method. The boxplot comparison results are illustrated in Fig. 3. Detailed improvement of each structure is presented in the supplementary material. Table 2 shows the results on CANDI. Please note all compared results except for nnU-Net in that table are directly copied from ACEnet paper [15] since CANDI has been explored more than JHU

Table 3. Analysis of the effect of OHSR on JHU Adult Atlas.

Method	DSC(%)
U-Net	73.82 ± 14.07
U-Net + OHSR (Type I)	77.34 ± 14.52 (+3.52)
U-Net + OHSR (Type II)	77.99 ± 12.93 (+4.17)
nnU-Net	82.97 ± 11.56
nnU-Net + OHSR (Type I)	83.49 ± 11.21 (+0.52)
nnU-Net + OHSR (Type II)	83.67 ± 11.26 (+0.70)

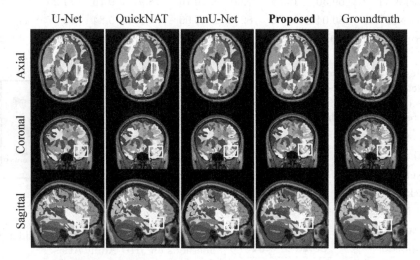

Fig. 4. Qualitative comparisons between our proposed method and SOTA. The regions of interest are highlighted in bounding boxes.

Adult Atlas. Our method achieves an average DSC of 88.23%, performing not only better than the hyperparameter-tuning method nnU-Net, but also better than ACEnet encoding anatomical context via an attention mechanism. These results also suggest that our method can leverage both simple (two-level) and sophisticated (five-level) hierarchies to enhance whole brain segmentation.

Ablation Studies. We first evaluate the effectiveness of incorporating OHSR into U-Net. The experiments are conducted on JHU Adult Atlas. As shown in Table 3, learning OHSR using a triplet loss with graph-based dynamic margin improves U-Net by 3.52% in DSC. This indicates our method can empower a relatively small network to segment fine-grained brain structures with considerable accuracy. Our method even outperforms nnU-Net by 0.70% in DSC, clearly demonstrating its superiority. We further compare the performance between two types of OHSR. As tabulated in Table 3, Type II ontology achieves better performance, indicating mimicking clinicians to understand brain hierarchy is of greater help.

Qualitative Results. Qualitative comparisons are demonstrated in Fig. 4. From the axial view, we can clearly see the external capsule is well-segmented by our proposed method, while other methods can hardly differentiate its boundary. From the coronal and sagittal views, we observe that our method can better capture and preserve the overall shape of the lateral frontal-orbital gyrus.

4 Conclusion

In this paper, we propose a novel approach to learn brain hierarchy based on ontology for whole brain segmentation. By introducing graph-based dynamic violate margin into the triplet loss, we encode multi-level ontology knowledge into a voxel-wise embedding space and mimic experts' hierarchical perception of the brain anatomy. We successfully demonstrate that our proposed method outperforms SOTA methods both quantitatively and qualitatively. We consider introducing hierarchical information into the output space as part of our future efforts.

Acknowledgement. This study was supported by the National Natural Science Foundation of China (62071210); the Shenzhen Science and Technology Program (RCYX20210609103056042); the Shenzhen Science and Technology Innovation Committee (KCXFZ2020122117340001); the Shenzhen Basic Research Program (JCYJ20200925153847004, JCYJ20190809120205578).

References

1. Ackerman, S.: Discovering the brain (1992)
2. Barz, B., Denzler, J.: Hierarchy-based image embeddings for semantic image retrieval. In: IEEE Winter Conference on Applications of Computer Vision, pp. 638–647 (2019)
3. Cardoso, M.J., et al.: MONAI: an open-source framework for deep learning in healthcare. arXiv preprint arXiv:2211.02701 (2022)
4. Ceritoglu, C., Tang, X., Chow, M., et al.: Computational analysis of LDDMM for brain mapping. Front. Neurosci. **7**, 151 (2013)
5. Chupin, M., Gérardin, E., Cuingnet, R., et al.: Fully automatic hippocampus segmentation and classification in Alzheimer's disease and mild cognitive impairment applied on data from ADNI. Hippocampus **19**(6), 579–587 (2009)
6. Coupé, P., et al.: AssemblyNet: a large ensemble of CNNs for 3D whole brain MRI segmentation. Neuroimage **219**, 117026 (2020)
7. Djamanakova, A., Tang, X., Li, X., et al.: Tools for multiple granularity analysis of brain MRI data for individualized image analysis. Neuroimage **101**, 168–176 (2014)
8. Erickson, K.I., Voss, M.W., Prakash, R.S., et al.: Exercise training increases size of hippocampus and improves memory. Proc. Natl. Acad. Sci. **108**(7), 3017–3022 (2011)

9. Ge, W.: Deep metric learning with hierarchical triplet loss. In: Proceedings of the European Conference on Computer Vision, pp. 269–285 (2018)

10. Gong, Y., Wu, H., Li, J., Wang, N., Liu, H., Tang, X.: Multi-granularity whole-brain segmentation based functional network analysis using resting-state fMRI. Front. Neurosci. **12**, 942 (2018)

11. Huo, Y., et al.: 3D whole brain segmentation using spatially localized atlas network tiles. Neuroimage **194**, 105–119 (2019)

12. Isensee, F., Jaeger, P.F., Kohl, S.A., Petersen, J., Maier-Hein, K.H.: nnU-Net: a self-configuring method for deep learning-based biomedical image segmentation. Nat. Methods **18**(2), 203–211 (2021)

13. Kennedy, D.N., Haselgrove, C., Hodge, S.M., Rane, P.S., Makris, N., Frazier, J.A.: CANDIShare: a resource for pediatric neuroimaging data. Neuroinformatics **10**, 319–322 (2012)

14. Li, L., Zhou, T., Wang, W., Li, J., Yang, Y.: Deep hierarchical semantic segmentation. In: Proceedings of the IEEE/CVF Conference on Computer Vision and Pattern Recognition, pp. 1246–1257 (2022)

15. Li, Y., Li, H., Fan, Y.: ACEnet: anatomical context-encoding network for neuroanatomy segmentation. Medi. Image Anal. **70**, 101991 (2021)

16. Lyu, J., Zhang, Y., Huang, Y., Lin, L., Cheng, P., Tang, X.: AADG: automatic augmentation for domain generalization on retinal image segmentation. IEEE Trans. Med. Imaging **41**(12), 3699–3711 (2022)

17. Mai, J.K., Majtanik, M., Paxinos, G.: Atlas of the Human Brain. Academic Press, Cambridge (2015)

18. Puelles, L., Harrison, M., Paxinos, G., Watson, C.: A developmental ontology for the mammalian brain based on the prosomeric model. Trends Neurosci. **36**(10), 570–578 (2013)

19. Tang, X., et al.: Bayesian parameter estimation and segmentation in the multi-atlas random orbit model. PLoS ONE **8**(6), e65591 (2013)

20. Ronneberger, O., Fischer, P., Brox, T.: U-Net: convolutional networks for biomedical image segmentation. In: Navab, N., Hornegger, J., Wells, W.M., Frangi, A.F. (eds.) MICCAI 2015. LNCS, vol. 9351, pp. 234–241. Springer, Cham (2015). https://doi.org/10.1007/978-3-319-24574-4_28

21. Roy, A.G., Conjeti, S., Navab, N., Wachinger, C.: Alzheimer's Disease Neuroimaging Initiative: QuickNAT: a fully convolutional network for quick and accurate segmentation of neuroanatomy. NeuroImage **186**, 713–727 (2019)

22. Roy, A.G., Navab, N., Wachinger, C.: Recalibrating fully convolutional networks with spatial and channel "squeeze and excitation" blocks. IEEE Trans. Med. Imaging **38**(2), 540–549 (2018)

23. Tang, X., Qin, Y., Wu, J., et al.: Shape and diffusion tensor imaging based integrative analysis of the hippocampus and the amygdala in Alzheimer's disease. Mag. Reson. Imaging **34**(8), 1087–1099 (2016)

24. Wang, H., Yushkevich, P.A.: Multi-atlas segmentation with joint label fusion and corrective learning-an open source implementation. Front. Neuroinf. **7**, 27 (2013)

25. Wu, D., et al.: Resource atlases for multi-atlas brain segmentations with multiple ontology levels based on T1-weighted MRI. Neuroimage **125**, 120–130 (2016)

26. Wu, J., Tang, X.: Brain segmentation based on multi-atlas and diffeomorphism guided 3D fully convolutional network ensembles. Pattern Recognit. **115**, 107904 (2021)

Fine-Grained Hand Bone Segmentation via Adaptive Multi-dimensional Convolutional Network and Anatomy-Constraint Loss

Bolun Zeng[1], Li Chen[2], Yuanyi Zheng[2], Ron Kikinis[3], and Xiaojun Chen[1(✉)]

[1] Institute of Biomedical Manufacturing and Life Quality Engineering, School of
Mechanical Engineering, Shanghai Jiao Tong University, Shanghai, China
`xiaojunchen@sjtu.edu.cn`
[2] Department of Ultrasound in Medicine, Shanghai Sixth People's Hospital Affiliated
to Shanghai Jiao Tong University School of Medicine, Shanghai, China
[3] The Surgical Planning Laboratory, Department of Radiology, Brigham and
Women's Hospital, Harvard Medical School, Boston, USA

Abstract. Ultrasound imaging is a promising tool for clinical hand examination due to its radiation-free and cost-effective nature. To mitigate the impact of ultrasonic imaging defects on accurate clinical diagnosis, automatic fine-grained hand bone segmentation is highly desired. However, existing ultrasound image segmentation methods face difficulties in performing this task due to the presence of numerous categories and insignificant inter-class differences. To address these challenges, we propose a novel Adaptive Multi-dimensional Convolutional Network (AMCNet) for fine-grained hand bone segmentation. It is capable of dynamically adjusting the weights of 2D and 3D convolutional features at different levels via an adaptive multi-dimensional feature fusion mechanism. We also design an anatomy-constraint loss to encourage the model to learn anatomical relationships and effectively mine hard samples. Experiments demonstrate that our method outperforms other comparison methods and effectively addresses the task of fine-grained hand bone segmentation in ultrasound volume. We have developed a user-friendly and extensible module on the 3D Slicer platform based on the proposed method and will release it globally to promote greater value in clinical applications. The source code is available at https://github.com/BL-Zeng/AMCNet.

Keywords: Hand bone segmentation · Adaptive convolution ·
Anatomical constraint · 3D Slicer · Ultrasound images

1 Introduction

Hand imaging examination is a standard clinical procedure commonly utilized for various medical purposes such as predicting biological bone age [16] and

---

Supplementary Information The online version contains supplementary material available at https://doi.org/10.1007/978-3-031-43901-8_38.

H. Greenspan et al. (Eds.): MICCAI 2023, LNCS 14223, pp. 395–404, 2023.
https://doi.org/10.1007/978-3-031-43901-8_38

diagnosing finger bone and joint diseases [4,8]. Ultrasound (US) is a promising alternative imaging modality for clinical examinations due to its radiation-free and cost-effective nature, especially the three-dimensional (3D) US volume, which is increasingly preferred for its intuitive visualization and comprehensive clinical information. However, the current US imaging technology is limited by low signal-to-noise ratio and inherent imaging artifacts, making the examination of hand with complex and delicate anatomical structure highly dependent on high-level expertise and experience.

To address this challenge, deep learning-based US image segmentation methods have been explored. For instance, Liu et al. [10] propose an attention-based network to segment seven key structures in the neonatal hip bone. Rahman et al. [14] present a graph convolutional network with orientation-guided supervision to segment bone surfaces. Studies such as [2,11] use the convolutional-based network for efficient bone surface segmentation. Additionally, some studies have focused on the automatic identification and segmentation of soft tissues, such as finger tendons and synovial sheaths [9,12]. Although these methods are effective in segmenting specific objects, they lack fine-grained analysis. Some studies have revealed that each hand bone has clinical analysis value [1], thus making fine-grained segmentation clinically significant. However, this is a challenging task. The hand comprises numerous structures, with a closely related anatomical relationship between the phalanges, metacarpal bones, and epiphysis. Moreover, different categories exhibit similar imaging features, with the epiphysis being particularly indistinguishable from the phalanges and metacarpal bones.

Fine-grained segmentation demands a model to maintain the inter-slice anatomical relationships while extracting intra-slice detailed features. The 2D convolution excels at capturing dense information but lacks inter-slice information, while the 3D convolution is complementary [6]. This motivates us to develop a proper adaptive fusion method. Previous studies used 2D/3D layouts to address data anisotropy problems. For example, Wang et al. [17] propose a 2.5D UNet incorporating 2D and 3D convolutions to improve the accuracy of MR image segmentation. Dong et al. [6] present a mesh network fusing multi-level features for better anisotropic feature extraction. However, the effectiveness of these methods in capturing fine-grained feature representations is limited due to their relatively fixed convolution distributions and feature fusion approaches. Moreover, the lack of supervision on complex anatomical relationships makes them inevitably suffer from anatomical errors such as missing or confusing categories.

To overcome the deficiencies of existing methods, this study proposes a novel Adaptive Multi-dimensional Convolutional Network (AMCNet) with an anatomy-constraint loss for fine-grained hand bone segmentation. Our contribution is three-fold. 1) First, to the best of our knowledge, this is the first work to address the challenge of automatic fine-grained hand bone segmentation in 3D US volume. We propose a novel multi-dimensional network to tackle the issue of multiple categories and insignificant feature differences. 2) Second, we propose an adaptive multi-dimensional feature fusion mechanism to dynamically adjust the weights of 2D and 3D convolutional feature layers according to different objectives, thus improving the fine-grained feature representation of the model.

3) Finally, we propose an anatomy-constraint loss that minimizes the anatomical error and mines hard samples, further improving the performance of the model.

2   Methods

2.1   Network Design

As shown in Fig. 1, the architecture of AMCNet consists of four down-sampling layers, four up-sampling layers, and four skip-connections. Each layer contains an adaptive 2D/3D convolutional module (ACM) which is proposed to dynamically balance inter-layer and intra-layer feature weight through adaptive 2D and 3D convolutions for better representations. In the encoder, each ACM block is followed by a max-pooling layer to compress features. In the decoder, the trilinear interpolate is used to up-sample features. The number of channels across each layer is empirically set to 64, 128, 256, and 512. The output layer uses the $1\times1\times1$ convolutional layer to obtain the segmentation map.

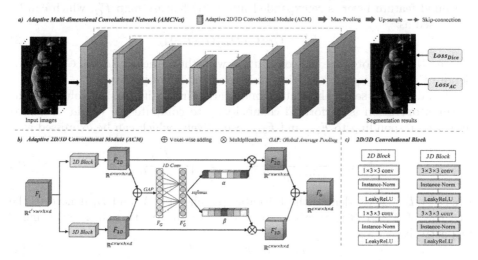

Fig. 1. An overview of the proposed AMCNet. The network consists of four down-sampling layers and four up-sampling layers. Each layer is composed of an ACM to fuse 2D and 3D convolutional features at different levels. $Loss_{Dice}$ denotes the Dice loss and $Loss_{AC}$ denotes the proposed anatomy-constraint loss.

2.2   Adaptive 2D/3D Convolutional Module (ACM)

To enable the model to capture the inter-layer anatomical connection and intra-layer dense semantic feature, the ACM is proposed to adaptively fuse the 2D and 3D convolution at different levels. Figure 1(b) illustrates the architecture of the ACM. Firstly, the feature map $F_i \in \mathbb{R}^{c' \times w \times h \times d}$ passes through 2D and 3D convolutional block respectively to obtain the 2D convolutional feature $F_{2D} \in \mathbb{R}^{c \times w \times h \times d}$ and 3D convolutional feature $F_{3D} \in \mathbb{R}^{c \times w \times h \times d}$. Figure 1(c) shows the details of the 2D and 3D convolutional block, which includes two $1 \times 3 \times 3$ or

$3 \times 3 \times 3$ convolution, instance normalization (Instance-Norm), and LeakyRuLU operations. The use of Instance-Norm considers the limitation of batch size in 3D medical image segmentation. Then, the F_{2D} and F_{3D} are performed the voxel-wise adding and the global average pooling (GAP) to generate channel-wise statistics $F_G \in \mathbb{R}^{c \times 1 \times 1 \times 1}$, which can be expressed as:

$$F_G = GAP(F_{2D} + F_{3D}) = \frac{1}{w \times h \times d} \sum_{i=1}^{w} \sum_{j=1}^{h} \sum_{k=1}^{d} (F_{2D} + F_{3D}) \qquad (1)$$

where w, h, and d are the width, height, and depth of the input feature map, respectively.

Further, a local cross-channel information interaction attention mechanism is applied for the fusion of multi-dimensional convolutional features. Specifically, the feature map F_G is squeezed to a one-dimension tensor of length c, which is the number of channels, and then a one-dimensional (1D) convolution with a kernel size of K is applied for information interaction between channels. The obtained feature layer is re-expanded into a 3D feature map F_G', which can be expressed as:

$$F_G' = G_U(C1D_K(G_S(F_G))) \qquad (2)$$

where $C1D_K$ denotes the 1D convolution with the kernel size of K, G_S and G_U denote the operation of squeezing and re-expanding respectively.

To adaptively select the feature information from different convolutions, the softmax operation is performed channel-wise to compute the weight vectors α and β corresponding to F_{2D} and F_{3D} respectively, which can be expressed as:

$$\alpha_i = \frac{e^{A_i F_G'}}{e^{A_i F_G'} + e^{B_i F_G'}}, \beta_i = \frac{e^{B_i F_G'}}{e^{A_i F_G'} + e^{B_i F_G'}} \qquad (3)$$

where $A, B \in \mathbb{R}^{c \times c}$ denote the learnable parameters, A_i and B_i denote to the i-th row of A and B respectively, α_i and β_i denote to i-th element of α and β respectively.

$$F_o = \alpha \cdot F_{2D} + \beta \cdot F_{3D} \qquad (4)$$

where F_o denotes the output feature map of the ACM.

2.3   Anatomy-Constraint Loss

Fine-grained hand bone segmentation places stringent demands on the anatomical relationships between categories, but the lack of supervision during model training renders it the primary source of segmentation error. For instance, the epiphysis is highly prone to neglect due to its small and imbalanced occupation, while the index and ring phalanx bones can easily be mistaken due to their symmetrical similarities. To address this issue, we propose the anatomy-constraint loss to facilitate the model's learning of anatomical relations.

Anatomical errors occur when pixels significantly deviate from their expected anatomical locations. Thus, we utilize the loss to compute and penalize these deviating pixels. Assume that the Y and P are the label and segmentation map respectively. First, the map representing anatomical errors is generated, where only pixels in the segmentation map that do not correspond to the anatomical relationship are activated. To mitigate subjective labeling errors caused by the unclear boundaries in US images, we perform morphological dilation on the segmentation map P. This operation expands the map and establishes an error tolerance, allowing us to disregard minor random errors and promote training stability. To make it differentiable, we implement this operation with a kernel=3 and stride=1 max-pooling operation. Subsequently, the anatomical error map F_E is computed by pixel-wise subtracting the Y and the expanded segmentation map P. The resulting difference map is then activated by ReLU, which ensures that only errors within the label region and beyond the anatomically acceptable range are penalized. The process can be expressed as:

$$F_E = \sigma_R(Y_{C_i} - G_{mp}(P_{C_i})) \tag{5}$$

where Y_{C_i} and P_{C_i} denote the i-th category maps of the label Y and segmentation map P respectively, $G_{mp}(\cdot)$ denotes the max-pooling operation, and $\sigma_R(\cdot)$ denotes the ReLU activation operation.

Next, we intersect F_E with the segmentation map P and label Y, respectively, based on which the cross entropy is computed, which is used to constrain the anatomical error:

$$Loss_{AC} = Loss_{CE}(P \odot F_E, \ Y \odot F_E) \tag{6}$$

where $Loss_{AC}(\cdot)$ denotes the proposed anatomy-constraint loss, $Loss_{CE}(\cdot)$ denotes the cross-entropy loss, and $\odot$ denotes the intersection operation.

To reduce the impact of class imbalance on model training and improve the stability of segmentation, we use a combination of Dice loss and anatomy-constraint loss function:

$$L = Loss_{Dice} + \gamma Loss_{AC} \tag{7}$$

where L is the overall loss, $Loss_{Dice}$ denotes the Dice loss, and γ denotes the weight-controlling parameter of the anatomy-constraint loss.

3   Experiments and Results

3.1   Dataset and Implementation

Our method is validated on an in-house dataset, which consists of 103 3D ultrasound volumes collected using device IBUS BE3 with a 12MHz linear transducer from pediatric hand examinations. The mean voxel resolution is $0.088 \times 0.130 \times 0.279 \, mm^3$ and the mean image size is $512 \times 1023 \times 609$. Two expert ultrasonographers manually annotated the data based on ITK-snap [18]. Each phalanx, metacarpal, and epiphysis were labeled with different categories, and there are a total of 39 categories including the background. The dataset was randomly spitted into 75% training set and 25% test set. All data were resized to 256×512 in the transverse plane and maintained the axial size. We extracted 256×512×16 voxels training patches from the resized images as the training samples.

The training and test phases of the network were implemented by PyTorch on an NVIDIA GeForce RTX 3090 GPU. The network was trained with Adam optimization with momentum of 0.9. The learning rate was set as 10e-3. For the hyperparameter, the kernel size K of 1D convolution in ACM was set as 3 and the weight-controlling parameter γ was set as 0.5. We used the Dice coefficient (DSC), Jaccard Similarity (Jaccard), Recall, F1-score, and Hausdorff Distance (HD95) as evaluation metrics.

3.2   Performance Comparison

We compared our network with recent and outstanding medical image segmentation methods, which contain UNet [15], UNet++ [19], 3D UNet [5], VNet [13], MNet [6], and the transformer baseline SwinUNet [3]. For a fair comparison, we used publicly available hyperparameters for each model. For the 3D network, the data processing method is consistent with ours, while for the 2D network, we slice the images along the axis to convert the 3D data into 2D.

Table 1 lists the results. Note that our method achieved the highest quantitative performance of DSC, Jaccard, Recall and F1-score, with values of 0.900, 0.819, 0.871, and 0.803, respectively. These results improved by 1.3%, 2.1%, 0.8%, and 1.3% compared to the best values of other methods. Note that our method outperformed the MNet that is a state-of-the-art (SOTA) method, which demonstrated the effectiveness of adaptive multi-dimensional feature fusion and anatomy-constraint for enhancing model performance.

Table 1. Quantitative comparison experiments between the proposed method and the outstanding segmentation methods. Dim denotes to dimension.

Dim	Methods	DSC↑	Jaccard↑	Recall↑	F1-score↑	HD95 (mm) ↓
2D	UNet [15]	0.855	0.747	0.837	0.736	1.215
	UNet++ [19]	0.875	0.779	0.763	0.697	3.328
	SwinUNet [3]	0.829	0.709	0.796	0.657	1.278
3D	3D UNet [5]	0.829	0.709	0.584	0.632	0.960
	VNet [13]	0.864	0.761	0.863	0.730	3.380
2D⊕3D	MNet [6]	0.887	0.798	0.830	0.790	**0.695**
	Ours	**0.900**	**0.819**	**0.871**	**0.803**	1.184

Figure 2 shows the visualization results of our method and comparative methods (Due to page limitations, we only presented the comparison results of our method with baseline and SOTA methods). Compared to other methods, our method has advantages in effectively mining difficult samples, particularly in accurately identifying and classifying clinically important but difficult-to-distinguish epiphysis. Additionally, it can effectively learn the anatomical relationships of different categories, reducing the occurrence of category confusion, particularly in the phalanges of the index finger, middle finger, and ring finger.

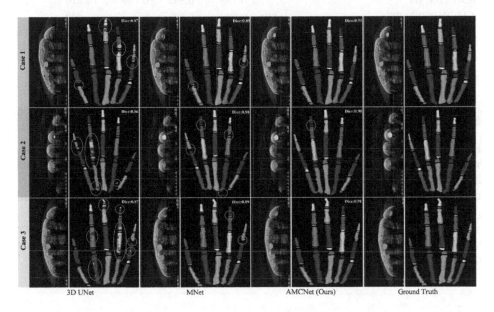

Fig. 2. The visualization results. Orange circles indicate obvious segmentation errors. (Color figure online)

3.3   Ablation Study

To validate the effect of anatomy-constraint loss, we compare the results of training with $Loss_{Dice}$ and the combination of $Loss_{Dice}$ and $Loss_{AC}$ on both 3D UNet and the proposed AMCNet. Table 2 lists the results. Note that compared with only $Loss_{Dice}$, UNet and our method have improved in various metrics after adding $Loss_{AC}$, boosting 0.6% in DSC and 1.1% in Jaccard. The results indicate that enforcing anatomical constraints to encourage the model to learn anatomical relationships improves the model's feature representation, resulting in better performance. Additionally, to verify the effect of the adaptive multi-dimensional feature fusion mechanism, we modified the ACM module to only 2D and 3D convolutional blocks, respectively. The results are shown in Table 2. Note that the 2D and 3D feature adaptive fusion mechanism improves model performance. Specifically, under $Loss_{Dice}$ and $Loss_{AC}$, it has resulted in an increase of 1.0% and 0.9% in DSC, 1.1% and 1.5% in Jaccard, respectively, compared to using only 2D or 3D convolution.

Table 2. Ablation study on the effect of $Loss_{AC}$ and ACM

Methods	DSC↑	Jaccard↑	Recall↑	F1-score↑	HD95 (mm) ↓
3D UNet [15]	0.829	0.709	0.584	0.632	0.960
3D UNet+$Loss_{AC}$	0.835	0.718	0.692	0.686	0.874
AMCNet 3D	0.887	0.797	0.787	0.648	2.705
AMCNet 2D	0.887	0.798	0.842	0.759	4.979
AMCNet 2D⊕3D	0.894	0.808	0.875	0.800	1.903
AMCNet 3D+$Loss_{AC}$	0.891	0.804	**0.877**	0.787	2.728
AMCNet 2D+$Loss_{AC}$	0.890	0.808	0.719	0.750	**0.748**
AMCNet 2D⊕3D +$Loss_{AC}$ (Ours)	**0.900**	**0.819**	0.871	**0.803**	1.184

3.4 Software Development and Application

Based on the method described above, a user-friendly and extensible module was developed on the 3D Slicer platform [7] to facilitate user access, as shown in Fig. 3. To use the module, users simply select the input and output file formats on the module interface and click the "apply" button. The software will then automatically perform image preprocessing, call the model for inference, and deliver the segmentation result within twenty seconds (see supplementary material 1). This plugin will be released globally to promote the greater value of the proposed method in clinical applications.

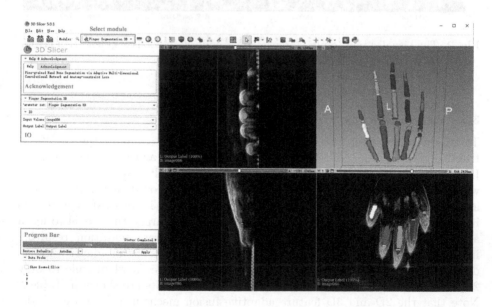

Fig. 3. The extensible module for our method based on the 3D Slicer platform

4 Conclusion

In this work, we have presented an adaptive multi-dimensional convolutional network, called AMCNet, to address the challenge of automatic fine-grained hand bone segmentation in 3D US volume. It adopts an adaptive multi-dimensional feature fusion mechanism to dynamically adjust the weights of 2D and 3D convolutional feature layers according to different objectives. Furthermore, an anatomy-constraint loss is designed to encourage the model to learn anatomical relationships and effectively mine hard samples. Experiments show that our proposed method outperforms other comparison methods and effectively addresses the task of fine-grained hand bone segmentation in ultrasound volume. The proposed method is general and could be applied to more medical segmentation scenarios in the future.

Acknowledgements. This work was supported by grants from the National Natural Science Foundation of China (81971709; M-0019; 82011530141), the Foundation of Science and Technology Commission of Shanghai Municipality (20490740700; 22Y11911700), Shanghai Jiao Tong University Foundation on Medical and Technological Joint Science Research (YG2021ZD21; YG2021QN72; YG2022QN056; YG2023ZD19; YG2023ZD15), the Funding of Xiamen Science and Technology Bureau (3502Z20221012).

References

1. Ahmed, O., Moore, D.D., Stacy, G.S.: Imaging diagnosis of solitary tumors of the phalanges and metacarpals of the hand. Am. J. Roentgenol. **205**(1), 106–115 (2015)
2. Alsinan, A.Z., Patel, V.M., Hacihaliloglu, I.: Automatic segmentation of bone surfaces from ultrasound using a filter-layer-guided CNN. Int. J. Comput. Assist. Radiol. Surg. **14**, 775–783 (2019)
3. Cao, H., et al.: Swin-Unet: Unet-like pure transformer for medical image segmentation. In: ECCV Workshops (2021)
4. Cecava, N.D., Kephart, D.A., Bui-Mansfield, L.T.: Bone lesions of the hand and wrist: systematic approach to imaging evaluation. Contemp. Diagn. Radiol. **43**(5), 1–7 (2020)
5. Çiçek, Ö., Abdulkadir, A., Lienkamp, S.S., Brox, T., Ronneberger, O.: 3D U-Net: learning dense volumetric segmentation from sparse annotation. In: Ourselin, S., Joskowicz, L., Sabuncu, M.R., Unal, G., Wells, W. (eds.) MICCAI 2016. LNCS, vol. 9901, pp. 424–432. Springer, Cham (2016). https://doi.org/10.1007/978-3-319-46723-8_49
6. Dong, Z., et al.: MNet: rethinking 2D/3D networks for anisotropic medical image segmentation. arXiv preprint arXiv:2205.04846 (2022)
7. Fedorov, A., et al.: 3D slicer as an image computing platform for the quantitative imaging network. Magn. Reson. Imaging **30**(9), 1323–1341 (2012)
8. Iagnocco, A., et al.: The reliability of musculoskeletal ultrasound in the detection of cartilage abnormalities at the metacarpo-phalangeal joints. Osteoarthritis Cartilage **20**(10), 1142–1146 (2012)
9. Kuok, C.P., et al.: Segmentation of finger tendon and synovial sheath in ultrasound image using deep convolutional neural network. Biomed. Eng. Online **19**(1), 24 (2020)

10. Liu, R., et al.: NHBS-Net: a feature fusion attention network for ultrasound neonatal hip bone segmentation. IEEE Trans. Med. Imaging **40**(12), 3446–3458 (2021)
11. Luan, K., Li, Z., Li, J.: An efficient end-to-end CNN for segmentation of bone surfaces from ultrasound. Comput. Med. Imaging Graph. **84**, 101766 (2020)
12. Martins, N., Sultan, S., Veiga, D., Ferreira, M., Teixeira, F., Coimbra, M.: A New active contours approach for finger extensor tendon segmentation in ultrasound images using prior knowledge and phase symmetry. IEEE J. Biomed. Health Inf. **22**(4), 1261–1268 (2018)
13. Milletari, F., Navab, N., Ahmadi, S.: V-Net: fully convolutional neural networks for volumetric medical image segmentation. In: 2016 Fourth International Conference on 3D Vision, pp. 565–571 (2016)
14. Rahman, A., Bandara, W.G.C., Valanarasu, J.M.J., Hacihaliloglu, I., Patel, V.M.: Orientation-guided graph convolutional network for bone surface segmentation. In: Wang, L., Dou, Q., Fletcher, P.T., Speidel, S., Li, S. (eds.) Medical Image Computing and Computer Assisted Intervention-MICCAI 2022. MICCAI 2022. Lecture Notes in Computer Science. vol. 13435. Springer, Cham (2022). https://doi.org/10.1007/978-3-031-16443-9_40
15. Ronneberger, O., Fischer, P., Brox, T.: U-Net: convolutional networks for biomedical image segmentation. In: Navab, N., Hornegger, J., Wells, W.M., Frangi, A.F. (eds.) MICCAI 2015. LNCS, vol. 9351, pp. 234–241. Springer, Cham (2015). https://doi.org/10.1007/978-3-319-24574-4_28
16. Spampinato, C., Palazzo, S., Giordano, D., Aldinucci, M., Leonardi, R.: Deep learning for automated skeletal bone age assessment in X-ray images. Med. Image Anal. **36**, 41–51 (2017)
17. Wang, G., et al.: Automatic segmentation of vestibular schwannoma from T2-weighted MRI by deep spatial attention with hardness-weighted loss. In: Shen, D., et al. (eds.) MICCAI 2019. LNCS, vol. 11765, pp. 264–272. Springer, Cham (2019). https://doi.org/10.1007/978-3-030-32245-8_30
18. Yushkevich, P.A., et al.: User-guided 3D active contour segmentation of anatomical structures: significantly improved efficiency and reliability. NeuroImage **31**(3), 1116–1128 (2006))
19. Zhou, Z., Siddiquee, M.M.R., Tajbakhsh, N.; Liang, J.:UNet++: redesigning skip connections to exploit multiscale features in image segmentation. IEEE Trans. Med. Imaging **39**(6), 1856–1867 (2020)

MedNeXt: Transformer-Driven Scaling of ConvNets for Medical Image Segmentation

Saikat Roy[1,3]([✉]), Gregor Koehler[1], Constantin Ulrich[1,5],
Michael Baumgartner[1,3,4], Jens Petersen[1], Fabian Isensee[1,4], Paul F. Jäger[4,6],
and Klaus H. Maier-Hein[1,2]

[1] Division of Medical Image Computing (MIC), German Cancer Research Center
(DKFZ), Heidelberg, Germany
[2] Pattern Analysis and Learning Group, Department of Radiation Oncology,
Heidelberg University Hospital, Heidelberg, Germany
[3] Faculty of Mathematics and Computer Science, Heidelberg University,
Heidelberg, Germany
saikat.roy@dkfz-heidelberg.de
[4] Helmholtz Imaging, German Cancer Research Center, Heidelberg, Germany
[5] National Center for Tumor Diseases (NCT), NCT Heidelberg, A Partnership
Between DKFZ and University Medical Center Heidelberg, Heidelberg, Germany
[6] Interactive Machine Learning Group, German Cancer Research Center, Heidelberg,
Germany

Abstract. There has been exploding interest in embracing Transformer-based architectures for medical image segmentation. However, the lack of large-scale annotated medical datasets make achieving performances equivalent to those in natural images challenging. Convolutional networks, in contrast, have higher inductive biases and consequently, are easily trainable to high performance. Recently, the ConvNeXt architecture attempted to modernize the standard ConvNet by mirroring Transformer blocks. In this work, we improve upon this to design a modernized and scalable convolutional architecture customized to challenges of data-scarce medical settings. We introduce MedNeXt, a *Transformer-inspired* large kernel segmentation network which introduces – 1) A *fully* ConvNeXt 3D Encoder-Decoder Network for medical image segmentation, 2) Residual ConvNeXt up and downsampling blocks to preserve semantic richness across scales, 3) A novel technique to iteratively increase kernel sizes by upsampling small kernel networks, to prevent performance saturation on limited medical data, 4) Compound scaling at multiple levels (depth, width, kernel size) of MedNeXt. This leads to state-of-the-art performance on 4 tasks on CT and MRI modalities and varying dataset sizes, representing a *modernized* deep architecture for medical image segmentation. Our code is made publicly available at: https://github.com/MIC-DKFZ/MedNeXt.

Keywords: Medical Image Segmentation · Transformers · MedNeXt · Large Kernels · ConvNeXt

H. Greenspan et al. (Eds.): MICCAI 2023, LNCS 14223, pp. 405–415, 2023.
https://doi.org/10.1007/978-3-031-43901-8_39

1   Introduction

Transformers [7,21,30] have seen wide-scale adoption in medical image segmentation as either components of hybrid architectures [2,3,8,9,31,33] or standalone techniques [15,25,34] for state-of-the-art performance. The ability to learn long-range spatial dependencies is one of the major advantages of the Transformer architecture in visual tasks. However, Transformers are plagued by the necessity of large annotated datasets to maximize performance benefits owing to their limited inductive bias. While such datasets are common to natural images (ImageNet-1k [6], ImageNet-21k [26]), medical image datasets usually suffer from the lack of abundant high quality annotations [19]. To retain the inherent inductive bias of convolutions while taking advantage of architectural improvements of Transformers, the ConvNeXt [22] was recently introduced to re-establish the competitive performance of convolutional networks for natural images. The ConvNeXt architecture uses an inverted bottleneck mirroring that of Transformers, composed of a depthwise layer, an expansion layer and a contraction layer (Sect. 2.1), in addition to large depthwise kernels to replicate their scalability and long-range representation learning. The authors paired large kernel ConvNeXt networks with enormous datasets to outperform erstwhile state-of-the-art Transformer-based networks. In contrast, the VGGNet [28] approach of stacking small kernels continues to be the predominant technique for designing ConvNets in medical image segmentation. Out-of-the-box data-efficient solutions such as nnUNet [13], using variants of a standard UNet [5], have still remained effective across a wide range of tasks.

The ConvNeXt architecture marries the scalability and long-range spatial representation learning capabilities of Vision [7] and Swin Transformers [21] with the inherent inductive bias of ConvNets. Additionally, the inverted bottleneck design allows us to scale width (increase channels) while not being affected by kernel sizes. Effective usage in medical image segmentation would allow benefits from – **1)** learning long-range spatial dependencies via large kernels, **2)** less intuitively, simultaneously scaling multiple network levels. To achieve this would require techniques to combat the tendency of large networks to overfit on limited training data. Despite this, there have been recent attempts to introduce large kernel techniques to the medical vision domain. In [18], a large kernel 3D-UNet [5] was used by decomposing the kernel into depthwise and depthwise dilated kernels for improved performance in organ and brain tumor segmentation – exploring kernel scaling, while using constant number of layers and channels. The ConvNeXt architecture itself was utilized in 3D-UX-Net [17], where the Transformer of SwinUNETR [8] was replaced with ConvNeXt blocks for high performance on multiple segmentation tasks. However, 3D-UX-Net only uses these blocks partially in a standard convolutional encoder, limiting their possible benefits.

In this work, we maximize the potential of a ConvNeXt design while uniquely addressing challenges of limited datasets in medical image segmentation. We present the first *fully* ConvNeXt 3D segmentation network, **MedNeXt**, which is a scalable Encoder-Decoder network, and make the following contributions:

- We utilize an architecture composed **purely of ConvNeXt blocks** which enables network-wide advantages of the ConvNeXt design. (Sect. 2.1)
- We introduce **Residual Inverted Bottlenecks** in place of regular up and downsampling blocks, to preserve contextual richness while resampling to benefit dense segmentation tasks. The modified residual connection in particular improves gradient flow during training. (Sec. 2.2)
- We introduce a simple but effective technique of iteratively increasing kernel size, **UpKern**, to prevent performance saturation on large kernel MedNeXts by initializing with trained upsampled small kernel networks. (Sect. 2.3)
- We propose applying **Compound Scaling** [29] of multiple network parameters owing to our network design, allowing orthogonality of width (*channels*), receptive field (*kernel size*) and depth (*number of layers*) scaling. (Sect. 2.4)

MedNeXt achieves state-of-the-art performance against baselines consisting of Transformer-based, convolutional and large kernel networks. We show performance benefits on 4 tasks of varying modality (CT, MRI) and sizes (ranging from 30 to 1251 samples), encompassing segmentation of organs and tumors. We propose MedNeXt as a strong and modernized alternative to standard ConvNets for building deep networks for medical image segmentation.

2 Proposed Method

2.1 Fully ConvNeXt 3D Segmentation Architecture

In prior work, ConvNeXt [22] distilled architectural insights from Vision Transformers [7] and Swin Transformers [21] into a convolutional architecture. The ConvNeXt block inherited a number of significant design choices from Transformers, designed to limit computation costs while scaling the network, which demonstrated performance improvements over standard ResNets [10]. In this work, we leverage these strengths by adopting the general design of ConvNeXt as the building block in a 3D-UNet-like [5] macro architecture to obtain the **MedNeXt**. We extend these blocks to up and downsampling layers as well (Sect. 2.2), resulting in the first fully ConvNeXt architecture for medical image segmentation. The macro architecture is illustrated in Fig. 1a. MedNeXt blocks (similar to ConvNeXt blocks) have 3-layers mirroring a Transformer block and are described for a C-channel input as follows:

1. **Depthwise Convolution Layer:** This layer contains a Depthwise Convolution with kernel size $k \times k \times k$, followed by normalization, with C output channels. We use channel-wise GroupNorm [32] for stability with small batches [27], instead of the original LayerNorm. The depthwise nature of convolutions allow large kernels in this layer to replicate a large attention window of Swin-Transformers, while simultaneously limiting compute and thus delegating the "heavy lifting" to the Expansion Layer.

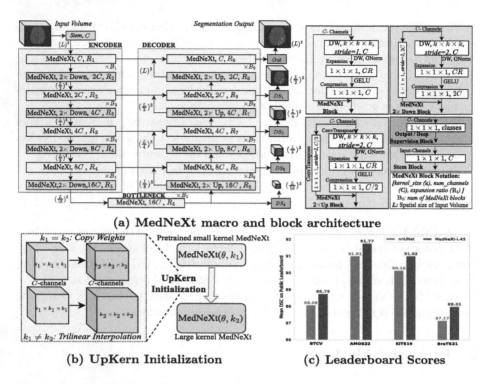

(a) MedNeXt macro and block architecture

(b) UpKern Initialization

(c) Leaderboard Scores

Fig. 1. (a) Architectural design of the MedNeXt. The network has 4 Encoder and Decoder layers each, with a bottleneck layer. MedNeXt blocks are present in Up and Downsampling layers as well. Deep Supervision is used at each decoder layer, with lower loss weights at lower resolutions. All residuals are *additive* while convolutions are padded to retain tensor sizes. **(b)** Upsampled Kernel (UpKern) initialization of a pair of MedNeXt architectures with similar configurations (θ) except kernel size (k_1, k_2). **(c)** MedNeXt-L ($5 \times 5 \times 5$) leaderboard performance.

2. **Expansion Layer:** Corresponding to a similar design in Transformers, this layer contains an overcomplete Convolution Layer with CR output channels, where R is the expansion ratio, followed by a GELU [12] activation. Large values of R allow the network to scale *width-wise* while $1 \times 1 \times 1$ kernel limits compute. It is important to note that this layer effectively decouples width scaling from receptive field (kernel size) scaling in the previous layer.

3. **Compression Layer:** Convolution layer with $1 \times 1 \times 1$ kernel and C output channels performing channel-wise compression of the feature maps.

MedNeXt is convolutional and retains the inductive bias inherent to Conv-Nets that allows easier training on sparse medical datasets. Our fully ConvNeXt architecture also enables width (more channels) and receptive field (larger kernels) scaling at both standard and up/downsampling layers. Alongside depth scaling (more layers), we explore these 3 orthogonal types of scaling to design a *compound scalable* MedNeXt for effective medical image segmentation (Sect. 2.4).

Table 1. (**Left**) MedNeXt configurations from scaling Block Counts (B) and Expansion Ratio (R) as in Fig. 1a. (**Right**) MedNext-B ablations (Sect. 4.1).

Config.	# Blocks (B)	Exp. Rat. (R)
S	$B_{all} = 2$	$R_{all} = 2$
B		$R_1 = R_9 = 2$
M	$B_1 = B_9 = 3$	$R_2 = R_8 = 3$
	$B_{2-8} = 4$	$R_{3-7} = 4$
	$B_1 = B_9 = 3$	$R_1 = R_9 = 3$
L	$B_2 = B_8 = 4$	$R_2 = R_8 = 4$
	$B_{3-7} = 8$	$R_{3-7} = 8$

Network Variants	BTCV		AMOS22	
	DSC	SDC	DSC	SDC
MedNeXt-B Resampling	**84.01**	**86.77**	**89.14**	**92.10**
Standard Up and Downsampling	83.13	85.64	88.96	91.86
MedNeXt-B ($5 \times 5 \times 5$) + UpKern	**84.23**	**87.06**	**89.38**	**92.36**
MedNeXt-B ($5 \times 5 \times 5$) from scratch	84.03	86.71	89.12	92.10
MedNeXt-B ($5 \times 5 \times 5$) + UpKern	**84.23**	**87.06**	**89.38**	**92.36**
MedNeXt-B ($3 \times 3 \times 3$) trained 2×	84.00	86.85	89.18	92.16
nnUNet *(non-MedNeXt baseline)*	83.56	86.07	88.88	91.70

2.2   Resampling with Residual Inverted Bottlenecks

The original ConvNeXt design utilizes separate downsampling layers which consist of standard strided convolutions. An equivalent upsampling block would be standard strided transposed convolutions. However, this design does not implicitly take advantage of width or kernel-based ConvNeXt scaling while resampling. We improve upon this by extending the Inverted Bottleneck to resampling blocks in MedNeXt. This is done by inserting the strided convolution or transposed convolution in the first *Depthwise Layer* for Downsampling and Upsampling MedNeXt blocks respectively. The corresponding channel reduction or increase is inserted in the last *compression* layer of our MedNeXt 2× Up or Down block design as in Fig. 1a. Additionally, to enable easier gradient flow, we add a residual connection with $1 \times 1 \times 1$ convolution or transposed convolution with *stride* of 2. In doing so, MedNeXt fully leverages the benefits from *Transformer-like* inverted bottlenecks to preserve rich semantic information in lower spatial resolutions in all its components, which should benefit dense medical image segmentation tasks.

2.3   UpKern: Large Kernel Convolutions Without Saturation

Large convolution kernels approximate the large attention windows in Transformers, but remain prone to performance saturation. ConvNeXt architectures in classification of natural images, despite the benefit of large datasets such as ImageNet-1k and ImageNet-21k, are seen to saturate at kernels of size $7 \times 7 \times 7$ [22]. Medical image segmentation tasks have significantly less data and performance saturation can be a problem in large kernel networks. To propose a solution, we borrow inspiration from Swin Transformer V2 [20] where a large-attention-window network is initialized with another network trained with a smaller attention window. Specifically, Swin Transformers use a bias matrix $\hat{B} \in \mathbb{R}^{(2M-1) \times (2M-1)}$ to store learnt relative positional embeddings, where M is the number of patches in an attention window. On increasing the window size, M increases and necessitates a larger $\hat{B}$. The authors proposed spatially interpolating an existing bias matrix to the larger size as a pretraining step, instead of training from scratch, which demonstrated improved performance. We propose a similar approach but customized to convolutions kernels, as seen in Fig. 1b, to overcome performance saturation. **UpKern** allows us to iteratively increase kernel size by initializing a large kernel network with a *compatible* pretrained small kernel network by *trilinearly upsampling* convolutional kernels (represented as tensors) of incompatible size. All other layers with identical tensor sizes (including normalization layers) are initialized by copying the unchanged pretrained weights. This leads to a simple but effective initialization technique for MedNeXt which helps large kernel networks overcome performance saturation in the comparatively limited data scenarios common to medical image segmentation.

2.4   Compound Scaling of Depth, Width and Receptive Field

Compound scaling [29] is the idea that simultaneous scaling on multiple levels (depth, width, receptive field, resolution etc.) offers benefits beyond that of scaling at one single level. The computational requirements of indefinitely scaling kernel sizes in 3D networks quickly becomes prohibitive and leads us to investigate simultaneous scaling at different levels. Keeping with Fig. 1a, our scaling is tested for block count (B), expansion ratio (R) and kernel size (k) – corresponding to depth, width and receptive field size. We use 4 model configurations of the MedNeXt to do so, as detailed in Table 1 **(Left)**. The basic functional design (MedNeXt-S) uses number of channels (C) as 32, $R = 2$ and $B = 2$. Further variants increase on just R (MedNeXt-B) or both R and B (MedNeXt-M). The largest 70-MedNext-block architecture uses high values of both R and B (MedNeXt-L) and is used to demonstrate the ability of MedNeXt to be significantly scaled depthwise (even at standard kernel sizes). We further explore large kernel sizes and experiment with $k = \{3, 5\}$ for each configuration, to maximize performance via *compound scaling* of the MedNeXt architecture.

3   Experimental Design

3.1   Configurations, Implementation and Baselines

We use PyTorch [24] for implementing our framework. We experiment with 4 configurations of the MedNeXt with 2 kernel sizes as detailed in Sect. 2.4. The GPU memory requirements of scaling are limited via – 1) Mixed precision training with `PyTorch AMP`, 2) Gradient Checkpointing. [4]. Our experimental framework uses the nnUNet [13] as a backbone - where the training schedule (epochs $= 1000$, batches per epoch $= 250$), inference (50% patch overlap) and data augmentation remain unchanged. All networks, except nnUNet, are trained with AdamW [23] as optimizer. The data is resampled to 1.0 mm isotropic spacing during training and inference (with results on original spacing), using input patch size of $128 \times 128 \times 128$ and 512×512, and batch size 2 and 14, for 3D and 2D networks respectively. The learning rate for all MedNeXt models is 0.001, except kernel:5 in KiTS19, which uses 0.0001 for stability. For baselines, all Swin models and 3D-UX-Net use 0.0025, while ViT models use 0.0001. We use Dice Similarity

Coefficient (DSC) and Surface Dice Similarity (SDC) at 1.0 mm tolerance for volumetric and surface accuracy. 5-fold cross-validation (CV) mean performance for supervised training using 80:20 splits for all models are reported. We also provide test set DSC scores for a 5-fold ensemble of MedNeXt-L (kernel: $5 \times 5 \times 5$) without postprocessing. Our extensive baselines consist of a high-performing convolutional network (nnUNet [13]), 4 convolution-transformer hybrid networks with transformers in the encoder (UNETR [9], SwinUNETR [8]) and in intermediate layers (TransBTS [31], TransUNet [3]), a fully transformer network (nnFormer [34]) as well as a partially ConvNeXt network (3D-UX-Net [17]). TransUNet is a 2D network while the rest are 3D networks. The uniform framework provides a common testbed for all networks, without incentivizing one over the other on aspects of patch size, spacing, augmentations, training and evaluation.

3.2 Datasets

We use 4 popular tasks, encompassing organ as well as tumor segmentation tasks, to comprehensively demonstrate the benefits of the MedNeXt architecture – 1) Beyond-the-Cranial-Vault (BTCV) Abdominal CT Organ Segmentation [16], 2) AMOS22 Abdominal CT Organ Segmentation [14] 3) Kidney Tumor Segmentation Challenge 2019 Dataset (KiTS19) [11], 4) Brain Tumor Segmentation Challenge 2021 (BraTS21) [1]. BTCV, AMOS22 and KiTS19 datasets contain 30, 200 and 210 CT volumes with 13, 15 and 2 classes respectively, while the BraTS21 dataset contains 1251 MRI volumes with 3 classes. This diversity shows the effectiveness of our methods across imaging modalities and training set sizes.

4 Results and Discussion

4.1 Performance Ablation of Architectural Improvements

We ablate the MedNeXt-B configuration on AMOS22 and BTCV datasets to highlight the efficacy of our improvements and demonstrate that a *vanilla* ConvNeXt is unable to compete with existing segmentation baselines such as nnUNet. The following are observed in ablation tests in Table 1 **(Right)** –

1. Residual Inverted Bottlenecks, specifically in Up and Downsampling layers, *functionally enables* MedNeXt (MedNeXt-B Resampling vs Standard Resampling) for medical image segmentation. In contrast, absence of these modified blocks lead to **considerably worse** performance. This is possibly owing to preservation of semantic richness in feature maps while resampling.
2. Training large kernel networks for medical image segmentation is a non-trivial task, with large kernel MedNeXts trained from scratch failing to perform in seen in MedNeXt-B (UpKern vs From Scratch). UpKern *improves performance* in kernel $5 \times 5 \times 5$ on both BTCV and AMOS22, whereas large kernel performance is **indistinguishable** from small kernels *without* it.

Table 2. 5-fold CV results of MedNeXt at kernel sizes: $\{3,5\}$ outperforming 7 baselines – consisting of convolutional, transformer and large kernel networks. [**val** (bold): better than or equal to $(\geq)$ top baseline, val (underline): better than $(>)$ *kernel:* 3 of same configuration]

Networks	Cat.	BTCV		AMOS22		KiTS19		BraTS21		AVG	
		DSC	SDC	DSC	SDC	DSC	SDC	DSC	SDC	DSC	SDC
nnUNet	*Baselines*	83.56	86.07	88.88	91.70	89.88	86.88	91.23	90.46	88.39	88.78
UNETR		75.06	75.00	81.98	82.65	84.10	78.05	89.65	88.28	82.36	81.00
TransUNet		76.72	76.64	85.05	86.52	80.82	72.90	89.17	87.78	82.94	80.96
TransBTS		82.35	84.33	86.52	88.84	87.03	83.53	90.66	89.71	86.64	86.60
nnFormer		80.76	82.37	84.20	86.38	89.09	85.08	90.42	89.83	86.12	85.92
SwinUNETR		80.95	82.43	86.83	89.23	87.36	83.09	90.48	89.56	86.41	86.08
3D-UX-Net		80.76	82.30	87.28	89.74	88.39	84.03	90.63	89.63	86.77	86.43
MedNeXt-S	*kernel: 3*	**83.90**	**86.60**	**89.03**	**91.97**	**90.45**	**87.80**	**91.27**	**90.46**	**88.66**	**89.21**
MedNeXt-B		**84.01**	**86.77**	**89.14**	**92.10**	**91.02**	**88.24**	**91.30**	**90.51**	**88.87**	**89.41**
MedNeXt-M		**84.31**	**87.34**	**89.27**	**92.28**	**90.78**	**88.22**	**91.57**	**90.78**	**88.98**	**89.66**
MedNeXt-L		**84.57**	**87.54**	**89.58**	**92.62**	**90.61**	**88.08**	**91.57**	**90.81**	**89.08**	**89.76**
MedNeXt-S	*kernel: 5*	83.92	86.80	89.27	92.26	**90.08**	**87.04**	91.40	90.57	88.67	**89.17**
MedNeXt-B		84.23	87.06	89.38	92.36	**90.30**	**87.40**	91.48	90.70	**88.85**	**89.38**
MedNeXt-M		84.41	87.48	89.58	92.65	**90.87**	**88.15**	91.49	90.67	89.09	89.74
MedNeXt-L		**84.82**	**87.85**	**89.87**	**92.95**	90.71	**87.85**	91.46	90.73	**89.22**	**89.85**

3. The performance boost in large kernels is seen to be due to the combination of UpKern with a larger kernel and not merely a longer *effective* training schedule (Upkern vs Trained $2\times$), as a trained MedNeXt-B with kernel $3\times3\times3$ retrained again is **unable to match** its large kernel counterpart.

This highlights that the MedNeXt modifications successfully translate the ConvNeXt architecture to medical image segmentation. We further establish the performance of the MedNeXt architecture against our baselines – comprising of convolutional, transformer-based and large kernel baselines – on all 4 datasets. We discuss the effectiveness of the MedNeXt on multiple levels.

4.2   Performance Comparison to Baselines

There are 2 levels at which MedNeXt successfully overcomes existing baselines - 5 fold CV and public testset performance. In 5-fold CV scores in Table 2, Med-NeXt, with $3 \times 3 \times 3$ kernels, takes advantage of depth and width scaling to provide state-of-the-art segmentation performance against **every baseline on all 4 datasets** with no additional training data. MedNeXt-L outperforms or is competitive with smaller variants despite task heterogeneity (brain and kidney tumors, organs), modality (CT, MRI) and training set size (BTCV: 18 samples vs BraTS21: 1000 samples), establishing itself as a powerful alternative to established methods such as nnUNet. With UpKern and $5 \times 5 \times 5$ kernels, MedNeXt takes advantage of full compound scaling to **improve further** on its own small kernel networks, comprehensively on organ segmentation (BTCV, AMOS22) and in a more limited fashion on tumor segmentation (KiTS19, BraTS21).

Furthermore, in leaderboard scores on official testsets (Fig. 1c), 5-fold ensembles for MedNeXt-L (kernel: $5 \times 5 \times 5$) and nnUNet, its strongest competitor are compared – **1) BTCV:** MedNeXt beats nnUNet and, to the best of our knowledge, is one of the leading methods with *only supervised training* and *no extra training data* (DSC: 88.76, HD95: 15.34), **2) AMOS22:** MedNeXt not only surpasses nnUNet, but is also **Rank 1** (date: 09.03.23) currently on the leaderboard (DSC: 91.77, NSD: 84.00), **3) KITS19:** MedNeXt exceeds nnUNet performance (DSC: 91.02), **4) BraTS21:** MedNeXt surpasses nnUNet in both volumetric and surface accuracy (DSC: 88.01, HD95: 10.69). MedNeXt attributes its performance solely to its architecture without leveraging techniques like transfer learning (3D-UX-Net) or repeated 5-fold ensembling (UNETR, SwinUNETR), thus establishing itself as the state-of-the-art for medical image segmentation.

5 Conclusion

In comparison to natural image analysis, medical image segmentation lacks architectures that benefit from scaling networks due to inherent domain challenges such as limited training data. In this work, MedNeXt is presented as a scalable *Transformer-inspired* fully-ConvNeXt 3D segmentation architecture customized for high performance on limited medical image datasets. We demonstrate MedNeXt's state-of-the-art performance across 4 challenging tasks against 7 strong baselines. Additionally, similar to ConvNeXt for natural images [22], we offer the *compound scalable* MedNeXt design as an effective modernization of standard convolution blocks for building deep networks for medical image segmentation.

References

1. Baid, U., et al.: The RSNA-ASNR-MICCAI BraTS 2021 benchmark on brain tumor segmentation and radiogenomic classification. arXiv preprint arXiv:2107.02314 (2021)
2. Cao, H., et al.: Swin-Unet: Unet-like pure transformer for medical image segmentation. arXiv preprint arXiv:2105.05537 (2021)
3. Chen, J., et al.: TransUNet: transformers make strong encoders for medical image segmentation. arXiv preprint arXiv:2102.04306 (2021)
4. Chen, T., Xu, B., Zhang, C., Guestrin, C.: Training deep nets with sublinear memory cost. arXiv preprint arXiv:1604.06174 (2016)
5. Çiçek, Ö., Abdulkadir, A., Lienkamp, S.S., Brox, T., Ronneberger, O.: 3D U-Net: learning dense volumetric segmentation from sparse annotation. In: Ourselin, S., Joskowicz, L., Sabuncu, M.R., Unal, G., Wells, W. (eds.) MICCAI 2016. LNCS, vol. 9901, pp. 424–432. Springer, Cham (2016). https://doi.org/10.1007/978-3-319-46723-8_49
6. Deng, J., Dong, W., Socher, R., Li, L.J., Li, K., Fei-Fei, L.: ImageNet: a large-scale hierarchical image database. In: 2009 IEEE Conference on Computer Vision and Pattern Recognition, pp. 248–255. IEEE (2009)

7. Dosovitskiy, A., et al.: An image is worth 16x16 words: Transformers for image recognition at scale. arXiv preprint arXiv:2010.11929 (2020)

8. Hatamizadeh, A., Nath, V., Tang, Y., Yang, D., Roth, H.R., Xu, D.: Swin UNETR: swin transformers for semantic segmentation of brain tumors in MRI images. In: Crimi, A., Bakas, S. (eds.) Brainlesion: Glioma, Multiple Sclerosis, Stroke and Traumatic Brain Injuries. BrainLes 2021. Lecture Notes in Computer Science. vol. 12962. Springer, Cham (2022). https://doi.org/10.1007/978-3-031-08999-2_22

9. Hatamizadeh, A., et al.: UNETR: transformers for 3D medical image segmentation. In: Proceedings of the IEEE/CVF Winter Conference on Applications of Computer Vision, pp. 574–584 (2022)

10. He, K., Zhang, X., Ren, S., Sun, J.: Deep residual learning for image recognition. In: Proceedings of the IEEE Conference on Computer Vision and Pattern Recognition, pp. 770–778 (2016)

11. Heller, N., et al.: The state of the art in kidney and kidney tumor segmentation in contrast-enhanced CT imaging: results of the KiTS19 challenge. Med. Image Anal. **67**, 101821 (2020)

12. Hendrycks, D., Gimpel, K.: Gaussian error linear units (gelus). arXiv preprint arXiv:1606.08415 (2016)

13. Isensee, F., Jaeger, P.F., Kohl, S.A., Petersen, J., Maier-Hein, K.H.: nnU-Net: a self-configuring method for deep learning-based biomedical image segmentation. Nat. Methods **18**(2), 203–211 (2021)

14. Ji, Y., et al.: AMOS: A large-scale abdominal multi-organ benchmark for versatile medical image segmentation. arXiv preprint arXiv:2206.08023 (2022)

15. Karimi, D., Vasylechko, S.D., Gholipour, A.: Convolution-free medical image segmentation using transformers. In: de Bruijne, M., et al. (eds.) MICCAI 2021. LNCS, vol. 12901, pp. 78–88. Springer, Cham (2021). https://doi.org/10.1007/978-3-030-87193-2_8

16. Landman, B., Xu, Z., Igelsias, J., Styner, M., Langerak, T., Klein, A.: Miccai multi-atlas labeling beyond the cranial vault-workshop and challenge. In: Proceedings of MICCAI Multi-Atlas Labeling Beyond Cranial Vault Challenge. vol. 5, p. 12 (2015)

17. Lee, H.H., Bao, S., Huo, Y., Landman, B.A.: 3D UX-Net: a large kernel volumetric convnet modernizing hierarchical transformer for medical image segmentation. arXiv preprint arXiv:2209.15076 (2022)

18. Li, H., Nan, Y., Del Ser, J., Yang, G.: Large-kernel attention for 3D medical image segmentation. Cognitive Computation, pp. 1–15 (2023)

19. Litjens, G., et al.: A survey on deep learning in medical image analysis. Med. Image Anal. **42**, 60–88 (2017)

20. Liu, Z., Hu, H., Lin, Y., Yao, Z., Xie, Z., Wei, Y., et al.: Swin transformer v2: scaling up capacity and resolution. In: Proceedings of the IEEE/CVF Conference on Computer Vision and Pattern Recognition, pp. 12009–12019 (2022)

21. Liu, Z., Lin, Y., Cao, Y., Hu, H., Wei, Y., Zhang, Z., et al.: Swin transformer: hierarchical vision transformer using shifted windows. In: Proceedings of the IEEE/CVF International Conference on Computer Vision, pp. 10012–10022 (2021)

22. Liu, Z., Mao, H., Wu, C.Y., Feichtenhofer, C., Darrell, T., Xie, S.: A convnet for the 2020s. In: Proceedings of the IEEE/CVF Conference on Computer Vision and Pattern Recognition, pp. 11976–11986 (2022)

23. Loshchilov, I., Hutter, F.: Decoupled weight decay regularization. arXiv preprint arXiv:1711.05101 (2017)

24. Paszke, A., et al.: PyTorch: an imperative style, high-performance deep learning library. In: Advances in Neural Information Processing Systems. vol. 32 (2019)

25. Peiris, H., Hayat, M., Chen, Z., Egan, G., Harandi, M.: A robust volumetric transformer for accurate 3D tumor segmentation. In: Wang, L., Dou, Q., Fletcher, P.T., Speidel, S., Li, S. (eds.) Medical Image Computing and Computer Assisted Intervention-MICCAI 2022. MICCAI 2022. Lecture Notes in Computer Science. vol. 13435. Springer, Cham (2022). https://doi.org/10.1007/978-3-031-16443-9_16
26. Ridnik, T., Ben-Baruch, E., Noy, A., Zelnik-Manor, L.: ImageNet-21k pretraining for the masses. arXiv preprint arXiv:2104.10972 (2021)
27. Roy, S., Kügler, D., Reuter, M.: Are 2.5 d approaches superior to 3D deep networks in whole brain segmentation? In: International Conference on Medical Imaging with Deep Learning, pp. 988–1004. PMLR (2022)
28. Simonyan, K., Zisserman, A.: Very deep convolutional networks for large-scale image recognition. arXiv preprint arXiv:1409.1556 (2014)
29. Tan, M., Le, Q.: EfficientNet: rethinking model scaling for convolutional neural networks. In: International Conference on Machine Learning, pp. 6105–6114. PMLR (2019)
30. Vaswani, A., et al.: Attention is all you need. In: Advances in neural Information Processing Systems. vol. 30 (2017)
31. Wang, W., Chen, C., Ding, M., Yu, H., Zha, S., Li, J.: TransBTS: multimodal brain tumor segmentation using transformer. In: de Bruijne, M., et al. (eds.) MICCAI 2021. LNCS, vol. 12901, pp. 109–119. Springer, Cham (2021). https://doi.org/10.1007/978-3-030-87193-2_11
32. Wu, Y., He, K.: Group normalization. In: Proceedings of the European Conference on Computer Vision (ECCV), pp. 3–19 (2018)
33. Xie, Y., Zhang, J., Shen, C., Xia, Y.: CoTr: efficiently bridging CNN and transformer for 3D medical image segmentation. In: de Bruijne, M., et al. (eds.) MICCAI 2021. LNCS, vol. 12903, pp. 171–180. Springer, Cham (2021). https://doi.org/10.1007/978-3-030-87199-4_16
34. Zhou, H.Y., Guo, J., Zhang, Y., Yu, L., Wang, L., Yu, Y.: nnFormer: Interleaved transformer for volumetric segmentation. arXiv preprint arXiv:2109.03201 (2021)

SwinUNETR-V2: Stronger Swin Transformers with Stagewise Convolutions for 3D Medical Image Segmentation

Yufan He$^{(\boxtimes)}$, Vishwesh Nath, Dong Yang, Yucheng Tang, Andriy Myronenko, and Daguang Xu

NVidia, Santa Clara, US
yufanh@nvidia.com

Abstract. Transformers for medical image segmentation have attracted broad interest. Unlike convolutional networks (CNNs), transformers use self-attentions that do not have a strong inductive bias. This gives transformers the ability to learn long-range dependencies and stronger modeling capacities. Although they, e.g. SwinUNETR, achieve state-of-the-art (SOTA) results on some benchmarks, the lack of inductive bias makes transformers harder to train, requires much more training data, and are sensitive to training recipes. In many clinical scenarios and challenges, transformers can still have inferior performances than SOTA CNNs like nnUNet. A transformer backbone and corresponding training recipe, which can achieve top performances under different medical image segmentation scenarios, still needs to be developed. In this paper, we enhance the SwinUNETR with convolutions, which results in a surprisingly stronger backbone, the SwinUNETR-V2, for 3D medical image segmentation. It achieves top performance on a variety of benchmarks of different sizes and modalities, including the Whole abdominal ORgan Dataset (WORD), MICCAI FLARE2021 dataset, MSD pancreas dataset, MSD prostate dataset, and MSD lung cancer dataset, all using the same training recipe (https://github.com/Project-MONAI/research-contributions/tree/main/SwinUNETR/BTCV, our training recipe is the same as that by SwinUNETR) with minimum changes across tasks.

Keywords: Swin transformer · Convolution · Hybrid model · Medical image segmentation

1 Introduction

Medical image segmentation is a core step for quantitative and precision medicine. In the past decade, Convolutional Neural Networks (CNNs) became the SOTA method to achieve accurate and fast medical image segmentation [10, 12, 21]. nn-UNet [12], which is based on UNet [21], has achieved top performances on over 20 medical segmentation challenges. Parallel to manually created networks such as nn-UNet, DiNTS [10], a CNN designed by automated neural network search, also achieved top performances in medical segmentation decathlon (MSD) [1] challenges. The convolution operation in CNN provides a strong inductive bias which is translational equivalent and efficient in capturing local features like boundary and texture. However, this inductive bias limits the representation power of CNN models which means a potentially lower

H. Greenspan et al. (Eds.): MICCAI 2023, LNCS 14223, pp. 416–426, 2023.
https://doi.org/10.1007/978-3-031-43901-8_40

performance ceiling on more challenging tasks [7]. Additionally, CNN has a local receptive field and are not able to capture long-range dependencies unlike transformers. Recently, vision transformers have been proposed, which adopt the transformers in natural language processing by splitting images into patches (tokens) [6], and use self-attention to learn features. The self-attention mechanism enables learning long-range dependencies between far-away tokens. This is intriguing and numerous works have been proposed to incorporate transformer attentions into medical image segmentation [2,3,9,23,24,30,32,35]. Among them, SwinUNETR [23] has achieved the new top performance in the MSD challenge and Beyond the Cranial Vault (BTCV) Segmentation Challenge by pretraining on large datasets. It has a U-shaped structure where the encoder is a Swin-Transformer [16].

Although transformers have achieved certain success in medical imaging, the lack of inductive bias makes them harder to be trained and requires much more training data to avoid overfitting. The self-attentions are good at learning complicated relational interactions for high-level concepts [5] but are also observed to be ignoring local feature details [5]. Unlike natural image segmentation benchmarks, e.g. ADE20k [34], where the challenge is in learning complex relationships and scene understanding from a large amount of labeled training images, many medical image segmentation networks need to be extremely focused on local boundary details while less in need of high-level relationships. Moreover, the number of training data is also limited. Hence in real clinical studies and challenges, CNNs can still achieve better results than transformers. For example, the top solutions in the last year MICCAI challenges HECTOR [19], FLARE [11], INSTANCE [15,22] and AMOS [13] are all CNN based. Besides lacking inductive bias and enough training data, one extra reason could be that transformers are computationally much expensive and harder to tune. More improvements and empirical evidence are needed before we say transformers are ready to replace CNNs for medical image segmentation.

In this paper, we try to develop a new "to-go" transformer for 3D medical image segmentation, which is expected to exhibit strong performance under different data situations and does not require extensive hyperparameter tuning. SwinUNETR reaches top performances on several large benchmarks, making itself the current SOTA, but without effective pretraining and excessive tuning, its performance on new datasets and challenges is not as high-performing as expected.

A straightforward direction to improve transformers is to combine the merits of both convolutions and self-attentions. Many methods have been proposed and most of them fall into two directions: 1) a new self-attention scheme to have convolution-like properties [5,7,16,25,26,29]. Swin-transformer [16] is a typical work in the first direction. It uses a local window instead of the whole image to perform self-attention. Although the basic operation is still self-attention, the local window and relative position embedding give self-attention a conv-like local receptive field and less computation cost. Another line in 1) is changing the self-attention operation directly. CoAtNet [5] unifies convolution and self-attention with relative attention, while ConViT [7] uses gated positional self-attention which is equipped with a soft convolutional inductive bias. Works in the second direction 2) employs both convolution and self-attention in the network [3,4,8,20,27,28,30,31,33,35]. For the works in this direction, we summarize them into three major categories as shown in Fig. 1: 2.a) dual branch feature fusion. MobileFormer [4], Conformer [20], and TransFuse [33] use a CNN branch and a transformer branch in parallel to fuse the features, thus the local details and global features are learned separately and fused altogether. However, this doubles the computation cost. Another line of works 2.b) focuses on the bottleneck design. The low-level features are extracted by convolution blocks and the bottleneck is the transformer, like the TransUNet [3], Cotr [30] and TransBTS [27]. The third direction 2.c) is a new block containing both convolution and self-attention. MOAT [31] removes the MLP in self-attention and uses a mobile convolution block at the front. The MOAT block is then used as the basic block in building the network. CvT [28] uses convolution as the embedding layer for key, value, and query. nnFormer [35] replaces the patch merging with convolution with stride.

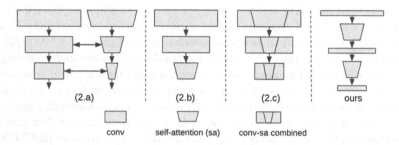

Fig. 1. Three major categories of methods combining convolution with transformers. (2.a): parallel branches with a CNN branch and a transformer branch [4,20,33]. (2.b): Using CNNs to extract local features in the lower level and use transformers in the bottleneck [3,27,30]. (2.c): New transformer blocks with convolution added [28,31]. Our SwinUNETR-V2 adds a convolution block at the beginning of each resolution stage.

Although those works showed strong performances, which works best and can be the "to go" transformer for 3D medical image segmentation is still unknown. For this purpose, we design the SwinUNETR-V2, which improves the current SOTA SwinnUNETR by introducing stage-wise convolutions into the backbone. Our network belongs to the second category, which employs convolution and self-attention directly. At each resolution level, we add a residual convolution (ResConv) block at the beginning, and the output is then used as input to the swin transformer blocks (contains a swin block and a shifted window swin block). MOAT [31] and CvT [28] add convolution before self-attention as a micro-level building block, and nnFormer has a similar design that uses convolution with stride to replace the patch merging layer for downsampling. Differently, our work only adds a ResConv block at the beginning of each stage, which is a macro-network level design. It is used to regularize the features for the following transformers. Although simple, we found it surprisingly effective for 3D medical image segmentation. The network is evaluated extensively on a variety of benchmarks and achieved top performances on the WORD [17], FLARE2021 [18], MSD prostate, MSD lung cancer, and MSD pancreas cancer datasets [1]. Compared to the original Swin-UNETR which needs extensive recipe tuning on a new dataset, we utilized the same training recipe with minimum changes across all benchmarks, showcasing the straight-forward applicability of SwinUNETR-V2 to reach state-of-the-art without extensive hyperparameter tuning or pretraining. We also experimented with four design variations inspired by existing works to justify the SwinUNETR-V2 design.

2   Method

Our SwinUNETR-V2 is based on the original SwinUNETR, and we focus on the trans-former encoder. The overall framework is shown in Fig. 2.

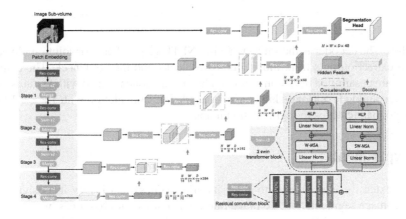

Fig. 2. The SwinUNETR-V2 architecture

Swin-Transformer. We briefly introduce the 3D swin-transformer as used in Swin-UNETR [23]. A patch embedding layer of 3D convolution (stride = 2,2,2, kernel size = 2,2,2) is used to embed the patch into tokens. Four stages of swin transformer block followed by patch merging are used to encode the input patches. Given an input tensor z^i of size (B, C, H, W, D) at swin block i, the swin transformer block splits the tensor into $(\lceil H/M \rceil, \lceil W/M \rceil, \lceil D/M \rceil)$ windows. It performs four operations

$$z^i = \text{W-MSA}(\text{LN}(z^{i-1})) + z^{i-1}; \quad z^i = \text{MLP}(\text{LN}(z^i)) + z^i$$
$$z^{i+1} = \text{SW-MSA}(\text{LN}(z^i)) + z^i; \quad z^{i+1} = \text{MLP}(\text{LN}(z^{i+1})) + z^{i+1}$$

W-MSA and SW-MSA represent regular window and shifted window multi-head self-attention, respectively. MLP and LN represent multilayer perceptron and layernorm, respectively. A patch merging layer is applied after every swin transformer block to reduce each spatial dimension by half.

Stage-Wise Convolution. Although Swin-transformer uses local window attention to introduce inductive bias like convolutions, self-attentions can still mess up with the local details. We experimented with multiple designs as in Fig. 3 and found that interleaved stage-wise convolution is the most effective for swin: convolution followed by swin blocks, then convolution goes on. At the beginning of each resolution level (stage), the input tokens are reshaped back to the original 3D volumes. A residual convolution (ResConv) block with two sets of $3 \times 3x3$ convolution, instance normalization, and leaky relu are used. The output then goes to a set of following swin transformer blocks (we use 2 in the paper). There are in total 4 ResConv blocks at 4 stages. We also tried inverted convolution blocks with depth-wise convolution like MOAT [31] or with original 3D convolution, they improve the performance but are worse than the ResConv block.

Decoder. The decoder is the same as SwinUNETR [23], where convolution blocks are used to extract outputs from those four swin blocks and the bottleneck. The extracted features are upsampled by deconvolutional layers and concatenated with features from a higher-resolution level(long-skip connection). A final convolution with $1 \times 1 \times 1$ kernel is used to map features to segmentation maps.

3   Experiments

We use extensive experiments to show its effectiveness and justify its design for 3D medical image segmentation. To make fair comparisons with baselines, we did not use any pre-trained weights.

Datasets. The network is validated on five datasets of different sizes, targets and modalities:

1) **The WORD dataset** [17] (large-scale Whole abdominal ORgan Dataset) contains 150 high-resolution abdominal CT volumes, each with 16 pixel-level organ annotations. A predefined data split of 100 training, 30 validation, and 20 test are provided. We use this split for our experiments.

2) **The MICCAI FLARE 2021 dataset** [18]. It provides 361 training scans with manual labels from 11 medical centers. Each scan is an abdominal 3D CT image with 4 organ annotations. We follow the test split in the 3D-UXNET[1] [14]: 20 hold-out test scans, and perform 5-fold 80%/20% train validation split on the rest 341 scans.

3) **MSD Task05 prostate, Task06 lung tumour and Task07 pancreas**. The Medical segmentation decathlon (MSD) [1] prostate dataset contains 32 labeled prostate MRI with two modalities for the prostate peripheral zone (PZ) and the transition zone (TZ). The challenges are the large inter-subject variability and limited training data. The lung tumor dataset contains 63 lung CT images with tumor annotations. The challenge comes from segmenting small tumors from large full 3D CT images. The pancreas dataset contains 281 3D CT scans with annotated pancreas and tumors

---

[1] https://github.com/MASILab/3DUX-Net.

(or cysts). The challenge is from the large label imbalances between the background, pancreas, and tumor structures. For all three MSD tasks, we perform 5-fold cross-validation with 70%/10%/20% train, validation, and test splits. These 20% test data will not overlap with other folds and cover all data by 5 folds.

Implementation Details
The training pipeline is based on the publicly available SwinUNETR codebase (https://github.com/Project-MONAI/research-contributions/tree/main/SwinUNETR/BT CV, our training recipe is the same as that by SwinUNETR). We changed the initial learning rate to 4e-4, and the training epoch is adapted to each task such that the total training iteration is about 40k. Random Gaussian smooth, Gaussian noise, and random gamma correction are also added as additional data augmentation. There are differences in data preprocessing across tasks. MSD data are resampled to $1 \times 1x1$ mm resolution and normalized to zero mean and standard deviation (CT images are firstly clipped by .5% and 99.5% foreground intensity percentile). For WORD and FLARE preprocessing, we use the default transforms in SwinUNETR codebase (https://github. com/Project-MONAI/research-contributions/tree/main/SwinUNETR/BTCV, our training recipe is the same as that by SwinUNETR) and 3D UXNet codebase (see footnote 1). Besides these, all other training hyperparameters are the same. We only made those minimal changes for different tasks and show surprisingly good generalizability of the SwinUNETR-V2 and the pipeline across tasks.

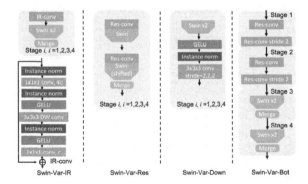

Fig. 3. Design variations for SwinUNETR-V2. Swin-Var-IR replaces the ResConv block in SwinUNETR-V2 with an inverted depth-wise convolution block. Swin-Var-Res added a ResConv block to every swin transformer block. Swin-Var-Down replaces the patch merging with a convolution block with stride 2. Swin-Var-Bot changes the top 2 stages of the encoder with ResConv blocks and only keeps transformer blocks in the higher stages.

Results

WORD Result. We follow the data split in [17] and report the test scores. All the baseline scores are from [17] except nnFormer and SwinUNETR. To make a fair comparison, we didn't use any test-time augmentation or model ensemble. The test set dice

Table 1. WORD test set Dice scores (%) and standard deviation in brackets. The best score is in bold. nnFormer, SwinUNETR and SwinUNETR-V2 results are from our codebase training, the rest is from the WORD paper [17].

Method	nnUNetV2 (2D)	ResUNet (2D)	AttUNet (3D)	nnUNet (3D)	nnUNetV2 (3D)	UNETR (3D)	nnFormer (3D)	CoTr (3D)	SwinUNETR (3D)	SwinUNETR-V2 (3D)
Liver	96.19	96.55	96.00	96.45	96.59	94.67	95.52	95.58	96.6	**96.65**(0.007)
Spleen	94.33	95.26	94.90	95.98	96.09	92.85	94.05	94.9	95.93	**96.16**(0.009)
Kidney (L)	91.29	95.63	94.65	95.40	95.63	91.49	92.8	93.26	94.93	**95.73**(0.009)
Kidney (R)	91.20	95.84	94.7	95.68	95.83	91.72	93.53	93.63	95.5	**95.91**(0.011)
Stomach	91.12	91.58	91.15	91.69	91.57	85.56	88.26	89.99	91.28	**92.31**(0.025)
Gallbladder	83.19	82.83	81.38	83.19	**83.72**	65.08	71.55	76.4	79.67	81.02(0.159)
Esophagus	77.79	77.17	76.87	**78.51**	77.36	67.71	58.89	74.37	77.68	78.36(0.122)
Pancreas	83.55	83.56	83.55	85.04	85.00	74.79	75.28	81.02	85.16	**85.51**(0.06)
Duodenum	64.47	66.67	67.68	68.31	67.73	57.56	58.76	63.58	68.11	**69.93**(0.152)
Colon	83.92	83.57	85.72	87.41	87.26	74.62	77.20	84.14	86.07	**87.46**(0.07)
Intestine	86.83	86.76	88.19	89.3	89.37	80.4	80.78	86.39	88.66	**89.71**(0.029)
Adrenal	70.0	70.9	70.23	72.38	**72.98**	60.76	57.13	69.06	70.58	71.75(0.09)
Rectum	81.49	82.16	80.47	82.41	82.32	74.06	73.42	80.0	81.73	**82.56**(0.05)
Bladder	90.15	91.0	89.71	**92.59**	92.11	85.42	86.97	89.27	91.79	91.56(0.11)
Head of Femur (L)	93.28	**93.39**	91.90	91.99	92.56	89.47	87.04	91.03	92.88	92.64(0.04)
Head of Femur (R)	**93.93**	93.88	92.43	92.74	92.49	90.17	86.87	91.87	92.77	92.9 (0.037)
Mean	85.80	86.67	86.21	87.44	87.41	79.77	79.88	84.66	86.83	**87.51**(0.062)

Table 2. WORD test set HD95 scores and standard deviation in brackets. The best score is in bold. nnFormer, SwinUNETR, and SwinUNETR-V2 are from our codebase training, the rest is from the WORD paper [17].

Method	nnUNetV2 (2D)	ResUNet (2D)	AttUNet (3D)	nnUNet (3D)	nnUNetV2 (3D)	UNETR (3D)	nnFormer (3D)	CoTr (3D)	SwinUNETR (3D)	SwinUNETR-V2 (3D)
Liver	7.34	4.64	3.61	3.31	3.17	8.36	3.95	7.47	2.63	**2.54**(1.36)
Spleen	9.53	8.7	2.74	2.15	2.12	14.84	3.02	8.14	1.78	**1.44**(0.48)
Kidney (L)	10.33	5.4	6.28	6.07	**2.46**	23.37	9.28	16.42	5.24	5.18 (18.98)
Kidney (R)	10.85	2.47	2.86	2.35	2.24	7.9	9.69	12.79	5.77	**1.58** (0.59)
Stomach	13.97	9.98	**8.23**	8.47	9.47	19.25	11.99	10.26	9.95	8.61 (7.86)
Gallbladder	7.91	9.48	**5.11**	5.24	6.04	12.72	6.58	11.32	6.46	5.29(7.54)
Esophagus	6.7	6.7	5.35	5.49	5.83	9.31	7.99	6.29	3.89	**3.32** (1.83)
Pancreas	7.82	7.82	6.96	6.84	6.87	10.66	7.96	8.88	4.84	**4.98** (5.66)
Duodenum	23.29	21.79	21.61	21.3	21.15	25.15	18.18	24.83	18.03	**17.13** (10.44)
Colon	15.68	17.41	10.21	9.99	10.42	20.32	15.38	12.41	9.93	**8.48** (9.28)
Intestine	8.96	9.54	5.68	5.14	5.27	12.62	8.82	7.96	5.33	**3.84** (2.33)
Adrenal	6.42	6.67	5.98	5.46	5.43	8.73	7.53	6.76	5.32	**4.81** (3.89)
Rectum	11.15	10.62	11.67	11.57	12.39	12.79	9.79	11.26	7.71	**7.16** (4.03)
Bladder	4.97	5.02	4.83	3.68	4.17	14.71	4.7	14.34	**2.38**	2.74 (4.15)
Head of Femur (L)	6.54	6.56	6.93	35.18	17.05	38.11	4.21	19.42	**2.78**	2.84 (2.45)
Head of Femur (R)	5.74	5.98	6.06	33.03	27.29	38.62	4.3	26.78	2.99	**2.79** (2.19)
Mean	9.88	8.6	7.13	10.33	8.84	17.34	8.34	12.83	5.94	**5.17** (5.19)

Table 3. FLARE 2021 5-fold cross-validation average test dice scores (on held-out test scans) and standard deviation in brackets. Baseline results from 3D UX-Net paper [14].

	3D U-Net	SegResNet	RAP-Net	nn-UNet	TransBTS	UNETR	nnFormer	SwinUNETR	3D UX-Net	SwinUNETR-V2
Spleen	0.911	0.963	0.946	0.971	0.964	0.927	0.973	0.979	**0.981**	0.980 (0.018)
Kidney	0.962	0.934	0.967	0.966	0.959	0.947	0.960	0.965	0.969	**0.973** (0.013)
Liver	0.905	0.965	0.940	0.976	0.974	0.960	0.975	0.980	0.982	**0.983** (0.008)
Pancreas	0.789	0.745	0.799	0.792	0.711	0.710	0.717	0.788	0.801	**0.851** (0.037)
Mean	0.892	0.902	0.913	0.926	0.902	0.886	0.906	0.929	0.934	**0.947** (0.019)

Table 4. MSD prostate, lung, and pancreas 5-fold cross-validation average test dice scores and standard deviation in brackets. The best score is in bold.

	Task05 Prostate			Task06 Lung	Task07 Pancreas		
	Peripheral zone	Transition zone	Avg.	Tumour (Avg.)	Pancreas	Tumour	Avg.
nnUNet2D	0.5838(0.1789)	0.8063(0.0902)	0.6950(0.1345)	–	–	–	–
nnUNet3D	0.5764(0.1697)	0.7922(0.0979)	0.6843(0.1338)	0.6067(0.2545)	0.7937(0.0882)	0.4507(0.3321)	0.6222(0.2101)
nnFormer	0.5666(0.1955)	0.7876(0.1228)	0.6771(0.1591)	0.4363(0.2080)	0.6405(0.1340)	0.3061(0.2687)	0.4733(0.2013)
UNETR	0.5440(0.1881)	0.7618(0.1213)	0.6529(0.1547)	0.2999(0.1785)	0.7262(0.1109)	0.2606(0.2732)	0.4934(0.1920)
3D-UXNet	0.6102(0.1760)	0.8410(0.0637)	0.7256(0.1198)	0.5999(0.2057)	0.7643(0.0987)	0.4374(0.2930)	0.6009(0.1959)
SwinUNETR	0.6167(0.1862)	**0.8498**(0.0518)	0.7332(0.1190)	0.5672(0.1968)	0.7546(0.0978)	0.3552(0.2514)	0.5549(0.1746)
SwinUNETR-V2	**0.6353**(0.1688)	0.8457(0.0567)	**0.7405**(0.1128)	**0.6203**(0.2012)	**0.8001**(0.0802)	**0.4805**(0.2973)	**0.6403**(0.1887)

Table 5. Dice and HD95 on WORD test set of the variations of SwinUNETR-V2.

	SwinUNETR	Swin-Var-Bot	Swin-Var-IR	Swin-Var-Res	Swin-Var-Down	SwinUNETR-V2
Dice (↑)	0.8683	0.8685	0.8713	0.8713	0.8687	0.8751
HD95 (↓)	5.94	5.18	6.64	5.3	14.04	5.17

score and 95% Hausdorff Distance (hd95) are shown in Table 1 and Table 2. We don't have the original baseline results for statistical testing (we reproduced some baseline results but the results are lower than reported), so we report the standard deviation of our methods. SwinUNETR has 62.5M parameters/295 GFlops and SwinUNETR-V2 has 72.8M parameters/320 GFlops. The baseline parameters/flops can be found in [14].

FLARE 2021 Result. We use the 5-fold cross-validation data split and baseline scores from [14]. Following [14], the five trained models are evaluated on 20 held-out test scans, and the average dice scores (not model ensemble) are shown in Table 3. We can see our SwinUNETR-V2 surpasses all the baseline methods by a large margin.

MSD Results. For MSD datasets, we perform 5-fold cross-validation and ran the baseline experiments with our codebase using exactly the same hyperparameters as mentioned. nnunet2D/3D baseline experiments are performed using nnunet's original codebase[2] since it has its own automatic hyperparameter selection. The test dice score and standard deviation (averaged over 5 fold) are shown in Table 4. We did not do any post-processing or model ensembling, thus there can be a gap between the test values and online MSD leaderboard values. We didn't compare with leaderboard results because the purpose of the experiments is to make fair comparisons, while not resorting to additional training data/pretraining, postprocessing, or model ensembling.

Variations of SwinUNetR-V2 In this section, we investigate other variations of adding convolutions into swin transformer. We follow Fig. 1 and investigate the (2.b) and (2.c) schemes, as well as the inverted convolution block. As for the (2.a) of parallel branches, it increases the GPU memory usage for 3D medical image too much and

---

[2] https://github.com/MIC-DKFZ/nnUNet.

we keep it for future investigation. As shown in Fig. 3, we investigate 1) Swin-Var-Bot (2.b scheme): Replacing the top 2 stages of swin transformer with ResConv block, and keeping the bottom two stages using swin blocks. 2) Swin-Var-IR: Using inverted residual blocks (with 3D depthwise convolution) instead of ResConv blocks. 3) Swin-Var-Res (2.c scheme): Instead of only adding Resconv blocks at the beginning of each stage, we create a new swin transformer block which all starts with this ResConv block, like the MOAT [31] work. 4) Swin-Var-Down: the patch merging is replaced by convolution with stride 2 like nnFormer [35]. We perform the study on the WORD dataset, and the mean test Dice and HD95 scores are shown in Table 5. We can see that adding convolution at different places does affect the performances, and the SwinUNETR-V2 design is the optimal on WORD test set.

4   Discussion and Conclusion

In this paper, we propose a new 3D medical image segmentation network SwinUNETR-V2. For some tasks, we found the original SwinUNETR with pure transformer backbones (or other ViT-based models) may have inferior performance and training stability than CNNs. To improve this, our core intuition is to combine convolution with window-based self-attention. Although existing window-based attention already has a convolution-like inductive bias, it is still not good enough for learning local details as convolutions. We tried multiple combination strategies as in Table 5 and found our current design most effective. By only adding one ResConv block at the beginning of each resolution level, the features can be well-regularized while not too constrained by the convolution inductive bias, and the computation cost will not increase by a lot. Extensive experiments are performed on a variety of challenging datasets, and SwinUNETR-V2 achieved promising improvements. The optimal combination of swin transformer and convolution still lacks a clear principle and theory, and we can only rely on trial and error in designing new architectures. We will apply the network to active challenges for more evaluation.

References

1. Antonelli, M., et al.: The medical segmentation decathlon. Nat. Commun. **13**(1), 1–13 (2022)
2. Cao, H., et al.: Swin-Unet: Unet-like pure transformer for medical image segmentation. arXiv preprint arXiv:2105.05537 (2021)
3. Chen, J., et al.: TransUNet: Transformers make strong encoders for medical image segmentation. arXiv preprint arXiv:2102.04306 (2021)
4. Chen, Y., et al.: Mobile-Former: Bridging mobileNet and transformer. In: Proceedings of the IEEE/CVF Conference on Computer Vision and Pattern Recognition, pp. 5270–5279 (2022)
5. Dai, Z., Liu, H., Le, Q.V., Tan, M.: CoAtNet: marrying convolution and attention for all data sizes. Adv. Neural Inf. Process. Syst. **34**, 3965–3977 (2021)
6. Dosovitskiy, A., et al.: An image is worth 16×16 words: Transformers for image recognition at scale. arXiv preprint arXiv:2010.11929 (2020)
7. d'Ascoli, S., Touvron, H., Leavitt, M.L., Morcos, A.S., Biroli, G., Sagun, L.: ConViT: improving vision transformers with soft convolutional inductive biases. In: International Conference on Machine Learning, pp. 2286–2296. PMLR (2021)

8. Guo, J., et al.: CMT: convolutional neural networks meet vision transformers. In: Proceedings of the IEEE/CVF Conference on Computer Vision and Pattern Recognition, pp. 12175–12185 (2022)
9. Hatamizadeh, A., et al.: UNETR: transformers for 3D medical image segmentation. In: Proceedings of the IEEE/CVF Winter Conference on Applications of Computer Vision, pp. 574–584 (2022)
10. He, Y., Yang, D., Roth, H., Zhao, C., Xu, D.: DiNTS: differentiable neural network topology search for 3D medical image segmentation. In: Proceedings of the IEEE/CVF Conference on Computer Vision and Pattern Recognition, pp. 5841–5850 (2021)
11. Huang, Z. et al.: Revisiting nnU-Net for iterative pseudo labeling and efficient sliding window inference. In: Ma, J., Wang, B. (eds.) Fast and Low-Resource Semi-supervised Abdominal Organ Segmentation. FLARE 2022. Lecture Notes in Computer Science. vol. 13816. Springer, Cham (2022). https://doi.org/10.1007/978-3-031-23911-3_16
12. Isensee, F., Jaeger, P.F., Kohl, S.A., Petersen, J., Maier-Hein, K.H.: nnU-Net: a self-configuring method for deep learning-based biomedical image segmentation. Nat. Methods **18**(2), 203–211 (2021)
13. Ji, Y., et al.: AMOS: A large-scale abdominal multi-organ benchmark for versatile medical image segmentation. arXiv preprint arXiv:2206.08023 (2022)
14. Lee, H.H., Bao, S., Huo, Y., Landman, B.A.: 3D UX-Net: A large kernel volumetric convnet modernizing hierarchical transformer for medical image segmentation. arXiv (2022)
15. Li, X., et al.: The state-of-the-art 3d anisotropic intracranial hemorrhage segmentation on non-contrast head CT: the instance challenge. arXiv preprint arXiv:2301.03281 (2023)
16. Liu, Z., et al.: Swin transformer: hierarchical vision transformer using shifted windows. In: Proceedings of the IEEE/CVF International Conference on Computer Vision, pp. 10012–10022 (2021)
17. Luo, X.: Word: a large scale dataset, benchmark and clinical applicable study for abdominal organ segmentation from CT image. Med. Image Anal. **82**, 102642 (2022)
18. Ma, J., et al.: Fast and low-GPU-memory abdomen CT organ segmentation: the flare challenge. Med. Image Anal. **82**, 102616 (2022)
19. Myronenko, A., Siddiquee, M.M.R., Yang, D., He, Y., Xu, D.: Automated head and neck tumor segmentation from 3D PET/CT. arXiv preprint arXiv:2209.10809 (2022)
20. Peng, Z., et al.: Conformer: local features coupling global representations for visual recognition. In: Proceedings of the IEEE/CVF International Conference on Computer Vision, pp. 367–376 (2021)
21. Ronneberger, O., Fischer, P., Brox, T.: U-Net: convolutional networks for biomedical image segmentation. In: Navab, N., Hornegger, J., Wells, W.M., Frangi, A.F. (eds.) MICCAI 2015. LNCS, vol. 9351, pp. 234–241. Springer, Cham (2015). https://doi.org/10.1007/978-3-319-24574-4_28
22. Siddiquee, M.M.R., Yang, D., He, Y., Xu, D., Myronenko, A.: Automated segmentation of intracranial hemorrhages from 3D CT. arXiv preprint arXiv:2209.10648 (2022)
23. Tang, Y., et al.: Self-supervised pre-training of swin transformers for 3D medical image analysis. In: Proceedings of the IEEE/CVF Conference on Computer Vision and Pattern Recognition, pp. 20730–20740 (2022)
24. Valanarasu, J.M.J., Oza, P., Hacihaliloglu, I., Patel, V.M.: Medical transformer: gated axial-attention for medical image segmentation. In: de Bruijne, M., et al. (eds.) MICCAI 2021. LNCS, vol. 12901, pp. 36–46. Springer, Cham (2021). https://doi.org/10.1007/978-3-030-87193-2_4
25. Vaswani, A., Ramachandran, P., Srinivas, A., Parmar, N., Hechtman, B., Shlens, J.: Scaling local self-attention for parameter efficient visual backbones. In: Proceedings of the IEEE/CVF Conference on Computer Vision and Pattern Recognition, pp. 12894–12904 (2021)

26. Wang, W., et al.: Pyramid vision transformer: a versatile backbone for dense prediction without convolutions. In: Proceedings of the IEEE/CVF International Conference on Computer Vision, pp. 568–578 (2021)

27. Wang, W., Chen, C., Ding, M., Yu, H., Zha, S., Li, J.: TransBTS: multimodal brain tumor segmentation using transformer. In: de Bruijne, M., et al. (eds.) MICCAI 2021. LNCS, vol. 12901, pp. 109–119. Springer, Cham (2021). https://doi.org/10.1007/978-3-030-87193-2_11

28. Wu, H., et al.: CvT: introducing convolutions to vision transformers. In: Proceedings of the IEEE/CVF International Conference on Computer Vision, pp. 22–31 (2021)

29. Xia, Z., Pan, X., Song, S., Li, L.E., Huang, G.: Vision transformer with deformable attention. In: Proceedings of the IEEE/CVF Conference on Computer Vision and Pattern Recognition, pp. 4794–4803 (2022)

30. Xie, Y., Zhang, J., Shen, C., Xia, Y.: CoTr: efficiently bridging CNN and transformer for 3D medical image segmentation. In: de Bruijne, M., et al. (eds.) MICCAI 2021. LNCS, vol. 12903, pp. 171–180. Springer, Cham (2021). https://doi.org/10.1007/978-3-030-87199-4_16

31. Yang, C., et al.: MOAT: alternating mobile convolution and attention brings strong vision models. arXiv preprint arXiv:2210.01820 (2022)

32. Yang, D., et al.: T-AutoML: automated machine learning for lesion segmentation using transformers in 3d medical imaging. In: Proceedings of the IEEE/CVF International Conference on Computer Vision, pp. 3962–3974 (2021)

33. Zhang, Y., Liu, H., Hu, Q.: TransFuse: fusing transformers and CNNs for medical image segmentation. In: de Bruijne, M., et al. (eds.) MICCAI 2021. LNCS, vol. 12901, pp. 14–24. Springer, Cham (2021). https://doi.org/10.1007/978-3-030-87193-2_2

34. Zhou, B., Zhao, H., Puig, X., Fidler, S., Barriuso, A., Torralba, A.: Scene parsing through ADE20K dataset. In: Proceedings of the IEEE/CVF Conference on Computer Vision and Pattern Recognition, pp. 633–641 (2017)

35. Zhou, H.Y., Guo, J., Zhang, Y., Yu, L., Wang, L., Yu, Y.: nnFormer: Interleaved transformer for volumetric segmentation. arXiv preprint arXiv:2109.03201 (2021)

Medical Boundary Diffusion Model for Skin Lesion Segmentation

Jiacheng Wang[1,2], Jing Yang[1], Qichao Zhou[2(✉)], and Liansheng Wang[1(✉)]

[1] Department of Computer Science at School of Informatics, Xiamen University,
Xiamen 361005, China
{jiachengw,jingy77}@stu.xmu.edu.cn, lswang@xmu.edu.cn
[2] Manteia Technologies Co., Ltd., Xiamen 361005, China
zhouqc@manteiatech.com

Abstract. Skin lesion segmentation in dermoscopy images has seen recent success due to advancements in multi-scale boundary attention and feature-enhanced modules. However, existing methods that rely on end-to-end learning paradigms, which directly input images and output segmentation maps, often struggle with extremely hard boundaries, such as those found in lesions of particularly small or large sizes. This limitation arises because the receptive field and local context extraction capabilities of any finite model are inevitably limited, and the acquisition of additional expert-labeled data required for larger models is costly. Motivated by the impressive advances of diffusion models that regard image synthesis as a parameterized chain process, we introduce a novel approach that formulates skin lesion segmentation as a boundary evolution process to thoroughly investigate the boundary knowledge. Specifically, we propose the Medical Boundary Diffusion Model (MB-Diff), which starts with a randomly sampled Gaussian noise, and the boundary evolves within finite times to obtain a clear segmentation map. First, we propose an efficient multi-scale image guidance module to constrain the boundary evolution, which makes the evolution direction suit our desired lesions. Second, we propose an evolution uncertainty-based fusion strategy to refine the evolution results and yield more precise lesion boundaries. We evaluate the performance of our model on two popular skin lesion segmentation datasets and compare our model to the latest CNN and transformer models. Our results demonstrate that our model outperforms existing methods in all metrics and achieves superior performance on extremely challenging skin lesions. The proposed approach has the potential to significantly enhance the accuracy and reliability of skin lesion segmentation, providing critical information for diagnosis and treatment. All resources will be publicly available at https://github.com/jcwang123/MBDiff.

Keywords: Skin lesion segmentation · Diffusion model

Supplementary Information The online version contains supplementary material available at https://doi.org/10.1007/978-3-031-43901-8_41.

1 Introduction

Segmentation of skin lesions from dermoscopy images is a critical task in disease diagnosis and treatment planning of skin cancers [17]. Manual lesion segmentation is time-consuming and prone to inter- and intra-observer variability. To improve the efficiency and accuracy of clinical workflows, numerous automated skin lesion segmentation models have been developed over the years [1,2,7,10,18,19,21]. These models have focused on enhancing feature representations using various techniques such as multi-scale feature fusion [10], attention mechanisms [1,7], self-attention mechanisms [18,19], and boundary-aware attention [2,18,19], resulting in significant improvements in skin lesion segmentation performance. Despite these advances, the segmentation of skin lesions with ambiguous boundaries, particularly at extremely challenging scales, remains a bottleneck issue that needs to be addressed. In such cases, even state-of-the-art segmentation models struggle to achieve accurate and consistent results.

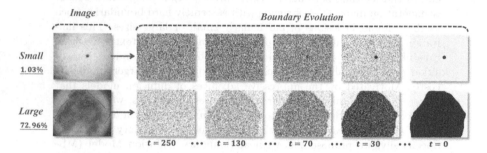

Fig. 1. The boundary evolution process. It could be seen that various lesions can be accurately segmented by splitting the segmentation into sequential timesteps (t), named as **boundary evolution** in this work.

Two representative boundaries are visualized in Fig. 1, where one extremely small lesion and one particularly large lesion are presented. The small one covers 1.03% in the image space and the large one covers 72.96%. As studied prior, solving the segmentation problems of such two types of lesions have different strategies. (1) For the small lesions, translating the features at a lower depth to the convolutional layers at a higher depth can avoid losing local contexts [10]. (2) For the large lesions, enlarging the receptive field by dilated convolution [1], and even global attention [18] can capture the long-range dependencies to improve the boundary decision. Besides the challenge of how to yield stable representations for various scales, multi-scale lesions will cause training fluctuation, that is, small lesions usually lead to large Dice loss. Feeding more boundary-aware supervision can reduce these negative effects to some degree [2,19]. The latest transformer, Xbound-Former, comprehensively addresses the multi-scale boundary problem through cross-scale boundary learning and exactly reaches higher performance on whatever small or large lesions.

However, current models for skin lesion segmentation are still struggling with extremely challenging cases, which are often encountered in clinical practice. While some approaches aim to optimize the model architecture by incorporating local and global contexts and multi-task supervision, and others seek to improve performance by collecting more labeled data and building larger models, both strategies are costly and can be limited by the inherent complexity of skin lesion boundaries. Therefore, we propose a novel approach that shifts the focus from merely segmenting lesion boundaries to predicting their evolution. Our approach is inspired by recent advances in image synthesis achieved by diffusion probabilistic models [6,9,14,15], which generate synthetic samples from a randomly sampled Gaussian distribution in a series of finite steps. We adapt this process to model the evolution of skin lesion boundaries as a parameterized chain process, starting from Gaussian noise and progressing through a series of denoising steps to yield a clear segmentation map with well-defined lesion boundaries. By predicting the next step in the chain process rather than the final segmentation map, our approach enables the more accurate segmentation of challenging lesions than previous models. We illustrate the process of boundary evolution in Fig. 1, where each row corresponds to a different step in the evolution process, culminating in a clear segmentation map with well-defined boundaries.

In this paper, we propose a **Medical Boundary Diff**usion model (**MB-Diff**) to improve the skin lesion segmentation, particularly in cases where the lesion boundaries are ambiguous and have extremely large or small sizes. The MB-Diff model follows the basic design of the plain diffusion model, using a sequential denoising process to generate the lesion mask. However, it also includes two key innovations: Firstly, we have developed an efficient multi-scale image guidance module, which uses a pretrained transformer encoder to extract multi-scale features from prior images. These features are then fused with the evolution features to constrain the direction of evolution. Secondly, we have implemented an evolution uncertainty-based fusion strategy, which takes into account the uncertainty of different initializations to refine the evolution results and obtain more precise lesion boundaries. We evaluate our model on two popular skin lesion segmentation datasets, ISIC-2016 and PH^2 datasets, and find that it performs significantly better than existing models.

2   Method

The key objective of MB-Diff is to improve the representation of ambiguous boundaries by learning boundary evolution through a cascaded series of steps, rather than a single step. In this section, we present the details of our cascaded boundary evolution learning process and the parameterized architecture of the evolution process. We also introduce our evolution-based uncertainty estimation and boundary ensemble techniques, which have significant potential for enhancing the precision and reliability of the evolved boundaries.

2.1   Boundary Evolution Process

We adopt a step-by-step denoising process to model boundary evolution in MB-Diff, drawing inspiration from recent diffusion probabilistic models (DPMs). Specifically, given the image and boundary mask distributions as $(\mathcal{X}, \mathcal{Y})$, assuming that the evolution consists of T steps in total, the boundary at T-th step (y_T) is the randomly initialized noise and the boundary at 0-th (y_0) step denotes the accurate result. We formulate the boundary evolution process as follows:

$$y_0 \sim p_\theta(y_0|x) := \int p_\theta(y_{0:T}|x) dy_{1:T} := p(y_T) \prod_{t=1}^{T} p_\theta(y_{t-1}|y_t, x), \tag{1}$$

where $p(y_T) = \mathcal{N}(y_T; 0, \mathbf{I})$ is the initialized Gaussian distribution and $p_\theta(y_{t-1}|y_t)$ is each learnable evolution step, formulated as the Gaussian transition, denoted as:

$$p_\theta(y_{t-1}|y_t) := \mathcal{N}(y_{t-1}; \mu_\theta(y_t, x, t), \textstyle\sum_\theta(y_t, x, t)). \tag{2}$$

Note that the prediction function takes the input image as a condition, enabling the evolving boundary to fit the corresponding lesion accurately. By modeling boundary evolution as a step-by-step denoising process, MB-Diff can effectively capture the complex structures of skin lesions with ambiguous boundaries, leading to superior performance in lesion segmentation.

To optimize the model parameters θ, we use the evolution target as an approximation of the posterior at each evolution step. Given the segmentation label y as y_0, the label is gradually added by a Gaussian noise as:

$$q(y_{1:T}|y_0) := \prod_{t=1}^{T} q(y_t|y_{t-1}) := \prod_{t=1}^{T} \mathcal{N}(y_t; \sqrt{1 - \beta_t} y_{t-1}, \beta_t \mathbf{I}), \tag{3}$$

where $\{\beta_t\}_{t=1}^{T}$ is a set of constants ranging from 0 to 1. After that, we compute the posterior $q(y_{t-1}|y_t, y_0)$ using Bayes' rule. The MSE loss function is utilized to measure the distance between the predicted mean and covariance of the Gaussian transition distribution and the evolution target $q(y_{t-1}|y_t, y_0)$.

2.2   Paramterized Architecture with Image Prior

The proposed model is a parameterized chain process that predicts the μ_{t-1}^* and $\sum_{t-1}^*$ at each evolution step t under the prior conditions of the image x and the prior evolution y_t^*. To capture the deep semantics of these conditions and perform efficient fusion, we adopt a basic U-Net [16] architecture inspired by the plain DPM and introduce novel designs for condition fusion, that is the efficient multi-scale image guidance module.

The architecture consists of a multi-level convolutional encoder and a symmetric decoder with short connection layers between them. To incorporate the variable t into the model, we first embed it into the latent space. Then, the prior evolution y_t^* is added to the latent t before each convolution. At the bottleneck layer, we fuse the evolution features with the image guidance to constrain the evolution and ensure that the final boundary suits the conditional image.

To achieve this, priors train a segmentation model concurrently with the evolution model and use an attention-based parser to translate the image features in the segmentation branch into the evolution branch [22]. Since the segmentation model is trained much faster than the evolution model, we adopt a pretrained pyramid vision transformer (PVT) [20] as the image feature extractor to obtain the multi-scale image features. Let $\{f_l\}_{l=1}^4$ denote the extracted features at four levels, with a 2x, 4x, 8x, 16x, smaller size of the original input. Each feature at the three lower levels is resized to match the scale of f_4 using Adaptive Averaging Pooling layers. After that, the four features are concatenated and fed into a full-connection layer to map the image feature space into the evolution space. We then perform a simple yet effective addition of the mapped image feature and the encoded prior evolution feature, similar to the fusion of time embeddings, to avoid redundant computation.

2.3   Evolution Uncertainty

Similar to typical evolutionary algorithms, the final results of boundary evolution are heavily influenced by the initialized population. As a stochastic chain process, the boundary evolution process may result in different endpoints due to the random Gaussian samples at each evolution step. This difference is particularly evident when dealing with larger ambiguity in boundary regions. The reason is that the image features in such ambiguous regions may not provide discriminative guidance for the evolution, resulting in significant variations in different evolution times. Instead of reducing the differences, we surprisingly find that these differences can represent segmentation uncertainty. Based on the evolution-based uncertainty estimation, the segmentation results become more accurate and trustworthy in practice [4,5,12].

Uncertainty Estimation: To estimate uncertainty, the model parameters θ are fixed, and the evolution starts with a randomly sampled Gaussian noise $y_T^* \sim \mathcal{N}(0, \mathbf{I})$. Let $\{y_T^{*,i}\}_{i=1}^n$ denote a total of n initializations. Once the evolution is complete, the obtained $\{\mu^{*,i}\}_{i=1}^n, \{\sum^{*,i}\}_{i=1}^n$ are used to the sample final lesion maps as: $y^{*,i} = \mu^{*,i} + exp(\frac{1}{2}\sum^{*,i})\mathcal{N}(0, \mathbf{I})$. Unlike traditional segmentation models that typically scale the prediction into the range of 0 to 1, the evolved maps generated by MB-Diff have unfixed distributions due to random sampling. Since the final result is primarily determined by the mean value μ, and the predicted $\sum$ has a limited range [6], we calculate the uncertainty as:

$$\delta = \sqrt{\frac{1}{n}\sum_{i=1}^n(\mu^{*,i} - \frac{1}{n}\sum_{j=1}^n \mu^{*,j})^2}, \tag{4}$$

Evolution Ensemble: Instead of training multiple networks or parameters to make the ensemble, MB-Diff allows running the inference multiple times and fusing the obtained evolutions. However, simply averaging the predicted identities from multiple evolutions is not effective, as the used MSE loss without activation constrains the predicted identities to be around 0 or 1, unlike the Sigmoid function which would limit the identities to a range between 0 and 1. Therefore,

Table 1. Comparison of skin lesion segmentation with different approaches on the ISIC-2016 and PH^2 datasets. The averaged scores of both sets are presented respectively.

Method	ISIC-2016 [8]				PH^2 [13]			
	IoU↑	Dice↑	ASSD↓	HD95↓	IoU↑	Dice↑	ASSD↓	HD95↓
U-Net++ [24]	81.84	88.93	15.01	44.83	81.26	88.99	15.97	46.66
CA-Net [7]	80.73	88.10	15.67	44.98	75.18	84.66	21.06	64.53
TransFuse [23]	86.19	92.03	10.04	30.33	82.32	89.75	15.00	39.98
TransUNet [3]	84.89	91.26	10.63	28.51	83.99	90.96	12.65	33.30
XBound-Former [18]	87.69	93.08	8.21	21.83	85.38	91.80	10.72	26.00
MedSegDiff [22]	83.39	89.85	12.38	31.23	82.21	89.73	13.53	36.59
MB-Diff (Ours)	**88.87**	**93.78**	**7.19**	**18.90**	**87.12**	**92.85**	**9.16**	**22.95**

we employ the max vote algorithm to obtain the final segmentation map. In this algorithm, each pixel is classified as a lesion only if its identity sum across all n evolutions is greater than a threshold value τ. Finally, the segmentation map is generated as $y^* = (\sum_{i=1}^{n} y^{*,i}) \geq \tau$.

3   Experiment

3.1   Datasets and Evaluation Metrics

Datasets: We use two publicly available skin lesion segmentation datasets from different institutions in our experiments: the ISIC-2016 dataset and the PH^2 dataset. The ISIC-2016 dataset [8] is provided by the International Skin Imaging Collaboration (ISIC) archive and consists of 900 samples in the public training set and 379 samples in the public validation set. As the annotation for its public test set is not currently available, we additionally collect the PH^2 dataset [13], which contains 200 labeled samples and is used to evaluate the generalization performance of our methods.

Evaluation Metrics: To comprehensively compare the segmentation results, particularly the boundary delineations, we employ four commonly used metrics to quantitatively evaluate the performance of our segmentation methods. These metrics include the *Dice* score, the *IoU* score, Average Symmetric Surface Distance (*ASSD*), and Hausdorff Distance of boundaries (95−th percentile; *HD*95). To ensure fair comparison, all labels and predictions are resized to (512×512) before computing these scores, following the approach of a previous study [18].

3.2   Implementation Details

For the diffusion model hyper-parameters, we use the default settings of the plain diffusion model, which can be found in the supplementary materials. Regarding

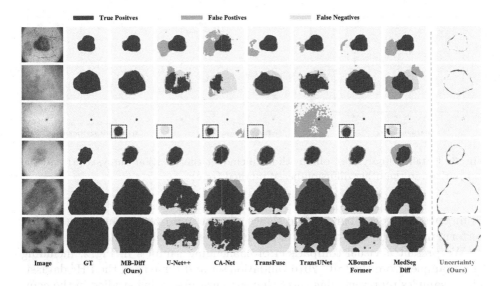

Fig. 2. Visual comparison of our method and the SOTAs. The first three rows are samples from the ISIC-2016 validation set and the last three rows are from the PH² dataset. We highlight the small lesions using dotted boxes in the third row.

the training parameters, we resize all images to (256×256) for efficient memory utilization and computation. We use a set of random augmentations, including vertical flipping, horizontal flipping, and random scale change (limited to $0.9 \sim 1.1$), to augment the training data. We set the batch size to 4 and train our model for a total of 200,000 iterations. During training, we use the AdamW optimizer with an initial learning rate of 1e-4. For the inference, we set $n = 4$ and $\tau = 2$ considering the speeds.

3.3   Comparison with State-of-the-Arts

We majorly compare our method to the latest skin lesion segmentation models, including the CNN-based and transformer-based models, i.e., U-Net++ [24], CA-Net [7], TransFuse [23], TransUNet [3], and especially the boundary-enhanced method, X-BoundFormer [18]. Additionally, we evaluate our method against MedSegDiff [22], a recently released diffusion-based model, which we re-trained for 200,000 steps to ensure a fair comparison.

The quantitative results are shown in Table 1, which reports four evaluation scores for two datasets. Though the parameters of CNNs and Transformers are selected with the best performance on ISIC-2016 validation set and the parameters of our method are selected by completing the 200,000 iterations, MB-Diff still achieves the 1.18% IoU improvement and 0.7% Dice improvement. Additionally, our predicted boundaries are closer to the annotations, as evidenced by the ASSD and HD95 metrics, which reduce by 1.02 and 1.93 pixels, respectively. When compared to MedSegDiff, MB-Diff significantly outperforms it in

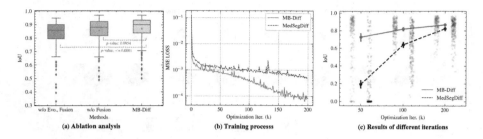

(a) Ablation analysis   (b) Training processs   (c) Results of different iterations

Fig. 3. Detailed analysis of our method, including the ablation analysis (a) and the comparison to the other diffusion-based method (b, c).

all metrics. Moreover, our method shows a larger improvement in generalization performance on the PH^2 dataset, indicating its better ability to handle new data.

We present a visual comparison of challenging samples in Fig. 2, including three samples from the ISIC-2016 validation set and three from the PH^2 dataset. These samples represent edge cases that are currently being studied in the community, including size variation, boundary ambiguity, and neighbor confusion. Our visual comparison reveals several key findings: (1) MB-Diff consistently achieves better segmentation performance on small and large lesions due to its thorough learning of boundary evolution, as seen in rows 3, 5, and 6. (2) MB-Diff is able to produce correct boundaries even in cases where they are nearly indistinguishable in human perception, eliminating the need for further manual adjustments and demonstrating significant practical value. (3) MB-Diff generates fewer false positive segmentation, resulting in cleaner predictions that enhance the user experience.

Furthermore, we provide a visualization of evolution uncertainties in Fig. 2, where deeper oranges indicate larger uncertainties. It is evident that most regions with high uncertainties correspond to false predictions. This information can be used to guide human refinement of the segmentation in practical applications, ultimately increasing the AI's trustworthiness.

3.4   Detailed Analysis of the Evolution

In this subsection, we make a comprehensive analysis to investigate the performance of each component in our method and compare it to the diffusion-based model, MedSegDiff. The results of our ablation study are presented in Fig. 3(a), where "w/o Evo" refers to using image features to directly train a segmentation model with FPN [11] architecture and "w/o Fusion" means no evolution fusion is used. To ensure a fair comparison, we average the scores of multiple evolutions to represent the performance of "w/o Fusion". The results demonstrate that our evolutionary approach can significantly improve performance, and the evolution uncertainty-based fusion strategy further enhances performance. Comparing our method to MedSegDiff, the training loss curve in Fig. 3(b) shows that our method converges faster and achieves smaller losses, indicating that our multi-scale image guidance is more effective than that of MedSegDiff. Furthermore, we evaluate our method's performance using parameters saved at different iterations, as shown in Fig. 3(c). Our results demonstrate that our method has competitive performance at 50k iterations versus MedSegDiff at 200k iterations and our method at 100k iterations has already outperformed well-trained MedSegDiff.

4 Conclusion

In this paper, we introduced the medical boundary diffusion (MB-Diff) model, which is a novel approach to segment skin lesions. Our proposed method formulates lesion segmentation as a boundary evolution process with finite timesteps, which allows for efficient and accurate segmentation of skin lesions. To guide the boundary evolution towards the lesions, we introduce an efficient multi-scale image guidance module. Additionally, we propose an evolution uncertainty-based fusion strategy to yield more accurate segmentation. Our method is evaluated on two well-known skin lesion segmentation datasets, and the results demonstrate superior performance and generalization ability in unseen domains. Through a detailed analysis of our training program, we find that our model has faster convergence and better performance compared to other diffusion-based models. Overall, our proposed MB-Diff model offers a promising solution to accurately segment skin lesions, and has the potential to be applied in a clinical setting.

Acknowledgement. This work is supported by the National Key Research and Development Program of China (2019YFE0113900).

References

1. Azad, R., Asadi-Aghbolaghi, M., Fathy, M., Escalera, S.: Attention Deeplabv3+: multi-level context attention mechanism for skin lesion segmentation. In: Bartoli, A., Fusiello, A. (eds.) ECCV 2020. LNCS, vol. 12535, pp. 251–266. Springer, Cham (2020). https://doi.org/10.1007/978-3-030-66415-2_16
2. Cao, W., et al.: ICL-Net: global and local inter-pixel correlations learning network for skin lesion segmentation. IEEE J. Biomed. Health Inf. **27**(1), 145–156 (2022)
3. Chen, J., et al.: TransUNet: Transformers make strong encoders for medical image segmentation. arXiv preprint arXiv:2102.04306 (2021)
4. Czolbe, S., Arnavaz, K., Krause, O., Feragen, A.: Is segmentation uncertainty useful? In: Feragen, A., Sommer, S., Schnabel, J., Nielsen, M. (eds.) IPMI 2021. LNCS, vol. 12729, pp. 715–726. Springer, Cham (2021). https://doi.org/10.1007/978-3-030-78191-0_55
5. DeVries, T., Taylor, G.W.: Leveraging uncertainty estimates for predicting segmentation quality. arXiv preprint arXiv:1807.00502 (2018)
6. Dhariwal, P., Nichol, A.: Diffusion models beat GANs on image synthesis. Adv. Neural Inf. Process. Syst. **34**, 8780–8794 (2021)
7. Gu, R., et al.: Ca-net: Comprehensive attention convolutional neural networks for explainable medical image segmentation. IEEE Trans. Med. Imaging **40**(2), 699–711 (2020)

8. Gutman, D., et al.: Skin lesion analysis toward melanoma detection: A challenge at the international symposium on biomedical imaging (ISBI) 2016, hosted by the international skin imaging collaboration (ISIC). arXiv preprint arXiv:1605.01397 (2016)
9. Ho, J., Jain, A., Abbeel, P.: Denoising diffusion probabilistic models. Adv. Neural Inf. Process. Syst. **33**, 6840–6851 (2020)
10. Li, H., et al.: Dense deconvolutional network for skin lesion segmentation. IEEE J. Biomed. Health Inf. **23**(2), 527–537 (2018)
11. Lin, T.Y., Dollár, P., Girshick, R., He, K., Hariharan, B., Belongie, S.: Feature pyramid networks for object detection. In: Proceedings of the IEEE/CVF Conference on Computer Vision and Pattern Recognition, pp. 2117–2125 (2017)
12. Mehrtash, A., Wells, W.M., Tempany, C.M., Abolmaesumi, P., Kapur, T.: Confidence calibration and predictive uncertainty estimation for deep medical image segmentation. IEEE Trans. Med. Imaging **39**(12), 3868–3878 (2020)
13. Mendonça, T., Ferreira, P.M., Marques, J.S., Marcal, A.R., Rozeira, J.: PH 2-A dermoscopic image database for research and benchmarking. In: 2013 35th Annual International Conference of the IEEE Engineering in Medicine and Biology Society (EMBC), pp. 5437–5440. IEEE (2013)
14. Nichol, A.Q., Dhariwal, P.: Improved denoising diffusion probabilistic models. In: International Conference on Machine Learning, pp. 8162–8171. PMLR (2021)
15. Rombach, R., Blattmann, A., Lorenz, D., Esser, P., Ommer, B.: High-resolution image synthesis with latent diffusion models. In: Proceedings of the IEEE/CVF Conference on Computer Vision and Pattern Recognition, pp. 10684–10695 (2022)
16. Ronneberger, O., Fischer, P., Brox, T.: U-Net: convolutional networks for biomedical image segmentation. In: Navab, N., Hornegger, J., Wells, W.M., Frangi, A.F. (eds.) MICCAI 2015. LNCS, vol. 9351, pp. 234–241. Springer, Cham (2015). https://doi.org/10.1007/978-3-319-24574-4_28
17. Siegel, R.L., Miller, K.D., Fuchs, H.E., Jemal, A.: Cancer statistics, 2022. CA: Cancer J. Clin. **72**(1), 7–33 (2022)
18. Wang, J., et al.: XBound-former: toward cross-scale boundary modeling in transformers. IEEE Trans. Med. Imaging **42**(6), 1735–1745 (2023)
19. Wang, J., Wei, L., Wang, L., Zhou, Q., Zhu, L., Qin, J.: Boundary-aware transformers for skin lesion segmentation. In: de Bruijne, M., et al. (eds.) MICCAI 2021. LNCS, vol. 12901, pp. 206–216. Springer, Cham (2021). https://doi.org/10.1007/978-3-030-87193-2_20
20. Wang, W., et al.: Pyramid vision transformer: a versatile backbone for dense prediction without convolutions. In: Proceedings of the IEEE/CVF International Conference on Computer Vision, pp. 568–578 (2021)
21. Wu, H., Pan, J., Li, Z., Wen, Z., Qin, J.: Automated skin lesion segmentation via an adaptive dual attention module. IEEE Trans. Med. Imaging **40**(1), 357–370 (2020)
22. Wu, J., Fang, H., Zhang, Y., Yang, Y., Xu, Y.: MedSegDiff: Medical image segmentation with diffusion probabilistic model. arXiv preprint arXiv:2211.00611 (2022)
23. Zhang, Y., Liu, H., Hu, Q.: TransFuse: fusing transformers and CNNs for medical image segmentation. In: de Bruijne, M., et al. (eds.) MICCAI 2021. LNCS, vol. 12901, pp. 14–24. Springer, Cham (2021). https://doi.org/10.1007/978-3-030-87193-2_2
24. Zhou, Z., Rahman Siddiquee, M.M., Tajbakhsh, N., Liang, J.: UNet++: a nested u-net architecture for medical image segmentation. In: Stoyanov, D., et al. (eds.) DLMIA/ML-CDS -2018. LNCS, vol. 11045, pp. 3–11. Springer, Cham (2018). https://doi.org/10.1007/978-3-030-00889-5_1

Instructive Feature Enhancement for Dichotomous Medical Image Segmentation

Lian Liu[1,2,3], Han Zhou[1,2,3], Jiongquan Chen[1,2,3], Sijing Liu[1,2,3], Wenlong Shi[4], Dong Ni[1,2,3], Deng-Ping Fan[5(✉)], and Xin Yang[1,2,3(✉)]

[1] National-Regional Key Technology Engineering Laboratory for Medical Ultrasound, School of Biomedical Engineering, Health Science Center, Shenzhen University, Shenzhen, China
xinyang@szu.edu.cn
[2] Medical Ultrasound Image Computing (MUSIC) Lab, Shenzhen University, Shenzhen, China
[3] Marshall Laboratory of Biomedical Engineering, Shenzhen University, Shenzhen, China
[4] Shenzhen RayShape Medical Technology Co., Ltd., Shenzhen, China
[5] Computer Vision Lab (CVL), ETH Zurich, Zurich, Switzerland
dengpfan@gmail.com

Abstract. Deep neural networks have been widely applied in dichotomous medical image segmentation (DMIS) of many anatomical structures in several modalities, achieving promising performance. However, existing networks tend to struggle with task-specific, heavy and complex designs to improve accuracy. They made little instructions to which feature channels would be more beneficial for segmentation, and that may be why the performance and universality of these segmentation models are hindered. In this study, we propose an instructive feature enhancement approach, namely **IFE**, to adaptively select feature channels with rich texture cues and strong discriminability to enhance raw features based on local curvature or global information entropy criteria. Being plug-and-play and applicable for diverse DMIS tasks, IFE encourages the model to focus on texture-rich features which are especially important for the ambiguous and challenging boundary identification, simultaneously achieving simplicity, universality, and certain interpretability. To evaluate the proposed IFE, we constructed the first large-scale DMIS dataset **Cosmos55k**, which contains 55,023 images from 7 modalities and 26 anatomical structures. Extensive experiments show that IFE can improve the performance of classic segmentation networks across different anatomies and modalities with only slight modifications. Code is available at https://github.com/yezi-66/IFE.

1 Introduction

Medical image segmentation (MIS) can provide important information regarding anatomical or pathological structural changes and plays a critical role in

---

L. Liu and H. Zhou—Contributed equally to this work.

© The Author(s), under exclusive license to Springer Nature Switzerland AG 2023
H. Greenspan et al. (Eds.): MICCAI 2023, LNCS 14223, pp. 437–447, 2023.
https://doi.org/10.1007/978-3-031-43901-8_42

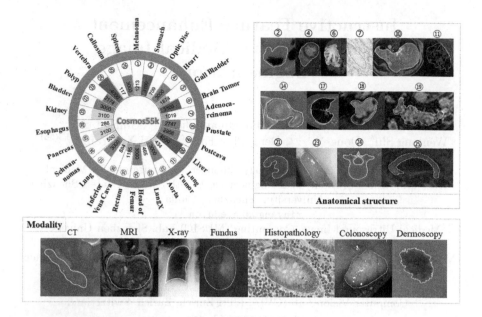

Fig. 1. Statistics, modalities, and examples of anatomical structures in Cosmos55k.

computer-aided diagnosis. With the rapid development of intelligent medicine, MIS involves an increasing number of imaging modalities, raising requirements for the accuracy and universality of deep models.

Medical images have diverse modalities owing to different imaging methods. Images from the same modality but various sites can exhibit high similarity in overall structure but diversity in details and textures. Models trained on specific modalities or anatomical structure datasets may not adapt to new datasets. Similar to dichotomous image segmentation tasks [26], MIS tasks typically input an image and output a binary mask of the object, which primarily relies on the dataset. To facilitate such dichotomous medical image segmentation (DMIS) task, we constructed Cosmos55k, a large-scale dataset of 55,023 challenging medical images covering 26 anatomical structures and 7 modalities (see Fig. 1).

Most current MIS architectures are carefully designed. Some increased the depth or width of the backbone network, such as UNet++ [36], which uses nested and dense skip connections, or DeepLabV3+ [9], which combines dilated convolutions and feature pyramid pooling with an effective decoder module. Others have created functional modules such as Inception and its variants [10,31], depthwise separable convolution [17], attention mechanism [16,33], and multi-scale feature fusion [8]. Although these modules are promising and can be used flexibly, they typically require repeated and handcrafted adaptation for diverse DMIS tasks. Alternatively, frameworks like nnUNet [18] developed an adaptive segmentation pipeline for multiple DMIS tasks by integrating key dataset attributes and achieved state-of-the-art. However, heavy design efforts are needed for nnUNet and the cost is very expensive. More importantly, previous DMIS networks often ignored the importance of identifying and enhancing the determining and instructive feature channels for segmentation, which potentially limits their performance in general DMIS tasks.

We observed that the texture-rich and sharp-edge cues in specific feature channels are crucial and instructive for accurately segmenting objects. Curvature [15] can represent the edge characteristics in images. Information entropy [1] can describe the texture and content complexity of images. In this study, we focus on exploring their roles in quantifying feature significance. Based on curvature and information entropy, we propose a simple approach to balance accuracy and universality with only minor modifications of the classic networks. Our contribution is three folds. First, we propose the novel 2D DMIS task and construct a large-scale dataset (*Cosmos55k*) to benchmark the goal and we will release the dataset to contribute to the community. Second, we propose a simple, generalizable, and effective instructive feature enhancement approach (*IFE*). With extensive experiments on Cosmos55k, IFE soundly proves its advantages in promoting various segmentation networks and tasks. Finally, we provide an interpretation of which feature channels are more beneficial to the final segmentation results. We believe IFE will benefit various DMIS tasks.

2   Methodology

Overview. Figure 2 and Fig. 3 illustrate two choices of the proposed IFE. It is general for different DMIS tasks. **1)** We introduce a feature quantization method based on either curvature (Fig. 2) or information entropy (Fig. 3), which characterizes the content abundance of each feature channel. The larger these parameters, the richer the texture and detail of the corresponding channel feature. **2)** We select a certain proportion of channel features with high curvature or information entropy and combine them with raw features. IFE improves performance with minor modifications to the segmentation network architecture.

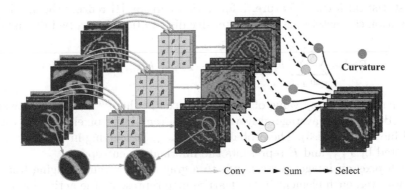

Fig. 2. Example of feature selection using curvature.

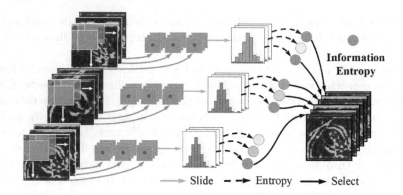

Fig. 3. Example of feature selection using 2D information entropy. (Color figure online)

2.1   Curvature-Based Feature Selection

For a two-dimensional surface embedded in Euclidean space R^3, two curvatures exist: Gaussian curvature and mean curvature. Compared with the Gaussian curvature, the mean curvature can better reflect the unevenness of the surface. Gong [15] proposed a calculation formula that only requires a simple linear convolution to obtain an approximate mean curvature, as shown below:

$$C = \begin{bmatrix} C_1 \ C_2 \ C_3 \end{bmatrix} * X, \tag{1}$$

where $C_1 = [\alpha, \beta, \alpha]^T$, $C_2 = [\beta, \gamma, \beta]^T$, $C_3 = [\alpha, \beta, \alpha]^T$, the values of α, β, and γ are $-1/16$, $5/16$, -1. $*$ denotes convolution, X represents the input image, and C is the mean curvature. Figure 2 illustrates the process, showing that the curvature image can effectively highlight the edge details in the features.

2.2   Information Entropy-Based Feature Selection

As a statistical form of feature, information entropy [1] reflects the spatial and aggregation characteristics of the intensity distribution. It can be formulated as:

$$E = -\sum_{i=0}^{255}\sum_{j=0}^{255} P_{i,j} log_2\left(P_{i,j}\right), \ P_{i,j} = f\left(i_n, j_n\right)/\left(H \times W\right), \tag{2}$$

i_n denotes the gray value of the center pixel in the nth 3×3 sliding window and j_n denotes the average gray value of the remaining pixels in that window (teal blue box in Fig. 3). The probability of (i_n, j_n) occurring in the entire image is denoted by $P_{i,j}$, and E represents the information entropy.

Each pixel on images corresponds to a gray or color value ranging from 0 to 255. However, each element in the feature map represents the activation level of the convolution filter at a particular position in the input image. Given an input feature F_x, as shown in Fig. 3, the tuples (i, j) obtained by the sliding windows are transformed to a histogram, representing the magnitude of the activation

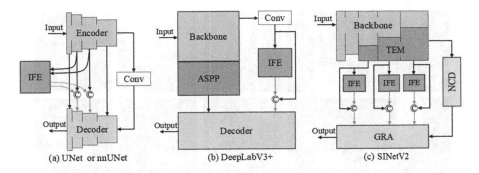

Fig. 4. Implementation of IFE in exemplar networks. ⓒ is concatenation.

level and distribution within the neighborhood. This involves rearranging the activation levels of the feature map. The histogram converting method *histc* and information entropy E are presented in Algorithm 1. Note that the probability $P_{hist(i,j)}$ will be used to calculate the information entropy.

2.3 Instructive Feature Enhancement

Although IFE can be applied to various deep neural networks, this study mainly focuses on widely used segmentation networks. Figure 4 shows the framework of IFE embedded in representative networks, *e.g.*, DeepLabV3+ [9], UNet [28], nnUNet [18], and SINetV2 [14]. The first three are classic segmentation networks. Because the MIS task is similar to the camouflage object detection, such as low contrast and blurred edges, we also consider SINetV2 [14]. According to [27], we implement IFE on the middle layers of UNet [28] and nnUNet [18], on the low-level features of DeepLabV3+ [9], and on the output of the TEM of SINetV2 [14].

While the input images are encoded into the feature space, the different channel features retain textures in various directions and frequencies. Notably, the information contained by the same channel may vary across different images, which can be seen in Fig. 5. For instance, the 15th channel of the lung CT feature map contains valuable texture and details, while the same channel in the aortic CT feature map may not provide significant informative content. However,

---

Algorithm 1. Information entropy of features with histogram.

Input:$F_x \in \mathbb{R}^{C \times \mathbb{H} \times W}, bins = 256, kernel\_size = 3$
Output:$E \in \mathbb{R}^C$
$z = unfold\,(F, kernel\_size)$ ▷ Sliding window operation.
$i = flatten\,(z)\,[(H \times W)\,//2]$
$j = (sum\,(z) - i)\,/((H \times W) - 1)$
$f_{hist(i,j)} = histc\,((i,j),bins)$ ▷ Compute the histogram of (i,j).
$ext\_k = kernel\_size//2$
$P_{hist(i,j)} = f_{hist(i,j)}/\,((H + ext_k) \times (W + ext_k))$
$E = sum\,(-P_{hist(i,j)} \times log_2\,(P_{hist(i,j)}))$

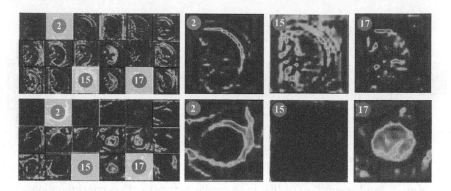

Fig. 5. Feature maps visualization of the lung (first row) and aortic (second row) CT images from stage 3 of SINetV2. 2, 15, and 17 are the indexes of channels. The information contained by the same channel may vary across different images.

their 2nd channel features both focus on the edge details. By preserving the raw features, the channel features that contribute more to the segmentation of the current object can be enhanced by dynamically selecting from the input features. Naturally, it is possible to explicitly increase the sensitivity of the model to the channel information. Specifically, for input image $\mathbf{X}$, the deep feature $\boldsymbol{F_x} = [f_1, f_2, f_3, \cdots, f_C] \in \mathbb{R}^{C*H*W}$ can be obtained by an encoder with the weights θ_x: $\boldsymbol{F_x} = Encoder\,(\mathbf{X}, \theta_x)$, our IFE can be expressed as:

$$\boldsymbol{F}_x^{'} = max\{S\,(\boldsymbol{F_x})\,, r\},\tag{3}$$

$\boldsymbol{F}_x^{'}$ is the selected feature map and S is the quantification method (see Fig. 2 or Fig. 3), and r is the selected proportion. As discussed in [6], enhancing the raw features through pixel-wise addition may introduce unwanted background noise and cause interference. In contrast, the concatenate operation directly joins the features, allowing the network to learn how to fuse them automatically, reducing the interference caused by useless background noises. Therefore, we used the concatenation and employed the concatenated features $\boldsymbol{F} = [\boldsymbol{F_x}, \boldsymbol{F}_x^{'}]$ as the input to the next stage of the network. Only the initialization channel number of the corresponding network layer needs to be modified.

3   Experimental Results

Cosmos55k. To construct the large-scale Cosmos55k, 30 publicly available datasets [3,4,11,19–24,29,30,32,35] were collected and processed with organizers' permission. The processing procedures included uniform conversion to PNG format, cropping, and removing mislabeled images. Cosmos55k (Fig. 1) offers 7 imaging modalities, including CT, MRI, X-ray, fundus, etc., covering 26 anatomical structures such as the liver, polyp, melanoma, and vertebra, among others.

Table 1. Quantitative comparison on IFE. +C, +E means curvature- or information entropy-based IFE. DLV3+ is DeepLabV3+. **Bolded** means the best group result. * denotes that DSC passed *t-test*, $p < 0.05$.

Method	Con(%)↑	DSC (%)↑	JC (%)↑	F1 (%)↑	HCE↓	MAE (%)↓	HD↓	ASD↓	RVD↓
UNet	84.410	93.695	88.690	**94.534**	1.988	**1.338**	11.464	2.287	7.145
+C	85.666	94.003*	89.154	94.528	1.777	1.449	**11.213**	2.222	6.658
+E	**86.664**	**94.233***	**89.526**	94.466	**1.610**	1.587	11.229	**2.177**	**6.394**
nnUNet	−53.979	92.617	87.267	92.336	2.548	0.257	19.963	3.698	9.003
+C	−30.908	**92.686**	**87.361**	**92.399**	2.521	0.257	19.840	**3.615**	8.919
+E	−34.750	92.641	87.313	92.367	**2.510**	0.257	**19.770**	3.637	**8.772**
SINetV2	80.824	93.292	88.072	93.768	2.065	1.680	11.570	2.495	7.612
+C	**84.883**	**93.635***	**88.525**	**94.152**	**1.971**	**1.573**	**11.122**	**2.402**	**7.125**
+E	83.655	93.423	88.205	93.978	2.058	1.599	11.418	2.494	7.492
DLV3+	88.566	94.899	90.571	95.219	1.289	**1.339**	9.113	2.009	5.603
+C	**88.943**	95.000*	90.738	**95.239**	1.274	1.369	**8.885**	**1.978**	**5.391**
+E	88.886	**95.002***	**90.741**	95.103	**1.257**	1.448	9.108	2.011	5.468

Images contain just one labeled object, reducing confusion from multiple objects with different structures.

Implementation Details. Cosmos55k comprises 55,023 images, with 31,548 images used for training, 5,884 for validation, and 17,591 for testing. We conducted experiments using Pytorch for UNet [28], DeeplabV3+ [9], SINetV2 [14], and nnUNet [18]. The experiments were conducted for 100 epochs on an RTX 3090 GPU. The batch sizes for the first three networks were 32, 64, and 64, respectively, and the optimizer used was Adam with an initial learning rate of 10^{-4}. Every 50 epochs, the learning rate decayed to $1/10$ of the former. Considering the large scale span of the images in Cosmos55k, the images were randomly resized to one of seven sizes (224, 256, 288, 320, 352, or 384) before being fed into the network for training. During testing, the images were resized to a fixed size of 224. Notably, the model set the hyperparameters for nnUNet [18] automatically.

Table 2. Ablation studies of UNet about ratio r. **Bolded** means the best result.

Method	Ratio	Con (%)↑	DSC (%)↑	JC (%)↑	F1 (%)↑	HCE↓	MAE (%)↓	HD↓	ASD↓	RVD↓
UNet	0,0	84.410	93.695	88.690	**94.534**	1.988	**1.338**	11.464	2.287	7.145
	1.0,1.0	84.950	93.787	88.840	94.527	1.921	1.376	11.417	2.271	6.974
+C	0.75,0.75	84.979	93.855	88.952	94.528	1.865	1.405	11.378	2.246	6.875
	0.75,0.50	85.666	94.003	89.154	94.528	1.777	1.449	**11.213**	2.222	6.658
	0.50,0.50	84.393	93.742	88.767	94.303	1.908	1.497	11.671	2.305	7.115
+E	0.75,0.75	85.392	93.964	89.106	94.290	1.789	1.597	11.461	2.252	6.712
	0.75,0.50	83.803	93.859	88.949	94.420	1.877	1.471	11.351	2.260	6.815
	0.50,0.50	**86.664**	**94.233**	**89.526**	94.466	**1.610**	1.587	11.229	**2.177**	**6.394**

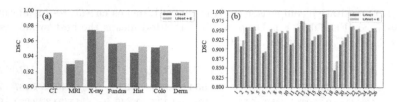

Fig. 6. DSC of UNet (blue) and UNet+E (pink) in modalities (a) and anatomic structures (b). Hist, Colo, and Derm are Histopathology, Colonoscopy, and Dermoscopy. (Color figure online)

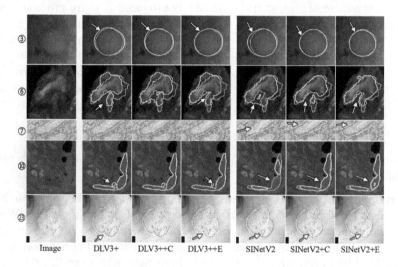

Fig. 7. Qualitative comparison of different models equipped with IFE. Red and green denote prediction and ground truth, respectively. (Color figure online)

Quantitative and Qualitative Analysis. To demonstrate the efficacy of IFE, we employ the following metrics: *Conformity (Con)* [7], *Dice Similarity Coefficient (DSC)* [5], *Jaccard Distance (JC)* [12], *F1* [2], *Human Correction Efforts (HCE)* [26], *Mean Absolute Error (MAE)* [25], *Hausdorff Distance (HD)* [34], *Average Symmetric Surface Distance (ASD)* [13], *Relative Volume Difference(RVD)* [13]. The quantitative results for UNet [28], DeeplabV3+ [9], SINetV2 [14], and nnUNet [18] are presented in Table 1. From the table, it can be concluded that IFE can improve the performance of networks on most segmentation metrics. Besides, Fig. 6 shows that IFE helps models perform better in most modalities and anatomical structures. Figure 7 presents a qualitative comparison. IFE aids in locating structures in an object that may be difficult to notice and enhances sensitivity to edge gray variations. IFE can substantially improve the segmentation accuracy of the base model in challenging scenes.

Ablation Studies. Choosing a suitable selection ratio r is crucial when applying IFE to different networks. Different networks' encoders are not equally capable of extracting features, and the ratio of channel features more favorable to the segmentation result varies. To analyze the effect of r, we conducted experiments using UNet [28]. As shown in Table 2, either too large or too small r will lead to a decline in the model performance.

4 Conclusion

In order to benchmark the general DMIS, we build a large-scale dataset called Cosmos55k. To balance universality and accuracy, we proposed an approach (IFE) that can select instructive feature channels to further improve the segmentation over strong baselines against challenging tasks. Experiments showed that IFE can improve the performance of classic models with slight modifications in the network. It is simple, universal, and effective. Future research will focus on extending this approach to 3D tasks.

Acknowledgements. The authors of this paper sincerely appreciate all the challenge organizers and owners for providing the public MIS datasets including AbdomenCT-1K, ACDC, AMOS 2022, BraTS20, CHAOS, CRAG, crossMoDA, EndoTect 2020, ETIS-Larib Polyp DB, iChallenge-AMD, iChallenge-PALM, IDRiD 2018, ISIC 2018, I2CVB, KiPA22, KiTS19& KiTS21, Kvasir-SEG, LUNA16, Multi-Atlas Labeling Beyond the Cranial Vault (Abdomen), Montgomery County CXR Set, M&Ms, MSD, NCI-ISBI 2013, PROMISE12, QUBIQ 2021, SIIM-ACR, SLIVER07, VerSe19 & VerSe20, Warwick-QU, and WORD.

This work was supported by the grant from National Natural Science Foundation of China (Nos. 62171290, 62101343), Shenzhen-Hong Kong Joint Research Program (No. SGDX20201103095613036), and Shenzhen Science and Technology Innovations Committee (No. 20200812143441001).

References

1. Abdel-Khalek, S., Ishak, A.B., Omer, O.A., Obada, A.S.: A two-dimensional image segmentation method based on genetic algorithm and entropy. Optik **131**, 414–422 (2017)
2. Achanta, R., Hemami, S., Estrada, F., Susstrunk, S.: Frequency-tuned salient region detection. In: IEEE CVPR (2009)
3. Bakas, S., et al.: Identifying the best machine learning algorithms for brain tumor segmentation, progression assessment, and overall survival prediction in the brats challenge. arXiv preprint arXiv:1811.02629 (2018)
4. Bernard, O., et al.: Deep learning techniques for automatic MRI cardiac multi-structures segmentation and diagnosis: is the problem solved? IEEE TMI **37**(11), 2514–2525 (2018)
5. Bilic, P., et al.: The liver tumor segmentation benchmark (LiTS). MIA **84**, 102680 (2023)
6. Cao, G., Xie, X., Yang, W., Liao, Q., Shi, G., Wu, J.: Feature-fused SSD: fast detection for small objects. In: SPIE ICGIP, vol. 10615 (2018)
7. Chang, H.H., Zhuang, A.H., Valentino, D.J., Chu, W.C.: Performance measure characterization for evaluating neuroimage segmentation algorithms. Neuroimage **47**(1), 122–135 (2009)

8. Chen, L.C., Papandreou, G., Kokkinos, I., Murphy, K., Yuille, A.L.: DeepLab: semantic image segmentation with deep convolutional nets, atrous convolution, and fully connected CRFs. IEEE TPAMI **40**(4), 834–848 (2017)

9. Chen, L.-C., Zhu, Y., Papandreou, G., Schroff, F., Adam, H.: Encoder-decoder with atrous separable convolution for semantic image segmentation. In: Ferrari, V., Hebert, M., Sminchisescu, C., Weiss, Y. (eds.) ECCV 2018. LNCS, vol. 11211, pp. 833–851. Springer, Cham (2018). https://doi.org/10.1007/978-3-030-01234-2_49

10. Chollet, F.: Xception: deep learning with depthwise separable convolutions. In: IEEE CVPR (2017)

11. Codella, N.C., et al.: Skin lesion analysis toward melanoma detection: a challenge at the 2017 international symposium on biomedical imaging (ISBI), hosted by the international skin imaging collaboration (ISIC). In: ISBI (2018)

12. Crum, W.R., Camara, O., Hill, D.L.: Generalized overlap measures for evaluation and validation in medical image analysis. IEEE TMI **25**(11), 1451–1461 (2006)

13. Dubuisson, M.P., Jain, A.K.: A modified Hausdorff distance for object matching. In: IEEE ICPR (1994)

14. Fan, D.P., Ji, G.P., Cheng, M.M., Shao, L.: Concealed object detection. IEEE TPAMI **44**(10), 6024–6042 (2021)

15. Gong, Y., Sbalzarini, I.F.: Curvature filters efficiently reduce certain variational energies. IEEE TIP **26**(4), 1786–1798 (2017)

16. Hou, Q., Zhou, D., Feng, J.: Coordinate attention for efficient mobile network design. In: IEEE CVPR (2021)

17. Howard, A., et al.: Searching for MobileNetV3. In: IEEE ICCV (2019)

18. Isensee, F., Jaeger, P.F., Kohl, S.A., Petersen, J., Maier-Hein, K.H.: nnU-Net: a self-configuring method for deep learning-based biomedical image segmentation. Nat. Meth. **18**(2), 203–211 (2021)

19. Jha, D., et al.: Kvasir-SEG: a segmented polyp dataset. In: Ro, Y.M., et al. (eds.) MMM 2020. LNCS, vol. 11962, pp. 451–462. Springer, Cham (2020). https://doi.org/10.1007/978-3-030-37734-2_37

20. Ji, W., et al.: Learning calibrated medical image segmentation via multi-rater agreement modeling. In: IEEE CVPR (2021)

21. Ji, Y., et al.: AMOS: a large-scale abdominal multi-organ benchmark for versatile medical image segmentation. arXiv preprint arXiv:2206.08023 (2022)

22. Kavur, A.E., et al.: Chaos challenge-combined (CT-MR) healthy abdominal organ segmentation. MIA **69**, 101950 (2021)

23. Litjens, G., et al.: Evaluation of prostate segmentation algorithms for MRI: the PROMISE12 challenge. MIA **18**(2), 359–373 (2014)

24. Luo, X., et al.: WORD: a large scale dataset, benchmark and clinical applicable study for abdominal organ segmentation from CT image. MIA **82**, 102642 (2022)

25. Perazzi, F., Krähenbühl, P., Pritch, Y., Hornung, A.: Saliency filters: contrast based filtering for salient region detection. In: IEEE CVPR (2012)

26. Qin, X., Dai, H., Hu, X., Fan, D.P., Shao, L., Van Gool, L.: Highly accurate dichotomous image segmentation. In: Avidan, S., Brostow, G., Cissé, M., Farinella, G.M., Hassner, T. (eds.) Computer Vision, ECCV 2022. LNCS, vol. 13678, pp. 38–56. Springer, Cham (2022). https://doi.org/10.1007/978-3-031-19797-0_3

27. Raghu, M., Unterthiner, T., Kornblith, S., Zhang, C., Dosovitskiy, A.: Do vision transformers see like convolutional neural networks? Adv. Neural. Inf. Process. Syst. **34**, 12116–12128 (2021)

28. Ronneberger, O., Fischer, P., Brox, T.: U-Net: convolutional networks for biomedical image segmentation. In: Navab, N., Hornegger, J., Wells, W.M., Frangi, A.F. (eds.) MICCAI 2015. LNCS, vol. 9351, pp. 234–241. Springer, Cham (2015). https://doi.org/10.1007/978-3-319-24574-4_28

29. Sekuboyina, A., et al.: Verse: a vertebrae labelling and segmentation benchmark for multi-detector CT images. MIA **73**, 102166 (2021)

30. Simpson, A.L., et al.: A large annotated medical image dataset for the development and evaluation of segmentation algorithms. arXiv preprint arXiv:1902.09063 (2019)

31. Szegedy, C., Ioffe, S., Vanhoucke, V., Alemi, A.: Inception-v4, inception-ResNet and the impact of residual connections on learning. In: AAAI (2017)

32. Zhao, Z., Chen, H., Wang, L.: A coarse-to-fine framework for the 2021 kidney and kidney tumor segmentation challenge. In: Heller, N., Isensee, F., Trofimova, D., Tejpaul, R., Papanikolopoulos, N., Weight, C. (eds.) KiTS 2021. LNCS, vol. 13168, pp. 53–58. Springer, Cham (2022). https://doi.org/10.1007/978-3-030-98385-7_8

33. Zhong, Z., et al.: Squeeze-and-attention networks for semantic segmentation. In: IEEE CVPR (2020)

34. Zhou, D., et al.: Iou loss for 2D/3D object detection. In: IEEE 3DV (2019)

35. Zhou, Y., et al.: Prior-aware neural network for partially-supervised multi-organ segmentation. In: IEEE ICCV (2019)

36. Zhou, Z., Siddiquee, M.M.R., Tajbakhsh, N., Liang, J.: UNet++: redesigning skip connections to exploit multiscale features in image segmentation. IEEE TMI **39**(6), 1856–1867 (2019)

MDViT: Multi-domain Vision Transformer for Small Medical Image Segmentation Datasets

Siyi Du[1](✉)⬤, Nourhan Bayasi[1]⬤, Ghassan Hamarneh[2]⬤, and Rafeef Garbi[1]⬤

[1] University of British Columbia, Vancouver, BC, Canada
{siyi,nourhanb,rafeef}@ece.ubc.ca
[2] Simon Fraser University, Burnaby, BC, Canada
hamarneh@sfu.ca

Abstract. Despite its clinical utility, medical image segmentation (MIS) remains a daunting task due to images' inherent complexity and variability. Vision transformers (ViTs) have recently emerged as a promising solution to improve MIS; however, they require larger training datasets than convolutional neural networks. To overcome this obstacle, data-efficient ViTs were proposed, but they are typically trained using a single source of data, which overlooks the valuable knowledge that could be leveraged from other available datasets. Naïvly combining datasets from different domains can result in negative knowledge transfer (NKT), i.e., a decrease in model performance on some domains with non-negligible inter-domain heterogeneity. In this paper, we propose MDViT, the first multi-domain ViT that includes domain adapters to mitigate data-hunger and combat NKT by adaptively exploiting knowledge in multiple small data resources (domains). Further, to enhance representation learning across domains, we integrate a mutual knowledge distillation paradigm that transfers knowledge between a universal network (spanning all the domains) and auxiliary domain-specific network branches. Experiments on 4 skin lesion segmentation datasets show that MDViT outperforms state-of-the-art algorithms, with superior segmentation performance and a fixed model size, at inference time, even as more domains are added. Our code is available at https://github.com/siyi-wind/MDViT.

Keywords: Vision Transformer · Data-efficiency · Multi-domain Learning · Medical Image Segmentation · Dermatology

1 Introduction

Medical image segmentation (MIS) is a crucial component in medical image analysis, which aims to partition an image into distinct regions (or segments) that are semantically related and/or visually similar. This process is essential for clinicians to, among others, perform qualitative and quantitative assessments

---

Supplementary Information The online version contains supplementary material available at https://doi.org/10.1007/978-3-031-43901-8_43.

of various anatomical structures or pathological conditions and perform image-guided treatments or treatment planning [2]. Vision transformers (ViTs), with their inherent ability to model long-range dependencies, have recently been considered a promising technique to tackle MIS. They process images as sequences of patches, with each patch having a global view of the entire image. This enables a ViT to achieve improved segmentation performance compared to traditional convolutional neural networks (CNNs) on plenty of segmentation tasks [16]. However, due to the lack of inductive biases, such as weight sharing and locality, ViTs are more data-hungry than CNNs, i.e., require more data to train [31]. Meanwhile, it is common to have access to multiple, diverse, yet small-sized datasets (100 s to 1000 ss of images per dataset) for the same MIS task, e.g., PH2 [25] and ISIC 2018 [11] in dermatology, LiTS [6] and CHAOS [18] in liver CT, or OASIS [24] and ADNI [17] in brain MRI. As each dataset alone is too small to properly train a ViT, the challenge becomes how to effectively leverage the different datasets.

Table 1. Related works on mitigating ViTs' data-hunger or multi-domain adaptive learning. **U** (universal) implies a model spans multiple domains. **F** means the model's size at inference time remains fixed even when more domains are added.

Method	ViT	Mitigate ViTs' data-hunger	U	F
[7, 22, 39]	√	√ by adding inductive bias	×	−
[31, 37]	√	√ by knowledge sharing	×	−
[34]	√	√ by increasing dataset size	×	−
[8]	√	√ by unsupervised pretraining	×	−
[28, 35]	×	×	√	√
[21, 26]	×	×	√	×
[32]	√	×	√	×
MDViT	√	√ by multi-domain learning	√	√

Various strategies have been proposed to address ViTs' data-hunger (Table 1), mainly: *Adding inductive bias* by constructing a hybrid network that fuses a CNN with a ViT [39], imitating CNNs' shifted filters and convolutional operations [7], or enhancing spatial information learning [22]; *sharing knowledge* by transferring knowledge from a CNN [31] or pertaining ViTs on multiple related tasks and then fine-tuning on a down-stream task [37]; *increasing data* via augmentation [34]; and *non-supervised pre-training* [8]. Nevertheless, one notable limitation in these approaches is that they are not universal, i.e., they rely on *separate training* for each dataset rather than incorporate valuable knowledge from related domains. As a result, they can incur additional training, inference, and memory costs, which is especially challenging when dealing with multiple small datasets in the context of MIS tasks. Multi-domain learning, which trains a single universal model to tackle all the datasets simultaneously, has been found promising for reducing computational demands while still leveraging information from multiple domains [1, 21]. To the best of our knowledge, multi-domain universal models have not yet been investigated for alleviating ViTs' data-hunger.

Given the inter-domain heterogeneity resulting from variations in imaging protocols, scanner manufacturers, etc. [4, 21], directly mixing all the datasets for training, i.e., *joint training*, may improve a model's performance on one dataset while degrading performance on other datasets with non-negligible unrelated domain-specific information, a phenomenon referred to as *negative knowledge transfer* (NKT) [1, 38]. A common strategy to mitigate NKT in computer vision is to introduce adapters aiding the model to adapt to different domains, i.e., *multi-domain adaptive training* (MAT), such as domain-specific mechanisms [21, 26, 32], and squeeze-excitation layers [28, 35] (Table 1). However, those MAT techniques are built based on CNN rather than ViT or are scalable, i.e., the models' size at the inference time increases linearly with the number of domains.

To address ViTs' data-hunger, in this work, we propose MDViT, a novel fixed-size multi-domain ViT trained to adaptively aggregate valuable knowledge from multiple datasets (domains) for improved segmentation. In particular, we introduce a domain adapter that adapts the model to different domains to mitigate negative knowledge transfer caused by inter-domain heterogeneity. Besides, for better representation learning across domains, we propose a novel mutual knowledge distillation approach that transfers knowledge between a universal network (spanning all the domains) and additional domain-specific network branches.

We summarize our contributions as follows: (1) To the best of our knowledge, we are the first to introduce multi-domain learning to alleviate ViTs' data-hunger when facing limited samples per dataset. (2) We propose a multi-domain ViT, MDViT, for medical image segmentation with a novel domain adapter to counteract negative knowledge transfer and with mutual knowledge distillation to enhance representation learning. (3) The experiments on 4 skin lesion segmentation datasets show that our multi-domain adaptive training outperforms separate and joint training (ST and JT), especially a 10.16% improvement in IOU on the skin cancer detection dataset compared to ST and that MDViT outperforms state-of-the-art data-efficient ViTs and multi-domain learning strategies.

2   Methodology

Let $X \in \mathbb{R}^{H \times W \times 3}$ be an input RGB image and $Y \in \{0, 1\}^{H \times W}$ be its ground-truth segmentation mask. Training samples $\{(X, Y)\}$ come from M datasets, each representing a domain. We aim to build and train a single ViT that performs well on all domain data and addresses the insufficiency of samples in any of the datasets. We first introduce our baseline (BASE), a ViT with hierarchical transformer blocks (Fig. 1-a). Our proposed MDViT extends BASE with 1) a domain adapter (DA) module inside the factorized multi-head self-attention (MHSA) to adapt the model to different domains (Fig. 1-b,c), and 2) a mutual knowledge distillation (MKD) strategy to extract more robust representations across domains (Fig. 1-d). We present the details of MDViT in Sect. 2.1.

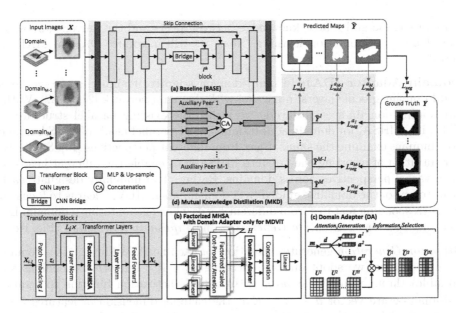

Fig. 1. Overall architecture of MDViT, which is trained on multi-domain data by optimizing two types of losses: L_{seg} and L_{mkd}. MDViT extends BASE (a) with DA inside factorized MHSA (b), which is detailed in (c), and MKD (d).

BASE is a U-shaped ViT based on the architecture of U-Net [27] and pyramid ViTs [7,19]. It contains encoding (the first four) and decoding (the last four) transformer blocks, a two-layer CNN bridge, and skip connections. As described in [19], the ith transformer block involves a convolutional patch embedding layer with a patch size of 3×3 and L_i transformer layers with factorized MHSA in linear complexity, the former of which converts a feature map X_{i-1} into a sequence of patch embeddings $z_i \in \mathbb{R}^{N_i \times C_i}$, where $N_i = \frac{H}{2^{i+1}} \frac{W}{2^{i+1}}$, $1 \le i \le 4$ is the number of patches and C_i is the channel dimension. We use the same position embedding as [19] and skip connections as [27]. To reduce computational complexity, following [19], we add two and one CNN layer before and after transformer blocks, respectively, enabling the 1st transformer block to process features starting from a lower resolution: $\frac{H}{4} \times \frac{W}{4}$. We do not employ integrated and hierarchical CNN backbones, e.g., ResNet, in BASE as data-efficient hybrid ViTs [33,39], to clearly evaluate the efficacy of multi-domain learning in mitigating ViTs' data-hunger.

2.1 MDViT

MDViT consists of a universal network (spanning M domains) and M auxiliary network branches, i.e., peers, each associated with one of the M domains. The universal network is the same as BASE, except we insert a domain adapter (DA) in each factorized MHSA to tackle negative knowledge transfer. Further, we employ a mutual knowledge distillation (MKD) strategy to transfer domain-specific and shared knowledge between peers and the universal network to enhance representation learning. Next, we will introduce DA and MKD in detail.

Domain Adapter (DA): In multi-domain adaptive training, some methods build domain-specific layers in parallel with the main network [21,26,32]. Without adding domain-specific layers, we utilize the existing parallel structure in ViTs, i.e., MHSA, for domain adaptation. The H parallel heads of MHSA mimic how humans examine the same object from different perspectives [10]. Similarly, our intuition of inserting the DA into MHSA is to enable the different heads to have varied perspectives across domains. Rather than manually designate each head to one of the domains, guided by a domain label, MDViT learns to focus on the corresponding features from different heads when encountering a domain. DA contains two steps: *Attention Generation* and *Information Selection* (Fig. 1-c).

Attention Generation generates attention for each head. We first pass a domain label vector m (we adopt one-hot encoding $m \in \mathbb{R}^M$ but other encodings are possible) through one linear layer with a ReLU activation function to acquire a domain-aware vector $d \in \mathbb{R}^{\frac{K}{r}}$. K is the channel dimension of features from the heads. We set the reduction ratio r to 2. After that, similar to [20], we calculate attention for each head: $a^h = \psi(W^h d) \in \mathbb{R}^K, h = 1, 2, ...H$, where ψ is a softmax operation across heads and $W^h \in \mathbb{R}^{K \times \frac{K}{r}}$.

Information Selection adaptively selects information from different heads. After getting the feature $U^h = [u_1^h, u_2^h, ..., u_K^h] \in \mathbb{R}^{N \times K}$ from the hth head, we utilize a^h to calibrate the information along the channel dimension: $\tilde{u}_k^h = a_k^h \cdot u_k^h$.

Mutual Knowledge Distillation (MKD): Distilling knowledge from domain-specific networks has been found beneficial for universal networks to learn more robust representations [21,40]. Moreover, mutual learning that transfers knowledge between teachers and students enables both to be optimized simultaneously [15]. To realize these benefits, we propose MKD that mutually transfers knowledge between auxiliary peers and the universal network. In Fig. 1-d, the mth auxiliary peer is only trained on the mth domain, producing output $\hat{Y}^m$, whereas the universal network's output is $\hat{Y}$. Similar to [21], we utilize a symmetric Dice loss $L_{mkd}^{a_m} = Dice(\hat{Y}, \hat{Y}^m)$ as the knowledge distillation loss. Each peer is an expert in a certain domain, guiding the universal network to learn domain-specific information. The universal network experiences all the domains and grasps the domain-shared knowledge, which is beneficial for peer learning.

Each *Auxiliary Peer* is trained on a small, individual dataset specific to that peer (Fig. 1-d). To achieve a rapid training process and prevent overfitting, particularly when working with numerous training datasets, we adapt a lightweight multilayer perception (MLP) decoder designed for ViT encoders [36] to our peers' architecture. Specifically, multi-level features from the encoding transformer blocks (Fig. 1-a) go through an MLP layer and an up-sample operation to unify the channel dimension and resolution to $\frac{H}{4} \times \frac{W}{4}$, which are then concatenated with the feature involving domain-shared information from the

Table 2. Segmentation results comparing BASE, MDViT, and SOTA methods. We report the models' parameter count at inference time in millions (M). **T** means training paradigms. † represents using domain-specific normalization.

Model	#Param. (millions) (M)	T	Segmentation Results in Test Sets (%)									
			Dice ↑					IOU ↑				
			ISIC	DMF	SCD	PH2	avg ± std	ISIC	DMF	SCD	PH2	avg ± std
(a) BASE												
BASE	27.8×	ST	90.18	90.68	86.82	93.41	90.27 ± 1.16	82.82	83.22	77.64	87.84	82.88 ± 1.67
BASE	27.8	JT	89.42	89.89	92.96	94.24	91.63 ± 0.42	81.68	82.07	87.03	89.36	85.04 ± 0.64
(b) Our Method												
MDViT	28.5	MAT	90.29	90.78	**93.22**	**95.53**	**92.45 ± 0.65**	82.99	83.41	**87.80**	**91.57**	**86.44 ± 0.94**
(c) Other Data-efficient MIS ViTs												
SwinUnet	41.4×	ST	89.25	90.69	88.58	94.13	90.66 ± 0.87	81.51	83.25	80.40	89.00	83.54 ± 1.27
SwinUnet	41.4	JT	89.64	90.40	92.98	94.86	91.97 ± 0.30	81.98	82.80	87.08	90.33	85.55 ± 0.50
UTNet	10.0×	ST	89.74	90.01	88.13	93.23	90.28 ± 0.62	82.16	82.13	79.87	87.60	82.94 ± 0.82
UTNet	10.0	JT	90.24	89.85	92.06	94.75	91.72 ± 0.63	82.92	82.00	85.66	90.17	85.19 ± 0.96
BAT	32.2×	ST	**90.45**	90.56	90.78	94.72	91.63 ± 0.68	83.04	82.97	83.66	90.03	84.92 ± 1.01
BAT	32.2	JT	90.06	90.06	92.66	93.53	91.58 ± 0.33	82.44	82.18	86.48	88.11	84.80 ± 0.53
TransFuse	26.3×	ST	90.43	**91.04**	91.37	94.93	91.94 ± 0.67	**83.18**	**83.86**	84.91	90.44	85.60 ± 0.95
TransFuse	26.3	JT	90.03	90.48	92.54	95.14	92.05 ± 0.36	82.56	82.97	86.50	90.85	85.72 ± 0.56
Swin UNETR	25.1×	ST	90.29	90.95	91.10	94.45	91.70 ± 0.51	82.93	83.69	84.16	89.59	85.09 ± 0.79
Swin UNETR	25.1	JT	89.81	90.87	92.29	94.73	91.93 ± 0.29	82.21	83.58	86.10	90.11	85.50 ± 0.44
(d) Other Multi-domain Learning Methods												
Rundo et al.	28.2	MAT	89.43	89.46	92.62	94.68	91.55 ± 0.64	81.73	81.40	86.71	90.12	84.99 ± 0.90
Wang et al.	28.1	MAT	89.46	89.62	92.62	94.47	91.55 ± 0.54	81.79	81.59	86.71	89.76	84.96 ± 0.74
BASE†	27.8(.02×)	MAT	90.22	90.61	**93.69**	95.55	92.52 ± 0.45	82.91	83.14	**88.28**	91.58	86.48 ± 0.74
MDViT†	28.6(.02×)	MAT	**90.24**	**90.71**	93.38	**95.90**	**92.56 ± 0.52**	82.97	**83.31**	88.06	**92.19**	**86.64 ± 0.76**

universal network's last transformer block. Finally, we pass the fused feature to an MLP layer and do an up-sample to obtain a segmentation map.

2.2 Objective Function

Similar to Combo loss [29], BASE's segmentation loss combines Dice and binary cross entropy loss: $L_{seg} = L_{Dice} + L_{bce}$. In MDViT, we use the same segmentation loss for the universal network and auxiliary peers, denoted as L_{seg}^{u} and L_{seg}^{a}, respectively. The overall loss is calculated as follows.

$$L_{total} = L_{seg}^{u}(\boldsymbol{Y}, \hat{\boldsymbol{Y}}) + \alpha \sum_{m=1}^{M} L_{seg}^{a_m}(\boldsymbol{Y}, \hat{\boldsymbol{Y}}^m) + \beta \sum_{m=1}^{M} L_{mkd}^{a_m}(\hat{\boldsymbol{Y}}, \hat{\boldsymbol{Y}}^m). \quad (1)$$

We set both α and β to 0.5. $L_{seg}^{a_m}$ does not optimize DA to avoid interfering with the domain adaptation learning. After training, we discard the auxiliary peers and only utilize the universal network for inference.

3 Experiments

Datasets and Evaluation Metrics: We study 4 skin lesion segmentation datasets collected from varied sources: ISIC 2018 (ISIC) [11], Dermofit Image Library (DMF) [3], Skin Cancer Detection (SCD) [14], and PH2 [25], which contain

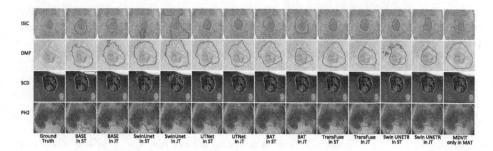

Fig. 2. Visual result comparison of MDViT, BASE and SOTA data-efficient MIS ViTs in ST and JT training paradigms on four datasets. The green and red contours present the ground truth and segmentation results, respectively. (Color figure online)

2594, 1300, 206, and 200 samples, respectively. To facilitate a fairer performance comparison across datasets, as in [4], we only use the 1212 images from DMF that exhibited similar lesion conditions as those in other datasets. We perform 5-fold cross-validation and utilize Dice and IOU metrics for evaluation as [33].

Implementation Details: We conduct 3 training paradigms: separate (ST), joint (JT), and multi-domain adaptive training (MAT), described in Sect. 1, to train all the models from scratch on the skin datasets. Images are resized to 256×256 and then augmented through random scaling, shifting, rotation, flipping, Gaussian noise, and brightness and contrast changes. The encoding transformer blocks' channel dimensions are [64, 128, 320, 512] (Fig. 1-a). We use two transformer layers in each transformer block and set the number of heads in MHSA to 8. The hidden dimensions of the CNN bridge and auxiliary peers are 1024 and 512. We deploy models on a single TITAN V GPU and train them for 200 epochs with the AdamW [23] optimizer, a batch size of 16, ensuring 4 samples from each dataset, and an initial learning rate of 1×10^{-4}, which changes through a linear decay scheduler whose step size is 50 and decay factor $\gamma = 0.5$.

Comparing Against BASE: In Table 2-a,b, compared with BASE in ST, BASE in JT improves the segmentation performance on small datasets (PH2 and SCD) but at the expense of diminished performance on larger datasets (ISIC and DMF). This is expected given the non-negligible inter-domain heterogeneity between skin lesion datasets, as found by Bayasi et al. [5]. The above results demonstrate that shared knowledge in related domains facilitates training a ViT on small datasets while, without a well-designed multi-domain algorithm, causing negative knowledge transfer (NKT) due to inter-domain heterogeneity, i.e., the model's performance decreases on other datasets. Meanwhile, MDViT fits all the domains without NKT and outperforms BASE in ST by a large margin; significantly increasing Dice and IOU on SCD by 6.4% and 10.16%, showing that MDViT smartly selects valuable knowledge when given data from a certain domain. Additionally, MDViT outperforms BASE in JT across all the domains, with average improvements of 0.82% on Dice and 1.4% on IOU.

Table 3. Ablation studies of MDViT and experiments of DA's plug-in capability. KD means general knowledge distillation, i.e., we only transfer knowledge from auxiliary peers to the universal network. ^{D} or ^{B} refers to using DeepLabv3's decoder or BASE's decoding layers as auxiliary peers.

Model	#Param. (M)	T	Dice ↑					IOU ↑				
			ISIC	DMF	SCD	PH2	avg ± std	ISIC	DMF	SCD	PH2	avg ± std
(a) Plug-in Capability of DA												
DosViT	14.6	JT	88.66	89.72	90.65	94.26	90.82 ± 0.43	80.45	81.73	83.29	89.26	83.68 ± 0.68
DosViT+DA	14.9	MAT	89.22	89.91	90.73	94.42	91.07 ± 0.32	81.28	82.00	83.44	89.57	84.07 ± 0.50
TransFuse	26.3	JT	90.03	90.48	92.54	95.14	92.05 ± 0.36	82.56	82.97	86.50	90.85	85.72 ± 0.56
TransFuse+DA	26.9	MAT	90.13	90.47	**93.62**	95.21	92.36 ± 0.38	82.80	82.94	**88.16**	90.97	86.22 ± 0.64
(b) Ablation Study for DA and MKD												
BASE	27.8	JT	89.42	89.89	92.96	94.24	91.63 ± 0.42	81.68	82.07	87.03	89.36	85.04 ± 0.64
BASE+DA	28.5	MAT	89.96	90.66	93.36	95.46	92.36 ± 0.51	82.52	83.24	87.98	91.43	86.29 ± 0.72
BASE+MKD	27.8	JT	89.27	89.53	92.66	94.83	91.57 ± 0.53	81.45	81.49	86.81	90.42	85.04 ± 0.74
BASE+DA+KD	28.5	MAT	90.03	90.59	93.26	**95.63**	92.38 ± 0.39	82.67	83.12	87.85	**91.72**	86.34 ± 0.51
(c) Ablation Study for Auxiliary Peers												
MDViT^D	28.5	MAT	89.64	90.25	92.24	95.36	91.87 ± 0.45	82.10	82.55	86.12	91.24	85.50 ± 0.67
MDViT^B	28.5	MAT	90.03	90.73	92.72	95.32	92.20 ± 0.50	82.66	83.35	87.01	91.17	86.05 ± 0.70
MDViT	28.5	MAT	**90.29**	**90.78**	93.22	95.53	**92.45 ± 0.65**	**82.99**	**83.41**	87.80	91.57	**86.44 ± 0.94**

Comparing Against State-of-the-Art (SOTA) Methods: We conduct experiments on SOTA data-efficient MIS ViTs and multi-domain learning methods. Previous MIS ViTs mitigated the data-hunger in one dataset by adding inductive bias, e.g., SwinUnet [7], UTNet [13], BAT [33], TransFuse [39], and Swin UNETR [30]. We implement ResNet-34 as the backbone of BAT for fair comparison (similar model size). As illustrated in Table 2-a,b,c, these SOTA models are superior to BASE in ST. This is expected since they are designed to reduce data requirements. Nevertheless, in JT, these models also suffer from NKT: They perform better than models in ST on some datasets, like SCD, and worse on others, like ISIC. Finally, MDViT achieves the best segmentation performance in average Dice and IOU without NKT and has the best results on SCD and PH2. Figure 2 shows MDViT's excellent performance on ISIC and DMF and that it achieves the closest results to ground truth on SCD and PH2. More segmentation results are presented in the supplementary material. Though BAT and TransFuse in ST have better results on some datasets like ISIC, they require extra compute resources to train M models as well as an M-fold increase in memory requirements. The above results indicate that domain-shared knowledge is especially beneficial for training relatively small datasets such as SCD.

We employ the two fixed-size (i.e., independent of M) multi-domain algorithms proposed by Rundo et al. [28] and Wang et al. [35] on BASE. We set the number of parallel SE adapters in [35] to 4. In Table 2-b,d, MDViT outperforms both of them on all the domains, showing the efficacy of MDViT and that multi-domain methods built on ViTs might not perform as well as on CNNs. We also apply the domain-specific normalization [21] to BASE and MDViT to get BASE† and MDViT†, respectively. In Table 2-d, BASE† confronts NKT, which lowers the performance on DMF compared with BASE in ST, whereas MDViT† not only addresses NKT but also outperforms BASE† on average Dice and IOU.

Ablation Studies and Plug-in Capability of DA: We conduct ablation studies to demonstrate the efficacy of DA, MKD, and auxiliary peers. Table 3-b reveals that using one-direction knowledge distillation (KD) or either of the critical components in MDViT, i.e., DA or MKD, but not together, could not achieve the best results. Table 3-c exemplifies that, for building the auxiliary peers, our proposed MLP architecture is more effective and has fewer parameters (1.6M) than DeepLabv3's decoder [9] (4.7M) or BASE's decoding layers (10.8M). Finally, we incorporate DA into two ViTs: TransFuse and DosViT (the latter includes the earliest ViT encoder [12] and a DeepLabv3's decoder). As shown in Table 3-a,b, DA can be used in various ViTs but is more advantageous in MDViT with more transformer blocks in the encoding and decoding process.

4   Conclusion

We propose a new algorithm to alleviate vision transformers (ViTs)' data-hunger in small datasets by aggregating valuable knowledge from multiple related domains. We constructed MDViT, a robust multi-domain ViT leveraging novel domain adapters (DAs) for negative knowledge transfer mitigation and mutual knowledge distillation (MKD) for better representation learning. MDViT is non-scalable, i.e., has a fixed model size at inference time even as more domains are added. The experiments on 4 skin lesion segmentation datasets show that MDViT outperformed SOTA data-efficient medical image segmentation ViTs and multi-domain learning methods. Our ablation studies and application of DA on other ViTs show the effectiveness of DA and MKD and DA's plug-in capability.

References

1. Adadi, A.: A survey on data-efficient algorithms in big data era. J. Big Data **8**(1), 24 (2021)
2. Asgari Taghanaki, S., Abhishek, K., Cohen, J.P., Cohen-Adad, J., Hamarneh, G.: Deep semantic segmentation of natural and medical images: a review. Artif. Intell. Rev. **54**, 137–178 (2021)
3. Ballerini, L., Fisher, R.B., Aldridge, B., Rees, J.: A color and texture based hierarchical K-NN approach to the classification of non-melanoma skin lesions. In: Celebi, M., Schaefer, G. (eds.) Color medical image analysis. LNCS, vol. 6, pp. 63–86. Springer, Dordrecht (2013). https://doi.org/10.1007/978-94-007-5389-1_4
4. Bayasi, N., Hamarneh, G., Garbi, R.: Culprit-prune-net: efficient continual sequential multi-domain learning with application to skin lesion classification. In: de Bruijne, M., et al. (eds.) MICCAI 2021. LNCS, vol. 12907, pp. 165–175. Springer, Cham (2021). https://doi.org/10.1007/978-3-030-87234-2_16
5. Bayasi, N., Hamarneh, G., Garbi, R.: BoosterNet: improving domain generalization of deep neural nets using culpability-ranked features. In: CVPR 2022, pp. 538–548 (2022)
6. Bilic, P., et al.: The liver tumor segmentation benchmark (LiTS). Med. Image Anal. **84**, 102680 (2023)

7. Cao, H., et al.: Swin-Unet: Unet-like pure transformer for medical image segmentation. In: Karlinsky, L., Michaeli, T., Nishino, K. (eds.) ECCV 2022. LNCS, vol. 13803, pp. 205–218. Springer, Cham (2023). https://doi.org/10.1007/978-3-031-25066-8_9

8. Cao, Y.H., Yu, H., Wu, J.: Training vision transformers with only 2040 images. arXiv preprint arXiv:2201.10728 (2022)

9. Chen, L.C., Papandreou, G., Schroff, F., Adam, H.: Rethinking atrous convolution for semantic image segmentation. arXiv preprint arXiv:1706.05587 (2017)

10. Clark, K., Khandelwal, U., Levy, O., Manning, C.D.: What does BERT look at? an analysis of BERT's attention. ACL **2019**, 276 (2019)

11. Codella, N., Rotemberg, V., Tschandl, P., Celebi, M.E., et al.: Skin lesion analysis toward melanoma detection 2018: A challenge hosted by the international skin imaging collaboration (ISIC). arXiv preprint arXiv:1902.03368 (2019)

12. Dosovitskiy, A., et al.: An image is worth 16x16 words: Transformers for image recognition at scale. In: ICLR 2020 (2020)

13. Gao, Y., Zhou, M., Metaxas, D.N.: UTNet: a hybrid transformer architecture for medical image segmentation. In: de Bruijne, M., et al. (eds.) MICCAI 2021. LNCS, vol. 12903, pp. 61–71. Springer, Cham (2021). https://doi.org/10.1007/978-3-030-87199-4_6

14. Glaister, J., Amelard, R., Wong, A., Clausi, D.A.: MSIM: multistage illumination modeling of dermatological photographs for illumination-corrected skin lesion analysis. IEEE Trans. Biomed. Eng. **60**(7), 1873–1883 (2013)

15. Gou, J., Yu, B., Maybank, S.J., Tao, D.: Knowledge distillation: a survey. Int. J. Comput. Vis. **129**, 1789–1819 (2021)

16. Han, K., Wang, Y., Chen, H., Chen, X., Guo, J., Liu, Z., et al.: A survey on vision transformer. IEEE Trans. Pattern Anal. Mach. Intell. **45**(1), 87–110 (2022)

17. Jack, C.R., Jr., et al.: The Alzheimer's disease neuroimaging initiative (ADNI): MRI methods. J. Mag. Reson. Imaging **27**(4), 685–691 (2008)

18. Kavur, A.E., Gezer, N.S., Barış, M., Aslan, S., Conze, P.H., Groza, V., et al.: CHAOS challenge-combined (CT-MR) healthy abdominal organ segmentation. Med. Image Anal. **69**, 101950 (2021)

19. Lee, Y., Kim, J., Willette, J., Hwang, S.J.: MPViT: multi-path vision transformer for dense prediction. In: CVPR 2022, pp. 7287–7296 (2022)

20. Li, X., Wang, W., et al.: Selective kernel networks. In: CVPR 2019, pp. 510–519 (2019)

21. Liu, Q., Dou, Q., Yu, L., Heng, P.A.: MS-Net: multi-site network for improving prostate segmentation with heterogeneous MRI data. IEEE Trans. Med. Imaging **39**(9), 2713–2724 (2020)

22. Liu, Y., Sangineto, E., Bi, W., Sebe, N., Lepri, B., Nadai, M.: Efficient training of visual transformers with small datasets. NeurIPS **2021**(34), 23818–23830 (2021)

23. Loshchilov, I., Hutter, F.: Decoupled weight decay regularization. arXiv preprint arXiv:1711.05101 (2017)

24. Marcus, D.S., Wang, T.H., Parker, J., et al.: Open access series of imaging studies (OASIS): cross-sectional MRI data in young, middle aged, nondemented, and demented older adults. J. Cogn. Neurosci. **19**(9), 1498–1507 (2007)

25. Mendonça, T., Ferreira, P.M., Marques, J.S., Marcal, A.R., Rozeira, J.: PH 2-A dermoscopic image database for research and benchmarking. In: EMBC 2013, pp. 5437–5440. IEEE (2013)

26. Rebuffi, S.A., Bilen, H., Vedaldi, A.: Efficient parametrization of multi-domain deep neural networks. In: CVPR 2018, pp. 8119–8127 (2018)

27. Ronneberger, O., Fischer, P., Brox, T.: U-Net: convolutional networks for biomedical image segmentation. In: Navab, N., Hornegger, J., Wells, W.M., Frangi, A.F. (eds.) MICCAI 2015. LNCS, vol. 9351, pp. 234–241. Springer, Cham (2015). https://doi.org/10.1007/978-3-319-24574-4_28

28. Rundo, L., et al.: USE-Net: incorporating squeeze-and-excitation blocks into u-net for prostate zonal segmentation of multi-institutional MRI datasets. Neurocomputing **365**, 31–43 (2019)

29. Taghanaki, S.A., Zheng, Y., Zhou, S.K., Georgescu, B., Sharma, P., Xu, D., et al.: Combo loss: handling input and output imbalance in multi-organ segmentation. Comput. Med. Imaging Graph. **75**, 24–33 (2019)

30. Tang, Y., et al.: Self-supervised pre-training of Swin transformers for 3d medical image analysis. In: CVPR 2022, pp. 20730–20740 (2022)

31. Touvron, H., Cord, M., Douze, M., Massa, F., Sablayrolles, A., Jégou, H.: Training data-efficient image transformers & distillation through attention. In: ICML 2021, pp. 10347–10357. PMLR (2021)

32. Wallingford, M., Li, H., Achille, A., Ravichandran, A., et al.: Task adaptive parameter sharing for multi-task learning. In: CVPR 2022, pp. 7561–7570 (2022)

33. Wang, J., Wei, L., Wang, L., Zhou, Q., Zhu, L., Qin, J.: Boundary-aware transformers for skin lesion segmentation. In: de Bruijne, M., et al. (eds.) MICCAI 2021. LNCS, vol. 12901, pp. 206–216. Springer, Cham (2021). https://doi.org/10.1007/978-3-030-87193-2_20

34. Wang, W., Zhang, J., Cao, Y., Shen, Y., Tao, D.: Towards data-efficient detection transformers. In: Avidan, S., Brostow, G., Cisse, M., Farinella, G.M., Hassner, T. (eds.) ECCV 2022. LNCS, vol. 13669, pp. 88–105. Springer, Cham (2022). https://doi.org/10.1007/978-3-031-20077-9_6

35. Wang, X., Cai, Z., Gao, D., Vasconcelos, N.: Towards universal object detection by domain attention. In: CVPR 2019, pp. 7289–7298 (2019)

36. Xie, E., Wang, W., Yu, Z., et al.: SegFormer: simple and efficient design for semantic segmentation with transformers. NeurIPS **2021**(34), 12077–12090 (2021)

37. Xie, Y., Zhang, J., et al.: UniMiSS: universal medical self-supervised learning via breaking dimensionality barrier. In: Avidan, S., Brostow, G., Cisse, M., Farinella, G.M., Hassner, T. (eds.) Computer Vision - ECCV 2022. LNCS, vol. 13681, pp. 558–575. Springer, Cham (2022). https://doi.org/10.1007/978-3-031-19803-8_33

38. Zhang, W., Deng, L., Zhang, L., Wu, D.: A survey on negative transfer. IEEE/CAA J. Automatica Sinica (2022)

39. Zhang, Y., Liu, H., Hu, Q.: TransFuse: fusing transformers and CNNs for medical image segmentation. In: de Bruijne, M., et al. (eds.) MICCAI 2021. LNCS, vol. 12901, pp. 14–24. Springer, Cham (2021). https://doi.org/10.1007/978-3-030-87193-2_2

40. Zhou, C., Wang, Z., He, S., Zhang, H., Su, J.: A novel multi-domain machine reading comprehension model with domain interference mitigation. Neurocomputing **500**, 791–798 (2022)

Semi-supervised Class Imbalanced Deep Learning for Cardiac MRI Segmentation

Yuchen Yuan[1], Xi Wang[1]([✉]) [ID], Xikai Yang[1], Ruijiang Li[2],
and Pheng-Ann Heng[1,3]

[1] Department of Computer Science and Engineering, The Chinese University
of Hong Kong, Shatin, Hong Kong
xiwang@cse.cuhk.edu.hk
[2] Department of Radiation Oncology, Stanford University School of Medicine,
Stanford, USA
[3] Institute of Medical Intelligence and XR, The Chinese University of Hong Kong,
Shatin, Hong Kong

Abstract. Despite great progress in semi-supervised learning (SSL) that leverages unlabeled data to improve the performance over fully supervised models, existing SSL approaches still fail to exhibit good results when faced with a severe class imbalance problem in medical image segmentation. In this work, we propose a novel Mean-teacher based class imbalanced learning framework for cardiac magnetic resonance imaging (MRI) segmentation, which can effectively conquer the problems of class imbalance and limited labeled data simultaneously. Specifically, in parallel to the traditional linear-based classifier, we additionally train a prototype-based classifier that makes dense predictions by matching test samples with a set of prototypes. The prototypes are iteratively updated by in-class features encoded in the entire sample set, which can better guide the model training by alleviating the class-wise bias exhibited in each individual sample. To reduce the noises in the pseudo labels, we propose a cascaded refining strategy by utilizing two multi-level tree filters that are built upon pairwise pixel similarity in terms of intensity values and semantic features. With the assistance of pixel affinities, soft pseudo labels are properly refined on-the-fly. Upon evaluation on ACDC and MMWHS, two cardiac MRI datasets with prominent class imbalance problem, the proposed method demonstrates the superiority compared to several state-of-the-art methods, especially in the case where few annotations are available (Code is available in https://github.com/IsYuchenYuan/SSCI).

Keywords: Semi-supervised learning · Medical image segmentation ·
Class imbalance · Prototype learning · Multi-level tree filters

1 Introduction

Recent progress in deep learning techniques, particularly convolutional neural networks (CNN), has shown enormous success in the field of 3D medical image

---

Supplementary Information The online version contains supplementary material available at https://doi.org/10.1007/978-3-031-43901-8_44.

segmentation [1]. However, these marvelous achievements are always accompanied by a high cost of substantial high-quality annotations, which are usually prohibitively difficult to obtain in the medical domain due to expensive time costs and highly-demanded expertise. Besides, class imbalance, caused by the huge variation of anatomical structure volumes, is another frequently occurring problem, further posing great challenges in automated and accurate 3D medical image segmentation with limited annotations.

To handle the label scarcity, a large bunch of semi-supervised learning (SSL) methods have been proposed by leveraging abundant unlabeled data to compensate for the annotation scarcity. Concretely, pseudo-labeling [2,3] and consistency regularization [4,5] are two effective SSL schemes. [6]. For example, Bai et al. [7] used the model trained on the labeled pairs to generate pseudo labels that serve as targets for the unlabeled data. To guarantee the quality of pseudo labels, Sedai et al. [8] used Bayesian deep learning [9] to measure the pixel-wise reliability of soft labels and suppressed regions with unreliable ones. Concurrently, Yu et al. designed a 3D uncertainty-aware semi-supervised MRI segmentation framework [10], where the uncertainty mechanism [9] is integrated with consistency regularization.

Despite their success, few of the SSL segmentation methods considered class imbalance that is naturally inherent in medical images. For example, in cardiac MRI images, *myocardium* is often missing or too small to be detected in apical slices [11], while other structures (e.g., the left ventricle blood cavity) usually have a large span in the whole volume. There are two possible hazards resulting from such severe class imbalance when annotations are limited. One is that the model can easily become biased towards the dominant classes. The class-wise bias substantially affects model convergence during training and the generalization on the test domain [11]. The other is that inadequate learning from the tail classes could introduce more noises to pseudo labels, which will dramatically accumulate during the training phase and destroy the model training consequently.

To solve the problems above, we propose a novel semi-supervised class imbalanced deep learning approach on top of the Mean-teacher framework [12] for MRI segmentation. Firstly, an extra prototype-based classifier is introduced into the student model in parallel to the traditional linear-based classifier. With prototype learning, each in-class features encoded in the entire training set can be leveraged as the prior knowledge to assist the learning of each individual, conducive to accurate segmentation for tail classes. Besides, two multi-level tree filters modeling the pairwise pixel similarity in terms of intensity and semantic features are employed to refine pseudo labels. Through a comprehensive evaluation on two 3D cardiac MRI datasets with different experimental settings, our method achieves state-of-the-art results. Notably, the outstanding superiority of our method is exhibited (2.2% and 10% improvement) in the case where only quite a few labels are available (1.25% for ACDC and 10% for MMWHS).

2   Methods

Problem Setting. Given limited labeled data $\mathcal{D}_L = \{X_i, Y_i\}_{i=1}^{L}$ and unlabeled data $\mathcal{D}_U = \{X_i\}_{i=L+1}^{L+U}$, $X_i \in \mathbb{R}^{H \times W \times D}$ represents the original image and $Y_i =$

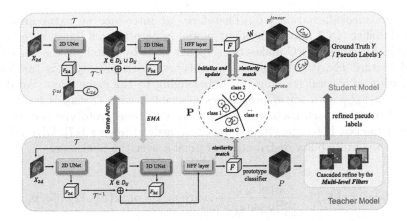

Fig. 1. Overview of the proposed semi-supervised class imbalance learning framework.

$\{0, 1, ..., C\}^{H \times W \times D}$ is the corresponding ground truth of C classes of objects in total. The given data are severely class imbalanced.

Overview. The overview of the proposed method is shown in Fig. 1. Our approach is built on top of the Mean-teacher [12] framework and adopts hybrid UNet (H-UNet) as backbone. Following [13], the networks learn through the student branch only, and the teacher model is update by using an Exponential Moving Average (EMA). The student model is trained with labeled pairs and pseudo labels generated by the teacher model. We train a linear- and a prototype-based classifier in parallel to alleviate class imbalance. Moreover, to guarantee the label quality, two multi-level tree filters (TFs) constructed based on the pixel affinity in terms of intensity and semantic features are employed to refine the teacher model's predictions from the prototype-based classifier.

2.1 H-UNet for 3D Medical Image Segmentation

The backbone model H-UNet is a variant of H-DenseUNet [14]. Given n samples in a training batch $X \in \mathbb{R}^{n \times 1 \times H \times W \times D}$, the 3D UNet produces features $F_{3d} \in \mathbb{R}^{n \times m \times H \times W \times D}$. Meanwhile, the 2D slices $X_{2d} \in \mathbb{R}^{nD \times 1 \times H \times W}$ obtained by performing the transformation operation $\mathcal{T}$ [14] on the original inputs are fed into the 2D UNet to generate features $F_{2d} \in \mathbb{R}^{nD \times m \times H \times W}$. By conducting inverse transformation operation, $\mathcal{T}^{-1}$, F_{2d} is aligned with F_{3d} before the two features are added and fed into the hybrid feature fusion (HFF) layer to yield the hybrid features F. We apply supervision on both 2D UNet and 3D UNet to train the H-UNet.

2.2 Class Imbalance Alleviation via Prototype-Based Classifier

Prototype-based classifier, leveraging prototypes instead of a parameterized predictor to make predictions, presents efficacy for semantic segmentation [15] and class imbalance alleviation [16]. Therefore, we introduce an extra prototype-based classifier (PC) into the student model in parallel to a linear-based classifier (LC). Specifically, PC makes dense predictions by matching the normalized hybrid feature F with the nearest prototypes. Each prototype is the feature aggregation of several training pixels belonging to the same class. We denote $\mathbf{P} = \{\boldsymbol{p}_{c,k}\}_{c,k=1}^{C,K}$ as a set of prototypes associated with C classes for segmentation. Note that each class may have more than one prototype (i.e., K) due to the intra-class heterogeneity exhibited in medical images [17]. Given X_i, PC produces the probability distribution of pixel a over the C classes:

$$p^{proto}(Y_i^{'}[a] = c|X_i) = \frac{exp(s_{a,c})}{\sum_{c'}^{C} exp(s_{a,c'})}, \text{with } s_{a,c} = max\{\mathcal{S}(\|F_i[a]\|_2, \boldsymbol{p}_{c,k})\}_{k=1}^{K},$$
(1)

where $s_{a,c} \in [-1,1]$ denotes the *pixel-class* similarity between pixel a and its closest prototype of class c. $\mathcal{S}(,)$ is the similarity measure (i.e., *cosine similarity*). $F_i[a]$ denotes the extracted features of pixel a, and $Y_i^{'}[a]$ denotes the predicted probability. $\|\cdot\|_2$ stands for the ℓ_2 normalization.

Meanwhile, the features F are also fed into LC parameterized by $W = [\mathbf{w}_1, \cdots, \mathbf{w}_C] \in \mathbb{R}^{C \times m}$; $\mathbf{w}_c \in \mathbb{R}^m$ is a learnable projection vector for class c. The probability distribution of pixel a estimated by LC is defined as:

$$p^{linear}(Y_i^{'}[a] = c|X_i) = \frac{exp(\mathbf{w}_c^T F_i[a])}{\sum_{c'=1}^{C} exp(\mathbf{w}_{c'}^T F_i[a])}.$$
(2)

These two classifiers complete each other during training. In the early training phase, LC dominates knowledge learning and provides PC with discriminative features to initialize and update the prototypes (See Sect. 2.4). With the addition of PC, the feature embedding space is further regularized, along with intra-class features being more compact and inter-class features being more separated, which in turn benefits the learning of LC.

2.3 Pseudo Label Refinement via Multi-level Tree Filters

Rather than directly using the teacher's high-confidence prediction to generate pseudo labels [13], we propose a cascaded refining strategy (CRS) to improve the label quality in a slice-by-slice manner on 3D volumes by using TFs [18], which proceeds with the following three steps as depicted in Fig. 2.

Graph Construction: First, we respectively represent the topology of the low-level original unlabeled image and the high-level hybrid features as two 4-connected planar graphs: $\mathcal{G}^* = \{\mathcal{V}^*, \mathcal{E}^*\}$ where $\mathcal{V}^*$ is the vertex set associated with each pixel and $\mathcal{E}^*$ is the edge set, and $* \in \{low, high\}$. The weight of the edge connecting two adjacent nodes a and b indicates their dissimilarity, which is defined by:

$$w_{a,b}^{low} = w_{b,a}^{low} = d(I[a], I[b]), \quad w_{a,b}^{high} = w_{b,a}^{high} = d(F[a], F[b]),$$
(3)

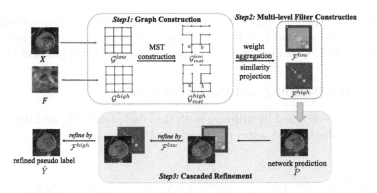

Fig. 2. The illustration of the cascaded refining strategy by using multi-level tree filters.

where $I[a] \in \mathbb{R}$ is the intensity of pixel a and $F[a] \in \mathbb{R}^m$ is the hybrid features. $d(\cdot, \cdot)$ denotes the squared Euclidean distance. Then, an edge pruning strategy is performed on $\mathcal{G}^{low}$ and $\mathcal{G}^{high}$ by sequentially removing the edges with substantial dissimilarity (i.e., larger distance) from $\mathcal{E}^*$ to generate two *minimum spanning trees* (MSTs) [19], $\mathcal{G}^{low}_{mst}$ and $\mathcal{G}^{high}_{mst}$, in which vertices are preferentially connected with similar ones.

Multi-level Filter Construction: Based on the two MSTs, we build the low-level TF $\mathcal{F}^{low}$ and the high-level TF $\mathcal{F}^{high}$. The filter weight $\mathcal{F}^*_{a,b}$ of any two nodes is obtained by aggregating the MST edge weights along the path between the two nodes [20]:

$$\mathcal{F}^*_{a,b} = \frac{1}{z_a} S_{\mathcal{G}^*_{mst}}(\mathbb{E}^*_{a,b}), \quad \text{where} \quad S_{\mathcal{G}^*_{mst}}(\mathbb{E}^*_{a,b}) = exp(-\sum_{\forall(q,o)\in\mathbb{E}^*_{a,b}} w^*_{q,o}). \quad (4)$$

Here, $\mathbb{E}^*_{a,b}$ is the edge set in the path from node a to node b. $S_{\mathcal{G}^*_{mst}}(\cdot)$ maps the distance of two vertices into a positive scalar which measures the pixel affinity. z_a is a normalization factor, which is the summation of the similarity between node a and all other nodes in the MST.

Cascaded Refinement: Lastly, we refine the teacher's prediction P with the two filters in a cascade manner to acquire high-quality pseudo labels $\hat{Y}$:

$$\hat{Y} = \mathsf{R}(\mathcal{F}^{high}, \mathsf{R}(\mathcal{F}^{low}, P)), \quad \text{with} \quad \mathsf{R}(\mathcal{F}^*, P_a) = \sum_{\forall b\in\mathcal{V}^*} \mathcal{F}^*_{a,b} P_b, \quad (5)$$

where $\mathsf{R}(\cdot, \cdot)$ is the refinement process where each unlabeled pixel can contribute to the refinement for other pixels with a contribution proportional to their similarity. By exploiting the multi-level complementary features, i.e., the object boundary information encoded in $\mathcal{F}^{low}$ and the semantic similarity encoded in $\mathcal{F}^{high}$, CRS shows superiority when a single TF fails in some cases. E.g., when two pixels of different classes have similar intensity values (pixel a and b in Fig. 2),

$\mathcal{F}^{low}$ will model them as affinity pairs, which is not expected. Fortunately, $\mathcal{F}^{high}$ can suppress the mutual interference between them according to their distinct semantic features, thus ensuring refinement efficacy (See Supplementary Fig. 1).

2.4   Network Training and Prototype Update

The framework is trained by utilizing both the labeled data D_L and the unlabeled data D_U. For the labeled data, the supervised loss is defined as:

$$
\mathcal{L}^l = \frac{1}{L} \sum_{i=1}^{L} \left(l(\widetilde{Y}_i^{2D}, Y_i^{2D}) + l(p^{proto}(Y_i^{'}|X_i), Y_i) + l(p^{linear}(Y_i^{'}|X_i), Y_i) \right), \quad (6)
$$

where $\widetilde{Y}^{2D}$ is the prediction generated by the 2D UNet in H-UNet, and Y^{2D} is the transformation of the 3D ground truth Y by conducting $\mathcal{T}$ (See Sect. 2.1). l is the weighted sum of cross entropy loss and dice loss.

For the unlabeled data, the student model is trained with the pseudo labels $\hat{Y}$ refined by the proposed CRS. The unsupervised loss is defined as:

$$
\mathcal{L}^u = \frac{1}{U} \sum_{i=L+1}^{L+U} \left(l(\widetilde{Y}_i^{2D}, \hat{Y}_i^{2D}) + l(p^{proto}(Y_i^{'}|X_i), \hat{Y}_i) + l(p^{linear}(Y_i^{'}|X_i), \hat{Y}_i) \right).
$$

$$(7)$$

Following [10], we use a time-dependent Gaussian warming up function to control the balance between the supervised and unsupervised loss.

Prototype Initialize and Update: The initialization of prototypes determines the PC performance. To get better prototypes, we first pretrain the network with the LC solely. Then, we collect pixelwise deep features and use K-means [21] to generate K subclusters for each class. Finally, initial prototypes of each class are obtained by averaging the features in each subcluster. Following [15], after each training iteration, we first conduct online clustering by few steps of Sinkhorn-Knopp iteration [22] and let the prototypes evolve continuously with the clustering features of both the labeled and unlabeled samples:

$$
p_{c,k} \leftarrow \alpha p_{c,k} + (1 - \alpha) \|F_{c,k}\|_2, \tag{8}
$$

where α is a momentum coefficient, i.e., 0.999. $\|F_{c,k}\|_2$ are the normalized features of pixels which belong to the c-th class and are closest to the k-th subcluster. Since the prototypes are iteratively updated by the in-class features encoded in the entire sample set, such prior knowledge can alleviate the class-wise bias exhibited in each individual sample, which can better guide the model training.

3 Experiments and Results

3.1 Datasets and Implementation Details

Datasets and Evaluation Metrics: The model is evaluated on two public cardiac MRI datasets: (1) The ACDC[1] [23] contains 100 patients' scans, with expert annotations for 3 structures: left ventricle (LV), myocardium (MYO), and right ventricle (RV); (2) The MMWHS[2] [24] consists of 20 3D cardiac MRIs with annotations for 7 structures: LV, RV, MYO, left atrium (LA), right atrium (RA), ascending aorta (AA), and pulmonary artery (PA). We adopt Dice Similarity Score (DSC) to evaluate model performance.

Pre-processing: Ahead of training, we removed the top 2% of the highest intensity value to reduce artifact impact for MMWHS dataset, followed by z-score normalization. As ACDC has a low through-plane resolution of 5–10 mm, we resample all 2D slices to a fixed in-plane resolution of 1×1 mm with a size of 256×256 as the training samples. For MMWHS, we randomly crop 3D cubes with a size of $80 \times 80 \times 80$ for training.

Implementation Details: We split the datasets into the training and validation set randomly at a ratio of 4:1 [11]. To assess the model performance on different label percentages, we respectively take 1.25%, 2.5%, and 10% data from the ACDC training set and 10%, 20%, and 40% data from the MMWHS training set as the labeled data and the rest as unlabeled. The training details for the two datasets are provided in Section A of the Supplementary material.

3.2 Comparison with State-of-the-Arts

Table 1. Comparison with state-of-art methods on the ACDC and MMWHS datasets

Methods	Avg DSC (ACDC)			Avg DSC (MMWHS)		
	$L=1.25\%$	$L=2.5\%$	$L=10\%$	$L=10\%$	$L=20\%$	$L=40\%$
Self train [7]	0.717	0.856	0.874	0.529	0.630	0.792
Data Aug [25]	0.657	0.833	0.865	0.513	0.679	0.791
Mixmatch [26]	0.616	0.812	0.852	0.550	0.689	0.792
Global+Local CL [27]	0.739	0.800	0.885	0.561	0.687	0.779
Class-wise Sampling [11]	0.746	0.842	0.889	0.626	**0.791**	0.815
Proposed	**0.768**	**0.869**	**0.907**	**0.732**	0.789	**0.840**

We compare our method with several state-of-the-art frameworks for semi-supervised medical image segmentation. The results are reported in Table 1.

---

[1] https://www.creatis.insa-lyon.fr/Challenge/acdc/databases.html.
[2] https://zmiclab.github.io/zxh/0/mmwhs/.

Among these, the proposed method outperforms competing methods across all datasets and label percentages, except for a slightly lower DSC than method Class-wise Sampling [11] by using 20% labeled MMWHS data. However, method [11] fails to produce satisfactory results using very few annotations (L = 1.25% in ACDC and L = 10% in MMWHS). In contrast, our method exhibits quite consistent performance on different label percentages and has substantial improvement when only few annotations are available, with more than 10% increase by using 10% labeled MMWHS. This indicates the benefit of fully leveraging unlabeled data via the proposed PC and CRS, in which every pixel in the unlabeled data can properly contribute to the model training and the refinement for pseudo labels. Methods like Data Aug [25] and contrastive learning of global and local features [27] do not explicitly address the class imbalance problem, thus they have inferior performance on both datasets. Also, these two methods both require a complex pretraining process, which is inefficient and time-consuming. Mixmatch [26], mixing labeled and unlabeled data using MixUp in SSL, cannot work very well for the datasets with a severe class imbalance problem, mainly because simply mixing two images could oppositely increase difficulty in the less-dominant classes learning.

3.3   Ablation Studies

Table 2. Ablation studies on the efficacy of different components in the proposed method. The mean and standard deviation of the testing results are reported.

Methods		LC	PC	CRS	Avg DSC	Dice of heart substructures ↑						
						MYO	LA	LV	RA	RV	AA	PA
3D UNet		✓			0.770	0.756(.08)	0.662(.08)	0.896(.02)	0.740(.13)	0.854(.04)	0.743(.06)	0.738(.02)
		✓	✓	✓	0.800	0.793(.04)	0.751(.02)	0.909(.02)	0.798(.06)	0.837(.09)	0.789(.06)	0.725(.04)
H-UNet	①	✓			0.782	0.757(.06)	0.758(.02)	0.897(.03)	0.730(.08)	0.805(.07)	0.775(.08)	0.756(.03)
	②	✓		✓	0.799	0.776(.04)	0.763(.01)	0.893(.02)	0.809(.07)	0.786(.05)	0.789(.08)	**0.775(.01)**
	③		✓		0.797	0.762(.06)	0.750(.02)	0.909(.02)	0.789(.08)	0.835(.04)	0.786(.05)	0.751(.03)
	④	✓	✓		0.825	0.800(.03)	**0.800(.02)**	0.910(.02)	0.843(.04)	0.864(.04)	0.789(.09)	0.766(.03)
	⑤	✓	✓	✓	**0.840**	**0.808(.07)**	0.794(.01)	**0.927(.01)**	**0.860(.05)**	**0.897(.03)**	**0.838(.04)**	0.755(.04)

We conducted ablation analyses on MMWHS by using 40% labeled data to investigate the contribution of the proposed key components (PC and CRS) and the choice of the backbone network. All the models are trained in the SSL way, and we use the thresholded pseudo-labeling [3] (threshold = 0.8) to generate pseudo labels for the models when CRS is absent. Table 2 presents the results of several variants by using different combinations of the proposed key components and two backbone networks. Compared with 3D UNet, H-UNet (model ①) shows more consistent results on different substructures and better performance regardless of the use of the proposed components, with a 1.2% and 4.0% increase in average DSC with and without PC and CRS respectively, indicating the merits of exploiting both intra-slice and inter-slice information for 3D medical image segmentation. When thresholded pseudo-labeling is replaced with CRS, model ② improves the average DSC by 1.7% with a smaller deviation for each heart substructure among all test samples, suggesting the effectiveness of CRS. PC alleviates the class imbalance problem by leveraging the prior knowledge of the entire sample set to assist individual learning. This is justified by the improvements of 1.5% on average DSC and better results on tail classes (e.g., RA, MYO) brought by substituting LC with PC (model ③). Moreover, the performance is further boosted when these two classifiers are simultaneously adopted (model ④), outperforming LC and PC by a margin of 4.3% and 2.8% respectively. This is because the mutual promotion between the two classifiers could provide a better regularized feature space for pixel predictions. With CRS integrated, PC and LC are trained with more reliable pseudo labels. The DSC is further improved by 1.5% (model ⑤), arriving at the highest value (0.840). Noticeably, when 3D UNet is equipped with all the components, the average DSC is improved by 3.0% with performance soar on tail classes. Such consistent efficacy indicates the potential of using the proposed PC and CRS to endow any segmentation backbones with the capability of addressing the class imbalance problem.

Please refer to Section B in the Supplementary material for more results in terms of average symmetric surface distance (ASSD) in voxel and the qualitative analyses of multi-level tree filters.

4 Conclusion

The scarcity of pixel-level annotations affects the performance of deep neural networks for medical image segmentation. Moreover, the class imbalanced problem existing in the medical data can further exacerbate the model degradation. To address the problems, we propose a novel semi-supervised class imbalanced learning approach by additionally introducing the prototype-based classifier into the student model and constructing two multi-level tree filters to refine the pseudo labels for more robust learning. Experiments conducted on two public cardiac MRI datasets demonstrate the superiority of the proposed method.

Acknowledgement. This work described in this paper was supported in part by the Shenzhen Portion of Shenzhen-Hong Kong Science and Technology Innovation Cooperation Zone under HZQB-KCZYB-20200089. The work was also partially supported by a grant from the Research Grants Council of the Hong Kong Special Administrative Region, China (Project Number: T45-401/22-N) and by a grant from the Hong Kong Innovation and Technology Fund (Project Number: MHP/085/21).

References

1. Hesamian, M.H., Jia, W., He, X., Kennedy, P.: Deep learning techniques for medical image segmentation: achievements and challenges. J. Digit. Imaging **32**, 582–596 (2019)
2. Lee, D.-H., et al.: Pseudo-label: the simple and efficient semi-supervised learning method for deep neural networks. In: Workshop on Challenges in Representation Learning. ICML, vol. 3, p. 896 (2013)
3. Zhang, B., et al.: Flexmatch: boosting semi-supervised learning with curriculum pseudo labeling. Adv. Neural. Inf. Process. Syst. **34**, 18408–18419 (2021)
4. Laine, S., Aila, T.: Temporal ensembling for semi-supervised learning, arXiv preprint arXiv:1610.02242 (2016)
5. Zhou, D., Bousquet, O., Lal, T., Weston, J., Schölkopf, B.: Learning with local and global consistency. In: Advances in Neural Information Processing Systems, vol. 16 (2003)
6. Tajbakhsh, N., Jeyaseelan, L., Li, Q., Chiang, J.N., Wu, Z., Ding, X.: Embracing imperfect datasets: a review of deep learning solutions for medical image segmentation. Med. Image Anal. **63**, 101693 (2020)
7. Bai, W., et al.: Semi-supervised learning for network-based cardiac MR image segmentation. In: Descoteaux, M., Maier-Hein, L., Franz, A., Jannin, P., Collins, D.L., Duchesne, S. (eds.) MICCAI 2017. LNCS, vol. 10434, pp. 253–260. Springer, Cham (2017). https://doi.org/10.1007/978-3-319-66185-8_29
8. Sedai, S., et al.: Uncertainty guided semi-supervised segmentation of retinal layers in OCT images. In: Shen, D., et al. (eds.) MICCAI 2019. LNCS, vol. 11764, pp. 282–290. Springer, Cham (2019). https://doi.org/10.1007/978-3-030-32239-7_32
9. Gal, Y., Ghahramani, Z.: Dropout as a Bayesian approximation: representing model uncertainty in deep learning. In: International Conference on Machine Learning, pp. 1050–1059. PMLR (2016)
10. Yu, L., Wang, S., Li, X., Fu, C.-W., Heng, P.-A.: Uncertainty-aware self-ensembling model for semi-supervised 3D left atrium segmentation. In: Shen, D., et al. (eds.) MICCAI 2019. LNCS, vol. 11765, pp. 605–613. Springer, Cham (2019). https://doi.org/10.1007/978-3-030-32245-8_67
11. Basak, H., Ghosal, S., Sarkar, R.: Addressing class imbalance in semi-supervised image segmentation: a study on cardiac MRI. In: Wang, L., Dou, Q., Fletcher, P.T., Speidel, S., Li, S. (eds.) MICCAI 2022. LNCS, vol. 13438, pp. 224–233. Springer, Cham (2022). https://doi.org/10.1007/978-3-031-16452-1_22
12. Tarvainen, A., Valpola, H.: Mean teachers are better role models: weight-averaged consistency targets improve semi-supervised deep learning results. In: Advances in Neural Information Processing Systems, vol. 30 (2017)
13. Sohn, K., et al.: Fixmatch: simplifying semi-supervised learning with consistency and confidence. Adv. Neural. Inf. Process. Syst. **33**, 596–608 (2020)
14. Li, X., Chen, H., Qi, X., Dou, Q., Fu, C.-W., Heng, P.-A.: H-DenseUNet: hybrid densely connected UNet for liver and tumor segmentation from CT volumes. IEEE Trans. Med. Imaging **37**(12), 2663–2674 (2018)
15. Zhou, T., Wang, W., Konukoglu, E., Van Gool, L.: Rethinking semantic segmentation: a prototype view. In: Proceedings of the IEEE/CVF Conference on Computer Vision and Pattern Recognition, pp. 2582–2593 (2022)
16. Huang, C., Wu, X., Zhang, X., Lin, S., Chawla, N.V.: Deep prototypical networks for imbalanced time series classification under data scarcity. In: Proceedings of the 28th ACM International Conference on Information and Knowledge Management, pp. 2141–2144 (2019)

17. Pham, D.L., Xu, C., Prince, J.L.: Current methods in medical image segmentation. Annu. Rev. Biomed. Eng. **2**(1), 315–337 (2000)
18. Song, L., Li, Y., Li, Z., Yu, G., Sun, H., Sun, J., Zheng, N.: Learnable tree filter for structure-preserving feature transform. In: Advances in Neural Information Processing Systems, vol. 32 (2019)
19. Kruskal, J.B.: On the shortest spanning subtree of a graph and the traveling salesman problem. Proc. Am. Math. Soc. **7**(1), 48–50 (1956)
20. Yang, Q.: Stereo matching using tree filtering. IEEE Trans. Pattern Anal. Mach. Intell. **37**(4), 834–846 (2014)
21. Krishna, K., Murty, M.N.: Genetic k-means algorithm. IEEE Trans. Syst. Man Cybern. Part B (Cybern.) **29**(3), 433–439 (1999)
22. Cuturi, M.: Sinkhorn distances: lightspeed computation of optimal transport. In: Advances in Neural Information Processing Systems, vol. 26 (2013)
23. Bernard, O., et al.: Deep learning techniques for automatic MRI cardiac multi-structures segmentation and diagnosis: is the problem solved? IEEE Trans. Med. Imaging **37**(11), 2514–2525 (2018)
24. Zhuang, X., Shen, J.: Multi-scale patch and multi-modality atlases for whole heart segmentation of MRI. Med. Image Anal. **31**, 77–87 (2016)
25. Chaitanya, K., Karani, N., Baumgartner, C.F., Becker, A., Donati, O., Konukoglu, E.: Semi-supervised and task-driven data augmentation. In: Chung, A.C.S., Gee, J.C., Yushkevich, P.A., Bao, S. (eds.) IPMI 2019. LNCS, vol. 11492, pp. 29–41. Springer, Cham (2019). https://doi.org/10.1007/978-3-030-20351-1_3
26. Berthelot, D., Carlini, N., Goodfellow, I., Papernot, N., Oliver, A., Raffel, C.A.: Mixmatch: a holistic approach to semi-supervised learning. In: Advances in Neural Information Processing Systems, vol. 32 (2019)
27. Chaitanya, K., Erdil, E., Karani, N., Konukoglu, E.: Contrastive learning of global and local features for medical image segmentation with limited annotations. Adv. Neural. Inf. Process. Syst. **33**, 12546–12558 (2020)

Collaborative Modality Generation and Tissue Segmentation for Early-Developing Macaque Brain MR Images

Xueyang Wu[1,2,3], Tao Zhong[1,2,3], Shujun Liang[1,2,3], Li Wang[4], Gang Li[4], and Yu Zhang[1,2,3(✉)]

[1] School of Biomedical Engineering, Southern Medical University, Guangzhou 510515, China
[2] Guangdong Provincial Key Laboratory of Medical Image Processing, Southern Medical University, Guangzhou 510515, China
[3] Guangdong Province Engineering Laboratory for Medical Imaging and Diagnostic Technology, Southern Medical University, Guangzhou 510515, China
yuzhang@smu.edu.cn
[4] Department of Radiology and BRIC, University of North Carolina at Chapel Hill, Chapel Hill, USA

Abstract. In neuroscience research, automatic segmentation of macaque brain tissues in magnetic resonance imaging (MRI) is crucial for understanding brain structure and function during development and evolution. Acquisition of multimodal information is a key enabler of accurate tissue segmentation, especially in early-developing macaques with extremely low contrast and dynamic myelination. However, many MRI scans of early-developing macaques are acquired only in a single modality. While various generative adversarial networks (GAN) have been developed to impute missing modality data, current solutions treat modality generation and image segmentation as two independent tasks, neglecting their inherent relationship and mutual benefits. To address these issues, this study proposes a novel Collaborative Segmentation-Generation Framework (CSGF) that enables joint missing modality generation and tissue segmentation of macaque brain MR images. Specifically, the CSGF consists of a modality generation module (MGM) and a tissue segmentation module (TSM) that are trained jointly by a cross-module feature sharing (CFS) and transferring generated modality. The training of the MGM under the supervision of the TSM enforces anatomical feature consistency, while the TSM learns multi-modality information related to anatomical structures from both real and synthetic multi-modality MR images. Experiments show that the CSGF outperforms the conventional independent-task mode on an early-developing macaque MRI dataset with 155 scans, achieving superior quality in both missing modality generation and tissue segmentation.

Keywords: Tissue segmentation · Missing-modality generation · Macaque · Early brain development · Multi-task collaboration

© The Author(s), under exclusive license to Springer Nature Switzerland AG 2023
H. Greenspan et al. (Eds.): MICCAI 2023, LNCS 14223, pp. 470–480, 2023.
https://doi.org/10.1007/978-3-031-43901-8_45

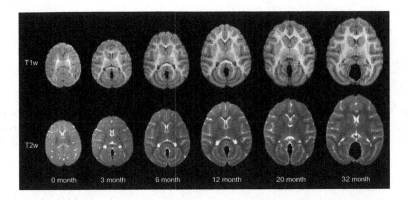

Fig. 1. T1w and T2w brain MR image of macaques at different ages.

1 Introduction

The precise processing and analysis of brain MR images are critical in advancing neuroscience research. As a widely used animal model with high systematic similarity to humans in various aspects, macaques play an indispensable role in understanding brain mechanisms in development, aging and evolution, exploring the pathogenesis of neurological diseases, and validating the effectiveness of clinical techniques and drugs [1]. In both humans and macaques, early brain development is a complex, dynamic, and regionally heterogeneous process that plays a crucial role in shaping later brain structural, functional, and cognitive outcomes [19]. While the human neuroimaging research is advancing rapidly, the macaque neuroimaging research lags behind, partly due to challenges in data acquisition, lack of tailored processing tools, and the inapplicability of human neuroimaging analysis tools to macaques [2]. Therefore, the macaque neuroimaging research is a rapidly evolving field that demands specialized tools and expertise. Of particular importance is brain tissue segmentation, a crucial prerequisite for quantitative volumetric analysis and surface-based studies, which aim to partition the brain into distinct regions such as white matter (WM), gray matter (GM), and cerebrospinal fluid (CSF) [3]. As illustrated in Fig. 1, the anatomical structures, age-related contrasts, and size of macaque brains undergo significant changes during early development [18], posing challenges for accurate tissue segmentation. Especially during the early postnatal months, the WM and GM show extremely low contrast, making their boundaries difficult to detect. The development of automatic and robust tissue segmentation algorithms for macaque brains during early developing stages is thus of great importance.

Deep learning has emerged as a powerful tool in medical image processing and analysis, with Convolutional Neural Networks (CNNs) showing remarkable performance in various applications, e.g., image segmentation [4], cross-modality generation [5,6] and multi-task learning [7,9] with state-of-the-art results. However, there is currently a lack of dedicated deep learning-based tools for brain tissue segmentation of macaque MRI data during early development, which can be

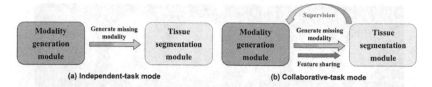

Fig. 2. A brief comparison of two modes: (a) Independent-task mode conducts image generation and tissue segmentation separately and step by step; (b) the proposed collaborative-task mode combines them through feature sharing and supervised learning.

attributed to several reasons. First, collecting macaque brain MRI data is challenging, and there is a dearth of publicly available large-scale labeled datasets for deep learning-based tissue segmentation. Second, brain tissue segmentation in early postnatal stages usually requires multi-modality images to provide more useful information than a single modality [8]. However, macaque datasets that are publicly available often only contain 3D T1-weighted (T1w) images and lack corresponding tissue labels and 3D T2-weighted (T2w) information. For example, the UW-Madison Rhesus MRI dataset [24] includes 592 macaque samples with T1w images only. To compensate for the missing modality information, generative adversarial networks (GAN) [13] have been applied to impute the missing modality in neuroimages, due to their potential for image-to-image translation [14–16]. Among them, pix2pix algorithm first proposes a solution for style transfer from image to image, which can be applied in the modality translation of medical images. Recently PTNet3D transforms the CNN-based frame into a Transformer-based frame, which reduces the computation amount required when processing high-resolution images. In these solutions, modality generation and downstream tasks are treated as two cascading independent tasks, as shown in Fig. 2(a). However, it is difficult to determine whether the generated missing modality has a natural positive effect on the downstream tasks [11]. In fact, both modality generation and tissue segmentation tasks require the feature extraction and learning of brain anatomical structures to achieve voxel-wise translation, showing a high task similarity and correlation. Essentially, the tissue map can guide the missing modality generation with anatomical information, while the generated missing modality can also provide useful complementary information for refining tissue segmentation [12]. To effectively capture and utilize the features and task correlation between two tasks, it is desirable to integrate tissue segmentation and modality generation into a unified framework. This can be achieved by imputing the missing modality image in a tissue-oriented manner, as illustrated in Fig. 2(b).

In this study, we present a 3D Collaborative Segmentation-Generation Framework (CSGF) for early-developing macaque MR images. The CSGF is designed in the form of multi-task collaboration and feature sharing, which enables it to complete the missing modality generation and tissue segmentation simultaneously. As depicted in Fig. 3(a), the CSGF comprises two modules:

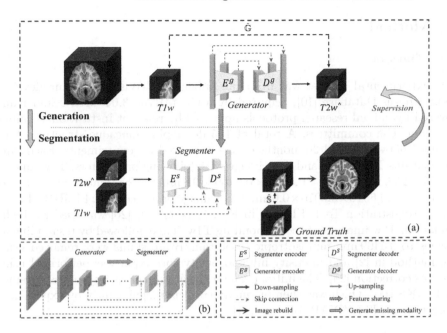

Fig. 3. Overview of the proposed 3D Collaborative Segmentation-Generation Framework (CSGF) (a). We use the T1w as the input of the modality generation module (MGM) to generate the missing modality T2w (T2w^). Meanwhile, the generated T2w (T2w^) and T1w are used as the input to the tissue segmentation module (TSM), and the results will be rebuilt into complete 3D MR images and represent the MGM loss and the TSM loss, respectively. In the structure diagram of cross-module feature sharing (CFS) (b), green blocks represent the decoder of the generator, and orange blocks represent the encoder of the segmenter. (Color figure online)

a modality generation module (MGM) and a tissue segmentation module (TSM), which are trained collaboratively through two forward information flows. Specifically, MGM and TSM will be linked by transferring generated missing-modality and cross-module feature sharing (CFS), ensuring that the MGM is trained under the supervision of the TSM, thereby imposing a constraint on anatomical feature consistency and providing multi-modality information.

The proposed CSGF offers several advantages over existing methods: (1) The collaborative learning of both MGM and TSM enables the missing modality to be imputed in a tissue-oriented manner, and hence the generated neuroimages are more consistent with real neuroimages from an anatomical point of view; (2) The CFS mechanism between the two modules provides more prosperous feature guidance for the TSM, thus also imposing constraints on the anatomical feature consistency of the MGM encoder; (3) The CSGF achieves improvements in both missing modality generation and brain tissue segmentation, especially for infant macaques with exceptionally low contrast between different tissues.

2   Method

2.1   Dataset

The experimental dataset was from the public UNC-Wisconsin Neurodevelopment Rhesus Database [10], acquired by a GE MR750 3.0T MRI scanner and covered by animal research protocols approved by relevant Institutional Animal Care and Use committees. A total of 155 developing macaque structural MRI samples between 0 and 36 months of age were used for experiments. Each sample contains T1w image and T2w image with following parameters: T1w, matrix = 256×256, resolution = $0.5469 \times 0.5469 \times 0.8\,\mathrm{mm}^3$; T2w, matrix = 256×256, resolution= $0.6016 \times 0.6016 \times 0.6\,\mathrm{mm}^3$. For image preprocessing, FMRIB's Linear Image Registration Tool (FLIRT) in FSL (version 5.0) [20] was used to rigidly align each T2w image to its corresponding T1w image, followed by resampling all images to $0.5469\,\mathrm{mm}$ mm isotropic resolution. Brain skulls were removed using the method in [2] and intensity inhomogeneity correction was performed using N4 bias correction [25]. The initial tissue segmentation of each scan was obtained by LINKS [17], which was then manually corrected significantly by experienced experts using ITK-SNAP software [21].

2.2   Problem Formulation

We construct this Collaborative Segmentation-Generation Framework (CSGF) based on the primary modality T1w (denoted as $\mathbf{T}_i^1$) and the auxiliary modality T2w (denoted as $\mathbf{T}_i^2$) data. Let $\{\mathbf{T}_i^1, \mathbf{T}_i^2, \mathbf{M}_i\}_{i=1 \sim N}$ be the dataset consisting of N scans, where $\mathbf{T}_i^1$ and $\mathbf{T}_i^2$ represent the T1w and T2w of ith scan, respectively; $\mathbf{M}_i$ represents the tissue label of ith scan. This framework can be formulated as

$$\hat{\mathbf{M}}_i = \mathbb{S}(\mathbf{T}_i^1, \mathbf{T}_i^2), \tag{1}$$

where $\hat{\mathbf{M}}_i$ is the estimated tissue label of the ith scan. However, in practical application, it is often the case that some s only contain the T1w modality. Therefore we construct a modality generation module (MGM) by a mapping function $\mathbb{G}$ to generate the missing T2w modality as

$$\hat{\mathbf{T}}_i^2 = \mathbb{G}(\mathbf{T}_i^1), \tag{2}$$

where $\hat{\mathbf{T}}_i^2$ is the generated T2w based on T1w modality for the ith scan. Then, the tissue segmentation module (TSM) constructs a mapping function $\mathbb{S}$ as

$$\hat{\mathbf{M}}_i = \mathbb{S}(\mathbf{T}_i^1, \mathbb{G}(\mathbf{T}_i^1)) = \mathbb{S}(\mathbf{T}_i^1, \hat{\mathbf{T}}_i^2) \approx \mathbb{S}(\mathbf{T}_i^1, \mathbf{T}_i^2). \tag{3}$$

2.3   Framework Overview and Collaborative Learning

The proposed CSGF consists of two key modules: the MGM and the TSM. These two modules are linked through the cross-module feature sharing (CFS), thus being trained collaboratively, as illustrated in Fig. 3(a). The CFS was created by a skip connection between the generator's down-sampling and the segmenter's up-sampling, as depicted in Fig. 3(b). This connection enables greater interaction and anatomy constraints between the two modules by additional back-propagating the gradient of the loss function.

The proposed MGM and TSM are trained collaboratively. Given a dataset $\{\mathbf{T}_i^1, \mathbf{T}_i^2, \mathbf{M}_i\}_{i=1 \sim N}$, the generator in MGM can be trained as

$$\hat{\mathbb{G}} = argmin \sum_{i=1}^{N} \{\|\mathbb{G}(\mathbf{T}_i^1) - \mathbf{T}_i^2\|_{GAN} + \|\mathbb{G}(\mathbf{T}_i^1) - \mathbf{T}_i^2\|_{MSE}\}, \qquad (4)$$

where $\| * \|_{GAN}$ represents the GAN loss and $\| * \|_{MSE}$ represents mean square error (MSE) loss. Meanwhile, the segmenter in TSM can be trained as

$$\hat{\mathbb{S}} = argmin \sum_{i=1}^{N} \{\|\mathbb{S}(\mathbf{T}_i^1, \mathbb{G}(\mathbf{T}_i^1)) - \mathbf{M}_i\|_{CE} + \|\mathbb{S}(\mathbf{T}_i^1, \mathbb{G}(\mathbf{T}_i^1)) - \mathbf{M}_i\|_{DICE}\}, \quad (5)$$

where $\| * \|_{CE}$ represents the cross-entropy loss and $\| * \|_{DICE}$ represents dice coefficient loss. During testing, for each input scan, if it has both T1w and T2w, we predict its tissue label as $\hat{\mathbf{M}}_i = \mathbb{S}(\mathbf{T}_i^1, \mathbf{T}_i^2)$. If it only has T1w modality, we first generate its T2w modality by $\hat{\mathbf{T}}_i^2 = \mathbb{G}(\mathbf{T}_i^1)$, and then predict its tissue label as $\hat{\mathbf{M}}_i = \mathbb{S}(\mathbf{T}_i^1, \mathbb{G}(\mathbf{T}_i^1))$.

3   Results and Discussion

3.1   Experimental Details

The total scans of MRI data used in this study was 155, divided into training and test sets in a 4:1 ratio. Before training, the data had the redundant background removed and cropped into a size of $160 \times 160 \times 160$, and then overlapping patches of size 64 were generated. The intensity values of all patches were normalized to $[-1, 1]$. Both MGM and TSM were optimized simultaneously during training. We employed the Adam optimizer with a learning rate of 0.0002 for the first 50 epochs, after which the learning rate gradually decreased to 0 over the following 50 epochs. In the testing phase, overlapping patches with stride 32 generated from testing volumes were input into the model, and the obtained results were rebuilt to a volume, during which the overlaps between adjacent patches were averaged. All experiments were implemented based on Pytorch 1.13.0 and conducted on NVIDIA RTX 4090 GPUs with 24GB VRAM in Ubuntu 18.04.

3.2   Evaluation of Framework

Totally 31 scans of early-developing macaques were used for framework evaluation. To provide a comparison baseline, we used PTNet3D (PTNet), one state-of-the-art algorithm for medical image translation, and U-Net [23], the most widely used segmentation model, to construct the CSGF. First, we trained two U-Net models as baseline models, one of which used only the T1w modality as input, while the other used both T1w and T2w modalities. Next, we trained a PTNet model, which generates corresponding T2w images based on the input T1w images. Finally, we constructed the CSGF framework using the PTNet and U-Net models described in Sect. 2. It is important to note that we constructed two versions of CSGF, one with the embedding of CFS to reflect the role of feature sharing and one without. According to the developmental stage of macaques [22], we divided the evaluation scans into three groups, including 7 in early infancy (age range of 0 to 6 months), 9 in infancy (age range of 7 to 12 months) and 15 in yearlings and juveniles (age range of 13 to 36 months). Compared with the other two stages, the brain in the early infancy stage from 0 to 6 months is in a state of vigorous development, especially the extremely low contrast between tissues from 0 to 2 months, which brings great challenges to this study.

Table 1. Quantitative comparison of different ablation settings. And 'ind.' means independent-task mode.

Method	Cerebrospinal fluid		Gray matter		White matter	
	Dice (%)	ASD (mm)	Dice (%)	ASD (mm)	Dice (%)	ASD (mm)
Early infancy (0–6 months)						
U-Net	74.82 ± 11.51	0.31 ± 0.19	90.59 ± 5.60	0.30 ± 0.21	79.57 ± 14.37	0.36 ± 0.28
PTNet+U-Net(ind.)	73.30 ± 11.23	0.49 ± 0.23	90.02 ± 5.46	0.35 ± 0.22	77.34 ± 14.43	0.41 ± 0.27
CSGF w\o CFS	76.90 ± 11.80	0.28 ± 0.19	90.97 ± 5.54	0.27 ± 0.19	80.21 ± 14.20	0.35 ± 0.28
CSGF with CFS	**77.54 ± 11.44**	**0.26 ± 0.18**	**90.99 ± 5.73**	**0.27 ± 0.19**	**80.57 ± 14.40**	**0.33 ± 0.27**
Infancy (7–12 months)						
U-Net	92.07 ± 3.21	0.06 ± 0.09	96.65 ± 0.73	0.06 ± 0.02	94.46 ± 1.21	0.08 ± 0.03
PTNet+U-Net(ind.)	91.11 ± 3.85	0.13 ± 0.09	96.30 ± 0.83	0.09 ± 0.05	93.82 ± 1.36	0.09 ± 0.03
CSGF w\o CFS	93.12 ± 3.10	0.06 ± 0.09	96.89 ± 0.75	0.05 ± 0.02	94.58 ± 1.19	**0.07 ± 0.02**
CSGF with CFS	**93.44 ± 3.01**	**0.05 ± 0.09**	**97.01 ± 0.76**	**0.05 ± 0.02**	**94.80 ± 1.27**	0.07 ± 0.03
Yearlings and juveniles (13–36 months)						
U-Net	92.71 ± 0.99	0.04 ± 0.01	96.69 ± 0.41	0.05 ± 0.01	94.70 ± 0.89	0.07 ± 0.02
PTNet+U-Net(ind.)	91.30 ± 2.01	0.11 ± 0.09	96.36 ± 0.49	0.07 ± 0.02	94.04 ± 1.01	0.08 ± 0.03
CSGF w±o CFS	93.65 ± 1.00	0.03 ± 0.01	96.95 ± 0.43	0.05 ± 0.01	94.81 ± 0.87	0.07 ± 0.02
CSGF with CFS	**93.79 ± 0.46**	**0.03 ± 0.01**	**97.08 ± 0.43**	**0.04 ± 0.01**	**95.03 ± 0.88**	**0.06 ± 0.02**

In the evaluation of tissue segmentation, we assumed the practical scenario that the test scans contain only T1w images. We compared four results, including U-Net with T1w input, U-Net with T1w and generated T2w input from PTNet (i.e. independent-task mode), CSGF without CFS embedding and the full CSGF framework. We quantitatively evaluate the results by using the average surface distance (ASD) and the dice coefficient, as reported in Table 1. The results show that compared to U-Net with a single T1w images input, U-Net with independent PTNet-generated T2w images as an additional modality input has worse results, especially for data during infancy. This may be due to the fact that the

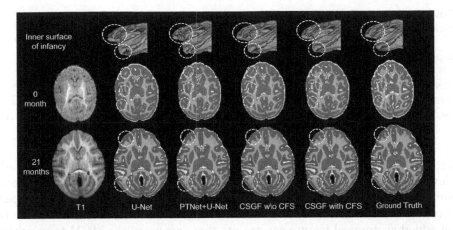

Fig. 4. Representative results of tissue segmentation. Red, green, and yellow colors represent cerebrospinal fluid (CSF), gray matter (GM), and white matter (WM), respectively. Blue dotted circles indicate significantly different results across different methods. Typical results of the reconstructed inner surface during infancy are also displayed on the top row, where red dotted circles indicate apparent differences. (Color figure online)

T2w images generated in the independent-task mode do not accurately express the anatomical structures of different tissues based on the real T1w images, thus introducing unnecessary noise for subsequent tissue segmentation. In comparison, the results of the two CSGF methods are superior to those of the above two, which may be due to the fact that under task-oriented supervised training, the T2w images generated by PTNet tend to retain anatomical structures that are conducive to subsequent tissue segmentation. Notably, the method in which CFS is embedded achieves the best results, showing that tissue segmentation also benefits from the deconstruction of features by the MGM during the encoding stage. Figure 4 presents representative visual results of tissue segmentation at two age stages. It can be seen that the result of CSGF is the closest to the ground truth, retaining more anatomical details.

In the evaluation of the modality generation, we compared generated T2w images from T1w images based on three methods, including PTNet, CSGF w/o CFS (i.e. PTNet under the supervision of U-Net) and the final CSGF framework. We quantitatively evaluated the results by using the peak signal to noise ratio (PSNR) and structural similarity index (SSIM), as reported in Table 2. It demonstrates that co-training supervised by subsequent tissue segmentation leads to PTNet in CSGF with improved generative results, as compared to a single PTNet. Especially in the infancy stage, the results of CSGF have been significantly improved. This improvement can be attributed to the fact that the generation module is constrained by task-oriented anatomical features, which encourage the preservation of complete tissue structures. This helps to address the challenge of low tissue contrast to some extent. The stable performance of

Table 2. Modality generation results on early infancy (0–6 months), infancy (7–12 months), yearlings and juveniles (13–36 months) macaques.

Method	Early infancy		Infancy		Yearlings and juveniles	
	PSNR(dB)	SSIM(%)	PSNR(dB)	SSIM(%)	PSNR(dB)	SSIM(%)
PTNet	25.74 ± 2.12	84.51 ± 6.19	28.07 ± 1.54	$\mathbf{92.22 \pm 1.92}$	$\mathbf{27.44 \pm 3.71}$	88.74 ± 12.53
CSGF w\o CFS	26.64 ± 1.90	84.97 ± 5.62	27.95 ± 1.39	92.02 ± 1.63	26.99 ± 3.46	88.72 ± 11.41
CSGF with CFS	$\mathbf{27.74 \pm 1.56}$	$\mathbf{85.92 \pm 5.38}$	$\mathbf{28.17 \pm 1.66}$	92.03 ± 1.60	27.37 ± 3.53	$\mathbf{88.97 \pm 11.57}$

the generative module suggests that anatomical feature consistency and task relevance can coexist and be co-optimized between MGM loss and TSM loss, leading to the preservation of image quality in the generative model. Furthermore, the incorporation of CFS enhances the generation results, possibly due to the added anatomical feature constraints that encourage the MGM to prioritize the encoding of anatomical information.

4   Conclusion

We propose the novel Collaborative Segmentation-Generation Framework (CSGF) to deal with the missing-modality brain MR images for tissue segmentation in early-developing macaques. Under this framework, the modality generation module (MGM) and tissue segmentation module (TSM) are jointly trained through cross-module feature sharing (CFS). The MGM is trained under the supervision of the TSM, while the TSM is trained with real and generated neuroimages. Comparative experiments on 155 scans of developing macaque data show that our CSGF outperforms conventional independent-task mode in both modality generation and tissue segmentation, showing its great potential in neuroimage research. Furthermore, as the proposed CSGF is a general framework, it can be easily extended to other types of modality generation, such as CT and PET, combined with other image segmentation tasks.

Acknowledgements. This work was supported in part by the National Natural Science Foundation of China #61971213, and #U22A20350, the National Natural Science Foundation of Youth Science Foundation project of China #62201246, and #62001206, and the Guangzhou Basic and Applied Basic Research Project #2023A04J2262.

References

1. Roelfsema, P.R., et al.: Basic neuroscience research with nonhuman primates: a small but indispensable component of biomedical research. Neuron **82**(6), 1200–1204 (2014)
2. Zhong, T., et al.: DIKA-Nets: Domain-invariant knowledge-guided attention networks for brain skull stripping of early developing macaques. Neuroimage **227**, 117649 (2021)
3. Li, G., et al.: Computational neuroanatomy of baby brains: a review. Neuroimage **185**, 906–925 (2019)

4. Isensee, F., et al.: nnU-Net: a self-configuring method for deep learning-based biomedical image segmentation. Nat. Methods **18**(2), 203–211 (2021)
5. Yu, B., et al.: Medical image synthesis via deep learning. Deep Learn. Med. Image Anal. 23–44 (2020)
6. Zhou, B., Liu, C., Duncan, J.S.: Anatomy-constrained contrastive learning for synthetic segmentation without ground-truth. In: de Bruijne, M., et al. (eds.) MICCAI 2021. LNCS, vol. 12901, pp. 47–56. Springer, Cham (2021). https://doi.org/10.1007/978-3-030-87193-2_5
7. Zheng, H., et al.: Phase Collaborative Network for Two-Phase Medical Imaging Segmentation. arXiv:1811.11814 (2018)
8. Zhou, T., et al.: A review: deep learning for medical image segmentation using multi-modality fusion. Array **3**, 100004 (2019)
9. Pan, Y., Chen, Y., Shen, D., Xia, Y.: Collaborative image synthesis and disease diagnosis for classification of neurodegenerative disorders with incomplete multimodal neuroimages. In: de Bruijne, M., et al. (eds.) MICCAI 2021. LNCS, vol. 12905, pp. 480–489. Springer, Cham (2021). https://doi.org/10.1007/978-3-030-87240-3_46
10. Young, J.T., et al.: The UNC-Wisconsin rhesus macaque neurodevelopment database: a structural MRI and DTI database of early postnatal development. Front. Neurosci. **11**, 29 (2017). https://pubmed.ncbi.nlm.nih.gov/28210206/
11. Xie, G., et al.: A Survey of Cross-Modality Brain Image Synthesis. arXiv preprint arXiv:2202.06997 (2022)
12. Yu, Z., Zhai, Y., Han, X., Peng, T., Zhang, X.-Y.: MouseGAN: GAN-based multiple MRI modalities synthesis and segmentation for mouse brain structures. In: de Bruijne, M., et al. (eds.) MICCAI 2021. LNCS, vol. 12901, pp. 442–450. Springer, Cham (2021). https://doi.org/10.1007/978-3-030-87193-2_42
13. Goodfellow, I., et al.: Generative adversarial networks. Commun. ACM **63**(11), 139–144 (2020)
14. Isola, P., et al.: Image-to-image translation with conditional adversarial networks. In: Proceedings of the IEEE Conference on Computer Vision and Pattern Recognition (CVPR 2017), pp. 1125–1134. IEEE(2017)
15. Zhu, J., et al.: Unpaired image-to-image translation using cycle-consistent adversarial networks. In: Proceedings of the IEEE Conference on Computer Vision and Pattern Recognition (CVPR 2017), pp. 2223–2232. IEEE (2017)
16. Zhang, X., et al.: PTNet3D: a 3D high-resolution longitudinal infant brain MRI synthesizer based on transformers. IEEE Trans. Med. Imaging **41**(10), 2925–2940 (2022)
17. Wang, L., et al.: Links: learning-based multi-source integration framework for segmentation of infant brain images. NeruoImage **108**, 160–172 (2015)
18. Zhong, T., et al.: Longitudinal brain atlases of early developing cynomolgus macaques from birth to 48 months of age. NeruoImage **247**, 118799 (2022)
19. Wang, F., et al.: Developmental topography of cortical thickness during infancy. Proc. Natl. Acad. Sci. **116**(32), 15855–15860 (2019)
20. Jenkinson, M., et al.: FSL. Neuroimage **62**(2), 782–790 (2012)
21. Yushkevich, P.A., et al.: User-guided 3D active contour segmentation of anatomical structures: significantly improved efficiency and reliability. Neuroimage **31**(3), 1116–1128 (2006)
22. National Centre for the Replacement, Reduction and Refinement of Animals in Research. https://macaques.nc3rs.org.uk/about-macaques/life-history

23. Ronneberger, O., Fischer, P., Brox, T.: U-Net: convolutional networks for biomedical image segmentation. In: Navab, N., Hornegger, J., Wells, W.M., Frangi, A.F. (eds.) MICCAI 2015. LNCS, vol. 9351, pp. 234–241. Springer, Cham (2015). https://doi.org/10.1007/978-3-319-24574-4_28
24. UW-Madison Rhesus MRI dataset. https://fcon_1000.projects.nitrc.org/indi/PRIME/uwmadison.html
25. Tustison, N.J., et at.: N4ITK: improved N3 bias correction. IEEE Trans. Med. Imaging **29**(6), 1310–1320 (2010)

EGE-UNet: An Efficient Group Enhanced UNet for Skin Lesion Segmentation

Jiacheng Ruan, Mingye Xie, Jingsheng Gao, Ting Liu, and Yuzhuo Fu[✉]

Shanghai Jiao Tong University, Shanghai, China
{jackchenruan,xiemingye,gaojingsheng,louisa_liu,yzfu}@sjtu.edu.cn

Abstract. Transformer and its variants have been widely used for medical image segmentation. However, the large number of parameter and computational load of these models make them unsuitable for mobile health applications. To address this issue, we propose a more efficient approach, the Efficient Group Enhanced UNet (**EGE-UNet**). We incorporate a Group multi-axis Hadamard Product Attention module (GHPA) and a Group Aggregation Bridge module (GAB) in a lightweight manner. The GHPA groups input features and performs Hadamard Product Attention mechanism (HPA) on different axes to extract pathological information from diverse perspectives. The GAB effectively fuses multi-scale information by grouping low-level features, high-level features, and a mask generated by the decoder at each stage. Comprehensive experiments on the ISIC2017 and ISIC2018 datasets demonstrate that EGE-UNet outperforms existing state-of-the-art methods. In short, compared to the TransFuse, our model achieves superior segmentation performance while reducing parameter and computation costs by **494x** and **160x**, respectively. Moreover, to our best knowledge, this is the first model with a parameter count limited to just **50KB**. Our code is available at https://github.com/JCruan519/EGE-UNet.

Keywords: Medical image segmentation · Light-weight model · mobile health

1 Introduction

Malignant melanoma is one of the most rapidly growing cancers in the world. As estimated by the American Cancer Society, there were approximately 100,350 new cases and over 6,500 deaths in 2020 [14]. Thus, an automated skin lesion segmentation system is imperative, as it can assist medical professionals in swiftly identifying lesion areas and facilitating subsequent treatment processes. To enhance the segmentation performance, recent studies tend to employ modules with larger parameter and computational complexity, such as incorporating self-attention mechanisms of Vision Transformer (ViT) [7]. For example, Swin-UNet [4], based on the Swin Transformer [11], leverages the feature extraction ability of self-attention mechanisms to improve segmentation performance.

This work was partially supported by the National Natural Science Foundation of China (Grant No. 61977045).

H. Greenspan et al. (Eds.): MICCAI 2023, LNCS 14223, pp. 481–490, 2023.
https://doi.org/10.1007/978-3-031-43901-8_46

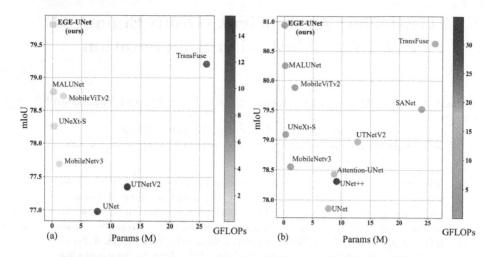

Fig. 1. (a) and (b) respectively show the visualization of comparative experimental results on the ISIC2017 and ISIC2018 datasets. The X-axis represents the number of parameters (lower is better), while Y-axis represents mIoU (higher is better). The color depth represents computational complexity (GFLOPs, lighter is better). (Color figure online)

TransUNet [5] has pioneered a serial fusion of CNN and ViT for medical image segmentation. TransFuse [26] employs a dual-path structure, utilizing CNN and ViT to capture local and global information, respectively. UTNetV2 [8] utilizes a hybrid hierarchical architecture, efficient bidirectional attention, and semantic maps to achieve global multi-scale feature fusion, combining the strengths of CNN and ViT. TransBTS [23] introduces self-attention into brain tumor segmentation tasks and uses it to aggregate high-level information.

Prior works have enhanced performance by introducing intricate modules, but neglected the constraint of computational resources in real medical settings. Hence, there is an urgent need to design a low-parameter and low-computational load model for segmentation tasks in mobile healthcare. Recently, UNeXt [22] has combined UNet [18] and MLP [21] to develop a lightweight model that attains superior performance, while diminishing parameter and computation. Furthermore, MALUNet [19] has reduced the model size by declining the number of model channels and introducing multiple attention modules, resulting in better performance for skin lesion segmentation than UNeXt. However, while MALUNet greatly reduces the number of parameter and computation, its segmentation performance is still lower than some large models, such as Trans-Fuse. Therefore, in this study, we propose EGE-UNet, a lightweight skin lesion segmentation model that achieves state-of-the-art while significantly reducing parameter and computation costs. Additionally, to our best knowledge, this is the first work to reduce parameter to approximately **50KB**.

To be specific, EGE-UNet leverages two key modules: the Group multi-axis Hadamard Product Attention module (GHPA) and Group Aggregation Bridge module (GAB). On the one hand, recent models based on ViT [7] have shown promise, owing to the multi-head self-attention mechanism (MHSA). MHSA divides the input into multiple heads and calculates self-attention in each head, which allows the model to obtain information from diverse perspectives, integrate different knowledge, and improve performance. Nonetheless, the quadratic complexity of MHSA enormously increases the model's size. Therefore, we present the Hadamard Product Attention mechanism (HPA) with linear complexity. HPA employs a learnable weight and performs a hadamard product operation with the input to obtain the output. Subsequently, inspired by the multi-head mode in MHSA, we propose GHPA, which divides the input into different groups and performs HPA in each group. However, it is worth noting that we perform HPA on different axes in different groups, which helps to further obtain information from diverse perspectives. On the other hand, for GAB, since the size and shape of segmentation targets in medical images are inconsistent, it is essential to obtain multi-scale information [19]. Therefore, GAB integrates high-level and low-level features with different sizes based on group aggregation, and additionally introduce mask information to assist feature fusion. Via combining the above two modules with UNet, we propose EGE-UNet, which achieves excellent segmentation performance with extremely low parameter and computation. Unlike previous approaches that focus solely on improving performance, our model also prioritizes usability in real-world environments. A clear comparison of EGE-UNet with others is shown in Fig. 1.

In summary, our contributions are threefold: (1) GHPA and GAB are proposed, with the former efficiently acquiring and integrating multi-perspective information and the latter accepting features at different scales, along with an auxiliary mask for efficient multi-scale feature fusion. (2) We propose EGE-UNet, an extremely lightweight model designed for skin lesion segmentation. (3) We conduct extensive experiments, which demonstrate the effectiveness of our methods in achieving state-of-the-art performance with significantly lower resource requirements.

2   EGE-UNet

The Overall Architecture. EGE-UNet is illustrated in Fig. 2, which is built upon the U-Shape architecture consisting of symmetric encoder-decoder parts. We take encoder part as an example. The encoder is composed of six stages, each with channel numbers of {8, 16, 24, 32, 48, 64}. While the first three stages employ plain convolutions with a kernel size of 3, the last three stages utilize the proposed GHPA to extract representation information from diverse perspectives. In contrast to the simple skip connections in UNet, EGE-UNet incorporates GAB for each stage between the encoder and decoder. Furthermore, our model leverages deep supervision [27] to generate mask predictions of varying scales, which are utilized for loss function and serve as one of the inputs to GAB. Via the integration of these advanced modules, EGE-UNet significantly reduces the parameter and computational load while enhancing the segmentation performance compared to prior approaches.

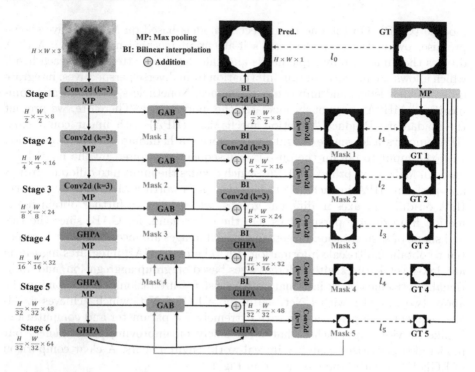

Fig. 2. The overview of EGE-UNet.

Algorithm 1: The Pytorch-style Pseudo-code for GHPA

 # Input: **X**, *the feature map with shape [B, C, H, W]*
 # Output: **Out**, *the feature map with shape [B, C, H, W]*
 # Params: ***a***, *the hyperparameter and default by 8 in this paper*
       b, *the hyperparameter and default by 8 in this paper*
       P_{xy}, *the randomly initialized tensor with shape [1, C//4, a, b]*
       P_{zx}, *the randomly initialized tensor with shape [1, 1, C//4, a]*
       P_{zy}, *the randomly initialized tensor with shape [1, 1, C//4, b]*
 # Operator: **DW**, *Depthwise Separable Convolution*
        LN, *LayerNorm*   **BI**, *Bilinear interpolation*

```
x1, x2, x3, x4 = torch.chunk(LN(X), 4, dim=1)
x1, x4 = x1 * DW(BI(Pxy))), DW(x4)
x2 = (x2.permute(0,3,1,2) * DW(BI(Pzx))).permute(0,2,3,1)
x3 = (x3.permute(0,2,1,3) * DW(BI(Pzy))).permute(0,2,1,3)
Out = DW(LN(torch.cat([x1,x2,x3,x4], dim=1)))
```

Group Multi-axis Hadamard Product Attention Module. To overcome the quadratic complexity issue posed by MHSA, we propose HPA with linear complexity. Given an input x and a randomly initialized learnable tensor p, bilinear interpolation is first utilized to resize p to match the size of x. Then, we employ depth-wise separable convolution (DW) [10,20] on p, followed by a hadamard product operation between x and p to obtain the output. However,

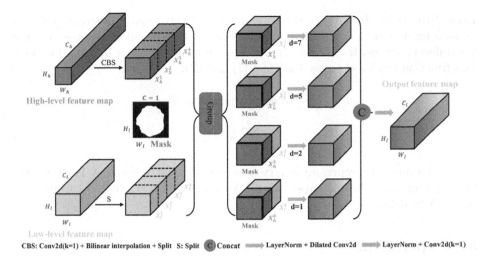

Fig. 3. The architecture of Group Aggregation Bridge module (GAB).

utilizing simple HPA alone is insufficient to extract information from multiple perspectives, resulting in unsatisfactory results. Motivated by the multi-head mode in MHSA, we introduce GHPA based on HPA, as illustrated in Algorithm 1. We divide the input into four groups equally along the channel dimension and perform HPA on the height-width, channel-height, and channel-width axes for the first three groups, respectively. For the last group, we only use DW on the feature map. Finally, we concatenate the four groups along the channel dimension and apply another DW to integrate the information from different perspectives. Note that all kernel size employed in DW are 3.

Group Aggregation Bridge Module. The acquisition of multi-scale information is deemed pivotal for dense prediction tasks, such as medical image segmentation. Hence, as shown in Fig. 3, we introduce GAB, which takes three inputs: low-level features, high-level features, and a mask. Firstly, depthwise separable convolution (DW) and bilinear interpolation are employed to adjust the size of high-level features, so as to match the size of low-level features. Secondly, we partition both feature maps into four groups along the channel dimension, and concatenate one group from the low-level features with one from the high-level features to obtain four groups of fused features. For each group of fused features, the mask is concatenated. Next, dilated convolutions [25] with kernel size of 3 and different dilated rates of {1, 2, 5, 7} are applied to the different groups, in order to extract information at different scales. Finally, the four groups are concatenated along the channel dimension, followed by the application of a plain convolution with the kernel size of 1 to enable interaction among features at different scales.

Loss Function. In this study, since different GAB require different scales of mask information, deep supervision [27] is employed to calculate the loss function for different stages, in order to generate more accurate mask information. Our loss function can be expressed as Eqs. (1) and (2).

$$l_i = Bce(y, \hat{y}) + Dice(y, \hat{y}) \tag{1}$$

$$\mathcal{L} = \sum_{i=0}^{5} \lambda_i \times l_i \tag{2}$$

where Bce and $Dice$ represent binary cross entropy and dice loss. λ_i is the weight for different stage. In this paper, we set λ_i to 1, 0.5, 0.4, 0.3, 0.2, 0.1 from $i = 0$ to $i = 5$ by default.

3   Experiments

Datasets and Implementation Details. To assess the efficacy of our model, we select two public skin lesion segmentation datasets, namely ISIC2017 [1,3] and ISIC2018 [2,6], containing 2150 and 2694 dermoscopy images, respectively. Consistent with prior research [19], we randomly partition the datasets into training and testing sets at a 7:3 ratio.

EGE-UNet is developed by Pytorch [17] framework. All experiments are performed on a single NVIDIA RTX A6000 GPU. The images are normalized and resized to 256×256. We apply various data augmentation, including horizontal flipping, vertical flipping, and random rotation. AdamW [13] is utilized as the optimizer, initialized with a learning rate of 0.001 and the CosineAnnealingLR [12] is employed as the scheduler with a maximum number of iterations of 50 and a minimum learning rate of 1e-5. A total of 300 epochs are trained with a batch size of 8. To evaluate our method, we employ Mean Intersection over Union (mIoU), Dice similarity score (DSC) as metrics, and we conduct 5 times and report the mean and standard deviation of the results for each dataset.

Comparative Results. The comparative experimental results presented in Table 1 reveal that our EGE-UNet exhibits a comprehensive state-of-the-art performance on the **ISIC2017** dataset. Specifically, in contrast to larger models, such as TransFuse, our model not only demonstrates superior performance, but also significantly curtails the number of parameter and computation by 494x and 160x, respectively. In comparison to other lightweight models, EGE-UNet surpasses UNeXt-S with a mIoU improvement of 1.55% and a DSC improvement of 0.97%, while exhibiting parameter and computation reductions of 17% and 72% of UNeXt-S. Furthermore, EGE-UNet outperforms MALUNet with a mIoU improvement of 1.03% and a DSC improvement of 0.64%, while reducing parameter and computation to 30% and 85% of MALUNet. For the **ISIC2018** dataset, the performance of our model also outperforms that of the best-performing model. Besides, it is noteworthy that EGE-UNet is the first lightweight model

Table 1. Comparative experimental results on the ISIC2017 and ISIC2018 dataset.

Dataset	Model	Params(M)↓	GFLOPs↓	mIoU(%)↑	DSC(%)↑
ISIC2017	UNet [18]	7.77	13.76	76.98	86.99
	UTNetV2 [8]	12.80	15.50	77.35	87.23
	TransFuse [26]	26.16	11.50	79.21	88.40
	MobileViTv2 [15]	1.87	0.70	78.72	88.09
	MobileNetv3 [9]	1.19	0.10	77.69	87.44
	UNeXt-S [22]	0.32	0.10	78.26	87.80
	MALUNet [19]	0.177	0.085	78.78	88.13
	EGE-UNet (Ours)	**0.053**	**0.072**	**79.81 ± 0.10**	**88.77 ± 0.06**
ISIC2018	UNet [18]	7.77	13.76	77.86	87.55
	UNet++ [27]	9.16	34.86	78.31	87.83
	Att-UNet [16]	8.73	16.71	78.43	87.91
	UTNetV2 [8]	12.80	15.50	78.97	88.25
	SANet [24]	23.90	5.96	79.52	88.59
	TransFuse [26]	26.16	11.50	80.63	89.27
	MobileViTv2 [15]	1.87	0.70	79.88	88.81
	MobileNetv3 [9]	1.19	0.10	78.55	87.98
	UNeXt-S [22]	0.32	0.10	79.09	88.33
	MALUNet [19]	0.177	0.085	80.25	89.04
	EGE-UNet (Ours)	**0.053**	**0.072**	**80.94 ± 0.11**	**89.46 ± 0.07**

Table 2. Ablation studies on the ISIC2017 dataset. (a) the macro ablation on two modules. (b) the micro ablation on GHPA. (c) the micro ablation on GAB.

Type	Model	Params(M)↓	GFLOPs↓	mIoU(%)↑	DSC(%)↑
(a)	Baseline	0.107	0.076	76.30	86.56
	Baseline + GHPA	0.034	0.058	78.82	88.16
	Baseline + GAB	0.126	0.086	78.78	88.13
(b)	w/o multi-axis grouping	0.074	0.074	79.13	88.35
	w/o DW for initialized tensor	0.050	0.072	79.03	88.29
(c)	w/o mask information	0.052	0.070	78.97	88.25
	w/o dilation rate of Conv2d	0.053	0.072	79.11	88.34

reducing parameter to about 50KB with excellent segmentation performance. Figure 1 presents a more clear visualization of the experimental findings and Fig. 4 shows some segmentation results.

Ablation Results. We conduct extensive ablation experiments to demonstrate the effectiveness of our proposed modules. The baseline utilized in our work is referenced from MALUNet [19], which employs a six-stage U-shaped architecture

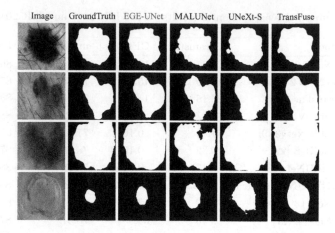

Fig. 4. Qualitative comparisons on the ISIC2018 dataset.

with symmetric encoder and decoder components. Each stage includes a plain convolution operation with a kernel size of 3, and the number of channels at each stage is set to {8, 16, 24, 32, 48, 64}. In Table 2(a), we conduct macro ablations on GHPA and GAB. Firstly, we replace the plain convolutions in the last three layers of baseline with GHPA. Due to the efficient multi-perspective feature acquisition of GHPA, it not only outperforms the baseline, but also greatly reduces the parameter and computation. Secondly, we substitute the skip-connection operation in baseline with GAB, resulting in further improved performance. Table 2(b) presents the ablations for GHPA. We replace the multi-axis grouping with single-branch and initialize the learnable tensors with only random values. It is evident that the removal of these two key designs leads to a marked drop. Table 2(c) illustrates the ablations for GAB. Initially, we omit the mask information, and mIoU metric even drops below 79%, thereby confirming once again the critical role of mask information in guiding feature fusion. Furthermore, we substitute the dilated convolutions in GAB with plain convolutions, which also leads to a reduction in performance.

4   Conclusions and Future Works

In this paper, we propose two advanced modules. Our GHPA uses a novel HPA mechanism to simplify the quadratic complexity of the self-attention to linear complexity. It also leverages grouping to fully capture information from different perspectives. Our GAB fuses low-level and high-level features and introduces a mask to integrate multi-scale information. Based on these modules, we propose EGE-UNet for skin lesion segmentation tasks. Experimental results demonstrate the effectiveness of our approach in achieving state-of-the-art performance with significantly lower resource requirements. We hope that our work can inspire further research on lightweight models for the medical image community.

Regarding limitations and future works, on the one hand, we mainly focus on how to greatly reduce the parameter and computation complexity while improving performance in this paper. Thus, we plan to deploy EGE-UNet in a real-world environment in the future work. On the other hand, EGE-UNet is currently designed only for the skin lesion segmentation task. Therefore, we will extend our lightweight design to other tasks.

References

1. https://challenge.isic-archive.com/data/#2017
2. https://challenge.isic-archive.com/data/#2018
3. Berseth, M.: ISIC 2017-skin lesion analysis towards melanoma detection. arXiv preprint arXiv:1703.00523 (2017)
4. Cao, H., et al.: Swin-unet: unet-like pure transformer for medical image segmentation. arXiv preprint arXiv:2105.05537 (2021)
5. Chen, J., et al.: Transunet: transformers make strong encoders for medical image segmentation. arXiv preprint arXiv:2102.04306 (2021)
6. Codella, N., et al.: Skin lesion analysis toward melanoma detection 2018: a challenge hosted by the international skin imaging collaboration (ISIC). arXiv preprint arXiv:1902.03368 (2019)
7. Dosovitskiy, A., et al.: An image is worth 16x16 words: transformers for image recognition at scale. arXiv preprint arXiv:2010.11929 (2020)
8. Gao, Y., Zhou, M., Liu, D., Metaxas, D.: A multi-scale transformer for medical image segmentation: architectures, model efficiency, and benchmarks. arXiv preprint arXiv:2203.00131 (2022)
9. Howard, A., et al.: Searching for mobilenetv3. In: Proceedings of the IEEE/CVF International Conference on Computer Vision, pp. 1314–1324 (2019)
10. Howard, A.G., et al.: Mobilenets: efficient convolutional neural networks for mobile vision applications. arXiv preprint arXiv:1704.04861 (2017)
11. Liu, Z., et al.: Swin transformer: hierarchical vision transformer using shifted windows. In: Proceedings of the IEEE/CVF International Conference on Computer Vision, pp. 10012–10022 (2021)
12. Loshchilov, I., Hutter, F.: SGDR: stochastic gradient descent with warm restarts. arXiv preprint arXiv:1608.03983 (2016)
13. Loshchilov, I., Hutter, F.: Decoupled weight decay regularization. arXiv preprint arXiv:1711.05101 (2017)
14. Mathur, P., et al.: Cancer statistics, 2020: report from national cancer registry programme, India. JCO Glob. Oncol. 6, 1063–1075 (2020)
15. Mehta, S., Rastegari, M.: Separable self-attention for mobile vision transformers. arXiv preprint arXiv:2206.02680 (2022)
16. Oktay, O., et al.: Attention U-Net: learning where to look for the pancreas. arXiv preprint arXiv:1804.03999 (2018)
17. Paszke, A., et al.: Pytorch: an imperative style, high-performance deep learning library. In: Advances in Neural Information Processing Systems, vol. 32 (2019)
18. Ronneberger, O., Fischer, P., Brox, T.: U-Net: convolutional networks for biomedical image segmentation. In: Navab, N., Hornegger, J., Wells, W.M., Frangi, A.F. (eds.) MICCAI 2015. LNCS, vol. 9351, pp. 234–241. Springer, Cham (2015). https://doi.org/10.1007/978-3-319-24574-4_28

19. Ruan, J., Xiang, S., Xie, M., Liu, T., Fu, Y.: MALUNet: a multi-attention and lightweight UNet for skin lesion segmentation. In: 2022 IEEE International Conference on Bioinformatics and Biomedicine (BIBM), pp. 1150–1156. IEEE (2022)

20. Sandler, M., Howard, A., Zhu, M., Zhmoginov, A., Chen, L.C.: Mobilenetv 2: inverted residuals and linear bottlenecks. In: Proceedings of the IEEE Conference on Computer Vision and Pattern Recognition, pp. 4510–4520 (2018)

21. Tolstikhin, I.O., et al.: MLP-mixer: an all-MLP architecture for vision. Adv. Neural. Inf. Process. Syst. **34**, 24261–24272 (2021)

22. Valanarasu, J.M.J., Patel, V.M.: UNeXt: MLP-based rapid medical image segmentation network. arXiv preprint arXiv:2203.04967 (2022)

23. Wang, W., Chen, C., Ding, M., Yu, H., Zha, S., Li, J.: TransBTS: multimodal brain tumor segmentation using transformer. In: de Bruijne, M., et al. (eds.) MICCAI 2021. LNCS, vol. 12901, pp. 109–119. Springer, Cham (2021). https://doi.org/10.1007/978-3-030-87193-2_11

24. Wei, J., Hu, Y., Zhang, R., Li, Z., Zhou, S.K., Cui, S.: Shallow attention network for polyp segmentation. In: de Bruijne, M., et al. (eds.) MICCAI 2021. LNCS, vol. 12901, pp. 699–708. Springer, Cham (2021). https://doi.org/10.1007/978-3-030-87193-2_66

25. Yu, F., Koltun, V.: Multi-scale context aggregation by dilated convolutions. arXiv preprint arXiv:1511.07122 (2015)

26. Zhang, Y., Liu, H., Hu, Q.: TransFuse: fusing transformers and CNNs for medical image segmentation. In: de Bruijne, M., et al. (eds.) MICCAI 2021. LNCS, vol. 12901, pp. 14–24. Springer, Cham (2021). https://doi.org/10.1007/978-3-030-87193-2_2

27. Zhou, Z., Rahman Siddiquee, M.M., Tajbakhsh, N., Liang, J.: UNet++: a nested U-net architecture for medical image segmentation. In: Stoyanov, D., et al. (eds.) DLMIA/ML-CDS -2018. LNCS, vol. 11045, pp. 3–11. Springer, Cham (2018). https://doi.org/10.1007/978-3-030-00889-5_1

BerDiff: Conditional Bernoulli Diffusion Model for Medical Image Segmentation

Tao Chen[1], Chenhui Wang[1], and Hongming Shan[1,2,3]($\boxtimes$) (iD)

[1] Institute of Science and Technology for Brain-Inspired Intelligence and MOE Frontiers Center for Brain Science, Fudan University, Shanghai, China
hmshan@fudan.edu.cn
[2] Key Laboratory of Computational Neuroscience and Brain-Inspired Intelligence (Fudan University), Ministry of Education, Shanghai, China
[3] Shanghai Center for Brain Science and Brain-Inspired Technology, Shanghai, China

Abstract. Medical image segmentation is a challenging task with inherent ambiguity and high uncertainty attributed to factors such as unclear tumor boundaries and multiple plausible annotations. The *accuracy* and *diversity* of segmentation masks are both crucial for providing valuable references to radiologists in clinical practice. While existing diffusion models have shown strong capacities in various visual generation tasks, it is still challenging to deal with discrete masks in segmentation. To achieve accurate and diverse medical image segmentation masks, we propose a novel conditional **Ber**noulli **Diff**usion model for medical image segmentation (**BerDiff**). Instead of using the Gaussian noise, we first propose to use the Bernoulli noise as the diffusion kernel to enhance the capacity of the diffusion model for binary segmentation tasks, resulting in more accurate segmentation masks. Second, by leveraging the stochastic nature of the diffusion model, our **BerDiff** randomly samples the initial Bernoulli noise and intermediate latent variables multiple times to produce a range of diverse segmentation masks, which can highlight salient regions of interest that can serve as a valuable reference for radiologists. In addition, our **BerDiff** can efficiently sample sub-sequences from the overall trajectory of the reverse diffusion, thereby speeding up the segmentation process. Extensive experimental results on two medical image segmentation datasets with different modalities demonstrate that our **BerDiff** outperforms other recently published state-of-the-art methods. Source code is made available at https://github.com/takimailto/BerDiff.

Keywords: Conditional diffusion · Bernoulli noise · Medical image segmentation

1 Introduction

Medical image segmentation plays a crucial role in enabling better diagnosis, surgical planning, and image-guided surgery [8]. The inherent ambiguity and high

---

Supplementary Information The online version contains supplementary material available at https://doi.org/10.1007/978-3-031-43901-8_47.

H. Greenspan et al. (Eds.): MICCAI 2023, LNCS 14223, pp. 491–501, 2023.
https://doi.org/10.1007/978-3-031-43901-8_47

uncertainty of medical images pose significant challenges [5] for accurate segmentation, attributed to factors such as unclear tumor boundaries in brain Magnetic Resonance Imaging (MRI) images and multiple plausible annotations of lung nodule in Computed Tomography (CT) images. Existing medical image segmentation methods typically provide a single, deterministic, most likely hypothesis mask, which may lead to misdiagnosis or sub-optimal treatment. Therefore, providing *accurate* and *diverse* segmentation masks as valuable references [17] for radiologists is crucial in clinical practice.

Recently, diffusion models [10] have shown strong capacities in various visual generation tasks [21, 22]. However, how to better deal with discrete segmentation tasks needs further consideration. Although many researches [1, 26] have combined diffusion model with segmentation tasks, all these methods do not take full account of the discrete characteristic of segmentation task and still use Gaussian noise as their diffusion kernel.

To achieve accurate and diverse segmentation masks, we propose a novel Conditional **Bernoulli Diff**usion model for medical image segmentation (`BerDiff`). Instead of using the Gaussian noise, we first propose to use the Bernoulli noise as the diffusion kernel to enhance the capacity of the diffusion model for segmentation, resulting in more accurate segmentation masks. Moreover, by leveraging the stochastic nature of the diffusion model, our `BerDiff` randomly samples the initial Bernoulli noise and intermediate latent variables multiple times to produce a range of diverse segmentation masks, highlighting salient regions of interest (ROI) that can serve as a valuable reference for radiologists. In addition, our `BerDiff` can efficiently sample sub-sequences from the overall trajectory of the reverse diffusion based on the rationale behind the Denoising Diffusion Implicit Models (DDIM) [25], thereby speeding up the segmentation process.

The contributions of this work are summarized as follows. 1) Instead of using the Gaussian noise, we propose a novel conditional diffusion model based on the Bernoulli noise for discrete binary segmentation tasks, achieving accurate and diverse medical image segmentation masks. 2) Our `BerDiff` can efficiently sample sub-sequences from the overall trajectory of the reverse diffusion, thereby speeding up the segmentation process. 3) Experimental results on LIDC-IDRI and BRATS 2021 datasets demonstrate that our `BerDiff` outperforms other state-of-the-art methods.

2   Methodology

In this section, we first describe the problem definitions, and then demonstrate the Bernoulli forward and diverse reverse processes of our `BerDiff`, as shown in Fig. 1. Finally, we provide an overview of the training and sampling procedures.

2.1   Problem Definition

Let us assume that $x \in \mathbb{R}^{H \times W \times C}$ denotes the input medical image with a spatial resolution of $H \times W$ and C channels. The ground-truth mask is represented

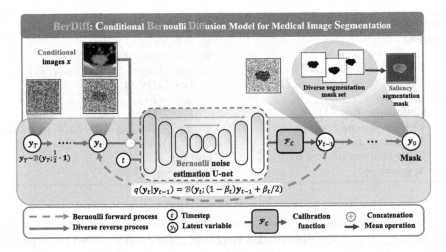

Fig. 1. Illustration of Bernoulli forward and diverse reverse processes of our `BerDiff`.

as $\boldsymbol{y}_0 \in \{0,1\}^{H \times W}$, where 0 represents background while 1 ROI. Inspired by diffusion-based models such as denoising diffusion probabilistic model (DDPM) and DDIM, we propose a novel conditional Bernoulli diffusion model, which can be represented as $p_\theta(\boldsymbol{y}_0|\boldsymbol{x}) := \int p_\theta(\boldsymbol{y}_{0:T}|\boldsymbol{x})\mathrm{d}\boldsymbol{y}_{1:T}$, where $\boldsymbol{y}_1,\ldots,\boldsymbol{y}_T$ are latent variables of the same size as the mask $\boldsymbol{y}_0$. For medical binary segmentation tasks, the diverse reverse process of our `BerDiff` starts from the initial Bernoulli noise $\boldsymbol{y}_T \sim \mathcal{B}(\boldsymbol{y}_T; \frac{1}{2}\mathbf{1})$ and progresses through intermediate latent variables constrained by the input medical image $\boldsymbol{x}$ to produce segmentation masks, where $\mathbf{1}$ denotes an all-ones matrix of the size $H \times W$.

2.2 Bernoulli Forward Process

In previous generation-related diffusion models, Gaussian noise is progressively added with increasing timestep t. However, for segmentation tasks, the ground-truth masks are represented by discrete values. To address this, our `BerDiff` gradually adds more Bernoulli noise using a noise schedule $\beta_1,\ldots,\beta_T$, as shown in Fig. 1. The Bernoulli forward process $q(\boldsymbol{y}_{1:T}|\boldsymbol{y}_0)$ of our `BerDiff` is a Markov chain, which can be represented as:

$$q\left(\boldsymbol{y}_{1:T} \mid \boldsymbol{y}_0\right) := \prod_{t=1}^{T} q\left(\boldsymbol{y}_t \mid \boldsymbol{y}_{t-1}\right), \tag{1}$$

$$q\left(\boldsymbol{y}_t \mid \boldsymbol{y}_{t-1}\right) := \mathcal{B}(\boldsymbol{y}_t; (1-\beta_t)\boldsymbol{y}_{t-1} + \beta_t/2), \tag{2}$$

where $\mathcal{B}$ denotes the Bernoulli distribution with the probability parameters $(1-\beta_t)\boldsymbol{y}_{t-1} + \beta_t/2$. Using the notation $\alpha_t = 1-\beta_t$ and $\bar{\alpha}_t = \prod_{\tau=1}^{t}\alpha_\tau$, we can efficiently sample $\boldsymbol{y}_t$ at an arbitrary timestep t in closed form:

$$q\left(\boldsymbol{y}_t \mid \boldsymbol{y}_0\right) = \mathcal{B}(\boldsymbol{y}_t; \bar{\alpha}_t\boldsymbol{y}_0 + (1-\bar{\alpha}_t)/2)). \tag{3}$$

Algorithm 1. Training	**Algorithm 2.** Sampling
repeat	$y_T \sim \mathcal{B}(y_T; \frac{1}{2} \cdot 1)$
$\quad (\boldsymbol{x}, \boldsymbol{y}_0) \sim q(\boldsymbol{x}, \boldsymbol{y}_0)$	**for** $t = T$ to 1 **do**
$\quad t \sim \text{Uniform}(\{1, \dots, T\})$	$\quad \hat{\mu}(\boldsymbol{y}_t, t, \boldsymbol{x}) = \mathcal{F}_C(\boldsymbol{y}_t, \hat{\epsilon}(\boldsymbol{y}_t, t, \boldsymbol{x}))$
$\quad \epsilon \sim \mathcal{B}(\epsilon; (1 - \bar{\alpha}_t)/2)$	$\quad$ **For DDPM:**
$\quad \boldsymbol{y}_t = \boldsymbol{y}_0 \oplus \epsilon$	$\quad \boldsymbol{y}_{t-1} \sim \mathcal{B}(\boldsymbol{y}_{t-1}; \hat{\mu}(\boldsymbol{y}_t, t, \boldsymbol{x}))$
$\quad$ **Calculate** Eq. (4)	$\quad$ **For DDIM:**
$\quad$ **Estimate** $\hat{\epsilon}(\boldsymbol{y}_t, t, \boldsymbol{x})$	$\quad \boldsymbol{y}_{t-1} \sim \mathcal{B}(\boldsymbol{y}_{t-1}; \sigma_t \boldsymbol{y}_t + (\bar{\alpha}_{t-1} - \sigma_t \bar{\alpha}_t) \vert \boldsymbol{y}_t -$
$\quad$ **Calculate** Eq. (6)	$\quad \hat{\epsilon}(\boldsymbol{y}_t, t, \boldsymbol{x}) \vert + ((1 - \bar{\alpha}_{t-1}) - (1 - \bar{\alpha}_t)\sigma_t)/2)$
$\quad$ **Take** gradient descent on $\nabla_\theta(\mathcal{L}_{\text{Total}})$	**end for**
until converged	**return** $\boldsymbol{y}_0$

To ensure that the objective function described in Sect. 2.4 is tractable and easy to compute, we use the sampled Bernoulli noise $\epsilon \sim \mathcal{B}(\epsilon; \frac{1 - \bar{\alpha}_t}{2} \cdot 1)$ to reparameterize $\boldsymbol{y}_t$ of Eq. (3) as $\boldsymbol{y}_0 \oplus \epsilon$, where $\oplus$ denotes the logical operation of "exclusive or (XOR)". Additionally, let $\odot$ denote elementwise product, and Norm($\cdot$) denote normalizing the input data along the channel dimension and then returning the second channel. The concrete Bernoulli posterior can be represented as:

$$q(\boldsymbol{y}_{t-1} \mid \boldsymbol{y}_t, \boldsymbol{y}_0) = \mathcal{B}(\boldsymbol{y}_{t-1}; \theta_{\text{post}}(\boldsymbol{y}_t, \boldsymbol{y}_0)), \tag{4}$$

where $\theta_{\text{post}}(\boldsymbol{y}_t, \boldsymbol{y}_0) = \text{Norm}([\alpha_t[1 - \boldsymbol{y}_t, \boldsymbol{y}_t] + \frac{1 - \alpha_t}{2}] \odot [\bar{\alpha}_{t-1}[1 - \boldsymbol{y}_0, \boldsymbol{y}_0] + \frac{1 - \bar{\alpha}_{t-1}}{2}])$.

2.3 Diverse Reverse Process

The diverse reverse process $p_\theta(\boldsymbol{y}_{0:T})$ can also be viewed as a Markov chain that starts from the Bernoulli noise $\boldsymbol{y}_T \sim \mathcal{B}(\boldsymbol{y}_T; \frac{1}{2} \cdot 1)$ and progresses through intermediate latent variables constrained by the input medical image $\boldsymbol{x}$ to produce diverse segmentation masks, as shown in Fig. 1. The concrete diverse reverse process of our `BerDiff` can be represented as:

$$p_\theta(\boldsymbol{y}_{0:T} \mid \boldsymbol{x}) := p(\boldsymbol{y}_T) \prod_{t=1}^{T} p_\theta(\boldsymbol{y}_{t-1} \mid \boldsymbol{y}_t, \boldsymbol{x}), \tag{5}$$

$$p_\theta(\boldsymbol{y}_{t-1} \mid \boldsymbol{y}_t, \boldsymbol{x}) := \mathcal{B}(\boldsymbol{y}_{t-1}; \hat{\mu}(\boldsymbol{y}_t, t, \boldsymbol{x})). \tag{6}$$

Specifically, we utilize the estimated Bernoulli noise $\hat{\epsilon}(\boldsymbol{y}_t, t, \boldsymbol{x})$ of $\boldsymbol{y}_t$ to parameterize $\hat{\mu}(\boldsymbol{y}_t, t, \boldsymbol{x})$ via a calibration function $\mathcal{F}_C$, as follows:

$$\hat{\mu}(\boldsymbol{y}_t, t, \boldsymbol{x}) = \mathcal{F}_C(\boldsymbol{y}_t, \hat{\epsilon}(\boldsymbol{y}_t, t, \boldsymbol{x})) = \theta_{\text{post}}(\boldsymbol{y}_t, \vert \boldsymbol{y}_t - \hat{\epsilon}(\boldsymbol{y}_t, t, \boldsymbol{x}) \vert), \tag{7}$$

where $\vert \cdot \vert$ denotes the absolute value operation. The calibration function aims to calibrate the latent variable $\boldsymbol{y}_t$ to a less noisy latent variable $\boldsymbol{y}_{t-1}$ in two steps: 1) estimating the segmentation mask $\boldsymbol{y}_0$ by computing the absolute deviation between $\boldsymbol{y}_t$ and the estimated noise $\hat{\epsilon}$; and 2) estimating the distribution of $\boldsymbol{y}_{t-1}$ by calculating the Bernoulli posterior, $p(\boldsymbol{y}_{t-1} \vert \boldsymbol{y}_t, \boldsymbol{y}_0)$, using Eq. (4).

2.4 Detailed Procedure

Here, we provide an overview of the training and sampling procedure in Algorithms 1 and 2. During the training phase, given an image and mask data pair $\{x, y_0\}$, we sample a random timestep t from a uniform distribution $\{1, \ldots, T\}$, which is used to sample the Bernoulli noise ϵ.

We then use ϵ to sample y_t from $q(y_t \mid y_0)$, which allows us to obtain the Bernoulli posterior $q(y_{t-1} \mid y_t, y_0)$. We pass the estimated Bernoulli noise $\hat{\epsilon}(y_t, t, x)$ through the calibration function $\mathcal{F}_C$ to parameterize $p_\theta(y_{t-1} \mid y_t, x)$. Based on the variational upper bound on the negative log-likelihood in previous diffusion models [3], we adopt Kullback-Leibler (KL) divergence and binary cross-entropy (BCE) loss to optimize our BerDiff as follows:

$$\mathcal{L}_{\text{KL}} = \mathbb{E}_{q(x, y_0)} \mathbb{E}_{q(y_t | y_0)} [D_{\text{KL}}[q(y_{t-1} \mid y_t, y_0) \| p_\theta(y_{t-1} \mid y_t, x)]], \qquad (8)$$

$$\mathcal{L}_{\text{BCE}} = -\mathbb{E}_{(\epsilon, \hat{\epsilon})} \sum_{i,j}^{H, W} [\epsilon_{i,j} \log \hat{\epsilon}_{i,j} + (1 - \epsilon_{i,j}) \log (1 - \hat{\epsilon}_{i,j})]. \qquad (9)$$

Finally, the overall objective function is presented as: $\mathcal{L}_{\text{Total}} = \mathcal{L}_{\text{KL}} + \lambda_{\text{BCE}} \mathcal{L}_{\text{BCE}}$, where λ_{BCE} is set to 1 in our experiments.

During the sampling phase, our BerDiff first samples the initial latent variable y_T, followed by iterative calculation of the probability parameters of y_{t-1} for different t. In Algorithm 2, we present two different sampling strategies from DDPM and DDIM for the latent variable y_{t-1}. Finally, our BerDiff is capable of producing diverse segmentation masks. By taking the mean of these masks, we can further obtain a saliency segmentation mask to highlight salient ROI that can serve as a valuable reference for radiologists. Note that our BerDiff has a novel parameterization technique, *i.e.* calibration function, to estimate the Bernoulli noise of y_t, which is different from previous works [3,11,24].

3 Experiment

3.1 Experimental Setup

Dataset and Preprocessing. The data used in this experiment are obtained from LIDC-IDRI [2,7] and BRATS 2021 [4] datasets. LIDC-IDRI contains 1,018 lung CT scans with plausible segmentation masks annotated by four radiologists. We adopt a standard preprocessing pipeline for lung CT scans and the train-validation-test partition as in previous work [5,15,23]. BRATS 2021 consists of four different sequence (T1, T2, FlAIR, T1CE) MRI images for each patient. All 3D scans are sliced into axial slices and discarded the bottom 80 and top 26 slices. Note that we treat the original four types of brain tumors as one type following previous work [25], converting the multi-target segmentation problem into binary. Our training set includes 55,174 2D images scanned from 1,126 patients, and the test set comprises 3,991 2D images scanned from 125 patients. Finally, the sizes of images from LIDC-IDRI and BRAST 2021 are resized to a resolution of 128×128 and 224×224, respectively.

Table 1. Ablation results of hyperparameters on LIDC-IDRI.

Loss	Estimation Target	GED				HM-IoU
		16	8	4	1	16
$\mathcal{L}_{KL}$	Bernoulli noise	0.332	0.365	0.430	0.825	0.517
$\mathcal{L}_{BCE}$	Bernoulli noise	<u>0.251</u>	<u>0.287</u>	<u>0.359</u>	<u>0.785</u>	<u>0.566</u>
$\mathcal{L}_{BCE} + \mathcal{L}_{KL}$	Bernoulli noise	**0.249**	**0.287**	**0.358**	**0.775**	**0.575**
$\mathcal{L}_{BCE} + \mathcal{L}_{KL}$	Ground-truth mask	0.277	0.317	0.396	0.866	0.509

♯ The best and second best results are highlighted in **bold** and <u>underlined</u>, respectively.

Table 2. Ablation results of diffusion kernel on LIDC-IDRI.

Training Iteration	Diffusion Kernel	GED				HM-IoU
		16	8	4	1	16
21,000	Gaussian	0.671	0.732	0.852	1.573	0.020
	Bernoulli	**0.252**	**0.287**	**0.358**	**0.775**	**0.575**
86,500	Gaussian	0.251	0.282	0.345	**0.719**	0.587
	Bernoulli	**0.238**	**0.271**	**0.340**	0.748	**0.596**

Implementation Details. We implement all the methods with the PyTorch library and train the models on NVIDIA V100 GPUs. All the networks are trained using the AdamW [19] optimizer with a mini-batch size of 32. The initial learning rate is set to 1×10^{-4} for BRATS 2021 and 5×10^{-5} for LIDC-IDRI. The Bernoulli noise estimation U-net network in Fig. 1 of our `BerDiff` is the same as previous diffusion-based models [20]. We employ a linear noise schedule for $T = 1000$ timesteps for all the diffusion models. And we use the sub-sequence sampling strategy of DDIM to accelerate the segmentation process. During mini-batch training of LIDC-IDRI, our `BerDiff` learns diverse expertise by randomly sampling one from four annotated segmentation masks for each image. Four metrics are used for performance evaluation, including Generalized Energy Distance (GED), Hungarian-Matched Intersection over Union (HM-IoU), Soft-Dice and Dice coefficient. We compute GED using varying numbers of segmentation samples (1, 4, 8, and 16), HM-IoU and Soft-Dice using 16 samples.

3.2   Ablation Study

We start by conducting ablation experiments to demonstrate the effectiveness of different losses and estimation targets, as shown in Table 1. All experiments are trained for 21,000 training iterations on LIDC-IDRI. We first explore the selection of losses in the top three rows. We find that the combination of KL divergence and BCE loss can achieve the best performance. Then, we explore the selection of estimation targets in the bottom two rows. We observe that estimating Bernoulli noise, instead of directly estimating the ground-truth mask, is

Table 3. Results on LIDC-IDRI.

Methods	GED 16	HM-IoU 16	Soft-Dice 16
Prob.U-net [15]	0.320 ± 0.03	0.500 ± 0.03	-
Hprob.U-net [16]	0.270 ± 0.01	0.530 ± 0.01	0.624 ± 0.01
CAR [14]	0.264 ± 0.00	0.592 ± 0.01	0.633 ± 0.00
JPro.U-net [29]	0.260 ± 0.00	0.585 ± 0.00	-
PixelSeg [28]	0.260 ± 0.00	0.587 ± 0.01	-
SegDiff [1]	0.248 ± 0.01	0.585 ± 0.00	0.637 ± 0.01
MedSegDiff [26]	0.420 ± 0.03	0.413 ± 0.03	0.453 ± 0.02
Zhang et al. [27]	0.400	0.534	0.599
Ji et al. [13]	0.658	0.447	0.616
Liao et al. [18]	0.593	0.453	0.587
BerDiff (Ours)	$\mathbf{0.238 \pm 0.01}$	$\mathbf{0.596 \pm 0.00}$	$\mathbf{0.644 \pm 0.00}$

Table 4. Results on BRATS 2021.

Methods	Dice
nnU-net [12]	88.2
TransU-net [6]	88.6
Swin UNETR [9]	89.0
U-net[♯]	89.2
SegDiff [1]	89.3
BerDiff (Ours)	**89.7**

♯ The U-net has the same architecture as the noise estimation network in our BerDiff and previous diffusion-based models.

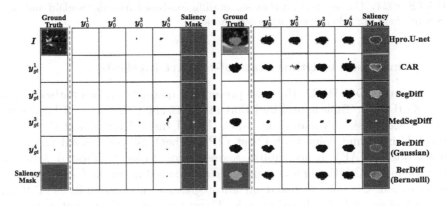

Fig. 2. Diverse segmentation masks and the corresponding saliency mask of two lung nodules randomly selected in LIDC-IDRI. y_0^i and y_{gt}^i refer to the i-th generated and ground-truth segmentation masks, respectively. Saliency Mask is the mean of diverse segmentation masks.

more suitable for our binary segmentation task. All of these findings are consistent with previous works [3,10].

Here, we conduct ablation experiments on our BerDiff with Gaussian or Bernoulli noise, and the results are shown in Table 2. For discrete segmentation tasks, we find that using Bernoulli noise can produce favorable results when training iterations are limited (e.g. 21,000 iterations) and even outperform using Gaussian noise when training iterations are sufficient (e.g. 86,500 iterations). We also provide a more detailed performance comparison between Bernoulli- and Gaussian-based diffusion models over training iterations in Fig. S3.

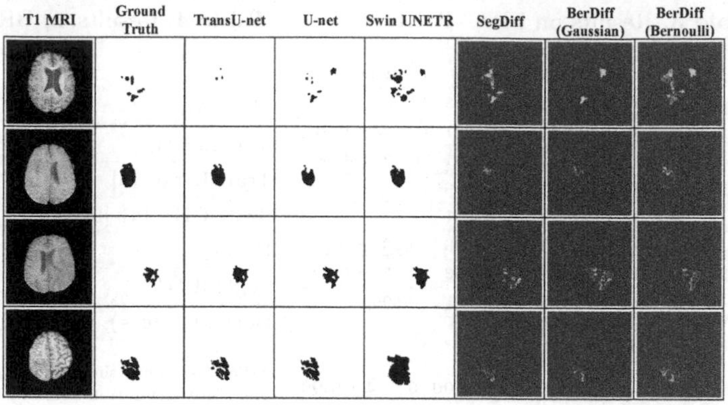

Fig. 3. Segmentation masks of four MRI images randomly selected in BRATS 2021. The segmentation masks of diffusion-based models (SegDiff and ours) presented here are saliency segmentation mask.

3.3 Comparison to Other State-of-the-Art Methods

Results on LIDC-IDRI. Here, we present the quantitative comparison results of LIDC-IDRI in Table 3, and find that our `BerDiff` performs well for discrete segmentation tasks. Probabilistic U-net (Prob.U-net), Hierarchical Prob.U-net (Hprob.U-net), and Joint Prob.U-net (JPro.U-net) use conditional variational autoencoder (cVAE) to accomplish segmentation tasks. Calibrated Adversarial Refinement (CAR) employs generative adversarial networks (GAN) to refine segmentation. PixelSeg is based on autoregressive models, while SegDiff and MedSegDiff are diffusion-based models. There are also methods that attempt to model multi-annotators explicitly [13,18,27]. We have the following three observations: 1) diffusion-based methods demonstrate significant superiority over traditional approaches based on VAE, GAN, and autoregression models for discrete segmentation tasks; 2) our `BerDiff` outperforms other diffusion-based models that use Gaussian noise as the diffusion kernel; and 3) our `BerDiff` also outperforms the methods that explicitly model the annotator, striking a good balance between diversity and accuracy. At the same time, we present comparison segmentation results in Fig. 2. Compared to other models, our `BerDiff` can effectively learn diverse expertise, resulting in more diverse and accurate segmentation masks. Especially for small nodules that can create ambiguity, such as the lung nodule on the left, our `BerDiff` approach produces segmentation masks that are more in line with the ground-truth masks.

Results on BRATS 2021. Here, we present the quantitative and qualitative results of BRATS 2021 in Table 4 and Fig. 3, respectively. We conducted a comparative analysis of our `BerDiff` with other models such as nnU-net, transformer-based models like TransU-net and Swin UNETR, as well as diffusion-based methods like SegDiff. First, we find that diffusion-based methods show superior performance compared to traditional U-net and transformer-based approaches. Besides, the high performance achieved by U-net, which shares the same architecture as our noise estimation network, highlights the effectiveness of the backbone design in diffusion-based models. Moreover, our proposed `BerDiff` surpasses other diffusion-based models that use Gaussian noise as the diffusion kernel. Finally, from Fig. 3, we find that our `BerDiff` segments more accurately on parts that are difficult to recognize by the human eye, such as the tumor in the 3rd row. At the same time, we can also generate diverse plausible segmentation masks to produce a saliency segmentation mask. We note that some of these masks may be false positives, as shown in the 1st row, but they can be filtered out due to low saliency. Please refer to Figs. S1 and S2 for more examples of diverse segmentation masks generated by our `BerDiff`.

4 Conclusion

In this paper, we proposed to use the Bernoulli noise as the diffusion kernel to enhance the capacity of the diffusion model for binary segmentation tasks, achieving accurate and diverse medical image segmentation results. Our `BerDiff` only focuses on binary segmentation tasks and takes much time during the iterative sampling process as other diffusion-based models; *e.g.* our `BerDiff` takes 0.4 s to segment one medical image, which is ten times of traditional U-net. In the future, we will extend our `BerDiff` to the multi-target segmentation problem and implement additional strategies for speeding up the segmentation process.

Acknowledgements. This work was supported in part by Natural Science Foundation of Shanghai (No. 21ZR1403600), National Natural Science Foundation of China (No. 62101136), Shanghai Municipal Science and Technology Major Project (No. 2018SHZDZX01) and ZJLab, Shanghai Municipal of Science and Technology Project (No. 20JC1419500), and Shanghai Center for Brain Science and Brain-inspired Technology.

References

1. Amit, T., Shaharbany, T., Nachmani, E., Wolf, L.: SegDiff: image segmentation with diffusion probabilistic models. arXiv preprint arXiv:2112.00390 (2021)
2. Armato, S.G., III., et al.: The lung image database consortium (LIDC) and image database resource initiative (IDRI): a completed reference database of lung nodules on CT scans. Med. Phys. **38**(2), 915–931 (2011)
3. Austin, J., Johnson, D.D., Ho, J., Tarlow, D., van den Berg, R.: Structured denoising diffusion models in discrete state-spaces. Adv. Neural. Inf. Process. Syst. **34**, 17981–17993 (2021)
4. Baid, U., et al.: The RSNA-ASNR-MICCAI BraTS 2021 benchmark on brain tumor segmentation and radiogenomic classification. arXiv preprint arXiv:2107.02314 (2021)
5. Baumgartner, C.F., et al.: PHiSeg: capturing uncertainty in medical image segmentation. In: Shen, D., et al. (eds.) MICCAI 2019. LNCS, vol. 11765, pp. 119–127. Springer, Cham (2019). https://doi.org/10.1007/978-3-030-32245-8_14

6. Chen, J., et al.: Transunet: transformers make strong encoders for medical image segmentation. arXiv preprint arXiv:2102.04306 (2021)

7. Clark, K., et al.: The cancer imaging archive (TCIA): maintaining and operating a public information repository. J. Digit. Imaging **26**(6), 1045–1057 (2013)

8. Haque, I.R.I., Neubert, J.: Deep learning approaches to biomedical image segmentation. Inform. Med. Unlocked **18**, 100297 (2020)

9. Hatamizadeh, A., Nath, V., Tang, Y., Yang, D., Roth, H.R., Xu, D.: Swin UNETR: swin transformers for semantic segmentation of brain tumors in MRI images. In: Crimi, A., Bakas, S. (eds.) BrainLes 2021. LNCS, vol. 12962, pp. 272–284. Springer, Cham (2021). https://doi.org/10.1007/978-3-031-08999-2_22

10. Ho, J., Jain, A., Abbeel, P.: Denoising diffusion probabilistic models. Adv. Neural. Inf. Process. Syst. **33**, 6840–6851 (2020)

11. Hoogeboom, E., Nielsen, D., Jaini, P., Forré, P., Welling, M.: Argmax flows and multinomial diffusion: Learning categorical distributions. Adv. Neural. Inf. Process. Syst. **34**, 12454–12465 (2021)

12. Isensee, F., Jaeger, P.F., Kohl, S.A., Petersen, J., Maier-Hein, K.H.: nnU-Net: a self-configuring method for deep learning-based biomedical image segmentation. Nat. Methods **18**(2), 203–211 (2021)

13. Ji, W., et al.: Learning calibrated medical image segmentation via multi-rater agreement modeling. In: 2021 IEEE/CVF Conference on Computer Vision and Pattern Recognition (CVPR), pp. 12336–12346 (2021)

14. Kassapis, E., Dikov, G., Gupta, D.K., Nugteren, C.: Calibrated adversarial refinement for stochastic semantic segmentation. In: 2021 IEEE/CVF International Conference on Computer Vision (ICCV), pp. 7037–7047 (2020)

15. Kohl, S., et al.: A probabilistic U-Net for segmentation of ambiguous images. In: Advances in Neural Information Processing Systems, vol. 31 (2018)

16. Kohl, S.A., et al.: A hierarchical probabilistic U-Net for modeling multi-scale ambiguities. arXiv preprint arXiv:1905.13077 (2019)

17. Lenchik, L., et al.: Automated segmentation of tissues using CT and MRI: a systematic review. Acad. Radiol. **26**(12), 1695–1706 (2019)

18. Liao, Z., Hu, S., Xie, Y., Xia, Y.: Modeling annotator preference and stochastic annotation error for medical image segmentation. arXiv preprint arXiv:2111.13410 (2021)

19. Loshchilov, I., Hutter, F.: Decoupled weight decay regularization. arXiv preprint arXiv:1711.05101 (2017)

20. Nichol, A.Q., Dhariwal, P.: Improved denoising diffusion probabilistic models. In: International Conference on Machine Learning, pp. 8162–8171. PMLR (2021)

21. Ramesh, A., Dhariwal, P., Nichol, A., Chu, C., Chen, M.: Hierarchical text-conditional image generation with clip latents. arXiv preprint arXiv:2204.06125 (2022)

22. Saharia, C., et al.: Photorealistic text-to-image diffusion models with deep language understanding. Adv. Neural. Inf. Process. Syst. **35**, 36479–36494 (2022)

23. Selvan, R., Faye, F., Middleton, J., Pai, A.: Uncertainty quantification in medical image segmentation with normalizing flows. In: Liu, M., Yan, P., Lian, C., Cao, X. (eds.) MLMI 2020. LNCS, vol. 12436, pp. 80–90. Springer, Cham (2020). https://doi.org/10.1007/978-3-030-59861-7_9

24. Sohl-Dickstein, J., Weiss, E., Maheswaranathan, N., Ganguli, S.: Deep unsupervised learning using nonequilibrium thermodynamics. In: International Conference on Machine Learning, pp. 2256–2265. PMLR (2015)

25. Wolleb, J., Sandkühler, R., Bieder, F., Valmaggia, P., Cattin, P.C.: Diffusion models for implicit image segmentation ensembles. In: International Conference on Medical Imaging with Deep Learning, pp. 1336–1348. PMLR (2022)
26. Wu, J., Fang, H., Zhang, Y., Yang, Y., Xu, Y.: MedSegDiff: medical image segmentation with diffusion probabilistic model. arXiv preprint arXiv:2211.00611 (2022)
27. Zhang, L., et al.: Disentangling human error from ground truth in segmentation of medical images. Adv. Neural. Inf. Process. Syst. **33**, 15750–15762 (2020)
28. Zhang, W., Zhang, X., Huang, S., Lu, Y., Wang, K.: PixelSeg: pixel-by-pixel stochastic semantic segmentation for ambiguous medical images. In: Proceedings of the 30-th ACM International Conference on Multimedia, pp. 4742–4750 (2022)
29. Zhang, W., Zhang, X., Huang, S., Lu, Y., Wang, K.: A probabilistic model for controlling diversity and accuracy of ambiguous medical image segmentation. In: Proceedings of the 30-th ACM International Conference on Multimedia, pp. 4751–4759 (2022)

DBTrans: A Dual-Branch Vision Transformer for Multi-Modal Brain Tumor Segmentation

Xinyi Zeng, Pinxian Zeng, Cheng Tang, Peng Wang, Binyu Yan, and Yan Wang[✉]

School of Computer Science, Sichuan University, Chengdu, China
wangyanscu@hotmail.com

Abstract. 3D Spatially Aligned Multi-modal MRI Brain Tumor Segmentation (SAMM-BTS) is a crucial task for clinical diagnosis. While Transformer-based models have shown outstanding success in this field due to their ability to model global features using the self-attention mechanism, they still face two challenges. First, due to the high computational complexity and deficiencies in modeling local features, the traditional self-attention mechanism is ill-suited for SAMM-BTS tasks that require modeling both global and local volumetric features within an acceptable computation overhead. Second, existing models only stack spatially aligned multi-modal data on the channel dimension, without any processing for such multi-channel data in the model's internal design. To address these challenges, we propose a Transformer-based model for the SAMM-BTS task, namely DBTrans, with dual-branch architectures for both the encoder and decoder. Specifically, the encoder implements two parallel feature extraction branches, including a local branch based on Shifted Window Self-attention and a global branch based on Shuffle Window Cross-attention to capture both local and global information with linear computational complexity. Besides, we add an extra global branch based on Shifted Window Cross-attention to the decoder, introducing the key and value matrices from the corresponding encoder block, allowing the segmented target to access a more complete context during up-sampling. Furthermore, the above dual-branch designs in the encoder and decoder are both integrated with improved channel attention mechanisms to fully explore the contribution of features at different channels. Experimental results demonstrate the superiority of our DBTrans model in both qualitative and quantitative measures. Codes will be released at https://github.com/Aru321/DBTrans.

Keywords: Spatially aligned multi-modal MRI (SAMM) · Brain tumor segmentation (BTS) · Transformer · Cross-Attention · Channel-Attention

1 Introduction

Glioma is one of the most common malignant brain tumors with varying degrees of invasiveness [1]. Brain Tumor Semantic segmentation of gliomas based on 3D spatially aligned Magnetic Resonance Imaging (SAMM-BTS) is crucial for accurate diagnosis

---

X. Zeng and P. Zeng—Contribute equally to this work.

© The Author(s), under exclusive license to Springer Nature Switzerland AG 2023
H. Greenspan et al. (Eds.): MICCAI 2023, LNCS 14223, pp. 502–512, 2023.
https://doi.org/10.1007/978-3-031-43901-8_48

and treatment planning. Unfortunately, radiologists suffer from spending several hours manually performing the SAMM-BTS task in clinical practice, resulting in low diagnostic efficiency. In addition, manual delineation requires doctors to have high professionalism. Therefore, it is necessary to design an efficient and accurate glioma lesion segmentation algorithm to effectively alleviate this problem and relieve doctors' workload and improve radiotherapy quality.

With the rise of deep learning, researchers have begun to study deep learning-based image analysis methods [2, 37]. Specifically, many convolutional neural network-based (CNN-based) models have achieved promising results [3–8]. Compared with natural images, medical image segmentation often requires higher accuracy to make subsequent treatment plans for patients. U-Net reaches an outstanding performance on medical image segmentation by combining the features from shallow and deep layers using skip-connection [9–11]. Based on U-Net, Brugger. et al. [12] proposed a partially reversible U-Net to reduce memory consumption while maintaining acceptable segmentation results. Pei et al. [13] explored the efficiency of residual learning and designed a 3D ResUNet for multi-modal brain tumor segmentation. However, due to the lack of global understanding of images for convolution operation, CNN-based methods struggle to model the dependencies between distant features and make full use of the contextual information [14]. But for semantic segmentation tasks whose results need to be predicted at pixel-level or voxel-level, both local spatial details and global dependencies are extremely important.

In recent years, models based on the self-attention mechanism, such as Transformer, have received widespread attention due to their excellent performance in Natural Language Processing (NLP) [15]. Compared with convolution operation, the self-attention mechanism is not restricted by local receptive fields and can capture long-range dependencies. Many works [16–19] have applied Transformers to computer vision tasks and achieved favorable results. For classification tasks, Vision Transformer (ViT) [19] was a groundbreaking innovation that first introduced pure Transformer layers directly across domains. And for semantic segmentation tasks, many methods, such as SETR [20] and Segformer [21], use ViT as the direct backbone network and combine it with a task-specific segmentation head for prediction results, reaching excellent performance on some 2D natural image datasets. For 3D medical image segmentation, Vision Transformer has also been preferred by researchers. A lot of robust variants based on Transformer have been designed to endow U-Net with the ability to capture contextual information in long-distance dependencies, further improving the semantic segmentation results of medical images [22–27]. Wang et al. [25] proposed a novel framework named TransBTS that embeds the Transformer in the bottleneck part of a 3D U-Net structure. Peiris et al. [26] introduced a 3D Swin-Transformer [28] to segmentation tasks and first incorporated the attention mechanism into skip-connection.

While Transformer-based models have shown effectiveness in capturing long-range dependencies, designing a Transformer architecture that performs well on the SAMM-BTS task remains challenging. First, modeling relationships between 3D voxel sequences is much more difficult than 2D pixel sequences. When applying 2D models, 3D images need to be sliced along one dimension. However, the data in each slice is related to three views, discarding any of them may lead to the loss of local information, which may cause the degradation of performance [29]. Second, most existing MRI segmentation methods still have difficulty capturing global interaction information while effectively encoding local information. Moreover, current methods just stack modalities and pass them through a network, which treats each modality equally along the channel dimension and may ignore the contribution of different modalities. To address the above limitations, we propose a novel encoder-decoder model, namely DBTrans, for multi-modal medical image segmentation. In the encoder, two types of window-based attention mechanisms, i.e., Shifted Window-based Multi-head Self Attention (Shifted-W-MSA) and Shuffle Window-based Multi-head Cross Attention (Shuffle-W-MCA), are introduced and applied in parallel to dual-branch encoder layers, while in the decoder, in addition to Shifted-W-MSA mechanism, Shifted Window-based Multi-head Cross Attention (Shifted-W-MCA) is designed for the dual-branch decoder layers. These mechanisms in the dual-branch architecture greatly enhance the ability of both local and global feature extraction. Notably, DBTrans is designed for 3D medical images, avoiding the information loss caused by data slicing.

The contributions of our proposed method can be described as follows: 1) Based on Transformer, we construct dual-branch encoder and decoder layers that assemble two attention mechanisms, being able to model close-window and distant-window dependencies without any extra computational cost. 2) In addition to the traditional skip-connection structure, in the dual-branch decoder, we also establish an extra path to facilitate the decoding process. We design a Shifted-W-MCA-based global branch to build a bridge between the decoder and encoder, maintaining affluent information of the segmentation target during the decoding process. 3) For the multi-modal data adopt in the task of SAMM-BTS, we improve the channel attention mechanism in SE-Net by applying SE-weights to features from both branches in the encoder and decoder layers. By this means, we implicitly consider the importance of multiple MRI modalities and two window-based attention branches, thereby strengthening the fusion effect of the multi-modal information from a global perspective.

2   Methodology

Figure 1 shows the overall structure of the proposed DBTrans. It is an end-to-end framework that has a 3D patch embedding along with a U-shaped model containing an encoder and a decoder. The model takes MRI data of $D \times H \times W \times C$ with four modalities stacked along channel dimensions as the input. The 3D patch embedding converts the input data to feature embedding $e^1 \in R^{D_1 \times H_1 \times W_1 \times C_1}$ which will be further processed by encoder layers. At the tail of the decoder, the segmentation head takes the output of the last layer and generates the final segmentation result of $D \times H \times W \times K$.

2.1   Dual-Branch in Encoder

As shown in Fig. 1(b), the encoder consists of four dual-branch encoder layers (one bottleneck included). Each encoder layer contains two consecutive encoder blocks and a 3D patch merging to down-sample the feature embedding. Note that there is only one encoder block in the bottleneck. The encoder block includes a dual-branch architecture with the local feature extraction branch, the global feature extraction branch, and

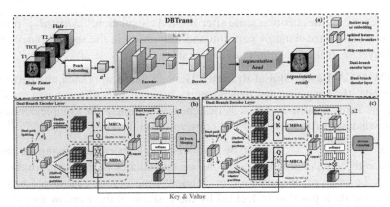

Fig. 1. The overall framework of the proposed DBTrans. The encoder contains dual-branch encoder layers (including a bottleneck), and the decoder contains dual-branch encoder layers. Skip connections based on cross attention are built between encoder and decoder.

the channel-attention-based dual-branch fusion module. After acquiring the embedding $e^i \in R^{D_i \times H_i \times W_i \times C_i}$ in the i-th encoder layer ($i \in [1, 4]$), we split it along the channel dimension, obtaining $e_1^i, e_2^i \in R^{D_i \times H_i \times W_i \times [C_i/2]}$ which are then separately fed into two branches.

Shifted-W-MSA-Based Local Branch. The image embedding $e_1^i \in R^{D_i \times H_i \times W_i \times [C_i/2]}$ is fed into the local branch in the encoder block. In the process of window partition (denoted as WP), e_1^i is split into non-overlapping windows after a layer normalization (LN) operation to obtain the window matrix m_1^i. Since we set M as 2, the input of $4 \times 4 \times 4$ size is uniformly divided into 8 windows of $2 \times 2 \times 2$. Following 3D Swin-Transformer [27], we introduce MSA based on the shifted-window partition to the second block in an encoder layer. During window partition of Shifted-W-MSA, the whole feature map is shifted by half of the window size, i.e., $\left(\frac{M}{2}, \frac{M}{2}, \frac{M}{2}\right)$. After the window partition, W-MSA is applied to calculate multi-head self-attention within each window. Specifically, for window matrix m_1^i, we first apply projection matrices W_Q^i, W_K^i, and W_V^i to obtain Q_1^i, K_1^i, and V_1^i matrices (the process is denoted as *Proj*). After projection, multi-head self-attention calculation is performed on Q_1^i, K_1^i, and V_1^i to get the attention score of every window. Finally, we rebuild the feature map from the windows, which serves as the inverse process of the window partition. After calculating the attention score, other basic components in Transformer are also employed, that is, layer normalization (LN), as well as a multi-layer perceptron (MLP) with two fully connected layers and a Gaussian Error Linear Unit (GELU). Residual connection is applied after each module. The whole process can be expressed as follows:

$$m_1^i = [\textbf{\textit{Shifted-}}]WP\left(LN\left(e_1^i\right)\right),$$
$$Q_1^i, K_1^i, V_1^i \in R^{\frac{D_i H_i W_i}{M^3} \times M^3 \times [C_i/2]} = Proj^i\left(m_1^i\right) = W_Q^i \cdot m_1^i, W_K^i \cdot m_1^i, W_V^i \cdot m_1^i, \quad (1)$$
$$\hat{z}_1^i = W\text{-}MSA\left(Q_1^i, K_1^i, V_1^i\right),$$
$$o_1^i = MLP^i\left(LN\left(\left(\hat{z}_1^i + e_1^i\right)\right)\right) + \left(\hat{z}_1^i + e_1^i\right),$$

where $\hat{z}_1^i$ represents the attention score after W-MSA calculation, "$[Shifted\text{-}]$" represents that we use the shifted-window partition and restoration in the second block of every encoder layer. o_1^i is the final output of the local branch.

Shuffle-W-MCA-Based Global Branch. Through the local branch, the network still cannot model the long-distance dependencies between non-adjacent windows in the same layer. Thus, Shuffle-W-MCA is designed to complete the complementary task. After the window-partition process that converts the embedding $e_2^i \in R^{D_i \times H_i \times W_i \times [C_i/2]}$ to $m_2^i \in R^{\frac{D_i H_i W_i}{M^3} \times M^3 \times [C_i/2]}$, inspired by ShuffleNet [35], instead of moving the channels, we propose to conduct shuffle operations on the patches in different windows. The patches at the same relative position in different windows are rearranged together in a window, and their position is decided by the position of the window they originally belong to. Then, this branch takes the query from the local branch, while generating keys and values from m_2^i to compute cross-attention scores. Such a design aims to enable the network to model the relationship between a window and other distant windows. Note that we adopt the projection function $Proj$ from the local branch, indicating that the weights of the two branches are shared. Through the shuffle operation, the network can generate both local and global feature representations without setting additional weight parameters. The whole process can be formulated as follows:

$$
\begin{aligned}
Q_2^i, K_2^i, V_2^i \in R^{\frac{D_i H_i W_i}{M^3} \times M^3 \times [C_i/2]} &= Proj^i\left(Shuffle(m_2^i)\right), \\
\hat{z}_2^i &= W\text{-}MCA\left(Q = Q_2^i, K = K_2^i, V = V_2^i\right), \\
o_2^i &= MLP^i\left(LN\left(\hat{z}_2^i + e_2^i\right)\right) + \left(\hat{z}_1^i + e_1^i\right).
\end{aligned}
\tag{2}
$$

In the second block, we get the final output o_2^i of the layer through the same process.

2.2 Dual-Branch in Decoder

As shown in Fig. 1(c), the decoder takes the output of the encoder as the input and generates the final prediction through three dual-branch decoder layers as well as a segmentation head. Each decoder layer consists of two consecutive decoder blocks and a 3D patch expanding layer. As for the j-th decoder layer ($j \in [1, 3]$), the embedding $d^j \in R^{D_j \times H_j \times W_j \times C_j}$ is also divided into the feature maps $d_1^j, d_2^j \in R^{D_j \times H_j \times W_j \times [C_j/2]}$. We further process d_1^j, d_2^j using a dual-branch architecture similar to that of the encoder, but with an additional global branch based on Shifted-W-MCA mechanism. Finally, the segmentation head generates the final segmentation result of $D \times H \times W \times K$, where K represents the number of classes. The local branch based on Shifted-W-MSA is the same as that in the encoder and will not be introduced in this section.

Shifted-W-MCA-Based Global Branch. Apart from employing Shifted-W-MSA to form the local branch of the decoder layer, we design a novel Shifted-W-MCA mechanism for the global branch to ease the information loss during the decoding process and take full advantage of the features from the encoder layers. The global branch receives the query matrix from the split feature map $d_2^j \in R^{D_j \times H_j \times W_j \times [C_j/2]}$, while receiving key and value matrices from the encoder block in the corresponding stage, denoted as

Q_2^j, K_{e_i}, and V_{e_i}. The process of Shifted-W-MCA can be formulated as follows:

$$Q_2^j, K_2^j, V_2^j = Proj^j\left([\textbf{Shifted-}]WP\left(LN\left(d_2^j\right)\right)\right),$$
$$\hat{z}_2^j = W\text{-}MCA\left(Q = Q_2^j, K = K_{e_1^{4-j}}, V = V_{e_1^{4-j}}\right),$$
$$o_2^j = MLP^i\left(LN\left(\hat{z}_2^j + d_2^j\right)\right) + \left(\hat{z}_1^j + d_1^j\right), \tag{3}$$

where $\hat{z}_2^j$ denotes the attention score after MCA calculation, o_2^j denotes the final output of the global branch.

2.3 Channel-Attention-Based Dual-Branch Fusion

As shown in Fig. 1(b) and Fig. 1(c), the dual-branch fusion module is based on the channel attention. For the block of the m-th ($m \in [1, 3]$) encoder or decoder layer, the dual-branch fusion module combines the features $o_1^m, o_2^m \in R^{D_m \times H_m \times W_m \times [C_m/2]}$ from the two extraction branches, obtaining a feature map filled with abundant multi-scale information among different modalities. Subsequently, the dependencies between the feature channels within the individual branches are implicitly modeled with the SE-Weight assignment first proposed in SE-Net [30]. Different from SE-Net, we dynamically assign weights for both dual-branch fusion and multi-modal fusion. The process of obtaining the attention weights can be represented as the formula (5) below:

$$Z_p = SE\_Weight\left(o_p^m\right), p = 1, 2, \tag{4}$$

where $Z_p \in R^{[C/2] \times 1 \times 1 \times 1}$ is the attention weight of a single branch. Then, the weight vectors of the two branches are re-calibrated using a Softmax function. Finally, the weighted channel attention is multiplied with the corresponding scale feature map to obtain the refined output feature map with richer multi-scale feature information:

$$attn_p = Softmax\left(Z_p\right) = \frac{\exp(Z_p)}{\sum_{p=1}^2 \exp(Z_p)},$$
$$Y_p = o_p^m \odot attn_p, p = 1, 2, \tag{5}$$
$$O = Cat([Y_1, Y_2]),$$

where "$\odot$" represents the operation of element-wise multiplication and "Cat" represents the concatenation. The concatenated output O, serving as the dual-branch output of a block in the encoder or decoder, implicitly integrates attention interaction information within individual branches across different channels/modalities, as well as across different branches.

2.4 Training Details

For the loss function, the widely used cross entropy (CE) loss and Dice loss [32] are introduced to train our DBTrans. Here we use parameter γ to balance the two loss parts. Our network is implemented based on PyTorch, and trained for 300 epochs using a single

RTX 3090 with 24G memory. The weights of the network were updated using the Adam optimizer, the batch size was set to 1, and the initial learning rate was set to 1×10^{-4}. A cosine decay scheduler was used as the adjustment strategy for the learning rate during training. We set the embedding dimensions C_0 as 144. Following previous segmentation methods, the parameter γ is set to 0.5.

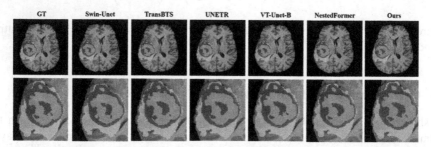

Fig. 2. Qualitative segmentation results on the test samples. The bottom row zooms-in the segmentation regions. Green, yellow, and red represent the peritumoral edema (ED), enhancing tumor (ET) and non-enhancing tumor/necrotic tumor (NET/NCR).

3 Experiments and Results

Datasets. We use the Multimodal Brain Tumor Segmentation Challenge (BraTS 2021 [33, 34, 38]) as the benchmark training set, validation set, and testing set. We divide the 1251 scans provided into 834, 208, and 209 (in a ratio of 2:1:1), respectively for training and testing. The ground truth labels of GBM segmentation necrotic/active tumor and edema are used to train the model. The BraTS 2021 dataset reflects real clinical diagnostic species and has four spatially aligned MRI modality data, namely T1, T1CE, T2, and Flair, which are obtained from different devices or according to different imaging protocols. The dataset contains three distinct sub-regions of brain tumors, namely peritumoral edema, enhancing tumor, and tumor core. The data augmentation includes random flipping, intensity scaling and intensity shifting on each axis with probabilities set to 0.5, 0.1 and 0.1, respectively.

Comparative Experiments. To evaluate the effectiveness of the proposed DBTrans, we compare it with the state-of-the-art brain tumor segmentation methods including six Transformer-based networks Swin-Unet [27], TransBTS [25], UNETR [22], nnFormer [23], VT-Unet-B [26], NestedFormer [36] as well as the most basic CNN network 3D U-Net [31] as the baseline. During inference, for any size of 3D images, we utilize the overlapping sliding windows technique to generate multi-class prediction results and take average values for the voxels in the overlapping region. The evaluation strategy adopted in this work is consistent with that of VT-Unet [26]. For other methods, we used the corresponding hyperparameter configuration mentioned in the original papers and reported the average metrics over 3 runs. Table 1 presents the Dice scores and 95% HDs of different methods for segmentation results on three different tumor regions (i.e.,

Table 1. Quantitative comparison with other state-of-the-arts methods in terms of dice score and 95% Hausdorff distance. Best results are bold, and second best are underlined.

Method	#param	FLOPS	Dice Score				95% Hausdorff Distance			
			ET	TC	WT	AVG	ET	TC	WT	AVG
3D U-Net[31]	11.9M	557.9G	83.39	86.28	89.59	86.42	6.15	**6.18**	11.49	7.94
Swin-Unet[27]	52.5M	93.17G	83.34	87.62	89.81	89.61	6.19	6.35	11.53	8.03
TransBTS[25]	33M	333G	80.35	85.35	89.25	84.99	7.83	8.21	15.12	10.41
UNETR[22]	102.5M	193.5G	79.78	83.66	90.10	84.51	9.72	10.01	15.99	11.90
nnFormer[23]	39.7M	110.7G	82.83	86.48	90.37	86.56	8.00	7.89	11.66	9.18
VT-Unet-B[26]	20.8M	165.0G	85.59	87.41	91.02	88.07	6.23	6.29	10.03	7.52
NestedFormer[36]	**10.48M**	**71.77G**	85.62	88.18	90.12	87.88	**6.08**	6.43	10.23	7.63
DBTrans(Ours)	24.6M	146.2G	**86.70**	**90.26**	**92.41**	**89.69**	6.13	6.24	**9.84**	**7.38**

ET, TC and WT) respectively, where a higher Dice score indicates better results and a lower 95% HD indicates better performance. By observation, our approach achieved the best performance on average among these methods in all three tumor regions. Compared with the state-of-the-art method VT-Unet, our method increased the average Dice score by 1.62 percentage points and achieved the lowest average 95% HD. Moreover, to verify the significance of our improvements, we calculate the variances of all results and conduct statistical tests (i.e., paired t-test). The results show that p-values on Dice and 95%HD are less than 0.05 in most comparison cases, indicating that the improvements are statistically significant. Compared with other methods, our DBTrans can not only capture more precise long-term dependencies between non-adjacent slices through Shuffle-W-MCA, but also accomplish dual-branch fusion and multi-modal fusion through dynamic SE-Weight assignment, obtaining better feature representations for segmentation.

Table 2. Quantitative comparison of ablation models in terms of average dice score.

Name	Index	DB-E	DB-D	Dual-branch fusion	Avg-Dice	Param
SwinUnet-1	(1)	✗	✗	✗	86.73	52.5M
SwinUnet-2	(2)	✓	✗	✗	87.52	43.2M
SwinUnet-3	(3)	✗	✓	✗	88.26	46.1M
SwinUnet-4	(4)	✓	✓	✗	88.86	27.7M
proposed	**(5)**	✓	✓	✓	**89.69**	**24.6M**

Figure 2 also shows the qualitative segmentation results on the test samples of test set patients, which further proves the feasibility and superiority of our DBTrans model. From the zoom-in area, we can observe that our model can more accurately segment tumor structures and delineate tumor boundaries compared with other methods.

Ablation Study. To further verify the contribution of each module, we establish the ablation models based on the modules introduced above. Note that, DP-E represents the dual-branch encoder layer, while DB-D represents the dual-branch decoder layer. When the dual-branch fusion is not included, we do not split the input, and simply fuse the features from the two branches using a convolution layer. In all, there are 5 models included in this ablation study: (1) SwinUnet-1 (baseline): We use Swin-Transformer layers without any dual-branch module. (2) SwinUnet-2: Based on (1), we add dual-branch encoder layers to the model. (3) SwinUnet-3: Based on (1), we add dual-branch decoder layers to the model. (4) SwinUnet-4: Based on (1), add both the encoder and decoder without the dual-branch fusion module. (5) Our proposed DBTrans model. As shown in Table 2, After applying our proposed dual-branch encoder and decoder layers to the baseline model, the average Dice score notably increased by 2.13. Subsequently, applying the dual-branch fusion module also prominently contributes to the performance of the model by an improvement of 0.83 on the Dice score. Notably, our dual-branch designs achieve higher performance while also reducing the number of parameters required. This is because we split the original feature embedding into two parts, thus the channel dimensions of features in two branches are halved.

4 Conclusion

In this paper, we innovatively proposed an end-to-end model named DBTrans for multi-modal medical image segmentation. In DBTrans, first, we well designed the dual-branch structures in encoder and decoder layers with Shifted-W-MSA, Shuffle-W-MCA, and Shifted-W-MCA mechanisms, facilitating feature extraction from both local and global views. Moreover, in the decoder, we establish a bridge between the query of the decoder and the key/value of the encoder to maintain the global context during the decoding process for the segmentation target. Finally, for the multi-modal superimposed data, we modify the channel attention mechanism in SE-Net, focusing on exploring the contribution of different modalities and branches to the effective information of feature maps. Experimental results demonstrate the superiority of DBTrans compared with the state-of-the-art medical image segmentation methods.

Acknowledgement. This work is supported by the National Natural Science Foundation of China (NSFC 62371325, 62071314), Sichuan Science and Technology Program 2023YFG0263, 2023YFG0025, 2023NSFSC0497, and Opening Foundation of Agile and Intelligent Computing Key Laboratory of Sichuan Province.

References

1. Gordillo, N., Montseny, E., et al.: State of the art survey on MRI brain tumor segmentation. Magn. Reson. Imaging **31**(8), 1426–1438 (2013)
2. Luo, Y., Zhou, L., Zhan, B., et al.: Adaptive rectification based adversarial network with spectrum constraint for high-quality PET image synthesis. Med. Image Anal. **77**, 102335 (2022)

3. Wang, K., Zhan, B., Zu, C., Wu, X., et al.: Semi-supervised medical image segmentation via a tripled-uncertainty guided mean teacher model with contrastive learning. Med. Image Anal. **79**, 102447 (2022)

4. Ma, Q., Zu, C., Wu, X., Zhou, J., Wang, Y.: Coarse-To-fine segmentation of organs at risk in nasopharyngeal carcinoma radiotherapy. In: de Bruijne, M., et al. (eds.) MICCAI 2021. LNCS, vol. 12901, pp. 358–368. Springer, Cham (2021). https://doi.org/10.1007/978-3-030-87193-2_34

5. Tang, P., Yang, P., Nie, D., et al.: Unified medical image segmentation by learning from uncertainty in an end-to-end manner. Knowl.-Based Syst. **241**, 108215 (2022)

6. Zhang, J., Zhang, Z., Wang, L., et al.: Kernel-based feature aggregation framework in point cloud networks. Pattern Recogn. **1**(1), 1–15 (2023)

7. Shi, Y., Zu, C., Yang, P., et al.: Uncertainty-weighted and relation-driven consistency training for semi-supervised head-and-neck tumor segmentation. Knowl.-Based Syst. **272**, 110598 (2023)

8. Wang, K., Wang, Y., Zhan, B., et al.: An efficient semi-supervised framework with multi-task and curriculum learning for medical image. Int. J. Neural Syst. **32**(09), 2250043 (2022)

9. Ronneberger, O., Fischer, P., Brox, T.: U-net: convolutional networks for biomedical image segmentation. In: Navab, N., Hornegger, J., Wells, W.M., Frangi, A.F. (eds.) MICCAI 2015. LNCS, vol. 9351, pp. 234–241. Springer, Cham (2015). https://doi.org/10.1007/978-3-319-24574-4_28

10. Zhou, T., Zhou, Y., He, K., et al.: Cross-level feature aggregation network for polyp segmentation. Pattern Recogn. **140**, 109555 (2023)

11. Du, G., Cao, X., Liang, J., et al.: Medical image segmentation based on u-net: a review. J. Imaging Sci. Technol. **64**(2), 020508-1–020508-12 (2020)

12. Brügger, R., Baumgartner, C.F., Konukoglu, E.: A partially reversible U-Net for memory-efficient volumetric image segmentation. In: Shen, D., et al. (eds.) MICCAI 2019. LNCS, vol. 11766, pp. 429–437. Springer, Cham (2019). https://doi.org/10.1007/978-3-030-32248-9_48

13. Pei, L., Liu, Y.: Multimodal brain tumor segmentation using a 3D ResUNet in BraTS 2021. In: 7th International Workshop, BrainLes 2021, Held in Conjunction with MICCAI 2021, Virtual Event, Revised Selected Papers, Part I, pp. 315–323. Springer, Cham (2022). https://doi.org/10.1007/978-3-031-08999-2_26

14. Zeng, P., Wang, Y., et al.: 3D CVT-GAN: a 3D convolutional vision transformer-GAN for PET reconstruction.In: Wang, L., et al. (eds.) MICCAI 2022, Proceedings, pp. 516-526. Springer, Cham (2022). https://doi.org/10.1007/978-3-031-16446-0_49

15. Vaswani, A., Shazeer, N., Parmar, N., et al.: Attention is all you need. Adv. Neural Inf. Process. Syst. **30** (2017)

16. Parmar, N., Vaswani, A,, Uszkoreit, J., et al.: Image transformer. In: International Conference on Machine Learning, pp. 4055–4064. PMLR (2018)

17. Luo, Y., et al.: 3D Transformer-GAN for high-quality PET reconstruction. In: de Bruijne, M., et al. (eds.) MICCAI 2021. LNCS, vol. 12906, pp. 276–285. Springer, Cham (2021). https://doi.org/10.1007/978-3-030-87231-1_27

18. Wen, L., Xiao, J., Tan, S., et al.: A transformer-embedded multi-task model for dose distribution prediction. Int. J. Neural Syst. **33**, 2350043–2350043 (2023)

19. Dosovitskiy, A., Beyer, L., Kolesnikov, A., et al.: An image is worth 16x16 words: transformers for image recognition at scale. In: Proceedings of International Conference on Learning Representations (2021)

20. Zheng, S., Lu, J., Zhao, H., et al.: Rethinking semantic segmentation from a sequence-to-sequence perspective with transformers. In: Proceedings of the IEEE/CVF Conference on Computer Vision and Pattern Recognition, pp. 6881–6890 (2021)

21. Xie, E., Wang, W., Yu, Z., et al.: SegFormer: simple and efficient design for semantic segmentation with transformers. Adv. Neural. Inf. Process. Syst. **34**, 12077–12090 (2021)
22. Hatamizadeh, A., Tang, Y., Nath, V., et al.: Unetr: transformers for 3d medical image segmentation. In: Proceedings of the IEEE/CVF Winter Conference on Applications of Computer Vision, pp. 574–584 (2022)
23. Zhou, H.Y., Guo, J., Zhang, Y., et al.: nnformer: Interleaved transformer for volumetric segmentation. arXiv preprint arXiv:2109.03201 (2021)
24. Lin, A., Chen, B., Xu, J., et al.: Ds-transunet: dual swin transformer u-net for medical image segmentation. IEEE Trans. Instrum. Meas. **71**, 1–15 (2022)
25. Wang, W., Chen, C., Ding, M., Yu, H., Zha, S., Li, J.: Transbts: multimodal brain tumor segmentation using transformer. In: de Bruijne, M., et al. (eds.) MICCAI 2021. LNCS, vol. 12901, pp. 109–119. Springer, Cham (2021). https://doi.org/10.1007/978-3-030-87193-2_11
26. Peiris, H., Hayat, M., Chen, Z., et al.: A robust volumetric transformer for accurate 3d tumor segmentation. In: Wang, L., et al. (eds.) MICCAI 2022, Proceedings, Part V, pp. 162–172. Springer, Cham (2022). https://doi.org/10.1007/978-3-031-16443-9_16
27. Cao, H., Wang, Y., Chen, J., et al.: Swin-unet: unet-like pure transformer for medical image segmentation. In: Computer Vision–ECCV 2022 Workshops. Proceedings, Part III, pp. 205–218. Springer, Cham (2023). https://doi.org/10.1007/978-3-031-25066-8_9
28. Liu, Z., Ning, J., Cao, Y., et al.: Video swin transformer. In: Proceedings of the IEEE/CVF Conference on Computer Vision and Pattern Recognition, pp. 3202–3211 (2022)
29. Gholami, A., et al.: A novel domain adaptation framework for medical image segmentation. In: Crimi, A., Bakas, S., Kuijf, H., Keyvan, F., Reyes, M., van Walsum, T. (eds.) BrainLes 2018. LNCS, vol. 11384, pp. 289–298. Springer, Cham (2019). https://doi.org/10.1007/978-3-030-11726-9_26
30. Hu, J., Shen, L., Sun, G.: Squeeze-and-excitation networks. In: Proceedings of the IEEE Conference on Computer Vision and Pattern Recognition, pp. 7132–7141 (2018)
31. Çiçek, Ö., Abdulkadir, A., Lienkamp, S.S., Brox, T., Ronneberger, O.: 3D U-Net: learning dense volumetric segmentation from sparse annotation. In: Ourselin, S., Joskowicz, L., Sabuncu, M.R., Unal, G., Wells, W. (eds.) MICCAI 2016. LNCS, vol. 9901, pp. 424–432. Springer, Cham (2016). https://doi.org/10.1007/978-3-319-46723-8_49
32. Jadon S.: A survey of loss functions for semantic segmentation. In: Proceedings of IEEE Conference on Computational Intelligence in Bioinformatics and Computational Biology (CIBCB), pp. 1–7. IEEE (2020)
33. Baid, U., et al.: The RSNA-ASNR-MICCAI BraTS 2021 benchmark on brain tumor segmentation and radiogenomic classification. arXiv:2107.02314 (2021)
34. Bakas, S., Akbari, H., Sotiras, A., Bilello, M., Rozycki, M., Kirby, J.S., et al.: Advancing the cancer genome atlas glioma mri collections with expert segmentation labels and radiomic features. Nat. Sci. Data **4**, 170117 (2017)
35. Zhang, X., Zhou, X., Lin, M, et al.: Shufflenet: an extremely efficient convolutional neural network for mobile devices. In: Proceedings of the IEEE Conference on Computer Vision and Pattern Recognition, pp. 6848–6856 (2018)
36. Xing, Z., Yu, L., Wan, L., et al.: NestedFormer: nested modality-aware transformer for brain tumor segmentation. In: Wang, L., et al. (eds.) MICCAI 2022, Proceedings, Part V, pp. 140–150. Springer, Cham (2022). https://doi.org/10.1007/978-3-031-16443-9_14
37. Wang, Y., Zhou, L., Yu, B., Wang, L., et al.: 3D auto-context-based locality adaptive multi-modality GANs for PET synthesis. IEEE Trans. Med. Imaging **38**(6), 1328–1339 (2019)
38. Menze, B.H., Jakab, A., Bauer, S., et al.: The multimodal brain tumor image segmentation benchmark (BRATS). IEEE Trans. Med. Imaging **34**(10), 1993–2024 (2014)

Memory Replay for Continual Medical Image Segmentation Through Atypical Sample Selection

Sutanu Bera$^{(\boxtimes)}$, Vinay Ummadi, Debashis Sen, Subhamoy Mandal, and Prabir Kumar Biswas

Indian Institute of Technology Kharagpur, Kharagpur, India
`sutanu.bera@iitkgp.ac.in`

Abstract. Medical image segmentation is critical for accurate diagnosis, treatment planning and disease monitoring. Existing deep learning-based segmentation models can suffer from catastrophic forgetting, especially when faced with varying patient populations and imaging protocols. Continual learning (CL) addresses this challenge by enabling the model to learn continuously from a stream of incoming data without the need to retrain from scratch. In this work, we propose a continual learning-based approach for medical image segmentation using a novel memory replay-based learning scheme. The approach uses a simple and effective algorithm for image selection to create the memory bank by ranking and selecting images based on their contribution to the learning process. We evaluate our proposed algorithm on three different problems and compare it with several baselines, showing significant improvements in performance. Our study highlights the potential of continual learning based algorithms for medical image segmentation and underscores the importance of efficient sample selection in creating memory banks.

Keywords: Memory Replay · Continual Learning · Medical Image Segmentation

1 Introduction

Medical image segmentation is an essential task in clinical practice, enabling accurate diagnosis, treatment planning, and disease monitoring. However, existing medical segmentation methods often encounter challenges related to changes in imaging protocols and variations in patient populations. These challenges can significantly impact the performance and generalizability of segmentation models. For instance, a segmentation model trained on MRI images from a specific

---

S. Bera and V. Ummadi—These authors contributed equally to this work and share the first authorship.

Supplementary Information The online version contains supplementary material available at https://doi.org/10.1007/978-3-031-43901-8_49.

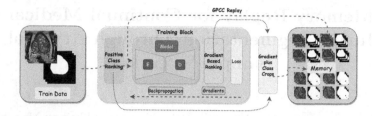

Fig. 1. Graphical summary of the concept of the proposed method. Positive Class Ranking and Gradient Based Ranking are computed online while training. Both rankings are used to create crops and stored in memory. The stored crops are to be used for replay while training on future datasets.

patient population may not perform well when applied to a different population with distinct demographic and clinical characteristics. Similarly, variations in imaging protocols, such as the use of different contrast agents or imaging parameters, can also affect the model's accuracy and reliability. To ensure accurate segmentation, it is necessary to retrain or fine-tune the model with current data before deploying it. However, this process often leads to catastrophic forgetting, where the model loses previously acquired knowledge while being trained on the current data. Catastrophic forgetting occurs due to the neural network's inability to learn from a continuous stream of data without disregarding previously learned information. Retraining the network using the complete training set, including both old and current data, is not always feasible or practical due to reasons such as the unavailability of old data or data privacy concerns. Moreover, training the network from scratch every time for every perturbation is a resource-intensive and time-sensitive process. Continual learning aims to address this limitation of catastrophic forgetting by enabling the model to learn continuously from a stream of incoming data without the need to retrain the model from scratch. Continual learning algorithms have gained significant interest lately for computer vision tasks like image denoising, super-resolution, and image classification. However, the development of efficient continual learning algorithms specifically designed for medical image segmentation has been largely overlooked in the literature. To address the above gap, our study proposes a continual learning based approach for medical image segmentation, which can be used to train any backbone network. In our approach, we leverage the recently proposed concept of the memory replay-based continual learning (MBCL) [6,15,17]. In MBCL, a memory buffer is used to store and replay previously learned data, enabling the model to retain important information while learning from new data. MBCL is, however, hampered by a few bottlenecks associated with medical images that pose a serious obstacle to its proper use in medical image segmentation. The efficiency of MBCL largely depends on the images stored in the memory bank [4,18], as the stored images must faithfully represent the previous task. It is known that medical image segmentation faces a major challenge of class imbalance. If an image with an under-representation of the positive class is stored in the memory bank, then it impedes the network from effectively remembering the previous task. In addition, not all medical images for training contribute equally to the learning process. So, images that pose greater challenges for the segmentation network should be saved in the memory bank. The importance of identifying atypical examples before creating memory banks cannot be overstated.

We propose a simple yet effective algorithm for image selection while creating the memory bank. Two different ranking mechanisms, which address the bottle-necks related to medical images discussed above, are proposed to rank all the images present in the training set. Then, images to be stored in the memory bank are selected using a combined ranking. Further, we suggest the cropping of the images around the organ of interest in order to minimize the size of the memory bank. An extensive evaluation is performed on three different problems, i.e., con-tinual prostate segmentation, continual hippocampus segmentation, and task incremental segmentation of the prostate, hippocampus and spleen. We consider several baselines including EWC [12], L2 regularization-based [10], and represen-tation learning-based [16]. Our method is found to outperform the conventional MBCL, and all the baseline mentioned above by a significant margin, creating a new benchmark for continual learning-based medical image segmentation.

2 Proposed Methodology

We are given a sequential stream of images from K sites, which are sequentially used to train a segmentation model. In a round $k \in [1, K]$ of this continual learning procedure, we can only obtain images and ground truths $\{(x_{k,i}, y_{k,i})\}_{i-1}^{n_k}$ from a new incoming site (dataset) D_k without access to old data from previous sites D_{k-1}. Due to catastrophic forgetting, this type of sequential learning results in a drop in performance for all the previous sites ($\leq D_{k-1}$) after training with images from D_k site as the parameters of the previous site or task are overwritten while learning a new task. In naive memory replay-based continual learning, a memory buffer, $\mathcal{M}$, is used to store a small number of examples of past sites ($\leq D_{k-1}$), which can be used to train the model along with the new data. Unlike other tasks like image classification or image restoration, for downstream tasks like medical image segmentation, the selection of images for storing in the $\mathcal{M}$ is very crucial. A medical image segmentation (like hippocampus segmentation) approach typically has a very small target organ. It is very likely that randomly selected images for storage in the $\mathcal{M}$ will have an under-representation of the positive (hippocampus) class. Further, the contribution of each training sample is not equal towards the learning, as a network usually learns more from examples that are challenging to segment. Based on the above observations, we propose two image ranking schemes to sort the images for storing in $\mathcal{M}$ (see Fig. 1).

2.1 Positive Class Based Ranking (PCR)

In this ranking scheme, we rank an input image volume according to the per-centage of voxels corresponding to the positive class available in the volume.

Let $cr_{k,i}$ be the positive class based ranking score for the sample input-ground truth pair $(x_{k,i}, y_{k,i})$ from the dataset D_k. We use the ground truth label $y_{k,i}$ to calculate the score of each volume. Let H, W and D respectively be the height, width and number of slices in the 3D volume of $y_{k,i}$. The voxel value at location (h, w, d) in the ground truth label $y_{k,i}$ is represented by $y_{k,i_{h,w,d}} \in 0, 1$. If the voxel value at a location (h, w, d) is equal to 1, then the voxel belongs to the positive class. Let us use $|y_{k,i}|$ to represent the total number of voxels in the 3D volume. For a sample pair $(x_{k,i}, y_{k,i})$, $cr_{k,i}$ is computed as follows:

$$cr_{k,i} = \frac{\sum_{h=0}^{H-1} \sum_{w=o}^{W-1} \sum_{d=0}^{D-1} (y_{k,i_{h,w,d}})}{|y_{k,i}|} \tag{1}$$

The rationale behind this ranking scheme is that by selecting volumes with a higher positive class occupancy, we can minimize the risk of underrepresentation of the positive class leading to a continuously trained network that remembers previous tasks more faithfully.

2.2 Gradient Based Ranking (GBR)

Here, we intend to identify the examples which the segmentation network finds hard to segment. In this ranking, we leverage the relationship between example difficulty and gradient variance, which has been previously observed in other studies [2]. Specifically, the neural network tends to encounter high gradients on examples that are hard to learn. With this motivation, we devise a simple method for calculating the score of every sample based on gradients, outlined in Algorithm 1. This algorithm enables the calculation of the gradient-based score during the network's training in real-time. In the algorithm, $f(\theta)$ refers to the image segmentation network with parameter θ and $L_k(\theta) = \frac{1}{n_k} \sum_{i=1}^{n_k} l(f(\theta; x_{k,i}), y_{k,i})$ denotes the loss function employed to train $f(\theta)$. $gr_{k,i}$ is the *gradient based score* assigned to a sample $D_{k,i}$, where $i \in [1, n_k]$ and $k \in [1, K]$. The corresponding

---

Algorithm 1. Gradient-based Sample Score

1: **for** $k = 1 \ldots K$ **do**	▷ Incoming dataset D_k
2: $n_k = |D_k|$	
3: $\theta \leftarrow \theta_0$	▷ Parameters randomly initialized
4: $pg \leftarrow 0$	▷ Prevoius gradients initially 0
5: **for** $epoch = 1 \ldots E$ **do**	▷ Traning for E epochs
6: **for** $i = 1 \ldots n_k$ **do**	
7: $cg = \frac{\partial L_k(f(\theta; x_{k,i}), y_{k,i})}{\partial \theta}$	▷ Loss function gradients
8: $gr_{k,i} \mathrel{+}= |pg_{k,i} - cg|$	▷ Absolute gradient difference for each sample
9: $pg_{k,i} \leftarrow cg$	▷ Set current gradients cg to previous gradients
10: $\theta \leftarrow \theta - \eta * cg$	▷ η is learning rate
11: **end for**	
12: **end for**	
13: **end for**	

gradients for the current sample are represented by cg, and $gr_{k,i}$ accumulates the absolute gradient difference between cg and previous gradients $pg_{k,i}$ for all training epochs. Essentially, a high value of $gr_{k,i}$ signifies a large gradient variance and implies that the example is difficult.

2.3   Gradient Plus Class Score (GPCS) Sampling for Memory

Once the score $gr_{k,i}$ and $cr_{k,i}$ are available for a dataset D_k, they are normalized to $[0,1]$ range and stored for future use. If p samples are to be selected for the replay memory $\mathcal{M}$, then $\frac{p}{2}$ will be selected using $gr_{k,i}$ and other $\frac{p}{2}$ using $cr_{k,i}$. However, we do not store entire image volumes in the memory. We propose a straightforward yet effective strategy to optimally utilize the memory bank size for continual medical segmentation tasks. Typically, the organ of interest in these tasks occupies only a small portion of the entire image, resulting in significant memory wastage when storing the complete volume. For example, the hippocampus, which is a common region of interest in medical image segmentation, occupies less than 0.5% of the total area. We propose to store only a volumetric crop of the sequence where the region of interest is present rather than the complete volume. This enables us to make more efficient use of the memory, allowing significant amounts of memory for storing additional crops within a given memory capacity. Thus, we can reduce memory wastage and optimize memory usage for medical image segmentation.

Consider that an image-label pair $(x_{k,i}, y_{k,i})$ from a dataset D_k has dimensions $H \times W \times D$ (height$\times$width$\times$sequence length). Instead of a complete image-label pair, a crop-sequence of dimension $h_c \times w_c \times d_c$ with $h_c \leq H, w_c \leq W, d_c \leq D$ is stored. We also use an additional hyperparameter, *foreground background ratio* $fbr \in [0,1]$ to store some background areas, as described below

$$fbr = \frac{Number\ of\ crops\ having\ RoI}{Number\ of\ crops\ not\ having\ RoI} \qquad (2)$$

Using only foreground regions in problems such as task incremental learning may result in a high degree of false positive segmentation. In such cases, having a few crops of background regions helps in reducing the forgetting in background regions as discussed in Sect. 3.3.

3   Experimental Details

3.1   Datasets

We conduct experiments to evaluate the effectiveness of our methods in two different types of incremental learning tasks. To this end, we use seven openly available datasets for binary segmentation tasks: four prostate datasets (Prostate158 [1], NCI-ISBI [3], Promise12 [14], and Decathlon [5]), two hippocampus datasets (Drayd [8] and HarP [7]), and one Spleen dataset from Decathlon [5]. The first set of experiments involved domain incremental prostate segmentation, with the datasets being trained in the following order: Prostate158 → ISBI → Promise12 → Decathlon. The second experiment involved domain incremental hippocampus segmentation, with the datasets being trained in the order HarP → Drayd. Finally, we conducted task incremental segmentation for three organs - prostate, spleen, and hippocampus - following the sequence: Promise12 (prostate) → MSD (spleen) → Drayd (hippocampus).

3.2 Evaluation Metrics

To evaluate the effectiveness of our proposed methods against baselines, we use the segmentation evaluation metric Dice Similarity Coefficient (DSC) along with standard continual learning (CL) metrics. The CL metrics [9] comprise Average Accuracy (ACC), Backward Transfer (BWT) [13], and Average Forgetting (AFGT) [19]. BWT measures the model's ability to apply newly learned knowledge to previously learned tasks, While AFGT measures the model's retention of previously learned knowledge after learning a new task.

3.3 Training Details

In this work, we consider a simple segmentation backbone network UNet, which is widely used in medical image segmentation. Both the proposed and baseline methods were used to train a Residual UNet [11]. All the methods for comparison (Tables 2, 3 and 4), except for Sequential (SGD), are trained using the Adam optimizer with a learning rate of 0.001, a momentum of 0.9, and a weight decay of 0.00001. Sequential(SGD) employs an SGD optimizer with the same hyperparameters as Adam. The training loss used is DiceCELoss, which combines Dice loss and Cross Entropy loss. During task incremental segmentation training using GPCC, the fbr parameter is set to 0.7, while the default value of 1.0 is used for other tasks. In parts of our GPCC experiments, where we examine continual prostate segmentation as well as continual prostate, spleen, and hippocampus segmentation, both the height (h_c) and width (w_c) of each volume are fixed to 160. In continual hippocampus segmentation experiments, these values are reduced to 128. Note that although we perform the experiments using UNet network, any sophisticated network like VNet, or DeepMedic can also be trained continually using our method.

4 Results and Discussion

Ablation Study of Our Atypical Sample Selection:[1] In order to evaluate the effectiveness of every module proposed in our work, an ablation study is conducted. The results of the analysis shown in Table 1 indicates that randomly storing samples leads to a significant decrease in performance during in the earlier trained domains due to insufficient representation of the domain distributions. Both PCR and GBR shows improvements in ACC over random replay

---

[1] All experimental values reported here are the average over four random seeds.

Table 1. Ablation study on continual prostate segmentation. The best DSC, ACC and AFGT are presented in bold. GPCC gives the best performance in general.

Learning Method	Dataset wise DSC scores (%)				CL Metrics		
	Prostate158 (↑)	NCI-ISBI (↑)	Promise12 (↑)	Decathlon (↑)	ACC (↑)	BWT (↑)	AFGT (↓)
Random Replay(3)	67.5 ± 3.3	85.7 ± 3.0	74.8 ± 10.3	85.7 ± 1.3	78.4 ± 2.8	−6.8 ± 3.2	7.7 ± 2.9
PCR Replay(3)	**82.9 ± 1.4**	81.6 ± 1.6	77.5 ± 3.8	82.8 ± 1.8	81.2 ± 2.5	−3.6 ± 1.3	4.6 ± 1.4
GBR Replay(3)	82.4 ± 1.2	83.9 ± 1.8	78.6 ± 3.4	81.6 ± 1.4	81.6 ± 2.2	−2.1 ± 0.9	2.7 ± 0.7
GPCC Replay(6)	82.1 ± 1.8	**85.9 ± 2.3**	**80.8 ± 5.2**	**86.3 ± 1.5**	**82.8 ± 1.3**	**−1.6 ± 0.7**	**2.3 ± 0.3**

by **2.8%** and **3.2%**, respectively. GPCC provides further enhancements across all domains, resulting in a final accuracy gain of **4.4%** over the random replay.

Comparison with Baselines: We consider several benchmark continual learning algorithms including L2 Regularization, EWC and Representation Replay as our baselines. First, we compare the performance of our method on continual prostate segmentation tasks. The objective comparison among different methods on this task is shown in Table 2. The proposed methods perform on par or better with joint learning, which is considered an upper bound in continual learning settings. Compared with Sequential (SGD) and Random Replay(3)[2], our method with GPCC(6) shows an ACC improvement of **9.2%** and **4.4%** respectively. With a memory footprint, GPCC with six crops of $160 \times 160 \times D$ is **65%** lighter compared to Random Replay (3) with an average image size of $384 \times 384 \times D$. Next, the objective comparison of different methods of continual hippocampus segmentation is shown in Table 3. This task poses a significant challenge due to the small region of interest (RoI) in whole brain MRI scans. Our proposed approach outperforms all other baseline methods in terms of CL metrics. When comparing GPCC Replay with six crops of $128 \times 128 \times D$ images to Random Replay(3) using full-size images of $233 \times 189 \times D$, we find that GPCC Replay is more memory-efficient consuming **26%** less memory while still achieving an **1.1%** ACC performance improvement. While some baselines showed higher DSC scores on the second domain, a high value of AFGT indicates their inability to

Table 2. Objective comparison of different methods on continual prostate segmentation. The best DSC (in continual), ACC and AFGT are presented in bold.

Learning Method	Dataset wise DSC scores(%)				CL Metrics		
	Prostate158 (↑)	NCI-ISBI (↑)	Promise12 (↑)	Decathlon (↑)	ACC (↑)	BWT (↑)	AFGT (↓)
Individual	82.3 ± 1.3	79.4 ± 1.6	87.5 ± 2.3	81.5 ± 2.1	−	−	−
Joint	83.5 ± 1.4	86.8 ± 1.8	82.6 ± 2.6	86.4 ± 1.5	−	−	−
Sequential(SGD)	61.8 ± 2.7	82.4 ± 2.4	66.4 ± 9.4	83.6 ± 1.5	73.6 ± 2.3	−10.6 ± 1.4	11.8 ± 1.2
L2 Regularization	36.7 ± 1.6	59.8 ± 5.2	59.3 ± 10.6	78.9 ± 4.8	58.7 ± 3.8	0.13 ± 0.0	**0.1 ± 0.0**
EWC	61.6 ± 3.2	82.6 ± 2.6	64.2 ± 10.9	85.3 ± 1.3	73.4 ± 3.1	−4.6 ± 1.7	6.07 ± 1.8
Representation Replay	60.2 ± 2.3	65.6 ± 3.2	58.3 ± 6.7	72.1 ± 2.6	64.0 ± 6.2	−10.3 ± 2.5	11.3 ± 3.4
Random Replay(3)	67.5 ± 3.3	85.7 ± 3.0	74.8 ± 10.3	85.7 ± 1.3	78.4 ± 2.8	−6.8 ± 3.2	7.7 ± 2.9
GPCC Replay(6)	**82.1 ± 1.8**	**85.9 ± 2.3**	**80.8 ± 5.2**	**86.3 ± 1.5**	**82.8 ± 1.3**	**−1.6 ± 0.7**	2.3 ± 0.3

---

[2] 3 (the number within brackets) is number of volumes stored in memory buffer.

Table 3. Objective comparison of different methods on continual hippocampus segmentation. The best DSC (in continual), ACC and AFGT are presented in bold.

Learning Method	Dataset wise DSC scores (%)		CL Metrics		
	HarP (↑)	Drayd (↑)	ACC (↑)	BWT (↑)	AFGT (↓)
Individual	80.5 ± 2.0	79.1 ± 1.1	-	-	-
Joint	82.1 ± 0.9	80.2 ± 0.8	-	-	-
Sequential(SGD)	55.2 ± 2.3	76.5 ± 1.2	63.4 ± 1.5	-15 ± 2.1	15.3 ± 1.9
EWC	71.3 ± 0.8	83.4 ± 0.7	77.3 ± 1.1	-7.6 ± 1.2	7.8 ± 1.3
Representation Replay	70.8 ± 1.2	$\mathbf{83.6 \pm 1.5}$	77.2 ± 1.4	-8.2 ± 1.3	8.3 ± 1.5
L2 Regularization	75.2 ± 0.6	79.1 ± 0.8	77.3 ± 0.7	-2.2 ± 0.9	2.2 ± 0.9
Random Replay(3)	$\mathbf{83.5 \pm 1.8}$	73.5 ± 1.7	78.7 ± 1.4	-4.5 ± 0.8	4.5 ± 0.9
GPCC Replay (6)	83.0 ± 0.7	77.1 ± 0.4	$\mathbf{79.8 \pm 0.6}$	-2.1 ± 0.5	$\mathbf{2.1 \pm 0.5}$

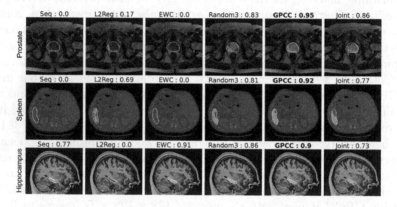

Fig. 2. Qualitative results for task incremental segmentation of prostate, spleen, and hippocampus using the methods in Table 4. Ground truths are in yellow borders and predictions are in peach. The bolded method has the highest DSC score.

retain previously learned knowledge, which suggests limitations in their ability to perform continual training of the backbone.

We finally assess the performance of our method on an even more challenging task of incremental learning segmentation. This involves continuously training a single model to accurately segment various organs while incorporating new organs as segmentation targets during each episode. Utilizing a single model for segmenting multiple organs offers potential advantages, particularly when there are constraints such as limited annotation data that hinder joint training. Task or class incremental learning becomes invaluable in such scenarios, as it allows us to incorporate new organs as segmentation targets without requiring a complete retraining process. To investigate the feasibility of this concept, we dedicate a section in our study to experimental analysis. The results of this analysis are presented in Table 4. We find that GPCC replay(12) shows substantial improvement by increasing the ACC score by **45.4%** over Sequential(SGD) and achieves

Table 4. Comparison of Baselines for Task Incremental Segmentation. The best DSC (in continual), ACC and AFGT are presented in bold.

Learning Method	Dataset wise DSC scores (%)			CL Metrics		
	Promise12 Prostate (↑)	MSD Spleen(↑)	Drayd Hippocampus(↑)	ACC (↑)	BWT (↑)	AFGT (↓)
Individual	81.5 ± 2.1	91.5 ± 1.6	79.1 ± 1.1	-	-	-
Joint	80.5 ± 0.7	80.5 ± 0.8	79.1 ± 0.6	-	-	-
Sequential(SGD)	0.0 ± 0.0	0.0 ± 0.0	67.5 ± 2.4	37.9 ± 1.7	-80 ± 1.8	80 ± 1.7
EWC	0.0 ± 0.0	0.0 ± 0.0	$\mathbf{80.8 \pm 1.4}$	42.6 ± 1.3	-87.3 ± 1.6	87.2 ± 1.7
Representation Replay	0.0 ± 0.0	0.0 ± 0.0	80.1 ± 2.4	42.2 ± 1.3	-86.3 ± 1.2	86.3 ± 1.0
L2 Regularization	49.7 ± 3.2	56.7 ± 2.5	15.2 ± 3.5	49.4 ± 3.3	-17.2 ± 2.6	17.2 ± 2.6
Random Replay(3)	66.5 ± 2.5	83.7 ± 2.3	80.1 ± 1.5	78.3 ± 1.3	-13.1 ± 1.1	13.1 ± 1.1
GPCC Replay (12)	$\mathbf{78.4 \pm 1.6}$	$\mathbf{87.9 \pm 1.4}$	80.5 ± 1.5	$\mathbf{83.3 \pm 1.3}$	-4.2 ± 1.2	$\mathbf{4.2 \pm 1.2}$

a **5%** increase in ACC compared to Random Replay(3), outperforming all the baselines. In this case, GPCC Replay is also lighter in memory consumption by up to **32%**.

Visually analyzing the predictions for a test sample in Fig. 2 from the task of incremental segmentation, L2 regression and Random Replay(3) are seen to produce only partial segmentation of RoI. On the other hand, GPCC predictions outperform joint learning and are very close to Ground Truths, with a DSC score $\geq$ **90%**. More visual comparison among different methods is given in the supplementary.

5   Conclusion

This paper proposes a novel approach to address the challenge of catastrophic forgetting in medical image segmentation using continual learning. The paper presents a memory replay-based continual learning paradigm that enables the model to learn continuously from a stream of incoming data without the need to retrain from scratch. The proposed algorithm includes an effective image selection method that ranks and selects images based on their contribution to the learning process and faithful representation of the task. The study evaluates the proposed algorithm on three different problems and demonstrates significant performance improvements compared to several relevant baselines.

References

1. Adams, L.C., et al.: Prostate158-an expert-annotated 3T MRI dataset and algorithm for prostate cancer detection. Comput. Biol. Med. **148**, 105817 (2022)
2. Agarwal, C., D'souza, D., Hooker, S.: Estimating example difficulty using variance of gradients. In: Proceedings of the IEEE/CVF Conference on Computer Vision and Pattern Recognition, pp. 10368–10378 (2022)

3. NCI-ISBI 2013 challenge: automated segmentation of prostate structures. https://wiki.cancerimagingarchive.net/display/Public/NCI-ISBI+2013+Challenge+-+Automated+Segmentation+of+Prostate+Structures

4. Aljundi, R., Lin, M., Goujaud, B., Bengio, Y.: Gradient based sample selection for online continual learning. In: Advances in Neural Information Processing Systems, vol. 32 (2019)

5. Antonelli, M., et al.: The medical segmentation decathlon. Nat. Commun. **13**(1), 4128 (2022)

6. Balaji, Y., Farajtabar, M., Yin, D., Mott, A., Li, A.: The effectiveness of memory replay in large scale continual learning. arXiv preprint arXiv:2010.02418 (2020)

7. Boccardi, M., et al.: Training labels for hippocampal segmentation based on the EADC-ADNI harmonized hippocampal protocol. Alzheimer's Dement. **11**(2), 175–183 (2015)

8. Denovellis, E., et al.: Data from: hippocampal replay of experience at real-world speeds (2021). https://doi.org/10.7272/Q61N7ZC3

9. Díaz-Rodríguez, N., Lomonaco, V., Filliat, D., Maltoni, D.: Don't forget, there is more than forgetting: new metrics for continual learning. arXiv preprint arXiv:1810.13166 (2018)

10. Hsu, Y.C., Liu, Y.C., Ramasamy, A., Kira, Z.: Re-evaluating continual learning scenarios: a categorization and case for strong baselines. arXiv preprint arXiv:1810.12488 (2018)

11. Kerfoot, E., Clough, J., Oksuz, I., Lee, J., King, A.P., Schnabel, J.A.: Left-ventricle quantification using residual U-Net. In: Pop, M., et al. (eds.) STACOM 2018. LNCS, vol. 11395, pp. 371–380. Springer, Cham (2019). https://doi.org/10.1007/978-3-030-12029-0_40

12. Kirkpatrick, J., et al.: Overcoming catastrophic forgetting in neural networks. Proc. Natl. Acad. Sci. **114**(13), 3521–3526 (2017)

13. Li, Z., Hoiem, D.: Learning without forgetting. IEEE Trans. Pattern Anal. Mach. Intell. **40**(12), 2935–2947 (2017)

14. Litjens, G., et al.: Evaluation of prostate segmentation algorithms for MRI: the promise12 challenge. Med. Image Anal. **18**(2), 359–373 (2014)

15. Lopez-Paz, D., Ranzato, M.: Gradient episodic memory for continual learning. In: Advances in Neural Information Processing Systems, vol. 30 (2017)

16. Pellegrini, L., Graffieti, G., Lomonaco, V., Maltoni, D.: Latent replay for real-time continual learning. In: 2020 IEEE/RSJ International Conference on Intelligent Robots and Systems (IROS), pp. 10203–10209 (2020). https://doi.org/10.1109/IROS45743.2020.9341460

17. Rolnick, D., Ahuja, A., Schwarz, J., Lillicrap, T., Wayne, G.: Experience replay for continual learning. In: Advances in Neural Information Processing Systems, vol. 32 (2019)

18. Tiwari, R., Killamsetty, K., Iyer, R., Shenoy, P.: GCR: gradient coreset based replay buffer selection for continual learning. In: Proceedings of the IEEE/CVF Conference on Computer Vision and Pattern Recognition, pp. 99–108 (2022)

19. Zenke, F., Poole, B., Ganguli, S.: Continual learning through synaptic intelligence. In: International Conference on Machine Learning, pp. 3987–3995. PMLR (2017)

Structure-Decoupled Adaptive Part Alignment Network for Domain Adaptive Mitochondria Segmentation

Rui Sun[1] , Huayu Mai[1] , Naisong Luo[1] , Tianzhu Zhang[1,2,3(✉)] ,
Zhiwei Xiong[1,2] , and Feng Wu[1,2]

[1] University of Science and Technology of China, Hefei, China
{issunrui,mai556,lns6}@mail.ustc.edu.cn,
{tzzhang,zwxiong,fengwu}@ustc.edu.cn
[2] Hefei Comprehensive National Science Center, Institute of Artificial Intelligence,
Hefei, China
[3] Deep Space Exploration Lab, Hefei, China

Abstract. Existing methods for unsupervised domain adaptive mitochondria segmentation perform feature alignment via adversarial learning, and achieve promising performance. However, these methods neglect the differences in structure of long-range sections. Besides, they fail to utilize the context information to merge the appropriate pixels to construct a part-level discriminator. To mitigate these limitations, we propose a Structure-decoupled Adaptive Part Alignment Network (SAPAN) including a structure decoupler and a part miner for robust mitochondria segmentation. The proposed SAPAN model enjoys several merits. First, the structure decoupler is responsible for modeling long-range section variation in structure, and decouple it from features in pursuit of domain invariance. Second, the part miner aims at absorbing the suitable pixels to aggregate diverse parts in an adaptive manner to construct part-level discriminator. Extensive experimental results on four challenging benchmarks demonstrate that our method performs favorably against state-of-the-art UDA methods.

Keywords: Mitochondria segmentation · Unsupervised domain adaptation · Electron microscopy images

1 Introduction

Automatic mitochondria segmentation from electron microscopy (EM) volume is a fundamental task, which has been widely applied to basic scientific research and

---

R. Sun and H. Mai—Equal Contribution.

Supplementary Information The online version contains supplementary material available at https://doi.org/10.1007/978-3-031-43901-8_50.

H. Greenspan et al. (Eds.): MICCAI 2023, LNCS 14223, pp. 523–533, 2023.
https://doi.org/10.1007/978-3-031-43901-8_50

clinical diagnosis [1,12,13,15,17]. Recent works like [6,7,16,21,22] have achieved conspicuous achievements attributed to the development of deep neural networks [10,23,26,27]. However, these approaches tend to suffer from severe performance degradation when evaluated on target volume (*i.e.*, target domain) that are sampled from a different distribution (caused by different devices used to image different organisms and tissues) compared to that of training volume (*i.e.*, source volume/domain). Therefore, how to alleviate this gap to empower the learned model generalization capability is very challenging.

In this work, we tackle the unsupervised domain adaptive (UDA) problem, where there are no accessible labels in the target volume. To alleviate this issue, representative and competitive methods [5,8,18,19,29,30] attempt to align feature distribution from source and target domains via adversarial training in a pixel-wise manner, that is, learning domain-invariant features against the substantial variances in data distribution. However, after an in-depth analysis of adversarial training, we find two key ingredients lacking in previous works. (1) **Structure-entangled feature.** In fact, there exists a large variation in the complex mitochondrial structure from different domains, as proven in [28]. However, previous methods only focus on domain gap either in each individual section or in adjacent sections, and neglect the differences in morphology and distribution (*i.e.*, structure) of long-range sections, leading to sub-optimal results in the form of hard-to-align features. (2) **Noisy discrimination.** Intuitively, humans can quickly distinguish domain of images with the same categories from cluttered backgrounds by automatically decomposing the foreground into multiple local parts, and then discriminate them in a fine-grained manner. Inspired by this, we believe that during adversarial learning, relying solely on context-limited pixel-level features to discriminate domains will inevitably introduce considerable noise, considering that the segmentation differences between the source and target domain are usually in local parts (*e.g.*, boundary). Thus, it is highly desirable to make full use of the context information to merge the appropriate neighboring pixel features to construct a part-level discriminator.

In this paper, we propose a **S**tructure-decoupled **A**daptive **P**art **A**lignment Network (**SAPAN**) including a structure decoupler and a part miner for robust mitochondria segmentation. (1) In the **structure decoupler**, we draw inspiration from [25] to model long-range section variation in distribution and morphology (*i.e.*, structural information), and decouple it from features in pursuit of domain invariance. In specific, we first prepend a spatial smoothing mechanism for each pixel in the current section to seek the corresponding location of other sections to attain the smoothed features, which are subsequently modulated to obtain decoupled features with easy-to-align properties. (2) Then, we devise a **part miner** as discriminator, which can dynamically absorb the suitable pixels to aggregate diverse parts against noise in an adaptive manner, thus the detailed local differences between the source and target domain can be accurately discriminated. Extensive experimental results on four challenging benchmarks demonstrate that our method performs favorably against SOTA UDA methods.

2   Related Work

Domain adaptation in EM volume [5,8,18,19,29,30] has attracted the attention of more and more researchers due to the difficulty in accessing manual annotation. In [29], they employ self-training paradigm while in [18] adversarial training is adopted and performs better. To further improve the domain generalization, [5] considering the inter-section gap. However, those methods neglect the differences in morphology and distribution (*i.e.*, structure) of long-range sections. In addition, existing adversarial training [4,5,18] adopt context-limited pixel-wise discriminator leading to sub-optimal results.

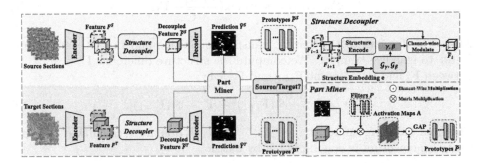

Fig. 1. Overview of our method. There are two main modules in SAPAN, *i.e.*, the structure decoupler for decoupling the feature from the domain-specific structure information and the part miner for adaptively discovering different parts.

3   Method

3.1   Overview

The domain adaptive mitochondria segmentation task aims at learning an accurate segmentation model in the target domain based on a labeled source domain EM volume $V^S = \{\mathbf{X}_i^S, \mathbf{Y}_i^S\}_{i=1}^N$ and an unlabeled target domain EM volume $V^T = \{\mathbf{X}_j^T\}_{j=1}^M$. As shown in Fig. 1, given an EM section $\mathbf{X}_i \in \mathbb{R}^{H \times W}$ (omit the superscript S/T for convenience) from either domain, the encoder of U-Net [20] first extracts its feature map $\mathbf{F}_i \in \mathbb{R}^{h \times w \times C}$, where h, w and C denote the height, width and channel number of the feature map respectively. A structure decoupler is applied to decouple the extracted feature map from domain-specific structure knowledge, which involves a channel-wise modulation mechanism. Subsequently, the decoupled feature map $\widetilde{\mathbf{F}}_i$ is fed into the decoder of U-Net which outputs prediction of the original size. In the end, we design a part miner to dynamically divide foreground and background into diverse parts in an adaptive manner, which are utilized to facilitate adversarial training. The details are as follows.

3.2   Structure Decoupler

In order to bridge the gap in mitochondrial structure between domains, a structure decoupler is designed to decouple the extracted feature map from domain-specific structure information, realized by a channel-wise modulation mechanism. To better model the structure information, we first apply attention-based spatial smoothing for adjacent sections. Concretely, given feature maps of adjacent sections $\mathbf{F}_i$ and $\mathbf{F}_{i+1}$, we define the spatial smoothing $\mathcal{S}_{i+1}(\cdot)$ w.r.t $\mathbf{F}_{i+1}$ as: each pixel $\mathbf{f}_{i,j} \in \mathbb{R}^{1 \times C}$ ($j = 1, 2, ..., hw$) in feature map $\mathbf{F}_i$ as query, the feature map of adjacent section $\mathbf{F}_{i+1} \in \mathbb{R}^{h \times w \times C}$ as keys and values. Formally,

$$\mathcal{S}_{i+1}(\mathbf{f}_{i,j}) = \text{softmax}(\frac{\mathbf{f}_{i,j}\mathbf{F}_{i+1}^{\mathsf{T}}}{\sqrt{C}})\mathbf{F}_{i+1}, \tag{1}$$

where the T refers to the matrix transpose operation and the $\sqrt{C}$ is a scaling factor to stabilize training. We compute the structure difference $\mathbf{D}_i \in \mathbb{R}^{h \times w \times C}$ between $\mathbf{F}_i$ and its adjacent $\mathbf{F}_{i+1}$ and $\mathbf{F}_{i-1}$ by:

$$\mathbf{D}_i = [\mathbf{F}_i - \mathcal{S}_i(\mathbf{F}_{i+1})] + [\mathbf{F}_i - \mathcal{S}_i(\mathbf{F}_{i-1})]. \tag{2}$$

The final structure embedding $\mathbf{e} \in \mathbb{R}^{1 \times C}$ for each domain is calculated by exponential momentum averaging batch by batch:

$$\mathbf{e}_b = \frac{1}{B} \sum_{i=1}^{B} \text{GAP}(\mathbf{D}_i), \ \mathbf{e} = \theta\mathbf{e} + (1 - \theta)\mathbf{e}_b, \tag{3}$$

where GAP$(\cdot)$ denotes the global average pooling, B denotes the batch size and $\theta \in [0, 1]$ is a momentum coefficient.

In this way, $\mathbf{e}^S$ and $\mathbf{e}^T$ condense the structure information for the whole volume of the corresponding domain. To effectively mitigate the discrepancy between different domains, we employ channel-wise modulation to decouple the feature from the domain-specific structure information. Taking source domain data $\mathbf{F}_i^S$ as example, we first produce the channel-wise modulation factor $\gamma^S \in \mathbb{R}^{1 \times C}$ and $\beta^S \in \mathbb{R}^{1 \times C}$ conditioned on $\mathbf{e}^S$:

$$\gamma^S = \mathcal{G}_\gamma(\mathbf{e}^S), \beta^S = \mathcal{G}_\beta(\mathbf{e}^S), \tag{4}$$

where $\mathcal{G}_\gamma(\cdot)$ and $\mathcal{G}_\beta(\cdot)$ shared by the source domain and target domain, are implemented by two linear layers and an activation layer. Then the decoupled source feature can be obtained by:

$$\widetilde{\mathbf{F}}_i^S = \gamma^S \odot \mathbf{F}_i^S + \beta^S, \tag{5}$$

where $\odot$ denotes element-wise multiplication. The decoupled target feature $\widetilde{\mathbf{F}}_i^T$ can be acquired in a similar way by Eq. (4) and Eq. (5). Subsequently, the structure decoupled feature is fed into the decoder of U-Net, which outputs the foreground-background probability map $\widetilde{\mathbf{Y}}_i \in \mathbb{R}^{2 \times H \times W}$. The finally prediction $\widehat{\mathbf{Y}}_i \in \mathbb{R}^{H \times W}$ can be obtained through simply apply argmax$(\cdot)$ operation on $\widetilde{\mathbf{Y}}_i$.

3.3   Part Miner

As far as we know, a good generator is inseparable from a powerful discriminator. To inspire the discriminator to focus on discriminative regions, we design a part miner to dynamically divide foreground and background into diverse parts in an adaptive manner, which are classified by the discriminator $\mathcal{D}_{part}$ subsequently.

To mine different parts, we first learn a set of part filters $\mathbf{P} = \{\mathbf{p}_k\}_{k=1}^{2K}$, each filter $\mathbf{p}_k$ is represented as a C-dimension vector to interact with the feature map $\widetilde{\mathbf{F}}$ (omit the subscript i for convenience). The first half $\{\mathbf{p}_k\}_{k=1}^{K}$ are responsible for dividing the foreground pixels into K groups and vice versa. Take the foreground filters for example. Before the interaction between $\mathbf{p}_k$ and $\widetilde{\mathbf{F}}$, we first filter out the pixels belonging to the background using downsampled prediction $\widehat{\mathbf{Y}}'$. Then we get K activation map $\mathbf{A}_i \in \mathbb{R}^{h \times w}$ by multiplying $\mathbf{p}_k$ with the masked feature map:

$$\mathbf{A}_i = \text{sigmoid}(\mathbf{p}_k(\widehat{\mathbf{Y}}' \odot \widetilde{\mathbf{F}})^\mathsf{T}). \tag{6}$$

In this way, the pixels with a similar pattern will be highlighted in the same activation map. And then, the foreground part-aware prototypes $\widetilde{\mathbf{P}} = \{\widetilde{\mathbf{p}}_k\}_{k=1}^{K}$ can be got by:

$$\widetilde{\mathbf{p}}_k = \text{GAP}(\mathbf{A}_i \odot \widetilde{\mathbf{F}}). \tag{7}$$

Substituting $\widehat{\mathbf{Y}}'$ with $(1 - \widehat{\mathbf{Y}}')$, we can get the background part-aware prototypes $\widetilde{\mathbf{P}} = \{\widetilde{\mathbf{p}}_k\}_{k=K+1}^{2K}$ in the same manner.

3.4   Training Objectives

During training, we calculate the supervise loss with the provided label $\mathbf{Y}_i^S$ of the source domain by:

$$\mathcal{L}_{sup} = \frac{1}{N} \sum_{i=1}^{N} \text{CE}(\widetilde{\mathbf{Y}}_i^S, \mathbf{Y}_i^S), \tag{8}$$

where the $\text{CE}(\cdot, \cdot)$ refers to the standard cross entropy loss.

Considering the part-aware prototypes may focus on the same, making the part miner degeneration, we impose a diversity loss to expand the discrepancy among part-aware prototypes. Formally,

$$\mathcal{L}_{div} = \sum_{i=1}^{2K} \sum_{j=1, i \neq j}^{2K} \cos(\widetilde{\mathbf{p}}_i, \widetilde{\mathbf{p}}_j), \tag{9}$$

where the $\cos(\cdot, \cdot)$ denotes cosine similarity between two vectors.

The discriminator $\mathcal{D}_{part}$ takes $\widetilde{\mathbf{p}}_k$ as input and outputs a scalar representing the probability that it belongs to the target domain. The loss function of $\mathcal{D}_{part}$ can be formulated as:

$$\mathcal{L}_{part} = \frac{1}{2K} \sum_{k=1}^{2K} [\text{CE}(\mathcal{D}_{part}(\widetilde{\mathbf{p}}_k^S), 0) + \text{CE}(\mathcal{D}_{part}(\widetilde{\mathbf{p}}_k^T), 1)]. \tag{10}$$

Table 1. Comparison with other SOTA methods on the Lucchi dataset. Note that "VNC III → Lucchi-Test" means training the model with VNC III as source domain and Lucchi training set as target domain and testing it on Lucchi testing set, and vice versa.

Methods	VNC III → Lucchi-Test				VNC III → Lucchi-Train			
	mAP	F1	MCC	IoU	mAP	F1	MCC	IoU
Oracle	97.5	92.9	92.3	86.8	99.1	94.2	93.7	88.8
NoAdapt	74.1	57.6	58.6	40.5	78.5	61.4	62.0	44.7
Y-Net [19]	-	68.2	-	52.1	-	71.8	-	56.4
DANN [2]	-	68.2	-	51.9	-	74.9	-	60.1
AdaptSegNet [24]	-	69.9	-	54.0	-	79.0	-	65.5
UALR [29]	80.2	72.5	71.2	57.0	87.2	78.8	77.7	65.2
DAMT-Net [18]	-	74.7	-	60.0	-	81.3	-	68.7
DA-VSN [4]	82.8	75.2	73.9	60.3	91.3	83.1	82.2	71.1
DA-ISC [5]	89.5	81.3	80.5	68.7	92.4	85.2	84.5	74.3
Ours	**91.1**	**84.1**	**83.5**	**72.8**	**94.4**	**86.7**	**86.1**	**77.1**

Table 2. Comparison with other SOTA methods on the MitoEM dataset. Note that "MitoEM-R → MitoEM-H" means training the model with MitoEM-R training set as the source domain and MitoEM-H training set as the target domain and testing it on MitoEM-H validation set, and vice versa.

Method	MitoEM-R → MitoEM-H				MitoEM-H → MitoEM-R			
	mAP	F1	MCC	IoU	mAP	F1	MCC	IoU
Oracle	97.0	91.6	91.2	84.5	98.2	93.2	92.9	87.3
NoAdapt	74.6	56.8	59.2	39.6	88.5	76.5	76.8	61.9
UALR [29]	90.7	83.8	83.2	72.2	92.6	86.3	85.5	75.9
DAMT-Net [18]	92.1	84.4	83.7	73.0	94.8	86.0	85.7	75.4
DA-VSN [4]	91.6	83.3	82.6	71.4	94.5	86.7	86.3	76.5
DA-ISC [5]	92.6	85.6	84.9	74.8	96.8	88.5	88.3	79.4
Ours	**93.9**	**86.1**	**85.5**	**75.6**	**97.0**	**89.2**	**88.8**	**80.6**

As a result, the overall objective of our SAPAN is as follows:

$$\mathcal{L} = \mathcal{L}_{sup} + \lambda_{div} \times \mathcal{L}_{div} - \lambda_{part} \times \mathcal{L}_{part}, \tag{11}$$

where the λ_{div} and λ_{part} are the trade-off weights. The segmentation network and the $\mathcal{D}_{part}$ are trained alternately by minimizing the $\mathcal{L}$ and the $\mathcal{L}_{part}$, respectively.

4   Experiments

4.1   Dataset and Evaluation Metric

Dataset. We evaluate our approach on three widely used EM datasets: the VNC III [3] dataset, the Lucchi dataset [9] and the MitoEM dataset [28]. These datasets exhibit significant diversity making the domain adaptation task challenging. The VNC III [3] dataset contains 20 sections of size 1024×1024. The training and testing set of Lucchi [9] dataset are both $165 \times 1024 \times 768$. The MitoEM dataset consists of two subsets of size $1000 \times 4096 \times 4096$, dubbed MitoEM-R and MitoEM-H respectively. The ground-truth of their training set (400) and validation set (100) are publicly available.

Metrics. Following [5, 29], four widely used metrics are used for evaluation, *i.e.*, mean Average Precision (mAP), F1 score, Mattews Correlation Coefficient (MCC) [14] and Intersection over Union (IoU).

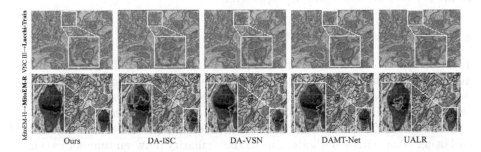

Fig. 2. Qualitative comparison with different methods. Note that the red and green contours denote the ground-truth and prediction. And we mark significant improvements using blue boxes. (Color figure online)

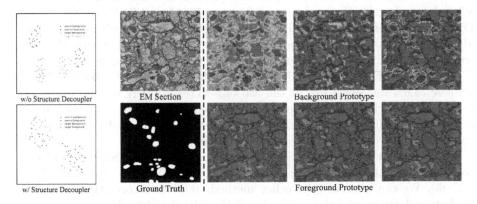

Fig. 3.   t-SNE visualization.   **Fig. 4.** Visualization of activation maps **A** for different prototypes in part miner.

Table 3. Ablation study on components.

SD	PM	mAP	F1	MCC	IoU
		85.3	80.3	79.4	67.5
✓		88.8	82.6	81.9	70.5
	✓	87.9	81.6	80.9	69.1
✓	✓	**91.1**	**84.1**	**83.5**	**72.8**

Table 4. Ablation study on smoothing operations (SMT).

	mAP	F1	MCC	IoU
w/o SMT	84.6	81.5	80.7	68.9
Conv. SMT	85.9	82.6	82.0	70.6
Atten. SMT	**91.1**	**84.1**	**83.5**	**72.8**

Table 5. Ablation on different ways for modeling part prototypes.

$fg.$	$bg.$	$\mathcal{L}_{div}$	mAP	F1	MCC	IoU
✓			75.3	75.5	74.7	61.2
✓		✓	88.4	82.2	81.4	69.9
✓	✓	✓	**91.1**	**84.1**	**83.5**	**72.8**

4.2   Implementation Details

We adopt a five-stage U-Net with feature channel number of $[16, 32, 48, 64, 80]$. During training, we randomly crop the original EM section into 512×512 with random augmentation including flip, transpose, rotate, resize and elastic transformation. All models are trained 20,000 iterations using Adam optimizer with batch size of 12, learning rate of 0.001 and β of (0.9, 0.99). The λ_{div}, λ_{part} and K are set as 0.01, 0.001 and 4, respectively.

4.3   Comparisons with SOTA Methods

Two groups of experiments are conducted to verify the effectiveness of our method. Note that "Oracle" means directly training the model on the target domain dataset with the ground truth and "NoAdapt" means training the model only using the source domain dataset without the target domain dataset.

Table 1 and Table 2 present the performance comparison between our method and other competitive methods. Due to the similarity between time series (video) and space series (EM volume), the SOTA domain adaptive video segmentation method [4] is also compared. We consistently observe that our method outperforms all other models on two groups of experiments, which strongly proves the effectiveness of our method. For example, our SAPAN enhances the IoU of VNC III → Lucchi-Test and Lucchi-Train to 72.8% and 77.1%, outperforming DA-ISC by significant margins of 4.1% and 2.8%. Compared with the pixel-wise alignment, the part-aware alignment can make the model fully utilize the global context information so as to perceive the mitochondria comprehensively.

On the MitoEM dataset with a larger structure discrepancy, our SAPAN can achieve 75.6% and 80.6% IoU on the two subsets respectively, outperforming DA-ISC by 0.8% and 1.2%. It demonstrates the remarkable generalization capacity of SAPAN, credited to the structure decoupler, which can effectively alleviate the domain gap caused by huge structural difference.

Figure 2 shows the qualitative comparison between our SAPAN and other competitive methods including UALR [29], DAMT-Net [18], DA-VSN [4] and ISC [5]. We can observe that other methods tend to incorrectly segment the background region or fail to activate all the mitochondria. We deem the main reason is that the large domain gap severely confuses the segmentation model. With the assistance of the structure decoupler and the part miner, SAPAN is more robust in the face of the large domain gap and generates a more accurate prediction.

4.4 Ablation Study and Analysis

To look deeper into our method, we perform a series of ablation studies on VNC III → Lucchi-Test to analyze each component of our SAPAN, including the **Structure Decoupler** (SD) and the **Part Miner** (PM). Note that we remove all modules except the U-Net and the pixel-wise discriminator as our baseline. Hyperparameters are discussed in the supplementary material.

Effectiveness of Components. As shown in Table 3, both structure decoupler and part miner bring a certain performance lift compared with the baseline. (1) With the utilization of SD, a 3.0% improvement of IoU can be observed, indicating that decoupling the feature from domain-specific information can benefit the domain adaptation task. In Fig. 3, we visualize the feature distribution with/without SD using t-SNE [11]. We can see that SD makes the feature distribution more compact and effectively alleviates the domain gap. (2) The introduction of PM achieves further accuracy gains, mainly ascribed to the adaptive part alignment mechanism. As shown in Fig. 4, the different prototypes focus on significant distinct areas. The discriminator benefits from the diverse parts-aware prototypes, which in turn promotes the segmentation network.

Effectiveness of the Attention-Based Smoothing. As shown in Table 4, abandoning the spatial smoothing operation makes the performance decrease. Compared with simply employing a convolution layer for smoothing, attention-based smoothing contributes to a remarkable performance (72.8% *vs.* 70.6% IoU), thanks to the long-range modeling capabilities of the attention mechanism.

Effectiveness of the Ways Modeling Part-Aware Prototypes. In Table 5, $fg.$ means only focusing on foreground and vice versa. Neglecting $\mathcal{L}_{div}$ leads to severe performance degradation, that is because $\mathcal{L}_{div}$ is able to prevent the prototypes from focusing on similar local semantic clues. And simultaneously modeling foreground/background prototypes brings further improvement, demonstrating there is a lot of discriminative information hidden in the background region.

5   Conclusion

We propose a structure decoupler to decouple the distribution and morphology, and a part miner to aggregate diverse parts for UDA mitochondria segmentation. Experiments show the effectiveness.

Acknowledgments. This work was partially supported by the National Nature Science Foundation of China (Grant 62022078, 62021001).

References

1. Duchen, M.: Mitochondria and Ca2+ in cell physiology and pathophysiology. Cell Calcium **28**(5–6), 339–348 (2000)
2. Ganin, Y., et al.: Domain-adversarial training of neural networks. J. Mach. Learn. Res. **17**(1) (2016). 2096-2030
3. Gerhard, S., Funke, J., Martel, J., Cardona, A., Fetter, R.: Segmented anisotropic sstem dataset of neural tissue. Figshare (2013)
4. Guan, D., Huang, J., Xiao, A., Lu, S.: Domain adaptive video segmentation via temporal consistency regularization. In: Proceedings of the IEEE/CVF International Conference on Computer Vision, pp. 8053–8064 (2021)
5. Huang, W., Liu, X., Cheng, Z., Zhang, Y., Xiong, Z.: Domain adaptive mitochondria segmentation via enforcing inter-section consistency. In: Wang, L., Dou, Q., Fletcher, P.T., Speidel, S., Li, S. (eds.) International Conference on Medical Image Computing and Computer-Assisted Intervention, pp. 89–98. Springer, Cham (2022). https://doi.org/10.1007/978-3-031-16440-8_9
6. Li, M., Chen, C., Liu, X., Huang, W., Zhang, Y., Xiong, Z.: Advanced deep networks for 3D mitochondria instance segmentation. In: 2022 IEEE 19th International Symposium on Biomedical Imaging (ISBI), pp. 1–5. IEEE (2022)
7. Li, Z., Chen, X., Zhao, J., Xiong, Z.: Contrastive learning for mitochondria segmentation. In: 2021 43rd Annual International Conference of the IEEE Engineering in Medicine & Biology Society (EMBC), pp. 3496–3500. IEEE (2021)
8. Liu, D., et al.: PDAM: a panoptic-level feature alignment framework for unsupervised domain adaptive instance segmentation in microscopy images. IEEE Trans. Med. Imaging **40**(1), 154–165 (2020)
9. Lucchi, A., Li, Y., Fua, P.: Learning for structured prediction using approximate subgradient descent with working sets. In: Proceedings of the IEEE Conference on Computer Vision and Pattern Recognition, pp. 1987–1994 (2013)
10. Luo, N., Pan, Y., Sun, R., Zhang, T., Xiong, Z., Wu, F.: Camouflaged instance segmentation via explicit de-camouflaging. In: Proceedings of the IEEE/CVF Conference on Computer Vision and Pattern Recognition, pp. 17918–17927 (2023)
11. Van der Maaten, L., Hinton, G.: Visualizing data using t-SNE. J. Mach. Learn. Res. **9**(11) (2008)
12. Mai, H., Sun, R., Zhang, T., Xiong, Z., Wu, F.: DualRel: semi-supervised mitochondria segmentation from a prototype perspective. In: Proceedings of the IEEE/CVF Conference on Computer Vision and Pattern Recognition, pp. 19617–19626 (2023)
13. Martin, L.J.: Biology of mitochondria in neurodegenerative diseases. Prog. Mol. Biol. Transl. Sci. **107**, 355–415 (2012)
14. Matthews, B.W.: Comparison of the predicted and observed secondary structure of T4 phage lysozyme. Biochimica et Biophysica Acta (BBA)-Protein Structure **405**(2), 442–451 (1975)
15. Newsholme, P., Gaudel, C., Krause, M.: Mitochondria and diabetes. An intriguing pathogenetic role. Adv. Mitochondrial Med., 235–247 (2012)
16. Nightingale, L., de Folter, J., Spiers, H., Strange, A., Collinson, L.M., Jones, M.L.: Automatic instance segmentation of mitochondria in electron microscopy data. BioRxiv, 2021–05 (2021)
17. Pan, Y., et al.: Adaptive template transformer for mitochondria segmentation in electron microscopy images. In: Proceedings of the IEEE/CVF International Conference on Computer Vision (2023)

18. Peng, J., Yi, J., Yuan, Z.: Unsupervised mitochondria segmentation in EM images via domain adaptive multi-task learning. IEEE J. Sel. Top. Sig. Process. **14**(6), 1199–1209 (2020)
19. Roels, J., Hennies, J., Saeys, Y., Philips, W., Kreshuk, A.: Domain adaptive segmentation in volume electron microscopy imaging. In: 2019 IEEE 16th International Symposium on Biomedical Imaging (ISBI 2019), pp. 1519–1522. IEEE (2019)
20. Ronneberger, O., Fischer, P., Brox, T.: U-Net: convolutional networks for biomedical image segmentation. In: Navab, N., Hornegger, J., Wells, W.M., Frangi, A.F. (eds.) MICCAI 2015. LNCS, vol. 9351, pp. 234–241. Springer, Cham (2015). https://doi.org/10.1007/978-3-319-24574-4_28
21. Sun, R., Li, Y., Zhang, T., Mao, Z., Wu, F., Zhang, Y.: Lesion-aware transformers for diabetic retinopathy grading. In: Proceedings of the IEEE/CVF Conference on Computer Vision and Pattern Recognition, pp. 10938–10947 (2021)
22. Sun, R., et al.: Appearance prompt vision transformer for connectome reconstruction. In: IJCAI (2023)
23. Sun, R., Wang, Y., Mai, H., Zhang, T., Wu, F.: Alignment before aggregation: trajectory memory retrieval network for video object segmentation. In: Proceedings of the IEEE/CVF International Conference on Computer Vision (2023)
24. Tsai, Y.H., Hung, W.C., Schulter, S., Sohn, K., Yang, M.H., Chandraker, M.: Learning to adapt structured output space for semantic segmentation. In: Proceedings of the IEEE Conference on Computer Vision and Pattern Recognition, pp. 7472–7481 (2018)
25. Wang, L., Tong, Z., Ji, B., Wu, G.: TDN: temporal difference networks for efficient action recognition. In: Proceedings of the IEEE/CVF Conference on Computer Vision and Pattern Recognition, pp. 1895–1904 (2021)
26. Wang, Y., Sun, R., Zhang, T.: Rethinking the correlation in few-shot segmentation: a buoys view. In: Proceedings of the IEEE/CVF Conference on Computer Vision and Pattern Recognition, pp. 7183–7192 (2023)
27. Wang, Y., Sun, R., Zhang, Z., Zhang, T.: Adaptive agent transformer for few-shot segmentation. In: Avidan, S., Brostow, G., Cissé, M., Farinella, G.M., Hassner, T. (eds.) European Conference on Computer Vision, pp. 36–52. Springer, Cham (2022). https://doi.org/10.1007/978-3-031-19818-2_3
28. Wei, D., et al.: MitoEM dataset: large-scale 3D mitochondria instance segmentation from EM images. In: Martel, A.L., et al. (eds.) MICCAI 2020. LNCS, vol. 12265, pp. 66–76. Springer, Cham (2020). https://doi.org/10.1007/978-3-030-59722-1_7
29. Wu, S., Chen, C., Xiong, Z., Chen, X., Sun, X.: Uncertainty-aware label rectification for domain adaptive mitochondria segmentation. In: de Bruijne, M., et al. (eds.) MICCAI 2021. LNCS, vol. 12903, pp. 191–200. Springer, Cham (2021). https://doi.org/10.1007/978-3-030-87199-4_18
30. Yi, J., Yuan, Z., Peng, J.: Adversarial-prediction guided multi-task adaptation for semantic segmentation of electron microscopy images. In: 2020 IEEE 17th International Symposium on Biomedical Imaging (ISBI), pp. 1205–1208. IEEE (2020)

Deep Probability Contour Framework for Tumour Segmentation and Dose Painting in PET Images

Wenhui Zhang[✉][iD] and Surajit Ray[iD]

University of Glasgow, Glasgow G12 8QQ, UK
W.Zhang.2@research.gla.ac.uk, Surajit.Ray@glasgow.ac.uk

Abstract. The use of functional imaging such as PET in radiotherapy (RT) is rapidly expanding with new cancer treatment techniques. A fundamental step in RT planning is the accurate segmentation of tumours based on clinical diagnosis. Furthermore, recent tumour control techniques such as intensity modulated radiation therapy (IMRT) dose painting requires the accurate calculation of multiple nested contours of intensity values to optimise dose distribution across the tumour. Recently, convolutional neural networks (CNNs) have achieved tremendous success in image segmentation tasks, most of which present the output map at a pixel-wise level. However, its ability to accurately recognize precise object boundaries is limited by the loss of information in the successive downsampling layers. In addition, for the dose painting strategy, there is a need to develop image segmentation approaches that reproducibly and accurately identify the high recurrent-risk contours. To address these issues, we propose a novel hybrid-CNN that integrates a kernel smoothing-based probability contour approach (KsPC) to produce contour-based segmentation maps, which mimic expert behaviours and provide accurate probability contours designed to optimise dose painting/IMRT strategies. Instead of user-supplied tuning parameters, our final model, named KsPC-Net, applies a CNN backbone to automatically learn the parameters and leverages the advantage of KsPC to simultaneously identify object boundaries and provide probability contour accordingly. The proposed model demonstrated promising performance in comparison to state-of-the-art models on the MICCAI 2021 challenge dataset (HECKTOR).

Keywords: Image Segmentation · PET imaging · Probability Contour · Dose painting · Deep learning

1 Introduction

Fluorodeoxyglucose Positron Emission Tomography (PET) is widely recognized as an essential tool in oncology [10], playing an important role in the stag-

Supplementary Information The online version contains supplementary material available at https://doi.org/10.1007/978-3-031-43901-8_51.

ing, monitoring, and follow-up radiotherapy (RT) planning [2,19]. Delineation of Region of Interest (ROI) is a crucial step in RT planning. It enables the extraction of semi-quantitative metrics such as standardized uptake values (SUVs), which normalize pixel intensities based on patient weight and radiotracer dose [20]. Manual delineation is a time-consuming and laborious task that is prone to poor reproducibility in medical imaging, and this is particularly true for PET, due to its low signal-to-noise ratio and limited spatial resolution [10]. In addition, manual delineation depends heavily on the expert's prior knowledge, which often leads to large inter-observer and intra-observer variations [8]. Therefore, there is an urgent need for developing accurate automatic segmentation algorithms in PET images which will reduce expert workload, speed up RT planning while reducing intra-observer variability.

In the last decade, CNNs have demonstrated remarkable achievements in medical image segmentation tasks. This is primarily due to their ability to learn informative hierarchical features directly from data. However, as illustrated in [9,23], it is rather difficult for CNNs to recognize the object boundary precisely due to the information loss in the successive downsampling layers. Despite the headway made in using CNNs, their applications have been restricted to the generation of pixel-wise segmentation maps instead of smooth contour. Although CNNs may yield satisfactory segmentation results, low values of the loss function may not always indicate a meaningful segmentation. For instance, a noisy result can create incorrect background contours and blurry object boundaries near the edge pixels [6]. To address this, a kernel smoothing-based probability contour (KsPC) approach was proposed in our previous work [22]. Instead of a pixel-wise analysis, we assume that the true SUVs come from a smooth underlying spatial process that can be modelled by kernel estimates. The KsPC provides a surface over images that naturally produces contour-based results rather than pixel-wise results, thus mimicking experts' hand segmentation. However, the performance of KsPC depends heavily on the tuning parameters of bandwidth and threshold in the model, and it lacks information from other patients.

Beyond tumour delineation, another important use of functional images, such as PET images is their use for designing IMRT dose painting (DP). In particular, dose painting uses functional images to paint optimised dose prescriptions based on the spatially varying radiation sensitivities of tumours, thus enhancing the efficacy of tumour control [14,18]. One of the popular DP strategies is dose painting by contours (DPBC), which assigns a homogeneous boost dose to the subregions defined by SUV thresholds. However, there is an urgent need to develop image segmentation approaches that reproducibly and accurately identify the high recurrent-risk contours [18]. Our previously proposed KsPC provides a clear framework to calculate the probability contours of the SUV values and can readily be used to define an objective strategy for segmenting tumours into subregions based on metabolic activities, which in turn can be used to design the IMRT DP strategy.

To address both tumour delineation and corresponding dose painting challenges, we propose to combine the expressiveness of deep CNNs with the versatility of KsPC in a unified framework, which we call KsPC-Net. In the proposed KsPC-Net, a CNN is employed to learn directly from the data to produce the pixel-wise bandwidth feature map and initial segmentation map, which are used to define the tuning parameters in the KsPC module. Our framework is completely automatic and differentiable. More specifically, we use the classic UNet [17] as the CNN backbone and evaluate our KsPC-Net on the publicly available MICCAI HECKTOR (HEad and neCK TumOR segmentation) challenge 2021 dataset. Our proposed KsPC-Net yields superior results in terms of both Dice similarity scores and Hausdorff distance compared to state-of-art models. Moreover, it can produce contour-based segmentation results which provide a more accurate delineation of object edges and provide probability contours as a byproduct, which can readily be used for DP planning.

2   Methods

2.1   Kernel Smoothing Based Probability Contour

Kernel-based method and follow up approach of modal clustering [13,16] have been used to cluster high-dimensional random variables and natural-scene image segmentation. In this work, we propose to model the pixel-specific SUV as a discretized version of the underlying unknown smooth process of "metabolic activity". The smooth process can then be estimated as kernel smoothed surface of the SUVs over the domain of the entire slice. In particular, let $Y = (y_1, y_2, ..., y_N)$ denote N pixel's SUV in a 2D PET image sequentially, and $\mathbf{x}_i = (x_{i1}, x_{i2}), i = 1, ..., N$ denote position vector with x_{i1} and x_{i2} being the position in 2D respectively. Note that $\mathbf{x}_i \in \mathbb{R}^d$ and $d = 2$ in our case. We assume that for each position vector $\mathbf{x}$, the SUV represents the frequency of $\mathbf{x}$ appearing in the corresponding grid. The SUV surface can therefore be modelled as kernel density estimate (KDE) [3,15] of an estimated point $\mathbf{x}$, which is defined generally as

$$\hat{f}(\mathbf{x}; \mathbf{H}) = \left(\sum_{i=1}^{N} y_i\right)^{-1} \sum_{t=1}^{y_1 + ... + y_N} K_{\mathbf{H}}(\mathbf{x} - \mathbf{x}_t), \tag{1}$$

where K is a kernel function and $\mathbf{H}$ is a symmetric, positive definite, $d \times d$ matrix of smoothing tuning parameters, called bandwidth which controls the orientation and amount of smoothing via the scaled kernel $K_{\mathbf{H}}(\mathbf{x}) = |\mathbf{H}|^{-1/2} K(|\mathbf{H}|^{-1/2} \mathbf{x})$. On the other hand, since $\mathbf{x}_t$ is counted y_i times at the same position, Eq. 1 can be further simplified as

$$\hat{f}(\mathbf{x}; \mathbf{H}) = \left(\sum_{i=1}^{N} y_i\right)^{-1} \sum_{i=1}^{N} K_{\mathbf{H}}(\mathbf{x} - \mathbf{x}_i) y_i. \tag{2}$$

A scaled kernel is positioned so that its mode coincides with each data point $\mathbf{x}_i$ which is expressed mathematically as $K_{\mathbf{H}}(\mathbf{x} - \mathbf{x}_i)$. In this paper, we have used a Guassian kernel which is denoted as:

$$K_{\mathbf{H}}(\mathbf{x} - \mathbf{x_i}) = (2\pi)^{-1/2} |\mathbf{H}|^{-1/2} \exp\left(-\frac{1}{2}(\mathbf{x} - \mathbf{x_i})^T \mathbf{H}^{-1}(\mathbf{x} - \mathbf{x_i})\right),$$

which is a normal distribution with mean $\mathbf{x}_i$ and variance-covariance matrix $\mathbf{H}$. Therefore, we can interpret $\hat{f}$ in Eq. (2) as the probability mass of the data point $\mathbf{x}$ which is estimated by smoothing the SUVs of the local neighbourhood using the Gaussian kernel. The resulting surface built by the KDE process can be visualized in Fig. 1(c). By placing a threshold plane, a contour-based segmentation map can naturally be obtained. Note that one can obtain a pixel-based segmentation map, by thresholding the surface at the observed grid points.

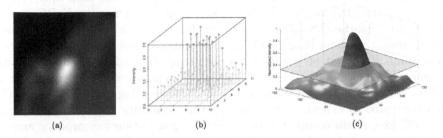

Fig. 1. A visualization example of how KsPC works: (a) An example of a PET image (b) Grid-level intensity values as observations (c) The resulting smoothed surface built by KsPC with a threshold plane.

After delineating the gross tumour volume, a follow-up application of the kernel smoothed surface is to construct probability contours. Mathematically, a $100\,\omega\%$ region of a density f is defined as the level set $\mathcal{L}(f_\omega) = \{f(\mathbf{x}) \geq f_\omega\}$ with its corresponding contour level f_ω such that $\mathcal{P}(\mathbf{x} \in \mathcal{L}(f_\omega) = 1 - \omega$, where $\mathbf{x}$ is a random variable and $\mathcal{L}(f_\omega)$ has a minimal hypervolume [11]. In other words, for any $\omega \in (0, 1)$, the $100\,\omega\%$ contour refers to the region with the smallest area which encompasses $100\,\omega\%$ of the probability mass of the density function [11]. In practice, f_ω can be estimated using the following result.

Result. The estimated probability contour level f_ω can be computed as the ω-th quantile of $\hat{f}_\omega$ of $\hat{f}(\mathbf{x}_1; \mathbf{H}), ..., \hat{f}(\mathbf{x}_n; \mathbf{H})$ (Proof in supplementary materials).

The primary advantage of utilizing probability contours is their ability to assign a clear probabilistic interpretation on the defined contours, which are scale-invariant [5]. This provides a robust definition of probability under the perturbation of the input data. In addition, these contours can be mapped to the IMRT dose painting contours, thus providing an alternative prescription strategy for IMRT. Examples of the application of probability contours will be demonstrated and explained in Sect. 4.2.

2.2 The KsPC-Net Architecture

In the KsPC module, the model performance heavily depends on the bandwidth matrix $\mathbf{H}$ and it is often assumed that each kernel shares the same scalar bandwidth parameter. However, one may want to use different amounts of smoothing in the kernel at different grid positions. The commonly used approach for bandwidth selection is cross-validation [4], which is rather time-consuming even in the simpler scalar situation. In this paper, we instead use the classic 2D-Unet [17] as our CNN backbone to compute the pixel-level bandwidth feature map, which informs the KsPc bandwidth. Additionally, we obtain the optimal threshold for constructing the KsPC contour from the initial segmentation map. As shown in Fig. 2 the proposed KsPC-Net integrates the KsPC approach with a CNN backbone (UNet) in an end-to-end differentiable manner. First, the initial segmentation map and pixel-level bandwidth parameter map $h(x_{i1}, x_{i2})$ of KsPC are learned from data by the CNN backbone. Then the KsPC module obtains the quantile threshold value for each image by identifying the quantile corresponding to the minimum SUV of the tumour class in the initial segmentation map. The next step involves transmitting the bandwidth map, quantile threshold, and raw image to KsPC module to generate the segmentation map and its corresponding probability contours. The resulting output from KsPC is then compared to experts' labels using a Dice similarity loss function, referred to KsPC loss. Additionally, the initial Unet segmentation can produce another loss function, called CNN loss, which serves as an auxiliary supervision for the CNN backbone. The final loss can then be constructed as the weighted sum of CNN loss and KsPC loss. By minimizing the final loss, the error can be backpropagated through the entire KsPC architecture to guide the weights updating the CNN backbone.

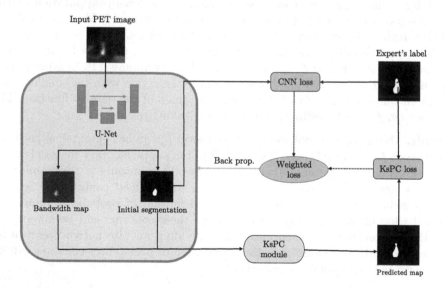

Fig. 2. The architecture of KsPC-Net. KsPC-Net is an end-to-end trainable framework with KsPC module. In contrast to the traditional kernel smoothing process, bandwidth parameters learned by the CNN backbone are pixel-wise functions $h(x_{i1}, x_{i2})$ rather than scalars.

2.3   Loss Function

The Dice similarity coefficient is widely employed to evaluate segmentation models. We utilize the Dice loss function to optimize the model performance during training, which is defined as:

$$\mathcal{L}_{dice}(y, \hat{y}) = 1 - \frac{2 \sum_i^N y_i \hat{y}_i}{\sum_i^N y_i + \sum_i^N \hat{y}_i + \epsilon},$$

where y_i is the label from experts and $\hat{y}_i$ is the predicted label of i-th pixel. N is the total number of pixels and ϵ is a small constant in case of zero division. As shown in Fig. 2, we construct the weighted Dice loss to train the model as follows:

$$\mathcal{L}_{final} = \alpha * \mathcal{L}_{CNN} + (1 - \alpha) * \mathcal{L}_{KsPC},$$

where $\mathcal{L}_{final}$ denotes the weighed dice loss while $\mathcal{L}_{CNN}$ and $\mathcal{L}_{KsPC}$ denotes the CNN loss and KsPC loss, respectively. In addition, α is a balancing parameter and is set to be 0.01 in this work.

3   Experiments

3.1   Dataset

The dataset is from the HECKTOR challenge in MICCAI 2021 (HEad and neCK TumOR segmentation challenge). The HECKTOR training dataset consists of 224 patients diagnosed with oropharyngeal cancer [1]. For each patient, FDG-PET input images and corresponding labels in binary description (0 s and 1 s) for the primary gross tumour volume are provided and co-registered to a size of 144 × 144 × 144 using bounding box information encompassing the tumour. Five-fold cross-validation is used to generalize the performance of models.

3.2   Implementation Details

We used Python and a trained network on a NVIDIA Dual Quadro RTX machine with 64 GB RAM using the PyTorch package. We applied a batch size of 12 and the Adam algorithm [12] with default parameters to minimize the dice loss function. All models were trained for 300 epochs. Each convolutional layer is followed by RELU activation and batch normalization.

4   Results

4.1   Results on HECKTOR 2021 Dataset

To evaluate the performance of our KsPC-Net, we compared it with the results of 5-fold cross-validation against three widely-used models namely, the standard 2D Unet, the 2D residual Unet and the 3D Unet. Additionally, we compare our performance against newly developed approaches MSA-Net [7] and CCUT-Net [21] which were reported in the HECKTOR 2021 challenges [1]. To quantify the performance, we report several metrics including Dice similarity scores, Precision, Recall, and Hausdorff distance. Table 1 shows the quantitative comparison of different approaches on HECKTOR dataset. It is worth mentioning that since our KsPC-Net is in a 2D Unet structure, the Hausdorff distance here was calculated on slice averages to use a uniform metric across all 2D and 3D segmentation models. However, the results of 2D Hausdorff distances of MSA-Net and CCUT-Net are not available and therefore they are omitted in the table of comparison.

Table 1. Mean segmentation results of different models and our proposed model. The model with best performance for each metric is indicated in **bold***.

Method	Dice Score	Hausdorff Dist	Precision	Recall
2D-Unet	0.740	0.561	0.797	0.873
Res-Unet	0.680	0.611	0.740	0.841
3D-Unet	0.764	0.546	**0.839***	0.797
MSA-Net	0.757	-	0.788	0.785
CCUT-Net	0.750	-	0.776	0.804
KsPC-Net(Ours)	**0.768***	**0.521***	0.793	**0.911***

The results clearly demonstrate that the proposed KsPC-Net is effective in segmenting H&N tumours, achieving a mean Dice score of 0.768. This represents a substantial improvement over alternative approaches, including 2D-UNet (0.740), 3D U-Net (0.764), Residual-Unet (0.680), MSA-Net (0.757) and CCUT-Net (0.750). While we acknowledge that there was no statistically significant improvement compared to other SOTA models, it is important to note that our main goal is to showcase the ability to obtain probability contours as a natural byproduct while preserving state-of-the-art accuracy levels. On the other hand, in comparison to the baseline 2D-Unet model, KsPC-Net yields a higher Recall (0.911) with a significant improvement (4.35%), indicating that KsPC-Net generates fewer false negatives (FN). Although the Precision of KsPC-Net is slightly lower than the best-performing method (3D Unet), it achieves a relatively high value of 0.793. In addition, the proposed KsPC-Net achieves the best performance on Hausdorff distance among the three commonly used Unet models (2D-Unet, Res-Unet and 3D-Unet), which indicates that KsPC-Net exhibits a stronger capacity for accurately localizing the boundaries of objects. This is consistent with the mechanisms of KsPC, which leverages neighbouring weights to yield outputs with enhanced smoothness.

4.2 Probability Contours

One of the byproducts of using the kernel-smoothed densities to model the SUVs is the associated probability contours, which can be readily used to develop a comprehensive inferential framework and uncertainty quantification. For example, Fig. 3 provides two examples of PET image segmentation maps by KsPC-Net and their corresponding probability contours in the last column. There are 5 contours in each case which is linear in probability space, in the sense that each contour encloses 10%, 30%, 50%, 70% and 90% probability mass respectively (from inner to outer), thus dividing the density surface into subregions with attached probability mass.

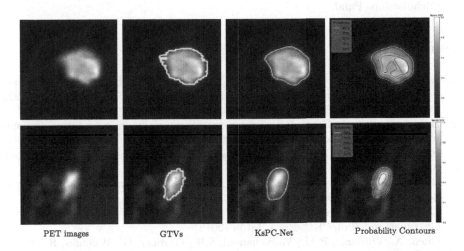

PET images   GTVs   KsPC-Net   Probability Contours

Fig. 3. Illustrations of the segmentation results and probability contours on two examples. The four columns are original PET images, ground truth provided by experts, segmentation maps from KsPC-Net and its probability contours (in 10%, 30%, 50%, 70%, 90% respectively).

These probability contours can provide a rigorous framework for designing the number and magnitude of SUV thresholds for designing optimal DP strategies. Since the SUVs are smoothed by the kernel density heights, the inner 10% probability contour corresponds to the subregion with relatively higher SUVs. In other words, there is an inverse mapping between the probability contours and the amount of dose boost assigned to subvolumes.

5   Conclusion

In this paper, we present a novel network, KsPC-Net, for the segmentation in 2D PET images, which integrates KsPC into the UNet architecture in an end-to-end differential manner. The KsPC-Net utilizes the benefits of KsPC to deliver both contour-based and grid-based segmentation outcomes, leading to improved precision in the segmentation of contours. Promising performance was achieved by our proposed KsPC-Net compared to the state-of-the-art approaches on the MICCAI 2021 challenge dataset (HECKTOR). It is worth mentioning that the architecture of our KsPC-Net is not limited to Head & Neck cancer type and can be broadcast to different cancer types. Additionally, a byproduct application of our KsPC-Net is to construct probability contours, which enables probabilistic interpretation of contours. The subregions created by probability contours allow for a strategy planning for the assigned dose boosts, which is a necessity for the treatment planning of radiation therapy for cancers.

Acknowledgement. This work was supported by the Carnegie Trust of Scotland PhD Scholarships Fund.

References

1. Andrearczyk, V., Oreiller, V., Hatt, M., Depeursinge, A.: Head and Neck Tumor Segmentation and Outcome Prediction. LNCS, vol. 13209. Springer, Cham (2022). https://doi.org/10.1007/978-3-030-98253-9
2. Bai, B., Bading, J., Conti, P.S.: Tumor quantification in clinical positron emission tomography. Theranostics **3**(10), 787 (2013)
3. Bowman, A., Foster, P.: Density based exploration of bivariate data. Stat. Comput. **3**, 171–177 (1993)
4. Bowman, A.W.: An alternative method of cross-validation for the smoothing of density estimates. Biometrika **71**(2), 353–360 (1984)
5. Chacón, J.E., Duong, T.: Multivariate Kernel Smoothing and Its Applications. Chapman and Hall/CRC, Boca Raton (2018)
6. Chen, X., Williams, B.M., Vallabhaneni, S.R., Czanner, G., Williams, R., Zheng, Y.: Learning active contour models for medical image segmentation. In: Proceedings of the IEEE/CVF Conference on Computer Vision and Pattern Recognition, pp. 11632–11640 (2019)
7. Cho, M., Choi, Y., Hwang, D., Yie, S.Y., Kim, H., Lee, J.S.: Multimodal spatial attention network for automatic head and neck tumor segmentation in FDG-PET and CT images. In: Andrearczyk, V., Oreiller, V., Hatt, M., Depeursinge, A. (eds.) HECKTOR 2021. LNCS, vol. 13209, pp. 75–82. Springer, Cham (2022). https://doi.org/10.1007/978-3-030-98253-9_6
8. Gudi, S., et al.: Interobserver variability in the delineation of gross tumour volume and specified organs-at-risk during IMRT for head and neck cancers and the impact of FDG-PET/CT on such variability at the primary site. J. Med. Imaging Radiat. Sci. **48**(2), 184–192 (2017)
9. Hatamizadeh, A., et al.: Deep active lesion segmentation. In: Suk, H.-I., Liu, M., Yan, P., Lian, C. (eds.) MLMI 2019. LNCS, vol. 11861, pp. 98–105. Springer, Cham (2019). https://doi.org/10.1007/978-3-030-32692-0_12
10. Hatt, M., et al.: The first MICCAI challenge on pet tumor segmentation. Med. Image Anal. **44**, 177–195 (2018)
11. Hyndman, R.J.: Computing and graphing highest density regions. Am. Stat. **50**(2), 120–126 (1996)
12. Kingma, D.P., Ba, J.: Adam: a method for stochastic optimization. arXiv preprint arXiv:1412.6980 (2014)
13. Li, J., Ray, S., Lindsay, B.G.: A nonparametric statistical approach to clustering via mode identification. J. Mach. Learn. Res. **8**(8) (2007)

14. Ling, C.C., et al.: Towards multidimensional radiotherapy (MD-CRT): biological imaging and biological conformality. Int. J. Radiat. Oncol. Biol. Phys. **47**(3), 551–560 (2000)

15. Parzen, E.: On estimation of a probability density function and mode. Ann. Math. Stat. **33**(3), 1065–1076 (1962)

16. Ray, S., Lindsay, B.G.: The topography of multivariate normal mixtures. Ann. Stat. **33**(5) (2005). https://doi.org/10.1214/009053605000000417

17. Ronneberger, O., Fischer, P., Brox, T.: U-Net: convolutional networks for biomedical image segmentation. In: Navab, N., Hornegger, J., Wells, W.M., Frangi, A.F. (eds.) MICCAI 2015. LNCS, vol. 9351, pp. 234–241. Springer, Cham (2015). https://doi.org/10.1007/978-3-319-24574-4_28

18. Shi, X., Meng, X., Sun, X., Xing, L., Yu, J.: PET/CT imaging-guided dose painting in radiation therapy. Cancer Lett. **355**(2), 169–175 (2014)

19. Vallieres, M., et al.: Radiomics strategies for risk assessment of tumour failure in head-and-neck cancer. Sci. Rep. **7**(1), 10117 (2017)

20. Visser, E.P., Boerman, O.C., Oyen, W.J.: SUV: from silly useless value to smart uptake value. J. Nucl. Med. **51**(2), 173–175 (2010)

21. Wang, J., Peng, Y., Guo, Y., Li, D., Sun, J.: CCUT-net: pixel-wise global context channel attention UT-net for head and neck tumor segmentation. In: Andrearczyk, V., Oreiller, V., Hatt, M., Depeursinge, A. (eds.) HECKTOR 2021. LNCS, vol. 13209, pp. 38–49. Springer, Cham (2022). https://doi.org/10.1007/978-3-030-98253-9_2

22. Zhang, W., Ray, S.: Kernel smoothing-based probability contours for tumour segmentation. In: 26th UK Conference on Medical Image Understanding and Analysis. Springer, Cham (2022)

23. Zhang, Y., Chung, A.C.S.: Deep supervision with additional labels for retinal vessel segmentation task. In: Frangi, A.F., Schnabel, J.A., Davatzikos, C., Alberola-López, C., Fichtinger, G. (eds.) MICCAI 2018. LNCS, vol. 11071, pp. 83–91. Springer, Cham (2018). https://doi.org/10.1007/978-3-030-00934-2_10

Annotator Consensus Prediction for Medical Image Segmentation with Diffusion Models

Tomer Amit, Shmuel Shichrur, Tal Shaharabany[(✉)], and Lior Wolf

Tel-Aviv University, Tel Aviv, Israel
{tomeramit1,shmuels1,shaharabany,wolf}@mail.tau.ac.il

Abstract. A major challenge in the segmentation of medical images is the large inter- and intra-observer variability in annotations provided by multiple experts. To address this challenge, we propose a novel method for multi-expert prediction using diffusion models. Our method leverages the diffusion-based approach to incorporate information from multiple annotations and fuse it into a unified segmentation map that reflects the consensus of multiple experts. We evaluate the performance of our method on several datasets of medical segmentation annotated by multiple experts and compare it with the state-of-the-art methods. Our results demonstrate the effectiveness and robustness of the proposed method. Our code is publicly available at https://github.com/tomeramit/Annotator-Consensus-Prediction.

Keywords: Multi annotator · Image segmentation · Diffusion Model

1 Introduction

Medical image segmentation is a challenging task that requires accurate delineation of structures and regions of interest in complex and noisy images. Multiple expert annotators are often employed to address this challenge, to provide binary segmentation annotations for the same image. However, due to differences in experience, expertise, and subjective judgments, annotations can vary significantly, leading to inter- and intra-observer variability. In addition, manual annotation is a time-consuming and costly process, which limits the scalability and applicability of segmentation methods.

To overcome these limitations, automated methods for multi-annotator prediction have been proposed, which aim to fuse the annotations from multiple annotators and generate an accurate and consistent segmentation result. Existing approaches for multi-annotator prediction include majority voting [7], label fusion [3], and label sampling [12].

In recent years, diffusion models have emerged as a promising approach for image segmentation, for example by using learned semantic features [2]. By modeling the diffusion of image intensity values over the iterations, diffusion models

H. Greenspan et al. (Eds.): MICCAI 2023, LNCS 14223, pp. 544–554, 2023.
https://doi.org/10.1007/978-3-031-43901-8_52

capture the underlying structure and texture of the images and can separate regions of interest from the background. Moreover, diffusion models can handle noise and image artifacts, and adapt to different image modalities.

In this work, we propose a novel method for multi-annotator prediction, using diffusion models for medical segmentation. The goal is to fuse multiple annotations of the same image from different annotators and obtain a more accurate and reliable segmentation result. In practice, we leverage the diffusion-based approach to create one map for each level of consensus. To obtain the final prediction, we average the obtained maps and obtain one soft map.

We evaluate the performance of the proposed method on a dataset of medical images annotated by multiple annotators. Our results demonstrate the effectiveness and robustness of the proposed method in handling inter- and intra-observer variability and achieving higher segmentation accuracy than the state-of-the-art methods. The proposed method could improve the efficiency and quality of medical image segmentation and facilitate the clinical decision-making process.

2   Related Work

Multi-annotator Strategies. Research attention has recently been directed towards the issues of multi-annotator labels [7,12]. During training, Jensen et al. [12] randomly sampled different labels per image. This method produced a more calibrated model. Guan et al. [7] predicted the gradings of each annotator individually and acquired the corresponding weights for the final prediction. Kohl et al. [15] used the same sampling strategy to train a probabilistic model, based on a U-Net combined with a conditional variational autoencoder. Another recent probabilistic approach [20] combines a diffusion model with KL divergence to capture the variability between the different annotators. In our work, we use consensus maps as the ground truth and compare them to other strategies.

Diffusion Probabilistic Models (DPM). [23] are a class of generative models based on a Markov chain, which can transform a simple distribution (e.g. Gaussian) to data sampled from a complex distribution. Diffusion models are capable of generating high-quality images that can compete with and even outperform the latest GAN methods [5,9,19,23]. A variational framework for the likelihood estimation of diffusion models was introduced by Huang et al. [11]. Subsequently, Kingma et al. [14] proposed a Variational Diffusion Model that produces state-of-the-art results in likelihood estimation for image density.

Conditional Diffusion Probabilistic Models. In our work, we use diffusion models to solve the image segmentation problem as conditional generation, given the image. Conditional generation with diffusion models includes methods for class-conditioned generation, which is obtained by adding a class embedding to the timestep embedding [19]. In [4], a method for guiding the generative process in DDPM is present. This method allows the generation of images based on a given reference image without any additional learning. In the domain

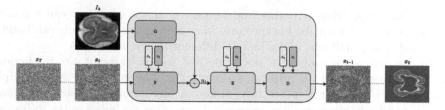

Fig. 1. The figure below illustrates our proposed method for multi-annotator segmentation. The input I_k image with the noisy segmentation map x_t is passed through our network iteratively T times in order to obtain an output segmentation map x_0. Each network receives the consensus level c as an embedding z_c as well as the time step data.

of super-resolution, the lower-resolution image is upsampled and then concatenated, channelwise, to the generated image at each iteration [10,21]. A similar approach passes the low-resolution images through a convolutional block [16] prior to the concatenation.

A previous study directly applied a diffusion model to generate a segmentation mask based on a conditioned input image [1]. Baranchuk et al. [2] extract features from a pretrained diffusion model for training a segmentation network, while our diffusion model generates the output mask. Compared to the diffusion-based image segmentation method of Wolleb et al. [26], our architecture differs in two main aspects: (i) the concatenation method of the condition signal, and (ii) an encoder that processes the conditioning signal. We also use a lower value of T, which reduces the running time.

3　Method

Our approach for binary segmentation with multi-annotators employs a diffusion model that is conditioned on the input image $I \in R^{W \times H}$, the step estimation t, and the consensus index c. The diffusion model updates its current estimate x_t iteratively, using the step estimation function ϵ_θ. See Fig. 1 for an illustration.

Given a set of C annotations $\{A_k^i\}_{i=1}^C$ associated with input sample I_k, we define the ground truth consensus map at level c to be

$$M_k^c[x,y] = \begin{cases} 1 & \sum_{i=1}^C A_k^i[x,y] \geq c, \\ 0 & \text{otherwise,} \end{cases} \tag{1}$$

During training, our algorithm iteratively samples a random level of the consensus $c \sim U[1, 2, ..., C]$ and an input-output pair (I_k, M_k^c). The iteration number $1 \leq t \leq T$ is sampled from a uniform distribution and X_T is sampled from a normal distribution.

We then compute x_t from X_T, M_k^c and t according to:

$$x_t = \sqrt{\bar{\alpha}_t} M_k^c + \sqrt{(1 - \bar{\alpha}_t)} X_T, X_T \sim N(0, I_{n \times n}). \tag{2}$$

Algorithm 1. Training Algorithm

Input T, $D = \{(I_k, M_k^1, ..., M_k^C)\}_k^K$
repeat
    sample $c \sim \{1, ..., C\}$
    sample $(I_k, M_k^c) \sim D'$
    sample $\epsilon \sim N(\mathbf{0}, \mathbf{I}_{n \times n})$
    sample $t \sim (\{1, ..., T\})$
    $z_c = LUT_c(c)$
    $z_t = LUT_t(t)$
    $\beta_t = \frac{10^{-4}(T-t)+2*10^{-2}(t-1)}{T-1}$
    $\alpha_t = 1 - \beta_t$
    $\bar{\alpha}_t = \prod_{s=0}^{t} \alpha_s$
    $x_t = \sqrt{\bar{\alpha}_t} M_k^c + \sqrt{1-\bar{\alpha}_t} \epsilon$
    $\nabla_\theta \|\epsilon - \epsilon_\theta(x_t, I_k, z_t, z_c)\|$
until convergence

Algorithm 2. Inference Algorithm

Input T, I
for $c = 1, ..., C$ **do**
    sample $x_{T_c} \sim N(\mathbf{0}, \mathbf{I}_{n \times n})$
    for $t = T, T-1, ..., 1$ **do**
        sample $z \sim N(\mathbf{0}, \mathbf{I}_{n \times n})$
        $z_c = LUT_c(c)$, $z_t = LUT_t(t)$
        $\beta_t = \frac{10^{-4}(T-t)+2*10^{-2}(t-1)}{T-1}$
        $\alpha_t = 1 - \beta_t$. $\bar{\alpha}_t = \prod_{s=0}^{t} \alpha_s$
        $\tilde{\beta}_t = \frac{1-\bar{\alpha}_{t-1}}{1-\bar{\alpha}_t} \beta_t$
        $\epsilon_t' = \frac{1-\alpha_t}{\sqrt{1-\bar{\alpha}_t}} \epsilon_\theta(\bar{x}_t, I, z_t, z_c)$
        $\bar{x}_{t-1_c} = \alpha_t^{-\frac{1}{2}} (x_t - \epsilon_t')$
        $x_{t-1_c} = \bar{x}_{t-1_c} + \mathbb{1}_{[t>1]} \tilde{\beta}_t^{\frac{1}{2}} z$
return $(\sum_{i=1}^{C} x_{0_i})/C$

where $\bar{\alpha}$ is a constant that defines the schedule of added noise.

The current step index t, and the consensus index c are integers that are translated to $z_t \in R^d$ and $z_c \in R^d$, respectively with a pair of lookup tables. The embeddings are passed to the different networks F, D and E.

In the next step, our algorithm encodes the input signal x_t with network F and encodes the condition image I_k with network G. We compute the conditioned signal $u_t = F(x_t, z_c, z_t) + G(I_k)$, and apply it to the networks E and D, where the output is the estimation of x_{t-1}.

$$\epsilon_\theta(x_t, I_k, z_t, z_c) = D(E(F(x_t, z_t, z_c) + G(I_k), z_t, z_c), z_t, z_c). \tag{3}$$

The loss function being minimized is:

$$E_{x_0, c, x_e, t, c}[\|\epsilon - \epsilon_\theta(x_t, I_k, z_t, z_c)\|^2]. \tag{4}$$

The training procedure is depicted in Algorithm 1. The total number of diffusion steps T is set by the user, and C is the number of different annotators in the dataset. Our model is trained using binary consensus maps (M_k^c) as the ground truth, where k is the sample id, and c is the consensus index.

The inference process is described in Algorithm 2. We sample our model for each consensus index, and then calculate the mean of all results to obtain our target, which is a soft-label map representing the annotator agreement. Mathematically, if the consensus maps are perfect, this is equivalent to assigning each image location with the fraction of annotations that consider this location to be part of the mask (if c annotators mark a pixel, it would appear in levels $1..c$). In Sect. 4, we compare our method with other variants and show that estimating the fraction map directly, using an identical diffusion model, is far inferior to estimating each consensus level separately and then averaging.

Employing Multiple Generations. Since calculating x_{t-1} during inference includes the addition of $\mathbb{1}_{[t>1]} \tilde{\beta}_t^{\frac{1}{2}} z$ where z is from a standard distribution, there is significant variability between different runs of the inference method on the same inputs, see Fig. 2(b).

In order to exploit this phenomenon, we run the inference algorithm multiple times, then average the results. This way, we stabilize the results of segmentation and improve performance, as demonstrated in Fig. 2(c). We use twenty-five generated instances in all experiments. In the ablation study, we quantify the gain of this averaging procedure.

Architecture. In this architecture, the U-Net's decoder D is conventional and its encoder is broken down into three networks: E, F, and G. The last encodes the input image, while F encodes the segmentation map of the current step x_t. The two processed inputs have the same spatial dimensionality and number of channels. Based on the success of residual connections [8], we sum these signals $F(x_t, z_t, z_c) + G(I)$. This sum then passes to the rest of the U-Net encoder E.

The input image encoder G is built from Residual in Residual Dense Blocks [24] (RRDBs), which combine multi-level residual connections without batch normalization layers. G has an input 2D-convolutional layer, an RRDB with a residual connection around it, followed by another 2D-convolutional layer, leaky RELU activation and a final 2D-convolutional output layer. F is a 2D-convolutional layer with a single-channel input and an output of L channels.

The encoder-decoder part of ϵ_θ, i.e., D and E, is based on U-Net, similarly to [19]. Each level is composed of residual blocks, and at resolution 16×16 and 8×8 each residual block is followed by an attention layer. The bottleneck contains two residual blocks with an attention layer in between. Each attention layer contains multiple attention heads.

The residual block is composed of two convolutional blocks, where each convolutional block contains group-norm, SiLU activation, and a 2D-convolutional layer. The residual block receives the time embedding through a linear layer, SiLU activation, and another linear layer. The result is then added to the output of the first 2D-convolutional block. Additionally, the residual block has a residual connection that passes all its content.

On the encoder side (network E), there is a downsample block after the residual blocks of the same depth, which is a 2D-convolutional layer with a stride of two. On the decoder side (network D), there is an upsample block after the residual blocks of the same depth, which is composed of the nearest interpolation that doubles the spatial size, followed by a 2D-convolutional layer. Each layer in the encoder has a skip connection to the decoder side.

4 Experiments

We conducted a series of experiments to evaluate the performance of our proposed method for multi-annotator prediction. Our experiments were carried out on datasets of the QUBIQ benchmark[1]. We compared the performance of our proposed method with several state-of-the-art methods.

---

[1] Quantification of Uncertainties in Biomedical Image Quantification Challenge in MICCAI20'- link.

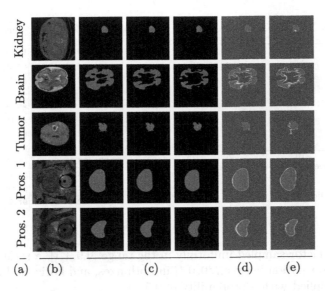

Fig. 2. Multiple segmentation results on all datasets of the QUBIQ benchmark. (a) dataset, (b) input image, (c) a subset of the obtained consensus maps for multiple runs with different consensus index on the same input, (d) average result, visualized by the 'bwr' color scale between 0 (blue) and 1 (red), and (e) ground truth. (Color figure online)

Datasets. The Quantification of Uncertainties in Biomedical Image Quantification Challenge (QUBIQ), is a recently available challenge dataset specifically for the evaluation of inter-rater variability. QUBIQ comprises four different segmentation datasets with CT and MRI modalities, including brain growth (one task, MRI, seven raters, 34 cases for training and 5 cases for testing), brain tumor (one task, MRI, three raters, 28 cases for training and 4 cases for testing), prostate (two subtasks, MRI, six raters, 33 cases for training and 15 cases for testing), and kidney (one task, CT, three raters, 20 cases for training and 4 cases for testing). Following [13], the evaluation is performed using the soft Dice coefficient with five threshold levels, set as (0.1, 0.3, 0.5, 0.7, 0.9).

Implementation Details. The number of diffusion steps in previous works was 1000 [9] and even 4000 [19]. The literature suggests that more is better [22]. In our experiments, we employ 100 diffusion steps, to reduce inference time.

The AdamW [18] optimizer is used in all our experiments. Based on the intuition that the more RRDB blocks, the better the results, we used as many blocks as we could fit on the GPU without overly reducing batch size.

Following [13], for all datasets of the QUBIQ benchmark the input image resolution, as well as the test image resolution, was 256×256. The experiments were performed with a batch size of four images and eight RRDB blocks. The network depth was seven, and the number of channels in each depth was $[L, L, L, 2L, 2L, 4L, 4L]$, with $L = 128$. The augmentations used were: random

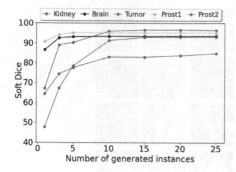

Fig. 3. Soft Dice vs. #generated images.

Table 1. QUBIQ soft Dice results.

Method	Kidney	Brain	Tumor	Prost1	Prost2
FCN	70.03	80.99	83.12	84.55	67.81
MCD	72.93	82.91	86.17	86.40	70.95
FPM	72.17	–	–	–	–
DAF	–	–	–	85.98	72.87
MV-UNet	70.65	81.77	84.03	85.18	68.39
LS-UNet	72.31	82.79	85.85	86.23	69.05
MH-UNet	73.44	83.54	86.74	87.03	75.61
MRNet	74.97	84.31	88.40	87.27	76.01
AMIS	68.53	74.09	92.95	91.64	21.91
DMISE	74.50	92.80	87.80	94.70	80.20
Ours	**96.58**	**93.81**	**93.16**	**95.21**	**84.62**

scaling by a factor sampled uniformly in the range $[0.9, 1.1]$, a rotation between 0 and 15°, translation between $[0, 0.1]$ in both axes, and horizontal and vertical flips, each applied with a probability of 0.5.

Results. We compare our method with FCN [17], MCD [6], FPM [27], DAF [25], MV-UNet [13], LS-UNet [12], MH-UNet [7], and MRNet [13].

We also compare with models that we train ourselves, using public code AMIS [20], and DMISE [26]. The first is trained in a scenario where each annotator is a different sample ("No annotator" variant of our ablation results below), and the second is trained on the consensus setting, similar to our method. As can be seen in Table 1, our method outperforms all other methods across all datasets of QUBIQ benchmark.

Ablation Study. We evaluate alternative training variants as an ablation study in Table 2. The "Annotator" variant, in which our model learns to produce each annotator binary segmentation map and then averages all the results to obtain the required soft-label map, achieves lower scores compared to the "Consensus" variant, which is our full method. The "No annotator" variant, where images were paired with random annotators without utilizing the annotator IDs, achieves a slightly lower average score compared to the "Annotator" variant. We also note that our "No annotator" variant outperforms the analog AMIS model in four out of five datasets, indicating that our architecture is somewhat preferable. In a third variant, our model learns to predict the soft-label map that denotes the fraction of annotators that mark each image location directly. Since this results in fewer generated images, we generate C times as many images per test sample. The score of this variant is also much lower than that of our method.

Next, we study the effect of the number of generated images on performance. The results can be seen in Fig. 3. In general, increasing the number of generated instances tends to improve performance. However, the number of runs required to reach optimal performance varies between classes. For example, for the Brain and the Prostate 1 datasets, optimal performance is achieved using 5 generated images, while on Prostate 2 the optimal performance is achieved using 25 gen-

Table 2. Ablation study showing soft Dice results for various alternative methods of training similar diffusion models.

Method	Kidney	Brain	Tumor	Prostate 1	Prostate 2
Annotator	96.13	89.88	92.51	93.89	76.89
No annotator	94.46	89.78	91.78	92.58	78.61
Soft-label	65.41	79.56	75.60	73.23	65.24
Consensus (our method)	**96.58**	**93.81**	**93.16**	**95.21**	**84.62**

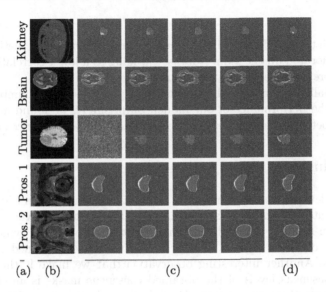

Fig. 4. Multiple segmentation results per number of generated images. (a) dataset, (b) input image, (c) results for 1, 5, 10, 25 generated images, and (d) ground truth.

erated images. Figure 4 depicts samples from multiple datasets and presents the progression as the number of generated images increases. As can be seen, as the number of generated images increases, the outcome becomes more and more similar to the target segmentation.

5 Discussion

In order to investigate the relationship between the annotator agreement and the performance of our model, we conducted an analysis by calculating the average Dice score between each pair of annotators across the entire dataset. The results of this pairwise Dice analysis can be found in Table 3, where higher mean-scores indicate a greater consensus among the annotators.

We observed that our proposed method demonstrated improved performance on datasets with higher agreement among annotators, specifically the kidney and

Table 3. Pairwise Dice scores per dataset.

Dataset	Mean score between pairs
Kidney	94.95
Brain	85.74
Tumor	90.65
Prostate 1	94.64
Prostate 2	89.91

prostate 1 datasets. Conversely, the performance of the other methods significantly deteriorated on the kidney dataset, leading to a lower correlation between the Dice score and the overall performance.

Additionally, we examined the relationship between the number of annotators and the performance of our model. Surprisingly, we found no significant correlation between the number of annotators and the performance of our model.

6 Conclusions

Shifting the level of consensus required to mark a region from very high to as low as one annotator, can be seen as creating a dynamic shift from a very conservative segmentation mask to a very liberal one. As it turns out, this dynamic is well-captured by diffusion models, which can be readily conditioned on the level of consensus. Another interesting observation that we make is that the mean (over the consensus level) of the obtained consensus masks is an effective soft mask. Applying these two elements together, we obtain state-of-the-art results on multiple binary segmentation tasks.

Acknowledgment. This project has received funding from the ISRAEL SCIENCE FOUNDATION (grant No. 2923/20) within the Israel Precision Medicine Partnership program. It was also supported by a grant from the Tel Aviv University Center for AI and Data Science (TAD).

References

1. Amit, T., Nachmani, E., Shaharbany, T., Wolf, L.: SegDiff: image segmentation with diffusion probabilistic models. arXiv preprint arXiv:2112.00390 (2021)
2. Baranchuk, D., Rubachev, I., Voynov, A., Khrulkov, V., Babenko, A.: Label-efficient semantic segmentation with diffusion models. arXiv preprint arXiv:2112.03126 (2021)
3. Chen, G., et al.: Automatic pathological lung segmentation in low-dose CT image using eigenspace sparse shape composition. IEEE Trans. Med. Imaging **38**(7), 1736–1749 (2019)
4. Choi, J., Kim, S., Jeong, Y., Gwon, Y., Yoon, S.: ILVR: conditioning method for denoising diffusion probabilistic models. arXiv preprint arXiv:2108.02938 (2021)

5. Dhariwal, P., Nichol, A.: Diffusion models beat GANs on image synthesis. In: Advances in Neural Information Processing Systems, vol. 34 (2021)
6. Gal, Y., Ghahramani, Z.: Dropout as a Bayesian approximation: representing model uncertainty in deep learning. In: International Conference on Machine Learning, pp. 1050–1059. PMLR (2016)
7. Guan, M., Gulshan, V., Dai, A., Hinton, G.: Who said what: modeling individual labelers improves classification. In: Proceedings of the AAAI Conference on Artificial Intelligence, vol. 32 (2018)
8. He, K., Zhang, X., Ren, S., Sun, J.: Deep residual learning for image recognition. In: Proceedings of the IEEE Conference on Computer Vision and Pattern Recognition, pp. 770–778 (2016)
9. Ho, J., Jain, A., Abbeel, P.: Denoising diffusion probabilistic models. Adv. Neural. Inf. Process. Syst. **33**, 6840–6851 (2020)
10. Ho, J., Saharia, C., Chan, W., Fleet, D.J., Norouzi, M., Salimans, T.: Cascaded diffusion models for high fidelity image generation. J. Mach. Learn. Res. **23**(47), 1–33 (2022)
11. Huang, C.W., Lim, J.H., Courville, A.C.: A variational perspective on diffusion-based generative models and score matching. In: Advances in Neural Information Processing Systems, vol. 34 (2021)
12. Jensen, M.H., Jørgensen, D.R., Jalaboi, R., Hansen, M.E., Olsen, M.A.: Improving uncertainty estimation in convolutional neural networks using inter-rater agreement. In: Shen, D., et al. (eds.) MICCAI 2019, Part IV. LNCS, vol. 11767, pp. 540–548. Springer, Cham (2019). https://doi.org/10.1007/978-3-030-32251-9_59
13. Ji, W., et al.: Learning calibrated medical image segmentation via multi-rater agreement modeling. In: Proceedings of the IEEE/CVF Conference on Computer Vision and Pattern Recognition, pp. 12341–12351 (2021)
14. Kingma, D.P., Salimans, T., Poole, B., Ho, J.: Variational diffusion models. arXiv preprint arXiv:2107.00630 (2021)
15. Kohl, S., et al.: A probabilistic U-Net for segmentation of ambiguous images. In: Advances in Neural Information Processing Systems, vol. 31 (2018)
16. Li, H., et al.: SRDIFF: single image super-resolution with diffusion probabilistic models. Neurocomputing **479**, 47–59 (2022)
17. Long, J., Shelhamer, E., Darrell, T.: Fully convolutional networks for semantic segmentation. In: Proceedings of the IEEE Conference on Computer Vision and Pattern Recognition, pp. 3431–3440 (2015)
18. Loshchilov, I., Hutter, F.: Decoupled weight decay regularization. arXiv preprint arXiv:1711.05101 (2017)
19. Nichol, A.Q., Dhariwal, P.: Improved denoising diffusion probabilistic models. In: International Conference on Machine Learning, pp. 8162–8171. PMLR (2021)
20. Rahman, A., Valanarasu, J.M.J., Hacihaliloglu, I., Patel, V.M.: Ambiguous medical image segmentation using diffusion models. In: Proceedings of the IEEE/CVF Conference on Computer Vision and Pattern Recognition, pp. 11536–11546 (2023)
21. Saharia, C., Ho, J., Chan, W., Salimans, T., Fleet, D.J., Norouzi, M.: Image super-resolution via iterative refinement. arXiv preprint arXiv:2104.07636 (2021)
22. San-Roman, R., Nachmani, E., Wolf, L.: Noise estimation for generative diffusion models. arXiv preprint arXiv:2104.02600 (2021)
23. Sohl-Dickstein, J., Weiss, E., Maheswaranathan, N., Ganguli, S.: Deep unsupervised learning using nonequilibrium thermodynamics. In: International Conference on Machine Learning, pp. 2256–2265. PMLR (2015)

24. Wang, X., et al.: ESRGAN: enhanced super-resolution generative adversarial networks. In: Proceedings of the European Conference on Computer Vision (ECCV) Workshops, pp. 0–0 (2018)
25. Wang, Y., et al.: Deep attentional features for prostate segmentation in ultrasound. In: Frangi, A.F., Schnabel, J.A., Davatzikos, C., Alberola-López, C., Fichtinger, G. (eds.) MICCAI 2018, Part IV. LNCS, vol. 11073, pp. 523–530. Springer, Cham (2018). https://doi.org/10.1007/978-3-030-00937-3_60
26. Wolleb, J., Sandkühler, R., Bieder, F., Valmaggia, P., Cattin, P.C.: Diffusion models for implicit image segmentation ensembles (2021)
27. Zhou, Y., Xie, L., Shen, W., Wang, Y., Fishman, E.K., Yuille, A.L.: A fixed-point model for pancreas segmentation in abdominal CT scans. In: Descoteaux, M., Maier-Hein, L., Franz, A., Jannin, P., Collins, D.L., Duchesne, S. (eds.) MICCAI 2017, Part I. LNCS, vol. 10433, pp. 693–701. Springer, Cham (2017). https://doi.org/10.1007/978-3-319-66182-7_79

Anti-adversarial Consistency Regularization for Data Augmentation: Applications to Robust Medical Image Segmentation

Hyuna Cho, Yubin Han, and Won Hwa Kim[✉]

Pohang University of Science and Technology (POSTECH), Pohang, South Korea
{hyunacho,yubin,wonhwa}@postech.ac.kr

Abstract. Modern deep learning methods for semantic segmentation require labor-intensive labeling for large-scale datasets with dense pixel-level annotations. Recent data augmentation methods such as dropping, mixing image patches, and adding random noises suggest effective ways to address the labeling issues for natural images. However, they can only be restrictively applied to medical image segmentation as they carry risks of distorting or ignoring the underlying clinical information of local regions of interest in an image. In this paper, we propose a novel data augmentation method for medical image segmentation without losing the semantics of the key objects (e.g., polyps). This is achieved by perturbing the objects with quasi-imperceptible adversarial noises and training a network to expand discriminative regions with a guide of anti-adversarial noises. Such guidance can be realized by a consistency regularization between the two contrasting data, and the strength of regularization is automatically and adaptively controlled considering their prediction uncertainty. Our proposed method significantly outperforms various existing methods with high sensitivity and Dice scores and extensive experiment results with multiple backbones on two datasets validate its effectiveness.

Keywords: Adversarial attack and defense · Data augmentation · Semantic segmentation

1 Introduction

Semantic segmentation aims to segment objects in an image by classifying each pixel into an object class. Training a deep neural network (DNN) for such a task is known to be data-hungry, as labeling dense pixel-level annotations requires laborious and expensive human efforts in practice [23, 32]. Furthermore, semantic segmentation in medical imaging suffers from privacy and data sharing issues [13, 35] and a lack of experts to secure accurate and clinically meaningful regions of interest (ROIs). This data shortage problem causes overfitting for training DNNs, resulting in the networks being biased by outliers and ignorant of unseen data.

To alleviate the sample size and overfitting issues, diverse data augmentations have been recently developed. For example, CutMix [31] and CutOut [4] augment images by dropping random-sized image patches or replacing the removed

H. Greenspan et al. (Eds.): MICCAI 2023, LNCS 14223, pp. 555–566, 2023.
https://doi.org/10.1007/978-3-031-43901-8_53

regions with a patch from another image. Random Erase [33] extracts noise from a uniform distribution and injects it into patches. Geometric transformations such as Elastic Transformation [26] warp images and deform the original shape of objects. Alternatively, feature perturbation methods augment data by perturbing data in feature space [7,22] and logit space [9].

Although these augmentation approaches have been successful for natural images, their usage for medical image semantic segmentation is quite restricted as objects in medical images contain non-rigid morphological characteristics that should be sensitively preserved. For example, basalioma (e.g., pigmented basal cell carcinoma) may look similar to malignant melanoma or mole in terms of color and texture [6,20], and early-stage colon polyps are mostly small and indistinguishable from background entrail surfaces [14]. In these cases, the underlying clinical features of target ROIs (e.g., polyp, tumor and cancer) can be distorted if regional colors and textures are modified with blur-based augmentations or geometric transformations. Also, cut-and-paste and crop-based methods carry risks of dropping or distorting key objects such that expensive pixel-level annotations could not be properly used. Considering the ROIs are usually small and under-represented compared to the backgrounds, the loss of information may cause a fatal class imbalance problem in semantic segmentation tasks.

In these regards, we tackle these issues with a novel augmentation method without distorting the semantics of objects in image space. This can be achieved by slightly but effectively perturbing target objects with adversarial noises at the object level. We first augment hard samples with adversarial attacks [18] that deceive a network and defend against such attacks with anti-adversaries. Specifically, multi-step adversarial noises are injected into ROIs to *maximize* loss and induce false predictions. Conversely, anti-adversaries are obtained with anti-adversarial perturbations that *minimize* a loss which eventually become easier samples to predict. We impose consistency regularization between these contrasting samples by evaluating their prediction ambiguities via supervised losses with true labels. With this regularization, the easier samples provide adaptive guidance to the misclassified data such that the difficult (but object-relevant) pixels can be gradually integrated into the correct prediction. From active learning perspective [12,19], as vague samples near the decision boundary are augmented and trained, improvement on a downstream prediction task is highly expected.

We summarize our main contributions as follows: **1)** We propose a novel online data augmentation method for semantic segmentation by imposing object-specific consistency regularization between anti-adversarial and adversarial data. **2)** Our method provides a flexible regularization between differently perturbed data such that a vulnerable network is effectively trained on challenging samples considering their ambiguities. **3)** Our method preserves underlying morphological characteristics of medical images by augmenting data with quasi-imperceptible perturbation. As a result, our method significantly improves sensitivity and Dice scores over existing augmentation methods on Kvasir-Seg [11] and ETIS-Larib Polyp DB [25] benchmarks for medical image segmentation.

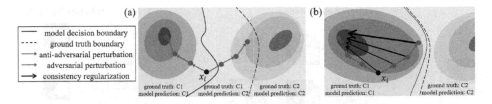

Fig. 1. (a) Conceptual illustration of the adversarial attack (red) and anti-adversarial perturbation (blue) in the latent feature space. Given a predicted sample embedding x_i, let its true label be a class 1 (C1). The adversarial attack sends the data point toward class 2 (C2) whereas the anti-adversarial perturbation increases its classification score. (b) Adaptive anti-adversarial consistency regularization (AAC) between the adversarially attacked data and the anti-adversary. (Color figure online)

2 Preliminary: Adversarial Attack and Anti-adversary

Adversarial attack is an input perturbation method that adds quasi-imperceptible noises into images to deceive a DNN. Given an image x, let μ be a noise bounded by l_∞-norm. While the difference between x and the perturbed sample $x' = x + \mu$ is hardly noticeable to human perception, a network $f_\theta(\cdot)$ can be easily fooled (i.e., $f_\theta(x) \neq f_\theta(x + \mu)$) as the μ pushes x' across the decision boundary.

To fool a DNN, Fast Gradient Sign Method (FGSM) [8] perturbs x toward *maximizing* a loss function L by defining a noise μ as the sign of loss derivative with respect to x as follows: $x' = x + \epsilon\, \text{sign}(\nabla_x L)$, where ϵ controls the magnitude of perturbation. The authors in [18] proposed an extension of FGSM, i.e., Projected Gradient Descent (PGD), which is an iterative adversarial attack that also finds x' with a higher loss. Given an iteratively perturbed sample x'_t at t-th perturbation where $x'_0 = x$, the x'_t of PGD is defined as $x'_t = \prod(x'_{t-1} + \epsilon\, \text{sign}(\nabla_x L))$ for T perturbation steps.

Recently, anti-adversarial methods were proposed for the benign purpose to defend against such attacks. The work in [15] used an anti-adversarial class activation map to identify objects and the authors in [1] proposed an anti-adversary layer to handle adversaries. In contrast to adversarial attacks, these works find μ that *minimizes* a loss to make easier samples to predict. Figure 1a shows multi-step adversarial and anti-adversarial perturbations in the latent space. To increase a classification score, the anti-adversarial noises move data away from the decision boundary, which is the opposite direction of the adversarial perturbations.

3 Method

Let $\{X_i\}_{i=1}^N$ be an image set with N samples each paired with corresponding ground truth pixel-level annotations Y_i. Our proposed method aims to 1) generate realistic images with adversarial attacks and 2) train a segmentation model $f_\theta(X_i) = Y_i$ for robust semantic segmentation with anti-adversarial consistency regularization (AAC). Figure 2 shows the overall training scheme with three phases: **1)** online data augmentation, **2)** computing adaptive AAC between

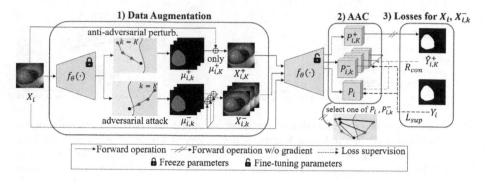

Fig. 2. An overview of our training scheme. Adversarial and anti-adversarial perturbations are iteratively performed for the objects of a given image X_i. Adversarial noise $\mu_{i,k}^-$ moves X_i across the decision boundary, whereas anti-adversarial noise $\mu_{i,k}^+$ pushes X_i away from the boundary. Downstream consistency regularization loss R_{con} minimizes the gap between adversaries $\{X_{i,k}^-\}_{k=1}^K$ and anti-adversary $X_{i,K}^+$.

differently perturbed samples, and **3)** updating the segmentation model using the loss from the augmented and original data. First, we generate plausible images with iterative adversarial and anti-adversarial perturbations. We separate the roles of perturbed data: adversaries are used as training samples and anti-adversaries are used to provide guidance (i.e., pseudo-labels) to learn the adversaries. Specifically, consistency regularization is imposed between these contrasting data by adaptively controlling the regularization magnitude in the next phase. Lastly, considering each sample's ambiguity, the network parameters θ are updated for learning the adversaries along with the given data so that discriminative regions are robustly expanded for challenging samples.

Data Augmentation with Object-Targeted Adversarial Attack. In many medical applications, false negatives (i.e., failing to diagnose a critical disease) are much more fatal than false positives. To deal with these false negatives, we mainly focus on training a network to learn diverse features at target ROIs (e.g., polyps) where disease-specific variations exist. To do so, we first exclude the background and perturb only the objects in the given image. Given o as the target object class and (p, q) as a pixel coordinate, a masked object is defined as $\hat{X}_i = \{(p, q) | X_{i_{(p,q)}} = o\}$.

As in PGD [18], we perform iterative perturbations on the $\hat{X}_i$ for K steps. Given $\hat{X}_{i,k}$ as a perturbed sample at k-th step ($k = 1, ..., K$), the adversarial and anti-adversarial perturbations use the same initial image as $X_{i,0}^- = \hat{X}_i$ and $X_{i,0}^+ = \hat{X}_i$, respectively. With this input, the iterative adversarial attack is defined as

$$X_{i,k+1}^- = X_{i,k}^- + \mu_{i,k}^- = X_{i,k}^- + \epsilon \, \text{sign}(\nabla_{X_{i,k}^-} L(f_\theta(X_{i,k}^-), Y_i)) \tag{1}$$

where $\mu_{i,k}^- = argmax_\mu L(f_\theta(X_{i,k+1}^-), Y_i)$ is a quasi-imperceptible adversarial noise that fools $f_\theta(\cdot)$ and ϵ is a perturbation magnitude that limits the noise (i.e., $|\mu_{(p,q)}| \le \epsilon$, s.t. $(p,q) \in \hat{X}_i$). Similarly, iterative anti-adversarial perturbation is defined as

$$X_{i,k+1}^+ = X_{i,k}^+ + \mu_{i,k}^+ = X_{i,k}^+ + \epsilon \, \text{sign}(\nabla_{X_{i,k}^+} L(f_\theta(X_{i,k}^+), Y_i)). \quad (2)$$

In contrast to the adversarial attack in Eq. 1, the anti-adversarial noise $\mu_{i,k}^+ = argmin_\mu L(f_\theta(X_{i,k+1}^+), Y_i)$ manipulates samples to increase the classification score.

Note that, generating noises and images are online and training-free as the loss derivatives are calculated with freezed network parameters. The adversaries $X_{i,1}^-, ..., X_{i,K}^-$ are used as additional training samples so that the network includes the non-discriminative yet object-relevant features for the prediction. On the other hand, as the anti-adversaries are sufficiently discriminative, we do not use them as training samples. Instead, only the K-th anti-adversary $X_{i,K}^+$ (i.e., the most perturbed sample with the lowest loss) is used for downstream consistency regularization to provide informative guidance to the adversaries.

Computing Adaptive Consistency Toward Anti-adversary. Let X_i' be either X_i or $X_{i,k}^-$. As shown in Fig. 1b, consistency regularization is imposed between the anti-adversary $X_{i,K}^+$ and X_i' to reduce the gap between samples with different prediction uncertainties. The weight of regularization between X_i' and $X_{i,K}^+$ is automatically determined by evaluating the gap in their prediction quality via supervised losses with ground truth Y_i as

$$w(X_i', X_{i,K}^+) = max(\frac{1}{2}, \frac{l(f_\theta(X_i'), Y_i)}{l(f_\theta(X_i'), Y_i) + l(f_\theta(X_{i,K}^+), Y_i)}) = max(\frac{1}{2}, \frac{l(P_i', Y_i)}{l(P_i', Y_i) + l(P_{i,K}^+, Y_i)}), \quad (3)$$

where $l(\cdot)$ is Dice loss [28] and P_i is the output of $f_\theta(\cdot)$ for X_i. Specifically, if X_i' is a harder sample to predict than $X_{i,K}^+$, i.e., $l(P_i', Y_i) > l(P_{i,K}^+, Y_i)$, the weight gets larger, and thus consistency regularization is intensified between the images.

Training a Segmentation Network. Let $\hat{Y}_{i,K}^+$ be a segmentation outcome, i.e., one-hot encoded pseudo-label from the network output $P_{i,K}^+$ of anti-adversary $X_{i,K}^+$. Given X_i and $\{X_{i,k}^-\}_{k=1}^K$ as training data, the supervised segmentation loss L_{sup} and the consistency regularization R_{con} are defined as

$$L_{sup} = \frac{1}{N} \sum_{i=1}^N l(P_i, Y_i) + \frac{1}{NK} \sum_{i=1}^N \sum_{k=1}^K l(P_{i,k}^-, Y_i) \quad \text{and} \quad (4)$$

$$R_{con} = \frac{1}{N} \sum_{i=1}^N w(X_i, X_{i,K}^+) l(P_i, \hat{Y}_{i,K}^+) + \frac{1}{NK} \sum_{i=1}^N \sum_{k=1}^K w(X_{i,k}^-, X_{i,K}^+) l(P_{i,k}^-, \hat{Y}_{i,K}^+). \quad (5)$$

Using the pseudo-label from anti-adversary as *a perturbation of the ground truth*, the network is supervised by diverse and realistic labels that contain auxiliary information that the originally given labels do not provide. With a hyperparameter α, the whole training loss $\mathcal{L} = L_{sup} + \alpha R_{con}$ is minimized via backpropagation to optimize the network parameters for semantic segmentation.

4   Experiments

4.1   Experimental Setup

Dataset. We conducted experiments on two representative public polyp segmentation datasets: Kvasir-SEG [11] and ETIS-Larib Polyp DB [25] (ETIS). Both are comprised of two classes: polyp and background. They provide 1000/196 (Kvasir-SEG/ETIS) input-label pairs in total and we split train/validation/test sets into 80%/10%/10% as in [5,10,24,27,29]. The images of Kvasir-SEG were resized to 512×608 ($H \times W$) and that of ETIS was set with 966×1255 resolution.

Implementation. We implemented our method on Pytorch framework with 4 NVIDIA RTX A6000 GPUs. Adam optimizer with learning rates of 4e-3/1e-4 (Kavsir-SEG/ETIS) were used for 200 epochs with a batch size of 16. We set the number of perturbation steps K as 10 and the magnitude of perturbation ϵ as 0.001. The weight α for R_{con} was set to 0.01.

Baselines. Along with conventional augmentation methods (i.e., random horizontal and vertical flipping denoted as 'Basic' in Table 1), recent methods such as CutMix [31], CutOut [4], Elastic Transform [26], Random Erase [33], DropBlock [7], Gaussian Noise Training (GNT) [22], Logit Uncertainty (LU) [9] and Tumor Copy-Paste (TumorCP) [30] were used as baselines. Their hyperparameters were adopted from the original papers. The Basic augmentation was used in all methods including ours by default. For the training, we used K augmented images with the given images for all baselines as in ours for a fair comparison.

Evaluation. To verify the effectiveness of our method, evaluations are conducted using various popular backbone architectures such as U-Net [21], U-Net++ [34], LinkNet [2], and DeepLabv3+ [3]. As the evaluation metric, mean Intersection over Union (mIoU) and mean Dice coefficient (mDice) are used for all experiments on test sets. Additionally, we provide recall and precision scores to offer a detailed analysis of class-specific misclassification performance.

4.2   Comparison with Existing Methods

As shown in Table 1, our method outperforms all baselines for all settings by at most 10.06%p and 5.98%p mIoU margin on Kvasir-SEG and ETIS, respectively. Moreover, in Fig. 3, our method with U-Net on Kvasir-SEG surpasses the baselines by ~8.2%p and ~7.2%p in precision and recall, respectively. Note that, all baselines showed improvements in most cases. However, our method

Table 1. Performance comparison with existing data augmentation methods.

Method	U-Net	U-Net++	LinkNet	DeepLabv3+	U-Net	U-Net++	LinkNet	DeepLabv3+
	mIoU				mDice			
Kvasir-SEG								
No Aug.	81.76	63.13	73.92	85.75	86.29	74.75	85.00	89.65
Basic	89.73	89.94	90.52	90.43	94.63	94.70	95.02	94.97
CutMix [31]	89.84	90.80	90.56	90.60	94.65	95.18	95.05	95.08
CutOut [4]	88.63	88.63	89.52	90.70	93.29	93.97	94.47	95.12
Elastic Trans. [26]	89.71	88.34	89.89	91.44	94.57	93.81	94.63	95.40
Random Erase [33]	88.73	89.45	90.72	90.94	94.03	94.43	95.14	95.25
DropBlock [7]	86.88	88.40	90.75	90.22	92.98	93.84	95.15	94.86
GNT [22]	82.37	88.71	90.32	90.88	90.36	94.02	94.91	95.22
LU [9]	89.51	90.84	87.71	90.52	94.46	95.20	93.45	95.02
TumorCP [30]	90.92	91.18	90.87	91.65	95.24	95.39	95.22	95.64
Ours	**92.43**	**91.43**	**91.51**	**92.42**	**96.07**	**95.52**	**95.57**	**96.06**
ETIS								
No Aug.	83.80	83.96	82.18	82.52	91.18	91.28	90.22	90.43
Basic	86.08	87.02	84.75	82.69	92.52	93.06	91.75	90.52
CutMix [31]	86.03	86.78	82.52	82.20	92.49	92.92	90.42	90.83
CutOut [4]	84.37	84.90	86.55	84.50	91.52	91.83	92.79	91.60
Elastic Trans. [26]	85.12	85.10	86.55	84.13	91.96	91.95	92.79	91.38
Random Erase [33]	85.20	84.12	82.52	83.63	92.01	91.37	90.43	90.83
DropBlock [7]	82.52	85.33	82.49	84.27	90.42	92.08	90.41	91.46
GNT [22]	85.36	84.94	84.19	84.55	92.10	91.86	91.42	91.63
LU [9]	82.52	82.52	82.43	84.33	90.43	90.42	90.37	91.50
TumorCP [30]	82.59	84.99	85.69	85.32	90.46	91.89	92.30	92.08
Ours	**88.35**	**87.62**	**88.41**	**85.58**	**93.81**	**93.40**	**93.85**	**92.23**

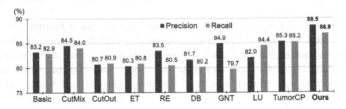

Fig. 3. Comparisons of precision and recall on the test set of Kvasir-SEG with U-Net.

performed better even compared with the TumorCP which uses seven differ-ent augmentations methods together for tumor segmentation. This is because our method *preserves the semantics of the key ROIs* with small but effective noises unlike geometric transformations [26,30], drop and cut-and-paste-based methods [4,7,30,31,33]. Also, as we augment uncertain samples that deliber-ately deceive a network as in Active Learning [12,16], our method is able to sensitively include the challenging (but ROI-relevant) features into prediction, unlike existing noise-based methods that extract noises from known distributions [9,22,30].

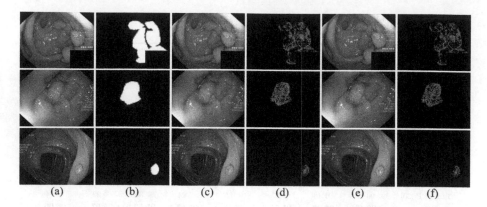

(a)   (b)   (c)   (d)   (e)   (f)

Fig. 4. (a) Input data (b) Ground truth label (c) Adversarially perturbed data (d) Adversarial noise (e) Anti-adversarially perturbed data (f) Anti-adversarial noise.

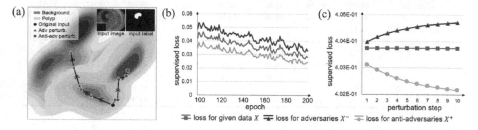

Fig. 5. (a) Density of pixel embeddings (orange and green), a sample (black) and its perturbations (red and blue) in 2D feature space via t-SNE. (b) and (c) show loss flow comparisons between a sample X and its perturbations X^- and X^+. Supervised losses are compared w.r.t. epochs and perturbation steps, and the anti-adversaries (blue) always demonstrate the lowest loss (i.e., closer to the ground truth). (Color figure online)

4.3   Analysis on Anti-adversaries and Adversaries

In Fig. 4, we visualize data augmentation results with (anti-) adversarial perturbations on Kvasir-SEG dataset. The perturbed data (c and e) are the addition of noise (d and f) to the given data (a), respectively. Interestingly, while adversaries (c) and anti-adversaries (e) are visually indistinguishable, they induce totally opposite model decisions towards different classes. In Fig. 5, we qualitatively and quantitatively compared their effects via visualizing perturbation trajectories in the feature space projected with t-SNE [17] and comparing their supervision losses. In Fig. 5a, the adversarial attacks send a pixel embedding of a polyp class to the background class, anti-adversarial perturbations push it towards the true class with a higher classification score. Also, loss comparisons in Fig. 5b and 5c demonstrate that the anti-adversaries (blue) are consistently easier to predict than the given data (grey) and adversaries (red) during the training and their differences get larger as the perturbations are iterated.

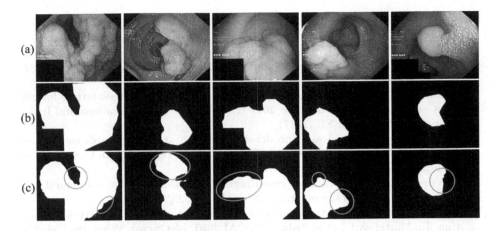

Fig. 6. Comparison of ground truths and pseudo labels from anti-adversaries. (a) Input data (b) Ground truth label (c) Pseudo-label $\hat{Y}_K^+$. (Color figure online)

These results confirm that the anti-adversaries send their pseudo label $\hat{Y}_K^+$ closer to the ground truth with a slight change. Therefore, they can be regarded as a perturbation of the ground truth that contain a potential to provide additional information to train a network on the adversaries. We empirically show that $\hat{Y}_K^+$ is able to provide such auxiliary information that the true labels do not provide, as our method performs better *with* R_{con} (i.e., $\mathcal{L} = L_{sup} + \alpha R_{con}$, 92.43% mIoU) than the case *without* R_{con} (i.e., $\mathcal{L} = L_{sup}$, 92.15% mIoU) using U-Net on Kvasir-SEG. Training samples in Fig. 6 show that the pseudo-labels $\hat{Y}_K^+$ can capture detailed abnormalities (marked in red circles) which are not included in the ground truths. Moreover, as the AAC considers sample-level ambiguity, the effect from $\hat{Y}_K^+$ is sensitively controlled and a network can selectively learn the under-trained yet object-relevant features from adversarial samples.

5   Conclusion

We present a novel data augmentation method for semantic segmentation using a flexible anti-adversarial consistency regularization. In particular, our method is tailored for medical images that contain small and underrepresented key objects such as a polyp and tumor. With object-level perturbations, our method effectively expands discriminative regions on challenging samples while preserving the morphological characteristics of key objects. Extensive experiments with various backbones and datasets confirm the effectiveness of our method.

Acknowledgement. This research was supported by IITP-2022-0-00290 (50%), IITP-2019-0-01906 (AI Graduate Program at POSTECH, 10%), IITP-2022-2020-0-01461 (ITRC, 10%) and NRF-2022R1A2C2092336 (30%) funded by the Korean government (MSIT).

References

1. Alfarra, M., Pérez, J.C., et al.: Combating adversaries with anti-adversaries. In: Proceedings of the AAAI Conference on Artificial Intelligence, vol. 36, pp. 5992–6000 (2022)
2. Chaurasia, A., Culurciello, E.: Linknet: exploiting encoder representations for efficient semantic segmentation. In: 2017 IEEE Visual Communications and Image Processing, pp. 1–4. IEEE (2017)
3. Chen, L.C., Zhu, Y., et al.: Encoder-decoder with atrous separable convolution for semantic image segmentation. In: Proceedings of the European Conference on Computer Vision, pp. 801–818 (2018)
4. DeVries, T., Taylor, G.W.: Improved regularization of convolutional neural networks with cutout. arXiv preprint arXiv:1708.04552 (2017)
5. Fan, D.-P., et al.: PraNet: parallel reverse attention network for polyp segmentation. In: Martel, A.L., et al. (eds.) MICCAI 2020. LNCS, vol. 12266, pp. 263–273. Springer, Cham (2020). https://doi.org/10.1007/978-3-030-59725-2_26
6. Felder, S., Rabinovitz, H., et al.: Dermoscopic pattern of pigmented basal cell carcinoma, blue-white variant. Dermatol. Surg. **32**(4), 569–570 (2006)
7. Ghiasi, G., Lin, T.Y., et al.: Dropblock: a regularization method for convolutional networks. In: Advances in Neural Information Processing Systems, vol. 31 (2018)
8. Goodfellow, I.J., Shlens, J., et al.: Explaining and harnessing adversarial examples. arXiv preprint arXiv:1412.6572 (2014)
9. Hu, Y., Zhong, Z., Wang, R., Liu, H., Tan, Z., Zheng, W.-S.: Data augmentation in logit space for medical image classification with limited training data. In: de Bruijne, M., et al. (eds.) MICCAI 2021. LNCS, vol. 12905, pp. 469–479. Springer, Cham (2021). https://doi.org/10.1007/978-3-030-87240-3_45
10. Jha, D., Riegler, M.A., et al.: Doubleu-net: a deep convolutional neural network for medical image segmentation. In: IEEE International Symposium on Computer-Based Medical Systems, pp. 558–564. IEEE (2020)
11. Jha, D., et al.: Kvasir-SEG: a segmented polyp dataset. In: Ro, Y.M., et al. (eds.) MMM 2020. LNCS, vol. 11962, pp. 451–462. Springer, Cham (2020). https://doi.org/10.1007/978-3-030-37734-2_37
12. Juszczak, P., Duin, R.P.: Uncertainty sampling methods for one-class classifiers. In: Proceedings of ICML-03, Workshop on Learning with Imbalanced Data Sets II, pp. 81–88 (2003)
13. Kaissis, G.A., Makowski, M.R., et al.: Secure, privacy-preserving and federated machine learning in medical imaging. Nat. Mach. Intell. **2**(6), 305–311 (2020)
14. Lee, J.Y., Jeong, J., et al.: Real-time detection of colon polyps during colonoscopy using deep learning: systematic validation with four independent datasets. Sci. Rep. **10**(1), 8379 (2020)
15. Lee, J., Kim, E., et al.: Anti-adversarially manipulated attributions for weakly and semi-supervised semantic segmentation. In: Proceedings of the IEEE/CVF Conference on Computer Vision and Pattern Recognition, pp. 4071–4080 (2021)
16. Lewis, D.D., Catlett, J.: Heterogeneous uncertainty sampling for supervised learning. In: Machine Learning Proceedings 1994, pp. 148–156. Elsevier (1994)
17. Van der Maaten, L., Hinton, G.: Visualizing data using t-SNE. J. Mach. Learn. Res. **9**(11) (2008)
18. Madry, A., Makelov, A., et al.: Towards deep learning models resistant to adversarial attacks. arXiv preprint arXiv:1706.06083 (2017)

19. Nguyen, V.L., Shaker, M.H., et al.: How to measure uncertainty in uncertainty sampling for active learning. Mach. Learn. **111**(1), 89–122 (2022)
20. Parmar, B., Talati, B.: Automated melanoma types and stages classification for dermoscopy images. In: 2019 Innovations in Power and Advanced Computing Technologies, vol. 1, pp. 1–7. IEEE (2019)
21. Ronneberger, O., Fischer, P., Brox, T.: U-Net: convolutional networks for biomedical image segmentation. In: Navab, N., Hornegger, J., Wells, W.M., Frangi, A.F. (eds.) MICCAI 2015. LNCS, vol. 9351, pp. 234–241. Springer, Cham (2015). https://doi.org/10.1007/978-3-319-24574-4_28
22. Rusak, E., et al.: A simple way to make neural networks robust against diverse image corruptions. In: Vedaldi, A., Bischof, H., Brox, T., Frahm, J.-M. (eds.) ECCV 2020. LNCS, vol. 12348, pp. 53–69. Springer, Cham (2020). https://doi.org/10.1007/978-3-030-58580-8_4
23. Saleh, F.S., Aliakbarian, M.S., et al.: Effective use of synthetic data for urban scene semantic segmentation. In: Proceedings of the European Conference on Computer Vision, pp. 84–100 (2018)
24. Sanderson, E., Matuszewski, B.J.: FCN-transformer feature fusion for polyp segmentation. In: Yang, G., Aviles-Rivero, A., Roberts, M., Schönlieb, C.B. (eds.) MIUA 2022. LNCS, vol. 13413, pp. 892–907. Springer, Cham (2022). https://doi.org/10.1007/978-3-031-12053-4_65
25. Silva, J., Histace, A., et al.: Toward embedded detection of polyps in WCE images for early diagnosis of colorectal cancer. Int. J. Comput. Assist. Radiol. Surg. **9**, 283–293 (2014)
26. Simard, P.Y., Steinkraus, D., et al.: Best practices for convolutional neural networks applied to visual document analysis. In: International Conference on Document Analysis and Recognition, vol. 3 (2003)
27. Srivastava, A., Jha, D., et al.: MSRF-Net: a multi-scale residual fusion network for biomedical image segmentation. IEEE J. Biomed. Health Inform. **26**(5), 2252–2263 (2021)
28. Sudre, C.H., Li, W., Vercauteren, T., Ourselin, S., Jorge Cardoso, M.: Generalised dice overlap as a deep learning loss function for highly unbalanced segmentations. In: Cardoso, M.J., et al. (eds.) DLMIA/ML-CDS -2017. LNCS, vol. 10553, pp. 240–248. Springer, Cham (2017). https://doi.org/10.1007/978-3-319-67558-9_28
29. Wang, J., Huang, Q., et al.: Stepwise feature fusion: Local guides global. In: Wang, L., Dou, Q., Fletcher, P.T., Speidel, S., Li, S. (eds.) MICCAI 2022. LNCS, vol. 13433, pp. 110–120. Springer, Cham (2022). https://doi.org/10.1007/978-3-031-16437-8_11
30. Yang, J., Zhang, Y., Liang, Y., Zhang, Y., He, L., He, Z.: TumorCP: a simple but effective object-level data augmentation for tumor segmentation. In: de Bruijne, M., et al. (eds.) MICCAI 2021. LNCS, vol. 12901, pp. 579–588. Springer, Cham (2021). https://doi.org/10.1007/978-3-030-87193-2_55
31. Yun, S., Han, D., et al.: Cutmix: regularization strategy to train strong classifiers with localizable features. In: Proceedings of the IEEE/CVF International Conference on Computer Vision, pp. 6023–6032 (2019)
32. Zhao, X., Vemulapalli, R., et al.: Contrastive learning for label efficient semantic segmentation. In: Proceedings of the IEEE/CVF International Conference on Computer Vision, pp. 10623–10633 (2021)
33. Zhong, Z., Zheng, L., et al.: Random erasing data augmentation. In: Proceedings of the AAAI Conference on Artificial Intelligence, vol. 34, pp. 13001–13008 (2020)

34. Zhou, Z., Rahman Siddiquee, M.M., Tajbakhsh, N., Liang, J.: UNet++: a nested
U-net architecture for medical image segmentation. In: Stoyanov, D., et al. (eds.)
DLMIA/ML-CDS -2018. LNCS, vol. 11045, pp. 3–11. Springer, Cham (2018).
https://doi.org/10.1007/978-3-030-00889-5_1

35. Ziller, A., Usynin, D., et al.: Medical imaging deep learning with differential privacy.
Sci. Rep. **11**(1), 1–8 (2021)

SimPLe: Similarity-Aware Propagation Learning for Weakly-Supervised Breast Cancer Segmentation in DCE-MRI

Yuming Zhong and Yi Wang[✉]

Smart Medical Imaging, Learning and Engineering (SMILE) Lab, Medical
UltraSound Image Computing (MUSIC) Lab, School of Biomedical Engineering,
Shenzhen University Medical School, Shenzhen University, Shenzhen, China
onewang@szu.edu.cn

Abstract. Breast dynamic contrast-enhanced magnetic resonance imaging (DCE-MRI) plays an important role in the screening and prognosis assessment of high-risk breast cancer. The segmentation of cancerous regions is essential useful for the subsequent analysis of breast MRI. To alleviate the annotation effort to train the segmentation networks, we propose a weakly-supervised strategy using extreme points as annotations for breast cancer segmentation. Without using any bells and whistles, our strategy focuses on fully exploiting the learning capability of the routine training procedure, i.e., the *train - fine-tune - retrain* process. The network first utilizes the pseudo-masks generated using the extreme points to *train* itself, by minimizing a contrastive loss, which encourages the network to learn more representative features for cancerous voxels. Then the trained network *fine-tunes* itself by using a similarity-aware propagation learning (SimPLe) strategy, which leverages feature similarity between unlabeled and positive voxels to propagate labels. Finally the network *retrains* itself by employing the pseudo-masks generated using previous fine-tuned network. The proposed method is evaluated on our collected DCE-MRI dataset containing 206 patients with biopsy-proven breast cancers. Experimental results demonstrate our method effectively fine-tunes the network by using the SimPLe strategy, and achieves a mean Dice value of 81%. *Our code is publicly available at* https://github.com/Abner228/SmileCode.

Keywords: Breast cancer · Weakly-supervised learning · Medical image segmentation · Contrastive learning · DCE-MRI

1 Introduction

Breast cancer is the most common cause of cancer-related deaths among women all around the world [8]. Early diagnosis and treatment is beneficial to improve the survival rate and prognosis of breast cancer patients. Mammography, ultrasonography, and magnetic resonance imaging (MRI) are routine imaging modalities for breast examinations [15]. Recent clinical studies have proven that

H. Greenspan et al. (Eds.): MICCAI 2023, LNCS 14223, pp. 567–577, 2023.
https://doi.org/10.1007/978-3-031-43901-8_54

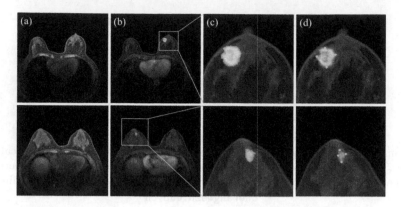

Fig. 1. Breast MRI and different annotations: (a) T1-weighted images, (b) corresponding contrast-enhanced images, (c) the cancer annotation with full segmentation masks, and (d) the cancer annotation using extreme points (*note that to facilitate the visualization, here we show the extreme points in 2D images, our method is based on 3D*).

dynamic contrast-enhanced (DCE)-MRI has the capability to reflect tumor morphology, texture, and kinetic heterogeneity [14], and is with the highest sensitivity for breast cancer screening and diagnosis among current clinical imaging modalities [17]. The basis for DCE-MRI is a dynamic T1-weighted contrast enhanced sequence (Fig. 1). T1-weighted acquisition depicts enhancing abnormalities after contrast material administration, that is, the cancer screening is performed by using the post-contrast images. Radiologists will analyze features such as texture, morphology, and then make the treatment plan or prognosis assessment. Computer-aided feature quantification and diagnosis algorithms have recently been exploited to facilitate radiologists analyze breast DCE-MRI [12,22], in which automatic cancer segmentation is the very first and important step.

To better support the radiologists with breast cancer diagnosis, various segmentation algorithms have been developed [20]. Early studies focused on image processing based approaches by conducting graph-cut segmentation [29] or analyzing low-level hand-crafted features [1,11,19]. These methods may encounter the issue of high computational complexity when analyzing volumetric data, and most of them require manual interactions. Recently, deep-learning-based methods have been applied to analyze breast MRI. Zhang *et al.* [28] proposed a mask-guided hierarchical learning framework for breast tumor segmentation via convolutional neural networks (CNNs), in which breast masks were also required to train one of CNNs. This framework achieved a mean Dice value of 72% on 48 testing T1-weighted scans. Li *et al.* [16] developed a multi-stream fusion mechanism to analyze T1/T2-weighted scans, and obtained a Dice result of 77% on 313 subjects. Gao *et al.* [7] proposed a 2D CNN architecture with designed attention modules, and got a Dice result of 81% on 87 testing samples. Zhou *et al.* [30] employed a 3D affinity learning based multi-branch ensemble network for the segmentation refinement and generated 78% Dice on 90 testing subjects. Wang *et al.* [24] integrated a combined 2D and 3D CNN and a contextual pyramid into U-net to obtain a Dice result of 76% on 90 subjects. Wang *et al.* [25] proposed a tumor-sensitive synthesis module to reduce false segmentation and obtained 78% Dice value. To reduce the huge annotation burden for the segmentation task, Zeng *et al.* [27] presented a semi-supervised strategy to segment the manually cropped DCE-MRI scans, and attained a Dice value of 78%.

Although [27] has been proposed to alleviate the annotation effort, to acquire the voxel-level segmentation masks is still time-consuming and laborious, see Fig. 1(c). Weakly-supervised learning strategies such as extreme points [5,21], bounding box [6] and scribbles [4] can be promising solutions. Roth *et al.* [21] utilized extreme points to generate scribbles to supervise the training of the segmentation network. Based on [21], Dorent *et al.* [5] introduced a regularized loss [4] derived from a Conditional Random Field (CRF) formulation to encourage the prediction consistency over homogeneous regions. Du *et al.* [6] employed bounding boxes to train the segmentation network for organs. However, the geometric prior used in [6] can not be an appropriate strategy for the segmentation of lesions with various shapes. To our knowledge, currently only one weakly-supervised work [18] has been proposed for breast mass segmentation in DCE-MRI. This method employed three partial annotation methods including single-slice, orthogonal-slice (i.e., 3 slices) and interval-slice ($\sim$6 slices) to alleviate the annotation cost, and then constrained segmentation by estimated volume using the partial annotation. The method obtained a Dice value of 83% using the interval-slice annotation, on a testing dataset containing only 28 patients.

In this study, we propose a simple yet effective weakly-supervised strategy, by using extreme points as annotations (see Fig. 1(d)) to segment breast cancer. Specifically, we attempt to optimize the segmentation network via the conventional *train - fine-tune - retrain* process. The initial training is supervised by a contrastive loss to pull close positive voxels in feature space. The fine-tune is conducted by using a similarity-aware propagation learning (SimPLe) strategy to update the pseudo-masks for the subsequent retrain. We evaluate our method on a collected DCE-MRI dataset containing 206 subjects. Experimental results show our method achieves competitive performance compared with fully supervision, demonstrating the efficacy of the proposed SimPLe strategy.

2   Method

The proposed SimPLe strategy and the *train - fine-tune - retrain* procedure is illustrated in Fig. 2. The extreme points are defined as the left-, right-, anterior-, posterior-, inferior-, and superior-most points of the cancerous region in 3D. The initial pseudo-masks are generated according to the extreme points by using the random walker algorithm. The segmentation network is firstly trained based on the initial pseudo-masks. Then SimPLe is employed to fine-tune the network and update the pseudo-masks. At last, the network is retrained from random initialization using the updated pseudo-masks.

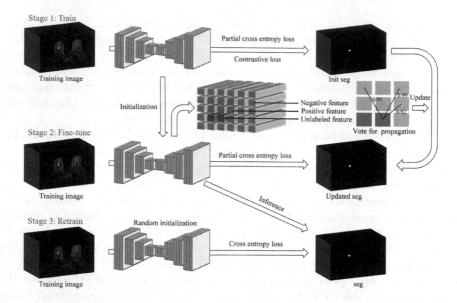

Fig. 2. The schematic illustration of the proposed similarity-aware propagation learning (SimPLe) and the *train - fine-tune - retrain* procedure for the breast cancer segmentation in DCE-MRI.

2.1   Generate Initial Pseudo-masks

We use the extreme points to generate pseudo-masks based on random walker algorithm [9]. To improve the performance of random walker, according to [21], we first generate scribbles by searching the shortest path on gradient magnitude map between each extreme point pair via the Dijkstra algorithm [3]. After generating the scribbles, we propose to dilate them to increase foreground seeds for random walker. Voxels outside the bounding box (note that once we have the six extreme points, we have the 3D bounding box of the cancer) are expected to be the background seeds. Next, the random walker algorithm is used to produce a foreground probability map $\widehat{Y} : \Omega \subset \mathbb{R}^3 \to [0, 1]$, where Ω is the spatial domain. To further increase the area of foreground, the voxel at location $\boldsymbol{k}$ is considered as new foreground seed if $\widehat{Y}(\boldsymbol{k})$ is greater than 0.8 and new background seed if $\widehat{Y}(\boldsymbol{k})$ is less than 0.1. Then we run the random walker algorithm repeatedly. After seven times iterations, we set foreground in the same way via the last output probability map. Voxels outside the bounding box are considered as background. The rest of voxels remain unlabeled. This is the way initial pseudo-masks $Y_{init} : \Omega \subset \mathbb{R}^3 \to \{0, 1, 2\}$ generated, where 0, 1 and 2 represent negative, positive and unlabeled.

2.2   Train Network with Initial Pseudo-masks

Let $X : \Omega \subset \mathbb{R}^3 \to \mathbb{R}$ denotes a training volume. Let f and θ be network and its parameters, respectively. A simple training approach is to minimize the partial cross entropy loss $\mathcal{L}_{pce}$, which is formulated as:

$$\mathcal{L}_{pce} = -\sum_{Y_{init}(k)=0} log(1 - f(X;\theta)(k)) - \sum_{Y_{init}(k)=1} log(f(X;\theta)(k)). \tag{1}$$

Moreover, supervised contrastive learning is employed to encourage voxels of the same label to gather around in feature space. It ensures the network to learn discriminative features for each category. Specifically, features corresponding to N negative voxels and N positive voxels are randomly sampled, then the contrastive loss $\mathcal{L}_{ctr}$ is minimized:

$$\mathcal{L}_{ctr} = -\frac{1}{2N-1} \sum_{k_n \in \mathcal{N}(k)} log(1 - \sigma(sim(\mathcal{Z}(k), \mathcal{Z}(k_n))/\tau))$$
$$-\frac{1}{2N-1} \sum_{k_p \in \mathcal{P}(k)} log(\sigma(sim(\mathcal{Z}(k), \mathcal{Z}(k_p))/\tau)), \tag{2}$$

where $\mathcal{P}(k)$ denotes the set of points with the same label as the voxel k and $\mathcal{N}(k)$ denotes the set of points with the different label. $\mathcal{Z}(k)$ denotes the feature vector of the voxel at location k. $sim(\cdot, \cdot)$ is the cosine similarity function. σ denotes sigmoid function. τ is a temperature parameter.

To summarize, we employ the sum of the partial cross entropy loss $\mathcal{L}_{pce}$ and the contrastive loss $\mathcal{L}_{ctr}$ to train the network with initial pseudo-masks:

$$\mathcal{L}_{train} = \mathcal{L}_{pce} + \mathcal{L}_{ctr}. \tag{3}$$

2.3 SimPLe-Based Fine-Tune and Retrain

The performance of the network trained by the incomplete initial pseudo-masks is still limited. We propose to fine-tune the entire network using the pre-trained weights as initialization. The fine-tune follows the SimPLe strategy which evaluates the similarity between unlabeled voxels and positive voxels to propagate labels to unlabeled voxels. Specifically, N positive voxels are randomly sampled as the referring voxel. For each unlabeled voxel k, we evaluate its similarity with all referring voxels:

$$\mathcal{S}(k) = \sum_{i=1}^{N} \mathbb{I}\{sim(\mathcal{Z}_k, \mathcal{Z}_i) > \lambda\}, \tag{4}$$

where $\mathbb{I}(\cdot)$ is the indicator function, which is equal to 1 if the cosine similarity is greater than λ and 0 if less. If $\mathcal{S}(k)$ is greater than αN, the voxel at location k is considered as positive. Then the network is fine-tuned using the partial cross entropy loss same as in the initial train stage. The loss function $\mathcal{L}_{finetune}$ is formulated as:

$$\mathcal{L}_{finetune} = \mathcal{L}_{pce} - w \cdot \sum_{\mathcal{S}(k)>\alpha N} log(f(X;\theta)(k)), \tag{5}$$

where w is the weighting coefficient that controls the influence of the pseudo labels. To reduce the influence of possible incorrect label propagation, pseudo labels for unlabeled voxels are valid only for the current iteration when they are generated.

After the fine-tune completed, the network generates binary pseudo-masks for every training data, which are expected to be similar to the ground-truths provided by radiologists. Finally the network is retrained from random initialization by minimizing the cross entropy loss with the binary pseudo-masks.

3   Experiments

Dataset. We evaluated our method on an in-house breast DCE-MRI dataset collected from the Cancer Center of Sun Yat-Sen University. In total, we collected 206 DCE-MRI scans with biopsy-proven breast cancers. All MRI scans were examined with 1.5T MRI scanner. The DCE-MRI sequences (TR/TE = 4.43 ms/1.50 ms, and flip angle = 10°) using gadolinium-based contrast agent were performed with the T1-weighted gradient echo technique, and injected 0.2 ml/kg intravenously at 2.0 ml/s followed by 20 ml saline. The DCE-MRI volumes have two kinds of resolution, $0.379 \times 0.379 \times 1.700$ mm^3 and $0.511 \times 0.511 \times 1.000$ mm^3.

All cancerous regions and extreme points were manually annotated by an experienced radiologist via ITK-SNAP [26] and further confirmed by another radiologist. We randomly divided the dataset into 21 scans for training and the remaining scans for testing[1]. Before training, we resampled all volumes into the same target spacing $0.600 \times 0.600 \times 1.000$ mm^3 and normalized all volumes as zero mean and unit variance.

Implementation Details. The framework was implemented in PyTorch, using a NVIDIA GeForce GTX 1080 Ti with 11GB of memory. We employed 3D U-net [2] as our network backbone.

- *Train*: The network was trained by stochastic gradient descent (SGD) for 200 epochs, with an initial learning rate $\eta = 0.01$. The ploy learning policy was used to adjust the learning rate, $(1 - \text{epoch}/200)^{0.9}$. The batch size was 2, consisting of a random foreground patch and a random background patch located via initial segmentation Y_{init}. Such setting can help alleviate class imbalance issue. The patch size was $128 \times 128 \times 96$. For the contrastive loss, we set N = 100, temperature parameter $\tau = 0.1$.
- *Fine-tune*: We initialized the network with the trained weights. We trained it by SGD for 100 iterations, with $\eta = 0.0001$. The ploy learning policy was also used. For the SimPLe strategy, we set $N = 100, \lambda = 0.96, \alpha = 0.96, w = 0.1$.

---

[1] We have tried different amount of training data to investigate the segmentation performance of the fully-supervised network. The results showed that when using 21, 42, 63 scans for training, the Dice results changed very little, within 0.3%. Therefore, to include more testing data, we chose to use 21 (10%) out of 206 scans for training.

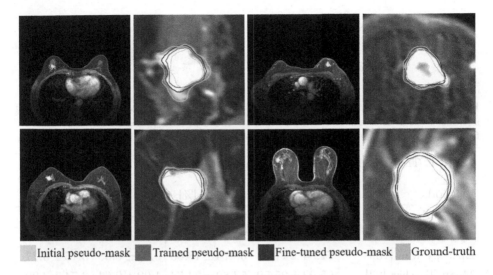

Initial pseudo-mask   Trained pseudo-mask   Fine-tuned pseudo-mask   Ground-truth

Fig. 3. The pseudo-masks (shown as boundaries) at different training stages, including the initial pseudo-mask generated by the random walker (purple), the trained network's output (green), the fine-tuned pseudo-mask using SimPLe (blue) and the ground-truth (red). Note that all images here are the training images. (Color figure online)

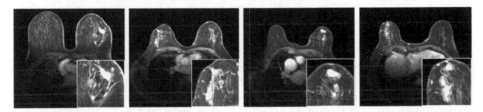

Fig. 4. The segmentation visualization in transversal slices. The blue and red contours are the segmented boundaries and the ground-truths, respectively. (Color figure online)

After fine-tuned, the last weights were used to generate the binary pseudo-masks.

- *Retrain*: The training strategy was the same as the initial train stage.
- *Inference*: A sliding window approach was used. The window size equaled the patch size used during training. Stride was set as half of a patch.

Quantitative and Qualitative Analysis. We first verified the efficacy of our SimPLe in the training stage. Figure 3 illustrates the pseudo-masks at different training stages. It is obvious that our SimPLe effectively updated the pseudo-masks to make them approaching the ground-truths. Therefore, such fune-tuned pseudo-masks could be used to retrain the network for better performance.

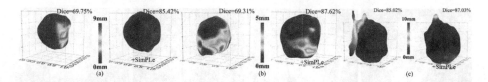

Fig. 5. Three cases of 3D visualization of the surface distance between segmented surface and ground-truth. Each case shows the $\mathcal{L}_{pce} + \mathcal{L}_{ctr}$ result and the SimPLe fine-tuned result. The proposed SimPLe consistently enhances the segmentation.

Table 1. The numerical results of different methods for breast cancer segmentation

Methods	Dice [%]	Jaccard [%]	ASD [mm]	95HD [mm]
$\mathcal{L}_{pce} + \mathcal{L}_{crf}$ [4]	64.97 ± 28.66	53.83 ± 27.26	1.01 ± 0.81	3.90 ± 3.91
$\mathcal{L}_{pce} + \mathcal{L}_{ctr}$	69.39 ± 24.09	57.36 ± 23.31	0.95 ± 0.66	3.60 ± 3.49
Entropy Min [10]	71.94 ± 17.86	58.74 ± 18.69	0.88 ± 0.58	3.16 ± 3.77
Mean Teacher [23]	65.92 ± 25.59	53.82 ± 24.74	1.02 ± 0.64	3.74 ± 3.29
Bounding Box [13]	77.02 ± 17.15	65.08 ± 17.92	0.89 ± 0.57	2.54 ± 5.20
$\mathcal{L}_{pce} + \mathcal{L}_{crf}+$ SimPLe	79.71 ± 17.72	68.99 ± 18.84	0.74 ± 0.55	2.48 ± 1.93
$\mathcal{L}_{pce} + \mathcal{L}_{ctr}+$ SimPLe	81.20 ± 13.28	70.01 ± 15.02	0.69 ± 0.44	2.40 ± 1.69
Fully Supervision	81.52 ± 19.40	72.10 ± 20.45	0.68 ± 0.63	2.40 ± 2.76

Figure 4 visualizes our cancer segmentation results on the testing data. Table 1 reports the quantitative Dice, Jaccard, average surface distance (ASD), and Hausdorff distance (95HD) results of different methods. We compared our method with an end-to-end approach [4] that proposed to optimize network via CRF-regularized loss $\mathcal{L}_{crf}$. Although our $\mathcal{L}_{ctr}$ supervised method outcompeted $\mathcal{L}_{crf}$ [4], the networks trained only using the initial pseudo-masks could not achieve enough high accuracy (Dice values<70%). In contrast, the proposed SimPLe largely boosted the performance of the basically trained networks, by +14.74% Dice and +15.16% Jaccard (*v.s.* $\mathcal{L}_{crf}$), +11.81% Dice and +12.65% Jaccard (*v.s.* $\mathcal{L}_{ctr}$). Table 1 also shows the comparison results of three general weakly-supervised strategies, including entropy minimization [10], mean teacher [23], and bounding box [13]. Our method consistently outperformed these strategies with respect to all evaluation metrics. Furthermore, our method achieved competitive Dice results compared with fully supervision, which again proves the efficacy of the proposed SimPLe strategy. Note that the average annotation time for extreme points and full masks were 31 s and 95 s per scan, respectively. Figure 5 visualizes the 3D distance map between the segmented surface and ground-truth. It can be observed that our SimPLe consistently enhanced the segmentation.

4   Conclusion

We introduce a simple yet effective weakly-supervised learning method for breast cancer segmentation in DCE-MRI. The primary attribute is to fully exploit the simple *train - fine-tune - retrain* process to optimize the segmentation network via only extreme point annotations. This is achieved by employing a similarity-aware propagation learning (SimPLe) strategy to update the pseudo-masks. Experimental results demonstrate the efficacy of the proposed SimPLe strategy for weakly-supervised segmentation.

Acknowledgements. This work was supported in part by the National Natural Science Foundation of China under Grants 62071305, 61701312 and 81971631, and in part by the Guangdong Basic and Applied Basic Research Foundation under Grant 2022A1515011241.

References

1. Ashraf, A.B., Gavenonis, S.C., Daye, D., Mies, C., Rosen, M.A., Kontos, D.: A multichannel markov random field framework for tumor segmentation with an application to classification of gene expression-based breast cancer recurrence risk. IEEE Trans. Med. Imaging **32**(4), 637–648 (2012)
2. Çiçek, Ö., Abdulkadir, A., Lienkamp, S.S., Brox, T., Ronneberger, O.: 3D U-Net: learning dense volumetric segmentation from sparse annotation. In: Ourselin, S., Joskowicz, L., Sabuncu, M.R., Unal, G., Wells, W. (eds.) MICCAI 2016. LNCS, vol. 9901, pp. 424–432. Springer, Cham (2016). https://doi.org/10.1007/978-3-319-46723-8_49
3. Dijkstra, E.: A note on two problems in connexion with graphs. Numerische Mathematik **1**, 269–271 (1959)
4. Dorent, R., et al.: Scribble-based domain adaptation via co-segmentation. In: Martel, A.L., et al. (eds.) MICCAI 2020. LNCS, vol. 12261, pp. 479–489. Springer, Cham (2020). https://doi.org/10.1007/978-3-030-59710-8_47
5. Dorent, R., et al.: Inter extreme points geodesics for end-to-end weakly supervised image segmentation. In: de Bruijne, M., et al. (eds.) MICCAI 2021. LNCS, vol. 12902, pp. 615–624. Springer, Cham (2021). https://doi.org/10.1007/978-3-030-87196-3_57
6. Du, H., Dong, Q., Xu, Y., Liao, J.: Weakly-supervised 3D medical image segmentation using geometric prior and contrastive similarity. arXiv preprint arXiv:2302.02125 (2023)
7. Gao, Y., Zhao, Y., Luo, X., Hu, X., Liang, C.: Dense encoder-decoder network based on two-level context enhanced residual attention mechanism for segmentation of breast tumors in magnetic resonance imaging. In: 2019 IEEE International Conference on Bioinformatics and Biomedicine, pp. 1123–1129. IEEE (2019)
8. Giaquinto, A.N., et al.: Breast cancer statistics, 2022. CA Cancer J. Clin. **72**(6), 524–541 (2022)
9. Grady, L.: Random walks for image segmentation. IEEE Trans. Pattern Anal. Mach. Intell. **28**(11), 1768–1783 (2006)
10. Grandvalet, Y., Bengio, Y.: Semi-supervised learning by entropy minimization. Adv. Neural Inf. Process. Syst. **17**, 1–8 (2004)

11. Gubern-Mérida, A., et al.: Automated localization of breast cancer in DCE-MRI. Med. Image Anal. **20**(1), 265–274 (2015)
12. Jiang, Y., Edwards, A.V., Newstead, G.M.: Artificial intelligence applied to breast MRI for improved diagnosis. Radiology **298**(1), 38–46 (2021)
13. Kervadec, H., Dolz, J., Wang, S., Granger, E., Ayed, I.B.: Bounding boxes for weakly supervised segmentation: global constraints get close to full supervision. In: Medical Imaging with Deep Learning, pp. 365–381 (2020)
14. Kim, J.Y., et al.: Kinetic heterogeneity of breast cancer determined using computer-aided diagnosis of preoperative MRI scans: relationship to distant metastasis-free survival. Radiology **295**(3), 517–526 (2020)
15. Lee, C.H., et al.: Breast cancer screening with imaging: recommendations from the society of breast imaging and the ACR on the use of mammography, breast MRI, breast ultrasound, and other technologies for the detection of clinically occult breast cancer. J. Am. Coll. Radiol. **7**(1), 18–27 (2010)
16. Li, C., Sun, H., Liu, Z., Wang, M., Zheng, H., Wang, S.: Learning cross-modal deep representations for multi-modal MR image segmentation. In: Shen, D., et al. (eds.) MICCAI 2019. LNCS, vol. 11765, pp. 57–65. Springer, Cham (2019). https://doi. org/10.1007/978-3-030-32245-8_7
17. Mann, R.M., Cho, N., Moy, L.: Breast MRI: state of the art. Radiology **292**(3), 520–536 (2019)
18. Meng, X., et al.: Volume-awareness and outlier-suppression co-training for weakly-supervised MRI breast mass segmentation with partial annotations. Knowl.-Based Syst. **258**, 109988 (2022)
19. Militello, C., et al.: Semi-automated and interactive segmentation of contrast-enhancing masses on breast DCE-MRI using spatial fuzzy clustering. Biomed. Signal Process. Control **71**, 103113 (2022)
20. Rezaei, Z.: A review on image-based approaches for breast cancer detection, segmentation, and classification. Expert Syst. Appl. **182**, 115204 (2021)
21. Roth, H.R., Yang, D., Xu, Z., Wang, X., Xu, D.: Going to extremes: weakly supervised medical image segmentation. Mach. Learn. Knowl. Extract. **3**(2), 507–524 (2021)
22. Sheth, D., Giger, M.L.: Artificial intelligence in the interpretation of breast cancer on MRI. J. Magn. Reson. Imaging **51**(5), 1310–1324 (2020)
23. Tarvainen, A., Valpola, H.: Mean teachers are better role models: weight-averaged consistency targets improve semi-supervised deep learning results. Adv. Neural Inf. Process. Syst. **30**, 1–10 (2017)
24. Wang, H., Cao, J., Feng, J., Xie, Y., Yang, D., Chen, B.: Mixed 2D and 3D convolutional network with multi-scale context for lesion segmentation in breast DCE-MRI. Biomed. Signal Process. Control **68**, 102607 (2021)
25. Wang, S., et al.: Breast tumor segmentation in DCE-MRI with tumor sensitive synthesis. IEEE Trans. Neural Netw. Learn. Syst. **34**, 4990–5001 (2021)
26. Yushkevich, P.A., Piven, J., Cody Hazlett, H., Gimpel Smith, R., Ho, S., Gee, J.C., Gerig, G.: User-guided 3D active contour segmentation of anatomical structures: significantly improved efficiency and reliability. Neuroimage **31**(3), 1116–1128 (2006)
27. Zeng, X., Huang, R., Zhong, Y., Xu, Z., Liu, Z., Wang, Y.: A reciprocal learning strategy for semisupervised medical image segmentation. Med. Phys. **50**(1), 163–177 (2023)
28. Zhang, J., Saha, A., Zhu, Z., Mazurowski, M.A.: Hierarchical convolutional neural networks for segmentation of breast tumors in MRI with application to radiogenomics. IEEE Trans. Med. Imaging **38**(2), 435–447 (2018)

29. Zheng, Y., Baloch, S., Englander, S., Schnall, M.D., Shen, D.: Segmentation and classification of breast tumor using dynamic contrast-enhanced MR images. In: Ayache, N., Ourselin, S., Maeder, A. (eds.) MICCAI 2007. LNCS, vol. 4792, pp. 393–401. Springer, Heidelberg (2007). https://doi.org/10.1007/978-3-540-75759-7_48

30. Zhou, L., Wang, S., Sun, K., Zhou, T., Yan, F., Shen, D.: Three-dimensional affinity learning based multi-branch ensemble network for breast tumor segmentation in MRI. Pattern Recogn. **129**, 108723 (2022)

Multi-shot Prototype Contrastive Learning and Semantic Reasoning for Medical Image Segmentation

Yuhui Song[1], Xiuquan Du[1], Yanping Zhang[1], and Chenchu Xu[1,2(✉)]

[1] Artificial Intelligence Institute, Anhui University, Hefei, China
cxu332@gmail.com
[2] Institute of Artificial Intelligence, Hefei Comprehensive National Science Center, Hefei, China

Abstract. Despite the remarkable achievements made by deep convolutional neural networks in medical image segmentation, the limitation that they rely heavily on high-precision and intensively annotated samples makes it difficult to adapt to novel classes that have not been seen before. Few-shot learning is introduced to solve these challenges by learning the generalized representation of a semantic class from very few annotated support samples that can be used as a reference for unannotated query samples. In this paper, instead of averaging multiple support prototypes, we propose a multi-shot prototype contrastive learning and semantic reasoning network (MPSNet) for medical image segmentation. The multi-shot learning network exists independently within the support set, obtains effective semantic features for support images and gives priority to training the core segmentation model of prototype contrastive learning. We also propose a semantic reasoning network that takes the prior semantic features and prior segmentation model learned from the support set as the immediate and necessary conditions for the query image to deduce its segmentation mask. The proposed method is verified to be superior to the state-of-the-art methods on three public datasets, revealing its powerful segmentation and generalization abilities. Code: https://github.com/H51705/FSS_MPSNet.

Keywords: Medical image segmentation · Multi-shot learning · Prototype contrastive learning · Semantic reasoning

1 Introduction

The successes of deep convolutional neural networks (CNNs) [9] in image tasks [12,20–22] rely heavily on a large number of intensively annotated training samples, and obtaining these annotated images is time-consuming, laborious and costly. In addition to the extreme lack of training data, these segmentation models also have low generalization ability due to the specificity of training classes.

© The Author(s), under exclusive license to Springer Nature Switzerland AG 2023
H. Greenspan et al. (Eds.): MICCAI 2023, LNCS 14223, pp. 578–588, 2023.
https://doi.org/10.1007/978-3-031-43901-8_55

This limitation is more significant in medical images. Therefore, few-shot learning (FSL) [11,14,17] emerged to deal with these challenges. By learning the generalized representation of a certain semantic class from a few labeled samples (i.e., support images) to guide the segmentation of that class in a large number of unlabeled images (i.e., query images).

The development of FSL has derived many effective medical image segmentation methods, including pure CNNs [2,10] and CNNs with a nonparametric learning module embedded [1], among which the prototype network is the most widely used [8,15,23]. For example, Yu et al. [23] proposed a location-sensitive local prototype network. Tang et al. [15] proposed a recurrent mask refinement algorithm based on a prototype network. In addition, Ouyang et al. [8] introduced a self-supervised super-pixel few-shot segmentation (FSS) method to get rid of the demand for labeled data in the training stage. Although FSS methods based on prototype networks have the ability to achieve image segmentation, when the prototype learned from the support image is used to judge the pixel category of the query image, it still cannot avoid the problem of insufficient prototype discriminability caused by the differences between the images as well as the complexity of the target. Furthermore, it is easy to lose important information when the target features are taken as a prototype by GAP (global average pooling). In this regard, Wang et al. [18] introduced a prototype alignment regularization between the support image and the query image to better utilize the information of the support set. Meanwhile, compared with one-shot mode, the multi-shot prototype strategy is widely used to improve segmentation performance [6,8,16]. However, the existing multi-shot strategies usually take the average of multiple support prototypes as the final prototype, which not only weakens the independent contribution of multiple support images but also destroys the semantic information of support prototypes. Especially for medical images with lower image intensity and contrast, prototype alignment regularization and the mean prototype strategy do not significantly increase the discriminant ability and prediction accuracy of prototypes. Especially for medical images with lower image intensity and contrast, prototype alignment regularization and mean prototype strategy may aggravate the confusion of target pixel categories. For pure prototype technology, Zhou et al. [25] explored a better non-parametric segmentation model based on non-learnable prototypes in their latest work and successfully applied it to the semantic segmentation of natural images, but this method has not been verified in medical images.

In this paper, we propose a prototype contrastive learning and semantic reasoning network based on multi-shot strategy (MPSNet). To our knowledge, this is a great improvement in medical image analysis. First, we design a multi-shot learning network (MLN) within the support set to generate prior semantic features and a prior segmentation model for the query image. Cross-validation mode is used to construct the support-query image pairs within the support set, and a complete FSS model is formed by prototype contrastive learning and supervised training. Second, we propose a prototype contrastive learning module (PCLM) to ascertain the positive contribution of multiple support images to the query image in segmentation guidance, leading to better segmentation performance of the query image. Third, we design a semantic reasoning network (SRN) that is convenient to directly transfer the prior semantic features and prior segmentation model to the query image to deduce its segmentation mask quickly. Our contributions can be summarized as follows:

1) A novel FSS method based on a multi-shot prototype strategy is proposed for the first time to replace the commonly used mean prototype method to improve the guidance ability of the support images to the query images in segmentation.
2) A multi-shot prototype contrastive learning network within the support set is constructed with supervised training to generate prior semantic features and a prior segmentation model, and transfer them to the query image to deduce its segmentation mask.
3) The proposed method achieves the latest performance on three public datasets that is superior to the state-of-the-art (SOTA) methods.

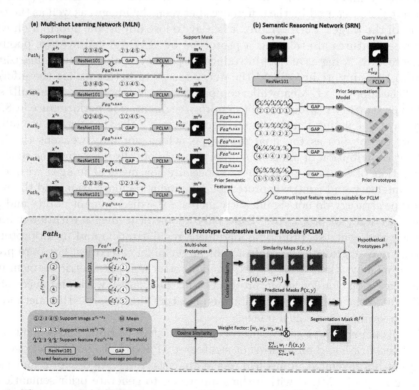

Fig. 1. The proposed MPSNet consists of three parts: (a) the multi-shot learning network (MLN), (b) the semantic reasoning network (SRN) and (c) the prototype contrastive learning module (PCLM). PCLM is a core segmentation module that is embedded in both MLN and SRN. MPSNet follows a 1-way 5-shot learning mode. For each episode, the input is 5 support image-mask pairs and a query image, and the output is the segmentation mask of the query image.

2 Method

2.1 Problem Setting

In FSS problem, a segmentation model is trained on a training set D_{train}, then the model is used to evaluate a test set D_{test}. Assumes that the semantic classes in D_{train} is C_{train} and in D_{test} is C_{test}, the particular definition of FSL is $C_{train} \cap C_{test} = \phi$. The training set and test set have a common intrinsic pattern that both contain a support set S and a query set Q. At training stage, we use episode mode to set N-way K-shot segmentation task, where N represents the number of semantic classes contained in each episode and K represents the number of images contained in each class. For the task of few-shot medical image segmentation, N is usually set to 1, that is, each episode only sets the support set and the query set for a specific class c. For one episode, suppose that the support set and the query set are $S_{train} = \{(I_k^s, M_k^s)\}_{k=1}^{K^s}$ and $Q_{train} = \{(I_n^q, M_n^q)\}_{n=1}^{N^q}$. The support images, masks and the query image constitute the model input $Input_{model} = \{I^s, M^s, I^q\}$, query mask M^q as the supervisory information. At testing stage, we set episodes with the same mode, that is, we use the trained FSS model to segment the query set under the function of the support set. Since our research is carried out based on multi-shot strategy, we mainly take 5-shot as an example to verify the effectiveness of the proposed method.

2.2 Multi-shot Learning Network

The framework of MLN is shown in Fig. 1(a), it is a supervised learning network that exists independently within the support set, mainly including 5 same structures but different input and output of 4-shot learning branch $\{Path_l\}_{l=1}^5$. When one support image is used as the fake query image, the remaining 4 support images are used as the fake support images. Therefore, in the support set, each support image can be used as a fake query image to form a 4-shot learning network. The specific implementation process can be expressed as follows:

$$\hat{m}^{s_i}/\hat{m}^{f_q} = PCLM(GAP(RN(\{x^{f_{s_i}}\}_{i=1}^4), \{m^{f_{s_i}}\}_{i=1}^4), RN(x^{f_q})), \quad (1)$$

where $RN(\cdot)$ refers to the shared feature extractor ResNet101, $GAP(\cdot)$ refers to global average pooling, $PCLM(\cdot)$ refers to the proposed prototype contrastive learning module. $x^{f_{s_i}}$, $m^{f_{s_i}}$, x^{f_q} represent the fake support image, mask and query image. We use cross-entropy (CE) loss to supervise the training process:

$$L_{seg}^s = \sum_{i=1}^5 L_{seg}^{s_i} = \sum_{i=1}^5 CE(\hat{m}^{s_i}, m^{s_i}). \quad (2)$$

As shown in the lower part of Fig. 1, we take $Path_1$ branch as an example in detail the specific structures. Fake support images $\{x^{f_{s_i}}\}_{i=1}^4 \in \mathbb{R}^{3 \times H \times W}$ and fake query image $x^{f_q} \in \mathbb{R}^{3 \times H \times W}$ are fed into ResNet101 for support features $\{Fea^{f_{s_i}}\}_{i=1}^4 \in \mathbb{R}^{C \times h \times w}$ and query features $Fea^{f_q} \in \mathbb{R}^{C \times h \times w}$. Under the function of support masks $\{m^{fs_i}\}_{i=1}^{4} \in \mathbb{R}^{1 \times H \times W}$, the support proto-types $P = \{P_i\}_{i=1}^{4}$ have been extracted from the support features, denoted as $P_i \in \mathbb{R}^{C \times 1 \times 1}$, this process is implemented by GAP:

$$P_i = \frac{\sum Fea^{fs_i} \odot m^{fs_i}}{\sum m^{fs_i}}, \qquad (3)$$

where support features Fea^{fs_i} are resized to the mask size (H, W), (h, w) denotes the shape of the feature maps, C denotes the channel number of the feature maps, $\odot$ denotes the Hadamard product. The obtained multi-shot prototypes P and query features Fea^{fq} are sent into PCLM to get the final segmentation of the query image.

2.3 Prototype Contrastive Learning Module

As shown in Fig. 1(c), PCLM is embedded in MLN and SRN as a core segmentation module, assumes the task of predicting the pixel-level categories of the query image by taking the multi-shot prototypes as evaluation criteria. Following Hansen et al. [3], we consider using a single foreground prototype as the feature compression of the target region, and adopt the anomaly score thresholding method to segment the query image. This method not only avoids the construction of the prototype for the background with more impurities, but also saves the calculation cost compared with the decoder. First of all, the similarity maps $\{S^i(x, y)\}_{i=1}^{4}$ are obtained by doing cosine similarity between the multi-shot prototypes P and the query features Fea^{fq}:

$$S^i(x, y) = -\alpha \frac{Fea^{fq}(x, y) \cdot P_i}{\left\| Fea^{fq}(x, y) \right\| \|P_i\|}. \qquad (4)$$

The foreground prediction mask is then obtained:

$$\widehat{m}_{fore_i}^{fq} = 1 - \delta\left(S^i(x, y) - T^{fq}\right) = 1 - \delta\left(S^i(x, y) - g_\theta\left(Fea^{fq}\right)\right), \qquad (5)$$

where (x, y) refers to the spatial location of the query image, and the prototype generates a similarity coefficient for each pixel. $\delta(\cdot)$ denotes the Sigmoid function with a steepness parameter $k = 0.5$. T^{fq} is the correlation threshold obtained by the query features Fea^{fq} through the full connection layers $g_\theta(\cdot)$. Then, we assume that the prediction mask is no different from the real mask, and GAP it with Fea^{fq} to obtain the hypothetical prototypes $P^h = \{P_i^h\}_{i=1}^{4}$:

$$P_i^h = \frac{\sum Fea^{fq} \odot \widehat{m}_{fore_i}^{fq}}{\sum \widehat{m}_{fore_i}^{fq}}. \qquad (6)$$

Theoretically, under the assumption that there is no pixel intensity and contrast difference between the query image and the support image, the hypothetical prototypes P^h should be similar to the multi-shot prototypes P. But in reality, it is difficult to make the query image and the support images have a consistent feature hierarchy. That is to say, the prior knowledge provided by multiple support images for the query image is not equally important. We need to assign a weight factor w_i to each support image to represent its positive contribution to query image segmentation. To this end, we determine the weight of each foreground prediction mask by comparing the hypothetical prototypes P^h with real support prototypes P:

$$w_i = \frac{P_i^h \cdot P_i}{\left\| P_i^h \right\| \left\| P_i \right\|}. \tag{7}$$

At this time, the foreground and background segmentation mask can be modified as:

$$\widehat{m}_{fore}^{fq} = \frac{\sum_{i=1}^{4} w_i \cdot \widehat{m}_{fore_i}^{fq}}{\sum_{i=1}^{4} w_i}, \tag{8}$$

$$\widehat{m}_{back}^{fq} = 1 - \widehat{m}_{fore}^{fq}, \tag{9}$$

$$\widehat{m}^{fq} = concatenate\left(\widehat{m}_{fore}^{fq}, \widehat{m}_{back}^{fq} \right), \tag{10}$$

we take $\widehat{m}^{fq}$ as the output of PCLM that is also the final segmentation mask.

2.4   Semantic Reasoning Network

In our 5-shot segmentation task, the query image needs to be accurately segmented under the guidance of the given 5 support images. Such an FSS task can be realized through the migration of the prior semantic features and prior segmentation model from MLN. Specifically, as shown in Fig. 1(b), we directly prototype the support features obtained from MLN with the corresponding support masks. It is worth noting that the support images in each supervised training path are gathered together to form the support feature-mask cluster to obtain the prototype cluster. The purpose of this is to make the FSS network designed for query images conform to the laws of prior semantic features and the prior segmentation model. At the same time, prior prototypes are obtained by internal averaging of the prototype cluster. On this basis, the implementation of the same PCLM module as MLN can output the segmentation mask $\widehat{m}^q$ of the query image, and use the query mask m^q to conduct supervised training:

$$\widehat{m}^q = PCLM(GAP(\{Fea^{s_i}\}_{i=1}^{5}), \{m^{s_i}\}_{i=1}^{5}), RN(x^q)), \tag{11}$$

$$L_{seg}^q = CE(\widehat{m}^q, m^q), \tag{12}$$

$$L_{seg} = L_{seg}^s + L_{seg}^q, \tag{13}$$

so far, the overall loss of the proposed segmentation network MPSNet is L_{seg}.

3   Experiments

3.1   Experimental Setup

Datasets: We evaluate the proposed MPSNet on three public datasets: the Combined (CT-MR) Healthy Abdominal Organ Segmentation (CHAOS) dataset [4] (Abd-MRI) with 20 3D T2-SPIR MRI scan images, the MICCAI 2015 Multi-Atlas Abdomen Labeling Challenge [5] (Abd-CT) with 30 3D CT scan images and the MICCAI 2019 Multi-sequence Cardiac MRI Segmentation Challenge (bSSFP fold) [26] (Cardiac-MRI) with 35 3D cardiac MRI scans. Our task is to segment the liver, right kidney (RK), left kidney (LK), and spleen from Abd-MRI and Abd-CT, and the left ventricle (LV), myocardium (Myo), and right ventricle (RV) from Cardiac MRI. In the preprocessing of these datasets, we follow Ouyang et al.'s scheme [8] to make a fair comparison with the SOTA methods. Similarly, we also use the super-pixels technology to generate the pseudo-labels for the preprocessed datasets.

Implementation Details: We use 5-fold cross validation to conduct training and testing. We chose the same allocation strategy for the support slices and the query slices as Roy et al. [10], and we randomly selected 1000 support-query image pairs to form the training set. All codes are based on PyTorch. We use the stochastic gradient descent (SGD) optimizer with a batch size of 1 and set the initial learning rate to 0.001, momentum to 0.9 and weight decay to 0.0005. The total number of iterations of the training is 30,000, and it takes an average of 9h to complete end-to-end training on an Nvidia RTX 3090 graphics card. To facilitate direct comparison with the SOTA methods, we follow the practice of most papers, using the pretrained ResNet101 framework as the feature encoder and the mean Sørensen-Dice coefficient (DSC) as the evaluation metric.

3.2   Superior Performance over the SOTA Methods

We compare the quantitative results of our method with the SOTA methods, including SE-Net [10], RP-Net [15], GCN-DE [13], ADNet [3], ALPNet [8] and RSCNet [19] are FSS methods specially proposed for medical images that have been verified on the datasets used by us. The remaining methods are classical FSS methods that are used in natural image. In particular, PANet [18] contains a prototype alignment strategy that is also applicable to medical images. As shown in Table 1, our method has obtained the best mean DSC of 82.01%, 74.72% and 77.90% for Abd-MRI, Abd-CT and Cardiac MRI respectively. Compared with the second-best method ALPNet [8], the mean DSC achieved by our method increased by 3.17%, 1.37% and 1%.

3.3   Ablation Studies

The specific implementation scheme of the ablation studies can be divided into three aspects: 1) MLN is removed from MPSNet to prove that the prior semantic

Table 1. Quantitative comparison between MPSNet and the SOTA methods.

Dataset	Abd-MRI					Abd-CT					Cardiac-MRI			
Methods	Liver	RK	LK	Spleen	Mean	Liver	RK	LK	Spleen	Mean	LV	Myo	RV	Mean
SE-Net [10]	27.43	61.32	62.11	51.80	50.66	38.20	23.60	32.70	32.53	31.76	58.04	25.18	12.86	32.03
RP-Net [15]	73.51	85.78	81.40	76.35	79.26	79.62	70.00	70.48	69.85	72.48	–	–	–	–
GCN-DE [13]	49.47	83.03	76.07	60.63	67.30	46.77	**75.50**	68.13	56.53	61.73	–	–	–	–
PANet [18]	66.59	42.18	44.32	54.59	51.90	69.16	36.42	38.83	42.40	46.70	70.62	46.03	67.16	61.27
PPNet [7]	73.12	71.78	62.13	66.57	68.40	–	–	–	–	–	67.78	42.61	60.80	57.06
CANet [24]	72.88	77.15	69.53	67.05	71.65	–	–	–	–	–	78.99	43.61	61.10	61.07
ADNet [3]	80.81	83.28	75.28	75.92	78.82	–	–	–	–	–	87.53	62.43	77.31	75.76
ALPNet [8]	76.10	85.18	81.92	72.18	78.84	78.29	71.81	**72.36**	70.96	73.35	83.99	**66.74**	79.96	76.90
RSCNet [19]	75.55	84.24	77.07	73.73	77.65	73.63	63.37	67.39	67.36	67.94	–	–	–	–
MPSNet (Ours)	**81.88**	**87.26**	**82.10**	**76.80**	**82.01**	**81.13**	69.56	70.26	**77.96**	**74.72**	**87.98**	65.08	**80.63**	**77.90**

features and the prior segmentation model are more effective than the independent segmentation based on the prototype network (w/o MLN). 2) We use PCLM to replace the mean prototype strategy to verify its improvement in segmentation performance (w/o PCLM). 3) The initial segmentation model based on the

Table 2. Quantitative results of the ablation studies.

Dataset	Abd-MRI					Abd-CT					Cardiac-MRI			
Methods	Liver	RK	LK	Spleen	Mean	Liver	RK	LK	Spleen	Mean	LV	Myo	RV	Mean
MPSNet	**81.88**	**87.26**	**82.10**	**76.80**	**82.01**	81.13	**69.56**	**70.26**	**77.96**	**74.72**	**87.98**	**65.08**	**80.63**	**77.90**
w/o MLN	79.18	78.88	70.62	66.90	73.90	82.25	57.21	58.48	66.28	66.05	84.04	51.83	68.47	68.12
w/o PCLM	80.52	86.51	81.26	74.72	80.75	80.89	69.29	68.29	75.37	73.46	86.01	62.51	77.19	75.24
PNet	81.13	78.63	74.35	68.81	75.73	**82.85**	62.48	64.20	67.43	69.24	83.42	50.08	66.76	66.75

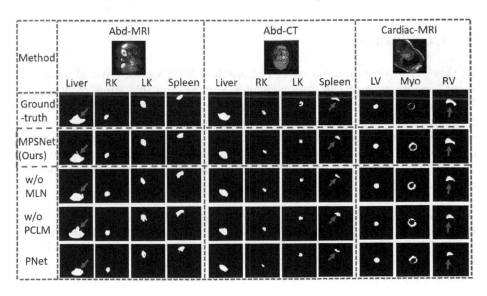

Fig. 2. Visualization of query image segmentation in test set.

prototype network is reverted, that is, MLN and PCLM are removed, and the multi-shot support prototypes are directly averaged for similarity measurement with the query features (PNet). Note that SRN cannot be implemented when MLN and PCLM are absent.

Table 2 shows the quantitative results of the ablation studies. In the absence of MLN, the average DSC is 8.11%, 8.67% and 9.78% lower than MPSNet. In the absence of PCLM, the average DSC is 1.26%, 1.26% and 2.66% lower than MPSNet. More significantly, our method has improved average DSC by 6.28%, 5.48% and 11.15% over PNet due to the inclusion of MLN and PCLM. Ablation studies fully demonstrate the effectiveness of the proposed method and the necessity of the internal parts. As shown in Fig. 2, one subject is randomly selected for each dataset to present the visualization results. In Fig. 3, we plot the training loss of the support images and the query image. It's obvious that the query loss decreases faster and becomes more stable than the support loss, indicating that our proposed MLN has a positive effect.

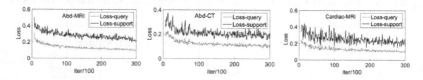

Fig. 3. Comparison of training loss between the support images and the query image. The value of the abscissa represents the loss measured every 100 iterations.

4   Conclusion

In this paper, we propose a novel FSS method MPSNet for medical image segmentation to replace the approach of averaging multiple support prototypes in the existing FSS methods. This is the first time that a multi-shot learning pattern is built within the support set and applied to the query image to significantly improve segmentation performance. Compared with the SOTA methods, our method is evaluated on three public datasets and achieves improvements in DSC of 3.17%, 1.37% and 1%, respectively. Relevant ablation studies also demonstrate the necessity and validity of our method.

Acknowledgment. This work was supported in part by The National Natural Science Foundation of China (62106001, U1908211), The University Synergy Innovation Program of Anhui Province (GXXT-2021-007, GXXT-2022-052), and The Anhui Provincial Natural Science Foundation (2208085Y19).

References

1. Feng, R., Zheng, X., Gao, T., Chen, J., Wu, J.: Interactive few-shot learning: Limited supervision, better medical image segmentation. IEEE Trans. Med. Imaging **40**, 2575–2588 (2021)
2. Guo, S., Xu, L., Feng, C., Xiong, H., Zhang, H.: Multi-level semantic adaptation for few-shot segmentation on cardiac image sequences. Med. Image Anal. **1**, 102170 (2021)
3. Hansen, S., Gautam, S., Jenssen, R., Kampffmeyer, M.: Anomaly detection-inspired few-shot medical image segmentation through self-supervision with super-voxels. Med. Image Anal. **78**, 102385 (2022)
4. Kavur, A., Gezer, N., Bar, M., Aslan, S., Conze, P.H.: Chaos challenge-combined (ct-mr) healthy abdominal organ segmentation. Med. Image Anal. **69**, 101950 (2021)
5. Landman, B., Xu, Z., Igelsias, J., Styner, M., Langerak, T., Klein, A.: Miccai multi atlas labeling beyond the cranial vault-workshop and challenge. In: MICCAI Multi-Atlas Labeling Beyond Cranial Vault-Workshop Challenge, vol. 5, no. 12 (2015)
6. Li, G., Jampani, V., Sevilla-Lara, L., Sun, D., Kim, J.: Adaptive prototype learning and allocation for few-shot segmentation. In: IEEE Conference on Computer Vision and Pattern Recognition, pp. 8334–8343 (2021)
7. Liu, Y., Zhang, X., Zhang, S., He, X.: Part-aware prototype network for few-shot semantic segmentation. In: Vedaldi, A., Bischof, H., Brox, T., Frahm, J.-M. (eds.) ECCV 2020. LNCS, vol. 12354, pp. 142–158. Springer, Cham (2020). https://doi.org/10.1007/978-3-030-58545-7_9
8. Ouyang, C., Biffi, C., Chen, C., Kart, T., Qiu, H., Rueckert, D.: Self-supervision with superpixels: training few-shot medical image segmentation without annotation, pp. 762–780 (2020)
9. Ronneberger, O., Fischer, P., Brox, T.: U-Net: convolutional networks for biomedical image segmentation. In: Navab, N., Hornegger, J., Wells, W.M., Frangi, A.F. (eds.) MICCAI 2015. LNCS, vol. 9351, pp. 234–241. Springer, Cham (2015). https://doi.org/10.1007/978-3-319-24574-4_28
10. Roy, A., Siddiqui, S., Polsterl, S., Navab, N., Wachinger, C.: 'squeeze & excite' guided few-shot segmentation of volumetric images. Med. Image Anal. **59**, 101587 (2020)
11. Snell, J., Swersky, K., Zemel, R.: Prototypical networks for few-shot learning. In: Neural Information Processing Systems Foundation, p. 4077–4087 (2017)
12. Song, Y., Du, X., Zhang, Y., Li, S.: Two-stage segmentation network with feature aggregation and multi-level attention mechanism for multi-modality heart images. Comput. Med. Imaging Graph. **97**, 102054 (2022)
13. Sun, L., Ding, X., Huang, Y., Wang, G., Yu, Y.: Few-shot medical image segmentation using a global correlation network with discriminative embedding. Comput. Biol. Med. **140**, 105067 (2022)
14. Sung, F., Yang, Y., Zhang, L., Xiang, T., Torr, P., Hospedales, T.: Learning to compare: relation network for few-shot learning. In: IEEE Conference on Computer Vision and Pattern Recognition, pp. 1199–1208 (2018)
15. Tang, H., Liu, X., Sun, S., Yan, X., Xie, X.: Recurrent mask refinement for few-shot medical image segmentation. In: IEEE/CVF International Conference on Computer Vision. p. 3918–3928 (2020)

16. Tian, Z., Zhao, H., Shu, M., Yang, Z., Jia, J.: Prior guided feature enrichment network for few-shot segmentation. IEEE Trans. Pattern Anal. Mach. Intell. **99**, 1–1 (2020)

17. Vinyals, O., Blundell, C., Lillicrap, T., Kavukcuoglu, K., Wierstra, D.: Matching networks for one shot learning. In: Neural Information Processing Systems Foundation, pp. 3630–3638 (2016)

18. Wang, K., Liew, J., Zou, Y., Zhou, D., Feng, J.: Panet: few-shot image semantic segmentation with prototype alignment. In: IEEE/CVF International Conference on Computer Vision, pp. 9197–9206 (2019)

19. Wang, R., Zhou, Q., Zheng, G.: Few-shot medical image segmentation regularized with self-reference and contrastive learning. In: Wang, L., Dou, Q., Fletcher, P.T., Speidel, S., Li, S. (eds.) MICCAI 202. LNCS, vol. 13434, pp. 514–523. Springer, Heidelberg (2022). https://doi.org/10.1007/978-3-031-16440-8_49

20. Xu, C., Gao, Z., Zhang, H., Li, S., Albuquerque, V.: Video salient object detection using dual-stream spatiotemporal attention. Appl. Soft Comput. **108**, 107433 (2021)

21. Xu, C., et al.: Bmanet: boundary mining with adversarial learning for semi-supervised 2d myocardial infarction segmentation. IEEE J. Biomed. Health Inf. **27**(1), 87–96 (2023)

22. Xu, C., Zhang, D., Chong, J., Chen, B., Li, S.: Synthesis of gadolinium-enhanced liver tumors on nonenhanced liver mr images using pixel-level graph reinforcement learning. Med. Image Anal. **69**, 101976 (2021)

23. Yu, Q., Dang, K., Tajbakhsh, N., Terzopoulos, D., Ding, X.: A location-sensitive local prototype network for few-shot medical image segmentation. In: IEEE International Symposium on Biomedical Imaging, p. 262–266 (2021)

24. Zhang, C., Lin, G., Liu, F., Yao, R., Shen, C.: Canet: class-agnostic segmentation networks with iterative refinement and attentive few-shot learning. In: IEEE Conference on Computer Vision and Pattern Recognition, pp. 5217–5226 (2019)

25. Zhou, T., Wang, W., Konukoglu, E., Gool, L.: Rethinking semantic segmentation: a prototype view, pp. 2582–2593 (2022)

26. Zhuang, X.: Multivariate mixture model for myocardial segmentation combining multi-source images. IEEE Trans. Pattern Anal. Mach. Intell. **41**(12), 2933–2946 (2018)

Self-supervised Learning via Inter-modal Reconstruction and Feature Projection Networks for Label-Efficient 3D-to-2D Segmentation

José Morano[1]([✉]) [iD], Guilherme Aresta[1] [iD], Dmitrii Lachinov[1] [iD], Julia Mai[2] [iD], Ursula Schmidt-Erfurth[2] [iD], and Hrvoje Bogunović[1,2] [iD]

[1] Christian Doppler Laboratory for Artificial Intelligence in Retina, Department of Ophthalmology and Optometry, Medical University of Vienna, Vienna, Austria
{jose.moranosanchez,hrvoje.bogunovic}@meduniwien.ac.at
[2] Lab for Ophthalmic Image Analysis, Department of Ophthalmology and Optometry, Medical University of Vienna, Vienna, Austria

Abstract. Deep learning has become a valuable tool for the automation of certain medical image segmentation tasks, significantly relieving the workload of medical specialists. Some of these tasks require segmentation to be performed on a subset of the input dimensions, the most common case being 3D→2D. However, the performance of existing methods is strongly conditioned by the amount of labeled data available, as there is currently no data efficient method, e.g. transfer learning, that has been validated on these tasks. In this work, we propose a novel convolutional neural network (CNN) and self-supervised learning (SSL) method for label-efficient 3D→2D segmentation. The CNN is composed of a 3D encoder and a 2D decoder connected by novel 3D→2D blocks. The SSL method consists of reconstructing image pairs of modalities with different dimensionality. The approach has been validated in two tasks with clinical relevance: the en-face segmentation of geographic atrophy and reticular pseudodrusen in optical coherence tomography. Results on different datasets demonstrate that the proposed CNN significantly improves the state of the art in scenarios with limited labeled data by up to 8% in Dice score. Moreover, the proposed SSL method allows further improvement of this performance by up to 23%, and we show that the SSL is beneficial regardless of the network architecture.

Keywords: image segmentation · self-supervised learning · OCT · retina

1 Introduction

Deep learning can significantly reduce the workload of medical specialists during image segmentation tasks, which are essential for patient diagnosis and

Supplementary Information The online version contains supplementary material available at https://doi.org/10.1007/978-3-031-43901-8_56.

follow-up management [8,14,16]. For most tasks, segmentation masks have the same dimensionality as the input. However, there are some tasks for which segmentation has to be performed in a subset of the dimensions of the data, e.g. 3D→2D [13,21]. This occurs, for example, for the segmentation of geographic atrophy (GA) in optical coherence tomography (OCT), where the segmentation is performed on the OCT projection. In recent years, several methods have been proposed for this type of tasks [9,11–13]. Li *et al.* [11] proposed an image projection network (IPN) that reduces the features to the target dimensionality using unidirectional pooling layers in the encoder. However, IPN follows a patch-based approach with fixed patch size, which prevents its direct application to full 3D volumes of varying size. Also, it does not have skip connections, which have proven to be highly useful for accurate segmentation. Later, Lachinov *et al.* [9] proposed a U-Net-like convolutional neural network (CNN) for 3D→2D segmentation that overcomes the limitations of IPN, which were also later overcome by the second version of IPN (IPNv2) [12]. However, they still require a large amount of labeled data to provide adequate performance. In addition, there are works that explore the use of CNNs for 3D→2D regression, where Seeböck *et al.* [20] proposed ReSensNet, a novel CNN based on Residual 3D U-Net [10], with a 3D encoder and a 2D decoder connected by 3D→2D blocks. However, ReSensNet only works at concrete input resolutions, and it is applied pixel-wise.

In general, one of the issues of these and other deep learning segmentation methods is that their performance strongly depends on the amount of annotated data [22], which hinders their deployment to real-world medical image analysis settings. Transfer learning from ImageNet is the standard approach to mitigate this issue [22]. However, specifically for segmentation, ImageNet pre-training has shown minimal performance gains [5,17], partially because it can only be performed on the encoder part of the very common encoder-decoder architectures.

A possible alternative is to pre-train the models using a self-supervised learning (SSL) paradigm [1,2,4,6,7,15,18]. However, only some of these approaches have the potential to be applied for 3D→2D segmentation, as many of them, such as image denoising [2], require input and output images to have the same dimensionality. Among the suitable approaches, multi-modal reconstruction pre-training (MMRP) shows great potential in multi-modal scenarios [7]. In this approach, models are trained to reconstruct pairs of images from different modalities, learning relevant patterns in the data without requiring manual annotations. MMRP, however, has only been proven useful for localizing non-pathological structures on 2D color fundus photography, using fluorescein angiography as the modality to reconstruct. Moreover, image pairs of these modalities have to be registered using a separate method.

Contributions. In this work, we propose a novel approach for label-efficient 3D→2D segmentation. In particular, our contributions are as follows: (1) As an alternative to state-of-the-art network architectures, we propose a 3D→2D segmentation CNN based on ReSensNet [20] that has a 3D encoder and a 2D decoder connected by novel 3D→2D projective blocks. (2) We propose a novel SSL strategy for 3D→2D models based on the reconstruction of modalities of

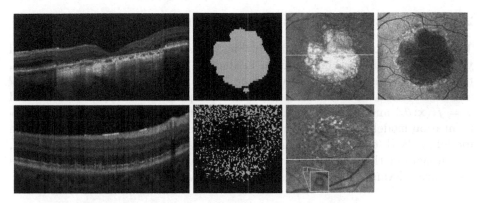

Fig. 1. From left to right: OCT slice (B-scan) with the corresponding ground truth annotations overlaid in green, ground truth, SLO with the location of the B-scan indicated in yellow and a zoom-in view in red, and FAF. Top: GA. Bottom: RPD. (Color figure online)

different dimensionality, and show that it significantly improves the performance of the models in the target segmentation tasks. This is the first data efficient method proposed for 3D→2D models and the first work exploring 3D→2D reconstruction. (3) Lastly, the performed experiments deepen the understanding of the proposed SSL paradigm, by exploring different settings with different image modalities. The proposed approach was validated on two clinically-relevant tasks: the en-face segmentation of GA and reticular pseudodrusen (RPD) in retinal OCT. The results demonstrate that the proposed approach clearly outperforms the state of the art in scenarios with scarce labeled data. Our code is publicly available on GitHub[1].

Clinical Background. *Geographic atrophy* (GA) is an advanced form of age-related macular degeneration (AMD) that corresponds to a progressive loss of retinal photoreceptors and leads to irreversible visual impairment. GA is typically assessed with OCT and/or fundus autofluorescence (FAF) imaging modalities [3,24]. In OCT, it is characterized by the loss of retinal pigment epithelium (RPE) tissue, accompanied by the contrast enhancement of the signal below the retina, and in FAF, by the loss of RPE autofluorescence [19] (see Fig. 1). In both cases, GA lesion is delineated as a 2D en-face area. Also, GA frequently appears brighter than the surrounding areas on scanning laser ophthalmoscopy (SLO) images due to its higher reflectance.

Reticular pseudodrusen (RPD) are accumulations of extracellular material that commonly occur in association with AMD. In OCT scans, these lesions are shown as granular hyperreflective deposits situated between the RPE layer and the ellipsoid zone. SLO visualizes RPD as a reticular pattern of iso-reflective round lesions surrounded by a hyporeflective border (see Fig. 1).

---

[1] https://github.com/j-morano/multimodal-ssl-fpn.

2   Methods and Experimental Setup

The proposed approach, illustrated in Fig. 2, is as follows. Let $\mathbf{x} \in \mathcal{X} \subset \mathbb{R}^n$ and $\mathbf{y} \in \mathcal{Y} \subset \mathbb{R}^{n-1}$ be two images from modalities $\mathcal{X}$ and $\mathcal{Y}$, and $\mathbf{z} \in \mathcal{Z} \subset \mathbb{R}^{n-1}$, their corresponding target segmentation mask. Let images and masks from $\mathcal{Y}$ and $\mathcal{Z}$ have related anatomical features. We optimize a reconstruction model $\mathbf{y} = f_r(\mathbf{x}; \theta_r)$ and transfer its knowledge by initializing the weights of the segmentation model $\mathbf{z} = f_s(\mathbf{x}; \theta_s)$ with the optimized weights of the reconstruction model f_r. With this approach, modality $\mathcal{Y}$ serves as a free source of supervision, and images of this modality serve as soft segmentation targets. Thus, models can learn relevant patterns in a self-supervised way.

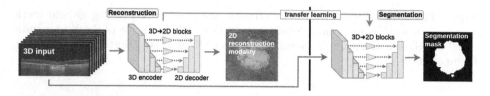

Fig. 2. Illustration of the proposed approach for 3D→2D segmentation. A novel 3D→2D model is trained for reconstructing image pairs of modalities with different dimensionality in a SSL setting, and then fine-tuned in the target segmentation task.

In this work, we propose a new CNN for the special case of 3D→2D segmentation, and we evaluate the proposed SSL approach for this case. In particular, we pre-train the new CNN to reconstruct SLO/FAF images from OCT (3D→2D), and then fine-tune it for GA/RPD segmentation. The advantage of reconstructing SLO over FAF is that several modern OCT devices allow to obtain co-registered OCT and SLO scans, providing the coordinates of each OCT slice within the SLO; thus, there is no need to use a separate registration method.

Network Architecture. The proposed network architecture (Fig. 3) is based on ReSensNet [20], and consists of a 3D encoder and a 2D decoder connected by novel 3D→2D feature projection blocks (FPBs). In the original work [20], training and inference are performed pixel-wise using fixed-size input patches. In contrast, we use full-size volumes of arbitrary resolution. To this end, we propose a novel type of FPB. In particular, all convolutions whose kernel size was equal to the expected feature size (calculated from the fixed size of the input patch) were replaced by $1 \times 1 \times 4$ convolutions. Then, to project 3D features at the output of each FPB to the 2D feature space, we add an adaptive average pooling of size 1 in the depth dimension at the end of each block. With this setting, feature selection and dimension reduction are performed at different scales, and the decoder processes only 2D features in the selected dimensions. This allows the model to learn the 3D structure of the data while being able to perform the segmentation in 2D. In addition, to overcome memory constraints and avoid overfitting, we reduce by half the number of kernels in each convolutional block.

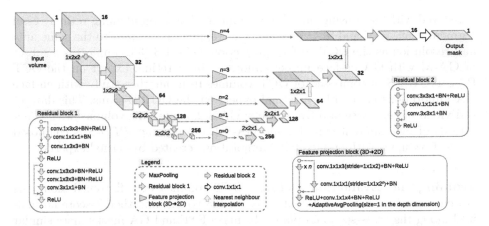

Fig. 3. Proposed 3D→2D CNN. Each residual encoder block has 8 3D convolutional layers, and each residual decoder block has 4 2D layers (number of feature maps also shown). The proposed feature projection block (FPB, in red) projects 3D features to the 2D feature space. FPBs have a variable number of $1 \times 1 \times 3$ convolutions followed by a $1 \times 1 \times 4$ convolution and a depth-wise adaptive average pooling of size 1. (Color figure online)

Table 1. Study data details. Dimensions: en-face height × en-face width × depth. FAF and SLO characteristics are after cropping to the same OCT en-face region projection.

Dataset	Scans	Patients (eyes)	Modality	Device	Area (mm)	Size (px)
GA-M	967	100 (184)	OCT	Spectralis	$6.68 \times 6.68 \times 1.92$	$49 \times 1024 \times 496$
			SLO	Spectralis	6.68×6.68	1024×1024
			FAF	Spectralis	6.68×6.68	1024×1024
GA-S	270	149 (166)	OCT	Spectralis	$6.02 \times 6.03 \times 1.92$	$49 \times 512 \times 496$
RPD-S	23	19 (23)	OCT	Spectralis	$5.73 \times 5.72 \times 1.92$	$97 \times 1024 \times 496$

Training Losses. As *reconstruction loss*, we use negative mean structural similarity index (NMSSIM) [23]. We empirically found that this loss performs equally or better than modern perceptual losses (e.g. LPIPS [25]) for our approach. NMSSIM loss can be defined as $\mathcal{L}_{\mathrm{NMSSIM}}(\mathbf{x}, \mathbf{y}) = -\frac{1}{HW} \sum_{h,w} \mathrm{SSIM}(\mathbf{x}_{hw}, \mathbf{y}_{hw})$, where $\mathbf{x}_{hw}$ and $\mathbf{y}_{hw}$ denote image patches of images $\mathbf{x}$ and $\mathbf{y}$ centered on the pixel with coordinates (h, w), $h \in H$ and $w \in W$, and $\mathrm{SSIM}(\mathbf{x}_{hw}, \mathbf{y}_{hw})$ is the SSIM map for those patches, as described in [23]. As *segmentation loss*, we use the direct sum of Dice loss and Binary Cross-Entropy. These two losses are standard for binary segmentation tasks [8,9,11,12,16].

Datasets. Experiments were performed using three datasets (Table 1). *GA-M* samples come from a clinical study on GA progression. OCT and SLO images were automatically co-registered by the imaging device, while FAF images were registered with SLO using an in-house pipeline based on aligning retinal vessel segmentation. FAF and SLO images were cropped and resized to the same area and resolution as the OCT en-face projection. GA-M-S (35 samples) is a subset of GA-M with GA en-face masks annotated by a retinal expert on the OCT B-Scans. *GA-S* is composed of OCT volumes from another study with en-face GA annotations created by a retinal expert on the OCT B-scans. This dataset is divided patient-wise into two subsets: GA-S-2, containing volumes with annotations of two different experts, and GA-S-1, of only one. *RPD-S* is composed of OCT volumes with en-face RPD annotations created by retinal experts.

Training and Evaluation Details. OCT volumes were flattened along the Bruch's membrane, rescaled depth-wise to 128 voxels, and then Z-score normalized along the cross-sectional plane. To make FAF and GA masks more similar and thus facilitate fine-tuning, FAF images were inverted. In all cases, models were trained for 800 epochs using SGD with a learning rate of 0.1 and a momentum of 0.9. Batch size was set to 4 for reconstruction and 8 for segmentation.

All datasets were split patient-wise into training (60%), validation (10%) and test (30%). For reconstruction, models were trained on GA-M. For GA segmentation, they were trained/fine-tuned on GA-S-1 and evaluated on GA-S-1, GA-S-2 and GA-M-S. For RPD segmentation, RPD-S was used. To evaluate the performance under label scarcity, we train with 5%, 10%, 20% and 100% of the data in GA-S, and 20% and 100%, in RPD-S. More details about the hardware used and the carbon footprint of our method are included in the Supplement.

To reduce inference variability, we average the predictions of the top-5 checkpoints of the models in terms of Dice (validation). Segmentations are evaluated via Dice and absolute area difference (Area diff.) of predicted and manual masks.

3   Results and Discussion

Baseline Comparison. We compared our approach to current state-of-the-art methods (IPN [11], IPNv2 [12], Lachinov *et al.* [9], and ReSensNet [20]), showing that we greatly improve the state of the art in GA segmentation in scenarios with limited labeled data (Fig. 4). When using only 5% of the data (40 samples), the mean Dice score was 23% higher than the best state-of-the-art approach. Even without SSL, the proposed CNN improves the Dice score by 8%. This gain is even greater in terms of Area diff. The improvement is also visible in the predicted segmentation masks (Fig. 5). When using our approach, the number of false positives and negatives is highly reduced. On the other hand, the improvement for RPD segmentation is more modest (in this case, we only compared with the current state-of-the-art-method: Lachinov et al. [9]). This can be explained by the greater visibility of GA features compared to RPD features in FAF and SLO (see Figs. 1 and 5). This suggests that SSL benefits from images with similar pathomorphological manifestations.

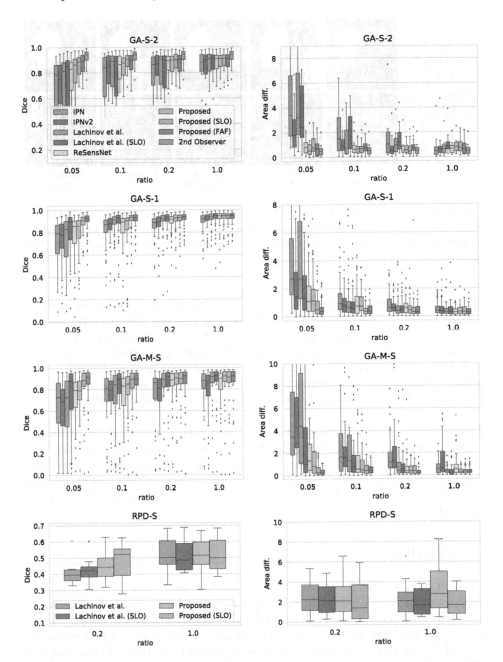

Fig. 4. Segmentation results of the models trained with different amounts of data. The title of each plot indicates the test dataset. If a model was pre-trained with SSL, the pre-training modality is shown in parentheses. A table with all means and standard deviations, as well as the results of a Wilcoxon signed rank test between our proposal and the others is included in the Supplement.

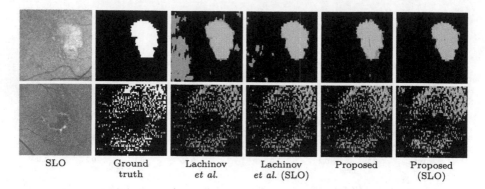

| SLO | Ground truth | Lachinov et al. | Lachinov et al. (SLO) | Proposed | Proposed (SLO) |

Fig. 5. Examples of GA (top) and RPD (bottom) segmentations from different models using the 5% and the 20% of the training data, respectively. True positives are depicted in green; true negatives, in black; false positives, in red; and false negatives, in blue. (Color figure online)

SSL Effect. To further assess the effect of the SSL, we also applied the strategy to the CNN by Lachinov et. al [9]. Figure 4 shows that SSL clearly improves the GA and RPD segmentation performance of both proposed and Lachinov *et al.* methods. These results are in line with the qualitative results in Fig. 5. This demonstrates that the SSL strategy is beneficial regardless of the architecture and the data. Notwithstanding, as discussed in the baseline comparison, the proposed SSL is more beneficial for GA segmentation than for RPD.

Reconstructed Modality Effect. We also conducted experiments to assess the effect of the reconstructed modality (SLO and FAF). Figure 4 shows that using FAF for SSL usually leads to better segmentation performance than using SLO. However, in multiple cases, the differences were not statistically significant. This is important because, unlike FAF, SLO does not require an external registration method.

4   Conclusions

Labeled data scarcity is one of the main limiting factors for the application of deep learning in medical imaging. In this work, we have proposed a new model and SSL strategy for label-efficient 3D→2D segmentation. The proposed approach was validated in two tasks with clinical relevance: the en-face segmentation of GA and RPD in OCT. The results demonstrate that: (1) the proposed CNN architecture clearly outperforms the state of the art when there is limited annotated data, (2) regardless of the architecture and the modality to be reconstructed, the proposed SSL strategy improves the performance of the models on the target tasks in those cases; (3) despite the greater diagnostic utility of FAF over SLO, SSL with FAF does not always result in a significant gain in model performance, with the advantage of the latter not requiring a supplementary registration method. On the other hand, although the proposed approach shows promising results in the en-face segmentation of RPD, further evaluation is needed.

Based on our findings, we believe that the proposed approach has the potential to be used in other common 3D→2D tasks, such as the prediction of retinal sensitivity in OCT, the segmentation of different structures in OCT-A, or the segmentation of intravascular ultrasound (IVUS). In addition, we also believe that the proposed SSL strategy could be easily extended to other imaging domains, such as magnetic resonance, where multi-modal data is widely used.

Acknowledgements. This work was supported in part by the Christian Doppler Research Association, Austrian Federal Ministry for Digital and Economic Affairs, the National Foundation for Research, Technology and Development, and Heidelberg Engineering.

References

1. Kalapos, A., Gyires-Tóth, B.: Self-supervised pretraining for 2D medical image segmentation. In: Karlinsky, L., Michaeli, T., Nishino, K. (eds.) ECCV 2022. LNCS, vol. 13807, pp. 472–484. Springer, Heidelberg (2022). https://doi.org/10.1007/978-3-031-25082-8_31
2. Brempong, E.A., Kornblith, S., Chen, T., Parmar, N., Minderer, M., Norouzi, M.: Denoising pretraining for semantic segmentation. In: Proceedings of the IEEE/CVF Conference on Computer Vision and Pattern Recognition, pp. 4175–4186 (2022)
3. Bui, P.T.A., et al.: Fundus autofluorescence and optical coherence tomography biomarkers associated with the progression of geographic atrophy secondary to age-related macular degeneration. Eye **36**(10), 2013–2019 (2021). https://doi.org/10.1038/s41433-021-01747-z
4. Grill, J.B., Strub, F., Altché, F., Tallec, C., Richemond, P.H., Buchatskaya, E., et al.: Bootstrap your own latent: a new approach to self-supervised learning. In: Proceedings of the 34th International Conference on Neural Information Processing Systems, NIPS 2020. Curran Associates Inc. (2020)
5. He, K., Girshick, R., Dollar, P.: Rethinking ImageNet pre-training. In: Proceedings of the IEEE/CVF International Conference on Computer Vision (ICCV) (2019)
6. Hervella, Á.S., Rouco, J., Novo, J., Ortega, M.: Retinal image understanding emerges from self-supervised multimodal reconstruction. In: Frangi, A.F., Schnabel, J.A., Davatzikos, C., Alberola-López, C., Fichtinger, G. (eds.) MICCAI 2018. LNCS, vol. 11070, pp. 321–328. Springer, Cham (2018). https://doi.org/10.1007/978-3-030-00928-1_37
7. Hervella, A.S., Rouco, J., Novo, J., Ortega, M.: Learning the retinal anatomy from scarce annotated data using self-supervised multimodal reconstruction. Appl. Soft Comput. **91**, 106210 (2020). https://doi.org/10.1016/j.asoc.2020.106210
8. Kavur, A.E., Gezer, N.S., Barış, M., Aslan, S., Conze, P.H., Groza, V., et al.: CHAOS challenge - combined (CT-MR) healthy abdominal organ segmentation. Med. Image Anal. **69**, 101950 (2021). https://doi.org/10.1016/j.media.2020.101950

9. Lachinov, D., Seeböck, P., Mai, J., Goldbach, F., Schmidt-Erfurth, U., Bogunovic, H.: Projective skip-connections for segmentation along a subset of dimensions in retinal OCT. In: de Bruijne, M., et al. (eds.) MICCAI 2021. LNCS, vol. 12901, pp. 431–441. Springer, Cham (2021). https://doi.org/10.1007/978-3-030-87193-2_41

10. Lee, K., Zung, J., Li, P., Jain, V., Seung, H.S.: Superhuman accuracy on the SNEMI3D connectomics challenge (2017). https://doi.org/10.48550/ARXIV.1706.00120

11. Li, M., et al.: Image projection network: 3D to 2D image segmentation in OCTA images. IEEE Trans. Med. Imaging **39**(11), 3343–3354 (2020). https://doi.org/10.1109/TMI.2020.2992244

12. Li, M., et al.: OCTA-500: a retinal dataset for optical coherence tomography angiography study (2022)

13. Liefers, B., González-Gonzalo, C., Klaver, C., van Ginneken, B., Sánchez, C.I.: Dense segmentation in selected dimensions: application to retinal optical coherence tomography. In: Cardoso, M.J., et al (eds.) Proceedings of The 2nd International Conference on Medical Imaging with Deep Learning. Proceedings of Machine Learning Research, vol. 102, pp. 337–346. PMLR (2019)

14. Menze, B.H., Jakab, A., Bauer, S., Kalpathy-Cramer, J., Farahani, K., Kirby, J., et al.: The multimodal brain tumor image segmentation benchmark (BRATS). IEEE Trans. Med. Imaging **34**(10), 1993–2024 (2015). https://doi.org/10.1109/TMI.2014.2377694

15. Morano, J., Álvaro S. Hervella, Barreira, N., Novo, J., Rouco, J.: Multimodal transfer learning-based approaches for retinal vascular segmentation. In: Giacomo, G.D., et al. (eds.) Proceedings of the 24th European Conference on Artificial Intelligence (ECAI 2020), pp. 1866–1873 (2020). https://doi.org/10.3233/FAIA200303

16. Orlando, J.I., Fu, H., Barbosa Breda, J., van Keer, K., Bathula, D.R., Diaz-Pinto, A., et al.: REFUGE challenge: a unified framework for evaluating automated methods for glaucoma assessment from fundus photographs. Med. Image Anal. **59**, 101570 (2020). https://doi.org/10.1016/j.media.2019.101570

17. Raghu, M., Zhang, C., Kleinberg, J., Bengio, S.: Transfusion: understanding transfer learning for medical imaging. In: Proceedings of the 33rd International Conference on Neural Information Processing Systems. Curran Associates Inc. (2019)

18. Ross, T., et al.: Exploiting the potential of unlabeled endoscopic video data with self-supervised learning. Int. J. Comput. Assist. Radiol. Surg. **13**(6), 925–933 (2018). https://doi.org/10.1007/s11548-018-1772-0

19. Schmitz-Valckenberg, S., et al.: Natural history of geographic atrophy progression secondary to age-related macular degeneration (geographic atrophy progression study). Ophthalmology **123**(2), 361–368 (2016). https://doi.org/10.1016/j.ophtha.2015.09.036

20. Seeböck, P., et al.: Linking function and structure with ReSensNet: predicting retinal sensitivity from OCT using deep learning. Ophthalmol. Retina **6**(6), 501–511 (2022). https://doi.org/10.1016/j.oret.2022.01.021

21. Sun, S., Sonka, M., Beichel, R.R.: Graph-based IVUS segmentation with efficient computer-aided refinement. IEEE Trans. Med. Imaging **32**(8), 1536–1549 (2013). https://doi.org/10.1109/TMI.2013.2260763

22. Tajbakhsh, N., Jeyaseelan, L., Li, Q., Chiang, J.N., Wu, Z., Ding, X.: Embracing imperfect datasets: a review of deep learning solutions for medical image segmentation. Med. Image Anal. **63**, 101693 (2020). https://doi.org/10.1016/j.media.2020.101693

23. Wang, Z., Bovik, A.C., Sheikh, H.R., Simoncelli, E.P.: Image quality assessment: from error visibility to structural similarity. IEEE Trans. Image Process. **13**(4), 600–612 (2004)
24. Wei, W., et al.: Two potentially distinct pathways to geographic atrophy in age-related macular degeneration characterized by quantitative fundus autofluorescence. Eye (2023). https://doi.org/10.1038/s41433-022-02332-8
25. Zhang, R., Isola, P., Efros, A.A., Shechtman, E., Wang, O.: The unreasonable effectiveness of deep features as a perceptual metric. In: 2018 IEEE/CVF Conference on Computer Vision and Pattern Recognition (CVPR), pp. 586–595 (2018). https://doi.org/10.1109/CVPR.2018.00068

RCS-YOLO: A Fast and High-Accuracy Object Detector for Brain Tumor Detection

Ming Kang⬛, Chee-Ming Ting(✉)⬛, Fung Fung Ting⬛,
and Raphaël C.-W. Phan⬛

School of Information Technology, Monash University, Malaysia Campus,
Subang Jaya, Malaysia
ting.cheeming@monash.edu

Abstract. With an excellent balance between speed and accuracy, cutting-edge YOLO frameworks have become one of the most efficient algorithms for object detection. However, the performance of using YOLO networks is scarcely investigated in brain tumor detection. We propose a novel YOLO architecture with Reparameterized Convolution based on channel Shuffle (RCS-YOLO). We present RCS and a One-Shot Aggregation of RCS (RCS-OSA), which link feature cascade and computation efficiency to extract richer information and reduce time consumption. Experimental results on the brain tumor dataset Br35H show that the proposed model surpasses YOLOv6, YOLOv7, and YOLOv8 in speed and accuracy. Notably, compared with YOLOv7, the precision of RCS-YOLO improves by 1%, and the inference speed by 60% at 114.8 images detected per second (FPS). Our proposed RCS-YOLO achieves state-of-the-art performance on the brain tumor detection task. The code is available at https://github.com/mkang315/RCS-YOLO.

Keywords: Medical image detection · YOLO · Reparameterized convolution · Channel shuffle · Computation efficiency

1 Introduction

Automatic detection of brain tumors from Magnetic Resonance Imaging (MRI) is complex, tedious, and time-consuming because there are a lot of missed, misinterpreted, and misleading tumor-like lesions in the images of the brain tumors [8]. Most of the current work focuses on brain tumor classification and segmentation from MRI and detection tasks are less explored [1,13,22]. While existing studies showed that various Convolutional Neural Networks (CNNs) are efficient for brain tumor detection, the performance of using You Only Look Once (YOLO) networks is scarcely investigated [12,20,23–25,27].

With the rapid development of CNNs, the accuracies of different visual tasks are constantly improved. However, the increasingly complex network architecture

Supplementary Information The online version contains supplementary material available at https://doi.org/10.1007/978-3-031-43901-8_57.

in CNN-based models, such as ResNet [6], DenseNet [9], Inception [28], etc. renders the inference speed slower. Though many advanced CNNs deliver higher accuracy, the complicated multi-branch designs (e.g., residual-addition in ResNet and branch-concatenation in Inception) make the models difficult to implement and customize, slowing down the inference and reducing memory utilization. The depth-wise separable convolutions used in MobileNets [7] also reduce the upper limit of the GPU inference speed. In addition, 3×3 regular convolution is highly optimized by some modern computing libraries. Consequently, VGG [26] is still heavily used for real-world applications in both research and industries.

RepVGG [2] is an extension of VGG via reparametrization to accelerate inference time. RepVGG uses a multi-branch topological architecture during the training phase, which is then reparameterized to a simplified single-branch architecture during the inference phase. In terms of the optimization strategy of network training, reparameterization was introduced in YOLOv6 [16], YOLOv7 [31], and YOLOv6 v3.0 [17]. YOLOv6 and YOLOv6 v3.0 employ reparameterization from RepVGG. RepConv, a RepVGG without an identity connection, is converted from RepVGG during inference time in YOLOv6, YOLOv6 v3.0, and YOLOv7 (named RepConvN in YOLOv7). Due to the removal of identity connections in RepConv, direct access to ResNet or the concatenation in DenseNet can provide more diversity of gradients for different feature maps. Grouped convolutions, which use a group of convolutions with multiple kernels per layer, like RepVGG, can also significantly reduce the computational complexity of the model, but there is no information communication between groups, which limits the ability of feature extraction of the convolution operator. In order to overcome the disadvantage of grouped convolutions, ShuffleNet V1 [34] and V2 [21] introduced the channel shuffle operation to facilitate information flows across different feature channels. In addition, when comparing Spatial Pyramid Pooling & Cross Stage Partial Network plus ConvBNSiLU (SPPCSPC) in YOLOv7 with Spatial Pyramid Pooling Fast (SPPF) in YOLOv5 [10] and YOLOv8 [11], it is found that more convolution layers in SPPCSPC architecture slow down the computation of the network. Nevertheless, SPP [4,5] module achieves the fusion of local features and global features by max-pooling different convolution kernels' sizes in the neck networks.

Aiming at a faster and high accuracy object detector for medical images, we propose a new YOLO architecture called RCS-YOLO by leveraging on the RepVGG/RepConv. The contributions of this work are summarized as follows:

1) We first develop a RepVGG/RepConv ShuffleNet (RCS) by combining the RepVGG/RepConv with a ShuffleNet which benefits from reparameterization to provide more feature information in the training stage and reduce inference time. Then, we build an RCS-based One-Shot Aggregation (RCS-OSA) module which allows not only low-cost memory consumption but also semantic information extraction.

2) We design new backbone and neck networks of YOLO architecture by incorporating the developed RCS-OSA and RepVGG/RepConv with path aggregation to shorten the information path between feature prediction layers.

This leads to fast propagation of accurate localization information to feature hierarchy in both the backbone and neck networks.

3) We apply the proposed RCS-YOLO model for a challenging task of brain tumor detection. To our best knowledge, this is the first work to leverage on YOLO-based model for fast brain tumor detection. Evaluation on a publicly available brain tumor detection annotated dataset shows superior detection accuracy and speed compared to other state-of-the-art YOLO architectures.

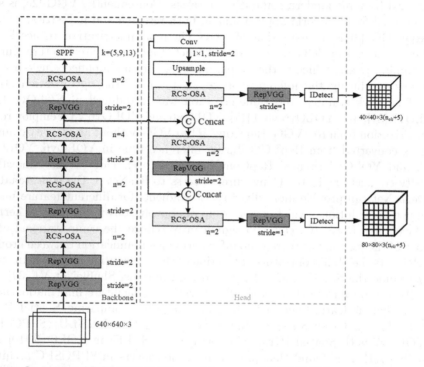

Fig. 1. Overview of RCS-YOLO. The architecture of RCS-YOLO is mainly comprised of the RCS-OSA (blue) and RepVGG (orange) modules. n represents the number of stacked RCS modules. n_{cls} represents the number of classes in detected objects. IDetect [29] from YOLOv7 denotes detection layers using 2D convolutional neural networks. (Color figure online)

2   Methods

The architecture of the proposed RCS-YOLO network is shown in Fig. 1. It incorporates a new module-RCS-OSA in the backbone and neck of the YOLO-based object detector.

2.1   RepVGG/RepConv ShuffleNet

Inspired by ShuffleNet, we design a structural reparameterized convolution based on channel shuffle. Figure 2 shows the structural schematic diagram of RCS.

Given that the feature dimensions of an input tensor are $C \times H \times W$, after the channel split operator, it is divided into two different channel-wise tensors with equal dimensions of $C \times H \times W$. For one of the tensors, we use the identity branch, 1×1 convolution, and 3×3 convolution to construct the training-time RCS. At the inference stage, the identity branch, 1×1 convolution, and 3×3 convolution are transformed to 3×3 RepConv by using structural reparameterization. The multi-branch topology architecture can learn abundant information about features during the training time, simplified single-branch architecture can save memory consumption during the inference time to achieve fast inference. After the multi-branch training of one of the tensors, it is concatenated to the other tensor in a channel-wise manner. The channel shuffle operator is also applied to enhance information fusion between two tensors so that the depth measurement between different channel features of the input can be realized with low computational complexity.

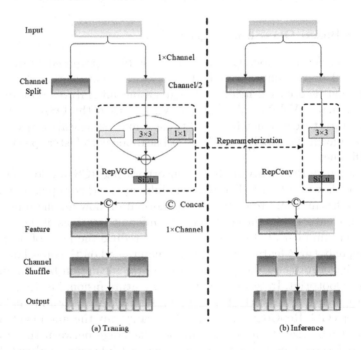

Fig. 2. The structure of RCS. (a) RepVGG at the training stage. (b) RepConv during model inference (or deployment). A rectangle with a black outer border represents the specific modular operation of the tensor; a rectangle with a gradient color represents the specific feature of the tensor, and the width of the rectangle represents the channel of the tensor.

When there is no channel shuffle, the output feature of each group only relates to the input feature within a group of grouped convolutions, and outputs from a certain group only relate to the input within the group. This blocks information flow between channel groups and weakens the ability of feature extraction. When channel shuffle is used, input and output features are fully related where one convolution group takes data from other groups, enabling more efficient feature information communication between different groups. The channel shuffle operates on stacked grouped convolutions and allows more informative feature representation. Moreover, assuming that the number of groups is g, for the same input feature, the computational complexity of channel shuffle is $\frac{1}{g}$ times that of a generic convolution.

Compared with the popular 3×3 convolution, during the inference stage, RCS uses the operators including channel split and channel shuffle to reduce the computational complexity by a factor of 2, while keeping the inter-channel information exchange. Moreover, using structural reparameterization enables deep representation learning from input features during the training stage, and reduction of inference-time memory consumption to achieve fast inference.

2.2 RCS-Based One-Shot Aggregation

The One-Shot Aggregation (OSA) module has been proposed to overcome the inefficiency of dense connections in DenseNet, by representing diversified features with multi-receptive fields and aggregating all features only once in the last feature maps. VoVNet V1 [14] and V2 [15] used the OSA module within its architecture to construct both lightweight and large-scale object detectors, which outperform the widely-used ResNet backbone with faster speed and better energy efficiency.

We develop an RCS-OSA module by incorporating RCS developed in Sect. 2.1 for OSA, as shown in Fig. 3. The RCS modules are stacked repeatedly to ensure the reuse of features and to enhance the information flow among different channels between features of adjacent layers. At different locations of the network, we set a different number of stacked modules. To reduce the level of network fragmentation, only three feature cascades are maintained on the one-shot aggregate path, which can mitigate the amount of network calculation burden and reduce the memory footprint. In terms of multi-scale feature fusion, inspired by the idea of Path Aggregation Network (PANet) [19], RCS-OSA + Upsampling and RCS-OSA + RepVGG/RepConv undersampling carry out the alignment of feature maps of different sizes to allow information exchange between the two prediction feature layers. This enables high-accuracy fast inference in object detection. Moreover, RCS-OSA maintains the same number of input channels and minimum output channels, thus reducing the memory access cost (MAC). For network building, we perpetuate max-pooling undersampling 32 times of YOLOv7 to construct a backbone network and adopt RepVGG/RepConv with a step of 2 to achieve undersampling. Due to the diversified feature representation of the RCS-OSA module and low-cost memory consumption, we use a different number of stacked RCS in RCS-OSA modules to achieve semantic information extraction during different stages of both backbone and neck networks.

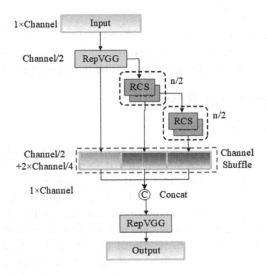

Fig. 3. The structure of RCS-OSA. n represents the number of stacked RCS modules.

The common evaluation metric of computation efficiency (or time complexity) is floating-point operations (FLOPs). FLOPs are only the indirect indicator to measure the speed of inference. However, the object detector with a DenseNet backbone shows rather slow speed and low energy efficiency because the linearly increasing number of channels by dense connection leads to heavy MAC, which causes considerable computation overhead. Given input features of dimension $M \times M$, the convolution kernel of size $K \times K$, number of input channels C_1, and the number of output channels C_2, FLOPs and MAC can be calculated as:

$$FLOPs = M^2 K^2 C_1 C_2 \tag{1}$$

$$MAC = M^2(C_1 + C_2) + K^2 C_1 C_2 \tag{2}$$

Assuming n to be 4, FLOPs of the proposed RCS-OSA and Efficient Layer Aggregation Networks (ELAN) [31,33] are $20.25C^2M^2$ and $40C^2M^2$ respectively. Compared with ELAN, FLOPs of RCS-OSA are reduced by nearly 50%. The MAC of RCS-OSA (i.e., $6CM^2 + 20.25C^2$) is also reduced compared to that of ELAN (i.e., $17CM^2 + 40C^2$).

2.3   Detection Head

To further reduce inference time, we decrease the number of detection heads comprised of RepVGG and IDetect from 3 to 2. The YOLOv5, YOLOv6, YOLOv7, and YOLOv8 have three detection heads. However, we use only two feature layers for prediction, reducing the number of original nine anchors with different scales to four and using the K-means unsupervised clustering method to regenerate anchors with different scales. The corresponding scales are (87, 90), (127,

139), (154, 171), (191,240). This not only reduces the number of convolution layers and computational complexity of RCS-YOLO but also reduces the overall computational requirements of the network during the inference stage and the computational time of postprocessing non-maximum suppression.

3 Experiments and Results

3.1 Dataset Details

To evaluate the proposed RCS-YOLO model, we used the brain tumor detection 2020 dataset (Br35H) [3], with a total of 701 images in the 'train' and 'val' two folders, 500 images of which are the 'train' folder were selected as the training set, while the other 201 images in the 'val' folder as the testing set. For the input size of 640×640 image, the actual corresponding size is 44×32. The small object is defined as the object whose pixel size is less than 32×32 defined by the MS COCO dataset [18], so there are no small objects in the brain tumor medical image data sets, and the scale change of the target boxes is smooth, almost square. The label boxes of the brain images were normalized (See supplementary material Sect. 1).

3.2 Implementation Details

For model training and inference, we used Ubuntu 18.04 LTS, Intel® Xeon® Gold 5218 CPU processor, CUDA 12.0, and cuDNN 8.2. GPU is GeForce RTX 3090 with 24G memory size. The networking development framework is Pytorch 1.9.1. The Integrated Development Environment (IDE) is PyCharm. We uniformly set epoch 150, the batch size as 8, image size as 640×640. Stochastic Gradient Descent (SGD) optimizer was used with an initial learning rate of 0.01 and weight decay of 0.0005.

3.3 Evaluation Metrics

In this paper, we choose precision, recall, AP_{50}, $AP_{50:95}$, FLOPs, and Frames Per Second (FPS) as comparative metrics of detection effect to determine the advantages and disadvantages of the model. Taking IoU = 0.5 as the standard, precision, and recall can be calculated by the following equations:

$$Precision = \frac{TP}{TP + FP} \tag{3}$$

$$Recall = \frac{TP}{(TP + PN)} \tag{4}$$

where TP represents the number of positive samples correctly identified as positive samples, FP represents the number of negative samples incorrectly identified as positive samples and FN represents the number of positive samples incorrectly identified as negative samples. AP_{50} is the area under the precision-recall (PR) curve formed by precision and recall. For $AP_{50:95}$, divide 10 IoU threshold of 0.5:0.05:0.95 to acquire the area under the PR curve, then average the results. FPS represents the number of images detected by the model per second.

3.4   Results

To highlight the accuracy and rapidity of the proposed model for the detection of brain tumor medical image data set, Table 1 shows the performance comparison between our proposed detector and other state-of-the-art object detectors. The time duration of FPS includes data preprocessing, forward model inference, and post-processing. The long border of the input images is set as 640 pixels. The short border adaptively scales without distortion, whilst keeping the grey filling with 32 times the pixels of the short border.

It can be seen that RCS-YOLO with the advantages of incorporating the RCS-OSA module performs well. Compared with YOLOv7, the FLOPs of the object detectors of this paper decrease by 8.8G, and the inference speed improves by 43.4 FPS. In terms of detection rate, precision improves by 0.024; AP_{50} increases by 0.01; $AP_{50:95}$ by 0.006. Also, RCS-YOLO is faster and more accurate than YOLOv6-L v3.0 and YOLOv8l. Although the $AP_{50:95}$ of RCS-YOLO equals that of YOLOv8l, it doesn't obscure the essential advantage of RCS-YOLO. The results clearly show the superior performance and efficiency of our method, compared to the state-of-the-art for brain tumor detection. As shown in supplementary material Fig. 2, brain tumor regions are accurately detected from MRI by using the proposed method.

Table 1. Quantitative results of different methods. The best results are shown in bold.

Model	Params	Precision	Recall	AP_{50}	$AP_{50:95}$	GFLOPs	FPS
YOLOv6-L [17]	59.6M	0.907	0.920	0.929	0.709	150.5	64.0
YOLOv7 [31]	36.9M	0.912	0.925	0.936	0.723	103.3	71.4
YOLOv8l [11]	43.9M	0.934	0.920	0.944	**0.729**	164.8	76.2
RCS-YOLO (Ours)	45.7M	**0.936**	**0.945**	**0.946**	**0.729**	**94.5**	**114.8**

Table 2. Ablation study on proposed RCS-OSA module. The best results are shown in bold.

Model	Params	Precision	Recall	AP_{50}	$AP_{50:95}$	GFLOPs	FPS
RepVGG-CSP (w/o RCS-OSA)	22.6M	0.926	0.930	0.933	0.689	**43.3**	6.1
RCS-YOLO (w/ RCS-OSA)	45.7M	**0.936**	**0.945**	**0.946**	**0.729**	94.5	**114.8**

3.5   Ablation Study

We demonstrate the effectiveness of the proposed RCS-OSA module in YOLO-based object detectors. The results of RepVGG-CSP in Table 2, where RCS-OSA in the RCS-YOLO is replaced with the Cross Stage Partial Network (CSP-Net) [32] used in existing YOLOv4-CSP [30] architecture, are decreased than RCS-YOLO except GFLOPs. Because the parameters of RepVGG-CSP (22.2M) are less than half those of RCS-YOLO (45.7M), the computation amount (i.e., GFLOPs) of RepVGG-CSP is accordingly smaller than RCS-YOLO. Nevertheless, RCS-YOLO still performs better in actual inference speed measured by FPS.

4 Conclusion

We developed an RCS-YOLO network for fast and accurate medical object detection, by leveraging the reparameterized convolution operator RCS based on channel shuffle in the YOLO architecture. We designed an efficient one-shot aggregation module RCS-OSA based on RCS, which serves as a computational unit in the backbone and neck of a new YOLO network. Evaluation of the brain MRI dataset shows superior performance for brain tumor detection in terms of both speed and precision, as compared to YOLOv6, YOLOv7, and YOLOv8 models.

References

1. Amin, J., Muhammad, S., Haldorai, A., Yasmin, M., Nayak, R.S.: Brain tumor detection and classification using machine learning: a comprehensive survey. Complex Intell. Syst. **8**, 3161–3183 (2022)
2. Ding, X., Zhang, X., Ma, N., Han, J., Ding, G., Sun, J.: RepVGG: making VGG-style ConvNets great again. In: 2021 IEEE/CVF Conference on Computer Vision and Pattern Recognition (CVPR), pp. 13733–13742. IEEE, Piscataway (2021)
3. Hamada, A.: Br35H : : brain tumor detection 2020. Kaggle (2020). https://www.kaggle.com/datasets/ahmedhamada0/brain-tumor-detection
4. He, K., Zhang, X., Ren, S., Sun, J.: Spatial pyramid pooling in deep convolutional networks for visual recognition. In: Fleet, D., Pajdla, T., Schiele, B., Tuytelaars, T. (eds.) ECCV 2014. LNCS, vol. 8691, pp. 346–361. Springer, Cham (2014). https://doi.org/10.1007/978-3-319-10578-9_23
5. He, K., Zhang, X., Ren, S., Sun, J.: Spatial pyramid pooling in deep convolutional networks for visual recognition. IEEE Trans. Pattern Anal. Mach. Intell. **37**(9), 1904–1916 (2015)
6. He, K., Zhang, X., Ren, S., Sun, J.: Deep residual learning for image recognition. In: 2016 IEEE Conference on Computer Vision and Pattern Recognition (CVPR), pp. 770–778. IEEE, Piscataway (2016)
7. Howard, A.G., et al.: MobileNets: efficient convolutional neural networks for mobile vision applications. arXiv preprint arXiv:1704.04861 (2017)
8. Huisman, T.A.: Tumor-like lesions of the brain. Cancer Imaging **9**(Special issue A), S10–S13 (2009)
9. Iandola, F., Moskewicz, M., Karayev, S., Girshick, R., Darrell, T., Keutzer, K.: DenseNet: implementing efficient ConvNet descriptor pyramids. arXiv preprint arXiv:1404.1869 (2014)

10. Jocher, G.: YOLOv5 (6.0/6.1) brief summary. GitHub (2022). https://github.com/ultralytics/yolov5/issues/6998
11. Jocher, G., Chaurasia, A., Qiu, J.: YOLO by Ultralytics (version 8.0.0). GitHub (2023). https://github.com/ultralytics/ultralytics
12. Kumar, V.V., Prince, P.G.K.: Brain lesion detection and analysis - a review. In: 2021 Fifth International Conference on I-SMAC (IoT in Social. Mobile, Analytics and Cloud) (I-SMAC), pp. 993–1001. Piscataway, IEEE (2021)
13. Lather, M., Singh, P.: Investigating brain tumor segmentation and detection techniques. Procedia Comput. Sci. **167**, 121–130 (2020)
14. Lee, Y., Hwang, J.-W., Lee, S., Bae, Y., Park, J.: An energy and GPU-computation efficient backbone network for real-time object detection. In: 2019 IEEE/CVF Conference on Computer Vision and Pattern Recognition Workshops (CVPRW), pp. 752–760. IEEE, Piscataway (2019)
15. Lee, Y., Park, J.: CenterMask: real-time anchor-free instance segmentation. In: 2020 IEEE/CVF Conference on Computer Vision and Pattern Recognition (CVPR), pp. 13903–13912. IEEE, Piscataway (2020)
16. Li, C., et al.: YOLOv6: a single-stage object detection framework for industrial applications. arXiv preprint arXiv:2209.02976 (2022)
17. Li, C., et al.: YOLOv6 v3.0: a full-scale reloading. arXiv preprint arXiv:2301.05586 (2023)
18. Lin, T.-Y., et al.: Microsoft COCO: common objects in context. In: Fleet, D., Pajdla, T., Schiele, B., Tuytelaars, T. (eds.) ECCV 2014. LNCS, vol. 8693, pp. 740–755. Springer, Cham (2014). https://doi.org/10.1007/978-3-319-10602-1_48
19. Liu, S., Qi, L., Qin, H., Shi, J., Jia, J.: Path aggregation network for instance segmentation. In: 2018 IEEE/CVF Conference on Computer Vision and Pattern Recognition (CVPR), pp. 8759–8768. IEEE, Piscataway (2018)
20. Lotlikar, V.S., Satpute, N., Gupta, A.: Brain tumor detection using machine learning and deep learning: a review. Curr. Med. Imaging **18**(6), 604–622 (2022)
21. Ma, N., Zhang, X., Zheng, H.-T., Sun, J.: ShuffleNet V2: practical guidelines for efficient CNN architecture design. In: Ferrari, V., Hebert, M., Sminchisescu, C., Weiss, Y. (eds.) ECCV 2018. LNCS, vol. 11218, pp. 122–138. Springer, Cham (2018). https://doi.org/10.1007/978-3-030-01264-9_8
22. Nazir, M., Shakil, S., Khurshid, K.: Role of deep learning in brain tumor detection and classification (2015 to 2020): a review. Comput. Med. Imaging Graph. **91**, 101940 (2021)
23. Rehman, A., Butt, M.A., Zaman, M.: A survey of medical image analysis using deep learning approaches. In: 2021 2nd International Conference on Smart Electronics and Communication (ICOSEC), pp. 1115–1120. IEEE, Piscataway (2021)
24. Shanishchara, P., Patel, V.D.: Brain tumor detection using supervised learning: a survey. In: 2022 Third International Conference on Intelligent Computing Instrumentation and Control Technologies (ICICICT), pp. 1159–1165. IEEE, Piscataway (2022)
25. Shirwaikar, R.D., Ramesh, K., Hiremath, A.: A survey on brain tumor detection using machine learning. In: 2021 International Conference on Forensics. Analytics, Big Data, Security (FABS), pp. 1–6. IEEE, Piscataway (2021)
26. Simonyan, K., Zisserman, A.: Very deep convolutional networks for large-scale image recognition. In: Bengio, Y., LeCun, Y. (eds.) 3rd International Conference on Learning Representations (ICLR) (2015)
27. Sravya, V., Malathi, S.: Survey on brain tumor detection using machine learning and deep learning. In: 2021 International Conference on Computer Communication and Informatics (ICCCI), pp. 1–3. IEEE, Piscataway (2021)

28. Szegedy, C., et al.: Going deeper with convolutions. In: 2015 IEEE Conference on Computer Vision and Pattern Recognition (CVPR), pp. 1–9. IEEE, Piscataway (2015)
29. Wang, C.-Y.: Yolov7.yaml. GitHub (2022). https://github.com/WongKinYiu/yolov7/blob/main/cfg/training/yolov7.yaml
30. Wang, C.-Y.: Yolov4-csp.yaml. GitHub (2022). https://github.com/WongKinYiu/yolov7/blob/main/cfg/baseline/yolov4-csp.yaml
31. Wang, C.-Y., Bochkovskiy, A., Liao, H.-Y.M.: YOLOv7: trainable bag-of-freebies sets new state-of-the-art for real-time object detectors. arXiv preprint arXiv:2207.02696 (2022)
32. Wang, C.-Y., Liao, H.-Y.M., Wu, Y.-H., Chen, P.-Y., Hsieh, J.-W., Yeh, I.-H.: CSPNet: a new backbone that can enhance learning capability of CNN. In: 2020 IEEE/CVF Conference on Computer Vision and Pattern Recognition Workshops (CVPRW), pp. 1571–1580. IEEE, Piscataway (2020)
33. Wang, C.-Y., Liao, H.-Y.M., Yeh, I.-H.: Designing network design strategies through gradient path analysis. arXiv preprint arXiv:2211.04800 (2022)
34. Zhang, X., Zhou, X., Lin, M., Sun, J.: ShuffleNet: an extremely efficient convolutional neural network for mobile devices. In: 2018 IEEE/CVF Conference on Computer Vision and Pattern Recognition (CVPR), pp. 6848–6856. IEEE, Piscataway (2018)

Certification of Deep Learning Models for Medical Image Segmentation

Othmane Laousy[1,2,3]([✉]), Alexandre Araujo[4], Guillaume Chassagnon[2], Nikos Paragios[5], Marie-Pierre Revel[2], and Maria Vakalopoulou[1,3]

[1] MICS, CentraleSupélec, Paris-Saclay University, Gif-sur-Yvette, France
`othmane.laousy@centralesupelec.fr`
[2] Hôpital Cochin, AP-HP, Paris-Cité University, Paris, France
[3] Inria Saclay, Gif-sur-Yvette, France
[4] New York University, New York, USA
[5] Therapanacea, Paris, France

Abstract. In medical imaging, segmentation models have known a significant improvement in the past decade and are now used daily in clinical practice. However, similar to classification models, segmentation models are affected by adversarial attacks. In a safety-critical field like healthcare, certifying model predictions is of the utmost importance. Randomized smoothing has been introduced lately and provides a framework to certify models and obtain theoretical guarantees. In this paper, we present for the first time a certified segmentation baseline for medical imaging based on randomized smoothing and diffusion models. Our results show that leveraging the power of denoising diffusion probabilistic models helps us overcome the limits of randomized smoothing. We conduct extensive experiments on five public datasets of chest X-rays, skin lesions, and colonoscopies, and empirically show that we are able to maintain high certified Dice scores even for highly perturbed images. Our work represents the first attempt to certify medical image segmentation models, and we aspire for it to set a foundation for future benchmarks in this crucial and largely uncharted area.

Keywords: Certified Robustness · Randomized Smoothing · Denoising Diffusion Models · Segmentation

1 Introduction

For the past decade, deep neural networks have dominated the computer vision community and provided near human performance on many different tasks, including classification [18], segmentation [24], and image generation [16]. Given these impressive results, convolutional neural networks are now used on a daily basis in fields like healthcare, self-driving cars, and robotics, to cite a few. In medical imaging, convolutional neural networks are particularly used to segment organs or regions of interest on different modalities such as X-rays, CT

Supplementary Information The online version contains supplementary material available at https://doi.org/10.1007/978-3-031-43901-8_58.

scans, MRIs, or ultrasound [36]. Indeed, segmentation techniques and variations of 2D and 3D U-Nets are currently the state-of-the-art to identify and isolate tumors, blood vessels, organs, or other structures within an image and provide crucial help to physicians for medical diagnosis, screening, and prognosis [32].

Nowadays, segmentation models are gaining widespread adoption in modern clinical practice and are being used with increasing frequency, making the results of these models critical for many patients. However, it is now commonly known that neural networks can be vulnerable to adversarial attacks [17,34], *i.e.*, small input perturbations invisible to humans crafted specifically such that the network performs errors. Over the past few years, a large body of work has devised empirical defenses against adversarial attacks for classification tasks [3,17,25], as well as segmentation tasks [37], including applications on medical imaging [27]. Although state-of-the-art empirical defenses provide significant robustness, these defenses do not guarantee *theoretical* robustness and stronger attacks can be crafted to break them [5]. Recently, *certified* defenses, for classification [2,11,26] and segmentation [15,23], have been proposed to guarantee the accuracy and reliability of neural networks. However, certified defenses for segmentation in the context of medical imaging are still lacking, even if models are getting market approvals (*e.g.*, FDA, CE) and are already adopted in clinical practice.

In this paper, we provide the first method for certified robustness in the context of segmentation for medical imaging. We leverage the *randomized smoothing* strategy [11,15], and the recent work on *diffusion models* [7] to achieve state-of-the-art certified robustness for segmentation models. Randomized smoothing consists in convolving the neural network with a Gaussian distribution (*i.e.*, by adding noise to the input) in order to obtain a smooth segmentation model. From the smoothness properties of the segmentation model, we can derive a robustness guarantee and compute a certified Dice score. We go even further by using diffusion models to first denoise the perturbed input and boost the certified robustness. By extension, we show that current diffusion models, trained on 'classical images' generalize well to medical datasets for denoising tasks. Extensive experiments on five public medical datasets of chest X-rays [21,31], skin lesions [10], and colonoscopies [6], and different popular segmentation models, prove the potential of our method. We hope that this study will provide the first step towards robustness guarantees for medical image segmentation.

2   Related Work

Since the discovery of adversarial attacks [17,34], numerous defenses [8,17,25] and attacks have been devised [8,25], demonstrating that neural networks are sensitive to small input perturbation and vulnerable to attacks. Adversarial training, which has been acknowledged as one of the most successful empirical defenses, consists in training a network directly on adversarial examples [25]. However, it is now known that even strong defenses can be bypassed by adaptive attacks [12]. Paschali et al. [27] were among the first to study adversarial attacks in the context of medical imaging. They conducted experiments using several neural network architectures [20,33] (*i.e.*, Inception V3, V4, MobileNet) and several attacks [17,25] to demonstrate that the vulnerability of neural networks is extended to medical images.

More specifically, in the context of classification, a previous work [4] has analyzed the robustness of neural networks for chest X-ray images and showed that gradient-based attacks were successful in fooling both machines and humans. In a similar line of work, Yao et al. [38] proposed an add-on to known attacks that bypasses state-of-the-art adversarial detectors making current defenses even less robust. On the other hand, several works have been focused on crafting defense strategies specifically in the context of medical imaging. For example, Almalik et al. [1] proposed a self-ensembling method to enhance the robustness of Vision Transformers in the presence of adversarial attacks. In the context of segmentation in medical imaging, [30] introduced a vector quantization approach by learning a discrete representation in a low dimensional embedding space and improving the robustness of a segmentation model. Finally, Daza et al. [13] proposed a lattice architecture that segments organs and lesions on MRI and CT scans and leveraged an efficient approach of adversarial training to defend against adversarial examples.

Although a large body of work has focused on constructing defenses for classification and segmentation tasks in the context of medical imaging, *certified* defenses are under-studied by the medical community. In this paper, we propose to leverage randomized smoothing and diffusion models for certified segmentation on medical datasets, setting the first baseline for this challenging problem and certifying popular segmentation architectures.

3   Randomized Smoothing

Randomized smoothing is a model agnostic technique, proposed by Cohen et al. [11], used to improve and certify the robustness of neural networks against adversarial attacks. This method consists in adding random noise (*e.g.*, noise generated from a Gaussian distribution) to the input data and then classifying the perturbed data using the neural network. Let $\mathcal{D} = \mathcal{X} \times \mathcal{Y}$ denote the data distribution where $\mathcal{X} \subset \mathbb{R}^d$ and $\mathcal{Y} = \{1, \ldots, k\}$ represent the input space and target space respectively and k is the number of classes. Let $f : \mathcal{X} \to \mathcal{Y}$ be a neural network such that for $(x, y) \in \mathcal{D}$, the classifier correctly classifies if $f(x) = y$. An adversarial attack is a small norm-bounded perturbation $\delta \in \mathbb{R}^d$ with $\|\delta\|_2 \leq \epsilon$ such that: $f(x + \delta) \neq y$. Randomized smoothing is a procedure to construct a new *smooth* classifier g given any base classifier f. Let $\mathcal{N}(0, \sigma^2 \mathbf{I})$ be a Gaussian distribution of mean 0 and variance σ, then, the smooth classifier g is defined as follows:

$$g(x) = \mathbb{P}_{\eta \sim \mathcal{N}(0,\sigma^2 \mathbf{I})} [f(x + \eta) = y]$$

Cohen et al. [11] have shown that if $R = \sigma \Phi^{-1}(g(x))$ where Φ is the cumulative distribution function of the standard Gaussian distribution and R can be considered the certified radius, then, $g(x + \delta) = y$ for all δ satisfying $\|\delta\|_2 \leq R$.

However, since it is not possible to compute g at x exactly, they proposed using Monte Carlo algorithms as an alternative approach for estimating $g(x)$ using random sampling. In order to obtain a reliable estimate of the probability $g(x)$, they also suggested a method that involves generating n samples of η from a normal distribution $\mathcal{N}(0, \sigma^2 \mathbf{I})$ and evaluating $f(x + \eta)$ for each sample. The resulting counts for each class in $\mathcal{Y}$ are then used to estimate probability p_y and the radius R with confidence $1 - \alpha$ (where α is a value between 0 and 1). If the confidence level cannot be achieved (for example, due to insufficient samples), the method will abstain from providing an estimate. More recently, Fischer et al. [15] built upon the work of [11] by introducing SEGCERTIFY, the first certified approach for image segmentation. The segmentation process involves assigning a segmentation class to every pixel in the image, which can be viewed as a form of classification at the pixel level. In the segmentation settings, the output space consists of regions or categories to be segmented, such as cars, roads, pedestrians, etc. The classifier function $f : \mathcal{X} \rightarrow \mathcal{Y}^d$ determines the class for each pixel and categorizes each component independently. In this context, the certification algorithm proposed by Cohen et al. [11] can be extended to accommodate the segmentation task.

To obtain a smooth classifier, it is necessary to add random noise to the input of the classifier. However, this creates a trade-off between accuracy and robustness. If low variance noise is added, accuracy won't be impacted significantly, but the certified radius will remain low. Conversely, adding high variance noise can improve certificates but at the expense of accuracy. To address this issue, Cohen et al. proposed a simple trick of training the network with noise injection during the training phase. While this method may reduce accuracy when evaluating the classifier with noise during the certification process, it can also help mitigate the trade-off between accuracy and robustness. One can note that during training, the network's objective is to learn to ignore the noise and classify at the same time. To improve the natural as well as the certified accuracy, Salman et al. [29] proposed to separate the two tasks with two networks trained separately. First, a network, $h : \mathcal{X} \rightarrow \mathcal{X}$, is trained to denoise the data such that for $\eta \sim \mathcal{N}(0, \sigma^2 \mathbf{I})$, we have $h(x + \eta) \approx x$, then, the output of the denoiser is given to the classifier.

In this paper, we leverage randomized smoothing and diffusion probabilistic models to obtain state-of-the-art results on certified segmentation for medical imaging. To the best of our knowledge, we are the first to propose a comprehensive study on certified segmentation for medical imaging.

4   Diffusion Probabilistic Models for Certification

The training of a Denoising Diffusion Probabilistic Model (DDPM) is an iterative process that involves adding a small amount of noise at every step of the diffusion process until random noise is reached. The reverse process then starts from random noise and generates a new image that conforms to the data distribution. Since DDPMs are inherently iterative denoising models, we can leverage this property for randomized smoothing. The idea would be to start the reverse process with a noisy image, rather than Gaussian noise, enabling the DDPM to output an image that resembles the original image.

Similar to Carlini et al. [7], our proposed pipeline is composed of two main steps: we denoise, then we certify. In order to complete the denoising, we need to first map between the noise model utilized in diffusion models and the one used in randomized smoothing. Randomized smoothing needs a data point that is enhanced with Gaussian noise added to it, given by $x_{rs} = x + \delta$ with $\delta \sim \mathcal{N}(x, \sigma^2 I)$. On the other hand, diffusion models suppose the noise model for $x_{DDPM} \sim \mathcal{N}(\sqrt{\alpha_t} x, (1 - \alpha_t) I)$. Programmatically, we start by adding Gaussian noise to an image x, obtaining x_{rs}. Then the timestep t^* on which we can use the diffusion model for randomized smoothing is defined. Depending on the scheduler of the denoiser, we compute t^* such that $\sigma^2 = \frac{1 - \alpha_{t^*}}{\alpha_{t^*}}$ (obtained by scaling x_{rs} with $\sqrt{\alpha_t}$ and pairing the variances). We then calculate $x_{DDPM} = \sqrt{\alpha_{t^*}}(x + \delta), \delta \sim \mathcal{N}(0, \sigma^2 I)$. After that, we apply a single-step denoiser and predict the completely denoised image. A single-step denoising involves directly predicting the image from t^* to $t = 0$. A multi-step denoising strategy implies iteratively predicting all images at $t^*, t^* - 1, \ldots$ until $t = 0$. Both techniques are explored in the next section and supplementary material.

Since randomized smoothing is applied to each pixel separately with a probability of $1 - \alpha$, considering the entire segmentation region would imply considering a union bound with significantly reduced confidence. Similar to Fischer et al. [15], we leverage the Holm-Bonferroni method [19] and perform multiple-testing corrections. For each image, we repeat this process $n = 100$ times, identifying pixels on which the model abstains, and computing the certified scores. We extend the work of Fischer et al. [15] to also compute a certified Dice score that is calculated ignoring the abstain class ($\oslash$). Our approach has a significant advantage compared to SEGCERTIFY since it leverages off-the-shelf and state-of-the-art pre-trained denoisers and segmentation models. SEGCERTIFY, on the other hand, relies on models trained with Gaussian noise.

5   Experiments and Results

Datasets: We perform experiments on 5 different publicly available datasets. All datasets were divided to 70% for training, 10% for validation, and 20% for testing. The testing set is the one used to compute certified results.

Chest X-rays Datasets: JSRT dataset [31] with annotations of lung, heart, and clavicles provided by [35] is used. This dataset contains 247 images. For lung segmentation only, we use both the Montgomery and Shenzen datasets [21]. Montgomery consists of 138 and Shenzen of 662 annotated images.

Skin Lesion: Skin images with their annotations provided by the ISIC 2018 boundary segmentation challenge [10] were used. This dataset consists of 2694 RGB dermatoscopy images.

Table 1. Comparison of our approach with three different model architectures on chest X-ray datasets. We report certified Dice, IoU and percentage of abstentions (%⊘) for different noise levels σ and radii R.

σ	R	JSRT							Montgomery			Schenzen		
		Lung		Heart		Clavicles			Lung			Lung		
		Dice	IoU	Dice	IoU	Dice	IoU	%⊘	Dice	IoU	%⊘	Dice	IoU	%⊘
UNet [28]														
0.25	0.17	0.94	0.91	0.88	0.79	0.75	0.63	0.07	0.93	0.89	0.07	0.95	0.90	0.05
0.50	0.34	0.90	0.83	0.88	0.79	0.61	0.45	0.09	0.89	0.80	0.07	0.93	0.90	0.02
1.00	0.67	0.87	0.79	0.84	0.75	0.23	0.15	0.15	0.88	0.80	0.14	0.89	0.83	0.10
ResUNet++ [22]														
0.25	0.17	0.95	0.91	0.93	0.87	0.78	0.65	0.05	0.96	0.93	0.02	0.95	0.91	0.01
0.50	0.34	0.94	0.88	0.91	0.83	0.63	0.48	0.08	0.94	0.89	0.03	0.93	0.90	0.02
1.00	0.67	0.90	0.82	0.87	0.77	0.28	0.19	0.12	0.89	0.83	0.07	0.90	0.85	0.06
DeeplabV2 [9]														
0.25	0.17	0.94	0.91	0.91	0.86	0.85	0.75	0.04	0.93	0.91	0.07	0.80	0.71	0.07
0.50	0.34	0.88	0.81	0.87	0.79	0.63	0.49	0.10	0.91	0.87	0.02	0.34	0.25	0.15
1.00	0.67	0.88	0.80	0.83	0.74	0.20	0.11	0.14	0.85	0.79	0.17	0.04	0.02	0.11

Colonoscopy Images: CVC-ClinicDB dataset [6] containing 612 colonoscopy images in RGB together with their annotations were utilized.

Implementation Details: We train three different segmentation models namely, a UNet [28], a ResUNet++ [22], and a DeeplabV2 [9] with and without noise. The models trained without noise are used exclusively with our method. The models trained with a Gaussian noise of 0.25 are used to compute SEGCERTIFY scores. All 6 models use an image input size of 512×512 for X-ray images, 384×512 for skin lesions, and 288×384 for colonoscopy. As a denoiser, we use an off-the-shelf denoising diffusion probabilistic model provided by [14]. We perform our experiments with the 256×256 class unconditional denoiser with a linear scheduler and without timestep respacing. For each noise level, our method follows the steps described in the previous section and uses $n_0 = 10$, n=100 for each image, and $\alpha = 0.001$, and $\tau = 0.75$. Our code is made publicly available at: https://github.com/othmanela/medical_cert_seg.

Results and Discussion: For all five datasets, we compute a certified Dice score and certified mean Intersection over Union (IoU). We also report the percentage of abstentions (%⊘) representing the mean number of pixels on which the model's prediction confidence was insufficient with respect to the radius R. The lower the percentage of abstentions the better the segmentation model is.

In Table 1, we compare our method using 3 different and popular architectures (UNet, ResUNet++, and DeeplabV2) on the chest X-rays datasets. We

Table 2. Certified segmentation results of our technique and SEGCERTIFY [15] on the chest X-ray JSRT dataset. We report Dice, IoU, and percentage of abstentions (%⊘) for each class.

Model	Trained with noise	σ	R	Lung Dice	Lung IoU	Heart Dice	Heart IoU	Clavicles Dice	Clavicles IoU	%⊘
ResUNet++ [22]	✗	0.00	0.00	0.97	0.94	0.94	0.91	0.93	0.91	0.00
	✓	0.25	0.00	0.91	0.90	0.89	0.87	0.84	0.79	0.00
SEGCERTIFY [15]	✓	0.25	0.17	**0.96**	**0.92**	**0.93**	**0.88**	**0.83**	**0.72**	0.04
	✓	0.50	0.34	0.89	0.84	0.85	0.79	0.58	0.43	0.13
	✓	1.00	0.67	0.07	0.04	0.02	0.01	0.00	0.00	0.24
	✗	0.25	0.17	0.95	0.91	**0.93**	0.87	0.78	0.65	0.05
Ours	✗	0.50	0.34	**0.94**	**0.88**	**0.91**	**0.83**	0.63	0.48	0.08
	✗	1.00	0.67	**0.90**	**0.82**	**0.87**	**0.77**	0.28	0.19	0.12

notice that our method maintains overall good results on all three model backbones. A similar table with SEGCERTIFY results is provided in Table S2 of the supplementary material. Overall, for both methods, ResUNet++ is the most robust architecture followed by UNet and then DeeplabV2 for all σ and R combinations. Moreover, certified metrics for lungs and heart remain high for our method, even with high levels of noise. However, the increasing level of noise affects the clavicles since these are smaller structures.

A comparison of our method and SEGCERTIFY using the ResUNet++ architecture is presented in Table 2 for the three chest X-ray datasets. We observe that we outperform SEGCERTIFY, especially for high sigma values. For $\sigma = 0.25$, SEGCERTIFY performs slightly better. This is due to the fact that the model used with SEGCERTIFY is trained with a noise level of 0.25. The main drawback however is that its Dice on unperturbed images drops considerably (*e.g.*, from 0.96 to 0.91 on lung segmentation). On the other hand, our pipeline does not require training a segmentation model with noise or even a denoising model. Our methodology relies only on off-the-shelf models. For the highest noise level of $\sigma = 1.0$, we notice that the certified Dice and IoU with SEGCERTIFY both drop to 0 whereas our proposed method is able to maintain high certified scores.

Qualitative results are provided in Fig. 1 for our proposed method and SEGCERTIFY for the different datasets and different levels of noise. Regarding the structures to segment, we notice that the abstentions around the clavicles (the smallest benchmarked region of interest on chest X-rays) get bigger. We also notice that the fine segmentation boundaries (*e.g.*, area around the skin lesion) may not be as sharp after denoising. As we increase the noise, the decision boundary is harder to find for all models. This may be due to the fact that fine details on the image are lost after the denoising step. However, our method is still able to segment the large majority of pixels properly on the image, contrary to its competitor, especially for high noise levels (third row on chest X-rays).

Table 3 reports certified segmentation results for skin lesions and colonoscopy on both techniques. We notice that our method is still performing better than

Table 3. Results on skin lesions [10] and CVC-ClinicDB [6] segmentation.

Model	Method	σ	R	Skin Lesions			CVC-ClinicDB		
				Dice	IoU	%⊘	Dice	IoU	%⊘
ResUNet++ [22]	SegCertify [15]	0.25	0.17	0.79	0.68	0.07	0.63	0.56	0.05
		0.50	0.34	0.41	0.27	0.06	0.15	0.10	0.01
		1.00	0.67	0.00	0.00	0.01	0.00	0.00	0.00
	Ours	0.25	0.17	0.85	0.77	0.03	0.65	0.57	0.04
		0.50	0.34	0.83	0.76	0.04	0.45	0.39	0.07
		1.00	0.67	0.77	0.69	0.06	0.26	0.23	0.14

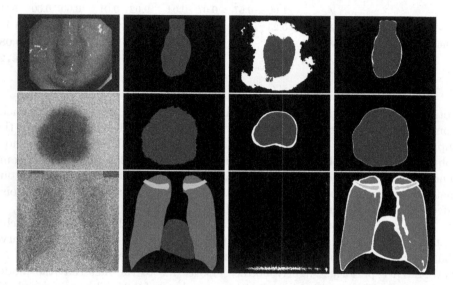

Fig. 1. Qualitative results of SegCertify and our method on colonoscopy, skin lesion, and chest X-ray images. From left to right: image with added noise, ground truth, SegCertify segmentation, our segmentation. White pixels denote abstention areas of the segmentation models. We increase the noise level from top to bottom: $\sigma = 0.25, 0.5$, and 1.0.

SegCertify. This supports our claim that DDPMs generalize quite well to medical images and that harnessing their potential boosts the state-of-the-art. Regarding the denoiser, we used a single-step denoising strategy, *i.e.*, we perform a single call to the DDPM to compute the denoised image from t^* to $t = 0$. Another strategy could be to iteratively denoise from t^*, $t^* - 1$, ... until $t = 0$. However, this implies predicting a denoised image multiple times and in the end, may result in images with unwanted artifacts. We perform multi-step denoising experiments and report results in Table S1 of the supplementary material. We note that the single-step denoising performs best since it relies more on the denoising power of DDPMs rather than their generative capabilities, and is also faster than the multi-step approach. Finally, we perform a comparison with another denoiser architecture. We train three UNet models (one for each noise level) on the JSRT dataset. We report results in Table S3 and notice that even with custom-trained denoisers, the DDPM outperforms the UNet denoising architecture. A comparison of denoised images is provided in Figure S1. We notice that the DDPM is able to keep high-fidelity images compared to the UNet and is therefore more relevant for certified medical image segmentation.

6 Conclusion

In this paper, we present the first work on certified segmentation for medical imaging, and extensively evaluate it on five different datasets and three deep learning segmentation models. Our technique leverages off-the-shelf denoising and segmentation models and provides the highest certified Dice and mIoU on multi-class and binary segmentation of five different datasets. With that, we are able to remove the overhead of having to train and fine-tune models specifically for robustness. This paradigm shift alleviates the dilemma of having to choose between highly accurate segmentation models or models robust to attacks. We hope that this work serves as a baseline for the unexplored yet critical topic of certified segmentation in medical imaging. Future work will involve extending our approach to 3D medical imaging modalities as well as exploring the realm of certified classification.

Acknowledgements. This work was granted access to the HPC resources of IDRIS under the allocation 2023-AD011013308R1 made by GENCI and it was partially supported by the ANR Hagnodice ANR-21-CE45-0007.

References

1. Almalik, F., Yaqub, M., Nandakumar, K.: Self-ensembling vision transformer (sevit) for robust medical image classification. In: Wang, L., Dou, Q., Fletcher, P.T., Speidel, S., Li, S. (eds.) MICCAI 2022. LNCS, vol. 13433, pp. 376–386. Springer, Heidelberg (2022). https://doi.org/10.1007/978-3-031-16437-8_36
2. Araujo, A., Havens, A., Delattre, B., Allauzen, A., Hu, B.: A unified algebraic perspective on lipschitz neural networks. In: ICLR (2023)
3. Araujo, A., Meunier, L., Pinot, R., Negrevergne, B.: Advocating for multiple defense strategies against adversarial examples. In: ECML (2020)
4. Asgari Taghanaki, S., Das, A., Hamarneh, G.: Vulnerability analysis of chest x-ray image classification against adversarial attacks. In: iMIMIC (2018)
5. Athalye, A., Carlini, N., Wagner, D.: Obfuscated gradients give a false sense of security: circumventing defenses to adversarial examples. In: ICML (2018)
6. Bernal, J., et al.: Wm-dova maps for accurate polyp highlighting in colonoscopy: validation vs. saliency maps from physicians. Comput. Med. Imaging Graph. **43**, 99–111 (2015)
7. Carlini, N., Tramer, F., Kolter, J.Z., et al.: (certified!!) adversarial robustness for free!. In: ICLR (2023)
8. Carlini, N., Wagner, D.: Towards evaluating the robustness of neural networks. In: IEEE on Security and Privacy (2017)

9. Chen, L.C., Papandreou, G., Kokkinos, I., Murphy, K., Yuille, A.L.: Deeplab: semantic image segmentation with deep convolutional nets, atrous convolution, and fully connected crfs. IEEE Pattern Anal. Mach. Intell. **40**, 834–848 (2016)
10. Codella, N.C.F., et al.: Skin lesion analysis toward melanoma detection 2018: a challenge hosted by the international skin imaging collaboration (ISIC). CoRR (2019)
11. Cohen, J., Rosenfeld, E., Kolter, Z.: Certified adversarial robustness via randomized smoothing. In: ICML (2019)
12. Croce, F., Hein, M.: Reliable evaluation of adversarial robustness with an ensemble of diverse parameter-free attacks. In: ICML (2020)
13. Daza, L., Pérez, J.C., Arbeláez, P.: Towards robust general medical image segmentation. In: de Bruijne, M., et al. (eds.) MICCAI 2021. LNCS, vol. 12903, pp. 3–13. Springer, Cham (2021). https://doi.org/10.1007/978-3-030-87199-4_1
14. Dhariwal, P., Nichol, A.: Diffusion models beat gans on image synthesis. In: NeurIPS (2021)
15. Fischer, M., Baader, M., Vechev, M.: Scalable certified segmentation via randomized smoothing. In: ICML (2021)
16. Goodfellow, I., et al.: Generative adversarial networks. ACM (2020)
17. Goodfellow, I.J., Shlens, J., Szegedy, C.: Explaining and harnessing adversarial examples. arXiv:1412.6572 (2014)
18. He, K., Zhang, X., Ren, S., Sun, J.: Deep residual learning for image recognition. In: CVPR (2016)
19. Holm, S.: A simple sequentially rejective multiple test procedure. J. Stat. **6**, 65–70 (1979)
20. Howard, A.G., et al.: Mobilenets: efficient convolutional neural networks for mobile vision applications. arXiv:1704.04861 (2017)
21. Jaeger, S., Candemir, S., Antani, S., Wáng, Y.X.J., Lu, P.X., Thoma, G.: Two public chest x-ray datasets for computer-aided screening of pulmonary diseases. Quant. Imaging Med. Surg. **4**, 475 (2014)
22. Jha, D., et al.: Resunet++: an advanced architecture for medical image segmentation. In: IEEE on Multimedia (2019)
23. Laousy, O., et al.: Towards better certified segmentation via diffusion models. In: UAI (2023)
24. Long, J., Shelhamer, E., Darrell, T.: Fully convolutional networks for semantic segmentation. In: CVPR (2015)
25. Madry, A., Makelov, A., Schmidt, L., Tsipras, D., Vladu, A.: Towards deep learning models resistant to adversarial attacks. arXiv:1706.06083 (2017)
26. Meunier, L., Delattre, B., Araujo, A., Allauzen, A.: A dynamical system perspective for lipschitz neural networks. In: ICML (2022)
27. Paschali, M., Conjeti, S., Navarro, F., Navab, N.: Generalizability *vs.* robustness: investigating medical imaging networks using adversarial examples. In: Frangi, A.F., Schnabel, J.A., Davatzikos, C., Alberola-López, C., Fichtinger, G. (eds.) MICCAI 2018. LNCS, vol. 11070, pp. 493–501. Springer, Cham (2018). https://doi.org/10.1007/978-3-030-00928-1_56
28. Ronneberger, O., Fischer, P., Brox, T.: U-Net: convolutional networks for biomedical image segmentation. In: Navab, N., Hornegger, J., Wells, W.M., Frangi, A.F. (eds.) MICCAI 2015. LNCS, vol. 9351, pp. 234–241. Springer, Cham (2015). https://doi.org/10.1007/978-3-319-24574-4_28
29. Salman, H., Sun, M., Yang, G., Kapoor, A., Kolter, J.Z.: Denoised smoothing: a provable defense for pretrained classifiers. In: NeurIPS (2020)

30. Santhirasekaram, A., Kori, A., Winkler, M., Rockall, A., Glocker, B.: Vector quantisation for robust segmentation. In: Wang, L., Dou, Q., Fletcher, P.T., Speidel, S., Li, S. (eds.) MICCAI 2022. LNCS, vol. 13434, pp. 663–672. Springer, Heidelberg (2022). https://doi.org/10.1007/978-3-031-16440-8_63

31. Shiraishi, J., et al.: Development of a digital image database for chest radiographs with and without a lung nodule: receiver operating characteristic analysis of radiologists' detection of pulmonary nodules. Am. J. Roentgenol. **174**, 71–74 (2000)

32. Siddique, N., Paheding, S., Elkin, C.P., Devabhaktuni, V.: U-net and its variants for medical image segmentation: a review of theory and applications. IEEE Access **9**, 82031–82057 (2021)

33. Szegedy, C., Ioffe, S., Vanhoucke, V., Alemi, A.: Inception-v4, inception-resnet and the impact of residual connections on learning. In: AAAI (2017)

34. Szegedy, C., et al.: Intriguing properties of neural networks. arXiv:1312.6199 (2013)

35. Van Ginneken, B., Stegmann, M.B., Loog, M.: Segmentation of anatomical structures in chest radiographs using supervised methods: a comparative study on a public database. Med. Image Anal. **10**, 19–40 (2006)

36. Wang, R., Lei, T., Cui, R., Zhang, B., Meng, H., Nandi, A.K.: Medical image segmentation using deep learning: a survey. arXiv:2009.13120v3 (2020)

37. Xie, C., Wang, J., Zhang, Z., Zhou, Y., Xie, L., Yuille, A.: Adversarial examples for semantic segmentation and object detection. In: CVPR (2017)

38. Yao, Q., He, Z., Lin, Y., Ma, K., Zheng, Y., Zhou, S.K.: A hierarchical feature constraint to camouflage medical adversarial attacks. In: de Bruijne, M., et al. (eds.) MICCAI 2021. LNCS, vol. 12903, pp. 36–47. Springer, Cham (2021). https://doi.org/10.1007/978-3-030-87199-4_4

Diffusion Transformer U-Net for Medical Image Segmentation

G. Jignesh Chowdary$^{(\boxtimes)}$ and Zhaozheng Yin

Stony Brook University, Stony Brook, NY, USA
jigneshchowdary@gmail.com

Abstract. Diffusion model has shown its power on various generation tasks. When applying the diffusion model in medical image segmentation, there are a few roadblocks to remove: the semantic features required for the conditioning of the diffusion process are not well aligned with the noise embedding; and the U-Net backbone employed in these diffusion models is not sensitive to contextual information that is essential during the reverse diffusion process for accurate pixel-level segmentation. To overcome these limitations, we present a cross-attention module to enhance the conditioning from source images, and a transformer based U-Net with multi-sized windows for the extraction of various scales of contextual information. Evaluated on five benchmark datasets with different imaging modalities including Kvasir-Seg, CVC Clinic DB, ISIC 2017, ISIC 2018, and Refuge, our diffusion transformer U-Net achieves great generalization ability and outperforms all the state-of-the-art models on these datasets.

Keywords: Diffusion model · Transformer · U-Net · Medical Image Segmentation

1 Introduction

Deep Learning (DL) methods like Convolutional Neural Networks (CNN) and Vision-Transformers (ViT) have been applied to medical image segmentation [7, 8, 17] with good performance. However, these DL methods have some inherent limitations on their network architectures. For example, CNNs are capable of extracting local features but not direct global features, whereas ViTs employ a fixed window which limit their capability to extract fine contextual details that are necessary for accurate pixel-level segmentation.

Recently, Denoising Diffusion Probabilistic Model (DDPM) [9] shows great performance in various conditional and unconditional generation tasks, and it is also applied to medical image segmentation [23, 24]. Despite of the success, there are a few shortcomings to overcome: (1) The semantic embedding extracted from the source image is not well aligned with the noise embedding in the diffusion

---

Supplementary Information The online version contains supplementary material available at https://doi.org/10.1007/978-3-031-43901-8_59.

process, leading to poor conditioning and subpar performance; and (2) The U-Net backbone in these DDPM-based methods is not sensitive to various scales of contextual information during the reverse diffusion (denoising) process, observed in CNNs and ViTs as well.

Motivated by the underlined limitations, we propose a Diffusion Transformer U-Net, with the following contributions:

- A conditional diffusion model with forward and backward processes is proposed to train segmentation networks. In the backward denoising process, the feature embedding of a noise image is aligned with that of the conditional source image by a new cross-attention module. Then, it is denoised into a segmentation mask of the source image by the segmentation network.
- A transformer-based U-Net with multi-sized windows, named as *MT U-Net*, is designed to extract both pixel-level and global contextual features for achieving good segmentation performance.
- The MT U-Net trained by the diffusion model has a great generalization capability on various imaging modalities, and outperforms all the current state-of-the-art on five benchmark datasets including polyp segmentation from colonoscopy images [1,10], skin lesion segmentation from dermoscopy images [4,5], and optic-cup segmentation from retinal fundus images [14].

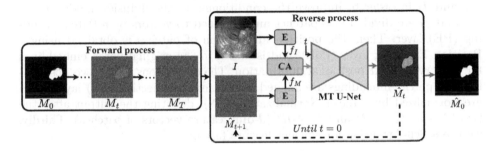

Fig. 1. Diffusion model with Cross-Attention (CA) to train the MT U-Net.

2 Method

2.1 Diffusion Model

The diffusion has two processes (Fig. 1): forward and reverse. In the forward process, the ground truth M_0 is transformed into noisy ground truth M_T by gradually adding Gaussian noise through T time steps. In the reverse process, first, the source image I and noise map $\hat{M}_{t+1}$ pass through an encoder E (two residual-inception blocks [18]) to obtain embedding $f_I \in R^{h \times w \times c_1}$ and $f_M \in R^{h \times w \times c_2}$ (subscripts I and M denote image and map), where h, w and c_1 (c_2) are the height, width and channels of the embeddings, respectively. Then, the two embeddings are aligned by a Cross-Attention (CA) module in the feature space. The aligned feature map is given as a noisy input to MT U-Net to recover

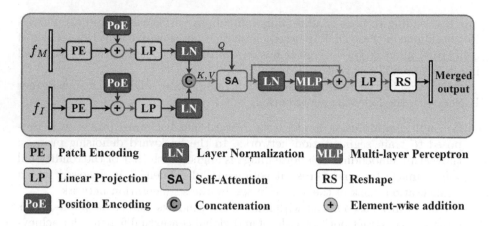

PE Patch Encoding   LN Layer Normalization   MLP Multi-layer Perceptron

LP Linear Projection   SA Self-Attention   RS Reshape

PoE Position Encoding   C Concatenation   + Element-wise addition

Fig. 2. Architecture of the Cross-Attention (CA) module.

$\hat{M}_t$. This reverse process iterates from $t = T - 1$ till $t = 0$ (i.e., the initial $\hat{M}_{t+1}$ when $t = T - 1$, $\hat{M}_T$, is set as M_T, and $\hat{M}_0$ is recovered eventually, which is expected to be identical to the ground truth M_0).

Figure 2 presents the architecture of our CA module, which is used to align f_M and f_I in order to improve the conditioning of the diffusion model. First, f_M and f_I are divided into patches and flattened to vectors by a Patch Encoding (PE) layer. Then, the position information of patches is obtained using a Position Encoding layer (PoE), and is added to the original patch embeddings for preserving their positional information. The dimensions of the two position-built-in patch embeddings are aligned using Linear Projection (LP) layers, and are normalized by a Layer Normalization (LN), denoting the output after the two LN's as $f_M^p \in R^d$ and $f_I^p \in R^d$ (d-dim feature vectors of patches). Thirdly, we use a Self-Attention for efficient feature fusion:

$$SA = Softmax(\frac{QK^\top}{\sqrt{d}})V \tag{1}$$

where f_M^p is the query (Q), the concatenation of f_M^p and f_I^p is the key (K) and value (V). $\top$ denotes the transpose. Fourthly, following [20], we encode the output of L_{SA} by a Layer Normalization (LN) and a two-layered Multi-Layer Perceptron (MLP) for extracting more contextual information. An auxiliary connection (residual) is used to enhance the information propagation. Lastly, we apply a Reshape (RS) layer to reshape and assemble the patches into the same size as f_M.

2.2   Multi-sized Transformer U-Net (MT U-Net)

Figure 3(a) presents the architecture of our MT U-Net, with the encoding and decoding parts. The encoding part consists of a Patch Partitioning layer, a Linear Embedding layer, a PoE and four Encoder blocks. The Patch Partitioning layer splits the input into non-overlapping patches with a patch size of 2×2.

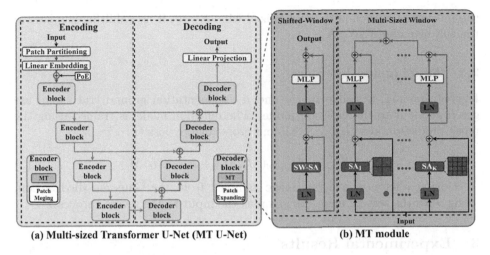

(a) Multi-sized Transformer U-Net (MT U-Net)   **(b) MT module**

Fig. 3. Architecture of the proposed MT U-Net, and the MT module. The time step embedding is not presented in the figure for clarity.

These patches, along with the time embedding are flattened into a $D \times 1$ dimension linear embedding using a Linear Embedding layer. Then positional information obtained from the PoE is added to the linear embedding before passing through the four Encoder blocks. Each Encoder block consists of a Multi-sized Transformer (MT) module and a Patch Merging layer, except the last encoder block which only contains the MT module. The MT module extracts multi-scale contextual features (to be elaborated later), and the Patch Merging layer down-samples the feature maps. With the inspiration from U-Net [15], a skip connection is employed for using the multi-scale contextual information from the encoder to overcome the loss of spatial information during down-sampling. Similar to the Encoder block, each Decoder block consists of an MT module and a patch-expanding layer, except the first decoder block which only contains the MT module. The patch-expanding layer performs the up sampling, and reshaping operation on feature maps. Finally, we employ a Linear Projection layer to obtain the pixel-level predictions.

The proposed Multi-sized Transformer (MT) module (Fig. 3(b)) is different from the conventional transformer [6]. The MT module consists of two parts: multi-sized window and shifted-window. The multi-sized window part extracts multi-scale contextual information, and the shifted-window part enriches the extracted information. The multi-sized window part has K parallel branches, with each branch consisting of a Layer Normalization (LN), multi-head Self-Attention (SA), auxiliary connection (residual), and a Multi-layer perceptron (MLP) with two layers followed by the GELU activation function. The window size used in the multi-head self-attention is varied to extract multi-scale contextual features. The output of these individual branches is combined, and is sent to the shifted-window part. The shifted-window part has a structure similar to the individual branch in the multi-sized window, but it uses shifting windows in the self attention (SW-SA).

2.3 Training and Inference

During training, a source image and its segmentation ground truth map are given as input to the diffusion model. The diffusion model is trained using the noise prediction loss (L_{Noise}) [12] and cross-entropy loss (L_{CE}).

$$L_{TOTAL} = L_{Noise}(M_t, \hat{M}_t) + L_{CE}(M_t, \hat{M}_t) \tag{2}$$

During inference, a noise image sampled from the Gaussian distribution, along with the testing image, is given as the input to the reverse process.

3   Experimental Results

3.1   Datasets and Evaluation Metrics

To evaluate the effectiveness and generalization ability of the proposed method, different medical image segmentation tasks are tested, including: (1) **Polyp segmentation from colonoscopy images** (Kvasir-SEG (KSEG) [10], CVC-Clinic DB (CVC) [1]), (2) **Skin lesion segmentation from dermoscopy images** (ISIC 2017 (IS17') [5], ISIC 2018 (IS18') [4,19]), and (3) **Optic-cup segmentation from retinal fundus images** (REFUGE (REF) [14]). Dice Coefficient (DC) and Intersection over Union (IoU) are used as evaluation metrics.

3.2   Implementation Details

The number of branches in the MT module is set to 3 by cross-validation, with window sizes as 4, 8, and 16 respectively. The diffusion transformer U-Net is trained for 40,000 iterations using SGD optimizer with a momentum of 0.6, with a batch size of 16, and the learning rate is set to 0.0005. In the diffusion, we use a linear noise scheduler with $T = 1000$ steps. For fair comparisons with the recent diffusion-based segmentation models [23,24], during inference an average ensemble of 25 predictions is considered as the final prediction. All the experiments are conducted using a NVIDIA Tesla V-100 GPU with 32 GB RAM.

3.3   Performance Comparison

First, we quantitatively compare our method with several well-known U-Net and/or Transformer-related segmentation models, including U-Net [15], U-Net++ [26], Attention U-Net [13], Swin U-Net [2], Trans U-Net [3], and Seg-Former [25]. With their source codes, these models are trained, and evaluated on the experiment datasets. For fair comparisons, all models use the same experimental protocol for each dataset. The quantitative results are shown in Table 1.

Table 1. Comparison with state-of-the-art methods related to U-Net and/or Transformers. '80 : 10 : 10' (data split on training:validation:testing) experimental protocol is employed on KSEG, CVC, IS18'; respective default splits are used on REF, and IS17'.

Metric	Datasets	Models						
		U-Net [15]	U-Net ++ [26]	Attention U-Net [13]	Swin U-Net [2]	Trans U-Net [3]	SegFormer [25]	Ours
DC	KSEG [10]	0.775	0.786	0.798	0.867	0.889	0.905	**0.946**
	CVC [1]	0.856	0.874	0.887	0.914	0.920	0.931	**0.954**
	IS18' [4,19]	0.813	0.833	0.844	0.869	0.904	0.914	**0.931**
	IS17' [5]	0.794	0.814	0.815	0.851	0.873	0.884	**0.935**
	REF [14]	0.769	0.779	0.792	0.815	0.821	0.843	**0.887**
IoU	KSEG [10]	0.714	0.725	0.752	0.854	0.863	0.874	**0.916**
	CVC [1]	0.805	0.821	0.845	0.874	0.889	0.905	**0.920**
	IS18' [4,19]	0.691	0.703	0.732	0.813	0.821	0.847	**0.879**
	IS17' [5]	0.659	0.663	0.682	0.721	0.745	0.768	**0.801**
	REF [14]	0.692	0.701	0.714	0.746	0.774	0.796	**0.815**

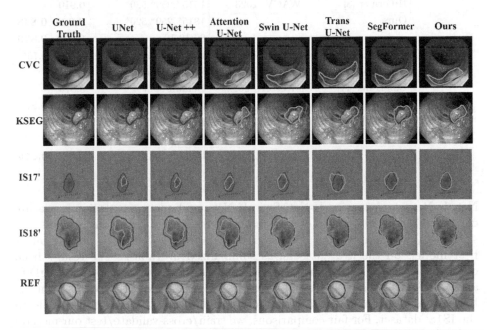

Fig. 4. Qualitative comparison with SOTA approaches on KSEG [10], CVC [1], IS18' [4,19], IS17' [5], and REF [14] datasets. The blue contours represent the ground truth, and the green contours represent the predicted results.

Our Diffusion Transformer U-Net outperforms all other U-Net or Transformer related models on the five datasets with various imaging modalities, validating its effectiveness and generalization capability.

Secondly, we qualitatively compare our Diffusion Transformer U-Net with other U-Net or Transformer related models. From the randomly sampled testing images in Fig. 4, we observe that the other models produce either over-segmented (e.g., Trans U-Net, SegFormer) or under-segmented results (e.g., U-

Table 2. Comparison with SOTA results. '-': No results reported. '*': number of images.

Dataset	Models	Publication, Year	Experimental protocol	DC	IoU
KSEG [10]	MSRF-Net [17]	IEEE JBHI, 2022	80:10:10	0.921	0.891
	Li-SegPNet [16]	IEEE TBE, 2022	80:10:10	0.905	0.828
	SSFormer [21]	MICCAI, 2022	80:10:10	0.935	0.890
	Ours		80:10:10	**0.946**	**0.916**
CVC [1]	MSRF-Net [17]	IEEE JBHI, 2022	80:10:10	0.942	0.904
	Li-SegPNet [16]	IEEE TBE, 2022	80:10:10	0.925	0.860
	SSFormer [21]	MICCAI, 2022	80:10:10	0.944	0.899
	Ours		80:10:10	**0.954**	**0.920**
IS18' [4,19]	MSRF-Net [17]	IEEE JBHI, 2022	80:10:10	0.882	0.837
	FAT-Net [22]	MedIA, 2022	1815*,259*,520*	0.890	0.820
	HiFormer [8]	WACV, 2023	1815*,259*,520*	0.910	-
	Ours		1815*,259*,520*	**0.924**	**0.843**
			80:10:10	**0.931**	**0.879**
IS17' [5]	FAT-Net [22]	MedIA, 2022	Default split	0.850	0.765
	HiFormer [8]	WACV, 2023	Default split	0.925	-
	ConTrans [11]	MICCAI, 2022	Default split	0.875	-
	Ours		Default split	**0.935**	**0.801**
REF [14]	MedSegDiff [23]	Arxiv, 2022	Default Split	0.863	0.782
	MedSegDiff-V2 [24]	Arxiv, 2023	Default Split	0.859	0.796
	Ours		Default Split	**0.887**	**0.815**

Net, U-Net++, Attention U-Net, Swin U-Net), and our segmentation masks are closest to the ground truth, demonstrating the effectiveness of our method.

Lastly, we compare our Diffusion Transformer U-Net with all the latest best models on the five datasets, as summarized in Table 2. The results from cited methods are copied from their papers directly, except for MedSegDiff, and MedSegDiff-V2. These two approaches are re-trained, and evaluated on the REF dataset. Note, since some methods use different experiment protocols on the IS18' dataset. For fair comparisons, we train/cross-validate/test our method using two different protocols, and compare ours with other methods with the same protocol. As shown in Table 2, our method consistently outperforms all the current best models on these five datasets, which again verifies its effectiveness and superiority.

3.4   Ablation Study

We perform a set of ablation studies to evaluate the contribution of each module in our Diffusion Transformer U-Net, as shown in Table 3:

- The original U-Net [15] is used as a baseline (row 1 in Table 3).

Table 3. Ablation on KSEG [10], CVC [1], IS18' [4,19], IS17' [5], and REF [14].

U-Net	Diff	CA	Vanila Trans	MT	IoU					DC				
					KSEG	CVC	IS17'	IS18'	REF	KSEG	CVC	IS17'	IS18'	REF
✓	✗	✗	✗	✗	0.714	0.805	0.659	0.691	0.692	0.775	0.856	0.794	0.813	0.769
✓	✓	✗	✗	✗	0.843	0.842	0.734	0.801	0.732	0.873	0.879	0.853	0.885	0.815
✓	✓	✓	✗	✗	0.875	0.875	0.763	0.841	0.763	0.905	0.932	0.889	0.901	0.831
✓	✓	✓	✓	✗	0.889	0.894	0.784	0.863	0.782	0.921	0.939	0.914	0.913	0.847
✓	✓	✓	✗	✓	**0.916**	**0.920**	**0.801**	**0.879**	**0.815**	**0.946**	**0.954**	**0.935**	**0.931**	**0.887**

- We replace the Cross-Attention (CA) in our diffusion model by a simple concatenation operation, and apply this simplified diffusion model to train the U-Net. Even this simplified diffusion model (row 2) can boost the U-Net performance in row 1, showing the effectiveness of diffusion.
- Using our diffusion model with the CA module (row 3), the performance is further improved, compared to the basic concatenation operation (row 2), which validates the contribution of the CA model for aligning feature embeddings during the denoising process of the diffusion model.
- Using our diffusion model with the CA module, we add the basic transformer units [6] without the multi-sized window into the U-Net (row 4). This also increases the segmentation performance, compared to row 3, which demonstrates that transformers can help the U-Net on segmentation.
- Based on the model from row 4, we add multi-sized windows into the transformer (i.e., our Diffusion Transformer U-Net, row 5). This gives the highest performance, compared to other configurations in the ablation studies.

4   Conclusion

A Diffusion Transformer U-Net is proposed for medical image segmentation. Instead of a standard U-Net in the diffusion model, we propose a transformer based U-Net with multi-sized windows for enhancing the contextual information extraction and reconstruction. We also design a cross-attention module to align feature embeddings, providing a better conditioning from the source image to the diffusion model. The evaluation on various datasets of different modalities shows the effectiveness and generalization ability of the proposed method.

References

1. Bernal, J., Sánchez, F.J., Fernández-Esparrach, G., Gil, D., Rodríguez, C., Vilariño, F.: Wm-dova maps for accurate polyp highlighting in colonoscopy: validation vs. saliency maps from physicians. Comput. Med. Imaging Graph. **43**, 99–111 (2015)
2. Cao, H., et al.: Swin-unet: Unet-like pure transformer for medical image segmentation. In: ECCV 2022, Part III, pp. 205–218. Springer, Cham (2023). https://doi.org/10.1007/978-3-031-25066-8_9

3. Chen, J., et al.: Transunet: transformers make strong encoders for medical image segmentation. arXiv preprint arXiv:2102.04306 (2021)
4. Codella, N., et al.: Skin lesion analysis toward melanoma detection 2018: a challenge hosted by the international skin imaging collaboration (ISIC). arXiv preprint arXiv:1902.03368 (2019)
5. Codella, N.C., et al.: Skin lesion analysis toward melanoma detection: a challenge at the 2017 international symposium on biomedical imaging (ISBI), hosted by the international skin imaging collaboration (isic). In: 2018 IEEE 15th International Symposium on Biomedical Imaging (ISBI 2018), pp. 168–172. IEEE (2018)
6. Dosovitskiy, A., et al.: An image is worth 16x16 words: transformers for image recognition at scale. arXiv preprint arXiv:2010.11929 (2020)
7. Gu, R., et al.: Ca-net: comprehensive attention convolutional neural networks for explainable medical image segmentation. IEEE Trans. Med. Imaging 40(2), 699–711 (2020)
8. Heidari, M., et al.: Hiformer: hierarchical multi-scale representations using transformers for medical image segmentation. In: Proceedings of the IEEE/CVF Winter Conference on Applications of Computer Vision, pp. 6202–6212 (2023)
9. Ho, J., Jain, A., Abbeel, P.: Denoising diffusion probabilistic models. Adv. Neural. Inf. Process. Syst. 33, 6840–6851 (2020)
10. Jha, D., et al.: Kvasir-SEG: a segmented polyp dataset. In: Ro, Y.M., Cheng, W.-H., Kim, J., Chu, W.-T., Cui, P., Choi, J.-W., Hu, M.-C., De Neve, W. (eds.) MMM 2020. LNCS, vol. 11962, pp. 451–462. Springer, Cham (2020). https://doi.org/10.1007/978-3-030-37734-2_37
11. Lin, A., Xu, J., Li, J., Lu, G.: Contrans: improving transformer with convolutional attention for medical image segmentation. In: Medical Image Computing and Computer Assisted Intervention-MICCAI 2022: 25th International Conference, Singapore, September 18–22, 2022, Proceedings, Part V. pp. 297–307. Springer, Cham (2022). https://doi.org/10.1007/978-3-031-16443-9_29
12. Nichol, A.Q., Dhariwal, P.: Improved denoising diffusion probabilistic models. In: International Conference on Machine Learning, pp. 8162–8171. PMLR (2021)
13. Oktay, O., et al.: Attention u-net: Learning where to look for the pancreas. arXiv preprint arXiv:1804.03999 (2018)
14. Orlando, J.I., et al.: Refuge challenge: a unified framework for evaluating automated methods for glaucoma assessment from fundus photographs. Med. Image Anal. 59, 101570 (2020)
15. Ronneberger, O., Fischer, P., Brox, T.: U-net: convolutional networks for biomedical image segmentation. In: Medical Image Computing and Computer-Assisted Intervention–MICCAI 2015: 18th International Conference, Munich, Germany, October 5-9, 2015, Proceedings, Part III 18. pp. 234–241. Springer, Cham (2015). https://doi.org/10.1007/978-3-319-24574-4_28
16. Sharma, P., Gautam, A., Maji, P., Pachori, R.B., Balabantaray, B.K.: Li-segpnet: encoder-decoder mode lightweight segmentation network for colorectal polyps analysis. IEEE Trans. Biomed. Eng. (2022)
17. Srivastava, A., et al.: Msrf-net: a multi-scale residual fusion network for biomedical image segmentation. IEEE J. Biomed. Health Inform. 26(5), 2252–2263 (2021)
18. Szegedy, C., Ioffe, S., Vanhoucke, V., Alemi, A.: Inception-v4, inception-resnet and the impact of residual connections on learning. In: Proceedings of the AAAI Conference on Artificial Intelligence, vol. 31 (2017)
19. Tschandl, P., Rosendahl, C., Kittler, H.: The ham10000 dataset, a large collection of multi-source dermatoscopic images of common pigmented skin lesions. Scientific data 5(1), 1–9 (2018)

20. Vaswani, A., Shazeer, N., Parmar, N., Uszkoreit, J., Jones, L., Gomez, A.N., Kaiser, L., Polosukhin, I.: Attention is all you need. Advances in neural information processing systems 30 (2017)
21. Wang, J., Huang, Q., Tang, F., Meng, J., Su, J., Song, S.: Stepwise feature fusion: Local guides global. In: Medical Image Computing and Computer Assisted Intervention-MICCAI 2022: 25th International Conference, Singapore, September 18–22, 2022, Proceedings, Part III. pp. 110–120. Springer (2022)
22. Wu, H., Chen, S., Chen, G., Wang, W., Lei, B., Wen, Z.: Fat-net: feature adaptive transformers for automated skin lesion segmentation. Med. Image Anal. **76**, 102327 (2022)
23. Wu, J., Fang, H., Zhang, Y., Yang, Y., Xu, Y.: Medsegdiff: medical image segmentation with diffusion probabilistic model. arXiv preprint arXiv:2211.00611 (2022)
24. Wu, J., Fu, R., Fang, H., Zhang, Y., Xu, Y.: Medsegdiff-v2: diffusion based medical image segmentation with transformer. arXiv preprint arXiv:2301.11798 (2023)
25. Xie, E., Wang, W., Yu, Z., Anandkumar, A., Alvarez, J.M., Luo, P.: Segformer: simple and efficient design for semantic segmentation with transformers. Adv. Neural. Inf. Process. Syst. **34**, 12077–12090 (2021)
26. Zhou, Z., Siddiquee, M.M.R., Tajbakhsh, N., Liang, J.: Unet++: redesigning skip connections to exploit multiscale features in image segmentation. IEEE Trans. Med. Imaging **39**(6), 1856–1867 (2019)

Scaling up 3D Kernels with Bayesian Frequency Re-parameterization for Medical Image Segmentation

Ho Hin Lee[1]([✉]), Quan Liu[1], Shunxing Bao[1], Qi Yang[1], Xin Yu[1], Leon Y. Cai[3], Thomas Z. Li[3], Yuankai Huo[1,2], Xenofon Koutsoukos[1], and Bennett A. Landman[1,2,3]

[1] Department of Computer Science, Vanderbilt University, Nashville, TN 37212, USA
ho.hin.lee@vanderbilt.edu
[2] Department of Electrical and Computer Engineering, Vanderbilt University, Nashville, TN 37212, USA
[3] Department of Biomedical Engineering, Vanderbilt University, Nashville, TN 37212, USA

Abstract. With the inspiration of vision transformers, the concept of depth-wise convolution revisits to provide a large Effective Receptive Field (ERF) using Large Kernel (LK) sizes for medical image segmentation. However, the segmentation performance might be saturated and even degraded as the kernel sizes scaled up (e.g., $21 \times 21 \times 21$) in a Convolutional Neural Network (CNN). We hypothesize that convolution with LK sizes is limited to maintain an optimal convergence for locality learning. While Structural Re-parameterization (SR) enhances the local convergence with small kernels in parallel, optimal small kernel branches may hinder the computational efficiency for training. In this work, we propose RepUX-Net, a pure CNN architecture with a simple large kernel block design, which competes favorably with current network state-of-the-art (SOTA) (e.g., 3D UX-Net, SwinUNETR) using 6 challenging public datasets. We derive an equivalency between kernel re-parameterization and the branch-wise variation in kernel convergence. Inspired by the spatial frequency in the human visual system, we extend to vary the kernel convergence into element-wise setting and model the spatial frequency as a Bayesian prior to re-parameterize convolutional weights during training. Specifically, a reciprocal function is leveraged to estimate a frequency-weighted value, which rescales the corresponding kernel element for stochastic gradient descent. From the experimental results, RepUX-Net consistently outperforms 3D SOTA benchmarks with internal validation (FLARE: 0.929 to 0.944), external validation (MSD: 0.901 to 0.932, KiTS: 0.815 to 0.847, LiTS: 0.933 to 0.949, TCIA: 0.736 to 0.779) and transfer learning (AMOS: 0.880 to 0.911) scenarios in Dice Score. Both codes and pre-trained models are available at: https://github.com/MASILab/RepUX-Net.

Supplementary Information The online version contains supplementary material available at https://doi.org/10.1007/978-3-031-43901-8_60.

Keywords: Bayesian Frequency Re-parameterization · Large Kernel Convolution · Medical Image Segmentation

1 Introduction

With the introduction of Vision Transformers (ViTs), CNNs have been greatly challenged as seen with the leading performance in multiple volumetric data benchmarks, especially for medical image segmentation [7,8,21,23]. The key contribution of ViTs is largely credited to the large Effective Receptive Field (ERF) with a multi-head self-attention mechanism [6]. Note the attention mechanism is computationally unscalable with respect to the input resolutions [17,18]. Therefore, the concept of depth-wise convolution is revisited to provide a scalable and efficient feature computation with large ERF using large kernel sizes (e.g., $7 \times 7 \times 7$) [14,18]. However, either from prior works or our experiments, the model performance becomes saturated or even degraded when the kernel size is scaled up in encoder blocks [4,16]. We hypothesize that scaling up the kernel size in convolution may limit the optimal learning convergences across local to global scales. Recently, the feasibility of leveraging large kernel convolutions (e.g., 31×31 [4], 51×51 [16]) has been shown with natural image domain with Structural Re-parameterization (SR), which adapts Constant-Scale Linear Addition (CSLA) block (Fig. 2b) and re-parameterizes the large kernel weights during inference [4]. As convolutions with small kernel sizes converge more easily, the convergence of small kernel regions enhances in the re-parameterized weight, as shown in Fig. 1a. With such observation, we further ask: **Can we adapt variable convergence across elements of the convolution kernel during training, instead of regional locality only?**

In this work, we first derive and extend the theoretical equivalency of the weight optimization in the CSLA block. We observe that the kernel weight of each branch can be optimized with variable convergence using branch-specific learning rates. Furthermore, the ERF with SR is visualized to be more widely distributed from the center element to the global surroundings [4], demonstrating a similar behavior to the spatial frequency in the human visual system [13].

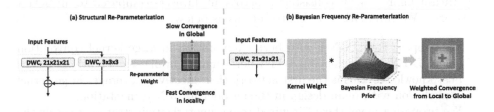

Fig. 1. With the fast convergence in small kernels, SR merges the branches weights and enhances the locality convergence with respect to the kernel size (deep blue region), while the global convergence is yet to be optimal (light blue region). By adapting BFR, the learning convergence can rescale in an element-wise setting and distribute the learning importance from local to global. (Color figure online)

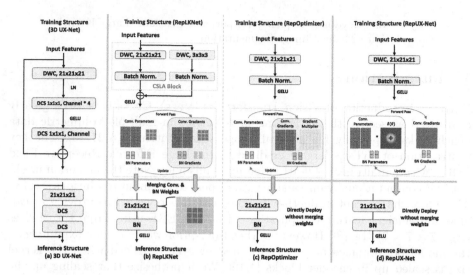

Fig. 2. Overview of RepUX-Net. Unlike performing SR to merge branches weight or performing GR within the optimizer, we propose to multiply a Bayesian function δ and scale the element-wise learning importance in each large kernel. We then put the scaled weights back into the convolution layer for training.

Inspired by the reciprocal characteristics of spatial frequency, we model the spatial frequency as a Bayesian prior to adapt variable convergence of each kernel element with stochastic gradient descent (Fig. 1b). Specifically, we compute a scaling factor with respect to the distance from the kernel center and multiply the corresponding element for re-parameterization during training. Furthermore, we simplify the encoder block design into a plain convolution block only to minimize the computation burden in training and achieve State-Of-The-Art (SOTA) performance. We propose RepUX-Net, a pure 3D CNN with the large kernel size (e.g., $21 \times 21 \times 21$) in encoder blocks, to compete favorably with current SOTA segmentation networks. We evaluate RepUX-Net on supervised multi-organ segmentation with 6 different public volumetric datasets. RepUX-Net demonstrates significant improvement consistently across all datasets compared to all SOTA networks. We summarize our contributions as below:

- We propose RepUX-Net with better adaptation in large kernel convolution than 3D UX-Net, achieving SOTA performance in 3D segmentation. To our best knowledge, this is the first network that effectively leverages large kernel convolution with plain design in the encoder for 3D segmentation.
- We propose a novel theory-inspired re-parameterization strategy to scale the element-wise learning convergence in large kernels with Bayesian prior knowledge. To our best knowledge, this is the first re-parameterization strategy to adapt 3D large kernels in the medical domain.
- We leverage six challenging public datasets to evaluate RepUX-Net in 1) direct training and 2) transfer learning scenarios with 3D multi-organ seg-

mentation. RepUX-Net achieves significant improvement consistently in both scenarios across all SOTA networks.

2 Related Works

Weights Re-parameterization: SR is a methodology of equivalently converting model structures via transforming the parameters in kernel weights. For example, RepVGG demonstrates to construct one extra ResNet-style shortcut as a 1×1 convolution, parallel to 3×3 convolution during training [5]. Such parallel branch design is claimed to enhance the learning efficiency during training, in which the 1×1 branch is then merged into the parallel 3×3 kernel via a series of linear transformation in the inference stage. OREPA further adds more parallel branches with linear scaling modules to enhance training efficiency [10]. Inspired by the parallel branches design, RepLKNet is proposed to scale up the 2D kernel size (e.g., 31×31) with a 3×3 convolution as the parallel branch [4]. SLaK further extends the kernel size to 51×51 by decomposing the large kernel into two rectangular parallel kernels with sparse groups and training the model with dynamic sparsity [16]. However, the proposed models' FLOPs remain at a high-level with the parallel branch design and demonstrates to have a trade-off between model performance and training efficiency. To tackle the trade-off, RepOptimizer provides an alternative to re-parameterize the back-propagate gradient, instead of the structural parameters of kernel weights, to enhance the training efficiency with plain convolution block design [3]. Significant efforts have been demonstrated to enlarge the 2D kernel size in the natural image domain, while limited studies have been proposed for 3D kernels in medical domain. As 3D kernels have a larger number of parameters than 2D, it is challenging to directly leverage the parallel branch design and maintain an optimal convergence of learning large kernel convolution without trading off the computation efficiency significantly.

3 Methods

Instead of changing the gradient dynamics during training [3], we introduce RepUX-Net, a pure 3D CNN architecture that performs element-wise scaling in large kernel weights to enhance the learning convergence and effectively adapts large receptive field for volumetric segmentation. To design such behavior, we adapt a two-step pipeline: 1) we define the theoretical equivalency of variable learning convergence in convolution branches; 2) we simulate the behavior of spatial frequency to re-weight the learning importance of each element in kernels for stochastic gradient descent. Note the theoretical derivation depends on the optimization with first-order gradient-driven optimizer (e.g., SGD, AdamW) [3].

3.1 Variable Learning Convergence in Multi-Branch Design

From Fig. 2b & 2c, previous re-parameterization strategies only demonstrate the benefits of the parallel branch design by either adding up the encoded outputs from both small and large kernels with SR (RepLKNet [4]) or performing Gradient Re-parameterization (GR) by multiplying with constant values (RepOptimizer [3]) in a Single Operator (SO) to enhance the locality learning in large kernels. Inspired by the concepts of SR and GR, we extend the theoretical equivalency proof in RepOptimizer to adapt variable learning convergence in branches. Here, we only showcase the conclusion with two convolutions and two constant scalars as the scaling factors for simplicity. The complete proof of equivalency is demonstrated in Supplementary 1.1. Let $\{\alpha_L, \alpha_S\}$ and $\{W_L, W_S\}$ be the two constant scalars and two convolution kernels (Large & Small) respectively. Let X and Y be the input and output features, the CSLA block is formulated as $Y_{CSLA} = \alpha_L(X \star W_L) + \alpha_S(X \star W_S)$, where $\star$ denotes as convolution. For SO blocks, we train the plain structure parameterized by W' and $Y_{SO} = X \star W'$. Let i be the number of training iterations, we ensure that $Y^{(i)}{}_{CSLA} = Y^{(i)}{}_{SO}, \forall i \geq 0$ and derive the stochastic gradient descent of parallel branches as follows:

$$\alpha_L W_{L(i+1)} + \alpha_S W_{S(i+1)} = \alpha_L W_{L(i)} - \lambda_L \alpha_L \frac{\partial \mathcal{L}}{\partial W_{L_i}} + \alpha_S W_{S(i)} - \lambda_S \alpha_S \frac{\partial \mathcal{L}}{\partial W_{S_i}}, \quad (1)$$

where $\mathcal{L}$ is the objective function; λ_L and λ_S are the Learning Rate (LR) of each branch respectively. We observe that the optimization of each branch can be different, which is feasible to control by adjusting the branch-specific LR. The locality convergence in large kernels enhance with the quick convergence in small kernels. Additionally from our experiments, a significant improvement is demonstrated with different branch-wise LR using SGD (Table 2). Building upon this insight, we further hypothesize that **the convergence of each large kernel element can be optimized differently by linear scaling with prior knowledge.**

3.2 Bayesian Frequency Re-parameterization (BFR)

With the visualization of ERF in RepLKNet [4], the diffused distribution (from local to global) in ERF demonstrates similar behavior with the spatial frequency in the human visual system [13]. High spatial frequency (small ERF) allows to refine and sharpen details with high acuity, while global details are demonstrated with low spatial frequency. Inspired by the reciprocal characteristics in spatial frequency, we first generate a Bayesian prior distribution to model the spatial frequency by computing a reciprocal distance function between each element and the central point of the kernel weight as follows:

$$d(x, y, z, c) = \sqrt{(x - c)^2 + (y - c)^2 + (z - c)^2}$$
$$\delta(x_k, y_k, z_k, c, \alpha) = \frac{\alpha}{d(x_k, y_k, z_k, c) + \alpha} \quad (2)$$

where k and c are the element and central index of the kernel weight, α is the hyperparameter to control the shape of the generated frequency distribution. Instead of adjusting the LR in parallel branches, we propose to re-parameterize the convolution weights by multiplying the scaling factor δ to each kernel element and apply a static LR λ for stochastic gradient descent in single operator setting as follows:

$$W'_{i+1} = \delta W'_i - \lambda \frac{\partial L}{\partial \delta W'_i} \tag{3}$$

With the multiplication with δ, each element in the kernel weight is rescaled with respect to the frequency level and allow to converge differently with a static LR in stochastic gradient descent. Such design demonstrates to influence the weighted convergence diffused from local to global in theory, thus tackling the limitation of enhancing the local convergence only in branch-wise setting.

3.3 Model Architecture

The backbone of RepUX-Net is based on 3D UX-Net [14], which comprises multiple volumetric convolution blocks that directly utilize 3D patches and leverage skip connections to transfer hierarchical multi-resolution features for end-to-end optimization. Inspired by [15], we choose a kernel size of $21 \times 21 \times 21$ for Depth-Wise Convolution (DWC-21) as the optimal choice without significant trade-off between model performance and computational efficiency in 3D. We further simplify the block design as a plain convolution block design to minimize the computational burden from additional modules. The encoder blocks in layers l and $l + 1$ are defined as follows:

$$\hat{z}^l = \text{GeLU}(\text{DWC-21}(\text{BN}(z^{l-1}))), \quad \hat{z}^{l+1} = \text{GeLU}(\text{DWC-21}(\text{BN}(z^l))) \tag{4}$$

where $\hat{z}_l$ and $\hat{z}_{l+1}$ are the outputs from the DWC layer in each depth level; BN denotes as the batch normalization layer.

4 Experimental Setup

Datasets. We perform experiments on six public datasets for volumetric segmentation, which comprise with 1) Medical Segmentation Decathlon (MSD) spleen dataset [1], 2) MICCAI 2017 LiTS Challenge dataset (LiTS) [2], 3) MICCAI 2019 KiTS Challenge dataset (KiTS) [9], 4) NIH TCIA Pancreas-CT dataset (TCIA) [20], 5) MICCAI 2021 FLARE Challenge dataset (FLARE) [19], and 6) MICCAI 2022 AMOS challenge dataset (AMOS) [12]. More details of each dataset (including data split for training and inference) are described in Supplementary Material (SM) Table 1.

Implementation. We evaluate RepUX-Net with three different scenarios: 1) internal validation with direct supervised learning, 2) external validation with the unseen datasets, and 3) transfer learning with pretrained weights. All preprocessing and training details including baselines, are followed with [14] for benchmarking. For external validations, we leverage the AMOS-pretrained weights to evaluate 4 unseen datasets. In summary, we evaluate the segmentation performance of RepUX-Net by comparing current SOTA networks in a fully-supervised

Table 1. Comparison of SOTA approaches on the five different testing datasets. (*: $p < 0.01$, with Paired Wilcoxon signed-rank test to all baseline networks)

| | | | Internal Testing | | | | | External Testing | | | |
| | | | FLARE | | | | | MSD | KiTS | LiTS | TCIA |
Methods	#Params	FLOPs	Spleen	Kidney	Liver	Pancreas	Mean	Spleen	Kidney	Liver	Pancreas
nn-UNet [11]	31.2M	743.3G	0.971	0.966	0.976	0.792	0.926	0.917	0.829	0.935	0.739
TransBTS [22]	31.6M	110.4G	0.964	0.959	0.974	0.711	0.902	0.881	0.797	0.926	0.699
UNETR [8]	92.8M	82.6G	0.927	0.947	0.960	0.710	0.886	0.857	0.801	0.920	0.679
nnFormer [23]	149.3M	240.2G	0.973	0.960	0.975	0.717	0.906	0.880	0.774	0.927	0.690
SwinUNETR [7]	62.2M	328.4G	0.979	0.965	0.980	0.788	0.929	0.901	0.815	0.933	0.736
3D UX-Net (k=7) [14]	53.0M	639.4G	0.981	0.969	0.982	0.801	0.934	0.926	0.836	0.939	0.750
3D UX-Net (k=21) [14]	65.9M	757.6G	0.980	0.968	0.979	0.795	0.930	0.908	0.808	0.929	0.720
RepOptimizer [3]	65.8M	757.4G	0.981	0.969	0.981	0.822	0.937	0.913	0.833	0.934	0.746
3D RepUX-Net (Ours)	65.8M	757.4G	**0.984**	**0.970**	**0.983**	**0.837**	**0.944***	**0.932***	**0.847***	**0.949***	**0.779***

Table 2. Ablation studies with quantitative Comparison on Block Designs with/out frequency modeling using different optimizer

Optimizer	Main Branch	Para. Branch	BFR	Train Steps	Main LR	Para. LR	Mean Dice
SGD	$21 \times 21 \times 21$	$\times$	$\times$	40000	0.0003	$\times$	0.898
AdamW	$21 \times 21 \times 21$	$\times$	$\times$	40000	0.0001	$\times$	0.906
SGD	$21 \times 21 \times 21$	$3 \times 3 \times 3$	$\times$	40000	0.0003	0.0006	0.917
AdamW	$21 \times 21 \times 21$	$3 \times 3 \times 3$	$\times$	40000	0.0001	0.0001	0.929
AdamW	$21 \times 21 \times 21$	$\times$	$\checkmark$	40000	0.0001	$\times$	**0.938**
SGD	$21 \times 21 \times 21$	$3 \times 3 \times 3$	$\times$	60000	0.0003	0.0006	0.930
AdamW	$21 \times 21 \times 21$	$3 \times 3 \times 3$	$\times$	60000	0.0001	0.0001	0.938
AdamW	$21 \times 21 \times 21$	$\times$	$\checkmark$	60000	0.0001	$\times$	**0.944**

setting. Furthermore, we perform ablation studies to investigate the effect on Bayesian frequency distribution with different scales generated by α and the variability of branch-wise learning rates with first-order gradient optimizers (e.g., SGD, AdamW) for volumetric segmentation. Dice similarity coefficient is leveraged as an evaluation metric to measure the overlapping regions between the model predictions and the manual ground-truth labels.

5   Results

Different Scenarios Evaluations. Table 1 shows the result comparison of current SOTA networks on medical image segmentation in a volumetric setting. With our designed convolutional blocks as the encoder backbone, RepUX-Net demonstrates the best performance across all segmentation task with significant improvement in Dice score (FLARE: 0.934 to 0.944, AMOS: 0.891 to 0.902). Furthermore, RepUX-Net demonstrates the best generalizability consistently with a significant boost in performance across 4 different external datasets (MSD: 0.926 to 0.932, KiTS: 0.836 to 0.847, LiTS: 0.939 to 0.949, TCIA: 0.750 to 0.779). For transfer learning scenario, the performance of RepUX-Net significantly outper-

Table 3. Evaluations on the AMOS testing split in different scenarios. (*: $p < 0.01$, with Paired Wilcoxon signed-rank test to all baseline networks)

Train From Scratch Scenario																
Methods	Spleen	R. Kid	L. Kid	Gall.	Eso.	Liver	Stom.	Aorta	IVC	Panc.	RAG	LAG	Duo.	Blad.	Pros.	Avg
nn-UNet	0.951	0.919	0.930	0.845	0.797	0.975	0.863	0.941	0.898	0.813	0.730	0.677	0.772	0.797	0.815	0.850
TransBTS	0.930	0.921	0.909	0.798	0.722	0.966	0.801	0.900	0.820	0.702	0.641	0.550	0.684	0.730	0.679	0.783
UNETR	0.925	0.923	0.903	0.777	0.701	0.964	0.759	0.887	0.821	0.687	0.688	0.543	0.629	0.710	0.707	0.740
nnFormer	0.932	0.928	0.914	0.831	0.743	0.968	0.820	0.905	0.838	0.725	0.678	0.578	0.677	0.737	0.596	0.785
SwinUNETR	0.956	0.957	0.949	0.891	0.820	0.978	0.880	0.939	0.894	0.818	0.800	0.730	0.803	0.849	0.819	0.871
3D UX-Net (k=7)	0.966	0.959	0.951	0.903	0.833	0.980	0.910	0.950	0.913	0.830	0.805	0.756	**0.846**	0.897	0.863	0.890
3D UX-Net (k=21)	0.963	0.959	0.953	**0.921**	0.848	0.981	0.903	0.953	0.910	0.828	0.815	0.754	0.824	0.900	0.878	0.891
RepOptimizer	0.968	**0.964**	0.953	0.903	0.857	0.981	0.915	0.950	0.915	0.826	0.802	0.756	0.813	0.906	0.867	0.892
RepUX-Net (Ours)	**0.972**	0.963	**0.964**	0.911	**0.861**	**0.982**	**0.921**	**0.956**	**0.924**	**0.837**	**0.818**	**0.777**	0.831	**0.916**	**0.879**	**0.902***

Transfer Learning Scenario																
Methods	Spleen	R. Kid	L. Kid	Gall.	Eso.	Liver	Stom.	Aorta	IVC	Panc.	RAG	LAG	Duo.	Blad.	Pros.	Avg
nn-UNet	0.965	0.959	0.951	0.889	0.820	0.980	0.890	0.948	0.901	0.821	0.785	0.739	0.806	0.869	0.839	0.878
TransBTS	0.885	0.931	0.916	0.817	0.744	0.969	0.837	0.914	0.855	0.724	0.630	0.566	0.704	0.741	0.722	0.792
UNETR	0.926	0.936	0.918	0.785	0.702	0.969	0.788	0.893	0.828	0.732	0.717	0.554	0.658	0.683	0.722	0.762
nnFormer	0.935	0.904	0.887	0.836	0.712	0.964	0.798	0.901	0.821	0.734	0.665	0.587	0.641	0.744	0.714	0.790
SwinUNETR	0.959	0.960	0.949	0.894	0.827	0.979	0.899	0.944	0.899	0.828	0.791	0.745	0.817	0.875	0.841	0.880
3D UX-Net (k=7)	0.970	0.967	0.961	0.923	0.832	0.984	0.920	0.951	0.914	0.856	0.825	0.739	0.853	0.906	0.876	0.900
3D UX-Net (k=21)	0.969	0.965	0.962	0.910	0.824	0.982	0.918	0.949	0.915	0.850	0.823	0.740	0.843	0.905	0.877	0.898
RepOptimizer	0.967	0.967	0.957	0.908	0.847	0.983	0.913	0.945	0.914	0.838	0.825	0.780	0.836	0.915	0.864	0.897
RepUX-Net	**0.973**	**0.968**	**0.965**	**0.933**	**0.865**	**0.985**	**0.930**	**0.960**	**0.923**	**0.859**	**0.829**	**0.793**	**0.869**	**0.918**	**0.891**	**0.911***

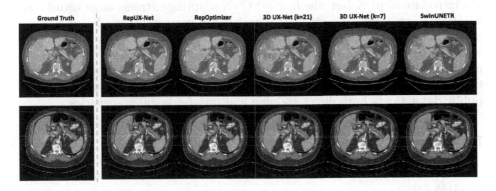

Fig. 3. Qualitative Representations of organ segmentation in LiTS and TCIA datasets

forms the current SOTA networks with mean Dice of 0.911 (1.22% enhancement), as shown in Table 2. RepUX-Net demonstrates its capabilities across the generalizability of unseen datasets and transfer learning ability. The qualitative representations (in Fig. 3) further provides additional confidence of the quality improvement in segmentation predictions with RepUX-Net (Table 3).

Ablation Studies with Block Designs & Optimizers. With the plain convolution design, a mean dice score of 0.906 is demonstrated with AdamW optimizer and perform slightly better than that with SGD. With the additional design of a parallel small kernel branch, the segmentation performance significantly improved (SGD: 0.898 to 0.917, AdamW: 0.906 to 0.929) with the optimized parallel branch LR using SR. The performance is further enhanced (SGD: 0.917

to 0.930, AdamW: 0.929 to 0.937) without being saturated with the increase of the training steps. By adapting BFR, the segmentation performance outperforms the parallel branch design significantly with a Dice score of 0.944.

Effectiveness on Different Frequency Distribution. From Fig. 1 in SM, RepUX-Net demonstrates the best performance when $\alpha = 1$, while comparable performance is demonstrated in both $\alpha = 0.5$ and $\alpha = 8$. A possible family of Bayesian distributions (different shapes) may need to further optimize the learning convergence of kernels across each channel.

Limitations. The shape of the generated Bayesian distribution is fixed across all kernel weights with an unlearnable distance function. Each channel in kernels is expected to extract variable features with different distributions. Exploring different families of distributions to rescale the element-wise convergence in kernels will be our potential future direction.

6 Conclusion

We introduce RepUX-Net, the first 3D CNN adapting extreme large kernel convolution in encoder network for medical image segmentation. We propose to model the spatial frequency in the human visual system as a reciprocal function, which generates a Bayesian prior to rescale the learning convergence of each element in kernel weights. By introducing the frequency-guided importance during training, RepUX-Net outperforms current SOTA networks on six challenging public datasets via both direct training and transfer learning scenarios.

References

1. Antonelli, M., et al.: The medical segmentation decathlon. Nat. Commun. **13**(1), 4128 (2022)
2. Bilic, P., et al.: The liver tumor segmentation benchmark (LITS). Med. Image Anal. **84**, 102680 (2023)
3. Ding, X., Chen, H., Zhang, X., Huang, K., Han, J., Ding, G.: Re-parameterizing your optimizers rather than architectures. arXiv preprint arXiv:2205.15242 (2022)
4. Ding, X., Zhang, X., Han, J., Ding, G.: Scaling up your kernels to 31x31: revisiting large kernel design in CNNs. In: Proceedings of the IEEE/CVF Conference on Computer Vision and Pattern Recognition, pp. 11963–11975 (2022)
5. Ding, X., Zhang, X., Ma, N., Han, J., Ding, G., Sun, J.: RepVGG: making VGG-style convnets great again. In: Proceedings of the IEEE/CVF Conference on Computer Vision and Pattern Recognition, pp. 13733–13742 (2021)
6. Dosovitskiy, A., et al.: An image is worth 16x16 words: transformers for image recognition at scale. arXiv preprint arXiv:2010.11929 (2020)
7. Hatamizadeh, A., Nath, V., Tang, Y., Yang, D., Roth, H.R., Xu, D.: Swin UNETR: swin transformers for semantic segmentation of brain tumors in MRI images. In: Crimi, A., Bakas, S. (eds.) BrainLes 2021. LNCS, vol. 12962, pp. 272–284. Springer, Cham (2022). https://doi.org/10.1007/978-3-031-08999-2_22

8. Hatamizadeh, A., et al.: UNETR: transformers for 3D medical image segmentation. In: Proceedings of the IEEE/CVF Winter Conference on Applications of Computer Vision, pp. 574–584 (2022)
9. Heller, N., et al.: An international challenge to use artificial intelligence to define the state-of-the-art in kidney and kidney tumor segmentation in CT imaging (2020)
10. Hu, M., et al.: Online convolutional re-parameterization. In: Proceedings of the IEEE/CVF Conference on Computer Vision and Pattern Recognition, pp. 568–577 (2022)
11. Isensee, F., Jaeger, P.F., Kohl, S.A., Petersen, J., Maier-Hein, K.H.: nnU-Net: a self-configuring method for deep learning-based biomedical image segmentation. Nat. Methods 18(2), 203–211 (2021)
12. Ji, Y., et al.: Amos: a large-scale abdominal multi-organ benchmark for versatile medical image segmentation. arXiv preprint arXiv:2206.08023 (2022)
13. Kulikowski, J.J., Marčelja, S., Bishop, P.O.: Theory of spatial position and spatial frequency relations in the receptive fields of simple cells in the visual cortex. Biol. Cybern. 43(3), 187–198 (1982)
14. Lee, H.H., Bao, S., Huo, Y., Landman, B.A.: 3D UX-Net: a large kernel volumetric convnet modernizing hierarchical transformer for medical image segmentation. arXiv preprint arXiv:2209.15076 (2022)
15. Li, H., Nan, Y., Del Ser, J., Yang, G.: Large-kernel attention for 3D medical image segmentation. arXiv preprint arXiv:2207.11225 (2022)
16. Liu, S., et al.: More convnets in the 2020s: scaling up kernels beyond 51x51 using sparsity. arXiv preprint arXiv:2207.03620 (2022)
17. Liu, Z., et al.: Swin transformer: hierarchical vision transformer using shifted windows. In: Proceedings of the IEEE/CVF International Conference on Computer Vision, pp. 10012–10022 (2021)
18. Liu, Z., Mao, H., Wu, C.Y., Feichtenhofer, C., Darrell, T., Xie, S.: A convnet for the 2020s. In: Proceedings of the IEEE/CVF Conference on Computer Vision and Pattern Recognition, pp. 11976–11986 (2022)
19. Ma, J., et al.: Abdomenct-1k: is abdominal organ segmentation a solved problem? IEEE Trans. Pattern Anal. Mach. Intell. (2021). https://doi.org/10.1109/TPAMI.2021.3100536
20. Roth, H.R., et al.: DeepOrgan: multi-level deep convolutional networks for automated pancreas segmentation. In: Navab, N., Hornegger, J., Wells, W.M., Frangi, A.F. (eds.) MICCAI 2015. LNCS, vol. 9349, pp. 556–564. Springer, Cham (2015). https://doi.org/10.1007/978-3-319-24553-9_68
21. Tang, Y., et al.: Self-supervised pre-training of swin transformers for 3D medical image analysis. In: Proceedings of the IEEE/CVF Conference on Computer Vision and Pattern Recognition, pp. 20730–20740 (2022)
22. Wang, W., Chen, C., Ding, M., Yu, H., Zha, S., Li, J.: TransBTS: multimodal brain tumor segmentation using transformer. In: de Bruijne, M., et al. (eds.) MICCAI 2021. LNCS, vol. 12901, pp. 109–119. Springer, Cham (2021). https://doi.org/10.1007/978-3-030-87193-2_11
23. Zhou, H.Y., Guo, J., Zhang, Y., Yu, L., Wang, L., Yu, Y.: nnFormer: interleaved transformer for volumetric segmentation. arXiv preprint arXiv:2109.03201 (2021)

ConvFormer: Plug-and-Play CNN-Style Transformers for Improving Medical Image Segmentation

Xian Lin[1], Zengqiang Yan[1(✉)], Xianbo Deng[2], Chuansheng Zheng[2], and Li Yu[1]

[1] School of Electronic Information and Communications, Huazhong University of Science and Technology, Wuhan, China
{xianlin,z_yan,hustlyu}@hust.edu.cn
[2] Department of Radiology, Union Hospital, Tongji Medical College, Huazhong University of Science and Technology,Wuhan, China
dengxianbo@hotmail.com, cszheng@hust.edu.cn

Abstract. Transformers have been extensively studied in medical image segmentation to build pairwise long-range dependence. Yet, relatively limited well-annotated medical image data makes transformers struggle to extract diverse global features, resulting in attention collapse where attention maps become similar or even identical. Comparatively, convolutional neural networks (CNNs) have better convergence properties on small-scale training data but suffer from limited receptive fields. Existing works are dedicated to exploring the combinations of CNN and transformers while ignoring attention collapse, leaving the potential of transformers under-explored. In this paper, we propose to build CNN-style Transformers (ConvFormer) to promote better attention convergence and thus better segmentation performance. Specifically, ConvFormer consists of pooling, CNN-style self-attention (CSA), and convolutional feed-forward network (CFFN) corresponding to tokenization, self-attention, and feed-forward network in vanilla vision transformers. In contrast to positional embedding and tokenization, ConvFormer adopts 2D convolution and max-pooling for both position information preservation and feature size reduction. In this way, CSA takes 2D feature maps as inputs and establishes long-range dependency by constructing self-attention matrices as convolution kernels with adaptive sizes. Following CSA, 2D convolution is utilized for feature refinement through CFFN. Experimental results on multiple datasets demonstrate the effectiveness of ConvFormer working as a plug-and-play module for consistent performance improvement of transformer-based frameworks. Code is available at https://github.com/xianlin7/ConvFormer.

Keywords: CNN-Style Transformers · Attention Collapse · Adaptive Self-Attention · Medical Image Segmentation

Supplementary Information The online version contains supplementary material available at https://doi.org/10.1007/978-3-031-43901-8_61.

1   Introduction

Benefiting from the prominent ability to model long-range dependency, transformers have become the de-facto standard for natural language processing [1]. Compared with convolutional neural networks (CNNs), which encourage locality, weight sharing, and translation equivariance, transformers build global dependency through self-attention layers, bringing more possibilities for feature exaction and breaking the performance ceiling of CNNs in return [2–6].

Inspired by this, transformers are introduced into medical image segmentation and arouse wide concerns [7–11]. In vision transformers, each medical image is first split into a series of patches and then projected into a 1D sequence of patch embeddings [4]. Through building pairwise interaction among patches/tokens, transformers are supposed to aggregate global information for robust feature exaction. However, learning well-convergence global dependency in transformers is highly data-intensive, making transformers less effective given relatively limited medical imaging data.

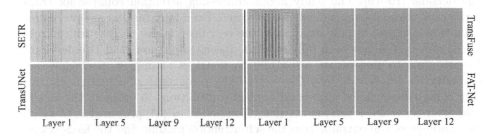

Fig. 1. Visualization of attention maps from the selected layers of the first head in different transformer frameworks. The darker the color, the closer the dependency.

To figure out how transformers work in medical image segmentation, we trained four state-of-the-art transformer-based models [5, 12–14] on the ACDC dataset and visualized the learned self-attention matrices across different layers as illustrated in Fig. 1. For all approaches, the attention matrices tend to become uniform among patches (*i.e.*, attention collapse [15]), especially in deeper layers. Attention collapse is more noticeable, especially in CNN-Transformer hybrid approaches (*i.e.*, TransUNet, TransFuse, and FAT-Net). On the one hand, insufficient training data would make transformers learn sub-optimal long-range dependency. On the other hand, directly combining CNNs with transformers would make the network biased to the learning of CNNs, as the convergence of CNNs is more achievable compared to transformers, especially on small-scale training data. Therefore, how to address attention collapse and improve the convergence of transformers is crucial for performance improvement.

In this work, we propose a plug-and-play module named ConvFormer to address attention collapse by constructing a kernel-scalable CNN-style transformer. In ConvFormer, 2D images can directly build sufficient long-range dependency without being split into 1D sequences. Specifically, corresponding to tokenization, self-attention, and feed-forward network in vanilla vision transformers, ConvFormer consists of pooling, CNN-style self-attention (CSA), and convolutional feed-forward network (CFFN) respectively. For an input image/feature map, its resolution is first reduced by applying convolution and max-pooling alternately. Then, CSA builds appropriate dependency for each pixel by adaptively generating a scalable convolutional, being smaller to include locality or being larger for long-range global interaction. Finally, CFFN refines the features of each pixel by applying continuous convolutions. Extensive experiments on three datasets across five state-of-the-art transformer-based methods validate the effectiveness of ConvFormer, outperforming existing solutions to attention collapse.

2   Related Work

Recent transformer-based approaches for medical image analysis mainly focus on introducing transformers for robust features exaction in the encoder, cross-scale feature interactive in skip connection, and multifarious feature fusion in the decoder [16–19]. The study about addressing attention collapse for transformers in medical imaging is under-explored. Even in natural image processing, attention collapse, usually existing in the very deep layers of deep transformer-based models, has not been fully studied. Specifically, Zhou et al. [15] developed Re-attention to re-generate self-attention matrices aiming at increasing their diversity on different layers. Zhou et al. [20] projected self-attention matrices into a high-dimensional space and applied convolutions to promote the locality and diversity of self-attention matrices. Touvron et al. [21] proposed to re-weight the channels of the outputs from the self-attention module and the feed-forward module to facilitate the convergence of transformers.

3   Method

The comparison between the vision transformer (ViT) and ConvFormer is illustrated in Fig. 2. The greatest difference is that our ConvFormer is conducted on 2D inputs while ViT is applied to 1D sequences. Specifically, the pooling module is utilized to replace tokenization in ViT, which well preserves locality and positional information without extra positional embeddings. The CNN-style self-attention (CSA) module, i.e. the core of ConvFormer, is developed to replace the self-attention (SA) module in ViT to build long-range dependency by constructing self-attention matrices in a similar way like convolutions with adaptive and scalable kernels. The convolutional feed-forward network (CFFN) is developed to refine the features for each pixel corresponding to the feed-forward network (FFN) in ViT. No upsampling procedure is adopted to resize the output of ConvFormer back to the input size as the pooling module can match the output size by adjusting the maxpooling times. It should be noticed that ConvFormer is realized based on convolutions, which eliminates the training tension between CNNs and transformers as analyzed in Sect. 1. Each module of ConvFormer is described in the following.

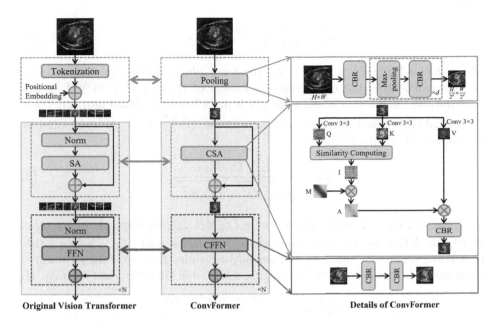

Fig. 2. Comparison between vanilla vision transformer and ConvFormer. CBR is short for the combination of convolution, batch normalization, and Relu. Multiple heads are omitted for simplicity.

3.1   Pooling vs. Tokenization

The pooling module is developed to realize the functions of tokenization (*i.e.*, making the input suitable to transformers in the channel dimension and shaping and reducing the input size when needed) while without losing details in the grid lines in tokenization. For an input $X_{in} \in \mathbb{R}^{c \times H \times W}$, convolution with a kernel size of 3×3 followed by batch normalization and Relu, is first applied to capture local features. Then, corresponding to each patch size S in ViT, total $d = \log_2 S$ downsampling operations are applied in the pooling module to produce the same resolutions. Here, each downsampling operation consists of a max-pooling with a kernel size of 2×2 and a combination of 3×3 convolution, batch normalization, and Relu. Finally, X_{in} becomes $X_1 \in \mathbb{R}^{c_m \times \frac{H}{2^d} \times \frac{W}{2^d}}$ through the pooling module where c_m is corresponding to the embedding dimension in ViT.

3.2   CNN-Style vs. Sequenced Self-attention

The building of long-range dependency in ConvFormer is relying on CNN-style self-attention, which creates an adaptive receptive field for each pixel by constructing a customized convolution kernel. Specifically, for each pixel $x_{i,j}$ of X_1, the convolution kernel $A^{i,j}$ is constructed based on two intermediate variables:

$$Q_{i,j} = \sum_{l=-1}^{1} \sum_{g=-1}^{1} E_{2+l,2+g}^{q} x_{i+l,j+g}, \tag{1}$$

$$K_{i,j} = \sum_{l=-1}^{1} \sum_{g=-1}^{1} E_{2+l,2+g}^{k} x_{i+l,j+g}, \tag{2}$$

where E^q and $E^k \in \mathbb{R}^{c_q \times c_m \times 3 \times 3}$ are the learnable projection matrices and c_q is corresponding to the embedding dimension of Q, K, and V in ViT, which incorporates the features of adjacent pixels in 3×3 neighborhood into $x_{i,j}$. Then, the initial customized convolutional kernel $I^{i,j} \in \mathbb{R}^{\frac{H}{2d} \times \frac{W}{2d}}$ for $x_{i,j}$ is calculated by computing the cosine similarity:

$$I_{m,n}^{i,j} = \frac{\sum_{l=0}^{c_q} Q_{i,j} K_{m,n}}{\sqrt{\sum_{l=0}^{c_q} Q_{i,j}^2} \sqrt{\sum_{l=0}^{c_q} K_{m,n}^2}}. \tag{3}$$

Here, $I_{m,n}^{i,j} \in [-1, 1]$ and seldom occurs $I_{m,n}^{i,j} = 0$. $I_{m,n}^{i,j}$ corresponds to attention score calculation in ViT (constrained to be positive while $I_{m,n}^{i,j}$ can be either positive or negative). Then, we dynamically determine the size of the customized convolution kernel for $x_{i,j}$ by introducing a learnable Gaussian distance map M:

$$M_{m,n}^{i,j} = e^{-\frac{(i-m)^2 (2^d/H)^2 + (j-n)^2 (2^d/W)^2}{2(\theta \times \alpha)^2}}, \tag{4}$$

where $\theta \in (0, 1)$ is a learnable network parameter to control the receptive field of A and α is a hyper-parameter to control the tendency of the receptive field. θ is proportional to the receptive field. For instance, under the typical setting $H = W = 256$, $d = 3$, and $\alpha = 1$, when $\theta = 0.003$, the receptive field only covers five adjacent pixels, when $\theta > 0.2$, the receptive field is global. The larger α is, the more likely A tends to have a global receptive field. Based on $I^{i,j}$ and $M^{i,j}$, $A^{i,j}$ is calculated by $A^{i,j} = I^{i,j} \times M^{i,j}$. In this way, every pixel $x^{i,j}$ has a customized size-scalable convolution kernel $A^{i,j}$. By multiplying A with V, CSA can build adaptive long-range dependency, where V can be formulated similarly according to Eq. (1). Finally, the combination of 1×1 convolution, batch normalization, and Relu is utilized to integrate features learned from long-range dependency.

3.3   Convolution Vs. Vanilla Feed-Forward Network

The convolution feed-forward network (CFFN) is to refine the features produced by CSA, just consisting of two combinations of 1×1 convolution, batch normalization, and Relu. By replacing linear projection and layer normalization in ViT, CFFN makes ConvFormer completely CNN-based, avoiding the combat between CNN and Transformer during training like CNN-Transformer hybrid approaches.

4   Experiments

4.1   Datasets and Implementation Details

ACDC[1]. A publicly-available dataset for the automated cardiac diagnosis challenge. Totally 100 scans with pixel-wise annotations of left ventricle (LV),

---

[1] https://www.creatis.insa-lyon.fr/Challenge/acdc/.

Table 1. Quantitative results in Dice (DSC) and Hausdorff Distance (HD).

| Method | DSC (%) ↑ | | | | | | HD (%) ↓ | | |
| | ACDC | | | | ISIC | ICH | ACDC | ISIC | ICH |
	Avg.	RV	MYO	LV			Avg.		
SETR [5]	87.14	83.95	83.89	93.59	89.03	80.17	17.24	22.33	14.90
+Re-attention [15]	85.91	82.52	82.27	92.93	88.00	79.08	18.21	24.57	14.50
+LayerScale [21]	85.74	81.54	82.70	92.98	87.98	78.91	17.88	23.94	14.45
+Refiner [20]	85.75	83.18	81.61	92.46	86.63	78.35	18.09	25.43	14.95
+ConvFormer	**91.00**	**89.26**	**88.60**	**95.15**	**90.41***	**81.56**	**14.08**	**21.68**	**13.50**
TransUNet [12]	90.80	89.59	87.81	94.99	88.75	78.52	14.58	25.11	15.90
+Re-attention [15]	91.25	89.91	88.61	95.22	88.35	77.50	13.79	**23.15**	**15.40**
+LayerScale [21]	91.30	89.37	88.79	**95.75**	88.75	75.60	**13.68**	23.32	15.50
+Refiner [20]	90.76	88.66	88.39	95.22	87.90	76.47	14.73	25.31	15.65
+ConvFormer	**91.42**	**90.17**	**88.84**	95.25	**89.40**	**80.66**	13.96	23.19	15.70
TransFuse [13]	89.10	87.85	85.73	93.73	89.28	75.11	14.98	23.08	18.60
+Re-attention [15]	88.48	87.05	85.37	93.00	88.28	73.74	16.20	24.56	18.45
+LayerScale [21]	88.85	87.81	85.24	93.50	89.00	74.18	13.53	23.96	20.00
+Refiner [20]	89.06	87.88	85.55	93.75	85.65	74.16	14.05	26.30	18.60
+ConvFormer	**89.88**	**88.85**	**86.50**	**94.30**	**90.56***	**75.56**	**12.84**	**21.30**	**17.60**
FAT-Net [14]	91.46	90.13	88.61	95.60	89.72	83.73	13.82	22.63	16.20
+Re-attention [15]	91.61*	89.99	89.18	95.64	89.84	84.42	14.00	22.54	14.20
+LayerScale [21]	91.71*	90.01	89.39	95.71	90.06	83.87	13.50	21.93	13.70
+Refiner [20]	91.94*	90.54	**89.70**	95.58	89.20	83.14	13.37	23.35	**13.25**
+ConvFormer	**92.18***	**90.69**	89.57	**96.28**	**90.36***	**84.97**	**11.32**	21.73	14.10
Patcher [27]	91.41	89.56	89.12	95.53	89.11	80.54	13.55	22.16	15.70
+Re-attention [15]	91.25	89.77	88.58	95.39	89.73	79.08	14.49	21.86	17.60
+LayerScale [21]	91.07	88.94	88.81	95.46	90.16	74.13	15.48	**21.78**	18.60
+Refiner [20]	91.26	89.65	88.58	95.57	68.92	79.66	13.93	56.62	15.60
+ConvFormer	**92.07***	**90.91**	**89.54**	**95.78**	**90.18**	**81.69**	**12.29**	21.88	**15.35**

★ Approaches outperforming the state-of-the-art 2D approaches on the publicly-available ACDC (*i.e.*, FAT-Net [14]: 91.46% in Avg. DSC) and ISIC (*i.e.*, Ms Red [28]: 90.25% in Avg. DSC) datasets respectively. *More comprehensive quantitative comparison results can be found in the supplemental materials.*

myocardium (MYO), and right ventricle (RV) are available [22]. Following [12,17,18], 70, 10, and 20 cases are used for training, validation, and testing respectively.

ISIC 2018[2]. A publicly-available dataset for skin lesion segmentation. Totally 2594 dermoscopic lesion images with pixel-level annotations are available [23,24]. Following [25,26], the dataset is randomly divided into 2076 images for training and 520 images for testing.

---

[2] https://challenge.isic-archive.com/data/.

ICH. A locally-collected dataset for hematoma segmentation. Totally 99 CT scans consisting of 2648 slices were collected and annotated by three radiologists. The dataset is randomly divided into the training, validation, and testing sets according to a ratio of 7:1:2.

Implementation Details. For a fair comparison, all the selected state-of-the-art transformer-based baselines were trained with or without ConvFormer under the same settings. All models were trained by an Adam optimizer with a learning rate of 0.0001 and a batch size of 4 for 400 rounds. Data augmentation includes random rotation, scaling, contrast augmentation, and gamma augmentation.

4.2 Results

ConvFormer can work as a plug-and-play module and replace the vanilla transformer blocks in transformer-based baselines. To evaluate the effectiveness of ConvFormer, five state-of-the-art transformer-based approaches are selected as backbones, including SETR [5], TransUNet [12], TransFuse [13], FAT-Net [14], and Patcher [27]. SETR and Patcher utilize pure-transformer encoders, while TransUNet, TransFuse, and FAT-Net adopt CNN-Transformer hybrid encoders. In addition, three state-of-the-art methods for addressing attention collapse, including Re-attention [15], LayerScale [21], and Refiner [20], are equipped with the above transformer-based baselines for comparison.

Quantitative Results. Quantitative results of ConvFormer embedded into various transformer-based baselines on the three datasets are summarized in Table 1. ConvFormer achieves consistent performance improvements on all five backbones. Compared to CNN-Transformer hybrid approaches (*i.e.*, TransUNet, TransFuse, and FAT-Net), ConvFormer is more beneficial on pure-transformer approaches (*i.e.*, SETR and Patcher). Specifically, with ConvFormer, SETR achieves an average increase of 3.86%, 1.38%, and 1.39% in Dice on the ACDC, ISIC, and ICH datasets respectively, while the corresponding performance improvements of Patcher are 0.66%, 1.07%, and 1.15% respectively. Comparatively, in CNN-Transformer hybrid approaches, as analyzed above, CNNs would be more dominating against transformers during training. Despite this, re-balancing CNNs and Transformers through ConvFormer can build better long-range dependency for consistent performance improvement.

Comparison with SOTA Approaches. Quantitative results compared with the state-of-the-art approaches to addressing attention collapse are summarized in Table 1. In general, given relatively limited training data, existing approaches designed for natural image processing are unsuitable for medical image segmentation, resulting in unstable performance across different backbones and datasets. Comparatively, ConvFormer consistently outperforms these approaches and brings stable performance improvements to various backbones across datasets, demonstrating the excellent generalizability of ConvFormer as a plug-and-play module.

Visualization of Self-Attention Matrices. To qualitatively evaluate the effectiveness of ConvFormer in addressing attention collapse and building efficient long-range dependency, we visualize the self-attention matrices with and

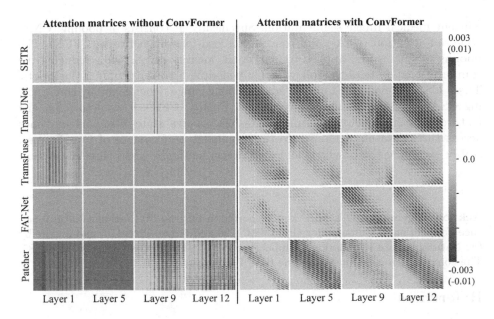

Fig. 3. Visualization of self-attention matrices by baselines w/ and w/o ConvFormer.

Table 2. Ablation study of hyper-parameter α on the ACDC dataset.

α	0.2	0.4	0.6	0.8	1.0
Dice (%)	90.71	**91.00**	90.76	90.66	90.45

without ConvFormer as illustrated in Fig. 3. By introducing ConvFormer, attention collapse is effectively alleviated. Compare to the self-attention matrices of baselines, the matrices learned by ConvFormer are more diverse. Specifically, the interactive range for each pixel is scalable, being small for locality preserving or being large for global receptive fields. Besides, dependency is no longer constrained to be positive like ViT, which is more consistent with convolution kernels. *Qualitative segmentation results of different approaches on the three datasets can be found in the supplemental materials.*

Ablation Study As described in Sec. 3.2, α is to control the receptive field tendency in ConvFormer, The larger the α, the more likely ConvFormer contains larger receptive fields. To validate this, we conduct an ablation study on α as summarized in Table 2. In general, using a large α does not necessarily lead to more performance improvements, which is consistent with our observation that not every pixel needs global information for segmentation.

5   Conclusions

In this paper, we construct the transformer as a kernel-scalable convolution to address the attention collapse and build diverse long-range dependencies for efficient medical image segmentation. Specifically, it consists of pooling, CNN-style self-attention (CSA), and convolution feed-forward network (CFFN). The pooling module is first applied to extract the locality details while reducing the computational costs of the following CSA module by downsampling the inputs. Then, CSA is developed to build adaptive long-range dependency by constructing CSA as a kernel-scalable convolution, Finally, CFFN is used to refine the features of each pixel. Experimental results on five state-of-the-art baselines across three datasets demonstrate the prominent performance of ConvFormer, stably exceeding the baselines and comparison methods across three datasets.

Acknowledgement. This work was supported in part by the National Natural Science Foundation of China under Grant 62271220 and Grant 62202179, and in part by the Natural Science Foundation of Hubei Province of China under Grant 2022CFB585. The computation is supported by the HPC Platform of HUST.

References

1. Vaswani, A., et al.: Attention is all you need. arXiv preprint arXiv:1706.03762 (2017)
2. Ronneberger, O., Fischer, P., Brox, T.: U-net: convolutional networks for biomedical image segmentation. In: Navab, N., Hornegger, J., Wells, W.M., Frangi, A.F. (eds.) MICCAI 2015, LNCS, vol. 9351, pp. 234–241. Springer, Cham (2015). https://doi.org/10.1007/978-3-319-24574-4_28
3. Schlemper, J., et al.: Attention gated networks: learning to leverage salient regions in medical images. Med. Image Anal. **53**, 197–207 (2019)
4. Dosovitskiy, A., et al.: An image is worth 16x16 words: transformers for image recognition at scale. arXiv preprint arXiv:2010.11929 (2020)
5. Zheng, S., et al.: Rethinking semantic segmentation from a sequence-to-sequence perspective with transformers. In: Proceedings of the IEEE/CVF International Conference on Computer Vision, pp. 6881–6890 (2021)
6. He, K., Chen, X., Xie, S., Li, Y., Dollár, P., Girshick, R.: Masked autoencoders are scalable vision learners. In: Proceedings of the IEEE/CVF International Conference on Computer Vision, pp. 16000–16009 (2022)
7. You, C., et al.: Class-aware generative adversarial transformers for medical image segmentation. arXiv preprint arXiv:2201.10737 (2022)
8. Karimi, D., Vasylechko, S.D., Gholipour, A.: Convolution-free medical image segmentation using transformers. In: de Bruijne, M., et al. (eds.) MICCAI 2021. LNCS, vol. 12901, pp. 78–88. Springer, Cham (2021). https://doi.org/10.1007/978-3-030-87193-2_8
9. Zhang, Y., et al.: mmFormer: multimodal medical transformer for incomplete multimodal learning of brain tumor segmentation. In: Wang, Li., Dou, Q., Fletcher, P.T., Speidel S., Li, S. (eds.) MICCAI 2022. LNCS, vol. 13431, pp. 107–117. Springer, Cham (2022). https://doi.org/10.1007/978-3-031-16443-9_11
10. Wang, Z., et al.: SMESwin unet: merging CNN and transformer for medical image segmentation. In: Wang, Li., Dou, Q., Fletcher, P.T., Speidel S., Li, S. (eds.) MICCAI 2022. LNCS, vol. 13431, pp. 517–526. Springer, Cham (2022). https://doi.org/10.1007/978-3-031-16443-9_50

11. Li, H., Chen, L., Han, H., Zhou, S.K.: SATr: slice attention with transformer for universal lesion detection. In: Wang, Li., Dou, Q., Fletcher, P.T., Speidel S., Li, S. (eds.) MICCAI 2022, LNCS, vol. 13431, pp. 163–174. Springer, Cham (2022). https://doi.org/10.1007/978-3-031-16437-8_16

12. Chen, J., et al. Transunet: transformers make strong encoders for medical image segmentation. arXiv preprint arXiv:2102.04306 (2021)

13. Zhang, Y., Liu, H., Hu, Q.: Transfuse: fusing transformers and CNNs for medical image segmentation. In: de Bruijne, M., et al. (eds.) MICCAI 2021. LNCS, vol. 12901, pp. 14–24. Springer, Cham (2021). https://doi.org/10.1007/978-3-030-87193-2_2

14. Wu, H., Chen, S., Chen, G., Wang, W., Lei, B., Wen, Z.: FAT-Net: feature adaptive transformers for automated skin lesion segmentation. Med. Image Anal. **76**, 102327 (2022)

15. Zhou, D., et al.: DeepViT: towards deeper vision transformer. arXiv preprint arXiv:2103.11886 (2021)

16. Huang, X., Deng, Z., Li, D., Yuan, X.: Missformer: an effective medical image segmentation transformer. arXiv preprint arXiv:2109.07162 (2021)

17. Cao, H., et al. Swin-unet: Unet-like pure transformer for medical image segmentation. arXiv preprint arXiv:2105.05537 (2021)

18. Xu, G., Wu, X., Zhang, X., He, X.: Levit-unet: make faster encoders with transformer for medical image segmentation. arXiv preprint arXiv:2107.08623 (2021)

19. Liu, W., et al.: Phtrans: parallelly aggregating global and local representations for medical image segmentation. In: Wang, Li., Dou, Q., Fletcher, P.T., Speidel S., Li, S. (eds.) MICCAI 2022. LNCS, vol. 13431, pp. 235–244. Springer, Cham (2022). https://doi.org/10.1007/978-3-031-16443-9_23

20. Zhou, D., et al.: Refiner: refining self-attention for vision transformers. arXiv preprint arXiv:2106.03714 (2021)

21. Touvron, H., Cord, M., Sablayrolles, A., Synnaeve, G., Jégou, H.: Going deeper with image transformers. In: Proceedings of the IEEE/CVF International Conference on Computer Vision, pp. 32–42 (2021)

22. Bernard, O., et al.: Deep learning techniques for automatic MRI cardiac multi-structures segmentation and diagnosis: is the problem solved? IEEE Trans. Med. Imag. **37**(11), 2514–2525 (2018)

23. Codella, N., et al.: Skin lesion analysis toward melanoma detection 2018: a challenge hosted by the international skin imaging collaboration (ISIC). arXiv preprint arXiv:1902.03368 (2019)

24. Tschandl, P., Rosendahl, C., Kittler, H.: The HAM10000 dataset, a large collection of multi-source dermatoscopic images of common pigmented skin lesions. Sci. Data. **5**(1), 1–9 (2018)

25. Lin, A., Chen, B., Xu, J., Zhang, Z., Lu, G., Zhang, D.: Ds-transunct: dual swin transformer u-net for medical image segmentation. IEEE Instrum. Meas. **71**, 1–15 (2022)

26. Chen, B., Liu, Y., Zhang, Z., Lu, G., Kong, A.W.K.: Transattunet: multi-level attention-guided u-net with transformer for medical image segmentation. arXiv preprint arXiv:2107.05274 (2021)

27. Ou, Y., et al.: Patcher: patch transformers with mixture of experts for precise medical image segmentation. In: Wang, Li., Dou, Q., Fletcher, P.T., Speidel S., Li, S. (eds.) MICCAI 2022. LNCS, vol. 13431, pp. 475–484. Springer, Cham (2022). https://doi.org/10.1007/978-3-031-16443-9_46

28. Dai D., et al.: Ms RED: a novel multi-scale residual encoding and decoding network for skin lesion segmentation. Med. Image Anal. **75**, 102293 (2022)

Edge-Aware Multi-task Network for Integrating Quantification Segmentation and Uncertainty Prediction of Liver Tumor on Multi-modality Non-contrast MRI

Xiaojiao Xiao, Qinmin Vivian Hu, and Guanghui Wang(✉)

Department of Computer Science, Toronto Metropolitan University, Toronto, Canada
wangcs@torontomu.ca

Abstract. Simultaneous multi-index quantification, segmentation, and uncertainty estimation of liver tumors on multi-modality non-contrast magnetic resonance imaging (NCMRI) are crucial for accurate diagnosis. However, existing methods lack an effective mechanism for multi-modality NCMRI fusion and accurate boundary information capture, making these tasks challenging. To address these issues, this paper proposes a unified framework, namely edge-aware multi-task network (EaMtNet), to associate multi-index quantification, segmentation, and uncertainty of liver tumors on the multi-modality NCMRI. The EaMt-Net employs two parallel CNN encoders and the Sobel filters to extract local features and edge maps, respectively. The newly designed edge-aware feature aggregation module (EaFA) is used for feature fusion and selection, making the network edge-aware by capturing long-range dependency between feature and edge maps. Multi-tasking leverages prediction discrepancy to estimate uncertainty and improve segmentation and quantification performance. Extensive experiments are performed on multi-modality NCMRI with 250 clinical subjects. The proposed model outperforms the state-of-the-art by a large margin, achieving a dice similarity coefficient of 90.01 ± 1.23 and a mean absolute error of $2.72 \pm 0.58\,\text{mm}$ for MD. The results demonstrate the potential of EaMtNet as a reliable clinical-aided tool for medical image analysis.

Keywords: Segmentation · Quantification · Liver tumor · Multi-modality · Uncertainty

1 Introduction

Simultaneous multi-index quantification (i.e., max diameter (MD), center point coordinates (X_o, Y_o), and Area), segmentation, and uncertainty prediction of liver tumor have essential significance for the prognosis and treatment of patients [6,16]. In clinical settings, segmentation and quantitation are manually performed by the clinicians through visually analyzing the contrast-enhanced MRI images (CEMRI) [9,10,18]. However, as shown in Fig. 1(b), Contrast-enhanced

H. Greenspan et al. (Eds.): MICCAI 2023, LNCS 14223, pp. 652–661, 2023.
https://doi.org/10.1007/978-3-031-43901-8_62

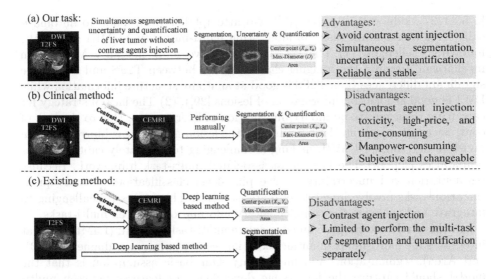

Fig. 1. Our method integrates segmentation and quantification of liver tumor using multi-modality NCMRI, which has the advantages of avoiding contrast agent injection, mutual promotion of multi-task, and reliability and stability.

MRI (CEMRI) has the drawbacks of being toxic, expensive, and time-consuming due to the need for contrast agents (CA) to be injected [2,4]. Moreover, manually annotating medical images is a laborious and tedious process that requires human expertise, making it manpower-intensive, subjective, and prone to variation [14]. Therefore, it is desirable to provide a reliable and stable tool for simultaneous segmentation, quantification, and uncertainty analysis, without requiring the use of contrast agents, as shown in Fig. 1(a).

Recently, an increasing number of works have been attempted on liver tumor segmentation or quantification [25,26,28,30]. As shown in Fig. 1(c), the work [26] attempted to use the T2FS for liver tumor segmentation, while it ignored the complementary information between multi-modality NCMRI of T2FS and DWI. In particular, there is evidence that diffusion-weighted imaging (DWI) helps to improve the detection sensitivity of focal lesions as these lesions typically have higher cell density and microstructure heterogeneity [20]. The study in [25,30] attempted to quantify the multi-index of liver tumor, however, the approach is limited to using multi-phase CEMRI that requires the injection of CA. In addition, all these works are limited to a single task and ignore the constraints and mutual promotion between multi-tasks. Available evidence suggests that uncertainty information regarding segmentation results is important as it guides clinical decisions and helps understand the reliability of the provided segmentation. However, current research on liver tumors tends to overlook this vital task.

To the best of our knowledge, although many works focus on the simultaneous quantization, segmentation, and uncertainty in medical images (i.e., heart

[3,5,11,27], kidney [17], polyp [13]). No attempt has been made to automatically liver tumor multi-task via integrating multi-modality NCMRI due to the following challenges: (1) The lack of an effective multi-modality MRI fusion mechanism. Because the imaging characteristics between T2FS and DWI have significant differences (i.e., T2FS is good at anatomy structure information while DWI is good at location information of lesions [29]). (2) The lack of strategy for capturing the accurate boundary information of liver tumors. Due to the lack of contrast agent injection, the boundary of the lesion may appear blurred or even invisible in a single NCMRI, making it challenging to accurately capture tumor boundaries [29]. (3) The lack of an associated multi-task framework. Because segmentation and uncertainty involve pixel-level classification, whereas quantification tasks involve image-level regression [11]. This makes it challenging to integrate and optimize the complementary information between multi-tasks.

In this study, we propose an edge-aware multi-task network (EaMtNet) that integrates the multi-index quantification (i.e., center point, max-diameter (MD), and Area), segmentation, and uncertainty. Our basic assumption is that the model should capture the long-range dependency of features between multi-modality and enhance the boundary information for quantification, segmentation, and uncertainty of liver tumors. The two parallel CNN encoders first extract local feature maps of multi-modality NCMRI. Meanwhile, to enhance the weight of tumor boundary information, the Sobel filters are employed to extract edge maps that are fed into edge-aware feature aggregation (EaFA) as prior knowledge. Then, the EaFA module is designed to select and fuse the information of multi-modality, making our EaMtNet edge-aware by capturing the long-range dependency of features maps and edge maps. Lastly, the proposed method estimates segmentation, uncertainty prediction, and multi-index quantification simultaneously by combining multi-task and cross-task joint loss.

The contributions of this work mainly include: (1) For the first time, multi-index quantification, segmentation, and uncertainty of the liver tumor on multi-modality NCMRI are achieved simultaneously, providing a time-saving, reliable, and stable clinical tool. (2) The edge information extracted by the Sobel filter enhances the weight of the tumor boundary by connecting the local feature as prior knowledge. (3) The novel EaFA module makes our EaMtNet edge-aware by capturing the long-range dependency of features maps and edge maps for feature fusion. The source code will be available on the author's website.

2   Method

The EaMtNet employs an innovative approach for simultaneous tumor multi-index quantification, segmentation, and uncertainty prediction on multi-modality NCMRI. As shown in Fig. 2, the EaMtNet inputs multi-modality NCMRI of T2FS and DWI for capturing the feature and outputs the multi-index quantification, segmentation, and uncertainty. Specifically, the proposed approach mainly consists of three steps: 1) The CNN encoders for capturing feature maps and the Sobel filters for extracting edge maps (Sect. 2.1); 2) The

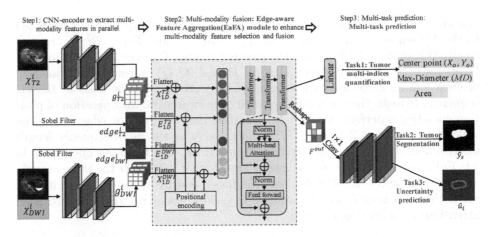

Fig. 2. Overview of the EaMtNet. It mainly consists of three steps: 1) CNN encoders for extracting local features of multi-modality NCMRI; 2) EaFA for enhancing multi-modality NCMRI feature selection and fusion; and 3) Multi-task prediction.

edge-aware feature aggregation (EaFA) for multi-modality feature selection and fusion via capturing the long-distance dependence (Sect. 2.2); and 3) Multi-task prediction module (Sect. 2.3).

2.1 CNN Encoder for Feature Extraction

In Step 1 of Fig. 2, the multi-modality NCMRI (i.e., $\chi^i_{T2} \in R^{H \times W}$, $\chi^i_{DWI} \in R^{H \times W}$) are fed into two parallel encoders and the Sobel filter to extract the feature maps (i.e., $g^i_{T2} \in R^{H \times W \times N}$, $g^i_{DWI} \in R^{H \times W \times N}$) and the corresponding edge maps (i.e., $edge^i_{T2} \in R^{H \times W}$, $edge^i_{DWI} \in R^{H \times W}$) respectively. Specifically, EaMtNet employs UNet as the backbone for segmentation because the CNN encoder has excellent capabilities in low-range semantic information extraction [15]. The two parallel CNN encoders have the same architecture where each encoder contains three shallow convolutional network blocks to capture features of adjacent slices. Each conv block consists of a convolutional layer, batch normalization, ReLU, and non-overlapping subsampling. At the same time, EaMt-Net utilizes the boundary information extracted by the Sobel filter [19] as prior knowledge to enhance the weight of tumor edge information to increase the awareness of the boundary.

2.2 Edge-Aware Feature Aggregation(EaFA) for Multi-modality Feature Selection and Fusion

In Step 2 of the proposed model, the feature maps (i.e., g^i_{T2}, g^i_{DWI}) and the edge maps (i.e., $edge^i_{T2}$, $edge^i_{DWI}$) are fed into EaFA for multi-modality feature fusion with edge-aware. In particular, the EaFA makes the EaMtNet edge-aware by using the Transformer to capture the long-range dependency of feature maps and edge maps. Specifically, the feature maps and edge maps are first flattened to the 1D sequence corresponding to $X_{1D} \in R^{N \times P^2}$ and $E_{1D} \in R^{2 \times Q^2}$, respectively. Where $N = 2 \times C$ means the channel number C of the last convolutional layer from the two parallel encoders. (P, P) and (Q, Q) represent the resolution of each feature map and each edge map, respectively. On the basis of the 1D sequence, to make the feature fusion with edge awareness, the operation of position encoding is performed not only on feature maps but also on edge maps. The yielded embeddings $Z \in R^{N \times P^2 + 2 \times Q^2}$ can serve as the input sequence length for the multi-head attention layer in Transformer. The following operations in our EaFA are similar to the traditional Transformer [22]. After the three cascade Transformer layers, the EaFA yields the fusion feature vector F for multi-task prediction. The specific computation of the self-attention matrix and multi-head attention are defined below [22]:

$$Attention(\mathcal{Q}, \mathcal{K}, \mathcal{V}) = softmax(\frac{\mathcal{Q}\mathcal{K}^T}{\sqrt{d_k}})(\mathcal{V}) \tag{1}$$

$$MultiHead(\mathcal{Q}, \mathcal{K}, \mathcal{V}) = Concat(head_1, ..., head_h)\mathcal{W}^{\mathcal{O}} \tag{2}$$

$$head_i = Attention(\mathcal{Q}\mathcal{W}_i^{\mathcal{Q}}, \mathcal{K}\mathcal{W}_i^{\mathcal{O}}, \mathcal{V}\mathcal{W}_i^{\mathcal{V}}) \tag{3}$$

where query $\mathcal{Q}$, key $\mathcal{K}$, and value $\mathcal{V}$ are all vectors of the flattened 1D sequences of X_{1D} and E_{1D}. $\mathcal{W}_i^{\mathcal{O}}$ is the projection matrix, and $\frac{1}{\sqrt{d_k}}$ is the scaling factor.

2.3   Multi-task Prediction

In Step 3 of Fig. 2, the EaMtNet outputs the multi-modality quantification $\hat{y}_Q$ (i.e., MD, X_o, Y_o and Area), segmentation result $\hat{y}_s$ and uncertainty map $\hat{u}_i$. Specifically, for the quantification path, $\hat{y}_Q$ is directly obtained by performing a linear layer to the feature F from EaFA. For the segmentation and uncertainty path, the output feature F from EaFA is first reshaped into a 2D feature map F^{out}. Then, to scale up to higher-resolution images, a 1×1 convolution layer is employed to change the channel number of F^{out} for feeding into the decoder. After upsampling by the CNN decoder, EaMtNet predicts the segmentation result $\hat{y}_s$ with $H \times W$ and uncertainty map $\hat{u}_i$ with $H \times W$. The CNN decoder contains three shallow deconv blocks, which consist of deconv layer, batch normalization, and ReLU. Inspired by [24], we select the entropy map as our uncertainty measure. Given the prediction probability after softmax, the entropy map is computed as follows:

$$H[x] = -\sum_{i=1}^{K} z_i(x) * log_2(z_i(x)) \tag{4}$$

where z_i is the probability of pixel x belonging to category i. When a pixel has high entropy, it means that the network is uncertain about its classification. Therefore, pixels with high entropy are more likely to be misclassified. In other words, its entropy will decrease when the network is confident in a pixel's label.

Under the constraints of uncertainty, the EaMtNet can effectively rectify the errors in tumor segmentation because the uncertainty estimation can avoid overconfidence and erroneous quantification [23]. Moreover, the EaMtNet novelly make represent different tasks in a unified framework, leading to beneficial interactions. Thus, the quantification performance is improved through back-propagation by the joint loss function $L_{multi-task}$. The function comprises segmentation loss L_{seg} and quantification loss L_{qua}, where the loss function L_{seg} is utilized for optimizing tumor segmentation, and L_{qua} is utilized for optimization of multi-index quantification. It can be defined as:

$$L_{Dice} = \frac{2\sum_i^N y_i \hat{y}_s}{\sum_i^S {y_i}^2 + \sum_i^N \hat{y}_s^2} \tag{5}$$

$$L_{qua}\left(\hat{y}_{task}^i, y_{task}^i\right) = \sum_{i=1} \left|y_{task}^i - \hat{y}_{task}^i\right| \tag{6}$$

where $\hat{y}_s$ represents the prediction, and y_i represents the ground truth label. The sum is performed on S pixels, $\hat{y}_{task}^i$ represents the predicted multi-index value, and y_{task}^i represents the ground truth of multi-index value, $task \in \{MD, X, Y, Area\}$.

3   Experimental Results and Discussion

For the first time, EaMtNet has achieved high performance with the dice similarity coefficient (DSC) up to $90.01 \pm 1.23\%$, and the mean absolute error (MAE) of the MD, X_o, Y_o and Area are down to 2.72 ± 0.58 mm, 1.87 ± 0.76 mm, 2.14 ± 0.93 mm and 15.76 ± 8.02 cm^2, respectively.

Dataset and Configuration. An axial dataset includes 250 distinct subjects, each underwent initial standard clinical liver MRI protocol examinations with corresponding pre-contrast images (T2FS [4mm]) and DWI [4mm]) was collected. The ground truth was reviewed by two abdominal radiologists with 10 and 22 years of experience in liver imaging, respectively. If any interpretations demonstrated discrepancies between the reviewers, they would re-evaluate the examinations together and reach a consensus. To align the paired images of T2 and DWI produced at different times. We set the T2 as the target image and the DWI as the source image to perform the pre-processing of non-rigid registration between T2 and DWI by using the Demons non-rigid registration method. It has been widely used in the field of medical image registration since it was proposed by Thirion [21]. We perform the Demons non-rigid registration on an open-source toolbox DIRART using Matlab 2017b.

Inspired by the work [22], we set the scaling factor d_k to 64 in equation (1). All experiments were assessed with a 5-fold cross-validation test. To quantitatively evaluate the segmentation results, we calculated the dice coefficient scores (DSC) metric that measures the overlapping between the segmentation prediction and ground truth [12]. To quantitatively evaluate the quantification results,

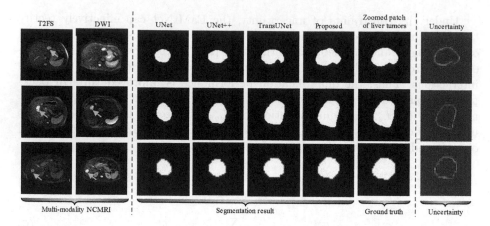

Fig. 3. The comparison of segmentation results between the proposed EaMtNet and three SOTA methods. The results show that our network yields high performance.

Table 1. Segmentation and quantification performance of the EaMtNet under different configurations. DSC is used to evaluate the segmentation performance. MAE is used to evaluate the quantification performance.

	DSC(%)	MD(mm)	$X_o(mm)$	$Y_o(mm)$	Area(cm^2)
No-EaFA	85.82 ± 1.92	3.49 ± 0.94	2.51 ± 1.43	3.12 ± 1.84	28.55 ± 9.75
No-Uncertainty	88.37 ± 2.71	3.25 ± 0.77	2.36 ± 0.92	2.78 ± 1.18	24.15 ± 9.19
Our method	90.01 ± 1.23	2.72 ± 0.58	1.87 ± 0.76	2.14 ± 0.93	15.76 ± 8.02

we calculated the mean absolute error (MAE). Our EaMtNet was implemented using Ubuntu 18.04 platform, Python v3.6, PyTorch v0.4.0, and running on two NVIDIA GTX 3090Ti GPUs.

Accurate Segmentation. The segmentation performance of EaMtNet has been validated and compared with three state-of-the-art (SOTA) segmentation methods (TransUNet [1], UNet [15], and UNet++ [31]). Furthermore, to ensure consistency in input modality, the channel number of the first convolution layer in the three comparison methods is set to 2. The visual examples of liver tumors are shown in Fig. 3, it is evident that our proposed EaMtNet outperforms the three SOTA methods. Some quantitative analysis results are shown in Table 1 and Table 2, our network achieves high performance with the DSC of $90.01 \pm 1.23\%$ (5.39% higher than the second-best). The results demonstrate that edge-aware, multi-modality fusion, and uncertainty prediction are essential for segmentation.

Ablation Study. To verify the contributions of edge-aware feature aggregation (EaFA) and uncertainty, we performed ablation study and compared and performance of different networks. First, we removed the EaFA and used concatenate,

Table 2. The quantitative evaluation of segmentation. DSC is used to evaluate the performance of our EaMtNet and three SOTA methods.

	UNet	UNet++	TransUNet	Our method
DSC(%)	76.59 ± 1.86	80.97 ± 2.37	84.62 ± 1.45	90.01 ± 1.23

Table 3. The quantitative evaluation of the multi-index quantification. The criteria of MAE is used to evaluate the performance of our EaMtNet and two SOTA methods.

	MD(mm)	X(mm)	Y(mm)	Area(cm^2)
ResNet-50	6.24 ± 2.81	3.15 ± 1.25	3.38 ± 1.27	31.32 ± 8.47
DenseNet	4.85 ± 1.67	2.73 ± 0.89	2.95 ± 1.15	25.37 ± 7.63
Our method	2.72 ± 0.58	1.87 ± 0.76	2.14 ± 0.93	15.76 ± 8.02

meaning we removed fusion multi-modality (No-EaFA). Then, we removed the uncertainty task (No-Uncertainty). The quantitative analysis results of these ablation studies are shown in Table 1. Our method exhibits high performance in both segmentation and quantification, indicating that each component of the EaMtNet plays a vital role in liver tumor segmentation and quantification.

Performance Comparison with State-of-the-Art. The EaMtNet has been validated and compared with three SOTA segmentation methods and two SOTA quantification methods (i.e., ResNet-50 [7] and DenseNet [8]). Furthermore, the channel number of the first convolution layer in the two quantification comparison methods is set to 2 to ensure the consistency of input modalities. The visual segmentation results are shown in Fig. 3. Moreover, the quantitative results (as shown in Table 2) corresponding to the visualization results (i.e., Fig. 3) obtained from the existing experiments further demonstrate that our method outperforms the three SOTA methods. Specifically, compared with the second-best approach, the DSC is boosted from 84.62 ± 1.45% to 90.01 ± 1.23%. The quantitative analysis results are shown in Table 3. It is evident that our method outperforms the two SOTA methods with a large margin in all metrics, owing to the proposed multi-modality fusing and multi-task association.

4   Conclusion

In this paper, we have proposed an EaMtNet for the simultaneous segmentation and multi-index quantification of liver tumors on multi-modality NCMRI. The new EaFA enhances edge awareness by utilizing boundary information as prior knowledge while capturing the long-range dependency of features to improve feature selection and fusion. Additionally, multi-task leverages the prediction discrepancy to estimate uncertainty, thereby improving segmentation and quantification performance. Extensive experiments have demonstrated the proposed model outperforms the SOTA methods in terms of DSC and MAE, with great potential to be a diagnostic tool for doctors.

Acknowledgements. This work is partly supported by the Natural Sciences and Engineering Research Council of Canada (NSERC) and TMU FOS Postdoctoral Fellowship.

References

1. Chen, J., et al.: Transunet: transformers make strong encoders for medical image segmentation. arXiv preprint arXiv:2102.04306 (2021)
2. Danet, I.M., Semelka, R.C., Braga, L.: Mr imaging of diffuse liver disease. Radiologic Clinics **41**(1), 67–87 (2003)
3. Du, X., Tang, R., Yin, S., Zhang, Y., Li, S.: Direct segmentation-based full quantification for left ventricle via deep multi-task regression learning network. IEEE J. Biomed. Health Inform. **23**(3), 942–948 (2018)
4. Fishbein, M., et al.: Hepatic MRI for fat quantitation: its relationship to fat morphology, diagnosis, and ultrasound. J. Clin. Gastroenterol. **39**(7), 619–625 (2005)
5. Ge, R., et al.: K-net: integrate left ventricle segmentation and direct quantification of paired echo sequence. IEEE Trans. Med. Imaging **39**(5), 1690–1702 (2019)
6. Gonzalez-Guindalini, F.D., et al.: Assessment of liver tumor response to therapy: role of quantitative imaging. Radiographics **33**(6), 1781–1800 (2013)
7. He, K., Zhang, X., Ren, S., Sun, J.: Deep residual learning for image recognition. In: Proceedings of the IEEE Conference on Computer Vision and Pattern Recognition, pp. 770–778 (2016)
8. Huang, G., Liu, Z., Van Der Maaten, L., Weinberger, K.Q.: Densely connected convolutional networks. In: Proceedings of the IEEE Conference on Computer Vision and Pattern Recognition, pp. 4700–4708 (2017)
9. Huo, J., Wu, J., Cao, J., Wang, G.: Supervoxel based method for multi-atlas segmentation of brain MR images. Neuroimage **175**, 201–214 (2018)
10. Lee, Y.J., et al.: Hepatocellular carcinoma: diagnostic performance of multidetector CT and MR imaging-a systematic review and meta-analysis. Radiology **275**(1), 97–109 (2015)
11. Luo, G., et al.: Commensal correlation network between segmentation and direct area estimation for bi-ventricle quantification. Med. Image Anal. **59**, 101591 (2020)
12. Milletari, F., Navab, N., Ahmadi, S.A.: V-net: fully convolutional neural networks for volumetric medical image segmentation. In: 2016 Fourth International Conference on 3D Vision (3DV), pp. 565–571. IEEE (2016)
13. Patel, K.B., Li, F., Wang, G.: Fuzzynet: a fuzzy attention module for polyp segmentation. In: NeurIPS'22 Workshop
14. Petitclerc, L., Sebastiani, G., Gilbert, G., Cloutier, G., Tang, A.: Liver fibrosis: review of current imaging and MRI quantification techniques. J. Magn. Reson. Imaging **45**(5), 1276–1295 (2017)
15. Ronneberger, O., Fischer, P., Brox, T.: U-Net: convolutional networks for biomedical image segmentation. In: Navab, N., Hornegger, J., Wells, W.M., Frangi, A.F. (eds.) MICCAI 2015. LNCS, vol. 9351, pp. 234–241. Springer, Cham (2015). https://doi.org/10.1007/978-3-319-24574-4_28
16. Rovira, À., León, A.: MR in the diagnosis and monitoring of multiple sclerosis: an overview. Eur. J. Radiol. **67**(3), 409–414 (2008)

17. Ruan, Y., et al.: MB-FSGAN: joint segmentation and quantification of kidney tumor on CT by the multi-branch feature sharing generative adversarial network. Med. Image Anal. **64**, 101721 (2020)
18. Sirlin, C.B., et al.: Consensus report from the 6th international forum for liver MRI using gadoxetic acid. J. Magn. Reson. Imaging **40**(3), 516–529 (2014)
19. Sobel, I., Feldman, G.: A 3x3 isotropic gradient operator for image processing. a talk at the Stanford Artificial Project in pp. 271–272 (1968)
20. Tang, L., Zhou, X.J.: Diffusion MRI of cancer: from low to high b-values. J. Magn. Reson. Imaging **49**(1), 23–40 (2019)
21. Thirion, J.P.: Non-rigid matching using demons. In: Proceedings CVPR IEEE Computer Society Conference on Computer Vision and Pattern Recognition, pp. 245–251. IEEE (1996)
22. Vaswani, A., et al.: Attention is all you need. In: Advances in Neural Information Processing Systems 30 (2017)
23. Wang, G., Li, W., Aertsen, M., Deprest, J., Ourselin, S., Vercauteren, T.: Aleatoric uncertainty estimation with test-time augmentation for medical image segmentation with convolutional neural networks. Neurocomputing **338**, 34–45 (2019)
24. Wang, H., Wang, Y., Zhang, Q., Xiang, S., Pan, C.: Gated convolutional neural network for semantic segmentation in high-resolution images. Remote Sens. **9**(5), 446 (2017)
25. Xiao, X., Zhao, J., Li, S.: Task relevance driven adversarial learning for simultaneous detection, size grading, and quantification of hepatocellular carcinoma via integrating multi-modality mri. Med. Image Anal. **81**, 102554 (2022)
26. Xiao, X., Zhao, J., Qiang, Y., Chong, J., Yang, X.T., Kazihise, N.G.-F., Chen, B., Li, S.: Radiomics-guided GAN for segmentation of liver tumor without contrast agents. In: Shen, D., Liu, T., Peters, T.M., Staib, L.H., Essert, C., Zhou, S., Yap, P.-T., Khan, A. (eds.) MICCAI 2019. LNCS, vol. 11765, pp. 237–245. Springer, Cham (2019). https://doi.org/10.1007/978-3-030-32245-8_27
27. Xu, C., Howey, J., Ohorodnyk, P., Roth, M., Zhang, H., Li, S.: Segmentation and quantification of infarction without contrast agents via spatiotemporal generative adversarial learning. Med. Image Anal. **59**, 101568 (2020)
28. Zhang, D., Chen, B., Chong, J., Li, S.: Weakly-supervised teacher-student network for liver tumor segmentation from non-enhanced images. Med. Image Anal. **70**, 102005 (2021)
29. Zhao, J., et al.: United adversarial learning for liver tumor segmentation and detection of multi-modality non-contrast MRI. Med. Image Anal. **73**, 102154 (2021)
30. Zhao, J., et al.: mfTrans-Net: quantitative measurement of hepatocellular carcinoma via multi-function transformer regression network. In: de Bruijne, M., Cattin, P.C., Cotin, S., Padoy, N., Speidel, S., Zheng, Y., Essert, C. (eds.) MICCAI 2021. LNCS, vol. 12905, pp. 75–84. Springer, Cham (2021). https://doi.org/10.1007/978-3-030-87240-3_8
31. Zhou, Z., Siddiquee, M.M.R., Tajbakhsh, N., Liang, J.: Unet++: redesigning skip connections to exploit multiscale features in image segmentation. IEEE Trans. Med. Imaging **39**(6), 1856–1867 (2019)

UPCoL: Uncertainty-Informed Prototype Consistency Learning for Semi-supervised Medical Image Segmentation

Wenjing Lu[1], Jiahao Lei[2], Peng Qiu[2], Rui Sheng[3], Jinhua Zhou[4], Xinwu Lu[2,5], and Yang Yang[1(✉)]

[1] Department of Computer Science and Engineering, Shanghai Jiao Tong University, Shanghai, China
yangyang@cs.sjtu.edu.cn
[2] Department of Vascular Surgery, Shanghai Ninth People's Hospital Affiliated to Shanghai Jiao Tong University, Shanghai, China
[3] Chaohu Clinical Medcial College, Anhui Medical University, Hefei, China
[4] School of Biomedical Engineering, Anhui Medical University, Hefei, China
[5] Shanghai Key Laboratory of Tissue Engineering, Shanghai Ninth People's Hospital, Shanghai Jiao Tong University School of Medicine, Shanghai, China

Abstract. Semi-supervised learning (SSL) has emerged as a promising approach for medical image segmentation, while its capacity has still been limited by the difficulty in quantifying the reliability of unlabeled data and the lack of effective strategies for exploiting unlabeled regions with ambiguous predictions. To address these issues, we propose an Uncertainty-informed Prototype Consistency Learning (UPCoL) framework, which learns fused prototype representations from labeled and unlabeled data judiciously by incorporating an entropy-based uncertainty mask. The consistency constraint enforced on prototypes leads to a more discriminative and compact prototype representation for each class, thus optimizing the distribution of hidden embeddings. We experiment with two benchmark datasets of two-class semi-supervised segmentation, left atrium and pancreas, as well as a three-class multi-center dataset of type B aortic dissection. For all three datasets, UPCoL outperforms the state-of-the-art SSL methods, demonstrating the efficacy of the uncertainty-informed prototype learning strategy (Code is available at https://github.com/VivienLu/UPCoL).

Keywords: Semi-supervised learning · Uncertainty assessment · Prototype learning · Medical image segmentation

1 Introduction

The field of medical image segmentation has been increasingly drawn to semi-supervised learning (SSL) due to the great difficulty and cost of data labeling.

Supplementary Information The online version contains supplementary material available at https://doi.org/10.1007/978-3-031-43901-8_63.

By utilizing both labeled and unlabeled data, SSL can significantly reduce the need for labeled training data and address inter-observer variability [1,8,14].

Typical SSL approaches involve techniques such as self-training, uncertainty estimation, and consistency regularization. Self-training aims to expand labeled training set by selecting the most confident predictions from the unlabeled data to augment the labeled data [23]. To obtain high-quality pseudo-labels, uncertainty estimation is often employed in self-training models. Various uncertainty estimation methods have been proposed to reduce the influence of ambiguous unlabeled data, e.g., Monte Carlo dropout [27] and ensemble-based methods [15,22]. Also, some metrics have been defined to quantify the degree of uncertainty. The most widely-used one is information entropy [18,27], where a threshold or a percentage is set to determine whether an unlabeled sample is reliable, i.e., its predicted label can be used as pseudo-label during the training phase. Besides pseudo-labeling, it is common to add a consistency regularization in loss function [5,11,28]. For instance, UA-MT [27] and CoraNet [15] impose consistency constraints on the teacher-student model for specific regions (certain/uncertain area or both). URPC [12] uses multilevel extraction of multi-scale uncertainty-corrected features to moderate the anomalous pixel of consistency loss.

Despite current progress, the performance of pseudo-labeling and consistency constraints has been limited for two reasons. First, defining an appropriate quantification criterion for reliability across various tasks can be challenging due to the inherent complexity of uncertainty. Second, most of the consistency constraints are imposed at decision space with the assumption that the decision boundary must be located at the low-density area, while the latent feature space of unlabeled data has not been fully exploited, and the low-density assumption may be incapable to guide model learning in the correct way.

Recently, prototype alignment has been introduced into SSL. Prototype-based methods have the potential of capturing underlying data structure including unlabeled information, and optimizing the distribution of feature embeddings across various categories [3,17]. Existing semi-supervised segmentation methods based on prototype learning aim to learn each class prototype from sample averaging and leverage consistency constraints to train the segmentation network. U²PL [18] distinguishes reliable samples among unlabeled data by uncertainty estimation, and constructs prototypes for the whole dataset by class averaging its features with labeled sample and reliable unlabeled features. CISC-R [19] queries a guiding labeled image that shares similar semantic information with an unlabeled image, then estimates pixel-level similarity between unlabeled features and labeled prototypes, thereby rectifying the pseudo labels with reliable pixel-level precision. CPCL [24] introduces a cyclic prototype consistency learning framework to exploit unlabeled data and enhance the prototype representation.

Overall, prototype learning has much room for improvement in semi-supervised segmentation. As voxel-level averaging is only reliable for labeled data, current prototype learning approaches rely on labeled data and a small amount of unlabeled data, or learn prototypes separately for labeled and unlabeled data. In this way, they may not fully represent the distribution of the embedding space. Here raises the question:

Can we capture the embedding distribution by considering all voxels, including both labeled and unlabeled, and exploit the knowledge of the entire dataset?

To answer it, we propose to learn fused prototypes through uncertainty-based attention pooling. The fused prototypes represent the most representative and informative examples from both the labeled and unlabeled data for each class. The main contributions of our work can be summarized as follows:

1) We develop a novel uncertainty-informed prototype consistency learning framework, UPCoL, by considering voxel-level consistency in both latent feature space (i.e., prototype) and decision space.
2) Different from previous studies, we design a fused prototype learning scheme, which jointly learns from labeled and unlabeled data embeddings.
3) For stable prototype learning, we propose a new entropy measure to qualify the reliability of unlabeled voxel and an attention-weighted strategy for fusion.
4) We apply UPCoL to two-class and three-class segmentation tasks. UPCoL outperforms the SOTA SSL methods by large margins.

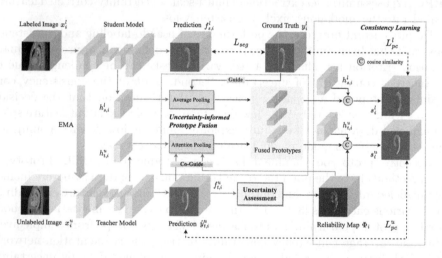

Fig. 1. Overview of UPCoL (using Aortic Dissection segmentation for illustration). The labeled images go through the student model for supervised learning, while the unlabeled images go through the teacher model for segmentation and uncertainty estimation. In Uncertainty-informed Prototype Fusion module, we utilize the reliability map to fuse prototypes learned from labeled and unlabeled embeddings. The similarity between fused prototypes and feature embeddings at each spatial location is then measured for consistency learning.

2   Methodology

Given a dataset $\mathcal{D} = \{\mathcal{D}^l, \mathcal{D}^u\}$, the labeled set $\mathcal{D}^l = \{x_i^l, y_i^l\}_{i=1}^N$ contains N samples, and the unlabeled set $\mathcal{D}^u = \{x_i^u\}_{i=N}^{N+M}$ contains M samples, where

$x_i^l, x_i^u \in \mathbb{R}^{H \times W \times D}$ represent the input with height H, width W, depth D, and $y_i^u \in \{0, 1, ..., C-1\}^{H \times W \times D}$. The proposed framework UPCoL includes a student model and a self-ensembling teacher model, each consisting of a representation head h and a segmentation head f. Figure 1 shows the overview.

2.1 Uncertainty Assessment

To assess the uncertainty at both voxel-level and geometry-level, we adopt the same ensemble of classifiers as [22] using different loss functions, i.e., cross-entropy loss, focal loss [9], Dice loss, and IoU loss. Unlike [22], which simply differentiates certain and uncertain regions based on the result discrepancy of four classifiers, we use the average of four prediction results and define an entropy-based measure to quantify the reliability for each voxel. Specifically, let $f_{t,i}^{(x,y,z)} \in \mathcal{R}^C$ denote the softmax probability for voxel at position (x, y, z) in i-th unlabeled image yielded by the segmentation head of the teacher model, where the segmentation result is the average over multiple classifiers (AMC), and C is the number of classes. The entropy is formulated in Eq. (1),

$$\mathcal{H}(f_{t,i}^{(x,y,z)}) = -\sum_{c=0}^{C-1} f_{t,i}^{(x,y,z)}(c) \log f_{t,i}^{(x,y,z)}(c). \tag{1}$$

Intuitively, voxels with high entropy are ambiguous. Thus, a reliability map can be defined accordingly, denoted by $\Phi_i^{(x,y,z)}$, which enables the model to assign varying degrees of importance to voxels,

$$\Phi_i^{(x,y,z)} = \frac{1}{H \times W \times D} \left(1 - \frac{\mathcal{H}(f_{t,i}^{(x,y,z)})}{\sum_{x,y,z} \mathcal{H}(f_{t,i}^{(x,y,z)})}\right). \tag{2}$$

2.2 Prototype Consistency Learning

Uncertainty-Informed Prototype Fusion. The prototypes from labeled and unlabeled data are first extracted separately. Both of them originate from the feature maps of the 3rd-layer decoder, which are upsampled to the same size as segmentation labels by trilinear interpolation. Let $h_{s,i}^l$ be the output feature by the representation head of the student model for the i-th labeled image, and $h_{t,i}^u$ be the hidden feature by the teacher representation head for the i-th unlabeled image. $\mathcal{B}_l$ and $\mathcal{B}_u$ denote the batch sizes of labeled and unlabeled set respectively, and (x, y, z) denotes voxel coordinate. For labeled prototype, the feature maps are masked directly using ground truth labels, and the prototype of class c is computed via masked average pooling [17, 29]:

$$p_c^l = \frac{1}{\mathcal{B}_l} \sum_{i=1}^{\mathcal{B}_l} \frac{\sum_{x,y,z} h_{s,i}^{l(x,y,z)} \mathbb{1}\left[y_i^{l(x,y,z)} = c\right]}{\sum_{x,y,z} \mathbb{1}\left[y_i^{l(x,y,z)} = c\right]}. \tag{3}$$

For unlabeled data, instead of simply averaging features from the same predicted class, UPCoL obtains the prototypes in an uncertainty-informed manner, i.e., using a masked attention pooling based on each voxel's reliability:

$$p_c^u = \frac{1}{\mathcal{B}_u} \sum_{i=1}^{\mathcal{B}_u} \frac{\sum_{x,y,z} h_{t,i}^{u(x,y,z)} \Phi_i^{(x,y,z)} \mathbb{1}\left[\hat{y}_i^{u(x,y,z)} = c\right]}{\sum_{x,y,z} \mathbb{1}\left[\hat{y}_i^{u(x,y,z)} = c\right]}. \tag{4}$$

Temporal Ensembling technique in Mean-Teacher architecture enhances model performance and augments the predictive label quality [16], leading to a progressive improvement in the reliability of predictive labels and the refinement of unlabeled prototypes throughout the training process. Thus, we adopt a nonlinear updating strategy to adjust the proportion of unlabeled prototypes for fusing the labeled prototypes and unlabeled prototypes, i.e.,

$$p_c = \lambda_{lab} p_c^l + \lambda_{unlab} p_c^u, \tag{5}$$

where $\lambda_{lab} = \frac{1}{1+\lambda_{con}}$, $\lambda_{unlab} = \frac{\lambda_{con}}{1+\lambda_{con}}$, and λ_{con} is the widely-used time-dependent Gaussian warming up function [16]. During the training process, the proportion of labeled prototypes decreases from 1 to 0.5, while the proportion of unlabeled prototypes increases from 0 to 0.5. This adjustment strategy ensures that labeled prototypes remain the primary source of information during training, even as the model gradually gives more attention to the unlabeled prototypes.

Consistency Learning. We adopt non-parametric metric learning to obtain representative prototypes for each semantic class. The feature-to-prototype similarity is employed to approximate the probability of voxels in each class,

$$s_i^{(x,y,z)} = \text{CosSim}\left(h_i^{(x,y,z)}, p_c\right) = \frac{h_i^{(x,y,z)} \cdot p_c}{\max(\left\|h_i^{(x,y,z)}\right\|_2 \cdot \|p_c\|_2, \epsilon)} \tag{6}$$

where the value of ϵ is fixed to $1e^{-8}$, and CosSim($\cdot$) denotes cosine similarity. To ensure the accuary of the prototype, prototype-based predictions for labeled voxels expect to close to ground truth. And the prototype-based predictions for unlabeled voxels expect to close to segmentor prediction since prototype predictions are considered to be reliable aid. Then the prototype consistency losses for labeled and unlabeled samples are defined respectively as:

$$\mathcal{L}_{pc}^l = \mathcal{L}_{CE}(s_i^l, y_i^l), \qquad \mathcal{L}_{pc}^u = \sum_{x,y,z} \Phi_i^{(x,y,z)} \mathcal{L}_{CE}(s_i^{u(x,y,z)}, \hat{y}_i^{u(x,y,z)}), \tag{7}$$

where $\hat{y}_i^{u(x,y,z)}$ is the student model prediction of the i-th unlabeled sample at (x, y, z). Equation (7) is a variant of expanding training set by pseudo labels of the unlabeled data commonly adopted in SSL, with the difference that we use the reliability-aware pseudo labels at the voxel level. Finally, the total loss of our UPCoL network is shown in Eq. (8),

$$\mathcal{L} = \mathcal{L}_{seg} + \mathcal{L}_{pc}^l + \lambda_{con}\mathcal{L}_{pc}^u. \tag{8}$$

3   Experimental Results

Datasets. We evaluate our approach on three datasets: the pancreas dataset (82 CTA scans), the left atrium dataset (100 MR images), and a multi-center dataset for type B aortic dissection (TBAD, 124 CTA scans). The pancreas and left atrium datasets are preprocessed following previous studies [7,10,21,22,27,28]. The TBAD dataset is well-annotated by experienced radiologists, with 100 scans for training and 24 for test, including both public data [25] and data collected by our team. The dataset was resampled to 1mm^3 and resized to $128 \times 128 \times 128$, in accordance with [2]. For all three datasets, we use only 20% of the training data with labels and normalize the voxel intensities to zero mean and unit variance.
Implementation Details. We adopt V-Net [13] as the backbone, and use the results of V-Nets trained with 20% and 100% labeled data as the lower and upper bounds, respectively. For the Mean-Teacher framework, the student network is trained for 10k iterations using Adam optimizer and learning rate 0.001, while the teacher is updated with exponential moving average (EMA) of the student's parameters. The batch size is 3, including 1 labeled and 2 unlabeled samples. Following [7,10,22], we randomly crop cubes of size $96 \times 96 \times 96$ and $112 \times 112 \times 80$ for the pancreas and left atrium datasets, respectively. In addition, we use 3-fold cross validation and apply data augmentation by rotation within the range $(-10°, 10°)$ and zoom factors within the range $(0.9, 1.1)$ for TBAD dataset during training, as proposed in [2,4]. Four performance metrics are adopted, i.e., Dice coefficient (Dice), Jaccard Index (Jac), 95% Hausdorff Distance (95HD), and Average Symmetric Surface Distance (ASD).
Results on the Left Atrium Dataset and Pancreas-CT Dataset. We compare UPCoL with nine SOTA SSL methods, including consistency-based (MT [16], MC-Net [21], ASE-Net [6], URPC [12]), uncertainty-based (UA-MT [27], MC-Net+ [20]), and divergence-based (DTC [10], SimCVD [26], CoraNet [15]). The results (Table 1) demonstrate a substantial performance gap between the lower and upper bounds due to the limited labeled data. Remarkably, our proposed framework outperforms the theoretical upper bound (the second row) in terms of Dice, Jac, 95HD on the left atrium dataset, and 95HD, ASD on the pancreas dataset, suggesting that the unlabeled knowledge extracted from deep features is reliable and well complements the information that is not captured in the fully supervised prediction phase.
Results on the Aortic Dissection Dataset. We compare with two SOTA SSL methods, the uncertainty-based method FUSSNet [22] and embedding-based method URPC [12], as well as two common SSL approaches, MT [16] and UA-MT [27]. As shown in Table 2, the proposed UPCoL obtains the best segmentation results and outperforms fully-supervised V-Net over 6% on Dice score, but

Table 1. Performance comparison on the Left Atrium and Pancreas datasets.

Method	Left Atrium						Pancreas-CT					
	Lb	Unlb	Dice	Jac	95HD	ASD	Lb	Unlb	Dice	Jac	95HD	ASD
V-Net	16	0	84.41	73.54	19.94	5.32	12	0	70.63	56.72	22.54	6.29
V-Net	80	0	91.42	84.27	5.15	1.50	62	0	82.60	70.81	5.61	1.33
MT [16] (NIPS'17)	16	64	86.00	76.27	9.75	2.80	12	50	75.85	61.98	12.59	3.40
UA-MT [27] (MICCAI'19)	16	64	88.88	80.21	7.32	2.26	12	50	77.26	63.28	11.90	3.06
DTC [10] (AAAI'21)	16	64	89.42	80.98	7.32	2.10	12	50	76.27	62.82	8.70	2.20
MC-Net [21] (MICCAI'21)	16	64	90.34	82.48	6.00	1.77	12	50	78.17	65.22	6.90	1.55
MC-Net+ [20] (MIA'22)	16	64	91.07	83.67	5.84	1.67	12	50	80.59	68.08	6.47	1.74
ASE-Net [6] (TMI'22)	16	64	90.29	82.76	7.18	1.64	-	-	-	-	-	-
CoraNet [15] (TMI'21)	-	-	-	-	-	-	12	50	79.49	67.10	11.10	3.06
SimCVD [26] (TMI'22)	16	64	90.85	83.80	6.03	1.86	-	-	-	-	-	-
URPC [12] (MIA'22)	-	-	-	-	-	-	12	50	80.02	67.30	8.51	1.98
UPCoL(Ours)	16	64	**91.69**	**84.69**	**4.87**	**1.56**	12	50	**81.78**	**69.66**	**3.78**	**0.63**

'Lb' and 'Unlb' denote the number of labeled samples and unlabeled samples, respectively.

Table 2. Performance comparison on the Aortic Dissection dataset.

Method	Lb	Unlb	Dice(%) ↑			Jaccard(%) ↑			95HD(voxel) ↓			ASD(voxel) ↓		
			TL	FL	Mean	TL	FL	Mean	TL	FL	Mean	TL	FL	Mean
V-Net	20	0	55.51	48.98	52.25	39.81	34.79	37.30	7.24	10.17	8.71	1.27	3.19	2.23
V-Net	100	0	75.98	64.02	70.00	61.89	50.05	55.97	3.16	7.56	5.36	0.48	2.44	1.46
MT [16]	20	80	57.62	49.95	53.78	41.57	35.52	38.54	6.00	8.98	7.49	0.97	2.77	1.87
UA-MT [27]	20	80	70.91	60.66	65.78	56.15	46.24	51.20	4.44	7.94	6.19	0.83	2.37	1.60
FUSSNet [22]	20	80	79.73	65.32	72.53	67.31	51.74	59.52	3.46	7.87	5.67	0.61	2.93	1.77
URPC [12]	20	80	81.84	69.15	75.50	70.35	57.00	63.68	4.41	9.13	6.77	0.93	**1.11**	**1.02**
UPCoL	20	80	**82.65**	**69.74**	**76.19**	**71.49**	**57.42**	**64.45**	**2.82**	**6.81**	**4.82**	**0.43**	2.22	1.33

only requires 20% labels. The accuracy of FUSSNet [22], URPC [12], and UPCoL surpassing the upper bound demonstrates the effectiveness of uncertainty and embedding-based approaches in exploiting the latent information of the data, particularly in challenging classification tasks. We further visualize the segmentation results on test data of different methods in Fig. 2. As can be seen, UPCoL achieves superior segmentation performance with fewer false positives and superior capability in capturing intricate geometric features, such as the vessel walls between True Lumen (TL) and False Lumen (FL), and effectively smoothing out rough portions of the manual annotation.

Ablation Study. Here we investigate the contribution of key components, including the mean-teacher architecture (MT), the average multi-classifier (AMC) (to yield segmentation results), and the prototype learning (PL) strategy. As shown in Table 3, the MT model, which enforces a consistency cost between the predictions of student model and teacher model, outperforms vanilla V-Net by a large margin (over 5% on Dice), and AMC can also enhance the MT's performance (over 2% on Dice). Compared to the consistency cost in original MT

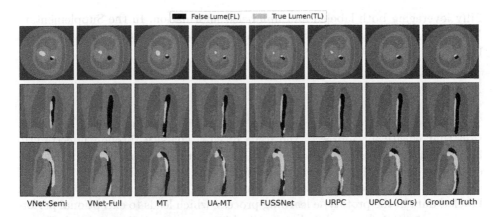

Fig. 2. Visualization of segmentation results on Type B Aortic Dissection dataset.

model [16] (used by MT and MT+AMC), the prototype consistency leads to better performance. Especially, we compare different prototype learning strategies in three methods, MT+PL, CPCL*, and UPCoL, which have the same backbone and AMC module. MT+PL performs prototype learning only for labeled data, CPCL* learns prototypes for labeled and unlabeled data separately using the same strategy proposed in [24], and UPCoL learns fused prototypes. As can be seen, prototype learning using unlabeled data is beneficial for performance improvement, but it requires a well-designed mechanism. Here CPCL* is slightly worse than using only labeled data for prototype learning, potentially due to the isolation of updating labeled and unlabeled prototypes, which may hinder their interaction and prevent the full utilization of knowledge. This highlights the importance of fusing labeled and unlabeled prototypes. UPCoL, on the other hand, successfully merges labeled and unlabeled prototypes through the use of reliability maps, resulting in SOTA performance. This demonstrates the effectiveness of uncertainty-based reliability assessment and prototype fusion in

Table 3. Ablation study results on the pancreas dataset.

Method		Component indication				Metric			
		MT	PL			Dice (%)	Jaccard (%)	95HD (voxel)	ASD (voxel)
			Lb	Unlb	Fuse				
W/o Consistency						70.63	56.72	22.54	6.29
Prediction Consistency	MT	✓				75.85	61.98	12.59	3.40
	MT + AMC	✓				77.97	64.46	9.80	2.52
Prototype Consistency	MT + PL	✓	✓			80.50	68.11	4.49	0.74
	CPCL*	✓	✓	✓		80.08	67.29	7.68	2.08
	UPCoL	✓	✓	✓	✓	**81.78**	**69.66**	**3.78**	**0.63**

'Lb', 'Unlb' and 'Fuse' denote the labeled, unlabeled, and fused prototypes, respectively.

fully leveraging both labeled and unlabeled information. In the Supplementary Materials, visualized results showcase improved predicted labels and unlabeled prototypes as training progresses.

4   Conclusion

This paper presents a novel framework, UPCoL, for semi-supervised segmentation that effectively addresses the issue of label sparsity through uncertainty-based prototype consistency learning. To better utilize unlabeled data, UPCoL employs a quantitative uncertainty measure at the voxel level to assign degrees of attention. UPCoL achieves a careful and effective fusion of unlabeled data with labeled data in the prototype learning process, which leads to exceptional performance on both 2-class and 3-class medical image segmentation tasks. As future work, a possible extension is to allow multiple prototypes for a class with diversified semantic concepts, and a memory-bank-like mechanism could be introduced to learn prototypes from large sample pools more efficiently.

Acknowledgements. This work was supported by the National Natural Science Foundation of China (Nos. 61972251 and 62272300).

References

1. Bai, Wenjia, et al.: Semi-supervised learning for network-based cardiac MR image segmentation. In: Descoteaux, Maxime, Maier-Hein, Lena, Franz, Alfred, Jannin, Pierre, Collins, D. Louis., Duchesne, Simon (eds.) MICCAI 2017. LNCS, vol. 10434, pp. 253–260. Springer, Cham (2017). https://doi.org/10.1007/978-3-319-66185-8_29
2. Cao, L., et al.: Fully automatic segmentation of type b aortic dissection from CTA images enabled by deep learning. Europ. J. Radiol. **121**, 108713 (2019)
3. Dong, N., Xing, E.P.: Few-shot semantic segmentation with prototype learning. In: BMVC, vol. 3 (2018)
4. Fantazzini, A., et al.: 3d automatic segmentation of aortic computed tomography angiography combining multi-view 2d convolutional neural networks. Cardiovascular Eng. Technol. **11**, 576–586 (2020)
5. Hang, Wenlong, et al.: Local and global structure-aware entropy regularized mean teacher model for 3D left atrium segmentation. In: Martel, Anne L.., Abolmaesumi, Purang, Stoyanov, Danail, Mateus, Diana, Zuluaga, Maria A.., Zhou, S. Kevin., Racoceanu, Daniel, Joskowicz, Leo (eds.) MICCAI 2020. LNCS, vol. 12261, pp. 562–571. Springer, Cham (2020). https://doi.org/10.1007/978-3-030-59710-8_55
6. Lei, T., Zhang, D., Du, X., Wang, X., Wan, Y., Nandi, A.K.: Semi-supervised medical image segmentation using adversarial consistency learning and dynamic convolution network. IEEE Trans. Med. Imaging (2022)
7. Li, Shuailin, Zhang, Chuyu, He, Xuming: Shape-aware semi-supervised 3d semantic segmentation for medical images. In: Martel, Anne L.., Abolmaesumi, Purang, Stoyanov, Danail, Mateus, Diana, Zuluaga, Maria A.., Zhou, S. Kevin., Racoceanu, Daniel, Joskowicz, Leo (eds.) MICCAI 2020. LNCS, vol. 12261, pp. 552–561. Springer, Cham (2020). https://doi.org/10.1007/978-3-030-59710-8_54

8. Li, X., Yu, L., Chen, H., Fu, C.W., Xing, L., Heng, P.A.: Transformation-consistent self-ensembling model for semisupervised medical image segmentation. IEEE Trans. Neural Networks Learn. Syst. **32**(2), 523–534 (2020)

9. Lin, T.Y., Goyal, P., Girshick, R., He, K., Dollár, P.: Focal loss for dense object detection. In: Proceedings of the IEEE International Conference on Computer Vision, pp. 2980–2988 (2017)

10. Luo, X., Chen, J., Song, T., Wang, G.: Semi-supervised medical image segmentation through dual-task consistency. In: Proceedings of the AAAI Conference on Artificial Intelligence. vol. 35, pp. 8801–8809 (2021)

11. Luo, X., Liao, W., Chen, J., Song, T., Chen, Y., Zhang, S., Chen, N., Wang, G., Zhang, S.: Efficient semi-supervised gross target volume of nasopharyngeal carcinoma segmentation via uncertainty rectified pyramid consistency. In: MICCAI 2021. pp. 318–329. Springer (2021)

12. Luo, X., Wang, G., Liao, W., Chen, J., Song, T., Chen, Y., Zhang, S., Metaxas, D.N., Zhang, S.: Semi-supervised medical image segmentation via uncertainty rectified pyramid consistency. Medical Image Analysis **80**, 102517 (2022)

13. Milletari, F., Navab, N., Ahmadi, S.A.: V-net: Fully convolutional neural networks for volumetric medical image segmentation. In: 2016 fourth international conference on 3D vision (3DV). pp. 565–571. Ieee (2016)

14. Nie, D., Gao, Y., Wang, L., Shen, D.: Asdnet: attention based semi-supervised deep networks for medical image segmentation. In: MICCAI 2018. pp. 370–378. Springer (2018)

15. Shi, Y., Zhang, J., Ling, T., Lu, J., Zheng, Y., Yu, Q., Qi, L., Gao, Y.: Inconsistency-aware uncertainty estimation for semi-supervised medical image segmentation. IEEE transactions on medical imaging **41**(3), 608–620 (2021)

16. Tarvainen, A., Valpola, H.: Mean teachers are better role models: Weight-averaged consistency targets improve semi-supervised deep learning results. Advances in neural information processing systems 30 (2017)

17. Wang, K., Liew, J.H., Zou, Y., Zhou, D., Feng, J.: Panet: Few-shot image semantic segmentation with prototype alignment. In: proceedings of the IEEE/CVF international conference on computer vision. pp. 9197–9206 (2019)

18. Wang, Y., Wang, H., Shen, Y., Fei, J., Li, W., Jin, G., Wu, L., Zhao, R., Le, X.: Semi-supervised semantic segmentation using unreliable pseudo-labels. In: Proceedings of the IEEE/CVF Conference on Computer Vision and Pattern Recognition. pp. 4248–4257 (2022)

19. Wu, L., Fang, L., He, X., He, M., Ma, J., Zhong, Z.: Querying labeled for unlabeled: Cross-image semantic consistency guided semi-supervised semantic segmentation. IEEE Transactions on Pattern Analysis and Machine Intelligence (2023)

20. Wu, Y., Ge, Z., Zhang, D., Xu, M., Zhang, L., Xia, Y., Cai, J.: Mutual consistency learning for semi-supervised medical image segmentation. Medical Image Analysis **81**, 102530 (2022)

21. Wu, Y., Xu, M., Ge, Z., Cai, J., Zhang, L.: Semi-supervised left atrium segmentation with mutual consistency training. In: MICCAI 2021. pp. 297–306. Springer (2021)

22. Xiang, J., Qiu, P., Yang, Y.: Fussnet: Fusing two sources of uncertainty for semi-supervised medical image segmentation. In: MICCAI 2022. pp. 481–491. Springer (2022)

23. Xie, Q., Luong, M.T., Hovy, E., Le, Q.V.: Self-training with noisy student improves imagenet classification. In: Proceedings of the IEEE/CVF conference on computer vision and pattern recognition. pp. 10687–10698 (2020)

24. Xu, Z., Wang, Y., Lu, D., Yu, L., Yan, J., Luo, J., Ma, K., Zheng, Y., Tong, R.K.y.: All-around real label supervision: Cyclic prototype consistency learning for semi-supervised medical image segmentation. IEEE Journal of Biomedical and Health Informatics 26(7), 3174–3184 (2022)

25. Yao, Z., Xie, W., Zhang, J., Dong, Y., Qiu, H., Yuan, H., Jia, Q., Wang, T., Shi, Y., Zhuang, J., et al.: Imagetbad: A 3d computed tomography angiography image dataset for automatic segmentation of type-b aortic dissection. Frontiers in Physiology p. 1611 (2021)

26. You, C., Zhou, Y., Zhao, R., Staib, L., Duncan, J.S.: Simcvd: Simple contrastive voxel-wise representation distillation for semi-supervised medical image segmentation. IEEE Transactions on Medical Imaging 41(9), 2228–2237 (2022)

27. Yu, L., Wang, S., Li, X., Fu, C.W., Heng, P.A.: Uncertainty-aware self-ensembling model for semi-supervised 3d left atrium segmentation. In: MICCAI 2019. pp. 605–613. Springer (2019)

28. Zeng, X., Huang, R., Zhong, Y., Sun, D., Han, C., Lin, D., Ni, D., Wang, Y.: Reciprocal learning for semi-supervised segmentation. In: MICCAI 2021. pp. 352–361. Springer (2021)

29. Zhang, X., Wei, Y., Yang, Y., Huang, T.S.: Sg-one: Similarity guidance network for one-shot semantic segmentation. IEEE transactions on cybernetics 50(9), 3855–3865 (2020)

A2FSeg: Adaptive Multi-modal Fusion Network for Medical Image Segmentation

Zirui Wang and Yi Hong[✉]

Department of Computer Science and Engineering, Shanghai Jiao Tong University,
Shanghai 200240, China
yi.hong@sjtu.edu.cn

Abstract. Magnetic Resonance Imaging (MRI) plays an important role in multi-modal brain tumor segmentation. However, missing modality is very common in clinical diagnosis, which will lead to severe segmentation performance degradation. In this paper, we propose a *simple* adaptive multi-modal fusion network for brain tumor segmentation, which has two stages of feature fusion, including a simple average fusion and an adaptive fusion based on an attention mechanism. Both fusion techniques are capable to handle the missing modality situation and contribute to the improvement of segmentation results, especially the adaptive one. We evaluate our method on the BraTS2020 dataset, achieving the state-of-the-art performance for the incomplete multi-modal brain tumor segmentation, compared to four recent methods. Our A2FSeg (Average and Adaptive Fusion Segmentation network) is simple yet effective and has the capability of handling any number of image modalities for incomplete multi-modal segmentation. Our source code is online and available at https://github.com/Zirui0623/A2FSeg.git.

Keywords: Modality-adaptive fusion · Missing modality · Brain tumor segmentation · Incomplete multi-modal segmentation

1 Introduction

Extracting brain tumors from medical image scans plays an important role in further analysis and clinical diagnosis. Typically, a brain tumor includes peritumoral edema, enhancing tumor, and non-enhancing tumor core. Since different modalities present different clarity of brain tumor components, we often use multi-modal image scans, such as T1, T1c, T2, and Flair, in the task of brain tumor segmentation [12]. Works have been done to handle brain tumor segmentation using image scans collected from all four modalities [11,15]. However, in practice, we face the challenge of collecting all modalities at the same time, with often one or more missing. Therefore, in this paper, we consider the problem of segmenting brain tumors with missing image modalities.

Current image segmentation methods for handling missing modalities can be divided into three categories, including: 1) brute-force methods: designing individual segmentation networks for each possible modality combination [18], 2)

H. Greenspan et al. (Eds.): MICCAI 2023, LNCS 14223, pp. 673–681, 2023.
https://doi.org/10.1007/978-3-031-43901-8_64

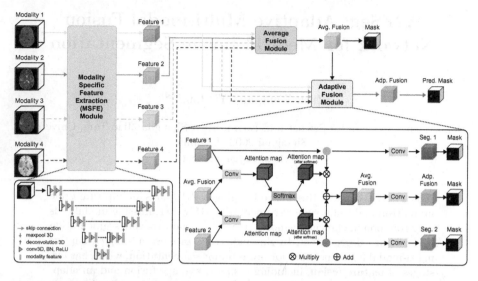

Fig. 1. Overview of our proposed adaptive multi-modal fusion network (A2FSeg, short for Average and Adaptive Fusion Segmentation network). The dashed lines indicate the possibility of missing some modalities. If so, both the average fusion module and the adaptive fusion module will ignore the missing ones. The final tumor mask is predicted based on feature maps after the adaptive fusion, indicated by the solid red arrows. (Best viewed in color) (Color figure online)

completion methods: synthesizing the missing modalities to complete all modalities required for conventional image segmentation methods [16], and 3) fusion-based methods: mapping images from different modalities into the same feature space for fusion and then segmenting brain tumors based on the fused features [10]. Methods in the first category have good segmentation performance; however, they are resource intensive and often require more training time. The performance of methods in the second category is limited by the synthesis quality of the missing modality. The third category often has one single network to take care of different scenarios of missing modalities, which is the most commonly used one in practice.

To handle various numbers of modal inputs, HeMIS [5] projects the image features of different modalities into the same feature space, by computing the mean and variance of the feature maps extracted from different modalities as the fused features. To improve the representation of feature fusion, HVED [3] treats the input of each modality as a Gaussian distribution, and fuses feature maps from different modalities through a Gaussian mixture model. RobustSeg [1], on the other hand, decomposes the modality features into modality-invariant content code and modality-specific appearance code, for more accurate fusion and segmentation. Considering the different clarity of brain tumor regions observed in different modalities, RFNet [2] introduces an attention mechanism to model the relations of modalities and tumor regions adaptively. Based on graph structure and attention mechanism, MFI [21] is proposed to learn adaptive complementary information between modalities in different missing situations.

Due to the complexity of current models, we tend to develop a simple model, which adopts a simple average fusion and attention mechanism. These two techniques are demonstrated to be effective in handling missing modalities and multimodal fusion [17]. Inspired by MAML [20], we propose a model called A2FSeg (Average and Adaptive Fusion Segmentation network, see Fig. 1), which has two fusion steps, i.e., an average fusion and an attention-based adaptive fusion, to integrate features from different modalities for segmentation. Although our fusion idea is quite simple, A2FSeg achieves state-of-the-art (SOTA) performance in the incomplete multimodal brain tumor image segmentation task on the BraTS2020 dataset. Our contributions in this paper are summarized below:

- We propose a *simple* multi-modal fusion network, A2FSeg, for brain tumor segmentation, which is general and can be extended to any number of modalities for incomplete image segmentation.
- We conduct experiments on the BraTS 2020 dataset and achieve the SOTA segmentation performance, having a mean Dice core of 89.79% for the whole tumor, 82.72% for the tumor core, and 66.71% for the enhancing tumor.

2   Method

Figure 1 presents the network architecture of our A2FSeg. It consists of four modality-specific sub-networks to extract features from each modality, an average fusion module to simply fuse features from available modalities at the first stage, and an adaptive fusion module based on an attention mechanism to adaptively fuse those features again at the second stage.

Modality-Specific Feature Extraction (MSFE) Module. Before fusion, we first extract features for every single modality, using the nnUNet model [7] as shown in Fig. 1. In particular, this MSFE model takes a 3D image scan from a specific modality m, i.e., $\mathbf{I}_m \in \mathbb{R}^{H \times W \times D}$ and $m \in \{$T1, T2, T1c, Flair$\}$, and outputs the corresponding image features $\mathbf{F}_m \in \mathbb{R}^{C \times H_f \times W_f \times D_f}$. Here, the number of channels is $C = 32$; H_f, W_f, and D_f are the height, width, and depth of feature maps $\mathbf{F}_m$, which share the same size as the input image. For every single modality, each MSFE module is supervised by the image segmentation mask to fasten its convergence and provide a good feature extraction for fusion later. All four MSFEs have the same architecture but with different weights.

Average Fusion Module. To aggregate image features from different modalities and handle the possibility of missing one or more modalities, we use the average of the available features from different modalities as the first fusion result. That is, we obtain a fused average feature $\bar{\mathbf{F}} = \frac{1}{N_m} \sum_{m=1}^{N_m} \mathbf{F}_m$. Here, N_m is the number of available modalities. For example, as shown in Fig. 1, if only the first two modalities are available at an iteration, then $N_m = 2$, and we will take the average of these two modalities, ignoring those missing ones.

Adaptive Fusion Module. Since each modality contributes differently to the final tumor segmentation, similar to MAML [20], we adopt the attention mechanism to measure the voxel-level contributions of each modality to the final segmentation. As shown in Fig. 1, to generate the attention map for a specific modality m, we take the concatenation of its feature extracted by the MSFE module $\mathbf{F}_m$ and the mean feature after the average fusion $\bar{\mathbf{F}}$, which is passed through a convolutional layer to generate the initial attention weights:

$$\mathbf{W}_m = \sigma\left(\mathcal{F}_m\left([\bar{\mathbf{F}}; \mathbf{F}_m]; \theta_m\right)\right), \quad m \in \{\text{T1, T1c, T2, Flair}\}. \tag{1}$$

Here, $\mathcal{F}_m$ is a convolutional layer for this specific modality m, and θ_m represents the parameters of this layer, and σ is a Sigmoid function. That is, we have an individual convolution layer $\mathcal{F}_m$ for each modality to generate different weights.

Due to the possibility of missing modalities, we will have different numbers of feature maps for fusion. To address this issue, we normalize the different attention weights by using a Softmax function:

$$\hat{\mathbf{W}}_m = \frac{\exp\left(\mathbf{W}_m\right)}{\sum_m^{N_m} \exp\left(\mathbf{W}_m\right)}. \tag{2}$$

That is, we only consider feature maps from those available modalities but normalize their contribution to the final fusion result, so that, the fused one has a consistent value range, no matter how many modalities are missing. Then, we perform voxel-wise multiplication of the attention weight with the corresponding modal feature maps. As a result, the adaptively fused feature maps $\hat{\mathbf{F}}$ is calculated by the weighted sum of each modal feature:

$$\hat{\mathbf{F}} = \sum_m \hat{\mathbf{W}}_m \otimes \mathbf{F}_m. \tag{3}$$

Here, $\otimes$ indicates the voxel-wise multiplication.

Loss Function. We have multiple segmentation heads, which are distributed in each module of A2FSeg. For each segmentation head, we use the combination of the cross-entropy and the soft dice score as the basic loss function, which is defined as

$$\mathcal{L}(\hat{y}, y) = \mathcal{L}_{CE}(\hat{y}, y) + \mathcal{L}_{Dice}(\hat{y}, y), \tag{4}$$

where $\hat{y}$ and y represent the segmentation prediction and the ground truth, respectively. Based on this basic one, we have the overall loss function defined as

$$\mathcal{L}_{total} = \sum_m \mathcal{L}_m(\hat{y}_m, y) + \mathcal{L}_{avg}(\hat{y}_{avg}, y) + \mathcal{L}_{adp}(\hat{y}_{adp}, y), \tag{5}$$

where the first term is the basic segmentation loss for each modality m after feature extraction; the second term is the loss for the segmentation output of the average fusion module; and the last term is the segmentation loss for the final output from the adaptive fusion module.

Table 1. Comparison among recent methods, including HeMIS [5], U-HVED [3], mmFormer [19], and MFI [21], and ours on BraTS2020 in terms of Dice%. Missing and available modalities are denoted by ○ and ●, respectively. F indicates Flair, HVED indicates U-HVED, and Former indicates mmFormer because of space issue.

Modalities				Complete					Core					Enhancing				
T_1	T_{1c}	T_2	F	Hemis	HVED	Former	MFI	Ours	Hemis	HVED	Former	MFI	Ours	Hemis	HVED	Former	MFI	Ours
○	○	○	●	87.76	86.49	90.08	90.60	91.48	66.56	64.42	71.13	75.59	76.21	44.95	43.32	48.25	51.96	53.80
○	○	●	○	85.53	85.14	87.00	88.38	88.82	65.55	64.87	72.85	75.38	76.40	43.77	43.31	50.18	52.72	54.46
○	○	●	●	90.51	89.87	91.19	91.65	91.95	70.82	70.55	75.18	77.42	77.83	48.32	47.86	52.51	54.77	56.10
○	●	○	○	72.83	74.31	80.00	80.16	83.11	83.59	83.96	85.29	85.35	86.95	75.54	77.34	76.17	76.91	78.01
○	●	○	●	91.29	90.45	91.51	92.36	92.42	86.27	85.78	87.05	87.96	88.67	76.30	76.29	76.99	77.26	78.07
○	●	●	○	86.32	86.82	88.79	89.53	89.90	85.61	85.11	87.41	87.83	88.75	75.57	75.68	77.46	76.56	77.85
○	●	●	●	91.82	91.46	91.93	92.38	92.72	86.63	86.06	87.87	88.56	87.96	76.25	75.47	76.15	76.69	76.98
●	○	○	○	75.02	76.64	81.20	79.91	83.67	61.18	62.78	71.36	72.36	75.52	37.55	39.46	46.65	50.40	52.58
●	○	○	●	90.29	88.81	91.29	91.45	91.89	71.95	70.18	76.01	78.22	78.07	48.16	46.53	51.20	55.05	54.00
●	○	●	○	86.66	87.13	88.22	88.03	89.40	67.67	70.21	75.00	75.85	77.39	44.86	46.95	51.37	54.39	54.58
●	○	●	●	90.85	90.34	91.61	91.67	92.23	72.75	73.22	77.05	78.30	78.64	48.48	49.45	52.51	55.44	55.34
●	●	○	○	77.42	79.40	82.53	82.50	84.81	84.76	84.94	86.03	86.52	87.40	75.43	76.56	76.84	76.76	77.80
●	●	○	●	91.65	90.97	91.95	92.24	92.29	86.79	86.61	87.44	88.84	87.75	76.44	75.79	76.91	77.06	76.75
●	●	●	○	86.75	87.72	89.19	88.81	89.49	86.11	85.36	87.30	87.22	87.16	75.16	75.62	76.37	76.52	77.69
●	●	●	●	92.00	91.62	92.26	92.33	92.71	87.67	86.46	88.13	88.60	88.74	75.39	75.66	76.08	76.66	76.70
Means				86.45	86.48	88.58	88.80	89.79	77.59	77.37	81.01	82.31	82.72	61.48	61.69	64.38	65.94	66.71

3 Experiments

3.1 Dataset

Our experiments are conducted on BraTS2020, which contains 369 multi-contrast MRI scans with four modalities: T1, T1c, T2, and Flair. These images went through a sequence of preprocessing steps, including co-registration to the same anatomical template, resampling to the same resolution ($1\,\mathrm{mm}^3$), and skull-stripping. The segmentation masks have three labels, including the whole tumor (abbreviated as Complete), tumor core (abbreviated as Core), and enhancing tumor (abbreviated as Enhancing). These annotations are manually provided by one to four radiologists according to the same annotation protocol.

3.2 Experimental Settings and Implementation Details

We implement our model with PyTorch [13] and perform experiments on an Nvidia RTX3090 GPU. We use the Adam optimizer [8], with an initial learning rate of 0.01. Since we use the method of exponential decay of learning rate, the initial learning rate is then multiplied by $(1 - \frac{\#\text{epoch}}{\#\text{max\_epoch}})^{0.9}$. Due to the limitation of GPU memory, each volume is randomly cropped into multiple patches with the size of $128 \times 128 \times 128$ for training. The network is trained for 400 epochs. In the inference stage, we use a sliding window to produce the final segmentation prediction of the input image.

3.3 Experimental Results and Comparison to Baseline Methods

To evaluate the performance of our model, we compare it with four recent models, HeMIS [5], U-HVED [3], mmFormer [19], and MFI [21]. The dataset is randomly split into 70% for training, 10% for validation, and 20% for testing, and all methods are evaluated on the same dataset and data splitting. We use the Dice score

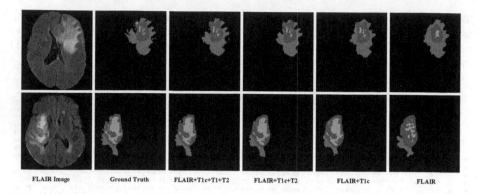

Fig. 2. Visualization of our A2FSeg results using a different number of modalities for brain tumor segmentation. Red: peritumoral edema; Blue: enhancing tumor; Green: the necrotic and non-enhancing tumor core. (Color figure online)

as the metric. As shown in Table 1, our method achieves the best result. For example, our method outperforms the current SOTA method MFI [21] in most missing-modality cases, including all cases for the whole/complete tumor, 8 out of 15 cases for the tumor core, 12 out of 15 cases for the enhancing tumor. Compared to MFI, for the whole tumor, tumor core, and enhancing tumor regions, we improve the average Dice scores by 0.99%, 0.41%, and 0.77%, respectively. Although the design of our model is quite simple, these results demonstrate its effectiveness for the incomplete multimodel segmentation task of brain tumors.

Figure 2 visualizes the segmentation results of samples from the BraTS2020 dataset. With only one Flair image available, the segmentation results of the tumor core and enhancing tumor are poor, because little information on these two regions is observed in the Flair image. With an additional T1c image, the segmentation results of these two regions are significantly improved and quite close to the ground truth. Although adding T1 and T2 images does not greatly improve the segmentation of the tumor core and the enhancing tumor, the boundary of the whole tumor is refined with their help.

Figure 3 visualizes the contribution to each tumor region from each modality. The numbers are the mean values of the attention maps computed for images in the test set. Overall, in our model, each modality has its contribution to the final segmentation, and no one dominates the result. This is because we have supervision on the segmentation branch of each modality, so that, each modality has the ability to segment each region to some extent. However, we still observe that Flair and T2 modalities have relatively larger contributions to the segmentation of all tumor regions, followed by T1c and then T1. This is probably because the whole tumor area is much clear in Flair and T2 compared to the other two modalities. Each modality shows its preference when segmenting different regions. Flair and T2 are more useful for extracting the peritumoral edema (ED) than the enhancing tumor (ET) and the non-enhancing tumor and

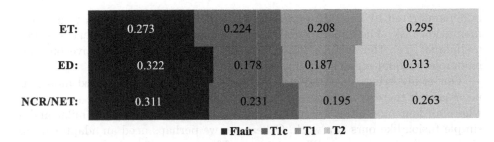

Fig. 3. Summary of the contribution of each modality to each tumor region from the estimated attention maps. ET: enhancing tumor, ED: the peritumoral edema, NCR/NET: non-enhancing tumor and necrosis.

Table 2. Ablation study of the adaptive fusion model in our method.

Methods	Complete	Core	Enhancing	Average
MFI	88.80	82.31	65.94	79.02
Baseline (Average fusion module only)	89.29	82.00	66.00	79.10
+Adaptive fusion module	**89.79**	**82.72**	**66.71**	**79.74**

necrosis (NCR/NET); while T1c and T1 are on the opposite and more helpful for extracting ET and NCR/NET.

3.4 Ablation Study

In this part, we investigate the effectiveness of the average fusion module and the adaptive fusion module, which are two important components of our method. Firstly, we set a baseline model without any modal interaction, that is, with the average fusion module only. Then, we add the adaptive fusion module to the baseline model. Table 2 reports this ablation study. With only adding the average fusion module, our method already obtains comparable performance with the current SOTA method MFI. By adding the adaptive fusion module, the dice scores of the three regions further increase by 0.50%, 0.72%, and 0.71%, respectively. This shows that both the average fusion module and the adaptive fusion module are effective in this brain tumor segmentation task.

4 Discussion and Conclusion

In this paper, we propose an average and adaptive fusion segmentation network (A2FSeg) for the incomplete multi-model brain tumor segmentation task. The essential components of our A2FSeg network are the two stages of feature fusion, including an average fusion and an adaptive fusion. Compare to existing complicated models, our model is much simpler and more effective, which is demonstrated by the best performance on the BraTS 2020 brain tumor segmentation task. The experimental results demonstrate the effectiveness of two techniques, i.e., the average fusion and the attention-based adaptive one, for incomplete modal segmentation tasks.

Our study brings up the question of whether having complicated models is necessary. If there is no huge gap between different modalities, like in our case where all four modalities are images, the image feature maps are similar and a simple fusion like ours can work. Otherwise, we perhaps need an adaptor or an alignment strategy to fuse different types of features, such as images and audio.

Also, we observe that a good feature extractor is essential for improving the segmentation results. In this paper, we only explore a reduced UNet for feature extraction. In future work, we will explore other feature extractors, such as Vision Transformer (ViT) or other pre-trained visual foundation models [4, 6, 14]. Recently, the segment anything model (SAM) [9] demonstrates its general ability to extract different regions of interest, which is promising to be adopted as a good starting point for brain tumor segmentation. Besides, our model is general for multi-modal segmentation and we will apply it to other multi-model segmentation tasks to evaluate its generalization on other applications.

Acknowledgements. This work was supported by NSFC 62203303 and Shanghai Municipal Science and Technology Major Project 2021SHZDZX0102.

References

1. Chen, C., Dou, Q., Jin, Y., Chen, H., Qin, J., Heng, P.-A.: Robust multimodal brain tumor segmentation via feature disentanglement and gated fusion. In: Shen, D., et al. (eds.) MICCAI 2019, Part III. LNCS, vol. 11766, pp. 447–456. Springer, Cham (2019). https://doi.org/10.1007/978-3-030-32248-9_50
2. Ding, Y., Yu, X., Yang, Y.: RFNET: region-aware fusion network for incomplete multi-modal brain tumor segmentation. In: Proceedings of the IEEE/CVF International Conference on Computer Vision, pp. 3975–3984 (2021)
3. Dorent, R., Joutard, S., Modat, M., Ourselin, S., Vercauteren, T.: Hetero-modal variational encoder-decoder for joint modality completion and segmentation. In: Shen, D., et al. (eds.) MICCAI 2019, Part II. LNCS, vol. 11765, pp. 74–82. Springer, Cham (2019). https://doi.org/10.1007/978-3-030-32245-8_9
4. Dosovitskiy, A., et al.: An image is worth 16x16 words: transformers for image recognition at scale (2021)
5. Havaei, M., Guizard, N., Chapados, N., Bengio, Y.: HeMIS: hetero-modal image segmentation. In: Ourselin, S., Joskowicz, L., Sabuncu, M.R., Unal, G., Wells, W. (eds.) MICCAI 2016, Part II. LNCS, vol. 9901, pp. 469–477. Springer, Cham (2016). https://doi.org/10.1007/978-3-319-46723-8_54
6. He, K., Chen, X., Xie, S., Li, Y., Dollár, P., Girshick, R.: Masked autoencoders are scalable vision learners. In: Proceedings of the IEEE/CVF Conference on Computer Vision and Pattern Recognition (CVPR), pp. 16000–16009 (2022)
7. Isensee, F., Jaeger, P.F., Kohl, S.A., Petersen, J., Maier-Hein, K.H.: NNU-Net: a self-configuring method for deep learning-based biomedical image segmentation. Nat. Methods **18**(2), 203–211 (2021)

8. Kingma, D.P., Ba, J.: Adam: a method for stochastic optimization. arXiv preprint arXiv:1412.6980 (2014)

9. Kirillov, A., et al.: Segment anything. arXiv preprint arXiv:2304.02643 (2023)

10. Li, R., et al.: Deep learning based imaging data completion for improved brain disease diagnosis. In: Golland, P., Hata, N., Barillot, C., Hornegger, J., Howe, R. (eds.) MICCAI 2014, Part III. LNCS, vol. 8675, pp. 305–312. Springer, Cham (2014). https://doi.org/10.1007/978-3-319-10443-0_39

11. Long, J., Shelhamer, E., Darrell, T.: Fully convolutional networks for semantic segmentation. In: Proceedings of the IEEE Conference on Computer Vision and Pattern Recognition, pp. 3431–3440 (2015)

12. Menze, B.H., et al.: The multimodal brain tumor image segmentation benchmark (brats). IEEE Trans. Med. Imaging **34**(10), 1993–2024 (2014)

13. Paszke, A., et al.:Automatic differentiation in pytorch (2017)

14. Radford, A., et al.: Learning transferable visual models from natural language supervision. In: Meila, M., Zhang, T. (eds.) Proceedings of the 38th International Conference on Machine Learning. In: Proceedings of Machine Learning Research, 18–24 July 2021, vol. 139, pp. 8748–8763. PMLR (2021), https://proceedings.mlr.press/v139/radford21a.html

15. Ronneberger, O., Fischer, P., Brox, T.: U-Net: convolutional networks for biomedical image segmentation. In: Navab, N., Hornegger, J., Wells, W.M., Frangi, A.F. (eds.) MICCAI 2015, Part III. LNCS, vol. 9351, pp. 234–241. Springer, Cham (2015). https://doi.org/10.1007/978-3-319-24574-4_28

16. van Tulder, G., de Bruijne, M.: Why does synthesized data improve multi-sequence classification? In: Navab, N., Hornegger, J., Wells, W.M., Frangi, A.F. (eds.) MICCAI 2015, Part I. LNCS, vol. 9349, pp. 531–538. Springer, Cham (2015). https://doi.org/10.1007/978-3-319-24553-9_65

17. Vaswani, A., et al.: Attention is all you need. In: Advances in Neural Information Processing Systems, vol. 30 (2017)

18. Wang, Y., et al.: ACN: adversarial co-training network for brain tumor segmentation with missing modalities. In: de Bruijne, M., et al. (eds.) MICCAI 2021, Part VII. LNCS, vol. 12907, pp. 410–420. Springer, Cham (2021). https://doi.org/10.1007/978-3-030-87234-2_39

19. Zhang, Y., et al.: mmFormer: multimodal medical transformer for incomplete multimodal learning of brain tumor segmentation. In: Wang, L., Dou, Q., Fletcher, P.T., Speidel, S., Li, S. (eds.) MICCAI 2022, Part V. LNCS, vol. 13435, pp. 107–117. Springer, Cham (2022). https://doi.org/10.1007/978-3-031-16443-9_11

20. Zhang, Y., et al.: Modality-aware mutual learning for multi-modal medical image segmentation. In: de Bruijne, M., et al. (eds.) MICCAI 2021, Part I. LNCS, vol. 12901, pp. 589–599. Springer, Cham (2021). https://doi.org/10.1007/978-3-030-87193-2_56

21. Zhao, Z., Yang, H., Sun, J.: Modality-adaptive feature interaction for brain tumor segmentation with missing modalities. In: Wang, L., Dou, Q., Fletcher, P.T., Speidel, S., Li, S. (eds.) MICCAI 2022, Part V. LNCS, vol. 13435, pp. 183–192. Springer, Cham (2022). https://doi.org/10.1007/978-3-031-16443-9_18

Learning Reliability of Multi-modality Medical Images for Tumor Segmentation via Evidence-Identified Denoising Diffusion Probabilistic Models

Jianfeng Zhao[1] and Shuo Li[2(✉)]

[1] School of Biomedical Engineering, Western University, London, ON, Canada
[2] School of Biomedical Engineering, Case Western Reserve University, Cleveland, OH, USA
slishuo@gmail.com

Abstract. Denoising diffusion probabilistic models (DDPM) for medical image segmentation are still a challenging task due to the lack of the ability to parse the reliability of multi-modality medical images. In this paper, we propose a novel evidence-identified DDPM (EI-DDPM) with contextual discounting for tumor segmentation by integrating multi-modality medical images. Advanced compared to previous work, the EI-DDPM deploys the DDPM-based framework for segmentation tasks under the condition of multi-modality medical images and parses the reliability of multi-modality medical images through contextual discounted evidence theory. We apply EI-DDPM on a BraTS 2021 dataset with 1251 subjects and a liver MRI dataset with 238 subjects. The extensive experiment proved the superiority of EI-DDPM, which outperforms the state-of-the-art methods.

Keywords: Denoising diffusion probabilistic models · Evidence theory · Multi-modality · Tumor segmentation

1 Introduction

Integrating multi-modality medical images for tumor segmentation is crucial for comprehensive diagnosis and surgical planning. In the clinic, the consistent information and complementary information in multi-modality medical images provide the basis for tumor diagnosis. For instance, the consistent anatomical structure information offers the location feature for tumor tracking [22], while the complementary information such as differences in lesion area among multi-modality medical images provides the texture feature for tumor characterization. Multi-modality machine learning aims to process and relate information from multiple modalities [4]. But it is still tricky to integrate multi-modality medical images due to the complexity of medical images.

Existing methods for multi-modality medical image integration can be categorized into three groups: (1) input-based integration methods that concatenate multi-modality images at the beginning of the framework to fuse them directly

H. Greenspan et al. (Eds.): MICCAI 2023, LNCS 14223, pp. 682–691, 2023.
https://doi.org/10.1007/978-3-031-43901-8_65

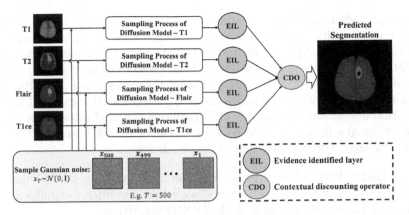

Fig. 1. The workflow of EI-DDPM for brain tumor segmentation of BraTS 2021 dataset. It mainly contains three parts: parallel DDPM path for multi-modality medical images feature extraction, EIL for preliminary multi-modality medical images integration, and CDO for parsing the reliability of multi-modality medical images.

[19,21], (2) feature-based fusion methods that incorporate a fusion module to merge feature maps [16,23], and (3) decision-based fusion methods that use weighted averaging to balance the weights of different modalities [11,15]. Essentially, these methods differ in their approach to modifying the number of channels, adding additional convolutional layers with a softmax layer for attention, or incorporating fixed modality-specific weights. However, there is no mechanism to evaluate the reliability of information from multi-modality medical images. Since the anatomical information of different modality medical images varies, the reliability provided by different modalities may also differ. Therefore, it remains challenging to consider the reliability of different modality medical images when combining multi-modality medical image information.

Dempster-Shafer theory (DST) [18], also known as evidence theory, is a powerful tool for modeling information, combining evidence, and making decisions by integrating uncertain information from various sources or knowledge [10].Some studies have attempted to apply DST to medical image processing [8]. However, using evidence theory alone does not enable us to weigh the different anatomical information from multi-modality medical images. To explore the reliability of different sources when using evidence theory, the work by Mercier et al. [13] proposed a contextual discounting mechanism to assign weights to different sources. Furthermore, for medical image segmentation, denoising diffusion probabilistic models (DDPM [7]) have shown remarkable performance [9,20]. Inspired by these studies, if DDPM can parse the reliability of multi-modality medical images to weigh the different anatomy information from them, it will provide a significant approach for tumor segmentation.

In this paper, we propose an evidence-identified DDPM (EI-DDPM) with contextual discounting for tumor segmentation via integrating multi-modality medical images. Our basic assumption is that we can learn the segmentation feature on single modality medical images using DDPM and parse the reliability of different modalities medical images by evidence theory with a contextual discounting mechanism. Specifically, the EI-DDPM first utilizes parallel conditional DDPM to learn the segmentation feature from a single modality image. Next, the evidence-identified layer (EIL) preliminarily integrates multi-modality images by comprehensively using the multi-modality uncertain information. Lastly, the contextual discounting operator (CDO) performs the final integration of multi-modality images by parsing the reliability of information from multi-modality medical images. The contributions of this work are:

- Our **EI-DDPM** achieves tumor segmentation by using DDPM under the guidance of evidence theory. It provides a solution to integrate multi-modality medical images when deploying the DDPM algorithm.
- The proposed EIL and CDO apply contextual discounting guided DST to parse the reliability of information from different modalities of medical images. This allows for the integration of multi-modality medical images with learned weights corresponding to their reliability.
- We conducted extensive experiments using the BraTS 2021 [12] dataset for brain tumor segmentation and a liver MRI dataset for liver tumor segmentation. Experimental results demonstrate the superiority of EI-DDPM over other State-of-The-Art (SoTA) methods.

2   Method

The EI-DDPM achieves tumor segmentation by parsing the reliability of multi-modality medical images. Specifically, as shown in Fig. 1, the EI-DDPM is fed with multi-modality medical images into the parallel DDPM path and performs the conditional sampling process to learn the segmentation feature from the single modality image (Sect. 2.1). Next, the EIL preliminary integrates multi-modality images by embedding the segmentation features from multi-modality images into the combination rule of DST (Sect. 2.2). Lastly, the CDO integrates multi-modality medical images for tumor segmentation by contextual discounting mechanism (Sect. 2.3).

2.1   Parallel DDPM Path for Segmentation Feature Learning

Background of DDPM: As an unconditional generative method, DDPM [7] has the form of $p_\theta(\mathbf{x}_0) := \int p_\theta(\mathbf{x}_{0:T}) d\mathbf{x}_{1:T}$, where $\mathbf{x}_1, ..., \mathbf{x}_T$ represent latents with the same dimensionality as the data $\mathbf{x}_0 \sim q(\mathbf{x}_0)$. It contains the forward process of diffusion and the reverse process of denoising. The forward process of diffusion that the approximate posterior $q(\mathbf{x}_{1:T}|\mathbf{x}_0)$, it is a Markov Chain by gradually adding Gaussian noise for converting the noise distribution to the data distribution according to the variance schedule $\beta_1, ..., \beta_T$:

$$q(\mathbf{x}_t|\mathbf{x}_{t-1}) := \mathcal{N}(\mathbf{x}_t; \sqrt{1-\beta_t}\mathbf{x}_{t-1}, \beta_t\mathbf{I}), \quad q(\mathbf{x}_t|\mathbf{x}_0) = \mathcal{N}(\mathbf{x}_t; \sqrt{\bar{\alpha}_t}\mathbf{x}_0, (1-\bar{\alpha}_t)\mathbf{I}) \quad (1)$$

The reverse process of denoising that the joint distribution $p_\theta(\mathbf{x}_{0:T})$, it can be defined as a Markov chain with learnt Gaussian transitions starting from $p(\mathbf{x}_T) = \mathcal{N}(\mathbf{x}_T; \mathbf{0}, \mathbf{I})$:

$$p_\theta(\mathbf{x}_{t-1}|\mathbf{x}_t) := \mathcal{N}(\mathbf{x}_{t-1}; \mu_\theta(\mathbf{x}_t, t), \sigma_t^2 \mathbf{I}), \quad p_\theta(\mathbf{x}_{0:T}) := p(\mathbf{x}_T) \prod_{t=1}^{T} p_\theta(\mathbf{x}_{t-1}|\mathbf{x}_t) \quad (2)$$

where $\alpha_t := 1 - \beta_t$, $\bar{\alpha}_t := \prod_{s=1}^{t} \alpha_s$, and $\sigma_t^2 = \frac{1-\bar{\alpha}_{t-1}}{1-\bar{\alpha}_t}\beta_t$.

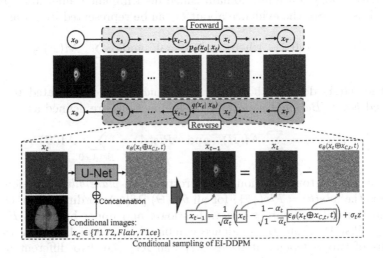

Fig. 2. The algorithm of multi-modality medical images conditioned DDPM.

Multi-modality Medical Images Conditioned DDPM: In Eq. 2 from DDPM [7], the unconditional prediction $\mathbf{x}_{t-1}$ at each step is obtained by subtracting the predicted noise from the previous $\mathbf{x}_t$, which can be defined as:

$$\mathbf{x}_{t-1} = \frac{1}{\sqrt{\alpha_t}}(\mathbf{x}_t - \frac{1-\alpha_t}{\sqrt{1-\bar{\alpha}_t}}\epsilon_\theta(\mathbf{x}_t, t)) + \sigma_t\mathbf{z}, \quad \mathbf{z} \sim \mathcal{N}(0, \mathbf{I}) \quad (3)$$

As shown in Fig. 2, to perform the conditional sampling in EI-DDPM, the prediction $\mathbf{x}_{t-1}$ at each step t is on the basis of the concatenation "$\oplus$" of previous $\mathbf{x}_t$ and the conditional image $\mathbf{x}_{C,t}$. Thus, the $\mathbf{x}_{t-1}$ here can be defined as:

$$\mathbf{x}_{t-1} = \frac{1}{\sqrt{\alpha_t}}(\mathbf{x}_t - \frac{1-\alpha_t}{\sqrt{1-\bar{\alpha}_t}}\epsilon_\theta(\mathbf{x}_t \oplus \mathbf{x}_{C,t}, t)) + \sigma_t\mathbf{z}, \quad \mathbf{z} \sim \mathcal{N}(0, \mathbf{I}) \quad (4)$$

where the conditional image $\mathbf{x}_C \in \{T1, T2, Flair, T1ce\}$ corresponding to the four parallel conditional DDPM path. And in each step t, the $\mathbf{x}_{C,t}$ was also performed the operation of adding Gaussian noise to convert the distribution:

$$\mathbf{x}_{C,t} = \mathbf{x}_{C,0} + \mathcal{N}(0, (1-\bar{\alpha}_t)\mathbf{I}) \quad (5)$$

where $\mathbf{x}_{C,0}$ presents the multi-modality medical images.

2.2   EIL Integrates Multi-modality Images

As shown in Fig. 1, the EIL is followed at the end of the parallel DDPM path. The hypothesis is to regard multi-modality images as independent and different sources knowledge. And then embedding multi-phase features into the combination rule of DST for evidence identification, which comprehensively parses the multi-phase uncertainty information for confident decision-making. The basic concepts of evidence identification come from DST [18]. It is assumed that $\Theta = \{\theta_1, \theta_2, ..., \theta_n\}$ is a finite domain called discriminant frame. *Mass function* is defined as $\mathcal{M}$. So, the evidence about Θ can be represented by $\mathcal{M}$ as:

$$\mathcal{M} : 2^\Theta \to [0,1] \quad \text{where} \quad \mathcal{M}(\emptyset) = 0 \quad \text{and} \quad \sum_{A \subseteq \Theta} \mathcal{M}(A) = 1 \qquad (6)$$

where the $\mathcal{M}(A)$ denotes the whole belief and evidence allocated to A. The associated *belief (Bel)* and *plausibility (Pls)* functions are defined as:

$$Bel(A) = \sum_{B \subseteq A} \mathcal{M}(B) \quad \text{and} \quad Pls(A) = \sum_{B \cap A \neq \emptyset} \mathcal{M}(B) \qquad (7)$$

And using the contour function pls to restrict the *plausibility* function Pls of singletons (i.e. $pls(\theta) = Pls(\{\theta\})$ for all $\theta \in \Theta$) [18]. According to the DST, the mass function $\mathcal{M}$ of a subset always has lower and upper bound as $Bel(A) \leq \mathcal{M}(A) \leq Pls(A)$. From the evidence combination rule of DST, for the mass functions of two independent items $\mathcal{M}_1$ and $\mathcal{M}_2$ (i.e. two different modality images), it can be calculated by a new mass function $\mathcal{M}_1 \oplus \mathcal{M}_2(A)$, which is the orthogonal sum of $\mathcal{M}_1$ and $\mathcal{M}_2$ as:

$$\mathcal{M}_1 \oplus \mathcal{M}_2(A) = \frac{1}{1 - \gamma} \sum_{B \cap C = A} \mathcal{M}_1(B)\mathcal{M}_2(C) \qquad (8)$$

where $\gamma = \sum_{B \cap C = \emptyset} \mathcal{M}_1(B)\mathcal{M}_2(C)$ is *conflict degree* between $\mathcal{M}_1$ and $\mathcal{M}_2$. And the combined contour function Pls_{12} corresponding to $\mathcal{M}_1 \oplus \mathcal{M}_2$ is:

$$Pls_{12} = \frac{Pls_1 Pls_2}{1 - \gamma} \qquad (9)$$

The specific evidence identification layer mainly contains three sub-layers (i.e. activation layer, mass function layer, and belief function layer). For activation layer, the activation of i unit can be defined as: $y_i = a_i exp(-\lambda_i ||x - w_i||^2)$. The w_i is the weight of i unit. $\lambda_i > 0$ and a_i are parameters. The mass function layer calculates the mass of each K classes using $\mathcal{M}_i(\{\theta_k\}) = u_{ik}y_i$, where $\sum_{k=1}^{K} u_{ik} = 1$. u_{ik} means the degree of i unit to class θ_k. Lastly, the third layer yields the final belief function about the class of each pixel using the combination rule of DST (Eq. 8).

U-Net   TransU-Net   SegDDPM   EI-DDPM   Ground truth

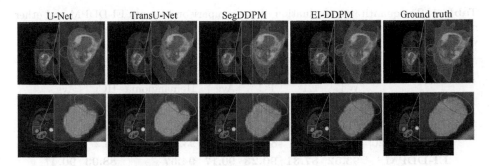

Fig. 3. Visualized segmentation results on BraTS 2021 dataset and liver MRI dataset. The first row presents tumor segmentation on BraTS 2021, where the label of green, yellow, and red represent the ED, ET, and NCR, respectively. The second row is HCC segmentation on liver MRI. The major segmentation differences between different methods are marked with blue circles. (Color figure online)

2.3 CDO Parses the Reliability of Multi-modality Medical Images

CDO performs the contextual discounting operation to EIL for parsing the reliability of multi-modality medical images on tumor segmentation. Firstly, define η as a coefficient in [0,1]. Then, transforming $\mathcal{M}$ using the discounting operation with discount rate $1 - \eta$ into a new mass function $^{\eta}\mathcal{M}$ [6]:

$$^{\eta}\mathcal{M} = \eta\mathcal{M} + (1 - \eta)\mathcal{M}_? \tag{10}$$

where $\mathcal{M}_?$ is a vacuous mass function defined by $\mathcal{M}(\Theta) = 1$, $^{\eta}\mathcal{M}$ is a mixture of $\mathcal{M}$ and $\mathcal{M}_?$. The coefficient η plays the role of weighting mass function $^{\eta}\mathcal{M}$. The corresponding contour function of $^{\eta}\mathcal{M}$ is:

$$^{\eta}pls(\{\theta_k\}) = 1 - \eta_k + \eta_k pls(\{\theta_k\}), \quad k = 1, ..., K \tag{11}$$

Advantage: *EIL and CDO parse the reliability of different modality medical images in different contexts.* For example, if we feed two modality medical images like $T1$ and $T2$ into the EI-DDPM, with the discount rate $1 - \eta_{T1}$ and $1 - \eta_{T2}$, we will have two contextual discounted contour functions $^{\eta_{T1}}pls_{T1}$ and $^{\eta_{T2}}pls_{T2}$. The combined contour function in Eq. 9 is proportional to the $^{\eta_{T1}}pls_{T1}{}^{\eta_{T2}}pls_{T2}$. In this situation, the η_{T1} and η_{T2} can be trained to weight the two modality medical images by parsing the reliability.

3   Experiment and Results

Dataset. We used two MRI datasets that BraTS 2021 [1,2,12] and a liver MRI dataset. BraTS 2021[1] contains 1251 subject with 4 aligned MRI modalities: T1, T2, Flair, and contrast-enhanced T1 (T1ce). The segmentation labels consist of

---

[1] http://braintumorsegmentation.org/.

Table 1. The quantitative evaluation of the comparison between EI-DDPM and other methods. The criteria of Dice evaluated the performance. Avg presents the average value in the whole dataset.

	Dice for BraTS 2021				Dice for Liver MRI		
	WT	ET	TC	Avg	Hemangioma	HCC	Avg
U-Net [17]	91.30	83.23	86.53	86.49	91.89	84.21	87.12
TransU-Net [5]	91.56	84.72	87.26	87.43	92.44	85.35	88.42
SegDDPM [20]	92.04	85.74	88.37	88.65	93.10	86.27	89.07
EI-DDPM	**93.52**	**87.31**	**90.23**	**90.17**	**94.57**	**88.03**	**90.47**

Table 2. Ablation studies for EIL and CDO. The criteria of Dice evaluated the performance. Avg presents the average value in the whole dataset.

	Dice for BraTS 2021				Dice for Liver MRI		
	WT	ET	TC	Avg	Hemangioma	HCC	Avg
No CDO	92.10	86.42	89.27	89.57	94.01	87.26	89.89
No EIL&CDO	92.33	85.83	88.50	89.03	93.24	86.33	89.25
EI-DDPM	**93.52**	**87.31**	**90.23**	**90.17**	**94.57**	**88.03**	**90.47**

GD-enhancing tumor (ET), necrotic tumor core (NCR), and peritumoral edematous (ED), which are combined into 3 nested subregions: Enhancing Tumor (ET) region, Tumor Core (TC) region (i.e. ET+NCR), and Whole Tumor (WT) region (i.e. ET+ED+NCR). The liver MRI dataset contains 238 subjects with 110 hemangioma subjects and 128 hepatocellular carcinoma (HCC) subjects. Each subject has corresponding 3 aligned MRI modalities: T1, T2, and T1ce (protocols were gadobutrol 0.1 mmol/kg on a 3T MRI scanner). The segmentation labels of hemangioma and HCC were performed by two radiologists with more than 5-year-experience. The resolution of the BraTS 2021 image and liver MRI image are 240×240 and 256×256, respectively. Both BraTS 2021 and liver MRI datasets were randomly divided into 3 groups following the ratio of training/validation/test as 7:1:2.

Implementation Details. For the number of DDPM paths, BraTS 2021 dataset is equal to 4 corresponding to the input 4 MRI modalities and the liver MRI dataset is equal to 3 corresponding to the input 3 MRI modalities. In the parallel DDPM path, the noise schedule followed the improved-DDPM [14], and the U-Net [17] was utilized as the denoising model with 300 sampling steps. In EIL, the initial values of a_i equal 0.5 and λ_i is equal to 0.01. For CDO, the initial of parameter η_k is equal to 0.5. With the Adam optimization algorithm, the denoising process was optimized using $\mathcal{L}_1$, and the EIL and CDO were optimized using Dice loss. The learning rate of EI-DDPM was set to 0.0001. The

Table 3. The learnt reliability coefficients η of BraTS 2021

η	ED	ET	NCR
T1	0.2725	0.3751	0.3318
T2	0.8567	0.9458	0.6912
Flair	**0.9581**	0.7324	0.5710
T1ce	0.7105	**0.9867**	**0.9879**

Table 4. The learnt reliability coefficients η of Liver MRI

η	Hemangioma	HCC
T1	0.8346	0.5941
T2	0.8107	0.5463
T1ce	**0.9886**	**0.9893**

framework was trained on Ubuntu 20.04 platform by using Pytorch and CUDA library, and it ran on an RTX 3090Ti graphics card with 24 GB memory.

Quantitative and Visual Evaluation. The performance of EI-DDPM is evaluated by comparing with three methods: a classical CNN-based method (U-Net [17]), a Transformer-based method (TransU-Net [5]), and a DDPM-based method for multi-modality medical image segmentation (SegDDPM [20]). The Dice score is used for evaluation criteria. Figure 3 shows the visualized segmentation results of EI-DDPM and compared methods. It shows some ambiguous area lost segmentation in three compared methods but can be segmented by our EI-DDPM. Table 1 reports the quantitative results of EI-DDPM and compared methods. Our EI-DDPM achieves highest Dice value on both BraTS dataset and liver MRI dataset. All these results proved EI-DDPM outperforms the three other methods.

Ablation Study. To prove the contribution from EIL and CDO, we performed 2 types of ablation studies: (1) removing CDO (i.e. No CDO) and (2) removing EIL and CDO (i.e. No EIL&CDO). Table 2 shows the quantitative results of ablation studies. Experimental results proved both EIL and CDO contribute to tumor segmentation on BraTS 2021 dataset and liver MRI dataset.

Discussion of Learnt Reliability Coefficients η. Table 3 and Table 4 show the learned reliability coefficients η on BraTS 2021 dataset and liver MRI dataset. The higher η value, the higher reliability of the corresponding region segmentation. As shown in Table 3, the Flair modality provides the highest reliability for ED segmentation. And both the T2 modality and T1ce modality provide relatively high reliability for ET and NCR segmentation. As shown in Table 4, the T1ce modality provides the highest reliability for hemangioma and HCC segmentation. These reliability values are the same as clinical experience [1,3].

4   Conclusion

In this paper, we proposed a novel DDPM-based framework for tumor segmentation under the condition of multi-modality medical images. The EIL and CDO enable our EI-DDPM to capture the reliability of different modality medical images with respect to different tumor regions. It provides a way of deploying contextual discounted DST to parse the reliability of multi-modality medical images. Extensive experiments prove the superiority of EI-DDPM for tumor segmentation on multi-modality medical images, which has great potential to aid in clinical diagnosis. The weakness of EI-DDPM is that it takes around 13 s to predict one segmentation image. In future work, we will focus on improving sampling steps in parallel DDPM paths to speed up EI-DDPM.

Acknowledgements. This work is partly supported by the China Scholarship Council (No. 202008370191)

References

1. Baid, U., et al.: The rsna-asnr-miccai brats 2021 benchmark on brain tumor segmentation and radiogenomic classification. arXiv preprint arXiv:2107.02314 (2021)
2. Bakas, S., et al.: Advancing the cancer genome atlas glioma MRI collections with expert segmentation labels and radiomic features. Sci. Data **4**(1), 1–13 (2017)
3. Balogh, J., et al.: Hepatocellular carcinoma: a review. Journal of hepatocellular carcinoma, pp. 41–53 (2016)
4. Baltrušaitis, T., Ahuja, C., Morency, L.P.: Multimodal machine learning: a survey and taxonomy. IEEE Trans. Pattern Anal. Mach. Intell. **41**(2), 423–443 (2018)
5. Chen, J., et al.: Transunet: transformers make strong encoders for medical image segmentation. arXiv preprint arXiv:2102.04306 (2021)
6. Denoeux, T., Kanjanatarakul, O., Sriboonchitta, S.: A new evidential k-nearest neighbor rule based on contextual discounting with partially supervised learning. Int. J. Approximate Reasoning **113**, 287–302 (2019)
7. Ho, J., Jain, A., Abbeel, P.: Denoising diffusion probabilistic models. Adv. Neural. Inf. Process. Syst. **33**, 6840–6851 (2020)
8. Huang, L., Ruan, S., Decazes, P., Denœux, T.: Lymphoma segmentation from 3d PET-CT images using a deep evidential network. Int. J. Approximate Reasoning **149**, 39–60 (2022)
9. Kim, B., Oh, Y., Ye, J.C.: Diffusion adversarial representation learning for self-supervised vessel segmentation. arXiv preprint arXiv:2209.14566 (2022)
10. Lian, C., Ruan, S., Denoeux, T., Li, H., Vera, P.: Joint tumor segmentation in PET-CT images using co-clustering and fusion based on belief functions. IEEE Trans. Image Process. **28**(2), 755–766 (2018)
11. Lim, K.Y., Mandava, R.: A multi-phase semi-automatic approach for multisequence brain tumor image segmentation. Expert Syst. Appl. **112**, 288–300 (2018)
12. Menze, B.H., et al.: The multimodal brain tumor image segmentation benchmark (brats). IEEE Trans. Med. Imaging **34**(10), 1993–2024 (2014)
13. Mercier, D., Quost, B., Denœux, T.: Refined modeling of sensor reliability in the belief function framework using contextual discounting. Inf. Fusion **9**(2), 246–258 (2008)
14. Nichol, A.Q., Dhariwal, P.: Improved denoising diffusion probabilistic models. In: International Conference on Machine Learning, pp. 8162–8171. PMLR (2021)
15. Qu, T., et al.: M3net: a multi-scale multi-view framework for multi-phase pancreas segmentation based on cross-phase non-local attention. Med. Image Anal. **75**, 102232 (2022)

16. Raju, A., et al.: Co-heterogeneous and adaptive segmentation from multi-source and multi-phase ct imaging data: a study on pathological liver and lesion segmentation. In: European Conference on Computer Vision. pp. 448–465. Springer (2020)
17. Ronneberger, O., Fischer, P., Brox, T.: U-Net: convolutional networks for biomedical image segmentation. In: Navab, N., Hornegger, J., Wells, W.M., Frangi, A.F. (eds.) MICCAI 2015. LNCS, vol. 9351, pp. 234–241. Springer, Cham (2015). https://doi.org/10.1007/978-3-319-24574-4_28
18. Shafer, G.: A mathematical theory of evidence, vol. 42. Princeton University Press (1976)
19. Wang, J., et al.: Tensor-based sparse representations of multi-phase medical images for classification of focal liver lesions. Pattern Recogn. Lett. **130**, 207–215 (2020)
20. Wolleb, J., Sandkühler, R., Bieder, F., Valmaggia, P., Cattin, P.C.: Diffusion models for implicit image segmentation ensembles. In: International Conference on Medical Imaging with Deep Learning, pp. 1336–1348. PMLR (2022)
21. Zhang, L., et al.: Robust pancreatic ductal adenocarcinoma segmentation with multi-institutional multi-phase partially-annotated CT scans. In: Martel, A.L., et al. (eds.) MICCAI 2020. LNCS, vol. 12264, pp. 491–500. Springer, Cham (2020). https://doi.org/10.1007/978-3-030-59719-1_48
22. Zhao, J., et al.: United adversarial learning for liver tumor segmentation and detection of multi-modality non-contrast MRI. Med. Image Anal. **73**, 102154 (2021)
23. Zhou, Y., et al.: Hyper-pairing network for multi-phase pancreatic ductal adenocarcinoma segmentation. In: Shen, D., et al. (eds.) MICCAI 2019. LNCS, vol. 11765, pp. 155–163. Springer, Cham (2019). https://doi.org/10.1007/978-3-030-32245-8_18

H-DenseFormer: An Efficient Hybrid Densely Connected Transformer for Multimodal Tumor Segmentation

Jun Shi[1], Hongyu Kan[1], Shulan Ruan[1], Ziqi Zhu[1], Minfan Zhao[1], Liang Qiao[1], Zhaohui Wang[1], Hong An[1(✉)], and Xudong Xue[2]

[1] University of Science and Technology of China, Hefei, China
shijun18@mail.ustc.edu.cn, han@ustc.edu.cn
[2] Hubei Cancer Hospital, Tongji Medical College, Huazhong University of Science and Technology, Wuhan, China

Abstract. Recently, deep learning methods have been widely used for tumor segmentation of multimodal medical images with promising results. However, most existing methods are limited by insufficient representational ability, specific modality number and high computational complexity. In this paper, we propose a hybrid densely connected network for tumor segmentation, named **H-DenseFormer**, which combines the representational power of the Convolutional Neural Network (CNN) and the Transformer structures. Specifically, H-DenseFormer integrates a Transformer-based Multi-path Parallel Embedding (**MPE**) module that can take an arbitrary number of modalities as input to extract the fusion features from different modalities. Then, the multimodal fusion features are delivered to different levels of the encoder to enhance multimodal learning representation. Besides, we design a lightweight Densely Connected Transformer (**DCT**) block to replace the standard Transformer block, thus significantly reducing computational complexity. We conduct extensive experiments on two public multimodal datasets, HECK-TOR21 and PI-CAI22. The experimental results show that our proposed method outperforms the existing state-of-the-art methods while having lower computational complexity. The source code is available at https://github.com/shijun18/H-DenseFormer.

Keywords: Tumor segmentation · Multimodal medical image · Transformer · Deep learning

1 Introduction

Accurate tumor segmentation from medical images is essential for quantitative assessment of cancer progression and preoperative treatment planning [3]. Tumor tissues usually present different features in different imaging modalities. For example, Computed Tomography (CT) and Positron Emission Tomography

J. Shi and H. Kan contributed equally. This study was supported by the Fundamental Research Funds for the Central Universities (No. YD2150002001).

(PET) are beneficial to represent morphological and metabolic information of tumors, respectively. In clinical practice, multimodal registered images, such as PET-CT images and Magnetic Resonance (MR) images with different sequences, are often utilized to delineate tumors to improve accuracy. However, manual delineation is time-consuming and error-prone, with a low inter-professional agreement [12]. These have prompted the demand for intelligent applications that can automatically segment tumors from multimodal images to optimize clinical procedures.

Recently, multimodal tumor segmentation has attracted the interest of many researchers. With the emergence of multimodal datasets (e.g., BRATS [25] and HECKTOR [1]), various deep-learning-based multimodal image segmentation methods have been proposed [3,10,13,27,29,31]. Overall, large efforts have been made on effectively fusing image features of different modalities to improve segmentation accuracy. According to the way of feature fusion, the existing methods can be roughly divided into three categories [15,36]: *input-level fusion, decision-level fusion*, and *layer-level fusion*. As a typical approach, input-level fusion [8,20,26,31,34] refers to concatenating multimodal images in the channel dimension as network input during the data processing or augmentation stage. This approach is suitable for most existing end-to-end models [6,32], such as U-Net [28] and U-Net++ [37]. However, the shallow fusion entangles the low-level features from different modalities, preventing the effective extraction of high-level semantics and resulting in limited performance gains. In contrast, [35] and [21] propose a solution based on decision-level fusion. The core idea is to train an independent segmentation network for each data modality and fuse the results in a specific way. These approaches can bring much extra computation at the same time, as the number of networks is positively correlated with the number of modalities. As a compromise alternative, layer-level fusion methods such as HyperDense-Net [10] advocate the cross-fusion of the multimodal features in the middle layer of the network.

In addition to the progress on the fusion of multimodal features, improving the model representation ability is also an effective way to boost segmentation performance. In the past few years, Transformer structure [11,24,30], centered on the multi-head attention mechanism, has been introduced to multimodal image segmentation tasks. Extensive studies [2,4,14,16] have shown that the Transformer can effectively model global context to enhance semantic representations and facilitate pixel-level prediction. Wang et al. [31] proposed TransBTS, a form of input-level fusion with a U-like structure, to segment brain tumors from multimodal MR images. TransBTS employs the Transformer as a bottleneck layer to wrap the features generated by the encoder, outperforming the traditional end-to-end models. Saeed et al. [29] adopted a similar structure in which the Transformer serves as the encoder rather than a wrapper, also achieving promising performance. Other works like [9] and [33], which combine the Transformer with the multimodal feature fusion approaches mentioned above, further demonstrate the potential of this idea for multimodal tumor segmentation.

Although remarkable performance has been accomplished with these efforts, there still exist several challenges to be resolved. Most existing methods are either limited to specific modality numbers due to the design of asymmetric connections or suffer from large computational complexity because of the huge amount of model parameters. Therefore, how to improve model ability while ensuring computational efficiency is the main focus of this paper.

To this end, we propose an efficient multimodal tumor segmentation solution named Hybrid Densely Connected Network (**H-DenseFormer**). First, our method leverages Transformer to enhance the global contextual information of different modalities. Second, H-DenseFormer integrates a Transformer-based Multi-path Parallel Embedding (**MPE**) module, which can extract and fuse multimodal image features as a complement to naive input-level fusion structure. Specifically, MPE assigns an independent encoding path to each modality, then merges the semantic features of all paths and feeds them to the encoder of the segmentation network. This decouples the feature representations of different modalities while relaxing the input constraint on the specific number of modalities. Finally, we design a lightweight, Densely Connected Transformer (**DCT**) module to replace the standard Transformer to ensure performance and computational efficiency. Extensive experimental results on two publicly available datasets demonstrate the effectiveness of our proposed method.

2   Method

2.1   Overall Architecture of H-DenseFormer

Figure 1 illustrates the overall architecture of our method. H-DenseFormer comprises a Multi-path Parallel Embedding (MPE) module and a U-shaped segmentation backbone network in form of input-level fusion. The former serves

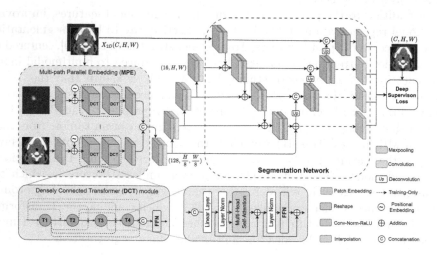

Fig. 1. Overall architecture of our proposed H-DenseFormer.

as the auxiliary extractor of multimodal fusion features, while the latter is used to generate predictions. Specifically, given a multimodal image input $X_{3D} \in \mathbb{R}^{C \times H \times W \times D}$ or $X_{2D} \in \mathbb{R}^{C \times H \times W}$ with a spatial resolution of $H \times W$, the depth dimension of D (number of slices) and C channels (number of modalities), we first utilize MPE to extract and fuse multimodal image features. Then, the obtained features are progressively upsampled and delivered to the encoder of the segmentation network to enhance the semantic representation. Finally, the segmentation network generates multi-scale outputs, which are used to calculate deep supervision loss as the optimization target.

2.2 Multi-path Parallel Embedding

Many methods [5,10,15] have proved that decoupling the feature representation of different modalities facilitates the extraction of high-quality multimodal features. Inspired by this, we design a Multip-path Parallel Embedding (MPE) module to enhance the representational ability of the network. As shown in Fig. 1, each modality has an independent encoding path consisting of a patch embedding module, stacked Densely Connected Transformer (DCT) modules, and a reshape operation. The independence of the different paths allows MPE to handle an **arbitrary** number of input modalities. Besides, the introduction of the Transformer provides the ability to model global contextual information. Given the input X_{3D}, after convolutional embedding and tokenization, the obtained feature of the i-th path is $\mathbf{F}_i \in \mathbb{R}^{l \times \frac{H}{p} \times \frac{W}{p} \times \frac{D}{p}}$, where $i \in [1, 2, ..., C]$, $p = 16$ and $l = 128$ denote the path size and embedding feature length respectively. First, we concatenate the features of all modalities and entangle them using a convolution operation. Then, interpolation upsampling is performed to obtain the multimodal fusion feature $\mathbf{F}_{out} \in \mathbb{R}^{k \times \frac{H}{8} \times \frac{W}{8} \times \frac{D}{8}}$, where $k = 128$ refers to the channel dimension. Finally, $\mathbf{F}_{out}$ is progressively upsampled to multiple scales and delivered to different encoder stages to enhance the learned representation.

2.3 Densely Connected Transformer

Standard Transformer structures [11] typically consist of dense linear layers with a computational complexity proportional to the feature dimension. Therefore, integrating the Transformer could lead to a mass of additional computation and memory requirements. Shortening the feature length can effectively reduce computation, but it also weakens the representation capability meanwhile. To address this problem, we propose the Densely Connected Transformer (DCT) module inspired by DenseNet [17] to balance computational cost and representation capability. Figure 1 details the DCT module, which consists of **four** Transformer layers and a feedforward layer. Each Transformer layer has a linear projection layer that reduces the input feature dimension to $g = 32$ to save computation. Different Transformer layers are connected densely to preserve representational power with lower feature dimensions. The feedforward layer at the end generates the fusion features of the different layers. Specifically, the output $\mathbf{z}_j$ of the j-th ($j \in [1, 2, 3, 4]$) Transformer layer can be calculated by:

$$\tilde{\mathbf{z}}_{j-1} = p(cat([\mathbf{z}_0; \mathbf{z}_1; ...; \mathbf{z}_{j-1}])), \tag{1}$$

$$\tilde{\mathbf{z}}_j = att(norm(\tilde{\mathbf{z}}_{j-1})) + \tilde{\mathbf{z}}_{j-1}, \tag{2}$$

$$\mathbf{z}_j = f(norm(\tilde{\mathbf{z}}_j)), \tag{3}$$

where $\mathbf{z}_0$ represents the original input, $cat(\cdot)$ and $p(\cdot)$ denote the concatenation operator and the linear layer, respectively. The $norm(\cdot)$, $att(\cdot)$, $f(\cdot)$ are the regular layer normalization, multi-head self-attention mechanism, and feedforward layer. The output of DCT is $\mathbf{z}_{out} = f(cat([\mathbf{z}_0; \mathbf{z}_1; ...; \mathbf{z}_4]))$. Table 1 shows that the stacked DCT has lower parameters and computational complexity than a standard Transformer structure with the same number of layers.

Table 1. Comparison of the computational complexity between the standard 12-layer Transformer structure and the stacked 3 (=12/4) DCT modules.

Feature Dimension	Resolution	Transformer		Stacked DCT (×3)	
		GFLOPs ↓	Params ↓	GFLOPs ↓	Params ↓
256	(512,512)	6.837	6.382M	**2.671**	**1.435M**
512	(512,512)	26.256	25.347M	**3.544**	**2.290M**

2.4   Segmentation Backbone Network

The H-DenseFormer adopts a U-shaped encoder-decoder structure as its backbone. As shown in Fig. 1, the encoder extracts features and reduces their resolution progressively. To preserve more details, we set the maximum downsampling factor to 8. The multi-level multimodal features from MPE are fused in a bitwise addition way to enrich the semantic information. The decoder is used to restore the resolution of the features, consisting of deconvolutional and convolutional layers with skip connections to the encoder. In particular, we employ Deep Supervision (**DS**) loss to improve convergence, which means that the multiscale output of the decoder is involved in the final loss computation.

Deep Supervision Loss. During training, the decoder has four outputs; for example, the i-th output of 2D H-DenseFormer is $\mathbf{O}^i \in \mathbb{R}^{c \times \frac{H}{2^i} \times \frac{W}{2^i}}$, where $i \in [0, 1, 2, 3]$, and $c = 2$ (tumor and background) represents the number of segmentation classes. To mitigate the pixel imbalance problem, we use a combined loss of Focal loss [23] and Dice loss as the optimization target, defined as follows:

$$\zeta_{FD} = 1 - \frac{2\sum_{t=1}^{N} p_t q_t}{\sum_{t=1}^{N} p_t + q_t} + \frac{1}{N} \sum_{t=1}^{N} -(1 - p_t)^\gamma log(p_t), \tag{4}$$

where N refers to the total number of pixels, p_t and q_t denote the predicted probability and ground truth of the t-th pixel, respectively, and $r = 2$ is the modulation factor. Thus, DS loss can be calculated as follows:

$$\zeta_{\mathbf{DS}} = \sum \alpha_i \cdot \zeta_{FD}(\mathbf{O}^i, \mathbf{G}^i), \alpha_i = 2^{-i}. \tag{5}$$

where $\mathbf{G}^i$ represents the ground truth after resizing and has the same size as $\mathbf{O}^i$. α is a weighting factor to control the proportion of loss corresponding to the output at different scales. This approach can improve the convergence speed and performance of the network.

3   Experiments

3.1   Dataset and Metrics

To validate the effectiveness of our proposed method, we performed extensive experiments on **HECKTOR21** [1] and **PI-CAI22**[1]. HECKTOR21 is a dual-modality dataset for head and neck tumor segmentation, containing 224 PET-CT image pairs. Each PET-CT pair is registered and cropped to a fixed size of (144,144,144). PI-CAI22 provides multimodal MR images of 220 patients with prostate cancer, including T2-Weighted imaging (T2W), high b-value Diffusion-Weighted imaging (DWI), and Apparent Diffusion Coefficient (ADC) maps. After standard resampling and center cropping, all images have a size of (24,384,384). We randomly select 180 samples for each dataset as the training set and the rest as the independent test set (44 cases for HECKTOR21 and 40 cases for PI-CAI22). Specifically, the training set is further randomly divided into five folds for cross-validation. For quantitative analysis, we use the Dice Similarity Coefficient (**DSC**), the Jaccard Index (**JI**), and the 95% Hausdorff Distance (**HD95**) as evaluation metrics for segmentation performance. A better segmentation will have a smaller HD95 and larger values for DSC and JI. We also conduct holistic t-tests of the overall performance for our method and all baseline models with the two-tailed $p < 0.05$.

3.2   Implementation Details

We use Pytorch to implement our proposed method and the baselines. For a fair comparison, all models are trained from scratch using two NVIDIA A100 GPUs and all comparison methods are implemented with open-source codes, following their original configurations. In particular, we evaluate the 3D and 2D H-DenseFormer on HECKTOR21 and PI-CAI22, respectively. During the training phase, the Adam optimizer is employed to minimize the loss with an initial learning rate of 10^{-3} and a weight decay of 10^{-4}. We use the PolyLR strategy [19] to control the learning rate change. We also use an early stopping strategy with a tolerance of 30 epochs to find the best model within 100 epochs. Online data augmentation, including random rotation and flipping, is performed to alleviate the overfitting problem.

---

[1] https://pi-cai.grand-challenge.org/.

3.3 Overall Performance

Table 2. Comparison with existing methods on independent test set. We show the mean ± std (standard deviation) scores of averaged over the 5 folds.

Methods (Year)	Params↓	GFLOPs↓	DSC(%) ↑	HD95(mm) ↓	JI(%) ↑
HECKTOR21, two modalities (CT and PET)					
3D U-Net (2016) [7]	12.95M	629.07	68.8 ± 1.4	14.9 ± 2.2	58.0 ± 1.4
UNETR (2022) [16]	95.76M	282.19	59.6 ± 2.5	23.7 ± 3.4	48.2 ± 2.6
Iantsen et al. (2021) [18]	38.66M	1119.75	72.4 ± 0.8	9.6 ± 1.0	60.5 ± 1.1
TransBTS (2021) [31]	30.62M	372.80	64.8 ± 1.0	20.9 ± 3.9	52.9 ± 1.2
3D H-DenseFormer	**3.64M**	**242.96**	**73.9 ± 0.5**	**8.1 ± 0.6**	**62.5 ± 0.5**
PI-CAI22, three modalities (T2W, DWI and ADC)					
Deeplabv3+ (2018) [6]	12.33M	**10.35**	47.4 ± 1.9	48.4 ± 14.3	35.4 ± 1.7
U-Net++ (2019) [37]	15.97M	36.08	49.7 ± 3.9	38.5 ± 6.7	36.9 ± 3.3
ITUNet (2022) [22]	18.13M	32.67	42.1 ± 2.3	67.6 ± 10.3	31.3 ± 1.6
Transunet (2021) [4]	93.23M	72.62	44.8 ± 3.0	59.3 ± 14.8	33.2 ± 2.5
2D H-DenseFormer	**4.25M**	31.46	**49.9 ± 1.2**	**35.9 ± 8.2**	**37.1 ± 1.2**

Table 2 compares the performance and computational complexity of our proposed method with the existing state-of-the-art methods on the independent **test** sets. For HECKTOR21, 3D H-DenseFormer achieves a DSC of 73.9%, HD95 of 8.1mm, and JI of 62.5%, which is a significant improvement ($p < 0.01$) over 3D U-Net [7], UNETR [16], and TransBTS [31]. It is worth noting that the performance of hybrid models such as UNETR is not as good as expected, even worse than 3D U-Net, perhaps due to the small size of the dataset. Moreover, compared to the champion solution of HECKTOR20 proposed by Iantsen et al. [18], our method has higher accuracy and about **10** and **5** times lower amount of network parameters and computational cost, respectively. For PI-CAI22, the 2D variant of H-DenseFormer also outperforms existing methods ($p < 0.05$), achieving a DSC of 49.9%, HD95 of 35.9 mm, and JI of 37.1%. Overall, H-DenseFormer reaches an effective balance of performance and computational cost compared to existing CNNs and hybrid structures. For qualitative analysis, we show a visual comparison of the different methods. It is evident from Fig. 2 that our approach can describe tumor contours more accurately while providing better segmentation accuracy for small-volume targets. These results further demonstrate the effectiveness of our proposed method in multimodal tumor segmentation tasks.

3.4 Parameter Sensitivity and Ablation Study

Impact of DCT Depth. As illustrated in Table 3, the network performance varies with the change in DCT depth. H-DenseFormer achieves the best performance at the DCT depth of **6**. An interesting finding is that although the depth

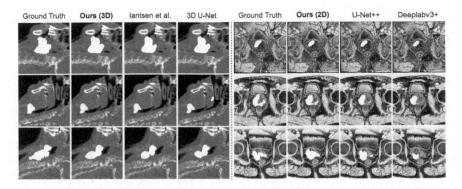

Ground Truth   **Ours (3D)**   Iantsen et al.   3D U-Net   Ground Truth   **Ours (2D)**   U-Net++   Deeplabv3+

Fig. 2. Visualizations of different models on HECKTOR21 (left) and PI-CAI22 (right).

Table 3. Parameter sensitivity analysis on DCT depth.

DCT Depth	Params↓	GFLOPs↓	DSC (%) ↑	HD95 (mm) ↓	JI (%) ↑
3	3.25M	242.38	73.5 ± 1.4	8.4 ± 0.7	62.2 ± 1.6
6	3.64M	242.96	**73.9 ± 0.5**	**8.1 ± 0.6**	**62.5 ± 0.5**
9	4.03M	243.55	72.7 ± 1.2	8.7 ± 0.6	61.2 ± 1.3

of the DCT has increased from 3 to 9, the performance does not improve or even worsen. We suspect that the reason is over-fitting due to over-parameterization. Therefore, choosing a proper DCT depth is crucial to improve accuracy.

Impact of Different Modules. The above results demonstrate the superiority of our method, but it is unclear which module plays a more critical role in performance improvement. Therefore, we perform ablation experiments on MPE, DCT and DS loss. Specifically, w/o MPE refers to keeping one embedding path, w/o DCT signifies using a standard 12-layer Transformer, and w/o DS loss denotes removing the deep supervision mechanism. As shown in Table 4, the performance decreases with varying degrees when removing them separately, which means all the modules are critical for H-DenseFormer. We can observe that DCT has a greater impact on overall performance than the others, further demonstrating its effectiveness. In particular, the degradation after removing the MPE also con-

Table 4. Ablation study of 3D H-DenseFormer, w/o denotes without.

Method	DSC (%) ↑	HD95 (mm) ↓	JI (%) ↑
3D H-DenseFormer **w/o MPE**	72.1 ± 0.8	10.8 ± 1.1	60.4 ± 0.8
3D H-DenseFormer **w/o DCT**	70.7 ± 1.8	11.9 ± 1.9	58.6 ± 2.1
3D H-DenseFormer **w/o DS loss**	72.2 ± 0.9	10.2 ± 1.0	60.1 ± 1.2
3D H-DenseFormer	**73.9 ± 0.5**	**8.1 ± 0.6**	**62.5 ± 0.5**

firms that decoupling the feature expression of different modalities helps obtain higher-quality multimodal features and improve segmentation performance.

4 Conclusion

In this paper, we proposed an efficient hybrid model (H-DenseFormer) that combines Transformer and CNN for multimodal tumor segmentation. Concretely, a Multi-path Parallel Embedding module and a Densely Connected Transformer block were developed and integrated to balance accuracy and computational complexity. Extensive experimental results demonstrated the effectiveness and superiority of our proposed H-DenseFormer. In future work, we will extend our method to more tasks and explore more efficient multimodal feature fusion methods to further improve computational efficiency and segmentation performance.

References

1. Andrearczyk, V., et al.: Overview of the HECKTOR challenge at MICCAI 2020: automatic head and neck tumor segmentation in PET/CT. In: Andrearczyk, V., Oreiller, V., Depeursinge, A. (eds.) HECKTOR 2020. LNCS, vol. 12603, pp. 1–21. Springer, Cham (2021). https://doi.org/10.1007/978-3-030-67194-5_1
2. Cao, H., et al.: Swin-UNet: UNet-like pure transformer for medical image segmentation. In: Karlinsky, L., Michaeli, T., Nishino, K. (eds.) Computer Vision – ECCV 2022 Workshops. ECCV 2022, Part III. LNCS, vol. 13803, pp. 205–218. Springer, Cham (2023). https://doi.org/10.1007/978-3-031-25066-8_9
3. Chen, C., Dou, Q., Jin, Y., Chen, H., Qin, J., Heng, P.-A.: Robust multimodal brain tumor segmentation via feature disentanglement and gated fusion. In: Shen, D., et al. (eds.) MICCAI 2019, Part III 22. LNCS, vol. 11766, pp. 447–456. Springer, Cham (2019). https://doi.org/10.1007/978-3-030-32248-9_50
4. Chen, J., et al.: TransUNet: transformers make strong encoders for medical image segmentation. arXiv preprint arXiv:2102.04306 (2021)
5. Chen, L., Wu, Y., DSouza, A.M., Abidin, A.Z., Wismüller, A., Xu, C.: MRI tumor segmentation with densely connected 3D CNN. In: Medical Imaging 2018: Image Processing, vol. 10574, pp. 357–364. SPIE (2018)
6. Chen, L.-C., Zhu, Y., Papandreou, G., Schroff, F., Adam, H.: Encoder-decoder with Atrous separable convolution for semantic image segmentation. In: Ferrari, V., Hebert, M., Sminchisescu, C., Weiss, Y. (eds.) ECCV 2018. LNCS, vol. 11211, pp. 833–851. Springer, Cham (2018). https://doi.org/10.1007/978-3-030-01234-2_49
7. Çiçek, Ö., Abdulkadir, A., Lienkamp, S.S., Brox, T., Ronneberger, O.: 3D U-Net: learning dense volumetric segmentation from sparse annotation. In: Ourselin, S., Joskowicz, L., Sabuncu, M.R., Unal, G., Wells, W. (eds.) MICCAI 2016. LNCS, vol. 9901, pp. 424–432. Springer, Cham (2016). https://doi.org/10.1007/978-3-319-46723-8_49
8. Cui, S., Mao, L., Jiang, J., Liu, C., Xiong, S.: Automatic semantic segmentation of brain gliomas from MRI images using a deep cascaded neural network. J. Healthc. Eng. **2018**, 4940593 (2018)

9. Dobko, M., Kolinko, D.I., Viniavskyi, O., Yelisieiev, Y.: Combining CNNs with transformer for multimodal 3D MRI brain tumor segmentation. In: Crimi, A., Bakas, S. (eds.) Brainlesion: Glioma, Multiple Sclerosis, Stroke and Traumatic Brain Injuries: 7th International Workshop, BrainLes 2021, Held in Conjunction with MICCAI 2021, Virtual Event, 27 September 2021, Revised Selected Papers, Part II, pp. 232–241. Springer, Cham (2022). https://doi.org/10.1007/978-3-031-09002-8_21

10. Dolz, J., Gopinath, K., Yuan, J., Lombaert, H., Desrosiers, C., Ayed, I.B.: HyperDense-Net: a hyper-densely connected CNN for multi-modal image segmentation. IEEE Trans. Med. Imag. **38**(5), 1116–1126 (2018)

11. Dosovitskiy, A., et al.: An image is worth 16 × 16 words: transformers for image recognition at scale. arXiv preprint arXiv:2010.11929 (2020)

12. Foster, B., Bagci, U., Mansoor, A., Xu, Z., Mollura, D.J.: A review on segmentation of positron emission tomography images. Comput. Bio. Med. **50**, 76–96 (2014)

13. Fu, X., Bi, L., Kumar, A., Fulham, M., Kim, J.: Multimodal spatial attention module for targeting multimodal PET-CT lung tumor segmentation. IEEE J. Biomed. Health Inform. **25**(9), 3507–3516 (2021)

14. Gao, Y., Zhou, M., Metaxas, D.N.: UTNet: a hybrid transformer architecture for medical image segmentation. In: de Bruijne, M., et al. (eds.) MICCAI 2021, Part III. LNCS, vol. 12903, pp. 61–71. Springer, Cham (2021). https://doi.org/10.1007/978-3-030-87199-4_6

15. Guo, Z., Li, X., Huang, H., Guo, N., Li, Q.: Deep learning-based image segmentation on multimodal medical imaging. IEEE Trans. Radiat. Plasma Med. Sci. **3**(2), 162–169 (2019)

16. Hatamizadeh, A., et al.: UNETR: transformers for 3D medical image segmentation. In: WACV 2022 Proceedings, pp. 574–584 (2022)

17. Huang, G., Liu, Z., Van Der Maaten, L., Weinberger, K.Q.: Densely connected convolutional networks. In: CVPR 2017 Proceedings, pp. 4700–4708 (2017)

18. Iantsen, A., Visvikis, D., Hatt, M.: Squeeze-and-excitation normalization for automated delineation of head and neck primary tumors in combined PET and CT images. In: Andrearczyk, V., Oreiller, V., Depeursinge, A. (eds.) HECKTOR 2020. LNCS, vol. 12603, pp. 37–43. Springer, Cham (2021). https://doi.org/10.1007/978-3-030-67194-5_4

19. Isensee, F., et al.: nnU-Net: a self-configuring method for deep learning-based biomedical image segmentation. Nat. Methods **18**(2), 203–211 (2021)

20. Kamnitsas, K., et al.: Efficient multi-scale 3D CNN with fully connected CRF for accurate brain lesion segmentation. Med. Image Anal. **36**, 61–78 (2017)

21. Kamnitsas, K., et al.: Ensembles of multiple models and architectures for robust brain tumour segmentation. In: Crimi, A., Bakas, S., Kuijf, H., Menze, B., Reyes, M. (eds.) BrainLes 2017. LNCS, vol. 10670, pp. 450–462. Springer, Cham (2018). https://doi.org/10.1007/978-3-319-75238-9_38

22. Kan, H., et al.: ITUnet: Integration of transformers and UNet for organs-at-risk segmentation. In: EMBC 2022, pp. 2123–2127. IEEE (2022)

23. Lin, T.Y., Goyal, P., Girshick, R., He, K., Dollár, P.: Focal loss for dense object detection. In: ICCV 2017 Proceedings, pp. 2980–2988 (2017)

24. Liu, Z., et al.: Swin transformer: hierarchical vision transformer using shifted windows. In: ICCV 2021 Proceedings, pp. 10012–10022 (2021)

25. Menze, B.H., et al.: The multimodal brain tumor image segmentation benchmark (BRATS). IEEE Trans. Med. Imag. **34**(10), 1993–2024 (2014)

26. Pereira, S., Pinto, A., Alves, V., Silva, C.A.: Brain tumor segmentation using convolutional neural networks in MRI images. IEEE Trans. Med. Imag. **35**(5), 1240–1251 (2016)
27. Rodríguez Colmeiro, R.G., Verrastro, C.A., Grosges, T.: Multimodal brain tumor segmentation using 3D convolutional networks. In: Crimi, A., Bakas, S., Kuijf, H., Menze, B., Reyes, M. (eds.) BrainLes 2017. LNCS, vol. 10670, pp. 226–240. Springer, Cham (2018). https://doi.org/10.1007/978-3-319-75238-9_20
28. Ronneberger, O., Fischer, P., Brox, T.: U-Net: convolutional networks for biomedical image segmentation. In: Navab, N., Hornegger, J., Wells, W.M., Frangi, A.F. (eds.) MICCAI 2015, Part III 18. LNCS, vol. 9351, pp. 234–241. Springer, Cham (2015). https://doi.org/10.1007/978-3-319-24574-4_28
29. Saeed, N., Sobirov, I., Al Majzoub, R., Yaqub, M.: TMSS: an end-to-end transformer-based multimodal network for segmentation and survival prediction. In: Wang, L., Dou, Q., Fletcher, P.T., Speidel, S., Li, S. (eds.) Medical Image Computing and Computer Assisted Intervention – MICCAI 2022. MICCAI 2022, Part VII. LNCS, vol. 13437, pp. 319–329. Springer, Cham (2022). https://doi.org/10.1007/978-3-031-16449-1_31
30. Vaswani, A., et al.: Attention is all you need. In: NIPS 2017, vol. 30 (2017)
31. Wang, W., Chen, C., Ding, M., Yu, H., Zha, S., Li, J.: TransBTS: multimodal brain tumor segmentation using transformer. In: de Bruijne, M., et al. (eds.) MICCAI 2021, Part I 24. LNCS, vol. 12901, pp. 109–119. Springer, Cham (2021). https://doi.org/10.1007/978-3-030-87193-2_11
32. Xiao, X., Lian, S., Luo, Z., Li, S.: Weighted Res-UNet for high-quality retina vessel segmentation. In: ITME 2018, pp. 327–331. IEEE (2018)
33. Zhang, Y., et al.: mmFormer: multimodal medical transformer for incomplete multimodal learning of brain tumor segmentation. In: Wang, L., Dou, Q., Fletcher, P.T., Speidel, S., Li, S. (eds.) MICCAI 2022 Proceedings, Part V, vol. 13435, pp. 107–117. Springer, Cham (2022). https://doi.org/10.1007/978-3-031-16443-9_11
34. Zhao, X., et al.: A deep learning model integrating FCNNs and CRFs for brain tumor segmentation. Med. Image Anal. **43**, 98–111 (2018)
35. Zhong, Z., et al.: 3D fully convolutional networks for co-segmentation of tumors on PET-CT images. In: ISBI 2018, pp. 228–231. IEEE (2018)
36. Zhou, T., Ruan, S., Canu, S.: A review: deep learning for medical image segmentation using multi-modality fusion. Array **3**, 100004 (2019)
37. Zhou, Z., Siddiquee, M.M.R., Tajbakhsh, N., Liang, J.: UNet++: redesigning skip connections to exploit multiscale features in image segmentation. IEEE Trans. Med. Imag. **39**(6), 1856–1867 (2019)

Evolutionary Normalization Optimization Boosts Semantic Segmentation Network Performance

Luisa Neubig$^{(\boxtimes)}$ and Andreas M. Kist

Department Artificial Intelligence in Biomedical Engineering,
Friedrich-Alexander-Universität Erlangen-Nürnberg (FAU), Erlangen, Germany
{luisa.e.neubig,andreas.kist}@fau.de

Abstract. Semantic segmentation is an important task in medical imaging. Typically, encoder-decoder architectures, such as the U-Net, are used in various variants to approach this task. Normalization methods, such as Batch or Instance Normalization are used throughout the architectures to adapt to data-specific noise. However, it is barely investigated which normalization method is most suitable for a given dataset and if a combination of those is beneficial for the overall performance. In this work, we show that by using evolutionary algorithms we can fully automatically select the best set of normalization methods, outperforming any competitive single normalization method baseline. We provide insights into the selection of normalization and how this compares across imaging modalities and datasets. Overall, we propose that normalization should be managed carefully during the development of the most recent semantic segmentation models as it has a significant impact on medical image analysis tasks, contributing to a more efficient analysis of medical data. Our code is openly available at https://github.com/neuluna/evoNMS.

Keywords: Semantic segmentation · Normalization · Evolutionary Algorithm

1 Introduction

Semantic segmentation, i.e., assigning a semantic label to each pixel in an image, is a common task in medical computer vision nowadays typically performed by fully convolutional encoder-decoder deep neural networks (DNNs). These DNNs usually incorporate some kind of normalization layers which are thought to reduce the impact of the internal covariate shift (ICS) [6]. This effect describes the adaption to small changes in the feature maps of deeper layers rather than learning the real representation of the target structures [21]. The understanding of Batch Normalization (BN) is very controversial, namely that its actual success is to use higher learning rates by smoothing the objective function instead of reducing the ICS [2]. Instance Normalization (IN), Layer Normalization (LN)

Supplementary Information The online version contains supplementary material available at https://doi.org/10.1007/978-3-031-43901-8_67.

and Group Normalization (GN) are examples of developments of BN to overcome its shortcomings, like reduced performance using smaller batch sizes [2]. As an alternative, Scaled Exponential Linear Units (SELUs) can act as self-normalization activation functions [9]. All normalization methods have different strengths and weaknesses, which influence the performance and generalizability of the network. Commonly, only a single normalization method is used throughout the network, and studies involving multiple normalization methods are rare.

Neural architecture search (NAS) is a strategy to tweak a neural architecture as such to discover efficient combinations of architectural building blocks for optimal performance on given datasets and tasks [12]. NAS strategies involve, for example, evolutionary algorithms to optimize an objective function by evaluating a set of candidate architectures and selecting the "fittest" architectures for breeding [12]. After several generations of training, selection, and breeding, the objective function of the evolutionary algorithm should be maximized.

In this study, we propose a novel evolutionary NAS approach to increase semantic segmentation performance by optimizing the spatiotemporal usage of normalization methods in a baseline U-Net [17]. Our study provides a uniquely and thorough analysis of the most effective layer-wise normalization configuration across medical datasets, rather than proposing a new normalization method. In the following, we refer to our proposed methodology as evoNMS (evolutionary Normalization Method Search).

We evaluated the performance of evoNMS on eleven biomedical segmentation datasets and compared it with a state-of-the-art semantic segmentation method (nnU-Net [7]) and U-Nets with constant normalization such as BN, IN, and no normalization (NoN). Our analysis demonstrates that evoNMS discovers very effective network architectures for semantic segmentation, achieves better or similar performance to state-of-the-art architectures, and guides the selection of the best normalization method for a specific semantic segmentation task. In addition, we analyze the normalization pattern across datasets and modalities and compare the normalization methods regarding their layer-specific contribution.

2   Related Works

To gain optimal performance in semantic segmentation tasks, it is important to optimize data preprocessing and architectural design. Popat et al. [15] and Wei et al. [19] concurrently developed an evolutionary approach to determine the best-performing U-Net architecture variant, considering the depth, filter size, pooling type, kernel type, and optimizer as hyperparameter-genes coding for a specific U-Net phenotype. When applied to retinal vessel segmentation, both showed that their approach finds a smaller U-Net configuration while achieving competitive performance with state-of-the-art architectures. Liu et al. [10] proved that not only the architecture has a huge impact on the generalizability of a neural network, but also the combination of normalization.

Various studies show that neural networks (NNs) benefit from normalization to enhance task performance, generalizability, and convergence behavior. Zhou et al. [22] showed the benefit of batch normalization, which focuses on the data bias in the latent space by introducing a dual normalization for better domain generalization. Dual normalization estimates the distribution from source-similar and source-dissimilar domains and achieves a more robust model for domain generalization. Domain-independent normalization also helps to improve unsupervised adversarial domain adaptation for improved generalization capability as shown in [16]. In [21], the authors analyzed the influence of different normalization methods, such as BN, IN, LN, and GN. Although many segmentation networks rely on BN, they recommend using normalization by dividing feature maps, such as GN (with a higher number of groups) or IN [21]. Normalization methods have been well discussed in the literature [3,5,9,18,20]. In a systematic review, Huang et al. [5] concluded that normalizing the activations is more efficient than normalizing the weights. In addition to the efficiency of normalizing the activations, Luo et al. [13] demonstrated a synergistic effect of their advantages by introducing switchable normalization (SN). SN alternates between the normalization strategies IN, LN, and BN according to their respective importance weights [13]. With an evolutionary approach, Liu et al. [11] also showed that the combination of normalization and activation functions improves the performance of the NN.

3   Methods

We investigated the impact of normalization on semantic segmentation using eleven different medical image datasets. Eight datasets were derived from the Medical Segmentation Decathlon (MSD) [1]. In this study, we selected subsets for segmenting the hippocampus, heart, liver, lung, colon, spleen, pancreas, and hepatic vessels. We only considered the datasets with 3D volumes of the MSD. In addition, we used the BAGLS [4] dataset (segmentation of glottal area), the Kvasir-SEG [8] dataset (gastrointestinal polyp images), and an in-house dataset for bolus segmentation in videofluoroscopic swallowing studies (VFSS). Table 1 lists the datasets used regarding their region of interest/organ, modality, number of segmented classes, and number of images. To minimize the differences between the datasets and to gain comparable results, all datasets were analyzed in 2D. The images were converted to grayscale images, resized, and cropped to a uniform size of 224×224 px. Their pixel intensity was normalized to a range from 0 to 1. The datasets were divided into training, validation, and test subsets, with percentages of 70%, 10%, and 20% for the BAGLS, Kvasir-SEG, and bolus datasets. If a test set was explicitly given, the split for training and validation was 80% and 20%, respectively.

We implemented our evolutionary optimization algorithm and the NN architectural design in TensorFlow 2.9.1/Keras 2.9.0 and executed our code on NVIDIA A100 GPUs. Each individual U-Net variant was trained for 20 epochs using the Adam optimizer, a constant learning rate of 1×10^{-3}, and a batch size of 64. All segmentation tasks were optimized using the Dice Loss (DL). To evaluate the performance of the trained network, we calculated the Dice Coefficient (DC), Intersection over Union (IoU) of the fitted bounding boxes (BBIoU), and Hausdorff Distance with a percentile of 95% (HD95) of the validation set after

Table 1. Overview of different datasets regarding their medical objective, number of segmentation labels, modality, and number of train, validation, test images, and the average evoNMS wall time (Time).

Dataset	ROI	Labels	Modality	Train	Validation	Test	Time
BAGLS	Glottal Area	1	Endoscope	16,277	2,325	4,650	96 h
Kvasir-SEG	Polyp	1	Endoscope	700	100	200	4 h
Bolus	Bolus	1	VFSS	7,931	1,133	2,266	39 h
Task02 MSD	Heart	1	MRI	1,215	135	1,297	7 h
Task03 MSD	Liver	2	CT	17,238	1,915	27,041	102 h
Task04 MSD	Hippocampus	2	MRI	5,960	662	4,499	40 h
Task06 MSD	Lung	1	CT	1,480	164	8,888	12 h
Task07 MSD	Pancreas	2	CT	7,907	878	13,544	49 h
Task08 MSD	Hepatic Vessel	2	CT	11,448	1,272	10,519	75 h
Task09 MSD	Spleen	1	CT	945	105	1,327	12 h
Task10 MSD	Colon	1	CT	1,152	128	6,616	8 h

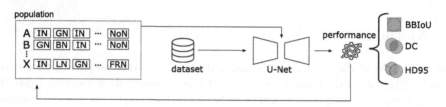

Fig. 1. Process of evolutionary optimization of normalization layers in a U-Net architecture. The sequence of normalization layers was randomly determined for a population of 20 individuals in the first generation. After training each U-Net variant on a given task/dataset, its performance was evaluated using a fitness function based on BBIoU, DC, and HD95. Based on the fitness value, normalization sequences were selected to breed the population of the next generation.

20 epochs. In addition, we included an early stopping criterion that is activated when the validation loss changes less than 0.1% to avoid unnecessary training time without further information gain. To compare our approach to state-of-the-art segmentation networks, we considered nnU-Net [7] as a baseline which was similarly trained as the above-mentioned U-Net variants.

Our proposed evoNMS approach is based on evolutionary optimization with leaderboard selection and is executed for 20 generations. Each generation's population consists of 20 individuals, i.e., U-Net variants, meaning that we train 400 variants for one evoNMS execution (duration 4 h (polyp) to 5 days (glottal area) on one A100 GPU). The first generation contains individuals with random sequences drawn from our gene pool containing either a BN, IN, FRN, GN2, GN4 layer or skips normalization (no normalization, NoN). Other U-Net-specific hyperparameters, such as initial filter size and activation functions were set across datasets to a fixed value (initial filter size of 16, and ReLU as activation function). In general, we kept all other hyperparameters fixed to focus only on the influence of normalization and infer whether it exhibits decoder/encoder dependence or even dependence on the underlying modality. After training each architecture, the fitness F_i (Eq. (1)) is evaluated for each individual i

$$F_i = \frac{1}{3} \cdot (DC_i + BBIoU_i + \frac{1}{HD95_i}) \quad , \text{where} \quad \frac{1}{HD95_i} \in [0,1], \quad (1)$$

where we compute the mean validation DC, validation IoU of the Bounding Box, and reciprocal of the validation HD95. We use the reciprocal of HD95 to balance the influence of each metric on the fitness value by a value ranging from 0 to 1. After each generation, the top ten individuals with the highest fitness F were bred. To breed a new individual, we selected two random individuals from the top ten candidates and combined them with a randomly selected crossing point across the normalization layer arrays of the two parental gene pools. Next, we applied mutations at a rate of 10%, which basically changes the normalization method of a random position to any normalization technique available in the gene pool.

Table 2. Overview of performance across datasets and normalization configurations. For multiclass segmentation, we define the highest validation DC as the mean across the segmentation labels. Each value is the mean value of five individual runs. For evoNMS, we selected the best normalization pattern w.r.t. the fitness value and trained this configuration five times. The ranking represents the average behavior of each network across the datasets (1-best, 6-worst). The bottom rows show the average behavior of each network across the eleven datasets regarding DC, BBIoU, and HD95 on the validation dataset.

Dataset	Labels	BN	IN	NoN	nnU-Net	Gen1 (Ours)	Gen20 (Ours)
glottal area	1	0.919	**0.933**	0.005	0.862	0.928	0.931
polyp	1	0.251	0.639	0.442	**0.804**	0.672	0.704
bolus	1	0.833	**0.841**	0.000	0.829	0.836	0.838
heart	1	0.016	0.844	0.029	**0.910**	0.243	0.885
liver	2	0.898	**0.932**	0.350	0.656	0.918	0.921
hippocampus	2	0.819	**0.829**	0.503	0.786	**0.829**	**0.829**
lung	1	0.701	0.827	0.008	0.672	0.663	**0.833**
pancreas	2	0.653	**0.722**	0.031	0.429	0.689	0.695
hepatic vessel	2	0.482	**0.721**	0.000	0.474	0.718	0.716
spleen	1	0.054	0.786	0.223	0.934	0.920	**0.953**
colon	1	0.022	0.717	0.074	0.695	0.735	**0.741**
Ranking		4.64	2.00	5.63	3.82	2.91	**1.72**
DC (avg)		0.514	0.799	0.151	0.732	0.741	**0.823**
BBIoU (avg)		0.446	0.770	0.099	0.686	0.691	**0.773**
HD95 (avg)		185.992	**4.121**	367.448	7.802	94.599	4.323

4   Results

We first evaluated the influence of different normalization methods on medical-image segmentation. We report the performance of neural architectures defined by our proposed evoNMS, which followed an evolutionary approach to determine the potentially best normalization method for each bottleneck layer, at generations 1 and 20. For each dataset, we evaluated the DC across different baselines of our default U-Net implementation with a fixed normalization method across layers (BN, IN, or NoN) and a state-of-the-art semantic segmentation network (nnU-Net) against our proposed evoNMS approach. Table 2 provides an overview of the mean validation DC as the main performance metric.

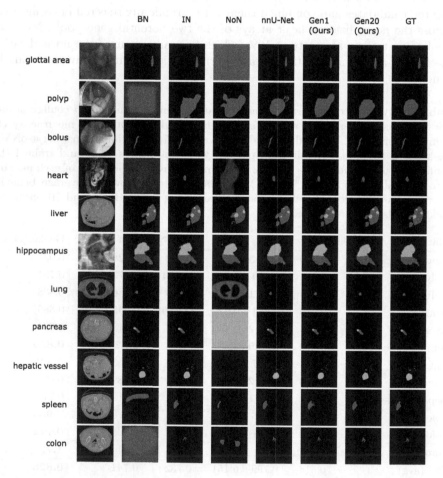

Fig. 2. Qualitative comparison of baseline architectures and the proposed evoNMS approach. Each horizontal set shows exemplary performance for a given dataset.

Overall, the architecture configurations with solely IN (six datasets), nnU-Net (two datasets), or our proposed evoNMS (one first generation, four in the last generation) achieved the highest mean validation DC across all datasets. Noteworthy, our evoNMS approach achieved competitive performance across all datasets, which is clearly shown at the ranking in Table 2, where the last generation of our approach achieved the best grade of 1.72. The results of our evoNMS demonstrate that our approach achieves superior performance in terms of the average DC and BBIoU scores on the validation dataset and yields comparable results in terms of HD95 to the U-Net trained with IN. In contrast, architectures that rely on BN or NoN consistently produced poor results due to exploding gradients across multiple datasets ([14]), questioning the broad application of BN. When comparing qualitative results of all baselines and evoNMS, we find that our approach accurately reflects the ground truth across datasets, especially after several generations of optimization (Fig. 2). We found that an evolutionary search without prior knowledge of the required hyperparameters and properties of the medical data can perform as well as, or in some cases, even better than, the best baseline of a U-net trained with IN or nnU-Net, showing the importance of normalization in semantic segmentation architectures.

We next were interested in the optimization behavior of evoNMS. We found that random initialization of normalization methods yielded poorly and highly variant converging behavior overall (Supplementary Fig. 1). However, after evolutionary optimization, the converging behavior is clearly improved in terms of convergence stability and lower variability. In Supplementary Fig. 2, we can exemplary show that evoNMS also improves all fitness-related metrics across generations. This highlights the ability of evoNMS to converge on multiple objectives. These findings suggest that our approach can also be used as a hyperparameter optimization problem to improve convergence behavior.

As we initialize the first population randomly, we determined whether the evoNMS algorithm converges to the same set of normalization methods for each dataset. We found for three independent evoNMS runs, that evoNMS converges for four out of eleven datasets on very similar patterns (Supplementary Fig. 3) across runs, with an overall average correlation of 36.3%, indicating that our algorithm is able to find relatively quickly a decent solution for a given dataset. In the case of the polyp dataset, top evoNMS-performing networks correlate with a striking 61.1% (Supplementary Table 1).

We next identified the final distribution of normalization methods across the encoder and decoder U-Net layers to determine dataset-specific normalization. In Fig. 3, we show the distribution of the top 10% performers in the last generation of evoNMS. We found consistent patterns across individuals especially in the last layer: in four out of eleven datasets, no normalization was preferred. In the colon dataset, the encoder was mainly dominated by IN, especially in the first encoding layer. In contrast, the decoder in the polyp, liver, and hippocampus datasets showed more consistency in the normalization methods suggesting that normalization methods could be encoder- and decoder-specific and are dataset dependent.

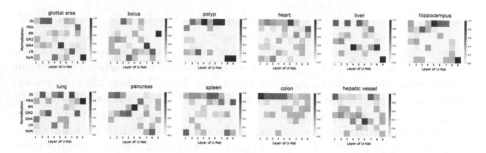

Fig. 3. Visualization of the relative frequency of selected normalizations over the U-Net layer for all datasets. This average was calculated for all individuals in the last generation of each dataset.

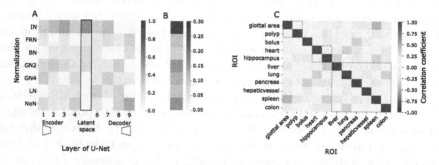

Fig. 4. Visualization of the best normalization pattern in the last generation and the selection for each individual. A) Distribution of normalizations across U-Net layers and datasets. B) Relative frequency of normalization patterns. C) shows the normalization pattern correlation of U-Net variants with the highest fitness F for each dataset in the last generation. The boxes indicated the same imaging modality.

To understand good normalization design across datasets in semantic segmentation tasks, we determined which normalization methods are more abundant at specific U-Net layers. IN is across layers a preferred choice by evoNMS except for the last two decoding layers (Fig. 4 A and B). Our results suggest that the latent space embedding heavily relies on IN (Fig. 4 A). Other normalization methods are less abundant in top-performing architectures, such as FRN, BN, and LN. However, NoN is mainly used in the last layer. FRN and especially BN seem to be inferior choices in semantic segmentation architectures.

Finally, we investigated if any set of normalization methods can be derived by the imaging modality, such as endoscopy, CT and MRI. In Fig. 4, we show the cross-correlation of all evoNMS top performers across datasets. In general, there are only weak correlations at the global level; a stronger correlation can be seen by correlating encoder and decoder separately (Supplementary Fig. 3). These results provide evidence, that *a priori* knowledge of the imaging modality does not hint towards a specific normalization pattern but rather has to be optimized for any given dataset.

5 Discussion and Conclusion

Our approach shows that using normalization methods wisely has a powerful impact on biomedical image segmentation across a variety of datasets acquired using different imaging modalities. Due to its inherent property of always finding an optimal solution, evoNMS is potentially capable of providing the best-performing set of normalization patterns for any given data set. For feasibility reasons, we considered only 20 epochs for each training set and 20 generations. However, we show that evoNMS with these constrained settings provides competitive results and outperforms or performs on par compared to all baselines.

Our results suggest the superior performance of IN and GN (Fig. 4) as overall very useful normalization strategies when used at their preferred location, in contrast to FRN, BN, and LN. State-of-the-art architectures, such as nnU-Net also rely on IN throughout the network [7], as well as evolutionary optimized U-Net architectures [19]. This established use of normalization confirms our findings for high-performing evoNMS networks and can be extrapolated to other semantic segmentation architectures that incorporate normalization methods. The advantage of evoNMS is its ability to also include non-learnable objectives, such as architectural scaling and reducing inference time, crucial for point-of-care solutions. In this study, we constrained the search space, but we will incorporate multiple hyperparameters in the future. On the other hand, approaches that use main building blocks and optimize mainly convolutional layer parameters and activation functions [15,19] would benefit from incorporating normalization strategies as well.

References

1. Antonelli, M., Reinke, A., Bakas, S., et al.: The medical segmentation decathlon. Nat. Commun. **13**(1), 4128 (2022). https://doi.org/10.1038/s41467-022-30695-9
2. Awais, M., Iqbal, M.T.B., Bae, S.H.: Revisiting internal covariate shift for batch normalization. IEEE Trans. Neural Networks Learn. Syst. **32**(11), 5082–5092 (2021). https://doi.org/10.1109/tnnls.2020.3026784
3. Ba, J.L., Kiros, J.R., Hinton, G.E.: Layer normalization (2016). https://doi.org/10.48550/ARXIV.1607.06450
4. Gómez, P., Kist, A.M., Schlegel, P., Berry, D.A., Chhetri, D.K., Dörr, S., Echternach, M., Johnson, A.M., Kniesburges, S., Kunduk, M., et al.: Bagls, a multihospital benchmark for automatic glottis segmentation. Sci. Data **7**(1), 186 (2020). https://doi.org/10.1038/s41597-020-0526-3
5. Huang, L., Qin, J., Zhou, Y., Zhu, F., Liu, L., Shao, L.: Normalization techniques in training DNNs: methodology, analysis and application (2020). https://doi.org/10.48550/ARXIV.2009.12836
6. Ioffe, S., Szegedy, C.: Batch normalization: accelerating deep network training by reducing internal covariate shift (2015). https://doi.org/10.48550/ARXIV.1502.03167
7. Isensee, F., Jaeger, P.F., Kohl, S.A.A., Petersen, J., Maier-Hein, K.H.: nnU-Net: a self-configuring method for deep learning-based biomedical image segmentation. Nat. Methods **18**(2), 203–211 (2020). https://doi.org/10.1038/s41592-020-01008-z

8. Jha, D., et al.: Kvasir-SEG: a segmented polyp dataset. In: Ro, Y.M., et al. (eds.) MMM 2020. LNCS, vol. 11962, pp. 451–462. Springer, Cham (2020). https://doi.org/10.1007/978-3-030-37734-2_37

9. Klambauer, G., Unterthiner, T., Mayr, A., Hochreiter, S.: Self-normalizing neural networks. In: Guyon, I., et al. (eds.) Advances in Neural Information Processing Systems, vol. 30. Curran Associates, Inc. (2017). https://proceedings.neurips.cc/paper/2017/file/5d44ee6f2c3f71b73125876103c8f6c4-Paper.pdf

10. Liu, C., et al.: Auto-DeepLab: hierarchical neural architecture search for semantic image segmentation, pp. 82–92 (2019). https://doi.org/10.1109/CVPR.2019.00017

11. Liu, H., Brock, A., Simonyan, K., Le, Q.: Evolving normalization-activation layers. **33**, 13539–13550 (2020). https://doi.org/10.48550/arXiv.2004.02967

12. Liu, Y., Sun, Y., Xue, B., Zhang, M., Yen, G.G., Tan, K.C.: A survey on evolutionary neural architecture search. IEEE Trans. Neural Networks Learn. Syst. **34**(2), 550–570 (2023). https://doi.org/10.1109/TNNLS.2021.3100554

13. Luo, P., Ren, J., Peng, Z., Zhang, R., Li, J.: Differentiable learning-to-normalize via switchable normalization (2018). https://doi.org/10.48550/ARXIV.1806.10779

14. Philipp, G., Song, D., Carbonell, J.G.: The exploding gradient problem demystified - definition, prevalence, impact, origin, tradeoffs, and solutions (2017). https://doi.org/10.48550/ARXIV.1712.05577

15. Popat, V., Mahdinejad, M., Cedeño, O., Naredo, E., Ryan, C.: GA-based U-Net architecture optimization applied to retina blood vessel segmentation. In: Proceedings of the 12th International Joint Conference on Computational Intelligence. SCITEPRESS - Science and Technology Publications (2020). https://doi.org/10.5220/0010112201920199

16. Romijnders, R., Meletis, P., Dubbelman, G.: A domain agnostic normalization layer for unsupervised adversarial domain adaptation, pp. 1866–1875, January 2019. https://doi.org/10.1109/WACV.2019.00203

17. Ronneberger, O., Fischer, P., Brox, T.: U-Net: Convolutional Networks for Biomedical Image Segmentation. In: Navab, N., Hornegger, J., Wells, W.M., Frangi, A.F. (eds.) MICCAI 2015. LNCS, vol. 9351, pp. 234–241. Springer, Cham (2015). https://doi.org/10.1007/978-3-319-24574-4_28

18. Singh, S., Krishnan, S.: Filter response normalization layer: eliminating batch dependence in the training of deep neural networks (2019). https://doi.org/10.48550/ARXIV.1911.09737

19. Wei, J., et al.: Genetic U-Net: automatically designed deep networks for retinal vessel segmentation using a genetic algorithm. IEEE Trans. Med. Imaging **41**(2), 292–307 (2022). https://doi.org/10.1109/TMI.2021.3111679

20. Wu, Y., He, K.: Group normalization. In: Ferrari, V., Hebert, M., Sminchisescu, C., Weiss, Y. (eds.) ECCV 2018. LNCS, vol. 11217, pp. 3–19. Springer, Cham (2018). https://doi.org/10.1007/978-3-030-01261-8_1

21. Zhou, X.Y., Yang, G.Z.: Normalization in training U-Net for 2-D biomedical semantic segmentation. IEEE Robot. Autom. Lett. **4**(2), 1792–1799 (2019). https://doi.org/10.1109/lra.2019.2896518

22. Zhou, Z., Qi, L., Yang, X., Ni, D., Shi, Y.: Generalizable cross-modality medical image segmentation via style augmentation and dual normalization. In: 2022 IEEE/CVF Conference on Computer Vision and Pattern Recognition (CVPR), pp. 20824–20833 (2022). https://doi.org/10.1109/CVPR52688.2022.02019

DOMINO++: Domain-Aware Loss Regularization for Deep Learning Generalizability

Skylar E. Stolte[1], Kyle Volle[2], Aprinda Indahlastari[3,4], Alejandro Albizu[3,5], Adam J. Woods[3,4,5], Kevin Brink[6], Matthew Hale[9], and Ruogu Fang[1,3,7,8(✉)]

[1] J. Crayton Pruitt Family Department of Biomedical Engineering, Herbert Wertheim College of Engineering, University of Florida (UF), Gainesville, USA
ruogu.fang@ufl.edu
[2] Torch Technologies, LLC, Shalimar, FL, USA
[3] Center for Cognitive Aging and Memory, McKnight Brain Institute, UF, Gainesville, USA
[4] Department of Clinical and Health Psychology, College of Public Health and Health Professions, UF, Gainesville, USA
[5] Department of Neuroscience, College of Medicine, UF, Gainesville, USA
[6] United States Air Force Research Laboratory, Eglin Air Force Base, Valparaiso, FL, USA
[7] Department of Electrical and Computer Engineering, Herbert Wertheim College of Engineering, UF, Gainesville, USA
[8] Department of Computer Information and Science and Engineering, Herbert Wertheim College of Engineering, UF, Gainesville, USA
[9] Department of Mechanical and Aerospace Engineering, Herbert Wertheim College of Engineering, UF, Gainesville, USA

Abstract. Out-of-distribution (OOD) generalization poses a serious challenge for modern deep learning (DL). OOD data consists of test data that is significantly different from the model's training data. DL models that perform well on in-domain test data could struggle on OOD data. Overcoming this discrepancy is essential to the reliable deployment of DL. Proper model calibration decreases the number of spurious connections that are made between model features and class outputs. Hence, calibrated DL can improve OOD generalization by only learning features that are truly indicative of the respective classes. Previous work proposed domain-aware model calibration (DOMINO) to improve DL calibration, but it lacks designs for model generalizability to OOD data. In this work, we propose DOMINO++, a dual-guidance and dynamic domain-aware loss regularization focused on OOD generalizability. DOMINO++ integrates expert-guided and data-guided knowledge in its regularization. Unlike DOMINO which imposed a fixed scaling and regularization rate, DOMINO++ designs a dynamic scaling factor and an adaptive regularization rate. Comprehensive evaluations compare DOMINO++ with DOMINO and the baseline model for head tissue segmentation from magnetic resonance images

Supplementary Information The online version contains supplementary material available at https://doi.org/10.1007/978-3-031-43901-8_68.

(MRIs) on OOD data. The OOD data consists of synthetic noisy and rotated datasets, as well as real data using a different MRI scanner from a separate site. DOMINO++'s superior performance demonstrates its potential to improve the trustworthy deployment of DL on real clinical data.

Keywords: Image Segmentation · Machine Learning Uncertainty · Model Calibration · Model Generalizability · Whole Head MRI

1   Introduction

Large open-access medical datasets are integral to the future of deep learning (DL) in medicine because they provide much-needed training data and a method of public comparison between researchers [20]. Researchers often curate their data for DL models; yet, even the selection process itself may contain inherent biases, confounding factors, and other "hidden" issues that cause failure on real clinical data [1]. In DL, out-of-distribution (OOD) generalizability refers to a model's ability to maintain its performance on data that is independent of the model's development [23]. OOD generalizability represents a critical issue in DL research since the point of artificial intelligence (AI) in medicine is to be capable of handling new patient cases. However, this important aspect of DL is often not considered. On the other hand, overcoming challenges such as scanner-induced variance are critical in the success of neuroimaging studies involving AI [5].

We hypothesize that adaptable domain-aware model calibration that combines expert-level and data-level knowledge can effectively generalize to OOD data. DL calibration is correlated with better OOD generalizability [21]. Calibrated models may accomplish this by learning less spurious connections between features and classes. This observation relates to how calibrated models reflect the true likelihood of a data point for a class. A calibrated model may let a confusing data point naturally lay closer to the class boundaries, rather than forcing tight decision boundaries that over-fit points. Calibration affects decision-making such that the models can better detect and handle OOD data [19].

In this work, we introduce DOMINO++, an adaptable regularization framework to calibrate DL models based on expert-guided and data-guided knowledge. DOMINO++ builds on the work DOMINO [18] with three important contributions: 1) combining expert-guided and data-guided regularization to fully exert the domain-aware regularization's potential. 2) Instead of using static scaling, DOMINO++ dynamically brings the domain-aware regularization term to the same order of magnitude as the base loss across epochs. 3) DOMINO++ adopts an adaptive regularization scheme by weighing the domain-aware regularization term in a progressive fashion. The strengths of DOMINO++'s regularization lie in its ability to take advantage of the benefits of both the semantic confusability derived from domain knowledge and data distribution, as well as its adaptive balance between the data term and the regularization strength. This work shows the advantages of DOMINO++ in a segmentation task from magnetic resonance (MR) images. DOMINO++ is tested in OOD datasets including synthesized noise additions, synthesized rotations, and a different MR scanner.

2   Dynamic Framework for DL Regularization

2.1   DL Backbone

U-Net transformer (UNETR) [10] serves as the DL backbone. UNETR is inspired by the awe-inspiring results of transformer modules in Natural Language Processing [22]. These modules use self-attention-based mechanisms to learn language range sequences better than traditional fully convolutional networks (FCNs). UNETR employs a transformer module as its encoder, whereas its decoder is an FCN like in the standard U-Net. This architecture learns three-dimensional (3D) volumes as sequences of one-dimensional (1D) patches. The FCN decoder receives the transformer's global information via skip connections and concatenates this information with local context that eventually recovers the original image dimensions. The baseline model does not include advanced calibration. However, basic principles to improve OOD generalizability are still incorporated for a more meaningful comparison. These principles include standard data augmentations like random Gaussian noise, rotations along each axis, and cropping. The model includes 12 attention heads and a feature size of 16.

2.2   DOMINO++ Loss Regularization

Derivation. The original DOMINO's loss regularization is as follows:

$$\mathcal{L}(y, \hat{y}) + \beta y^T W \hat{y}, \text{where } W = W_{HC} \text{ or } W_{CM} \tag{1}$$

where $\mathcal{L}$ can be any uncalibrated loss function (e.g., $DiceCE$ which is a hybrid of cross-entropy and Dice score [12]). y and $\hat{y}$ are the true labels and model output scores, respectively. β is an empirical static regularization rate that ranges between 0–1, and s is a pre-determined fixed scaling factor to balance the data term and the regularization term. The penalty matrix W has dimensions $N \times N$, where N is the number of classes. W_{HC} and W_{CM} represent the hierarchical clustering (HC)-based and confusion matrix (CM)-based penalty matrices.

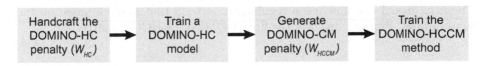

Fig. 1. The flowchart for the DOMINO++-HCCM pipeline

We improved its loss function to DOMINO++'s dual-guidance penalty matrix with adaptive scaling and regularization rate as follows:

$$(1 - \beta)\mathcal{L}(y, \hat{y}) + \beta y^T (s W_{HCCM})\hat{y} \tag{2}$$

where β dynamically changes over epochs. s is adaptively updated to balance the data and regularization terms. W_{HCCM} is the dual-guidance penalty matrix.

Combining Expert-Guided and Data-Guided Regularization. DOMINO-HC regularizes classes by arranging them into hierarchical groupings based on domain. DOMINO-HC is data-independent and thus immune to noise. Yet, it becomes less useful without clear hierarchical groups. DOMINO-CM calculates class penalties using the performance of an uncalibrated model on a held-out dataset. The CM method does not require domain knowledge, but it can be more susceptible to messy data. Overall, DOMINO-HC is expert-crafted and DOMINO-CM is data-driven. These approaches have complementary advantages and both perform very well on medical image segmentation [18]. Hence, this work combines these methods to learn from experts and data. Fig. 1 shows the process for creating this matrix term.

The combined regulation (a.k.a. DOMINO-HCCM) requires first replicating DOMINO-HC. For this step, we recreate the exact hierarchical groupings from the DOMINO paper [18]. A confusion matrix is generated using DOMINO-HC on an additional validation set for matrix penalty. Next, the confusion matrix is normalized by the number of true pixels in each class. The normalized terms are subtracted from the identity matrix. Finally, all diagonals are set to 0's. Next, a second DL model trains using the resulting penalty matrix in its regularization. This process differs from DOMINO-CM because DOMINO-HC was used to generate the final model's matrix penalty. The uncalibrated model may produce a matrix penalty that is susceptible to variable quality depending on the model's training data. In comparison, the initial regularization term adds an inductive bias in the first model that encodes more desirable qualities about the class mappings [13]. Namely, the initial model contains information about the hierarchical class groupings that drives the generation of the second model's matrix penalty. The final model can now use a regularization term that is more based on task than dataset. Figure 2 displays the final DOMINO++-HCCM matrix.

Dynamic Scaling Term. DOMINO++ adds a domain-aware regularization term to any standard loss. The resulting loss function combines the standard loss's goal of increasing accuracy with DOMINO++'s goal of reweighing the importance of different class mix-ups when incorrect. DL models are at risk of being dominated by a specific loss during training if the losses are of different scales [14]. DOMINO [18] neglects to account for this and provides a static scaling for the regularization term based on the first epoch standard loss. In comparison, DOMINO++ updates the scaling on the regularization term to be within the same scale as the current epoch standard loss. Specifically, the scaling is computed based on the closest value to the baseline loss on the log scale. For example, an epoch with $\mathcal{L} = 13$ will have a scaling factor $S = 10$.

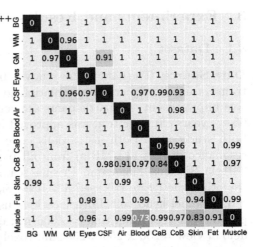

Fig. 2. Raw matrix penalty (W) for the combined method DOMINO-HCCM. Abbreviations - BG: Background, WM: White Matter, GM: Grey Matter, CSF: Cerebrospinal Fluid, CaB: Cancellous Bone, CoB: Cortical Bone.

2.3 Adaptive Regularization Weighting

Multiple loss functions must be balanced properly [6]. Most studies linearly balance the separate loss terms using hyper-parameters. Hyper-parameter selection is nontrivial and can greatly alter performance. Indeed, the timing of regularization during training is critical to the final performance [8]. Hence, the current work investigates the role of regularization timing in the final model performance. Equation 2 is similar to the original DOMINO equation [18]; however, the equation is modified to include a weighting term (e.g., $1 - \beta$) on the standard loss function. In DOMINO, the β term was simply set at a constant value of 1. As shown in Eq. 3, DOMINO++ weighs the loss regularization to decay (β) across epochs, while the standard loss is scaled reversely with regard to the regularization term (see Eq. 2).

$$\beta = 1 - \frac{CurrentEpoch}{TotalEpochs} \tag{3}$$

3   Experiments and Results

3.1   Dataset

Data Source. The data in this study is from a Phase III clinical trial that tests transcranial direct current stimulation to augment cognitive training for cognitive improvement. All participants are cognitively healthy older adults between 65–89 years old. The trial was approved by the Institutional Review Boards at both study sites. Both institutions collected structural T1-weighted magnetic resonance images (T1-MRIs) from all participants. One site ("Site A") used a 3-Tesla Siemens Magnetom Prisma scanner with a 64-channel head coil and the other site ("Site B") used a 3-Tesla Siemens Magnetom Skyra scanner with a 32-channel head coil. Both locations used the following MPRAGE sequence parameters: repetition time = 1800 ms; echo time = 2.26 ms; flip angle = 8°; field of view = $256 \times 256 \times 256$ mm; voxel size = 1 mm^3. The proper permissions were received for use of this dataset in this work. A total of 133 participants were included, including 123 from Site A and 10 participants from Site B.

Reference Segmentations. The T1 MRIs are segmented into 11 different tissues, which include grey matter (GM), white matter (WM), cerebrospinal fluid (CSF), eyes, muscle, cancellous bone, cortical bone, skin, fat, major artery (blood), and air. Trained labelers performed a combination of automated segmentation and manual correction. Initially, base segmentations for WM, GM, and bone are obtained using Headreco [16], while air is generated in the Statistical Parametric Mapping toolbox [2]. Afterward, these automated outputs are manually corrected using ScanIP Simpleware™. Thresholding and morphological operations are employed to differentiate between the bone compartments. Eyes, muscle, skin, fat, and blood are manually segmented in Simpleware. Finally, CSF is generated by subtracting the ten other tissues from the whole head.

Out-of-Domain (OOD) Testing Data Most DL work selects a testing set by splitting a larger dataset into training and testing participants. This work also incorporates "messy" or fully independent data. Thus, three additional testing datasets are used along with the traditional testing data (Site A - Clean).

Site A Noisy - MRI noise may be approximated as Gaussian for a signal-to-noise ratio (SNR) greater than 2 [9]. Therefore, this work simulates noisy MRI images using Gaussian noise of 0 mean with a variance of 0.01.

Site A Rotated - Rotated MRI data simulates other further disturbances or irregularities (e.g., head tilting) during scanning. The rotation dataset includes random rotation of 5- to 45° clockwise or counter-clockwise with respect to each 3D axis. The rotation angles are based on realistic scanner rotation [15].

Site B - Site A uses a 64-channel head coil and Site B uses a 32-channel head coil. The maximum theoretical SNR of an MRI increases with the number of channels [17]. Hence, this work seeks to test the performance of a model trained exclusively on a higher channel scanner on a lower channel testing dataset. Thus, the Site A data serves as the exclusive source of the training and validation data, and Site B serves as a unique and independent testing dataset.

3.2 Implementation Details

This study implements UNETR using the Medical Open Network for Artificial Intelligence (MONAI-1.1.0) in Pytorch 1.10.0 [4]. The Site A data is split from 123 MRIs into 93 training/10 validation/10 held-out validation (matrix penalty)/10 testing. 10 images from Site B serve as an additional testing dataset. Each DL model requires 1 GPU, 4 CPUs, and 30 GB of memory. Each model is trained for 25,000 iterations with evaluation at 500 intervals. The models are trained on $256 \times 256 \times 256$ images with batch sizes of 1 image. The optimization consists Adam optimization using stochastic gradient descent. All models segment a single head in 3–4 s during inference.

3.3 Analysis Approach

This section compares the results of 11 tissue head segmentation on each of the datasets using the baseline model, the best performing DOMINO approach, and the best performing DOMINO++ approach. The results are evaluated using Dice score [3] and Hausdorff Distance [7,11]. The Dice score represents the overlap of the model outputs with the true labels. It is better when greater and is optimally 1. Hausdorff distance represents the distance between the model outputs with the true labels. It is better when lesser and is optimally 0.

3.4 Segmentation Results

Qualitative Comparisons. Figure 3 features segmentation of one example slice from the Site A MRI dataset with Gaussian Noise. DOMINO substantially exaggerates the blood regions in this slice. In addition, DOMINO entirely misses a section of white matter near the eyes. However, DOMINO can also capture certain regions of the white matter, particularly in the back of the head, better than the baseline model. In general, all outputs have noisy regions where there appear to be "specks" of an erroneous tissue. For instance, grey matter is incorrectly identified as specks within the white matter. This issue is far more common in the DOMINO output compared to the baseline or DOMINO++ outputs.

Quantitative Comparisons. Tables 1 and 2 show that DOMINO++ achieves the best Dice scores and Hausdorff Distances across all test sets, respectively. As such, DOMINO++ produces the most accurate overall segmentation across tissue types. The supplementary material provides individual results across every dataset and tissue type. So far, DOMINO++ improves the model generalizability to the noisy and rotated datasets the most. These improvements are important in combating realistic MR issues such as motion artifacts. Future work will

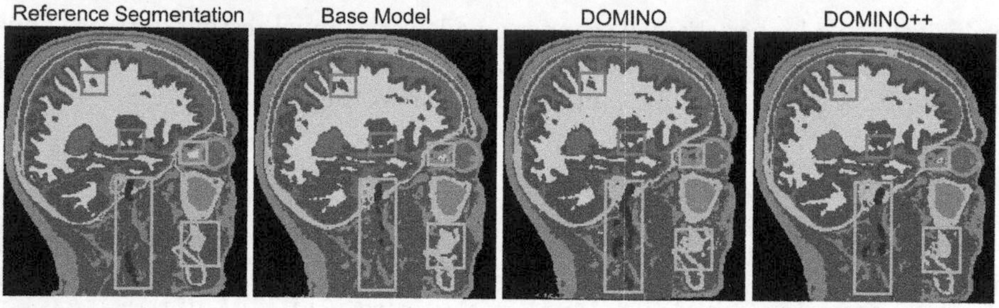

Fig. 3. Visual comparison of segmentation performance on a noisy MRI image from Site A. The yellow rectangular regions show areas where DOMINO++ improves the segmentation. The orange regions show areas that DOMINO and DOMION++ improve the segmentation over the baseline model. (Color figure online)

Table 1. Average Dice Scores. The data is written as mean ± standard deviation. * Denotes Significance using multiple comparisons tests.

Method	Site A clean	Site A noisy	Site A rotated	Site B
Base	0.808 ± 0.014	0.781 ± 0.015	0.727 ± 0.041	0.730 ± 0.028
DOMINO	0.826 ± 0.014	0.791 ± 0.018	0.777 ± 0.023	0.750 ± 0.026
DOMINO++	**0.842 ± 0.012***	**0.812 ± 0.016***	**0.789 ± 0.23***	**0.765 ± 0.027**

Table 2. Average Hausdorff Distances. The data is written as mean ± standard deviation. * Denotes Significance using multiple comparisons tests.

Method	Site A clean	Site A noisy	Site A rotated	Site B
Base	0.651 ± 0.116	0.669 ± 0.085	2.266 ± 1.373	1.699 ± 0.414
DOMINO	0.525 ± 0.090	0.565 ± 0.149	1.284 ± 0.500	1.782 ± 0.669
DOMINO++	**0.461 ± 0.077***	**0.457 ± 0.076***	**1.185 ± 0.411***	**1.228 ± 0.414**

build off of DOMINO++'s improvements on different scanner data to yield even better results. Table 3 displays the Hausdorff Distances for every tissue across Site B's test data. Site B is highlighted since that this is real-world OOD data. DOMINO++ performs better in most tissues and the overall segmentation. GM, cortical bone, and blood show the most significant differences with DOMINO++. This is highly relevant to T1 MRI segmentation. Bone is difficult to differentiate from CSF with only T1 scans due to similar contrast. Available automated segmentation tools use young adult heads as reference, whereas the bone structure between older and younger adults is very different (e.g., more porous in older adults). Hence, DOMINO++ is an important step in developing automated segmentation tools that are better suited for older adult heads.

Table 3. Hausdorff Distances on Site B data. The data is written as mean ± standard deviation. * Denotes Significance using multiple comparisons tests. Abbreviations - CaB: Cancellous Bone. CoB: Cortical Bone.

Tissue	WM	GM	Eyes	CSF	Air	Blood	CaB	CoB	Skin	Fat	Muscle
Base	0.215± 0.063	0.266± 0.107	1.089± 0.881	0.506± 0.226	2.281± 1.426	8.534± 2.564	1.581± 0.626	2.203± 1.279	0.610± 0.424	**0.958± 0.751**	0.363± 0.159
DOMINO	0.260± 0.117	0.221± 0.048	1.600± 4.108	0.564± 0.198	2.070± 1.380	9.934± 4.025	**1.456± 0.672**	1.331± 0.827	0.811± 0.838	1.040± 0.987	0.320± 0.234
DOMINO++	**0.189± 0.042**	**0.171± 0.036***	**0.149± 0.047**	**0.462± 0.106**	**1.446± 1.057**	**6.260± 2.195***	1.996± 0.960	**0.950± 0.705***	**0.570± 0.369**	1.060± 0.935	**0.308± 0.165**

Ablation Testing. The supplementary material provides the results of ablation testing on DOMINO++. These results compare how W_{HCCM}, s, and β individually contribute to the results. Interestingly, different individual terms cause the model to perform stronger in specific datasets. Yet, the combined DOMINO++ still performs the best across the majority of datasets and metrics. These observations suggest that each term has strengths on different data types that can strengthen the overall performance.

Training Time Analysis. DOMINO-HC took about 12 h to train whereas DOMINO-CM and DOMINO++ took about 24 h to train. All models took 3–4 s per MR volume at the inference time. A task that has very clear hierarchical groups may still favor DOMINO-HC for the convenient training time. This might include a task with well-documented taxonomic levels (e.g., animal classification). However, medical data is often not as clear, which is why models that can learn from the data are valuable. DOMINO++ makes up for the longer training time by learning more specific class similarities from the data. Tasks that benefit from DOMINO++ over DOMINO-HC are those that only have loosely-defined categories. Tissue segmentation falls under this domain because tissues largely occur in similar anatomical locations (strength of DOMINO-HC) but the overall process is still variable with individual heads (strength of DOMINO-CM).

4   Conclusions

DOMINO [18] established a framework for calibrating DL models using the semantic confusability and hierarchical similarity between classes. In this work, we proposed the DOMINO++ model which builds upon DOMINO's framework with important novel contributions: the integration of data-guided and expert-guided knowledge, better adaptability, and dynamic learning. DOMINO++ surpasses the equivalent uncalibrated DL model and DOMINO in 11-tissue segmentation on both standard and OOD datasets. OOD data is unavoidable and remains a pivotal challenge for the use of artificial intelligence in clinics, where there is great variability between different treatment sites and patient populations. Overall, this work indicates that DOMINO++ has great potential to improve the trustworthiness and reliability of DL models in real-life clinical data.

We will release DOMINO++ code to the community to support open science research.

Acknowledgements. This work was supported by the National Institutes of Health/National Institute on Aging, USA (NIA RF1AG071469, NIA R01AG054077), the National Science Foundation, USA (1908299), the Air Force Research Laboratory Munitions Directorate, USA (FA8651-08-D-0108 TO48), and NSF-AFRL INTERN Supplement to NSF IIS-1908299, USA.

References

1. Arjovsky, M., Bottou, L., Gulrajani, I., Lopez-Paz, D.: Invariant risk minimization (2019). https://doi.org/10.48550/ARXIV.1907.02893, https://arxiv.org/abs/1907.02893
2. Ashburner, J.: SPM: a history. Neuroimage **62**(2), 791–800 (2012)
3. Bertels, J., et al.: Optimizing the dice score and Jaccard index for medical image segmentation: theory and practice. In: Shen, D., et al. (eds.) MICCAI 2019. LNCS, vol. 11765, pp. 92–100. Springer, Cham (2019). https://doi.org/10.1007/978-3-030-32245-8_11
4. Consortium, M.: MONAI: Medical open network for AI, March 2020. https://doi.org/10.5281/zenodo.6114127, If you use this software, please cite it using these metadata
5. Dinsdale, N.K., Bluemke, E., Sundaresan, V., Jenkinson, M., Smith, S.M., Namburete, A.I.: Challenges for machine learning in clinical translation of big data imaging studies. Neuron **110**, 3866–3881 (2022)
6. Dosovitskiy, A., Djolonga, J.: You only train once: loss-conditional training of deep networks. In: International Conference on Learning Representations (2020)
7. Dubuisson, M.P., Jain, A.K.: A modified Hausdorff distance for object matching. In: Proceedings of 12th International Conference on Pattern Recognition, vol. 1, pp. 566–568. IEEE (1994)
8. Golatkar, A.S., Achille, A., Soatto, S.: Time matters in regularizing deep networks: weight decay and data augmentation affect early learning dynamics, matter little near convergence. In: Advances in Neural Information Processing Systems, vol. 32 (2019)
9. Gudbjartsson, H., Patz, S.: The Rician distribution of noisy MRI data. Magn. Reson. Med. **34**(6), 910–914 (1995)
10. Hatamizadeh, A., et al.: UNETR: transformers for 3D medical image segmentation. In: Proceedings of the IEEE/CVF Winter Conference on Applications of Computer Vision, pp. 574–584 (2022)
11. Huttenlocher, D.P., Klanderman, G.A., Rucklidge, W.J.: Comparing images using the Hausdorff distance. IEEE Trans. Pattern Anal. Mach. Intell. **15**(9), 850–863 (1993)
12. Jadon, S.: A survey of loss functions for semantic segmentation. In: 2020 IEEE Conference on Computational Intelligence in Bioinformatics and Computational Biology (CIBCB), pp. 1–7. IEEE (2020)
13. Kukačka, J., Golkov, V., Cremers, D.: Regularization for deep learning: a taxonomy. arXiv preprint arXiv:1710.10686 (2017)
14. Lee, J.H., Lee, C., Kim, C.S.: Learning multiple pixelwise tasks based on loss scale balancing. In: Proceedings of the IEEE/CVF International Conference on Computer Vision, pp. 5107–5116 (2021)

15. Runge, V.M., Osborne, M.A., Wood, M.L., Wolpert, S.M., Kwan, E., Kaufman, D.M.: The efficacy of tilted axial MRI of the CNS. Magn. Reson. Imaging **5**(6), 421–430 (1987)
16. Saturnino, G.B., Puonti, O., Nielsen, J.D., Antonenko, D., Madsen, K.H., Thielscher, A.: SimNIBS 2.1: a comprehensive pipeline for individualized electric field modelling for transcranial brain stimulation. In: Makarov, S., Horner, M., Noetscher, G. (eds.) Brain and Human Body Modeling, pp. 3–25. Springer, Cham (2019). https://doi.org/10.1007/978-3-030-21293-3_1
17. Sobol, W.T.: Recent advances in MRI technology: implications for image quality and patient safety. Saudi J. Ophthalmol. **26**(4), 393–399 (2012)
18. Stolte, S.E., et al.: DOMINO: domain-aware model calibration in medical image segmentation. In: Wang, L., Dou, Q., Fletcher, P.T., Speidel, S., Li, S. (eds.) Medical Image Computing and Computer Assisted Intervention-MICCAI 2022: 25th International Conference, Singapore, 18–22 September 2022, Proceedings, Part V, pp. 454–463. Springer, Cham (2022). https://doi.org/10.1007/978-3-031-16443-9_44
19. Tom, G., Hickman, R.J., Zinzuwadia, A., Mohajeri, A., Sanchez-Lengeling, B., Aspuru-Guzik, A.: Calibration and generalizability of probabilistic models on low-data chemical datasets with DIONYSUS. arXiv preprint arXiv:2212.01574 (2022)
20. Torralba, A., Efros, A.A.: Unbiased look at dataset bias. In: CVPR 2011, pp. 1521–1528 (2011). https://doi.org/10.1109/CVPR.2011.5995347
21. Wald, Y., Feder, A., Greenfeld, D., Shalit, U.: On calibration and out-of-domain generalization. Adv. Neural. Inf. Process. Syst. **34**, 2215–2227 (2021)
22. Wolf, T., et al.: HuggingFace's transformers: state-of-the-art natural language processing (2020)
23. Yang, J., Soltan, A.A., Clifton, D.A.: Machine learning generalizability across healthcare settings: insights from multi-site COVID-19 screening. npj Digit. Med. **5**(1), 69 (2022)

Ariadne's Thread: Using Text Prompts to Improve Segmentation of Infected Areas from Chest X-ray Images

Yi Zhong, Mengqiu Xu, Kongming Liang, Kaixin Chen, and Ming Wu[✉]

Beijing University of Posts and Telecommunications, Beijing, China
{xiliang2017,xumengqiu,liangkongming,chenkaixin,wuming}@bupt.edu.cn

Abstract. Segmentation of the infected areas of the lung is essential for quantifying the severity of lung disease like pulmonary infections. Existing medical image segmentation methods are almost uni-modal methods based on image. However, these image-only methods tend to produce inaccurate results unless trained with large amounts of annotated data. To overcome this challenge, we propose a language-driven segmentation method that uses text prompt to improve to the segmentation result. Experiments on the QaTa-COV19 dataset indicate that our method improves the Dice score by 6.09% at least compared to the uni-modal methods. Besides, our extended study reveals the flexibility of multi-modal methods in terms of the information granularity of text and demonstrates that multi-modal methods have a significant advantage over image-only methods in terms of the size of training data required.

Keywords: Multi-modal Learning · Medical Image Segmentation

1 Introduction

Radiology plays an important role in the diagnosis of some pulmonary infectious diseases, such as the COVID-19 pneumonia outbreak in late 2019 [1]. With the development of deep learning, deep neural networks are more and more used to process radiological images for assisted diagnosis, such as disease classification, lesion detection and segmentation, etc. With the fast processing of radiological images by deep neural networks, some diagnoses can be obtained immediately, such as the classification of bacterial or viral pneumonia and the segmentation mask for pulmonary infections, which is important for quantifying the severity of the disease as well as its progression [2]. Besides, these diagnoses given by the AI allow doctors to predict risks and prognostics in a "patient-specific" way [3]. Radiologists usually take more time to complete lesion annotation than AI, and annotation results can be influenced by individual bias and clinical experience [4].

Ariadne's thread, the name comes from ancient Greek myth, tells of Theseus walking out of the labyrinth with the help of Ariadne's golden thread.

© The Author(s), under exclusive license to Springer Nature Switzerland AG 2023
H. Greenspan et al. (Eds.): MICCAI 2023, LNCS 14223, pp. 724–733, 2023.
https://doi.org/10.1007/978-3-031-43901-8_69

Therefore, it is of importance to design automatic medical image segmentation algorithms to assist clinicians in developing accurate and fast treatment plans.

Most of the biomedical segmentation methods [5–9] are improved based on U-Net [10]. However, the performance of these image-only methods is constrained by the training data, which is also a dilemma in the medical image field. Radford et al. proposed CLIP [11] in 2021, where they used 4M image-text pairs for contrastive learning. With the rise of multi-modal learning in the recent years, there are also methods [12–16] that focus on vision-language pretraining/processing and applying them on local tasks. Li et al. proposed a language-driven medical image segmentation method LViT [16], using a hybrid CNN-Transformer structure to fuse text and image features. However, LViT uses an early fusion approach and the information contained in the text is not well represented. In this paper, we propose a multi-modal segmentation method that using independent text encoder and image encoder, and design a GuideDecoder to fuse the features of both modalities at decoding stage. Our main contributions are summarized as follow:

- We propose a language-driven segmentation method for segmenting infected areas from lung x-ray images. Source code of our method see: https://github.com/Junelin2333/LanGuideMedSeg-MICCAI2023
- The designed GuideDecoder in our method can adaptively propagate sufficient semantic information of the text prompts into pixel-level visual features, promoting consistency between two modalities.
- We have cleaned the errors contained in the text annotations of QaTa-COV19 [17] and contacted the authors of LViT to release a new version.
- Our extended study reveals the impact of information granularity in text prompts on the segmentation performance of our method, and demonstrates the significant advantage of multi-modal method over image-only methods in terms of the size of training data required.

2   Method

The overview of our proposed method is shown in Fig. 1(a). The model consists of three main components: Image Encoder, Text Encoder and GuideDecoder that enables multi-modal information fusion. As you can see, our proposed method uses a modular design. Compared to early stage fusion in LViT, our proposed method in modular design is more flexible. For example, when our method is used for brain MRI images, thanks to the modular design, we could first load pre-trained weights trained on the corresponding data to separate visual and text encoders, and then only need to train GuideDecoders.

Visual Encoder and Text Encoder. The Visual Encoder used in the model is ConvNeXt-Tiny [18]. For an input image $I \in \mathbb{R}^{H \times W \times 1}$, we extract multiple visual features from the four stages of ConvNeXt-Tiny, which are defined as $f_4 \in \mathbb{R}^{\frac{H}{4} \times \frac{W}{4} \times C_1}$, $f_8 \in \mathbb{R}^{\frac{H}{8} \times \frac{W}{8} \times C_2}$, $f_{16} \in \mathbb{R}^{\frac{H}{16} \times \frac{W}{16} \times C_3}$ and $f_{32} \in \mathbb{R}^{\frac{H}{32} \times \frac{W}{32} \times C_4}$,

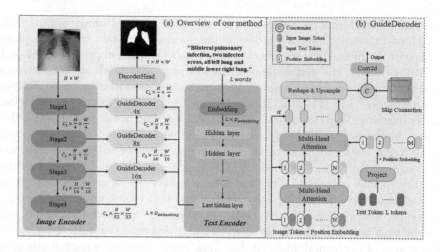

Fig. 1. The overview of the our proposed method (a) and the detail of the GuideDecoder in out method. Our proposed approach uses a modular design where the model consists mainly of an image encoder, a text encoder and several GuideDecoders. The GuideDecoder is used to adaptively propagate semantic information from textual features to visual features and output decoded visual features.

Note that C is the feature dimension, H and W are the height and width of the original image. For an input text prompt $T \in \mathbb{R}^L$, We adopt the CXR-BERT [19] to extract text features $g_t \in \mathbb{R}^{L \times C}$. Note that C is the feature dimension, L is the length of the text prompt.

GuideDecoder. Due to our modular design, visual features and textual features are encoded independently by different encoders. Therefore, the design of the decoder is particularly important, as we can only fuse multi-modal features from different encoders in post stage. The structure of GuideDecoder is shown in Fig. 1(b). The GuideDecoder first processes the input textual features and visual features before performing multi-modal interaction.

The input textual features first go through a projection module (i.e. Project in the figure) that aligns the dimensionality of the text token with that of the image token and reduces the number of text tokens. The projection process is shown in Eq. 1.

$$f_t = \sigma(Conv(TW_T)) \tag{1}$$

where W_T is a learnable matrix, $Conv(\cdot)$ denotes a 1×1 convolution layer, and $\sigma(\cdot)$ denotes the ReLU activation function. Given an input feature $T \in \mathbb{R}^{L \times D}$, the output projected features is $f_t \in \mathbb{R}^{M \times C_1}$, where M is the number of tokens after projection and C_1 is the dimension of the projected features, consistent with the dimension of the image token.

For the input visual features $I \in \mathbb{R}^{H \times W \times C_1}$, after adding the position encoding we use self-attention to enhance the visual information in them to obtain the evolved visual features. The process is shown in Eq. 2.

$$f_i = I + LN(MHSA(I)) \tag{2}$$

where $MHSA(\cdot)$ denotes Multi-Head Self-Attention layer, $LN(\cdot)$ denotes Layer Normalization, and finally the evolved visual features $f_i \in \mathbb{R}^{H \times W \times C_1}$ with residuals could be obtained.

After those, the multi-head cross-attention layer is adopted to propagate fine-grained semantic information into the evolved image features. To obtain the multi-modal feature $f_c \in \mathbb{R}^{H \times W \times C_1}$, the output further computed by layer normalization and residual connection:

$$f_c = f_i + \alpha(LN(MHCA(f_i, f_t))) \tag{3}$$

where $MHCA(\cdot)$ denotes multi-head cross-attention and α is a learnable parameter to control the weight of the residual connection.

Then, the multi-modal feature $f_c \in \mathbb{R}^{(H \times W) \times C_1}$ would be reshaped and upsampling to obtain $f_c' \in \mathbb{R}^{H' \times W' \times C_1}$. Finally the f_c' is concatenated with $f_s \in \mathbb{R}^{H' \times W' \times C_2}$ on the channel dimension, where f_s is the low-level visual feature obtained from visual encoder via skip connection. The concatenated features are processed through a convolution layer and a ReLU activation function to obtain the final decoded output $f_o \in \mathbb{R}^{H' \times W' \times C_2}$

$$f_c' = Upsample(Reshape(f_c))$$
$$f_o = \sigma(Conv([f_c', f_s'])) \tag{4}$$

where $[\cdot, \cdot]$ represents the concatenate operation on the channel dimension.

3   Experiments

3.1   Dataset

The dataset used to evaluate our method performance is the QaTa-COV19 dataset [17], which is compiled by researchers from Qatar University and Tampere University. It consists of 9258 COVID-19 chest radiographs with pixel-level manual annotations of infected lung areas, of which 7145 are in the training set and 2113 in the test set. However, the original QaTa-COV19 dataset does not contain any matched text annotations.

Li et al. [16] have made significant contributions by extending the text annotations of the dataset, their endeavors are worthy of commendation. We conducted a revisitation of the text annotations and found several notable features. Each sentence consists of three parts, containing position information at different granularity. However, these sentences cannot be considered as medical reports for lacking descriptions of the disease, we consider them as a kind of "text prompt" just as the title of the paper states.

Besides, we found some obvious errors (e.g. misspelled words, grammatical errors and unclear referents) in the extended text annotations. We have fixed these identified errors and contacted the authors of LViT to release a new version of the dataset. Dataset see Github link: https://github.com/HUANGLIZI/LViT.

3.2 Experiment Settings

Following the file name of the subjects in the original train set, we split the training set and the validation set uniformly in the ratio of 80% and 20%. Therefore, the training set has a total of 5716 samples, the validation set has 1429 samples and the test set has 2113 samples. All images are cropped to 224×224 and the data is augmented using a random zoom with 10% probability.

We used a number of open source libraries including but not limited to PyTorch, MONAI [20] and Transformers [21] to implement our method and baseline approach. We use PyTorch Lightning for the final training and inference wrapper. All the methods are training on one NVIDIA Tesla V100 SXM3 32 GB VRAM GPU. We use the Dice loss plus Cross-entropy loss as the loss function, and train the network using AdamW optimization with a batch size of 32. We utilize the cosine annealing learning rate policy, the initial learning rate is set to 3e−4 and the minimal learning rate is set to 1e−6.

We used three metrics to evaluate the segmentation results objectively: Accuracy, Dice coefficient and Jaccard coefficient. Both Dice and Jaccard coefficient calculate the intersection regions over the union regions of the given predicted mask and ground truth, where the Dice coefficient is more indicative of the segmentation performance of small targets.

3.3 Comparison Experiments

We compared our method with common mono-modal medical image segmentation methods and with the LViT previously proposed by Li et al. The quantitative results of the experiment are shown in Table 1. UNet++ achieves the best performance of the mono-modal approach. Comparing to UNet++, our method improves accuracy by 1.44%, Dice score by 6.09% and Jaccard score by 9.49%. Our method improves accuracy by 1.28%, Dice score by 4.86% and Jaccard coefficient by 7.66% compared to the previous multi-modal method LViT. In general, using text prompts could significantly improve segmentation performance.

Table 1. Comparisons with some mono-modal methods and previous multi-modal method on QaTa-COV19 test set. '*' denotes these methods use text prompt and CXR-BERT as the text embedding encoder.

Method		Acc	Dice	Jaccard
Mono-Modal	Unet	0.9584	0.8299	0.7092
	Unet++	**0.9608**	**0.8369**	**0.7196**
	Attention Unet	0.9567	0.8240	0.7006
	Swin UNETR	0.9511	0.8002	0.6669
Multi-Modal*	LViT	0.9624	0.8492	0.7379
	Our Method	**0.9752**	**0.8978**	**0.8145**

The results of the qualitative experiment are shown in Fig. 2. The image-only mono-modal methods tend to generate some over-segmentation, while the multi-modal approach refers to the specific location of the infected region through text prompts to make the segmentation results more accurate.

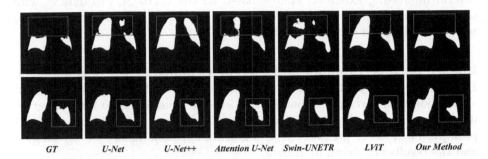

GT   U-Net   U-Net++   Attention U-Net   Swin-UNETR   LViT   Our Method

Fig. 2. Qualitative results on QaTa-COV19. The sample image in the first row is from 'sub-S09345_ses-E16849_run-1_bp-chest_vp-ap_dx.png' and The sample image in the second row is from 'sub-S09340_ses-E17282_run-1_bp-chest_vp-ap_cr.png'.

3.4   Ablation Study

Our proposed method introduces semantic information of text in the decoding process of image features and designs the GuideDecoder to let the semantic information in the text guide the generation of the final segmentation mask. We performed an ablation study on the number of GuideDecoder used in the model and the results are shown in the Table 2.

Table 2. Ablation studies on QaTa-COV19 test set. We used different numbers (0–3) of GuideDecoders in the model to verify the effectiveness of the GuideDecoder. Note that the GuideDecoder in the model is replaced in turn by the Decoder in the UNet, 'w/o text' means without text and the model use UNet Decoders only.

Method	Acc	Dice	Jaccard
w/o text	0.9610	0.8414	0.7262
1 layer	0.9735	0.8920	0.8050
2 layers	0.9748	0.8963	0.8132
3 layers	**0.9752**	**0.8978**	**0.8144**

As can be seen from the Table 2, the segmentation performance of the model improves as the number of GuideDecoders used in the model increases. The effectiveness of GuideDecoder could be proved by these results.

3.5   Extended Study

Considering the application of the algorithm in clinical scenarios, we conducted several interesting extension studies based on the QaTa-COV19 dataset with the text annotations. It is worth mentioning that the following extended studies were carried out on our proposed method.

Impact of Text Prompts at Different Granularity on Segmentation Performance. In Sect. 3.1 we mention that each sample is extended to a text annotation with three parts containing positional information at different granularity, as shown in the Fig. 3. Therefore we further explored the impact of text prompts at different granularity on segmentation performance of our method and the results are shown in Table 3.

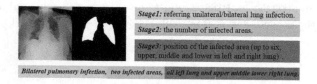

Fig. 3. The Split Example of Different Stages in Text Annotation. The chest x-ray, mask and text annotation correspond to 'covid_1.png'. We divide the sentence into three stages and distinguished them with a background colour, the darker the background colour the more detailed the stage has position information.

Table 3. Study of text prompts at different granularity and segmentation performance. The term *w/o text* in the table means *without text*, while *Stage1*, *Stage2*, and *Stage3* represent the three parts of each text annotation. See Fig. 3 for examples of the different Stages.

Stage of Text Prompt	Acc	Dice	Jaccard
w/o text	0.9610	0.8414	0.7262
stage1 + stage2	0.9648	0.8557	0.7479
stage3	**0.9753**	**0.8981**	**0.8151**
stage1 + stage2 + stage3	0.9752	0.8978	0.8145

The results in the table show that the segmentation performance of our proposed method is driven by the granularity of the position information contained in the text prompt. Our proposed method achieved better segmentation performance when given a text prompt with more detailed position information. Meanwhile, we observed that the performance of our method is almost identical when using two types of text prompts, i.e. *Stage3* alone and *Stage1 + Stage2 + Stage3*. It means the most detailed position information in the text prompt plays the most significant role in improving segmentation performance. But this does not mean that other granularity of position information in the text prompt does not contribute to the improvement in segmentation performance. Even when the input text prompts contain only the coarsest location information (*Stage1 + Stage2* items in the Table 3), our proposed method yielded a 1.43% higher Dice score than the method without text prompt.

Impact of the Size of Training Data on Segmentation Performance. As shown in Table 4, our proposed method demonstrates highly competitive performance even with a reduced amount of training data. With only a quarter of the training data, our proposed method achieves a 2.69% higher Dice score than UNet++, which is the best performing mono-modal model trained on the full dataset. This provides sufficient evidence for the superiority of multimodal approaches and the fact that suitable text prompts could significantly help improve the segmentation performance.

Table 4. Study of the size of training data and segmentation of performance. Note that the performance of UNet++ is used as a comparative reference.

Method	Acc	Dice	Jaccard
Unet++ (using 10% training data)	0.9425	0.7706	0.6268
Ours (using 10% training data)	0.9574	0.8312	0.7111
Unet++ (using 100% training data)	0.9608	0.8369	0.7196
Ours (using 15% training data)	0.9636	0.8503	0.7395
Ours (using 25% training data)	0.9673	0.8638	0.7602
Ours (using 50% training data)	0.9719	0.8821	0.7891
Ours (using 100% training data)	**0.9752**	**0.8978**	**0.8145**

We observed that when the training data was reduced to 10%, our method only began to exhibit inferior performance compared to UNet++, which was trained with all available data. Similar experiments could be found in the LViT paper. Therefore, it can be argued that multi-modal approaches require only a small amount of data (less than 15% in the case of our method) to achieve performance equivalent to that of mono-modal methods.

4 Conclusion

In this paper, we propose a language-driven method for segmenting infected areas from lung x-ray images. The designed GuideDecoder in our method can adaptively propagate sufficient semantic information of the text prompts into pixel-level visual features, promoting consistency between two modalities. The experimental results on the QaTa-COV19 dataset indicate that the multi-modal segmentation method based on text-image could achieve better performance compared to the image-only segmentation methods. Besides, we have conducted several extended studies on the information granularity of the text prompts and the size of the training data, which reveals the flexibility of multi-modal methods in terms of the information granularity of text and demonstrates that multi-modal methods have a significant advantage over image-only methods in terms of the size of training data required.

Acknowledgements. This work was supported by NSFC under Grant 62076093 and MoE-CMCC "Artificial Intelligence" Project No. MCM20190701.

References

1. Yin, S., Deng, H., Xu, Z., Zhu, Q., Cheng, J.: SD-UNet: a novel segmentation framework for CT images of lung infections. Electronics **11**(1), 130 (2022)
2. Oulefki, A., Agaian, S., Trongtirakul, T., Laouar, A.K.: Automatic COVID-19 lung infected region segmentation and measurement using CT-scans images. Pattern Recogn. **114**, 114 (2021)
3. Mu, N., Wang, H., Zhang, Y., Jiang, J., Tang, J.: Progressive global perception and local polishing network for lung infection segmentation of COVID-19 CT images. Pattern Recogn. **120**, 108168 (2021)
4. Fan, D.P., et al.: Inf-Net: automatic COVID-19 lung infection segmentation from CT images. IEEE Trans. Med. Imaging **39**(8), 2626–2637 (2020). https://doi.org/10.1109/TMI.2020.2996645
5. Cao, H., et al.: Swin-Unet: Unet-like pure transformer for medical image segmentation. In: Karlinsky, L., Michaeli, T., Nishino, K. (eds.) Computer Vision-ECCV 2022 Workshops: Tel Aviv, Israel, 23–27 October 2022, Proceedings, Part III, pp. 205–218. Springer, Cham (2023). https://doi.org/10.1007/978-3-031-25066-8_9
6. Hatamizadeh, A., Nath, V., Tang, Y., Yang, D., Roth, H.R., Xu, D.: Swin UNETR: Swin transformers for semantic segmentation of brain tumors in MRI images. In: Crimi, A., Bakas, S. (eds.) Brainlesion: Glioma, Multiple Sclerosis, Stroke and Traumatic Brain Injuries: 7th International Workshop, BrainLes 2021, Held in Conjunction with MICCAI 2021, Virtual Event, 27 September 2021, Revised Selected Papers, Part I, pp. 272–284. Springer, Cham (2022). https://doi.org/10.1007/978-3-031-08999-2_22
7. Zhou, Z., Siddiquee, M.M.R., Tajbakhsh, N., Liang, J.: UNet++: redesigning skip connections to exploit multiscale features in image segmentation. IEEE Trans. Med. Imaging **39**(6), 1856–1867 (2019)
8. Oktay, O., et al.: Attention U-Net: learning where to look for the pancreas. CoRR abs/1804.03999 (2018). http://arxiv.org/abs/1804.03999
9. Hatamizadeh, A., et al.: UNETR: transformers for 3D medical image segmentation. In: Proceedings of the IEEE/CVF Winter Conference on Applications of Computer Vision, pp. 574–584 (2022)

10. Ronneberger, O., Fischer, P., Brox, T.: U-Net: convolutional networks for biomedical image segmentation. In: Navab, N., Hornegger, J., Wells, W., Frangi, A. (eds.) Medical Image Computing and Computer-Assisted Intervention-MICCAI 2015: 18th International Conference, Munich, Germany, 5–9 October 2015, Proceedings, Part III 18, pp. 234–241. Springer, Cham (2015). https://doi.org/10.1007/978-3-319-24574-4_28

11. Radford, A., et al.: Learning transferable visual models from natural language supervision. In: International Conference on Machine Learning, pp. 8748–8763. PMLR (2021)

12. Wang, Z., et al.: CRIS: clip-driven referring image segmentation. In: Proceedings of the IEEE/CVF Conference on Computer Vision and Pattern Recognition, pp. 11686–11695 (2022)

13. Rao, Y., et al.: DenseCLIP: language-guided dense prediction with context-aware prompting. In: Proceedings of the IEEE/CVF Conference on Computer Vision and Pattern Recognition, pp. 18082–18091 (2022)

14. Bhalodia, R., et al.: Improving Pneumonia localization via cross-attention on medical images and reports. In: de Bruijne, M., et al. (eds.) Medical Image Computing and Computer Assisted Intervention-MICCAI 2021: 24th International Conference, Strasbourg, France, 27 September–1 October 2021, Proceedings, Part II 24, pp. 571–581. Springer, Cham (2021). https://doi.org/10.1007/978-3-030-87196-3_53

15. Müller, P., Kaissis, G., Zou, C., Rueckert, D.: Radiological reports improve pre-training for localized imaging tasks on chest X-rays. In: Wang, L., Dou, Q., Fletcher, P.T., Speidel, S., Li, S. (eds.) Medical Image Computing and Computer Assisted Intervention-MICCAI 2022: 25th International Conference, Singapore, 18–22 September 2022, Proceedings, Part V, pp. 647–657. Springer, Cham (2022). https://doi.org/10.1007/978-3-031-16443-9_62

16. Li, Z., et al.: LViT: language meets vision transformer in medical image segmentation. arXiv preprint arXiv:2206.14718 (2022)

17. Degerli, A., Kiranyaz, S., Chowdhury, M.E., Gabbouj, M.: OSegNet: operational segmentation network for Covid-19 detection using chest X-ray images. In: 2022 IEEE International Conference on Image Processing (ICIP), pp. 2306–2310. IEEE (2022)

18. Liu, Z., Mao, H., Wu, C.Y., Feichtenhofer, C., Darrell, T., Xie, S.: A ConvNet for the 2020s. In: Proceedings of the IEEE/CVF Conference on Computer Vision and Pattern Recognition, pp. 11976–11986 (2022)

19. Boecking, B., et al.: Making the most of text semantics to improve biomedical vision-language processing. In: Avidan, S., Brostow, G., Cissé, M., Farinella, G.M., Hassner, T. (eds.) Computer Vision-ECCV 2022: 17th European Conference, Tel Aviv, Israel, 23–27 October 2022, Proceedings, Part XXXVI, pp. 1–21. Springer, Cham (2022). https://doi.org/10.1007/978-3-031-20059-5_1

20. Cardoso, M.J., et al.: MONAI: an open-source framework for deep learning in healthcare. arXiv preprint arXiv:2211.02701 (2022)

21. Wolf, T., et al.: Transformers: state-of-the-art natural language processing. In: Proceedings of the 2020 Conference on Empirical Methods in Natural Language Processing: System Demonstrations, pp. 38–45. Association for Computational Linguistics, Online, October 2020. https://www.aclweb.org/anthology/2020.emnlp-demos.6

NISF: Neural Implicit Segmentation Functions

Nil Stolt-Ansó[1,2]([✉]) (iD), Julian McGinnis[2] (iD), Jiazhen Pan[3] (iD),
Kerstin Hammernik[2,4] (iD), and Daniel Rueckert[1,2,3,4] (iD)

[1] Munich Center for Machine Learning, Technical University of Munich,
Munich, Germany
nil.stolt@tum.de
[2] School of Computation, Information and Technology, Technical University of
Munich, Munich, Germany
[3] School of Medicine, Klinikum Rechts der Isar, Technical University of Munich,
Munich, Germany
[4] Department of Computing, Imperial College London, London, UK

Abstract. Segmentation of anatomical shapes from medical images has
taken an important role in the automation of clinical measurements.
While typical deep-learning segmentation approaches are performed on
discrete voxels, the underlying objects being analysed exist in a real-
valued continuous space. Approaches that rely on convolutional neural
networks (CNNs) are limited to grid-like inputs and not easily applicable
to sparse or partial measurements. We propose a novel family of image
segmentation models that tackle many of CNNs' shortcomings: Neural
Implicit Segmentation Functions (NISF). Our framework takes inspiration from the field of neural implicit functions where a network learns a
mapping from a real-valued coordinate-space to a shape representation.
NISFs have the ability to segment anatomical shapes in high-dimensional
continuous spaces. Training is not limited to voxelized grids, and covers
applications with sparse and partial data. Interpolation between observations is learnt naturally in the training procedure and requires no post-
processing. Furthermore, NISFs allow the leveraging of learnt shape priors to make predictions for regions outside of the original image plane.
We go on to show the framework achieves dice scores of 0.87 ± 0.045 on
a (3D+t) short-axis cardiac segmentation task using the UK Biobank
dataset. We also provide a qualitative analysis on our frameworks ability
to perform segmentation and image interpolation on unseen regions of
an image volume at arbitrary resolutions.

1 Introduction

Image segmentation is a core task in domains where the area, volume or surface of
an object is of interest. The principle of segmentation involves assigning a class to
every presented point in the input space. Typically, the input is presented in the
form of images: aligned pixel (or voxel) grids, with the intention to obtain a class
label for each. In this context, the application of deep learning to the medical

H. Greenspan et al. (Eds.): MICCAI 2023, LNCS 14223, pp. 734–744, 2023.
https://doi.org/10.1007/978-3-031-43901-8_70

imaging domain has shown great promise in recent years. With the advent of the U-Net [20], Convolutional Neural Networks (CNN) have been successfully applied to a multitude of imaging domains and achieved (or even surpassed) human performance [11]. The convolution operation make CNNs an obvious choice for dealing with inputs in the form of 2D pixel- or 3D voxel-grids.

Despite their efficacy, CNNs suffer from a range of limitations that lead to incompatibilities for some imaging domains. CNNs are restricted to data in the form of grids, and cannot easily handle sparse or partial inputs. Moreover, due to the CNN's segmentation output also being confined to a grid, obtaining smooth object surfaces requires post-processing heuristics. Predicting a high resolution segmentations also has implications on the memory and compute requirements in high-dimensional domains. Finally, the learning of long-distance spatial correlations requires deep stacks of layers, which may pose too taxing in low resource domains.

We introduce a novel approach to image segmentation that circumvents these shortcomings: Neural Implicit Segmentation Functions (NISF). Inspired by ongoing research in the field of neural implicit functions (NIF), a neural network is taught to learn a mapping from a coordinate space to any arbitrary real-valued space, such as segmentation, distance function, or image intensity. While CNNs employ the image's pixel or voxel intensities as an input, NISF's input is a real-valued vector $c \in \mathbb{R}^N$ for a single N-dimensional coordinate, alongside a subject-specific latent representation vector $h \in \mathbb{R}^d$. Given c and h, the network is taught to predict image intensity and segmentation value pairs. The space $\mathcal{H}$ over all possible latent vectors h serves as a learnable prior over all possible subject representations.

In this paper, we describe an auto-decoder process by which a previously unseen subject's pairs of coordinate-image intensity values (c, i) may be used to approximate that subject's latent representation h. Given a latent code, the intensity and segmentation predictions from any arbitrary coordinates in the volume may be sampled. We evaluate the proposed framework's segmentation scores and investigate its generalization properties on the UK-Biobank cardiac magnetic resonance imaging (MRI) short-axis dataset. We make the source code publicly available[1].

2   Related Work

Cardiac MRI. Cardiac magnetic resonance imaging (MRI) is often the preferred imaging modality for the assessment of function and structure of the cardiovascular system. This is in equal parts due to its non-invasive nature, and due to its high spatial and temporal resolution capabilities. The short-axis (SAX) view is a (3+t)-dimensional volume made up of stacked cross-sectional (2D+t) acquisitions which lay orthogonal to the ventricle's long axis (see Fig. 1). Spatial resolution is highest in-plane (typically <$3 \, mm^2$), with a much lower inter-slice resolution (10 mm), and a temporal resolution of $\leq 45 \, ms$ [15]. On the other hand,

---

[1] Code repository: https://github.com/NILOIDE/Implicit_segmentation.

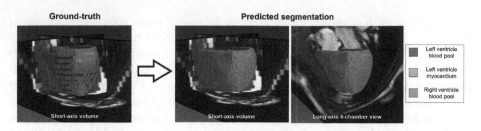

Fig. 1. Short axis volumes have low resolution along the ventricle's long axis. Given a short axis image volume, a NISF can produce arbitrary resolution segmentations along the long axis.

long-axis (LAX) views are (2D+t) acquisitions orthogonal to the SAX plane and provide high resolution along the ventricle's long axis.

Image Segmentation. The capabilities of the CNN has caused it to become the predominant choice for image segmentation tasks [8,20]. However, a pitfall of these models is their poor generalization to certain input transformations. One such transformation is scaling. This drawback limits the use of CNNs on domains with large variations in pixel spacings. Past works have attempted to mitigate this issue by accounting for dataset characteristics [11], building resilience through augmentations [29], or using multi-scale feature extractors [5].

Additionally, segmentation performed by fully convolutional model is restricted to predicting in pixel (or voxel) grids. This requires post-processing heuristics to extract smooth object surfaces. Works such as [12,19] try to mitigate this issue through point-wise decoders that operate on interpolated convolutional features. Alternatives to binarized segmentation have been recently proposed such as soft segmentations [7] and distance field predictions [6,24]. Smoothness can also be improved by predicting at higher resolutions. This is however limited by the exponential increase of memory that comes with high-dimensional data. Partitioning of the input can make memory requirements manageable [3,9], but doing so disallows the ability to learn long-distance spatial correlations.

Neural Implicit Functions. In recent years, NIFs have achieved notable milestones in the field of shape representations [17,18]. NIFs have multiple advantages over classical voxelized approaches that makes them remarkably interesting for applications in the medical imaging domain [10,28]. First, NIFs can sample shapes at any points in space at arbitrary resolutions. This makes them particularly fit for working with sparse, partial, or non-uniform data. Implicit functions thus remove the need for traditional interpolation as high-resolution shapes are learnt implicitly by the network [1]. This is specially relevant to the medical imaging community, where scans may have complex sampling strategies, have missing or unusable regions, or have highly anisotropic voxel sizes. These properties may further vary across scanners and acquisition protocols, making generalization across datasets a challenge. Additionally, the ability to process each point independently allows implicit functions to have flexible optimization strategies, making entire volumes be optimizable holistically.

Image Priors. The typical application of a NIF involves the training of a multi-layer perceptron (MLP) on a *single* scene. Although generalization still occurs in generating novel views of the target scene, the introduction of prior knowledge and conditioning of the MLP is subject to ongoing research [1,14,16,18,22,23]. Approaches such as [1,18] opt for auto-decoder architectures where the network is modulated by latent code at the input level. At inference time, the latent code of the target scene is optimized by backpropagation. Works such as [16] choose to instead modulate the network at its activation functions. Other frameworks obtain the latent code in a single-shot fashion through the use of an encoder network [14,16,22,23]. This latent code is then used by a hyper-network [14, 16,23] or a meta-learning approach [22] to generate the weights of a decoder network.

3 Methods

Shared Prior. In order to generalize to unseen subjects, we attempt to build a shared prior $\mathcal{H}$ over all subjects. This is done by conditioning the classifier with a latent vector $h \in \mathbb{R}^d$ at the input level. Each individual subject j in a population X, can be thought of having a distinct h_j that serves as a latent code of their unique features. Following [1,18], we initialize a matrix $H \in \mathbb{R}^{Xd}$, where each row is a latent vector h_j corresponding to a single subject j in the dataset. The latent vector h_j of a subject is fed to the MLP alongside a point's coordinate and can be optimized through back-propagation. This allows $\mathcal{H}$ to be optimized to capture useful inter-patient features.

Model Architecture. The architecture is composed of a segmentation function f_θ and a reconstruction function f_ϕ. At each continuous-valued coordinate $c \in \mathbb{R}^N$, function f_θ models the shape's segmentation probability s_c for all M classes, and function f_ϕ models the image intensity i_c. The functions are conditioned by a latent vector h at the input level as follows:

$$f_\theta : \left(c \in \mathbb{R}^N\right) \times \left(h \in \mathbb{R}^d\right) \to s_c \in [0,1]^M, \quad \sum_{i=1}^{M} s_c^i = 1 \tag{1}$$

$$f_\phi : \left(c \in \mathbb{R}^N\right) \times \left(h \in \mathbb{R}^d\right) \to i_c \in [0,1] \tag{2}$$

In order to improve local agreement between the segmentation and reconstruction functions, we jointly model f_θ and f_ϕ by a unique multi-layer perceptron (MLP) with two output heads (Fig. 2). We employ Gabor wavelet activation functions [21] which are known to be more expressive than Fourier Features combined with ReLU [26] or sinusoidal activation functions [23].

Prior Training. Following the setup described in [1], we randomly initialize the matrix H consisting of a trainable latent vector $h_j \sim \mathcal{N}\left(0, 10^{-2}\right)$ for each subject in the training set. On each training sample, the parameters of the MLP are jointly optimized with the subject's h_j. We select a training batch by uniformly sampling a time frame t and using all points within that 3D volume. Each voxel in the sample is processed in parallel along the batch dimension. Coordinates are normalized to the range $[0, 1]$ based on the voxel's relative position.

The difference in image reconstruction from the ground-truth voxel intensities is supervised using binary cross-entropy (BCE). This is motivated by our data's voxel intensity distribution being heavily skewed towards the extremes. The segmentation loss is a sum of a BCE loss component and a Dice loss component. We found that adding a weighting factor of $\alpha = 10$ to the image reconstruction loss component yielded inference-time improvements on both image reconstruction and segmentation metrics. Additionally, L2 regularization is applied to the latent vector h_j and the MLP's parameters. The full loss is summarized as follows:

$$
\begin{aligned}
\mathcal{L}_{train}(\theta, \phi, h_j) &= \mathcal{L}_{BCE}\left(f_\theta(c, h_j), s_c\right) + \mathcal{L}_{Dice}\left(f_\theta(c, h_j), s_c\right) \\
&+ \alpha\, \mathcal{L}_{BCE}\left(f_\phi(c, h_j), i_c\right) + \mathcal{L}_{L2}(\theta) + \mathcal{L}_{L2}(\phi) + \mathcal{L}_{L2}(h_j)
\end{aligned}
\tag{3}
$$

Inference. Once the segmentation function f_θ has learnt a mapping from the population prior $\mathcal{H}$ to the segmentation space S, inference becomes a task of finding a latent code h within $\mathcal{H}$ that correctly models the new subject's features. The ground-truth segmentation of a new subject is obviously not available at inference, and it is thus not possible to use f_θ to optimize h. However, since both functions f_ϕ (image reconstruction) and f_θ (segmentation) have been jointly trained by consistently using the same latent vector h, we make the following assumption: *A latent code h optimized for image reconstruction under f_ϕ will also produce accurate segmentations under f_θ.* This assumption makes it possible to use the image reconstruction function f_ϕ alone to find a latent code h for an unseen image in order to generalize segmentation predictions using f_θ.

For this task, a new $h \sim \mathcal{N}\left(0, 10^{-4}\right)$ is initialized. The weights of the MLP are frozen, such that the only tuneable parameters are those of h. Optimization is performed exclusively on the image reconstruction loss (dashed green line in Fig. 2):

$$
\mathcal{L}_{infer}(h_j) = \mathcal{L}_{BCE}\left(f_\phi(c, h_j), i_c\right) + \mathcal{L}_{L2}(h_j)
\tag{4}
$$

Due to the loss being composed exclusively by the image reconstruction term, h is expected to eventually overfit to f_ϕ. Special care should be taken to find a step-number hyperparameter that stops the optimization of h at the optimal segmentation performance. In our experiments, we chose this parameter based on the Dice score of the best validation run.

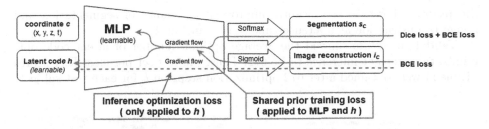

Fig. 2. Training and inference setups. During the prior's training, the MLP and the input latent code h are jointly optimized on image reconstruction and segmentation losses (solid blue line). At inference, solely the latent code h is optimized exclusively on the image reconstruction (dashed green line). (Color figure online)

4   Experiments and Results

Data Overview. The dataset consists of a random subset of 1150 subjects from the UK Biobank's short-axis cardiac MRI acquisitions [25]. An overview of the UK Biobank cohort's baseline statistics can be found in their showcase website [27]. The dataset split included 1000 subjects for the prior training, 50 for validation, and 100 for testing. The (3D+t) short-axis volumes are anisotropic in nature and have a wide range of shapes and pixel spacings along the spatial dimensions. No form of preprocessing was performed on the images except for an intensity normalization to the range $[0, 1]$ as performed in similar literature [2]. The high dimensionality of (3D+t) volumes makes manual annotation prohibitively time consuming. Due to this, we make use of synthetic segmentation as ground truth shapes created using a trained state of the art segmentation CNN provided by [2]. The object of interest in each scan is composed of three distinct, mutually exclusive sub-regions: The left ventricle (LV) blood pool, LV myocardium, and right ventricle (RV) blood pool (see Fig. 1).

Implementation Details. The architecture consists of 8 residual layers, each with 128 hidden units. The subject latent codes had 128 learnable parameters. The model was implemented using Pytorch and trained on an NVIDIA A40 GPU for 1000 epochs, lasting approximately 9 days. Inference optimization lasted 3–7 minutes per subject depending on volume dimensions. Losses are minimized using the ADAM optimizer [13] using a learning rate of 10^{-4} during the prior training and 10^{-4} during inference.

Results. As the latent code is optimized during inference, segmentation metrics follow an overfitting pattern (see Fig. 3). This is an expected consequence of the inference process optimizing solely on the image reconstruction loss. Early stopping should be employed to obtain the best performing latent code state.

The benefits of training a prior over the population is investigated by tracking inference-time Dice scores obtained from spaced-out validation runs. Training of the prior is shown to significantly improve performance of segmentation and image reconstruction at inference-time as seen in Fig. 4.

Validation results showed the average optimal number of latent code optimimization steps at inference to be 672. Thus, the test set per-class Dice scores (Table 1) were obtained after 672 optimization steps on h for each test subject.

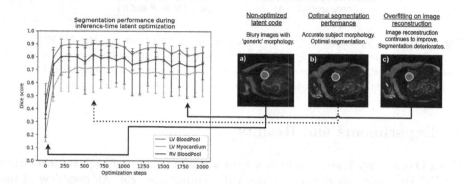

Fig. 3. Segmentation Dice trend during a subject's inference. Early stopping is important to prevent overfitting on reconstruction task. a) Non-optimized latent code creates blurry images with 'generic' morphology. b) As the latent code is optimized, subject morphology begins to be accurately reconstructed. Segmentation performance reaches an optimum. c) Reconstruction continues to improve, but segmentation deteriorates.

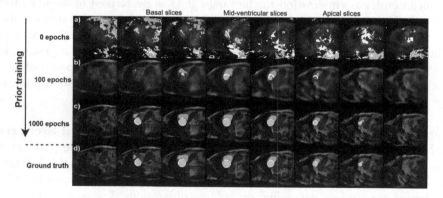

Fig. 4. Inference-time segmentation and image reconstruction at various stages of the prior's training process. a) Prior has not been trained. Inference can roughly reconstruct the image outline. Segmentation fails. b) Early on, reconstructed images are blurry. Segmentation is poor, but at the correct region. c) Eventually images are reconstructed with great detail and segmentations are accurate. d) Ground truth.

Table 1. Class Dice scores for the 100 subject test dataset.

Class	Classes average	LV blood Pool	LV myocardium	RV blood Pool
Dice score	0.87 ± 0.045	0.90 ± 0.037	0.82 ± 0.075	0.88 ± 0.063

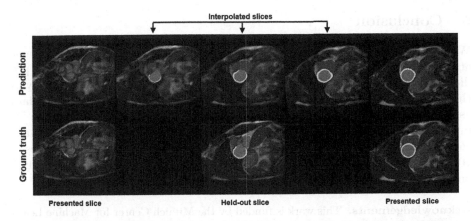

Fig. 5. Interpolation predictions for a held-out basal slice. Top row: Predicted segmentation overlayed on predicted image. Bottom row: Ground truth segmentation overlayed on original image. Middle column is never shown to network during inference. Black slices don't exist in original image volume. The model appears to understand how the ventricles come into view as we descend down the slice dimension.

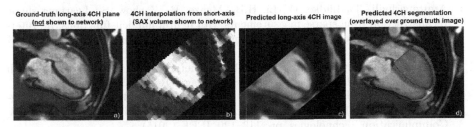

Fig. 6. Segmentation of a held-out long-axis 4-chamber plane from SAX image data. a) Ground-truth long-axis 4-chamber view (not presented to network). b) Nearest-neighbour interpolation of 4-chamber view from SAX volume. c) Predicted 4-chamber image plane. d) Predicted 4-chamber view segmentation.

Further investigation is performed on the generalization capabilities of the subject prior by producing segmentations for held-out sections of the image volume. First, the subject's latent code is optimized using the inference process. Then, the model's output is sampled at the held-out region's coordinates.

Right ventricle segmentation in basal slices is notoriously challenging to manually annotate due to the delineation of the atrial and ventricular cavity combined with the sparsity of the resolution along the long axis [4]. Nonetheless, as seen in Fig. 5, our approach is capable of capturing smooth and plausible morphology of these regions despite not having access to the image information.

We go on to show NISF's ability to generate high-resolution segmentation for out-of-plane views. We optimize on a short-axis volume at inference and subsequently sample coordinates corresponding to long-axis views. Despite *never* presenting a ground-truth long-axis image, the model reconstructs an interpolated view and provides an accurate segmentation along its plane (Fig. 6).

5 Conclusion

We present a novel family of image segmentation models that can model shapes at arbitrary resolutions. The approach is able to leverage priors to make predictions for regions not present in the original image data. Working directly on the coordinate space has the benefit of accepting high-dimensional sparse data, as well as not being affected by variations in image shapes and resolutions. We implement a simple version of this framework and evaluate it on a short-axis cardiac MRI segmentation task using the UK Biobank. Reported Dice scores on 100 unseen subjects average 0.87 ± 0.045. We also perform a qualitative analysis on the framework's ability to predict held-out sections of image volumes.

Acknowledgements. This work is funded by the Munich Center for Machine Learning and European Research Council (ERC) project Deep4MI (884622). This research has been conducted using the UK Biobank Resource under Application Number 87802.

References

1. Amiranashvili, T., Lüdke, D., Li, H.B., Menze, B., Zachow, S.: Learning shape reconstruction from sparse measurements with neural implicit functions. In: International Conference on Medical Imaging with Deep Learning, pp. 22–34. PMLR (2022)
2. Bai, W., et al.: Automated cardiovascular magnetic resonance image analysis with fully convolutional networks. J. Cardiovasc. Magn. Reson. **20**(1), 1–12 (2018)
3. Bali, A., Singh, S.N.: A review on the strategies and techniques of image segmentation. In: 2015 Fifth International Conference on Advanced Computing & Communication Technologies, pp. 113–120. IEEE (2015)
4. Budai, A., et al.: Fully automatic segmentation of right and left ventricle on short-axis cardiac MRI images. Comput. Med. Imaging Graph. **85**, 101786 (2020)
5. Chen, L.C., Yang, Y., Wang, J., Xu, W., Yuille, A.L.: Attention to scale: scale-aware semantic image segmentation. In: Proceedings of the IEEE Conference on Computer Vision and Pattern Recognition, pp. 3640–3649 (2016)
6. Dai, A., Ruizhongtai Qi, C., Nießner, M.: Shape completion using 3D-encoder-predictor CNNs and shape synthesis. In: Proceedings of the IEEE Conference on Computer Vision and Pattern Recognition, pp. 5868–5877 (2017)
7. Gros, C., Lemay, A., Cohen-Adad, J.: SoftSeg: advantages of soft versus binary training for image segmentation. Med. Image Anal. **71**, 102038 (2021)
8. He, K., Zhang, X., Ren, S., Sun, J.: Deep residual learning for image recognition. In: Proceedings of the IEEE Conference on Computer Vision and Pattern Recognition, pp. 770–778 (2016)
9. Hou, L., Samaras, D., Kurc, T.M., Gao, Y., Davis, J.E., Saltz, J.H.: Efficient multiple instance convolutional neural networks for gigapixel resolution image classification. arXiv preprint arXiv:1504.07947, vol. 7, pp. 174–182 (2015)
10. Huang, W., Li, H., Cruz, G., Pan, J., Rueckert, D., Hammernik, K.: Neural implicit k-Space for binning-free non-cartesian cardiac MR imaging. arXiv preprint arXiv:2212.08479 (2022)
11. Isensee, F., Jaeger, P.F., Kohl, S.A., Petersen, J., Maier-Hein, K.H.: nnU-Net: a self-configuring method for deep learning-based biomedical image segmentation. Nat. Methods **18**(2), 203–211 (2021)

12. Khan, M.O., Fang, Y.: Implicit neural representations for medical imaging segmentation. In: Wang, L., Dou, Q., Fletcher, P.T., Speidel, S., Li, S. (eds.) International Conference on Medical Image Computing and Computer-Assisted Intervention, vol. 13435, pp. 433–443. Springer, Cham (2022). https://doi.org/10.1007/978-3-031-16443-9_42

13. Kingma, D.P., Ba, J.: Adam: a method for stochastic optimization. In: 3rd International Conference on Learning Representations, ICLR 2015, San Diego, CA, USA, 7–9 May 2015, Conference Track Proceedings (2015)

14. Klocek, S., Maziarka, Ł., Wołczyk, M., Tabor, J., Nowak, J., Śmieja, M.: Hypernetwork functional image representation. In: Tetko, I., Kurková, V., Karpov, P., Theis, F. (eds.) Artificial Neural Networks and Machine Learning-ICANN 2019: Workshop and Special Sessions: 28th International Conference on Artificial Neural Networks, Munich, Germany, 17–19 September 2019, Proceedings 28, pp. 496–510. Springer, Cham (2019). https://doi.org/10.1007/978-3-030-30493-5_48

15. Kramer, C.M., Barkhausen, J., Bucciarelli-Ducci, C., Flamm, S.D., Kim, R.J., Nagel, E.: Standardized cardiovascular magnetic resonance imaging (CMR) protocols: 2020 update. J. Cardiovasc. Magn. Reson. **22**(1), 1–18 (2020)

16. Mehta, I., Gharbi, M., Barnes, C., Shechtman, E., Ramamoorthi, R., Chandraker, M.: Modulated periodic activations for generalizable local functional representations. In: Proceedings of the IEEE/CVF International Conference on Computer Vision, pp. 14214–14223 (2021)

17. Mildenhall, B., Srinivasan, P.P., Tancik, M., Barron, J.T., Ramamoorthi, R., Ng, R.: NeRF: representing scenes as neural radiance fields for view synthesis. Commun. ACM **65**(1), 99–106 (2021)

18. Park, J.J., Florence, P., Straub, J., Newcombe, R., Lovegrove, S.: DeepSDF: learning continuous signed distance functions for shape representation, pp. 165–174 (2019)

19. Peng, S., Niemeyer, M., Mescheder, L., Pollefeys, M., Geiger, A.: Convolutional occupancy networks. In: Vedaldi, A., Bischof, H., Brox, T., Frahm, J.M. (eds.) Computer Vision - ECCV 2020. LNCS, 12348, vol. 3, pp. 523–540. Springer, Cham (2020). https://doi.org/10.1007/978-3-030-58580-8_31

20. Ronneberger, O., Fischer, P., Brox, T.: U-Net: convolutional networks for biomedical image segmentation. In: Navab, N., Hornegger, J., Wells, W., Frangi, A. (eds.) Medical Image Computing and Computer-Assisted Intervention-MICCAI 2015: 18th International Conference, Munich, Germany, 5–9 October 2015, Proceedings, Part III 18, pp. 234–241. Springer, Cham (2015). https://doi.org/10.1007/978-3-319-24574-4_28

21. Saragadam, V., LeJeune, D., Tan, J., Balakrishnan, G., Veeraraghavan, A., Baraniuk, R.G.: WIRE: wavelet implicit neural representations. arXiv preprint arXiv:2301.05187 (2023)

22. Sitzmann, V., Chan, E., Tucker, R., Snavely, N., Wetzstein, G.: MetaSDF: meta-learning signed distance functions. Adv. Neural. Inf. Process. Syst. **33**, 10136–10147 (2020)

23. Sitzmann, V., Martel, J., Bergman, A., Lindell, D., Wetzstein, G.: Implicit neural representations with periodic activation functions. Adv. Neural. Inf. Process. Syst. **33**, 7462–7473 (2020)

24. Stutz, D., Geiger, A.: Learning 3D shape completion from laser scan data with weak supervision. In: Proceedings of the IEEE Conference on Computer Vision and Pattern Recognition, pp. 1955–1964 (2018)

25. Sudlow, C., et al.: UK Biobank: an open access resource for identifying the causes of a wide range of complex diseases of middle and old age. PLoS Med. **12**(3), e1001779 (2015)
26. Tancik, M., et al.: Fourier features let networks learn high frequency functions in low dimensional domains. Adv. Neural. Inf. Process. Syst. **33**, 7537–7547 (2020)
27. UK Biobank: Data showcase. https://biobank.ndph.ox.ac.uk/showcase/. Accessed 7 Mar 2023
28. Wolterink, J.M., Zwienenberg, J.C., Brune, C.: Implicit neural representations for deformable image registration. In: International Conference on Medical Imaging with Deep Learning, pp. 1349–1359. PMLR (2022)
29. Zhao, A., Balakrishnan, G., Durand, F., Guttag, J.V., Dalca, A.V.: Data augmentation using learned transformations for one-shot medical image segmentation. In: Proceedings of the IEEE/CVF Conference on Computer Vision and Pattern Recognition, pp. 8543–8553 (2019)

Multimodal CT and MR Segmentation of Head and Neck Organs-at-Risk

Gašper Podobnik[1](✉)(iD), Primož Strojan[2](iD), Primož Peterlin[2](iD),
Bulat Ibragimov[1,3](iD), and Tomaž Vrtovec[1](iD)

[1] Faculty of Electrical Engineering, University of Ljubljana, Ljubljana, Slovenia
gasper.podobnik@fe.uni-lj.si
[2] Institute of Oncology Ljubljana, Ljubljana, Slovenia
[3] Department of Computer Science, University of Copenhagen, Copenhagen, Denmark

Abstract. Radiotherapy (RT) is a standard treatment modality for head and neck (HaN) cancer that requires accurate segmentation of target volumes and nearby healthy organs-at-risk (OARs) to optimize radiation dose distribution. However, computed tomography (CT) imaging has low image contrast for soft tissues, making accurate segmentation of soft tissue OARs challenging. Therefore, magnetic resonance (MR) imaging has been recommended to enhance the segmentation of soft tissue OARs in the HaN region. Based on our two empirical observations that deformable registration of CT and MR images of the same patient is inherently imperfect and that concatenating such images at the input layer of a deep learning network cannot optimally exploit the information provided by the MR modality, we propose a novel modality fusion module (MFM) that learns to spatially align MR-based feature maps before fusing them with CT-based feature maps. The proposed MFM can be easily implemented into any existing multimodal backbone network. Our implementation within the nnU-Net framework shows promising results on a dataset of CT and MR image pairs from the same patients. Furthermore, the evaluation on a clinically realistic scenario with the missing MR modality shows that MFM outperforms other state-of-the-art multimodal approaches.

Keywords: Multimodal segmentation · Head and neck · Organs-at-risk · Computed tomography · Magnetic resonance · nnU-Net

1 Introduction

Head and neck (HaN) cancer is a prevalent type of cancer [3] with a yearly incidence of above 1 million cases and prevalence of above 4 million cases worldwide, accounting for around 5% of all cancer sites [17]. Radiotherapy (RT) is a standard

Supplementary Information The online version contains supplementary material available at https://doi.org/10.1007/978-3-031-43901-8_71.

treatment modality for HaN cancer, which aims to deliver high doses of radiation to cancerous cells while sparing nearby healthy organs-at-risk (OARs) [21]. To optimize radiation dose distribution, accurate three-dimensional (3D) segmentation of target volumes and OARs is required. Computed tomography (CT) is the primary imaging modality used for RT planning due to its ability to provide information about electron density, however, its low image contrast for soft tissues, including tumors, makes accurate segmentation of soft tissue OARs challenging. Therefore, the integration of complementary imaging modalities, such as magnetic resonance (MR), has been strongly recommended in clinical practice to enhance the segmentation of several soft tissue OARs in the HaN region [1]. This naturally poses a question of whether automatic OAR segmentation can benefit from the MR image modality. Our study therefore aims to evaluate the impact of MR integration on the quality and robustness of automatic OAR segmentation in the HaN region, therefore contributing to the growing body of research on multimodal methods for medical image analysis.

Related Work. A literature review by Zhang et al. [24] divides deep learning (DL)-based multimodal segmentation methods into three fusion strategy groups: *early*, *late* and *hybrid* (also named *layer*) fusion. The first two groups of methods are most commonly applied; early fusion comprises simple concatenation of modalities along the channel dimension before feeding them into the deep neural network. Additionally, concatenating feature maps (FMs) from separate modality encoders can also be considered as early fusion [7]. Late fusion, on the other hand, employs separate branches for each input modality and then fuses the output features by either plain concatenation or by weighing the contributions of separate branches at the decision level. For example, Zhang et al. [23] proposed an attention mechanism to fuse FMs from two separate U-Nets that accepted contrast-enhanced arterial and venous phase CT images. The third group, hybrid fusion, aims to combine the strengths of early and late fusion [24] by employing two or more separate encoders (i.e. one for each modality) and a single decoder, where features from different resolution levels of the encoder are fused and fed into the decoder that produces the final full-resolution segmentation. Such hybrid or multi-level fusion along with the adaptive fusion method represents the current trend in computer vision [24], with the self-supervised model adaptation method as a prime example [18]. One important aspect is also the missing modality scenario, meaning that the multimodal model should produce satisfactory results even if only one input modality is available. Nevertheless, the optimal fusion strategy remains an open question in need of further exploration. Similar conclusions were reached in a review of multimodal segmentation methods in the medical imaging community by Zhou et al. [25]. Most methods implement either early or late fusion, however, the layer fusion strategy was identified as a better choice, since dense connections among layers can exploit more complex and complementary information to enhance training. The highlight is HyperDenseNet, a dual-path 3D network proposed by Dolz et al. [4] that employs dense connections between two convolutional paths, and achieves improvements compared to other fusion strategies and single modality variants. However, other studies have shown that the best fusion strategy depends on the specific nature of the problem, e.g. Yan et al. [22] demonstrated that the late fusion outperforms the other two approaches for the longitudinal detection of diabetic retinopathy. Relevant to the field of multimodal segmentation are also developments on unpaired multimodal segmentation, where cross-modality learning is employed to take advantage of different image modalities covering the same anatomy, but without the constraint to collect images from the same patients [5,10,19]. Although the methodologies comprising CycleGANs and/or multiple segmentation networks [10,19] seem promising, they can be excessively complex for the task of HaN OAR segmentation where both CT and MR image modalities from the same patient are often available. Consequently, our primary focus is the paired multimodal segmentation problem, including the missing modality scenario.

Motivation. When segmenting OARs in the HaN region for the purpose of RT planning, a multimodal segmentation model that can leverage the information from CT and MR images of the same patient might be beneficial compared to separate single-modal models. Firstly, as intuition suggests, such a model would rely on the CT image for bone structures and on the MR image for soft tissues, and therefore improve the overall segmentation quality by exploiting the complementary information from both modalities. Secondly, a multimodal model would facilitate cross-modality learning by extracting knowledge from one and applying that knowledge to the other modality, potentially improving the segmentation accuracy. Several studies indicated that such an approach is feasible, for example, for improving video classification by training a model on an auxiliary audio reconstruction task [12], or for audio-based detection by using the multimodal knowledge distillation concept, where teacher networks trained on RGB, depth and thermal images improve a student network trained only on audio data [20]. Finally, from the DL infrastructure maintenance perspective, it is easier to maintain a single model that can handle both modalities than two separate models for each modality. However, clinical practice differs considerably from theory, meaning that a number of considerations must be taken into account. Firstly, although MR image acquisition is recommended, it is not always feasible due to time constraints, scanner occupancy and financial aspects. Consequently, automatic OAR multimodal segmentation is required to handle the missing modality scenario, and provide a similar segmentation quality as a single-modality system. Secondly, because CT and MR images are not acquired simultaneously and with the same acquisition parameters (e.g. resolution), there is an inherent misalignment between both modalities. This can be mitigated with image registration, but not completely, mainly due to different patient positioning that especially affects the deformation of soft tissues, and various modality-specific artifacts (e.g. motion, implants, partial volume effect, etc.).

Contributions. To tackle these considerations, we propose a mechanism named modality fusion module (MFM) that can generally be applied to any network architecture that learns features from multiple modalities, and shows promising performance also in the missing modality scenario. The advantages of the proposed MFM are the following: 1) it enables the spatial alignment of FMs from one with FMs from the other modality to further reduce errors that persist after deformable registration of input images, and enrich the FMs to improve the final OAR segmentation, 2) it significantly improves the performance of the missing modality scenario compared to other baseline fusion approaches, and 3) it performs well also on single modality out-of-distribution data, therefore facilitating cross-modality learning and contributing to better model generalizability.

2   Methods

Backbone Architecture. Our chosen backbone network is based on nnU-Net, a publicly available framework for DL-based segmentation [8] that builds on the U-Net architecture [16], adds self-configurable pre-processing, augmentation and post-processing, and employs efficient training strategies. However, nnU-Net, which uses an early fusion strategy by concatenating input images or patches before feeding them to the first network layer, may not be the optimal strategy for multimodal segmentation. Recent studies have shown that this approach does not allow the network to learn meaningful high-level features from each modality before their fusion, resulting in only simple relationships between intensities from each input modality [4,23]. This is particularly problematic when fusing CT and MR images, which differ in several aspects, such as the type and location of artifacts, acquisition parameters, and visibility of soft tissues and bone structures. While MR images can help to improve the delineations of OARs that are poorly visible in CT images, the primary delineation is always performed on CT images with the help of registered MR images. An important repercussion is that image registration errors propagate into OAR delineations, which is particularly salient in the HaN region. To address these challenges, we propose an upgraded nnU-Net network with two separate encoders, one for each modality, and a common decoder that fuses FMs using the proposed MFM that learns to infer affine transformation parameters in a single forward pass. This approach efficiently *pseudo-registers* FMs from the MR encoder with those from the CT encoder, mitigating the effects of registration errors caused by non-rigid deformation of OARs and imaging artifacts.

Modality Fusion Module. The proposed MFM draws inspiration from the work of Jaderberg et al. [9], who introduced a spatial transformer network (STN) that learns to infer transformation parameters in a single forward pass, and then uses them to transform images and/or FMs. The fundamental idea is that STN can learn meaningful features that are spatially invariant to characteristics of the input data, without the need for extra supervision, thereby enhancing task

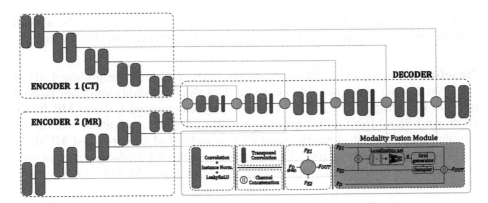

Fig. 1. The proposed backbone architecture is based on nnU-Net but with separate encoders for the computed tomography (CT) and magnetic resonance (MR) image, and with the proposed modality fusion module.

performance. While it was demonstrated that complete spatial invariance cannot be achieved with STNs [6], the work of Jaderberg et al. is crucial in showing that STNs can be implemented as differentiable modules, enabling the loss to be propagated through the sampling (interpolation) mechanism. The same underlying principle of STNs has also been leveraged in *optical flow* and its derivative work *semantic flow*, where the flow alignment module was proposed to resample low-resolution FMs and align them with high-resolution FMs [2]. We capitalize on the same principle to register FMs from MR images to those from CT images. Notably, MFM is different from semantic flow, as it takes two FMs of the same resolution but from different modalities, and aligns FMs from the auxiliary modality to FMs of the primary modality. We propose to use MFM at each resolution level of the nnU-Net backbone, which is schematically presented in Fig. 1, and consists of three blocks: *localization network, grid generator* and *sampler*. The localization network is a regressor network that accepts concatenated FMs from both encoders and applies four blocks of strided convolutions followed by the ReLU activation to reduce their spatial dimensions. The final FMs are flattened and fed into a simple two-layer fully connected network, which outputs 12 affine 3D transformation parameters that are then passed to the grid generator. The generated sampling grid is used by the sampler to resample FMs from the second encoder, which are then concatenated with the untouched FMs from the first encoder and the decoder (right before the bottleneck, only the first two are concatenated, as there are no decoder FMs at that level). Both the grid generator and the sampler and readily implemented in the PyTorch library [9], and because they are both differentiable, no special optimization is needed for the localization network, allowing localization parameters to be optimized with the main (segmentation) loss function. Since there is no additional supervision that would assure perfect registration, we refer to this process as *pseudo-registration.* The purpose of this architecture is to align FMs from both modalities and improve their fusion, leading to better segmentation results.

Baseline Comparison. We evaluate the performance of the proposed MFM nnU-Net against three baseline networks: 1) a single modality nnU-Net trained only on CT images, 2) a nnU-Net trained on concatenated CT and MR image pairs, and 3) a model with separate encoders for both modalities, but with a simple concatenation along the channel axis instead of the proposed MFM. In addition, we compare our model with the state-of-the-art *modality-aware mutual learning nnU-Net* (MAML) that was presented at MICCAI 2021 [23].

3   Experiments and Results

Image Datasets. The proposed methodology was evaluated on two publicly available datasets: our recently released HaN-Seg dataset [14] and the PDDCA dataset [15]. The HaN-Seg dataset comprises CT and T1-weighted MR images of 56 patients, which were deformably registered with the SimpleElastix registration tool, and corresponding curated manual delineations of 30 OARs (for details, please refer to [14]). Although only a subset of images is publicly available[1] due to the ongoing HaN-Seg challenge[2], both the publicly available training as well as the privately withheld test images were used in our 4-fold cross-validation experiments. On the other hand, to evaluate the generalization ability of our method, we also conducted experiments on the CT-only PDDCA dataset (for details, please refer to [15]), from which we collected 15 images from the off- and on-site test sets of the corresponding challenge for our evaluation. As this dataset is widely used for evaluating the performance of automatic HaN OAR segmentation methods, it serves as a valuable benchmark for comparison with other state-of-the-art methods. Note that none of the images from the CT-only PDDCA dataset were used for training, and as our model expects two inputs, we substituted the missing MR modality with an empty matrix (i.e. zeros).

Implementation Details. All models were trained for all OARs using the 3d_fullres configuration of nnU-Net, with the only modification that we reduced rotation around the axial axis and disabled image flipping along the sagittal plane, which eliminated segmentation errors that were previously observed for the paired (left and right) OARs. The same modification was also used with the MAML model. To ensure a fair model comparison, we set the number of filters in the encoder of the single modality baseline model to match the number of filters of the entry-level concatenation encoder. We also halved the number of filters in networks that have separate encoders so that the overall number of parameters in the proposed model and the baselines remains approximately the same (excluding the parameters in the localization part of MFM

---

[1] https://doi.org/10.5281/zenodo.7442914.
[2] https://hanseg2023.grand-challenge.org.

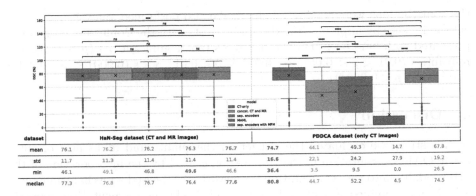

dataset	HaN-Seg dataset (CT and MR images)					PDDCA dataset (only CT images)			
mean	76.1	76.2	76.2	76.3	76.7	74.7	44.1	49.3	14.7	67.8
std	11.7	11.3	11.4	11.4	11.4	16.6	22.1	24.2	27.9	19.2
min	46.1	49.1	46.8	49.6	46.6	36.4	3.5	9.5	0.0	26.5
median	77.3	76.8	76.7	76.4	77.6	80.8	44.7	52.2	4.5	74.5

Fig. 2. The results in terms of the Dice similarity coefficient (DSC) for the HaN-Seg (left) and PDDCA (right) dataset.

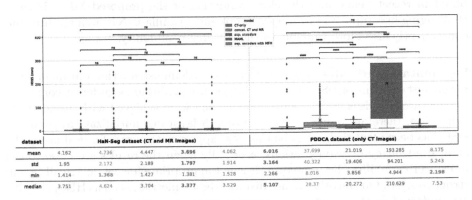

dataset	HaN-Seg dataset (CT and MR images)					PDDCA dataset (only CT images)			
mean	4.162	4.736	4.447	3.696	4.062	6.016	37.699	21.019	193.285	8.175
std	1.95	2.172	2.189	1.797	1.914	3.164	40.322	19.406	94.201	5.243
min	1.414	1.368	1.427	1.381	1.528	2.266	8.016	3.856	4.944	2.198
median	3.751	4.624	3.704	3.377	3.529	5.107	28.37	20.272	210.629	7.53

Fig. 3. The results in terms of the 95th-percentile Hausdorff distance (HD$_{95}$) for the HaN-Seg (left) and PDDCA (right) dataset. (Note: Infinite values of HD$_{95}$ were replaced with a maximal value over all data.)

block). Note that the MAML model, which is composed of two U-Nets, had a considerably higher number of parameters. To address the challenge of a relatively small dataset, we adopted a 4-fold cross-validation strategy without using any external training images. All models were trained until convergence, i.e. when the validation loss plateaued, and we selected the model with the best validation loss for inference.

Results. The quality of the obtained OAR segmentation masks was evaluated by computing the Dice similarity coefficient (DSC) and the 95th-percentile Hausdorff distance (HD$_{95}$) against reference manual delineations, and the results for all OARs are presented in Figs. 2 and 3, respectively. Since not all images contain all 30 OARs (due to a different field-of-view), we first calculated the mean metric for each OAR and then the overall mean across all OARs to ensure that the contributions were equally weighted. We also performed analysis of statistical significance by applying paired sample t-tests with the Bonferroni correction, presented with bars on top of the box plots (non-significant: ns ($p > 0.05$), significant: $*$ ($0.01 < p < 0.05$), $**$ ($0.001 < p < 0.01$), $***$ ($0.0001 < p < 0.001$) and $****$ ($p < 0.0001$)).

4 Discussion

In this study, we evaluated the impact on the quality and robustness of automatic OAR segmentation in the HaN region caused by the incorporation of the MR modality into the segmentation framework. We devised a mechanism named MFM and combined it with nnU-Net as our backbone segmentation network. The choice of using nnU-Net as the backbone was based on the rationale that nnU-Net already incorporates numerous state-of-the-art DL innovations proposed in recent years, and therefore validation of the proposed MFM is more challenging in comparison to simply improving a vanilla U-Net architecture, and consequently also more valuable to the research community.

Segmentation Results. The obtained results demonstrate that our model performs best in terms of DSC (Fig. 2). The resulting gains are significant compared to separate encoders and CT-only models, and were achieved with 4-fold cross-validation, therefore reducing the chance of a favorable initialization. However, DSC has been identified not to be the most appropriate metric for evaluating the clinical adequacy of segmentations, especially when the results are close to the interrater variability [13], moreover, it is not appropriate for volumetrically small structures [11]. On the other hand, distance-based metrics, such as HD_{95} (Fig. 3), are preferred as they better measure the shape consistency between the reference and predicted segmentations. Although MAML achieved the best results in terms of HD_{95}, indicating that late fusion can efficiently merge the information from both modalities, it should be noted that MAML has a considerate advantage due to having two decoders and an additional attention fusion block compared to the baseline nnU-Net with separate encoders and a single decoder. On the other hand, our approach based on separate encoders with MFM is not far behind, with a mean HD_{95} of 4.06 mm, which is more than 15% better than the early concatenation fusion. The comparison to the baseline nnU-Net with separate encoders offers the most direct evaluation of the proposed MFM. An approximate 10% improvement in HD_{95} suggests that MFM allows the network to learn more informative FMs that lead to a better overall performance.

Missing Modality Scenario. The overall good performance on the HaN-Seg dataset suggests that all models are close to the maximal performance, which is bounded by the quality of reference segmentations. However, the performance on the PDDCA dataset that consists only of CT images allows us to test how the models handle the missing modality scenario and perform on an out-of-distribution dataset, as images from this dataset were not used for training. As expected, the CT-only model performed best in its regular operating scenario, with a mean DSC of 74.7% (Fig. 2) and HD_{95} of 6.02 mm (Fig. 3). However, significant differences can be observed between multimodal methods, where the proposed model outperformed MAML and other baselines by a large margin in both metrics. The MAML model with a mean DSC of less than 15% and HD_{95} of more than 190 mm was not able to handle the missing modality scenario, whereas the MFM model performed almost as good as the CT-only model, with a mean DSC of 67.8% and HD_{95} of 8.18 mm. It should be noted that we did not employ any training strategies to improve handling of missing modalities, such as swapping input images or intensity augmentations. A possible explanation is that the proposed MFM facilitates cross-modality learning, enabling nnU-Net to extract better FMs from CT images even in such extreme scenarios.

5 Conclusions

In this study, we introduced MFM, a fusion module that aligns FMs from an auxiliary modality (e.g. MR) to FMs from the primary modality (e.g. CT). The proposed MFM is versatile, as it can be applied to any multimodal segmentation network. However, it has to be noted that it is not symmetrical, and therefore requires the user to specify the primary modality, which is typically the same as the primary modality used in manual delineation (i.e. in our case CT). We evaluated the performance of MFM combined with the nnU-Net backbone for segmentation of OARs in the HaN region, an important task in RT cancer treatment planning. The obtained results indicate that the performance of MFM is similar to other state-of-the-art methods, but it outperforms other multimodal methods in scenarios with one missing modality.

Acknowledgments. This work was supported by the Slovenian Research Agency (ARRS) under grants J2-1732, P2-0232 and P3-0307, and partially by the Novo Nordisk Foundation under grant NFF20OC0062056.

References

1. Brouwer, C.L., Steenbakkers, R.J., Bourhis, J., et al.: CT-based delineation of organs at risk in the head and neck region: DAHANCA, EORTC, GORTEC, HKN-PCSG, NCIC CTG, NCRI, NRG oncology and TROG consensus guidelines. Radiother. Oncol. **117**, 83–90 (2015). https://doi.org/10.1016/j.radonc.2015.07.041
2. Chen, J., Zhan, Y., Xu, Y., Pan, X.: FAFNet: fully aligned fusion network for RGBD semantic segmentation based on hierarchical semantic flows. IET Image Process **17**, 32–41 (2023). https://doi.org/10.1049/ipr2.12614
3. Chow, L.Q.M.: Head and neck cancer. N. Engl. J. Med. **382**, 60–72 (2020). https://doi.org/10.1056/NEJMRA1715715
4. Dolz, J., Gopinath, K., Yuan, J., Lombaert, H., Desrosiers, C., Ben Ayed, I.: HyperDense-Net: a hyper-densely connected CNN for multi-modal image segmentation. IEEE Trans. Med. Imaging **38**, 1116–1126 (2019). https://doi.org/10.1109/TMI.2018.2878669

5. Dou, Q., Liu, Q., Heng, P.A., Glocker, B.: Unpaired multi-modal segmentation via knowledge distillation. IEEE Trans. Med. Imaging **39**, 2415–2425 (2020). https://doi.org/10.1109/TMI.2019.2963882

6. Finnveden, L., Jansson, Y., Lindeberg, T.: Understanding when spatial transformer networks do not support invariance, and what to do about it. In: 25th International Conference on Pattern Recognition - ICPR 2020, Milan, Italy, pp. 3427–3434. IEEE (2020). https://doi.org/10.1109/ICPR48806.2021.9412997

7. Hu, X., Yang, K., Fei, L., Wang, K.: ACNet: attention based network to exploit complementary features for RGBD semantic segmentation. In: 26th International Conference on Image Processing - ICIP 2019, Taipei, Taiwan, pp. 1440–1444. IEEE (2019). https://doi.org/10.1109/ICIP.2019.8803025

8. Isensee, F., Jaeger, P.F., Kohl, S.A., Petersen, J., Maier-Hein, K.H.: nnU-Net: a self-configuring method for deep learning-based biomedical image segmentation. Nat. Methods **18**, 203–211 (2021). https://doi.org/10.1038/s41592-020-01008-z

9. Jaderberg, M., Simonyan, K., Zisserman, A., Kavukcuoglu, K.: Spatial transformer networks. In: Advances in Neural Information Processing Systems - NIPS 2015, vol. 28. Curran Associates, Montréal, QC, Canada (2015). https://doi.org/10.48550/arxiv.1506.02025

10. Jiang, J., Rimner, A., Deasy, J.O., Veeraraghavan, H.: Unpaired cross-modality educed distillation (CMEDL) for medical image segmentation. IEEE Trans. Med. Imaging **41**, 1057–1068 (2022). https://doi.org/10.1109/TMI.2021.3132291

11. Maier-Hein, L., Reinke, A., Kozubek, M., et al.: BIAS: transparent reporting of biomedical image analysis challenges. Med. Image Anal. **66**, 101796 (2020). https://doi.org/10.1016/j.media.2020.101796

12. Ngiam, J., Khosla, A., Kim, M., Nam, J., Lee, H., Ng, A.Y.: Multimodal deep learning. In: 28th International Conference on International Conference on Machine Learning - ICML 2011, Bellevue, WA, USA, pp. 689–696. Omnipress (2011)

13. Nikolov, S., Blackwell, S., Zverovitch, A., et al.: Clinically applicable segmentation of head and neck anatomy for radiotherapy: deep learning algorithm development and validation study. J. Med. Internet Res. **23**, e26151 (2021). https://doi.org/10.2196/26151

14. Podobnik, G., Strojan, P., Peterlin, P., Ibragimov, B., Vrtovec, T.: HaN-Seg: the head and neck organ-at-risk CT and MR segmentation dataset. Med. Phys. **50**, 1917–1927 (2023). https://doi.org/10.1002/mp.16197

15. Raudaschl, P.F., Zaffino, P., Sharp, G.C., et al.: Evaluation of segmentation methods on head and neck CT: auto-segmentation challenge 2015. Med. Phys. **44**, 2020–2036 (2017). https://doi.org/10.1002/mp.12197

16. Ronneberger, O., Fischer, P., Brox, T.: U-Net: convolutional networks for biomedical image segmentation. In: Navab, N., Hornegger, J., Wells, W.M., Frangi, A.F. (eds.) MICCAI 2015. LNCS, vol. 9351, pp. 234–241. Springer, Cham (2015). https://doi.org/10.1007/978-3-319-24574-4_28

17. Sung, H., Ferlay, J., Siegel, R.L., et al.: Global cancer statistics 2020: GLOBOCAN estimates of incidence and mortality worldwide for 36 cancers in 185 countries. CA Cancer J. Clin. **71**, 209–249 (2021). https://doi.org/10.3322/caac.21660

18. Valada, A., Mohan, R., Burgard, W.: Self-supervised model adaptation for multimodal semantic segmentation. Int. J. Comput. Vis. **128**, 1239–1285 (2018). https://doi.org/10.1007/s11263-019-01188-y

19. Valindria, V.V., Pawlowski, N., Rajchl, M., et al.: Multi-modal learning from unpaired images: application to multi-organ segmentation in CT and MRI. In: 2018 IEEE Winter Conference on Applications of Computer Vision - WACV 2018, Lake Tahoe, NV, USA, pp. 547–556. IEEE (2018). https://doi.org/10.1109/WACV.2018.00066

20. Valverde, F.R., Hurtado, J.V., Valada, A.: There is more than meets the eye: self-supervised multi-object detection and tracking with sound by distilling multimodal knowledge. In: 2021 IEEE/CVF Conference on Computer Vision and Pattern Recognition - CVPR 2021, Nashville, TN, USA, pp. 11607–11616. IEEE (2021). https://doi.org/10.1109/CVPR46437.2021.01144

21. Yan, F., Knochelmann, H.M., Morgan, P.F., et al.: The evolution of care of cancers of the head and neck region: state of the science in 2020. Cancers **12**, 1543 (2020). https://doi.org/10.3390/cancers12061543

22. Yan, Y., et al.: Longitudinal detection of diabetic retinopathy early severity grade changes using deep learning. In: Fu, H., Garvin, M.K., MacGillivray, T., Xu, Y., Zheng, Y. (eds.) OMIA 2021. LNCS, vol. 12970, pp. 11–20. Springer, Cham (2021). https://doi.org/10.1007/978-3-030-87000-3_2

23. Zhang, Y., et al.: Modality-aware mutual learning for multi-modal medical image segmentation. In: de Bruijne, M., et al. (eds.) MICCAI 2021. LNCS, vol. 12901, pp. 589–599. Springer, Cham (2021). https://doi.org/10.1007/978-3-030-87193-2_56

24. Zhang, Y., Sidibé, D., Morel, O., Mériaudeau, F.: Deep multimodal fusion for semantic image segmentation: a survey. Image Vis. Comput. **105**, 104042 (2021). https://doi.org/10.1016/j.imavis.2020.104042

25. Zhou, T., Ruan, S., Canu, S.: A review: deep learning for medical image segmentation using multi-modality fusion. Array **3–4**, 100004 (2019). https://doi.org/10.1016/j.array.2019.100004

Conditional Diffusion Models for Weakly Supervised Medical Image Segmentation

Xinrong Hu[1], Yu-Jen Chen[2], Tsung-Yi Ho[3], and Yiyu Shi[1(✉)]

[1] University of Notre Dame, Notre Dame, IN, USA
{xhu7,yshi4}@nd.edu
[2] National Tsing Hua University, Hsinchu, Taiwan
[3] The Chinese University of Hong Kong, Shatin, Hong Kong

Abstract. Recent advances in denoising diffusion probabilistic models have shown great success in image synthesis tasks. While there are already works exploring the potential of this powerful tool in image semantic segmentation, its application in weakly supervised semantic segmentation (WSSS) remains relatively under-explored. Observing that conditional diffusion models (CDM) is capable of generating images subject to specific distributions, in this work, we utilize category-aware semantic information underlied in CDM to get the prediction mask of the target object with only image-level annotations. More specifically, we locate the desired class by approximating the derivative of the output of CDM w.r.t the input condition. Our method is different from previous diffusion model methods with guidance from an external classifier, which accumulates noises in the background during the reconstruction process. Our method outperforms state-of-the-art CAM and diffusion model methods on two public medical image segmentation datasets, which demonstrates that CDM is a promising tool in WSSS. Also, experiment shows our method is more time-efficient than existing diffusion model methods, making it practical for wider applications. The codes are available at https://github.com/xhu248/cond_ddpm_wsss.

Keywords: weakly supervised semantic segmentation · diffusion models · brain tumor · magnetic resonance imaging

1 Introduction

Medical image segmentation is always a critical task as it can be used for disease diagnosis, treatment planning, and anomaly monitoring. Weakly supervised semantic segmentation attracts significant attention from medical image community since it greatly reduces the cost of dense pixel-wise labeling to get segmentation mask. In WSSS, the training labels are usually easier and faster to obtain, like image-level tags, bounding boxes, scribbles, or point annotations. This work only focuses on WSSS with image-level tags, like whether a tumor

---

Supplementary Information The online version contains supplementary material available at https://doi.org/10.1007/978-3-031-43901-8_72.

presents or not. In this field, previous WSSS works [10,14] are dominated by class activation map (CAM) [25] and its variants [4,18,21], which was firstly introduced as a tool to visualize saliency maps when making a class prediction.

Meanwhile, denoising diffusion models [6,9] demonstrate superior performance in image synthesis than other generative models. Also, there are several works exploring the application of diffusion models to semantic segmentation in natural images [2,8]and medical images [15,22–24]. To the best of our knowledge, Wolleb et al. [22] is the only work that introduces diffusion models to pixel-wise anomaly detection with only classification labels. They achieve this by utilizing an external classifier trained with image-level annotations to guide the reverse Markov chain. By passing the gradient of the classifier, the diffusion model gradually removes the anomaly areas during the denoising process and then obtains the anomaly map by comparing the reconstructed and original images However, this approach is based on the hypothesis that the classifier can accurately locate the target objects and that the background is not changed when removing the noise. This assumption does not always hold, especially when the distribution of positive images is diverse, and the reconstruction error can also be accumulated after hundreds of steps. As shown in Fig. 1, the reconstructed images guided by the gradient of non-kidney not only remove the kidney area but also change the content in the background. Another limitation of this method is the long inference time required for a single image, as hundreds of iterations are needed to restore the image to its original noise level. In contrast, CAM approaches need only one inference to get the saliency maps. Therefore, there is ample room for improvement in using diffusion models for WSSS task.

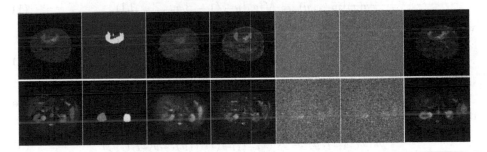

Fig. 1. Intuition behind our CDM based method. From left to right, each column represents original images, ground truth, reconstructed images with negative guidance, saliency map generated by guided diffusion model, images with one step of denoising conditioned on positive label and negative label, and saliency map generated by our method. Images are from BraTS and CHAOS dataset.

In this work, we propose a novel WSSS framework with conditional diffusion models (CDM) as we observe that the predicted noises on different condition show difference. Instead of completely removing the noises from images, we calculate the derivative of the predicted noise after a few stages with respect to conditions so that the related objects are highlighted in the gradient map with less background misidentified. As the output of diffusion model is not differentiable with respect to the discrete condition input, we adopt the finite difference method, i.e., perturbing the condition embedding by a small amplitude and logging the change of the output with DDIM [19] generative process. In addition, our method does not require the full reverse denoising process for the noised images and may only need one or a few iterations. Thus the inference time of our method is comparable to that of CAM-based approaches. We evaluate our methods on two different tasks, brain tumor segmentation and kidney segmentation, and provide the quantitative results of both CAM based and diffusion model based methods as comparison. Our approach achieves state-of-the-art performance on both datasets, demonstrating the effectiveness of the proposed framework. We also conduct extensive ablation studies to analyze the impact of various components in our framework and provide reasoning for each design.

2 Methods

2.1 Training Conditional Denoising Diffusion Models

Suppose that we have a sample x_0 from distribution $D(x|y)$, and y is the condition. The condition y can be various, like different modality [20], inpainting [3] and low resolution images [17]. In this work, $y \in \{y_0, y_1\}$ indicates the binary classification label, like brain CT scans without tumor vs. with tumor. We then gradually add Guassian noise to the original image sample with different level $t \in \{0, 1, ..., T\}$ as

$$q(x_t|x_{t-1}, y) := \mathcal{N}(x_t|y; \sqrt{1 - \beta_t}x_{t-1}|y, \beta_t \mathbf{I}) \tag{1}$$

With fixed variance $\{\beta_1, \beta_2, ..., \beta_t\}$, x_t can be explicitly expressed by x_0,

$$q(x_t|x_0, y) := \mathcal{N}(x_t|y; \sqrt{\bar{\alpha}_t}x_0|y, (1 - \bar{\alpha}_t)\mathbf{I}) \tag{2}$$

where $\alpha_t := 1 - \beta_t$, $\bar{\alpha}_t := \prod_{s=1}^{t} \alpha_s$.

Then a conditional U-Net [16] $\epsilon_\theta(x, t, y)$ is trained to approximate the reverse denoising process,

$$p_\theta(x_{t-1}|x_t, y) := \mathcal{N}(x_{t-1}; \mu_\theta(x_t, t, y), \Sigma_\theta(x_t, t, y)) \tag{3}$$

The variance μ_σ can be learnable parameters or a fixed set of scalars, and both settings achieve comparable results in [9]. As for the mean, after reparameterization with $x_t = \sqrt{\bar{\alpha}_t}x_0 + \sqrt{1 - \bar{\alpha}_t}\epsilon$ for $\epsilon \sim \mathcal{N}(\mathbf{0}, I)$, the loss function can be simplified as:

$$L := \mathbb{E}_{x_0, \epsilon} \left\| \epsilon - \epsilon_\theta(\sqrt{\bar{\alpha}_t}x_0 + \sqrt{1 - \bar{\alpha}_t}\epsilon, t, y) \right\| \tag{4}$$

As for how to infuse binary condition y in the U-Net, we follow the strategy in [6], using a embedding projection function $e = f(y), f \in \mathbb{R} \to \mathbb{R}^n$, with n being the embedding dimension. Then the condition embedding is added to feature maps in different blocks. After training the denoising model, Tashiro et al. [3] proved that the network can yield the desired conditional distribution $D(x|y)$ given condition y.

Algorithm 1. Generation of WSSS prediction mask using differentiate conditional model with DDIM sampling

Input: input image x with label y_1, noise level Q, inference stage R, noise predictor ϵ_θ, τ
Output: prediction mask of label y_1
for all t from 1 to Q **do**
$\quad x_t \leftarrow \sqrt{1 - \beta_t} x_{t-1} + \beta_t * \mathcal{N}(0, \mathbf{I})$
end for
$a \leftarrow \text{zeros}(x.\text{shape})$
$x'_t \leftarrow x_t.\text{copy}()$
for all t from Q to Q-R **do**
$\quad \hat{\epsilon}_1 \leftarrow \epsilon_\theta(x_t, t, f(y_1)), \quad \hat{\epsilon}_0 \leftarrow \epsilon_\theta(x'_t, t, (1 - \tau)f(y_1) + \tau f(y_0))$
$\quad x_{t-1} \leftarrow \sqrt{\bar{\alpha}_{t-1}}(\frac{x_t - \sqrt{1 - \bar{\alpha}_t}\hat{\epsilon}_1}{\sqrt{\bar{\alpha}_t}}) + \sqrt{1 - \bar{\alpha}_{t-1}}\hat{\epsilon}_1$
$\quad x'_{t-1} \leftarrow \sqrt{\bar{\alpha}_{t-1}}(\frac{x'_t - \sqrt{1 - \bar{\alpha}_t}\hat{\epsilon}_0}{\sqrt{\bar{\alpha}_t}}) + \sqrt{1 - \bar{\alpha}_{t-1}}\hat{\epsilon}_0$
$\quad a \leftarrow a + \frac{1}{\tau}(x_{t-1} - x'_{t-1})$
end for
return a

2.2 Gradient Map w.r.t Condition

Inspired by the finding in [2] that the denoising model extracts semantic information when performing reverse diffusion process, we aim to get segmentation mask from the sample generated by single or just several reverse Markov steps with DDIM [19]. The reason for using DDIM is that one can generate a sample x_{t-1} from x_t deterministically if removing the random noise term via:

$$x_{t-1}(x_t, t, y) = \sqrt{\bar{\alpha}_{t-1}}(\frac{x_t - \sqrt{1 - \bar{\alpha}_t}\hat{\epsilon}_\theta(x_t, y)}{\sqrt{\bar{\alpha}_t}}) + \sqrt{1 - \bar{\alpha}_{t-1}}\hat{\epsilon}_\theta(x_t, y) \quad (5)$$

When given the same images at noise level Q, but with different conditions, the noises predicted by the network ϵ_θ are supposed to reflect the localization of target objects, that is equivalently $\|x_{t-1}(x_t, t, y_1) - x_{t-1}(x_t, t, y_0)\|$. This idea is quite similar to [22] if we keep sampling x_{t-1} until x_0. However, it is not guaranteed that the condition y_0 does not change background areas besides the object needed to be localized. Therefore, in order to minimize the error caused by the generation process, we propose to visualize the sensitivity of $x_{t-1}(x_t, t, y_1)$ w.r.t condition y_1, that is $\frac{\partial x_{t-1}}{\partial y}$. Since the embedding projection function $f(\cdot)$ is not differentiable, we approximate the gradient using the finite difference method:

$$\frac{\partial x_{t-1}(x_t, t, y)}{\partial y}\bigg|_{y=y1} = \lim_{\tau \to 1} \frac{x_{t-1}(x_t, t, f(y_1)) - x_{t-1}(x_t, t, (1 - \tau)f(y_1) + \tau f(y_0))}{\tau} \quad (6)$$

in which, τ is the moving weight from $f(y_1)$ towards $f(y_0)$. The weight τ can not be too close to 1, otherwise there is no noticeable gap between x_{t-1} and x'_{t-1}, and we find $\tau = 0.95$ gives the best performance. Algorithm 1 shows the detailed workflow of obtaining the segmentation mask of samples with label y_1. Notice that we iterate the process (5) for R steps, and the default R in this work is set as 10, much smaller than $Q = 400$. The purpose of R is to amplify the change of x_{t-1} since the condition does not change the predicted noise a lot in one step.

In addition, we find that the guidance of additional classifier can further boost the WSSS task, by passing the gradient $\hat{\epsilon} \leftarrow \epsilon_\theta(x_t) - s\sqrt{1 - \bar{\alpha}_t}\nabla_{x_t}logp_\phi(y|x_t)$, where p_ϕ is the classifier and s is the gradient scale. Then, in Algorithm 1, $\hat{\epsilon}_1$ and $\hat{\epsilon}_0$ have additional terms guided by gradient of y_1 and y_0, respectively. The ablation studies of related hyperparameters can be seen in Sect. 3.2.

3   Experiments and Results

Brain Tumor Segmentation. BraTS (Brain Tumor Segmentation challenge) [1] contains 2,000 cases of 3D brain scans, each of which includes four different MRI modalities as well as tumor segmentation ground truth. In this work, we only use the FLAIR channel and treat all types of tumor as one single class. We divide the official training set into 9:1 for training and validation purpose, and evaluate our method on the official validation set. For preprocessing, we slice each volume into 2D images, following the setting in [5]. Then the total number of slices in training set is 193,905, and we report metrics on the 5802 positive samples in the test set

Kidney Segmentation. This task is conducted on dataset from ISBI 2019 CHAOS Challenge [12], which contains 20 volumes of T2-SPIR MR abdominal scans. CHAOS provides pixel-wise annotation for several organs, but we focus on the kidney. We split the 20 volumes into four folds for cross-validation, and then decompose 3D volumes to 2D slices in every fold. In the test stage, we remove slices with area of interest taking up less than 5% of the total area in the slice, in order to avoid the influence of extreme cases on the average results.

Only classification labels are used during training the diffusion models, and segmentation masks are used for evaluation in the test stage. For both datasets, we repeat the evaluation protocols for four times and report the average metrics and their standard deviation on test set.

3.1   Implementation Details

As for model architecture, we use the same setting as in [8]. The diffusion model is based on U-Net with encoder and decoder consisting of resnet blocks. We implement two different versions of the proposed method, one without classifier guidance and one with it. To differentiate the two in the experiments, we continue to call them former CDM, and the latter CG-CDM. The classifier used in CG-CDM is the same as the encoder of the diffusion model. We stop training the diffusion model after 50,000 iterations or 7 days, and the classifier is trained for 20,000 iterations. We choose AdamW as the optimizer with learning rate being 1e-5 and 5e-5 for diffusion model and classifier. The batch sizes for both datasets are 2. The implementation of all methods in this work is based on PyTorch library, and all experiments are run on a single NVIDIA RTX 2080Ti.

Table 1. Comparisons with state-of-the-art WSSS methods on BraTS dataset. "CAM" refers to CAM-based methods, "DM" means methods based on diffusion models, and "FSL" is short for fully supervised learning.

Types	Methods	Dice↑	mIoU↑	HD95↓	infer time
CAM	GradCAM [18]	0.235 ± 0.075	0.149 ± 0.051	44.4 ± 8.9	3.79 s
	GradCAM++ [4]	0.281 ± 0.084	0.187 ± 0.059	32.6 ± 6.0	3.59 s
	ScoreCAM [21]	0.303 ± 0.053	0.202 ± 0.039	32.7 ± 2.1	27.0 s
	LayerCAM [11]	0.276 ± 0.082	0.184 ± 0.058	30.4 ± 6.6	4.07 s
DM	CG-Diff [22]	0.456 ± 0.043	0.325 ± 0.036	43.4 ± 3.0	116 s
	CDM	0.525 ± 0.057	0.407 ± 0.051	26.0 ± 5.2	1.61 s
	CG-CDM	$\mathbf{0.563 \pm 0.023}$	$\mathbf{0.450 \pm 0.012}$	$\mathbf{19.2 \pm 3.2}$	2.78 s
FSL	N/A	0.902 ± 0.028	0.814 ± 0.023	8.1 ± 1.9	0.31 s

Table 2. Comparison with state-of-the-art WSSS methods on CHAOS dataset.

Types	Methods	Dice↑	mIoU↑	HD95↓
CAM	GradCAM [18]	0.105 ± 0.017	0.059 ± 0.010	33.9 ± 5.1
	GradCAM++ [4]	0.147 ± 0.016	0.085 ± 0.010	28.5 ± 4.5
	ScoreCAM [21]	0.135 ± 0.024	0.078 ± 0.015	32.1 ± 6.7
	LayerCAM [11]	0.194 ± 0.022	0.131 ± 0.018	29.7 ± 8.1
DM	CG-Diff [22]	0.235 ± 0.025	0.152 ± 0.020	27.1 ± 3.2
	CDM	0.263 ± 0.028	0.167 ± 0.042	26.6 ± 2.4
	CG-CDM	$\mathbf{0.311 \pm 0.018}$	$\mathbf{0.186 \pm 0.014}$	$\mathbf{23.3 \pm 3.4}$
FSL	N/A	0.847 ± 0.011	0.765 ± 0.023	3.6 ± 1.7

3.2 Results

Comparison with State of the Arts. We benchmark our methods against previous WSSS works on two datasets in Table 1 & 2, in terms of dice score, mean intersection over union (mIoU), and Hausdorff distance (HD95). For CAM based methods, we include the classical GradCAM [18] and GradCAM++ [4], as well as two more recent methods, ScoreCAM [21] and LayerCAM [11]. The implementation of these CAM approaches is based on the repository [7]. For diffusion based methods, we include the only diffusion model for medical image segmentation in the WSSS literature, namely CG-Diff [22]. We follow the default setting in [22], setting noise level $Q = 400$ and gradient scale $s = 100$. We also present the results under the fully supervised learning setting, which is the upper bond of all WSSS methods (Fig. 2).

From the results, we can make several key observations. Firstly, our proposed method, even without classifier guidance, outperform all other WSSS methods including the classifier guided diffusion model CG-Diff on both datasets for all

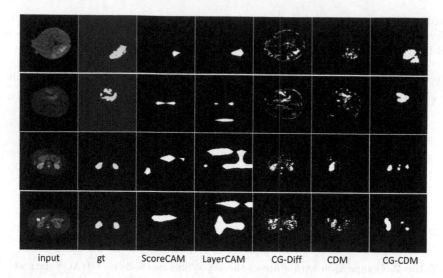

Fig. 2. Visualization of WSSS segmentation masks using different methods on both BraTS and CHAOS dataset. The threshold is decided by average Otsu thresholding [13].

three metrics. When classifier guidance is provided, the improvement gets even bigger, and CG-CDM can beat other methods regarding segmentation accuracy. Secondly, all WSSS methods have performance drop on kidney dataset compared with BraTS dataset. This demonstrates that the kidney segmentation task is a more challenging task for WSSS than brain tumor task, which may be caused by the small training size and diverse appearance across slices in the CHAOS dataset.

Time Efficiency. Regarding inference time for different methods, as shown in Table 1, both CDM and CG-CDM are much faster than CG-Diff. The default noise level Q is set as 400 for all diffusion model approaches, and our methods run 10 iterations during the denoising steps. For all CAM-based approaches, we add augmentation smooth and eigen smooth suggested in [7] to reduce noise in the prediction mask. This post-processing greatly increases the inference time. Without the two smooth methods, the inference time for GradCAM is 0.031 s, but the segmentation accuracy is significantly degraded. Therefore, considering both inference time and performance, our method is a better option than CAM for WSSS.

Ablation Studies. There are several important hyperparameters in our framework, noise level Q, number of iterations R, moving weight τ, and gradient scale s. The default setting is CG-CDM on BraTS dataset with $Q = 400$, $R = 10$, $\tau = 0.95$, and $s = 10$. We evaluate the influence of one hyperparameter at a time by keeping other parameters at their default values. As illustrated in Fig. 3, a few observations can be made: (1) Either too large or too small noise level can

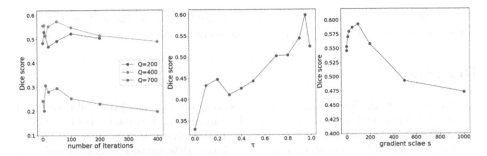

Fig. 3. Ablation studies of hyper-parameters on BraTS dataset, including noise level Q, number of iterations R, moving weight τ, and gradient scale s.

negatively influence the performance. When Q is small, most spatial information is still kept in x_t and the predicted noise by diffusion model contains no
semantic knowledge. When Q is large, most of the spatial information is lost
and the predicted noise can be distracted from original structure. Meanwhile,
larger number of iterations can lightly improve the dice score at the beginning.
When R gets too high, the error in the background is also accumulated after
too many iterations. (2) We try different τ in the range (0, 1.0). Small τ leads
to more noises in the background when calculating the difference in different
conditions. On the other hand, as τ gets close to 1, the difference between x_{t-1}
and x'_{t-1} becomes minor, and the gradient map mainly comes from the guidance
of the classifier, making localization not so accurate. Thus, $\tau = 0.95$ becomes the
optimal choice for this task. (3) As for gradient scale, Fig. 3 shows that before
$s = 100$, larger gradient scale can boost the CDM, because at this time, the
gradient from the classifier is at the same magnitude as the difference caused by
the changed condition embedding. When the guidance of the classifier becomes
dominant, the dice score gets lower as the background is distorted by too large
gradients.

4   Conclusion

In this paper, we present a novel weakly supervised semantic segmentation framework based on conditional diffusion models. Fundamentally, the essence of generative approaches on WSSS is maximizing the change in class-related areas while
minimizing the noise in the background. Our methods are designed around this
rule to enhance the state-of-the-art. First, existing work that utilizes a trained
classifier to remove target objects leads to unpredictable distortion in other areas,
thus we decide to iterate the reverse denoising process for as few steps as possible. Second, to amplify the difference caused by different conditions, we extract
the semantic information from gradient of the noise predicted by the diffusion
model. Finally, this rule also applies to all other designs and choice of hyperparameters in our framework. When compared with latest WSSS methods on

two public medical image segmentation datasets, our method shows superior performance regarding both segmentation accuracy and inference efficiency.

References

1. Bakas, S., et al.: Advancing the cancer genome atlas glioma MRI collections with expert segmentation labels and radiomic features. Sci. Data **4**(1), 1–13 (2017)
2. Baranchuk, D., Voynov, A., Rubachev, I., Khrulkov, V., Babenko, A.: Label-efficient semantic segmentation with diffusion models. In: International Conference on Learning Representations (2021)
3. Batzolis, G., Stanczuk, J., Schönlieb, C.B., Etmann, C.: Conditional image generation with score-based diffusion models. arXiv preprint arXiv:2111.13606 (2021)
4. Chattopadhay, A., Sarkar, A., Howlader, P., Balasubramanian, V.N.: Grad-CAM++: generalized gradient-based visual explanations for deep convolutional networks. In: 2018 IEEE Winter Conference on Applications of Computer Vision (WACV), pp. 839–847. IEEE (2018)
5. Dey, R., Hong, Y.: ASC-Net: adversarial-based selective network for unsupervised anomaly segmentation. In: de Bruijne, M., et al. (eds.) MICCAI 2021. LNCS, vol. 12905, pp. 236–247. Springer, Cham (2021). https://doi.org/10.1007/978-3-030-87240-3_23
6. Dhariwal, P., Nichol, A.: Diffusion models beat GANs on image synthesis. Adv. Neural. Inf. Process. Syst. **34**, 8780–8794 (2021)
7. Gildenblat, J., contributors: Pytorch library for cam methods (2021). https://github.com/jacobgil/pytorch-grad-cam
8. Graikos, A., Malkin, N., Jojic, N., Samaras, D.: Diffusion models as plug-and-play priors. arXiv preprint arXiv:2206.09012 (2022)
9. Ho, J., Jain, A., Abbeel, P.: Denoising diffusion probabilistic models. Adv. Neural. Inf. Process. Syst. **33**, 6840–6851 (2020)
10. Izadyyazdanabadi, M., et al.: Weakly-supervised learning-based feature localization for confocal laser endomicroscopy glioma images. In: Frangi, A.F., Schnabel, J.A., Davatzikos, C., Alberola-López, C., Fichtinger, G. (eds.) MICCAI 2018. LNCS, vol. 11071, pp. 300–308. Springer, Cham (2018). https://doi.org/10.1007/978-3-030-00934-2_34
11. Jiang, P.T., Zhang, C.B., Hou, Q., Cheng, M.M., Wei, Y.: Layercam: exploring hierarchical class activation maps for localization. IEEE Trans. Image Process. **30**, 5875–5888 (2021)
12. Kavur, A.E., et al.: CHAOS challenge - combined (CT-MR) healthy abdominal organ segmentation. Med. Image Anal. **69**, 101950 (2021). https://doi.org/10.1016/j.media.2020.101950. http://www.sciencedirect.com/science/article/pii/S1361841520303145
13. Otsu, N.: A threshold selection method from gray-level histograms. IEEE Trans. Syst. Man Cybern. **9**(1), 62–66 (1979)
14. Patel, G., Dolz, J.: Weakly supervised segmentation with cross-modality equivariant constraints. Med. Image Anal. **77**, 102374 (2022)
15. Pinaya, W.H., et al.: Fast unsupervised brain anomaly detection and segmentation with diffusion models. In: Wang, L., Dou, Q., Fletcher, P.T., Speidel, S., Li, S. (eds.) MICCAI 2022. LNCS, vol. 13438, pp. 705–714. Springer, Cham (2022). https://doi.org/10.1007/978-3-031-16452-1_67

16. Ronneberger, O., Fischer, P., Brox, T.: U-Net: convolutional networks for biomedical image segmentation. In: Navab, N., Hornegger, J., Wells, W.M., Frangi, A.F. (eds.) MICCAI 2015. LNCS, vol. 9351, pp. 234–241. Springer, Cham (2015). https://doi.org/10.1007/978-3-319-24574-4_28

17. Saharia, C., Ho, J., Chan, W., Salimans, T., Fleet, D.J., Norouzi, M.: Image super-resolution via iterative refinement. IEEE Trans. Pattern Anal. Mach. Intell. **45**(4), 4713–4726 (2022)

18. Selvaraju, R.R., Cogswell, M., Das, A., Vedantam, R., Parikh, D., Batra, D.: Gradcam: visual explanations from deep networks via gradient-based localization. In: Proceedings of the IEEE International Conference on Computer Vision, pp. 618–626 (2017)

19. Song, J., Meng, C., Ermon, S.: Denoising diffusion implicit models. In: International Conference on Learning Representations (2020)

20. Tashiro, Y., Song, J., Song, Y., Ermon, S.: CSDI: conditional score-based diffusion models for probabilistic time series imputation. Adv. Neural. Inf. Process. Syst. **34**, 24804–24816 (2021)

21. Wang, H., et al.: Score-CAM: score-weighted visual explanations for convolutional neural networks. In: Proceedings of the IEEE/CVF Conference on Computer Vision and Pattern Recognition Workshops, pp. 24 25 (2020)

22. Wolleb, J., Bieder, F., Sandkühler, R., Cattin, P.C.: Diffusion models for medical anomaly detection. In: Wang, L., Dou, Q., Fletcher, P.T., Speidel, S., Li, S. (eds.) MICCAI 2022. LNCS, vol. 13438, pp. 35–45. Springer, Cham (2022). https://doi.org/10.1007/978-3-031-16452-1_4

23. Wolleb, J., Sandkühler, R., Bieder, F., Valmaggia, P., Cattin, P.C.: Diffusion models for implicit image segmentation ensembles. In: International Conference on Medical Imaging with Deep Learning, pp. 1336–1348. PMLR (2022)

24. Wyatt, J., Leach, A., Schmon, S.M., Willcocks, C.G.: Anoddpm: anomaly detection with denoising diffusion probabilistic models using simplex noise. In: Proceedings of the IEEE/CVF Conference on Computer Vision and Pattern Recognition, pp. 650–656 (2022)

25. Zhou, B., Khosla, A., Lapedriza, A., Oliva, A., Torralba, A.: Learning deep features for discriminative localization. In: Proceedings of the IEEE Conference on Computer Vision and Pattern Recognition, pp. 2921–2929 (2016)

From Sparse to Precise: A Practical Editing Approach for Intracardiac Echocardiography Segmentation

Ahmed H. Shahin[1,2(✉)], Yan Zhuang[1,3], and Noha El-Zehiry[1,4]

[1] Siemens Healthineers, New Jersey, USA
ahmedhshahen@gmail.com
[2] University College London, London, UK
[3] National Institutes of Health Clinical Center, Maryland, USA
[4] Wipro, New Jersey, USA

Abstract. Accurate and safe catheter ablation procedures for atrial fibrillation require precise segmentation of cardiac structures in Intracardiac Echocardiography (ICE) imaging. Prior studies have suggested methods that employ 3D geometry information from the ICE transducer to create a sparse ICE volume by placing 2D frames in a 3D grid, enabling the training of 3D segmentation models. However, the resulting 3D masks from these models can be inaccurate and may lead to serious clinical complications due to the sparse sampling in ICE data, frames misalignment, and cardiac motion. To address this issue, we propose an interactive editing framework that allows users to edit segmentation output by drawing scribbles on a 2D frame. The user interaction is mapped to the 3D grid and utilized to execute an editing step that modifies the segmentation in the vicinity of the interaction while preserving the previous segmentation away from the interaction. Furthermore, our framework accommodates multiple edits to the segmentation output in a sequential manner without compromising previous edits. This paper presents a novel loss function and a novel evaluation metric specifically designed for editing. Cross-validation and testing results indicate that, in terms of segmentation quality and following user input, our proposed loss function outperforms standard losses and training strategies. We demonstrate quantitatively and qualitatively that subsequent edits do not compromise previous edits when using our method, as opposed to standard segmentation losses. Our approach improves segmentation accuracy while avoiding undesired changes away from user interactions and without compromising the quality of previously edited regions, leading to better patient outcomes.

Keywords: Interactive editing · Ultrasound · Echocardiography

Work done while the authors were employed by Siemens Healthineers.
Disclaimer: The concepts and information presented in this paper are based on research results that are not commercially available. Future commercial availability cannot be guaranteed.

Supplementary Information The online version contains supplementary material available at https://doi.org/10.1007/978-3-031-43901-8_73.

1   Introduction

Atrial Fibrillation (AFib) is a prevalent cardiac arrhythmia affecting over 45 million individuals worldwide as of 2016 [7]. Catheter ablation, which involves the elimination of affected cardiac tissue, is a widely used treatment for AFib. To ensure procedural safety and minimize harm to healthy tissue, Intracardiac Echocardiography (ICE) imaging is utilized to guide the intervention.

Intracardiac Echocardiography imaging utilizes an ultrasound probe attached to a catheter and inserted into the heart to obtain real-time images of its internal structures. In ablation procedures for Left Atrium (LA) AFib treatment, the ICE ultrasound catheter is inserted in the right atrium to image the left atrial structures. The catheter is rotated clockwise to capture image frames that show the LA body, the LA appendage and the pulmonary veins [12]. Unlike other imaging modalities, such as transesophageal echocardiography, ICE imaging does not require general anesthesia [3]. Therefore, it is a safer and more convenient option for cardiac interventions using ultrasound imaging.

The precise segmentation of cardiac structures, particularly the LA, is crucial for the success and safety of catheter ablation. However, segmentation of the LA is challenging due to the constrained spatial resolution of 2D ICE images and the manual manipulation of the ICE transducer. Additionally, the sparse sampling of ICE frames makes it difficult to train automatic segmentation models. Consequently, there is a persistent need to develop interactive editing tools to help experts modify the automatic segmentation to reach clinically satisfactory accuracy.

During a typical ICE imaging scan, a series of sparse 2D ICE frames is captured and a Clinical Application Specialist (CAS) annotates the boundaries of the desired cardiac structure in each frame[1] (Fig. 1a). To construct dense 3D masks for training segmentation models, Liao et al. utilized the 3D geometry information from the ICE transducer, to project the frames and their annotations onto a 3D grid [8]. They deformed a 3D template of the LA computed from 414 CT scans to align as closely as possible with the CAS contours, producing a 3D mesh to train a segmentation model [8]. However, the resulting mesh may not perfectly align with the original CAS contours due to factors such as frames misalignment and cardiac motion (Fig. 1b). Consequently, models trained with such 3D mesh as ground truth do not produce accurate enough segmentation results, which can lead to serious complications (Fig. 1c).

A natural remedy is to allow clinicians to edit the segmentation output and create a model that incorporates and follows these edits. In the case of ICE data, the user interacts with the segmentation output by drawing a scribble on one of the 2D frames (Fig. 1d). Ideally, the user interaction should influence the segmentation in the neighboring frames while preserving the original segmentation in the rest of the volume. Moreover, the user may make multiple edits to

---

[1] Annotations typically take the form of contours instead of masks, as the structures being segmented appear with open boundaries in the frames.

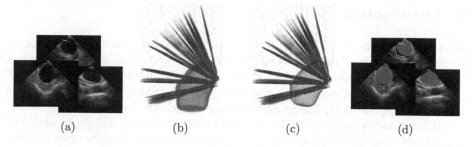

(a)   (b)   (c)   (d)

Fig. 1. Volumetric segmentation of ICE data. (a) 2D ICE frames with CAS contours outlining LA boundaries. (b) 2D frames (black shades) and CAS contours projected onto a 3D grid. The blue mesh represents the 3D segmentation mask obtained by deforming a CT template to fit the contours as closely as possible [8]. Note the sparsity of frames. (c) Predicted 3D segmentation mask generated by a model trained with masks from (b). (d) Predicted mask projected onto 2D (green) and compared with the original CAS contours. Note the misalignment between the mask and CAS contours in some frames. Yellow indicates an example of a user-corrective edit. (Color figure online)

the segmentation output, which must be incorporated in a sequential manner without compromising the previous edits.

In this paper, we present a novel interactive editing framework for the ICE data. This is the first study to address the specific challenges of interactive editing with ICE data. Most of the editing literature treats editing as an interactive segmentation problem and does not provide a clear distinction between interactive segmentation and interactive editing. We provide a novel method that is specifically designed for editing. The novelty of our approach is two-fold: 1) We introduce an editing-specific novel loss function that guides the model to incorporate user edits *while preserving the original segmentation in unedited areas*. 2) We present a novel evaluation metric that best reflects the editing formulation. Comprehensive evaluations of the proposed method on ICE data demonstrate that the presented loss function achieves superior performance compared to traditional interactive segmentation losses and training strategies, as evidenced by the experimental data.

2   Interactive Editing of ICE Data

2.1   Problem Definition

The user is presented first with an ICE volume, $x \in R^{H \times W \times D}$, and its initial imperfect segmentation, $y_{\text{init}} \in R^{H \times W \times D}$, where H, W and D are the dimensions of the volume. To correct inaccuracies in the segmentation, the user draws a scribble on one of the 2D ICE frames. Our goal is to use this 2D interaction to provide a 3D correction to y_{init} in the vicinity of the user interaction. We project the user interaction from 2D to 3D and encode it as a 3D Gaussian heatmap,

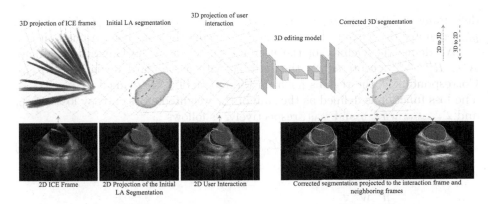

Fig. 2. The proposed interactive editing framework involves user interaction with the segmentation output by drawing a scribble on one of the 2D frames. The editing model is trained to incorporate user interaction while preserving the initial segmentation in unedited areas. Cyan shade: initial segmentation. Green contour: corrected segmentation. Yellow contour: user interaction. (Color figure online)

$u \in R^{H \times W \times D}$, centered on the scribble with a standard deviation of σ_{enc} [9]. The user iteratively interacts with the output until they are satisfied with the quality of the segmentation.

We train an editing model f to predict the corrected segmentation output $\hat{y}^t \in R^{H \times W \times D}$ given x, y_{init}^t, and u^t, where t is the iteration number. The goal is for $\hat{y}^t$ to accurately reflect the user's correction near their interaction while preserving the initial segmentation elsewhere. Since $y_{init}^{t+1} \equiv \hat{y}^t$, subsequent user inputs $u^{\{t+1,\dots,T\}}$ should not corrupt previous corrections $u^{\{0,\dots,t\}}$ (Fig. 2).

2.2 Loss Function

Most interactive segmentation methods aim to incorporate user guidance to enhance the overall segmentation [2,5,9]. However, in our scenario, this approach may undesirably modify previously edited areas and may not align with clinical expectations since the user has corrected these areas, and the changes are unexpected. To address the former issue, Bredell et al. proposed an iterative training strategy in which user edits are synthesized and accumulated over a fixed number of steps with every training iteration [1]. However, this approach comes with a significant increase in training time and does not explicitly instruct the model to preserve regions away from the user input.

We propose an **editing-specific** loss function $\mathcal{L}$ that encourages the model to preserve the initial segmentation while incorporating user input. The proposed loss function incentivizes the model to match the prediction $\hat{y}$ with the ground truth y in the vicinity of the user interaction. In regions further away from the user interaction, the loss function encourages the model to match the initial segmentation y_{init}, instead. Here, y represents the 3D mesh, which is created by deforming a CT template to align with the CAS contours y_{cas} [8]. Meanwhile, y_{init} denotes the output of a segmentation model that has been trained on y.

We define the vicinity of the user interaction as a 3D Gaussian heatmap, $A \in R^{H \times W \times D}$, centered on the scribble with a standard deviation of σ_{edit}. Correspondingly, the regions far from the interaction are defined as $\bar{A} = 1 - A$. The loss function is defined as the sum of the weighted cross entropy losses $\mathcal{L}_{edit}$ and $\mathcal{L}_{preserve}$ w.r.t y and y_{init}, respectively, as follows

$$\mathcal{L} = \mathcal{L}_{edit} + \mathcal{L}_{preserve} \tag{1}$$

where

$$\mathcal{L}_{edit} = -\sum_{i=1}^{H}\sum_{j=1}^{W}\sum_{k=1}^{D} A_{i,j,k} \left[y_{i,j,k} \log \hat{y}_{i,j,k} + (1 - y_{i,j,k}) \log(1 - \hat{y}_{i,j,k}) \right] \tag{2}$$

$$\mathcal{L}_{preserve} = -\sum_{i=1}^{H}\sum_{j=1}^{W}\sum_{k=1}^{D} \bar{A}_{i,j,k} \left[y_{init_{i,j,k}} \log \hat{y}_{i,j,k} + (1 - y_{init_{i,j,k}}) \log(1 - \hat{y}_{i,j,k}) \right]$$
$$\tag{3}$$

The Gaussian heatmaps facilitate a gradual transition between the edited and unedited areas, resulting in a smooth boundary between the two regions.

2.3   Evaluation Metric

The evaluation of segmentation quality typically involves metrics such as the Dice coefficient and the Jaccard index, which are defined for binary masks, or distance-based metrics, which are defined for contours [13]. In our scenario, where the ground truth is CAS contours, we use distance-based metrics[2]. However, standard utilization of these metrics computes the distance between the predicted and ground truth contours, which misleadingly incentivizes alignment with the ground truth contours in **all regions**. This approach incentivizes changes in the unedited regions, which is undesirable from a user perspective, as users want to see changes only in the vicinity of their edit. Additionally, this approach incentivizes the corruption of previous edits.

We propose a novel editing-specific evaluation metric that assesses how well the prediction $\hat{y}$ matches the CAS contours y_{cas} in the vicinity of the user interaction, and the initial segmentation y_{init} in the regions far from the interaction.

$$\mathcal{D} = \mathcal{D}_{edit} + \mathcal{D}_{preserve} \tag{4}$$

where, $\forall (i,j,k) \in \{1,\ldots,H\} \times \{1,\ldots,W\} \times \{1,\ldots,D\}$, $\mathcal{D}_{edit}$ is the distance from y_{cas} to $\hat{y}$ in the vicinity of the user edit, as follows

$$\mathcal{D}_{edit} = \mathbb{1}_{(y_{cas_{i,j,k}}=1)} \cdot A_{i,j,k} \cdot d(y_{cas_{i,j,k}}, \hat{y}) \tag{5}$$

---

[2] Contours are inferred from the predicted mask $\hat{y}$.

where d is the minimum Manhattan distance from $y_{cas_{i,j,k}}$ to any point on $\hat{y}$. For $\mathcal{D}_{\text{preserve}}$, we compute the average symmetric distance between y_{init} and $\hat{y}$, since the two contours are of comparable length. The average symmetric distance is defined as the average of the minimum Manhattan distance from each point on y_{init} contour to $\hat{y}$ contour and vice versa, as follows

$$\mathcal{D}_{\text{preserve}} = \frac{\bar{A}}{2} \cdot \left[\mathbb{1}_{(y_{\text{init}_{i,j,k}}=1)} \cdot d(y_{\text{init}_{i,j,k}}, \hat{y}) + \mathbb{1}_{(\hat{y}_{i,j,k}=1)} \cdot d(\hat{y}_{i,j,k}, y_{\text{init}}) \right] \quad (6)$$

The resulting $\mathcal{D}$ represents a distance map $\in R^{H \times W \times D}$ with defined values only on the contours $y_{\text{cas}}, y_{\text{init}}, \hat{y}$. Statistics such as the 95th percentile and mean can be computed on the corresponding values of these contours on the distance map.

3   Experiments

3.1   Dataset

Our dataset comprises ICE scans for 712 patients, each with their LA CAS contours y_{cas} and the corresponding 3D meshes y generated by [8]. Scans have an average of 28 2D frames. Using the 3D geometry information, frames are projected to a 3D grid with a resolution of $128 \times 128 \times 128$ and voxel spacing of $1.1024 \times 1.1024 \times 1.1024$ mm. We performed five-fold cross-validation on 85% of the dataset (605 patients) and used the remaining 15% (107 patients) for testing.

3.2   Implementation Details

To obtain the initial imperfect segmentation y_{init}, a U-Net model [11] is trained on the 3D meshes y using a Cross-Entropy (CE) loss. The same U-Net architecture is used for the editing model. The encoding block consists of two 3D convolutional layers followed by a max pooling layer. Each convolutional layer is followed by batch normalization and ReLU non-linearity layers [4]. The number of filters in the segmentation model convolutional layers are 16, 32, 64, and 128 for each encoding block, and half of them for the editing model. The decoder follows a similar architecture.

The input of the editing model consists of three channels: the input ICE volume x, the initial segmentation y_{init}, and the user input u. During training, the user interaction is synthesized on the frame with maximum error between y_{init} and y.[3] The region of maximum error is selected and a scribble is drawn on the boundary of the ground truth in that region to simulate the user interaction. During testing, the real contours of the CAS are used and the contour with the maximum distance from the predicted segmentation is chosen as the user interaction. The values of σ_{enc} and σ_{edit} are set to 20, chosen by cross-validation. Adam optimizer is used with a learning rate of 0.005 and a batch size of 4 to train the editing model for 100 epochs [6].

---

[3] We do not utilize the CAS contours during training and only use them for testing because the CAS contours do not align with the segmentation meshes y.

Table 1. Results on Cross-Validation (CV) and test set. We use the editing evaluation metric $\mathcal{D}$ and report the 95th percentile of the overall editing error, the error near the user input, and far from the user input (mm). The near and far regions are defined by thresholding A at 0.5. For the CV results, we report the mean and standard deviation over the five folds. The statistical significance is computed for the difference with InterCNN. †: p-value < 0.01, ‡: p-value < 0.001.

Method	CV			Test		
	Overall ↓	Near ↓	Far ↓	Overall ↓	Near ↓	Far ↓
No Editing	3.962 ± 0.148	-	-	4.126	-	-
CE Loss	1.164 ± 0.094	0.577 ± 0.024	0.849 ± 0.105	1.389	0.6	1.073
Dice Loss	1.188 ± 0.173	0.57 ± 0.089	0.892 ± 0.155	1.039	**0.46**	0.818
InterCNN	0.945 ± 0.049	**0.517 ± 0.052**	0.561 ± 0.006	0.94	0.509	0.569
Editing Loss	**0.809 ± 0.05**‡	0.621 ± 0.042	**0.182 ± 0.01**‡	**0.844**†	0.662	**0.184**‡

3.3   Results

We use the editing evaluation metric $\mathcal{D}$ (Sect. 2.3) for the evaluation of the different methods. For better interpretability of the results, we report the overall error, the error near the user input, and the error far from the user input. We define near and far regions by thresholding the Gaussian heatmap A at 0.5.

We evaluate our loss (editing loss) against the following baselines: (1) **No Editing**: the initial segmentation y_{init} is used as the final segmentation $\hat{y}$, and the overall error in this case is the distance from the CAS contours to y_{init}. This should serve as an upper bound for error. (2) **CE Loss**: an editing model trained using the standard CE segmentation loss w.r.t y. (3) **Dice Loss** [10]: an editing model trained using Dice segmentation loss w.r.t y. (4) **InterCNN** [1]: for every training sample, simulated user edits based on the prediction are accumulated with any previous edits and re-input to the model for 10 iterations, trained using CE loss. We report the results after a single edit (the furthest CAS contour from $\hat{y}$) in Table 1. A single training epoch takes ≈ 3 min for all models except InterCNN, which takes ≈ 14 min, on a single NVIDIA Tesla V100 GPU. The inference time through our model is ≈ 20 milliseconds per volume.

Our results demonstrate that the proposed loss outperforms all baselines in terms of overall error. Although all the editing methods exhibit comparable performance in the near region, in the far region where the error is calculated relative to y_{init}, our proposed loss outperforms all the baselines by a significant margin. This can be attributed to the fact that the baselines are trained using loss functions which aim to match the ground truth globally, resulting in deviations from the initial segmentation in the far region. In contrast, our loss takes into account user input in its vicinity and maintains the initial segmentation elsewhere.

Sequential Editing. We also investigate the scenario in which the user iteratively performs edits on the segmentation multiple times. We utilized the same models that were used in the single edit experiment and simulated 10 editing iterations. At each iteration, we selected the furthest CAS contour from $\hat{y}$,

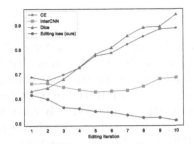

Fig. 3. 95th percentile of the distance from the CAS contours to the prediction.

ensuring that the same edit was not repeated twice. For the interCNN model, we aggregated the previous edits and input them into the model, whereas for all other models, we input a single edit per iteration. We assessed the impact of the number of edits on the overall error. In Fig. 3, we calculated the distance from all the CAS contours to the predicted segmentation and observed that the editing loss model improved with more edits. In contrast, the CE and Dice losses degraded with more edits due to compromising the previous corrections, while InterCNN had only marginal improvements.

Furthermore, in Fig. 4, we present a qualitative example to understand the effect of follow-up edits on the first correction. Edits after the first one are on other frames and not shown in the figure. We observe that the CE and InterCNN methods did not preserve the first correction, while the editing loss model maintained it. This is a crucial practical advantage of our loss, which allows the user to make corrections without compromising the previous edits.

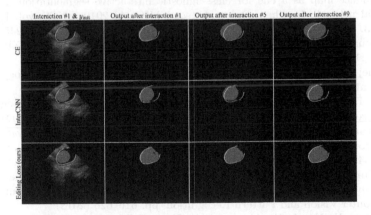

Fig. 4. Impact of follow-up edits on the first correction. Yellow: first user edit. Green: output after each edit. Our editing loss maintains the integrity of the previous edits, while in the other methods the previous edits are compromised. (Color figure online)

4 Conclusion

We presented an interactive editing framework for challenging clinical applications. We devised an editing-specific loss function that penalizes the deviation from the ground truth near user interaction and penalizes deviation from the initial segmentation away from user interaction. Our novel editing algorithm is more robust as it does not compromise previously corrected regions. We demonstrate the performance of our method on the challenging task of volumetric segmentation of sparse ICE data. However, our formulation can be applied to other editing tasks and different imaging modalities.

References

1. Bredell, G., Tanner, C., Konukoglu, E.: Iterative interaction training for segmentation editing networks. In: Shi, Y., Suk, H.I., Liu, M. (eds.) Machine Learning in Medical Imaging, pp. 363–370 (2018)
2. Dorent, R., et al.: Inter extreme points geodesics for end-to-end weakly supervised image segmentation. In: de Bruijne, M., et al. (eds.) MICCAI 2021. LNCS, vol. 12902, pp. 615–624. Springer, Cham (2021). https://doi.org/10.1007/978-3-030-87196-3_57
3. Enriquez, A., et al.: Use of intracardiac echocardiography in interventional cardiology: working with the anatomy rather than fighting it. Circulation 137(21), 2278–2294 (2018)
4. Ioffe, S., Szegedy, C.: Batch normalization: accelerating deep network training by reducing internal covariate shift. In: International Conference on Machine Learning, pp. 448–456 (2015)
5. Khan, S., Shahin, A.H., Villafruela, J., Shen, J., Shao, L.: Extreme points derived confidence map as a cue for class-agnostic interactive segmentation using deep neural network. In: Shen, D., et al. (eds.) MICCAI 2019. LNCS, vol. 11765, pp. 66–73. Springer, Cham (2019). https://doi.org/10.1007/978-3-030-32245-8_8
6. Kingma, D.P., Ba, J.: Adam: a method for stochastic optimization. arXiv preprint arXiv:1412.6980 (2014)
7. Kornej, J., Börschel, C.S., Benjamin, E.J., Schnabel, R.B.: Epidemiology of atrial fibrillation in the 21st century. Circ. Res. 127(1), 4–20 (2020)
8. Liao, H., Tang, Y., Funka-Lea, G., Luo, J., Zhou, S.K.: More knowledge is better: cross-modality volume completion and 3D+2D segmentation for intracardiac echocardiography contouring. In: Frangi, A.F., Schnabel, J.A., Davatzikos, C., Alberola-López, C., Fichtinger, G. (eds.) MICCAI 2018. LNCS, vol. 11071, pp. 535–543. Springer, Cham (2018). https://doi.org/10.1007/978-3-030-00934-2_60
9. Maninis, K.K., Caelles, S., Pont-Tuset, J., Van Gool, L.: Deep extreme cut: from extreme points to object segmentation. In: Proceedings of the IEEE Conference on Computer Vision and Pattern Recognition, pp. 616–625 (2018)
10. Milletari, F., Navab, N., Ahmadi, S.A.: V-net: fully convolutional neural networks for volumetric medical image segmentation. In: 2016 Fourth International Conference on 3D Vision (3DV), pp. 565–571 (2016)
11. Ronneberger, O., Fischer, P., Brox, T.: U-Net: convolutional networks for biomedical image segmentation. In: Navab, N., Hornegger, J., Wells, W.M., Frangi, A.F. (eds.) MICCAI 2015. LNCS, vol. 9351, pp. 234–241. Springer, Cham (2015). https://doi.org/10.1007/978-3-319-24574-4_28

12. Russo, A.D., et al.: Role of intracardiac echocardiography in atrial fibrillation ablation. J. Atr. Fibrillation **5**(6) (2013)
13. Taha, A.A., Hanbury, A.: Metrics for evaluating 3D medical image segmentation: analysis, selection, and tool. BMC Med. Imaging **15**(1), 1–28 (2015)

Ischemic Stroke Segmentation from a Cross-Domain Representation in Multimodal Diffusion Studies

Santiago Gómez[1]ⓘ, Daniel Mantilla[2]ⓘ, Brayan Valenzuela[1]ⓘ,
Andres Ortiz[2]ⓘ, Daniela D Vera[2]ⓘ, Paul Camacho[2]ⓘ,
and Fabio Martínez[1(✉)]ⓘ

[1] BIVL2ab, Universidad Industrial de Santander, Bucaramanga, Santander,
Colombia
famarcar@saber.uis.edu.co
[2] Clínica FOSCAL, Floridablanca, Santander, Colombia

Abstract. Localization and delineation of ischemic stroke are crucial for diagnosis and prognosis. Diffusion-weighted MRI studies allow to associate hypoperfused brain tissue with stroke findings, observed from ADC and DWI parameters. However, this process is expensive, time-consuming, and prone to expert observational bias. To address these challenges, currently, deep representations are based on deep autoencoder representations but are limited to learning from only ADC observations, biased also for one expert delineation. This work introduces a multimodal and multi-segmentation deep autoencoder that recovers ADC and DWI stroke segmentations. The proposed approach learns independent ADC and DWI convolutional branches, which are further fused into an embedding representation. Then, decoder branches are enriched with cross-attention mechanisms and adjusted from ADC and DWI findings. In this study, we validated the proposed approach from 82 ADC and DWI sequences, annotated by two interventional neuroradiologists. The proposed approach achieved higher mean dice scores of 55.7% and 57.7% for the ADC and DWI annotations by the training reference radiologist, out-performing models that only learn from one modality. Notably, it also demonstrated a proper generalization capability, obtaining mean dice scores of 60.5% and 61.0% for the ADC and DWI annotations of a second radiologist. This study highlights the effectiveness of modality-specific pattern learning in producing cross-domain embeddings that enhance ischemic stroke lesion estimations and generalize well over annotations by other radiologists.

Keywords: Ischemic stroke · Difussion-weighted MRI · Cross-domain learning · Multi-segmentation strategies

---

Ministry of science, technology and innovation of Colombia (MINCIENCIAS).

H. Greenspan et al. (Eds.): MICCAI 2023, LNCS 14223, pp. 776–785, 2023.
https://doi.org/10.1007/978-3-031-43901-8_74

1   Introduction

Stroke is the second leading cause of death worldwide, with over twelve million new cases reported annually. Projections estimate that one in four individuals will experience a stroke in their lifetime [3]. Ischemic stroke, caused by blood vessel occlusion, is the most common stroke, accounting for approximately 80% of all cases. These strokes carry a high risk of morbidity, emphasizing the need for timely and accurate diagnosis and treatment [4].

Lesion delineation and quantification are critical components of stroke diagnosis, guiding clinical intervention and informing patient prognosis [7]. Diffusion-weighted MRI studies are the primary tool for observing and characterizing stroke lesions. Among these sequences, the B-1000 (DWI) and the apparent diffusion coefficient (ADC) are biomarkers of cytotoxic injury that predicts edema formation and outcome after ischemic stroke [1]. Typically, the standard lesion analysis requires neuroradiologists to observe both modalities to derive a comprehensive lesion characterization. However, this observational task is challenging, time-consuming (taking approximately 15 min per case), and susceptible to biased errors [9,11].

To address these challenges, computational approaches have supported stroke lesion segmentation, including multi-context approaches from convolutional architectures and standard training schemes to achieve a latent lesion representation [5]. Other strategies have tackled the class imbalance between lesion regions and healthy tissue by using mini-batches of image patches, ensuring stratified information but overlooking global lesion references [2]. More recent architectures have included attention modules that integrate local and contextual information to enhance stroke-related patterns [8,12]. These approaches in general are however adjusted from only one modality reference, losing complementary radiological findings. Furthermore, many of these approaches are calibrated to one-expert annotations, introducing the possibility of learning expert bias.

To the best of our knowledge, this work is the first effort to explore and model complementary lesion annotations from DWI and ADC modalities, utilizing a novel multimodal cross-attentional autoencoder. The proposed architecture employs modality-specific encoders to extract stroke patterns and generate a cross-domain embedding that enhances lesion segmentations for both modalities. To mitigate noise propagation, cross-attention mechanisms were integrated with skip connections between encoder and decoder blocks with similar spatial dimensions. The approach was evaluated on 82 ADC and DWI sequences, annotated by two neuroradiologists, evidencing generalization capabilities and outperforming typical unimodal models. The proposed method makes significant contributions to the field, as outlined below.

- A multi-context and multi-segmentation deep attentional architecture that retrieves lesion annotations from ADC and DWI, sequences.
- An architecture generalization study by training the model with one reference radiologist but evaluating with respect to two radiologist annotations.

2   Proposed Approach

This work introduces a multi-path encoder-decoder representation with the capability to recover stroke segmentations from ADC and DWI parameters. Independent branches are encoded for each modality and fused into an embedding representation. Then, independent decoder paths are evolved to recover stroke lesions, supported by cross-attention modules from the respective encoders. A multi-segmentation task is herein implemented to update the proposed deep representation. The general description of the proposed approach is illustrated in Fig. 1.

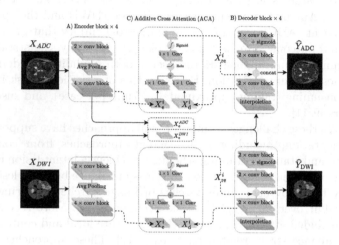

Fig. 1. Graphical representation of the proposed multimodal network for accurate segmentation of lesions in ADC and DWI modalities. The architecture comprises three main parts: A) the encoder, which characterizes the most relevant information in the input image and transforms it into a low-dimensional embedding; B) the decoder, which recovers high-dimensional information captured in the embedded representation; and C) the attention module, which filters and relates the most important features in both modalities.

2.1   Multimodal Attentional Mechanisms

In this work, a dual path encoder-decoder is introduced to recover stroke shape findings, observed from ADC and DWI. From the decoder section, two independent convolutional paths build a deep representation of each considered parameter. Both paths project studies into a low-dimensional embedding space that code stroke associated findings (see Fig. 1a).

Then, a single and complementary latent representation is achieved by the concatenation of both encoders. We hypothesize that such late fusion from embedding space allows a more robust integration of salient features related to stroke findings in both modalities.

Hence, independent decoder branches, for each modality, are implemented from the fused embedding representation. These decoder paths are updated at each level from skip connections, implemented as cross-attentional mechanisms. Particularly, these connections control the contribution of the encoder X_e^i into the decoder X_d^i representation, at each convolutional level i. The refined encoder features are computed as:

$$X_{re}^i = \sigma_2(W_{re}^T \sigma_1(W_e^T X_e^i + W_d^T X_d^i)) \cdot X_e^i$$

where σ_1 and σ_2 are ReLU and Sigmoid activations, respectively (see in Fig. 1c). Consequently, the refined features X_{re}^i preserve highly correlated features between encoder and decoder branches, at the same level of processing. These cross-attention mechanisms enhance the quality of lesion estimations as a result of complementing the decoder representation with early representations of the encoder.

2.2 Multi-segmentation Minimization

A multi-segmentation loss function is here introduced to capture complementary textural patterns from both MRI parameters and improve lesion segmentation. This function induces the learning of highly correlated features present in ADC and DWI images. Specifically, the loss function is defined as:

$$\mathcal{L} = -\alpha(1 - \hat{Y}_{ADC})^\gamma \log(\hat{Y}_{ADC})\mathbf{C} - \alpha(1 - \hat{Y}_{DWI})^\gamma \log(\hat{Y}_{DWI})\mathbf{C} \quad (1)$$

Here, γ is a focusing parameter, α is a weight balancing factor, and $\hat{Y}_{ADC}$ and $\hat{Y}_{DWI}$ are the annotations for ADC and DWI, respectively. It should be noted, that ischemic stroke segmentation is highly imbalanced, promoting background reconstruction. To overcome this issue, each modality loss is weighted by a set of class weight maps ($\mathbf{C} \in \mathbb{N}^{H \times W}$) that increase the contribution of lesion voxels and counteract the negative impact of class imbalance. From this multi-segmentation loss function, the model can more effectively capture complementary information from both ADC and DWI, leading to improved lesion segmentation.

2.3 ADC-DWI Data and Experimental Setup

A retrospective study was conducted to validate the capability of the proposed approach, estimating ischemic stroke lesion segmentation. The study collected 82 studies of patients with stroke symptoms at a clinical center between October 2021 and September 2022. Each study has between 20 to 26 slices with resolutions ranging in each axis as x, y $= [0.83 - 0.94]$ and z $= [5.50 - 7.20]$ mm. After reviewing imaging studies, the studies were categorized into control (n $= 7$) or ischemic stroke (n $= 75$) studies. Control patients with stroke symptoms were included to diversify tissue samples, potentially enhancing our models' ability to segment stroke lesion tissue. Each study included ADC and DWI sequences obtained using a Toshiba Vantage Titan MRI scanner. Two neuro-interventional radiologists, each with more than five years of experience, individually annotated each medical sequence based on clinical records using the MRIcroGL software [6]. To meet the inclusion criteria, patients had to be older than 18 years and show no signs of a cerebral hemorrhage. Patients who showed evidence of partial reperfusion were not excluded from the study.

The dataset was stratified into two subsets for training and testing. The training set consisted of 51 studies with annotations from a single expert (R1), while the test set contained 31 studies with manual delineations from two experts. In addition, a validation set was created from 11 studies of the training set. This validation set was used to measure the model's capability to segment ischemic stroke lesions on unseen data during training and to save the best weights before predicting on the test set. The test set was specifically designed to evaluate the model's performance on quantifying differences between lesion annotations. It allowed for a comparison of annotations between modalities of the same radiologist and between radiologists with the same modality. Furthermore, the test set was used to assess the generalizability of the model when compared to annotations made by the two radiologists. Segmentation capability was measured from the standard Dice score (DSC), precision, and recall.

Regarding the model, the proposed symmetric architecture consists of five convolutional levels (32, 64, 128, 256, and 512 filters for both networks and 1024 filters in the bottleneck) with residual convolutional blocks (3×3, batch normalization, and ReLU activation). For training, the architecture was updated over 300 epochs with a batch size of 16, and a Focal loss with an AdamW optimizer with $\gamma = 2$ and $\alpha = 0.25$. A linear warm-up strategy was used in the first 90 epochs (learning rate from $5e-3$ and decay $1e-5$), followed by 210 epochs of cosine decay. The class weight map values were set to 1 for non-lesion voxels and 3 for lesion voxels. The whole studies were normalized and resized to 224×224 pixels. Registration and skull stripping were not considered. Furthermore, data augmentation such as random brightness and contrast, flips, rotations, random elastic transformations, and random grid and optical distortions were applied to the slices. As a baseline, the proposed architecture was trained with a single modality (ADC, DWI). The code for our work can be found in https://gitlab.com/bivl2ab/research/2023-cross-domain-stroke-segmentation.

3   Evaluation and Results

From a retrospective study, we conducted an analysis to establish radiologist agreement in both ADC and DWI, following the Kappa coefficient from lesion differences. In Fig. 2a and Fig. 2b are summarized the distribution agreement between radiologist annotations, observing the ADC and DWI studies, respectively. In Fig. 2c is reported the agreement distribution between modalities, observed for each radiologist.

As expected, a substantial level of agreement among radiologists was found in both ADC and DWI sequences, with an average agreement of 65% and 70%,

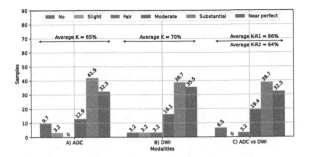

Fig. 2. Percentage of samples and average kappa values stratified in the different levels of agreement. The kappa values for inter-modality are given for both radiologists.

respectively. It should be noted however that in 25% of studies there exist between fair and moderate agreement. These discrepancies may be associated with acute strokes, where radiological findings are negligible and challenging even for expert radiologists. In such a sense, it should be noted that computational approaches may be biased for expert annotations, considering the substantial agreement. Also, among modalities is reported a kappa value of around 65% that evidences the importance to complement findings to achieve more precise stroke delineations.

Hence, an ablation study was then carried out to measure the capabilities of the proposed approach to recover segmentation, regarding multi-segmentation segmentation and the contribution of cross-attention mechanisms. The capability of architecture was measured with respect to the two radiologist delineations and with respect to each modality. Figure 3 illustrates four different variations of the proposed approach. Firstly, a unimodal configuration was implemented from decoder cross-attention mechanisms but adjusted with only one modality. The Dual V1 configuration corresponds to multi-segmentation proposed strategy that uses cross-attention mechanisms at decoder paths. In this version, each cross-attention is learned from the encoder and previous decoder levels. The version (Dual V2) implements cross-attention mechanisms into the encoder path, fusing both modalities at each processing level. Conversely, the Dual V3 integrates encoder and decoder cross-attention modules to force multimodal fusion at different levels of architecture.

Table 1 summarizes the achieved results for each architecture variation and with respect to each radiologist reference at ADC and DWI modalities. Interestingly, the Dual V1 architecture, following an embedding fusion, achieves consistent dice scores in both ADC (V1 = 0.58 vs V3 = 0.53) and DWI (V1 = 0.60 vs V3 = 0.62) for both radiologists, taking advantage of multi-segmentation segmentation learning. Notably, the Dual V1 outperforms most of the other Dual models by up to 10% and the typical approximation by up to 30%. It should be noted that the score achieved for unimodal architecture is coherent with the achieved by other strategies in public datasets [5, 10]. The other dual versions, despite of multilevel fusion, have a trend to filter stroke features and reduce the

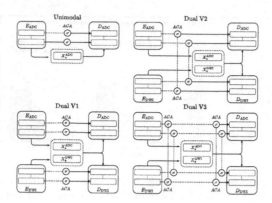

Fig. 3. Baseline unimodal models and different variations of the proposed approach.

quality of the estimated segmentations. Also, the proposed approach achieves remarkable results regarding expert radiologists, even with respect to R2 annotations, unseen during training. These results evidence the potential capabilities of the architecture to generalize segmentations, avoiding expert bias.

Table 1. Estimated segmentation metrics for the four proposed models. The best results are highlighted in gray.

Modality	Model	Radiologist 1			Radiologist 2		
		Dice ↑	Prec ↑	Sens ↑	Dice ↑	Prec ↑	Sens ↑
ADC	Unimodal	0.28 ± 0.30	0.66 ± 0.43	0.15 ± 0.18	0.32 ± 0.30	0.60 ± 0.46	0.17 ± 0.19
	Dual V1	0.56 ± 0.28	0.71 ± 0.28	0.44 ± 0.30	0.60 ± 0.29	0.69 ± 0.33	0.51 ± 0.32
	Dual V2	0.49 ± 0.29	0.68 ± 0.32	0.36 ± 0.27	0.53 ± 0.29	0.66 ± 0.37	0.41 ± 0.28
	Dual V3	0.51 ± 0.28	0.71 ± 0.29	0.43 ± 0.29	0.55 ± 0.29	0.69 ± 0.33	0.49 ± 0.30
DWI	Unimodal	0.45 ± 0.34	0.57 ± 0.37	0.45 ± 0.37	0.45 ± 0.34	0.59 ± 0.38	0.40 ± 0.34
	Dual V1	0.58 ± 0.30	0.61 ± 0.32	0.54 ± 0.36	0.61 ± 0.29	0.67 ± 0.33	0.51 ± 0.31
	Dual V2	0.52 ± 0.29	0.62 ± 0.30	0.56 ± 0.34	0.53 ± 0.28	0.67 ± 0.30	0.52 ± 0.28
	Dual V3	0.62 ± 0.26	0.67 ± 0.29	0.55 ± 0.34	0.61 ± 0.28	0.71 ± 0.31	0.50 ± 0.31

Taking advantage of the substantial agreement between the annotations of both radiologists in our study, we carried out a t-test analysis between different versions of the proposed approach, finding statistically significant differences between Unimodal and Dual models, while the distributions among different Dual models have a similar statistical performance. This analysis was extended to compare the masks of the models with respect to expert radiologists. In this study, the masks of the Dual V1 model have the same distribution ($p > 0.05$) as the reference dice scores over both ADC and DWI. Similarly, the Dual V3 model masks follow the same distribution as the reference dice scores, but only on DWI. Complementary, from Bland-Altman plots in Fig. 4 it can be seen that Dual models exhibit small and clustered differences around the mean, with a mean difference closer to zero. In contrast, single models show more spread-out differences and a higher mean difference compared to Dual models.

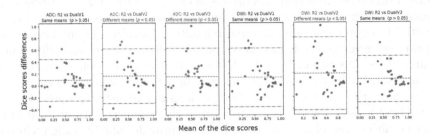

Fig. 4. Bland-Altmans between reference dice scores (R1 vs R2) and dice scores from the proposed dual models predictions.

Figure 5 illustrated the segmentations generated in each of the experiments for the proposed approach and the unimodal baseline. In top, the 3D brain rendering illustrates the stroke segmentations achieved by the dual architecture (red contours), the unimodal baseline (blue areas) and the reference of a radiologist (green areas). At the bottom, the slices illustrate each estimated segmentation as red contours, while the radiologist delineation is represented as the blue area. As expected, the proposed approach reports a remarkable capability to correctly define the limits of the ischemic lesion. Additionally, the proposed approach achieves adequate localization without overestimating the lesion, which is a key issue in identifying treatments and estimating patient diagnosis.

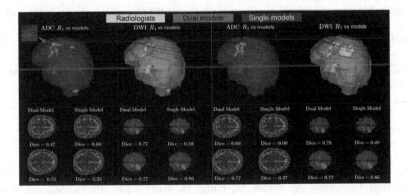

Fig. 5. Comparison of the ischemic stroke lesion predictions for the Dual V1 and unimodal baseline models against the two radiologists. (Color figure online)

4   Conclusions and Perspectives

The proposed approach demonstrates the potential of a multi-context and multi-segmentation cross-attention encoder-decoder architecture for recovering stroke segmentations from DWI and ADC images. The proposed approach has the potential to improve the accuracy and reliability of stroke segmentation in clinical practice, nevertheless, it introduces potential challenges in scalability. Future research directions include the development of attention mechanisms that can fuse additional MRI modalities to improve the accuracy of stroke segmentation. Furthermore, the study could benefit from a more exhaustive preprocessing for the medical images. Finally, an extension of the agreement study to include a larger number of radiologists and studies could provide valuable insights into the reliability and consistency of lesion annotation in MRI studies.

Acknowledgements. The authors thank Ministry of science, technology and innovation of Colombia (MINCIENCIAS) for supporting this research work by the project "Mecanismos computacionales de aprendizaje profundo para soportar tareas de localización, segmentación y pronóstico de lesiones asociadas con accidentes cerebrovasculares isquémicos", with code 91934.

References

1. Bevers, M.B., et al.: Apparent diffusion coefficient signal intensity ratio predicts the effect of revascularization on ischemic cerebral edema. Cerebrovasc. Dis. **45**(3–4), 93–100 (2018)
2. Clèrigues, A., et al.: Acute and sub-acute stroke lesion segmentation from multimodal MRI. Comput. Meth. Programs Biomed. **194**, 105521 (2020). https://doi.org/10.1016/j.cmpb.2020.105521
3. Feigin, V.L., et al.: World stroke organization (WSO): global stroke fact sheet 2022. Int. J. Stroke **17**(1), 18–29 (2022)
4. Hinkle, J.L., Guanci, M.M.: Acute ischemic stroke review. J. Neurosci. Nurs. **39**(5), 285–293 (2007)
5. Hu, X., et al.: Brain SegNet: 3D local refinement network for brain lesion segmentation. BMC Med. Imaging **20**(1), 1–10 (2020). https://doi.org/10.1186/s12880-020-0409-2
6. Li, X., et al.: The first step for neuroimaging data analysis: DICOM to NIfTI conversion. J. Neurosci. Meth. **264**, 47–56 (2016)
7. Liew, S.L., et al.: A large, open source dataset of stroke anatomical brain images and manual lesion segmentations. Sci. Data **5**(1), 1–11 (2018)
8. Liu, P.: Stroke lesion segmentation with 2D novel CNN pipeline and novel loss function. In: Crimi, A., Bakas, S., Kuijf, H., Keyvan, F., Reyes, M., van Walsum, T. (eds.) BrainLes 2018. LNCS, vol. 11383, pp. 253–262. Springer, Cham (2019). https://doi.org/10.1007/978-3-030-11723-8_25
9. Martel, A.L., Allder, S.J., Delay, G.S., Morgan, P.S., Moody, A.R.: Measurement of infarct volume in stroke patients using adaptive segmentation of diffusion weighted MR images. In: Taylor, C., Colchester, A. (eds.) MICCAI 1999. LNCS, vol. 1679, pp. 22–31. Springer, Heidelberg (1999). https://doi.org/10.1007/10704282_3

10. Pinto, A., et al.: Enhancing clinical MRI perfusion maps with data-driven maps of complementary nature for lesion outcome prediction. In: Frangi, A.F., Schnabel, J.A., Davatzikos, C., Alberola-López, C., Fichtinger, G. (eds.) MICCAI 2018. LNCS, vol. 11072, pp. 107–115. Springer, Cham (2018). https://doi.org/10.1007/978-3-030-00931-1_13

11. Rana, A.K., Wardlaw, J.M., Armitage, P.A., Bastin, M.E.: Apparent diffusion coefficient (ADC) measurements may be more reliable and reproducible than lesion volume on diffusion-weighted images from patients with acute ischaemic stroke-implications for study design. Magn. Reson. Imaging **21**(6), 617–624 (2003)

12. Wang, G., et al.: Automatic ischemic stroke lesion segmentation from computed tomography perfusion images by image synthesis and attention-based deep neural networks. Med. Image Anal. **65**, 101787 (2020). https://doi.org/10.1016/j.media.2020.101787

Pre-operative Survival Prediction of Diffuse Glioma Patients with Joint Tumor Subtyping

Zhenyu Tang[1], Zhenyu Zhang[2], Huabing Liu[1], Dong Nie[3], and Jing Yan[4(✉)]

[1] School of Computer Science and Engineering, Beihang University, Beijing 100191, China
tangzhenyu@buaa.edu.cn
[2] Department of Neurosurgery, The First Affiliated Hospital of Zhengzhou University, Zhengzhou 480082, Henan, China
[3] Alibaba Inc., Hangzhou 311121, Zhejiang, China
[4] Department of MRI, The First Affiliated Hospital of Zhengzhou University, Zhengzhou 480082, Henan, China
fccyanj@zzu.edu.cn

Abstract. Pre-operative survival prediction for diffuse glioma patients is desired for personalized treatment. Clinical findings show that tumor types are highly correlated with the prognosis of diffuse glioma. However, the tumor types are unavailable before craniotomy and cannot be used in pre-operative survival prediction. In this paper, we propose a new deep learning based pre-operative survival prediction method. Besides the common survival prediction backbone, a tumor subtyping network is integrated to provide tumor-type-related features. Moreover, a novel ordinal manifold mixup is presented to enhance the training of the tumor subtyping network. Unlike the original manifold mixup, which neglects the feature distribution, the proposed method forces the feature distribution of different tumor types in the order of risk grade, by which consistency between the augmented features and labels can be strengthened. We evaluate our method on both in-house and public datasets comprising 1936 patients and demonstrate up to a 10% improvement in concordance-index compared with the state-of-the-art methods. Ablation study further confirms the effectiveness of the proposed tumor subtyping network and the ordinal manifold mixup. Codes are available at https://github.com/ginobilinie/osPred.

Keywords: Brain tumor · Survival prediction · Subtyping · Ordinal mixup

1 Introduction

Diffuse glioma is a common malignant tumor with highly variable prognosis across individuals. To improve survival outcomes, many pre-operative survival prediction methods have been proposed with success. Based on the prediction results, personalized treatment can be achieved. For instance, Isensee et al. [1]

proposed a random forest model [2], which adopts the radiomics features [3] of the brain tumor images, to predict the overall survival (OS) time of diffuse glioma patients. Nie et al. [4] developed a multi-channel 3D convolutional neural network (CNN) [5] to learn features from multimodal MR brain images and classify OS time as long or short using a support vector machine (SVM) model [6]. In [7], an end-to-end CNN-based method was presented that uses multimodal MR brain images and clinical features such as Karnofsky performance score [8] to predict OS time. In [9], an imaging phenotype and genotype based survival prediction method (PGSP) was proposed, which integrates tumor genotype information to enhance prediction accuracy.

Despite the promising results of existing pre-operative survival prediction methods, they often overlook clinical knowledge that could aid in improving the prediction accuracy. Notably, tumor types have been found to be strongly correlated with the prognosis of diffuse glioma [10]. Unfortunately, tumor type information is unavailable before craniotomy. To address this limitation, we propose a new pre-operative survival prediction method that integrates a tumor subtyping network into the survival prediction backbone. The subtyping network is responsible for learning tumor-type-related features from pre-operative multimodal MR brain images. Concerning the inherent issue of imbalanced tumor types in the training data collected in clinic, a novel ordinal manifold mixup based feature augmentation is presented and applied in the training stage of the tumor subtyping network. Unlike the original manifold mixup [11], which ignores the feature distribution of different classes, in the proposed ordinal manifold mixup, feature distribution of different tumor types is encouraged to be in the order of risk grade, and the augmented features are produced between neighboring risk grades. In this way, inconsistency between the augmented features and the corresponding labels can be effectively reduced.

Our method is evaluated using pre-operative multimodal MR brain images of 1726 diffuse glioma patients collected from cooperation hospitals and a public dataset BraTS2019 [12] containing multimodal MR brain images of 210 patients. Our method achieves the highest prediction accuracy of all state-of-the-art methods under evaluation. In addition, ablation study further confirms the effectiveness of the proposed tumor subtyping network and the ordinal manifold mixup.

2   Methods

Diffuse glioma can be classified into three histological types: the oligodendroglioma, the astrocytoma, and the glioblastoma [10]. The median survival times (in months) are 119 (oligodendroglioma), 36 (astrocytoma), and 8 (glioblastoma) [13]. So the tumor types have strong correlation with the prognosis of diffused glioma. Based on this observation, we propose a new pre-operative survival prediction method (see Fig. 1). Our network is composed of two parts: the survival prediction backbone and the tumor subtyping network. The survival prediction backbone is a deep Cox proportional hazard model [14] which takes the multimodal MR brain images of diffuse glioma patients as inputs and

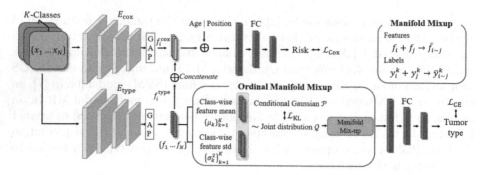

Fig. 1. Structure of our survival prediction network. It is composed of a survival prediction backbone (top) and a tumor subtyping network (bottom). To solve the imbalanced tumor type issue, a novel ordinal manifold mixup is introduced in the training stage of the tumor subtyping network.

predicts the corresponding risks. The tumor subtyping network is a classification network, which classifies the patient tumor types and feeds the learned tumor-type-related features to the backbone to enhance the survival prediction performance.

The tumor subtyping network is trained independently before being integrated into the backbone. To solve the inherent issue of imbalanced tumor type in the training data collected in clinic, a novel ordinal manifold mixup based feature augmentation is applied in the training of the tumor subtyping network. It is worth noting that the ground truth of tumor types, which is determined after craniotomy, is available in the training data, while for the testing data, tumor types are not required, because tumor-type-related features can be learned from the pre-operative multimodal MR brain images.

2.1   The Survival Prediction Backbone

The architecture of the survival prediction backbone, depicted in Fig. 1 top, consists of an encoder E_{cox} with four ResBlocks [15], a global average pooling layer (GAP), and three fully connected (FC) layers. Assume that $D = \{x_1, ..., x_N\}$ is the dataset containing pre-operative multimodal MR brain images of diffuse glioma patients, and N is the number of patients. The backbone is responsible for deriving features from x_i to predict the risk of the patient. Moreover, after the GAP of the backbone, the learned feature $f_i^{\mathrm{cox}} \in \mathbb{R}^M$ is concatenated with the tumor-type-related feature $f_i^{\mathrm{type}} \in \mathbb{R}^M$ learned from the tumor subtyping network (discussed later), and M is the vector dimension which is set to 128. As f_i^{type} has strong correlation with prognosis, the performance of the backbone can be improved. In addition, information of patient age and tumor position is also used. To encode the tumor position, the brain is divided into $3 \times 3 \times 3$ blocks, and the tumor position is represented by 27 binary values (0 or 1) with each value for one block. If a block contains tumors, then the corresponding binary value is 1, otherwise is 0. The backbone is based on the deep Cox proportional hazard model, and the loss function is defined as:

$$\mathcal{L}_{\text{Cox}} = -\sum_{i=1}^{N} \delta_i \left(h_\theta(x_i) - \log \sum_{j \in R(t_i)} e^{h_\theta(x_j)} \right), \tag{1}$$

where $h_\theta(x_i)$ represents the risk of the i-th patient predicted by the backbone, θ stands for the parameters of the backbone, x_i is the input multimodal MR brain images of the i-th patient, $R(t_i)$ is the risk group at time t_i, which contains all patients who are still alive before time t_i, t_i is the observed time (time of death happened) of x_i, and $\delta_i = 0/1$ for censored/non-censored patient.

2.2   The Tumor Subtyping Network

The tumor subtyping network has almost the same structure as the backbone. It is responsible for learning tumor-type-related features from each input pre-operative multimodal MR brain image x_i and classifying the tumor into oligo-dendroglioma, astrocytoma, or glioblastoma. The cross entropy is adopted as the loss function of the tumor subtyping network, which is defined as:

$$\mathcal{L}_{\text{CE}} = -\sum_{i=1}^{N} \sum_{k=1}^{3} y_i^k \log(p_i^k), \tag{2}$$

where y_i^k and p_i^k are the ground truth (0 or 1) and the prediction (probability) of the k-th tumor type ($k = 1, 2, 3$) of the i-th patient, respectively. The learned tumor-type-related feature $f_i^{\text{type}} \in \mathbb{R}^M$ is fed to the survival prediction backbone and concatenated with f_i^{cox} learned in the backbone to predict the risk.

In the in-house dataset, the proportions of the three tumor types are 20.9% (oligodendroglioma), 28.7% (astrocytoma), and 50.4% (glioblastoma), which is consistent with the statistical report in [13]. To solve the imbalance issue of tumor types in the training of the tumor subtyping network, a novel ordinal manifold mixup based feature augmentation is presented.

2.3   The Ordinal Manifold Mixup

In the original manifold mixup [11], features and the corresponding labels are augmented using linear interpolation on two randomly selected features (e.g., f_i and f_j). Specifically, the augmented feature $\bar{f}_{i \sim j}$ and label $\bar{y}_{i \sim j}$ is defined as:

$$\begin{aligned} \bar{f}_{i \sim j} &= \lambda f_i + (1 - \lambda) f_j \\ \bar{y}_{i \sim j}^k &= \lambda y_i^k + (1 - \lambda) y_j^k, \end{aligned} \tag{3}$$

where y_i^k and y_j^k stand for the labels of the k-th tumor type of the i-th and j-th patients, respectively, and $\lambda \in [0, 1]$ is a weighting factor. For binary classification, the original manifold mixup can effectively enhance the network performance, however, for the classification of more than two classes, e.g., tumor types, there exists a big issue.

As shown in Fig. 2 left, assume that f_i and f_j are features of oligoden-droglioma (green) and astrocytoma (yellow) learned in the tumor subtyping network, respectively. The augmented feature $\bar{f}_{i\sim j}$ (red) produced from the lin-ear interpolation between f_i and f_j has the corresponding label $\bar{y}_{i\sim j}$ with high probabilities for the tumor types of oligodendroglioma and astrocytoma. How-ever, since these is no constraint imposed on the feature distribution of different tumor types, $\bar{f}_{i\sim j}$ could fall into the distribution of glioblastoma (blue) as shown in Fig. 2 left. In this case, $\bar{f}_{i\sim j}$ and $\bar{y}_{i\sim j}$ are inconsistent, which could influence the training and degrade the performance of the tumor subtyping network.

Fig. 2. Illustration of feature augmentation using original manifold mixup (left) and the proposed ordinal manifold mixup (right). (Color figure online)

As aforementioned, the survival time of patients with different tumor types varies largely (oligodendroglioma > astrocytoma > glioblastoma), so the tumor types can be regarded as risk grade, which are ordered rather than categorical. Based on this assertion and inspired by [16], we impose an ordinal constraint on the tumor-type-related features to make the feature distribution of different tumor types in the order of risk grade. In this way, the manifold mixup strategy can be applied between each two neighboring tumor types to produce augmented features with consistent labels that reflect reasonable risk (see Fig. 2 right).

Normally, the feature distribution of each tumor type is assumed to be inde-pendent normal distribution, so their joint distribution is given by:

$$\mathcal{Q}(F^1, F^2, F^3) = \prod_{k=1}^{3} \mathcal{N}(\mu_k, \sigma_k^2), \tag{4}$$

where $F^k, k = 1, 2, 3$ represents the feature set of oligodendroglioma, astrocy-toma, and glioblastoma, respectively, μ_k and σ_k^2 are mean and variance of F^k.

To impose the ordinal constraint, we define the desired feature distribution of each tumor type as $\mathcal{N}(\hat{\mu}_1, \hat{\sigma}_1^2)$ for $k = 1$, and $\mathcal{N}(\hat{\mu}_{k-1} + \Delta_k, \hat{\sigma}_k^2)$ for $k = 2$ and 3. In this way, the feature distribution of each tumor type depends on its predecessor, and the mean feature of each tumor type $\hat{\mu}_k$ (except $\hat{\mu}_1$) is equal to the mean feature of its predecessor $\hat{\mu}_{k-1}$ shifted by Δ_k. Note that Δ_k is set to be larger than $3 \times \hat{\sigma}_k$ to ensure the desired ordering [16]. In this way, the conditional distribution under the ordinal constraint is defined as:

$$\mathcal{P}(F^1, F^2, F^3) = \mathcal{P}(F^1)\mathcal{P}(F^2|F^1)\mathcal{P}(F^3|F^2), \tag{5}$$

which can be represented as:

$$\mathcal{N}\left(\begin{bmatrix} \hat{\mu}_1 \\ \hat{\mu}_1 + \Delta_2 \\ \hat{\mu}_1 + \Delta_2 + \Delta_3 \end{bmatrix}, \begin{bmatrix} \hat{\sigma}_1^2 & \hat{\sigma}_1^2 & \hat{\sigma}_1^2 \\ \hat{\sigma}_1^2 & \hat{\sigma}_1^2 + \hat{\sigma}_2^2 & \hat{\sigma}_1^2 + \hat{\sigma}_2^2 \\ \hat{\sigma}_1^2 & \hat{\sigma}_1^2 + \hat{\sigma}_2^2 & \hat{\sigma}_1^2 + \hat{\sigma}_2^2 + \hat{\sigma}_3^2 \end{bmatrix}\right), \tag{6}$$

where $\hat{\mu}_1$ and $\hat{\sigma}_k, k = 1, 2, 3$ can be learned by the tumor subtyping network. Finally, the ordinal loss, which is in the form of KL divergence, is defined as:

$$\mathcal{L}_{\mathrm{KL}} = \mathrm{KL}(\prod_{k=1}^{3} \mathcal{N}(\mu_k, \sigma_k^2) \,\|\, \mathcal{P}(F^1, F^2, F^3)). \tag{7}$$

In our method, μ_k and σ_k^2 are calculated by

$$\mu_k = \frac{1}{N_k} \sum_{x_i \in D^k} \mathcal{G}(\Phi_\theta(x_i))$$

$$\sigma_k^2 = \frac{\sum_{x_i \in D^k}(x_i - \mu_k)^2}{N_k - 1}, \tag{8}$$

where Φ_θ and $\mathcal{G}$ are the encoder and GAP of the tumor subtyping network, respectively, θ is the parameter set of the encoder, $D^k = \{x_1, ..., x_{N_k}\}, k = 1, 2, 3$ stands for the subset containing the pre-operative multimodal MR brain images of the patients with the k-th tumor type, N_k is the patient number in D^k. So we impose the ordinal loss $\mathcal{L}_{\mathrm{KL}}$ to the features after the GAP of the tumor subtyping network as shown in Fig. 1. Since the ordinal constraint, feature distribution of different tumor types learned in the subtyping network is encouraged to be in the ordered of risk grade, and features can be augmented between neighboring tumor type. In this way, inconsistency between the resulting augmented features and labels can be effectively reduced.

The tumor subtyping network is first trained before being integrated into the survival prediction backbone. In the training stage of the tumor subtyping network, each input batch contains pre-operative multimodal MR brain images of N patients and can be divided into $K = 3$ subsets according to their corresponding tumor types, i.e., $D^k, k = 1, 2, 3$. With the ordinal constrained feature distribution, high consistent features can be augmented between neighboring tumor types. Based on the original and augmented features, the performance of the tumor subtyping network can be enhanced.

Once the tumor subtyping network has been trained, it is then integrated into the survival prediction backbone, which is trained under the constraint of the cox proportional hazard loss $\mathcal{L}_{\mathrm{Cox}}$.

3   Results

In our experiment, both in-house and public datasets are used to evaluate our method. Specifically, the in-house dataset collected in cooperation hospitals contains pre-operative multimodal MR images, including T1, T1 contrast enhanced

(T1c), T2, and FLAIR, of 1726 patients (age 49.7 ± 13.1) with confirmed diffuse glioma types. The patient number of each tumor type is 361 (oligodendroglioma), 495 (astrocytoma), and 870 (glioblastoma), respectively. In the 1726 patients, 743 have the corresponding overall survival time (dead, non-censored), and 983 patients have the last visiting time (alive, censored). Besides the in-house dataset, a public dataset BraTS2019, including pre-operative multimodal MR images of 210 non-censored patients (age 61.4 ± 12.2), is adopted as the external independent testing dataset. All images of the in-house and BraTS2019 datasets go through the same pre-processing stage, including image normalization and affine transformation to MNI152 [17]. Based on the tumor mask of each image, tumor bounding boxes can be calculated. According to the bounding boxes of all 1936 patients, the size of input 3D image patch is set to $96 \times 96 \times 64$ voxels, which can cover the entire tumor of every patient.

Besides our method, four state-of-the-art methods, including random forest based method (RF) [18], deep convolutional survival model (deepConvSurv) [19], multi-channel survival prediction method (MCSP) [20], and imaging phenotype and genotype based survival prediction method (PGSP) [9], are evaluated. It is worth noting that in the RF method, 100 decision trees and 390 handcrafted radiomics features are used. The output of RF, MCSP, and PGSP is the overall survival (OS) times in days, while for deepConvSurv and our method, the output is the risk (deep Cox proportional hazard models). Concordance index (C-index) is adopted to quantify the prediction accuracy:

$$\text{C-index}(D) = \frac{\sum_{i,j \in D} \mathbf{1}_{T_i < T_j} \cdot \mathbf{1}_{R_i < R_j} \cdot \delta_i}{\sum_{i,j \in D} \mathbf{1}_{T_i < T_j} \cdot \delta_i}, \tag{9}$$

where $D = \{x_1, ..., x_N\}$ is the dataset containing all patients, T_i and T_j are ground truth of survival times of the i-th and j-th patients, R_i and R_j are the days predicted by RF, MCSP, and PGSP or risks predicted by the deep Cox proportional hazard models (i.e., deepConvSurv and our method), $\mathbf{1}_{x < y} = 1$ if $x < y$, else 0, and $\delta_i = 0$ or 1 when the i-th patient is censored or non-censored.

As RF, MCSP, and PGSP cannot use the censored data in the in-house dataset, 80% of the non-censored data (594 patients) are randomly selected as the training data, and the rest 20% non-censored data (149 patients) are for testing. While deepConvSurv and our method are deep Cox models, both censored and non-censored patients can be utilized. So besides the 80% non-censored patients, all censored data (983 patients) are also included in the training data.

Table 1 shows the evaluation results of the in-house and the external independent (BraTS2019) testing datasets using all methods under evaluation. Our method achieves the highest C-index of all the methods under evaluation. Moreover, comparing with deepConvSurv, our method can improve the prediction accuracy up to 10% (in-house) and 8% (BraTS2019).

3.1   Ablation Study of Survival Prediction

To show the effect of the tumor subtyping network and the ordinal manifold mixup in survival prediction, our method without the tumor subtyping network

Table 1. Evaluation of the prediction results (C-Index) using the in-house and external independent (BraTS2019) testing datasets.

	RF	deepConvSurv	MCSP	PGSP	Ours
In-house	0.708	0.726	0.715	0.776	**0.798**
BraTS2019	0.619	0.705	0.700	0.738	**0.762**

(Baseline-1) and our method with the tumor subtyping network (using original manifold mixup instead, Baseline-2) are evaluated. For the in-house dataset, the resulting C-indices are 0.744 (Baseline-1) and 0.735 (Baseline-2). So our method make the improvement of C-index more than 8% comparing with Baseline-2. For the external independent testing dataset BraTS2019, the resulting C-indices are 0.738 (Baseline-1) and 0.714 (Baseline-2), and our method still has more than 6% improvement comparing with Baseline-2.

Figure 3 shows the distributions of tumor-type-related features (after the GAP) of the in-house testing data in Baseline-2, and our method. Principal component analysis [21] is used to project features to a 2D plane. Since the ordinal constraint, the feature distribution of different tumor types is in the order of risk grade using our method, which cannot be observed in Baseline-2.

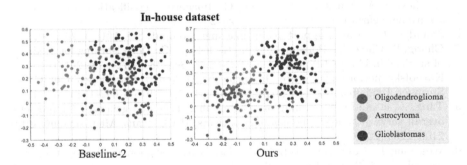

Fig. 3. Feature distributions of the in-house testing data in Baseline-2, and Ours.

4 Conclusions

We proposed a new method for pre-operative survival prediction of diffuse glioma patients, where a tumor subtyping network is integrated into the prediction backbone. Based on the tumor subtyping network, tumor type information, which are only available after craniotomy, can be derived from the pre-operative multimodal MR images to boost the survival prediction performance. Moreover, a novel ordinal manifold mixup was presented, where ordinal constraint is imposed to make feature distribution of different tumor types in the order of risk grade, and feature augmentation only takes place between neighboring tumor types. In this way, inconsistency between the augmented features and corresponding labels can be effectively reduced. Both in-house and public datasets containing 1936 patients were used in the experiment. Our method outperformed the state-of-the-art methods in terms of the concordance-index.

Acknowledgements. This work is supported in part by National Natural Science Foundation of China No. 62073012, 82102149, 82273493, Beijing Municipal Natural Science Foundation No. 7222307, and Natural Science Foundation of Henan Province for Excellent Young Scholars No. 232300421057.

References

1. Isensee, F., Kickingereder, P., Wick, W., Bendszus, M., Maier-Hein, K.H.: Brain tumor segmentation and radiomics survival prediction: contribution to the brats 2017 challenge. In: International MICCAI Brainlesion Workshop (2017)
2. Breiman, L.: Random forests. Mach. Learn. **45**, 5–32 (2001)
3. Gillies, R.J., Kinahan, P.E., Hricak, H.: Radiomics: images are more than pictures, they are data. Radiology **278**(2), 563–577 (2016)
4. Nie, D., Zhang, H., Ehsan, A., Liu, A., Shen, D.: 3D deep learning for multi-modal imaging-guided survival time prediction of brain tumor patients. In: MICCAI (2016)
5. Krizhevsky, A., Sutskever, I., Hinton, G.: ImageNet classification with deep convolutional neural networks. In: NIPS (2012)
6. Saunders, C., et al.: Support vector machine. Comput. Sci. **1**(4), 1–28 (2002)
7. Chang, P., Chow, D., Poisson, L., Jain, R., Filippi, C.: Deep learning for prediction of survival in idh wild-type gliomas. J. Neurol. Sci. **381**, 172–173 (2017)
8. Karnofsky performance score. In: Schwab, M. (eds.) Encyclopedia of Cancer. Springer, Heidelberg (2011). https://doi.org/10.1007/978-3-642-16483-5_3198
9. Tang, Z., et al.: Deep learning of imaging phenotype and genotype for predicting overall survival time of glioblastoma patients. IEEE Trans. Med. Imaging **39**(6), 2100–2109 (2020)
10. Wesseling, P., Capper, D.: WHO 2016 classification of gliomas. Neuropathol. Appl. Neurobiol. **44**(2), 139–150 (2019)
11. Verma, V., et al.: Manifold mixup: better representations by interpolating hidden states. In: International Conference on Machine Learning (ICML), pp. 6438–6447 (2019)
12. Menze, B.H., et al.: The multimodal brain tumor image segmentation benchmark (BRATS). IEEE Trans. Med. Imaging **34**(10), 1993–2024 (2014)
13. Ostrom, Q.T., Patil, G., Cioffi, N., Waite, K., Kruchko, C., Barnholtz-Sloan, J.S.: CBTRUS statistical report: primary brain and other central nervous system tumors diagnosed in the united states in 2013–2017. Neuro-Oncol. **22**(12), iv1–iv96 (2020)
14. Katzman, J., Shaham, U., Bates, U., Cloninger, A., Jiang, T., Kluger, Y.: Deep survival: a deep cox proportional hazards network. arXiv:1606.00931 (2016)
15. He, K., Zhang, X., Ren, X., Sun, J.: Deep residual learning for image recognition. In: CVPR, pp. 770–778 (2016)
16. Liu, X., Li, S., Ge, Y., Ye, P., You, J., Lu, J.: Recursively conditional gaussian for ordinal unsupervised domain adaptation. In: Proceedings of the IEEE/CVF International Conference on Computer Vision, pp. 764–773 (2021)

17. Fonov, V.S., Evans, A.C., Botteron, K., Almli, C.R., McKinstry, R.C., Collins, D.L.: BDCG: unbiased average age-appropriate atlases for pediatric studies. NeuroImage **54**(1) (2011)
18. Liaw, A., Wiener, M.: Classification and regression by randomForest. R News **23**(23), 18–22 (2002)
19. Zhu, X., Yao, J., Zhu, F., Huang, J.: WSISA: making survival prediction from whole slide histopathological images. In: CVPR, pp. 6855–6863 (2017)
20. Nie, D., et al.: Multi-channel 3D deep feature learning for survival time prediction of brain tumor patients using multi-modal neuroimages. Sci. Rep. **9**(1), 1103 (2019)
21. Kurita, T.: Principal Component Analysis (PCA), pp. 1–4. Springer, Cham (2019). https://doi.org/10.1007/978-3-030-03243-2_649-1

Author Index

H. Greenspan et al. (Eds.): MICCAI 2023, LNCS 14223, pp. 797–801, 2023.
https://doi.org/10.1007/978-3-031-43901-8

Printed in the United States
by Baker & Taylor Publisher Services

Printed in the United States
by Baker & Taylor Publisher Services